Calculus with Applications

TENTH EDITION

Calculus with Applications

TENTH EDITION

Margaret L. Lial
American River College

Raymond N. Greenwell
Hofstra University

Nathan P. Ritchey
Youngstown State University

PEARSON

Boston Columbus Indianapolis New York San Francisco Upper Saddle River
Amsterdam Cape Town Dubai London Madrid Milan Munich Paris Montréal Toronto
Delhi Mexico City São Paulo Sydney Hong Kong Seoul Singapore Taipei Tokyo

Editor in Chief: Deirdre Lynch
Executive Editor: Jennifer Crum
Executive Content Editor: Christine O'Brien
Senior Project Editor: Rachel S. Reeve
Editorial Assistant: Joanne Wendelken
Senior Managing Editor: Karen Wernholm
Senior Production Project Manager: Patty Bergin
Associate Director of Design, USHE North and West: Andrea Nix
Senior Designer: Heather Scott
Digital Assets Manager: Marianne Groth
Media Producer: Jean Choe
Software Development: Mary Durnwald and Bob Carroll
Executive Marketing Manager: Jeff Weidenaar
Marketing Coordinator: Caitlin Crain
Senior Author Support/Technology Specialist: Joe Vetere
Rights and Permissions Advisor: Michael Joyce
Image Manager: Rachel Youdelman
Senior Manufacturing Buyer: Carol Melville
Senior Media Buyer: Ginny Michaud
Production Coordination and Composition: Nesbitt Graphics, Inc.
Illustrations: Nesbitt Graphics, Inc., Network Graphics, and IllustraTech
Cover Design: Heather Scott
Cover Image: Lightpoet/Shutterstock

Credits appear on page C-1, which constitutes a continuation of the copyright page.

Many of the designations used by manufacturers and sellers to distinguish their products are claimed as trademarks. Where those designations appear in this book, and Pearson was aware of a trademark claim, the designations have been printed in initial caps or all caps.

Library of Congress Cataloging-in-Publication Data
Lial, Margaret L.
  Calculus with applications — 10th ed. / Margaret L. Lial, Raymond N.
Greenwell, Nathan P. Ritchey.
    p.cm.
Includes bibliographical references and index.
  ISBN-13: 978-0-321-74900-0 (student ed.)
  ISBN-10: 0-321-74900-6 (student ed.)
1. Calculus—Textbooks.   I. Greenwell, Raymond N.   II. Ritchey, Nathan P.   III. Title.
QA303.2.L53 2012
515—dc22
                                                        2010030933

5 6 7 8 9 10—RRDW—15

www.pearsonhighered.com

ISBN-10: 0-321-74900-6
ISBN-13: 978-0-321-74900-0

Contents

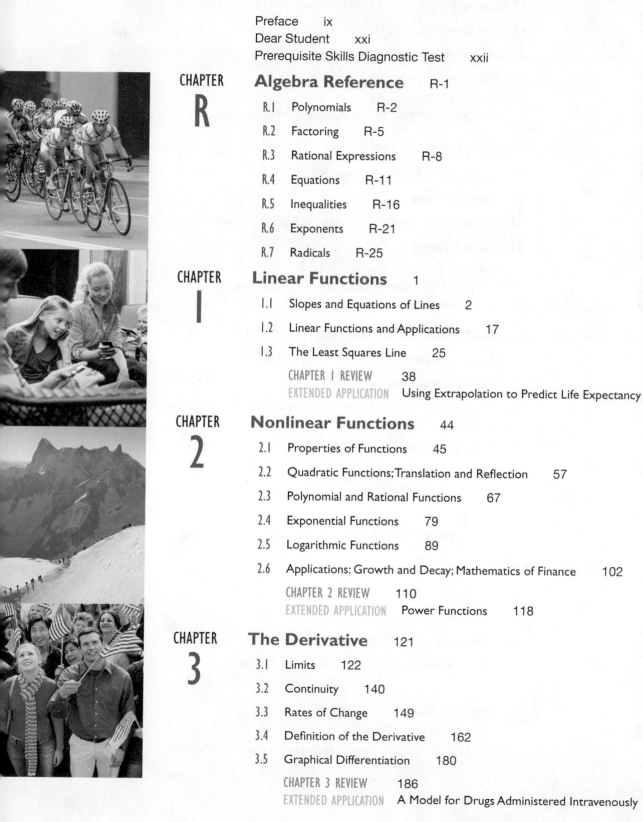

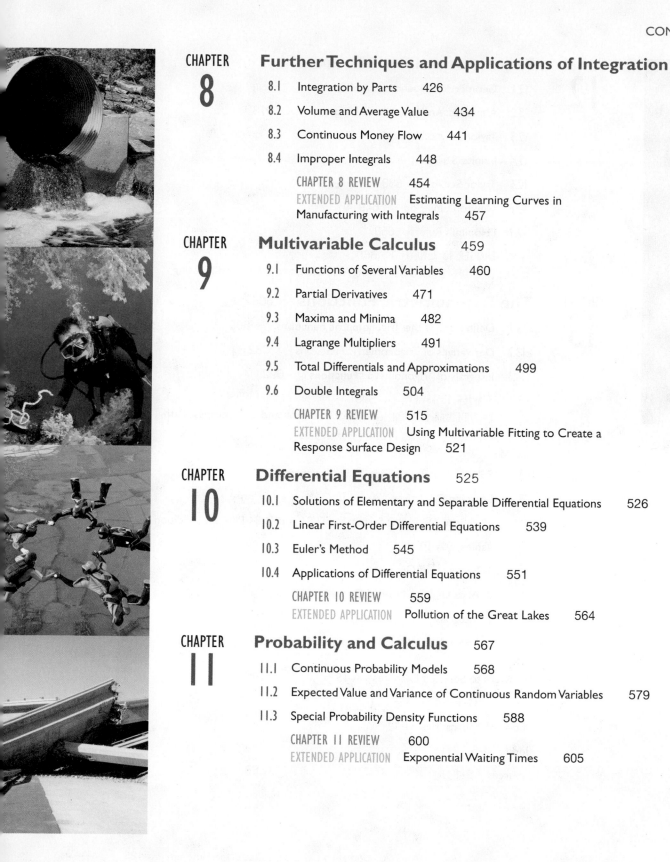

Preface

Calculus with Applications is a thorough, application-oriented text for students majoring in business, management, economics, or the life or social sciences. In addition to its clear exposition, this text consistently connects the mathematics to career and everyday-life situations. A prerequisite of two years of high school algebra is assumed. A renewed focus on quick and effective assessments, new applications and exercises, as well as other new learning tools make this 10th edition an even richer learning resource for students.

Our Approach

Our main goal is to present applied calculus in a concise and meaningful way so that students can understand the full picture of the concepts they are learning and apply it to real-life situations. This is done through a variety of ways.

Focus on Applications Making this course meaningful to students is critical to their success. Applications of the mathematics are integrated throughout the text in the exposition, the examples, the exercise sets, and the supplementary resources. *Calculus with Applications* presents students with a myriad of opportunities to relate what they're learning to career situations through the *Apply It* questions, the applied examples, and the *Extended Applications*. To get a sense of the breadth of applications presented, look at the Index of Applications in the back of the book or the extended list of sources of real-world data on www.pearsonhighered .com/mathstatsresources.

Pedagogy to Support Students Students need careful explanations of the mathematics along with examples presented in a clear and consistent manner. Additionally students and instructors should have a means to assess the basic prerequisite skills. This can now be done with the *Prerequisite Skills Diagnostic Test* located just before Chapter R. In addition, the students need a mechanism to check their understanding as they go and resources to help them remediate if necessary. *Calculus with Applications* has this support built into the pedagogy of the text through fully developed and annotated examples, *Your Turn* exercises, *For Review* references, and supplementary material.

Beyond the Textbook Students today take advantage of a variety of resources and delivery methods for instruction. As such, we have developed a robust MyMathLab course for *Calculus with Applications*. MyMathLab has a well-established and well-documented track record of helping students succeed in mathematics. The MyMathLab online course for *Calculus with Applications* contains over 2000 exercises to challenge students and provides help when they need it. Students who learn best by seeing and hearing can view section- and example-level videos within MyMathLab or on the book-specific DVD-Rom. These and other resources are available to students as a unified and reliable tool for their success.

New to the Tenth Edition

Based on the authors' experience in the classroom along with feedback from many instructors across the country, the focus of this revision is to improve the clarity of the presentation and provide students with more opportunities to learn, practice, and apply what they've learned on their own. This is done in both the presentation of the content and in new features added to the text.

New and Revised Content

- **Chapter R** The flow of the material was improved by reordering some exercises and examples. Exercises were added to Section R.1 (on performing algebraic operations) and Section R.5 (on solving inequalities).

- **Chapter 1** Changes in the presentation were made throughout to increase clarity, including adding some examples and rewriting others. Terminology in Section 1.2 was adjusted to be more consistent with usage in economics.

- **Chapter 2** The material in Section 2.1 on the Dow Jones Average was updated. Material on even and odd functions was added. Material on identifying the degree of a polynomial has been rewritten as an example to better highlight the concept. The discussion of the Rule of 70 and the Rule of 72 was improved. A new Extended Application on Power Functions has been added.

- **Chapter 3** In Section 3.1, the introduction of limits was completely revised. The opening discussion and example were transformed into a series of examples that progress through different limit scenarios: a function defined at the limit, a function undefined at the limit (a hole in the graph), a function defined at the limit but with a different value than the limit (a piecewise function), and then finally, finding a limit when one does not exist. New figures were added to illustrate the different scenarios. In Section 3.2 the definition and example of continuity has been revised using a simple process to test for continuity. The opening discussion of Section 3.5, showing how to sketch the graph of the derivative given the graph of the original function, was rewritten as an example.

- **Chapter 4** The introduction to the chain rule was rewritten as an example in Section 4.3. Exercise topics were revised to cover subjects such as worldwide Internet users, online learning, and the Gateway arch.

- **Chapter 5** In Section 5.1 the definition of increasing/decreasing functions has been moved to the beginning of the chapter, followed by the discussion of using derivatives to determine where the function increases and decreases. The determination of where a function is increasing or decreasing is divided into three examples: when the critical numbers are found by setting the derivative equal to zero, when the critical numbers are found by determining where the derivative is undefined, and when the function has no critical numbers.

- **Chapter 6** Changes in the presentation were made throughout to increase clarity and exercise sets were rearranged to improve progression and parity.

- **Chapter 7** The social sciences category of exercises was added to Section 7.1, including the topics of bachelor's degrees and the number of females earning degrees in dentistry. Color was added to the introduction and first example of substitution in Section 7.2 to enable students to follow the substitution more easily.

- **Chapter 8** In addition to exercises based on real data being updated, examples in this chapter were changed for pedagogical reasons.

- **Chapter 9** Graphs generated by Maple™ were added to Examples 2 and 4 in Section 9.3 to assist students in visualizing the concept of relative extrema. Material covering utility functions was added to Section 9.4. Many of the figures of three-dimensional surfaces were improved to make them clearer and more attractive.

- **Chapter 10** The notation in Section 10.1 was changed to improve clarity. Additionally, exercises and examples in this section were modified to emphasize checking that solutions satisfy the original differential equation.

• **Chapter 11** In Section 11.2, an example on how to calculate the probability within one standard deviation of the mean (which is required in many of the exercises) was added. The Social Sciences category was added to the exercise set, with exercises on calculating the median, expected value, and standard deviation. Topics include the time it takes to learn a task and the age of users of a social network.

• **Chapter 12** Examples were added on calculating depreciation with a geometric sequence and illustrating how to find the sum of a geometric sequence when the sequence is written in summation notation, similar to several of the exercises in Section 12.1. A new example, and corresponding exercises, was added to 12.4, illustrating how to solve a problem first using algebra and then using a geometric series. Four basic exercises finding the Taylor series were added to the beginning of the exercise set in Section 12.5, and five exercises that require l'Hôpital's rule were added to Section 12.7.

• **Chapter 13** Material was added to Section 13.1 to clarify the meaning of the sine and cosine. Example 7, along with related exercises, were added which explore where the trigonometric functions take on specific values.

Prerequisite Skills Diagnostic Test

The Prerequisite Skills Diagnostic Test gives students and instructors a means to assess the basic prerequisite skills needed to be successful in this course. In addition, the answers to the test include references to specific content in Chapter R as applicable so students can zero in on where they need improvement. Solutions to the questions in this test are in Appendix A.

More Applications and Exercises

This text is used in large part because of the enormous amounts of real data used in examples and exercises throughout the text. This 10th edition will not disappoint in this area. We have added or updated nearly 20% of the applications and 37% of the examples throughout the text and added or updated over 340 exercises.

Reference Tables for Exercises

The answers to odd-numbered exercises in the back of the textbook now contain a table referring students to a specific example in the section for help with most exercises. For the review exercises, the table refers to the section in the chapter where the topic of that exercise is first discussed.

Annotated Instructor's Edition

The annotated instructor's edition is filled with valuable teaching tips in the margins for those instructors who are new to teaching this course. In addition, answers to most exercises are provided directly on the exercise set page to make assigning and checking homework easier. In addition, answers to most exercises are provided directly on the exercise set page along with + symbol next to the most challenging exercises to make assigning and checking homework easier.

New to MyMathLab

Available now with *Calculus with Applications* are the following resources within MyMathLab that will benefit students in this course.

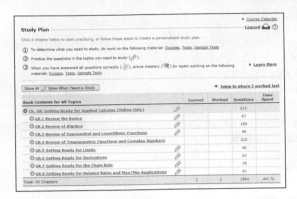

- "Getting Ready for Applied Calculus" chapter covers basic prerequisite skills
- Personalized Homework allows you to create homework assignments based on the results of student assessments
- Videos with extensive section coverage
- Hundreds more assignable exercises than the previous edition of the text
- Application labels within exercise sets (e.g., "Bus/Econ") make it easy for you to find types of applications appropriate to your students
- Additional graphing calculator and Excel spreadsheet help

A detailed description of the overall capabilities of MyMathLab is provided on page xvii.

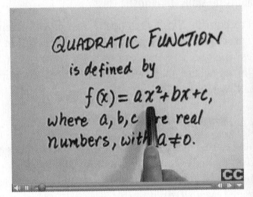

Source Lines

Sources for the exercises are now written in an abbreviated format within the actual exercise so that students immediately see that the problem comes from, or pulls data from, actual research or industry. The complete references are available at www.pearsonhighered.com/mathstatresources as well as on page S-1.

Other New Features

We have worked hard to meet the needs of today's students through this revision. In addition to the new content and resources listed above, there are many new features to this 10th edition including **new and enhanced examples, Your Turn** exercises, the inclusion of and instruction for **new technology**, and **new and updated Extended Applications**. You can view these new features in context in the following *Quick Walk-Through of Calculus with Applications, 10e.*

A Quick Walk-Through of *Calculus with Applications, 10e*

3 The Derivative

The population of the United States has been increasing since 1790, when the first census was taken. Over the past few decades, the population has not only been increasing, but the level of diversity has also been increasing. This fact is important to school districts, businesses, and government officials. Using examples in the third section of this chapter, we explore two rates of change related to the increase in minority population. In the first example, we calculate an average rate of change; in the second, we calculate the rate of change at a particular time. This latter rate is an example of a derivative, the subject of this chapter.

◀ **Chapter Opener**

Each chapter opens with a quick introduction that relates to an application presented in the chapter. In addition, a section-level table of contents is included.

Apply It ▶

An **Apply It** question, typically at the start of a section, asks students to consider how to solve a real-life situation related to the math they are about to learn. The **Apply It** question is answered in an application within the section or the exercise set. ("Apply It" was labeled "Think About It" in the previous edition.)

NEW!
Teaching Tips ▶

Teaching Tips are provided in the margins of the Annotated Instructor's Edition for those who are new to teaching this course. In addition, answers to most exercises are provided directly on the exercise set page making it easier to assign and check homework.

Apply It ▶
continued

The solution to the **Apply It** question often falls in the body of the text where it can be seen in context with the mathematics.

3.5 Graphical Differentiation

APPLY IT   Given a graph of the production function, how can we find the graph of the marginal production function?
We will answer this question in Example 1 using graphical differentiation.

In the previous section, we estimated the derivative at various points of a graph by estimating the slope of the tangent line at those points. We will now extend this process to show how to sketch the graph of the derivative given the graph of the original function. This is important because, in many applications, a graph is all we have, and it is easier to find the derivative graphically than to find a formula that fits the graph and take the derivative of that formula.

EXAMPLE 1   **Production of Landscape Mulch**

In Figure 45(a), the graph shows total production (TP), measured in cubic yards of landscape mulch per week, as a function of labor used, measured in workers hired by a small business. The graph in Figure 45(b) shows the marginal production curve (MP_L), which is the derivative of the total production function. Verify that the graph of the marginal production curve (MP_L) is the graph of the derivative of the total production curve (TP).

Teaching Tip: Graphical differentiation gets students to focus on the concept of the derivative rather than the mechanics. This topic is difficult for many students because there are no formulas to rely on. One must thoroughly understand what's going on to do anything. On the other hand, we have seen students who are weak in algebra but who possess a good intuitive grasp of geometry find this topic quite simple.

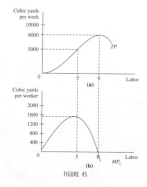

FIGURE 45

APPLY IT

SOLUTION   Let q refer to the quantity of labor. We begin by choosing a point where estimating the derivative of TP is simple. Observe that when $q = 8$, TP has a horizontal tangent line, so its derivative is 0. This explains why the graph of MP_L equals 0 when $q = 8$.

Now, observe in Figure 45(a) that when $q < 8$, the tangent lines of TP have positive slope and the slope is steepest when $q = 5$. This means that the derivative should be positive for $q < 8$ and largest when $q = 5$. Verify that the graph of MP_L has this property.

Finally, as Figure 45(a) shows, the tangent lines of TP have negative slope when $q > 8$, so its derivative, represented by the graph of MP_L, should also be negative there. Verify that the graph of MP_L, in Figure 45(b), has this property as well.

Caution ▶

Caution boxes provide students with a quick "heads-up" to common difficulties and errors.

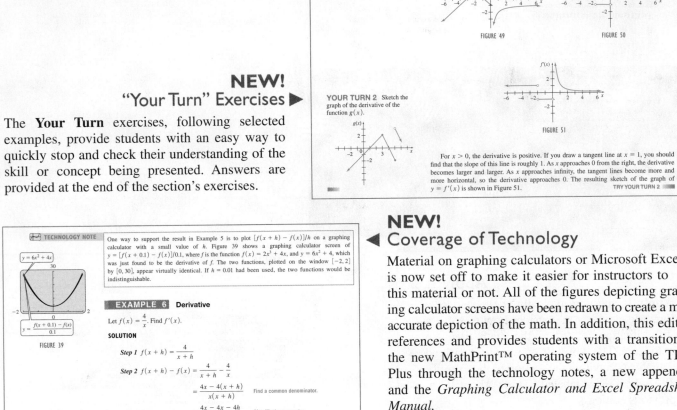

CAUTION Remember that when you graph the derivative, you are graphing the *slope* of the original function. Do not confuse the slope of the original function with the *y*-value of the original function. In fact, the slope of the original function is equal to the *y*-value of its derivative.

Sometimes the original function is not smooth or even continuous, so the graph of the derivative may also be discontinuous.

EXAMPLE 3   Graphing a Derivative

Sketch the graph of the derivative of the function shown in Figure 49.

SOLUTION Notice that when $x < -2$, the slope is 1, and when $-2 < x < 0$, the slope is -1. At $x = -2$, the derivative does not exist due to the sharp corner in the graph. The derivative also does not exist at $x = 0$ because the function is discontinuous there. Using this information, the graph of $f'(x)$ on $x < 0$ is shown in Figure 50.

FIGURE 49

FIGURE 50

YOUR TURN 2 Sketch the graph of the derivative of the function $g(x)$.

FIGURE 51

For $x > 0$, the derivative is positive. If you draw a tangent line at $x = 1$, you should find that the slope of this line is roughly 1. As x approaches 0 from the right, the derivative becomes larger and larger. As x approaches infinity, the tangent lines become more and more horizontal, so the derivative approaches 0. The resulting sketch of the graph of $y = f'(x)$ is shown in Figure 51.   TRY YOUR TURN 2

NEW!
"Your Turn" Exercises ▶

The **Your Turn** exercises, following selected examples, provide students with an easy way to quickly stop and check their understanding of the skill or concept being presented. Answers are provided at the end of the section's exercises.

TECHNOLOGY NOTE One way to support the result in Example 5 is to plot $[f(x + h) - f(x)]/h$ on a graphing calculator with a small value of h. Figure 39 shows a graphing calculator screen of $y = [f(x + 0.1) - f(x)]/0.1$, where f is the function $f(x) = 2x^3 + 4x$, and $y = 6x^2 + 4$, which was just found to be the derivative of f. The two functions, plotted on the window $[-2, 2]$ by $[0, 30]$, appear virtually identical. If $h = 0.01$ had been used, the two functions would be indistinguishable.

$y = 6x^2 + 4x$

$y = \dfrac{f(x + 0.1) - f(x)}{0.1}$

FIGURE 39

EXAMPLE 6   Derivative

Let $f(x) = \dfrac{4}{x}$. Find $f'(x)$.

SOLUTION

Step 1 $f(x + h) = \dfrac{4}{x + h}$

Step 2 $f(x + h) - f(x) = \dfrac{4}{x + h} - \dfrac{4}{x}$

$\quad = \dfrac{4x - 4(x + h)}{x(x + h)}$ Find a common denominator.

$\quad = \dfrac{4x - 4x - 4h}{x(x + h)}$ Simplify the numerator.

$\quad = \dfrac{-4h}{x(x + h)}$

Step 3 $\dfrac{f(x + h) - f(x)}{h} = \dfrac{\dfrac{-4h}{x(x + h)}}{h}$

$\quad = \dfrac{-4h}{x(x + h)} \cdot \dfrac{1}{h}$ Invert and multiply.

$\quad = \dfrac{-4}{x(x + h)}$

Step 4 $f'(x) = \displaystyle\lim_{h \to 0} \dfrac{f(x + h) - f(x)}{h}$

$\quad = \displaystyle\lim_{h \to 0} \dfrac{-4}{x(x + h)}$

NEW!
◀ Coverage of Technology

Material on graphing calculators or Microsoft Excel™ is now set off to make it easier for instructors to use this material or not. All of the figures depicting graphing calculator screens have been redrawn to create a more accurate depiction of the math. In addition, this edition references and provides students with a transition to the new MathPrint™ operating system of the TI-84 Plus through the technology notes, a new appendix, and the *Graphing Calculator and Excel Spreadsheet Manual*.

NEW!
◀ Enhanced Examples

Most learning from a textbook takes place within the examples of the text. The authors have taken advantage of this by adding more detailed annotations to the already well-developed examples to guide students through new concepts and skills.

For Review ▶

For Review boxes are provided in the margin as appropriate, giving students just-in-time help with skills they should already know but may have forgotten. **For Review** comments sometimes include an explanation while others refer students back to earlier parts of the book for a more thorough review.

FOR REVIEW

In Section 1.1, we saw that the equation of a line can be found with the point-slope form $y - y_1 = m(x - x_1)$, if the slope m and the coordinates (x_1, y_1) of a point on the line are known. Use the point-slope form to find the equation of the line with slope 3 that goes through the point $(-1, 4)$.

Let $m = 3$, $x_1 = -1$, $y_1 = 4$. Then

$y - y_1 = m(x - x_1)$
$y - 4 = 3(x - (-1))$
$y - 4 = 3x + 3$
$y = 3x + 7.$

Exercises ▶

Skill-based problems are followed by **application exercises**, which are grouped by subject with subheads indicating the specific topic (e.g. Business and Economics).

Writing exercises, labeled with 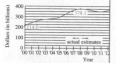 provide students with an opportunity to explain important mathematical ideas.

Technology exercises are labeled with [graph icon] for graphing calculator and [spreadsheet icon] for spreadsheet.

---

a. Find the average rate of change per year of the total amount in the account for the first five years of the investment (from $t = 0$ to $t = 5$).

b. Find the average rate of change per year of the total amount in the account for the second five years of the investment (from $t = 5$ to $t = 10$).

c. Estimate the instantaneous rate of change for $t = 5$.

30. Sales The graph shows annual sales (in thousands of dollars) of a Nintendo game at a particular store. Find the average annual rate of change in sales for the following changes in years.

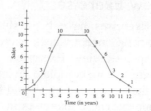

a. 1 to 4   **b.** 4 to 7   **c.** 7 to 12

d. What do your answers for parts a–c tell you about the sales of this product?

e. Give an example of another product that might have such a sales curve.

31. Gasoline Prices In 2008, the price of gasoline in the United States inexplicably spiked and then dropped. The average monthly price (in cents) per gallon of unleaded regular gasoline for 2008 is shown in the following chart. Find the average rate of change per month in the average price per gallon for each time period. *Source: U.S. Energy Information Administration.*

a. From January to July (the peak)

b. From July to December

c. From January to December

32. Medicare Trust Fund The graph shows in the Medicare Trust Fund at the end of th *Social Security Administration.*

Medicare Trust Fund

[graph: 378.1, 214.0, actual estimates, Year '00 '01 '02 '03 '04 '05 '06 '07 '08 '09 '10 '11 '12, Dollars in billions 400 300 200 100 0]

Using the Consumer Price Index for Urban Wage Earners and Clerical Wo

Find the approximate average rate of cha for each time period.

a. From 2000 to 2008 (the peak)

b. From 2008 to 2018

Life Sciences

33. Flu Epidemic Epidemiologists in Col estimate that t days after the flu begins t percent of the population infected by the f

$$p(t) = t^2 + t$$

for $0 \le t \le 5$.

a. Find the average rate of change of p with respect to t over the interval from 1 to 4 days.

b. Find and interpret the instantaneous rate of change of p with respect to t at $t = 3$.

34. World Population Growth The future size of the world population depends on how soon it reaches replacement-level fertility, the point at which each woman bears on average about 2.1 children. The graph shows projections for reaching that

Ultimate World Population Size Under Different Assumptions

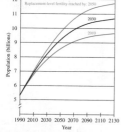

3.3 EXERCISES

Find the average rate of change for each function over the given interval.

1. $y = x^2 + 2x$ between $x = 1$ and $x = 3$

2. $y = -4x^2 - 6$ between $x = 2$ and $x = 6$

3. $y = -3x^3 + 2x^2 - 4x + 1$ between $x = -2$ and $x = 1$

4. $y = 2x^3 - 4x^2 + 6x$ between $x = -1$ and $x = 4$

5. $y = \sqrt{x}$ between $x = 1$ and $x = 4$

6. $y = \sqrt{3x - 2}$ between $x = 1$ and $x = 2$

7. $y = e^x$ between $x = -2$ and $x = 0$

8. $y = \ln x$ between $x = 2$ and $x = 4$

Suppose the position of an object moving in a straight line is given by $s(t) = t^2 + 5t + 2$. Find the instantaneous velocity at each time.

9. $t = 6$ **10.** $t = 1$

Suppose the position of an object moving in a straight line is given by $s(t) = 5t^2 - 2t - 7$. Find the instantaneous velocity at each time.

11. $t = 2$ **12.** $t = 3$

Suppose the position of an object moving in a straight line is given by $s(t) = t^3 + 2t + 9$. Find the instantaneous velocity at each time.

13. $t = 1$ **14.** $t = 4$

Find the instantaneous rate of change for each function at the given value.

15. $f(x) = x^2 + 2x$ at $x = 0$

16. $s(t) = -4t^2 - 6$ at $t = 2$

17. $g(t) = 1 - t^2$ at $t = -1$

18. $F(x) = x^2 + 2$ at $x = 0$

[icon] Use the formula for instantaneous rate of change, approximating the limit by using smaller and smaller values of h, to find the instantaneous rate of change for each function at the given value.

19. $f(x) = x^x$ at $x = 2$ **20.** $f(x) = x^x$ at $x = 3$

21. $f(x) = x^{\ln x}$ at $x = 2$ **22.** $f(x) = x^{\ln x}$ at $x = 3$

23. Explain the difference between the average rate of change of y as x changes from a to b, and the instantaneous rate of change of y at $x = a$.

24. If the instantaneous rate of change of $f(x)$ with respect to x is positive when $x = 1$, is f increasing or decreasing there?

APPLICATIONS

Business and Economics

25. Profit Suppose that the total profit in hundreds of dollars from selling x items is given by

$$P(x) = 2x^2 - 5x + 6.$$

Find the average rate of change of profit for the following changes in x.

a. 2 to 4   **b.** 2 to 3

c. Find and interpret the instantaneous rate of change of profit with respect to the number of items produced when $x = 2$. (This number is called the *marginal profit* at $x = 2$.)

d. Find the marginal profit at $x = 4$.

26. Revenue The revenue (in thousands of dollars) from producing x units of an item is

$$R(x) = 10x - 0.002x^2.$$

a. Find the average rate of change of revenue when production is increased from 1000 to 1001 units.

b. Find and interpret the instantaneous rate of change of revenue with respect to the number of items produced when 1000 units are produced. (This number is called the *marginal revenue* at $x = 1000$.)

c. Find the additional revenue if production is increased from 1000 to 1001 units.

d. Compare your answers for parts a and c. What do you find? How do these answers compare with your answer to part b?

27. Demand Suppose customers in a hardware store are willing to buy $N(p)$ boxes of nails at p dollars per box, as given by

$$N(p) = 80 - 5p^2, \quad 1 \le p \le 4.$$

a. Find the average rate of change of demand for a change in price from \$2 to \$3.

b. Find and interpret the instantaneous rate of change of demand when the price is \$2.

c. Find the instantaneous rate of change of demand when the price is \$3.

d. As the price is increased from \$2 to \$3, how is demand changing? Is the change to be expected? Explain.

28. Interest If \$1000 is invested in an account that pays 5% compounded annually, the total amount, $A(t)$, in the account after t years is

$$A(t) = 1000(1.05)^t.$$

a. Find the average rate of change per year of the total amount in the account for the first five years of the investment (from $t = 0$ to $t = 5$).

b. Find the average rate of change per year of the total amount in the account for the second five years of the investment (from $t = 5$ to $t = 10$).

c. Estimate the instantaneous rate of change for $t = 5$.

29. Interest If \$1000 is invested in an account that pays 5% compounded continuously, the total amount, $A(t)$, in the account after t years is

$$A(t) = 1000e^{0.05t}.$$

Connection exercises integrate topics presented in different sections or chapters and are indicated with ⊞.

Exercises that are particularly **challenging** are denoted with + in the Annotated Instructor's Edition only.

---

of the derivative to find the derivative of the

56. $y = 5x^2 - 6x + 7$

d 58, find the derivative of the function at a) by approximating the definition of the all values of h and (b) by using a graphing in on the function until it appears to be a hen finding the slope of that line.

; $x_0 = 3$ **58.** $f(x) = x^{\ln x}$; $x_0 = 2$

of the derivative for each function shown.

[graph $f(x)$]

[graph $f(x)$]

61. Let f and g be differentiable functions such that

$$\lim_{x \to \infty} f(x) = c$$
$$\lim_{x \to \infty} g(x) = d$$

where $c \ne d$. Determine

$$\lim_{x \to \infty} \frac{cf(x) - dg(x)}{f(x) - g(x)}.$$

(Choose one of the following.) *Source: Society of Actuaries.*

a. 0   **b.** $\dfrac{cf'(0) - dg'(0)}{f'(0) - g'(0)}$

c. $f'(0) - g'(0)$ **d.** $c - d$ **e.** $c + d$

APPLICATIONS

Business and Economics

62. Revenue Waverly Products has found that its revenue is related to advertising expenditures by the function

---

⊞ 63. Cost Analysis A company charges \$1.50 per lb when a certain chemical is bought in lots of 125 lb or less, with a price per pound of \$1.35 if more than 125 lb are purchased. Let $C(x)$ represent the cost of x lb. Find the cost for the following numbers of pounds.

a. 100   **b.** 125   **c.** 140   **d.** Graph $y = C(x)$.

e. Where is C discontinuous?

Find the average cost per pound if the following number of pounds are bought.

f. 100   **g.** 125   **h.** 140

Find and interpret the marginal cost (that is, the instantaneous rate of change of the cost) for the following numbers of pounds.

i. 100   **j.** 140

⊞ 64. Marginal Analysis Suppose the profit (in cents) from selling x lb of potatoes is given by

$$P(x) = 15x + 25x^2.$$

Find the average rate of change in profit from selling each of the following amounts.

a. 6 lb to 7 lb   **b.** 6 lb to 6.5 lb   **c.** 6 lb to 6.1 lb

Find the marginal profit (that is, the instantaneous rate of change of the profit) from selling the following amounts.

d. 6 lb   **e.** 20 lb   **f.** 30 lb

g. What is the domain of x?

h. Is it possible for the marginal profit to be negative here? What does this mean?

i. Find the average profit function. (Recall that average profit is given by total profit divided by the number produced, or $\overline{P}(x) = P(x)/x$.)

j. Find the marginal average profit function (that is, the function giving the instantaneous rate of change of the average profit function).

k. Is it possible for the marginal average profit to vary here? What does this mean?

l. Discuss whether this function describes a realistic situation.

⊞ 65. Average Cost The graph on the next page shows the total cost $C(x)$ to produce x tons of cement. (Recall that average cost is given by total cost divided by the number produced, or $\overline{C}(x) = C(x)/x$.)

End-of-Chapter Summary ▶

End-of-Chapter Summary provides students with a quick summary of the key ideas of the chapter followed by a list of key definitions, terms, and examples.

---

3 CHAPTER REVIEW

SUMMARY

In this chapter we introduced the ideas of limit and continuity of functions and then used these ideas to explore calculus. We saw that the difference quotient can represent

- the average rate of change,
- the slope of the secant line, and
- the average velocity.

We also learned how to estimate the value of the derivative using graphical differentiation. In the next chapter, we will take a closer look at the definition of the derivative to develop a set of rules to

We saw that the derivative can represent

- the instantaneous rate of change,
- the slope of the tangent line, and
- the instantaneous velocity.

quickly and easily calculate the derivative of a wide range of functions without the need to directly apply the definition of the derivative each time.

Limit of a Function Let f be a function and let a and L be real numbers. If

1. as x takes values closer and closer (but not equal) to a on both sides of a, the corresponding values of $f(x)$ get closer and closer (and perhaps equal) to L; and
2. the value of $f(x)$ can be made as close to L as desired by taking values of x close enough to a;

then L is the limit of $f(x)$ as x approaches a, written

$$\lim_{x \to a} f(x) = L.$$

ce of Limits The limit of f as x approaches a may not exist.

1. If $f(x)$ becomes infinitely large in magnitude (positive or negative) as x approaches the number a from either side, we write $\lim_{x \to a} f(x) = \infty$ or $\lim_{x \to a} f(x) = -\infty$. In either case, the limit does not exist.
2. If $f(x)$ becomes infinitely large in magnitude (positive) as x approaches a from one side and infinitely large in magnitude (negative) as x approaches a from the other side, then $\lim_{x \to a} f(x)$ does not exist.

KEY TERMS

To understand the concepts presented in this chapter, you should know the meaning and use of the following terms.
For easy reference, the section in the chapter where a word (or expression) was first used is provided.

3.1
limit
limit from the left/right
one-/two-sided limit
piecewise function
limit at infinity

3.2
continuous
discontinuous
removable discontinuity
continuous on an open/closed
  interval
continuous from the right/left
Intermediate Value Theorem

3.3
average rate of change
difference quotient
instantaneous rate of change
velocity

3.4
secant line
tangent line
slope of the curve
derivative
differentiable
differentiation

REVIEW EXERCISES

CONCEPT CHECK

Determine whether each of the following statements is true or false, and explain why.

1. The limit of a product is the product of the limits when each of the limits exists.
2. The limit of a function may not exist at a point even though the function is defined there.
3. If a rational function has a polynomial in the denominator of higher degree than the polynomial in the numerator, then the limit at infinity must equal zero.
4. If the limit of a function exists at a point, then the function is continuous there.
5. A polynomial function is continuous everywhere.
6. A rational function is continuous everywhere.
7. The derivative gives the average rate of change of a function.
8. The derivative gives the instantaneous rate of change of a function.
9. The instantaneous rate of change is a limit.
10. The derivative is a function.
11. The slope of the tangent line gives the average rate of change.
12. The derivative of a function exists wherever the function is continuous.

PRACTICE AND EXPLORATIONS

13. Is a derivative always a limit? Is a limit always a derivative? Explain.
14. Is every continuous function differentiable? Is every differentiable function continuous? Explain.
15. Describe how to tell when a function is discontinuous at the real number $x = a$.
16. Give two applications of the derivative

$$f'(x) = \lim_{h \to 0} \frac{f(x + h) - f(x)}{h}.$$

Decide whether the limits in Exercises 17–34 exist. If a limit exists, find its value.

17. **a.** $\lim_{x \to -3^-} f(x)$ **b.** $\lim_{x \to -3^+} f(x)$ **c.** $\lim_{x \to -3} f(x)$ **d.** $f(-3)$

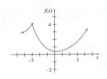

18. **a.** $\lim_{x \to -1^-} g(x)$ **b.** $\lim_{x \to -1^+} g(x)$ **c.** $\lim_{x \to -1} g(x)$ **d.** $g(-1)$

19. **a.** $\lim_{x \to 4^-} f(x)$ **b.** $\lim_{x \to 4^+} f(x)$ **c.** $\lim_{x \to 4} f(x)$ **d.** $f(4)$

◀ Chapter Review Exercises

Chapter Review Exercises have been slightly reorganized so that the Concept Check exercises fall within the Chapter Review Exercises. This provides students with a more complete review of both the skills and the concepts they should have mastered in this chapter. These exercises in their entirety provide a comprehensive review for a chapter-level exam.

EXTENDED APPLICATION

A MODEL FOR DRUGS ADMINISTERED INTRAVENOUSLY

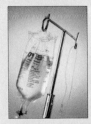

When a drug is administered intravenously it enters the bloodstream immediately, producing an immediate effect for the patient. The drug can be either given as a single rapid injection or given at a constant drip rate. The latter is commonly referred to as an intravenous (IV) infusion. Common drugs administered intravenously include morphine for pain, diazepam (or Valium) to control a seizure, and digoxin for heart failure.

SINGLE RAPID INJECTION

With a single rapid injection, the amount of drug in the bloodstream reaches its peak immediately and then the body eliminates the drug exponentially. The larger the amount of drug there is in the body, the faster the body eliminates it. If a lesser amount of drug is in the body, it is eliminated more slowly.

Since the half-life of this drug is 4 hours,

$$k = -\frac{\ln 2}{4} \approx -0.17.$$

Therefore, the model is

$$A(t) = 35e^{-0.17t}.$$

The graph of $A(t)$ is given in Figure 55.

Rapid IV Injection

$A(t) = 35e^{-0.17t}$

Hours since dose was administered

◀ Extended Applications

Extended Applications are provided now at the end of *every* chapter as in-depth applied exercises to help stimulate student interest. These activities can be completed individually or as a group project.

Supplements

STUDENT RESOURCES	**INSTRUCTOR RESOURCES**

Student Edition

- ISBN 0-321-74900-6 / 978-0-321-74900-0

Student's Solutions Manual

- Provides detailed solutions to all odd-numbered text exercises and sample chapter tests with answers.
- Authored by Elka Block and Frank Purcell
- ISBN 0-321-75790-4 / 978-0-321-75790-6

Graphing Calculator and Excel Spreadsheet Manual

- Provides instructions and keystroke operations for the TI-83/84 Plus, the TI-84 Plus with the new operating system featuring MathPrint™, and the TI-89 as well as for the Excel spreadsheet program.
- Authored by GEX Publishing Services
- ISBN 0-321-70966-7 / 978-0-321-70966-0

Video Lectures on DVD-ROM with Optional Captioning

- Complete set of digitized videos, with extensive section coverage, for student use at home or on campus
- Ideal for distance learning or supplemental instruction
- ISBN 0-321-74612-0 / 978-0-321-74612-2

Supplementary Content

- Additional Extended Applications
- Comprehensive source list
- Available at the Downloadable Student Resources site, www.pearsonhighered.com/mathstatsresources, and to qualified instructors within MyMathLab or through the Pearson Instructor Resource Center, www.pearsonhighered.com/irc

Annotated Instructor's Edition

- Numerous teaching tips
- Includes all the answers, usually on the same page as the exercises, for quick reference
- ISBN 0-321-73329-0 / 978-0-321-73329-0
- More challenging exercises are indicated with a + symbol

Instructor's Resource Guide and Solutions Manual (download only)

- Provides complete solutions to all exercises, two versions of a pre-test and final exam, and teaching tips.
- Authored by Elka Block and Frank Purcell
- Available to qualified instructors within MyMathLab or through the Pearson Instructor Resource Center, www.pearsonhighered.com/irc

PowerPoint Lecture Presentation

- Newly revised and greatly improved
- Classroom presentation slides are geared specifically to the sequence and philosophy of this textbook.
- Includes lecture content and key graphics from the book
- Available to qualified instructors within MyMathLab or through the Pearson Instructor Resource Center, www.pearsonhighered.com/irc
- Authored by Dr. Sharda K. Gudehithlu, Wilbur Wright College

Media Resources www.mymathlab.com

MyMathLab® Online Course (access code required)
www.mymathlab.com

MyMathLab delivers **proven results** in helping individual students succeed.

- MyMathLab has a consistently positive impact on the quality of learning in higher education math instruction. MyMathLab can be successfully implemented in any environment—lab-based, hybrid, fully online, traditional—and demonstrates the quantifiable difference that integrated usage has on student retention, subsequent success, and overall achievement.

- MyMathLab's comprehensive online gradebook automatically tracks your students' results on tests, quizzes, homework, and in the study plan. You can use the gradebook to quickly intervene if your students have trouble, or to provide positive feedback on a job well done. The data within MyMathLab is easily exported to a variety of spreadsheet programs, such as Microsoft Excel. You can determine which points of data you want to export, and then analyze the results to determine success.

MyMathLab provides **engaging experiences** that personalize, stimulate, and measure learning for each student.

- **Tutorial Exercises:** The homework and practice exercises in MyMathLab and MyStatLab are correlated to the exercises in the textbook, and they regenerate algorithmically to give students unlimited opportunity for practice and mastery. The software offers immediate, helpful feedback when students enter incorrect answers.

- **Multimedia Learning Aids:** Exercises include guided solutions, sample problems, animations, videos, and eText clips for extra help at point-of-use.

- **Expert Tutoring:** Although many students describe the whole of MyMathLab as "like having your own personal tutor," students using MyMathLab and MyStatLab do have access to live tutoring from Pearson, from qualified math and statistics instructors who provide tutoring sessions for students via MyMathLab and MyStatLab.

And, MyMathLab comes from a **trusted partner** with educational expertise and an eye on the future.

Knowing that you are using a Pearson product means knowing that you are using quality content. That means that our eTexts are accurate, that our assessment tools work, and that our questions are error-free. And whether you are just getting started with MyMathLab, or have a question along the way, we're here to help you learn about our technologies and how to incorporate them into your course.

To learn more about how MyMathLab combines proven learning applications with powerful assessment, visit **www.mymathlab.com** or contact your Pearson representative.

Math XL MathXL® Online Course (access code required)
www.mathxl.com

MathXL® is the homework and assessment engine that runs MyMathLab. (MyMathLab is MathXL plus a learning management system.) With MathXL, instructors can:

- Create, edit, and assign online homework and tests using algorithmically generated exercises correlated at the objective level to the textbook.

- Create and assign their own online exercises and import TestGen tests for added flexibility.

- Maintain records of all student work tracked in MathXL's online gradebook.

With MathXL, students can:

- Take chapter tests in MathXL and receive personalized study plans and/or personalized homework assignments based on their test results.

- Use the study plan and/or the homework to link directly to tutorial exercises for the objectives they need to study.

- Access supplemental animations and video clips directly from selected exercises.

MathXL is available to qualified adopters. For more information, visit our website at **www.mathxl.com**, or contact your Pearson representative.

InterAct Math Tutorial Website
www.interactmath.com

Get practice and tutorial help online! This interactive tutorial website provides algorithmically generated practice exercises that correlate directly to the exercises in the textbook. Students can retry an exercise as many times as they would like with new values each time for unlimited practice and mastery. Every exercise is accompanied by an interactive guided solution that provides helpful feedback for incorrect answers, and students can view a worked-out sample problem that guides them through an exercise similar to the one in which they're working.

TestGen®
www.pearsoned.com/testgen

TestGen® enables instructors to build, edit, print, and administer tests using a computerized bank of questions developed to cover all the objectives of the text. TestGen is algorithmically based, allowing instructors to create multiple but equivalent versions of the same question or test with the click of a button. Instructors can also modify test bank questions or add new questions. The software and testbank are available for download from Pearson Education's online catalog.

Acknowledgments

We wish to thank the following professors for their contributions in reviewing portions of this text:

John Alford, *Sam Houston State University*
Robert David Borgersen, *University of Manitoba*
C.T. Bruns, *University of Colorado, Boulder*
Nurit Budinsky, *University of Massachusetts—Dartmouth*
Martha Morrow Chalhoub, *Collin College*
Karabi Dattta, *Northern Illinois University*
James "Rob" Ely, *Blinn College—Bryan Campus*
Sam Evers, *The University of Alabama*
Chris Ferbrache, *Fresno City College*
Pete Gomez, *Houston Community College, Northwest*
Sharda K. Gudehithlu, *Wilbur Wright College*
Mary Beth Headlee, *State College of Florida*
David L. Jones, *University of Kansas*
Karla Karstens, *University of Vermont*
Lynette J. King, *Gadsden State Community College*
Jason Knapp, *University of Virginia*
Mark C. Lammers, *University of North Carolina, Wilmington*
Rebecca E. Lynn, *Colorado State University*
Rodolfo Maglio, *Northeastern Illinois University*
Cyrus Malek, *Collin College*
Javad Namazi, *Fairleigh Dickinson University*
Dana Nimic, *Southeast Community College—Lincoln*
Lisa Nix, *Shelton State Community College*
Sam Northshield, *SUNY, Plattsburgh*
Susan Ojala, *University of Vermont*
Brooke Quinlan, *Hillsborough Community College*
Candace Rainer, *Meridian Community College*
Theresa Rushing, *The University of Tennessee at Martin*
Katherine E. Schultz, *Pensacola Junior College*
Barbara Dinneen Sehr, *Indiana University, Kokomo*
Gordon H. Shumard, *Kennesaw State University*
Walter Sizer, *Minnesota State University, Moorhead*
Jennifer Strehler, *Oakton Community College*
Antonis P. Stylianou, *University of Missouri—Kansas City*

Jason Terry, *Central New Mexico Community College*
Yan Tian, *Palomar College*
Sara Van Asten, *North Hennepin Community College*
Amanda Wheeler, *Amarillo College*
Roger Zarnowski, *Angelo State University*

We also thank Elka Block and Frank Purcell of Twin Prime Editorial for doing an excellent job updating the *Student's Solutions Manual* and *Instructor's Resource Guide and Solutions Manual*, an enormous and time-consuming task. Further thanks go to our accuracy checkers Nathan Kidwell, John Samons, and Lauri Semarne. We are very thankful for the work of William H. Kazez, Theresa Laurent, and Richard McCall, in writing Extended Applications for the book. We are grateful to Karla Harby and Mary Ann Ritchey for their editorial assistance. We especially appreciate the staff at Pearson, whose contributions have been very important in bringing this project to a successful conclusion.

Margaret L. Lial
Raymond N. Greenwell
Nathan P. Ritchey

Dear Student,

Hello! The fact that you're reading this preface is good news. One of the keys to success in a math class is to read the book. Another is to answer all the questions correctly on your professor's tests. You've already started doing the first; doing the second may be more of a challenge, but by reading this book and working out the exercises, you'll be in a much stronger position to ace the tests. One last essential key to success is to go to class and actively participate.

You'll be happy to discover that we've provided the answers to the odd-numbered exercises in the back of the book. As you begin the exercises, you may be tempted to immediately look up the answer in the back of the book, and then figure out how to get that answer. It is an easy solution that has a consequence—you won't learn to do the exercises without that extra hint. Then, when you take a test, you will be forced to answer the questions without knowing what the answer is. Believe us, this is a lot harder! The learning comes from figuring out the exercises. Once you have an answer, look in the back and see if your answer agrees with ours. If it does, you're on the right path. If it doesn't, try to figure out what you did wrong. Once you've discovered your error, continue to work out more exercises to master the concept and skill.

Equations are a mathematician's way of expressing ideas in concise shorthand. The problem in reading mathematics is unpacking the shorthand. One useful technique is to read with paper and pencil in hand so you can work out calculations as you go along. When you are baffled, and you wonder, "How did they get that result?" try doing the calculation yourself and see what you get. You'll be amazed (or at least mildly satisfied) at how often that answers your question. Remember, math is not a spectator sport. You don't learn math by passively reading it or watching your professor. You learn mathematics by doing mathematics.

Finally, if there is anything you would like to see changed in the book, feel free to write to us at matrng@hofstra.edu or npritchey@ysu.edu. We're constantly trying to make this book even better. If you'd like to know more about us, we have Web sites that we invite you to visit: http://people.hofstra.edu/rgreenwell and http://people.ysu.edu/~npritchey.

Marge Lial
Ray Greenwell
Nate Ritchey

Prerequisite Skills Diagnostic Test

Below is a very brief test to help you recognize which, if any, prerequisite skills you may need to remediate in order to be successful in this course. After completing the test, check your answers in the back of the book. In addition to the answers, we have also provided the solutions to these problems in Appendix A. These solutions should help remind you how to solve the problems. For problems 5-26, the answers are followed by references to sections within Chapter R where you can find guidance on how to solve the problem and/or additional instruction. Addressing any weak prerequisite skills now will make a positive impact on your success as you progress through this course.

1. What percent of 50 is 10?

2. Simplify $\dfrac{13}{7} - \dfrac{2}{5}$.

3. Let x be the number of apples and y be the number of oranges. Write the following statement as an algebraic equation: "The total number of apples and oranges is 75."

4. Let s be the number of students and p be the number of professors. Write the following statement as an algebraic equation: "There are at least four times as many students as professors."

5. Solve for k: $7k + 8 = -4(3 - k)$.

6. Solve for x: $\dfrac{5}{8}x + \dfrac{1}{16}x = \dfrac{11}{16} + x$.

7. Write in interval notation: $-2 < x \le 5$.

8. Using the variable x, write the following interval as an inequality: $(-\infty, -3]$.

9. Solve for y: $5(y - 2) + 1 \le 7y + 8$.

10. Solve for p: $\dfrac{2}{3}(5p - 3) > \dfrac{3}{4}(2p + 1)$.

11. Carry out the operations and simplify: $(5y^2 - 6y - 4) - 2(3y^2 - 5y + 1)$.

12. Multiply out and simplify $(r^2 - 2r + 3)(r + 1)$.

13. Multiply out and simplify $(a - 2b)^2$.

14. Factor $3pq + 6p^2q + 9pq^2$.

15. Factor $3x^2 - x - 10$.

16. Perform the operation and simplify: $\dfrac{a^2 - 6a}{a^2 - 4} \cdot \dfrac{a - 2}{a}$.

17. Perform the operation and simplify: $\dfrac{x + 3}{x^2 - 1} + \dfrac{2}{x^2 + x}$.

18. Solve for x: $3x^2 + 4x = 1$.

19. Solve for z: $\dfrac{8z}{z + 3} \le 2$.

20. Simplify $\dfrac{4^{-1}(x^2 y^3)^2}{x^{-2} y^5}$.

21. Simplify $\dfrac{4^{1/4}(p^{2/3} q^{-1/3})^{-1}}{4^{-1/4}\, p^{4/3} q^{4/3}}$.

22. Simplify as a single term without negative exponents: $k^{-1} - m^{-1}$.

23. Factor $(x^2 + 1)^{-1/2}(x + 2) + 3(x^2 + 1)^{1/2}$.

24. Simplify $\sqrt[3]{64b^6}$.

25. Rationalize the denominator: $\dfrac{2}{4 - \sqrt{10}}$.

26. Simplify $\sqrt{y^2 - 10y + 25}$.

R

Algebra Reference

In this chapter, we will review the most important topics in algebra. Knowing algebra is a fundamental prerequisite to success in higher mathematics. This algebra reference is designed for self-study; study it all at once or refer to it when needed throughout the course. Since this is a review, answers to all exercises are given in the answer section at the back of the book.

R.1 Polynomials

An expression such as $9p^4$ is a **term**; the number 9 is the **coefficient**, p is the **variable**, and 4 is the **exponent**. The expression p^4 means $p \cdot p \cdot p \cdot p$, while p^2 means $p \cdot p$, and so on. Terms having the same variable and the same exponent, such as $9x^4$ and $-3x^4$, are **like terms**. Terms that do not have both the same variable and the same exponent, such as m^2 and m^4, are **unlike terms**.

   A **polynomial** is a term or a finite sum of terms in which all variables have whole number exponents, and no variables appear in denominators. Examples of polynomials include

$$5x^4 + 2x^3 + 6x, \qquad 8m^3 + 9m^2n - 6mn^2 + 3n^3, \qquad 10p, \qquad \text{and} \qquad -9.$$

Order of Operations
Algebra is a language, and you must be familiar with its rules to correctly interpret algebraic statements. The following order of operations have been agreed upon through centuries of usage.

- Expressions in **parentheses** are calculated first, working from the inside out. The numerator and denominator of a fraction are treated as expressions in parentheses.
- **Powers** are performed next, going from left to right.
- **Multiplication** and **division** are performed next, going from left to right.
- **Addition** and **subtraction** are performed last, going from left to right.

For example, in the expression $(6(x + 1)^2 + 3x - 22)^2$, suppose x has the value of 2. We would evaluate this as follows:

$$
\begin{aligned}
(6(2 + 1)^2 + 3(2) - 22)^2 &= (6(3)^2 + 3(2) - 22)^2 & \text{Evaluate the expression in the innermost parentheses.}\\
&= (6(9) + 3(2) - 22)^2 & \text{Evaluate 3 raised to a power.}\\
&= (54 + 6 - 22)^2 & \text{Perform the multiplication.}\\
&= (38)^2 & \text{Perform the addition and subtraction from left to right.}\\
&= 1444 & \text{Evaluate the power.}
\end{aligned}
$$

In the expression $\dfrac{x^2 + 3x + 6}{x + 6}$, suppose x has the value of 2. We would evaluate this as follows:

$$
\frac{2^2 + 3(2) + 6}{2 + 6} = \frac{16}{8} \qquad \text{Evaluate the numerator and the denominator.}
$$

$$
= 2 \qquad \text{Simplify the fraction.}
$$

Adding and Subtracting Polynomials
The following properties of real numbers are useful for performing operations on polynomials.

Properties of Real Numbers

For all real numbers a, b, and c:

1. $a + b = b + a$;   **Commutative properties**
   $ab = ba$;

2. $(a + b) + c = a + (b + c)$;   **Associative properties**
   $(ab)c = a(bc)$;

3. $a(b + c) = ab + ac$.   **Distributive property**

EXAMPLE 1 Properties of Real Numbers

(a) $2 + x = x + 2$ Commutative property of addition

(b) $x \cdot 3 = 3x$ Commutative property of multiplication

(c) $(7x)x = 7(x \cdot x) = 7x^2$ Associative property of multiplication

(d) $3(x + 4) = 3x + 12$ Distributive property

One use of the distributive property is to add or subtract polynomials. Only like terms may be added or subtracted. For example,

$$12y^4 + 6y^4 = (12 + 6)y^4 = 18y^4,$$

and

$$-2m^2 + 8m^2 = (-2 + 8)m^2 = 6m^2,$$

but the polynomial $8y^4 + 2y^5$ cannot be further simplified. To subtract polynomials, we use the facts that $-(a + b) = -a - b$ and $-(a - b) = -a + b$. In the next example, we show how to add and subtract polynomials.

EXAMPLE 2 Adding and Subtracting Polynomials

Add or subtract as indicated.

(a) $(8x^3 - 4x^2 + 6x) + (3x^3 + 5x^2 - 9x + 8)$

    SOLUTION Combine like terms.

$$(8x^3 - 4x^2 + 6x) + (3x^3 + 5x^2 - 9x + 8)$$
$$= (8x^3 + 3x^3) + (-4x^2 + 5x^2) + (6x - 9x) + 8$$
$$= 11x^3 + x^2 - 3x + 8$$

(b) $2(-4x^4 + 6x^3 - 9x^2 - 12) + 3(-3x^3 + 8x^2 - 11x + 7)$

    SOLUTION Multiply each polynomial by the coefficient in front of the polynomial, and then combine terms as before.

$$2(-4x^4 + 6x^3 - 9x^2 - 12) + 3(-3x^3 + 8x^2 - 11x + 7)$$
$$= -8x^4 + 12x^3 - 18x^2 - 24 - 9x^3 + 24x^2 - 33x + 21$$
$$= -8x^4 + 3x^3 + 6x^2 - 33x - 3$$

(c) $(2x^2 - 11x + 8) - (7x^2 - 6x + 2)$

YOUR TURN 1 Perform the operation $3(x^2 - 4x - 5) - 4(3x^2 - 5x - 7)$.

    SOLUTION Distributing the minus sign and combining like terms yields

$$(2x^2 - 11x + 8) + (-7x^2 + 6x - 2)$$
$$= -5x^2 - 5x + 6.$$ **TRY YOUR TURN 1**

Multiplying Polynomials
The distributive property is also used to multiply polynomials, along with the fact that $a^m \cdot a^n = a^{m+n}$. For example,

$$x \cdot x = x^1 \cdot x^1 = x^{1+1} = x^2 \qquad \text{and} \qquad x^2 \cdot x^5 = x^{2+5} = x^7.$$

EXAMPLE 3 Multiplying Polynomials

Multiply.

(a) $8x(6x - 4)$

    SOLUTION Using the distributive property yields

$$8x(6x - 4) = 8x(6x) - 8x(4)$$
$$= 48x^2 - 32x.$$

(b) $(3p - 2)(p^2 + 5p - 1)$

SOLUTION   Using the distributive property yields

$$(3p - 2)(p^2 + 5p - 1)$$
$$= 3p(p^2 + 5p - 1) - 2(p^2 + 5p - 1)$$
$$= 3p(p^2) + 3p(5p) + 3p(-1) - 2(p^2) - 2(5p) - 2(-1)$$
$$= 3p^3 + 15p^2 - 3p - 2p^2 - 10p + 2$$
$$= 3p^3 + 13p^2 - 13p + 2.$$

(c) $(x + 2)(x + 3)(x - 4)$

SOLUTION   Multiplying the first two polynomials and then multiplying their product by the third polynomial yields

$$(x + 2)(x + 3)(x - 4)$$
$$= [(x + 2)(x + 3)](x - 4)$$
$$= (x^2 + 2x + 3x + 6)(x - 4)$$
$$= (x^2 + 5x + 6)(x - 4)$$
$$= x^3 + 5x^2 + 6x - 4x^2 - 20x - 24$$
$$= x^3 + x^2 - 14x - 24. \qquad \textsf{TRY YOUR TURN 2}$$

YOUR TURN 2 Perform the operation $(3y + 2)(4y^2 - 2y - 5)$.

A **binomial** is a polynomial with exactly two terms, such as $2x + 1$ or $m + n$. When two binomials are multiplied, the FOIL method (First, Outer, Inner, Last) is used as a memory aid.

EXAMPLE 4   Multiplying Polynomials

Find $(2m - 5)(m + 4)$ using the FOIL method.

SOLUTION

$$
(2m - 5)(m + 4) = \overset{\text{F}}{(2m)(m)} + \overset{\text{O}}{(2m)(4)} + \overset{\text{I}}{(-5)(m)} + \overset{\text{L}}{(-5)(4)}
$$
$$= 2m^2 + 8m - 5m - 20$$
$$= 2m^2 + 3m - 20$$

EXAMPLE 5   Multiplying Polynomials

Find $(2k - 5m)^3$.

SOLUTION   Write $(2k - 5m)^3$ as $(2k - 5m)(2k - 5m)(2k - 5m)$. Then multiply the first two factors using FOIL.

$$(2k - 5m)(2k - 5m) = 4k^2 - 10km - 10km + 25m^2$$
$$= 4k^2 - 20km + 25m^2$$

Now multiply this last result by $(2k - 5m)$ using the distributive property, as in Example 3(b).

$$(4k^2 - 20km + 25m^2)(2k - 5m)$$
$$= 4k^2(2k - 5m) - 20km(2k - 5m) + 25m^2(2k - 5m)$$
$$= 8k^3 - 20k^2m - 40k^2m + 100km^2 + 50km^2 - 125m^3$$
$$= 8k^3 - 60k^2m + 150km^2 - 125m^3 \quad \textsf{Combine like terms.}$$

Notice in the first part of Example 5, when we multiplied $(2k - 5m)$ by itself, that the product of the square of a binomial is the square of the first term, $(2k)^2$, plus twice the product of the two terms, $(2)(2k)(-5m)$, plus the square of the last term, $(-5k)^2$.

| CAUTION | Avoid the common error of writing $(x + y)^2 = x^2 + y^2$. As the first step of Example 5 shows, the square of a binomial has three terms, so

$$(x + y)^2 = x^2 + 2xy + y^2.$$

Furthermore, higher powers of a binomial also result in more than two terms. For example, verify by multiplication that

$$(x + y)^3 = x^3 + 3x^2y + 3xy^2 + y^3.$$

Remember, for any value of $n \neq 1$,

$$(x + y)^n \neq x^n + y^n.$$

R.1 EXERCISES

Perform the indicated operations.

1. $(2x^2 - 6x + 11) + (-3x^2 + 7x - 2)$

2. $(-4y^2 - 3y + 8) - (2y^2 - 6y - 2)$

3. $-6(2q^2 + 4q - 3) + 4(-q^2 + 7q - 3)$

4. $2(3r^2 + 4r + 2) - 3(-r^2 + 4r - 5)$

5. $(0.613x^2 - 4.215x + 0.892) - 0.47(2x^2 - 3x + 5)$

6. $0.5(5r^2 + 3.2r - 6) - (1.7r^2 - 2r - 1.5)$

7. $-9m(2m^2 + 3m - 1)$

8. $6x(-2x^3 + 5x + 6)$

9. $(3t - 2y)(3t + 5y)$

10. $(9k + q)(2k - q)$

11. $(2 - 3x)(2 + 3x)$

12. $(6m + 5)(6m - 5)$

13. $\left(\dfrac{2}{5}y + \dfrac{1}{8}z\right)\left(\dfrac{3}{5}y + \dfrac{1}{2}z\right)$

14. $\left(\dfrac{3}{4}r - \dfrac{2}{3}s\right)\left(\dfrac{5}{4}r + \dfrac{1}{3}s\right)$

15. $(3p - 1)(9p^2 + 3p + 1)$

16. $(3p + 2)(5p^2 + p - 4)$

17. $(2m + 1)(4m^2 - 2m + 1)$

18. $(k + 2)(12k^3 - 3k^2 + k + 1)$

19. $(x + y + z)(3x - 2y - z)$

20. $(r + 2s - 3t)(2r - 2s + t)$

21. $(x + 1)(x + 2)(x + 3)$

22. $(x - 1)(x + 2)(x - 3)$

23. $(x + 2)^2$

24. $(2a - 4b)^2$

25. $(x - 2y)^3$

26. $(3x + y)^3$

YOUR TURN ANSWERS

1. $-9x^2 + 8x + 13$ **2.** $12y^3 + 2y^2 - 19y - 10$

R.2 Factoring

Multiplication of polynomials relies on the distributive property. The reverse process, where a polynomial is written as a product of other polynomials, is called **factoring**. For example, one way to factor the number 18 is to write it as the product $9 \cdot 2$; both 9 and 2 are **factors** of 18. Usually, only integers are used as factors of integers. The number 18 can also be written with three integer factors as $2 \cdot 3 \cdot 3$.

The Greatest Common Factor
To factor the algebraic expression $15m + 45$, first note that both $15m$ and 45 are divisible by 15; $15m = 15 \cdot m$ and $45 = 15 \cdot 3$. By the distributive property,

$$15m + 45 = 15 \cdot m + 15 \cdot 3 = 15(m + 3).$$

Both 15 and $m + 3$ are factors of $15m + 45$. Since 15 divides into both terms of $15m + 45$ (and is the largest number that will do so), 15 is the **greatest common factor** for the polynomial $15m + 45$. The process of writing $15m + 45$ as $15(m + 3)$ is often called **factoring out** the greatest common factor.

EXAMPLE 1 Factoring

Factor out the greatest common factor.

(a) $12p - 18q$

SOLUTION Both $12p$ and $18q$ are divisible by 6. Therefore,

$$12p - 18q = 6 \cdot 2p - 6 \cdot 3q = 6(2p - 3q).$$

(b) $8x^3 - 9x^2 + 15x$

SOLUTION Each of these terms is divisible by x.

$$8x^3 - 9x^2 + 15x = (8x^2) \cdot x - (9x) \cdot x + 15 \cdot x$$
$$= x(8x^2 - 9x + 15) \quad \text{or} \quad (8x^2 - 9x + 15)x$$

TRY YOUR TURN 1

> **YOUR TURN 1** Factor $4z^4 + 4z^3 + 18z^2$.

One can always check factorization by finding the product of the factors and comparing it to the original expression.

> **CAUTION** When factoring out the greatest common factor in an expression like $2x^2 + x$, be careful to remember the 1 in the second term.
>
> $$2x^2 + x = 2x^2 + 1x = x(2x + 1), \quad \text{not } x(2x).$$

Factoring Trinomials

A polynomial that has no greatest common factor (other than 1) may still be factorable. For example, the polynomial $x^2 + 5x + 6$ can be factored as $(x + 2)(x + 3)$. To see that this is correct, find the product $(x + 2)(x + 3)$; you should get $x^2 + 5x + 6$. A polynomial such as this with three terms is called a **trinomial**. To factor the trinomial $x^2 + 5x + 6$, where the coefficient of x^2 is 1, we use FOIL backwards.

EXAMPLE 2 Factoring a Trinomial

Factor $y^2 + 8y + 15$.

SOLUTION Since the coefficient of y^2 is 1, factor by finding two numbers whose *product* is 15 and whose *sum* is 8. Since the constant and the middle term are positive, the numbers must both be positive. Begin by listing all pairs of positive integers having a product of 15. As you do this, also form the sum of each pair of numbers.

Products	Sums
$15 \cdot 1 = 15$	$15 + 1 = 16$
$5 \cdot 3 = 15$	$5 + 3 = 8$

The numbers 5 and 3 have a product of 15 and a sum of 8. Thus, $y^2 + 8y + 15$ factors as

$$y^2 + 8y + 15 = (y + 5)(y + 3).$$

The answer also can be written as $(y + 3)(y + 5)$.

If the coefficient of the squared term is *not* 1, work as shown below.

EXAMPLE 3 Factoring a Trinomial

Factor $4x^2 + 8xy - 5y^2$.

SOLUTION The possible factors of $4x^2$ are $4x$ and x or $2x$ and $2x$; the possible factors of $-5y^2$ are $-5y$ and y or $5y$ and $-y$. Try various combinations of these factors until one works (if, indeed, any work). For example, try the product $(x + 5y)(4x - y)$.

$$(x + 5y)(4x - y) = 4x^2 - xy + 20xy - 5y^2$$
$$= 4x^2 + 19xy - 5y^2$$

This product is not correct, so try another combination.

$$(2x - y)(2x + 5y) = 4x^2 + 10xy - 2xy - 5y^2$$
$$= 4x^2 + 8xy - 5y^2$$

YOUR TURN 2 Factor $6a^2 + 5ab - 4b^2$.

Since this combination gives the correct polynomial,

$$4x^2 + 8xy - 5y^2 = (2x - y)(2x + 5y).$$ **TRY YOUR TURN 2**

Special Factorizations

Four special factorizations occur so often that they are listed here for future reference.

Special Factorizations

$x^2 - y^2 = (x + y)(x - y)$	**Difference of two squares**
$x^2 + 2xy + y^2 = (x + y)^2$	**Perfect square**
$x^3 - y^3 = (x - y)(x^2 + xy + y^2)$	**Difference of two cubes**
$x^3 + y^3 = (x + y)(x^2 - xy + y^2)$	**Sum of two cubes**

A polynomial that cannot be factored is called a **prime polynomial**.

EXAMPLE 4 Factoring Polynomials

Factor each polynomial, if possible.

(a) $64p^2 - 49q^2 = (8p)^2 - (7q)^2 = (8p + 7q)(8p - 7q)$ Difference of two squares

(b) $x^2 + 36$ is a prime polynomial.

(c) $x^2 + 12x + 36 = (x + 6)^2$ Perfect square

(d) $9y^2 - 24yz + 16z^2 = (3y - 4z)^2$ Perfect square

(e) $y^3 - 8 = y^3 - 2^3 = (y - 2)(y^2 + 2y + 4)$ Difference of two cubes

(f) $m^3 + 125 = m^3 + 5^3 = (m + 5)(m^2 - 5m + 25)$ Sum of two cubes

(g) $8k^3 - 27z^3 = (2k)^3 - (3z)^3 = (2k - 3z)(4k^2 + 6kz + 9z^2)$ Difference of two cubes

(h) $p^4 - 1 = (p^2 + 1)(p^2 - 1) = (p^2 + 1)(p + 1)(p - 1)$ Difference of two squares

CAUTION In factoring, always look for a common factor first. Since $36x^2 - 4y^2$ has a common factor of 4,

$$36x^2 - 4y^2 = 4(9x^2 - y^2) = 4(3x + y)(3x - y).$$

It would be incomplete to factor it as

$$36x^2 - 4y^2 = (6x + 2y)(6x - 2y),$$

since each factor can be factored still further. To *factor* means to factor completely, so that each polynomial factor is prime.

R.2 EXERCISES

Factor each polynomial. If a polynomial cannot be factored, write *prime*. Factor out the greatest common factor as necessary.

1. $7a^3 + 14a^2$

2. $3y^3 + 24y^2 + 9y$

3. $13p^4q^2 - 39p^3q + 26p^2q^2$

4. $60m^4 - 120m^3n + 50m^2n^2$

5. $m^2 - 5m - 14$

6. $x^2 + 4x - 5$

7. $z^2 + 9z + 20$

8. $b^2 - 8b + 7$

9. $a^2 - 6ab + 5b^2$

10. $s^2 + 2st - 35t^2$

11. $y^2 - 4yz - 21z^2$

12. $3x^2 + 4x - 7$

13. $3a^2 + 10a + 7$

14. $15y^2 + y - 2$

15. $21m^2 + 13mn + 2n^2$

16. $6a^2 - 48a - 120$

17. $3m^3 + 12m^2 + 9m$

18. $4a^2 + 10a + 6$

19. $24a^4 + 10a^3b - 4a^2b^2$

20. $24x^4 + 36x^3y - 60x^2y^2$

21. $x^2 - 64$

22. $9m^2 - 25$

23. $10x^2 - 160$

24. $9x^2 + 64$

25. $z^2 + 14zy + 49y^2$

26. $s^2 - 10st + 25t^2$

27. $9p^2 - 24p + 16$

28. $a^3 - 216$

29. $27r^3 - 64s^3$

30. $3m^3 + 375$

31. $x^4 - y^4$

32. $16a^4 - 81b^4$

YOUR TURN ANSWERS

1. $2z^2(2z^2 + 2z + 9)$ **2.** $(2a - b)(3a + 4b)$

R.3 Rational Expressions

Many algebraic fractions are **rational expressions**, which are quotients of polynomials with nonzero denominators. Examples include

$$\frac{8}{x - 1}, \qquad \frac{3x^2 + 4x}{5x - 6}, \qquad \text{and} \qquad \frac{2y + 1}{y^2}.$$

Next, we summarize properties for working with rational expressions.

Properties of Rational Expressions

For all mathematical expressions P, Q, R, and S, with $Q \neq 0$ and $S \neq 0$:

$$\frac{P}{Q} = \frac{PS}{QS} \qquad \qquad \textbf{Fundamental property}$$

$$\frac{P}{Q} + \frac{R}{Q} = \frac{P + R}{Q} \qquad \qquad \textbf{Addition}$$

$$\frac{P}{Q} - \frac{R}{Q} = \frac{P - R}{Q} \qquad \qquad \textbf{Subtraction}$$

$$\frac{P}{Q} \cdot \frac{R}{S} = \frac{PR}{QS} \qquad \qquad \textbf{Multiplication}$$

$$\frac{P}{Q} \div \frac{R}{S} = \frac{P}{Q} \cdot \frac{S}{R} \quad (R \neq 0) \qquad \textbf{Division}$$

When writing a rational expression in lowest terms, we may need to use the fact that $\dfrac{a^m}{a^n} = a^{m-n}$. For example,

$$\frac{x^4}{3x} = \frac{1x^4}{3x} = \frac{1}{3} \cdot \frac{x^4}{x} = \frac{1}{3} \cdot x^{4-1} = \frac{1}{3}x^3.$$

EXAMPLE 1   Reducing Rational Expressions

Write each rational expression in lowest terms, that is, reduce the expression as much as possible.

(a) $\dfrac{8x + 16}{4} = \dfrac{8(x + 2)}{4} = \dfrac{4 \cdot 2(x + 2)}{4} = 2(x + 2)$

Factor both the numerator and denominator in order to identify any common factors, which have a quotient of 1. The answer could also be written as $2x + 4$.

YOUR TURN 1 Write in lowest terms

$$\frac{z^2 + 5z + 6}{2z^2 + 7z + 3}.$$

(b) $\dfrac{k^2 + 7k + 12}{k^2 + 2k - 3} = \dfrac{(k + 4)(k + 3)}{(k - 1)(k + 3)} = \dfrac{k + 4}{k - 1}$

The answer cannot be further reduced.

TRY YOUR TURN 1

CAUTION One of the most common errors in algebra involves incorrect use of the fundamental property of rational expressions. Only common *factors* may be divided or "canceled." It is essential to factor rational expressions before writing them in lowest terms. In Example 1(b), for instance, it is not correct to "cancel" k^2 (or cancel k, or divide 12 by -3) because the additions and subtraction must be performed first. Here they cannot be performed, so it is not possible to divide. After factoring, however, the fundamental property can be used to write the expression in lowest terms.

EXAMPLE 2 Combining Rational Expressions

Perform each operation.

(a) $\dfrac{3y + 9}{6} \cdot \dfrac{18}{5y + 15}$

SOLUTION Factor where possible, then multiply numerators and denominators and reduce to lowest terms.

$$\frac{3y + 9}{6} \cdot \frac{18}{5y + 15} = \frac{3(y + 3)}{6} \cdot \frac{18}{5(y + 3)}$$

$$= \frac{3 \cdot 18(y + 3)}{6 \cdot 5(y + 3)}$$

$$= \frac{3 \cdot \cancel{6} \cdot 3\cancel{(y + 3)}}{\cancel{6} \cdot 5\cancel{(y + 3)}} = \frac{3 \cdot 3}{5} = \frac{9}{5}$$

(b) $\dfrac{m^2 + 5m + 6}{m + 3} \cdot \dfrac{m}{m^2 + 3m + 2}$

SOLUTION Factor where possible.

$$\frac{(m + 2)(m + 3)}{m + 3} \cdot \frac{m}{(m + 2)(m + 1)}$$

$$= \frac{m\cancel{(m + 2)}\cancel{(m + 3)}}{\cancel{(m + 3)}\cancel{(m + 2)}(m + 1)} = \frac{m}{m + 1}$$

(c) $\dfrac{9p - 36}{12} \div \dfrac{5(p - 4)}{18}$

SOLUTION Use the division property of rational expressions.

$$\frac{9p - 36}{12} \cdot \frac{18}{5(p - 4)}$$ *Invert and multiply.*

$$= \frac{9\cancel{(p - 4)}}{\cancel{6} \cdot 2} \cdot \frac{\cancel{6} \cdot 3}{5\cancel{(p - 4)}} = \frac{27}{10}$$ *Factor and reduce to lowest terms.*

(d) $\dfrac{4}{5k} - \dfrac{11}{5k}$

SOLUTION As shown in the list of properties, to subtract two rational expressions that have the same denominators, subtract the numerators while keeping the same denominator.

$$\frac{4}{5k} - \frac{11}{5k} = \frac{4 - 11}{5k} = -\frac{7}{5k}$$

(e) $\dfrac{7}{p} + \dfrac{9}{2p} + \dfrac{1}{3p}$

SOLUTION These three fractions cannot be added until their denominators are the same. A **common denominator** into which p, $2p$, and $3p$ all divide is $6p$. Note that $12p$ is also a common denominator, but $6p$ is the **least common denominator**. Use the fundamental property to rewrite each rational expression with a denominator of $6p$.

$$\frac{7}{p} + \frac{9}{2p} + \frac{1}{3p} = \frac{6 \cdot 7}{6 \cdot p} + \frac{3 \cdot 9}{3 \cdot 2p} + \frac{2 \cdot 1}{2 \cdot 3p}$$

$$= \frac{42}{6p} + \frac{27}{6p} + \frac{2}{6p}$$

$$= \frac{42 + 27 + 2}{6p}$$

$$= \frac{71}{6p}$$

(f) $\dfrac{x + 1}{x^2 + 5x + 6} - \dfrac{5x - 1}{x^2 - x - 12}$

SOLUTION To find the least common denominator, we first factor each denominator. Then we change each fraction so they all have the same denominator, being careful to multiply only by quotients that equal 1.

$$\frac{x + 1}{x^2 + 5x + 6} - \frac{5x - 1}{x^2 - x - 12}$$

$$= \frac{x + 1}{(x + 2)(x + 3)} - \frac{5x - 1}{(x + 3)(x - 4)}$$

$$= \frac{x + 1}{(x + 2)(x + 3)} \cdot \frac{(x - 4)}{(x - 4)} - \frac{5x - 1}{(x + 3)(x - 4)} \cdot \frac{(x + 2)}{(x + 2)}$$

$$= \frac{(x^2 - 3x - 4) - (5x^2 + 9x - 2)}{(x + 2)(x + 3)(x - 4)}$$

$$= \frac{-4x^2 - 12x - 2}{(x + 2)(x + 3)(x - 4)}$$

$$= \frac{-2(2x^2 + 6x + 1)}{(x + 2)(x + 3)(x - 4)}$$

YOUR TURN 2 Perform each of the following operations.

(a) $\dfrac{z^2 + 5z + 6}{2z^2 - 5z - 3} \cdot \dfrac{2z^2 - z - 1}{z^2 + 2z - 3}.$

(b) $\dfrac{a - 3}{a^2 + 3a + 2} + \dfrac{5a}{a^2 - 4}.$

Because the numerator cannot be factored further, we leave our answer in this form. We could also multiply out the denominator, but factored form is usually more useful.

TRY YOUR TURN 2

R.3 EXERCISES

Write each rational expression in lowest terms.

1. $\dfrac{5v^2}{35v}$

2. $\dfrac{25p^3}{10p^2}$

3. $\dfrac{8k + 16}{9k + 18}$

4. $\dfrac{2(t - 15)}{(t - 15)(t + 2)}$

5. $\dfrac{4x^3 - 8x^2}{4x^2}$

6. $\dfrac{36y^2 + 72y}{9y}$

7. $\dfrac{m^2 - 4m + 4}{m^2 + m - 6}$

8. $\dfrac{r^2 - r - 6}{r^2 + r - 12}$

9. $\dfrac{3x^2 + 3x - 6}{x^2 - 4}$

10. $\dfrac{z^2 - 5z + 6}{z^2 - 4}$

11. $\dfrac{m^4 - 16}{4m^2 - 16}$

12. $\dfrac{6y^2 + 11y + 4}{3y^2 + 7y + 4}$

Perform the indicated operations.

13. $\dfrac{9k^2}{25} \cdot \dfrac{5}{3k}$

14. $\dfrac{15p^3}{9p^2} \div \dfrac{6p}{10p^2}$

15. $\dfrac{3a + 3b}{4c} \cdot \dfrac{12}{5(a + b)}$

16. $\dfrac{a - 3}{16} \div \dfrac{a - 3}{32}$

17. $\dfrac{2k - 16}{6} \div \dfrac{4k - 32}{3}$

18. $\dfrac{9y - 18}{6y + 12} \cdot \dfrac{3y + 6}{15y - 30}$

19. $\dfrac{4a + 12}{2a - 10} \div \dfrac{a^2 - 9}{a^2 - a - 20}$

20. $\dfrac{6r - 18}{9r^2 + 6r - 24} \cdot \dfrac{12r - 16}{4r - 12}$

21. $\dfrac{k^2 + 4k - 12}{k^2 + 10k + 24} \cdot \dfrac{k^2 + k - 12}{k^2 - 9}$

22. $\dfrac{m^2 + 3m + 2}{m^2 + 5m + 4} \div \dfrac{m^2 + 5m + 6}{m^2 + 10m + 24}$

23. $\dfrac{2m^2 - 5m - 12}{m^2 - 10m + 24} \div \dfrac{4m^2 - 9}{m^2 - 9m + 18}$

24. $\dfrac{4n^2 + 4n - 3}{6n^2 - n - 15} \cdot \dfrac{8n^2 + 32n + 30}{4n^2 + 16n + 15}$

25. $\dfrac{a + 1}{2} - \dfrac{a - 1}{2}$

26. $\dfrac{3}{p} + \dfrac{1}{2}$

27. $\dfrac{6}{5y} - \dfrac{3}{2}$

28. $\dfrac{1}{6m} + \dfrac{2}{5m} + \dfrac{4}{m}$

29. $\dfrac{1}{m - 1} + \dfrac{2}{m}$

30. $\dfrac{5}{2r + 3} - \dfrac{2}{r}$

31. $\dfrac{8}{3(a - 1)} + \dfrac{2}{a - 1}$

32. $\dfrac{2}{5(k - 2)} + \dfrac{3}{4(k - 2)}$

33. $\dfrac{4}{x^2 + 4x + 3} + \dfrac{3}{x^2 - x - 2}$

34. $\dfrac{y}{y^2 + 2y - 3} - \dfrac{1}{y^2 + 4y + 3}$

35. $\dfrac{3k}{2k^2 + 3k - 2} - \dfrac{2k}{2k^2 - 7k + 3}$

36. $\dfrac{4m}{3m^2 + 7m - 6} - \dfrac{m}{3m^2 - 14m + 8}$

37. $\dfrac{2}{a + 2} + \dfrac{1}{a} + \dfrac{a - 1}{a^2 + 2a}$

38. $\dfrac{5x + 2}{x^2 - 1} + \dfrac{3}{x^2 + x} - \dfrac{1}{x^2 - x}$

▰ **YOUR TURN ANSWERS**

1. $(z + 2)/(2z + 1)$

2a. $(z + 2)/(z - 3)$

2b. $6(a^2 + 1)/[(a - 2)(a + 2)(a + 1)]$

R.4 Equations

Linear Equations

Equations that can be written in the form $ax + b = 0$, where a and b are real numbers, with $a \neq 0$, are **linear equations**. Examples of linear equations include $5y + 9 = 16$, $8x = 4$, and $-3p + 5 = -8$. Equations that are *not* linear include absolute value equations such as $|x| = 4$. The following properties are used to solve linear equations.

Properties of Equality

For all real numbers a, b, and c:

1. If $a = b$, then $a + c = b + c$.   **Addition property of equality**
  (The same number may be added to both sides of an equation.)

2. If $a = b$, then $ac = bc$.   **Multiplication property of equality**
  (Both sides of an equation may be multiplied by the same number.)

EXAMPLE 1   Solving Linear Equations

Solve the following equations.

(a) $x - 2 = 3$

SOLUTION   The goal is to isolate the variable. Using the addition property of equality yields

$$x - 2 + 2 = 3 + 2, \qquad \text{or} \qquad x = 5.$$

(b) $\dfrac{x}{2} = 3$

SOLUTION   Using the multiplication property of equality yields

$$2 \cdot \dfrac{x}{2} = 2 \cdot 3, \quad \text{or} \quad x = 6.$$

The following example shows how these properties are used to solve linear equations. The goal is to isolate the variable. The solutions should always be checked by substitution in the original equation.

EXAMPLE 2   Solving a Linear Equation

Solve $2x - 5 + 8 = 3x + 2(2 - 3x)$.

SOLUTION

$2x - 5 + 8 = 3x + 4 - 6x$	Distributive property
$2x + 3 = -3x + 4$	Combine like terms.
$5x + 3 = 4$	Add $3x$ to both sides.
$5x = 1$	Add -3 to both sides.
$x = \dfrac{1}{5}$	Multiply both sides by $\frac{1}{5}$.

YOUR TURN 1 Solve $3x - 7 = 4(5x + 2) - 7x$.

Check by substituting in the original equation. The left side becomes $2(1/5) - 5 + 8$ and the right side becomes $3(1/5) + 2[2 - 3(1/5)]$. Verify that both of these expressions simplify to $17/5$.   **TRY YOUR TURN 1**

Quadratic Equations
An equation with 2 as the highest exponent of the variable is a *quadratic equation*. A **quadratic equation** has the form $ax^2 + bx + c = 0$, where a, b, and c are real numbers and $a \neq 0$. A quadratic equation written in the form $ax^2 + bx + c = 0$ is said to be in **standard form**.

The simplest way to solve a quadratic equation, but one that is not always applicable, is by factoring. This method depends on the **zero-factor property**.

Zero-Factor Property
If a and b are real numbers, with $ab = 0$, then either

$$a = 0 \quad \text{or} \quad b = 0 \quad \text{(or both)}.$$

EXAMPLE 3   Solving a Quadratic Equation

Solve $6r^2 + 7r = 3$.

SOLUTION   First write the equation in standard form.

$$6r^2 + 7r - 3 = 0$$

Now factor $6r^2 + 7r - 3$ to get

$$(3r - 1)(2r + 3) = 0.$$

By the zero-factor property, the product $(3r - 1)(2r + 3)$ can equal 0 if and only if

$$3r - 1 = 0 \quad \text{or} \quad 2r + 3 = 0.$$

YOUR TURN 2 Solve $2m^2 + 7m = 15$.

Solve each of these equations separately to find that the solutions are $1/3$ and $-3/2$. Check these solutions by substituting them in the original equation.   **TRY YOUR TURN 2**

| **CAUTION** | Remember, the zero-factor property requires that the product of two (or more) factors be equal to *zero*, not some other quantity. It would be incorrect to use the zero-factor property with an equation in the form $(x + 3)(x - 1) = 4$, for example. |

If a quadratic equation cannot be solved easily by factoring, use the *quadratic formula*. (The derivation of the quadratic formula is given in most algebra books.)

Quadratic Formula
The solutions of the quadratic equation $ax^2 + bx + c = 0$, where $a \neq 0$, are given by

$$x = \frac{-b \pm \sqrt{b^2 - 4ac}}{2a}.$$

EXAMPLE 4 **Quadratic Formula**

Solve $x^2 - 4x - 5 = 0$ by the quadratic formula.

SOLUTION The equation is already in standard form (it has 0 alone on one side of the equal sign), so the values of a, b, and c from the quadratic formula are easily identified. The coefficient of the squared term gives the value of a; here, $a = 1$. Also, $b = -4$ and $c = -5$. (Be careful to use the correct signs.) Substitute these values into the quadratic formula.

$$x = \frac{-(-4) \pm \sqrt{(-4)^2 - 4(1)(-5)}}{2(1)} \qquad a = 1, b = -4, c = -5$$

$$x = \frac{4 \pm \sqrt{16 + 20}}{2} \qquad\qquad (-4)^2 = (-4)(-4) = 16$$

$$x = \frac{4 \pm 6}{2} \qquad\qquad\qquad \sqrt{16 + 20} = \sqrt{36} = 6$$

The $\pm$ sign represents the two solutions of the equation. To find both of the solutions, first use $+$ and then use $-$.

$$x = \frac{4 + 6}{2} = \frac{10}{2} = 5 \qquad \text{or} \qquad x = \frac{4 - 6}{2} = \frac{-2}{2} = -1$$

The two solutions are 5 and -1.

| **CAUTION** | Notice in the quadratic formula that the square root is added to or subtracted from the value of $-b$ *before* dividing by $2a$. |

EXAMPLE 5 **Quadratic Formula**

Solve $x^2 + 1 = 4x$.

SOLUTION First, add $-4x$ on both sides of the equal sign in order to get the equation in standard form.

$$x^2 - 4x + 1 = 0$$

Now identify the letters a, b, and c. Here $a = 1$, $b = -4$, and $c = 1$. Substitute these numbers into the quadratic formula.

$$x = \frac{-(-4) \pm \sqrt{(-4)^2 - 4(1)(1)}}{2(1)}$$

$$= \frac{4 \pm \sqrt{16 - 4}}{2}$$

$$= \frac{4 \pm \sqrt{12}}{2}$$

Simplify the solutions by writing $\sqrt{12}$ as $\sqrt{4 \cdot 3} = \sqrt{4} \cdot \sqrt{3} = 2\sqrt{3}$. Substituting $2\sqrt{3}$ for $\sqrt{12}$ gives

$$x = \frac{4 \pm 2\sqrt{3}}{2}$$

$$= \frac{2(2 \pm \sqrt{3})}{2} \qquad \text{Factor } 4 \pm 2\sqrt{3}.$$

$$= 2 \pm \sqrt{3}. \qquad \text{Reduce to lowest terms.}$$

The two solutions are $2 + \sqrt{3}$ and $2 - \sqrt{3}$.

The exact values of the solutions are $2 + \sqrt{3}$ and $2 - \sqrt{3}$. The $\sqrt{}$ key on a calculator gives decimal approximations of these solutions (to the nearest thousandth):

YOUR TURN 3 Solve $z^2 + 6 = 8z$.

$$2 + \sqrt{3} \approx 2 + 1.732 = 3.732*$$

$$2 - \sqrt{3} \approx 2 - 1.732 = 0.268 \qquad \text{TRY YOUR TURN 3}$$

NOTE Sometimes the quadratic formula will give a result with a negative number under the radical sign, such as $3 \pm \sqrt{-5}$. A solution of this type is a complex number. Since this text deals only with real numbers, such solutions cannot be used.

Equations with Fractions
When an equation includes fractions, first eliminate all denominators by multiplying both sides of the equation by a common denominator, a number that can be divided (with no remainder) by each denominator in the equation. When an equation involves fractions with variable denominators, it is *necessary* to check all solutions in the original equation to be sure that no solution will lead to a zero denominator.

> **EXAMPLE 6** **Solving Rational Equations**

Solve each equation.

(a) $\dfrac{r}{10} - \dfrac{2}{15} = \dfrac{3r}{20} - \dfrac{1}{5}$

SOLUTION The denominators are 10, 15, 20, and 5. Each of these numbers can be divided into 60, so 60 is a common denominator. Multiply both sides of the equation by 60 and use the distributive property. (If a common denominator cannot be found easily, all the denominators in the problem can be multiplied together to produce one.)

$$\frac{r}{10} - \frac{2}{15} = \frac{3r}{20} - \frac{1}{5}$$

$$60\left(\frac{r}{10} - \frac{2}{15}\right) = 60\left(\frac{3r}{20} - \frac{1}{5}\right) \qquad \text{Multiply by the common denominator.}$$

$$60\left(\frac{r}{10}\right) - 60\left(\frac{2}{15}\right) = 60\left(\frac{3r}{20}\right) - 60\left(\frac{1}{5}\right) \qquad \text{Distributive property}$$

$$6r - 8 = 9r - 12$$

*The symbol $\approx$ means "is approximately equal to."

Add $-9r$ and 8 to both sides.

$$6r - 8 + (-9r) + 8 = 9r - 12 + (-9r) + 8$$
$$-3r = -4$$
$$r = \frac{4}{3} \qquad \text{Multiply each side by } -\frac{1}{3}.$$

Check by substituting into the original equation.

(b) $\dfrac{3}{x^2} - 12 = 0$

SOLUTION Begin by multiplying both sides of the equation by x^2 to get $3 - 12x^2 = 0$. This equation could be solved by using the quadratic formula with $a = -12$, $b = 0$, and $c = 3$. Another method that works well for the type of quadratic equation in which $b = 0$ is shown below.

$$3 - 12x^2 = 0$$
$$3 = 12x^2 \qquad \text{Add } 12x^2.$$
$$\frac{1}{4} = x^2 \qquad \text{Multiply by } \frac{1}{12}.$$
$$\pm \frac{1}{2} = x \qquad \text{Take square roots.}$$

Verify that there are two solutions, $-1/2$ and $1/2$.

(c) $\dfrac{2}{k} - \dfrac{3k}{k+2} = \dfrac{k}{k^2 + 2k}$

SOLUTION Factor $k^2 + 2k$ as $k(k+2)$. The least common denominator for all the fractions is $k(k+2)$. Multiplying both sides by $k(k+2)$ gives the following:

$$k(k+2) \cdot \left(\frac{2}{k} - \frac{3k}{k+2} \right) = k(k+2) \cdot \frac{k}{k^2+2k}$$
$$2(k+2) - 3k(k) = k$$
$$2k + 4 - 3k^2 = k \qquad \text{Distributive property}$$
$$-3k^2 + k + 4 = 0 \qquad \text{Add } -k; \text{ rearrange terms.}$$
$$3k^2 - k - 4 = 0 \qquad \text{Multiply by } -1.$$
$$(3k - 4)(k + 1) = 0 \qquad \text{Factor.}$$
$$3k - 4 = 0 \qquad \text{or} \qquad k + 1 = 0$$
$$k = \frac{4}{3} \qquad k = -1$$

YOUR TURN 4 Solve $\dfrac{1}{x^2 - 4} + \dfrac{2}{x - 2} = \dfrac{1}{x}$.

Verify that the solutions are $4/3$ and -1. **TRY YOUR TURN 4**

> **CAUTION** It is possible to get, as a solution of a rational equation, a number that makes one or more of the denominators in the original equation equal to zero. That number is not a solution, so it is *necessary* to check all potential solutions of rational equations. These introduced solutions are called **extraneous solutions**.

EXAMPLE 7 **Solving a Rational Equation**

Solve $\dfrac{2}{x-3} + \dfrac{1}{x} = \dfrac{6}{x(x-3)}$.

SOLUTION The common denominator is $x(x - 3)$. Multiply both sides by $x(x - 3)$ and solve the resulting equation.

$$x(x - 3) \cdot \left(\frac{2}{x - 3} + \frac{1}{x} \right) = x(x - 3) \cdot \left[\frac{6}{x(x - 3)} \right]$$

$$2x + x - 3 = 6$$

$$3x = 9$$

$$x = 3$$

Checking this potential solution by substitution in the original equation shows that 3 makes two denominators 0. Thus, 3 cannot be a solution, so there is no solution for this equation. ▄

R.4 EXERCISES

Solve each equation.

1. $2x + 8 = x - 4$

2. $5x + 2 = 8 - 3x$

3. $0.2m - 0.5 = 0.1m + 0.7$

4. $\frac{2}{3}k - k + \frac{3}{8} = \frac{1}{2}$

5. $3r + 2 - 5(r + 1) = 6r + 4$

6. $5(a + 3) + 4a - 5 = -(2a - 4)$

7. $2[3m - 2(3 - m) - 4] = 6m - 4$

8. $4[2p - (3 - p) + 5] = -7p - 2$

Solve each equation by factoring or by using the quadratic formula. If the solutions involve square roots, give both the exact solutions and the approximate solutions to three decimal places.

9. $x^2 + 5x + 6 = 0$

10. $x^2 = 3 + 2x$

11. $m^2 = 14m - 49$

12. $2k^2 - k = 10$

13. $12x^2 - 5x = 2$

14. $m(m - 7) = -10$

15. $4x^2 - 36 = 0$

16. $z(2z + 7) = 4$

17. $12y^2 - 48y = 0$

18. $3x^2 - 5x + 1 = 0$

19. $2m^2 - 4m = 3$

20. $p^2 + p - 1 = 0$

21. $k^2 - 10k = -20$

22. $5x^2 - 8x + 2 = 0$

23. $2r^2 - 7r + 5 = 0$

24. $2x^2 - 7x + 30 = 0$

25. $3k^2 + k = 6$

26. $5m^2 + 5m = 0$

Solve each equation.

27. $\dfrac{3x - 2}{7} = \dfrac{x + 2}{5}$

28. $\dfrac{x}{3} - 7 = 6 - \dfrac{3x}{4}$

29. $\dfrac{4}{x - 3} - \dfrac{8}{2x + 5} + \dfrac{3}{x - 3} = 0$

30. $\dfrac{5}{p - 2} - \dfrac{7}{p + 2} = \dfrac{12}{p^2 - 4}$

31. $\dfrac{2m}{m - 2} - \dfrac{6}{m} = \dfrac{12}{m^2 - 2m}$

32. $\dfrac{2y}{y - 1} = \dfrac{5}{y} + \dfrac{10 - 8y}{y^2 - y}$

33. $\dfrac{1}{x - 2} - \dfrac{3x}{x - 1} = \dfrac{2x + 1}{x^2 - 3x + 2}$

34. $\dfrac{5}{a} + \dfrac{-7}{a + 1} = \dfrac{a^2 - 2a + 4}{a^2 + a}$

35. $\dfrac{5}{b + 5} - \dfrac{4}{b^2 + 2b} = \dfrac{6}{b^2 + 7b + 10}$

36. $\dfrac{2}{x^2 - 2x - 3} + \dfrac{5}{x^2 - x - 6} = \dfrac{1}{x^2 + 3x + 2}$

37. $\dfrac{4}{2x^2 + 3x - 9} + \dfrac{2}{2x^2 - x - 3} = \dfrac{3}{x^2 + 4x + 3}$

YOUR TURN ANSWERS

1. $-3/2$

2. $3/2, -5$

3. $4 \pm \sqrt{10}$

4. $-1, -4$

R.5 Inequalities

To write that one number is greater than or less than another number, we use the following symbols.

Inequality Symbols

$<$ means *is less than* $\leq$ means *is less than or equal to*

$>$ means *is greater than* $\geq$ means *is greater than or equal to*

Linear Inequalities

An equation states that two expressions are equal; an **inequality** states that they are unequal. A **linear inequality** is an inequality that can be simplified to the form $ax < b$. (Properties introduced in this section are given only for $<$, but they are equally valid for $>$, $\leq$, or $\geq$.) Linear inequalities are solved with the following properties.

Properties of Inequality

For all real numbers a, b, and c:

1. If $a < b$, then $a + c < b + c$.
2. If $a < b$ and if $c > 0$, then $ac < bc$.
3. If $a < b$ and if $c < 0$, then $ac > bc$.

Pay careful attention to property 3; it says that if both sides of an inequality are multiplied by a negative number, the direction of the inequality symbol must be reversed.

EXAMPLE 1 Solving a Linear Inequality

Solve $4 - 3y \leq 7 + 2y$.

SOLUTION Use the properties of inequality.

$$4 - 3y + (-4) \leq 7 + 2y + (-4) \qquad \text{Add } -4 \text{ to both sides.}$$
$$-3y \leq 3 + 2y$$

Remember that *adding* the same number to both sides never changes the direction of the inequality symbol.

$$-3y + (-2y) \leq 3 + 2y + (-2y) \qquad \text{Add } -2y \text{ to both sides.}$$
$$-5y \leq 3$$

Multiply both sides by $-1/5$. Since $-1/5$ is negative, change the direction of the inequality symbol.

$$-\frac{1}{5}(-5y) \geq -\frac{1}{5}(3)$$

$$y \geq -\frac{3}{5}$$

YOUR TURN 1 Solve $3z - 2 > 5z + 7$.

TRY YOUR TURN 1

CAUTION It is a common error to forget to reverse the direction of the inequality sign when multiplying or dividing by a negative number. For example, to solve $-4x \leq 12$, we must multiply by $-1/4$ on both sides *and* reverse the inequality symbol to get $x \geq -3$.

The solution $y \geq -3/5$ in Example 1 represents an interval on the number line. **Interval notation** often is used for writing intervals. With interval notation, $y \geq -3/5$ is written as $[-3/5, \infty)$. This is an example of a **half-open interval**, since one endpoint, $-3/5$, is included. The **open interval** $(2, 5)$ corresponds to $2 < x < 5$, with neither endpoint included. The **closed interval** $[2, 5]$ includes both endpoints and corresponds to $2 \leq x \leq 5$.

The **graph** of an interval shows all points on a number line that correspond to the numbers in the interval. To graph the interval $[-3/5, \infty)$, for example, use a solid circle at $-3/5$, since $-3/5$ is part of the solution. To show that the solution includes all real numbers greater than or equal to $-3/5$, draw a heavy arrow pointing to the right (the positive direction). See Figure 1.

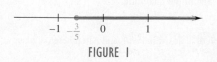

FIGURE 1

EXAMPLE 2   Graphing a Linear Inequality

Solve $-2 < 5 + 3m < 20$. Graph the solution.

SOLUTION   The inequality $-2 < 5 + 3m < 20$ says that $5 + 3m$ is *between* -2 and 20. Solve this inequality with an extension of the properties given above. Work as follows, first adding -5 to each part.

$$-2 + (-5) < 5 + 3m + (-5) < 20 + (-5)$$
$$-7 < 3m < 15$$

Now multiply each part by $1/3$.

$$-\frac{7}{3} < m < 5$$

A graph of the solution is given in Figure 2; here open circles are used to show that $-7/3$ and 5 are *not* part of the graph.*

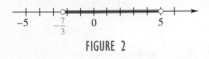

FIGURE 2

Quadratic Inequalities   A **quadratic inequality** has the form $ax^2 + bx + c > 0$ (or $<$, or $\leq$, or $\geq$). The highest exponent is 2. The next few examples show how to solve quadratic inequalities.

EXAMPLE 3   Solving a Quadratic Inequality

Solve the quadratic inequality $x^2 - x < 12$.

SOLUTION   Write the inequality with 0 on one side, as $x^2 - x - 12 < 0$. This inequality is solved with values of x that make $x^2 - x - 12$ negative (<0). The quantity $x^2 - x - 12$ changes from positive to negative or from negative to positive at the points where it equals 0. For this reason, first solve the *equation* $x^2 - x - 12 = 0$.

$$x^2 - x - 12 = 0$$
$$(x - 4)(x + 3) = 0$$
$$x = 4 \quad \text{or} \quad x = -3$$

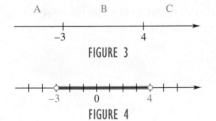

FIGURE 3

FIGURE 4

YOUR TURN 2 Solve $3y^2 \leq 16y + 12$.

Locating -3 and 4 on a number line, as shown in Figure 3, determines three intervals A, B, and C. Decide which intervals include numbers that make $x^2 - x - 12$ negative by substituting any number from each interval in the polynomial. For example,

choose -4 from interval A: $(-4)^2 - (-4) - 12 = 8 > 0$;

choose 0 from interval B: $0^2 - 0 - 12 = -12 < 0$;

choose 5 from interval C: $5^2 - 5 - 12 = 8 > 0$.

Only numbers in interval B satisfy the given inequality, so the solution is $(-3, 4)$. A graph of this solution is shown in Figure 4.   **TRY YOUR TURN 2**

EXAMPLE 4   Solving a Polynomial Inequality

Solve the inequality $x^3 + 2x^2 - 3x \geq 0$.

SOLUTION   This is not a quadratic inequality because of the x^3 term, but we solve it in a similar way by first factoring the polynomial.

$$x^3 + 2x^2 - 3x = x(x^2 + 2x - 3) \qquad \text{Factor out the common factor.}$$
$$= x(x - 1)(x + 3) \qquad \text{Factor the quadratic.}$$

*Some textbooks use brackets in place of solid circles for the graph of a closed interval, and parentheses in place of open circles for the graph of an open interval.

Now solve the corresponding equation.

$$x(x - 1)(x + 3) = 0$$

$$x = 0 \quad \text{or} \quad x - 1 = 0 \quad \text{or} \quad x + 3 = 0$$

$$x = 1 \qquad\qquad x = -3$$

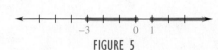

FIGURE 5

These three solutions determine four intervals on the number line: $(-\infty, -3)$, $(-3, 0)$, $(0, 1)$, and $(1, \infty)$. Substitute a number from each interval into the original inequality to determine that the solution consists of the numbers between -3 and 0 (including the endpoints) and all numbers that are greater than or equal to 1. See Figure 5. In interval notation, the solution is

$$[-3, 0] \cup [1, \infty).*$$

Inequalities with Fractions

Inequalities with fractions are solved in a similar manner as quadratic inequalities.

EXAMPLE 5 **Solving a Rational Inequality**

Solve $\dfrac{2x - 3}{x} \geq 1$.

SOLUTION First solve the corresponding equation.

$$\frac{2x - 3}{x} = 1$$

$$2x - 3 = x$$

$$x = 3$$

The solution, $x = 3$, determines the intervals on the number line where the fraction may change from greater than 1 to less than 1. This change also may occur on either side of a number that makes the denominator equal 0. Here, the x-value that makes the denominator 0 is $x = 0$. Test each of the three intervals determined by the numbers 0 and 3.

$$\text{For } (-\infty, 0), \text{ choose } -1: \frac{2(-1) - 3}{-1} = 5 \geq 1.$$

$$\text{For } (0, 3), \quad \text{choose} \quad 1: \frac{2(1) - 3}{1} = -1 \not\geq 1.$$

$$\text{For } (3, \infty), \quad \text{choose} \quad 4: \frac{2(4) - 3}{4} = \frac{5}{4} \geq 1.$$

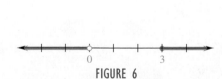

FIGURE 6

The symbol $\not\geq$ means "is *not* greater than or equal to." Testing the endpoints 0 and 3 shows that the solution is $(-\infty, 0) \cup [3, \infty)$, as shown in Figure 6.

CAUTION A common error is to try to solve the inequality in Example 5 by multiplying both sides by x. The reason this is wrong is that we don't know in the beginning whether x is positive or negative. If x is negative, the $\geq$ would change to $\leq$ according to the third property of inequality listed at the beginning of this section.

---

*The symbol $\cup$ indicates the *union* of two sets, which includes all elements in either set.

EXAMPLE 6 Solving a Rational Inequality

Solve $\dfrac{(x-1)(x+1)}{x} \leq 0$.

SOLUTION We first solve the corresponding equation.

$$\frac{(x-1)(x+1)}{x} = 0$$

$$(x-1)(x+1) = 0 \qquad \text{Multiply both sides by } x.$$

$$x = 1 \qquad \text{or} \qquad x = -1 \qquad \text{Use the zero-factor property.}$$

Setting the denominator equal to 0 gives $x = 0$, so the intervals of interest are $(-\infty, -1)$, $(-1, 0)$, $(0, 1)$, and $(1, \infty)$. Testing a number from each region in the original inequality and checking the endpoints, we find the solution is

$$(-\infty, -1] \cup (0, 1],$$

as shown in Figure 7.

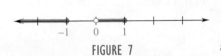

FIGURE 7

CAUTION Remember to solve the equation formed by setting the *denominator* equal to zero. Any number that makes the denominator zero always creates two intervals on the number line. For instance, in Example 6, substituting $x = 0$ makes the denominator of the rational inequality equal to 0, so we know that there may be a sign change from one side of 0 to the other (as was indeed the case).

EXAMPLE 7 Solving a Rational Inequality

Solve $\dfrac{x^2 - 3x}{x^2 - 9} < 4$.

SOLUTION Solve the corresponding equation.

$$\frac{x^2 - 3x}{x^2 - 9} = 4$$

$$x^2 - 3x = 4x^2 - 36 \qquad \text{Multiply by } x^2 - 9.$$

$$0 = 3x^2 + 3x - 36 \qquad \text{Get 0 on one side.}$$

$$0 = x^2 + x - 12 \qquad \text{Multiply by } \tfrac{1}{3}.$$

$$0 = (x + 4)(x - 3) \qquad \text{Factor.}$$

$$x = -4 \qquad \text{or} \qquad x = 3$$

Now set the denominator equal to 0 and solve that equation.

$$x^2 - 9 = 0$$

$$(x - 3)(x + 3) = 0$$

$$x = 3 \qquad \text{or} \qquad x = -3$$

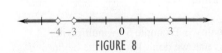

FIGURE 8

The intervals determined by the three (different) solutions are $(-\infty, -4)$, $(-4, -3)$, $(-3, 3)$, and $(3, \infty)$. Testing a number from each interval in the given inequality shows that the solution is

$$(-\infty, -4) \cup (-3, 3) \cup (3, \infty),$$

as shown in Figure 8. For this example, none of the endpoints are part of the solution because $x = 3$ and $x = -3$ make the denominator zero and $x = -4$ produces an equality.

YOUR TURN 3 Solve $\dfrac{k^2 - 35}{k} \geq 2$.

TRY YOUR TURN 3

R.5 EXERCISES

Write each expression in interval notation. Graph each interval.

1. $x < 4$

2. $x \geq -3$

3. $1 \leq x < 2$

4. $-2 \leq x \leq 3$

5. $-9 > x$

6. $6 \leq x$

Using the variable x, write each interval as an inequality.

7. $[-7, -3]$

8. $[4, 10)$

9. $(-\infty, -1]$

10. $(3, \infty)$

11.

12.

13.

14.

Solve each inequality and graph the solution.

15. $6p + 7 \leq 19$

16. $6k - 4 < 3k - 1$

17. $m - (3m - 2) + 6 < 7m - 19$

18. $-2(3y - 8) \geq 5(4y - 2)$

19. $3p - 1 < 6p + 2(p - 1)$

20. $x + 5(x + 1) > 4(2 - x) + x$

21. $-11 < y - 7 < -1$

22. $8 \leq 3r + 1 \leq 13$

23. $-2 < \dfrac{1 - 3k}{4} \leq 4$

24. $-1 \leq \dfrac{5y + 2}{3} \leq 4$

25. $\dfrac{3}{5}(2p + 3) \geq \dfrac{1}{10}(5p + 1)$

26. $\dfrac{8}{3}(z - 4) \leq \dfrac{2}{9}(3z + 2)$

Solve each quadratic inequality. Graph each solution.

27. $(m - 3)(m + 5) < 0$

28. $(t + 6)(t - 1) \geq 0$

29. $y^2 - 3y + 2 < 0$

30. $2k^2 + 7k - 4 > 0$

31. $x^2 - 16 > 0$

32. $2k^2 - 7k - 15 \leq 0$

33. $x^2 - 4x \geq 5$

34. $10r^2 + r \leq 2$

35. $3x^2 + 2x > 1$

36. $3a^2 + a > 10$

37. $9 - x^2 \leq 0$

38. $p^2 - 16p > 0$

39. $x^3 - 4x \geq 0$

40. $x^3 + 7x^2 + 12x \leq 0$

41. $2x^3 - 14x^2 + 12x < 0$

42. $3x^3 - 9x^2 - 12x > 0$

Solve each inequality.

43. $\dfrac{m - 3}{m + 5} \leq 0$

44. $\dfrac{r + 1}{r - 1} > 0$

45. $\dfrac{k - 1}{k + 2} > 1$

46. $\dfrac{a - 5}{a + 2} < -1$

47. $\dfrac{2y + 3}{y - 5} \leq 1$

48. $\dfrac{a + 2}{3 + 2a} \leq 5$

49. $\dfrac{2k}{k - 3} \leq \dfrac{4}{k - 3}$

50. $\dfrac{5}{p + 1} > \dfrac{12}{p + 1}$

51. $\dfrac{2x}{x^2 - x - 6} \geq 0$

52. $\dfrac{8}{p^2 + 2p} > 1$

53. $\dfrac{z^2 + z}{z^2 - 1} \geq 3$

54. $\dfrac{a^2 + 2a}{a^2 - 4} \leq 2$

YOUR TURN ANSWERS

1. $z < -9/2$ **2.** $[-2/3, 6]$ **3.** $[-5, 0) \cup [7, \infty)$

R.6 Exponents

Integer Exponents

Recall that $a^2 = a \cdot a$, while $a^3 = a \cdot a \cdot a$, and so on. In this section, a more general meaning is given to the symbol a^n.

Definition of Exponent

If n is a natural number, then

$$a^n = a \cdot a \cdot a \cdot \cdots \cdot a,$$

where a appears as a factor n times.

In the expression a^n, the power n is the **exponent** and a is the **base**. This definition can be extended by defining a^n for zero and negative integer values of n.

Zero and Negative Exponents

If a is any nonzero real number, and if n is a positive integer, then

$$a^0 = 1 \qquad \text{and} \qquad a^{-n} = \frac{1}{a^n}.$$

(The symbol 0^0 is meaningless.)

EXAMPLE 1 Exponents

(a) $6^0 = 1$

(b) $(-9)^0 = 1$

(c) $3^{-2} = \dfrac{1}{3^2} = \dfrac{1}{9}$

(d) $9^{-1} = \dfrac{1}{9^1} = \dfrac{1}{9}$

(e) $\left(\dfrac{3}{4}\right)^{-1} = \dfrac{1}{(3/4)^1} = \dfrac{1}{3/4} = \dfrac{4}{3}$

The following properties follow from the definitions of exponents given above.

Properties of Exponents

For any integers m and n, and any real numbers a and b for which the following exist:

1. $a^m \cdot a^n = a^{m+n}$ 4. $(ab)^m = a^m \cdot b^m$

2. $\dfrac{a^m}{a^n} = a^{m-n}$ 5. $\left(\dfrac{a}{b}\right)^m = \dfrac{a^m}{b^m}$

3. $(a^m)^n = a^{mn}$

Note that $(-a)^n = a^n$ if n is an even integer, but $(-a)^n = -a^n$ if n is an odd integer.

EXAMPLE 2 Simplifying Exponential Expressions

Use the properties of exponents to simplify each expression. Leave answers with positive exponents. Assume that all variables represent positive real numbers.

(a) $7^4 \cdot 7^6 = 7^{4+6} = 7^{10}$ (or 282,475,249)   Property 1

(b) $\dfrac{9^{14}}{9^6} = 9^{14-6} = 9^8$ (or 43,046,721)   Property 2

(c) $\dfrac{r^9}{r^{17}} = r^{9-17} = r^{-8} = \dfrac{1}{r^8}$ Property 2

(d) $(2m^3)^4 = 2^4 \cdot (m^3)^4 = 16m^{12}$ Properties 3 and 4

(e) $(3x)^4 = 3^4 \cdot x^4 = 81x^4$ Property 4

(f) $\left(\dfrac{x^2}{y^3}\right)^6 = \dfrac{(x^2)^6}{(y^3)^6} = \dfrac{x^{2\cdot6}}{y^{3\cdot6}} = \dfrac{x^{12}}{y^{18}}$ Properties 3 and 5

(g) $\dfrac{a^{-3}b^5}{a^4b^{-7}} = \dfrac{b^{5-(-7)}}{a^{4-(-3)}} = \dfrac{b^{5+7}}{a^{4+3}} = \dfrac{b^{12}}{a^7}$ Property 2

(h) $p^{-1} + q^{-1} = \dfrac{1}{p} + \dfrac{1}{q} = \dfrac{1}{p} \cdot \dfrac{q}{q} + \dfrac{1}{q} \cdot \dfrac{p}{p} = \dfrac{q}{pq} + \dfrac{p}{pq} = \dfrac{p+q}{pq}$

(i) $\dfrac{x^{-2} - y^{-2}}{x^{-1} - y^{-1}} = \dfrac{\dfrac{1}{x^2} - \dfrac{1}{y^2}}{\dfrac{1}{x} - \dfrac{1}{y}}$ Definition of a^{-n}

$= \dfrac{\dfrac{y^2 - x^2}{x^2 y^2}}{\dfrac{y - x}{xy}}$ Get common denominators and combine terms.

$= \dfrac{y^2 - x^2}{x^2 y^2} \cdot \dfrac{xy}{y - x}$ Invert and multiply.

$= \dfrac{(y - x)(y + x)}{x^2 y^2} \cdot \dfrac{xy}{y - x}$ Factor.

$= \dfrac{x + y}{xy}$ Simplify.

YOUR TURN 1
Simplify
$\left(\dfrac{y^2 z^{-4}}{y^{-3} z^4}\right)^{-2}$.

TRY YOUR TURN 1

CAUTION If Example 2(e) were written $3x^4$, the properties of exponents would not apply. When no parentheses are used, the exponent refers only to the factor closest to it. Also notice in Examples 2(c), 2(g), 2(h), and 2(i) that a negative exponent does *not* indicate a negative number.

Roots
For *even* values of n and nonnegative values of a, the expression $a^{1/n}$ is defined to be the **positive nth root** of a or the **principal nth root** of a. For example, $a^{1/2}$ denotes the positive second root, or **square root**, of a, while $a^{1/4}$ is the positive fourth root of a. When n is *odd*, there is only one nth root, which has the same sign as a. For example, $a^{1/3}$, the **cube root** of a, has the same sign as a. By definition, if $b = a^{1/n}$, then $b^n = a$. On a calculator, a number is raised to a power using a key labeled x^y, y^x, or $\wedge$. For example, to take the fourth root of 6 on a TI-84 Plus calculator, enter $6 \wedge (1/4)$ to get the result 1.56508458.

EXAMPLE 3 **Calculations with Exponents**

(a) $121^{1/2} = 11$, since 11 is positive and $11^2 = 121$.

(b) $625^{1/4} = 5$, since $5^4 = 625$.

(c) $256^{1/4} = 4$

(d) $64^{1/6} = 2$

(e) $27^{1/3} = 3$

(f) $(-32)^{1/5} = -2$

(g) $128^{1/7} = 2$

(h) $(-49)^{1/2}$ is not a real number.

Rational Exponents
In the following definition, the domain of an exponent is extended to include all rational numbers.

Definition of $a^{m/n}$

For all real numbers a for which the indicated roots exist, and for any rational number m/n,

$$a^{m/n} = (a^{1/n})^m.$$

EXAMPLE 4   Calculations with Exponents

(a) $27^{2/3} = (27^{1/3})^2 = 3^2 = 9$

(b) $32^{2/5} = (32^{1/5})^2 = 2^2 = 4$

(c) $64^{4/3} = (64^{1/3})^4 = 4^4 = 256$

(d) $25^{3/2} = (25^{1/2})^3 = 5^3 = 125$

NOTE $27^{2/3}$ could also be evaluated as $(27^2)^{1/3}$, but this is more difficult to perform without a calculator because it involves squaring 27 and then taking the cube root of this large number. On the other hand, when we evaluate it as $(27^{1/3})^2$, we know that the cube root of 27 is 3 without using a calculator, and squaring 3 is easy.

All the properties for integer exponents given in this section also apply to any rational exponent on a nonnegative real-number base.

EXAMPLE 5   Simplifying Exponential Expressions

(a) $\dfrac{y^{1/3}y^{5/3}}{y^3} = \dfrac{y^{1/3+5/3}}{y^3} = \dfrac{y^2}{y^3} = y^{2-3} = y^{-1} = \dfrac{1}{y}$

(b) $m^{2/3}(m^{7/3} + 2m^{1/3}) = m^{2/3+7/3} + 2m^{2/3+1/3} = m^3 + 2m$

(c) $\left(\dfrac{m^7 n^{-2}}{m^{-5}n^2}\right)^{1/4} = \left(\dfrac{m^{7-(-5)}}{n^{2-(-2)}}\right)^{1/4} = \left(\dfrac{m^{12}}{n^4}\right)^{1/4} = \dfrac{(m^{12})^{1/4}}{(n^4)^{1/4}} = \dfrac{m^{12/4}}{n^{4/4}} = \dfrac{m^3}{n}$

In calculus, it is often necessary to factor expressions involving fractional exponents.

EXAMPLE 6   Simplifying Exponential Expressions

Factor out the smallest power of the variable, assuming all variables represent positive real numbers.

(a) $4m^{1/2} + 3m^{3/2}$

SOLUTION   The smallest exponent is $1/2$. Factoring out $m^{1/2}$ yields

$$4m^{1/2} + 3m^{3/2} = m^{1/2}(4m^{1/2-1/2} + 3m^{3/2-1/2})$$
$$= m^{1/2}(4 + 3m).$$

Check this result by multiplying $m^{1/2}$ by $4 + 3m$.

(b) $9x^{-2} - 6x^{-3}$

SOLUTION   The smallest exponent here is -3. Since 3 is a common numerical factor, factor out $3x^{-3}$.

$$9x^{-2} - 6x^{-3} = 3x^{-3}(3x^{-2-(-3)} - 2x^{-3-(-3)}) = 3x^{-3}(3x - 2)$$

Check by multiplying. The factored form can be written without negative exponents as

$$\frac{3(3x - 2)}{x^3}.$$

(c) $(x^2 + 5)(3x - 1)^{-1/2}(2) + (3x - 1)^{1/2}(2x)$

SOLUTION   There is a common factor of 2. Also, $(3x - 1)^{-1/2}$ and $(3x - 1)^{1/2}$ have a common factor. Always factor out the quantity to the *smallest* exponent. Here $-1/2 < 1/2$, so the common factor is $2(3x - 1)^{-1/2}$ and the factored form is

$$2(3x - 1)^{-1/2}[(x^2 + 5) + (3x - 1)x] = 2(3x - 1)^{-1/2}(4x^2 - x + 5).$$

TRY YOUR TURN 2

YOUR TURN 2 Factor $5z^{1/3} + 4z^{-2/3}$.

R.6 EXERCISES

Evaluate each expression. Write all answers without exponents.

1. 8^{-2}

2. 3^{-4}

3. 5^0

4. $\left(-\dfrac{3}{4}\right)^0$

5. $-(-3)^{-2}$

6. $-(-3^{-2})$

7. $\left(\dfrac{1}{6}\right)^{-2}$

8. $\left(\dfrac{4}{3}\right)^{-3}$

Simplify each expression. Assume that all variables represent positive real numbers. Write answers with only positive exponents.

9. $\dfrac{4^{-2}}{4}$

10. $\dfrac{8^9 \cdot 8^{-7}}{8^{-3}}$

11. $\dfrac{10^8 \cdot 10^{-10}}{10^4 \cdot 10^2}$

12. $\left(\dfrac{7^{-12} \cdot 7^3}{7^{-8}}\right)^{-1}$

13. $\dfrac{x^4 \cdot x^3}{x^5}$

14. $\dfrac{y^{10} \cdot y^{-4}}{y^6}$

15. $\dfrac{(4k^{-1})^2}{2k^{-5}}$

16. $\dfrac{(3z^2)^{-1}}{z^5}$

17. $\dfrac{3^{-1} \cdot x \cdot y^2}{x^{-4} \cdot y^5}$

18. $\dfrac{5^{-2}m^2y^{-2}}{5^2m^{-1}y^{-2}}$

19. $\left(\dfrac{a^{-1}}{b^2}\right)^{-3}$

20. $\left(\dfrac{c^3}{7d^{-2}}\right)^{-2}$

Simplify each expression, writing the answer as a single term without negative exponents.

21. $a^{-1} + b^{-1}$

22. $b^{-2} - a$

23. $\dfrac{2n^{-1} - 2m^{-1}}{m + n^2}$

24. $\left(\dfrac{m}{3}\right)^{-1} + \left(\dfrac{n}{2}\right)^{-2}$

25. $(x^{-1} - y^{-1})^{-1}$

26. $(x \cdot y^{-1} - y^{-2})^{-2}$

Write each number without exponents.

27. $121^{1/2}$

28. $27^{1/3}$

29. $32^{2/5}$

30. $-125^{2/3}$

31. $\left(\dfrac{36}{144}\right)^{1/2}$

32. $\left(\dfrac{64}{27}\right)^{1/3}$

33. $8^{-4/3}$

34. $625^{-1/4}$

35. $\left(\dfrac{27}{64}\right)^{-1/3}$

36. $\left(\dfrac{121}{100}\right)^{-3/2}$

Simplify each expression. Write all answers with only positive exponents. Assume that all variables represent positive real numbers.

37. $3^{2/3} \cdot 3^{4/3}$

38. $27^{2/3} \cdot 27^{-1/3}$

39. $\dfrac{4^{9/4} \cdot 4^{-7/4}}{4^{-10/4}}$

40. $\dfrac{3^{-5/2} \cdot 3^{3/2}}{3^{7/2} \cdot 3^{-9/2}}$

41. $\left(\dfrac{x^6 y^{-3}}{x^{-2} y^5}\right)^{1/2}$

42. $\left(\dfrac{a^{-7} b^{-1}}{b^{-4} a^2}\right)^{1/3}$

43. $\dfrac{7^{-1/3} \cdot 7r^{-3}}{7^{2/3} \cdot (r^{-2})^2}$

44. $\dfrac{12^{3/4} \cdot 12^{5/4} \cdot y^{-2}}{12^{-1} \cdot (y^{-3})^{-2}}$

45. $\dfrac{3k^2 \cdot (4k^{-3})^{-1}}{4^{1/2} \cdot k^{7/2}}$

46. $\dfrac{8p^{-3} \cdot (4p^2)^{-2}}{p^{-5}}$

47. $\dfrac{a^{4/3} \cdot b^{1/2}}{a^{2/3} \cdot b^{-3/2}}$

48. $\dfrac{x^{3/2} \cdot y^{4/5} \cdot z^{-3/4}}{x^{5/3} \cdot y^{-6/5} \cdot z^{1/2}}$

49. $\dfrac{k^{-3/5} \cdot h^{-1/3} \cdot t^{2/5}}{k^{-1/5} \cdot h^{-2/3} \cdot t^{1/5}}$

50. $\dfrac{m^{7/3} \cdot n^{-2/5} \cdot p^{3/8}}{m^{-2/3} \cdot n^{3/5} \cdot p^{-5/8}}$

Factor each expression.

51. $3x^3(x^2 + 3x)^2 - 15x(x^2 + 3x)^2$

52. $6x(x^3 + 7)^2 - 6x^2(3x^2 + 5)(x^3 + 7)$

53. $10x^3(x^2 - 1)^{-1/2} - 5x(x^2 - 1)^{1/2}$

54. $9(6x + 2)^{1/2} + 3(9x - 1)(6x + 2)^{-1/2}$

55. $x(2x + 5)^2(x^2 - 4)^{-1/2} + 2(x^2 - 4)^{1/2}(2x + 5)$

56. $(4x^2 + 1)^2(2x - 1)^{-1/2} + 16x(4x^2 + 1)(2x - 1)^{1/2}$

YOUR TURN ANSWERS

1. z^{16}/y^{10}

2. $z^{-2/3}(5z + 4)$

R.7 Radicals

We have defined $a^{1/n}$ as the positive or principal nth root of a for appropriate values of a and n. An alternative notation for $a^{1/n}$ uses radicals.

Radicals

If n is an even natural number and $a > 0$, or n is an odd natural number, then

$$a^{1/n} = \sqrt[n]{a}.$$

The symbol $\sqrt[n]{}$ is a **radical sign**, the number a is the **radicand**, and n is the **index** of the radical. The familiar symbol $\sqrt{a}$ is used instead of $\sqrt[2]{a}$.

EXAMPLE 1 Radical Calculations

(a) $\sqrt[4]{16} = 16^{1/4} = 2$

(b) $\sqrt[5]{-32} = -2$

(c) $\sqrt[3]{1000} = 10$

(d) $\sqrt[6]{\dfrac{64}{729}} = \dfrac{2}{3}$

With $a^{1/n}$ written as $\sqrt[n]{a}$, the expression $a^{m/n}$ also can be written using radicals.

$$a^{m/n} = (\sqrt[n]{a})^m \qquad \text{or} \qquad a^{m/n} = \sqrt[n]{a^m}$$

The following properties of radicals depend on the definitions and properties of exponents.

Properties of Radicals

For all real numbers a and b and natural numbers m and n such that $\sqrt[n]{a}$ and $\sqrt[n]{b}$ are real numbers:

1. $(\sqrt[n]{a})^n = a$

2. $\sqrt[n]{a^n} = \begin{cases} |a| & \text{if } n \text{ is even} \\ a & \text{if } n \text{ is odd} \end{cases}$

3. $\sqrt[n]{a} \cdot \sqrt[n]{b} = \sqrt[n]{ab}$

4. $\dfrac{\sqrt[n]{a}}{\sqrt[n]{b}} = \sqrt[n]{\dfrac{a}{b}} \qquad (b \neq 0)$

5. $\sqrt[m]{\sqrt[n]{a}} = \sqrt[mn]{a}$

Property 3 can be used to simplify certain radicals. For example, since $48 = 16 \cdot 3$,

$$\sqrt{48} = \sqrt{16 \cdot 3} = \sqrt{16} \cdot \sqrt{3} = 4\sqrt{3}.$$

To some extent, simplification is in the eye of the beholder, and $\sqrt{48}$ might be considered as simple as $4\sqrt{3}$. In this textbook, we will consider an expression to be simpler when we have removed as many factors as possible from under the radical.

EXAMPLE 2 Radical Calculations

(a) $\sqrt{1000} = \sqrt{100 \cdot 10} = \sqrt{100} \cdot \sqrt{10} = 10\sqrt{10}$

(b) $\sqrt{128} = \sqrt{64 \cdot 2} = 8\sqrt{2}$

(c) $\sqrt{2} \cdot \sqrt{18} = \sqrt{2 \cdot 18} = \sqrt{36} = 6$

(d) $\sqrt[3]{54} = \sqrt[3]{27 \cdot 2} = \sqrt[3]{27} \cdot \sqrt[3]{2} = 3\sqrt[3]{2}$

(e) $\sqrt{288m^5} = \sqrt{144 \cdot m^4 \cdot 2m} = 12m^2\sqrt{2m}$

(f) $2\sqrt{18} - 5\sqrt{32} = 2\sqrt{9 \cdot 2} - 5\sqrt{16 \cdot 2}$

$$= 2\sqrt{9} \cdot \sqrt{2} - 5\sqrt{16} \cdot \sqrt{2}$$

$$= 2(3)\sqrt{2} - 5(4)\sqrt{2} = -14\sqrt{2}$$

YOUR TURN 1
Simplify $\sqrt{28x^9y^5}$.

(g) $\sqrt{x^5} \cdot \sqrt[3]{x^5} = x^{5/2} \cdot x^{5/3} = x^{5/2 + 5/3} = x^{25/6} = \sqrt[6]{x^{25}} = x^4\sqrt[6]{x}$ **TRY YOUR TURN 1**

When simplifying a square root, keep in mind that $\sqrt{x}$ is nonnegative by definition. Also, $\sqrt{x^2}$ is not x, but $|x|$, the **absolute value of x**, defined as

$$|x| = \begin{cases} x & \text{if } x \geq 0 \\ -x & \text{if } x < 0. \end{cases}$$

For example, $\sqrt{(-5)^2} = |-5| = 5$. It is correct, however, to simplify $\sqrt{x^4} = x^2$. We need not write $|x^2|$ because x^2 is always nonnegative.

EXAMPLE 3   Simplifying by Factoring

Simplify $\sqrt{m^2 - 4m + 4}$.

SOLUTION   Factor the polynomial as $m^2 - 4m + 4 = (m - 2)^2$. Then by property 2 of radicals and the definition of absolute value,

$$\sqrt{(m-2)^2} = |m - 2| = \begin{cases} m - 2 & \text{if } m - 2 \geq 0 \\ -(m - 2) = 2 - m & \text{if } m - 2 < 0. \end{cases}$$

CAUTION   Avoid the common error of writing $\sqrt{a^2 + b^2}$ as $\sqrt{a^2} + \sqrt{b^2}$. We must add a^2 and b^2 *before* taking the square root. For example, $\sqrt{16 + 9} = \sqrt{25} = 5$, *not* $\sqrt{16} + \sqrt{9} = 4 + 3 = 7$. This idea applies as well to higher roots. For example, in general,

$$\sqrt[3]{a^3 + b^3} \neq \sqrt[3]{a^3} + \sqrt[3]{b^3},$$

$$\sqrt[4]{a^4 + b^4} \neq \sqrt[4]{a^4} + \sqrt[4]{b^4}.$$

Also,

$$\sqrt{a + b} \neq \sqrt{a} + \sqrt{b}.$$

Rationalizing Denominators   The next example shows how to *rationalize* (remove all radicals from) the denominator in an expression containing radicals.

EXAMPLE 4   Rationalizing Denominators

Simplify each expression by rationalizing the denominator.

(a) $\dfrac{4}{\sqrt{3}}$

SOLUTION   To rationalize the denominator, multiply by $\sqrt{3}/\sqrt{3}$ (or 1) so that the denominator of the product is a rational number.

$$\frac{4}{\sqrt{3}} \cdot \frac{\sqrt{3}}{\sqrt{3}} = \frac{4\sqrt{3}}{3} \qquad \sqrt{3} \cdot \sqrt{3} = \sqrt{9} = 3$$

(b) $\dfrac{2}{\sqrt[3]{x}}$

SOLUTION   Here, we need a perfect cube under the radical sign to rationalize the denominator. Multiplying by $\sqrt[3]{x^2}/\sqrt[3]{x^2}$ gives

$$\frac{2}{\sqrt[3]{x}} \cdot \frac{\sqrt[3]{x^2}}{\sqrt[3]{x^2}} = \frac{2\sqrt[3]{x^2}}{\sqrt[3]{x^3}} = \frac{2\sqrt[3]{x^2}}{x}.$$

(c) $\dfrac{1}{1 - \sqrt{2}}$

SOLUTION The best approach here is to multiply both numerator and denominator by the number $1 + \sqrt{2}$. The expressions $1 + \sqrt{2}$ and $1 - \sqrt{2}$ are conjugates,* and their product is $1^2 - (\sqrt{2})^2 = 1 - 2 = -1$. Thus,

$$\frac{1}{1 - \sqrt{2}} = \frac{1(1 + \sqrt{2})}{(1 - \sqrt{2})(1 + \sqrt{2})} = \frac{1 + \sqrt{2}}{1 - 2} = -1 - \sqrt{2}.$$

YOUR TURN 2 Rationalize the denominator in
$$\frac{5}{\sqrt{x} - \sqrt{y}}.$$

TRY YOUR TURN 2

Sometimes it is advantageous to rationalize the *numerator* of a rational expression. The following example arises in calculus when evaluating a *limit*.

EXAMPLE 5 Rationalizing Numerators

Rationalize each numerator.

(a) $\dfrac{\sqrt{x} - 3}{x - 9}$.

SOLUTION Multiply the numerator and denominator by the conjugate of the numerator, $\sqrt{x} + 3$.

$$\frac{\sqrt{x} - 3}{x - 9} \cdot \frac{\sqrt{x} + 3}{\sqrt{x} + 3} = \frac{(\sqrt{x})^2 - 3^2}{(x - 9)(\sqrt{x} + 3)} \qquad (a - b)(a + b) = a^2 - b^2$$

$$= \frac{x - 9}{(x - 9)(\sqrt{x} + 3)}$$

$$= \frac{1}{\sqrt{x} + 3}$$

(b) $\dfrac{\sqrt{3} + \sqrt{x + 3}}{\sqrt{3} - \sqrt{x + 3}}$

SOLUTION Multiply the numerator and denominator by the conjugate of the numerator, $\sqrt{3} - \sqrt{x + 3}$.

$$\frac{\sqrt{3} + \sqrt{x + 3}}{\sqrt{3} - \sqrt{x + 3}} \cdot \frac{\sqrt{3} - \sqrt{x + 3}}{\sqrt{3} - \sqrt{x + 3}} = \frac{3 - (x + 3)}{3 - 2\sqrt{3}\sqrt{x + 3} + (x + 3)}$$

$$= \frac{-x}{6 + x - 2\sqrt{3(x + 3)}}$$

R.7 EXERCISES

Simplify each expression by removing as many factors as possible from under the radical. Assume that all variables represent positive real numbers.

1. $\sqrt[3]{125}$

2. $\sqrt[4]{1296}$

3. $\sqrt[5]{-3125}$

4. $\sqrt{50}$

5. $\sqrt{2000}$

6. $\sqrt{32y^5}$

7. $\sqrt{27} \cdot \sqrt{3}$

8. $\sqrt{2} \cdot \sqrt{32}$

9. $7\sqrt{2} - 8\sqrt{18} + 4\sqrt{72}$

10. $4\sqrt{3} - 5\sqrt{12} + 3\sqrt{75}$

11. $4\sqrt{7} - \sqrt{28} + \sqrt{343}$

*If a and b are real numbers, the *conjugate* of $a + b$ is $a - b$.

12. $3\sqrt{28} - 4\sqrt{63} + \sqrt{112}$

13. $\sqrt[3]{2} - \sqrt[3]{16} + 2\sqrt[3]{54}$

14. $2\sqrt[3]{5} - 4\sqrt[3]{40} + 3\sqrt[3]{135}$

15. $\sqrt{2x^3y^2z^4}$ **16.** $\sqrt{160r^7s^9t^{12}}$

17. $\sqrt[3]{128x^3y^8z^9}$ **18.** $\sqrt[4]{x^8y^7z^{11}}$

19. $\sqrt{a^3b^5} - 2\sqrt{a^7b^3} + \sqrt{a^3b^9}$

20. $\sqrt{p^7q^3} - \sqrt{p^5q^9} + \sqrt{p^9q}$

21. $\sqrt{a} \cdot \sqrt[3]{a}$

22. $\sqrt{b^3} \cdot \sqrt[4]{b^3}$

Simplify each root, if possible.

23. $\sqrt{16 - 8x + x^2}$

24. $\sqrt{9y^2 + 30y + 25}$

25. $\sqrt{4 - 25z^2}$

26. $\sqrt{9k^2 + h^2}$

Rationalize each denominator. Assume that all radicands represent positive real numbers.

27. $\dfrac{5}{\sqrt{7}}$ **28.** $\dfrac{5}{\sqrt{10}}$

29. $\dfrac{-3}{\sqrt{12}}$ **30.** $\dfrac{4}{\sqrt{8}}$

31. $\dfrac{3}{1 - \sqrt{2}}$ **32.** $\dfrac{5}{2 - \sqrt{6}}$

33. $\dfrac{6}{2 + \sqrt{2}}$ **34.** $\dfrac{\sqrt{5}}{\sqrt{5} + \sqrt{2}}$

35. $\dfrac{1}{\sqrt{r} - \sqrt{3}}$ **36.** $\dfrac{5}{\sqrt{m} - \sqrt{5}}$

37. $\dfrac{y - 5}{\sqrt{y} - \sqrt{5}}$ **38.** $\dfrac{\sqrt{z} - 1}{\sqrt{z} - \sqrt{5}}$

39. $\dfrac{\sqrt{x} + \sqrt{x + 1}}{\sqrt{x} - \sqrt{x + 1}}$ **40.** $\dfrac{\sqrt{p} + \sqrt{p^2 - 1}}{\sqrt{p} - \sqrt{p^2 - 1}}$

Rationalize each numerator. Assume that all radicands represent positive real numbers.

41. $\dfrac{1 + \sqrt{2}}{2}$ **42.** $\dfrac{3 - \sqrt{3}}{6}$

43. $\dfrac{\sqrt{x} + \sqrt{x + 1}}{\sqrt{x} - \sqrt{x + 1}}$ **44.** $\dfrac{\sqrt{p} - \sqrt{p - 2}}{\sqrt{p}}$

YOUR TURN ANSWERS

1. $2x^4y^2\sqrt{7xy}$ **2.** $5(\sqrt{x} + \sqrt{y})/(x - y)$

Linear Functions

Over short time intervals, many changes in the economy are well modeled by linear functions. In an exercise in the first section of this chapter, we will examine a linear model that predicts the number of cellular telephone users in the United States. Such predictions are important tools for cellular telephone company executives and planners.

Stockbyte/Getty Images

Before using mathematics to solve a real-world problem, we must usually set up a **mathematical model,** a mathematical description of the situation. In this chapter we look at some mathematics of *linear* models, which are used for data whose graphs can be approximated by straight lines. Linear models have an immense number of applications, because even when the underlying phenomenon is not linear, a linear model often provides an approximation that is sufficiently accurate and much simpler to use.

1.1   Slopes and Equations of Lines

APPLY IT   How fast has tuition at public colleges been increasing in recent years, and how well can we predict tuition in the future?
In Example 14 of this section, we will answer these questions using the equation of a line.

There are many everyday situations in which two quantities are related. For example, if a bank account pays 6% simple interest per year, then the interest I that a deposit of P dollars would earn in one year is given by

$$I = 0.06 \cdot P, \quad \text{or} \quad I = 0.06P.$$

The formula $I = 0.06P$ describes the relationship between interest and the amount of money deposited.

Using this formula, we see, for example, that if $P = \$100$, then $I = \$6$, and if $P = \$200$, then $I = \$12$. These corresponding pairs of numbers can be written as **ordered pairs,** $(100, 6)$ and $(200, 12)$, whose order is important. The first number denotes the value of P and the second number the value of I.

Ordered pairs are graphed with the perpendicular number lines of a **Cartesian coordinate system**, shown in Figure 1.* The horizontal number line, or x-**axis**, represents the first components of the ordered pairs, while the vertical or y-**axis** represents the second components. The point where the number lines cross is the zero point on both lines; this point is called the **origin**.

Each point on the xy-plane corresponds to an ordered pair of numbers, where the x-value is written first. From now on, we will refer to the point corresponding to the ordered pair (x, y) as "the point (x, y)."

Locate the point $(-2, 4)$ on the coordinate system by starting at the origin and counting 2 units to the left on the horizontal axis and 4 units upward, parallel to the vertical axis. This point is shown in Figure 1, along with several other sample points. The number -2 is the x-**coordinate** and the number 4 is the y-**coordinate** of the point $(-2, 4)$.

The x-axis and y-axis divide the plane into four parts, or **quadrants**. For example, quadrant I includes all those points whose x- and y-coordinates are both positive. The quadrants are numbered as shown in Figure 1. The points on the axes themselves belong to no quadrant. The set of points corresponding to the ordered pairs of an equation is the **graph** of the equation.

The x- and y-values of the points where the graph of an equation crosses the axes are called the x-**intercept** and y-**intercept**, respectively.** See Figure 2.

---

*The name "Cartesian" honors René Descartes (1596–1650), one of the greatest mathematicians of the seventeenth century. According to legend, Descartes was lying in bed when he noticed an insect crawling on the ceiling and realized that if he could determine the distance from the bug to each of two perpendicular walls, he could describe its position at any given moment. The same idea can be used to locate a point in a plane.
**Some people prefer to define the intercepts as ordered pairs, rather than as numbers.

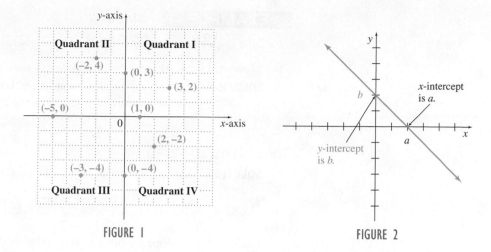

FIGURE 1

FIGURE 2

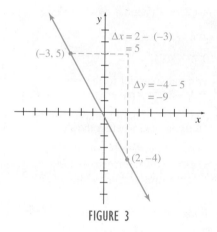

FIGURE 3

Slope of a Line

An important characteristic of a straight line is its *slope*, a number that represents the "steepness" of the line. To see how slope is defined, look at the line in Figure 3. The line goes through the points $(x_1, y_1) = (-3, 5)$ and $(x_2, y_2) = (2, -4)$. The difference in the two x-values,

$$x_2 - x_1 = 2 - (-3) = 5$$

in this example, is called the **change in x**. The symbol Δx (read "delta x") is used to represent the change in x. In the same way, Δy represents the **change in y**. In our example,

$$\Delta y = y_2 - y_1$$
$$= -4 - 5$$
$$= -9.$$

These symbols, Δx and Δy, are used in the following definition of slope.

Slope of a Line

The **slope** of a line is defined as the vertical change (the "rise") over the horizontal change (the "run") as one travels along the line. In symbols, taking two different points (x_1, y_1) and (x_2, y_2) on the line, the slope is

$$m = \frac{\text{Change in } y}{\text{Change in } x} = \frac{\Delta y}{\Delta x} = \frac{y_2 - y_1}{x_2 - x_1},$$

where $x_1 \neq x_2$.

By this definition, the slope of the line in Figure 3 is

$$m = \frac{\Delta y}{\Delta x} = \frac{-4 - 5}{2 - (-3)} = -\frac{9}{5}.$$

The slope of a line tells how fast y changes for each unit of change in x.

NOTE Using similar triangles, it can be shown that the slope of a line is independent of the choice of points on the line. That is, the same slope will be obtained for *any* choice of two different points on the line.

EXAMPLE 1 Slope

Find the slope of the line through each pair of points.

(a) $(7, 6)$ and $(-4, 5)$

SOLUTION Let $(x_1, y_1) = (7, 6)$ and $(x_2, y_2) = (-4, 5)$. Use the definition of slope.

$$m = \frac{\Delta y}{\Delta x} = \frac{5 - 6}{-4 - 7} = \frac{-1}{-11} = \frac{1}{11}$$

(b) $(5, -3)$ and $(-2, -3)$

SOLUTION Let $(x_1, y_1) = (5, -3)$ and $(x_2, y_2) = (-2, -3)$. Then

$$m = \frac{-3 - (-3)}{-2 - 5} = \frac{0}{-7} = 0.$$

Lines with zero slope are horizontal (parallel to the x-axis).

(c) $(2, -4)$ and $(2, 3)$

SOLUTION Let $(x_1, y_1) = (2, -4)$ and $(x_2, y_2) = (2, 3)$. Then

$$m = \frac{3 - (-4)}{2 - 2} = \frac{7}{0},$$

which is undefined. This happens when the line is vertical (parallel to the y-axis).

TRY YOUR TURN 1

YOUR TURN 1 Find the slope of the line through $(1, 5)$ and $(4, 6)$.

CAUTION The phrase "no slope" should be avoided; specify instead whether the slope is zero or undefined.

In finding the slope of the line in Example 1(a), we could have let $(x_1, y_1) = (-4, 5)$ and $(x_2, y_2) = (7, 6)$. In that case,

$$m = \frac{6 - 5}{7 - (-4)} = \frac{1}{11},$$

the same answer as before. The order in which coordinates are subtracted does not matter, as long as it is done consistently.

Figure 4 shows examples of lines with different slopes. Lines with positive slopes go up from left to right, while lines with negative slopes go down from left to right.

It might help you to compare slope with the percent grade of a hill. If a sign says a hill has a 10% grade uphill, this means the slope is 0.10, or 1/10, so the hill rises 1 foot for every 10 feet horizontally. A 15% grade downhill means the slope is −0.15.

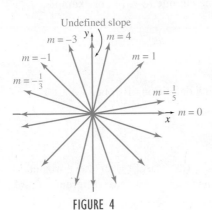

FIGURE 4

Equations of a Line

An equation in two first-degree variables, such as $4x + 7y = 20$, has a line as its graph, so it is called a **linear equation**. In the rest of this section, we consider various forms of the equation of a line.

Suppose a line has a slope m and y-intercept b. This means that it goes through the point $(0, b)$. If (x, y) is any other point on the line, then the definition of slope tells us that

$$m = \frac{y - b}{x - 0}.$$

We can simplify this equation by multiplying both sides by x and adding b to both sides. The result is

$$y = mx + b,$$

which we call the *slope-intercept* form of a line. This is the most common form for writing the equation of a line.

-FOR REVIEW-

For review on solving a linear equation, see Section R.4.

Slope-Intercept Form

If a line has slope m and y-intercept b, then the equation of the line in **slope-intercept form** is

$$y = mx + b.$$

When $b = 0$, we say that y is **proportional** to x.

EXAMPLE 2   Equation of a Line

Find an equation in slope-intercept form for each line.

(a) Through $(0, -3)$ with slope $3/4$

SOLUTION   We recognize $(0, -3)$ as the y-intercept because it's the point with 0 as its x-coordinate, so $b = -3$. The slope is $3/4$, so $m = 3/4$. Substituting these values into $y = mx + b$ gives

$$y = \frac{3}{4}x - 3.$$

(b) With x-intercept 7 and y-intercept 2

SOLUTION   Notice that $b = 2$. To find m, use the definition of slope after writing the x-intercept as $(7, 0)$ (because the y-coordinate is 0 where the line crosses the x-axis) and the y-intercept as $(0, 2)$:

$$m = \frac{0 - 2}{7 - 0} = -\frac{2}{7}.$$

YOUR TURN 2 Find the equation of the line with x-intercept -4 and y-intercept 6.

Substituting these values into $y = mx + b$, we have

$$y = -\frac{2}{7}x + 2.$$

TRY YOUR TURN 2

EXAMPLE 3   Finding the Slope

Find the slope of the line whose equation is $3x - 4y = 12$.

SOLUTION   To find the slope, solve the equation for y.

$$3x - 4y = 12$$
$$-4y = -3x + 12 \qquad \text{Subtract } 3x \text{ from both sides.}$$
$$y = \frac{3}{4}x - 3 \qquad \text{Divide both sides by } -4.$$

YOUR TURN 3 Find the slope of the line whose equation is $8x + 3y = 5$.

The coefficient of x is $3/4$, which is the slope of the line. Notice that this is the same line as in Example 2 (a).

TRY YOUR TURN 3

The slope-intercept form of the equation of a line involves the slope and the y-intercept. Sometimes, however, the slope of a line is known, together with one point (perhaps *not* the y-intercept) that the line goes through. The *point-slope form* of the equation of a line is used to find the equation in this case. Let (x_1, y_1) be any fixed point on the line, and let (x, y) represent any other point on the line. If m is the slope of the line, then by the definition of slope,

$$\frac{y - y_1}{x - x_1} = m,$$

or

$$y - y_1 = m(x - x_1). \qquad \text{Multiply both sides by } x - x_1.$$

> **Point-Slope Form**
> If a line has slope m and passes through the point (x_1, y_1), then an equation of the line is given by
> $$y - y_1 = m(x - x_1),$$
> the **point-slope form** of the equation of a line.

EXAMPLE 4 Point-Slope Form

Find an equation of the line that passes through the point $(3, -7)$ and has slope $m = 5/4$.

SOLUTION Use the point-slope form.

$$y - y_1 = m(x - x_1)$$

$$y - (-7) = \frac{5}{4}(x - 3) \qquad y_1 = -7, m = \tfrac{5}{4}, x_1 = 3$$

$$y + 7 = \frac{5}{4}(x - 3)$$

$$4y + 28 = 5(x - 3) \qquad \text{Multiply both sides by 4.}$$

$$4y + 28 = 5x - 15 \qquad \text{Distribute.}$$

$$4y = 5x - 43 \qquad \text{Combine constants.}$$

$$y = \frac{5}{4}x - \frac{43}{4} \qquad \text{Divide both sides by 4.}$$

FOR REVIEW

See Section R.4 for details on eliminating denominators in an equation.

The equation of the same line can be given in many forms. To avoid confusion, the linear equations used in the rest of this section will be written in slope-intercept form, $y = mx + b$, which is often the most useful form.

The point-slope form also can be useful to find an equation of a line if we know two different points that the line goes through, as in the next example.

EXAMPLE 5 Using Point-Slope Form to Find an Equation

Find an equation of the line through $(5, 4)$ and $(-10, -2)$.

SOLUTION Begin by using the definition of slope to find the slope of the line that passes through the given points.

$$\text{Slope} = m = \frac{-2 - 4}{-10 - 5} = \frac{-6}{-15} = \frac{2}{5}$$

Either $(5, 4)$ or $(-10, -2)$ can be used in the point-slope form with $m = 2/5$. If $(x_1, y_1) = (5, 4)$, then

$$y - y_1 = m(x - x_1)$$

$$y - 4 = \frac{2}{5}(x - 5) \qquad y_1 = 4, m = \tfrac{2}{5}, x_1 = 5$$

$$5y - 20 = 2(x - 5) \qquad \text{Multiply both sides by 5.}$$

$$5y - 20 = 2x - 10 \qquad \text{Distributive property}$$

$$5y = 2x + 10 \qquad \text{Add 20 to both sides.}$$

$$y = \frac{2}{5}x + 2 \qquad \text{Divide by 5 to put in slope-intercept form.}$$

YOUR TURN 4 Find the equation of the line through $(2, 9)$ and $(5, 3)$. Put your answer in slope-intercept form.

Check that the same result is found if $(x_1, y_1) = (-10, -2)$.   **TRY YOUR TURN 4**

EXAMPLE 6 **Horizontal Line**

Find an equation of the line through $(8, -4)$ and $(-2, -4)$.

SOLUTION Find the slope.

$$m = \frac{-4 - (-4)}{-2 - 8} = \frac{0}{-10} = 0$$

Choose, say, $(8, -4)$ as (x_1, y_1).

$$y - y_1 = m(x - x_1)$$
$$y - (-4) = 0(x - 8) \qquad y_1 = -4, m = 0, x_1 = 8$$
$$y + 4 = 0 \qquad\qquad 0(x - 8) = 0$$
$$y = -4$$

Plotting the given ordered pairs and drawing a line through the points show that the equation $y = -4$ represents a horizontal line. See Figure 5(a). Every horizontal line has a slope of zero and an equation of the form $y = k$, where k is the y-value of all ordered pairs on the line.

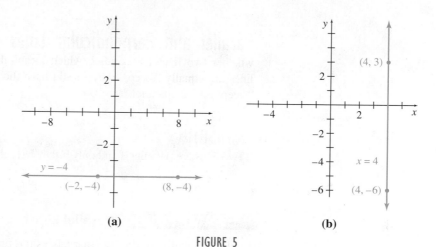

(a)   **(b)**

FIGURE 5

EXAMPLE 7 **Vertical Line**

Find an equation of the line through $(4, 3)$ and $(4, -6)$.

SOLUTION The slope of the line is

$$m = \frac{-6 - 3}{4 - 4} = \frac{-9}{0},$$

which is undefined. Since both ordered pairs have x-coordinate 4, the equation is $x = 4$. Because the slope is undefined, the equation of this line cannot be written in the slope-intercept form.

Again, plotting the given ordered pairs and drawing a line through them show that the graph of $x = 4$ is a vertical line. See Figure 5(b).

Slope of Horizontal and Vertical Lines

The slope of a horizontal line is 0.

The slope of a vertical line is undefined.

The different forms of linear equations discussed in this section are summarized below. The slope-intercept and point-slope forms are equivalent ways to express the equation of a nonvertical line. The slope-intercept form is simpler for a final answer, but you may find the point-slope form easier to use when you know the slope of a line and a point through which the line passes. The slope-intercept form is often considered the standard form. Any line that is not vertical has a unique slope-intercept form but can have many point-slope forms for its equation.

Equations of Lines	
Equation	**Description**
$y = mx + b$	**Slope-intercept form:** slope m, y-intercept b
$y - y_1 = m(x - x_1)$	**Point-slope form:** slope m, line passes through (x_1, y_1)
$x = k$	**Vertical line:** x-intercept k, no y-intercept (except when $k = 0$), undefined slope
$y = k$	**Horizontal line:** y-intercept k, no x-intercept (except when $k = 0$), slope 0

Parallel and Perpendicular Lines

One application of slope involves deciding whether two lines are parallel, which means that they never intersect. Since two parallel lines are equally "steep," they should have the same slope. Also, two lines with the same "steepness" are parallel.

Parallel Lines
Two lines are **parallel** if and only if they have the same slope, or if they are both vertical.

EXAMPLE 8   Parallel Line

Find the equation of the line that passes through the point $(3, 5)$ and is parallel to the line $2x + 5y = 4$.

SOLUTION   The slope of $2x + 5y = 4$ can be found by writing the equation in slope-intercept form.

$$2x + 5y = 4$$

$$y = -\frac{2}{5}x + \frac{4}{5} \qquad \text{Subtract } 2x \text{ from both sides and divide both sides by 5.}$$

This result shows that the slope is $-2/5$. Since the lines are parallel, $-2/5$ is also the slope of the line whose equation we want. This line passes through $(3, 5)$. Substituting $m = -2/5$, $x_1 = 3$, and $y_1 = 5$ into the point-slope form gives

$$y - y_1 = m(x - x_1)$$

$$y - 5 = -\frac{2}{5}(x - 3) = -\frac{2}{5}x + \frac{6}{5}$$

$$y = -\frac{2}{5}x + \frac{6}{5} + 5$$

$$y = -\frac{2}{5}x + \frac{31}{5}. \qquad \text{Multiply 5 by 5/5 to get a common denominator.}$$

YOUR TURN 5   Find the equation of the line that passes through the point $(4, 5)$ and is parallel to the line $3x - 6y = 7$. Put your answer in slope-intercept form.

TRY YOUR TURN 5

As already mentioned, two nonvertical lines are parallel if and only if they have the same slope. Two lines having slopes with a product of -1 are perpendicular. A proof of this fact, which depends on similar triangles from geometry, is given as Exercise 43 in this section.

> ### Perpendicular Lines
> Two lines are **perpendicular** if and only if the product of their slopes is -1, or if one is vertical and the other horizontal.

EXAMPLE 9   Perpendicular Line

Find the equation of the line L passing through the point $(3, 7)$ and perpendicular to the line having the equation $5x - y = 4$.

SOLUTION   To find the slope, write $5x - y = 4$ in slope-intercept form:

$$y = 5x - 4.$$

The slope is 5. Since the lines are perpendicular, if line L has slope m, then

$$5m = -1$$

$$m = -\frac{1}{5}.$$

Now substitute $m = -1/5$, $x_1 = 3$, and $y_1 = 7$ into the point-slope form.

$$y - 7 = -\frac{1}{5}(x - 3)$$

$$y - 7 = -\frac{1}{5}x + \frac{3}{5}$$

$$y = -\frac{1}{5}x + \frac{3}{5} + 7 \cdot \frac{5}{5} \qquad \text{Add 7 to both sides and get a common denominator.}$$

$$y = -\frac{1}{5}x + \frac{38}{5}$$

TRY YOUR TURN 6

YOUR TURN 6   Find the equation of the line passing through the point $(3, 2)$ and perpendicular to the line having the equation $2x + 3y = 4$.

The next example uses the equation of a line to analyze real-world data. In this example, we are looking at how one variable changes over time. To simplify the arithmetic, we will *rescale* the variable representing time, although computers and calculators have made rescaling less important than in the past. Here it allows us to work with smaller numbers, and, as you will see, find the y-intercept of the line more easily. We will use rescaling on many examples throughout this book. When we do, it is important to be consistent.

EXAMPLE 10   Prevalence of Cigarette Smoking

In recent years, the percentage of the U.S. population age 18 and older who smoke has decreased at a roughly constant rate, from 24.1% in 1998 to 20.6% in 2008. *Source: Centers for Disease Control and Prevention.*

(a) Find the equation describing this linear relationship.

SOLUTION   Let t represent time in years, with $t = 0$ representing 1990. With this rescaling, the year 1998 corresponds to $t = 8$ and the year 2008 corresponds to $t = 2008 - 1990 = 18$. Let y represent the percentage of the population who smoke. The two ordered pairs representing the given information are then $(8, 24.1)$ and $(18, 20.6)$. The slope of the line through these points is

$$m = \frac{20.6 - 24.1}{18 - 8} = \frac{-3.5}{10} = -0.35.$$

This means that, on average, the percentage of the adult population who smoke is decreasing by about 0.35% per year.

Using $m = -0.35$ in the point-slope form, and choosing $(t_1, y_1) = (8, 24.1)$, gives the required equation.

$$y - 24.1 = -0.35(t - 8)$$
$$y - 24.1 = -0.35t + 2.8$$
$$y = -0.35t + 26.9$$

We could have used the other point $(18, 20.6)$ and found the same answer. Instead, we'll use this to check our answer by observing that $-0.35(18) + 26.9 = 20.6$, which agrees with the y-value at $t = 18$.

(b) One objective of Healthy People 2010 (a campaign of the U.S. Department of Health and Human Services) was to reduce the percentage of U.S. adults who smoke to 12% or less by the year 2010. If this decline in smoking continued at the same rate, did they meet this objective?

SOLUTION Using the same rescaling, $t = 20$ corresponds to the year 2010. Substituting this value into the above equation gives

$$y = -0.35(20) + 26.9 = 19.9.$$

Continuing at this rate, an estimated 19.9% of the adult population still smoked in 2010, and the objective of Healthy People 2010 was not met.

Notice that if the formula from part (a) of Example 10 is valid for all nonnegative t, then eventually y becomes 0:

$$-0.35t + 26.9 = 0$$
$$-0.35t = -26.9 \qquad \text{Subtract 26.9 from both sides.}$$
$$t = \frac{-26.9}{-0.35} = 76.857 \approx 77^*, \qquad \text{Divide both sides by } -0.35.$$

which indicates that 77 years from 1990 (in the year 2067), 0% of the U.S. adult population will smoke. Of course, it is still possible that in 2067 there will be adults who smoke; the trend of recent years may not continue. Most equations are valid for some specific set of numbers. It is highly speculative to extrapolate beyond those values.

On the other hand, people in business and government often need to make some prediction about what will happen in the future, so a tentative conclusion based on past trends may be better than no conclusion at all. There are also circumstances, particularly in the physical sciences, in which theoretical reasons imply that the trend will continue.

Graph of a Line

We can graph the linear equation defined by $y = x + 1$ by finding several ordered pairs that satisfy the equation. For example, if $x = 2$, then $y = 2 + 1 = 3$, giving the ordered pair $(2, 3)$. Also, $(0, 1)$, $(4, 5)$, $(-2, -1)$, $(-5, -4)$, $(-3, -2)$, among many others, satisfy the equation.

To graph $y = x + 1$, we begin by locating the ordered pairs obtained above, as shown in Figure 6(a). All the points of this graph appear to lie on a straight line, as in Figure 6(b). This straight line is the graph of $y = x + 1$.

It can be shown that every equation of the form $ax + by = c$ has a straight line as its graph, assuming a and b are not both 0. Although just two points are needed to determine a line, it is a good idea to plot a third point as a check. It is often convenient to use the x- and y-intercepts as the two points, as in the following example.

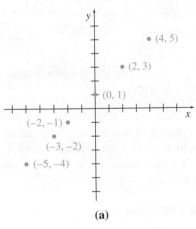

(a)

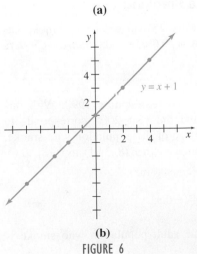

(b)

FIGURE 6

*The symbol $\approx$ means "is approximately equal to."

EXAMPLE 11   Graph of a Line

Graph $3x + 2y = 12$.

SOLUTION   To find the y-intercept, let $x = 0$.

$$3(0) + 2y = 12$$
$$2y = 12 \quad \text{Divide both sides by 2.}$$
$$y = 6$$

Similarly, find the x-intercept by letting $y = 0$, which gives $x = 4$. Verify that when $x = 2$, the result is $y = 3$. These three points are plotted in Figure 7(a). A line is drawn through them in Figure 7(b).

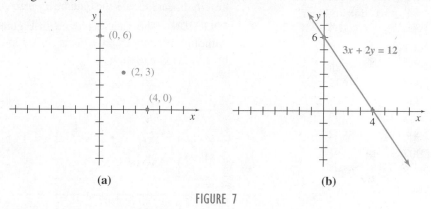

(a)   **(b)**

FIGURE 7

Not every line has two distinct intercepts; the graph in the next example does not cross the x-axis, and so it has no x-intercept.

EXAMPLE 12   Graph of a Horizontal Line

Graph $y = -3$.

SOLUTION   The equation $y = -3$, or equivalently, $y = 0x - 3$, always gives the same y-value, -3, for any value of x. Therefore, no value of x will make $y = 0$, so the graph has no x-intercept. As we saw in Example 6, the graph of such an equation is a horizontal line parallel to the x-axis. In this case the y-intercept is -3, as shown in Figure 8.

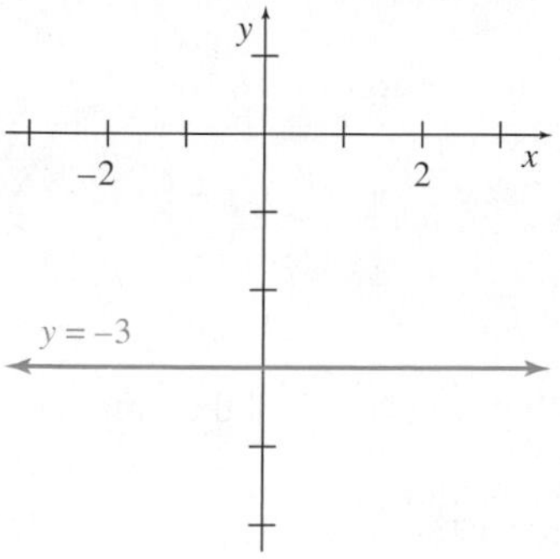

The graph in Example 12 has only one intercept. Another type of linear equation with coinciding intercepts is graphed in Example 13.

EXAMPLE 13   Graph of a Line Through the Origin

Graph $y = -3x$.

SOLUTION   Begin by looking for the x-intercept. If $y = 0$, then

$$y = -3x$$
$$0 = -3x \quad \text{Let } y = 0.$$
$$0 = x. \quad \text{Divide both sides by } -3.$$

We have the ordered pair $(0, 0)$. Starting with $x = 0$ gives exactly the same ordered pair, $(0, 0)$. Two points are needed to determine a straight line, and the intercepts have led to only one point. To get a second point, we choose some other value of x (or y). For example, if $x = 2$, then

$$y = -3x = -3(2) = -6, \quad \text{Let } x = 2.$$

giving the ordered pair $(2, -6)$. These two ordered pairs, $(0, 0)$ and $(2, -6)$, were used to get the graph shown in Figure 9.

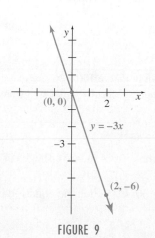

FIGURE 8

FIGURE 9

Linear equations allow us to set up simple mathematical models for real-life situations. In almost every case, linear (or any other reasonably simple) equations provide only approximations to real-world situations. Nevertheless, these are often remarkably useful approximations.

EXAMPLE 14 Tuition

The table on the left lists the average annual cost (in dollars) of tuition and fees at public four-year colleges for selected years. *Source: The College Board.*

Cost of Public College

Year	Tuition and Fees
2000	3508
2001	3766
2002	4098
2003	4645
2004	5126
2005	5492
2006	5804
2007	6191
2008	6591
2009	7020

(a) Plot the cost of public colleges by letting $t = 0$ correspond to 2000. Are the data *exactly* linear? Could the data be *approximated* by a linear equation?

SOLUTION The data is plotted in Figure 10(a) in a figure known as a **scatterplot**. Although it is not exactly linear, it is approximately linear and could be approximated by a linear equation.

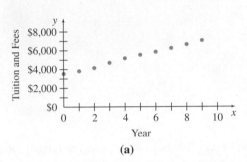

(a)

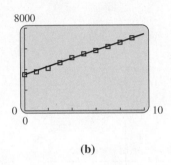

(b)

FIGURE 10

(b) Use the points (0, 3508) and (9, 7020) to determine an equation that models the data.

SOLUTION We first find the slope of the line as follows:

$$m = \frac{7020 - 3508}{9 - 0} = \frac{3512}{9} \approx 390.2.$$

We have rounded to four digits, noting that we cannot expect more accuracy in our answer than in our data, which is accurate to four digits. Using the slope-intercept form of the line, $y = mt + b$, with $m = 390.2$ and $b = 3508$, gives

$$y = 390.2t + 3508.$$

TECHNOLOGY NOTE A graphing calculator plot of this line and the data points are shown in Figure 10(b). Notice that the points closely fit the line. More details on how to construct this graphing calculator plot are given at the end of this example.

(c) Discuss the accuracy of using this equation to estimate the cost of public colleges in the year 2030.

SOLUTION The year 2030 corresponds to the year $t = 30$, for which the equation predicts a cost of

$$y = 390.2(30) + 3508 = 15{,}214, \quad \text{or} \quad \$15{,}214.$$

The year 2030 is many years in the future, however. Many factors could affect the tuition, and the actual figure for 2030 could turn out to be very different from our prediction. ▬

TECHNOLOGY NOTE

You can plot data with a TI-84 Plus graphing calculator using the following steps.

1. Store the data in lists.
2. Define the stat plot.
3. Turn off $Y =$ functions (unless you also want to graph a function).
4. Turn on the plot you want to display.
5. Define the viewing window.
6. Display the graph.

Consult the calculator's instruction booklet or the *Graphing Calculator and Excel Spreadsheet Manual*, available with this book, for specific instructions. See the calculator-generated graph in Figure 10(b), which includes the points and line from Example 14. Notice how the line closely approximates the data.

1.1 EXERCISES

Find the slope of each line.

1. Through $(4, 5)$ and $(-1, 2)$

2. Through $(5, -4)$ and $(1, 3)$

3. Through $(8, 4)$ and $(8, -7)$

4. Through $(1, 5)$ and $(-2, 5)$

5. $y = x$ **6.** $y = 3x - 2$

7. $5x - 9y = 11$ **8.** $4x + 7y = 1$

9. $x = 5$ **10.** The x-axis

11. $y = 8$ **12.** $y = -6$

13. A line parallel to $6x - 3y = 12$

14. A line perpendicular to $8x = 2y - 5$

In Exercises 15–24, find an equation in slope-intercept form for each line.

15. Through $(1, 3)$, $m = -2$

16. Through $(2, 4)$, $m = -1$

17. Through $(-5, -7)$, $m = 0$

18. Through $(-8, 1)$, with undefined slope

19. Through $(4, 2)$ and $(1, 3)$

20. Through $(8, -1)$ and $(4, 3)$

21. Through $(2/3, 1/2)$ and $(1/4, -2)$

22. Through $(-2, 3/4)$ and $(2/3, 5/2)$

23. Through $(-8, 4)$ and $(-8, 6)$

24. Through $(-1, 3)$ and $(0, 3)$

In Exercises 25–34, find an equation for each line in the form $ax + by = c$, where a, b, and c are integers with no factor common to all three and $a \geq 0$.

25. x-intercept -6, y-intercept -3

26. x-intercept -2, y-intercept 4

27. Vertical, through $(-6, 5)$

28. Horizontal, through $(8, 7)$

29. Through $(-4, 6)$, parallel to $3x + 2y = 13$

30. Through $(2, -5)$, parallel to $2x - y = -4$

31. Through $(3, -4)$, perpendicular to $x + y = 4$

32. Through $(-2, 6)$, perpendicular to $2x - 3y = 5$

33. The line with y-intercept 4 and perpendicular to $x + 5y = 7$

34. The line with x-intercept $-2/3$ and perpendicular to $2x - y = 4$

35. Do the points $(4, 3)$, $(2, 0)$, and $(-18, -12)$ lie on the same line? Explain why or why not. (*Hint:* Find the slopes between the points.)

36. Find k so that the line through $(4, -1)$ and $(k, 2)$ is
   a. parallel to $2x + 3y = 6$,
   b. perpendicular to $5x - 2y = -1$.

37. Use slopes to show that the quadrilateral with vertices at $(1, 3)$, $(-5/2, 2)$, $(-7/2, 4)$, and $(2, 1)$ is a parallelogram.

38. Use slopes to show that the square with vertices at $(-2, 5)$, $(4, 5)$, $(4, -1)$, and $(-2, -1)$ has diagonals that are perpendicular.

For the lines in Exercises 39 and 40, which of the following is closest to the slope of the line? (a) 1 (b) 2 (c) 3 (d) 21 (e) 22 (f) −3

39.

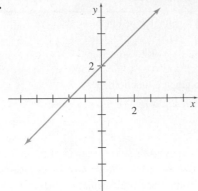

40.

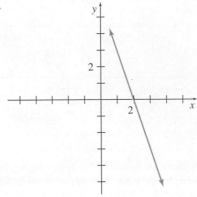

In Exercises 41 and 42, estimate the slope of the lines.

41.

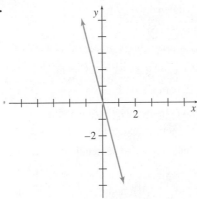

42.

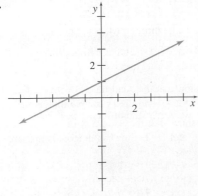

43. To show that two perpendicular lines, neither of which is vertical, have slopes with a product of −1, go through the following steps. Let line L_1 have equation $y = m_1x + b_1$, and let L_2 have equation $y = m_2x + b_2$, with $m_1 > 0$ and $m_2 < 0$. Assume that L_1 and L_2 are perpendicular, and use right triangle *MPN* shown in the figure. Prove each of the following statements.

a. *MQ* has length m_1.

b. *QN* has length $-m_2$.

c. Triangles *MPQ* and *PNQ* are similar.

d. $m_1/1 = 1/(-m_2)$ and $m_1m_2 = -1$.

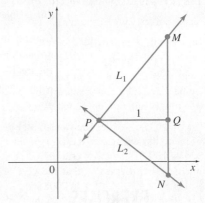

44. Consider the equation $\dfrac{x}{a} + \dfrac{y}{b} = 1$.

a. Show that this equation represents a line by writing it in the form $y = mx + b$.

b. Find the *x*- and *y*-intercepts of this line.

c. Explain in your own words why the equation in this exercise is known as the intercept form of a line.

Graph each equation.

45. $y = x - 1$

46. $y = 4x + 5$

47. $y = -4x + 9$

48. $y = -6x + 12$

49. $2x - 3y = 12$

50. $3x - y = -9$

51. $3y - 7x = -21$

52. $5y + 6x = 11$

53. $y = -2$

54. $x = 4$

55. $x + 5 = 0$

56. $y + 8 = 0$

57. $y = 2x$

58. $y = -5x$

59. $x + 4y = 0$

60. $3x - 5y = 0$

APPLICATIONS

Business and Economics

61. Sales The sales of a small company were $27,000 in its second year of operation and $63,000 in its fifth year. Let *y* represent sales in the *x*th year of operation. Assume that the data can be approximated by a straight line.

a. Find the slope of the sales line, and give an equation for the line in the form $y = mx + b$.

b. Use your answer from part a to find out how many years must pass before the sales surpass $100,000.

62. Use of Cellular Telephones The following table shows the subscribership of cellular telephones in the United States (in millions) for even-numbered years between 2000 and 2008. *Source: Time Almanac 2010.*

Year	2000	2002	2004	2006	2008
Subscribers (in millions)	109.48	140.77	182.14	233.04	270.33

a. Plot the data by letting $t = 0$ correspond to 2000. Discuss how well the data fit a straight line.

b. Determine a linear equation that approximates the number of subscribers using the points $(0, 109.48)$ and $(8, 270.33)$.

c. Repeat part b using the points $(2, 140.77)$ and $(8, 270.33)$.

d. Discuss why your answers to parts b and c are similar but not identical.

e. Using your equations from parts b and c, approximate the number of cellular phone subscribers in the year 2007. Compare your result with the actual value of 255.40 million.

63. Consumer Price Index The Consumer Price Index (CPI) is a measure of the change in the cost of goods over time. The index was 100 for the three-year period centered on 1983. For simplicity, we will assume that the CPI was exactly 100 in 1983. Then the CPI of 215.3 in 2008 indicates that an item that cost $1.00 in 1983 would cost $2.15 in 2008. The CPI has been increasing approximately linearly over the last few decades. *Source: Time Almanac 2010.*

a. Use this information to determine an equation for the CPI in terms of t, which represents the years since 1980.

b. Based on the answer to part a, what was the predicted value of the CPI in 2000? Compare this estimate with the actual CPI of 172.2.

c. Describe the rate at which the annual CPI is changing.

Life Sciences

64. HIV Infection The time interval between a person's initial infection with HIV and that person's eventual development of AIDS symptoms is an important issue. The method of infection with HIV affects the time interval before AIDS develops. One study of HIV patients who were infected by intravenous drug use found that 17% of the patients had AIDS after 4 years, and 33% had developed the disease after 7 years. The relationship between the time interval and the percentage of patients with AIDS can be modeled accurately with a linear equation. *Source: Epidemiologic Review.*

a. Write a linear equation $y = mt + b$ that models this data, using the ordered pairs $(4, 0.17)$ and $(7, 0.33)$.

b. Use your equation from part a to predict the number of years before half of these patients will have AIDS.

65. Exercise Heart Rate To achieve the maximum benefit for the heart when exercising, your heart rate (in beats per minute) should be in the target heart rate zone. The lower limit of this zone is found by taking 70% of the difference between 220 and your age. The upper limit is found by using 85%. *Source: Physical Fitness.*

a. Find formulas for the upper and lower limits (u and l) as linear equations involving the age x.

b. What is the target heart rate zone for a 20-year-old?

c. What is the target heart rate zone for a 40-year-old?

d. Two women in an aerobics class stop to take their pulse and are surprised to find that they have the same pulse. One woman is 36 years older than the other and is working at the upper limit of her target heart rate zone. The younger woman is working at the lower limit of her target heart rate zone. What are the ages of the two women, and what is their pulse?

e. Run for 10 minutes, take your pulse, and see if it is in your target heart rate zone. (After all, this is listed as an exercise!)

66. Ponies Trotting A 1991 study found that the peak vertical force on a trotting Shetland pony increased linearly with the pony's speed, and that when the force reached a critical level, the pony switched from a trot to a gallop. For one pony, the critical force was 1.16 times its body weight. It experienced a force of 0.75 times its body weight at a speed of 2 meters per second and a force of 0.93 times its body weight at 3 meters per second. At what speed did the pony switch from a trot to a gallop? *Source: Science.*

67. Life Expectancy Some scientists believe there is a limit to how long humans can live. One supporting argument is that during the last century, life expectancy from age 65 has increased more slowly than life expectancy from birth, so eventually these two will be equal, at which point, according to these scientists, life expectancy should increase no further. In 1900, life expectancy at birth was 46 yr, and life expectancy at age 65 was 76 yr. In 2004, these figures had risen to 77.8 and 83.7, respectively. In both cases, the increase in life expectancy has been linear. Using these assumptions and the data given, find the maximum life expectancy for humans. *Source: Science.*

Social Sciences

68. Child Mortality Rate The mortality rate for children under 5 years of age around the world has been declining in a roughly linear fashion in recent years. The rate per 1000 live births was 90 in 1990 and 65 in 2008. *Source: World Health Organization.*

a. Determine a linear equation that approximates the mortality rate in terms of time t, where t represents the number of years since 1900.

b. If this trend continues, in what year will the mortality rate first drop to 50 or below per 1000 live births?

69. Health Insurance The percentage of adults in the United States without health insurance increased at a roughly linear rate from 1999, when it was 17.2%, to 2008, when it was 20.3%. *Source: The New York Times.*

a. Determine a linear equation that approximates the percentage of adults in the United States without health insurance in terms of time t, where t represents the number of years since 1990.

b. If this trend were to continue, in what year would the percentage of adults without health insurance be at least 25%?

70. Marriage The following table lists the U.S. median age at first marriage for men and women. The age at which both groups marry for the first time seems to be increasing at a roughly linear rate in recent decades. Let t correspond to the number of years since 1980. *Source: U.S. Census Bureau.*

Age at First Marriage						
Year	1980	1985	1990	1995	2000	2005
Men	24.7	25.5	26.1	26.9	26.8	27.1
Women	22.0	23.3	23.9	24.5	25.1	25.3

a. Find a linear equation that approximates the data for men, using the data for the years 1980 and 2005.

b. Repeat part a using the data for women.

c. Which group seems to have the faster increase in median age at first marriage?

d. In what year will the men's median age at first marriage reach 30?

e. When the men's median age at first marriage is 30, what will the median age be for women?

71. Immigration In 1950, there were 249,187 immigrants admitted to the United States. In 2008, the number was 1,107,126. *Source: 2008 Yearbook of Immigration Statistics.*

a. Assuming that the change in immigration is linear, write an equation expressing the number of immigrants, y, in terms of t, the number of years after 1900.

b. Use your result in part a to predict the number of immigrants admitted to the United States in 2015.

c. Considering the value of the y-intercept in your answer to part a, discuss the validity of using this equation to model the number of immigrants throughout the entire 20th century.

Physical Sciences

72. Global Warming In 1990, the Intergovernmental Panel on Climate Change predicted that the average temperature on Earth would rise 0.3°C per decade in the absence of international controls on greenhouse emissions. Let t measure the time in years since 1970, when the average global temperature was 15°C. *Source: Science News.*

a. Find a linear equation giving the average global temperature in degrees Celsius in terms of t, the number of years since 1970.

b. Scientists have estimated that the sea level will rise by 65 cm if the average global temperature rises to 19°C. According to your answer to part a, when would this occur?

73. Galactic Distance The table lists the distances (in megaparsecs where 1 megaparsec $\approx 3.1 \times 10^{19}$ km) and velocities (in kilometers per second) of four galaxies moving rapidly away from Earth. *Source: Astronomical Methods and Calculations, and Fundamental Astronomy.*

Galaxy	Distance	Velocity
Virga	15	1600
Ursa Minor	200	15,000
Corona Borealis	290	24,000
Bootes	520	40,000

a. Plot the data points letting x represent distance and y represent velocity. Do the points lie in an approximately linear pattern?

b. Write a linear equation $y = mx$ to model this data, using the ordered pair (520, 40,000).

c. The galaxy Hydra has a velocity of 60,000 km per sec. Use your equation to approximate how far away it is from Earth.

d. The value of m in the equation is called the *Hubble constant*. The Hubble constant can be used to estimate the age of the universe A (in years) using the formula

$$A = \frac{9.5 \times 10^{11}}{m}.$$

Approximate A using your value of m.

General Interest

74. News/Talk Radio From 2001 to 2007, the number of stations carrying news/talk radio increased at a roughly linear rate, from 1139 in 2001 to 1370 in 2007. *Source: State of the Media.*

a. Find a linear equation expressing the number of stations carrying news/talk radio, y, in terms of t, the years since 2000.

b. Use your answer from part a to predict the number of stations carrying news/talk radio in 2008. Compare with the actual number of 2046. Discuss how the linear trend from 2001 to 2007 might have changed in 2008.

75. Tuition The table lists the annual cost (in dollars) of tuition and fees at private four-year colleges for selected years. (See Example 14.) *Source: The College Board.*

Year	Tuition and Fees
2000	16,072
2002	18,060
2004	20,045
2006	22,308
2008	25,177
2009	26,273

a. Sketch a graph of the data. Do the data appear to lie roughly along a straight line?

b. Let $t = 0$ correspond to the year 2000. Use the points (0, 16,072) and (9, 26,273) to determine a linear equation that models the data. What does the slope of the graph of the equation indicate?

c. Discuss the accuracy of using this equation to estimate the cost of private college in 2025.

YOUR TURN ANSWERS

1. 1/3
2. $y = (3/2)x + 6$
3. $-8/3$
4. $y = -2x + 13$
5. $y = (1/2)x + 3$
6. $y = (3/2)x - 5/2$

1.2 Linear Functions and Applications

APPLY IT   **How many units must be sold for a firm to break even?**
In Example 6 in this section, this question will be answered using a linear function.

As we saw in the previous section, many situations involve two variables related by a linear equation. For such a relationship, when we express the variable y in terms of x, we say that y is a **linear function** of x. This means that for any allowed value of x (the **independent variable**), we can use the equation to find the corresponding value of y (the **dependent variable**). Examples of equations defining linear functions include $y = 2x + 3$, $y = -5$, and $2x - 3y = 7$, which can be written as $y = (2/3)x - (7/3)$. Equations in the form $x = k$, where k is a constant, do not define linear functions. All other linear equations define linear functions.

$f(x)$ **Notation**   Letters such as f, g, or h are often used to name functions. For example, f might be used to name the function defined by

$$y = 5 - 3x.$$

To show that this function is named f, it is common to replace y with $f(x)$ (read "f of x") to get

$$f(x) = 5 - 3x.$$

By choosing 2 as a value of x, $f(x)$ becomes $5 - 3 \cdot 2 = 5 - 6 = -1$, written

$$f(2) = -1.$$

The corresponding ordered pair is $(2, -1)$. In a similar manner,

$$f(-4) = 5 - 3(-4) = 17, \qquad f(0) = 5, \qquad f(-6) = 23,$$

and so on.

EXAMPLE 1   Function Notation

Let $g(x) = -4x + 5$. Find $g(3)$, $g(0)$, $g(-2)$, and $g(b)$.

SOLUTION   To find $g(3)$, substitute 3 for x.

$$g(3) = -4(3) + 5 = -12 + 5 = -7$$

Similarly,

$$g(0) = -4(0) + 5 = 0 + 5 = 5,$$
$$g(-2) = -4(-2) + 5 = 8 + 5 = 13,$$

and

$$g(b) = -4b + 5.$$ **TRY YOUR TURN 1**

YOUR TURN 1
Calculate $g(-5)$.

We summarize the discussion below.

> **Linear Function**
> A relationship f defined by
>
> $$y = f(x) = mx + b,$$
>
> for real numbers m and b, is a **linear function**.

Supply and Demand   Linear functions are often good choices for supply and demand curves. Typically, as the price of an item increases, consumers are less likely to buy an increasingly expensive item, and so the demand for the item decreases. On the other hand, as the price of an item increases, producers are more likely to see a profit in selling the item, and so the supply of the item increases. The increase in the quantity supplied and decrease in the quantity demanded can eventually result in a surplus, which causes the price to fall. These countervailing trends tend to move the price, as well as the quantity supplied and demanded toward an equilibrium value.

For example, during the late 1980s and early 1990s, the consumer demand for cranberries (and all of their healthy benefits) soared. The quantity demanded surpassed the quantity supplied, causing a shortage, and cranberry prices rose dramatically. As prices increased, growers wanted to increase their profits, so they planted more acres of cranberries. Unfortunately, cranberries take 3 to 5 years from planting until they can first be harvested. As growers waited and prices increased, consumer demand decreased. When the cranberries were finally harvested, the supply overwhelmed the demand and a huge surplus occurred, causing the price of cranberries to drop in the late 1990s. *Source: Agricultural Marketing Resource Center.* Other factors were involved in this situation, but the relationship between price, supply, and demand was nonetheless typical.

Although economists consider price to be the independent variable, they have the unfortunate habit of plotting price, usually denoted by p, on the vertical axis, while everyone else graphs the independent variable on the horizontal axis. This custom was started by the English economist Alfred Marshall (1842–1924). In order to abide by this custom, we will write p, the price, as a function of q, the quantity produced, and plot p on the vertical axis. But remember, it is really *price* that determines how much consumers demand and producers supply, not the other way around.

Supply and demand functions are not necessarily linear, the simplest kind of function. Yet most functions are approximately linear if a small enough piece of the graph is taken, allowing applied mathematicians to often use linear functions for simplicity. That approach will be taken in this chapter.

EXAMPLE 2   Supply and Demand

Suppose that Greg Tobin, manager of a giant supermarket chain, has studied the supply and demand for watermelons. He has noticed that the demand increases as the price decreases. He has determined that the quantity (in thousands) demanded weekly, q, and the price (in dollars) per watermelon, p, are related by the linear function

$$p = D(q) = 9 - 0.75q. \quad \text{\small Demand function}$$

(a) Find the quantity demanded at a price of \$5.25 per watermelon and at a price of \$3.75 per watermelon.

SOLUTION   To find the quantity demanded at a price of \$5.25 per watermelon, replace p in the demand function with 5.25 and solve for q.

$$5.25 = 9 - 0.75q$$
$$-3.75 = -0.75q \qquad \text{\small Subtract 9 from both sides.}$$
$$5 = q \qquad \text{\small Divide both sides by } -0.75.$$

Thus, at a price of \$5.25, the quantity demanded is 5000 watermelons.

Similarly, replace p with 3.75 to find the demand when the price is \$3.75. Verify that this leads to $q = 7$. When the price is lowered from \$5.25 to \$3.75 per watermelon, the quantity demanded increases from 5000 to 7000 watermelons.

(b) Greg also noticed that the quantity of watermelons supplied decreased as the price decreased. Price p and supply q are related by the linear function

$$p = S(q) = 0.75q. \quad \text{\small Supply function}$$

Find the quantity supplied at a price of \$5.25 per watermelon and at a price of \$3.00 per watermelon.

SOLUTION Substitute 5.25 for p in the supply function, $p = 0.75q$, to find that $q = 7$, so the quantity supplied is 7000 watermelons. Similarly, replacing p with 3 in the supply equation gives a quantity supplied of 4000 watermelons. If the price decreases from $5.25 to $3.00 per watermelon, the quantity supplied also decreases, from 7000 to 4000 watermelons.

(c) Graph both functions on the same axes.

YOUR TURN 2 Find the quantity of watermelon demanded and supplied at a price of $3.30 per watermelon.

SOLUTION The results of part (a) are written as the ordered pairs (5, 5.25) and (7, 3.75). The line through those points is the graph of the demand function, $p = 9 - 0.75q$, shown in red in Figure 11(a). We used the ordered pairs (7, 5.25) and (4, 3) from the work in part (b) to graph the supply function, $p = 0.75q$, shown in blue in Figure 11(a).

TRY YOUR TURN 2

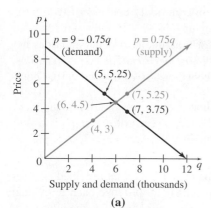

(a)

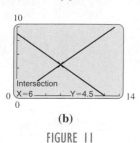

(b)

FIGURE 11

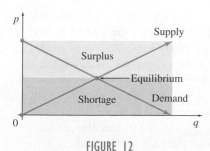

FIGURE 12

YOUR TURN 3 Repeat Example 3 using the demand equation $D(q) = 10 - 0.85q$ and the supply equation $S(q) = 0.4q$.

TECHNOLOGY NOTE A calculator-generated graph of the lines representing the supply and demand functions in Example 2 is shown in Figure 11(b). To get this graph, the equation of each line, using x and y instead of q and p, was entered, along with an appropriate window. A special menu choice gives the coordinates of the intersection point, as shown at the bottom of the graph.

NOTE Not all supply and demand problems will have the same scale on both axes. It helps to consider the intercepts of both the supply graph and the demand graph to decide what scale to use. For example, in Figure 11, the y-intercept of the demand function is 9, so the scale should allow values from 0 to at least 9 on the vertical axis. The x-intercept of the demand function is 12, so the values on the x-axis must go from 0 to 12.

As shown in the graphs of Figure 11, both the supply graph and the demand graph pass through the point (6, 4.5). If the price of a watermelon is more than $4.50, the quantity supplied will exceed the quantity demanded and there will be a **surplus** of watermelons. At a price less than $4.50, the quantity demanded will exceed the quantity supplied and there will be a **shortage** of watermelons. Only at a price of $4.50 will quantity demanded and supplied be equal. For this reason, $4.50 is called the *equilibrium price*. When the price is $4.50, quantity demanded and supplied both equal 6000 watermelons, the *equilibrium quantity*. In general, the **equilibrium price** of the commodity is the price found at the point where the supply and demand graphs for that commodity intersect. The **equilibrium quantity** is the quantity demanded and supplied at that same point. Figure 12 illustrates a general supply and demand situation.

EXAMPLE 3 **Equilibrium Quantity**

Use algebra to find the equilibrium quantity and price for the watermelons in Example 2.

SOLUTION The equilibrium quantity is found when the prices from both supply and demand are equal. Set the two expressions for p equal to each other and solve.

$$9 - 0.75q = 0.75q$$
$$9 = 1.5q \qquad \text{Add 0.75}q \text{ to both sides.}$$
$$6 = q$$

The equilibrium quantity is 6000 watermelons, the same answer found earlier.

The equilibrium price can be found by plugging the value of $q = 6$ into either the demand or the supply function. Using the demand function,

$$p = D(6) = 9 - 0.75(6) = 4.5.$$

The equilibrium price is $4.50, as we found earlier. Check your work by also plugging $q = 6$ into the supply function.

TRY YOUR TURN 3

⊞ **TECHNOLOGY NOTE** You may prefer to find the equilibrium quantity by solving the equation with your calculator. Or, if your calculator has a TABLE feature, you can use it to find the value of q that makes the two expressions equal.

Another important issue is how, in practice, the equations of the supply and demand functions can be found. Data need to be collected, and if they lie perfectly along a line, then the equation can easily be found with any two points. What usually happens, however, is that the data are scattered, and there is no line that goes through all the points. In this case we must find a line that approximates the linear trend of the data as closely as possible (assuming the points lie approximately along a line) as in Example 14 in the previous section. This is usually done by the *method of least squares*, also referred to as *linear regression*. We will discuss this method in Section 1.3.

Cost Analysis
The cost of manufacturing an item commonly consists of two parts. The first is a **fixed cost** for designing the product, setting up a factory, training workers, and so on. Within broad limits, the fixed cost is constant for a particular product and does not change as more items are made. The second part is a *cost per item* for labor, materials, packing, shipping, and so on. The total value of this second cost *does* depend on the number of items made.

EXAMPLE 4 Cost Analysis

A small company decides to produce video games. The owners find that the fixed cost for creating the game is $5000, after which they must spend $12 to produce each individual copy of the game. Find a formula $C(x)$ for the cost as a linear function of x, the number of games produced.

SOLUTION Notice that $C(0) = 5000$, since $5000 must be spent even if no games are produced. Also, $C(1) = 5000 + 12 = 5012$, and $C(2) = 5000 + 2 \cdot 12 = 5024$. In general,

$$C(x) = 5000 + 12x,$$

because every time x increases by 1, the cost should increase by $12. The number 12 is also the slope of the graph of the cost function; the slope gives us the cost to produce one additional item.

In economics, **marginal cost** is the rate of change of cost $C(x)$ at a level of production x and is equal to the slope of the cost function at x. It approximates the cost of producing one additional item. In fact, some books define the marginal cost to be the cost of producing one additional item. With *linear functions*, these two definitions are equivalent, and the marginal cost, which is equal to the slope of the cost function, is *constant*. For instance, in the video game example, the marginal cost of each game is $12. For other types of functions, these two definitions are only approximately equal. Marginal cost is important to management in making decisions in areas such as cost control, pricing, and production planning.

The work in Example 4 can be generalized. Suppose the total cost to make x items is given by the linear cost function $C(x) = mx + b$. The fixed cost is found by letting $x = 0$:

$$C(0) = m \cdot 0 + b = b;$$

thus, the fixed cost is b dollars. The additional cost of each additional item, the marginal cost, is m, the slope of the line $C(x) = mx + b$.

> ### Linear Cost Function
> In a cost function of the form $C(x) = mx + b$, the m represents the marginal cost and b the fixed cost. Conversely, if the fixed cost of producing an item is b and the marginal cost is m, then the **linear cost function** $C(x)$ for producing x items is $C(x) = mx + b$.

EXAMPLE 5   Cost Function

The marginal cost to make x batches of a prescription medication is $10 per batch, while the cost to produce 100 batches is $1500. Find the cost function $C(x)$, given that it is linear.

SOLUTION   Since the cost function is linear, it can be expressed in the form $C(x) = mx + b$. The marginal cost is $10 per batch, which gives the value for m. Using $x = 100$ and $C(x) = 1500$ in the point-slope form of the line gives

$$C(x) - 1500 = 10(x - 100)$$
$$C(x) - 1500 = 10x - 1000$$
$$C(x) = 10x + 500. \qquad \text{Add 1500 to both sides.}$$

YOUR TURN 4   Repeat Example 5, using a marginal cost of $15 per batch and a cost of $1930 to produce 80 batches.

The cost function is given by $C(x) = 10x + 500$, where the fixed cost is $500.

TRY YOUR TURN 4

Break-Even Analysis

The **revenue** $R(x)$ from selling x units of an item is the product of the price per unit p and the number of units sold (demand) x, so that

$$R(x) = px.$$

The corresponding **profit** $P(x)$ is the difference between revenue $R(x)$ and cost $C(x)$. That is,

$$P(x) = R(x) - C(x).$$

A company can make a profit only if the revenue received from its customers exceeds the cost of producing and selling its goods and services. The number of units at which revenue just equals cost is the **break-even quantity**; the corresponding ordered pair gives the **break-even point**.

EXAMPLE 6   Break-Even Analysis

APPLY IT

A firm producing poultry feed finds that the total cost $C(x)$ in dollars of producing and selling x units is given by

$$C(x) = 20x + 100.$$

Management plans to charge $24 per unit for the feed.

(a) How many units must be sold for the firm to break even?

SOLUTION   The firm will break even (no profit and no loss) as long as revenue just equals cost, or $R(x) = C(x)$. From the given information, since $R(x) = px$ and $p = \$24$,

$$R(x) = 24x.$$

Substituting for $R(x)$ and $C(x)$ in the equation $R(x) = C(x)$ gives

$$24x = 20x + 100,$$

from which $x = 25$. The firm breaks even by selling 25 units, which is the break-even quantity. The graphs of $C(x) = 20x + 100$ and $R(x) = 24x$ are shown in Figure 13.

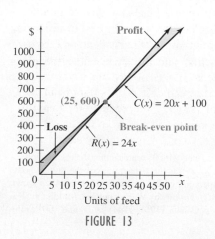

FIGURE 13

The break-even point (where $x = 25$) is shown on the graph. If the company sells more than 25 units (if $x > 25$), it makes a profit. If it sells fewer than 25 units, it loses money.

(b) What is the profit if 100 units of feed are sold?

SOLUTION   Use the formula for profit $P(x)$.

$$\begin{aligned} P(x) &= R(x) - C(x) \\ &= 24x - (20x + 100) \\ &= 4x - 100 \end{aligned}$$

Then $P(100) = 4(100) - 100 = 300$. The firm will make a profit of $300 from the sale of 100 units of feed.

(c) How many units must be sold to produce a profit of $900?

SOLUTION   Let $P(x) = 900$ in the equation $P(x) = 4x - 100$ and solve for x.

$$\begin{aligned} 900 &= 4x - 100 \\ 1000 &= 4x \\ x &= 250 \end{aligned}$$

Sales of 250 units will produce $900 profit.   **TRY YOUR TURN 5**

YOUR TURN 5   Repeat Example 6(c), using a cost function $C(x) = 35x + 250$, a charge of $58 per unit, and a profit of $8030.

Temperature   One of the most common linear relationships found in everyday situations deals with temperature. Recall that water freezes at 32° Fahrenheit and 0° Celsius, while it boils at 212° Fahrenheit and 100° Celsius.* The ordered pairs $(0, 32)$ and $(100, 212)$ are graphed in Figure 14 on axes showing Fahrenheit (F) as a function of Celsius (C). The line joining them is the graph of the function.

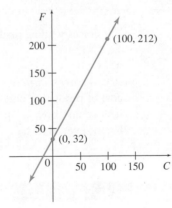

FIGURE 14

EXAMPLE 7   **Temperature**

Derive an equation relating F and C.

SOLUTION   To derive the required linear equation, first find the slope using the given ordered pairs, $(0, 32)$ and $(100, 212)$.

$$m = \frac{212 - 32}{100 - 0} = \frac{9}{5}$$

*Gabriel Fahrenheit (1686–1736), a German physicist, invented his scale with 0° representing the temperature of an equal mixture of ice and ammonium chloride (a type of salt), and 96° as the temperature of the human body. (It is often said, erroneously, that Fahrenheit set 100° as the temperature of the human body. Fahrenheit's own words are quoted in *A History of the Thermometer and Its Use in Meteorology* by W. E. Knowles, Middleton: The Johns Hopkins Press, 1966, p. 75.) The Swedish astronomer Anders Celsius (1701–1744) set 0° and 100° as the freezing and boiling points of water.

The F-intercept of the graph is 32, so by the slope-intercept form, the equation of the line is

$$F = \frac{9}{5}C + 32.$$

With simple algebra this equation can be rewritten to give C in terms of F:

$$C = \frac{5}{9}(F - 32).$$

1.2 EXERCISES

For Exercises 1–10, let $f(x) = 7 - 5x$ and $g(x) = 2x - 3$. Find the following.

1. $f(2)$ **2.** $f(4)$

3. $f(-3)$ **4.** $f(-1)$

5. $g(1.5)$ **6.** $g(2.5)$

7. $g(-1/2)$ **8.** $g(-3/4)$

9. $f(t)$ **10.** $g(k^2)$

In Exercises 11–14, decide whether the statement is true or false.

11. To find the x-intercept of the graph of a linear function, we solve $y = f(x) = 0$, and to find the y-intercept, we evaluate $f(0)$.

12. The graph of $f(x) = -5$ is a vertical line.

13. The slope of the graph of a linear function cannot be undefined.

14. The graph of $f(x) = ax$ is a straight line that passes through the origin.

15. Describe what fixed costs and marginal costs mean to a company.

16. In a few sentences, explain why the price of a commodity not already at its equilibrium price should move in that direction.

17. Explain why a linear function may not be adequate for describing the supply and demand functions.

18. In your own words, describe the break-even quantity, how to find it, and what it indicates.

Write a linear cost function for each situation. Identify all variables used.

19. A Lake Tahoe resort charges a snowboard rental fee of $10 plus $2.25 per hour.

20. An Internet site for downloading music charges a $10 registration fee plus 99 cents per downloaded song.

21. A parking garage charges 2 dollars plus 75 cents per half-hour.

22. For a one-day rental, a car rental firm charges $44 plus 28 cents per mile.

Assume that each situation can be expressed as a linear cost function. Find the cost function in each case.

23. Fixed cost: $100; 50 items cost $1600 to produce.

24. Fixed cost: $35; 8 items cost $395 to produce.

25. Marginal cost: $75; 50 items cost $4300 to produce.

26. Marginal cost: $120; 700 items cost $96,500 to produce.

APPLICATIONS

Business and Economics

27. Supply and Demand Suppose that the demand and price for a certain model of a youth wristwatch are related by

$$p = D(q) = 16 - 1.25q,$$

where p is the price (in dollars) and q is the quantity demanded (in hundreds). Find the price at each level of demand.

a. 0 watches   **b.** 400 watches   **c.** 800 watches

Find the quantity demanded for the watch at each price.

d. $8   **e.** $10   **f.** $12

g. Graph $p = 16 - 1.25q$.

Suppose the price and supply of the watch are related by

$$p = S(q) = 0.75q,$$

where p is the price (in dollars) and q is the quantity supplied (in hundreds) of watches. Find the quantity supplied at each price.

h. $0   **i.** $10   **j.** $20

k. Graph $p = 0.75q$ on the same axis used for part g.

l. Find the equilibrium quantity and the equilibrium price.

28. Supply and Demand Suppose that the demand and price for strawberries are related by

$$p = D(q) = 5 - 0.25q,$$

where p is the price (in dollars) and q is the quantity demanded (in hundreds of quarts). Find the price at each level of demand.

a. 0 quarts   **b.** 400 quarts   **c.** 840 quarts

Find the quantity demanded for the strawberries at each price.

d. $4.50   **e.** $3.25   **f.** $2.40

g. Graph $p = 5 - 0.25q$.

Suppose the price and supply of strawberries are related by

$$p = S(q) = 0.25q,$$

where p is the price (in dollars) and q is the quantity supplied (in hundreds of quarts) of strawberries. Find the quantity supplied at each price.

h. $0   **i.** $2   **j.** $4.50

k. Graph $p = 0.75q$ on the same axis used for part g.

l. Find the equilibrium quantity and the equilibrium price.

29. Supply and Demand Let the supply and demand functions for butter pecan ice cream be given by

$$p = S(q) = \frac{2}{5}q \quad \text{and} \quad p = D(q) = 100 - \frac{2}{5}q,$$

where p is the price in dollars and q is the number of 10-gallon tubs.

a. Graph these on the same axes.

b. Find the equilibrium quantity and the equilibrium price. (*Hint:* The way to divide by a fraction is to multiply by its reciprocal.)

30. Supply and Demand Let the supply and demand functions for sugar be given by

$$p = S(q) = 1.4q - 0.6 \quad \text{and}$$
$$p = D(q) = -2q + 3.2,$$

where p is the price per pound and q is the quantity in thousands of pounds.

a. Graph these on the same axes.

b. Find the equilibrium quantity and the equilibrium price.

31. Supply and Demand Suppose that the supply function for honey is $p = S(q) = 0.3q + 2.7$, where p is the price in dollars for an 8-oz container and q is the quantity in barrels. Suppose also that the equilibrium price is $4.50 and the demand is 2 barrels when the price is $6.10. Find an equation for the demand function, assuming it is linear.

32. Supply and Demand Suppose that the supply function for walnuts is $p = S(q) = 0.25q + 3.6$, where p is the price in dollars per pound and q is the quantity in bushels. Suppose also that the equilibrium price is $5.85, and the demand is 4 bushels when the price is $7.60. Find an equation for the demand function, assuming it is linear.

33. T-Shirt Cost Joanne Wendelken sells silk-screened T-shirts at community festivals and crafts fairs. Her marginal cost to produce one T-shirt is $3.50. Her total cost to produce 60 T-shirts is $300, and she sells them for $9 each.

a. Find the linear cost function for Joanne's T-shirt production.

b. How many T-shirts must she produce and sell in order to break even?

c. How many T-shirts must she produce and sell to make a profit of $500?

34. Publishing Costs Alfred Juarez owns a small publishing house specializing in Latin American poetry. His fixed cost to produce a typical poetry volume is $525, and his total cost to produce 1000 copies of the book is $2675. His books sell for $4.95 each.

a. Find the linear cost function for Alfred's book production.

b. How many poetry books must he produce and sell in order to break even?

c. How many books must he produce and sell to make a profit of $1000?

35. Marginal Cost of Coffee The manager of a restaurant found that the cost to produce 100 cups of coffee is $11.02, while the cost to produce 400 cups is $40.12. Assume the cost $C(x)$ is a linear function of x, the number of cups produced.

a. Find a formula for $C(x)$.

b. What is the fixed cost?

c. Find the total cost of producing 1000 cups.

d. Find the total cost of producing 1001 cups.

e. Find the marginal cost of the 1001st cup.

f. What is the marginal cost of *any* cup and what does this mean to the manager?

36. Marginal Cost of a New Plant In deciding whether to set up a new manufacturing plant, company analysts have decided that a linear function is a reasonable estimation for the total cost $C(x)$ in dollars to produce x items. They estimate the cost to produce 10,000 items as $547,500, and the cost to produce 50,000 items as $737,500.

a. Find a formula for $C(x)$.

b. Find the fixed cost.

c. Find the total cost to produce 100,000 items.

d. Find the marginal cost of the items to be produced in this plant and what does this mean to the manager?

37. Break-Even Analysis Producing x units of tacos costs $C(x) = 5x + 20$; revenue is $R(x) = 15x$, where $C(x)$ and $R(x)$ are in dollars.

a. What is the break-even quantity?

b. What is the profit from 100 units?

c. How many units will produce a profit of $500?

38. Break-Even Analysis To produce x units of a religious medal costs $C(x) = 12x + 39$. The revenue is $R(x) = 25x$. Both $C(x)$ and $R(x)$ are in dollars.

a. Find the break-even quantity.

b. Find the profit from 250 units.

c. Find the number of units that must be produced for a profit of $130.

Break-Even Analysis You are the manager of a firm. You are considering the manufacture of a new product, so you ask the accounting department for cost estimates and the sales department for sales estimates. After you receive the data, you must decide whether to go ahead with production of the new product. Analyze the data in Exercises 39–42 (find a break-even quantity) and then decide what you would do in each case. Also write the profit function.

39. $C(x) = 85x + 900$; $R(x) = 105x$; no more than 38 units can be sold.

40. $C(x) = 105x + 6000$; $R(x) = 250x$; no more than 400 units can be sold.

41. $C(x) = 70x + 500$; $R(x) = 60x$ (*Hint*: What does a negative break-even quantity mean?)

42. $C(x) = 1000x + 5000$; $R(x) = 900x$

43. Break-Even Analysis Suppose that the fixed cost for a product is $400 and the break-even quantity is 80. Find the marginal profit (the slope of the linear profit function).

44. Break-Even Analysis Suppose that the fixed cost for a product is $650 and the break-even quantity is 25. Find the marginal profit (the slope of the linear profit function).

Physical Sciences

45. Temperature Use the formula for conversion between Fahrenheit and Celsius derived in Example 7 to convert each temperature.

a. 58°F to Celsius

b. −20°F to Celsius

c. 50°C to Fahrenheit

46. Body Temperature You may have heard that the average temperature of the human body is 98.6°. Recent experiments show that the actual figure is closer to 98.2°. The figure of 98.6 comes from experiments done by Carl Wunderlich in 1868. But Wunderlich measured the temperatures in degrees Celsius and rounded the average to the nearest degree, giving 37°C as the average temperature. *Source: Science News.*

a. What is the Fahrenheit equivalent of 37°C?

b. Given that Wunderlich rounded to the nearest degree Celsius, his experiments tell us that the actual average human body temperature is somewhere between 36.5°C and 37.5°C. Find what this range corresponds to in degrees Fahrenheit.

47. Temperature Find the temperature at which the Celsius and Fahrenheit temperatures are numerically equal.

General Interest

48. Education Cost The 2009–2010 budget for the California State University system projected a fixed cost of $486,000 at each of five off-campus centers, plus a marginal cost of $1140 per student. *Source: California State University.*

a. Find a formula for the cost at each center, $C(x)$, as a linear function of x, the number of students.

b. The budget projected 500 students at each center. Calculate the total cost at each center.

c. Suppose, due to budget cuts, that each center is limited to $1 million. What is the maximum number of students that each center can then support?

YOUR TURN ANSWERS

1. 25   **2.** 7600 and 4400
3. 8000 watermelons and $3.20 per watermelon
4. $C(x) = 15x + 730$ **5.** 360

1.3 The Least Squares Line

APPLY IT How has the accidental death rate in the United States changed over time?
In Example 1 in this section, we show how to answer such questions using the method of least squares.

We use past data to find trends and to make tentative predictions about the future. The only assumption we make is that the data are related linearly—that is, if we plot pairs of data, the resulting points will lie close to some line. This method cannot give exact answers. The best we can expect is that, if we are careful, we will get a reasonable approximation.

The table lists the number of accidental deaths per 100,000 people in the United States through the past century. *Source: National Center for Health Statistics.* If you were a manager at an insurance company, these data could be very important. You might need to make some predictions about how much you will pay out next year in accidental death benefits, and even a very tentative prediction based on past trends is better than no prediction at all.

The first step is to draw a scatterplot, as we have done in Figure 15 on the next page. Notice that the points lie approximately along a line, which means that a linear function may give a good approximation of the data. If we select two points and find the line that passes through them, as we did in Section 1.1, we will get a different line for each pair of points, and in some cases the lines will be very different. We want to draw one line that is simultaneously close to all the points on the graph, but many such lines are possible, depending upon how we define the phrase "simultaneously close to all the points." How do we decide on the best possible line? Before going on, you might want to try drawing the line you think is best on Figure 15.

Accidental Death Rate	
Year	**Death Rate**
1910	84.4
1920	71.2
1930	80.5
1940	73.4
1950	60.3
1960	52.1
1970	56.2
1980	46.5
1990	36.9
2000	34.0

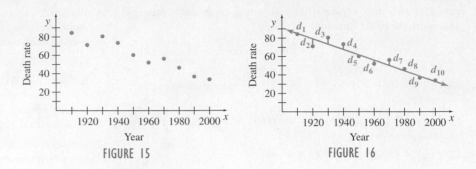

FIGURE 15   FIGURE 16

The line used most often in applications is that in which the sum of the squares of the vertical distances from the data points to the line is as small as possible. Such a line is called the **least squares line**. The least squares line for the data in Figure 15 is drawn in Figure 16. How does the line compare with the one you drew on Figure 15? It may not be exactly the same, but should appear similar.

In Figure 16, the vertical distances from the points to the line are indicated by d_1, d_2, and so on, up through d_{10} (read "d-sub-one, d-sub-two, d-sub-three," and so on). For n points, corresponding to the n pairs of data, the least squares line is found by minimizing the sum $(d_1)^2 + (d_2)^2 + (d_3)^2 + \cdots + (d_n)^2$.

We often use **summation notation** to write the sum of a list of numbers. The Greek letter sigma, Σ, is used to indicate "the sum of." For example, we write the sum $x_1 + x_2 + \cdots + x_n$, where n is the number of data points, as

$$x_1 + x_2 + \cdots + x_n = \Sigma x.$$

Similarly, Σxy means $x_1 y_1 + x_2 y_2 + \cdots + x_n y_n$, and so on.

> **CAUTION** Note that Σx^2 means $x_1^2 + x_2^2 + \cdots + x_n^2$, which is *not* the same as squaring Σx. When we square Σx, we write it as $(\Sigma x)^2$.

For the least squares line, the sum of the distances we are to minimize, $d_1^2 + d_2^2 + \cdots + d_n^2$, is written as

$$d_1^2 + d_2^2 + \cdots + d_n^2 = \Sigma d^2.$$

To calculate the distances, we let $(x_1, y_1), (x_2, y_2), \cdots, (x_n, y_n)$ be the actual data points and we let the least squares line be $Y = mx + b$. We use Y in the equation instead of y to distinguish the predicted values (Y) from the y-value of the given data points. The predicted value of Y at x_1 is $Y_1 = mx_1 + b$, and the distance, d_1, between the actual y-value y_1 and the predicted value Y_1 is

$$d_1 = |Y_1 - y_1| = |mx_1 + b - y_1|.$$

Likewise,

$$d_2 = |Y_2 - y_2| = |mx_2 + b - y_2|,$$

and

$$d_n = |Y_n - y_n| = |mx_n + b - y_n|.$$

The sum to be minimized becomes

$$\Sigma d^2 = (mx_1 + b - y_1)^2 + (mx_2 + b - y_2)^2 + \cdots + (mx_n + b - y_n)^2$$
$$= \Sigma(mx + b - y)^2,$$

where $(x_1, y_1), (x_2, y_2), \ldots, (x_n, y_n)$ are known and m and b are to be found.

The method of minimizing this sum requires advanced techniques and is not given here. To obtain the equation for the least squares line, a system of equations must be solved, producing the following formulas for determining the slope m and y-intercept b.*

Least Squares Line
The **least squares line** $Y = mx + b$ that gives the best fit to the data points $(x_1, y_1), (x_2, y_2), \ldots, (x_n, y_n)$ has slope m and y-intercept b given by

$$m = \frac{n(\Sigma xy) - (\Sigma x)(\Sigma y)}{n(\Sigma x^2) - (\Sigma x)^2} \quad \text{and} \quad b = \frac{\Sigma y - m(\Sigma x)}{n}.$$

EXAMPLE 1 Least Squares Line

APPLY IT

Calculate the least squares line for the accidental death rate data.

SOLUTION

Method 1
Calculating by Hand

To find the least squares line for the given data, we first find the required sums. To reduce the size of the numbers, we rescale the year data. Let x represent the years since 1900, so that, for example, $x = 10$ corresponds to the year 1910. Let y represent the death rate. We then calculate the values in the xy, x^2, and y^2 columns and find their totals. (The column headed y^2 will be used later.) Note that the number of data points is $n = 10$.

		Least Squares Calculations		
x	y	xy	x^2	y^2
10	84.4	844	100	7123.36
20	71.2	1424	400	5069.44
30	80.5	2415	900	6480.25
40	73.4	2936	1600	5387.56
50	60.3	3015	2500	3636.09
60	52.1	3126	3600	2714.41
70	56.2	3934	4900	3158.44
80	46.5	3720	6400	2162.25
90	36.9	3321	8100	1361.61
100	34.0	3400	10,000	1156.00
$\Sigma x = 550$	$\Sigma y = 595.5$	$\Sigma xy = 28,135$	$\Sigma x^2 = 38,500$	$\Sigma y^2 = 38,249.41$

Putting the column totals into the formula for the slope m, we get

$$m = \frac{n(\Sigma xy) - (\Sigma x)(\Sigma y)}{n(\Sigma x^2) - (\Sigma x)^2} \qquad \text{Formula for } m$$

$$= \frac{10(28,135) - (550)(595.5)}{10(38,500) - (550)^2} \qquad \text{Substitute from the table.}$$

$$= \frac{281,350 - 327,525}{385,000 - 302,500} \qquad \text{Multiply.}$$

$$= \frac{-46,175}{82,500} \qquad \text{Subtract.}$$

$$= -0.5596970 \approx -0.560.$$

*See Exercise 9 at the end of this section.

The significance of m is that the death rate per 100,000 people is tending to drop (because of the negative) at a rate of 0.560 per year.

Now substitute the value of m and the column totals in the formula for b:

$$b = \frac{\Sigma y - m(\Sigma x)}{n}$$ Formula for b

$$= \frac{595.5 - (-0.559697)(550)}{10}$$ Substitute.

$$= \frac{595.5 - (-307.83335)}{10}$$ Multiply.

$$= \frac{903.33335}{10} = 90.333335 \approx 90.3$$

Substitute m and b into the least squares line, $Y = mx + b$; the least squares line that best fits the 10 data points has equation

$$Y = -0.560x + 90.3.$$

This gives a mathematical description of the relationship between the year and the number of accidental deaths per 100,000 people. The equation can be used to predict y from a given value of x, as we will show in Example 2. As we mentioned before, however, caution must be exercised when using the least squares equation to predict data points that are far from the range of points on which the equation was modeled.

CAUTION In computing m and b, we rounded the final answer to three digits because the original data were known only to three digits. It is important, however, *not* to round any of the intermediate results (such as Σx^2) because round-off error may have a detrimental effect on the accuracy of the answer. Similarly, it is important not to use a rounded-off value of m when computing b.

Method 2
Graphing Calculator

The calculations for finding the least squares line are often tedious, even with the aid of a calculator. Fortunately, many calculators can calculate the least squares line with just a few keystrokes. For purposes of illustration, we will show how the least squares line in the previous example is found with a TI-84 Plus graphing calculator.

We begin by entering the data into the calculator. We will be using the first two lists, called L_1 and L_2. Choosing the STAT menu, then choosing the fourth entry ClrList, we enter L_1, L_2, to indicate the lists to be cleared. Now we press STAT again and choose the first entry EDIT, which brings up the blank lists. As before, we will only use the last two digits of the year, putting the numbers in L_1. We put the death rate in L_2, giving the two screens shown in Figure 17.

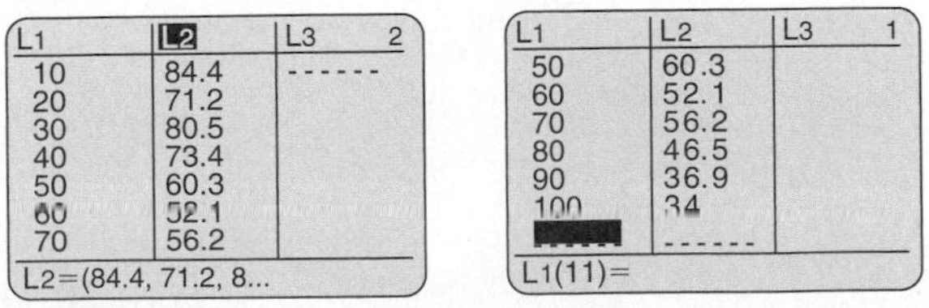

L1	**L2**	L3	2
10	84.4	------	
20	71.2		
30	80.5		
40	73.4		
50	60.3		
60	52.1		
70	56.2		
L2=(84.4, 71.2, 8...			

L1	L2	L3	1
50	60.3		
60	52.1		
70	56.2		
80	46.5		
90	36.9		
100	34		

L1(11)=			

FIGURE 17

Quit the editor, press STAT again, and choose CALC instead of EDIT. Then choose item 4 LinReg($ax + b$) to get the values of a (the slope) and b (the y-intercept) for the least squares line, as shown in Figure 18. With a and b rounded to three decimal places, the least squares line is $Y = -0.560x + 90.3$. A graph of the data points and the line is shown in Figure 19.

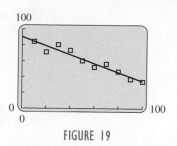

FIGURE 18   FIGURE 19

For more details on finding the least squares line with a graphing calculator, see the *Graphing Calculator and Excel Spreadsheet Manual* available with this book.

Method 3
Spreadsheet

YOUR TURN 1 Repeat Example 1, deleting the last pair of data (100, 34.0) and changing the second to last pair to (90, 40.2).

Many computer spreadsheet programs can also find the least squares line. Figure 20 shows the scatterplot and least squares line for the accidental death rate data using an Excel spreadsheet. The scatterplot was found using the Marked Scatter chart from the Gallery and the line was found using the Add Trendline command under the Chart menu. For details, see the *Graphing Calculator and Excel Spreadsheet Manual* available with this book.

TRY YOUR TURN 1

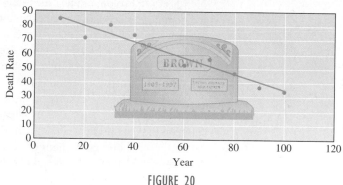

FIGURE 20

EXAMPLE 2 **Least Squares Line**

What do we predict the accidental death rate to be in 2012?

SOLUTION Use the least squares line equation given above with $x = 112$.

$$Y = -0.560x + 90.3$$
$$= -0.56(112) + 90.3$$
$$= 27.6$$

The accidental death rate in 2012 is predicted to be about 27.6 per 100,000 population. In this case, we will have to wait until the 2012 data become available to see how accurate our prediction is. We have observed, however, that the accidental death rate began to go up after 2000 and was 40.6 per 100,000 population in 2006. This illustrates the danger of extrapolating beyond the data.

EXAMPLE 3 **Least Squares Line**

In what year is the death rate predicted to drop below 26 per 100,000 population?

SOLUTION Let $Y = 26$ in the equation above and solve for x.

$$26 = -0.560x + 90.3$$

$$-64.3 = -0.560x \qquad \text{Subtract 90.3 from both sides.}$$

$$x = 115 \qquad \text{Divide both sides by } -0.560.$$

This corresponds to the year 2015 (115 years after 1900), when our equation predicts the death rate to be $-0.560(115) + 90.3 = 25.9$ per 100,000 population. ▬

Correlation

Although the least squares line can always be found, it may not be a good model. For example, if the data points are widely scattered, no straight line will model the data accurately. One measure of how well the original data fits a straight line is the **correlation coefficient,** denoted by r, which can be calculated by the following formula.

Correlation Coefficient

$$r = \frac{n(\Sigma xy) - (\Sigma x)(\Sigma y)}{\sqrt{n(\Sigma x^2) - (\Sigma x)^2} \cdot \sqrt{n(\Sigma y^2) - (\Sigma y)^2}}$$

Although the expression for r looks daunting, remember that each of the summations, Σx, Σy, Σxy, and so on, are just the totals from a table like the one we prepared for the data on accidental deaths. Also, with a calculator, the arithmetic is no problem! Furthermore, statistics software and many calculators can calculate the correlation coefficient for you.

The correlation coefficient measures the strength of the linear relationship between two variables. It was developed by statistics pioneer Karl Pearson (1857–1936). The correlation coefficient r is between 1 and -1 or is equal to 1 or -1. Values of exactly 1 or -1 indicate that the data points lie *exactly* on the least squares line. If $r = 1$, the least squares line has a positive slope; $r = -1$ gives a negative slope. If $r = 0$, there is no linear correlation between the data points (but some *nonlinear* function might provide an excellent fit for the data). A correlation coefficient of zero may also indicate that the data fit a horizontal line. To investigate what is happening, it is always helpful to sketch a scatterplot of the data. Some scatterplots that correspond to these values of r are shown in Figure 21.

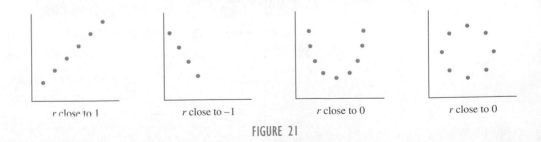

r close to 1 r close to -1 r close to 0 r close to 0

FIGURE 21

A value of r close to 1 or -1 indicates the presence of a linear relationship. The exact value of r necessary to conclude that there is a linear relationship depends upon n, the number of data points, as well as how confident we want to be of our conclusion. For details, consult a text on statistics.*

---

*For example, see *Introductory Statistics*, 8th edition, by Neil A. Weiss, Boston, Mass.: Pearson, 2008.

EXAMPLE 4   **Correlation Coefficient**

Find r for the data on accidental death rates in Example 1.

SOLUTION

Method 1
Calculating by Hand

From the table in Example 1,

$\Sigma x = 550$, $\Sigma y = 595.5$, $\Sigma xy = 28,135$, $\Sigma x^2 = 38,500$,

$\Sigma y^2 = 38,249.41$,   and $n = 10$.

Substituting these values into the formula for r gives

$$r = \frac{n(\Sigma xy) - (\Sigma x)(\Sigma y)}{\sqrt{n(\Sigma x^2) - (\Sigma x)^2} \cdot \sqrt{n(\Sigma y^2) - (\Sigma y)^2}}$$ Formula for r

$$= \frac{10(28,135) - (550)(595.5)}{\sqrt{10(38,500) - (550)^2} \cdot \sqrt{10(38,249.41) - (595.5)^2}}$$ Substitute.

$$= \frac{281,350 - 327,525}{\sqrt{385,000 - 302,500} \cdot \sqrt{382,494.1 - 354,620.25}}$$ Multiply.

$$= \frac{-46,175}{\sqrt{82,500} \cdot \sqrt{27,873.85}}$$ Subtract.

$$= \frac{-46,175}{47,954.06787}$$ Take square roots and multiply.

$$= -0.9629005849 \approx -0.963.$$

This is a high correlation, which agrees with our observation that the data fit a line quite well.

Method 2
Graphing Calculator

Most calculators that give the least squares line will also give the correlation coefficient. To do this on the TI-84 Plus, press the second function CATALOG and go down the list to the entry DiagnosticOn. Press ENTER at that point, then press STAT, CALC, and choose item 4 to get the display in Figure 22. The result is the same as we got by hand. The command DiagnosticOn need only be entered once, and the correlation coefficient will always appear in the future.

```
LinReg
 y=ax+b
 a=-.5596969697
 b=90.33333333
 r²=.9271775365
 r=-.962900585
■
```

FIGURE 22

Method 3
Spreadsheet

Many computer spreadsheet programs have a built-in command to find the correlation coefficient. For example, in Excel, use the command "= CORREL(A1:A10,B1:B10)" to find the correlation of the 10 data points stored in columns A and B. For more details, see the *Graphing Calculator and Excel Spreadsheet Manual* available with this text.

YOUR TURN 2 Repeat Example 4, deleting the last pair of data (100, 34.0) and changing the second to last pair to (90, 40.2).

TRY YOUR TURN 2

The square of the correlation coefficient gives the fraction of the variation in y that is explained by the linear relationship between x and y. Consider Example 4, where $r^2 = (-0.963)^2 = 0.927$. This means that 92.7% of the variation in y is explained by the linear relationship found earlier in Example 1. The remaining 7.3% comes from the scattering of the points about the line.

EXAMPLE 5 **Average Expenditure per Pupil Versus Test Scores**

Many states and school districts debate whether or not increasing the amount of money spent per student will guarantee academic success. The following scatterplot shows the average eighth grade reading score on the National Assessment of Education Progress (NAEP) for the 50 states and the District of Columbia in 2007 plotted against the average expenditure per pupil in 2007. Explore how the correlation coefficient is affected by the inclusion of the District of Columbia, which spent $14,324 per pupil and had a NAEP score of 241. *Source: U.S. Census Bureau and National Center for Education Statistics.*

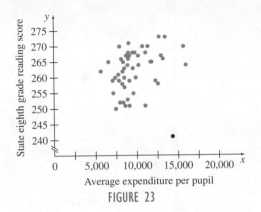

FIGURE 23

SOLUTION A spreadsheet was used to create a plot of the points shown in Figure 23. Washington D.C. corresponds to the red point in the lower right, which is noticeably separate from all the other points. Using the original data, the correlation coefficient when Washington D.C. is included is 0.1981, indicating that there is not a strong linear correlation. Excluding Washington D.C. raises the correlation coefficient to 0.3745, which is a somewhat stronger indication of a linear correlation. This illustrates that one extreme data point that is separate from the others, known as an **outlier**, can have a strong effect on the correlation coefficient.

Even if the correlation between average expenditure per pupil and reading score in Example 5 was high, this would not prove that spending more per pupil causes high reading scores. To prove this would require further research. It is a common statistical fallacy to assume that correlation implies causation. Perhaps the correlation is due to a third underlying variable. In Example 5, perhaps states with wealthier families spend more per pupil, and the students read better because wealthier families have greater access to reading material. Determining the truth requires careful research methods that are beyond the scope of this textbook.

1.3 EXERCISES

1. Suppose a positive linear correlation is found between two quantities. Does this mean that one of the quantities increasing causes the other to increase? If not, what does it mean?

2. Given a set of points, the least squares line formed by letting x be the independent variable will not necessarily be the same as the least squares line formed by letting y be the independent variable. Give an example to show why this is true.

3. For the following table of data,

x	1	2	3	4	5	6	7	8	9	10
y	0	0.5	1	2	2.5	3	3	4	4.5	5

a. draw a scatterplot.

b. calculate the correlation coefficient.

c. calculate the least squares line and graph it on the scatterplot.

d. predict the y-value when x is 11.

The following problem is reprinted from the November 1989 *Actuarial Examination on Applied Statistical Methods*. Source: *Society of Actuaries*.

4. You are given

X	6.8	7.0	7.1	7.2	7.4
Y	0.8	1.2	0.9	0.9	1.5

Determine r^2, the coefficient of determination for the regression of Y on X. Choose one of the following. (*Note:* The coefficient of determination is defined as the square of the correlation coefficient.)

a. 0.3   b. 0.4   c. 0.5   d. 0.6   e. 0.7

5. Consider the following table of data.

x	1	1	2	2	9
y	1	2	1	2	9

a. Calculate the least squares line and the correlation coefficient.

b. Repeat part a, but this time delete the last point.

c. Draw a graph of the data, and use it to explain the dramatic difference between the answers to parts a and b.

6. Consider the following table of data.

x	1	2	3	4	9
y	1	2	3	4	−20

a. Calculate the least squares line and the correlation coefficient.

b. Repeat part a, but this time delete the last point.

c. Draw a graph of the data, and use it to explain the dramatic difference between the answers to parts a and b.

7. Consider the following table of data.

x	1	2	3	4
y	1	1	1	1.1

a. Calculate the correlation coefficient.

b. Sketch a graph of the data.

c. Based on how closely the data fits a straight line, is your answer to part a surprising? Discuss the extent to which the correlation coefficient describes how well the data fit a horizontal line.

8. Consider the following table of data.

x	0	1	2	3	4
y	4	1	0	1	4

a. Calculate the least squares line and the correlation coefficient.

b. Sketch a graph of the data.

c. Comparing your answers to parts a and b, does a correlation coefficient of 0 mean that there is no relationship between the x and y values? Would some curve other than a line fit the data better? Explain.

9. The formulas for the least squares line were found by solving the system of equations

$$nb + (\Sigma x)m = \Sigma y$$
$$(\Sigma x)b + (\Sigma x^2)m = \Sigma xy.$$

Solve the above system for b and m to show that

$$m = \frac{n(\Sigma xy) - (\Sigma x)(\Sigma y)}{n(\Sigma x^2) - (\Sigma x)^2} \quad \text{and}$$

$$b = \frac{\Sigma y - m(\Sigma x)}{n}.$$

APPLICATIONS

Business and Economics

10. **Consumer Durable Goods**   The total value of consumer durable goods has grown at an approximately linear rate in recent years. The annual data for the years 2002 through 2008 can be summarized as follows, where x represents the years since 2000 and y the total value of consumer durable goods in trillions of dollars. *Source: Bureau of Economic Analysis.*

$n = 7$	$\Sigma x = 35$	$\Sigma x^2 = 203$
$\Sigma y = 28.4269$	$\Sigma y^2 = 116.3396$	$\Sigma xy = 147.1399$

a. Find an equation for the least squares line.

b. Use your result from part **a** to predict the total value of consumer durable goods in the year 2015.

c. If this growth continues linearly, in what year will the total value of consumer durable goods first reach at least 6 trillion dollars?

d. Find and interpret the correlation coefficient.

11. **Decrease in Banks**   The number of banks in the United States has been dropping steadily since 1984, and the trend in recent years has been roughly linear. The annual data for the years 1999 through 2008 can be summarized as follows, where x represents the years since 1990 and y the number of banks, in thousands, in the United States. *Source: FDIC.*

$n = 10$	$\Sigma x = 235$	$\Sigma x^2 = 5605$
$\Sigma y = 77.564$	$\Sigma y^2 = 603.60324$	$\Sigma xy = 1810.095$

a. Find an equation for the least squares line.

b. Use your result from part a to predict the number of U.S. banks in the year 2020.

c. If this trend continues linearly, in what year will the number of U.S. banks drop below 6000?

d. Find and interpret the correlation coefficient.

12. **Digital Cable Subscribers**   The number of subscribers to digital cable television has been growing steadily, as shown by the

following table. *Source: National Cable and Telecommunications Association.*

Year	2000	2002	2004	2006	2008
Customers (in millions)	8.5	19.3	25.4	32.6	40.4

a. Find an equation for the least squares line, letting x equal the number of years since 2000.

b. Based on your answer to part a, at approximately what rate is the number of subscribers to digital cable television growing per year?

c. Use your result from part a to predict the number of digital cable subscribers in the year 2012.

d. If this trend continues linearly, in what year will the number of digital cable subscribers first exceed 70 million?

e. Find and interpret the correlation coefficient.

13. **Consumer Credit** The total amount of consumer credit has been increasing steadily in recent years. The following table gives the total U.S. outstanding consumer credit. *Source: Federal Reserve.*

Year	2004	2005	2006	2007	2008
Consumer credit (in billions of dollars)	2219.5	2319.8	2415.0	2551.9	2592.1

a. Find an equation for the least squares line, letting x equal the number of years since 2000.

b. Based on your answer to part a, at approximately what rate is the consumer credit growing per year?

c. Use your result from part a to predict the amount of consumer credit in the year 2015.

d. If this trend continues linearly, in what year will the total debt first exceed $4000 billion?

e. Find and interpret the correlation coefficient.

14. **New Car Sales** New car sales have increased at a roughly linear rate. Sales, in millions of vehicles, from 2000 to 2007, are given in the table below. *Source: National Automobile Dealers Association.* Let x represent the number of years since 2000.

Year	Sales
2000	17.3
2001	17.1
2002	16.8
2003	16.6
2004	16.9
2005	16.9
2006	16.5
2007	16.1

a. Find the equation of the least squares line and the correlation coefficient.

b. Find the equation of the least squares line using only the data for every other year starting with 2000, 2002, and so on. Find the correlation coefficient.

c. Compare your results for parts a and b. What do you find? Why do you think this happens?

15. **Air Fares** In 2006, for passengers who made early reservations, American Airlines offered lower prices on one-way fares from New York to various cities. Fourteen of the cities are listed in the following table, with the distances from New York to the cities included. *Source: American Airlines.*

a. Plot the data. Do the data points lie in a linear pattern?

b. Find the correlation coefficient. Combining this with your answer to part a, does the cost of a ticket tend to go up with the distance flown?

c. Find the equation of the least squares line, and use it to find the approximate marginal cost per mile to fly.

d. For similar data in a January 2000 *New York Times* ad, the equation of the least squares line was $Y = 113 + 0.0243x$. *Source: The New York Times.* Use this information and your answer to part b to compare the cost of flying American Airlines for these two time periods.

e. Identify the outlier in the scatterplot. Discuss the reason why there would be a difference in price to this city.

City	Distance (x) (miles)	Price (y) (dollars)
Boston	206	95
Chicago	802	138
Denver	1771	228
Kansas City	1198	209
Little Rock	1238	269
Los Angeles	2786	309
Minneapolis	1207	202
Nashville	892	217
Phoenix	2411	109
Portland	2885	434
Reno	2705	399
St. Louis	948	206
San Diego	2762	239
Seattle	2815	329

Life Sciences

16. **Bird Eggs** The average length and width of various bird eggs are given in the following table. *Source: National Council of Teachers of Mathematics.*

Bird Name	Width (cm)	Length (cm)
Canada goose	5.8	8.6
Robin	1.5	1.9
Turtledove	2.3	3.1
Hummingbird	1.0	1.0
Raven	3.3	5.0

a. Plot the points, putting the length on the *y*-axis and the width on the *x*-axis. Do the data appear to be linear?

b. Find the least squares line, and plot it on the same graph as the data.

c. Suppose there are birds with eggs even smaller than those of hummingbirds. Would the equation found in part b continue to make sense for all positive widths, no matter how small? Explain.

d. Find the correlation coefficient.

17. Size of Hunting Parties In the 1960s, the famous researcher Jane Goodall observed that chimpanzees hunt and eat meat as part of their regular diet. Sometimes chimpanzees hunt alone, while other times they form hunting parties. The following table summarizes research on chimpanzee hunting parties, giving the size of the hunting party and the percentage of successful hunts. *Source: American Scientist and Mathematics Teacher.*

Number of Chimps in Hunting Party	Percentage of Successful Hunts
1	20
2	30
3	28
4	42
5	40
6	58
7	45
8	62
9	65
10	63
12	75
13	75
14	78
15	75
16	82

a. Plot the data. Do the data points lie in a linear pattern?

b. Find the correlation coefficient. Combining this with your answer to part a, does the percentage of successful hunts tend to increase with the size of the hunting party?

c. Find the equation of the least squares line, and graph it on your scatterplot.

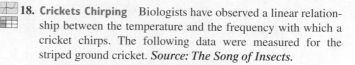

18. Crickets Chirping Biologists have observed a linear relationship between the temperature and the frequency with which a cricket chirps. The following data were measured for the striped ground cricket. *Source: The Song of Insects.*

Temperature °F (x)	Chirps Per Second (y)
88.6	20.0
71.6	16.0
93.3	19.8
84.3	18.4
80.6	17.1
75.2	15.5
69.7	14.7
82.0	17.1
69.4	15.4
83.3	16.2
79.6	15.0
82.6	17.2
80.6	16.0
83.5	17.0
76.3	14.4

a. Find the equation for the least squares line for the data.

b. Use the results of part a to determine how many chirps per second you would expect to hear from the striped ground cricket if the temperature were 73°F.

c. Use the results of part a to determine what the temperature is when the striped ground crickets are chirping at a rate of 18 times per sec.

d. Find the correlation coefficient.

Social Sciences

19. Pupil-Teacher Ratios The following table gives the national average pupil-teacher ratio in public schools over selected years. *Source: National Center for Education Statistics.*

Year	Ratio
1990	17.4
1994	17.7
1998	16.9
2002	16.2
2006	15.8

a. Find the equation for the least squares line. Let *x* correspond to the number of years since 1990 and let *y* correspond to the average number of pupils per 1 teacher.

b. Use your answer from part a to predict the pupil-teacher ratio in 2020. Does this seem realistic?

c. Calculate and interpret the correlation coefficient.

20. Poverty Levels The following table lists how poverty level income cutoffs (in dollars) for a family of four have changed over time. *Source: U.S. Census Bureau.*

Year	Income
1980	8414
1985	10,989
1990	13,359
1995	15,569
2000	17,604
2005	19,961
2008	22,207

Let x represent the year, with $x = 0$ corresponding to 1980 and y represent the income in thousands of dollars.

a. Plot the data. Do the data appear to lie along a straight line?

b. Calculate the correlation coefficient. Does your result agree with your answer to part a?

c. Find the equation of the least squares line.

d. Use your answer from part c to predict the poverty level in the year 2018.

21. Ideal Partner Height In an introductory statistics course at Cornell University, 147 undergraduates were asked their own height and the ideal height for their ideal spouse or partner. For this exercise, we are including the data for only a representative sample of 10 of the students, as given in the following table. All heights are in inches. *Source: Chance.*

Height	Ideal Partner's Height
59	66
62	71
66	72
68	73
71	75
67	63
70	63
71	67
73	66
75	66

a. Find the regression line and correlation coefficient for this data. What strange phenomenon do you observe?

b. The first five data pairs are for female students and the second five for male students. Find the regression line and correlation coefficient for each set of data.

c. Plot all the data on one graph, using different types of points to distinguish the data for the males and for the females. Using this plot and the results from part b, explain the strange phenomenon that you observed in part a.

22. SAT Scores At Hofstra University, all students take the math SAT before entrance, and most students take a mathematics placement test before registration. Recently, one professor collected the following data for 19 students in his Finite Mathematics class:

Math SAT	Placement Test	Math SAT	Placement Test	Math SAT	Placement Test
540	20	580	8	440	10
510	16	680	15	520	11
490	10	560	8	620	11
560	8	560	13	680	8
470	12	500	14	550	8
600	11	470	10	620	7
540	10				

a. Find an equation for the least squares line. Let x be the math SAT and y be the placement test score.

b. Use your answer from part a to predict the mathematics placement test score for a student with a math SAT score of 420.

c. Use your answer from part a to predict the mathematics placement test score for a student with a math SAT score of 620.

d. Calculate the correlation coefficient.

e. Based on your answer to part d, what can you conclude about the relationship between a student's math SAT and mathematics placement test score?

Physical Sciences

23. Length of a Pendulum Grandfather clocks use pendulums to keep accurate time. The relationship between the length of a pendulum L and the time T for one complete oscillation can be determined from the data in the table.* *Source: Gary Rockswold.*

L (ft)	T (sec)
1.0	1.11
1.5	1.36
2.0	1.57
2.5	1.76
3.0	1.92
3.5	2.08
4.0	2.22

a. Plot the data from the table with L as the horizontal axis and T as the vertical axis.

b. Find the least squares line equation and graph it simultaneously, if possible, with the data points. Does it seem to fit the data?

c. Find the correlation coefficient and interpret it. Does it confirm your answer to part b?

24. Air Conditioning While shopping for an air conditioner, Adam Bryer consulted the following table, which gives a machine's BTUs and the square footage (ft²) that it would cool.

*The actual relationship is $L = 0.81T^2$, which is not a linear relationship. This illustration that even if the relationship is not linear, a line can give a good approximation.

ft² (x)	BTUs (y)
150	5000
175	5500
215	6000
250	6500
280	7000
310	7500
350	8000
370	8500
420	9000
450	9500

a. Find the equation for the least squares line for the data.

b. To check the fit of the data to the line, use the results from part a to find the BTUs required to cool a room of 150 ft², 280 ft², and 420 ft². How well does the actual data agree with the predicted values?

c. Suppose Adam's room measures 230 ft². Use the results from part a to decide how many BTUs it requires. If air conditioners are available only with the BTU choices in the table, which would Adam choose?

d. Why do you think the table gives ft² instead of ft³, which would give the volume of the room?

General Interest

25. Football The following data give the expected points for a football team with first down and 10 yards to go from various points on the field. *Source: Operations Research.* (*Note:* $\sum x = 500$, $\sum x^2 = 33{,}250$, $\sum y = 20.668$, $\sum y^2 = 91.927042$, $\sum xy = 399.16$.)

Yards from Goal (x)	Expected Points (y)
5	6.041
15	4.572
25	3.681
35	3.167
45	2.392
55	1.538
65	0.923
75	0.236
85	−0.637
95	−1.245

a. Calculate the correlation coefficient. Does there appear to be a linear correlation?

b. Find the equation of the least squares line.

c. Use your answer from part a to predict the expected points when a team is at the 50-yd line.

26. Athletic Records The table shows the men's and women's outdoor world records (in seconds) in the 800-m run. *Source: Nature, Track and Field Athletics, Statistics in Sports, and The World Almanac and Book of Facts.*

Year	Men's Record	Women's Record
1905	113.4	—
1915	111.9	—
1925	111.9	144
1935	109.7	135.6
1945	106.6	132
1955	105.7	125
1965	104.3	118
1975	103.7	117.48
1985	101.73	113.28
1995	101.11	113.28
2005	101.11	113.28

Let x be the year, with $x = 0$ corresponding to 1900.

a. Find the equation for the least squares line for the men's record (y) in terms of the year (x).

b. Find the equation for the least squares line for the women's record.

c. Suppose the men's and women's records continue to improve as predicted by the equations found in parts a and b. In what year will the women's record catch up with the men's record? Do you believe that will happen? Why or why not?

d. Calculate the correlation coefficient for both the men's and the women's record. What do these numbers tell you?

e. Draw a plot of the data, and discuss to what extent a linear function describes the trend in the data.

27. Running If you think a marathon is a long race, consider the Hardrock 100, a 100.5 mile running race held in southwestern Colorado. The chart at right lists the times that the 2008 winner, Kyle Skaggs, arrived at various mileage points along the way. *Source: www.run100s.com.*

a. What was Skagg's average speed?

b. Graph the data, plotting time on the x-axis and distance on the y-axis. You will need to convert the time from hours and minutes into hours. Do the data appear to lie approximately on a straight line?

c. Find the equation for the least squares line, fitting distance as a linear function of time.

d. Calculate the correlation coefficient. Does it indicate a good fit of the least squares line to the data?

e. Based on your answer to part d, what is a good value for Skagg's average speed? Compare this with your answer to part a. Which answer do you think is better? Explain your reasoning.

Time (hr:min)	Miles
0	0
2:19	11.5
3:43	18.9
5:36	27.8
7:05	32.8
7:30	36.0
8:30	43.9
10:36	51.5
11:56	58.4
15:14	71.8
17:49	80.9
18:58	85.2
20:50	91.3
23:23	100.5

YOUR TURN ANSWERS

1. $Y = -0.535x + 89.5$ **2.** -0.949

CHAPTER REVIEW

SUMMARY

In this chapter we studied linear functions, whose graphs are straight lines. We developed the slope-intercept and point-slope formulas, which can be used to find the equation of a line, given a point and the slope or given two points. We saw that lines have many applications in virtually every discipline. Lines are used through the rest of this book, so fluency in their use is important. We concluded the chapter by introducing the method of least squares, which is used to find an equation of the line that best fits a given set of data.

Slope of a Line The slope of a line is defined as the vertical change (the "rise") over the horizontal change (the "run") as one travels along the line. In symbols, taking two different points (x_1, y_1) and (x_2, y_2) on the line, the slope is

$$m = \frac{y_2 - y_1}{x_2 - x_1},$$

where $x_1 \neq x_2$.

Equations of Lines

Equation	Description
$y = mx + b$	Slope intercept form: slope m and y-intercept b.
$y - y_1 = m(x - x_1)$	Point-slope form: slope m and line passes through (x_1, y_1).
$x = k$	Vertical line: x-intercept k, no y-intercept (except when $k = 0$), undefined slope.
$y = k$	Horizontal line: y-intercept k, no x-intercept (except when $k = 0$), slope 0.

Parallel Lines Two lines are parallel if and only if they have the same slope, or if they are both vertical.

Perpendicular Lines Two lines are perpendicular if and only if the product of their slopes is -1, or if one is vertical and the other horizontal.

Linear Function   A relationship f defined by

$$y = f(x) = mx + b,$$

for real numbers m and b, is a linear function.

Linear Cost Function   In a cost function of the form $C(x) = mx + b$, the m represents the marginal cost and b represents the fixed cost.

Least Squares Line   The least squares line $Y = mx + b$ that gives the best fit to the data points $(x_1, y_1), (x_2, y_2), \ldots,$ (x_n, y_n) has slope m and y-intercept b given by the equations

$$m = \frac{n(\Sigma xy) - (\Sigma x)(\Sigma y)}{n(\Sigma x^2) - (\Sigma x)^2}$$

$$b = \frac{\Sigma y - m(\Sigma x)}{n}$$

Correlation Coefficient

$$r = \frac{n(\Sigma xy) - (\Sigma x)(\Sigma y)}{\sqrt{n(\Sigma x^2) - (\Sigma x)^2} \ \sqrt{n(\Sigma y^2) - (\Sigma y)^2}}$$

KEY TERMS

To understand the concepts presented in this chapter, you should know the meaning and use of the following terms. For easy reference, the section in the chapter where a word (or expression) was first used is provided.

mathematical model

1.1
ordered pair
Cartesian coordinate system
axes
origin
coordinates
quadrants
graph
intercepts

slope
linear equation
slope-intercept form
proportional
point-slope form
parallel
perpendicular
scatterplot

1.2
linear function

independent variable
dependent variable
surplus
shortage
equilibrium price
equilibrium quantity
fixed cost
marginal cost
linear cost function
revenue

profit
break-even quantity
break-even point

1.3
least squares line
summation notation
correlation coefficient
outlier

REVIEW EXERCISES

CONCEPT CHECK

Determine whether each statement is true or false, and explain why.

1. A given line can have more than one slope.

2. The equation $y = 3x + 4$ represents the equation of a line with slope 4.

3. The line $y = -2x + 5$ intersects the point $(3, -1)$.

4. The line that intersects the points $(2, 3)$ and $(2, 5)$ is a horizontal line.

5. The line that intersects the points $(4, 6)$ and $(5, 6)$ is a horizontal line.

6. The x-intercept of the line $y = 8x + 9$ is 9.

7. The function $f(x) = \pi x + 4$ represents a linear function.

8. The function $f(x) = 2x^2 + 3$ represents a linear function.

9. The lines $y = 3x + 17$ and $y = -3x + 8$ are perpendicular.

10. The lines $4x + 3y = 8$ and $4x + y = 5$ are parallel.

11. A correlation coefficient of zero indicates a perfect fit with the data.

12. It is not possible to get a correlation coefficient of -1.5 for a set of data.

PRACTICE AND EXPLORATIONS

13. What is marginal cost? Fixed cost?

14. What six quantities are needed to compute a correlation coefficient?

Find the slope for each line that has a slope.

15. Through $(-3, 7)$ and $(2, 12)$

16. Through $(4, -1)$ and $(3, -3)$

17. Through the origin and $(11, -2)$

18. Through the origin and $(0, 7)$

19. $4x + 3y = 6$

20. $4x - y = 7$

21. $y + 4 = 9$

22. $3y - 1 = 14$

23. $y = 5x + 4$

24. $x = 5y$

Find an equation in the form $y = mx + b$ for each line.

25. Through $(5, -1)$; slope $= 2/3$

26. Through $(8, 0)$; slope $= -1/4$

27. Through $(-6, 3)$ and $(2, -5)$

28. Through $(2, -3)$ and $(-3, 4)$

29. Through $(2, -10)$, perpendicular to a line with undefined slope

30. Through $(-2, 5)$; slope $= 0$

Find an equation for each line in the form $ax + by = c$, where a, b, and c are integers with no factor common to all three and $a \geq 0$.

31. Through $(3, -4)$, parallel to $4x - 2y = 9$

32. Through $(0, 5)$, perpendicular to $8x + 5y = 3$

33. Through $(-1, 4)$; undefined slope

34. Through $(7, -6)$, parallel to a line with undefined slope

35. Through $(3, -5)$, parallel to $y = 4$

36. Through $(-3, 5)$, perpendicular to $y = -2$

Graph each linear equation defined as follows.

37. $y = 4x + 3$

38. $y = 6 - 2x$

39. $3x - 5y = 15$

40. $4x + 6y = 12$

41. $x - 3 = 0$

42. $y = 1$

43. $y = 2x$

44. $x + 3y = 0$

APPLICATIONS

Business and Economics

45. Profit To manufacture x thousand computer chips requires fixed expenditures of $352 plus $42 per thousand chips. Receipts from the sale of x thousand chips amount to $130 per thousand.

  a. Write an expression for expenditures.

  b. Write an expression for receipts.

  c. For profit to be made, receipts must be greater than expenditures. How many chips must be sold to produce a profit?

46. Supply and Demand The supply and demand for crabmeat in a local fish store are related by the equations

$$\text{Supply: } p = S(q) = 6q + 3$$

and

$$\text{Demand: } p = D(q) = 19 - 2q,$$

where p represents the price in dollars per pound and q represents the quantity of crabmeat in pounds per day. Find the quantity supplied and demanded at each of the following prices.

  a. $10   **b.** $15   **c.** $18

  d. Graph both the supply and the demand functions on the same axes.

  e. Find the equilibrium price.

  f. Find the equilibrium quantity.

47. Supply For a new diet pill, 60 pills will be supplied at a price of $40, while 100 pills will be supplied at a price of $60. Write a linear supply function for this product.

48. Demand The demand for the diet pills in Exercise 47 is 50 pills at a price of $47.50 and 80 pills at a price of $32.50. Determine a linear demand function for these pills.

49. Supply and Demand Find the equilibrium price and quantity for the diet pills in Exercises 47 and 48.

Cost In Exercises 50–53, find a linear cost function.

50. Eight units cost $300; fixed cost is $60.

51. Fixed cost is $2000; 36 units cost $8480.

52. Twelve units cost $445; 50 units cost $1585.

53. Thirty units cost $1500; 120 units cost $5640.

54. Break-Even Analysis The cost of producing x cartons of CDs is $C(x)$ dollars, where $C(x) = 200x + 1000$. The CDs sell for $400 per carton.

  a. Find the break-even quantity.

  b. What revenue will the company receive if it sells just that number of cartons?

55. Break-Even Analysis The cost function for flavored coffee at an upscale coffeehouse is given in dollars by $C(x) = 3x + 160$, where x is in pounds. The coffee sells for $7 per pound.

  a. Find the break-even quantity.

  b. What will the revenue be at that point?

56. U.S. Imports from China The United States is China's largest export market. Imports from China have grown from about 102 billion dollars in 2001 to 338 billion dollars in 2008. This growth has been approximately linear. Use the given data pairs to write a linear equation that describes this growth in imports over the years. Let $t = 1$ represent 2001 and $t = 8$ represent 2008. *Source: TradeStats Express™.*

57. U.S. Exports to China U.S. exports to China have grown (although at a slower rate than imports) since 2001. In 2001, about 19.1 billion dollars of goods were exported to China. By 2008, this amount had grown to 69.7 billion dollars. Write a linear equation describing the number of exports each year, with $t = 1$ representing 2001 and $t = 8$ representing 2008. *Source: TradeStats Express™.*

58. Median Income The U.S. Census Bureau reported that the median income for all U.S. households in 2008 was $50,303. In 1988, the median income (in 2008 dollars) was $47,614. The median income is approximately linear and is a function of time. Find a formula for the median income, I, as a function of the year t, where t is the number of years since 1900. *Source: U.S Census Bureau.*

59. New Car Cost The average new car cost (in dollars) for selected years from 1980 to 2005 is given in the table. *Source: Chicago Tribune and National Automobile Dealers Association.*

Year	1980	1985	1990	1995	2000	2005
Cost	7500	12,000	16,000	20,450	24,900	28,400

a. Find a linear equation for the average new car cost in terms of x, the number of years since 1980, using the data for 1980 and 2005.

b. Repeat part a, using the data for 1995 and 2005.

c. Find the equation of the least squares line using all the data.

d. Use a graphing calculator to plot the data and the three lines from parts a-c.

e. Discuss which of the three lines found in parts a–c best describes the data, as well as to what extent a linear model accurately describes the data.

f. Calculate the correlation coefficient.

Life Sciences

60. World Health In general, people tend to live longer in countries that have a greater supply of food. Listed below is the 2003–2005 daily calorie supply and 2005 life expectancy at birth for 10 randomly selected countries. *Source: Food and Agriculture Organization.*

Country	Calories (x)	Life Expectancy (y)
Belize	2818	75.4
Cambodia	2155	59.4
France	3602	80.4
India	2358	62.7
Mexico	3265	75.5
New Zealand	3235	79.8
Peru	2450	72.5
Sweden	3120	80.5
Tanzania	2010	53.7
United States	3826	78.7

a. Find the correlation coefficient. Do the data seem to fit a straight line?

b. Draw a scatterplot of the data. Combining this with your results from part a, do the data seem to fit a straight line?

c. Find the equation of the least squares line.

d. Use your answer from part c to predict the life expectancy in the United Kingdom, which has a daily calorie supply of 3426. Compare your answer with the actual value of 79.0 years.

e. Briefly explain why countries with a higher daily calorie supply might tend to have a longer life expectancy. Is this trend likely to continue to higher calorie levels? Do you think that an American who eats 5000 calories a day is likely to live longer than one who eats 3600 calories? Why or why not?

f. (For the ambitious!) Find the correlation coefficient and least squares line using the data for a larger sample of countries, as found in an almanac or other reference. Is the result in general agreement with the previous results?

61. Blood Sugar and Cholesterol Levels The following data show the connection between blood sugar levels and cholesterol levels for eight different patients.

Patient	Blood Sugar Level (x)	Cholesterol Level (y)
1	130	170
2	138	160
3	142	173
4	159	181
5	165	201
6	200	192
7	210	240
8	250	290

For the data given in the preceding table, $\sum x = 1394$, $\sum y = 1607$, $\sum xy = 291{,}990$, $\sum x^2 = 255{,}214$, and $\sum y^2 = 336{,}155$.

a. Find the equation of the least squares line.

b. Predict the cholesterol level for a person whose blood sugar level is 190.

c. Find the correlation coefficient.

Social Sciences

62. Beef Consumption The per capita consumption of beef in the United States decreased from 115.7 lb in 1974 to 92.9 lb in 2007. Assume a linear function describes the decrease. Write a linear equation defining the function. Let t represent the number of years since 1950 and y represent the number of pounds of red meat consumed. *Source: U.S. Department of Agriculture.*

63. Marital Status More people are staying single longer in the United States. In 1995, the number of never married adults, age 15 and over, was 55.0 million. By 2009, it was 72.1 million. Assume the data increase linearly, and write an equation that defines a linear function for this data. Let t represent the number of years since 1990. *Source: U.S. Census Bureau.*

64. Poverty The following table gives the number of families under the poverty level in the U.S. in recent years. *Source: U.S. Census Bureau.*

Year	Families Below Poverty Level (in thousands)
2000	6400
2001	6813
2002	7229
2003	7607
2004	7623
2005	7657
2006	7668
2007	7835
2008	8147

a. Find a linear equation for the number of families below poverty level (in thousands) in terms of x, the number of years since 2000, using the data for 2000 and 2008.

b. Repeat part a, using the data for 2004 and 2008.

c. Find the equation of the least squares line using all the data. Then plot the data and the three lines from parts a–c on a graphing calculator.

d. Discuss which of the three lines found in parts a–c best describes the data, as well as to what extent a linear model accurately describes the data.

e. Calculate the correlation coefficient.

65. Governors' Salaries In general, the larger a state's population, the more the governor earns. Listed in the table below are the estimated 2008 populations (in millions) and the salary of the governor (in thousands of dollars) for eight randomly selected states. *Source: U.S. Census Bureau and Alaska Department of Administration.*

State	AZ	DE	MD	MA	NY	PA	TN	WY
Population (x)	6.50	0.88	5.54	6.45	19.30	12.39	5.92	0.53
Governor's Salary (y)	95	133	150	141	179	170	160	105

a. Find the correlation coefficient. Do the data seem to fit a straight line?

b. Draw a scatterplot of the data. Compare this with your answer from part a.

c. Find the equation for the least squares line.

d. Based on your answer to part c, how much does a governor's salary increase, on average, for each additional million in population?

e. Use your answer from part c to predict the governor's salary in your state. Based on your answers from parts a and b, would this prediction be very accurate? Compare with the actual salary, as listed in an almanac or other reference.

f. (For the ambitious!) Find the correlation coefficient and least squares line using the data for all 50 states, as found in an almanac or other reference. Is the result in general agreement with the previous results?

66. Movies A mathematician exploring the relationship between ratings of movies, their year of release, and their length discovered a paradox. Rather than list the data set of 100 movies in the original research, we have created a sample of size 10 that captures the properties of the original dataset. In the following table, the rating is a score from 1 to 10, and the length is in minutes. *Source: Journal of Statistics Education.*

Year	Rating	Length
2001	10	120
2003	5	85
2004	3	100
2004	6	105
2005	4	110
2005	8	115
2006	6	135
2007	2	105
2007	5	125
2008	6	130

a. Find the correlation coefficient between the years since 2000 and the length.

b. Find the correlation coefficient between the length and the rating.

c. Given that you found a positive correlation between the year and the length in part a, and a positive correlation between the length and the rating in part b, what would you expect about the correlation between the year and the rating? Calculate this correlation. Are you surprised?

d. Discuss the paradoxical result in part c. Write out in words what each correlation tells you. Try to explain what is happening. You may want to look at a scatterplot between the year and the rating, and consider which points on the scatterplot represent movies of length no more than 110 minutes, and which represent movies of length 115 minutes or more.

EXTENDED APPLICATION

USING EXTRAPOLATION TO PREDICT LIFE EXPECTANCY

One reason for developing a mathematical model is to make predictions. If your model is a least squares line, you can predict the y-value corresponding to some new x by substituting this x into an equation of the form $Y = mx + b$. (We use a capital Y to remind us that we're getting a predicted value rather than an actual data value.) Data analysts distinguish between two very different kinds of prediction, *interpolation*, and *extrapolation*. An interpolation uses a new x inside the x range of your original data. For example, if you have inflation data at 5-year intervals from 1950 to 2000, estimating the rate of inflation in 1957 is an interpolation problem. But if you use the same data to estimate what the inflation rate was in 1920, or what it will be in 2020, you are extrapolating.

In general, interpolation is much safer than extrapolation, because data that are approximately linear over a short interval may be nonlinear over a larger interval. One way to detect nonlinearity is to look at *residuals*, which are the differences between the actual data values and the values predicted by the line of best fit. Here is a simple example:

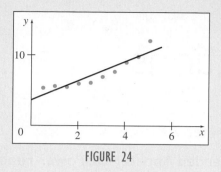

FIGURE 24

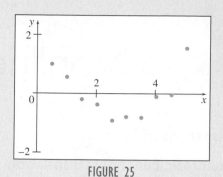

FIGURE 25

The regression equation for the linear fit in Figure 24 is $Y = 3.431 + 1.334x$. Since the r-value for this regression line is 0.93, our linear model fits the data very well. But we might notice that the predictions are a bit low at the ends and high in the middle. We can get a better look at this pattern by plotting the residuals. To find them, we put each value of the independent variable into the regression equation, calculate

the predicted value Y, and subtract it from the actual y-value. The residual plot is shown in Figure 25, with the vertical axis rescaled to exaggerate the pattern. The residuals indicate that our data have a nonlinear, U-shaped component that is not captured by the linear fit. Extrapolating from this data set is probably not a good idea; our linear prediction for the value of y when x is 10 may be much too low.

EXERCISES

The following table gives the life expectancy at birth of females born in the United States in various years from 1970 to 2005. **Source: National Center for Health Statistics.**

Year of Birth	Life Expectancy (years)
1970	74.7
1975	76.6
1980	77.4
1985	78.2
1990	78.8
1995	78.9
2000	79.3
2005	79.9

1. Find an equation for the least squares line for these data, using year of birth as the independent variable.

2. Use your regression equation to guess a value for the life expectancy of females born in 1900.

3. Compare your answer with the actual life expectancy for females born in 1900, which was 48.3 years. Are you surprised?

4. Find the life expectancy predicted by your regression equation for each year in the table, and subtract it from the actual value in the second column. This gives you a table of residuals. Plot your residuals as points on a graph.

5. Now look at the residuals as a fresh data set, and see if you can sketch the graph of a smooth function that fits the residuals well. How easy do you think it will be to predict the life expectancy at birth of females born in 2015?

6. What will happen if you try linear regression on the *residuals*? If you're not sure, use your calculator or software to find the regression equation for the residuals. Why does this result make sense?

7. Since most of the females born in 1995 are still alive, how did the Public Health Service come up with a life expectancy of 78.9 years for these women?

8. Go to the website WolframAlpha.com and enter: "linear fit {1970,74.7}, {1975,76.6}, etc.," putting in all the data from the table. Discuss how the solution compares with the solutions provided by a graphing calculator and by Microsoft Excel.

DIRECTIONS FOR GROUP PROJECT

Assume that you and your group (3–5 students) are preparing a report for a local health agency that is interested in using linear regression to predict life expectancy. Using the questions above as a guide, write a report that addresses the spirit of each question and any issues related to that question. The report should be mathematically sound, grammatically correct, and professionally crafted. Provide recommendations as to whether the health agency should proceed with the linear equation or whether it should seek other means of making such predictions.

2

Nonlinear Functions

There are fourteen mountain peaks over 8000 meters on the Earth's surface. At these altitudes climbers face the challenge of "thin air," since atmospheric pressure is about one third of the pressure at sea level. An exercise in Section 4 of this chapter shows how the change in atmospheric pressure with altitude can be modeled with an exponential function.

Figure 1 below shows the average price of gold for each year in the last few decades. *Source: finfacts.ie.* The graph is not a straight line and illustrates a function that, unlike those studied in Chapter 1, is *nonlinear.* Linear functions are simple to study, and they can be used to approximate many functions over short intervals. But most functions exhibit behavior that, in the long run, does not follow a straight line. In this chapter we will study some of the most common nonlinear functions.

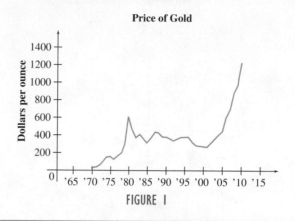

FIGURE 1

2.1 Properties of Functions

APPLY IT   How has the world's use of different energy sources changed over time?
We will analyze this question in Exercise 78 in this section, after developing the concept of nonlinear functions.

As we saw in Chapter 1, the linear cost function $C(x) = 12x + 5000$ for video games is related to the number of items produced. The number of games produced is the independent variable and the total cost is the dependent variable because it depends on the number produced. When a specific number of games (say 1000) is substituted for x, the cost $C(x)$ has one specific value $(12 \cdot 1000 + 5000)$. Because of this, the variable $C(x)$ is said to be a *function* of x.

> **Function**
>
> A **function** is a rule that assigns to each element from one set exactly one element from another set.

In most cases in this book, the "rule" mentioned in the box is expressed as an equation, such as $C(x) = 12x + 5000$, and each set mentioned in the definition will ordinarily be the real numbers or some subset of the reals. When an equation is given for a function, we say that the equation *defines* the function. Whenever x and y are used in this book to define a function, x represents the independent variable and y the dependent variable. Of course, letters other than x and y could be used and are often more meaningful. For example, if the independent variable represents the number of items sold for $4 each and the dependent variable represents revenue, we might write $R = 4s$.

The independent variable in a function can take on any value within a specified set of values called the *domain*.

Domain and Range

The set of all possible values of the independent variable in a function is called the **domain** of the function, and the resulting set of possible values of the dependent variable is called the **range**.

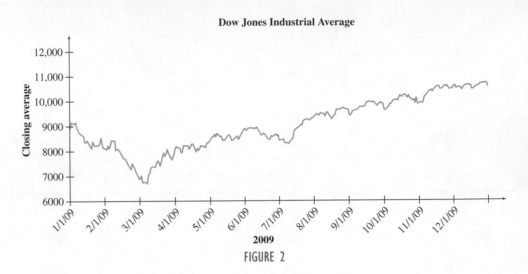

Dow Jones Industrial Average

FIGURE 2

An important function to investors around the world is the Dow Jones industrial average, a performance measure of the stock market. Figure 2 shows how this average varied over the year 2009. *Source: Yahoo! Finance.* Let us label this function $y = f(x)$, where y is the Dow Jones industrial average and x is the time in days from the beginning of 2009. Notice that the function increases and decreases during the year, so it is not linear, although a linear function could be used as a very rough approximation. Such a function, whose graph is not a straight line, is called a *nonlinear function*.

The concepts you learned in the section on linear functions apply to this and other nonlinear functions as well. The independent variable here is x, the time in days; the dependent variable is y, the average at any time. The domain is $\{x \mid 0 \le x \le 365\}$, or $[0, 365]$; $x = 0$ corresponds to the beginning of the day on January 1, and $x = 365$ corresponds to the end of the day on December 31. By looking for the lowest and highest values of the function, we estimate the range to be approximately $\{y \mid 6500 \le y \le 10{,}500\}$, or $[6500, 10{,}500]$. As with linear functions, the domain is mapped along the horizontal axis and the range along the vertical axis.

We do not have a formula for $f(x)$. (If we had possessed such a formula at the beginning of 2009, we could have made a lot of money!) Instead, we can use the graph to estimate values of the function. To estimate $f(10)$, for example, we draw a vertical line from January 10, as shown in Figure 3(a). The y-coordinate seems to be roughly 8600, so we estimate $f(10) \approx 8600$. Similarly, if we wanted to solve the equation $f(x) = 7000$, we would look for points on the graph that have a y-coordinate of 7000. As Figure 3(b) shows, this occurs at two points. The first time is around February 27 (the 58th day of the year), and the last time is around March 11 (the 70th day of the year). Thus $f(x) = 7000$ when $x = 58$ and $x = 70$.

This function can also be given as a table. The table in the margin shows the value of the function for several values of x.

Notice from the table that $f(0) = f(1) = 8772.25$. The stock market was closed for the first day of 2009, so the Dow Jones average did not change. This illustrates an important

Dow Jones Industrial Average	
Day (x)	Close (y)
0	8772.25
1	8772.25
2	9034.69
3	9034.69
4	9034.69
5	8952.89
6	9015.10
7	8769.70

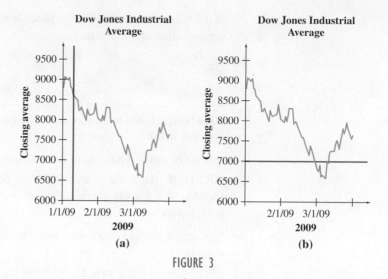

FIGURE 3

property of functions: Several different values of the independent variable can have the same value for the dependent variable. On the other hand, we cannot have several different y-values corresponding to the same value of x; if we did, this would not be a function.

What is $f(5.5)$? We do not know. When the stock market closed on January 5, the Dow Jones industrial average was 8952.89. The closing value the following day was 9015.10. We do not know what happened in between, although this information is recorded by the New York Stock Exchange.

Functions arise in numerous applications, and an understanding of them is critical for understanding calculus. The following example shows some of the ways functions can be represented and will help you in determining whether a relationship between two variables is a function or not.

EXAMPLE 1   Functions

Which of the following are functions?

(a)

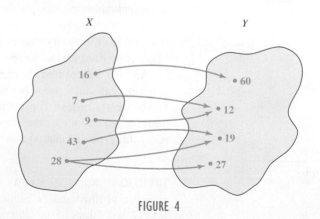

FIGURE 4

SOLUTION   Figure 4 shows that an x-value of 28 corresponds to *two* y-values, 19 and 27. In a function, each x must correspond to exactly one y, so this correspondence is not a function.

(b) The x^2 key on a calculator

SOLUTION   This correspondence between input and output is a function because the calculator produces just one x^2 (one y-value) for each x-value entered. Notice also that two x-values, such as 3 and -3, produce the same y-value of 9, but this does not violate the definition of a function.

(c)

x	1	1	2	2	3	3
y	3	-3	5	-5	8	-8

SOLUTION Since at least one x-value corresponds to more than one y-value, this table does not define a function.

(d) The set of ordered pairs with first elements mothers and second elements their children

SOLUTION Here the mother is the independent variable and the child is the dependent variable. For a given mother, there may be several children, so this correspondence is not a function.

(e) The set of ordered pairs with first elements children and second elements their birth mothers

SOLUTION In this case the child is the independent variable and the mother is the dependent variable. Since each child has only one birth mother, this is a function. ▬

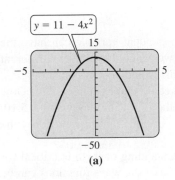

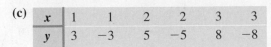

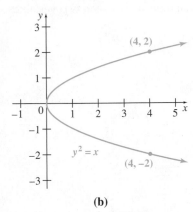

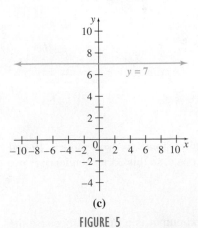

FIGURE 5

EXAMPLE 2 Functions

Decide whether each equation or graph represents a function. (Assume that x represents the independent variable here, an assumption we shall make throughout this book.) Give the domain and range of any functions.

(a) $y = 11 - 4x^2$

SOLUTION For a given value of x, calculating $11 - 4x^2$ produces exactly one value of y. (For example, if $x = -7$, then $y = 11 - 4(-7)^2 = -185$, so $f(-7) = -185$.) Since one value of the independent variable leads to exactly one value of the dependent variable, $y = 11 - 4x^2$ meets the definition of a function.

Because x can take on any real-number value, the domain of this function is the set of all real numbers. Finding the range is more difficult. One way to find it would be to ask what possible values of y could come out of this function. Notice that the value of y is 11 minus a quantity that is always 0 or positive, since $4x^2$ can never be negative. There is no limit to how large $4x^2$ can be, so the range is $(-\infty, 11]$.

Another way to find the range would be to examine the graph. Figure 5(a) shows a graphing calculator view of this function, and we can see that the function takes on y-values of 11 or less. The calculator cannot tell us, however, whether the function continues to go down past the viewing window, or turns back up. To find out, we need to study this type of function more carefully, as we will do in the next section.

(b) $y^2 = x$

SOLUTION Suppose $x = 4$. Then $y^2 = x$ becomes $y^2 = 4$, from which $y = 2$ or $y = -2$, as illustrated in Figure 5(b). Since one value of the independent variable can lead to two values of the dependent variable, $y^2 = x$ does not represent a function.

(c) $y = 7$

SOLUTION No matter what the value of x, the value of y is always 7. This is indeed a function; it assigns exactly one element, 7, to each value of x. Such a function is known as a **constant function**. The domain is the set of all real numbers, and the range is the set $\{7\}$. Its graph is the horizontal line that intersects the y-axis at $y = 7$, as shown in Figure 5(c). Every constant function has a horizontal line for its graph.

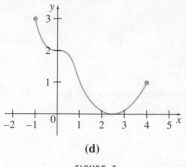

(d)

FIGURE 5

(d) The graph in Figure 5(d).

SOLUTION For each value of x, there is only one value of y. For example, the point $(-1, 3)$ on the graph shows that $f(-1) = 3$. Therefore, the graph represents a function. From the graph, we see that the values of x go from -1 to 4, so the domain is $[-1, 4]$. By looking at the values of y, we see that the range is $[0, 3]$.

The following agreement on domains is customary.

> **Agreement on Domains**
> Unless otherwise stated, assume that the domain of all functions defined by an equation is the largest set of real numbers that are meaningful replacements for the independent variable.

For example, suppose

$$y = \frac{-4x}{2x - 3}.$$

Any real number can be used for x except $x = 3/2$, which makes the denominator equal 0. By the agreement on domains, the domain of this function is the set of all real numbers except 3/2, which we denote $\{x \mid x \neq 3/2\}$, $\{x \neq 3/2\}$, or $(-\infty, 3/2) \cup (3/2, \infty)$.*

> **CAUTION** When finding the domain of a function, there are two operations to avoid: (1) dividing by zero; and (2) taking the square root (or any even root) of a negative number. Later sections will present other functions, such as logarithms, which require further restrictions on the domain. For now, just remember these two restrictions on the domain.

EXAMPLE 3 **Domain and Range**

Find the domain and range for each function defined as follows.

(a) $f(x) = x^2$

SOLUTION Any number may be squared, so the domain is the set of all real numbers, written $(-\infty, \infty)$. Since $x^2 \geq 0$ for every value of x, the range is $[0, \infty)$.

(b) $y = x^2$, with the domain specified as $\{-2, -1, 0, 1, 2\}$.

SOLUTION With the domain specified, the range is the set of values found by applying the function to the domain. Since $f(0) = 0$, $f(-1) = f(1) = 1$, and $f(-2) = f(2) = 4$, the range is $\{0, 1, 4\}$. The graph of the set of ordered pairs is shown in Figure 6.

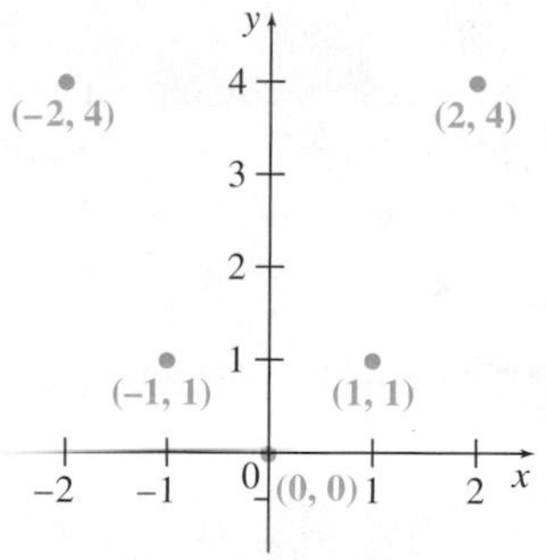

(c) $y = \sqrt{6 - x}$

SOLUTION For y to be a real number, $6 - x$ must be nonnegative. This happens only when $6 - x \geq 0$, or $6 \geq x$, making the domain $(-\infty, 6]$. The range is $[0, \infty)$ because $\sqrt{6 - x}$ is always nonnegative.

(d) $y = \sqrt{2x^2 + 5x - 12}$

SOLUTION The domain includes only those values of x satisfying $2x^2 + 5x - 12 \geq 0$. Using the methods for solving a quadratic inequality produces the domain

$$(-\infty, -4] \cup [3/2, \infty).$$

As in part (c), the range is $[0, \infty)$.

---

y 4 — (−2, 4) ... (2, 4)

3 —

2 —

(−1, 1) 1 — (1, 1)

−2 −1 0 (0, 0) 1 2 x

FIGURE 6

FOR REVIEW

Section R.5 demonstrates the method for solving a quadratic inequality. To solve $2x^2 + 5x - 12 \geq 0$, factor the quadratic to get $(2x - 3)(x + 4) \geq 0$. Setting each factor equal to 0 gives $x = 3/2$ or $x = -4$, leading to the intervals $(-\infty, -4]$, $[-4, 3/2]$, and $[3/2, \infty)$. Testing a number from each interval shows that the solution is $(-\infty, -4] \cup [3/2, \infty)$.

---

*The *union* of sets A and B, written $A \cup B$, is defined as the set of all elements in A or B or both.

(e) $y = \dfrac{2}{x^2 - 9}$

SOLUTION Since the denominator cannot be zero, $x \neq 3$ and $x \neq -3$. The domain is

$$(-\infty, -3) \cup (-3, 3) \cup (3, \infty).$$

Because the numerator can never be zero, $y \neq 0$. The denominator can take on any real number except for 0, allowing y to take on any value except for 0, so the range is $(-\infty, 0) \cup (0, \infty)$. **TRY YOUR TURN 1**

YOUR TURN 1 Find the domain and range for the function $y = \dfrac{1}{\sqrt{x^2 - 4}}$.

To understand how a function works, think of a function f as a machine—for example, a calculator or computer—that takes an input x from the domain and uses it to produce an output $f(x)$ (which represents the y-value), as shown in Figure 7. In the Dow Jones example, when we put 4 into the machine, we get out 9034.69, since $f(4) = 9034.69$.

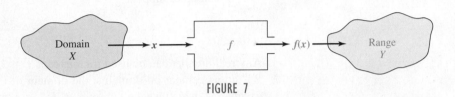

FIGURE 7

EXAMPLE 4 Evaluating Functions

Let $g(x) = -x^2 + 4x - 5$. Find the following.

(a) $g(3)$

SOLUTION Replace x with 3.

$$g(3) = -3^2 + 4 \cdot 3 - 5 = -9 + 12 - 5 = -2$$

(b) $g(a)$

SOLUTION Replace x with a to get $g(a) = -a^2 + 4a - 5$.

This replacement of one variable with another is important in later chapters.

(c) $g(x + h)$

SOLUTION Replace x with the expression $x + h$ and simplify.

$$\begin{aligned}
g(x + h) &= -(x + h)^2 + 4(x + h) - 5 \\
&= -(x^2 + 2xh + h^2) + 4(x + h) - 5 \\
&= -x^2 - 2xh - h^2 + 4x + 4h - 5
\end{aligned}$$

(d) $g\left(\dfrac{2}{r}\right)$

SOLUTION Replace x with $2/r$ and simplify.

$$g\left(\frac{2}{r}\right) = -\left(\frac{2}{r}\right)^2 + 4\left(\frac{2}{r}\right) - 5 = -\frac{4}{r^2} + \frac{8}{r} - 5$$

(e) Find all values of x such that $g(x) = -12$.

SOLUTION Set $g(x)$ equal to -12, and then add 12 to both sides to make one side equal to 0.

$$-x^2 + 4x - 5 = -12$$
$$-x^2 + 4x + 7 = 0$$

This equation does not factor, but can be solved with the quadratic formula, which says that if $ax^2 + bx + c = 0$, where $a \neq 0$, then

$$x = \frac{-b \pm \sqrt{b^2 - 4ac}}{2a}.$$

In this case, with $a = -1$, $b = 4$, and $c = 7$, we have

$$x = \frac{-4 \pm \sqrt{16 - 4(-1)7}}{2(-1)}$$

$$= \frac{-4 \pm \sqrt{44}}{-2}$$

$$= 2 \pm \sqrt{11}$$

$$\approx -1.317 \quad \text{or} \quad 5.317. \qquad \textbf{TRY YOUR TURN 2}$$

YOUR TURN 2 Given the function $f(x) = 2x^2 - 3x - 4$, find each of the following.
(a) $f(x + h)$ **(b)** All values of x such that $f(x) = -5$.

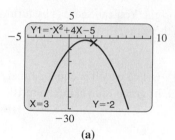

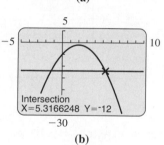

(a)

(b)

FIGURE 8

TECHNOLOGY NOTE

We can verify the results of parts (a) and (e) of the previous example using a graphing calculator. In Figure 8(a), after graphing $f(x) = -x^2 + 4x - 5$, we have used the "value" feature on the TI-84 Plus to support our answer from part (a). In Figure 8(b) we have used the "intersect" feature to find the intersection of $y = g(x)$ and $y = -12$. The result is $x = 5.3166248$, which is one of our two answers to part (e). The graph clearly shows that there is another answer on the opposite side of the y-axis.

CAUTION Notice from Example 4(c) that $g(x + h)$ is *not* the same as $g(x) + h$, which equals $-x^2 + 4x - 5 + h$. There is a significant difference between applying a function to the quantity $x + h$ and applying a function to x and adding h afterward.

If you tend to get confused when replacing x with $x + h$, as in Example 4(c), you might try replacing the x in the original function with a box, like this:

$$g\left(\boxed{}\right) = -\left(\boxed{}\right)^2 + 4\left(\boxed{}\right) - 5$$

Then, to compute $g(x + h)$, just enter $x + h$ into the box:

$$g\left(\boxed{x + h}\right) = -\left(\boxed{x + h}\right)^2 + 4\left(\boxed{x + h}\right) - 5$$

and proceed as in Example 4(c).

Notice in the Dow Jones example that to find the value of the function for a given value of x, we drew a vertical line from the value of x and found where it intersected the graph. If a graph is to represent a function, each value of x from the domain must lead to exactly one value of y. In the graph in Figure 9, the domain value x_1 leads to *two* y-values, y_1 and y_2. Since the given x-value corresponds to two different y-values, this is not the graph of a function. This example suggests the **vertical line test** for the graph of a function.

Vertical Line Test

If a vertical line intersects a graph in more than one point, the graph is not the graph of a function.

A graph represents a function if and only if every vertical line intersects the graph in no more than one point.

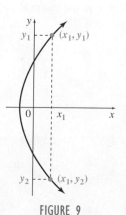

FIGURE 9

EXAMPLE 5 Vertical Line Test

Use the vertical line test to determine which of the graphs in Example 2 represent functions.

SOLUTION Every vertical line intersects the graphs in Figure 5(a), (c), and (d) in at most one point, so these are the graphs of functions. It is possible for a vertical line to intersect the graph in Figure 5(b) twice, so this is not a function.

A function f is called an **even function** if $f(-x) = f(x)$. This means that the graph is symmetric about the y-axis, so the left side is a mirror image of the right side. The function f is called an **odd function** if $f(-x) = -f(x)$. This means that the graph is symmetric about the origin, so the left side of the graph can be found by rotating the right side by $180°$ about the origin.

EXAMPLE 6 Even and Odd Functions

Determine whether each of the following functions is even, odd, or neither.

(a) $f(x) = x^4 - x^2$

─FOR REVIEW─
Recall from Sec. R.6 that
$(-a)^n = a^n$ if n is an even integer, and $(-a)^n = -a^n$ if n is an odd integer.

SOLUTION   Calculate $f(-x) = (-x)^4 - (-x)^2 = x^4 - x^2 = f(x)$, so the function is even. Its graph, shown in Figure 10(a), is symmetric about the y-axis.

(b) $f(x) = \dfrac{x}{x^2 + 1}$

SOLUTION   Calculate $f(-x) = \dfrac{(-x)}{(-x)^2 + 1} = \dfrac{-x}{x^2 + 1} = -f(x)$, so the function is odd.

Its graph, shown in Figure 10(b), is symmetric about the origin. You can see this by turning the book upside down and observing that the graph looks the same.

(c) $f(x) = x^4 - 4x^3$

SOLUTION   Calculate $f(-x) = (-x)^4 - 4(-x)^3 = x^4 + 4x^3$, which is equal neither to $f(x)$ nor to $-f(x)$. So the function is neither even nor odd. Its graph, shown in Figure 10(c), has no symmetry.

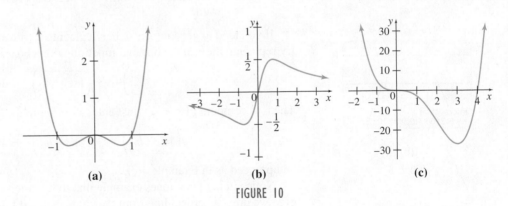

(a)   (b)   (c)

FIGURE 10

EXAMPLE 7 Delivery Charges

An overnight delivery service charges \$25 for a package weighing up to 2 lb. For each additional pound, or portion thereof, there is an additional charge of \$3. Let $D(w)$ represent the cost to send a package weighing w lb. Graph $D(w)$ for w in the interval $(0, 6]$.

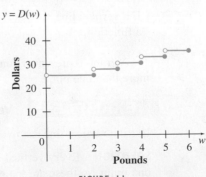

FIGURE 11

SOLUTION For w in the interval $(0, 2]$, the shipping cost is $y = 25$. For w in $(2, 3]$, the shipping cost is $y = 25 + 3 = 28$. For w in $(3, 4]$, the shipping cost is $y = 28 + 3 = 31$, and so on. The graph is shown in Figure 11.

The function discussed in Example 7 is called a **step function**. Many real-life situations are best modeled by step functions. Additional examples are given in the exercises.

In Chapter 1 you saw several examples of linear models. In Example 8, we use a quadratic equation to model the area of a lot.

EXAMPLE 8 Area

A fence is to be built against a brick wall to form a rectangular lot, as shown in Figure 12. Only three sides of the fence need to be built, because the wall forms the fourth side. The contractor will use 200 m of fencing. Let the length of the wall be l and the width w, as shown in Figure 12.

(a) Find the area of the lot as a function of the length l.

   SOLUTION The area formula for a rectangle is area = length × width, or

$$A = lw.$$

We want the area as a function of the length only, so we must eliminate the width. We use the fact that the total amount of fencing is the sum of the three sections, one length and two widths, so $200 = l + 2w$. Solve this for w:

$$200 = l + 2w$$
$$200 - l = 2w \qquad \text{Subtract } l \text{ from both sides.}$$
$$100 - l/2 = w. \qquad \text{Divide both sides by 2.}$$

Substituting this into the formula for area gives

$$A = l(100 - l/2).$$

(b) Find the domain of the function in part (a).

   SOLUTION The length cannot be negative, so $l \geq 0$. Similarly, the width cannot be negative, so $100 - l/2 \geq 0$, from which we find $l \leq 200$. Therefore, the domain is $[0, 200]$.

(c) Sketch a graph of the function in part (a).

   SOLUTION The result from a graphing calculator is shown in Figure 13. Notice that at the endpoints of the domain, when $l = 0$ and $l = 200$, the area is 0. This makes sense: If the length or width is 0, the area will be 0 as well. In between, as the length increases from 0 to 100 m, the area gets larger, and seems to reach a peak of 5000 m² when $l = 100$ m. After that, the area gets smaller as the length continues to increase because the width is becoming smaller.

In the next section, we will study this type of function in more detail and determine exactly where the maximum occurs.

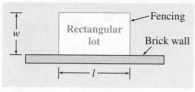

FIGURE 12

$A = l(100 - l/2)$

5000

0   200
0

FIGURE 13

2.1   EXERCISES

Which of the following rules define y as a function of x?

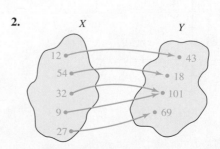

1.

2.

3.

x	y
3	9
2	4
1	1
0	0
−1	1
−2	4
−3	9

4.

x	y
9	3
4	2
1	1
0	0
1	−1
4	−2
9	−3

5. $y = x^3 + 2$

7. $x = |y|$

6. $y = \sqrt{x}$

8. $x = y^2 + 4$

List the ordered pairs obtained from each equation, given {−2, −1, 0, 1, 2, 3} as the domain. Graph each set of ordered pairs. Give the range.

9. $y = 2x + 3$

10. $y = -3x + 9$

11. $2y - x = 5$

12. $6x - y = -1$

13. $y = x(x + 2)$

14. $y = (x - 2)(x + 2)$

15. $y = x^2$

16. $y = -4x^2$

Give the domain of each function defined as follows.

17. $f(x) = 2x$

18. $f(x) = 2x + 3$

19. $f(x) = x^4$

20. $f(x) = (x + 3)^2$

21. $f(x) = \sqrt{4 - x^2}$

22. $f(x) = |3x - 6|$

23. $f(x) = (x - 3)^{1/2}$

24. $f(x) = (3x + 5)^{1/2}$

25. $f(x) = \dfrac{2}{1 - x^2}$

26. $f(x) = \dfrac{-8}{x^2 - 36}$

27. $f(x) = -\sqrt{\dfrac{2}{x^2 - 16}}$

28. $f(x) = -\sqrt{\dfrac{5}{x^2 + 36}}$

29. $f(x) = \sqrt{x^2 - 4x - 5}$

30. $f(x) = \sqrt{15x^2 + x - 2}$

31. $f(x) = \dfrac{1}{\sqrt{3x^2 + 2x - 1}}$

32. $f(x) = \sqrt{\dfrac{x^2}{3 - x}}$

Give the domain and the range of each function. Where arrows are drawn, assume the function continues in the indicated direction.

33.

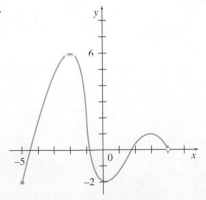

34.

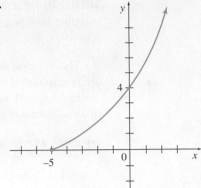

35.

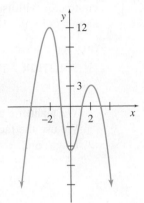

36.

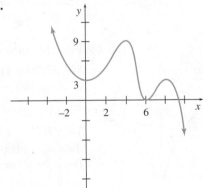

In Exercises 37–40, give the domain and range. Then, use each graph to find (a) $f(-2)$, (b) $f(0)$, (c) $f(1/2)$, and (d) any values of x such that $f(x) = 1$.

37.

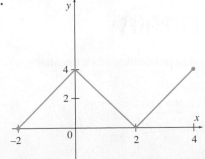

38.

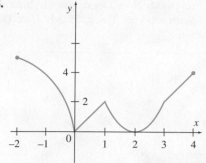

39.

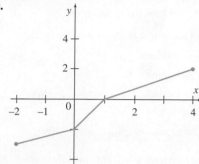

40.

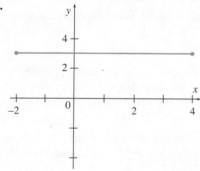

For each function, find (a) $f(4)$, (b) $f(-1/2)$, (c) $f(a)$, (d) $f(2/m)$, and (e) any values of x such that $f(x) = 1$.

41. $f(x) = 3x^2 - 4x + 1$ **42.** $f(x) = (x + 3)(x - 4)$

43. $f(x) = \begin{cases} \dfrac{2x + 1}{x - 4} & \text{if } x \neq 4 \\ \\ 7 & \text{if } x = 4 \end{cases}$

44. $f(x) = \begin{cases} \dfrac{x - 4}{2x + 1} & \text{if } x \neq -\dfrac{1}{2} \\ \\ 10 & \text{if } x = -\dfrac{1}{2} \end{cases}$

Let $f(x) = 6x^2 - 2$ and $g(x) = x^2 - 2x + 5$ to find the following values.

45. $f(t + 1)$ **46.** $f(2 - r)$

47. $g(r + h)$ **48.** $g(z - p)$

49. $g\left(\dfrac{3}{q}\right)$ **50.** $g\left(-\dfrac{5}{z}\right)$

For each function defined as follows, find (a) $f(x + h)$, (b) $f(x + h) - f(x)$, and (c) $[f(x + h) - f(x)]/h$.

51. $f(x) = 2x + 1$ **52.** $f(x) = x^2 - 3$

53. $f(x) = 2x^2 - 4x - 5$ **54.** $f(x) = -4x^2 + 3x + 2$

55. $f(x) = \dfrac{1}{x}$ **56.** $f(x) = -\dfrac{1}{x^2}$

Decide whether each graph represents a function.

57.

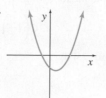

58.

59.

60.

61.

62.

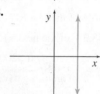

Classify each of the functions in Exercises 63–70 as even, odd, or neither.

63. $f(x) = 3x$ **64.** $f(x) = 5x$

65. $f(x) = 2x^2$ **66.** $f(x) = x^2 - 3$

67. $f(x) = \dfrac{1}{x^2 + 4}$ **68.** $f(x) = x^3 + x$

69. $f(x) = \dfrac{x}{x^2 - 9}$ **70.** $f(x) = |x - 2|$

APPLICATIONS

Business and Economics

71. Saw Rental A chain-saw rental firm charges $28 per day or fraction of a day to rent a saw, plus a fixed fee of $8 for re-sharpening the blade. Let $S(x)$ represent the cost of renting a saw for x days. Find the following.

a. $S\left(\dfrac{1}{2}\right)$ **b.** $S(1)$ **c.** $S\left(1\dfrac{1}{4}\right)$

d. $S\left(3\dfrac{1}{2}\right)$ **e.** $S(4)$ **f.** $S\left(4\dfrac{1}{10}\right)$

g. What does it cost to rent a saw for $4\frac{9}{10}$ days?

h. A portion of the graph of $y = S(x)$ is shown here. Explain how the graph could be continued.

i. What is the independent variable?

j. What is the dependent variable?

k. Is S a linear function? Explain.

l. Write a sentence or two explaining what part f and its answer represent.

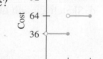

m. We have left $x = 0$ out of the graph. Discuss why it should or shouldn't be included. If it were included, how would you define $S(0)$?

72. Rental Car Cost The cost to rent a mid-size car is $54 per day or fraction of a day. If the car is picked up in Pittsburgh and dropped off in Cleveland, there is a fixed $44 drop-off charge. Let $C(x)$ represent the cost of renting the car for x days, taking it from Pittsburgh to Cleveland. Find the following.

a. $C(3/4)$ **b.** $C(9/10)$ **c.** $C(1)$ **d.** $C\left(1\dfrac{5}{8}\right)$

e. Find the cost of renting the car for 2.4 days.

f. Graph $y = C(x)$.

 g. Is C a function? Explain.

h. Is C a linear function? Explain.

i. What is the independent variable?

j. What is the dependent variable?

73. Attorney Fees According to Massachusetts state law, the maximum amount of a jury award that attorneys can receive is:

40% of the first $150,000,

33.3% of the next $150,000,

30% of the next $200,000, and

24% of anything over $500,000.

Let $f(x)$ represent the maximum amount of money that an attorney in Massachusetts can receive for a jury award of size x. Find each of the following, and describe in a sentence what the answer tells you. **Source: The New Yorker.**

a. $f(250,000)$ **b.** $f(350,000)$ **c.** $f(550,000)$

d. Sketch a graph of $f(x)$.

74. Tax Rates In New York state in 2010, the income tax rates for a single person were as follows:

4% of the first $8000 earned,

4.5% of the next $3000 earned,

5.25% of the next $2000 earned,

5.9% of the next $7000 earned,

6.85% of the next $180,000 earned,

7.85% of the next $300,000 earned, and

8.97% of any amount earned over $500,000.

Let $f(x)$ represent the amount of tax owed on an income of x dollars. Find each of the following, and explain in a sentence what the answer tells you. **Source: New York State.**

a. $f(10,000)$

b. $f(12,000)$

c. $f(18,000)$

d. Sketch a graph of $f(x)$.

Life Sciences

75. Whales Diving The figure in the next column shows the depth of a diving sperm whale as a function of time, as recorded by researchers at the Woods Hole Oceanographic Institution in Massachusetts. **Source: Peter Tyack, Woods Hole Oceanographic Institution.**

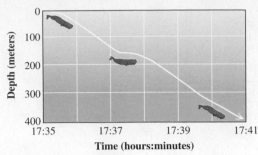

Find the depth of the whale at the following times.

a. 17 hours and 37 minutes

b. 17 hours and 39 minutes

76. Metabolic Rate The basal metabolic rate (in kcal/day) for large anteaters is given by

$$y = f(x) = 19.7x^{0.753},$$

where x is the anteater's weight in kilograms.* **Source: Wildlife Feeding and Nutrition.**

a. Find the basal metabolic rate for anteaters with the following weights.

   i. 5 kg   **ii.** 25 kg

b. Suppose the anteater's weight is given in pounds rather than kilograms. Given that 1 lb = 0.454 kg, find a function $x = g(z)$ giving the anteater's weight in kilograms if z is the animal's weight in pounds.

c. Write the basal metabolic rate as a function of the weight in pounds in the form $y = az^b$ by calculating $f(g(z))$.

77. Swimming Energy The energy expenditure (in kcal/km) for animals swimming at the surface of the water is given by

$$y = f(x) = 0.01x^{0.88},$$

where x is the animal's weight in grams. **Source: Wildlife Feeding and Nutrition.**

a. Find the energy for the following animals swimming at the surface of the water.

   i. A muskrat weighing 800 g

   ii. A sea otter weighing 20,000 g

b. Suppose the animal's weight is given in kilograms rather than grams. Given that 1 kg = 1000 g, find a function $x = g(z)$ giving the animal's weight in grams if z is the animal's weight in kilograms.

c. Write the energy expenditure as a function of the weight in kilograms in the form $y = az^b$ by calculating $f(g(z))$.

---

*Technically, kilograms are a measure of mass, not weight. Weight is a measure of the force of gravity, which varies with the distance from the center of Earth. For objects on the surface of Earth, weight and mass are often used interchangeably, and we will do so in this text.

GENERAL INTEREST

78. APPLY IT Energy Consumption Over the last century, the world has shifted from using high-carbon sources of energy such as wood to lower carbon fuels such as oil and natural gas, as shown in the figure. *Source: The New York Times.* The rise in carbon emissions during this time has caused concern because of its suspected contribution to global warming.

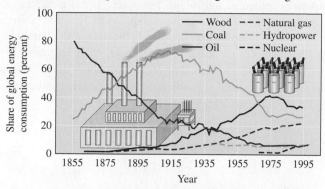

a. In what year were the percent of wood and coal use equal? What was the percent of each used in that year?

b. In what year were the percent of oil and coal use equal? What was the percent of each used in that year?

79. Perimeter A rectangular field is to have an area of 500 m².

    a. Write the perimeter, P, of the field as a function of the width, w.

    b. Find the domain of the function in part a.

    c. Use a graphing calculator to sketch the graph of the function in part a.

    d. Describe what the graph found in part c tells you about how the perimeter of the field varies with the width.

80. Area A rectangular field is to have a perimeter of 6000 ft.

    a. Write the area, A, of the field as a function of the width, w.

    b. Find the domain of the function in part a.

    c. Use a graphing calculator to sketch the graph of the function in part a.

    d. Describe what the graph found in part c tells you about how the area of the field varies with the width.

YOUR TURN ANSWERS

1. $(-\infty, -2) \cup (2, \infty), (0, \infty)$
2. (a) $2x^2 + 4xh + 2h^2 - 3x - 3h - 4$ (b) 1 and 1/2

2.2 Quadratic Functions; Translation and Reflection

APPLY IT How much should a company charge for its seminars? When Power and Money, Inc., charges $600 for a seminar on management techniques, it attracts 1000 people. For each $20 decrease in the fee, an additional 100 people will attend the seminar. The managers wonder how much to charge for the seminar to maximize their revenue.

In Example 6 in this section we will see how knowledge of quadratic functions *will help provide an answer to the question above.*

    A linear function is defined by

$$f(x) = ax + b,$$

for real numbers a and b. In a *quadratic function* the independent variable is squared. A quadratic function is an especially good model for many situations with a maximum or a minimum function value. Quadratic functions also may be used to describe supply and demand curves; cost, revenue, and profit; as well as other quantities. Next to linear functions, they are the simplest type of function, and well worth studying thoroughly.

FOR REVIEW

In this section you will need to know how to solve a quadratic equation by factoring and by the quadratic formula, which are covered in Sections R.2 and R.4. Factoring is usually easiest; when a polynomial is set equal to zero and factored, then a solution is found by setting any one factor equal to zero. But factoring is not always possible. The quadratic formula will provide the solution to *any* quadratic equation.

Quadratic Function

A **quadratic function** is defined by

$$f(x) = ax^2 + bx + c,$$

where a, b, and c are real numbers, with $a \neq 0$.

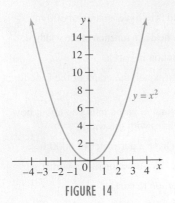

FIGURE 14

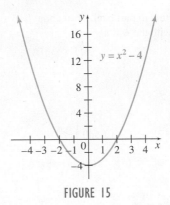

FIGURE 15

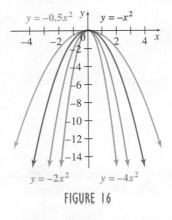

FIGURE 16

The simplest quadratic function has $f(x) = x^2$, with $a = 1$, $b = 0$, and $c = 0$. This function describes situations where the dependent variable y is proportional to the *square* of the independent variable x. The graph of this function is shown in Figure 14. This graph is called a **parabola**. Every quadratic function has a parabola as its graph. The lowest (or highest) point on a parabola is the **vertex** of the parabola. The vertex of the parabola in Figure 14 is $(0, 0)$.

If the graph in Figure 14 were folded in half along the y-axis, the two halves of the parabola would match exactly. This means that the graph of a quadratic function is *symmetric* with respect to a vertical line through the vertex; this line is the **axis** of the parabola.

There are many real-world instances of parabolas. For example, cross sections of spotlight reflectors or radar dishes form parabolas. Also, a projectile thrown in the air follows a parabolic path. For such applications, we need to study more complicated quadratic functions than $y = x^2$, as in the next several examples.

EXAMPLE 1   Graphing a Quadratic Function

Graph $y = x^2 - 4$.

SOLUTION   Each value of y will be 4 less than the corresponding value of y in $y = x^2$. The graph of $y = x^2 - 4$ has the same shape as that of $y = x^2$ but is 4 units lower. See Figure 15. The vertex of the parabola (on this parabola, the *lowest* point) is at $(0, -4)$. The x-intercepts can be found by letting $y = 0$ to get

$$0 = x^2 - 4,$$

from which $x = 2$ and $x = -2$ are the x-intercepts. The axis of the parabola is the vertical line $x = 0$.

Example 1 suggests that the effect of c in $ax^2 + bx + c$ is to lower the graph if c is negative and to raise the graph if c is positive. This is true for any function; the movement up or down is referred to as a **vertical translation** of the function.

EXAMPLE 2   Graphing Quadratic Functions

Graph $y = ax^2$ with $a = -0.5$, $a = -1$, $a = -2$, and $a = -4$.

SOLUTION   Figure 16 shows all four functions plotted on the same axes. We see that since a is negative, the graph opens downward. When a is between -1 and 1 (that is, when $a = -0.5$), the graph is wider than the original graph, because the values of y are smaller in magnitude. On the other hand, when a is greater than 1 or less than -1, the graph is steeper.

Example 2 shows that the sign of a in $ax^2 + bx + c$ determines whether the parabola opens upward or downward. Multiplying $f(x)$ by a negative number flips the graph of f upside down. This is called a **vertical reflection** of the graph. The magnitude of a determines how steeply the graph increases or decreases.

EXAMPLE 3   Graphing Quadratic Functions

Graph $y = (x - h)^2$ for $h = 3, 0$, and -4.

SOLUTION   Figure 17 shows all three functions on the same axes. Notice that since the number is subtracted *before* the squaring occurs, the graph does not move up or down but instead moves left or right. Evaluating $f(x) = (x - 3)^2$ at $x = 3$ gives the same result as evaluating $f(x) = x^2$ at $x = 0$. Therefore, when we subtract the positive number 3 from x, the graph shifts 3 units to the right, so the vertex is at $(3, 0)$. Similarly, when we subtract

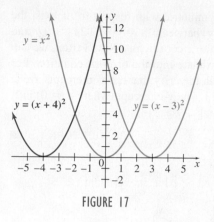

FIGURE 17

the negative number -4 from x—in other words, when the function becomes $f(x) = (x + 4)^2$—the graph shifts to the left 4 units.

The left or right shift of the graph illustrated in Figure 17 is called a **horizontal translation** of the function.

If a quadratic equation is given in the form $ax^2 + bx + c$, we can identify the translations and any vertical reflection by rewriting it in the form

$$y = a(x - h)^2 + k.$$

In this form, we can identify the vertex as (h, k). A quadratic equation not given in this form can be converted by a process called **completing the square**. The next example illustrates the process.

EXAMPLE 4   Graphing a Quadratic Function

Graph $y = -3x^2 - 2x + 1$.

Method 1
Completing the Square

SOLUTION   To begin, factor -3 from the x-terms so the coefficient of x^2 is 1:

$$y = -3\left(x^2 + \frac{2}{3}x\right) + 1.$$

Next, we make the expression inside the parentheses a perfect square by adding the square of one-half of the coefficient of x, which is $\left(\frac{1}{2} \cdot \frac{2}{3}\right)^2 = \frac{1}{9}$. Since there is a factor of -3 outside the parentheses, we are actually adding $-3 \cdot \left(\frac{1}{9}\right)$. To make sure that the value of the function is not changed, we must also add $3 \cdot \left(\frac{1}{9}\right)$ to the function. Actually, we are simply adding $-3 \cdot \left(\frac{1}{9}\right) + 3 \cdot \left(\frac{1}{9}\right) = 0$, and not changing the function. To summarize our steps,

$$y = -3\left(x^2 + \frac{2}{3}x\right) + 1 \qquad \text{Factor out } -3.$$

$$= -3\left(x^2 + \frac{2}{3}x + \frac{1}{9}\right) + 1 + 3\left(\frac{1}{9}\right) \qquad \begin{array}{l}\text{Add and subtract } -3 \text{ times} \\ \left(\frac{1}{2} \text{ the coefficient of } x\right)^2.\end{array}$$

$$= -3\left(x + \frac{1}{3}\right)^2 + \frac{4}{3}. \qquad \text{Factor and combine terms.}$$

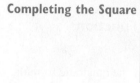

$$y = -3x^2 - 2x + 1$$

FIGURE 18

The function is now in the form $y = a(x - h)^2 + k$. Since $h = -1/3$ and $k = 4/3$, the graph is the graph of the parabola $y = x^2$ translated 1/3 unit to the left and 4/3 units upward. This puts the vertex at $(-1/3, 4/3)$. Since a $= -3$ is negative, the graph will be flipped upside down. The 3 will cause the parabola to be stretched vertically by a factor of 3. These results are shown in Figure 18.

Method 2
The Quadratic Formula

Instead of completing the square to find the vertex of the graph of a quadratic function given in the form $y = ax^2 + bx + c$, we can develop a formula for the vertex. By the quadratic formula, if $ax^2 + bx + c = 0$, where $a \neq 0$, then

$$x = \frac{-b \pm \sqrt{b^2 - 4ac}}{2a}.$$

Notice that this is the same as

$$x = \frac{-b}{2a} \pm \frac{\sqrt{b^2 - 4ac}}{2a} = \frac{-b}{2a} \pm Q,$$

YOUR TURN 1 For the function $y = 2x^2 - 6x - 1$,
(a) complete the square, (b) find the y-intercept, (c) find the x-intercepts, (d) find the vertex, and (e) sketch the graph.

where $Q = \sqrt{b^2 - 4ac}/(2a)$. Since a parabola is symmetric with respect to its axis, the vertex is halfway between its two roots. Halfway between $x = -b/(2a) + Q$ and $x = -b/(2a) - Q$ is $x = -b/(2a)$. Once we have the x-coordinate of the vertex, we can easily find the y-coordinate by substituting the x-coordinate into the original equation. For the function in this example, use the quadratic formula to verify that the x-intercepts are at $x = -1$ and $x = 1/3$, and the vertex is at $(-1/3, 4/3)$. The y-intercept (where $x = 0$) is 1. The graph is in Figure 18.

TRY YOUR TURN 1

> ### Graph of the Quadratic Function
> The graph of the quadratic function $f(x) = ax^2 + bx + c$ has its vertex at
> $$\left(\frac{-b}{2a}, f\!\left(\frac{-b}{2a} \right) \right).$$
> The graph opens upward if $a > 0$ and downward if $a < 0$.

Another situation that may arise is the absence of any x-intercepts, as in the next example.

EXAMPLE 5   Graphing a Quadratic Function

Graph $y = x^2 + 4x + 6$.

SOLUTION   This does not appear to factor, so we'll try the quadratic formula.

$$x = \frac{-b \pm \sqrt{b^2 - 4ac}}{2a} \qquad a = 1, b = 4, c = 6$$

$$= \frac{-4 \pm \sqrt{4^2 - 4(1)(6)}}{2(1)} = \frac{-4 \pm \sqrt{-8}}{2}$$

As soon as we see the negative value under the square root sign, we know the solutions are complex numbers. Therefore, there are no x-intercepts. Nevertheless, the vertex is still at

$$x = \frac{-b}{2a} = \frac{-4}{2} = -2.$$

Substituting this into the equation gives

$$y = (-2)^2 + 4(-2) + 6 = 2.$$

The y-intercept is at $(0, 6)$, which is 2 units to the right of the parabola's axis $x = -2$. Using the symmetry of the figure, we can also plot the mirror image of this point on the opposite side of the parabola's axis: at $x = -4$ (2 units to the left of the axis), y is also equal to 6. Plotting the vertex, the y-intercept, and the point $(-4, 6)$ gives the graph in Figure 19.

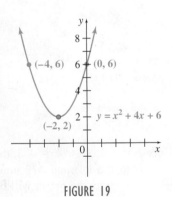

FIGURE 19

We now return to the question with which we started this section.

EXAMPLE 6   Management Science

APPLY IT

When Power and Money, Inc., charges \$600 for a seminar on management techniques, it attracts 1000 people. For each \$20 decrease in the fee, an additional 100 people will attend the seminar. The managers are wondering how much to charge for the seminar to maximize their revenue.

SOLUTION   Let x be the number of \$20 decreases in the price. Then the price charged per person will be

$$\text{Price per person} = 600 - 20x,$$

and the number of people in the seminar will be

$$\text{Number of people} = 1000 + 100x.$$

The total revenue, $R(x)$, is given by the product of the price and the number of people attending, or

$$R(x) = (600 - 20x)(1000 + 100x)$$
$$= 600{,}000 + 40{,}000x - 2000x^2.$$

We see by the negative in the x^2-term that this defines a parabola opening downward, so the maximum revenue is at the vertex. The x-coordinate of the vertex is

$$x = \frac{-b}{2a} = \frac{-40{,}000}{2(-2000)} = 10.$$

The y-coordinate is then

$$y = 600{,}000 + 40{,}000(10) - 2000(10^2)$$
$$= 800{,}000.$$

YOUR TURN 2 Solve Example 6 with the following changes: a $1650 price attracts 900 people, and each $40 decrease in the price attracts an additional 80 people.

Therefore, the maximum revenue is $800,000, which is achieved by charging $600 - 20x = 600 - 20(10) = \$400$ per person.　　　**TRY YOUR TURN 2**

　　　Notice in this last example that the maximum revenue was achieved by charging less than the current price of $600, which was more than made up for by the increase in sales. This is typical of many applications. Mathematics is a powerful tool for solving such problems, since the answer is not always what one might have guessed intuitively.

　　　To solve problems such as Example 6, notice the following:

1. The key step after reading and understanding the problem is identifying a useful variable.
2. Revenue is always price times the number sold.
3. The expressions for the price and for the number of people are both linear functions of x.
4. We know the constant term in each linear function because we know what happens when $x = 0$.
5. We know how much both the price and the number of people change each time x increases by 1, which gives us the slope of each linear function.
6. The maximum or minimum of a quadratic function occurs at its vertex.

The concept of maximizing or minimizing a function is important in calculus, as we shall see in future chapters.

　　　In the next example, we show how the calculation of profit can involve a quadratic function.

EXAMPLE 7　Profit

A deli owner has found that his revenue from producing x pounds of vegetable cream cheese is given by $R(x) = -x^2 + 30x$, while the cost in dollars is given by $C(x) = 5x + 100$.

(a) Find the minimum break-even quantity.

SOLUTION　Notice from the graph in Figure 20 that the revenue function is a parabola opening downward and the cost function is a linear function that crosses the revenue function at two points. To find the minimum break-even quantity, we find where the two functions are equal.

$$
\begin{aligned}
R(x) &= C(x) \\
-x^2 + 30x &= 5x + 100 \\
0 &= x^2 - 25x + 100 \quad \text{Subtract } -x^2 + 30x \text{ from both sides.} \\
&= (x - 5)(x - 20) \quad \text{Factor.}
\end{aligned}
$$

The two graphs cross when $x = 5$ and $x = 20$. The minimum break-even point is at $x = 5$. The deli owner must sell at least 5 lb of cream cheese to break even.

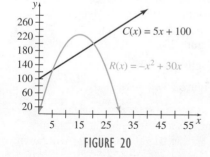

FIGURE 20

(b) Find the maximum revenue.

SOLUTION By factoring the revenue function, $R(x) = -x^2 + 30x = x(-x + 30)$, we can see that it has two roots, $x = 0$ and $x = 30$. The maximum is at the vertex, which has a value of x halfway between the two roots, or $x = 15$. (Alternatively, we could use the formula $x = -b/(2a) = -30/(-2) = 15$.) The maximum revenue is $R(15) = -15^2 + 30(15) = 225$, or $225.

(c) Find the maximum profit.

SOLUTION The profit is the difference between the revenue and the cost, or

$$\begin{aligned} P(x) &= R(x) - C(x) \\ &= (-x^2 + 30x) - (5x + 100) \\ &= -x^2 + 25x - 100. \end{aligned}$$

YOUR TURN 3 Suppose the revenue in dollars is given by $R(x) = -x^2 + 40x$ and the cost is given by $C(x) = 8x + 192$. Find **(a)** the minimum break-even quantity, **(b)** the maximum revenue, and **(c)** the maximum profit.

This is just the negative of the expression factored in part (a), where we found the roots to be $x = 5$ and $x = 20$. The value of x at the vertex is halfway between these two roots, or $x = (5 + 20)/2 = 12.5$. (Alternatively, we could use the formula $x = -b/(2a) = -25/(-2) = 12.5$.) The value of the function here is $P(12.5) = -12.5^2 + 25(12.5) - 100 = 56.25$. It is clear that this is a maximum, not only from Figure 20, but also because the profit function is a quadratic with a negative x^2-term. A maximum profit of $56.25 is achieved by selling 12.5 lb of cream cheese. **TRY YOUR TURN 3**

Below and on the next page, we provide guidelines for sketching graphs that involve translations and reflections.

Translations and Reflections of Functions

Let f be any function, and let h and k be positive constants (Figure 21).

The graph of $y = f(x) + k$ is the graph of $y = f(x)$ translated upward by an amount k (Figure 22).

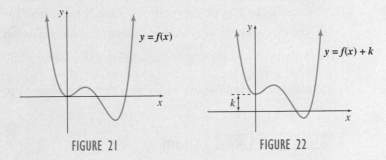

FIGURE 21   FIGURE 22

The graph of $y = f(x) - k$ is the graph of $y = f(x)$ translated downward by an amount k (Figure 23).

The graph of $y = f(x - h)$ is the graph of $y = f(x)$ translated to the right by an amount h (Figure 24).

The graph of $y = f(x + h)$ is the graph of $y = f(x)$ translated to the left by an amount h (Figure 25).

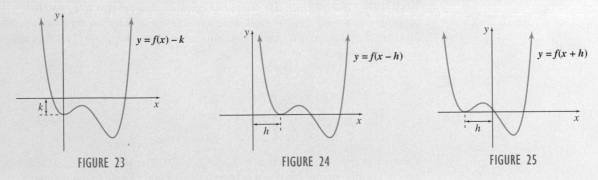

FIGURE 23   FIGURE 24   FIGURE 25

Translations and Reflections of Functions

The graph of $y = -f(x)$ is the graph of $y = f(x)$ reflected vertically across the x-axis, that is, turned upside down (Figure 26). The graph of $y = f(-x)$ is the graph of $y = f(x)$ reflected horizontally across the y-axis, that is, its mirror image (Figure 27).

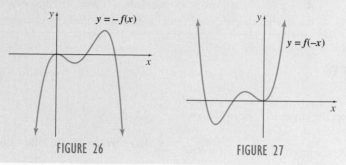

FIGURE 26   FIGURE 27

Notice in Figure 27 another type of reflection, known as a **horizontal reflection**. Multiplying x or $f(x)$ by a constant a, to get $y = f(ax)$ or $y = a \cdot f(x)$ does not change the general appearance of the graph, except to compress or stretch it. When a is negative, it also causes a reflection, as shown in the last two figures in the summary for $a = -1$. Also see Exercises 39–42 in this section.

EXAMPLE 8   Translations and Reflections of a Graph

Graph $f(x) = -\sqrt{4 - x} + 3$.

SOLUTION   Begin with the simplest possible function, then add each variation in turn. Start with the graph of $f(x) = \sqrt{x}$. As Figure 28 reveals, this is just one-half of the graph of $f(x) = x^2$ lying on its side.

Now add another component of the original function, the negative in front of the x, giving $f(x) = \sqrt{-x}$. This is a horizontal reflection of the $f(x) = \sqrt{x}$ graph, as shown in Figure 29. Next, include the 4 under the square root sign. To get $4 - x$ into the form $f(x - h)$ or $f(x + h)$, we need to factor out the negative: $\sqrt{4 - x} = \sqrt{-(x - 4)}$. Now the 4 is subtracted, so this function is a translation to the right of the function $f(x) = \sqrt{-x}$ by 4 units, as Figure 30 indicates.

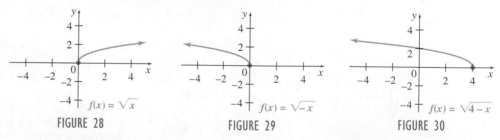

FIGURE 28   FIGURE 29   FIGURE 30

The effect of the negative in front of the radical is a vertical reflection, as in Figure 31, which shows the graph of $f(x) = -\sqrt{4 - x}$. Finally, adding the constant 3 raises the entire graph by 3 units, giving the graph of $f(x) = -\sqrt{4 - x} + 3$ in Figure 32(a).

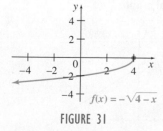

FIGURE 31

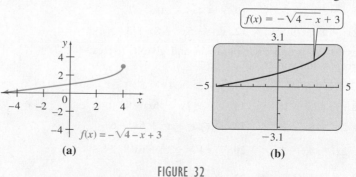

(a)   (b)

FIGURE 32

⊞ **TECHNOLOGY NOTE** │ If you viewed a graphing calculator image such as Figure 32(b), you might think the function continues to go up and to the right. By realizing that $(4, 3)$ is the vertex of the sideways parabola, we see
that this is the rightmost point on the graph. Another approach is to find the domain of f by setting
$4 - x \geq 0$, from which we conclude that $x \leq 4$. This demonstrates the importance of knowing
the algebraic techniques in order to interpret a graphing calculator image correctly.

2.2 EXERCISES

1. How does the value of a affect the graph of $y = ax^2$? Discuss the case for $a \geq 1$ and for $0 \leq a \leq 1$.

2. How does the value of a affect the graph of $y = ax^2$ if $a \leq 0$?

In Exercises 3–8, match the correct graph A–F to the function without using your calculator. Then, if you have a graphing calculator, use it to check your answers. Each graph in this group shows x and y in $[-10, 10]$.

3. $y = x^2 - 3$

4. $y = (x - 3)^2$

5. $y = (x - 3)^2 + 2$

6. $y = (x + 3)^2 + 2$

7. $y = -(3 - x)^2 + 2$

8. $y = -(x + 3)^2 + 2$

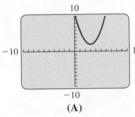
(A)

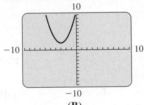
(B)

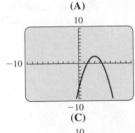
(C)

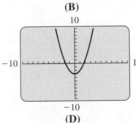
(D)

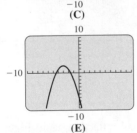
(E)

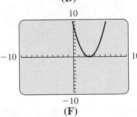
(F)

Complete the square and determine the vertex for each of the following.

9. $y = 3x^2 + 9x + 5$

10. $y = 4x^2 - 20x - 7$

11. $y = -2x^2 + 8x - 9$

12. $y = -5x^2 - 8x + 3$

In Exercises 13–24, graph each parabola and give its vertex, axis, x-intercepts, and y-intercept.

13. $y = x^2 + 5x + 6$

14. $y = x^2 + 4x - 5$

15. $y = -2x^2 - 12x - 16$

16. $y = -3x^2 - 6x + 4$

17. $f(x) = 2x^2 + 8x - 8$

18. $f(x) = -x^2 + 6x - 6$

19. $f(x) = 2x^2 - 4x + 5$

20. $f(x) = \frac{1}{2}x^2 + 6x + 24$

21. $f(x) = -2x^2 + 16x - 21$

22. $f(x) = \frac{3}{2}x^2 - x - 4$

23. $f(x) = \frac{1}{3}x^2 - \frac{8}{3}x + \frac{1}{3}$

24. $f(x) = -\frac{1}{2}x^2 - x - \frac{7}{2}$

In Exercises 25–30, follow the directions for Exercises 3–8.

25. $y = \sqrt{x + 2} - 4$

26. $y = \sqrt{x - 2} - 4$

27. $y = \sqrt{-x + 2} - 4$

28. $y = \sqrt{-x - 2} - 4$

29. $y = -\sqrt{x + 2} - 4$

30. $y = -\sqrt{x - 2} - 4$

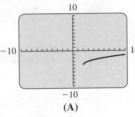
(A)

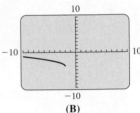
(B)

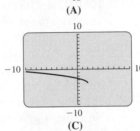
(C)

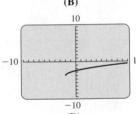
(D)

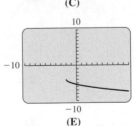
(E)

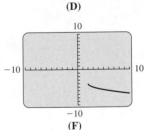
(F)

Given the following graph, sketch by hand the graph of the function described, giving the new coordinates for the three points labeled on the original graph.

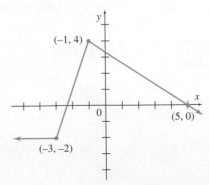

31. $y = -f(x)$

32. $y = f(x - 2) + 2$

33. $y = f(-x)$

34. $y = f(2 - x) + 2$

Use the ideas in this section to graph each function without a calculator.

35. $f(x) = \sqrt{x - 2} + 2$ **36.** $f(x) = \sqrt{x + 2} - 3$

37. $f(x) = -\sqrt{2 - x} - 2$ **38.** $f(x) = -\sqrt{2 - x} + 2$

Using the graph of $f(x)$ in Figure 21, show the graph of $f(ax)$ where a satisfies the given condition.

39. $0 < a < 1$ **40.** $1 < a$

41. $-1 < a < 0$ **42.** $a < -1$

Using the graph of $f(x)$ in Figure 21, show the graph of $af(x)$ where a satisfies the given condition.

43. $0 < a < 1$ **44.** $1 < a$

45. $-1 < a < 0$ **46.** $a < -1$

47. If r is an x-intercept of the graph of $y = f(x)$, what is an x-intercept of the graph of each of the following?

    a. $y = -f(x)$ **b.** $y = f(-x)$
    c. $y = -f(-x)$

48. If b is the y-intercept of the graph of $y = f(x)$, what is the y-intercept of the graph of each of the following?

    a. $y = -f(x)$ **b.** $y = f(-x)$
    c. $y = -f(-x)$

APPLICATIONS

Business and Economics

Profit In Exercises 49–52, let $C(x)$ be the cost to produce x batches of widgets, and let $R(x)$ be the revenue in thousands of dollars. For each exercise, (a) graph both functions, (b) find the minimum break-even quantity, (c) find the maximum revenue, and (d) find the maximum profit.

49. $R(x) = -x^2 + 8x$, $C(x) = 2x + 5$

50. $R(x) = -\dfrac{x^2}{2} + 5x$, $C(x) = \dfrac{3}{2}x + 3$

51. $R(x) = -\dfrac{4}{5}x^2 + 10x$, $C(x) = 2x + 15$

52. $R(x) = -4x^2 + 36x$, $C(x) = 16x + 24$

53. Maximizing Revenue The revenue of a charter bus company depends on the number of unsold seats. If the revenue $R(x)$ is given by

$$R(x) = 8000 + 70x - x^2,$$

where x is the number of unsold seats, find the maximum revenue and the number of unsold seats that corresponds to maximum revenue.

54. Maximizing Revenue A charter flight charges a fare of $200 per person plus $4 per person for each unsold seat on the plane. The plane holds 100 passengers. Let x represent the number of unsold seats.

    a. Find an expression for the total revenue received for the flight $R(x)$. (*Hint:* Multiply the number of people flying, $100 - x$, by the price per ticket.)

    b. Graph the expression from part a.

    c. Find the number of unsold seats that will produce the maximum revenue.

    d. What is the maximum revenue?

    e. Some managers might be concerned about the empty seats, arguing that it doesn't make economic sense to leave any seats empty. Write a few sentences explaining why this is not necessarily so.

55. Maximizing Revenue The demand for a certain type of cosmetic is given by

$$p = 500 - x,$$

where p is the price in dollars when x units are demanded.

    a. Find the revenue $R(x)$ that would be obtained at a price p. (*Hint:* Revenue = Demand × Price)

    b. Graph the revenue function $R(x)$.

    c. Find the price that will produce maximum revenue.

    d. What is the maximum revenue?

56. Revenue The manager of a peach orchard is trying to decide when to arrange for picking the peaches. If they are picked now, the average yield per tree will be 100 lb, which can be sold for 80¢ per pound. Past experience shows that the yield per tree will increase about 5 lb per week, while the price will decrease about 4¢ per pound per week.

    a. Let x represent the number of weeks that the manager should wait. Find the income per pound.

    b. Find the number of pounds per tree.

    c. Find the total revenue from a tree.

    d. When should the peaches be picked in order to produce maximum revenue?

    e. What is the maximum revenue?

57. Income The manager of an 80-unit apartment complex is trying to decide what rent to charge. Experience has shown that at a rent of $800, all the units will be full. On the average, one additional unit will remain vacant for each $25 increase in rent.

    a. Let x represent the number of $25 increases. Find an expression for the rent for each apartment.

    b. Find an expression for the number of apartments rented.

    c. Find an expression for the total revenue from all rented apartments.

    d. What value of x leads to maximum revenue?

    e. What is the maximum revenue?

58. Advertising A study done by an advertising agency reveals that when x thousands of dollars are spent on advertising, it results in a sales increase in thousands of dollars given by the function

$$S(x) = -\frac{1}{4}(x - 10)^2 + 40, \quad \text{for } 0 \le x \le 10.$$

    a. Find the increase in sales when no money is spent on advertising.

    b. Find the increase in sales when $10,000 is spent on advertising.

    c. Sketch the graph of $S(x)$ without a calculator.

Life Sciences

59. Length of Life   According to recent data from the Teachers Insurance and Annuity Association (TIAA), the survival function for life after 65 is approximately given by

$$S(x) = 1 - 0.058x - 0.076x^2,$$

where x is measured in decades. This function gives the probability that an individual who reaches the age of 65 will live at least x decades ($10x$ years) longer. *Source: Ralph DeMarr.*

a. Find the median length of life for people who reach 65, that is, the age for which the survival rate is 0.50.

b. Find the age beyond which virtually nobody lives. (There are, of course, exceptions.)

60. Tooth Length   The length (in mm) of the mesiodistal crown of the first molar for human fetuses can be approximated by

$$L(t) = -0.01t^2 + 0.788t - 7.048,$$

where t is the number of weeks since conception. *Source: American Journal of Physical Anthropology.*

a. What does this formula predict for the length at 14 weeks? 24 weeks?

b. What does this formula predict for the maximum length, and when does that occur? Explain why the formula does not make sense past that time.

61. Splenic Artery Resistance   Blood flow to the fetal spleen is of research interest because several diseases are associated with increased resistance in the splenic artery (the artery that goes to the spleen). Researchers have found that the index of splenic artery resistance in the fetus can be described by the function

$$y = 0.057x - 0.001x^2,$$

where x is the number of weeks of gestation. *Source: American Journal of Obstetrics and Gynecology.*

a. At how many weeks is the splenic artery resistance a maximum?

b. What is the maximum splenic artery resistance?

c. At how many weeks is the splenic artery resistance equal to 0, according to this formula? Is your answer reasonable for this function? Explain.

62. Cancer   From 1975 to 2007, the age-adjusted incidence rate of invasive lung and bronchial cancer among women can be closely approximated by

$$f(t) = -0.040194t^2 + 2.1493t + 23.921,$$

where t is the number of years since 1975. *Source: National Cancer Institute.* Based on this model, in what year did the incidence rate reach a maximum? On what years was the rate increasing? Decreasing?

Social Sciences

63. Age of Marriage   The following table gives the median age at their first marriage of women in the United States for some selected years. *Source: U.S. Census Bureau.*

Year	Age
1940	21.5
1950	20.3
1960	20.3
1970	20.8
1980	22.0
1990	23.9
2000	25.1

a. Plot the data using $x = 40$ for 1940, and so on.

b. Would a linear or quadratic function best model this data? Explain.

c. If your graphing calculator has a regression feature, find the quadratic function that best fits the data. Graph this function on the same calculator window as the data. (On a TI-84 Plus calculator, press the STAT key, and then select the CALC menu. QuadReg is item 5. The command QuadReg L_1,L_2,Y_1 finds the quadratic regression equation for the data in L_1 and L_2 and stores the function in Y_1.)

d. Find a quadratic function defined by $f(x) = a(x - h)^2 + k$ that models the data using $(60, 20.3)$ as the vertex and then choosing $(100, 25.1)$ as a second point to determine the value of a.

e. Graph the function from part d on the same calculator window as the data and function from part c. Do the graphs of the two functions differ by much?

64. Gender Ratio   The number of males per 100 females, age 65 or over, in the United States for some recent years is shown in the following table. *Source: The New York Times 2010 Almanac.*

Year	Males per 100 Females
1960	82.8
1970	72.1
1980	67.6
1990	67.2
2000	70.0
2007	72.9

a. Plot the data, letting x be the years since 1900.

b. Would a linear or quadratic function best model this data? Explain.

c. If your graphing calculator has a quadratic regression feature, find the quadratic function that best fits the data. Graph this function on the same calculator window as the data. (See Exercise 63(c).)

d. Choose the lowest point in the table above as the vertex and $(60, 82.8)$ as a second point to find a quadratic function defined by $f(x) = a(x - h)^2 + k$ that models the data.

e. Graph the function from part d on the same calculator window as the data and function from part c. Do the graphs of the two functions differ by much?

f. Predict the number of males per 100 females in 2004 using the two functions from parts c and d, and compare with the actual figure of 71.7.

65. Accident Rate According to data from the National Highway Traffic Safety Administration, the accident rate as a function of the age of the driver in years x can be approximated by the function

$$f(x) = 60.0 - 2.28x + 0.0232x^2$$

for $16 \leq x \leq 85$. Find the age at which the accident rate is a minimum and the minimum rate. *Source: Ralph DeMarr.*

Physical Sciences

66. Maximizing the Height of an Object If an object is thrown upward with an initial velocity of 32 ft/second, then its height after t seconds is given by

$$h = 32t - 16t^2.$$

a. Find the maximum height attained by the object.

b. Find the number of seconds it takes the object to hit the ground.

67. Stopping Distance According to data from the National Traffic Safety Institute, the stopping distance y in feet of a car traveling x mph can be described by the equation $y = 0.056057x^2 + 1.06657x$. *Source: National Traffic Safety Institute.*

a. Find the stopping distance for a car traveling 25 mph.

b. How fast can you drive if you need to be certain of stopping within 150 ft?

General Interest

68. Maximizing Area Glenview Community College wants to construct a rectangular parking lot on land bordered on one side by a highway. It has 380 ft of fencing to use along the other three sides. What should be the dimensions of the lot if the enclosed

area is to be a maximum? (*Hint:* Let x represent the width of the lot, and let $380 - 2x$ represent the length.)

69. Maximizing Area What would be the maximum area that could be enclosed by the college's 380 ft of fencing if it decided to close the entrance by enclosing all four sides of the lot? (See Exercise 68.)

In Exercises 70 and 71, draw a sketch of the arch or culvert on coordinate axes, with the horizontal and vertical axes through the vertex of the parabola. Use the given information to label points on the parabola. Then give the equation of the parabola and answer the question.

70. Parabolic Arch An arch is shaped like a parabola. It is 30 m wide at the base and 15 m high. How wide is the arch 10 m from the ground?

71. Parabolic Culvert A culvert is shaped like a parabola, 18 ft across the top and 12 ft deep. How wide is the culvert 8 ft from the top?

YOUR TURN ANSWERS

1. (a) $y = 2(x - 3/2)^2 - 11/2$ **(b)** -1
(c) $(3 \pm \sqrt{11})/2$ **(d)** $(3/2, -11/2)$
(e)

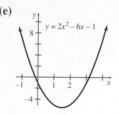

2. Charge \$1050 for a maximum revenue of \$2,205,000.
3. (a) 8   **(b)** \$400   **(c)** \$64

2.3 Polynomial and Rational Functions

APPLY IT How does the revenue collected by the government vary with the tax rate?

In Exercises 48–50 in this section, we will explore this question using polynomial *and* rational functions.

Polynomial Functions
Earlier, we discussed linear and quadratic functions and their graphs. Both of these functions are special types of *polynomial functions.*

> **Polynomial Function**
>
> A **polynomial function** of **degree** n, where n is a nonnegative integer, is defined by
>
> $$f(x) = a_n x^n + a_{n-1}x^{n-1} + \cdots + a_1 x + a_0,$$
>
> where $a_n, a_{n-1}, \ldots, a_1$, and a_0 are real numbers, called **coefficients**, with $a_n \neq 0$. The number a_n is called the **leading coefficient**.

For $n = 1$, a polynomial function takes the form

$$f(x) = a_1 x + a_0,$$

a linear function. A linear function, therefore, is a polynomial function of degree 1. (Note, however, that a linear function of the form $f(x) = a_0$ for a real number a_0 is a polynomial function of degree 0, the constant function.) A polynomial function of degree 2 is a quadratic function.

Accurate graphs of polynomial functions of degree 3 or higher require methods of calculus to be discussed later. Meanwhile, a graphing calculator is useful for obtaining such graphs, but care must be taken in choosing a viewing window that captures the significant behavior of the function.

The simplest polynomial functions of higher degree are those of the form $f(x) = x^n$. Such a function is known as a **power function**. Figure 33 below shows the graphs of $f(x) = x^3$ and $f(x) = x^5$, as well as tables of their values. These functions are simple enough that they can be drawn by hand by plotting a few points and connecting them with a smooth curve. An important property of all polynomials is that their graphs are smooth curves.

The graphs of $f(x) = x^4$ and $f(x) = x^6$, shown in Figure 34 along with tables of their values, can be sketched in a similar manner. These graphs have symmetry about the y-axis, as does the graph of $f(x) = ax^2$ for a nonzero real number a. As with the graph of $f(x) = ax^2$, the value of a in $f(x) = ax^n$ affects the direction of the graph. When $a > 0$, the graph has the same general appearance as the graph of $f(x) = x^n$. However, if $a < 0$, the graph is reflected vertically. Notice that $f(x) = x^3$ and $f(x) = x^5$ are odd functions, while $f(x) = x^4$ and $f(x) = x^6$ are even functions.

$f(x) = x^3$		$f(x) = x^5$	
x	$f(x)$	x	$f(x)$
-2	-8	-1.5	-7.6
-1	-1	-1	-1
0	0	0	0
1	1	1	1
2	8	1.5	7.6

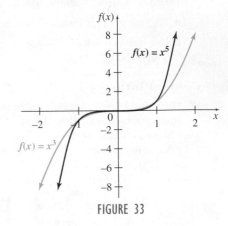

FIGURE 33

$f(x) = x^4$		$f(x) = x^6$	
x	$f(x)$	x	$f(x)$
-2	16	-1.5	11.4
-1	1	-1	1
0	0	0	0
1	1	1	1
2	16	1.5	11.4

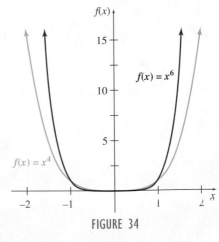

FIGURE 34

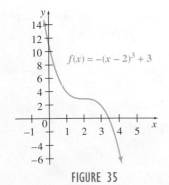

FIGURE 35

YOUR TURN 1 Graph $f(x) = 64 - x^6$.

EXAMPLE 1 Translations and Reflections

Graph $f(x) = -(x - 2)^3 + 3$.

SOLUTION Using the principles of translation and reflection from the previous section, we recognize that this is similar to the graph of $y = x^3$, but reflected vertically (because of the negative in front of $(x - 2)^3$), and with its center moved 2 units to the right and 3 units up. The result is shown in Figure 35. **TRY YOUR TURN 1**

A polynomial of degree 3, such as that in the previous example and in the next, is known as a **cubic polynomial.** A polynomial of degree 4, such as that in Example 3, is known as a **quartic polynomial.**

EXAMPLE 2 Graphing a Polynomial

Graph $f(x) = 8x^3 - 12x^2 + 2x + 1$.

SOLUTION Figure 36 shows the function graphed on the x- and y-intervals $[-0.5, 0.6]$ and $[-2, 2]$. In this view, it appears similar to a parabola opening downward. Zooming out to $[-1, 2]$ by $[-8, 8]$, we see in Figure 37 that the graph goes upward as x gets large. There are also two **turning points** near $x = 0$ and $x = 1$. (In a later chapter, we will introduce another term for such turning points: *relative extrema*.) By zooming in with the graphing calculator, we can find these turning points to be at approximately $(0.09175, 1.08866)$ and $(0.90825, -1.08866)$.

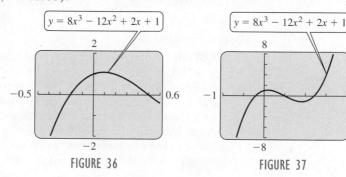

FIGURE 36   FIGURE 37

Zooming out still further, we see the function on $[-10, 10]$ by $[-300, 300]$ in Figure 38. From this viewpoint, we don't see the turning points at all, and the graph seems similar in shape to that of $y = x^3$. This is an important point: when x is large in magnitude, either positive or negative, $8x^3 - 12x^2 + 2x + 1$ behaves a lot like $8x^3$, because the other terms are small in comparison with the cubic term. So this viewpoint tells us something useful about the function, but it is less useful than the previous graph for determining the turning points.

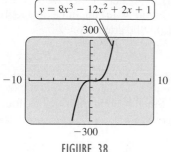

FIGURE 38

After the previous example, you may wonder how to be sure you have the viewing window that exhibits all the important properties of a function. We will find an answer to this question in later chapters using the techniques of calculus. Meanwhile, let us consider one more example to get a better idea of what polynomials look like.

EXAMPLE 3 Graphing a Polynomial

Graph $f(x) = -3x^4 + 14x^3 - 54x + 3$.

SOLUTION Figure 39 shows a graphing calculator view on $[-3, 5]$ by $[-50, 50]$. If you have a graphing calculator, we recommend that you experiment with various viewpoints and verify for yourself that this viewpoint captures the important behavior of the function. Notice that it has three turning points. Notice also that as $|x|$ gets large, the graph turns downward. This is because as $|x|$ becomes large, the x^4-term dominates the other terms, which are small in comparison, and the x^4-term has a negative coefficient.

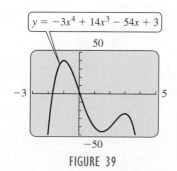

FIGURE 39

As suggested by the graphs above, the domain of a polynomial function is the set of all real numbers. The range of a polynomial function of odd degree is also the set of all real numbers. Some typical graphs of polynomial functions of odd and even degree are shown in Figure 40 on the next page. The first two graphs suggest that for every polynomial function f of odd degree, there is at least one real value of x for which $f(x) = 0$. Such a value of x is called a **real zero** of f; these values are also the x-intercepts of the graph.

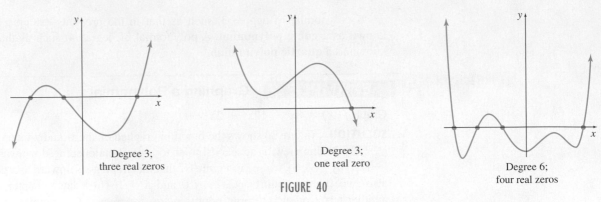

FIGURE 40

EXAMPLE 4   **Identifying the Degree of a Polynomial**

Identify the degree of the polynomial in each of the figures, and give the sign (+ or −) for the leading coefficient.

(a) Figure 41(a)

SOLUTION   Notice that the polynomial has a range $[k, \infty)$. This must be a polynomial of even degree, because if the highest power of x is an odd power, the polynomial can take on all real numbers, positive and negative. Notice also that the polynomial becomes a large positive number as x gets large in magnitude, either positive or negative, so the leading coefficient must be positive. Finally, notice that it has three turning points. Observe from the previous examples that a polynomial of degree n has at most $n − 1$ turning points. In a later chapter, we will use calculus to see why this is true. So the polynomial graphed in Figure 41(a) might be degree 4, although it could also be of degree 6, 8, etc. We can't be sure from the graph alone.

(b) Figure 41(b)

SOLUTION   Because the range is $(-\infty, \infty)$, this must be a polynomial of odd degree. Notice also that the polynomial becomes a large negative number as x becomes a large positive number, so the leading coefficient must be negative. Finally, notice that it has four turning points, so it might be degree 5, although it could also be of degree 7, 9, etc. ▪

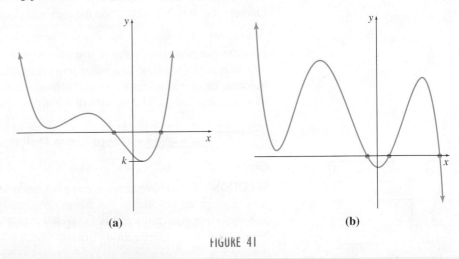

(a)   (b)

FIGURE 41

Properties of Polynomial Functions

1. A polynomial function of degree n can have at most $n − 1$ turning points. Conversely, if the graph of a polynomial function has n turning points, it must have degree at least $n + 1$.

2. In the graph of a polynomial function of even degree, both ends go up or both ends go down. For a polynomial function of odd degree, one end goes up and one end goes down.

3. If the graph goes up as x becomes a large positive number, the leading coefficient must be positive. If the graph goes down as x becomes a large positive number, the leading coefficient is negative.

Rational Functions

Many situations require mathematical models that are quotients. A common model for such situations is a *rational function*.

Rational Function

A **rational function** is defined by

$$f(x) = \frac{p(x)}{q(x)},$$

where $p(x)$ and $q(x)$ are polynomial functions and $q(x) \neq 0$.

Since any values of x such that $q(x) = 0$ are excluded from the domain, a rational function often has a graph with one or more breaks.

EXAMPLE 5   Graphing a Rational Function

Graph $y = \dfrac{1}{x}$.

SOLUTION   This function is undefined for $x = 0$, since 0 is not allowed as the denominator of a fraction. For this reason, the graph of this function will not intersect the vertical line $x = 0$, which is the y-axis. Since x can take on any value except 0, the values of x can approach 0 as closely as desired from either side of 0.

Values of 1/x for Small x								
				x approaches 0.				
x	-0.5	-0.2	-0.1	-0.01	0.01	0.1	0.2	0.5
$y = \dfrac{1}{x}$	-2	-5	-10	-100	100	10	5	2
					$\lvert y \rvert$ gets larger and larger.			

The table above suggests that as x gets closer and closer to 0, $\lvert y \rvert$ gets larger and larger. This is true in general: as the denominator gets smaller, the fraction gets larger. Thus, the graph of the function approaches the vertical line $x = 0$ (the y-axis) without ever touching it.

As $\lvert x \rvert$ gets larger and larger, $y = 1/x$ gets closer and closer to 0, as shown in the table below. This is also true in general: as the denominator gets larger, the fraction gets smaller.

Values of 1/x for Large |x|							
x	-100	-10	-4	-1	1	4	10	100
$y = \dfrac{1}{x}$	-0.01	-0.1	-0.25	-1	1	0.25	0.1	0.01

The graph of the function approaches the horizontal line $y = 0$ (the x-axis). The information from both tables supports the graph in Figure 42.

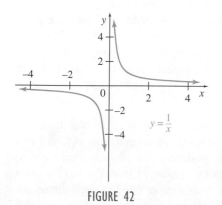

FIGURE 42

In Example 5, the vertical line $x = 0$ and the horizontal line $y = 0$ are *asymptotes*, defined as follows.

Asymptotes

If a function gets larger and larger in magnitude without bound as x approaches the number k, then the line $x = k$ is a **vertical asymptote**.

If the values of y approach a number k as $\lvert x \rvert$ gets larger and larger, the line $y = k$ is a **horizontal asymptote**.

There is an easy way to find any vertical asymptotes of a rational function. First, find the roots of the denominator. If a number k makes the denominator 0 but does not make the numerator 0, then the line $x = k$ is a vertical asymptote. If, however, a number k makes both the denominator and the numerator 0, then further investigation will be necessary, as we will see in the next example. In the next chapter we will show another way to find asymptotes using the concept of a *limit*.

EXAMPLE 6   Graphing a Rational Function

Graph the following rational functions:

(a) $y = \dfrac{x^2 + 3x + 2}{x + 1}$.

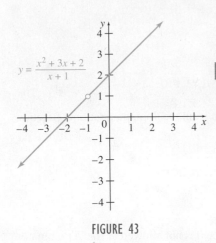

$y = \dfrac{x^2 + 3x + 2}{x + 1}$

FIGURE 43

SOLUTION   The value $x = -1$ makes the denominator 0, and so -1 is not in the domain of this function. Note that the value $x = -1$ also makes the numerator 0. In fact, if we factor the numerator and simplify the function, we get

$$y = \frac{x^2 + 3x + 2}{x + 1} = \frac{(x + 2)(x + 1)}{(x + 1)} = x + 2 \quad \text{for} \quad x \neq -1.$$

The graph of this function, therefore, is the graph of $y = x + 2$ with a hole at $x = -1$, as shown in Figure 43.

(b) $y = \dfrac{3x + 2}{2x + 4}$.

SOLUTION   The value $x = -2$ makes the denominator 0, but not the numerator, so the line $x = -2$ is a vertical asymptote. To find a horizontal asymptote, let x get larger and larger, so that $3x + 2 \approx 3x$ because the 2 is very small compared with $3x$. Similarly, for x very large, $2x + 4 \approx 2x$. Therefore, $y = (3x + 2)/(2x + 4) \approx (3x)/(2x) = 3/2$. This means that the line $y = 3/2$ is a horizontal asymptote. (A more precise way of approaching this idea will be seen in the next chapter when limits at infinity are discussed.)

The intercepts should also be noted. When $x = 0$, the y-intercept is $y = 2/4 = 1/2$. To make a fraction 0, the numerator must be 0; so to make $y = 0$, it is necessary that $3x + 2 = 0$. Solve this for x to get $x = -2/3$ (the x-intercept). We can also use these values to determine where the function is positive and where it is negative. Using the techniques described in Chapter R, verify that the function is negative on $(-2, -2/3)$ and positive on $(-\infty, -2) \cup (-2/3, \infty)$. With this information, the two asymptotes to guide us, and the fact that there are only two intercepts, we suspect the graph is as shown in Figure 44. A graphing calculator can support this.   **TRY YOUR TURN 2**

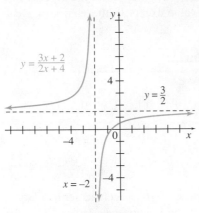

$y = \dfrac{3x + 2}{2x + 4}$

$y = \dfrac{3}{2}$

$x = -2$

FIGURE 44

YOUR TURN 2

Graph $y = \dfrac{4x - 6}{x - 3}$.

Rational functions occur often in practical applications. In many situations involving environmental pollution, much of the pollutant can be removed from the air or water at a fairly reasonable cost, but the last small part of the pollutant can be very expensive to remove. Cost as a function of the percentage of pollutant removed from the environment can be calculated for various percentages of removal, with a curve fitted through the resulting data points. This curve then leads to a mathematical model of the situation. Rational functions are often a good choice for these **cost-benefit models** because they rise rapidly as they approach a vertical asymptote.

EXAMPLE 7   Cost-Benefit Analysis

Suppose a cost-benefit model is given by

$$y = \frac{18x}{106 - x},$$

where y is the cost (in thousands of dollars) of removing x percent of a certain pollutant. The domain of x is the set of all numbers from 0 to 100 inclusive; any amount of pollutant from 0% to 100% can be removed. Find the cost to remove the following amounts of the pollutant: 100%, 95%, 90%, and 80%. Graph the function.

SOLUTION   Removal of 100% of the pollutant would cost

$$y = \frac{18(100)}{106 - 100} = 300,$$

or $300,000. Check that 95% of the pollutant can be removed for $155,000, 90% for $101,000, and 80% for $55,000. Using these points, as well as others obtained from the function, gives the graph shown in Figure 45.

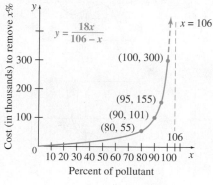

FIGURE 45

If a cost function has the form $C(x) = mx + b$, where x is the number of items produced, m is the marginal cost per item and b is the fixed cost, then the **average cost** per item is given by

$$\overline{C}(x) = \frac{C(x)}{x} = \frac{mx + b}{x}.$$

Notice that this is a rational function with a vertical asymptote at $x = 0$ and a horizontal asymptote at $y = m$. The vertical asymptote reflects the fact that, as the number of items produced approaches 0, the average cost per item becomes infinitely large, because the fixed costs are spread over fewer and fewer items. The horizontal asymptote shows that, as the number of items becomes large, the fixed costs are spread over more and more items, so most of the average cost per item is the marginal cost to produce each item. This is another example of how asymptotes give important information in real applications.

2.3   EXERCISES

1. Explain how translations and reflections can be used to graph $y = -(x - 1)^4 + 2$.

2. Describe an asymptote, and explain when a rational function will have (a) a vertical asymptote and (b) a horizontal asymptote.

Use the principles of the previous section with the graphs of this section to sketch a graph of the given function.

3. $f(x) = (x - 2)^3 + 3$

4. $f(x) = (x + 1)^3 - 2$

5. $f(x) = -(x + 3)^4 + 1$

6. $f(x) = -(x - 1)^4 + 2$

In Exercises 7–15, match the correct graph A–I to the function without using your calculator. Then, after you have answered all of them, if you have a graphing calculator, use your calculator to check your answers. Each graph is plotted on $[-6, 6]$ by $[-50, 50]$.

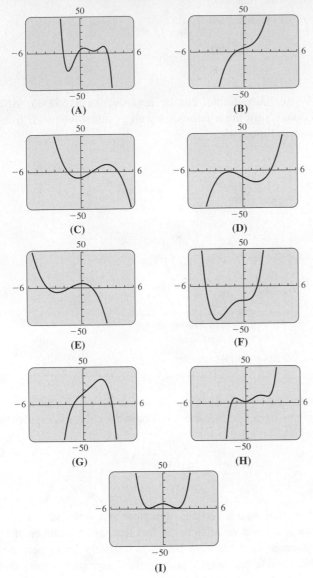

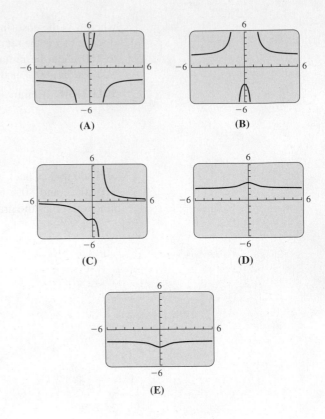

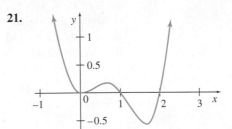

16. $y = \dfrac{2x^2 + 3}{x^2 - 1}$ **17.** $y = \dfrac{2x^2 + 3}{x^2 + 1}$

18. $y = \dfrac{-2x^2 - 3}{x^2 - 1}$ **19.** $y = \dfrac{-2x^2 - 3}{x^2 + 1}$

20. $y = \dfrac{2x^2 + 3}{x^3 - 1}$

Each of the following is the graph of a polynomial function. Give the possible values for the degree of the polynomial, and give the sign ($+$ or $-$) for the leading coefficient.

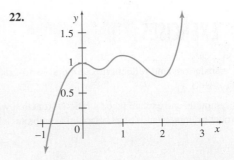

7. $y = x^3 - 7x - 9$ **8.** $y = -x^3 + 4x^2 + 3x - 8$

9. $y = -x^3 - 4x^2 + x + 6$ **10.** $y = 2x^3 + 4x + 5$

11. $y = x^4 - 5x^2 + 7$ **12.** $y = x^4 + 4x^3 - 20$

13. $y = -x^4 + 2x^3 + 10x + 15$

14. $y = 0.7x^5 - 2.5x^4 - x^3 + 8x^2 + x + 2$

15. $y = -x^5 + 4x^4 + x^3 - 16x^2 + 12x + 5$

In Exercises 16–20, match the correct graph A–E to the function without using your calculator. Then, after you have answered all of them, if you have a graphing calculator, use your calculator to check your answers. Each graph in this group is plotted on $[-6, 6]$ by $[-6, 6]$. Hint: Consider the asymptotes. (If you try graphing these graphs in Connected mode rather than Dot mode, you will see some lines that are not part of the graph, but the result of the calculator connecting disconnected parts of the graph.)

23.

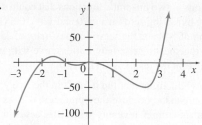

24.

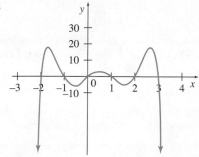

25.

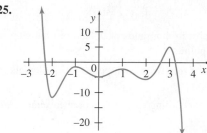

26.

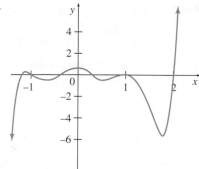

Find any horizontal and vertical asymptotes and any holes that may exist for each rational function. Draw the graph of each function, including any x- and y-intercepts.

27. $y = \dfrac{-4}{x + 2}$

28. $y = \dfrac{-1}{x + 3}$

29. $y = \dfrac{2}{3 + 2x}$

30. $y = \dfrac{8}{5 - 3x}$

31. $y = \dfrac{2x}{x - 3}$

32. $y = \dfrac{4x}{3 - 2x}$

33. $y = \dfrac{x + 1}{x - 4}$

34. $y = \dfrac{x - 4}{x + 1}$

35. $y = \dfrac{3 - 2x}{4x + 20}$

36. $y = \dfrac{6 - 3x}{4x + 12}$

37. $y = \dfrac{-x - 4}{3x + 6}$

38. $y = \dfrac{-2x + 5}{x + 3}$

39. $y = \dfrac{x^2 + 7x + 12}{x + 4}$

40. $y = \dfrac{9 - 6x + x^2}{3 - x}$

41. Write an equation that defines a rational function with a vertical asymptote at $x = 1$ and a horizontal asymptote at $y = 2$.

42. Write an equation that defines a rational function with a vertical asymptote at $x = -2$ and a horizontal asymptote at $y = 0$.

43. Consider the polynomial functions defined by $f(x) = (x - 1)(x - 2)(x + 3)$, $g(x) = x^3 + 2x^2 - x - 2$, and $h(x) = 3x^3 + 6x^2 - 3x - 6$.

   a. What is the value of $f(1)$?

   b. For what values, other than 1, is $f(x) = 0$?

   c. Verify that $g(-1) = g(1) = g(-2) = 0$.

   d. Based on your answer from part c, what do you think is the factored form of $g(x)$? Verify your answer by multiplying it out and comparing with $g(x)$.

   e. Using your answer from part d, what is the factored form of $h(x)$?

   f. Based on what you have learned in this exercise, fill in the blank: If f is a polynomial and $f(a) = 0$ for some number a, then one factor of the polynomial is

   _____.

44. Consider the function defined by

$$f(x) = \frac{x^7 - 4x^5 - 3x^4 + 4x^3 + 12x^2 - 12}{x^7}.$$

Source: The Mathematics Teacher.

   a. Graph the function on $[-6, 6]$ by $[-6, 6]$. From your graph, estimate how many x-intercepts the function has and what their values are.

   b. Now graph the function on $[-1.5, -1.4]$ by $[-10^{-4}, 10^{-4}]$ and on $[1.4, 1.5]$ by $[-10^{-5}, 10^{-5}]$. From your graphs, estimate how many x-intercepts the function has and what their values are.

   c. From your results in parts a and b, what advice would you give a friend on using a graphing calculator to find x-intercepts?

45. Consider the function defined by

$$f(x) = \frac{1}{x^5 - 2x^3 - 3x^2 + 6}.$$

Source: The Mathematics Teacher.

   a. Graph the function on $[-3.4, 3.4]$ by $[-3, 3]$. From your graph, estimate how many vertical asymptotes the function has and where they are located.

   b. Now graph the function on $[-1.5, -1.4]$ by $[-10, 10]$ and on $[1.4, 1.5]$ by $[-1000, 1000]$. From your graphs, estimate how many vertical asymptotes the function has and where they are located.

   c. From your results in parts a and b, what advice would you give a friend on using a graphing calculator to find vertical asymptotes?

APPLICATIONS

Business and Economics

46. Average Cost Suppose the average cost per unit $\overline{C}(x)$, in dollars, to produce x units of yogurt is given by

$$\overline{C}(x) = \frac{600}{x + 20}.$$

a. Find $\overline{C}(10), \overline{C}(20), \overline{C}(50), \overline{C}(75),$ and $\overline{C}(100)$.

b. Which of the intervals $(0, \infty)$ and $[0, \infty)$ would be a more reasonable domain for $\overline{C}$? Why?

c. Give the equations of any asymptotes. Find any intercepts.

d. Graph $y = \overline{C}(x)$.

47. Cost Analysis In a recent year, the cost per ton, y, to build an oil tanker of x thousand deadweight tons was approximated by

$$\overline{C}(x) = \frac{220{,}000}{x + 475}$$

for $x > 0$.

a. Find $\overline{C}(25), \overline{C}(50), \overline{C}(100), \overline{C}(200), \overline{C}(300),$ and $\overline{C}(400)$.

b. Find any asymptotes.

c. Find any intercepts.

d. Graph $y = \overline{C}(x)$.

APPLY IT **Tax Rates** Exercises 48–50 refer to the *Laffer curve*, originated by the economist Arthur Laffer. An idealized version of this curve is shown here. According to this curve, decreasing a tax rate, say from x_2 percent to x_1 percent on the graph, can actually lead to an increase in government revenue. The theory is that people will work harder and earn more money if they are taxed at a lower rate, so the government ends up with more revenue than it would at a higher tax rate. All economists agree on the endpoints—0 revenue at tax rates of both 0% and 100%—but there is much disagreement on the location of the tax rate x_1 that produces the maximum revenue.

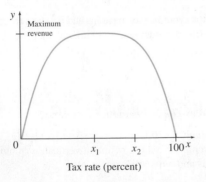

48. A function that might describe the entire Laffer curve is

$$y = x(100 - x)(x^2 + 500),$$

where y is government revenue in hundreds of thousands of dollars from a tax rate of x percent, with the function valid for $0 \le x \le 100$. Find the revenue from the following tax rates.

a. 10%   b. 40%   c. 50%   d. 80%

e. Graph the function.

49. Find the equations of two quadratic functions that could describe the Laffer curve by having zeros at $x = 0$ and $x = 100$. Give the first a maximum of 100 and the second a maximum of 250, then multiply them together to get a new Laffer curve with a maximum of 25,000. Plot the resulting function.

50. An economist might argue that the models in the two previous exercises are unrealistic because they predict that a tax rate of 50% gives the maximum revenue, while the actual value is probably less than 50%. Consider the function

$$y = \frac{300x - 3x^2}{5x + 100},$$

where y is government revenue in millions of dollars from a tax rate of x percent, where $0 \le x \le 100$. *Source: Dana Lee Ling.*

a. Graph the function, and discuss whether the shape of the graph is appropriate.

b. Use a graphing calculator to find the tax rate that produces the maximum revenue. What is the maximum revenue?

51. Cost-Benefit Model Suppose a cost-benefit model is given by

$$y = \frac{6.7x}{100 - x},$$

where y is the cost in thousands of dollars of removing x percent of a given pollutant.

a. Find the cost of removing each percent of pollutants: 50%; 70%; 80%; 90%; 95%; 98%; 99%.

b. Is it possible, according to this function, to remove *all* the pollutant?

c. Graph the function.

52. Cost-Benefit Model Suppose a cost-benefit model is given by

$$y = \frac{6.5x}{102 - x},$$

where y is the cost in thousands of dollars of removing x percent of a certain pollutant.

a. Find the cost of removing each percent of pollutants: 0%; 50%; 80%; 90%; 95%; 99%; 100%.

b. Graph the function.

Life Sciences

53. Contact Lenses The strength of a contact lens is given in units known as diopters, as well as in mm of arc. The following is taken from a chart used by optometrists to convert diopters to mm of arc. *Source: Bausch & Lomb.*

Diopters	mm of Arc
36.000	9.37
36.125	9.34
36.250	9.31
36.375	9.27
36.500	9.24
36.625	9.21
36.750	9.18
36.875	9.15
37.000	9.12

where t is the number of years since 1998. **Source: Association of American Medical Colleges.**

a. Graph the number of applicants on $0 \leq t \leq 11$.

b. Based on the graph in part a, during what years did the number of medical school applicants increase?

57. Population Biology The function

$$f(x) = \frac{\lambda x}{1 + (ax)^b}$$

is used in population models to give the size of the next generation $(f(x))$ in terms of the current generation (x). **Source: Models in Ecology.**

a. What is a reasonable domain for this function, considering what x represents?

b. Graph this function for $\lambda = a = b = 1$.

c. Graph this function for $\lambda = a = 1$ and $b = 2$.

d. What is the effect of making b larger?

58. Growth Model The function

$$f(x) = \frac{Kx}{A + x}$$

is used in biology to give the growth rate of a population in the presence of a quantity x of food. This is called Michaelis-Menten kinetics. **Source: Mathematical Models in Biology.**

a. What is a reasonable domain for this function, considering what x represents?

b. Graph this function for $K = 5$ and $A = 2$.

c. Show that $y = K$ is a horizontal asymptote.

d. What do you think K represents?

e. Show that A represents the quantity of food for which the growth rate is half of its maximum.

a. Notice that as the diopters increase, the mm of arc decrease. Find a value of k so the function $a = f(d) = k/d$ gives a, the mm of arc, as a function of d, the strength in diopters. (Round k to the nearest integer. For a more accurate answer, average all the values of k given by each pair of data.)

b. An optometrist wants to order 40.50 diopter lenses for a patient. The manufacturer needs to know the strength in mm of arc. What is the strength in mm of arc?

54. Cardiac Output A technique for measuring cardiac output depends on the concentration of a dye after a known amount is injected into a vein near the heart. In a normal heart, the concentration of the dye at time x (in seconds) is given by the function

$$g(x) = -0.006x^4 + 0.140x^3 - 0.053x^2 + 1.79x.$$

a. Graph $g(x)$ on $[0, 6]$ by $[0, 20]$.

b. In your graph from part a, notice that the function initially increases. Considering the form of $g(x)$, do you think it can keep increasing forever? Explain.

c. Write a short paragraph about the extent to which the concentration of dye might be described by the function $g(x)$.

55. Alcohol Concentration The polynomial function

$$A(x) = 0.003631x^3 - 0.03746x^2 + 0.1012x + 0.009$$

gives the approximate blood alcohol concentration in a 170-lb woman x hours after drinking 2 oz of alcohol on an empty stomach, for x in the interval $[0, 5]$. **Source: Medical Aspects of Alcohol Determination in Biological Specimens.**

a. Graph $A(x)$ on $0 \leq x \leq 5$.

b. Using the graph from part a, estimate the time of maximum alcohol concentration.

c. In many states, a person is legally drunk if the blood alcohol concentration exceeds 0.08%. Use the graph from part a to estimate the period in which this 170-lb woman is legally drunk.

56. Medical School For the years 1998 to 2009, the number of applicants to U.S. medical schools can be closely approximated by

$$A(t) = -6.7615t^4 + 114.7t^3 - 240.1t^2 - 2129t + 40,966,$$

59. Brain Mass The mass (in grams) of the human brain during the last trimester of gestation and the first two years after birth can be approximated by the function

$$m(c) = \frac{c^3}{100} - \frac{1500}{c},$$

where c is the circumference of the head in centimeters. **Source: Early Human Development.**

a. Find the approximate mass of brains with a head circumference of 30, 40, or 50 cm.

b. Clearly the formula is invalid for any values of c yielding negative values of w. For what values of c is this true?

c. Use a graphing calculator to sketch this graph on the interval $20 \leq c \leq 50$.

d. Suppose an infant brain has mass of 700 g. Use features on a graphing calculator to find what the circumference of the head is expected to be.

Social Sciences

60. Head Start The enrollment in Head Start for some recent years is included in the table. *Source: Administration for Children & Families.*

Year	Enrollment
1966	733,000
1970	477,400
1980	376,300
1990	540,930
1995	750,696
2000	857,664
2005	906,993

a. Plot the points from the table using 0 for 1960, and so on.

b. Use the quadratic regression feature of a graphing calculator to get a quadratic function that approximates the data. Graph the function on the same window as the scatterplot.

c. Use cubic regression to get a cubic function that approximates the data. Graph the function on the same window as the scatterplot.

d. Which of the two functions in part b and c appears to be a better fit for the data? Explain your reasoning.

Physical Sciences

61. Length of a Pendulum A simple pendulum swings back and forth in regular time intervals. Grandfather clocks use pendulums to keep accurate time. The relationship between the length of a pendulum L and the period (time) T for one complete oscillation can be expressed by the function $L = kT^n$, where k is a constant and n is a positive integer to be determined. The data below were taken for different lengths of pendulums.* *Source: Gary Rockswold.*

T (sec)	L (ft)
1.11	1.0
1.36	1.5
1.57	2.0
1.76	2.5
1.92	3.0
2.08	3.5
2.22	4.0

a. Find the value of k for $n = 1, 2$, and 3, using the data for the 4-ft pendulum.

b. Use a graphing calculator to plot the data in the table and to graph the function $L = kT^n$ for the three values of k (and their corresponding values of n) found in part a. Which function best fits the data?

*See Exercise 23, Section 1.3.

c. Use the best-fitting function from part a to predict the period of a pendulum having a length of 5 ft.

d. If the length of pendulum doubles, what happens to the period?

e. If you have a graphing calculator or computer program with a quadratic regression feature, use it to find a quadratic function that approximately fits the data. How does this answer compare with the answer to part b?

62. Coal Consumption The table gives U.S. coal consumption for selected years. *Source: U.S. Department of Energy.*

Year	Millions of Short Tons
1950	494.1
1960	398.1
1970	523.2
1980	702.7
1985	818.0
1990	902.9
1995	962.1
2000	1084.1
2005	1128.3

a. Draw a scatterplot, letting $x = 0$ represent 1950.

b. Use the quadratic regression feature of a graphing calculator to get a quadratic function that approximates the data.

c. Graph the function from part b on the same window as the scatterplot.

d. Use cubic regression to get a cubic function that approximates the data.

e. Graph the cubic function from part d on the same window as the scatterplot.

f. Which of the two functions in parts b and d appears to be a better fit for the data? Explain your reasoning.

YOUR TURN ANSWERS

1.

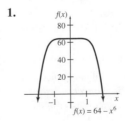

$f(x) = 64 - x^6$

2.

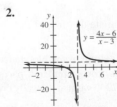

$y = \dfrac{4x - 6}{x - 3}$

2.4 Exponential Functions

APPLY IT **How much interest will an investment earn? What is the oxygen consumption of yearling salmon?**
Later in this section, in Examples 5 and 6, we will see that the answers to these questions depend on exponential functions.

In earlier sections we discussed functions involving expressions such as x^2, $(2x + 1)^3$, or x^{-1}, where the variable or variable expression is the base of an exponential expression, and the exponent is a constant. In an exponential function, the variable is in the exponent and the base is a constant.

> **Exponential Function**
> An **exponential function** with base a is defined as
> $$f(x) = a^x, \quad \text{where } a > 0 \text{ and } a \neq 1.$$

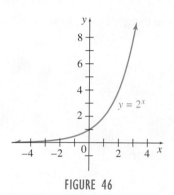

FIGURE 46

(If $a = 1$, the function is the constant function $f(x) = 1$.)

Exponential functions may be the single most important type of functions used in practical applications. They are used to describe growth and decay, which are important ideas in management, social science, and biology.

Figure 46 shows a graph of the exponential function defined by $f(x) = 2^x$. You could plot such a curve by hand by noting that $2^{-2} = 1/4$, $2^{-1} = 1/2$, $2^0 = 1$, $2^1 = 2$, and $2^2 = 4$, and then drawing a smooth curve through the points $(-2, 1/4)$, $(-1, 1/2)$, $(0, 1)$, $(1, 2)$, and $(2, 4)$. This graph is typical of the graphs of exponential functions of the form $y = a^x$, where $a > 1$. The y-intercept is $(0, 1)$. Notice that as x gets larger and larger, the function also gets larger. As x gets more and more negative, the function becomes smaller and smaller, approaching but never reaching 0. Therefore, the x-axis is a horizontal asymptote, but the function only approaches the left side of the asymptote. In contrast, rational functions approach both the left and right sides of the asymptote. The graph suggests that the domain is the set of all real numbers and the range is the set of all positive numbers.

FOR REVIEW

To review the properties of exponents used in this section, see Section R.6.

EXAMPLE 1 Graphing an Exponential Function

Graph $f(x) = 2^{-x}$.

SOLUTION The graph, shown in Figure 47, is the horizontal reflection of the graph of $f(x) = 2^x$ given in Figure 46. Since $2^{-x} = 1/2^x = (1/2)^x$, this graph is typical of the graphs of exponential functions of the form $y = a^x$ where $0 < a < 1$. The domain includes all real numbers and the range includes all positive numbers. The y-intercept is $(0, 1)$. Notice that this function, with $f(x) = 2^{-x} = (1/2)^x$, is decreasing over its domain.

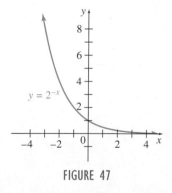

FIGURE 47

In the definition of an exponential function, notice that the base a is restricted to positive values, with negative or zero bases not allowed. For example, the function $y = (-4)^x$ could not include such numbers as $x = 1/2$ or $x = 1/4$ in the domain because the y-values would not be real numbers. The resulting graph would be at best a series of separate points having little practical use.

FOR REVIEW

Recall from Section 2.2 that the graph of $f(-x)$ is the reflection of the graph of $f(x)$ about the y-axis.

EXAMPLE 2 Graphing an Exponential Function

Graph $f(x) = -2^x + 3$.

SOLUTION The graph of $y = -2^x$ is the vertical reflection of the graph of $y = 2^x$, so this is a decreasing function. (Notice that -2^x is not the same as $(-2)^x$. In -2^x, we raise 2 to the x power and then take the negative.) The 3 indicates that the graph should be translated

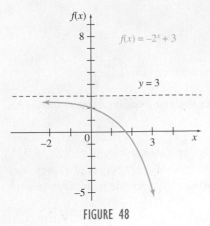

FIGURE 48

vertically 3 units, as compared to the graph of $y = -2^x$. Since $y = -2^x$ would have y-intercept $(0, -1)$, this function has y-intercept $(0, 2)$, which is up 3 units. For negative values of x, the graph approaches the line $y = 3$, which is a horizontal asymptote. The graph is shown in Figure 48.

Exponential Equations

In Figures 46 and 47, which are typical graphs of exponential functions, a given value of x leads to exactly one value of a^x. Because of this, an equation with a variable in the exponent, called an **exponential equation**, often can be solved using the following property.

> If $a > 0$, $a \neq 1$, and $a^x = a^y$, then $x = y$.

The value $a = 1$ is excluded, since $1^2 = 1^3$, for example, even though $2 \neq 3$. To solve $2^{3x} = 2^7$ using this property, work as follows.

$$2^{3x} = 2^7$$
$$3x = 7$$
$$x = \frac{7}{3}$$

EXAMPLE 3   Solving Exponential Equations

(a) Solve $9^x = 27$.

SOLUTION   First rewrite both sides of the equation so the bases are the same. Since $9 = 3^2$ and $27 = 3^3$,

$$9^x = 27$$
$$(3^2)^x = 3^3$$
$$3^{2x} = 3^3 \quad \text{Multiply exponents.}$$
$$2x = 3$$
$$x = \frac{3}{2}.$$

FOR REVIEW
Recall from Section R.6 that
$(a^m)^n = a^{mn}$.

(b) Solve $32^{2x-1} = 128^{x+3}$.

SOLUTION   Since the bases must be the same, write 32 as 2^5 and 128 as 2^7, giving

$$32^{2x-1} = 128^{x+3}$$
$$(2^5)^{2x-1} = (2^7)^{x+3}$$
$$2^{10x-5} = 2^{7x+21}. \quad \text{Multiply exponents.}$$

Now use the property from above to get

$$10x - 5 = 7x + 21$$
$$3x = 26$$
$$x = \frac{26}{3}.$$

YOUR TURN 1   Solve $25^{x/2} = 125^{x+3}$.

Verify this solution in the original equation.   **TRY YOUR TURN 1**

Compound Interest

The calculation of compound interest is an important application of exponential functions. The cost of borrowing money or the return on an investment is called **interest**. The amount borrowed or invested is the **principal**, P. The **rate of interest** r is given as a percent per year, and t is the **time**, measured in years.

Simple Interest
The product of the principal P, rate r, and time t gives **simple interest**, I:

$$I = Prt.$$

With **compound interest**, interest is charged (or paid) on interest as well as on the principal. To find a formula for compound interest, first suppose that P dollars, the principal, is deposited at a rate of interest r per year. The interest earned during the first year is found using the formula for simple interest.

$$\text{First-year interest} = P \cdot r \cdot 1 = Pr.$$

At the end of one year, the amount on deposit will be the sum of the original principal and the interest earned, or

$$P + Pr = P(1 + r). \tag{1}$$

If the deposit earns compound interest, the interest earned during the second year is found from the total amount on deposit at the end of the first year. Thus, the interest earned during the second year (again found by the formula for simple interest), is

$$[P(1 + r)](r)(1) = P(1 + r)r, \tag{2}$$

so the total amount on deposit at the end of the second year is the sum of amounts from (1) and (2) above, or

$$P(1 + r) + P(1 + r)r = P(1 + r)(1 + r) = P(1 + r)^2.$$

In the same way, the total amount on deposit at the end of three years is

$$P(1 + r)^3.$$

After t years, the total amount on deposit, called the *compound amount*, is $P(1 + r)^t$.

When interest is compounded more than once a year, the compound interest formula is adjusted. For example, if interest is to be paid quarterly (four times a year), 1/4 of the interest rate is used each time interest is calculated, so the rate becomes $r/4$, and the number of compounding periods in t years becomes $4t$. Generalizing from this idea gives the following formula.

Compound Amount
If P dollars is invested at a yearly rate of interest r per year, compounded m times per year for t years, the **compound amount** is

$$A = P\left(1 + \frac{r}{m}\right)^{tm} \text{ dollars.}$$

EXAMPLE 4   Compound Interest

Inga Moffitt invests a bonus of $9000 at 6% annual interest compounded semiannually for 4 years. How much interest will she earn?

SOLUTION   Use the formula for compound interest with $P = 9000$, $r = 0.06$, $m = 2$, and $t = 4$.

$$A = P\left(1 + \frac{r}{m}\right)^{tm}$$

$$= 9000\left(1 + \frac{0.06}{2}\right)^{4(2)}$$

$$= 9000(1.03)^8$$

$$\approx 11,400.93 \qquad \text{Use a calculator.}$$

YOUR TURN 2   Find the interest earned on $4400 at 3.25% interest compounded quarterly for 5 years.

The investment plus the interest is $11,400.93. The interest amounts to $11,400.93 − $9000 = $2400.93.

TRY YOUR TURN 2

NOTE When using a calculator to compute the compound interest, store each partial result in the calculator and avoid rounding off until the final answer.

The Number e Perhaps the single most useful base for an exponential function is the number e, an irrational number that occurs often in practical applications. The famous Swiss mathematician Leonhard Euler (pronounced "oiler") (1707–1783) was the first person known to have referred to this number as e, and the notation has continued to this day. To see how the number e occurs in an application, begin with the formula for compound interest,

$$P\left(1 + \frac{r}{m}\right)^{tm}.$$

Suppose that a lucky investment produces annual interest of 100%, so that $r = 1.00 = 1$. Suppose also that you can deposit only \$1 at this rate, and for only one year. Then $P = 1$ and $t = 1$. Substituting these values into the formula for compound interest gives

$$P\left(1 + \frac{r}{m}\right)^{t(m)} = 1\left(1 + \frac{1}{m}\right)^{1(m)} = \left(1 + \frac{1}{m}\right)^{m}.$$

As interest is compounded more and more often, m gets larger and the value of this expression will increase. For example, if $m = 1$ (interest is compounded annually),

$$\left(1 + \frac{1}{m}\right)^{m} = \left(1 + \frac{1}{1}\right)^{1} = 2^1 = 2,$$

so that your \$1 becomes \$2 in one year. Using a graphing calculator, we produced Figure 49 (where m is represented by X and $(1 + 1/m)^m$ by Y_1) to see what happens as m becomes larger and larger. A spreadsheet can also be used to produce this table.

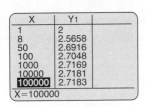

X	Y1
1	2
8	2.5658
50	2.6916
100	2.7048
1000	2.7169
10000	2.7181
100000	2.7183

X=100000

FIGURE 49

The table suggests that as m increases, the value of $(1 + 1/m)^m$ gets closer and closer to a fixed number, called e. As we shall see in the next chapter, this is an example of a limit.

> **Definition of e**
>
> As m becomes larger and larger, $\left(1 + \dfrac{1}{m}\right)^{m}$ becomes closer and closer to the number e, whose approximate value is 2.718281828.

The value of e is approximated here to 9 decimal places. Euler approximated e to 18 decimal places. Many calculators give values of e^x, usually with a key labeled e^x. Some require two keys, either INV LN or 2nd LN. (We will define $\ln x$ in the next section.) In Figure 50, the functions $y = 2^x$, $y = e^x$, and $y = 3^x$ are graphed for comparison. Notice that e^x is between 2^x and 3^x, because e is between 2 and 3. For $x > 0$, the graphs show that $3^x > e^x > 2^x$. All three functions have y-intercept $(0, 1)$. It is difficult to see from the graph, but $3^x < e^x < 2^x$ when $x < 0$.

The number e is often used as the base in an exponential equation because it provides a good model for many natural, as well as economic, phenomena. In the exercises for this section, we will look at several examples of such applications.

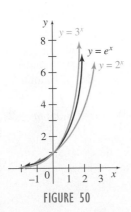

FIGURE 50

Continuous Compounding In economics, the formula for **continuous compounding** is a good example of an exponential growth function. Recall the formula for compound amount

$$A = P\left(1 + \frac{r}{m}\right)^{tm},$$

where m is the number of times annually that interest is compounded. As m becomes larger and larger, the compound amount also becomes larger but not without bound. Recall that as m becomes larger and larger, $(1 + 1/m)^m$ becomes closer and closer to e. Similarly,

$$\left(1 + \frac{1}{(m/r)}\right)^{m/r}$$

becomes closer and closer to e. Let us rearrange the formula for compound amount to take advantage of this fact.

$$A = P\left(1 + \frac{r}{m}\right)^{tm}$$

$$= P\left(1 + \frac{1}{(m/r)}\right)^{tm}$$

$$= P\left[\left(1 + \frac{1}{(m/r)}\right)^{m/r}\right]^{rt} \qquad \frac{m}{r} \cdot rt = tm$$

This last expression becomes closer and closer to Pe^{rt} as m becomes larger and larger, which describes what happens when interest is compounded continuously. Essentially, the number of times annually that interest is compounded becomes infinitely large. We thus have the following formula for the compound amount when interest is compounded continuously.

Continuous Compounding

If a deposit of P dollars is invested at a rate of interest r compounded continuously for t years, the compound amount is

$$A = Pe^{rt} \text{ dollars.}$$

EXAMPLE 5   Continuous Compound Interest

APPLY IT If $600 is invested in an account earning 2.75% compounded continuously, how much would be in the account after 5 years?

SOLUTION   In the formula for continuous compounding, let $P = 600$, $t = 5$, and $r = 0.0275$ to get

$$A = 600e^{5(0.0275)} \approx 688.44,$$

or $688.44.

YOUR TURN 3 Find the amount after 4 years if $800 is invested in an account earning 3.15% compounded continuously.

TRY YOUR TURN 3

In situations that involve growth or decay of a population, the size of the population at a given time t often is determined by an exponential function of t. The next example illustrates a typical application of this kind.

EXAMPLE 6   Oxygen Consumption

APPLY IT Biologists studying salmon have found that the oxygen consumption of yearling salmon (in appropriate units) increases exponentially with the speed of swimming according to the function defined by

$$f(x) = 100e^{0.6x},$$

where x is the speed in feet per second. Find the following.

(a) The oxygen consumption when the fish are still

FOR REVIEW

Refer to the discussion on linear regression in Section 1.3. A similar process is used to fit data points to other types of functions. Many of the functions in this chapter's applications were determined in this way, including that given in Example 6.

SOLUTION When the fish are still, their speed is 0. Substitute 0 for x:

$$f(0) = 100e^{(0.6)(0)} = 100e^0$$
$$= 100 \cdot 1 = 100. \qquad e^0 = 1$$

When the fish are still, their oxygen consumption is 100 units.

(b) The oxygen consumption at a speed of 2 ft per second

SOLUTION Find $f(2)$ as follows.

$$f(2) = 100e^{(0.6)(2)} = 100e^{1.2} \approx 332$$

At a speed of 2 ft per second, oxygen consumption is about 332 units rounded to the nearest integer. Because the function is only an approximation of the real situation, further accuracy is not realistic.

EXAMPLE 7 Food Surplus

A magazine article argued that the cause of the obesity epidemic in the United States is the decreasing cost of food (in real terms) due to the increasing surplus of food. *Source: The New York Times Magazine.* As one piece of evidence, the following table was provided, which we have updated, showing U.S. corn production (in billions of bushels) for selected years.

Year	Production (billions of bushels)
1930	1.757
1940	2.207
1950	2.764
1960	3.907
1970	4.152
1980	6.639
1990	7.934
2000	9.968
2005	11.112

(a) Plot the data. Does the production appear to grow linearly or exponentially?

SOLUTION Figure 51 shows a graphing calculator plot of the data, which suggests that corn production is growing exponentially.

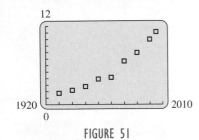

FIGURE 51

(b) Find an exponential function in the form of $p(x) = p_0 a^{x-1930}$ that models this data, where x is the year and $p(x)$ is the production of corn. Use the data for 1930 and 2005.

SOLUTION Since $p(1930) = p_0 a^0 = p_0$, we have $p_0 = 1.757$. Using $x = 2005$, we have

$$p(2005) = 1.757a^{2005-1930} = 1.757a^{75} = 11.112$$

$$a^{75} = \frac{11.112}{1.757} \qquad \text{Divide by 1.757.}$$

$$a = \left(\frac{11.112}{1.757}\right)^{1/75} \qquad \text{Take the 75th root.}$$

$$\approx 1.0249.$$

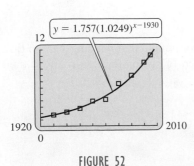

$y = 1.757(1.0249)^{x-1930}$

FIGURE 52

Thus $p(x) = 1.757(1.0249)^{x-1930}$. Figure 52 shows that this function fits the data well.

(c) Determine the expected annual percentage increase in corn production during this time period.

SOLUTION Since a is 1.0249, the production of corn each year is 1.0249 times its value the previous year, for a rate of increase of $0.0249 = 2.49\%$ per year.

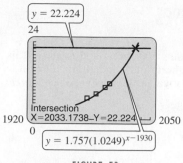

FIGURE 53

(d) Graph p and estimate the year when corn production will be double what it was in 2005.

SOLUTION   Figure 53 shows the graphs of $p(x)$ and $y = 2 \cdot 11.112 = 22.224$ on the same coordinate axes. (Note that the scale in Figure 53 is different than the scale in Figures 51 and 52 so that larger values of x and $p(x)$ are visible.) Their graphs intersect at approximately 2033, which is thus the year when corn production will be double its 2005 level. In the next section, we will see another way to solve such problems that does not require the use of a graphing calculator.

Another way to check whether an exponential function fits the data is to see if points whose x-coordinates are equally spaced have y-coordinates with a constant ratio. This must be true for an exponential function because if $f(x) = a \cdot b^x$, then $f(x_1) = a \cdot b^{x_1}$ and $f(x_2) = a \cdot b^{x_2}$, so

$$\frac{f(x_2)}{f(x_1)} = \frac{a \cdot b^{x_2}}{a \cdot b^{x_1}} = b^{x_2 - x_1}.$$

This last expression is constant if $x_2 - x_1$ is constant, that is, if the x-coordinates are equally spaced.

In the previous example, all data points but the last have x-coordinates 10 years apart, so we can compare the ratios of corn production for any of these first pairs of years. Here are the ratios for 1930–1940 and for 1990–2000:

$$\frac{2.207}{1.757} = 1.256$$

$$\frac{9.968}{7.934} = 1.256$$

These ratios are identical to 3 decimal places, so an exponential function fits the data very well. Not all ratios are this close; using the values at 1970 and 1980, we have $6.639/4.152 = 1.599$. From Figure 52, we can see that this is because the 1970 value is below the exponential curve and the 1980 value is above the curve.

TECHNOLOGY NOTE   Another way to find an exponential function that fits a set of data is to use a graphing calculator or computer program with an exponential regression feature. This fits an exponential function through a set of points using the least squares method, introduced in Section 1.3 for fitting a line through a set of points. On a TI-84 Plus, for example, enter the year into the list L_1 and the corn production into L_2. For simplicity, subtract 1930 from each year, so that 1930 corresponds to $x = 0$. Selecting `ExpReg` from the `STAT CALC` menu yields $y = 1.728(1.0254)^x$, which is close to the function we found in Example 7(b).

2.4 EXERCISES

A ream of 20-lb paper contains 500 sheets and is about 2 in. high. Suppose you take one sheet, fold it in half, then fold it in half again, continuing in this way as long as possible. *Source: The AMATYC Review.*

1. Complete the table.

Number of Folds	1	2	3	4	5	...	10	...	50
Layers of Paper									

2. After folding 50 times (if this were possible), what would be the height (in miles) of the folded paper?

For Exercises 3–11, match the correct graph A–F to the function without using your calculator. Notice that there are more functions than graphs; some of the functions are equivalent. After you have answered all of them, use a graphing calculator to check your answers. Each graph in this group is plotted on the window $[-2, 2]$ by $[-4, 4]$.

3. $y = 3^x$

4. $y = 3^{-x}$

5. $y = \left(\frac{1}{3}\right)^{1-x}$

6. $y = 3^{x+1}$

7. $y = 3(3)^x$

8. $y = \left(\frac{1}{3}\right)^x$

9. $y = 2 - 3^{-x}$

10. $y = -2 + 3^{-x}$

11. $y = 3^{x-1}$

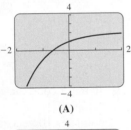

(A)

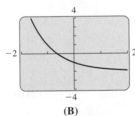

(B)

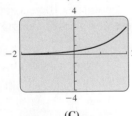

(C)

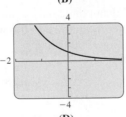

(D)

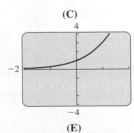

(E)

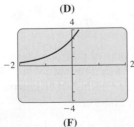

(F)

12. In Exercises 3–11, there were more formulas for functions than there were graphs. Explain how this is possible.

Solve each equation.

13. $2^x = 32$

14. $4^x = 64$

15. $3^x = \frac{1}{81}$

16. $e^x = \frac{1}{e^5}$

17. $4^x = 8^{x+1}$

18. $25^x = 125^{x+2}$

19. $16^{x+3} = 64^{2x-5}$

20. $(e^3)^{-2x} = e^{-x+5}$

21. $e^{-x} = (e^4)^{x+3}$

22. $2^{|x|} = 8$

23. $5^{-|x|} = \frac{1}{25}$

24. $2^{x^2-4x} = \left(\frac{1}{16}\right)^{x-4}$

25. $5^{x^2+x} = 1$

26. $8^{x^2} = 2^{x+4}$

27. $27^x = 9^{x^2+x}$

28. $e^{x^2+5x+6} = 1$

Graph each of the following.

29. $y = 5e^x + 2$

30. $y = -2e^x - 3$

31. $y = -3e^{-2x} + 2$

32. $y = 4e^{-x/2} - 1$

33. In our definition of exponential function, we ruled out negative values of a. The author of a textbook on mathematical economics, however, obtained a "graph" of $y = (-2)^x$ by plotting the following points and drawing a smooth curve through them.

x	-4	-3	-2	-1	0	1	2	3
y	$1/16$	$-1/8$	$1/4$	$-1/2$	1	-2	4	-8

The graph oscillates very neatly from positive to negative values of y. Comment on this approach. (This exercise shows the dangers of point plotting when drawing graphs.)

34. Explain why the exponential equation $4^x = 6$ cannot be solved using the method described in Example 3.

35. Explain why $3^x > e^x > 2^x$ when $x > 0$, but $3^x < e^x < 2^x$ when $x < 0$.

36. A friend claims that as x becomes large, the expression $1 + 1/x$ gets closer and closer to 1, and 1 raised to any power is still 1. Therefore, $f(x) = (1 + 1/x)^x$ gets closer and closer to 1 as x gets larger. Use a graphing calculator to graph f on $0.1 \leq x \leq 50$. How might you use this graph to explain to the friend why $f(x)$ does not approach 1 as x becomes large? What does it approach?

APPLICATIONS

Business and Economics

37. Interest Find the interest earned on $10,000 invested for 5 years at 4% interest compounded as follows.

a. Annually   **b.** Semiannually (twice a year)

c. Quarterly   **d.** Monthly   **e.** Continuously

38. Interest Suppose $26,000 is borrowed for 4 years at 6% interest. Find the interest paid over this period if the interest is compounded as follows.

a. Annually   **b.** Semiannually

c. Quarterly   **d.** Monthly   **e.** Continuously

39. Interest Ron Hampton needs to choose between two investments: One pays 6% compounded annually, and the other pays 5.9% compounded monthly. If he plans to invest $18,000 for 2 years, which investment should he choose? How much extra interest will he earn by making the better choice?

40. Interest Find the interest rate required for an investment of $5000 to grow to $7500 in 5 years if interest is compounded as follows.

a. Annually   **b.** Quarterly

41. Inflation Assuming continuous compounding, what will it cost to buy a $10 item in 3 years at the following inflation rates?

a. 3%   **b.** 4%   **c.** 5%

42. Interest Ali Williams invests a $25,000 inheritance in a fund paying 5.5% per year compounded continuously. What will be the amount on deposit after each time period?

a. 1 year   **b.** 5 years   **c.** 10 years

43. Interest Leigh Jacks plans to invest $500 into a money market account. Find the interest rate that is needed for the money to grow to $1200 in 14 years if the interest is compounded quarterly.

44. Interest Kristi Perez puts $10,500 into an account to save money to buy a car in 12 years. She expects the car of her dreams to cost $30,000 by then. Find the interest rate that is necessary if the interest is computed using the following methods.

a. Compounded quarterly   **b.** Compounded monthly

45. Inflation If money loses value at the rate of 8% per year, the value of $1 in t years is given by

$$y = (1 - 0.08)^t = (0.92)^t.$$

a. Use a calculator to help complete the following table.

t	0	1	2	3	4	5	6	7	8	9	10
y	1					0.66					0.43

b. Graph $y = (0.92)^t$.

c. Suppose a house costs $165,000 today. Use the results of part a to estimate the cost of a similar house in 10 years.

d. Find the cost of a $50 textbook in 8 years.

46. Interest On January 1, 2000, Jack deposited $1000 into Bank X to earn interest at the rate of j per annum compounded semiannually. On January 1, 2005, he transferred his account to Bank Y to earn interest at the rate of k per annum compounded quarterly. On January 1, 2008, the balance at Bank Y was $1990.76. If Jack could have earned interest at the rate of k per annum compounded quarterly from January 1, 2000, through January 1, 2008, his balance would have been $2203.76. Which of the following represents the ratio k/j? *Source: Society of Actuaries.*

a. 1.25   **b.** 1.30   **c.** 1.35   **d.** 1.40   **e.** 1.45

Life Sciences

47. Population Growth Since 1960, the growth in world population (in millions) closely fits the exponential function defined by
$$A(t) = 3100e^{0.0166t},$$
where t is the number of years since 1960. *Source: United Nations.*

a. World population was about 3686 million in 1970. How closely does the function approximate this value?

b. Use the function to approximate world population in 2000. (The actual 2000 population was about 6115 million.)

c. Estimate world population in the year 2015.

48. Growth of Bacteria Salmonella bacteria, found on almost all chicken and eggs, grow rapidly in a nice warm place. If just a few hundred bacteria are left on the cutting board when a chicken is cut up, and they get into the potato salad, the population begins compounding. Suppose the number present in the potato salad after x hours is given by

$$f(x) = 500 \cdot 2^{3x}.$$

a. If the potato salad is left out on the table, how many bacteria are present 1 hour later?

b. How many were present initially?

c. How often do the bacteria double?

d. How quickly will the number of bacteria increase to 32,000?

49. Minority Population According to the U.S. Census Bureau, the United States is becoming more diverse. Based on U.S. Census population projections for 2000 to 2050, the projected Hispanic population (in millions) can be modeled by the exponential function

$$h(t) = 37.79(1.021)^t,$$

where $t = 0$ corresponds to 2000 and $0 \le t \le 50$. *Source: U.S. Census Bureau.*

a. Find the projected Hispanic population for 2005. Compare this to the actual value of 42.69 million.

b. The U.S. Asian population is also growing exponentially, and the projected Asian population (in millions) can be modeled by the exponential function

$$a(t) = 11.14(1.023)^t,$$

where $t = 0$ corresponds to 2000 and $0 \le t \le 50$. Find the projected Asian population for 2005, and compare this to the actual value of 12.69 million.

c. Determine the expected annual percentage increase for Hispanics and for Asians. Which minority population, Hispanic or Asian, is growing at a faster rate?

d. The U.S. black population is growing at a linear rate, and the projected black population (in millions) can be modeled by the linear function

$$b(t) = 0.5116t + 35.43,$$

where $t = 0$ corresponds to 2000 and $0 \le t \le 50$. Find the projected black population for 2005 and compare this projection to the actual value of 37.91 million.

e. Graph the projected population function for Hispanics and estimate when the Hispanic population will be double its actual value for 2005. Then do the same for the Asian and black populations. Comment on the accuracy of these numbers.

50. Physician Demand The demand for physicians is expected to increase in the future, as shown in the table on the following page. *Source: Association of American Medical Colleges.*

Year	Demand for Physicians (in thousands)
2006	680.5
2015	758.6
2020	805.8
2025	859.3

a. Plot the data, letting $t = 0$ correspond to 2000. Does fitting an exponential curve to the data seem reasonable?

b. Use the data for 2006 and 2015 to find a function of the form $f(x) = Ce^{kt}$ that goes through these two points.

c. Use your function from part c to predict the demand for physicians in 2020 and 2025. How well do these predictions fit the data?

d. If you have a graphing calculator or computer program with an exponential regression feature, use it to find an exponential function that approximately fits the data. How does this answer compare with the answer to part b?

Physical Sciences

51. **Carbon Dioxide** The table gives the estimated global carbon dioxide (CO_2) emissions from fossil-fuel burning, cement production, and gas flaring over the last century. The CO_2 estimates are expressed in millions of metric tons. *Source: U.S. Department of Energy.*

Year	CO_2 Emissions (millions of metric tons)
1900	534
1910	819
1920	932
1930	1053
1940	1299
1950	1630
1960	2577
1970	4076
1980	5330
1990	6143
2000	6672

a. Plot the data, letting $x = 0$ correspond to 1900. Do the emissions appear to grow linearly or exponentially?

b. Find an exponential function in the form of $f(x) = f_0 a^x$ that fits this data at 1900 and 2000, where x is the number of years since 1900 and $f(x)$ is the CO_2 emissions.

c. Approximate the average annual percentage increase in CO_2 emissions during this time period.

d. Graph $f(x)$ and estimate the first year when emissions will be at least double what they were in 2000.

52. **Radioactive Decay** Suppose the quantity (in grams) of a radioactive substance present at time t is

$$Q(t) = 1000(5^{-0.3t}),$$

where t is measured in months.

a. How much will be present in 6 months?

b. How long will it take to reduce the substance to 8 g?

53. **Atmospheric Pressure** The atmospheric pressure (in millibars) at a given altitude (in meters) is listed in the table. *Source: Elements of Meteorology.*

Altitude	Pressure
0	1013
1000	899
2000	795
3000	701
4000	617
5000	541
6000	472
7000	411
8000	357
9000	308
10,000	265

a. Find functions of the form $P = ae^{kx}$, $P = mx + b$, and $P = 1/(ax + b)$ that fit the data at $x = 0$ and $x = 10,000$, where P is the pressure and x is the altitude.

b. Plot the data in the table and graph the three functions found in part a. Which function best fits the data?

c. Use the best-fitting function from part b to predict pressure at 1500 m and 11,000 m. Compare your answers to the true values of 846 millibars and 227 millibars, respectively.

d. If you have a graphing calculator or computer program with an exponential regression feature, use it to find an exponential function that approximately fits the data. How does this answer compare with the answer to part b?

54. **Computer Chips** The power of personal computers has increased dramatically as a result of the ability to place an increasing number of transistors on a single processor chip. The following table lists the number of transistors on some popular computer chips made by Intel. *Source: Intel.*

Year	Chip	Transistors (in millions)
1985	386	0.275
1989	486	1.2
1993	Pentium	3.1
1997	Pentium II	7.5
1999	Pentium III	9.5
2000	Pentium 4	42
2005	Pentium D	291
2007	Penryn	820
2009	Nehalem	1900

a. Let t be the year, where $t = 0$ corresponds to 1985, and y be the number of transistors (in millions). Find functions of the form $y = mt + b$, $y = at^2 + b$, and $y = ab^t$ that fit the data at 1985 and 2009.

b. Use a graphing calculator to plot the data in the table and to graph the three functions found in part a. Which function best fits the data?

c. Use the best-fitting function from part b to predict the number of transistors on a chip in the year 2015.

d. If you have a graphing calculator or computer program with an exponential regression feature, use it to find an exponential function that approximately fits the data. How does this answer compare with the answer to part b?

e. In 1965 Gordon Moore wrote a paper predicting how the power of computer chips would grow in the future. Moore's law says that the number of transistors that can be put on a chip doubles roughly every 18 months. Discuss the extent to which the data in this exercise confirms or refutes Moore's law.

55. Wind Energy The following table gives the total world wind energy capacity (in megawatts) in recent years. *Source: World Wind Energy Association.*

Year	Capacity (MW)
2001	24,322
2002	31,181
2003	39,295
2004	47,693
2005	58,024
2006	74,122
2007	93,930
2008	120,903
2009	159,213

a. Let t be the number of years since 2000, and C the capacity (in MW). Find functions of the form $C = mt + b$, $C = at^2 + b$, and $C = ab^t$ that fit the data at 2001 and 2009.

b. Use a graphing calculator to plot the data in the table and to graph the three functions found in part a. Which function best fits the data?

c. If you have a graphing calculator or computer program with an exponential regression feature, use it to find an exponential function that approximately fits the data in the table. How does this answer compare with the answer to part b?

d. Using the three functions from part b and the function from part c, predict the total world wind capacity in 2010. Compare these with the World Wind Energy Association's prediction of 203,500.

YOUR TURN ANSWERS

1. $-9/2$ **2.** 772.97

3. 907.43

2.5 Logarithmic Functions

APPLY IT   With an inflation rate averaging 5% per year, how long will it take for prices to double?

The number of years it will take for prices to double under given conditions is called the **doubling time**. For $1 to double (become $2) in t years, assuming 5% annual compounding, means that

$$A = P\left(1 + \frac{r}{m}\right)^{mt}$$

becomes

$$2 = 1\left(1 + \frac{0.05}{1}\right)^{1(t)}$$

or

$$2 = (1.05)^t.$$

This equation would be easier to solve if the variable were not in the exponent. **Logarithms** are defined for just this purpose. In Example 8, we will use logarithms to answer the question posed above.

Logarithm

For $a > 0$, $a \neq 1$, and $x > 0$,

$$y = \log_a x \qquad \text{means} \qquad a^y = x.$$

(Read $y = \log_a x$ as "y is the logarithm of x to the base a.") For example, the exponential statement $2^4 = 16$ can be translated into the logarithmic statement $4 = \log_2 16$. Also, in the problem discussed above, $(1.05)^t = 2$ can be rewritten with this definition as $t = \log_{1.05} 2$. A logarithm is an exponent: $\log_a x$ **is the exponent used with the base a to get x.**

EXAMPLE 1 Equivalent Expressions

This example shows the same statements written in both exponential and logarithmic forms.

Exponential Form	*Logarithmic Form*
(a) $3^2 = 9$	$\log_3 9 = 2$
(b) $(1/5)^{-2} = 25$	$\log_{1/5} 25 = -2$
(c) $10^5 = 100{,}000$	$\log_{10} 100{,}000 = 5$
(d) $4^{-3} = 1/64$	$\log_4(1/64) = -3$
(e) $2^{-4} = 1/16$	$\log_2(1/16) = -4$
(f) $e^0 = 1$	$\log_e 1 = 0$

YOUR TURN 1 Write the equation $5^{-2} = 1/25$ in logarithmic form.

TRY YOUR TURN 1

EXAMPLE 2 Evaluating Logarithms

Evaluate each of the following logarithms.

(a) $\log_4 64$

SOLUTION We seek a number x such that $4^x = 64$. Since $4^3 = 64$, we conclude that $\log_4 64 = 3$.

(b) $\log_2(-8)$

SOLUTION We seek a number x such that $2^x = -8$. Since 2^x is positive for all real numbers x, we conclude that $\log_2(-8)$ is undefined. (Actually, $\log_2(-8)$ can be defined if we use complex numbers, but in this textbook, we restrict ourselves to real numbers.)

(c) $\log_5 80$

SOLUTION We know that $5^2 = 25$ and $5^3 = 125$, so $\log_5 25 = 2$ and $\log_5 125 = 3$. Therefore, $\log_5 80$ must be somewhere between 2 and 3. We will find a more accurate answer in Example 4.

YOUR TURN 2
Evaluate $\log_3(1/81)$.

TRY YOUR TURN 2

Logarithmic Functions

For a given positive value of x, the definition of logarithm leads to exactly one value of y, so $y = \log_a x$ defines the *logarithmic function* of base a (the base a must be positive, with $a \neq 1$).

Logarithmic Function

If $a > 0$ and $a \neq 1$, then the **logarithmic function** of base a is defined by

$$f(x) = \log_a x$$

for $x > 0$.

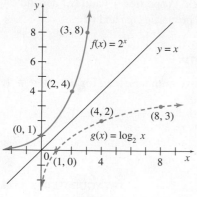

FIGURE 54

The graphs of the exponential function with $f(x) = 2^x$ and the logarithmic function with $g(x) = \log_2 x$ are shown in Figure 54. The graphs show that $f(3) = 2^3 = 8$, while $g(8) = \log_2 8 = 3$. Thus, $f(3) = 8$ and $g(8) = 3$. Also, $f(2) = 4$ and $g(4) = 2$. In fact, for any number m, if $f(m) = p$, then $g(p) = m$. Functions related in this way are called **inverse functions** of each other. The graphs also show that the domain of the exponential function (the set of real numbers) is the range of the logarithmic function. Also, the range of the exponential function (the set of positive real numbers) is the domain of the logarithmic function. Every logarithmic function is the inverse of some exponential function. This means that we can graph logarithmic functions by rewriting them as exponential functions using the definition of logarithm. The graphs in Figure 54 show a characteristic of a pair of inverse functions: their graphs are mirror images about the line $y = x$. Therefore, since exponential functions go through the point $(0, 1)$, logarithmic functions go through the point $(1, 0)$. Notice that because the exponential function has the x-axis as a horizontal asymptote, the logarithmic function has the y-axis as a vertical asymptote. A more complete discussion of inverse functions is given in most standard intermediate algebra and college algebra books.

The graph of $\log_2 x$ is typical of logarithms with bases $a > 1$. When $0 < a < 1$, the graph is the vertical reflection of the logarithm graph in Figure 54. Because logarithms with bases less than 1 are rarely used, we will not explore them here.

CAUTION   The domain of $\log_a x$ consists of all $x > 0$. In other words, you cannot take the logarithm of zero or a negative number. This also means that in a function such as $g(x) = \log_a(x - 2)$, the domain is given by $x - 2 > 0$, or $x > 2$.

Properties of Logarithms

The usefulness of logarithmic functions depends in large part on the following **properties of logarithms**.

Properties of Logarithms

Let x and y be any positive real numbers and r be any real number. Let a be a positive real number, $a \neq 1$. Then

a. $\log_a xy = \log_a x + \log_a y$

b. $\log_a \dfrac{x}{y} = \log_a x - \log_a y$

c. $\log_a x^r = r \log_a x$

d. $\log_a a = 1$

e. $\log_a 1 = 0$

f. $\log_a a^r = r$.

To prove property (a), let $m = \log_a x$ and $n = \log_a y$. Then, by the definition of logarithm,

$$a^m = x \qquad \text{and} \qquad a^n = y.$$

Hence,

$$a^m a^n = xy.$$

By a property of exponents, $a^m a^n = a^{m+n}$, so

$$a^{m+n} = xy.$$

Now use the definition of logarithm to write

$$\log_a xy = m + n.$$

Since $m = \log_a x$ and $n = \log_a y$,

$$\log_a xy = \log_a x + \log_a y.$$

Proofs of properties (b) and (c) are left for the exercises. Properties (d) and (e) depend on the definition of a logarithm. Property (f) follows from properties (c) and (d).

EXAMPLE 3   Properties of Logarithms

If all the following variable expressions represent positive numbers, then for $a > 0$, $a \neq 1$, the statements in (a)–(c) are true.

(a) $\log_a x + \log_a(x - 1) = \log_a x(x - 1)$

(b) $\log_a \dfrac{x^2 - 4x}{x + 6} = \log_a(x^2 - 4x) - \log_a(x + 6)$

(c) $\log_a(9x^5) = \log_a 9 + \log_a(x^5) = \log_a 9 + 5 \cdot \log_a x$ TRY YOUR TURN 3 ▬

YOUR TURN 3 Write the expression $\log_a (x^2/y^3)$ as a sum, difference, or product of simpler logarithms.

Evaluating Logarithms

The invention of logarithms is credited to John Napier (1550–1617), who first called logarithms "artificial numbers." Later he joined the Greek words *logos* (ratio) and *arithmos* (number) to form the word used today. The development of logarithms was motivated by a need for faster computation. Tables of logarithms and slide rule devices were developed by Napier, Henry Briggs (1561–1631), Edmund Gunter (1581–1626), and others.

For many years logarithms were used primarily to assist in involved calculations. Current technology has made this use of logarithms obsolete, but logarithmic functions play an important role in many applications of mathematics. Since our number system has base 10, logarithms to base 10 were most convenient for numerical calculations and so base 10 logarithms were called **common logarithms.** Common logarithms are still useful in other applications. For simplicity,

$$\log_{10} x \text{ is abbreviated } \log x.$$

Most practical applications of logarithms use the number e as base. (Recall that to 7 decimal places, $e = 2.7182818$.) Logarithms to base e are called **natural logarithms,** and

$$\log_e x \text{ is abbreviated } \ln x$$

(read "el-en x"). A graph of $f(x) = \ln x$ is given in Figure 55.

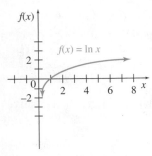

FIGURE 55

NOTE Keep in mind that $\ln x$ is a logarithmic function. Therefore, all of the properties of logarithms given previously are valid when a is replaced with e and $\log_e$ is replaced with $\ln$.

Although common logarithms may seem more "natural" than logarithms to base e, there are several good reasons for using natural logarithms instead. The most important reason is discussed later, in the section on Derivatives of Logarithmic Functions.

A calculator can be used to find both common and natural logarithms. For example, using a calculator and 4 decimal places, we get the following values.

$$\log 2.34 = 0.3692, \quad \log 594 = 2.7738, \quad \text{and} \quad \log 0.0028 = -2.5528.$$

$$\ln 2.34 = 0.8502, \quad \ln 594 = 6.3869, \quad \text{and} \quad \ln 0.0028 = -5.8781.$$

Notice that logarithms of numbers less than 1 are negative when the base is greater than 1. A look at the graph of $y = \log_2 x$ or $y = \ln x$ will show why.

Sometimes it is convenient to use logarithms to bases other than 10 or e. For example, some computer science applications use base 2. In such cases, the following theorem is useful for converting from one base to another.

Change-of-Base Theorem for Logarithms

If x is any positive number and if a and b are positive real numbers, $a \neq 1$, $b \neq 1$, then

$$\log_a x = \frac{\log_b x}{\log_b a}.$$

To prove this result, use the definition of logarithm to write $y = \log_a x$ as $x = a^y$ or $x = a^{\log_a x}$ (for positive x and positive a, $a \neq 1$). Now take base b logarithms of both sides of this last equation.

$$\log_b x = \log_b a^{\log_a x}$$
$$\log_b x = (\log_a x)(\log_b a), \quad \log_a x^r = r \log_a x$$
$$\log_a x = \frac{\log_b x}{\log_b a} \qquad \text{Solve for } \log_a x.$$

If the base b is equal to e, then by the change-of-base theorem,

$$\log_a x = \frac{\log_e x}{\log_e a}.$$

Using $\ln x$ for $\log_e x$ gives the special case of the theorem using natural logarithms.

> For any positive numbers a and x, $a \neq 1$,
>
> $$\log_a x = \frac{\ln x}{\ln a}.$$

The change-of-base theorem for logarithms is useful when graphing $y = \log_a x$ on a graphing calculator for a base a other than e or 10. For example, to graph $y = \log_2 x$, let $y = \ln x / \ln 2$. The change-of-base theorem is also needed when using a calculator to evaluate a logarithm with a base a other than e or 10.

EXAMPLE 4   Evaluating Logarithms

Evaluate $\log_5 80$.

SOLUTION   As we saw in Example 2, this number is between 2 and 3. Using the second form of the change-of-base theorem for logarithms with $x = 80$ and $a = 5$ gives

$$\log_5 80 = \frac{\ln 80}{\ln 5} \approx \frac{4.3820}{1.6094} \approx 2.723.$$

YOUR TURN 4
Evaluate $\log_3 50$.

To check, use a calculator to verify that $5^{2.723} \approx 80$.   **TRY YOUR TURN 4**

> **CAUTION**   As mentioned earlier, when using a calculator, do not round off intermediate results. Keep all numbers in the calculator until you have the final answer. In Example 4, we showed the rounded intermediate values of $\ln 80$ and $\ln 5$, but we used the unrounded quantities when doing the division.

Logarithmic Equations

Equations involving logarithms are often solved by using the fact that exponential functions and logarithmic functions are inverses, so a logarithmic equation can be rewritten (with the definition of logarithm) as an exponential equation. In other cases, the properties of logarithms may be useful in simplifying a **logarithmic equation**.

EXAMPLE 5   Solving Logarithmic Equations

Solve each equation.

(a) $\log_x \dfrac{8}{27} = 3$

SOLUTION Using the definition of logarithm, write the expression in exponential form. To solve for x, take the cube root on both sides.

$$x^3 = \frac{8}{27}$$

$$x = \frac{2}{3}$$

(b) $\log_4 x = \frac{5}{2}$

SOLUTION In exponential form, the given statement becomes

$$4^{5/2} = x$$

$$(4^{1/2})^5 = x$$

$$2^5 = x$$

$$32 = x.$$

(c) $\log_2 x - \log_2(x - 1) = 1$

SOLUTION By a property of logarithms,

$$\log_2 x - \log_2(x - 1) = \log_2 \frac{x}{x - 1},$$

so the original equation becomes

$$\log_2 \frac{x}{x - 1} = 1.$$

Now write this equation in exponential form, and solve.

$$\frac{x}{x - 1} = 2^1 = 2$$

Solve this equation.

$$\frac{x}{x - 1}(x - 1) = 2(x - 1) \qquad \text{Multiply both sides by } x-1.$$

$$x = 2(x - 1)$$

$$x = 2x - 2$$

$$-x = -2$$

$$x = 2$$

(d) $\log x + \log(x - 3) = 1$

SOLUTION Similar to part (c), we have

$$\log x + \log(x - 3) = \log[x(x - 3)] = 1.$$

Since the logarithm base is 10, this means that

$$x(x - 3) = 10 \qquad 10^1 = 10$$

$$x^2 - 3x - 10 = 0 \qquad \text{Subtract 10 from both sides.}$$

$$(x - 5)(x + 2) = 0. \qquad \text{Factor.}$$

YOUR TURN 5 Solve for x: $\log_2 x + \log_2(x + 2) = 3$.

This leads to two solutions: $x = 5$ and $x = -2$. But notice that -2 is not a valid value for x in the original equation, since the logarithm of a negative number is undefined. The only solution is, therefore, $x = 5$.　　**TRY YOUR TURN 5**

CAUTION It is important to check solutions when solving equations involving logarithms because $\log_a u$, where u is an expression in x, has domain given by $u > 0$.

Exponential Equations

In the previous section exponential equations like $(1/3)^x = 81$ were solved by writing each side of the equation as a power of 3. That method cannot be used to solve an equation such as $3^x = 5$, however, since 5 cannot easily be written as a power of 3. Such equations can be solved approximately with a graphing calculator, but an algebraic method is also useful, particularly when the equation involves variables such as a and b rather than just numbers such as 3 and 5. A general method for solving these equations is shown in the following example.

EXAMPLE 6 Solving Exponential Equations

Solve each equation.

(a) $3^x = 5$

SOLUTION Taking natural logarithms (logarithms to any base could be used) on both sides gives

$$\ln 3^x = \ln 5$$
$$x \ln 3 = \ln 5 \qquad \ln u^r = r \ln u$$
$$x = \frac{\ln 5}{\ln 3} \approx 1.465$$

(b) $3^{2x} = 4^{x+1}$

SOLUTION Taking natural logarithms on both sides gives

$$\ln 3^{2x} = \ln 4^{x+1}$$
$$2x \ln 3 = (x + 1) \ln 4 \qquad \ln u^r = r \ln u$$
$$(2 \ln 3)x = (\ln 4)x + \ln 4$$
$$(2 \ln 3)x - (\ln 4)x = \ln 4 \qquad \text{Subtract } (\ln 4)x \text{ from both sides.}$$
$$(2 \ln 3 - \ln 4)x = \ln 4 \qquad \text{Factor } x.$$
$$x = \frac{\ln 4}{2 \ln 3 - \ln 4}. \qquad \text{Divide both sides by } 2 \ln 3 - \ln 4.$$

Use a calculator to evaluate the logarithms, then divide, to get

$$x \approx \frac{1.3863}{2(1.0986) - 1.3863} \approx 1.710.$$

(c) $5e^{0.01x} = 9$

SOLUTION

$$e^{0.01x} = \frac{9}{5} = 1.8 \qquad \text{Divide both sides by 5.}$$
$$\ln e^{0.01x} = \ln 1.8 \qquad \text{Take natural logarithms on both sides.}$$
$$0.01x = \ln 1.8 \qquad \ln e^u = u$$
$$x = \frac{\ln 1.8}{0.01} \approx 58.779 \qquad \text{TRY YOUR TURN 6}$$

YOUR TURN 6 Solve for x: $2^{x+1} = 3^x$.

Just as $\log_a x$ can be written as a base e logarithm, any exponential function $y = a^x$ can be written as an exponential function with base e. For example, there exists a real number k such that

$$2 = e^k.$$

Raising both sides to the power x gives

$$2^x = e^{kx},$$

so that powers of 2 can be found by evaluating appropriate powers of e. To find the necessary number k, solve the equation $2 = e^k$ for k by first taking logarithms on both sides.

$$2 = e^k$$
$$\ln 2 = \ln e^k$$
$$\ln 2 = k \ln e$$
$$\ln 2 = k \qquad \ln e = 1$$

Thus, $k = \ln 2$. In the section on Derivatives of Exponential Functions, we will see why this change of base is useful. A general statement can be drawn from this example.

> **Change-of-Base Theorem for Exponentials**
> For every positive real number a,
> $$a^x = e^{(\ln a)x}.$$

Another way to see why the change-of-base theorem for exponentials is true is to first observe that $e^{\ln a} = a$. Combining this with the fact that $e^{ab} = (e^a)^b$, we have $e^{(\ln a)x} = (e^{\ln a})^x = a^x$.

EXAMPLE 7 Change-of-Base-Theorem

(a) Write 7^x using base e rather than base 7.

SOLUTION According to the change-of-base theorem,

$$7^x = e^{(\ln 7)x}.$$

Using a calculator to evaluate $\ln 7$, we could also approximate this as $e^{1.9459x}$.

(b) Approximate the function $f(x) = e^{2x}$ as $f(x) = a^x$ for some base a.

SOLUTION We do not need the change-of-base theorem here. Just use the fact that

$$e^{2x} = (e^2)^x \approx 7.389^x,$$

where we have used a calculator to approximate e^2.   **TRY YOUR TURN 7**

YOUR TURN 7 Approximate $e^{0.025x}$ in the form a^x.

EXAMPLE 8 Doubling Time

Complete the solution of the problem posed at the beginning of this section.

APPLY IT **SOLUTION** Recall that if prices will double after t years at an inflation rate of 5%, compounded annually, t is given by the equation

$$2 = (1.05)^t.$$

We solve this equation by first taking natural logarithms on both sides.

$$\ln 2 = \ln(1.05)^t$$
$$\ln 2 = t \ln 1.05 \qquad \ln x^r = r \ln x$$
$$t = \frac{\ln 2}{\ln 1.05} \approx 14.2$$

It will take about 14 years for prices to double.

The problem solved in Example 8 can be generalized for the compound interest equation

$$A = P(1 + r)^t.$$

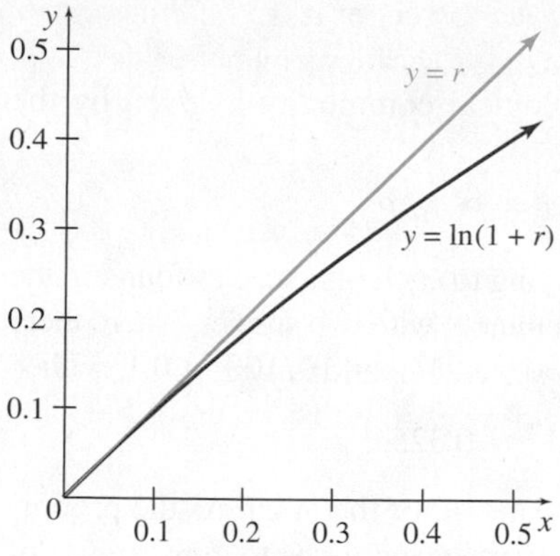

Solving for t as in Example 8 (with $A = 2$ and $P = 1$) gives the doubling time in years as

$$t = \frac{\ln 2}{\ln(1 + r)}.$$

If r is small, $\ln(1 + r) \approx r$, as Figure 56 shows, so that

$$t = \frac{\ln 2}{\ln(1 + r)} \approx \frac{\ln 2}{r} \approx \frac{0.693}{r}.$$

Notice from Figure 56 that the actual value of r is larger than $\ln(1 + r)$, causing the above formula to give a doubling time that is too small. By changing 0.693 to 0.70, the formula becomes reasonably accurate for $0.001 \le r < 0.05$. By increasing the numerator further to 0.72, the formula becomes reasonably accurate for $0.05 \le r < 0.12$. This leads to two useful rules for estimating the doubling time. The **rule of 70** says that for $0.001 \le r < 0.05$, the value of $70/100r$ gives a good approximation of t. The **rule of 72** says that for $0.05 \le r \le 0.12$, the value of $72/100r$ approximates t quite well.

Figure 57 shows the three functions $y = \ln 2/\ln(1 + r)$, $y = 70/100r$, and $y = 72/100r$ graphed on the same axes. Observe how close to each other they are. In an exercise we will ask you to explore the relationship between these functions further.

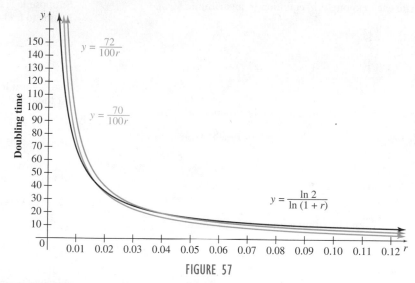

FIGURE 56

FIGURE 57

![Doubling time graph]

EXAMPLE 9   Rules of 70 and 72

Approximate the years to double at an interest rate of 6% using first the rule of 70, then the rule of 72.

SOLUTION   By the rule of 70, money will double at 6% interest after

$$\frac{70}{100r} = \frac{70}{100(0.06)} = \frac{70}{6} = 11.67 \left(\text{or } 11\frac{2}{3}\right)$$

years.

Using the rule of 72 gives

$$\frac{72}{100r} = \frac{72}{6} = 12$$

years doubling time. Since a more precise answer is given by

$$\frac{\ln 2}{\ln(1 + r)} = \frac{\ln 2}{\ln(1.06)} \approx \frac{0.693}{0.058} \approx 11.9,$$

the rule of 72 gives a better approximation than the rule of 70. This agrees with the statement that the rule of 72 works well for values of r where $0.05 \le r \le 0.12$, since $r = 0.06$ falls into this category.

> **EXAMPLE 10** Index of Diversity
>
> One measure of the diversity of the species in an ecological community is given by the **index of diversity** H, where
>
> $$H = -[P_1 \ln P_1 + P_2 \ln P_2 + \cdots + P_n \ln P_n],$$
>
> and $P_1, P_2, \ldots, P_n$ are the proportions of a sample belonging to each of n species found in the sample. **Source: Statistical Ecology.** For example, in a community with two species, where there are 90 of one species and 10 of the other, $P_1 = 90/100 = 0.9$ and $P_2 = 10/100 = 0.1$, with
>
> $$H = -[0.9 \ln 0.9 + 0.1 \ln 0.1] \approx 0.325.$$
>
> Verify that $H \approx 0.673$ if there are 60 of one species and 40 of the other. As the proportions of n species get closer to $1/n$ each, the index of diversity increases to a maximum of $\ln n$.

2.5 EXERCISES

Write each exponential equation in logarithmic form.

1. $5^3 = 125$

2. $7^2 = 49$

3. $3^4 = 81$

4. $2^7 = 128$

5. $3^{-2} = \dfrac{1}{9}$

6. $\left(\dfrac{5}{4}\right)^{-2} = \dfrac{16}{25}$

Write each logarithmic equation in exponential form.

7. $\log_2 32 = 5$

8. $\log_3 81 = 4$

9. $\ln \dfrac{1}{e} = -1$

10. $\log_2 \dfrac{1}{8} = -3$

11. $\log 100{,}000 = 5$

12. $\log 0.001 = -3$

Evaluate each logarithm without using a calculator.

13. $\log_8 64$

14. $\log_9 81$

15. $\log_4 64$

16. $\log_3 27$

17. $\log_2 \dfrac{1}{16}$

18. $\log_3 \dfrac{1}{81}$

19. $\log_2 \sqrt[3]{\dfrac{1}{4}}$

20. $\log_8 \sqrt[4]{\dfrac{1}{2}}$

21. $\ln e$

22. $\ln e^3$

23. $\ln e^{5/3}$

24. $\ln 1$

25. Is the "logarithm to the base 3 of 4" written as $\log_4 3$ or $\log_3 4$?

26. Write a few sentences describing the relationship between e^x and $\ln x$.

Use the properties of logarithms to write each expression as a sum, difference, or product of simpler logarithms. For example, $\log_2(\sqrt{3}x) = \frac{1}{2}\log_2 3 + \log_2 x$.

27. $\log_5(3k)$

28. $\log_9(4m)$

29. $\log_3 \dfrac{3p}{5k}$

30. $\log_7 \dfrac{15p}{7y}$

31. $\ln \dfrac{3\sqrt{5}}{\sqrt[3]{6}}$

32. $\ln \dfrac{9\sqrt[3]{5}}{\sqrt[4]{3}}$

Suppose $\log_b 2 = a$ and $\log_b 3 = c$. Use the properties of logarithms to find the following.

33. $\log_b 32$

34. $\log_b 18$

35. $\log_b(72b)$

36. $\log_b(9b^2)$

Use natural logarithms to evaluate each logarithm to the nearest thousandth.

37. $\log_5 30$

38. $\log_{12} 210$

39. $\log_{1.2} 0.95$

40. $\log_{2.8} 0.12$

Solve each equation in Exercises 41–64. Round decimal answers to four decimal places.

41. $\log_x 36 = -2$

42. $\log_9 27 = m$

43. $\log_8 16 = z$

44. $\log_y 8 = \dfrac{3}{4}$

45. $\log_r 5 = \dfrac{1}{2}$

46. $\log_4(5x + 1) = 2$

47. $\log_5(9x - 4) = 1$

48. $\log_4 x - \log_4(x + 3) = -1$

49. $\log_9 m - \log_9(m - 4) = -2$

50. $\log(x + 5) + \log(x + 2) = 1$

51. $\log_3(x - 2) + \log_3(x + 6) = 2$

52. $\log_3(x^2 + 17) - \log_3(x + 5) = 1$

53. $\log_2(x^2 - 1) - \log_2(x + 1) = 2$

54. $\ln(5x + 4) = 2$

55. $\ln x + \ln 3x = -1$

56. $\ln(x + 1) - \ln x = 1$

57. $2^x = 6$

58. $5^x = 12$

59. $e^{k-1} = 6$

60. $e^{2y} = 15$

61. $3^{x+1} = 5^x$

62. $2^{x+1} = 6^{x-1}$

63. $5(0.10)^x = 4(0.12)^x$

64. $1.5(1.05)^x = 2(1.01)^x$

Write each expression using base e rather than base 10.

65. 10^{x+1}

66. 10^{x^2}

Approximate each expression in the form a^x without using e.

67. e^{3x}

68. e^{-4x}

Find the domain of each function.

69. $f(x) = \log(5 - x)$

70. $f(x) = \ln(x^2 - 9)$

71. Lucky Larry was faced with solving

$$\log(2x + 1) - \log(3x - 1) = 0.$$

Larry just dropped the logs and proceeded:

$$(2x + 1) - (3x - 1) = 0$$
$$-x + 2 = 0$$
$$x = 2.$$

Although Lucky Larry is wrong in dropping the logs, his procedure will always give the correct answer to an equation of the form

$$\log A - \log B = 0,$$

where A and B are any two expressions in x. Prove that this last equation leads to the equation $A - B = 0$, which is what you get when you drop the logs. *Source: The AMATYC Review.*

72. Find all errors in the following calculation.

$$(\log(x + 2))^2 = 2 \log(x + 2)$$
$$= 2(\log x + \log 2)$$
$$= 2(\log x + 100)$$

73. Prove: $\log_a\left(\dfrac{x}{y}\right) = \log_a x - \log_a y$.

74. Prove: $\log_a x^r = r \log_a x$.

APPLICATIONS

Business and Economics

75. Inflation Assuming annual compounding, find the time it would take for the general level of prices in the economy to double at the following annual inflation rates.

a. 3%   **b.** 6%   **c.** 8%

d. Check your answers using either the rule of 70 or the rule of 72, whichever applies.

76. Interest Mary Klingman invests $15,000 in an account paying 7% per year compounded annually.

a. How many years are required for the compound amount to at least double? (Note that interest is only paid at the end of each year.)

b. In how many years will the amount at least triple?

c. Check your answer to part a using either the rule of 70 or the rule of 72, whichever applies.

77. Interest Leigh Jacks plans to invest $500 into a money market account. Find the interest rate that is needed for the money to grow to $1200 in 14 years if the interest is compounded continuously. (Compare with Exercise 43 in the previous section.)

78. Rule of 72 Complete the following table, and use the results to discuss when the rule of 70 gives a better approximation for the doubling time, and when the rule of 72 gives a better approximation.

r	0.001	0.02	0.05	0.08	0.12
$(\ln 2)/\ln(1 + r)$					
$70/100r$					
$72/100r$					

79. Pay Increases You are offered two jobs starting July 1, 2013. Humongous Enterprises offers you $45,000 a year to start, with a raise of 4% every July 1. At Crabapple Inc. you start at $30,000, with an annual increase of 6% every July 1. On July 1 of what year would the job at Crabapple Inc. pay more than the job at Humongous Enterprises? Use the algebra of logarithms to solve this problem, and support your answer by using a graphing calculator to see where the two salary functions intersect.

Life Sciences

80. Insect Species An article in *Science* stated that the number of insect species of a given mass is proportional to $m^{-0.6}$, where m is the mass in grams. *Source: Science.* A graph accompanying the article shows the common logarithm of the mass on the horizontal axis and the common logarithm of the number of species on the vertical axis. Explain why the graph is a straight line. What is the slope of the line?

Index of Diversity For Exercises 81–83, refer to Example 10.

81. Suppose a sample of a small community shows two species with 50 individuals each.

a. Find the index of diversity H.

b. What is the maximum value of the index of diversity for two species?

c. Does your answer for part a equal $\ln 2$? Explain why.

82. A virgin forest in northwestern Pennsylvania has 4 species of large trees with the following proportions of each: hemlock, 0.521; beech, 0.324; birch, 0.081; maple, 0.074. Find the index of diversity H.

83. Find the value of the index of diversity for populations with n species and $1/n$ of each if

a. $n = 3$;   **b.** $n = 4$.

c. Verify that your answers for parts a and b equal $\ln 3$ and $\ln 4$, respectively.

84. Allometric Growth The allometric formula is used to describe a wide variety of growth patterns. It says that $y = nx^m$, where x and y are variables, and n and m are constants. For example, the famous biologist J. S. Huxley used this formula to relate the weight of the large claw of the fiddler crab to the weight of the body without the claw. *Source: Problems of Relative Growth.* Show that if x and y are given by the allometric formula, then

$X = \log_b x$, $Y = \log_b y$, and $N = \log_b n$ are related by the linear equation

$$Y = mX + N.$$

85. **Drug Concentration** When a pharmaceutical drug is injected into the bloodstream, its concentration at time t can be approximated by $C(t) = C_0 e^{-kt}$, where C_0 is the concentration at $t = 0$. Suppose the drug is ineffective below a concentration C_1 and harmful above a concentration C_2. Then it can be shown that the drug should be given at intervals of time T, where

$$T = \frac{1}{k} \ln \frac{C_2}{C_1}.$$

Source: Applications of Calculus to Medicine.

A certain drug is harmful at a concentration five times the concentration below which it is ineffective. At noon an injection of the drug results in a concentration of 2 mg per liter of blood. Three hours later the concentration is down to 1 mg per liter. How often should the drug be given?

The graph for Exercise 86 is plotted on a logarithmic scale where differences between successive measurements are not always the same. Data that do not plot in a linear pattern on the usual Cartesian axes often form a linear pattern when plotted on a logarithmic scale. Notice that on the horizontal scale, the distance from 5 to 10 is not the same as the distance from 10 to 15, and so on. This is characteristic of a graph drawn on logarithmic scales.

86. **Metabolism Rate** The accompanying graph shows the basal metabolism rate (in cm³ of oxygen per gram per hour) for marsupial carnivores, which include the Tasmanian devil. This rate is inversely proportional to body mass raised to the power 0.25. *Source: The Quarterly Review of Biology.*

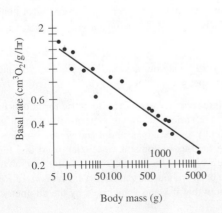

a. Estimate the metabolism rate for a marsupial carnivore with body mass of 10 g. Do the same for one with body mass of 1000 g.

b. Verify that if the relationship between x and y is of the form $y = ax^b$, then there will be a linear relationship between $\ln x$ and $\ln y$. (*Hint:* Apply ln to both sides of $y = ax^b$.)

c. If a function of the form $y = ax^b$ contains the points (x_1, y_1) and (x_2, y_2), then values for a and b can be found by dividing $y_1 = ax_1^b$ by $y_2 = ax_2^b$, solving the resulting equation for b, and putting the result back into either equation to solve for a. Use this procedure and the results from part a to find an equation of the form $y = ax^b$ that gives the basal metabolism rate as a function of body mass.

d. Use the result of part c to predict the basal metabolism rate of a marsupial carnivore whose body mass is 100 g.

87. **Minority Population** The U.S. Census Bureau has reported that the United States is becoming more diverse. In Exercise 49 of the previous section, the projected Hispanic population (in millions) was modeled by the exponential function

$$h(t) = 37.79 \, (1.021)^t$$

where $t = 0$ corresponds to 2000 and $0 \le t \le 50$. *Source: U.S. Census Bureau.*

a. Estimate in what year the Hispanic population will double the 2005 population of 42.69 million. Use the algebra of logarithms to solve this problem.

b. The projected U.S. Asian population (in millions) was modeled by the exponential function

$$h(t) = 11.14 \, (1.023)^t,$$

where $t = 0$ corresponds to 2000 and $0 \le t \le 50$. Estimate in what year the Asian population will double the 2005 population of 12.69 million.

Social Sciences

88. **Evolution of Languages** The number of years $N(r)$ since two independently evolving languages split off from a common ancestral language is approximated by

$$N(r) = -5000 \, \ln r,$$

where r is the proportion of the words from the ancestral language that are common to both languages now. Find the following.

a. $N(0.9)$ b. $N(0.5)$ c. $N(0.3)$

d. How many years have elapsed since the split if 70% of the words of the ancestral language are common to both languages today?

e. If two languages split off from a common ancestral language about 1000 years ago, find r.

Physical Sciences

89. Communications Channel According to the Shannon-Hartley theorem, the capacity of a communications channel in bits per second is given by

$$C = B \log_2\left(\frac{s}{n} + 1\right),$$

where B is the frequency bandwidth of the channel in hertz and s/n is its signal-to-noise ratio. **Source: Scientific American.** It is physically impossible to exceed this limit. Solve the equation for the signal-to-noise ratio s/n.

For Exercises 90–93, recall that log x represents the common (base 10) logarithm of x.

90. Intensity of Sound The loudness of sounds is measured in a unit called a *decibel*. To do this, a very faint sound, called the *threshold sound*, is assigned an intensity I_0. If a particular sound has intensity I, then the decibel rating of this louder sound is

$$10 \log \frac{I}{I_0}.$$

Find the decibel ratings of the following sounds having intensities as given. Round answers to the nearest whole number.

a. Whisper, $115I_0$

b. Busy street, $9{,}500{,}000I_0$

c. Heavy truck, 20 m away, $1{,}200{,}000{,}000I_0$

d. Rock music concert, $895{,}000{,}000{,}000I_0$

e. Jetliner at takeoff, $109{,}000{,}000{,}000{,}000I_0$

f. In a noise ordinance instituted in Stamford, Connecticut, the threshold sound I_0 was defined as 0.0002 microbars. **Source: The New York Times.** Use this definition to express the sound levels in parts c and d in microbars.

91. Intensity of Sound A story on the National Public Radio program *All Things Considered*, discussed a proposal to lower the noise limit in Austin, Texas, from 85 decibels to 75 decibels. A manager for a restaurant was quoted as saying, "If you cut from 85 to 75, . . . you're basically cutting the sound down in half." Is this correct? If not, to what fraction of its original level is the sound being cut? **Source: National Public Radio.**

92. Earthquake Intensity The magnitude of an earthquake, measured on the Richter scale, is given by

$$R(I) = \log \frac{I}{I_0},$$

where I is the amplitude registered on a seismograph located 100 km from the epicenter of the earthquake, and I_0 is the amplitude of a certain small size earthquake. Find the Richter scale ratings of earthquakes with the following amplitudes.

a. $1{,}000{,}000I_0$ **b.** $100{,}000{,}000I_0$

c. On June 15, 1999, the city of Puebla in central Mexico was shaken by an earthquake that measured 6.7 on the Richter scale. Express this reading in terms of I_0. **Source: Exploring Colonial Mexico.**

d. On September 19, 1985, Mexico's largest recent earthquake, measuring 8.1 on the Richter scale, killed about 10,000 people. Express the magnitude of an 8.1 reading in terms of I_0. **Source: History.com.**

e. Compare your answers to parts c and d. How much greater was the force of the 1985 earthquake than the 1999 earthquake?

f. The relationship between the energy E of an earthquake and the magnitude on the Richter scale is given by

$$R(E) = \frac{2}{3} \log\left(\frac{E}{E_0}\right),$$

where E_0 is the energy of a certain small earthquake. Compare the energies of the 1999 and 1985 earthquakes.

g. According to a newspaper article, "Scientists say such an earthquake of magnitude 7.5 could release 15 times as much energy as the magnitude 6.7 trembler that struck the Northridge section of Los Angeles" in 1994. **Source: The New York Times.** Using the formula from part f, verify this quote by computing the magnitude of an earthquake with 15 times the energy of a magnitude 6.7 earthquake.

93. Acidity of a Solution A common measure for the acidity of a solution is its pH. It is defined by pH $= -\log[H^+]$, where H^+ measures the concentration of hydrogen ions in the solution. The pH of pure water is 7. Solutions that are more acidic than pure water have a lower pH, while solutions that are less acidic (referred to as basic solutions) have a higher pH.

a. Acid rain sometimes has a pH as low as 4. How much greater is the concentration of hydrogen ions in such rain than in pure water?

b. A typical mixture of laundry soap and water for washing clothes has a pH of about 11, while black coffee has a pH of about 5. How much greater is the concentration of hydrogen ions in black coffee than in the laundry mixture?

94. Music Theory A music theorist associates the fundamental frequency of a pitch f with a real number defined by

$$p = 69 + 12 \log_2 (f/440).$$

Source: Science.

a. Standard concert pitch for an A is 440 cycles per second. Find the associated value of p.

b. An A one octave higher than standard concert pitch is 880 cycles per second. Find the associated value of p.

▰**YOUR TURN ANSWERS**

1. $\log_5 (1/25) = -2$ **2.** -4

3. $2 \log_a x - 3 \log_a y$ **4.** 3.561

5. 2   **6.** $(\ln 2)/\ln(3/2) \approx 1.7095$

7. 1.0253^x

2.6 Applications: Growth and Decay; Mathematics of Finance

APPLY IT What interest rate will cause $5000 to grow to $7250 in 6 years if money is compounded continuously?

This question, which will be answered in Example 7, is one of many situations that occur in biology, economics, and the social sciences, in which a quantity changes at a rate proportional to the amount of the quantity present.

In cases such as continuous compounding described above, the amount present at time *t* is a function of *t*, called the **exponential growth and decay function**. (The derivation of this equation is presented in a later section on Differential Equations.)

> **Exponential Growth and Decay Function**
>
> Let y_0 be the amount or number of some quantity present at time $t = 0$. The quantity is said to grow or decay exponentially if for some constant k, the amount present at time t is given by
>
> $$y = y_0 e^{kt}.$$

If $k > 0$, then k is called the **growth constant**; if $k < 0$, then k is called the **decay constant**. A common example is the growth of bacteria in a culture. The more bacteria present, the faster the population increases.

EXAMPLE 1 Yeast Production

Yeast in a sugar solution is growing at a rate such that 1 g becomes 1.5 g after 20 hours. Find the growth function, assuming exponential growth.

SOLUTION The values of y_0 and k in the exponential growth function $y = y_0 e^{kt}$ must be found. Since y_0 is the amount present at time $t = 0$, $y_0 = 1$. To find k, substitute $y = 1.5$, $t = 20$, and $y_0 = 1$ into the equation.

$$y = y_0 e^{kt}$$
$$1.5 = 1e^{k(20)}$$

Now take natural logarithms on both sides and use the power rule for logarithms and the fact that $\ln e = 1$.

$$1.5 = e^{20k}$$
$$\ln 1.5 = \ln e^{20k} \qquad \text{Take ln of both sides.}$$
$$\ln 1.5 = 20k \qquad \ln e^x = x$$
$$\frac{\ln 1.5}{20} = k \qquad \text{Divide both sides by 20.}$$
$$k \approx 0.02 \text{ (to the nearest hundredth)}$$

YOUR TURN 1 Find the growth function if 5 g grows exponentially to 18 g after 16 hours.

The exponential growth function is $y = e^{0.02t}$, where y is the number of grams of yeast present after t hours. **TRY YOUR TURN 1**

The decline of a population or decay of a substance may also be described by the exponential growth function. In this case the decay constant k is negative, since an increase in time leads to a decrease in the quantity present. Radioactive substances provide a good example of exponential decay. By definition, the **half-life** of a radioactive substance is the time it takes for exactly half of the initial quantity to decay.

EXAMPLE 2 Carbon Dating

Carbon-14 is a radioactive form of carbon that is found in all living plants and animals. After a plant or animal dies, the carbon-14 disintegrates. Scientists determine the age of the remains by comparing its carbon-14 with the amount found in living plants and animals. The amount of carbon-14 present after t years is given by the exponential equation

$$A(t) = A_0 e^{kt},$$

with $k = -[(\ln 2)/5600]$.

(a) Find the half-life of carbon-14.

SOLUTION Let $A(t) = (1/2)A_0$ and $k = -[(\ln 2)/5600]$.

$$\frac{1}{2}A_0 = A_0 e^{-[(\ln 2)/5600]t}$$

$$\frac{1}{2} = e^{-[(\ln 2)/5600]t} \qquad \text{Divide by } A_0.$$

$$\ln\frac{1}{2} = \ln e^{-[(\ln 2)/5600]t} \qquad \text{Take ln of both sides.}$$

$$\ln\frac{1}{2} = -\frac{\ln 2}{5600}t \qquad \ln e^x = x$$

$$-\frac{5600}{\ln 2}\ln\frac{1}{2} = t \qquad \text{Multiply by } -\frac{5600}{\ln 2}.$$

$$-\frac{5600}{\ln 2}(\ln 1 - \ln 2) = t \qquad \ln\frac{x}{y} = \ln x - \ln y$$

$$-\frac{5600}{\ln 2}(-\ln 2) = t \qquad \ln 1 = 0$$

$$5600 = t$$

The half-life is 5600 years.

(b) Charcoal from an ancient fire pit on Java had 1/4 the amount of carbon-14 found in a living sample of wood of the same size. Estimate the age of the charcoal.

SOLUTION Let $A(t) = (1/4)A_0$ and $k = -[(\ln 2)/5600]$.

$$\frac{1}{4}A_0 = A_0 e^{-[(\ln 2)/5600]t}$$

$$\frac{1}{4} = e^{-[(\ln 2)/5600]t}$$

$$\ln\frac{1}{4} = \ln e^{-[(\ln 2)/5600]t}$$

$$\ln\frac{1}{4} = -\frac{\ln 2}{5600}t$$

$$-\frac{5600}{\ln 2}\ln\frac{1}{4} = t$$

$$t = 11{,}200$$

YOUR TURN 2 Estimate the age of a sample with 1/10 the amount of carbon-14 as a live sample.

The charcoal is about 11,200 years old.

TRY YOUR TURN 2

By following the steps in Example 2, we get the general equation giving the half-life T in terms of the decay constant k as

$$T = -\frac{\ln 2}{k}.$$

For example, the decay constant for potassium-40, where t is in billions of years, is approximately -0.5545 so its half-life is

$$T = -\frac{\ln 2}{(-0.5545)}$$
$$\approx 1.25 \text{ billion years.}$$

We can rewrite the growth and decay function as

$$y = y_0 e^{kt} = y_0(e^k)^t = y_0 a^t,$$

where $a = e^k$. This is sometimes a helpful way to look at an exponential growth or decay function.

EXAMPLE 3 Radioactive Decay

Rewrite the function for radioactive decay of carbon-14 in the form $A(t) = A_0 a^{f(t)}$.

SOLUTION   From the previous example, we have

$$A(t) = A_0 e^{kt} = A_0 e^{-[(\ln 2)/5600]t}$$
$$= A_0 (e^{\ln 2})^{-t/5600}$$
$$= A_0 2^{-t/5600} = A_0 (2^{-1})^{t/5600} = A_0 \left(\frac{1}{2}\right)^{t/5600}.$$

This last expression shows clearly that every time t increases by 5600 years, the amount of carbon-14 decreases by a factor of $1/2$.

Effective Rate

We could use a calculator to see that $1 at 8% interest (per year) compounded semiannually is $1(1.04)^2 = 1.0816$ or $1.0816. The actual increase of $0.0816 is 8.16% rather than the 8% that would be earned with interest compounded annually. To distinguish between these two amounts, 8% (the annual interest rate) is called the **nominal** or **stated** interest rate, and 8.16% is called the **effective** interest rate. We will continue to use r to designate the stated rate and we will use r_E for the effective rate.

Effective Rate for Compound Interest
If r is the annual stated rate of interest and m is the number of compounding periods per year, the effective rate of interest is

$$r_E = \left(1 + \frac{r}{m}\right)^m - 1.$$

Effective rate is sometimes called *annual yield*.

With continuous compounding, $1 at 8% for 1 year becomes $(1)e^{1(0.08)} = e^{0.08} \approx 1.0833$. The increase is 8.33% rather than 8%, so a stated interest rate of 8% produces an effective rate of 8.33%.

Effective Rate for Continuous Compounding
If interest is compounded continuously at an annual stated rate of r, the effective rate of interest is

$$r_E = e^r - 1.$$

EXAMPLE 4 Effective Rate

Find the effective rate corresponding to each stated rate.

(a) 6% compounded quarterly

SOLUTION Using the formula, we get

$$\left(1 + \frac{0.06}{4}\right)^4 - 1 = (1.015)^4 - 1 \approx 0.0614.$$

The effective rate is 6.14%.

YOUR TURN 3 Find the effective rate corresponding to each stated rate. **(a)** 4.25% compounded monthly **(b)** 3.75% compounded continuously.

(b) 6% compounded continuously

SOLUTION The formula for continuous compounding gives

$$e^{0.06} - 1 \approx 0.0618,$$

so the effective rate is 6.18%. **TRY YOUR TURN 3**

The formula for interest compounded m times a year, $A = P(1 + r/m)^{tm}$, has five variables: A, P, r, m, and t. If the values of any four are known, then the value of the fifth can be found.

EXAMPLE 5 Interest

Meghan Moreau has received a bonus of $25,000. She invests it in an account earning 7.2% compounded quarterly. Find how long it will take for her $25,000 investment to grow to $40,000.

SOLUTION Here $P = \$25,000$, $r = 0.072$, and $m = 4$. We also know the amount she wishes to end up with, $A = \$40,000$. Substitute these values into the compound interest formula and solve for time, t.

$$40,000 = 25,000\left(1 + \frac{0.072}{4}\right)^{4t}$$

$$40,000 = 25,000(1.018)^{4t}$$

$$1.6 = 1.018^{4t} \qquad \text{Divide both sides by 25,000.}$$

$$\ln 1.6 = \ln(1.018)^{4t} \qquad \text{Take ln of both sides.}$$

$$\ln 1.6 = 4t \cdot \ln 1.018$$

YOUR TURN 4 Find the time needed for $30,000 to grow to $50,000 when invested in an account that pays 3.15% compounded quarterly.

$$t = \frac{\ln 1.6}{4 \ln 1.018} \approx 6.586 \qquad \text{Divide both sides by 4 ln 1.018.}$$

Note that the interest is calculated quarterly and is added only at the *end* of each quarter. Therefore, we need to round up to the nearest quarter. She will have $40,000 in 6.75 years. **TRY YOUR TURN 4**

> **CAUTION** When calculating the time it takes for an investment to grow, take into account that interest is added only at the *end* of each compounding period. In Example 5, interest is added quarterly. At the end of the second quarter of the sixth year ($t = 6.5$), she will have only $39,754.13, but at the end of the third quarter of that year ($t = 6.75$), she will have $40,469.70.

If A, the amount of money we wish to end up with, is given as well as r, m, and t, then P can be found using the formula for compounded interest. Here P is the amount that should be deposited today to produce A dollars in t years. The amount P is called the **present value** of A dollars.

EXAMPLE 6 Present Value

Tom Shaffer has a balloon payment of $100,000 due in 3 years. What is the present value of that amount if the money earns interest at 4% annually?

SOLUTION Here P in the compound interest formula is unknown, with $A = 100,000$, $r = 0.04$, $t = 3$, and $m = 1$. Substitute the known values into the formula to get $100,000 = P(1.04)^3$. Solve for P, using a calculator to find $(1.04)^3$.

$$P = \frac{100,000}{(1.04)^3} \approx 88,889.64$$

The present value of $100,000 in 3 years at 4% per year is $88,889.64.

In general, to find the present value for an interest rate r compounded m times per year for t years, solve the equation

$$A = P\left(1 + \frac{r}{m}\right)^{tm}$$

for the variable P. To find the present value for an interest rate r compounded continuously for t years, solve the equation

$$A = Pe^{rt}$$

for the variable P.

EXAMPLE 7 Continuous Compound Interest

Find the interest rate that will cause $5000 to grow to $7250 in 6 years if the money is compounded continuously.

APPLY IT

SOLUTION Use the formula for continuous compounding, $A = Pe^{rt}$, with $A = 7250$, $P = 5000$, and $t = 6$. Solve first for e^{rt}, then for r.

$$A = Pe^{rt}$$
$$7250 = 5000e^{6r}$$
$$1.45 = e^{6r} \qquad \text{Divide by 5000.}$$
$$\ln 1.45 = \ln e^{6r} \qquad \text{Take ln of both sides.}$$
$$\ln 1.45 = 6r \qquad \ln e^x = x$$
$$r = \frac{\ln 1.45}{6}$$
$$r \approx 0.0619$$

YOUR TURN 5 Find the interest rate that will cause $3200 to grow to $4500 in 7 years if the money is compounded continuously.

The required interest rate is 6.19%. **TRY YOUR TURN 5**

Limited Growth Functions
The exponential growth functions discussed so far all continued to grow without bound. More realistically, many populations grow exponentially for a while, but then the growth is slowed by some external constraint that eventually limits the growth. For example, an animal population may grow to the point where its habitat can no longer support the population and the growth rate begins to dwindle until a stable population size is reached. Models that reflect this pattern are called **limited growth functions.** The next example discusses a function of this type that occurs in industry.

EXAMPLE 8 Employee Turnover

Assembly-line operations tend to have a high turnover of employees, forcing companies to spend much time and effort in training new workers. It has been found that a worker new to a task on the line will produce items according to the function defined by

$$P(x) = 25 - 25e^{-0.3x},$$

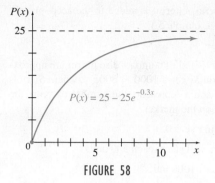

FIGURE 58

where $P(x)$ items are produced by the worker on day x.

(a) What happens to the number of items a worker can produce as x gets larger and larger?

SOLUTION As x gets larger, $e^{-0.3x}$ becomes closer to 0, so $P(x)$ approaches 25. This represents the limit on the number of items a worker can produce in a day. Note that this limit represents a horizontal asymptote on the graph of P, shown in Figure 58.

(b) How many days will it take for a new worker to produce at least 20 items in a day?

SOLUTION Let $P(x) = 20$ and solve for x.

$$P(x) = 25 - 25e^{-0.3x}$$
$$20 = 25 - 25e^{-0.3x}$$
$$-5 = -25e^{-0.3x}$$
$$0.2 = e^{-0.3x}$$

Now take natural logarithms of both sides and use properties of logarithms.

$$\ln 0.2 = \ln e^{-0.3x}$$
$$\ln 0.2 = -0.3x \qquad \ln e^u = u$$
$$x = \frac{\ln 0.2}{-0.3} \approx 5.4$$

This means that 5 days are not quite enough; on the fifth day, a new worker produces $P(5) = 25 - 25e^{-0.3(5)} \approx 19.4$ items. It takes 6 days, and on the sixth day, a new worker produces $P(6) = 25 - 25e^{-0.3(5)} \approx 20.9$ items.

Graphs such as the one in Figure 58 are called **learning curves**. According to such a graph, a new worker tends to learn quickly at first; then learning tapers off and approaches some upper limit. This is characteristic of the learning of certain types of skills involving the repetitive performance of the same task.

2.6 EXERCISES

 1. What is the difference between stated interest rate and effective interest rate?

2. In the exponential growth or decay function $y = y_0 e^{kt}$, what does y_0 represent? What does k represent?

 3. In the exponential growth or decay function, explain the circumstances that cause k to be positive or negative.

 4. What is meant by the half-life of a quantity?

5. Show that if a radioactive substance has a half-life of T, then the corresponding constant k in the exponential decay function is given by $k = -(\ln 2)/T$.

6. Show that if a radioactive substance has a half-life of T, then the corresponding exponential decay function can be written as $y = y_0(1/2)^{(t/T)}$.

APPLICATIONS

Business and Economics

Effective Rate Find the effective rate corresponding to each nominal rate of interest.

7. 4% compounded quarterly

8. 6% compounded monthly

9. 8% compounded continuously

10. 5% compounded continuously

Present Value Find the present value of each amount.

11. $10,000 if interest is 6% compounded quarterly for 8 years

12. $45,678.93 if interest is 7.2% compounded monthly for 11 months

13. $7300 if interest is 5% compounded continuously for 3 years

14. $25,000 if interest is 4.6% compounded continuously for 8 years

15. **Effective Rate** Tami Dreyfus bought a television set with money borrowed from the bank at 9% interest compounded semiannually. What effective interest rate did she pay?

16. **Effective Rate** A firm deposits some funds in a special account at 6.2% compounded quarterly. What effective rate will they earn?

17. **Effective Rate** Robin Kim deposits $7500 of lottery winnings in an account paying 6% interest compounded monthly. What effective rate does the account earn?

18. **Present Value** Frank Steed must make a balloon payment of $20,000 in 4 years. Find the present value of the payment if it includes annual interest of 6.5% compounded monthly.

19. **Present Value** A company must pay a $307,000 settlement in 3 years.

    a. What amount must be deposited now at 6% compounded semiannually to have enough money for the settlement?

    b. How much interest will be earned?

    c. Suppose the company can deposit only $200,000 now. How much more will be needed in 3 years?

    d. Suppose the company can deposit $200,000 now in an account that pays interest continuously. What interest rate would they need to accumulate the entire $307,000 in 3 years?

20. **Present Value** A couple wants to have $40,000 in 5 years for a down payment on a new house.

    a. How much should they deposit today, at 6.4% compounded quarterly, to have the required amount in 5 years?

    b. How much interest will be earned?

    c. If they can deposit only $20,000 now, how much more will they need to complete the $40,000 after 5 years?

    d. Suppose they can deposit $20,000 now in an account that pays interest continuously. What interest rate would they need to accumulate the entire $40,000 in 5 years?

21. **Interest** Christine O'Brien, who is self-employed, wants to invest $60,000 in a pension plan. One investment offers 8% compounded quarterly. Another offers 7.75% compounded continuously.

    a. Which investment will earn the most interest in 5 years?

    b. How much more will the better plan earn?

    c. What is the effective rate in each case?

    d. If Ms. O'Brien chooses the plan with continuous compounding, how long will it take for her $60,000 to grow to $80,000?

    e. How long will it take for her $60,000 to grow to at least $80,000 if she chooses the plan with quarterly compounding? (Be careful; interest is added to the account only every quarter. See Example 5.)

22. **Interest** Greg Tobin wishes to invest a $5000 bonus check into a savings account that pays 6.3% interest. Find how many years it will take for the $5000 to grow to at least $11,000 if interest is compounded

a. quarterly. (Be careful; interest is added to the account only every quarter. See Example 5.)

b. continuously.

23. **Sales** Sales of a new model of compact disc player are approximated by the function $S(x) = 1000 - 800e^{-x}$, where $S(x)$ is in appropriate units and x represents the number of years the disc player has been on the market.

    a. Find the sales during year 0.

    b. In how many years will sales reach 500 units?

    c. Will sales ever reach 1000 units?

    d. Is there a limit on sales for this product? If so, what is it?

24. **Sales** Sales of a new model of digital camera are approximated by

    $$S(x) = 5000 - 4000e^{-x},$$

    where x represents the number of years that the digital camera has been on the market, and $S(x)$ represents sales in thousands of dollars.

    a. Find the sales in year 0.

    b. When will sales reach $4,500,000?

    c. Find the limit on sales.

Life Sciences

25. **Population Growth** The population of the world in the year 1650 was about 500 million, and in the year 2010 was 6756 million. *Source: U.S. Census Bureau.*

    a. Assuming that the population of the world grows exponentially, find the equation for the population $P(t)$ in millions in the year t.

    b. Use your answer from part a to find the population of the world in the year 1.

    c. Is your answer to part b reasonable? What does this tell you about how the population of the world grows?

26. **Giardia** When a person swallows giardia cysts, stomach acids and pancreatic enzymes cause the cysts to release trophozoites, which divide every 12 hours. *Source: The New York Times.*

    a. Suppose the number of trophozoites at time $t = 0$ is y_0. Write a function in the form $y = y_0 e^{kt}$ giving the number after t hours.

    b. Write the function from part a in the form $y = y_0 2^{f(t)}$.

    c. The article cited above said that a single trophozoite can multiply to a million in just 10 days and a billion in 15 days. Verify this fact.

27. **Growth of Bacteria** A culture contains 25,000 bacteria, with the population increasing exponentially. The culture contains 40,000 bacteria after 10 hours.

    a. Write a function in the form $y = y_0 e^{kt}$ giving the number of bacteria after t hours.

    b. Write the function from part a in the form $y = y_0 a^t$.

    c. How long will it be until there are 60,000 bacteria?

28. **Decrease in Bacteria** When an antibiotic is introduced into a culture of 50,000 bacteria, the number of bacteria decreases exponentially. After 9 hours, there are only 20,000 bacteria.

a. Write an exponential equation to express the growth function y in terms of time t in hours.

b. In how many hours will half the number of bacteria remain?

29. Growth of Bacteria The growth of bacteria in food products makes it necessary to time-date some products (such as milk) so that they will be sold and consumed before the bacteria count is too high. Suppose for a certain product that the number of bacteria present is given by

$$f(t) = 500e^{0.1t},$$

under certain storage conditions, where t is time in days after packing of the product and the value of $f(t)$ is in millions.

a. If the product cannot be safely eaten after the bacteria count reaches 3000 million, how long will this take?

b. If $t = 0$ corresponds to January 1, what date should be placed on the product?

30. Cancer Research An article on cancer treatment contains the following statement: A 37% 5-year survival rate for women with ovarian cancer yields an estimated annual mortality rate of 0.1989. The authors of this article assume that the number of survivors is described by the exponential decay function given at the beginning of this section, where y is the number of survivors and k is the mortality rate. Verify that the given survival rate leads to the given mortality rate. *Source: American Journal of Obstetrics and Gynecology.*

31. Chromosomal Abnormality The graph below shows how the risk of chromosomal abnormality in a child rises with the age of the mother. *Source: downsyndrome.about.com.*

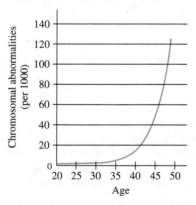

a. Read from the graph the risk of chromosomal abnormality (per 1000) at ages 20, 35, 42, and 49.

b. Assuming the graph to be of the form $y = Ce^{kt}$, find k using $t = 20$ and $t = 35$.

c. Still assuming the graph to be of the form $y = Ce^{kt}$, find k using $t = 42$ and $t = 49$.

d. Based on your results from parts a–c, is it reasonable to assume the graph is of the form $y = Ce^{kt}$? Explain.

e. In situations such as parts a–c, where an exponential function does not fit because different data points give different values for the growth constant k, it is often appropriate to describe the data using an equation of the form $y = Ce^{kt^n}$. Parts b and c show that $n = 1$ results in a smaller constant using the interval $[20, 35]$ than using the interval $[42, 49]$.

Repeat parts b and c using $n = 2, 3$, etc., until the interval $[20, 35]$ yields a larger value of k than the interval $[42, 49]$, and then estimate what n should be.

Physical Sciences

32. Carbon Dating Refer to Example 2. A sample from a refuse deposit near the Strait of Magellan had 60% of the carbon-14 found in a contemporary living sample. How old was the sample?

Half-Life Find the half-life of each radioactive substance. See Example 2.

33. Plutonium-241; $A(t) = A_0 e^{-0.053t}$

34. Radium-226; $A(t) = A_0 e^{-0.00043t}$

35. Half-Life The half-life of plutonium-241 is approximately 13 years.

a. How much of a sample weighing 4 g will remain after 100 years?

b. How much time is necessary for a sample weighing 4 g to decay to 0.1 g?

36. Half-Life The half-life of radium-226 is approximately 1620 years.

a. How much of a sample weighing 4 g will remain after 100 years?

b. How much time is necessary for a sample weighing 4 g to decay to 0.1 g?

37. Radioactive Decay 500 g of iodine-131 is decaying exponentially. After 3 days 386 g of iodine-131 is left.

a. Write a function in the form $y = y_0 e^{kt}$ giving the number of grams of iodine-131 after t days.

b. Write the function from part a in the form $y = y_0(386/500)^{f(t)}$.

c. Use your answer from part a to find the half-life of iodine-131.

38. Radioactive Decay 25 g of polonium-210 is decaying exponentially. After 50 days 19.5 g of polonium-210 is left.

a. Write a function in the form $y = y_0 e^{kt}$ giving the number of grams of polonium-210 after t days.

b. Write the function from part a in the form $y = y_0 a^{t/50}$.

c. Use your answer from part a to find the half-life of polonium-210.

39. Nuclear Energy Nuclear energy derived from radioactive isotopes can be used to supply power to space vehicles. The output of the radioactive power supply for a certain satellite is given by the function $y = 40e^{-0.004t}$, where y is in watts and t is the time in days.

a. How much power will be available at the end of 180 days?

b. How long will it take for the amount of power to be half of its original strength?

c. Will the power ever be completely gone? Explain.

40. Botany A group of Tasmanian botanists have claimed that a King's holly shrub, the only one of its species in the world, is also the oldest living plant. Using carbon-14 dating of charcoal found along with fossilized leaf fragments, they arrived at an age of 43,000 years for the plant, whose exact location in southwest Tasmania is being kept a secret. What percent of the original carbon-14 in the charcoal was present? *Source: Science.*

41. **Decay of Radioactivity** A large cloud of radioactive debris from a nuclear explosion has floated over the Pacific Northwest, contaminating much of the hay supply. Consequently, farmers in the area are concerned that the cows who eat this hay will give contaminated milk. (The tolerance level for radioactive iodine in milk is 0.) The percent of the initial amount of radioactive iodine still present in the hay after t days is approximated by $P(t)$, which is given by the mathematical model

$$P(t) = 100e^{-0.1t}.$$

   a. Find the percent remaining after 4 days.

   b. Find the percent remaining after 10 days.

   c. Some scientists feel that the hay is safe after the percent of radioactive iodine has declined to 10% of the original amount. Solve the equation $10 = 100e^{-0.1t}$ to find the number of days before the hay may be used.

   d. Other scientists believe that the hay is not safe until the level of radioactive iodine has declined to only 1% of the original level. Find the number of days that this would take.

42. **Chemical Dissolution** The amount of chemical that will dissolve in a solution increases exponentially as the temperature is increased. At 0°C, 10 g of the chemical dissolves, and at 10°C, 11 g dissolves.

   a. Write an equation to express the amount of chemical dissolved, y, in terms of temperature, t, in degrees Celsius.

   b. At what temperature will 15 g dissolve?

Newton's Law of Cooling Newton's law of cooling says that the rate at which a body cools is proportional to the difference in temperature between the body and an environment into which it is introduced. This leads to an equation where the temperature $f(t)$ of the body at time t after being introduced into an environment having constant temperature T_0 is

$$f(t) = T_0 + Ce^{-kt},$$

where C and k are constants. Use this result in Exercises 43–45.

43. Find the temperature of an object when $t = 9$ if $T_0 = 18$, $C = 5$, and $k = 0.6$.

44. If $C = 100$, $k = 0.1$, and t is time in minutes, how long will it take a hot cup of coffee to cool to a temperature of 25°C in a room at 20°C?

45. If $C = -14.6$ and $k = 0.6$ and t is time in hours, how long will it take a frozen pizza to thaw to 10°C in a room at 18°C?

YOUR TURN ANSWERS

1. $y = 5e^{0.08t}$ 2. 18,600 years old

3. **(a)** 4.33%   **(b)** 3.82%   4. 16.5 years

5. 4.87%

2 CHAPTER REVIEW

SUMMARY

In this chapter we defined functions and studied some of their properties. In particular, we studied several families of functions including quadratic, polynomial, rational, exponential, and logarithmic functions. By knowing the properties of a family of functions, we can immediately apply that knowledge to any member of the family we encounter, giving us valuable information about the domain and the behavior of the function. Furthermore, this knowledge can help us to choose an appropriate function for an application. Exponential functions have so many important applications that we highlighted some of them in the last section of the chapter. In the next chapters, we see how calculus gives us even more information about the behavior of functions.

Function	A function is a rule that assigns to each element from one set exactly one element from another set.
Domain and Range	The set of all possible values of the independent variable in a function is called the domain of the function, and the resulting set of possible values of the dependent variable is called the range.
Vertical Line Test	A graph represents a function if and only if every vertical line intersects the graph in no more than one point.
Quadratic Function	A quadratic function is defined by $$f(x) = ax^2 + bx + c,$$ where a, b, and c are real numbers, with $a \neq 0$.
Graph of a Quadratic Function	The graph of the quadratic function $f(x) = ax^2 + bx + c$ is a parabola with its vertex at $$\left(\frac{-b}{2a}, f\left(\frac{-b}{2a}\right)\right).$$ The graph opens upward if $a > 0$ and downward if $a < 0$.

Polynomial Function A polynomial function of degree n, where n is a nonnegative integer, is defined by

$$f(x) = a_n x^n + a_{n-1} x^{n-1} + \cdots + a_1 x + a_0,$$

where $a_n, a_{n-1}, \ldots, a_1$ and a_0 are real numbers, called coefficients, with $a_n \neq 0$. The number a_n is called the leading coefficient.

Properties of Polynomial Functions

1. A polynomial function of degree n can have at most $n - 1$ turning points. Conversely, if the graph of a polynomial function has n turning points, it must have degree at least $n + 1$.
2. In the graph of a polynomial function of even degree, both ends go up or both ends go down. For a polynomial function of odd degree, one end goes up and one end goes down.
3. If the graph goes up as x becomes large, the leading coefficient must be positive. If the graph goes down as x becomes large, the leading coefficient is negative.

Rational Function A rational function is defined by

$$f(x) = \frac{p(x)}{q(x)},$$

where $p(x)$ and $q(x)$ are polynomial functions and $q(x) \neq 0$.

Asymptotes If a function gets larger and larger in magnitude without bound as x approaches the number k, then the line $x = k$ is a vertical asymptote.

If the values of y approach a number k as $|x|$ gets larger and larger, the line $y = k$ is a horizontal asymptote.

Exponential Function An exponential function with base a is defined as
$$f(x) = a^x, \text{ where } a > 0 \text{ and } a \neq 1.$$

Simple Interest If P dollars is invested at a yearly simple interest rate r per year for time t (in years), the interest I is given by
$$I = Prt.$$

Math of Finance Formulas If P is the principal or present value, r is the annual interest rate, t is time in years, and m is the number of compounding periods per year:

	Compounded m Times per Year	Compounded Continuously
Compound amount	$A = P\left(1 + \dfrac{r}{m}\right)^{tm}$	$A = Pe^{rt}$
Effective rate	$r_E = \left(1 + \dfrac{r}{m}\right)^{m} - 1$	$r_E = e^r - 1$

Definition of e As m becomes larger and larger, $\left(1 + \dfrac{1}{m}\right)^m$ becomes closer and closer to the number e, whose approximate value is 2.718281828.

Logarithm For $a > 0$, $a \neq 1$, and $x > 0$,
$$y = \log_a x \text{ means } a^y = x.$$

Logarithmic Function If $a > 0$ and $a \neq 1$, then the logarithmic function of base a is defined by
$$f(x) = \log_a x,$$
for $x > 0$.

Properties of Logarithms Let x and y be any positive real numbers and r be any real number. Let a be a positive real number, $a \neq 1$. Then

a. $\log_a xy = \log_a x + \log_a y$
b. $\log_a \dfrac{x}{y} = \log_a x - \log_a y$
c. $\log_a x^r = r \log_a x$
d. $\log_a a = 1$
e. $\log_a 1 = 0$
f. $\log_a a^r = r.$

Change-of-Base Theorem for Logarithms   If x is any positive number and if a and b are positive real numbers, $a \neq 1, b \neq 1$, then

$$\log_a x = \frac{\log_b x}{\log_b a} = \frac{\ln x}{\ln a}.$$

Change-of-Base Theorem for Exponentials   For every positive real number a

$$a^x = e^{(\ln a)x}.$$

Exponential Growth and Decay Function   Let y_0 be the amount or number of some quantity present at time $t = 0$. The quantity is said to grow or decay exponentially if, for some constant k, the amount present at any time t is given by

$$y = y_0 e^{kt}.$$

Graphs of Basic Functions

Quadratic

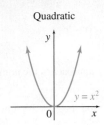

$y = x^2$

Absolute Value

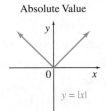

$y = |x|$

Square Root

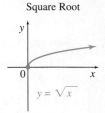

$y = \sqrt{x}$

Rational

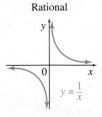

$y = \frac{1}{x}$

Exponential

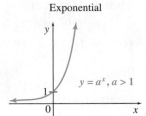

$y = a^x, a > 1$

Logarithmic

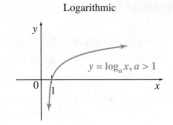

$y = \log_a x, a > 1$

KEY TERMS

To understand the concepts presented in this chapter, you should know the meaning and use of the following terms.
For easy reference, the section in the chapter where a word (or expression) was first used is provided.

2.1
function
domain
range
constant function
vertical line test
even function
odd function
step function

2.2
quadratic function
parabola
vertex
axis
vertical translation
vertical reflection
horizontal translation
completing the square
horizontal reflection

2.3
polynomial function
degree
coefficient
leading coefficient
power function
cubic polynomial
quartic polynomial
turning point
real zero
rational function
vertical asymptote
horizontal asymptote
cost–benefit model
average cost

2.4
exponential function
exponential equation
interest

principal
rate of interest
time
simple interest
compound interest
compound amount
e
continuous compounding

2.5
doubling time
logarithm
logarithmic function
inverse function
properties of logarithms
common logarithms
natural logarithms
change-of-base theorem for
  logarithms
logarithmic equation

change-of-base theorem for
  exponentials
rule of 70
rule of 72
index of diversity

2.6
exponential growth and decay
  function
growth constant
decay constant
half-life
nominal (stated) rate
effective rate
present value
limited growth function
learning curve

REVIEW EXERCISES

CONCEPT CHECK

Determine whether each of the following statements is true or false, and explain why.

1. A linear function is an example of a polynomial function.

2. A rational function is an example of an exponential function.

3. The function $f(x) = 3x^2 - 75x + 2$ is a quadratic function.

4. The function $f(x) = x^2 - 6x + 4$ has a vertex at $x = 3$.

5. The function $g(x) = x^\pi$ is an exponential function.

6. The function $f(x) = \dfrac{1}{x - 6}$ has a vertical asymptote at $y = 6$.

7. Since $3^{-2} = \dfrac{1}{9}$ we can conclude that $\log_3 \dfrac{1}{9} = -2$.

8. The domain of the function $f(x) = \dfrac{1}{x^2 - 4}$ includes all real numbers except $x = 2$.

9. The amount of money after two years if \$2000 is invested in an account that is compounded monthly with an annual rate of 4% is $A = 2000\left(1 + \dfrac{4}{12}\right)^{24}$ dollars.

10. $\log_1 1 = 0$

11. $\ln(5 + 7) = \ln 5 + \ln 7$

12. $(\ln 3)^4 = 4 \ln 3$

13. $\log_{10} 0 = 1$

14. $e^{\ln 2} = 2$

15. $e^{\ln(-2)} = -2$

16. $\dfrac{\ln 4}{\ln 8} = \ln 4 - \ln 8$

17. The function $g(x) = e^x$ grows faster than the function $f(x) = \ln x$.

18. The half-life of a radioactive substance is the time required for half of the initial quantity to decay.

PRACTICE AND EXPLORATIONS

19. What is a function? A linear function? A quadratic function? A rational function?

20. How do you find a vertical asymptote? A horizontal asymptote?

21. What can you tell about the graph of a polynomial function of degree n before you plot any points?

22. Describe in words what a logarithm is.

List the ordered pairs obtained from the following if the domain of x for each exercise is $\{-3, -2, -1, 0, 1, 2, 3\}$. Graph each set of ordered pairs. Give the range.

23. $y = (2x - 1)(x + 1)$

24. $y = \dfrac{x}{x^2 + 1}$

25. Let $f(x) = 5x^2 - 3$ and $g(x) = -x^2 + 4x + 1$. Find the following.

 a. $f(-2)$ b. $g(3)$ c. $f(-k)$ d. $g(3m)$

 e. $f(x + h)$ f. $g(x + h)$ g. $\dfrac{f(x + h) - f(x)}{h}$

 h. $\dfrac{g(x + h) - g(x)}{h}$

26. Let $f(x) = 2x^2 + 5$ and $g(x) = 3x^2 + 4x - 1$. Find the following.

 a. $f(-3)$ b. $g(2)$ c. $f(3m)$ d. $g(-k)$

 e. $f(x + h)$ f. $g(x + h)$ g. $\dfrac{f(x + h) - f(x)}{h}$

 h. $\dfrac{g(x + h) - g(x)}{h}$

Find the domain of each function defined as follows.

27. $y = \dfrac{3x - 4}{x}$

28. $y = \dfrac{\sqrt{x - 2}}{2x + 3}$

29. $y = \ln(x + 7)$

30. $y = \ln(x^2 - 16)$

Graph the following by hand.

31. $y = 2x^2 + 3x - 1$

32. $y = -\dfrac{1}{4}x^2 + x + 2$

33. $y = -x^2 + 4x + 2$

34. $y = 3x^2 - 9x + 2$

35. $f(x) = x^3 - 3$

36. $f(x) = 1 - x^4$

37. $y = -(x - 1)^4 + 4$

38. $y = -(x + 2)^3 - 2$

39. $f(x) = \dfrac{8}{x}$

40. $f(x) = \dfrac{2}{3x - 6}$

41. $f(x) = \dfrac{4x - 2}{3x + 1}$

42. $f(x) = \dfrac{6x}{x + 2}$

43. $y = 4^x$

44. $y = 4^{-x} + 3$

45. $y = \left(\dfrac{1}{5}\right)^{2x-3}$

46. $y = \left(\dfrac{1}{2}\right)^{x-1}$

47. $y = \log_2(x - 1)$

48. $y = 1 + \log_3 x$

49. $y = -\ln(x + 3)$

50. $y = 2 - \ln x^2$

Solve each equation.

51. $2^{x+2} = \dfrac{1}{8}$

52. $\left(\dfrac{9}{16}\right)^x = \dfrac{3}{4}$

53. $9^{2y+3} = 27^y$

54. $\dfrac{1}{2} = \left(\dfrac{b}{4}\right)^{1/4}$

Write each equation using logarithms.

55. $3^5 = 243$

56. $5^{1/2} = \sqrt{5}$

57. $e^{0.8} = 2.22554$

58. $10^{1.07918} = 12$

Write each equation using exponents.

59. $\log_2 32 = 5$

60. $\log_9 3 = \dfrac{1}{2}$

61. $\ln 82.9 = 4.41763$

62. $\log 3.21 = 0.50651$

Evaluate each expression without using a calculator. Then support your work using a calculator and the change-of-base theorem for logarithms.

63. $\log_3 81$

64. $\log_{32} 16$

65. $\log_4 8$

66. $\log_{100} 1000$

Simplify each expression using the properties of logarithms.

67. $\log_5 3k + \log_5 7k^3$

68. $\log_3 2y^3 - \log_3 8y^2$

69. $4 \log_3 y - 2 \log_3 x$

70. $3 \log_4 r^2 - 2 \log_4 r$

Solve each equation. If necessary, round each answer to the nearest thousandth.

71. $6^p = 17$

72. $3^{z-2} = 11$

73. $2^{1-m} = 7$

74. $12^{-k} = 9$

75. $e^{-5-2x} = 5$

76. $e^{3x-1} = 14$

77. $\left(1 + \dfrac{m}{3}\right)^5 = 15$

78. $\left(1 + \dfrac{2p}{5}\right)^2 = 3$

79. $\log_k 64 = 6$

80. $\log_3(2x + 5) = 5$

81. $\log(4p + 1) + \log p = \log 3$

82. $\log_2(5m - 2) - \log_2(m + 3) = 2$

83. Give the following properties of the exponential function $f(x) = a^x; a > 0, a \neq 1$.

  a. Domain   **b.** Range   **c.** y-intercept

  d. Asymptote(s)

  e. Increasing if a is _____

  f. Decreasing if a is _____

84. Give the following properties of the logarithmic function $f(x) = \log_a x; a > 0, a \neq 1$.

  a. Domain   **b.** Range   **c.** x-intercept

  d. Asymptote(s)

  e. Increasing if a is _____

  f. Decreasing if a is _____

85. Compare your answers for Exercises 83 and 84. What similarities do you notice? What differences?

APPLICATIONS

Business and Economics

86. Car Rental To rent a mid-size car from one agency costs $60 per day or fraction of a day. If you pick up the car in Boston and drop it off in Utica, there is a fixed $40 charge. Let $C(x)$ represent the cost of renting the car for x days and taking it from Boston to Utica. Find the following.

  a. $C\left(\dfrac{3}{4}\right)$ **b.** $C\left(\dfrac{9}{10}\right)$ **c.** $C(1)$

  d. $C\left(1\dfrac{5}{8}\right)$ **e.** $C\left(2\dfrac{1}{9}\right)$

  f. Graph the function defined by $y = C(x)$ for $0 < x \leq 5$.

g. What is the independent variable?

h. What is the dependent variable?

87. Pollution The cost to remove x percent of a pollutant is

$$y = \frac{7x}{100 - x},$$

in thousands of dollars. Find the cost of removing the following percents of the pollutant.

  a. 80%   **b.** 50%   **c.** 90%

  d. Graph the function.

  e. Can all of the pollutant be removed?

Interest Find the amount of interest earned by each deposit.

88. $6902 if interest is 6% compounded semiannually for 8 years

89. $2781.36 if interest is 4.8% compounded quarterly for 6 years

Interest Find the compound amount if $12,104 is invested at 6.2% compounded continuously for each period.

90. 2 years   **91.** 4 years

Interest Find the compound amounts for the following deposits if interest is compounded continuously.

92. $1500 at 6% for 9 years

93. $12,000 at 5% for 8 years

94. How long will it take for $1000 deposited at 6% compounded semiannually to double? To triple?

95. How long will it take for $2100 deposited at 4% compounded quarterly to double? To triple?

Effective Rate Find the effective rate to the nearest hundredth for each nominal interest rate.

96. 7% compounded quarterly

97. 6% compounded monthly

98. 5% compounded continuously

Present Value Find the present value of each amount.

99. $2000 if interest is 6% compounded annually for 5 years

100. $10,000 if interest is 8% compounded semiannually for 6 years

101. Interest To help pay for college expenses, Julie Davis borrowed $10,000 at 7% interest compounded semiannually for 8 years. How much will she owe at the end of the 8-year period?

102. Inflation How long will it take for $1 to triple at an annual inflation rate of 8% compounded continuously?

103. Interest Find the interest rate needed for $6000 to grow to $8000 in 3 years with continuous compounding.

104. Present Value Frank Steed wants to open a camera shop. How much must he deposit now at 6% interest compounded monthly to have $25,000 at the end of 3 years?

105. Revenue A concert promoter finds she can sell 1000 tickets at $50 each. She will not sell the tickets for less than $50, but she finds that for every $1 increase in the ticket price above $50, she will sell 10 fewer tickets.

a. Express n, the number of tickets sold, as a function of p, the price.

b. Express R, the revenue, as a function of p, the price.

c. Find the domain of the function found in part b.

d. Express R, the revenue, as a function of n, the number sold.

e. Find the domain of the function found in part d.

f. Find the price that produces the maximum revenue.

g. Find the number of tickets sold that produces the maximum revenue.

h. Find the maximum revenue.

i. Sketch the graph of the function found in part b.

j. Describe what the graph found in part i tells you about how the revenue varies with price.

106. Cost Suppose the cost in dollars to produce x posters is given by

$$C(x) = \frac{5x + 3}{x + 1}.$$

a. Sketch a graph of $C(x)$.

b. Find a formula for $C(x + 1) - C(x)$, the cost to produce an additional poster when x posters are already produced.

c. Find a formula for $A(x)$, the average cost per poster.

d. Find a formula for $A(x + 1) - A(x)$, the change in the average cost per poster when one additional poster is produced. (This quantity is approximately equal to the marginal average cost, which will be discussed in the chapter on the derivative.)

107. Cost Suppose the cost in dollars to produce x hundreds of nails is given by

$$C(x) = x^2 + 4x + 7.$$

a. Sketch a graph of $C(x)$.

b. Find a formula for $C(x + 1) - C(x)$, the cost to produce an additional hundred nails when x hundred are already produced. (This quantity is approximately equal to the marginal cost.)

c. Find a formula for $A(x)$, the average cost per hundred nails.

d. Find a formula for $A(x + 1) - A(x)$, the change in the average cost per nail when one additional batch of 100 nails is produced. (This quantity is approximately equal to the marginal average cost, which will be discussed in the chapter on the derivative.)

108. Consumer Price Index The U.S. consumer price index (CPI, or cost of living index) has risen over the years, as shown in the table in the next column, using an index in which the average over the years 1982 to 1984 is set to 100. *Source: Bureau of Labor Statistics.*

a. Letting t be the years since 1960, write an exponential function in the form $y = a^t$ that fits the data at 1960 and 2005.

b. If your calculator has an exponential regression feature, find the best fitting exponential function for the data.

Year	CPI
1960	29.6
1970	38.8
1980	82.4
1990	130.7
1995	152.4
2000	172.2
2005	195.3

c. Use a graphing calculator to plot the answers to parts a and b on the same axes as the data. Are the answers to parts a and b close to each other?

d. If your calculator has a quadratic and cubic regression feature, find the best-fitting quadratic and cubic functions for the data.

e. Use a graphing calculator to plot the answers to parts b and d on the same window as the data. Discuss the extent to which any one of these functions models the data better than the others.

Life Sciences

109. Fever A certain viral infection causes a fever that typically lasts 6 days. A model of the fever (in °F) on day x, $1 \le x \le 6$, is

$$F(x) = -\frac{2}{3}x^2 + \frac{14}{3}x + 96.$$

According to the model, on what day should the maximum fever occur? What is the maximum fever?

110. Sunscreen An article in a medical journal says that a sunscreen with a sun protection factor (SPF) of 2 provides 50% protection against ultraviolet B (UVB) radiation, an SPF of 4 provides 75% protection, and an SPF of 8 provides 87.5% protection (which the article rounds to 87%). *Source: Family Practice.*

a. 87.5% protection means that 87.5% of the UVB radiation is screened out. Write as a fraction the amount of radiation that is let in, and then describe how this fraction, in general, relates to the SPF rating.

b. Plot UVB percent protection (y) against x, where $x = 1/\text{SPF}$.

c. Based on your graph from part b, give an equation relating UVB protection to SPF rating.

d. An SPF of 8 has double the chemical concentration of an SPF 4. Find the increase in the percent protection.

e. An SPF of 30 has double the chemical concentration of an SPF 15. Find the increase in the percent protection.

f. Based on your answers from parts d and e, what happens to the increase in the percent protection as the SPF continues to double?

111. HIV in Infants The following table lists the reported number of cases of infants born in the United States with HIV in recent years because their mother was infected.* *Source: Centers for Disease Control and Prevention.*

Year	Cases
1995	295
1997	166
1999	109
2001	115
2003	94
2005	107
2007	79

a. Plot the data on a graphing calculator, letting $t = 0$ correspond to the year 1995.

b. Using the regression feature on your calculator, find a quadratic, a cubic, and an exponential function that models this data.

c. Plot the three functions with the data on the same coordinate axes. Which function or functions best capture the behavior of the data over the years plotted?

d. Find the number of cases predicted by all three functions for 2015. Which of these are realistic? Explain.

112. Respiratory Rate Researchers have found that the 95th percentile (the value at which 95% of the data is at or below) for respiratory rates (in breaths per minute) during the first 3 years of infancy are given by

$$y = 10^{1.82411 - 0.0125995x + 0.00013401x^2}$$

for awake infants and

$$y = 10^{1.72858 - 0.0139928x + 0.00017646x^2}$$

for sleeping infants, where x is the age in months. *Source: Pediatrics.*

a. What is the domain for each function?

b. For each respiratory rate, is the rate decreasing or increasing over the first 3 years of life? (*Hint:* Is the graph of the quadratic in the exponent opening upward or downward? Where is the vertex?)

c. Verify your answer to part b using a graphing calculator.

d. For a 1-year-old infant in the 95th percentile, how much higher is the waking respiratory rate than the sleeping respiratory rate?

113. Polar Bear Mass One formula for estimating the mass (in kg) of a polar bear is given by

$$m(g) = e^{0.02 + 0.062g - 0.000165g^2},$$

where g is the axillary girth in centimeters. It seems reasonable that as girth increases, so does the mass. What is the largest girth for which this formula gives a reasonable answer? What is the predicted mass of a polar bear with this girth? *Source: Wildlife Management.*

*These data include only those infants born in the 25 states with confidential name-based HIV infection reporting.

114. Population Growth A population of 15,000 small deer in a specific region has grown exponentially to 17,000 in 4 years.

a. Write an exponential equation to express the population growth y in terms of time t in years.

b. At this rate, how long will it take for the population to reach 45,000?

115. Population Growth In 1960 in an article in *Science* magazine, H. Van Forester, P. M. Mora, and W. Amiot predicted that world population would be infinite in the year 2026. Their projection was based on the rational function defined by

$$p(t) = \frac{1.79 \times 10^{11}}{(2026.87 - t)^{0.99}},$$

where $p(t)$ gives population in year t. This function has provided a relatively good fit to the population until very recently. *Source: Science.*

a. Estimate world population in 2010 using this function, and compare it with the estimate of 6.909 billion. *Source: United Nations.*

b. What does the function predict for world population in 2020? 2025?

c. Discuss why this function is not realistic, despite its good fit to past data.

116. Intensity of Light The intensity of light (in appropriate units) passing through water decreases exponentially with the depth it penetrates beneath the surface according to the function

$$I(x) = 10e^{-0.3x},$$

where x is the depth in meters. A certain water plant requires light of an intensity of 1 unit. What is the greatest depth of water in which it will grow?

117. Drug Concentration The concentration of a certain drug in the bloodstream at time t (in minutes) is given by

$$c(t) = e^{-t} - e^{-2t}.$$

Use a graphing calculator to find the maximum concentration and the time when it occurs.

118. Glucose Concentration When glucose is infused into a person's bloodstream at a constant rate of c grams per minute, the glucose is converted and removed from the bloodstream at a rate proportional to the amount present. The amount of glucose in grams in the bloodstream at time t (in minutes) is given by

$$g(t) = \frac{c}{a} + \left(g_0 - \frac{c}{a}\right)e^{-at},$$

where a is a positive constant. Assume $g_0 = 0.08$, $c = 0.1$, and $a = 1.3$.

a. At what time is the amount of glucose a maximum? What is the maximum amount of glucose in the bloodstream?

b. When is the amount of glucose in the bloodstream 0.1 g?

c. What happens to the amount of glucose in the bloodstream after a very long time?

119. Species Biologists have long noticed a relationship between the area of a piece of land and the number of species found there. The following data shows a sample of the British Isles and how many vascular plants are found on each. *Source: Journal of Biogeography.*

Isle	Area (km²)	Species
Ailsa	0.8	75
Fair	5.2	174
Iona	9.1	388
Man	571.6	765
N. Ronaldsay	7.3	131
Skye	1735.3	594
Stronsay	35.2	62
Wight	380.7	1008

a. One common model for this relationship is logarithmic. Using the logarithmic regression feature on a graphing calculator, find a logarithmic function that best fits the data.

b. An alternative to the logarithmic model is a power function of the form $S = b(A^c)$. Using the power regression feature on a graphing calculator, find a power function that best fits the data.

c. Graph both functions from parts a and b along with the data. Give advantages and drawbacks of both models.

d. Use both functions to predict the number of species found on the isle of Shetland, with an area of 984.2 km². Compare with the actual number of 421.

e. Describe one or more situations where being able to predict the number of species could be useful.

Physical Sciences

120. Oil Production The production of an oil well has decreased exponentially from 128,000 barrels per year 5 years ago to 100,000 barrels per year at present.

a. Letting $t = 0$ represent the present time, write an exponential equation for production y in terms of time t in years.

b. Find the time it will take for production to fall to 70,000 barrels per year.

121. Dating Rocks Geologists sometimes measure the age of rocks by using "atomic clocks." By measuring the amounts of potassium-40 and argon-40 in a rock, the age t of the specimen (in years) is found with the formula

$$t = (1.26 \times 10^9)\frac{\ln[1 + 8.33(A/K)]}{\ln 2},$$

where A and K, respectively, are the numbers of atoms of argon-40 and potassium-40 in the specimen.

a. How old is a rock in which $A = 0$ and $K > 0$?

b. The ratio A/K for a sample of granite from New Hampshire is 0.212. How old is the sample?

c. Let $A/K = r$. What happens to t as r gets larger? Smaller?

122. Planets The following table contains the average distance D from the sun for the eight planets and their period P of revolution around the sun in years. *Source: The Natural History of the Universe.*

Planet	Distance (D)	Period (P)
Mercury	0.39	0.24
Venus	0.72	0.62
Earth	1	1
Mars	1.52	1.89
Jupiter	5.20	11.9
Saturn	9.54	29.5
Uranus	19.2	84.0
Neptune	30.1	164.8

The distances are given in astronomical units (A.U.); 1 A.U. is the average distance from Earth to the sun. For example, since Jupiter's distance is 5.2 A.U., its distance from the sun is 5.2 times farther than Earth's.

a. Find functions of the form $P = kD^n$ for $n = 1$, 1.5, and 2 that fit the data at Neptune.

b. Use a graphing calculator to plot the data in the table and to graph the three functions found in part a. Which function best fits the data?

c. Use the best-fitting function from part b to predict the period of Pluto (which was removed from the list of planets in 2006), which has a distance from the sun of 39.5 A.U. Compare your answer to the true value of 248.5 years.

d. If you have a graphing calculator or computer program with a power regression feature, use it to find a power function (a function of the form $P = kD^n$) that approximately fits the data. How does this answer compare with the answer to part b?

EXTENDED APPLICATION

POWER FUNCTIONS

In this chapter we have seen several applications of power functions, which have the general form $y = ax^b$. Power functions are so named because the independent variable is raised to a power. (These should not be confused with exponential functions, in which the independent variable appears in the power.) We explored some special cases of power functions, such as $b = 2$ (a simple quadratic function) and $b = 1/2$ (a square root function). But applications of power functions vary greatly and are not limited to these special cases.

For example, in Exercise 86 in Section 2.5, we saw that the basal metabolism rate of marsupial carnivores is a power function of the body mass. In that exercise we also saw a way to verify that empirical data can be modeled with a power function. By taking the natural logarithm of both sides of the equation for a power function,

$$y = ax^b, \tag{1}$$

we obtain the equation

$$\ln y = \ln a + b \ln x. \tag{2}$$

Letting $Y = \ln y$, $X = \ln x$, and $A = \ln a$ results in the linear equation

$$Y = A + bX. \tag{3}$$

Plotting the logarithm of the original data reveals whether a straight line approximates the data well. If it does, then a power function is a good fit to the original data.

Here is another example. In an attempt to measure how the pace of city life is related to the population of the city, two researchers estimated the average speed of pedestrians in 15 cities by measuring the mean time it took them to walk 50 feet. Their results are shown in the table in the next column. *Source: Nature.*

Figure 59(a) shows the speed (stored in the list L_2 on a TI-84 Plus) plotted against the population (stored in L_1). The natural logarithm of the data was then calculated and stored using the commands $\ln(L_1) \rightarrow L_3$ and $\ln(L_2) \rightarrow L_4$. A plot of the data in L_3 and L_4 is shown in Figure 59(b). Notice that the data lie fairly closely along a straight line, confirming that a power function is an appropriate model for the original data. (These calculations and plots could also be carried out on a spreadsheet.)

A power function that best fits the data according to the least squares principle of Section 1.3 is found with the TI-84 Plus command PwrReg L_1, L_2, Y_1. The result is

$$y = 1.363x^{0.09799}, \tag{4}$$

City	Population (x)	Speed (ft/sec) (y)
Brno, Czechoslovakia	341,948	4.81
Prague, Czechoslovakia	1,092,759	5.88
Corte, Corsica	5491	3.31
Bastia, France	49,375	4.90
Munich, Germany	1,340,000	5.62
Psychro, Crete	365	2.67
Itea, Greece	2500	2.27
Iráklion, Greece	78,200	3.85
Athens, Greece	867,023	5.21
Safed, Israel	14,000	3.70
Dimona, Israel	23,700	3.27
Netanya, Israel	70,700	4.31
Jerusalem, Israel	304,500	4.42
New Haven, U.S.A.	138,000	4.39
Brooklyn, U.S.A.	2,602,000	5.05

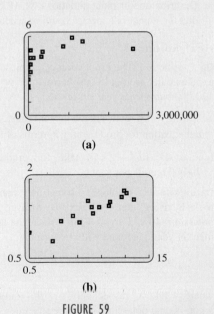

(a)

(b)

FIGURE 59

with a correlation coefficient of $r = 0.9081$. (For more on the correlation coefficient, see Section 1.3.) Because this value is close to 1, it indicates a good fit. Similarly, we can find a line that fits the data in Figure 59(b) with the command LinReg(ax + b) L_3, L_4, Y_2.

The result is

$$Y = 0.30985 + 0.09799X, \tag{5}$$

again with $r = 0.9081$. The identical correlation coefficient is not a surprise, since the two commands accomplish essentially the same thing. Comparing Equations (4) and (5) with Equations (1), (2), and (3), notice that $b = 0.09799$ in both Equations (4) and (5), and that $A = 0.30985 \approx \ln a = \ln 1.363$. (The slight difference is due to rounding.) Equations (4) and (5) are plotted on the same window as the data in Figure 60(a) and (b), respectively.

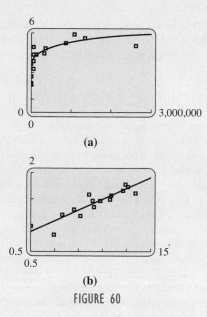

(a)

(b)

FIGURE 60

These results raise numerous questions worth exploring. What does this analysis tell you about the connection between the pace of city life and the population of a city? What might be some reasons for this connection?

A third example was considered in Review Exercise 119, where we explored the relationship between the area of each of the British isles and the number of species of vascular plants on the isles. In Figure 61(a) we have plotted the complete set of data from the original article (except for the Isle of Britain, whose area is so large that it doesn't fit on a graph that shows the other data in detail). *Source: Journal of Biogeography.* In Figure 61(b) we have plotted the natural logarithm of the data (again leaving out Britain). Notice that despite the large amount of scatter in the data, there is a linear trend. Figure 61(a) includes the best-fitting power function

$$y = 125.9x^{0.2088},\qquad (6)$$

while Figure 61 (b) includes the best-fitting line

$$Y = 4.836 + 0.2088X.\qquad (7)$$

(We have included the data for Britain in both of these calculations.) Notice as before that the exponent of the power function equals the slope of the linear function, and that $A = 4.836 \approx \ln a = \ln 125.9$. The correlation is 0.6917 for both, indicating that there is a trend, but that the data is somewhat scattered about the best-fitting curve or line.

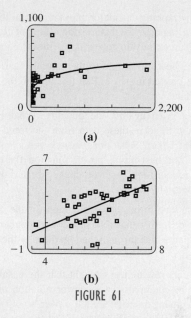

(a)

(b)

FIGURE 61

There are many more examples of data that fit a power function. Bohorquez et al. have found that the frequency of attacks in a war is a power function of the number of people killed in the attacks. *Source: Nature.* Amazingly, this holds true for a wide variety of different wars, with the average value of the exponent as $b \approx 2.5$. For a fragmented, fluid enemy with more groups, the value of b tends to be larger, and for a robust, stronger enemy with fewer groups, the value of b tends to be smaller.* You will explore other applications of power functions in the exercises.

EXERCISES

1. Gwartney et al. listed the following data relating the price of a monthly cellular bill (in dollars) and the demand (in millions of subscribers).[†] *Source: Economics: Private and Public Choice.*

Quantity (in millions)	Price (in dollars)
2.1	123
3.5	107
5.3	92
7.6	79
11.0	73
16.0	63
24.1	56

*For a TED video on this phenomenon, see http://www.ted.com/talks/sean_gourley_on_the_mathematics_of_war.html.
[†] The authors point out that these are actual prices and quantities annually for 1988 to 1994. If they could assume that other demand determinants, such as income, had remained constant during that period, this would give an accurate measurement of the demand function.

119

a. Using a graphing calculator, plot the natural logarithm of the price against the natural logarithm of the quantity. Does the relationship appear to be linear?

b. Find the best-fitting line to the natural logarithm of the data, as plotted in part a. Plot this line on the same axes as the data.

c. Plot the price against the quantity. What is different about the trend in these data from the trend in Figures 59(a) and 61(a)? What does this tell you about the exponent of the best-fitting power function for these data? What conclusions can you make about how demand varies with the price?

d. Find the best-fitting power function for the data plotted in part c. Verify that this function is equivalent to the least squares line through the logarithm of the data found in part b.

2. For many years researchers thought that the basal metabolic rate (BMR) in mammals was a power function of the mass, with disagreement on whether the power was 0.67 or 0.75. More recently, White et al. proposed that the power may vary for mammals with different evolutionary lineages. *Source: Evolution.* The following table shows a portion of their data containing the natural logarithm of the mass (in grams) and the natural logarithm of the BMR (in mL of oxygen per hour) for 12 species from the genus Peromyscus, or deer mouse.

Species	ln(mass)	ln(BMR)
Peromyscus boylii	3.1442	3.9943
Peromyscus californicus	3.8618	3.9506
Peromyscus crinitus	2.7663	3.2237
Peromyscus eremicus	3.0681	3.4998
Peromyscus gossypinus	3.0681	3.6104
Peromyscus leucopus	3.1355	3.8111
Peromyscus maniculatus	3.0282	3.6835
Peromyscus megalops	4.1927	4.5075
Peromyscus oreas	3.2019	3.7729
Peromyscus polionotus	2.4849	3.0671
Peromyscus sitkensis	3.3439	3.8447
Peromyscus truei	3.5041	4.0378

a. Plot ln(BMR) against ln(mass) using a graphing calculator. Does the relationship appear to be linear?

b. Find the least squares line for the data plotted in part a. Plot the line on the same axes as the data.

c. Calculate the mass and BMR for each species, and then find the best-fitting power function for these data. Plot this function on the same axes as the mass and BMR data.

d. What would you conclude about whether the deer mouse BMR can be modeled as a power function of the mass? What seems to be an approximate value of the power?

DIRECTIONS FOR GROUP PROJECT

Go to the section on "build your own tables" of the Human Development Reports website at http://hdrstats.undp.org/en/buildtables. *Select a group of countries, as well as two indicators that you think might be related by a power function. For example, you might choose "GDP per capita" and "Population not using an improved water source (%)" Click on "Display indicators in Row" and then "Show results." Then click on "Export to Excel." From the Excel spreadsheet, create a scatterplot of the original data, as well as a scatterplot of the natural logarithm of the data. Find data for which the natural logarithm is roughly a straight line, and find the least squares line. Then convert this to a power function modeling the original data. Present your results in a report, describing in detail what your analysis tells you about the countries under consideration and any other conclusions that you can make.*

3 The Derivative

The population of the United States has been increasing since 1790, when the first census was taken. Over the past few decades, the population has not only been increasing, but the level of diversity has also been increasing. This fact is important to school districts, businesses, and government officials. Using examples in the third section of this chapter, we explore two rates of change related to the increase in minority population. In the first example, we calculate an average rate of change; in the second, we calculate the rate of change at a particular time. This latter rate is an example of a derivative, the subject of this chapter.

The algebraic problems considered in earlier chapters dealt with *static* situations:

- What is the revenue when 100 items are sold?
- How much interest is earned in three years?
- What is the equilibrium price?

Calculus, on the other hand, deals with *dynamic* situations:

- At what rate is the demand for a product changing?
- How fast is a car moving after 2 hours?
- When does the growth of a population begin to slow down?

The techniques of calculus allow us to answer these questions, which deal with rates of change.

The key idea underlying calculus is the concept of limit, so we will begin by studying limits.

3.1 Limits

APPLY IT **What happens to the demand of an essential commodity as its price continues to increase?**
We will find an answer to this question in Exercise 82 using the concept of limit.

The limit is one of the tools that we use to describe the behavior of a function as the values of x approach, or become closer and closer to, some particular number.

EXAMPLE 1 Finding a Limit

What happens to $f(x) = x^2$ when x is a number *very close* to (but not equal to) 2?

SOLUTION We can construct a table with x values getting closer and closer to 2 and find the corresponding values of $f(x)$.

	x approaches 2 from left →					← *x* approaches 2 from right			
x	1.9	1.99	1.999	1.9999	2	2.0001	2.001	2.01	2.1
$f(x)$	3.61	3.9601	3.996001	3.99960001	4	4.00040001	4.004001	4.0401	4.41

$f(x)$ approaches 4 → ↑ ← $f(x)$ approaches 4

The table suggests that, as x gets closer and closer to 2 from either side, $f(x)$ gets closer and closer to 4. In fact, you can use a calculator to show the values of $f(x)$ can be made as close as you want to 4 by taking values of x close enough to 2. This is not surprising since the value of the function at $x = 2$ is $f(x) = 4$. We can observe this fact by looking at the graph $y = x^2$, as shown in Figure 1. In such a case, we say "the limit of $f(x)$ as x approaches 2 is 4," which is written as

$$\lim_{x \to 2} f(x) = 4.$$

YOUR TURN 1
Find $\lim_{x \to 1} (x^2 + 2)$.

TRY YOUR TURN 1

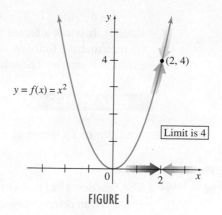

FIGURE 1

The phrase "x approaches 2 from the left" is written $x \to 2^-$. Similarly, "x approaches 2 from the right" is written $x \to 2^+$. These expressions are used to write **one-sided limits**. The **limit from the left** (as x approaches 2 from the negative direction) is written

$$\lim_{x \to 2^-} f(x) = 4,$$

and shown in red in Figure 1. The **limit from the right** (as x approaches 2 from the positive direction) is written

$$\lim_{x \to 2^+} f(x) = 4,$$

and shown in blue in Figure 1. A **two-sided limit**, such as

$$\lim_{x \to 2} f(x) = 4,$$

exists only if both one-sided limits exist and are the same; that is, if $f(x)$ approaches the same number as x approaches a given number from *either* side.

> **CAUTION** Notice that $\lim_{x \to a^-} f(x)$ does not mean to take negative values of x, nor does it mean to choose values of x to the right of a and then move in the negative direction. It means to use values less than a ($x < a$) that get closer and closer to a.

The previous example suggests the following informal definition.

Limit of a Function

Let f be a function and let a and L be real numbers. If

1. as x takes values closer and closer (but not equal) to a on both sides of a, the corresponding values of $f(x)$ get closer and closer (and perhaps equal) to L; and

2. the value of $f(x)$ can be made as close to L as desired by taking values of x close enough to a;

then L is the **limit** of $f(x)$ as x approaches a, written

$$\lim_{x \to a} f(x) = L.$$

This definition is informal because the expressions "closer and closer to" and "as close as desired" have not been defined. A more formal definition would be needed to prove the rules for limits given later in this section.*

---

*The limit is the key concept from which all the ideas of calculus flow. Calculus was independently discovered by the English mathematician Isaac Newton (1642–1727) and the German mathematician Gottfried Wilhelm Leibniz (1646–1716). For the next century, supporters of each accused the other of plagiarism, resulting in a lack of communication between mathematicians in England and on the European continent. Neither Newton nor Leibniz developed a mathematically rigorous definition of the limit (and we have no intention of doing so here). More than 100 years passed before the French mathematician Augustin-Louis Cauchy (1789–1857) accomplished this feat.

The definition of a limit describes what happens to $f(x)$ when x is near, but not at, the value a. It is not affected by how (or even whether) $f(a)$ is defined. Also the definition implies that the function values cannot approach two different numbers, so that if a limit exists, it is unique. These ideas are illustrated in the following examples.

EXAMPLE 2   Finding a Limit

Find $\lim\limits_{x \to 2} g(x)$, where $g(x) = \dfrac{x^3 - 2x^2}{x - 2}$.

Method 1
Using a Table

SOLUTION

The function $g(x)$ is undefined when $x = 2$ since the value $x = 2$ makes the denominator 0. However, in determining the limit as x approaches 2 we are concerned only with the values of $g(x)$ when x is close to but *not equal to* 2. To determine if the limit exists, consider the value of g at some numbers close to but not equal to 2, as shown in the following table.

		x approaches 2 from left				x approaches 2 from right			
x	1.9	1.99	1.999	1.9999	2	2.0001	2.001	2.01	2.1
$g(x)$	3.61	3.9601	3.996001	3.99960001		4.00040001	4.004001	4.0401	4.41

$f(x)$ approaches 4 $f(x)$ approaches 4

Undefined

Notice that this table is almost identical to the previous table, except that g is undefined at $x = 2$. This suggests that $\lim\limits_{x \to 2} g(x) = 4$, in spite of the fact that the function g does not exist at $x = 2$.

Method 2
Using Algebra

A second approach to this limit is to analyze the function. By factoring the numerator,

$$x^3 - 2x^2 = x^2(x - 2),$$

$g(x)$ simplifies to

$$g(x) = \frac{x^2(x - 2)}{x - 2} = x^2, \qquad \text{provided } x \neq 2.$$

The graph of $g(x)$, as shown in Figure 2, is almost the same as the graph of $y = x^2$, except that it is undefined at $x = 2$ (illustrated by the "hole" in the graph).

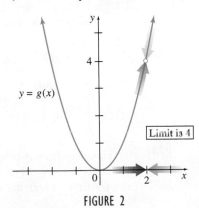

FIGURE 2

Since we are looking at the limit as x approaches 2, we look at values of the function for x close to but not equal to 2. Thus, the limit is

$$\lim_{x \to 2} g(x) = \lim_{x \to 2} x^2 = 4.$$

TRY YOUR TURN 2

YOUR TURN 2

Find $\lim\limits_{x \to 2} \dfrac{x^2 - 4}{x - 2}$.

⟨⊞⟩ **TECHNOLOGY NOTE**

We can use the TRACE feature on a graphing calculator to determine the limit. Figure 3 shows the graph of the function in Example 2 drawn with a graphing calculator. Notice that the function has a small gap at the point (2, 4), which agrees with our previous observation that the function is undefined at $x = 2$, where the limit is 4. (Due to the limitations of the graphing calculator, this gap may vanish when the viewing window is changed very slightly.)

The result after pressing the TRACE key is shown in Figure 4. The cursor is already located at $x = 2$; if it were not, we could use the right or left arrow key to move the cursor there. The calculator does not give a y-value because the function is undefined at $x = 2$. Moving the cursor back a step gives $x = 1.98$, $y = 3.92$. Moving the cursor forward two steps gives $x = 2.02$, $y = 4.09$. It seems that as x approaches 2, y approaches 4, or at least something close to 4. Zooming in on the point (2, 4) (such as using the window [1.9, 2.1] by [3.9, 4.1]) allows the limit to be estimated more accurately and helps ensure that the graph has no unexpected behavior very close to $x = 2$.

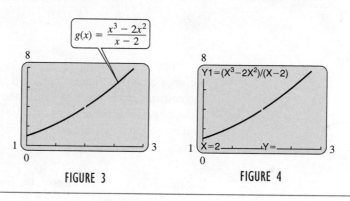

$$g(x) = \frac{x^3 - 2x^2}{x - 2}$$

FIGURE 3   **FIGURE 4**

EXAMPLE 3 **Finding a Limit**

Determine $\lim_{x \to 2} h(x)$ for the function h defined by

$$h(x) = \begin{cases} x^2, & \text{if } x \neq 2, \\ 1, & \text{if } x = 2. \end{cases}$$

YOUR TURN 3
Find $\lim_{x \to 3} f(x)$ if
$$f(x) = \begin{cases} 2x - 1 & \text{if } x \neq 3 \\ 1 & \text{if } x = 3. \end{cases}$$

SOLUTION A function defined by two or more cases is called a **piecewise function**. The domain of h is all real numbers, and its graph is shown in Figure 5. Notice that $h(2) = 1$, but $h(x) = x^2$ when $x \neq 2$. To determine the limit as x approaches 2, we are concerned only with the values of $h(x)$ when x is close but not equal to 2. Once again,

$$\lim_{x \to 2} h(x) = \lim_{x \to 2} x^2 = 4.$$

TRY YOUR TURN 3

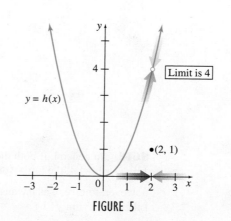

FIGURE 5

EXAMPLE 4 **Finding a Limit**

Find $\lim\limits_{x \to -2} f(x)$, where

$$f(x) = \frac{3x + 2}{2x + 4}.$$

SOLUTION The graph of the function is shown in Figure 6. A table with the values of $f(x)$ as x gets closer and closer to -2 is given below.

	x approaches −2 from left				x approaches −2 from right			
x	−2.1	−2.01	−2.001	−2.0001	−1.9999	−1.999	−1.99	−1.9
f(x)	21.5	201.5	2001.5	20,001.5	−19,998.5	−1998.5	−198.5	−18.5

Both the graph and the table suggest that as x approaches -2 from the left, $f(x)$ becomes larger and larger without bound. This happens because as x approaches -2, the denominator approaches 0, while the numerator approaches -4, and -4 divided by a smaller and smaller number becomes larger and larger. When this occurs, we say that "the limit as x approaches -2 from the left is infinity," and we write

$$\lim_{x \to -2^-} f(x) = \infty.$$

Because ∞ is not a real number, the limit in this case does not exist.

In the same way, the behavior of the function as x approaches -2 from the right is indicated by writing

$$\lim_{x \to -2^+} f(x) = -\infty,$$

since $f(x)$ becomes more and more negative without bound. Since there is no real number that $f(x)$ approaches as x approaches -2 (from either side), nor does $f(x)$ approach either ∞ or $-\infty$, we simply say

$$\lim_{x \to -2} \frac{3x + 2}{2x + 4} \quad \text{does not exist.}$$

TRY YOUR TURN 4

YOUR TURN 4

Find $\lim\limits_{x \to 0} \dfrac{2x - 1}{x}$.

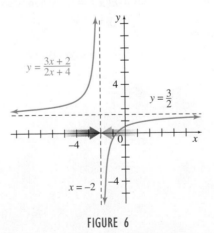

FIGURE 6

NOTE In general, if both the limit from the left and from the right approach ∞, so that $\lim\limits_{x \to a} f(x) = \infty$, the limit would not exist because ∞ is not a real number. It is customary, however, to give ∞ as the answer since it describes how the function is behaving near $x = a$. Likewise, if $\lim\limits_{x \to a} f(x) = -\infty$, we give $-\infty$ as the answer.

We have shown three methods for determining limits: (1) using a table of numbers, (2) using algebraic simplification, and (3) tracing the graph on a graphing calculator. Which method you choose depends on the complexity of the function and the accuracy required by the application. Algebraic simplification gives the exact answer, but it can be difficult or even impossible to use in some situations. Calculating a table of numbers or tracing the graph may be easier when the function is complicated, but be careful, because the results could be inaccurate, inconclusive, or misleading. A graphing calculator does not tell us what happens between or beyond the points that are plotted.

EXAMPLE 5 Finding a Limit

Find $\lim\limits_{x \to 0} \dfrac{|x|}{x}$.

SOLUTION

Method 1
Algebraic Approach

The function $f(x) = |x|/x$ is not defined when $x = 0$. When $x > 0$, the definition of absolute value says that $|x| = x$, so $f(x) = |x|/x = x/x = 1$. When $x < 0$, then $|x| = -x$ and $f(x) = |x|/x = -x/x = -1$. Therefore,

$$\lim_{x \to 0^+} f(x) = 1 \qquad \text{and} \qquad \lim_{x \to 0^-} f(x) = -1.$$

Since the limits from the left and from the right are different, the limit does not exist.

Method 2
Graphing Calculator Approach

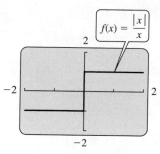

FIGURE 7

A calculator graph of f is shown in Figure 7.

As x approaches 0 from the right, x is always positive and the corresponding value of $f(x)$ is 1, so

$$\lim_{x \to 0^+} f(x) = 1.$$

But as x approaches 0 from the left, x is always negative and the corresponding value of $f(x)$ is -1, so

$$\lim_{x \to 0^-} f(x) = -1.$$

As in the algebraic approach, the limits from the left and from the right are different, so the limit does not exist.

The discussion up to this point can be summarized as follows.

Existence of Limits

The limit of f as x approaches a may not exist.

1. If $f(x)$ becomes infinitely large in magnitude (positive or negative) as x approaches the number a from either side, we write $\lim\limits_{x \to a} f(x) = \infty$ or $\lim\limits_{x \to a} f(x) = -\infty$. In either case, the limit does not exist.

2. If $f(x)$ becomes infinitely large in magnitude (positive) as x approaches a from one side and infinitely large in magnitude (negative) as x approaches a from the other side, then $\lim\limits_{x \to a} f(x)$ does not exist.

3. If $\lim\limits_{x \to a^-} f(x) = L$ and $\lim\limits_{x \to a^+} f(x) = M$, and $L \neq M$, then $\lim\limits_{x \to a} f(x)$ does not exist.

Figure 8 illustrates these three facts.

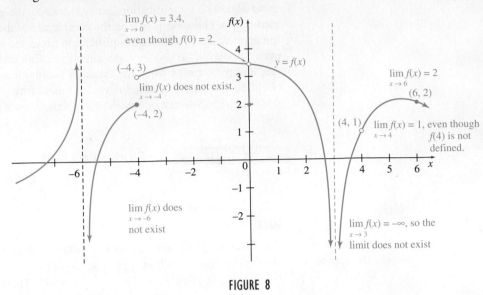

FIGURE 8

Rules for Limits

As shown by the preceding examples, tables and graphs can be used to find limits. However, it is usually more efficient to use the rules for limits given below. (Proofs of these rules require a formal definition of limit, which we have not given.)

Rules for Limits

Let a, A, and B be real numbers, and let f and g be functions such that

$$\lim_{x \to a} f(x) = A \qquad \text{and} \qquad \lim_{x \to a} g(x) = B.$$

1. If k is a constant, then $\lim_{x \to a} k = k$ and $\lim_{x \to a} [k \cdot f(x)] = k \cdot \lim_{x \to a} f(x) = k \cdot A$.

2. $\lim_{x \to a} [f(x) \pm g(x)] = \lim_{x \to a} f(x) \pm \lim_{x \to a} g(x) = A \pm B$

   (The limit of a sum or difference is the sum or difference of the limits.)

3. $\lim_{x \to a} [f(x) \cdot g(x)] = \left[\lim_{x \to a} f(x)\right] \cdot \left[\lim_{x \to a} g(x)\right] = A \cdot B$

   (The limit of a product is the product of the limits.)

4. $\lim_{x \to a} \dfrac{f(x)}{g(x)} = \dfrac{\lim_{x \to a} f(x)}{\lim_{x \to a} g(x)} = \dfrac{A}{B}$ if $B \neq 0$

   (The limit of a quotient is the quotient of the limits, provided the limit of the denominator is not zero.)

5. If $p(x)$ is a polynomial, then $\lim_{x \to a} p(x) = p(a)$.

6. For any real number k, $\lim_{x \to a} [f(x)]^k = \left[\lim_{x \to a} f(x)\right]^k = A^k$, provided this limit exists.*

7. $\lim_{x \to a} f(x) = \lim_{x \to a} g(x)$ if $f(x) = g(x)$ for all $x \neq a$.

8. For any real number $b > 0$, $\lim_{x \to a} b^{f(x)} = b^{[\lim_{x \to a} f(x)]} = b^A$.

9. For any real number b such that $0 < b < 1$ or $1 < b$,

   $\lim_{x \to a} [\log_b f(x)] = \log_b [\lim_{x \to a} f(x)] = \log_b A$ if $A > 0$.

*This limit does not exist, for example, when $A < 0$ and $k = 1/2$, or when $A = 0$ and $k \leq 0$.

This list may seem imposing, but these limit rules, once understood, agree with common sense. For example, Rule 3 says that if $f(x)$ becomes close to A as x approaches a, and if $g(x)$ becomes close to B, then $f(x) \cdot g(x)$ should become close to $A \cdot B$, which seems plausible.

EXAMPLE 6 Rules for Limits

Suppose $\lim\limits_{x \to 2} f(x) = 3$ and $\lim\limits_{x \to 2} g(x) = 4$. Use the limit rules to find the following limits.

(a) $\lim\limits_{x \to 2} [f(x) + 5g(x)]$

SOLUTION

$$\lim_{x \to 2} [f(x) + 5g(x)] = \lim_{x \to 2} f(x) + \lim_{x \to 2} 5g(x) \quad \text{Rule 2}$$

$$= \lim_{x \to 2} f(x) + 5 \lim_{x \to 2} g(x) \quad \text{Rule 1}$$

$$= 3 + 5(4)$$

$$= 23$$

(b) $\lim\limits_{x \to 2} \dfrac{[f(x)]^2}{\ln g(x)}$

SOLUTION

$$\lim_{x \to 2} \frac{[f(x)]^2}{\ln g(x)} = \frac{\lim\limits_{x \to 2} [f(x)]^2}{\lim\limits_{x \to 2} \ln g(x)} \quad \text{Rule 4}$$

$$= \frac{[\lim\limits_{x \to 2} f(x)]^2}{\ln[\lim\limits_{x \to 2} g(x)]} \quad \text{Rule 6 and Rule 9}$$

$$= \frac{3^2}{\ln 4}$$

$$\approx \frac{9}{1.38629} \approx 6.492$$

YOUR TURN 5 Find $\lim\limits_{x \to 2} [f(x) + g(x)]^2$.

TRY YOUR TURN 5

EXAMPLE 7 Finding a Limit

Find $\lim\limits_{x \to 3} \dfrac{x^2 - x - 1}{\sqrt{x + 1}}$.

SOLUTION

$$\lim_{x \to 3} \frac{x^2 - x - 1}{\sqrt{x + 1}} = \frac{\lim\limits_{x \to 3} (x^2 - x - 1)}{\lim\limits_{x \to 3} \sqrt{x + 1}} \quad \text{Rule 4}$$

$$= \frac{\lim\limits_{x \to 3} (x^2 - x - 1)}{\sqrt{\lim\limits_{x \to 3} (x + 1)}} \quad \text{Rule 6 } (\sqrt{a} = a^{1/2})$$

$$= \frac{3^2 - 3 - 1}{\sqrt{3 + 1}} \quad \text{Rule 5}$$

$$= \frac{5}{\sqrt{4}}$$

$$= \frac{5}{2}$$

As Examples 6 and 7 suggest, the rules for limits actually mean that many limits can be found simply by evaluation. This process is valid for polynomials, rational functions, exponential functions, logarithmic functions, and roots and powers, as long as this does not involve an illegal operation, such as division by 0 or taking the logarithm of a negative number. Division by 0 presents particular problems that can often be solved by algebraic simplification, as the following example shows.

EXAMPLE 8 Finding a Limit

Find $\lim\limits_{x \to 2} \dfrac{x^2 + x - 6}{x - 2}$.

SOLUTION Rule 4 cannot be used here, since

$$\lim_{x \to 2}(x - 2) = 0.$$

The numerator also approaches 0 as x approaches 2, and $0/0$ is meaningless. For $x \neq 2$, we can, however, simplify the function by rewriting the fraction as

$$\frac{x^2 + x - 6}{x - 2} = \frac{(x + 3)(x - 2)}{x - 2} = x + 3.$$

Now Rule 7 can be used.

YOUR TURN 6 Find $\lim\limits_{x \to -3} \dfrac{x^2 - x - 12}{x + 3}$.

$$\lim_{x \to 2} \frac{x^2 + x - 6}{x - 2} = \lim_{x \to 2}(x + 3) = 2 + 3 = 5$$

TRY YOUR TURN 6

NOTE Mathematicians often refer to a limit that gives $0/0$, as in Example 8, as an *indeterminate form*. This means that when the numerator and denominator are polynomials, they must have a common factor, which is why we factored the numerator in Example 8.

EXAMPLE 9 Finding a Limit

Find $\lim\limits_{x \to 4} \dfrac{\sqrt{x} - 2}{x - 4}$.

SOLUTION As $x \to 4$, the numerator approaches 0 and the denominator also approaches 0, giving the meaningless expression $0/0$. In an expression such as this involving square roots, rather than trying to factor, you may find it simpler to use algebra to rationalize the numerator by multiplying both the numerator and the denominator by $\sqrt{x} + 2$. This gives

$$\frac{\sqrt{x} - 2}{x - 4} \cdot \frac{\sqrt{x} + 2}{\sqrt{x} + 2} = \frac{(\sqrt{x})^2 - 2^2}{(x - 4)(\sqrt{x} + 2)} \qquad (a - b)(a + b) = a^2 - b^2$$

$$= \frac{x - 4}{(x - 4)(\sqrt{x} + 2)} = \frac{1}{\sqrt{x} + 2}$$

if $x \neq 4$. Now use the rules for limits.

YOUR TURN 7 Find $\lim\limits_{x \to 1} \dfrac{\sqrt{x} - 1}{x - 1}$.

$$\lim_{x \to 4} \frac{\sqrt{x} - 2}{x - 4} = \lim_{x \to 4} \frac{1}{\sqrt{x} + 2} = \frac{1}{\sqrt{4} + 2} = \frac{1}{2 + 2} = \frac{1}{4}$$

TRY YOUR TURN 7

CAUTION | Simply because the expression in a limit is approaching 0/0, as in Examples 8 and 9, does *not* mean that the limit is 0 or that the limit does not exist. For such a limit, try to simplify the expression using the following principle: **To calculate the limit of $f(x)/g(x)$ as x approaches a, where $f(a) = g(a) = 0$, you should attempt to factor $x - a$ from both the numerator and the denominator**.

EXAMPLE 10 Finding a Limit

Find $\lim\limits_{x \to 1} \dfrac{x^2 - 2x + 1}{(x - 1)^3}$.

Method 1
Algebraic Approach

SOLUTION

Again, Rule 4 cannot be used, since $\lim\limits_{x \to 1} (x - 1)^3 = 0$. If $x \neq 1$, the function can be rewritten as

$$\frac{x^2 - 2x + 1}{(x - 1)^3} = \frac{(x - 1)^2}{(x - 1)^3} = \frac{1}{x - 1}.$$

Then

$$\lim_{x \to 1} \frac{x^2 - 2x + 1}{(x - 1)^3} = \lim_{x \to 1} \frac{1}{x - 1}$$

by Rule 7. None of the rules can be used to find

$$\lim_{x \to 1} \frac{1}{x - 1},$$

but as x approaches 1, the denominator approaches 0 while the numerator stays at 1, making the result larger and larger in magnitude. If $x > 1$, both the numerator and denominator are positive, so $\lim\limits_{x \to 1^+} 1/(x - 1) = \infty$. If $x < 1$, the denominator is negative, so $\lim\limits_{x \to 1^-} 1/(x - 1) = -\infty$. Therefore,

$$\lim_{x \to 1} \frac{x^2 - 2x + 1}{(x - 1)^3} = \lim_{x \to 1} \frac{1}{x - 1} \text{ does not exist.}$$

Method 2
Graphing Calculator Approach

Using the TABLE feature on a TI-84 Plus, we can produce the table of numbers shown in Figure 9, where Y_1 represents the function $y = 1/(x - 1)$. Figure 10 shows a graphing calculator view of the function on $[0, 2]$ by $[-10, 10]$. The behavior of the function indicates a vertical asymptote at $x = 1$, with the limit approaching $-\infty$ from the left and ∞ from the right, so

$$\lim_{x \to 1} \frac{x^2 - 2x + 1}{(x - 1)^3} = \lim_{x \to 1} \frac{1}{x - 1} \text{ does not exist.}$$

Both the table and the graph can be easily generated using a spreadsheet. Consult the *Graphing Calculator and Excel Spreadsheet Manual*, available with this text, for details.

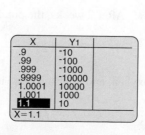

FIGURE 9

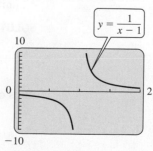

FIGURE 10

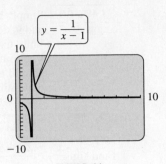

$$y = \frac{1}{x-1}$$

FIGURE 11

| CAUTION | A graphing calculator can give a deceptive view of a function. Figure 11 shows the result if we graph the previous function on $[0, 10]$ by $[-10, 10]$. Near $x = 1$, the graph appears to be a steep line connecting the two pieces. The graph in Figure 10 is more representative of the function near $x = 1$. When using a graphing calculator, you may need to experiment with the viewing window, guided by what you have learned about functions and limits, to get a good picture of a function. On many calculators, extraneous lines connecting parts of the graph can be avoided by using DOT mode rather than CONNECTED mode. |

NOTE Another way to understand the behavior of the function in the previous example near $x = 1$ is to recall from the section on Polynomial and Rational Functions that a rational function often has a vertical asymptote at a value of x where the denominator is 0, although it may not if the numerator there is also 0. In this example, we see after simplifying that the function has a vertical asymptote at $x = 1$ because that would make the denominator of $1/(x - 1)$ equal to 0, while the numerator is 1.

Limits at Infinity

Sometimes it is useful to examine the behavior of the values of $f(x)$ as x gets larger and larger (or more and more negative). The phrase "x approaches infinity," written $x \to \infty$, expresses the fact that x becomes larger without bound. Similarly, the phrase "x approaches negative infinity" (symbolically, $x \to -\infty$) means that x becomes more and more negative without bound (such as -10, -1000, $-10,000$, etc.). The next example illustrates a **limit at infinity**.

EXAMPLE 11 Oxygen Concentration

Suppose a small pond normally contains 12 units of dissolved oxygen in a fixed volume of water. Suppose also that at time $t = 0$ a quantity of organic waste is introduced into the pond, with the oxygen concentration t weeks later given by

$$f(t) = \frac{12t^2 - 15t + 12}{t^2 + 1}.$$

As time goes on, what will be the ultimate concentration of oxygen? Will it return to 12 units?

SOLUTION After 2 weeks, the pond contains

$$f(2) = \frac{12 \cdot 2^2 - 15 \cdot 2 + 12}{2^2 + 1} = \frac{30}{5} = 6$$

units of oxygen, and after 4 weeks, it contains

$$f(4) = \frac{12 \cdot 4^2 - 15 \cdot 4 + 12}{4^2 + 1} \approx 8.5$$

units. Choosing several values of t and finding the corresponding values of $f(t)$, or using a graphing calculator or computer, leads to the table and graph in Figure 12.

The graph suggests that, as time goes on, the oxygen level gets closer and closer to the original 12 units. If so, the line $y = 12$ is a horizontal asymptote. The table suggests that

$$\lim_{t \to \infty} f(t) = 12.$$

Thus, the oxygen concentration will approach 12, but it will never be *exactly* 12.

x	$f(x)$
10	10.515
100	11.85
1000	11.985
10,000	11.9985
100,000	11.99985
(a)	

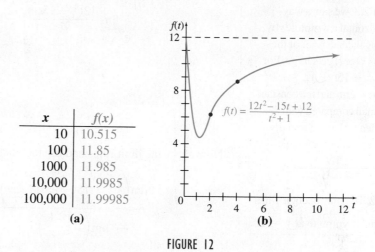

(b)

FIGURE 12

As we saw in the previous example, *limits at infinity or negative infinity, if they exist, correspond to horizontal asymptotes of the graph of the function.* In the previous chapter, we saw one way to find horizontal asymptotes. We will now show a more precise way, based upon some simple limits at infinity. The graphs of $f(x) = 1/x$ (in red) and $g(x) = 1/x^2$ (in blue) shown in Figure 13, as well as the table there, indicate that $\lim_{x \to \infty} 1/x = 0$, $\lim_{x \to -\infty} 1/x = 0$, $\lim_{x \to \infty} 1/x^2 = 0$, and $\lim_{x \to -\infty} 1/x^2 = 0$, suggesting the following rule.

x	$\dfrac{1}{x}$	$\dfrac{1}{x^2}$
-100	-0.01	0.0001
-10	-0.1	0.01
-1	-1	1
1	1	1
10	0.1	0.01
100	0.01	0.0001

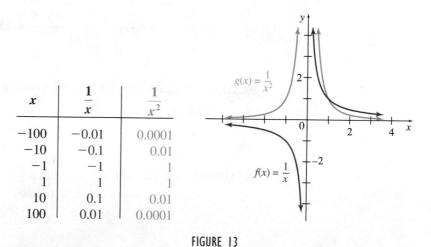

FIGURE 13

Limits at Infinity

For any positive real number n,

$$\lim_{x \to \infty} \frac{1}{x^n} = 0 \qquad \text{and} \qquad \lim_{x \to -\infty} \frac{1}{x^n} = 0.^*$$

---

*If x is negative, x^n does not exist for certain values of n, so the second limit is undefined for those values of n.

The rules for limits given earlier remain unchanged when a is replaced with ∞ or $-\infty$.

To evaluate the limit at infinity of a rational function, divide the numerator and denominator by the largest power of the variable that appears in the denominator, t^2 here, and then use these results. In the previous example, we find that

$$\lim_{t \to \infty} \frac{12t^2 - 15t + 12}{t^2 + 1} = \lim_{t \to \infty} \frac{\dfrac{12t^2}{t^2} - \dfrac{15t}{t^2} + \dfrac{12}{t^2}}{\dfrac{t^2}{t^2} + \dfrac{1}{t^2}}$$

$$= \lim_{t \to \infty} \frac{12 - 15 \cdot \dfrac{1}{t} + 12 \cdot \dfrac{1}{t^2}}{1 + \dfrac{1}{t^2}}.$$

Now apply the limit rules and the fact that $\lim\limits_{t \to \infty} 1/t^n = 0$.

$$\frac{\lim\limits_{t \to \infty}\left(12 - 15 \cdot \dfrac{1}{t} + 12 \cdot \dfrac{1}{t^2}\right)}{\lim\limits_{t \to \infty}\left(1 + \dfrac{1}{t^2}\right)}$$

$$= \frac{\lim\limits_{t \to \infty} 12 - \lim\limits_{t \to \infty} 15 \cdot \dfrac{1}{t} + \lim\limits_{t \to \infty} 12 \cdot \dfrac{1}{t^2}}{\lim\limits_{t \to \infty} 1 + \lim\limits_{t \to \infty} \dfrac{1}{t^2}} \qquad \text{Rules 4 and 2}$$

$$= \frac{12 - 15\left(\lim\limits_{t \to \infty} \dfrac{1}{t}\right) + 12\left(\lim\limits_{t \to \infty} \dfrac{1}{t^2}\right)}{1 + \lim\limits_{t \to \infty} \dfrac{1}{t^2}} \qquad \text{Rule 1}$$

$$= \frac{12 - 15 \cdot 0 + 12 \cdot 0}{1 + 0} = 12. \qquad \text{Limits at infinity}$$

EXAMPLE 12 **Limits at Infinity**

Find each limit.

(a) $\lim\limits_{x \to \infty} \dfrac{8x + 6}{3x - 1}$

SOLUTION We can use the rule $\lim\limits_{x \to \infty} 1/x^n = 0$ to find this limit by first dividing the numerator and denominator by x, as follows.

$$\lim_{x \to \infty} \frac{8x + 6}{3x - 1} = \lim_{x \to \infty} \frac{\dfrac{8x}{x} + \dfrac{6}{x}}{\dfrac{3x}{x} - \dfrac{1}{x}} = \lim_{x \to \infty} \frac{8 + 6 \cdot \dfrac{1}{x}}{3 - \dfrac{1}{x}} = \frac{8 + 0}{3 - 0} = \frac{8}{3}$$

(b) $\lim\limits_{x \to \infty} \dfrac{3x + 2}{4x^3 - 1} = \lim\limits_{x \to \infty} \dfrac{3 \cdot \dfrac{1}{x^2} + 2 \cdot \dfrac{1}{x^3}}{4 - \dfrac{1}{x^3}} = \dfrac{0 + 0}{4 - 0} = \dfrac{0}{4} = 0$

Here, the highest power of x in the denominator is x^3, which is used to divide each term in the numerator and denominator.

(c) $\displaystyle\lim_{x\to\infty} \frac{3x^2 + 2}{4x - 3} = \lim_{x\to\infty} \frac{3x + \dfrac{2}{x}}{4 - \dfrac{3}{x}}$

The highest power of x in the denominator is x (to the first power). There is a higher power of x in the numerator, but we don't divide by this. Notice that the denominator approaches 4, while the numerator becomes infinitely large, so

$$\lim_{x\to\infty} \frac{3x^2 + 2}{4x - 3} = \infty.$$

(d) $\displaystyle\lim_{x\to\infty} \frac{5x^2 - 4x^3}{3x^2 + 2x - 1} = \lim_{x\to\infty} \frac{5 - 4x}{3 + \dfrac{2}{x} - \dfrac{1}{x^2}}$

YOUR TURN 8 Find
$$\lim_{x\to\infty} \frac{2x^2 + 3x - 4}{6x^2 - 5x + 7}.$$

The highest power of x in the denominator is x^2. The denominator approaches 3, while the numerator becomes a negative number that is larger and larger in magnitude, so

$$\lim_{x\to\infty} \frac{5x^2 - 4x^3}{3x^2 + 2x - 1} = -\infty.$$

TRY YOUR TURN 8

The method used in Example 12 is a useful way to rewrite expressions with fractions so that the rules for limits at infinity can be used.

> **Finding Limits at Infinity**
>
> If $f(x) = p(x)/q(x)$, for polynomials $p(x)$ and $q(x)$, $q(x) \neq 0$, $\displaystyle\lim_{x\to -\infty} f(x)$ and $\displaystyle\lim_{x\to\infty} f(x)$ can be found as follows.
>
> 1. Divide $p(x)$ and $q(x)$ by the highest power of x in $q(x)$.
> 2. Use the rules for limits, including the rules for limits at infinity,
>
> $$\lim_{x\to\infty} \frac{1}{x^n} = 0 \qquad \text{and} \qquad \lim_{x\to -\infty} \frac{1}{x^n} = 0,$$
>
> to find the limit of the result from step 1.

For an alternate approach to finding limits at infinity, see Exercise 81.

3.1 EXERCISES

In Exercises 1–4, choose the best answer for each limit.

1. If $\displaystyle\lim_{x\to 2^-} f(x) = 5$ and $\displaystyle\lim_{x\to 2^+} f(x) = 6$, then $\displaystyle\lim_{x\to 2} f(x)$
   - **a.** is 5.
   - **b.** is 6.
   - **c.** does not exist.
   - **d.** is infinite.

2. If $\displaystyle\lim_{x\to 2^-} f(x) = \lim_{x\to 2^+} f(x) = -1$, but $f(2) = 1$, then $\displaystyle\lim_{x\to 2} f(x)$
   - **a.** is -1.
   - **b.** does not exist.
   - **c.** is infinite.
   - **d.** is 1.

3. If $\displaystyle\lim_{x\to 4^-} f(x) = \lim_{x\to 4^+} f(x) = 6$, but $f(4)$ does not exist, then $\displaystyle\lim_{x\to 4} f(x)$
   - **a.** does not exist.
   - **b.** is 6.
   - **c.** is $-\infty$.
   - **d.** is ∞.

4. If $\displaystyle\lim_{x\to 1^-} f(x) = -\infty$ and $\displaystyle\lim_{x\to 1^+} f(x) = -\infty$, then $\displaystyle\lim_{x\to 1} f(x)$
   - **a.** is ∞.
   - **b.** is $-\infty$.
   - **c.** does not exist.
   - **d.** is 1.

Decide whether each limit exists. If a limit exists, estimate its value.

5. a. $\lim\limits_{x \to 3} f(x)$ **b.** $\lim\limits_{x \to 0} f(x)$

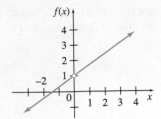

6. a. $\lim\limits_{x \to 2} F(x)$ **b.** $\lim\limits_{x \to -1} F(x)$

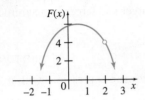

7. a. $\lim\limits_{x \to 0} f(x)$ **b.** $\lim\limits_{x \to 2} f(x)$

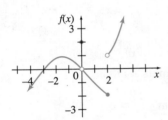

8. a. $\lim\limits_{x \to 3} g(x)$ **b.** $\lim\limits_{x \to 5} g(x)$

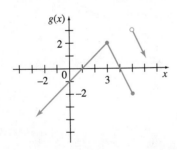

In Exercises 9 and 10, use the graph to find (i) $\lim\limits_{x \to a^-} f(x)$, (ii) $\lim\limits_{x \to a^+} f(x)$, (iii) $\lim\limits_{x \to a} f(x)$, and (iv) $f(a)$ if it exists.

9. a. $a = -2$ **b.** $a = -1$

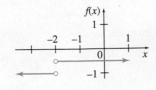

10. a. $a = 1$ **b.** $a = 2$

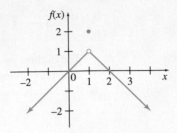

Decide whether each limit exists. If a limit exists, find its value.

11. $\lim\limits_{x \to \infty} f(x)$

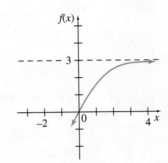

12. $\lim\limits_{x \to -\infty} g(x)$

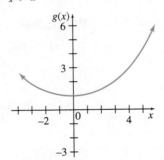

13. Explain why $\lim\limits_{x \to 2} F(x)$ in Exercise 6 exists, but $\lim\limits_{x \to -2} f(x)$ in Exercise 9 does not.

14. In Exercise 10, why does $\lim\limits_{x \to 1} f(x) = 1$, even though $f(1) = 2$?

15. Use the table of values to estimate $\lim\limits_{x \to 1} f(x)$.

x	0.9	0.99	0.999	0.9999	1.0001	1.001	1.01	1.1
$f(x)$	3.9	3.99	3.999	3.9999	4.0001	4.001	4.01	4.1

Complete the tables and use the results to find the indicated limits.

16. If $f(x) = 2x^2 - 4x + 7$, find $\lim\limits_{x \to 1} f(x)$.

x	0.9	0.99	0.999		1.001	1.01	1.1
$f(x)$			5.000002		5.000002		

17. If $k(x) = \dfrac{x^3 - 2x - 4}{x - 2}$, find $\lim\limits_{x \to 2} k(x)$.

x	1.9	1.99	1.999	2.001	2.01	2.1
$k(x)$						

18. If $f(x) = \dfrac{2x^3 + 3x^2 - 4x - 5}{x + 1}$, find $\lim\limits_{x \to -1} f(x)$.

x	-1.1	-1.01	-1.001	-0.999	-0.99	-0.9
$f(x)$						

19. If $h(x) = \dfrac{\sqrt{x} - 2}{x - 1}$, find $\lim\limits_{x \to 1} h(x)$.

x	0.9	0.99	0.999	1.001	1.01	1.1
$h(x)$						

20. If $f(x) = \dfrac{\sqrt{x} - 3}{x - 3}$, find $\lim\limits_{x \to 3} f(x)$.

x	2.9	2.99	2.999	3.001	3.01	3.1
$f(x)$						

Let $\lim\limits_{x \to 4} f(x) = 9$ and $\lim\limits_{x \to 4} g(x) = 27$. **Use the limit rules to find each limit.**

21. $\lim\limits_{x \to 4}[f(x) - g(x)]$

22. $\lim\limits_{x \to 4}[g(x) \cdot f(x)]$

23. $\lim\limits_{x \to 4} \dfrac{f(x)}{g(x)}$

24. $\lim\limits_{x \to 4} \log_3 f(x)$

25. $\lim\limits_{x \to 4} \sqrt{f(x)}$

26. $\lim\limits_{x \to 4} \sqrt[3]{g(x)}$

27. $\lim\limits_{x \to 4} 2^{f(x)}$

28. $\lim\limits_{x \to 4}[1 + f(x)]^2$

29. $\lim\limits_{x \to 4} \dfrac{f(x) + g(x)}{2g(x)}$

30. $\lim\limits_{x \to 4} \dfrac{5g(x) + 2}{1 - f(x)}$

Use the properties of limits to help decide whether each limit exists. If a limit exists, find its value.

31. $\lim\limits_{x \to 3} \dfrac{x^2 - 9}{x - 3}$

32. $\lim\limits_{x \to -2} \dfrac{x^2 - 4}{x + 2}$

33. $\lim\limits_{x \to 1} \dfrac{5x^2 - 7x + 2}{x^2 - 1}$

34. $\lim\limits_{x \to -3} \dfrac{x^2 - 9}{x^2 + x - 6}$

35. $\lim\limits_{x \to -2} \dfrac{x^2 - x - 6}{x + 2}$

36. $\lim\limits_{x \to 5} \dfrac{x^2 - 3x - 10}{x - 5}$

37. $\lim\limits_{x \to 0} \dfrac{1/(x + 3) - 1/3}{x}$

38. $\lim\limits_{x \to 0} \dfrac{-1/(x + 2) + 1/2}{x}$

39. $\lim\limits_{x \to 25} \dfrac{\sqrt{x} - 5}{x - 25}$

40. $\lim\limits_{x \to 36} \dfrac{\sqrt{x} - 6}{x - 36}$

41. $\lim\limits_{h \to 0} \dfrac{(x + h)^2 - x^2}{h}$

42. $\lim\limits_{h \to 0} \dfrac{(x + h)^3 - x^3}{h}$

43. $\lim\limits_{x \to \infty} \dfrac{3x}{7x - 1}$

44. $\lim\limits_{x \to -\infty} \dfrac{8x + 2}{4x - 5}$

45. $\lim\limits_{x \to -\infty} \dfrac{3x^2 + 2x}{2x^2 - 2x + 1}$

46. $\lim\limits_{x \to \infty} \dfrac{x^2 + 2x - 5}{3x^2 + 2}$

47. $\lim\limits_{x \to \infty} \dfrac{3x^3 + 2x - 1}{2x^4 - 3x^3 - 2}$

48. $\lim\limits_{x \to \infty} \dfrac{2x^2 - 1}{3x^4 + 2}$

49. $\lim\limits_{x \to \infty} \dfrac{2x^3 - x - 3}{6x^2 - x - 1}$

50. $\lim\limits_{x \to \infty} \dfrac{x^4 - x^3 - 3x}{7x^2 + 9}$

51. $\lim\limits_{x \to \infty} \dfrac{2x^2 - 7x^4}{9x^2 + 5x - 6}$

52. $\lim\limits_{x \to \infty} \dfrac{-5x^3 - 4x^2 + 8}{6x^2 + 3x + 2}$

53. Let $f(x) = \begin{cases} x^3 + 2 & \text{if } x \neq -1 \\ 5 & \text{if } x = -1 \end{cases}$. Find $\lim\limits_{x \to -1} f(x)$.

54. Let $g(x) = \begin{cases} 0 & \text{if } x = -2 \\ \frac{1}{2}x^2 - 3 & \text{if } x \neq -2 \end{cases}$. Find $\lim\limits_{x \to -2} g(x)$.

55. Let $f(x) = \begin{cases} x - 1 & \text{if } x < 3 \\ 2 & \text{if } 3 \leq x \leq 5 \\ x + 3 & \text{if } x > 5 \end{cases}$.

  a. Find $\lim\limits_{x \to 3} f(x)$.   **b.** Find $\lim\limits_{x \to 5} f(x)$.

56. Let $g(x) = \begin{cases} 5 & \text{if } x < 0 \\ x^2 - 2 & \text{if } 0 \leq x \leq 3 \\ 7 & \text{if } x > 3 \end{cases}$.

  a. Find $\lim\limits_{x \to 0} g(x)$.   **b.** Find $\lim\limits_{x \to 3} g(x)$.

In Exercises 57–60, calculate the limit in the specified exercise, using a table such as in Exercises 15–20. Verify your answer by using a graphing calculator to zoom in on the point on the graph.

57. Exercise 31

58. Exercise 32

59. Exercise 33

60. Exercise 34

61. Let $F(x) = \dfrac{3x}{(x + 2)^3}$.

  a. Find $\lim\limits_{x \to -2} F(x)$.

  b. Find the vertical asymptote of the graph of $F(x)$.

  c. Compare your answers for parts a and b. What can you conclude?

62. Let $G(x) = \dfrac{-6}{(x - 4)^2}$.

  a. Find $\lim\limits_{x \to 4} G(x)$.

  b. Find the vertical asymptote of the graph of $G(x)$.

  c. Compare your answers for parts a and b. Are they related? How?

63. How can you tell that the graph in Figure 10 is more representative of the function $f(x) = 1/(x - 1)$ than the graph in Figure 11?

64. A friend who is confused about limits wonders why you investigate the value of a function closer and closer to a point, instead of just finding the value of a function at the point. How would you respond?

65. Use a graph of $f(x) = e^x$ to answer the following questions.

a. Find $\lim_{x \to -\infty} e^x$.

b. Where does the function e^x have a horizontal asymptote?

66. Use a graphing calculator to answer the following questions.

a. From a graph of $y = xe^{-x}$, what do you think is the value of $\lim_{x \to \infty} xe^{-x}$? Support this by evaluating the function for several large values of x.

b. Repeat part a, this time using the graph of $y = x^2e^{-x}$.

c. Based on your results from parts a and b, what do you think is the value of $\lim_{x \to \infty} x^ne^{-x}$, where n is a positive integer? Support this by experimenting with other positive integers n.

67. Use a graph of $f(x) = \ln x$ to answer the following questions.

a. Find $\lim_{x \to 0^+} \ln x$.

b. Where does the function $\ln x$ have a vertical asymptote?

68. Use a graphing calculator to answer the following questions.

a. From a graph of $y = x \ln x$, what do you think is the value of $\lim_{x \to 0^+} x \ln x$? Support this by evaluating the function for several small values of x.

b. Repeat part a, this time using the graph of $y = x(\ln x)^2$.

c. Based on your results from parts a and b, what do you think is the value of $\lim_{x \to 0^+} x(\ln x)^n$, where n is a positive integer? Support this by experimenting with other positive integers n.

69. Explain in your own words why the rules for limits at infinity should be true.

70. Explain in your own words what Rule 4 for limits means.

Find each of the following limits (a) by investigating values of the function near the x-value where the limit is taken, and (b) using a graphing calculator to view the function near that value of x.

71. $\lim_{x \to 1} \dfrac{x^4 + 4x^3 - 9x^2 + 7x - 3}{x - 1}$

72. $\lim_{x \to 2} \dfrac{x^4 + x - 18}{x^2 - 4}$

73. $\lim_{x \to -1} \dfrac{x^{1/3} + 1}{x + 1}$

74. $\lim_{x \to 4} \dfrac{x^{3/2} - 8}{x + x^{1/2} - 6}$

Use a graphing calculator to graph the function. (a) Determine the limit from the graph. (b) Explain how your answer could be determined from the expression for $f(x)$.

75. $\lim_{x \to \infty} \dfrac{\sqrt{9x^2 + 5}}{2x}$

76. $\lim_{x \to -\infty} \dfrac{\sqrt{9x^2 + 5}}{2x}$

77. $\lim_{x \to -\infty} \dfrac{\sqrt{36x^2 + 2x + 7}}{3x}$

78. $\lim_{x \to \infty} \dfrac{\sqrt{36x^2 + 2x + 7}}{3x}$

79. $\lim_{x \to \infty} \dfrac{(1 + 5x^{1/3} + 2x^{5/3})^3}{x^5}$

80. $\lim_{x \to -\infty} \dfrac{(1 + 5x^{1/3} + 2x^{5/3})^3}{x^5}$

81. Explain why the following rules can be used to find $\lim_{x \to \infty} [p(x)/q(x)]$:

a. If the degree of $p(x)$ is less than the degree of $q(x)$, the limit is 0.

b. If the degree of $p(x)$ is equal to the degree of $q(x)$, the limit is A/B, where A and B are the leading coefficients of $p(x)$ and $q(x)$, respectively.

c. If the degree of $p(x)$ is greater than the degree of $q(x)$, the limit is ∞ or $-\infty$.

APPLICATIONS

Business and Economics

82. APPLY IT Consumer Demand When the price of an essential commodity (such as gasoline) rises rapidly, consumption drops slowly at first. If the price continues to rise, however, a "tipping" point may be reached, at which consumption takes a sudden substantial drop. Suppose the accompanying graph shows the consumption of gasoline, $G(t)$, in millions of gallons, in a certain area. We assume that the price is rising rapidly. Here t is time in months after the price began rising. Use the graph to find the following.

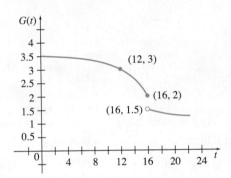

a. $\lim_{t \to 12} G(t)$

b. $\lim_{t \to 16} G(t)$

c. $G(16)$

d. The tipping point (in months)

83. Sales Tax Officials in California tend to raise the sales tax in years in which the state faces a budget deficit and then cut the tax when the state has a surplus. The graph below shows the California state sales tax since it was first established in 1933. Let $T(x)$ represent the sales tax per dollar spent in year x. Find the following. *Source: California State.*

a. $\lim_{x \to 53} T(x)$

b. $\lim_{x \to 09^-} T(x)$

c. $\lim_{x \to 09^+} T(x)$

d. $\lim_{x \to 09} T(x)$

e. $T(09)$

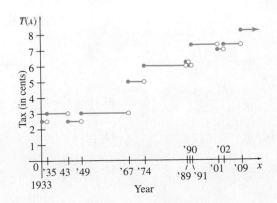

84. Postage The graph below shows how the postage required to mail a letter in the United States has changed in recent years. Let $C(t)$ be the cost to mail a letter in the year t. Find the following. *Source: United States Postal Service.*

a. $\lim_{t \to 2009^-} C(t)$

b. $\lim_{t \to 2009^+} C(t)$

c. $\lim_{t \to 2009} C(t)$

d. $C(2009)$

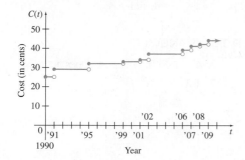

85. Average Cost The cost (in dollars) for manufacturing a particular DVD is

$$C(x) = 15{,}000 + 6x,$$

where x is the number of DVDs produced. Recall from the previous chapter that the average cost per DVD, denoted by $\overline{C}(x)$, is found by dividing $C(x)$ by x. Find and interpret $\lim_{x \to \infty} \overline{C}(x)$.

86. Average Cost In Chapter 1, we saw that the cost to fly x miles on American Airlines could be approximated by the equation

$$C(x) = 0.0738x + 111.83.$$

Recall from the previous chapter that the average cost per mile, denoted by $\overline{C}(x)$, is found by dividing $C(x)$ by x. Find and interpret $\lim_{x \to \infty} \overline{C}(x)$. *Source: American Airlines.*

87. Employee Productivity A company training program has determined that, on the average, a new employee produces $P(s)$ items per day after s days of on-the-job training, where

$$P(s) = \frac{63s}{s + 8}.$$

Find and interpret $\lim_{s \to \infty} P(s)$.

88. Preferred Stock In business finance, an annuity is a series of equal payments received at equal intervals for a finite period of time. The *present value* of an n-period annuity takes the form

$$P = R \left[\frac{1 - (1 + i)^{-n}}{i} \right],$$

where R is the amount of the periodic payment and i is the fixed interest rate per period. Many corporations raise money by issuing preferred stock. Holders of the preferred stock, called a *perpetuity*, receive payments that take the form of an annuity in that the amount of the payment never changes. However, normally the payments for preferred stock do not end but theoretically continue forever. Find the limit of this present value equation as n approaches infinity to derive a formula for the present value of a share of preferred stock paying a periodic dividend R. *Source: Robert D. Campbell.*

89. Growing Annuities For some annuities encountered in business finance, called *growing annuities*, the amount of the periodic payment is not constant but grows at a constant periodic rate. Leases with escalation clauses can be examples of growing annuities. The present value of a growing annuity takes the form

$$P = \frac{R}{i - g} \left[1 - \left(\frac{1 + g}{1 + i} \right)^n \right],$$

where

$R =$ amount of the next annuity payment,

$g =$ expected constant annuity growth rate,

$i =$ required periodic return at the time the annuity is evaluated,

$n =$ number of periodic payments.

A corporation's common stock may be thought of as a claim on a growing annuity where the annuity is the company's annual dividend. However, in the case of common stock, these payments have no contractual end but theoretically continue forever. Compute the limit of the expression above as n approaches infinity to derive the Gordon–Shapiro Dividend Model popularly used to estimate the value of common stock. Make the reasonable assumption that $i > g$. (*Hint:* What happens to a^n as $n \to \infty$ if $0 < a < 1$?) *Source: Robert D. Campbell.*

Life Sciences

90. Alligator Teeth Researchers have developed a mathematical model that can be used to estimate the number of teeth $N(t)$ at time t (days of incubation) for *Alligator mississippiensis*, where

$$N(t) = 71.8e^{-8.96e^{-0.0685t}}.$$

Source: Journal of Theoretical Biology.

a. Find $N(65)$, the number of teeth of an alligator that hatched after 65 days.

b. Find $\lim_{t \to \infty} N(t)$ and use this value as an estimate of the number of teeth of a newborn alligator. (*Hint:* See Exercise 65.) Does this estimate differ significantly from the estimate of part a?

91. Sediment To develop strategies to manage water quality in polluted lakes, biologists must determine the depths of sediments and the rate of sedimentation. It has been determined that the depth of sediment $D(t)$ (in centimeters) with respect to time (in years before 1990) for Lake Coeur d'Alene, Idaho, can be estimated by the equation

$$D(t) = 155(1 - e^{-0.0133t}).$$

Source: Mathematics Teacher.

a. Find $D(20)$ and interpret.

b. Find $\lim_{t \to \infty} D(t)$ and interpret.

92. Drug Concentration The concentration of a drug in a patient's bloodstream h hours after it was injected is given by

$$A(h) = \frac{0.17h}{h^2 + 2}.$$

Find and interpret $\lim\limits_{h \to \infty} A(h)$.

Social Sciences

93. Legislative Voting Members of a legislature often must vote repeatedly on the same bill. As time goes on, members may change their votes. Suppose that p_0 is the probability that an individual legislator favors an issue before the first roll call vote, and suppose that p is the probability of a change in position from one vote to the next. Then the probability that the legislator will vote "yes" on the nth roll call is given by

$$p_n = \frac{1}{2} + \left(p_0 - \frac{1}{2}\right)(1 - 2p)^n.$$

For example, the chance of a "yes" on the third roll call vote is

$$p_3 = \frac{1}{2} + \left(p_0 - \frac{1}{2}\right)(1 - 2p)^3.$$

Source: Mathematics in the Behavioral and Social Sciences.

Suppose that there is a chance of $p_0 = 0.7$ that Congressman Stephens will favor the budget appropriation bill before the first roll call, but only a probability of $p = 0.2$ that he will change his mind on the subsequent vote. Find and interpret the following.

a. p_2 **b.** p_4

c. p_8 **d.** $\lim\limits_{n \to \infty} p_n$

YOUR TURN ANSWERS

1. 3   **2.** 4   **3.** 5   **4.** Does not exist.   **5.** 49

6. -7 **7.** 1/2   **8.** 1/3

3.2 Continuity

APPLY IT How does the average cost per day of a rental car change with the number of days the car is rented?

We will answer this question in Exercise 38.

In 2009, Congress passed legislation raising the federal minimum wage for the third time in three years. Figure 14 below shows how that wage has varied since it was instituted in 1938. We will denote this function by $f(t)$, where t is the year. *Source: U.S. Department of Labor.*

Notice from the graph that $\lim\limits_{t \to 1997^-} f(t) = 4.75$ and that $\lim\limits_{t \to 1997^+} f(t) = 5.15$, so that $\lim\limits_{t \to 1997} f(t)$ does not exist. Notice also that $f(1997) = 5.15$. A point such as this, where a function has a sudden sharp break, is a point where the function is *discontinuous*. In this case, the discontinuity is caused by the jump in the minimum wage from $4.75 per hour to $5.15 per hour in 1997.

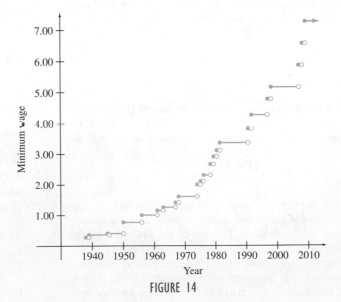

FIGURE 14

Intuitively speaking, a function is *continuous* at a point if you can draw the graph of the function in the vicinity of that point without lifting your pencil from the paper. As we already mentioned, this would not be possible in Figure 14 if it were drawn correctly; there would be a break in the graph at $t = 1997$, for example. Conversely, a function is discontinuous at any x-value where the pencil *must* be lifted from the paper in order to draw the graph on both sides of the point. A more precise definition is as follows.

Continuity at $x = c$

A function f is **continuous** at $x = c$ if the following three conditions are satisfied:

1. $f(c)$ is defined,

2. $\lim\limits_{x \to c} f(x)$ exists, and

3. $\lim\limits_{x \to c} f(x) = f(c)$.

If f is not continuous at c, it is **discontinuous** there.

The following example shows how to check a function for continuity at a specific point. We use a three-step test, and if any step of the test fails, the function is not continuous at that point.

EXAMPLE 1   Continuity

Determine if each function is continuous at the indicated x-value.

(a) $f(x)$ in Figure 15 at $x = 3$

  SOLUTION

  Step 1 Does the function exist at $x = 3$?

  The open circle on the graph of Figure 15 at the point where $x = 3$ means that $f(x)$ does not exist at $x = 3$. Since the function does not pass the first test, it is discontinuous at $x = 3$, and there is no need to proceed to Step 2.

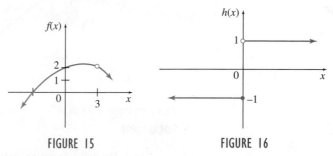

FIGURE 15   FIGURE 16

(b) $h(x)$ in Figure 16 at $x = 0$

  SOLUTION

  Step 1 Does the function exist at $x = 0$?

  According to the graph in Figure 16, $h(0)$ exists and is equal to -1.

  Step 2 Does the limit exist at $x = 0$?

  As x approaches 0 from the left, $h(x)$ is -1. As x approaches 0 from the right, however, $h(x)$ is 1. In other words,

$$\lim_{x \to 0^-} h(x) = -1,$$

  while

$$\lim_{x \to 0^+} h(x) = 1.$$

Since no single number is approached by the values of $h(x)$ as x approaches 0, the limit $\lim\limits_{x \to 0} h(x)$ does not exist. Since the function does not pass the second test, it is discontinuous at $x = 0$, and there is no need to proceed to Step 3.

(c) $g(x)$ in Figure 17 at $x = 4$

SOLUTION

Step 1 Is the function defined at $x = 4$?

In Figure 17, the heavy dot above 4 shows that $g(4)$ is defined. In fact, $g(4) = 1$.

Step 2 Does the limit exist at $x = 4$?

The graph shows that

$$\lim_{x \to 4^-} g(x) = -2, \text{ and } \lim_{x \to 4^+} g(x) = -2.$$

Therefore, the limit exists at $x = 4$ and

$$\lim_{x \to 4} g(x) = -2.$$

Step 3 Does $g(4) = \lim\limits_{x \to 4} g(x)$?

Using the results of Step 1 and Step 2, we see that $g(4) \neq \lim\limits_{x \to 4} g(x)$.
Since the function does not pass the third test, it is discontinuous at $x = 4$.

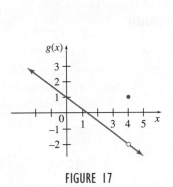

FIGURE 17

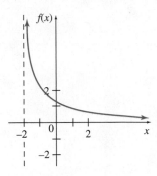

FIGURE 18

(d) $f(x)$ in Figure 18 at $x = -2$.

SOLUTION

Step 1 Does the function exist at $x = -2$?

The function f graphed in Figure 18 is not defined at $x = -2$. Since the function does not pass the first test, it is discontinuous at $x = -2$. (Function f is continuous at any value of x greater than -2, however.)　■

Notice that the function in part (a) of Example 1 could be made continuous simply by defining $f(3) = 2$. Similarly, the function in part (c) could be made continuous by redefining $g(4) = -2$. In such cases, when the function can be made continuous at a specific point simply by defining or redefining it at that point, the function is said to have a **removable discontinuity**.

A function is said to be **continuous on an open interval** if it is continuous at every x-value in the interval. Continuity on a closed interval is slightly more complicated because we must decide what to do with the endpoints. We will say that a function f is **continuous from the right** at $x = c$ if $\lim_{x \to c^+} f(x) = f(c)$. A function f is **continuous from the left** at $x = c$ if $\lim_{x \to c^-} f(x) = f(c)$. With these ideas, we can now define continuity on a closed interval.

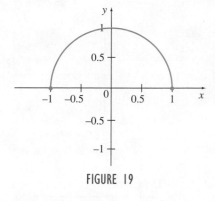

FIGURE 19

> ### Continuity on a Closed Interval
> A function is **continuous on a closed interval** $[a, b]$ if
>
> **1.** it is continuous on the open interval (a, b),
>
> **2.** it is continuous from the right at $x = a$, and
>
> **3.** it is continuous from the left at $x = b$.

For example, the function $f(x) = \sqrt{1 - x^2}$, shown in Figure 19, is continuous on the closed interval $[-1, 1]$. By defining continuity on a closed interval in this way, we need not worry about the fact that $\sqrt{1 - x^2}$ does not exist to the left of $x = -1$ or to the right of $x = 1$.

The table below lists some key functions and tells where each is continuous.

Continuous Functions		
Type of Function	**Where It Is Continuous**	**Graphic Example**
Polynomial Function $y = a_n x^n + a_{n-1} x^{n-1} + \cdots + a_1 x + a_0$, where a_n $a_{n-1}, \ldots, a_1, a_0$ are real numbers, not all 0	For all x	
Rational Function $y = \dfrac{p(x)}{q(x)}$, where $p(x)$ and $q(x)$ are polynomials, with $q(x) \neq 0$	For all x where $q(x) \neq 0$	
Root Function $y = \sqrt{ax + b}$, where a and b are real numbers, with $a \neq 0$ and $ax + b \geq 0$	For all x where $ax + b \geq 0$	

(continued)

Continuous Functions (cont.)		
Type of Function	**Where It Is Continuous**	**Graphic Example**
Exponential Function $y = a^x$ where $a > 0$	For all x	
Logarithmic Function $y = \log_a x$ where $a > 0$, $\quad a \neq 1$	For all $x > 0$	

Continuous functions are nice to work with because finding $\lim\limits_{x \to c} f(x)$ is simple if f is continuous: just evaluate $f(c)$.

When a function is given by a graph, any discontinuities are clearly visible. When a function is given by a formula, it is usually continuous at all x-values except those where the function is undefined or possibly where there is a change in the defining formula for the function, as shown in the following examples.

EXAMPLE 2   Continuity

Find all values $x = a$ where the function is discontinuous.

(a) $f(x) = \dfrac{4x - 3}{2x - 7}$

SOLUTION   This rational function is discontinuous wherever the denominator is zero. There is a discontinuity when $a = 7/2$.

(b) $g(x) = e^{2x - 3}$

SOLUTION   This exponential function is continuous for all x.   TRY YOUR TURN 1

YOUR TURN 1 Find all values $x = a$ where the function is discontinuous.

$f(x) = \sqrt{5x + 3}$

EXAMPLE 3   Continuity

Find all values of x where the following piecewise function is discontinuous.

$$f(x) = \begin{cases} x + 1 & \text{if } x < 1 \\ x^2 - 3x + 4 & \text{if } 1 \leq x \leq 3. \\ 5 - x & \text{if } x > 3 \end{cases}$$

SOLUTION   Since each piece of this function is a polynomial, the only x-values where f might be discontinuous here are 1 and 3. We investigate at $x = 1$ first. From the left, where x-values are less than 1,

$$\lim_{x \to 1^-} f(x) = \lim_{x \to 1^-} (x + 1) = 1 + 1 = 2.$$

From the right, where x-values are greater than 1,

$$\lim_{x \to 1^+} f(x) = \lim_{x \to 1^+} (x^2 - 3x + 4) = 1^2 - 3 + 4 = 2.$$

Furthermore, $f(1) = 1^2 - 3 + 4 = 2$, so $\lim_{x \to 1} f(x) = f(1) = 2$. Thus f is continuous at $x = 1$ since $f(1) = \lim_{x \to 1} f(x)$.

Now let us investigate $x = 3$. From the left,

$$\lim_{x \to 3^-} f(x) = \lim_{x \to 3^-} (x^2 - 3x + 4) = 3^2 - 3(3) + 4 = 4.$$

From the right,

$$\lim_{x \to 3^+} f(x) = \lim_{x \to 3^+} (5 - x) = 5 - 3 = 2.$$

Because $\lim_{x \to 3^-} f(x) \neq \lim_{x \to 3^+} f(x)$, the limit $\lim_{x \to 3} f(x)$ does not exist, so f is discontinuous at $x = 3$, regardless of the value of $f(3)$.

The graph of $f(x)$ can be drawn by considering each of the three parts separately. In the first part, the line $y = x + 1$ is drawn including only the section of the line to the left of $x = 1$. The other two parts are drawn similarly, as illustrated in Figure 20. We can see by the graph that the function is continuous at $x = 1$ and discontinuous at $x = 3$, which confirms our solution above.

TRY YOUR TURN 2

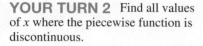

YOUR TURN 2 Find all values of x where the piecewise function is discontinuous.

$$f(x) = \begin{cases} 5x - 4 & \text{if } x < 0 \\ x^2 & \text{if } 0 \leq x \leq 3 \\ x + 6 & \text{if } x > 3 \end{cases}$$

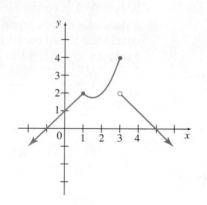

FIGURE 20

TECHNOLOGY NOTE

Some graphing calculators have the ability to draw piecewise functions. On the TI-84 Plus, letting

$$Y_1 = (X + 1)(X < 1) + (X^2 - 3X + 4)(1 \leq X)(X \leq 3) + (5 - X)(X > 3)$$

produces the graph shown in Figure 21(a).

CAUTION It is important here that the graphing mode be set on DOT rather than CONNECTED. Otherwise, the calculator will show a line segment at $x = 3$ connecting the parabola to the line, as in Figure 21(b), although such a segment does not really exist.

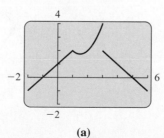

(a)

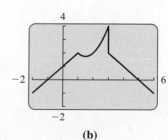

(b)

FIGURE 21

EXAMPLE 4   Cost Analysis

A trailer rental firm charges a flat $8 to rent a hitch. The trailer itself is rented for $22 per day or fraction of a day. Let $C(x)$ represent the cost of renting a hitch and trailer for x days.

(a) Graph C.

SOLUTION   The charge for one day is $8 for the hitch and $22 for the trailer, or $30. In fact, if $0 < x \leq 1$, then $C(x) = 30$. To rent the trailer for more than one day, but not more than two days, the charge is $8 + 2 \cdot 22 = 52$ dollars. For any value of x satisfying $1 < x \leq 2$, the cost is $C(x) = 52$. Also, if $2 < x \leq 3$, then $C(x) = 74$. These results lead to the graph in Figure 22.

(b) Find any values of x where C is discontinuous.

SOLUTION   As the graph suggests, C is discontinuous at $x = 1, 2, 3, 4$, and all other positive integers.

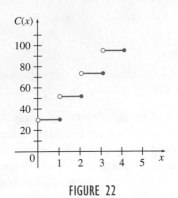

FIGURE 22

One application of continuity is the **Intermediate Value Theorem**, which says that if a function is continuous on a closed interval $[a, b]$, the function takes on every value between $f(a)$ and $f(b)$. For example, if $f(1) = -3$ and $f(2) = 5$, then f must take on every value between -3 and 5 as x varies over the interval $[1, 2]$. In particular (in this case), there must be a value of x in the interval $(1, 2)$ such that $f(x) = 0$. If f were discontinuous, however, this conclusion would not necessarily be true. This is important because, if we are searching for a solution to $f(x) = 0$ in $[1, 2]$, we would like to know that a solution exists.

3.2   EXERCISES

In Exercises 1–6, find all values $x = a$ where the function is discontinuous. For each point of discontinuity, give (a) $f(a)$ if it exists, (b) $\lim_{x \to a^-} f(x)$, (c) $\lim_{x \to a^+} f(x)$, (d) $\lim_{x \to a} f(x)$, and (e) identify which conditions for continuity are not met. Be sure to note when the limit doesn't exist.

1.

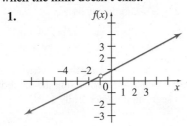

2.

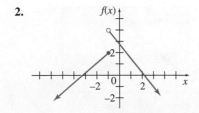

3.

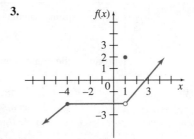

4.

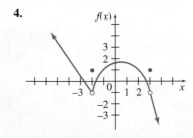

5.

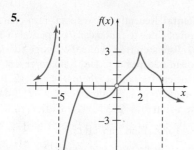

6.

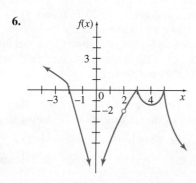

Find all values $x = a$ where the function is discontinuous. For each value of x, give the limit of the function as x approaches a. Be sure to note when the limit doesn't exist.

7. $f(x) = \dfrac{5 + x}{x(x - 2)}$

8. $f(x) = \dfrac{-2x}{(2x + 1)(3x + 6)}$

9. $f(x) = \dfrac{x^2 - 4}{x - 2}$

10. $f(x) = \dfrac{x^2 - 25}{x + 5}$

11. $p(x) = x^2 - 4x + 11$

12. $q(x) = -3x^3 + 2x^2 - 4x + 1$

13. $p(x) = \dfrac{|x + 2|}{x + 2}$

14. $r(x) = \dfrac{|5 - x|}{x - 5}$

15. $k(x) = e^{\sqrt{x-1}}$

16. $j(x) = e^{1/x}$

17. $r(x) = \ln\left|\dfrac{x}{x - 1}\right|$

18. $j(x) = \ln\left|\dfrac{x + 2}{x - 3}\right|$

In Exercises 19–24, (a) graph the given function, (b) find all values of x where the function is discontinuous, and (c) find the limit from the left and from the right at any values of x found in part b.

19. $f(x) = \begin{cases} 1 & \text{if } x < 2 \\ x + 3 & \text{if } 2 \le x \le 4 \\ 7 & \text{if } x > 4 \end{cases}$

20. $f(x) = \begin{cases} x - 1 & \text{if } x < 1 \\ 0 & \text{if } 1 \le x \le 4 \\ x - 2 & \text{if } x > 4 \end{cases}$

21. $g(x) = \begin{cases} 11 & \text{if } x < -1 \\ x^2 + 2 & \text{if } -1 \le x \le 3 \\ 11 & \text{if } x > 3 \end{cases}$

22. $g(x) = \begin{cases} 0 & \text{if } x < 0 \\ x^2 - 5x & \text{if } 0 \le x \le 5 \\ 5 & \text{if } x > 5 \end{cases}$

23. $h(x) = \begin{cases} 4x + 4 & \text{if } x \le 0 \\ x^2 - 4x + 4 & \text{if } x > 0 \end{cases}$

24. $h(x) = \begin{cases} x^2 + x - 12 & \text{if } x \le 1 \\ 3 - x & \text{if } x > 1 \end{cases}$

In Exercises 25–28, find the value of the constant k that makes the function continuous.

25. $f(x) = \begin{cases} kx^2 & \text{if } x \le 2 \\ x + k & \text{if } x > 2 \end{cases}$

26. $g(x) = \begin{cases} x^3 + k & \text{if } x \le 3 \\ kx - 5 & \text{if } x > 3 \end{cases}$

27. $g(x) = \begin{cases} \dfrac{2x^2 - x - 15}{x - 3} & \text{if } x \ne 3 \\ kx - 1 & \text{if } x = 3 \end{cases}$

28. $h(x) = \begin{cases} \dfrac{3x^2 + 2x - 8}{x + 2} & \text{if } x \ne -2 \\ 3x + k & \text{if } x = -2 \end{cases}$

29. Explain in your own words what the Intermediate Value Theorem says and why it seems plausible.

30. Explain why $\lim\limits_{x \to 2}(3x^2 + 8x)$ can be evaluated by substituting $x = 2$.

In Exercises 31–32, (a) use a graphing calculator to tell where the rational function $P(x)/Q(x)$ is discontinuous, and (b) verify your answer from part (a) by using the graphing calculator to plot $Q(x)$ and determine where $Q(x) = 0$. You will need to choose the viewing window carefully.

31. $f(x) = \dfrac{x^2 + x + 2}{x^3 - 0.9x^2 + 4.14x - 5.4}$

32. $f(x) = \dfrac{x^2 + 3x - 2}{x^3 - 0.9x^2 + 4.14x + 5.4}$

33. Let $g(x) = \dfrac{x + 4}{x^2 + 2x - 8}$. Determine all values of x at which g is discontinuous, and for each of these values of x, define g in such a manner so as to remove the discontinuity, if possible. Choose one of the following. *Source: Society of Actuaries.*

 a. g is discontinuous only at -4 and 2.
   Define $g(-4) = -\frac{1}{6}$ to make g continuous at -4.
   $g(2)$ cannot be defined to make g continuous at 2.

 b. g is discontinuous only at -4 and 2.
   Define $g(-4) = -\frac{1}{6}$ to make g continuous at -4.
   Define $g(2) = 6$ to make g continuous at 2.

 c. g is discontinuous only at -4 and 2.
   $g(-4)$ cannot be defined to make g continuous at -4.
   $g(2)$ cannot be defined to make g continuous at 2.

 d. g is discontinuous only at 2.
   Define $g(2) = 6$ to make g continuous at 2.

 e. g is discontinuous only at 2.
   $g(2)$ cannot be defined to make g continuous at 2.

34. Tell at what values of x the function $f(x)$ in Figure 8 from the previous section is discontinuous. Explain why it is discontinuous at each of these values.

APPLICATIONS

Business and Economics

35. Production The graph shows the profit from the daily production of x thousand kilograms of an industrial chemical. Use the graph to find the following limits.

   a. $\lim\limits_{x \to 6} P(x)$ **b.** $\lim\limits_{x \to 10^-} P(x)$ **c.** $\lim\limits_{x \to 10^+} P(x)$

   d. $\lim\limits_{x \to 10} P(x)$

   e. Where is the function discontinuous? What might account for such a discontinuity?

   f. Use the graph to estimate the number of units of the chemical that must be produced before the second shift is as profitable as the first.

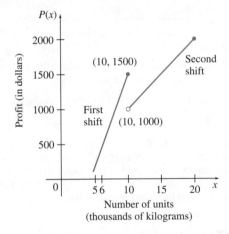

36. Cost Analysis The cost to transport a mobile home depends on the distance, x, in miles that the home is moved. Let $C(x)$ represent the cost to move a mobile home x miles. One firm charges as follows.

Cost per Mile	Distance in Miles
$4.00	$0 < x \le 150$
$3.00	$150 < x \le 400$
$2.50	$400 < x$

Find the cost to move a mobile home the following distances.

   a. 130 miles   **b.** 150 miles   **c.** 210 miles

   d. 400 miles   **e.** 500 miles

   f. Where is C discontinuous?

37. Cost Analysis A company charges $1.25 per lb for a certain fertilizer on all orders 100 lb. or less, and $1 per lb for orders over 100 lb. Let $F(x)$ represent the cost for buying x lb of the fertilizer. Find the cost of buying the following.

   a. 80 lb   **b.** 150 lb   **c.** 100 lb

   d. Where is F discontinuous?

38. APPLY IT Car Rental Recently, a car rental firm charged $36 per day or portion of a day to rent a car for a period of 1 to 5 days. Days 6 and 7 were then free, while the charge for days 8 through 12 was again $36 per day. Let $A(t)$ represent the average cost to rent the car for t days, where $0 < t \le 12$. Find the average cost of a rental for the following number of days.

   a. 4   **b.** 5   **c.** 6   **d.** 7   **e.** 8

   f. Find $\lim\limits_{x \to 5^-} A(t)$.   **g.** Find $\lim\limits_{x \to 5^+} A(t)$.

   h. Where is A discontinuous on the given interval?

39. Postage To send international first class mail (large envelopes) from the United States to Australia in 2010, it cost $1.24 for the first ounce, $0.84 for each additional ounce up to a total of 8 oz, and $1.72 for each additional four ounces after that up to a total of 64 oz. Let $C(x)$ be the cost to mail x ounces. Find the following. *Source: U.S. Postal Service.*

   a. $\lim\limits_{x \to 3^-} C(x)$ **b.** $\lim\limits_{x \to 3^+} C(x)$ **c.** $\lim\limits_{x \to 3} C(x)$

   d. $C(3)$ **e.** $\lim\limits_{x \to 14^+} C(x)$ **f.** $\lim\limits_{x \to 14^-} C(x)$

   g. $\lim\limits_{x \to 14} C(x)$ **h.** $C(14)$

   i. Find all values on the interval $(0, 64)$ where the function C is discontinuous.

Life Sciences

40. Pregnancy A woman's weight naturally increases during the course of a pregnancy. When she delivers, her weight immediately decreases by the approximate weight of the child. Suppose that a 120-lb woman gains 27 lb during pregnancy, delivers a 7-lb baby, and then, through diet and exercise, loses the remaining weight during the next 20 weeks.

   a. Graph the weight gain and loss during the pregnancy and the 20 weeks following the birth of the baby. Assume that the pregnancy lasts 40 weeks, that delivery occurs immediately after this time interval, and that the weight gain/loss before and after birth is linear.

   b. Is this a continuous function? If not, then find the value(s) of t where the function is discontinuous.

41. Poultry Farming Researchers at Iowa State University and the University of Arkansas have developed a piecewise function that can be used to estimate the body weight (in grams) of a male broiler during the first 56 days of life according to

$$W(t) = \begin{cases} 48 + 3.64t + 0.6363t^2 + 0.00963t^3 & \text{if } 1 \le t \le 28, \\ -1004 + 65.8t & \text{if } 28 < t \le 56, \end{cases}$$

where t is the age of the chicken (in days). *Source: Poultry Science.*

   a. Determine the weight of a male broiler that is 25 days old.

   b. Is $W(t)$ a continuous function?

   c. Use a graphing calculator to graph $W(t)$ on $[1, 56]$ by $[0, 3000]$. Comment on the accuracy of the graph.

   d. Comment on why researchers would use two different types of functions to estimate the weight of a chicken at various ages.

YOUR TURN ANSWERS

1. Discontinuous when $a < -3/5$.

2. Discontinuous at $x = 0$.

3.3 Rates of Change

APPLY IT   How does the manufacturing cost of a DVD change as the number of DVDs manufactured changes?

This question will be answered in Example 4 of this section as we develop a method for finding the rate of change of one variable with respect to a unit change in another variable.

Average Rate of Change   One of the main applications of calculus is determining how one variable changes in relation to another. A marketing manager wants to know how profit changes with respect to the amount spent on advertising, while a physician wants to know how a patient's reaction to a drug changes with respect to the dose.

For example, suppose we take a trip from San Francisco driving south. Every half-hour we note how far we have traveled, with the following results for the first three hours.

Distance Traveled							
Time in Hours	0	0.5	1	1.5	2	2.5	3
Distance in Miles	0	30	55	80	104	124	138

If s is the function whose rule is

$$s(t) = \text{Distance from San Francisco at time } t,$$

then the table shows, for example, that $s(0) = 0$, $s(1) = 55$, $s(2.5) = 124$, and so on. The distance traveled during, say, the second hour can be calculated by $s(2) - s(1) = 104 - 55 = 49$ miles.

Distance equals time multiplied by rate (or speed); so the distance formula is $d = rt$. Solving for rate gives $r = d/t$, or

$$\text{Average speed} = \frac{\text{Distance}}{\text{Time}}.$$

For example, the average speed over the time interval from $t = 0$ to $t = 3$ is

$$\text{Average speed} = \frac{s(3) - s(0)}{3 - 0} = \frac{138 - 0}{3} = 46,$$

or 46 mph. We can use this formula to find the average speed for any interval of time during the trip, as shown below.

FOR REVIEW

Recall from Section 1.1 the formula for the slope of a line through two points (x_1, y_1) and (x_2, y_2):

$$\frac{y_2 - y_1}{x_2 - x_1}.$$

Find the slopes of the lines through the following points.

$(0.5, 30)$ and $(1, 55)$
$(0.5, 30)$ and $(1.5, 80)$
$(1, 55)$ and $(2, 104)$

Compare your answers to the average speeds shown in the table.

Average Speed	
Time Interval	Average Speed $= \dfrac{\text{Distance}}{\text{Time}}$
$t = 0.5$ to $t = 1$	$\dfrac{s(1) - s(0.5)}{1 - 0.5} = \dfrac{25}{0.5} = 50$
$t = 0.5$ to $t = 1.5$	$\dfrac{s(1.5) - s(0.5)}{1.5 - 0.5} = \dfrac{50}{1} = 50$
$t = 1$ to $t = 2$	$\dfrac{s(2) - s(1)}{2 - 1} = \dfrac{49}{1} = 49$
$t = 1$ to $t = 3$	$\dfrac{s(3) - s(1)}{3 - 1} = \dfrac{83}{2} = 41.5$
$t = a$ to $t = b$	$\dfrac{s(b) - s(a)}{b - a}$

The analysis of the average speed or *average rate of change* of distance s with respect to t can be extended to include any function defined by $f(x)$ to get a formula for the average rate of change of f with respect to x.

Average Rate of Change

The **average rate of change** of $f(x)$ with respect to x for a function f as x changes from a to b is

$$\frac{f(b) - f(a)}{b - a}.$$

NOTE
The formula for the average rate of change is the same as the formula for the slope of the line through $(a, f(a))$ and $(b, f(b))$. This connection between slope and rate of change will be examined more closely in the next section.

In Figure 23 we have plotted the distance vs. time for our trip from San Francisco, connecting the points with straight line segments. Because the change in y gives the change in distance, and the change in x gives the change in time, the slope of each line segment gives the average speed over that time interval:

$$\text{Slope} = \frac{\text{Change in } y}{\text{Change in } x} = \frac{\text{Change in distance}}{\text{Change in time}} = \text{Average speed.}$$

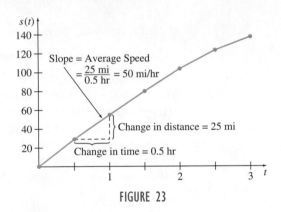

FIGURE 23

EXAMPLE 1   Minority Population

The United States population is becoming more diverse. Based on the U.S. Census population projections for 2000 to 2050, the projected Hispanic population (in millions) can be modeled by the exponential function

$$H(t) = 37.791(1.021)^t,$$

where $t = 0$ corresponds to 2000 and $0 \le t \le 50$. Use H to estimate the average rate of change in the Hispanic population from 2000 to 2010. *Source: U.S. Census Bureau.*

SOLUTION   On the interval from $t = 0$ (2000) to $t = 10$ (2010), the average rate of change is

$$\frac{H(10) - H(0)}{10 - 0} = \frac{37.791(1.021)^{10} - 37.791(1.021)^0}{10}$$

$$\approx \frac{46.521 - 37.791}{10} = \frac{8.73}{10}$$

$$= 0.873,$$

YOUR TURN 1   The projected U.S. Asian population (in millions) for this same time period is $A(t) = 11.14(1.023)^t$. Use A to estimate the average rate of change from 2000 to 2010.

or 0.873 million. Based on this model, it is estimated that the Hispanic population in the United States increased, on average, at a rate of about 873,000 people per year between 2000 and 2010.   **TRY YOUR TURN 1**

EXAMPLE 2 Household Telephones

Some U.S. households are substituting wireless telephones for traditional landline telephones. The graph in Figure 24 shows the percent of households in the United States with landline telephones for the years 2005 to 2009. Find the average rate of change in the percent of households with a landline between 2005 and 2009. *Source: Centers for Disease Control and Prevention.*

SOLUTION   Let $L(t)$ be the percent of U.S. households with landlines in the year t. Then the average rate of change between 2005 and 2009 was

$$\frac{L(2009) - L(2005)}{2009 - 2005} = \frac{73.5 - 89.7}{4} = \frac{-16.2}{4} = -4.05,$$

or -4.05%. On average, the percent of U.S. households with landline telephones decreased by about 4.05% per year during this time period. **TRY YOUR TURN 2**

YOUR TURN 2 In Example 2, find the average rate of change in percent of households with a landline between 2007 and 2009.

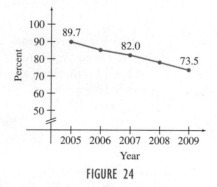

FIGURE 24

Instantaneous Rate of Change

Suppose a car is stopped at a traffic light. When the light turns green, the car begins to move along a straight road. Assume that the distance traveled by the car is given by the function

$$s(t) = 3t^2, \quad \text{for} \quad 0 \le t \le 15,$$

where t is the time in seconds and $s(t)$ is the distance in feet. We have already seen how to find the *average* speed of the car over any time interval. We now turn to a different problem, that of determining the exact speed of the car at a particular instant, say $t = 10$.

The intuitive idea is that the exact speed at $t = 10$ is very close to the average speed over a very short time interval near $t = 10$. If we take shorter and shorter time intervals near $t = 10$, the average speeds over these intervals should get closer and closer to the exact speed at $t = 10$. In other words, the exact speed at $t = 10$ is the limit of the average speeds over shorter and shorter time intervals near $t = 10$. The following chart illustrates this idea. The values in the chart are found using $s(t) = 3t^2$, so that, for example, $s(10) = 3(10)^2 = 300$ and $s(10.1) = 3(10.1)^2 = 306.03$.

Approximation of Speed at 10 Seconds	
Interval	**Average Speed**
$t = 10$ to $t = 10.1$	$\dfrac{s(10.1) - s(10)}{10.1 - 10} = \dfrac{306.03 - 300}{0.1} = 60.3$
$t = 10$ to $t = 10.01$	$\dfrac{s(10.01) - s(10)}{10.01 - 10} = \dfrac{300.6003 - 300}{0.01} = 60.03$
$t = 10$ to $t = 10.001$	$\dfrac{s(10.001) - s(10)}{10.001 - 10} = \dfrac{300.060003 - 300}{0.001} = 60.003$

The results in the chart suggest that the exact speed at $t = 10$ is 60 ft/sec. We can confirm this by computing the average speed from $t = 10$ to $t = 10 + h$, where h is a small, but nonzero, number that represents a small change in time. (The chart does this for $h = 0.1$, $h = 0.01$, and $h = 0.001$.) The average speed from $t = 10$ to $t = 10 + h$ is then

$$\frac{s(10 + h) - s(10)}{(10 + h) - 10} = \frac{3(10 + h)^2 - 3 \cdot 10^2}{h}$$

$$= \frac{3(100 + 20h + h^2) - 300}{h}$$

$$= \frac{300 + 60h + 3h^2 - 300}{h}$$

$$= \frac{60h + 3h^2}{h}$$

$$= \frac{h(60 + 3h)}{h}$$

$$= 60 + 3h,$$

where h is not equal to 0. Saying that the time interval from 10 to $10 + h$ gets shorter and shorter is equivalent to saying that h gets closer and closer to 0. Therefore, the exact speed at $t = 10$ is the limit, as h approaches 0, of the average speed over the interval from $t = 10$ to $t = 10 + h$; that is,

$$\lim_{h \to 0} \frac{s(10 + h) - s(10)}{h} = \lim_{h \to 0} (60 + 3h)$$

$$= 60 \text{ ft/sec.}$$

This example can be easily generalized to any function f. Let a be a specific x-value, such as 10 in the example. Let h be a (small) number, which represents the distance between the two values of x, namely, a and $a + h$. The average rate of change of f as x changes from a to $a + h$ is

$$\frac{f(a + h) - f(a)}{(a + h) - a} = \frac{f(a + h) - f(a)}{h},$$

which is often called the **difference quotient**. Observe that the difference quotient is equivalent to the average rate of change formula, which can be verified by letting $b = a + h$ in the average rate of change formula. Furthermore, the exact rate of change of f at $x = a$, called the *instantaneous rate of change of f at $x = a$*, is the limit of this difference quotient.

Instantaneous Rate of Change

The **instantaneous rate of change** for a function f when $x = a$ is

$$\lim_{h \to 0} \frac{f(a + h) - f(a)}{h},$$

provided this limit exists.

CAUTION Remember that $f(x + h) \neq f(x) + f(h)$. To find $f(x + h)$, replace x with $(x + h)$ in the expression for $f(x)$. For example, if $f(x) = x^2$,

$$f(x + h) = (x + h)^2 = x^2 + 2xh + h^2,$$

but

$$f(x) + f(h) = x^2 + h^2.$$

In the example just discussed, with the car starting from the traffic light, we saw that the instantaneous rate of change gave the speed of the car. But speed is always positive, while instantaneous rate of change can be positive or negative. Therefore, we will refer to **velocity** when we want to consider not only how fast something is moving but also in what direction it is moving. In any motion along a straight line, one direction is arbitrarily labeled as positive, so when an object moves in the opposite direction, its velocity is negative. In general, velocity is the same as the instantaneous rate of change of a function that gives position in terms of time.

In Figure 25, we have plotted the function $s(t) = 3t^2$, giving distance as a function of time. We have also plotted in green a line through the points $(10, s(10))$ and $(15, s(15))$. As we observed earlier, the slope of this line is the same as the average speed between $t = 10$ and $t = 15$. Finally, in red, we have plotted the line that results when the second point, $(15, s(15))$, moves closer and closer to the first point until the two coincide. The slope of this line corresponds to the instantaneous velocity at $t = 10$. We will explore these ideas further in the next section. Meanwhile, you might think about how to calculate the equations of these lines.

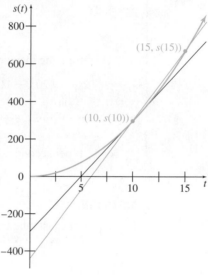

FIGURE 25

An alternate, but equivalent, approach is to let $a + h = b$ in the definition for instantaneous rate of change, so that $h = b - a$. This makes the instantaneous rate of change formula look more like the average rate of change formula.

> **Instantaneous Rate of Change (Alternate Form)**
>
> The **instantaneous rate of change** for a function f when $x = a$ can be written as
>
> $$\lim_{b \to a} \frac{f(b) - f(a)}{b - a},$$
>
> provided this limit exists.

EXAMPLE 3 Velocity

The distance in feet of an object from a starting point is given by $s(t) = 2t^2 - 5t + 40$, where t is time in seconds.

(a) Find the average velocity of the object from 2 seconds to 4 seconds.

SOLUTION The average velocity is

$$\frac{s(4) - s(2)}{4 - 2} = \frac{52 - 38}{2} = \frac{14}{2} = 7$$

ft per second.

(b) Find the instantaneous velocity at 4 seconds.

Method 1
Standard Form

SOLUTION

For $t = 4$, the instantaneous velocity is

$$\lim_{h \to 0} \frac{s(4 + h) - s(4)}{h}$$

ft per second. We first calculate $s(4 + h)$ and $s(4)$, that is,

$$
\begin{aligned}
s(4 + h) &= 2(4 + h)^2 - 5(4 + h) + 40 \\
&= 2(16 + 8h + h^2) - 20 - 5h + 40 \\
&= 32 + 16h + 2h^2 - 20 - 5h + 40 \\
&= 2h^2 + 11h + 52,
\end{aligned}
$$

and

$$s(4) = 2(4)^2 - 5(4) + 40 = 52.$$

Therefore, the instantaneous velocity at $t = 4$ is

$$\lim_{h \to 0} \frac{(2h^2 + 11h + 52) - 52}{h} = \lim_{h \to 0} \frac{2h^2 + 11h}{h} = \lim_{h \to 0} \frac{h(2h + 11)}{h}$$

$$= \lim_{h \to 0} (2h + 11) = 11,$$

or 11 ft per second.

Method 2
Alternate Form

SOLUTION

For $t = 4$, the instantaneous velocity is

$$\lim_{b \to 4} \frac{s(b) - s(4)}{b - 4}$$

ft per second. We first calculate $s(b)$ and $s(4)$, that is,

$$s(b) = 2b^2 - 5b + 40$$

and

$$s(4) = 2(4)^2 - 5(4) + 40 = 52.$$

The instantaneous rate of change is then

$$\lim_{b \to 4} \frac{2b^2 - 5b + 40 - 52}{b - 4} = \lim_{b \to 4} \frac{2b^2 - 5b - 12}{b - 4} \qquad \text{Simplify the numerator.}$$

$$= \lim_{b \to 4} \frac{(2b + 3)(b - 4)}{b - 4} \qquad \text{Factor the numerator.}$$

$$= \lim_{b \to 4} (2b + 3) \qquad \text{Cancel the } h - 4$$

$$= 11, \qquad \text{Calculate the limit.}$$

YOUR TURN 3 For the function in Example 3, find the instantaneous velocity at 2 seconds.

or 11 ft per second.

TRY YOUR TURN 3

EXAMPLE 4 Manufacturing

APPLY IT

A company determines that the cost in dollars to manufacture x cases of the DVD "Mathematicians Caught in Embarrassing Moments" is given by

$$C(x) = 100 + 15x - x^2 \quad (0 \le x \le 7).$$

(a) Find the average rate of change of cost per case for manufacturing between 1 and 5 cases.

SOLUTION Use the formula for average rate of change.

The average rate of change of cost is

$$\frac{C(5) - C(1)}{5 - 1} = \frac{150 - 114}{4} = 9.$$

Thus, on average, the cost increases at the rate of $9 per case when production increases from 1 to 5 cases.

(b) Find the additional cost when production is increased from 1 to 2 cases.

SOLUTION The additional cost can be found by calculating the cost to produce 2 cases, and subtracting the cost to produce 1 case; that is,

$$C(2) - C(1) = 126 - 114 = 12.$$

The additional cost to produce the second case is $12.

(c) Find the instantaneous rate of change of cost with respect to the number of cases produced when just one case is produced.

SOLUTION The instantaneous rate of change for $x = 1$ is given by

$$\lim_{h \to 0} \frac{C(1 + h) - C(1)}{h}$$

$$= \lim_{h \to 0} \frac{[100 + 15(1 + h) - (1 + h)^2] - [100 + 15(1) - 1^2]}{h}$$

$$= \lim_{h \to 0} \frac{100 + 15 + 15h - 1 - 2h - h^2 - 114}{h}$$

$$= \lim_{h \to 0} \frac{13h - h^2}{h} \qquad \text{Combine terms.}$$

$$= \lim_{h \to 0} \frac{h(13 - h)}{h} \qquad \text{Factor.}$$

$$= \lim_{h \to 0} (13 - h) \qquad \text{Divide by } h.$$

$$= 13. \qquad \text{Calculate the limit.}$$

When 1 case is manufactured, the cost is increasing at the rate of $13 per case. Notice that this is close to the value calculated in part (b). ▬

As we mentioned in Chapter 1, economists sometimes define the marginal cost as the cost of producing one additional item and sometimes as the instantaneous rate of change of the cost function. These definitions are considered to be essentially equivalent. If a company (or an economy) produces millions of items, it makes little difference whether we let $h = 1$ or take the limit as h goes to 0, because 1 is very close to 0 when production is in the millions. The advantage of taking the instantaneous rate of change point of view is that it allows all the power of calculus to be used, including the Fundamental Theorem of Calculus, which is discussed later in this book.

Throughout this textbook, *we define the marginal cost to be the instantaneous rate of change of the cost function.* It can then be interpreted as the approximate cost of producing one additional item. For simplicity, we will make this interpretation even when production numbers are fairly small.

EXAMPLE 5 Manufacturing

For the cost function in the previous example, find the instantaneous rate of change of cost when 5 cases are made.

SOLUTION The instantaneous rate of change for $x = 5$ is given by

$$\lim_{h \to 0} \frac{C(5 + h) - C(5)}{h}$$

$$= \lim_{h \to 0} \frac{[100 + 15(5 + h) - (5 + h)^2] - [100 + 15(5) - 5^2]}{h}$$

$$= \lim_{h \to 0} \frac{100 + 75 + 15h - 25 - 10h - h^2 - 150}{h}$$

$$= \lim_{h \to 0} \frac{5h - h^2}{h} \qquad \text{Combine terms.}$$

$$= \lim_{h \to 0} \frac{h(5 - h)}{h} \qquad \text{Factor.}$$

$$= \lim_{h \to 0} (5 - h) \qquad \text{Divide by } h.$$

$$= 5. \qquad \text{Calculate the limit.}$$

YOUR TURN 4 If the cost function is given by $C(x) = x^2 - 2x + 12$, find the instantaneous rate of change of cost when $x = 4$.

When 5 cases are manufactured, the cost is increasing at the rate of \$5 per case; that is, the marginal cost when $x = 5$ is \$5. Notice that as the number of items produced goes up, the marginal cost goes down, as might be expected. **TRY YOUR TURN 4**

EXAMPLE 6 Minority Population

Estimate the instantaneous rate of change in 2010 in the Hispanic population of the United States.

SOLUTION We saw in Example 1 that the U.S. Hispanic population is approximately given by $H(t) = 37.791(1.021)^t$, where $t = 0$ corresponds to 2000. Unlike the previous example, in which the function was a polynomial, the function in this example is an exponential, making it harder to compute the limit directly using the formula for instantaneous rate of change at $t = 10$ (the year 2010):

YOUR TURN 5 Estimate the instantaneous rate of change in 2010 in the Asian population of the United States. An estimate of the U.S. Asian population is given by $A(t) = 11.14(1.023)^t$, where $t = 0$ corresponds to 2000.

$$\lim_{h \to 0} \frac{37.791(1.021)^{10+h} - 37.791(1.021)^{10}}{h}.$$

Instead, we will approximate the instantaneous rate of change by using smaller and smaller values of h. See the following table. The limit seems to be approaching 0.96682 (million). Thus, the instantaneous rate of change in the U.S. Hispanic population is about 966,820 people per year in 2010. **TRY YOUR TURN 5**

	Limit Calculations
	$\dfrac{37.791(1.021)^{10+h} - 37.791(1.021)^{10}}{h}$
h	h
1	0.97693
0.1	0.96782
0.01	0.96692
0.001	0.96683
0.0001	0.96682
0.00001	0.96682

The table could be created using the TABLE feature on a TI-84 Plus calculator by entering Y_1 as the function from Example 6, and $Y_2 = (Y_1(10 + X) - Y_1(10))/X$. (The calculator requires us to use X in place of h in the formula for instantaneous rate of change.) The result is shown in Figure 26. This table can also be generated using a spreadsheet.

X	Y₂
1	.97693
.1	.96782
.01	.96692
.001	.96683
1E⁻4	.96682
1E⁻5	.96682
X=	

FIGURE 26

EXAMPLE 7 Velocity

One day Musk, the friendly pit bull, escaped from the yard and ran across the street to see a neighbor, who was 50 ft away. An estimate of the distance Musk ran as a function of time is given by the following table.

Distance Traveled				
t (sec)	0	1	2	3	4
s (ft)	0	10	25	42	50

(a) Find Musk's average velocity during her 4-second trip.

SOLUTION The total distance she traveled is 50 ft, and the total time is 4 seconds, so her average velocity is $50/4 = 12.5$ ft per second.

(b) Estimate Musk's velocity at 2 seconds.

SOLUTION We could estimate her velocity by taking the short time interval from 2 to 3 seconds, for which the velocity is

$$\frac{42 - 25}{1} = 17 \text{ ft per second.}$$

Alternatively, we could estimate her velocity by taking the short time interval from 1 to 2 seconds, for which the velocity is

$$\frac{25 - 10}{1} = 15 \text{ ft per second.}$$

A better estimate is found by averaging these two values to get

$$\frac{17 + 15}{2} = 16 \text{ ft per second.}$$

Another way to get this same answer is to take the time interval from 1 to 3 seconds, for which the velocity is

$$\frac{42 - 10}{2} = 16 \text{ ft per second.}$$

This answer is reasonable if we assume Musk's velocity changes at a fairly steady rate and does not increase or decrease drastically from one second to the next. It is impossible to calculate Musk's exact velocity without knowing her position at times arbitrarily close to 2 seconds, or without a formula for her position as a function of time, or without a radar gun or speedometer on her. (In any case, she was very happy when she reached the neighbor.)

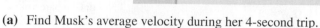

3.3 EXERCISES

Find the average rate of change for each function over the given interval.

1. $y = x^2 + 2x$ between $x = 1$ and $x = 3$

2. $y = -4x^2 - 6$ between $x = 2$ and $x = 6$

3. $y = -3x^3 + 2x^2 - 4x + 1$ between $x = -2$ and $x = 1$

4. $y = 2x^3 - 4x^2 + 6x$ between $x = -1$ and $x = 4$

5. $y = \sqrt{x}$ between $x = 1$ and $x = 4$

6. $y = \sqrt{3x - 2}$ between $x = 1$ and $x = 2$

7. $y = e^x$ between $x = -2$ and $x = 0$

8. $y = \ln x$ between $x = 2$ and $x = 4$

Suppose the position of an object moving in a straight line is given by $s(t) = t^2 + 5t + 2$. Find the instantaneous velocity at each time.

9. $t = 6$ **10.** $t = 1$

Suppose the position of an object moving in a straight line is given by $s(t) = 5t^2 - 2t - 7$. Find the instantaneous velocity at each time.

11. $t = 2$ **12.** $t = 3$

Suppose the position of an object moving in a straight line is given by $s(t) = t^3 + 2t + 9$. Find the instantaneous velocity at each time.

13. $t = 1$ **14.** $t = 4$

Find the instantaneous rate of change for each function at the given value.

15. $f(x) = x^2 + 2x$ at $x = 0$

16. $s(t) = -4t^2 - 6$ at $t = 2$

17. $g(t) = 1 - t^2$ at $t = -1$

18. $F(x) = x^2 + 2$ at $x = 0$

Use the formula for instantaneous rate of change, approximating the limit by using smaller and smaller values of h, to find the instantaneous rate of change for each function at the given value.

19. $f(x) = x^x$ at $x = 2$ **20.** $f(x) = x^x$ at $x = 3$

21. $f(x) = x^{\ln x}$ at $x = 2$ **22.** $f(x) = x^{\ln x}$ at $x = 3$

23. Explain the difference between the average rate of change of y as x changes from a to b, and the instantaneous rate of change of y at $x = a$.

24. If the instantaneous rate of change of $f(x)$ with respect to x is positive when $x = 1$, is f increasing or decreasing there?

APPLICATIONS

Business and Economics

25. Profit Suppose that the total profit in hundreds of dollars from selling x items is given by
$$P(x) = 2x^2 - 5x + 6.$$

Find the average rate of change of profit for the following changes in x.

a. 2 to 4   **b.** 2 to 3

c. Find and interpret the instantaneous rate of change of profit with respect to the number of items produced when $x = 2$. (This number is called the *marginal profit* at $x = 2$.)

d. Find the marginal profit at $x = 4$.

26. Revenue The revenue (in thousands of dollars) from producing x units of an item is
$$R(x) = 10x - 0.002x^2.$$

a. Find the average rate of change of revenue when production is increased from 1000 to 1001 units.

b. Find and interpret the instantaneous rate of change of revenue with respect to the number of items produced when 1000 units are produced. (This number is called the *marginal revenue* at $x = 1000$.)

c. Find the additional revenue if production is increased from 1000 to 1001 units.

d. Compare your answers for parts a and c. What do you find? How do these answers compare with your answer to part b?

27. Demand Suppose customers in a hardware store are willing to buy $N(p)$ boxes of nails at p dollars per box, as given by
$$N(p) = 80 - 5p^2, \quad 1 \le p \le 4.$$

a. Find the average rate of change of demand for a change in price from \$2 to \$3.

b. Find and interpret the instantaneous rate of change of demand when the price is \$2.

c. Find the instantaneous rate of change of demand when the price is \$3.

d. As the price is increased from \$2 to \$3, how is demand changing? Is the change to be expected? Explain.

28. Interest If \$1000 is invested in an account that pays 5% compounded annually, the total amount, $A(t)$, in the account after t years is
$$A(t) = 1000(1.05)^t.$$

a. Find the average rate of change per year of the total amount in the account for the first five years of the investment (from $t = 0$ to $t = 5$).

b. Find the average rate of change per year of the total amount in the account for the second five years of the investment (from $t = 5$ to $t = 10$).

c. Estimate the instantaneous rate of change for $t = 5$.

29. Interest If \$1000 is invested in an account that pays 5% compounded continuously, the total amount, $A(t)$, in the account after t years is
$$A(t) = 1000e^{0.05t}.$$

a. Find the average rate of change per year of the total amount in the account for the first five years of the investment (from $t = 0$ to $t = 5$).

b. Find the average rate of change per year of the total amount in the account for the second five years of the investment (from $t = 5$ to $t = 10$).

c. Estimate the instantaneous rate of change for $t = 5$.

30. Sales The graph shows annual sales (in thousands of dollars) of a Nintendo game at a particular store. Find the average annual rate of change in sales for the following changes in years.

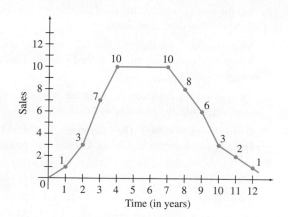

a. 1 to 4   **b.** 4 to 7   **c.** 7 to 12

d. What do your answers for parts a–c tell you about the sales of this product?

e. Give an example of another product that might have such a sales curve.

31. Gasoline Prices In 2008, the price of gasoline in the United States inexplicably spiked and then dropped. The average monthly price (in cents) per gallon of unleaded regular gasoline for 2008 is shown in the following chart. Find the average rate of change per month in the average price per gallon for each time period. *Source: U.S. Energy Information Administration.*

a. From January to July (the peak)

b. From July to December

c. From January to December

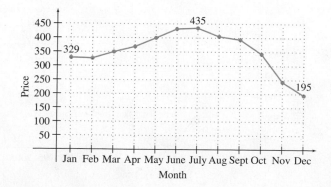

32. Medicare Trust Fund The graph shows the money remaining in the Medicare Trust Fund at the end of the fiscal year. *Source: Social Security Administration.*

Medicare Trust Fund

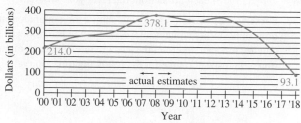

Using the Consumer Price Index for Urban Wage Earners and Clerical Workers

Find the approximate average rate of change in the trust fund for each time period.

a. From 2000 to 2008 (the peak)   **b.** From 2008 to 2018

Life Sciences

33. Flu Epidemic Epidemiologists in College Station, Texas, estimate that t days after the flu begins to spread in town, the percent of the population infected by the flu is approximated by

$$p(t) = t^2 + t$$

for $0 \le t \le 5$.

a. Find the average rate of change of p with respect to t over the interval from 1 to 4 days.

b. Find and interpret the instantaneous rate of change of p with respect to t at $t = 3$.

34. World Population Growth The future size of the world population depends on how soon it reaches replacement-level fertility, the point at which each woman bears on average about 2.1 children. The graph shows projections for reaching that point in different years. *Source: Population Reference Bureau.*

Ultimate World Population Size Under Different Assumptions

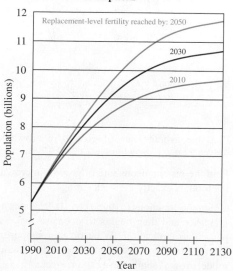

a. Estimate the average rate of change in population for each projection from 1990 to 2050. Which projection shows the smallest rate of change in world population?

b. Estimate the average rate of change in population from 2090 to 2130 for each projection. Interpret your answer.

35. Bacteria Population The graph shows the population in millions of bacteria t minutes after an antibiotic is introduced into a culture. Find and interpret the average rate of change of population with respect to time for the following time intervals.

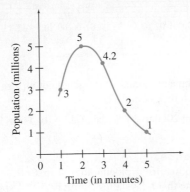

Population (millions) / Time (in minutes)

a. 1 to 2   **b.** 2 to 3   **c.** 3 to 4   **d.** 4 to 5

e. How long after the antibiotic was introduced did the population begin to decrease?

f. At what time did the rate of decrease of the population slow down?

36. Molars The crown length (as shown below) of first molars in fetuses is related to the postconception age of the tooth as

$$L(t) = -0.01t^2 + 0.788t - 7.048,$$

where $L(t)$ is the crown length, in millimeters, of the molar t weeks after conception. *Source: American Journal of Physical Anthropology.*

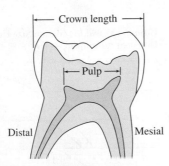

a. Find the average rate of growth in crown length during weeks 22 through 28.

b. Find the instantaneous rate of growth in crown length when the tooth is exactly 22 weeks of age.

c. Graph the function on $[0, 50]$ by $[0, 9]$. Does a function that increases and then begins to decrease make sense for this particular application? What do you suppose is happening during the first 11 weeks? Does this function accurately model crown length during those weeks?

37. Thermic Effect of Food The metabolic rate of a person who has just eaten a meal tends to go up and then, after some time has passed, returns to a resting metabolic rate. This phenomenon is known as the thermic effect of food. Researchers have indicated that the thermic effect of food (in kJ/hr) for a particular person is

$$F(t) = -10.28 + 175.9te^{-t/1.3},$$

where t is the number of hours that have elapsed since eating a meal. *Source: American Journal of Clinical Nutrition.*

a. Graph the function on $[0, 6]$ by $[-20, 100]$.

b. Find the average rate of change of the thermic effect of food during the first hour after eating.

c. Use a graphing calculator to find the instantaneous rate of change of the thermic effect of food exactly 1 hour after eating.

d. Use a graphing calculator to estimate when the function stops increasing and begins to decrease.

38. Mass of Bighorn Yearlings The body mass of yearling bighorn sheep on Ram Mountain in Alberta, Canada, can be estimated by

$$M(t) = 27.5 + 0.3t - 0.001t^2$$

where $M(t)$ is measured in kilograms and t is days since May 25. *Source: Canadian Journal of Zoology.*

a. Find the average rate of change of the weight of a bighorn yearling between 105 and 115 days past May 25.

b. Find the instantaneous rate of change of weight for a bighorn yearling sheep whose age is 105 days past May 25.

c. Graph the function $M(t)$ on $[5, 125]$ by $[25, 65]$.

d. Does the behavior of the function past 125 days accurately model the mass of the sheep? Why or why not?

Social Sciences

39. Immigration The following graph shows how immigration (in thousands) to the United States has varied over the past century. *Source: Homeland Security.*

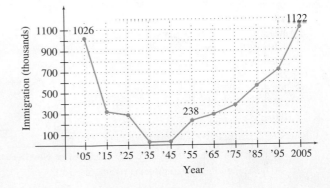

Immigration (thousands) / Year

a. Find the average annual rate of change in immigration for the first half of the century (from 1905 to 1955).

b. Find the average annual rate of change in immigration for the second half of the century (from 1955 to 2005).

c. Find the average annual rate of change in immigration for the entire century (from 1905 to 2005).

d. Average your answers to parts a and b, and compare the result with your answer from part c. Will these always be equal for any two time periods?

e. If the annual average rate of change for the entire century continues, predict the number of immigrants in 2009. Compare your answer to the actual number of 1,130,818 immigrants.

40. Drug Use The following chart shows how the percentage of eighth graders, tenth graders, and twelfth graders who have used marijuana in their lifetime has varied in recent years. *Source: National Institute of Health.*

a. Find the average annual rate of change in the percent of eighth graders who have used marijuana in their lifetime over the three-year period 2003–2006 and the three-year period 2006–2009. Then calculate the annual rate of change for 2003–2009.

b. Repeat part a for tenth graders.

c. Repeat part a for twelfth graders.

d. Discuss any similarities and differences between your answers to parts a through c, as well as possible reasons for these differences and similarities.

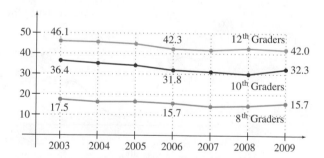

Physical Sciences

41. Temperature The graph shows the temperature T in degrees Celsius as a function of the altitude h in feet when an inversion layer is over Southern California. (An inversion layer is formed when air at a higher altitude, say 3000 ft, is warmer than air at sea level, even though air normally is cooler with increasing altitude.) Estimate and interpret the average rate of change in temperature for the following changes in altitude.

a. 1000 to 3000 ft   **b.** 1000 to 5000 ft

c. 3000 to 9000 ft   **d.** 1000 to 9000 ft

e. At what altitude at or below 7000 ft is the temperature highest? Lowest? How would your answer change if 7000 ft is changed to 10,000 ft?

f. At what altitude is the temperature the same as it is at 1000 ft?

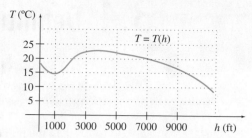

42. Velocity A car is moving along a straight test track. The position in feet of the car, $s(t)$, at various times t is measured, with the following results.

t (sec)	0	2	4	6	8	10
$s(t)$ (ft)	0	10	14	20	30	36

Find and interpret the average velocities for the following changes in t.

a. 0 to 2 seconds   **b.** 2 to 4 seconds

c. 4 to 6 seconds   **d.** 6 to 8 seconds

e. Estimate the instantaneous velocity at 4 seconds

    i. by finding the average velocity between 2 and 6 seconds, and

    ii. by averaging the answers for the average velocity in the two seconds before and the two seconds after (that is, the answers to parts b and c).

f. Estimate the instantaneous velocity at 6 seconds using the two methods in part e.

g. Notice in parts e and f that your two answers are the same. Discuss whether this will always be the case, and why or why not.

43. Velocity Consider the example at the beginning of this section regarding the car traveling from San Francisco.

a. Estimate the instantaneous velocity at 1 hour. Assume that the velocity changes at a steady rate from one half-hour to the next.

b. Estimate the instantaneous velocity at 2 hours.

44. Velocity The distance of a particle from some fixed point is given by

$$s(t) = t^2 + 5t + 2,$$

where t is time measured in seconds. Find the average velocity of the particle over the following intervals.

a. 4 to 6 seconds

b. 4 to 5 seconds

c. Find the instantaneous velocity of the particle when $t = 4$.

YOUR TURN ANSWERS

1. Increase, on average, by 284,000 people per year
2. Decrease, on average, of 4.25% per year
3. 3 ft per second
4. $6 per unit
5. About 0.318 million, or 318,000 people per year

3.4 Definition of the Derivative

How does the risk of chromosomal abnormality in a child change with the mother's age?
We will answer this question in Example 3, using the concept of the derivative.

In the previous section, the formula

$$\lim_{h \to 0} \frac{f(a+h) - f(a)}{h}$$

was used to calculate the instantaneous rate of change of a function f at the point where $x = a$. Now we will give a geometric interpretation of this limit.

The Tangent Line

In geometry, a *tangent line* to a circle is defined as a line that touches the circle at only one point, as at the point P in Figure 27 (which shows the top half of a circle). If you think of this half-circle as part of a curving road on which you are driving at night, then the tangent line indicates the direction of the light beam from your headlights as you pass through the point P. (We are not considering the new type of headlights on some cars that follow the direction of the curve.) Intuitively, the tangent line to an arbitrary curve at a point P on the curve should touch the curve at P, but not at any points nearby, and should indicate the direction of the curve. In Figure 28, for example, the lines through P_1 and P_3 are tangent lines, while the lines through P_2 and P_5 are not. The tangent lines just touch the curve and indicate the direction of the curve, while the other lines pass through the curve heading in some other direction. To decide about the line at P_4, we need to define the idea of a tangent line to the graph of a function more carefully.

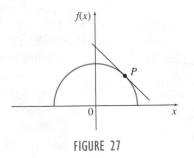

FIGURE 27

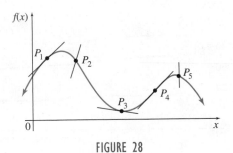

FIGURE 28

To see how we might define the slope of a line tangent to the graph of a function f at a given point, let R be a fixed point with coordinates $(a, f(a))$ on the graph of a function $y = f(x)$, as in Figure 29 on the next page. Choose a different point S on the graph and draw the line through R and S; this line is called a **secant line.** If S has coordinates $(a + h, f(a + h))$, then by the definition of slope, the slope of the secant line RS is given by

$$\text{Slope of secant} = \frac{\Delta y}{\Delta x} = \frac{f(a+h) - f(a)}{a+h-a} = \frac{f(a+h) - f(a)}{h}.$$

This slope corresponds to the average rate of change of y with respect to x over the interval from a to $a + h$. As h approaches 0, point S will slide along the curve, getting closer and closer to the fixed point R. See Figure 30, which shows successive positions S_1, S_2, S_3, and

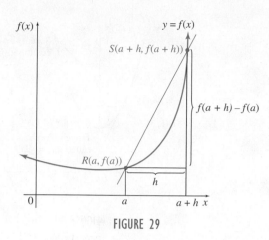

FIGURE 29

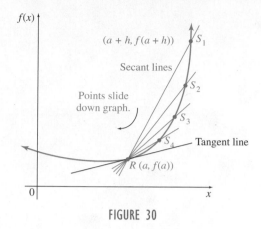

FIGURE 30

S_4 of the point S. If the slopes of the corresponding secant lines approach a limit as h approaches 0, then this limit is defined to be the slope of the tangent line at point R.

Slope of the Tangent Line

The **tangent line** of the graph of $y = f(x)$ at the point $(a, f(a))$ is the line through this point having slope

$$\lim_{h \to 0} \frac{f(a + h) - f(a)}{h},$$

provided this limit exists. If this limit does not exist, then there is no tangent at the point.

Notice that the definition of the slope of the tangent line is identical to that of the instantaneous rate of change discussed in the previous section and is calculated by the same procedure.

The slope of the tangent line at a point is also called the **slope of the curve** at the point and corresponds to the instantaneous rate of change of y with respect to x at the point. It indicates the direction of the curve at that point.

FOR REVIEW

In Section 1.1, we saw that the equation of a line can be found with the point-slope form $y - y_1 = m(x - x_1)$, if the slope m and the coordinates (x_1, y_1) of a point on the line are known. Use the point-slope form to find the equation of the line with slope 3 that goes through the point $(-1, 4)$.

Let $m = 3$, $x_1 = -1$, $y_1 = 4$. Then

$$y - y_1 = m(x - x_1)$$
$$y - 4 = 3(x - (-1))$$
$$y - 4 = 3x + 3$$
$$y = 3x + 7.$$

EXAMPLE 1 Tangent Line

Consider the graph of $f(x) = x^2 + 2$.

(a) Find the slope and equation of the secant line through the points where $x = -1$ and $x = 2$.

SOLUTION Use the formula for slope as the change in y over the change in x, where y is given by $f(x)$. Since $f(-1) = (-1)^2 + 2 = 3$ and $f(2) = 2^2 + 2 = 6$, we have

$$\text{Slope of secant line} = \frac{f(2) - f(-1)}{2 - (-1)} = \frac{6 - 3}{3} = 1.$$

The slope of the secant line through $(-1, f(-1)) = (-1, 3)$ and $(2, f(2)) = (2, 6)$ is 1.

The equation of the secant line can be found with the point-slope form of the equation of a line from Chapter 1. We'll use the point $(-1, 3)$, although we could have just as well used the point $(2, 6)$.

$$y - y_1 = m(x - x_1)$$
$$y - 3 = 1[x - (-1)]$$
$$y - 3 = x + 1$$
$$y = x + 4$$

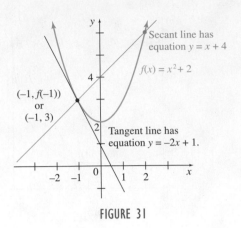

FIGURE 31

Figure 31 shows a graph of $f(x) = x^2 + 2$, along with a graph of the secant line (in green) through the points where $x = -1$ and $x = 2$.

(b) Find the slope and equation of the tangent line at $x = -1$.

SOLUTION Use the definition given previously, with $f(x) = x^2 + 2$ and $a = -1$. The slope of the tangent line is given by

$$\text{Slope of tangent} = \lim_{h \to 0} \frac{f(a + h) - f(a)}{h}$$

$$= \lim_{h \to 0} \frac{[(-1 + h)^2 + 2] - [(-1)^2 + 2]}{h}$$

$$= \lim_{h \to 0} \frac{[1 - 2h + h^2 + 2] - [1 + 2]}{h}$$

$$= \lim_{h \to 0} \frac{-2h + h^2}{h}$$

$$= \lim_{h \to 0} \frac{h(-2 + h)}{h}$$

$$= \lim_{h \to 0} (-2 + h) = -2.$$

The slope of the tangent line at $(-1, f(-1)) = (-1, 3)$ is -2.

The equation of the tangent line can be found with the point-slope form of the equation of a line from Chapter 1.

$$y - y_1 = m(x - x_1)$$
$$y - 3 = -2[x - (-1)]$$
$$y - 3 = -2(x + 1)$$
$$y - 3 = -2x - 2$$
$$y = -2x + 1$$

YOUR TURN 1 For the graph of $f(x) = x^2 - x$, (a) find the equation of the secant line through the points where $x = -2$ and $x = 1$, and (b) find the equation of the tangent line at $x = -2$.

The tangent line at $x = -1$ is shown in red in Figure 31. **TRY YOUR TURN 1**

Figure 32 shows the result of zooming in on the point $(-1, 3)$ in Figure 31. Notice that in this closeup view, the graph and its tangent line appear virtually identical. This gives us another interpretation of the tangent line. Suppose, as we zoom in on a function, the graph appears to become a straight line. Then this line is the tangent line to the graph at that point. In other words, the tangent line captures the behavior of the function very close to the point under consideration. (This assumes, of course, that the function when viewed close up is approximately a straight line. As we will see later in this section, this may not occur.)

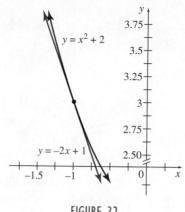

FIGURE 32

If it exists, the tangent line at $x = a$ is a good approximation of the graph of a function near $x = a$.

Consequently, another way to approximate the slope of the curve is to zoom in on the function using a graphing calculator until it appears to be a straight line (the tangent line). Then find the slope using any two points on that line.

TECHNOLOGY

EXAMPLE 2 **Slope (Using a Graphing Calculator)**

Use a graphing calculator to find the slope of the graph of $f(x) = x^x$ at $x = 1$.

SOLUTION The slope would be challenging to evaluate algebraically using the limit definition. Instead, using a graphing calculator on the window $[0, 2]$ by $[0, 2]$, we see the graph in Figure 33.

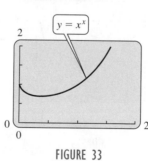

FIGURE 33   **FIGURE 34**

Zooming in gives the view in Figure 34. Using the TRACE key, we find two points on the line to be $(1, 1)$ and $(1.0021277, 1.0021322)$. Therefore, the slope is approximately

$$\frac{1.0021322 - 1}{1.0021277 - 1} \approx 1.$$

TECHNOLOGY NOTE

In addition to the method used in Example 2, there are other ways to use a graphing calculator to determine slopes of tangent lines and estimate instantaneous rates. Some alternate methods are listed below.

1. The Tangent command (under the DRAW menu) on a TI-84 Plus allows the tangent line to be drawn on a curve, giving an easy way to generate a graph with its tangent line similar to Figure 31.

2. Rather than using a graph, we could use a TI-84 Plus to create a table, as we did in the previous section, to estimate the instantaneous rate of change. Letting $Y_1 = X^X$ and

$Y_2 = (Y_1(1 + X) - Y_1(1))/X$, along with specific table settings, results in the table shown in Figure 35. Based on this table, we estimate that the slope of the graph of $f(x) = x^x$ at $x = 1$ is 1.

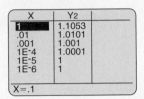

FIGURE 35

3. An even simpler method on a TI-84 Plus is to use the dy/dx command (under the CALC menu) or the $nDeriv$ command (under the MATH menu). We will use this method in Example 4(b). But be careful, because sometimes these commands give erroneous results. For an example, see the Caution at the end of this section. For more details on the dy/dx command or the $nDeriv$ command, see the *Graphing Calculator and Excel Spreadsheet Manual* available with this book.

EXAMPLE 3 Genetics

Figure 36 shows how the risk of chromosomal abnormality in a child increases with the age of the mother. Find the rate that the risk is rising when the mother is 40 years old. *Source: downsyndrome.about.com.*

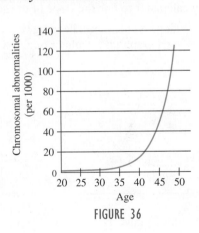

FIGURE 36

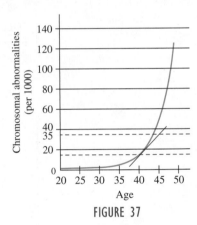

FIGURE 37

APPLY IT

SOLUTION In Figure 37, we have added the tangent line to the graph at the point where the age of the mother is 40. At that point, the risk is approximately 15 per 1000. Extending the line, we estimate that when the age is 45, the y-coordinate of the line is roughly 35. Thus, the slope of the line is

$$\frac{35 - 15}{45 - 40} = \frac{20}{5} = 4.$$

Therefore, at the age of 40, the risk of chromosomal abnormality in the child is increasing at the rate of about 4 per 1000 for each additional year of the mother's age.

The Derivative

If $y = f(x)$ is a function and a is a number in its domain, then we shall use the symbol $f'(a)$ to denote the special limit

$$\lim_{h \to 0} \frac{f(a + h) - f(a)}{h},$$

provided that it exists. This means that for each number a we can assign the number $f'(a)$ found by calculating this limit. This assignment defines an important new function.

Derivative

The **derivative** of the function f at x is defined as

$$f'(x) = \lim_{h \to 0} \frac{f(x + h) - f(x)}{h},$$

provided this limit exists.

NOTE

The derivative is a *function of x*, since $f'(x)$ varies as x varies. This differs from both the slope of the tangent line and the instantaneous rate of change, either of which is represented by the number $f'(a)$ that corresponds to a number a. Otherwise, the formula for the derivative is identical to the formula for the slope of the tangent line given earlier in this section and to the formula for instantaneous rate of change given in the previous section.

The notation $f'(x)$ is read "f-prime of x." The function $f'(x)$ is called the derivative of f with respect to x. If x is a value in the domain of f and if $f'(x)$ exists, then f is **differentiable** at x. The process that produces f' is called **differentiation**.

The derivative function has several interpretations, two of which we have discussed.

1. The function $f'(x)$ represents the *instantaneous rate of change* of $y = f(x)$ with respect to x. This instantaneous rate of change could be interpreted as marginal cost, revenue, or profit (if the original function represented cost, revenue, or profit) or velocity (if the original function described displacement along a line). From now on we will use *rate of change* to mean *instantaneous* rate of change.

2. The function $f'(x)$ represents the *slope* of the graph of $f(x)$ at any point x. If the derivative is evaluated at the point $x = a$, then it represents the slope of the curve, or the slope of the tangent line, at that point.

The following table compares the different interpretations of the difference quotient and the derivative.

The Difference Quotient and the Derivative	
Difference Quotient	**Derivative**
$\dfrac{f(x + h) - f(x)}{h}$	$\lim\limits_{h \to 0} \dfrac{f(x + h) - f(x)}{h}$
▪ Slope of the secant line	▪ Slope of the tangent line
▪ Average rate of change	▪ Instantaneous rate of change
▪ Average velocity	▪ Instantaneous velocity
▪ Average rate of change in cost, revenue, or profit	▪ Marginal cost, revenue, or profit

Just as we had an alternate definition in the previous section by using b instead of $a + h$, we now have an alternate definition by using b in place of $x + h$.

Derivative (Alternate Form)

The **derivative** of function f at x can be written as

$$f'(x) = \lim_{b \to x} \frac{f(b) - f(x)}{b - x},$$

provided this limit exists.

The next few examples show how to use the definition to find the derivative of a function by means of a four-step procedure.

> **EXAMPLE 4** **Derivative**

Let $f(x) = x^2$.

(a) Find the derivative.

Method 1
Original Definition

SOLUTION By definition, for all values of x where the following limit exists, the derivative is given by

$$f'(x) = \lim_{h \to 0} \frac{f(x + h) - f(x)}{h}.$$

Use the following sequence of steps to evaluate this limit.

Step 1 Find $f(x + h)$.

Replace x with $x + h$ in the equation for $f(x)$. Simplify the result.

$$f(x) = x^2$$

$$f(x + h) = (x + h)^2$$

$$= x^2 + 2xh + h^2$$

(Note that $f(x + h) \neq f(x) + h$, since $f(x) + h = x^2 + h$.)

Step 2 Find $f(x + h) - f(x)$.

Since $f(x) = x^2$,

$$f(x + h) - f(x) = (x^2 + 2xh + h^2) - x^2 = 2xh + h^2.$$

Step 3 Find and simplify the quotient $\dfrac{f(x + h) - f(x)}{h}$. We find that

$$\frac{f(x + h) - f(x)}{h} = \frac{2xh + h^2}{h} = \frac{h(2x + h)}{h} = 2x + h,$$

except that $2x + h$ is defined for all real numbers h, while $[f(x + h) - f(x)]/h$ is not defined at $h = 0$. But this makes no difference in the limit, which ignores the value of the expression at $h = 0$.

Step 4 Finally, find the limit as h approaches 0. In this step, h is the variable and x is fixed.

$$f'(x) = \lim_{h \to 0} \frac{f(x + h) - f(x)}{h}.$$

$$= \lim_{h \to 0} (2x + h)$$

$$= 2x + 0 = 2x$$

Method 2
Alternate Form

SOLUTION Use

$$f(b) = b^2$$

and

$$f(x) = x^2.$$

We apply the alternate definition of the derivative as follows.

$$\lim_{b \to x} \frac{f(b) - f(x)}{b - x} = \lim_{b \to x} \frac{b^2 - x^2}{b - x}$$

$$= \lim_{b \to x} \frac{(b + x)(b - x)}{b - x} \qquad \text{Factor the numerator.}$$

$$= \lim_{b \to x} (b + x) \qquad \text{Divide by } b - x.$$

$$= x + x \qquad \text{Calculate the limit.}$$

$$= 2x$$

The alternate method appears shorter here because factoring $b^2 - x^2$ may seem simpler than calculating $f(x + h) - f(x)$. In other problems, however, factoring may be harder, in which case the first method may be preferable. Thus, from now on, we will use only the first method.

(b) Calculate and interpret $f'(3)$.

Method 1
Algebraic Method

    SOLUTION Since $f'(x) = 2x$, we have

$$f'(3) = 2 \cdot 3 = 6.$$

The number 6 is the slope of the tangent line to the graph of $f(x) = x^2$ at the point where $x = 3$, that is, at $(3, f(3)) = (3, 9)$. See Figure 38(a).

Method 2
Graphing Calculator

As we mentioned earlier, some graphing calculators can calculate the value of the derivative at a given x-value. For example, the TI-84 Plus uses the `nDeriv` command as shown in Figure 38(b), with the expression for $f(x)$, the variable, and the value of a entered to find $f'(3)$ for $f(x) = x^2$.

YOUR TURN 2 Let $f(x) = x^2 - x$. Find the derivative, and then find $f'(-2)$.

TRY YOUR TURN 2

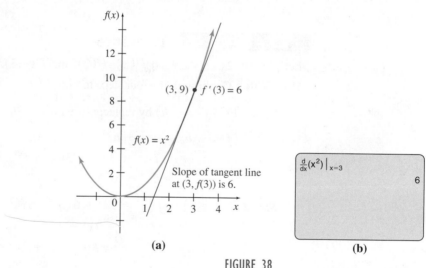

(a)   (b)

FIGURE 38

CAUTION **1.** In Example 4(a) notice that $f(x + h)$ is *not* equal to $f(x) + h$. In fact,

$$f(x + h) = (x + h)^2 = x^2 + 2xh + h^2,$$

but

$$f(x) + h = x^2 + h.$$

2. In Example 4(b), do not confuse $f(3)$ and $f'(3)$. The value $f(3)$ is the y-value that corresponds to $x = 3$. It is found by substituting 3 for x in $f(x)$; $f(3) = 3^2 = 9$. On the other hand, $f'(3)$ is the slope of the tangent line to the curve at $x = 3$; as Example 4(b) shows, $f'(3) = 2 \cdot 3 = 6$.

Finding $f'(x)$ from the Definition of Derivative

The four steps used to find the derivative $f'(x)$ for a function $y = f(x)$ are summarized here.

1. Find $f(x + h)$.
2. Find and simplify $f(x + h) - f(x)$.
3. Divide by h to get $\dfrac{f(x + h) - f(x)}{h}$.
4. Let $h \to 0$; $f'(x) = \displaystyle\lim_{h \to 0} \dfrac{f(x + h) - f(x)}{h}$, if this limit exists.

We now have four equivalent expressions for the change in x, but each has its uses, as the following box shows. We emphasize that these expressions all represent the same concept.

Equivalent Expressions for the Change in x

$x_2 - x_1$	Useful for describing the equation of a line through two points
$b - a$	A way to write $x_2 - x_1$ without the subscripts
Δx	Useful for describing slope without referring to the individual points
h	A way to write Δx with just one symbol

EXAMPLE 5 Derivative

Let $f(x) = 2x^3 + 4x$. Find $f'(x)$, $f'(2)$, and $f'(-3)$.

SOLUTION Go through the four steps to find $f'(x)$.

Step 1 Find $f(x + h)$ by replacing x with $x + h$.

$$
\begin{aligned}
f(x + h) &= 2(x + h)^3 + 4(x + h) \\
&= 2(x^3 + 3x^2 h + 3xh^2 + h^3) + 4(x + h) \\
&= 2x^3 + 6x^2 h + 6xh^2 + 2h^3 + 4x + 4h
\end{aligned}
$$

Step 2 $\begin{aligned}[t] f(x + h) - f(x) &= 2x^3 + 6x^2 h + 6xh^2 + 2h^3 + 4x + 4h \\ &\quad - 2x^3 - 4x \\ &= 6x^2 h + 6xh^2 + 2h^3 + 4h \end{aligned}$

Step 3 $\begin{aligned}[t] \dfrac{f(x + h) - f(x)}{h} &= \dfrac{6x^2 h + 6xh^2 + 2h^3 + 4h}{h} \\ &= \dfrac{h(6x^2 + 6xh + 2h^2 + 4)}{h} \\ &= 6x^2 + 6xh + 2h^2 + 4 \end{aligned}$

Step 4 Now use the rules for limits to get

$$
\begin{aligned}
f'(x) &= \lim_{h \to 0} \dfrac{f(x + h) - f(x)}{h} \\
&= \lim_{h \to 0} (6x^2 + 6xh + 2h^2 + 4) \\
&= 6x^2 + 6x(0) + 2(0)^2 + 4 \\
f'(x) &= 6x^2 + 4.
\end{aligned}
$$

YOUR TURN 3 Let $f(x) = x^3 - 1$. Find $f'(x)$ and $f'(-1)$.

Use this result to find $f'(2)$ and $f'(-3)$.

$$f'(2) = 6 \cdot 2^2 + 4 = 28$$
$$f'(-3) = 6 \cdot (-3)^2 + 4 = 58$$

TRY YOUR TURN 3

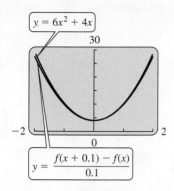

TECHNOLOGY NOTE

$y = 6x^2 + 4x$

30

−2 ⌞_____⌟ 2

0

$y = \dfrac{f(x + 0.1) - f(x)}{0.1}$

FIGURE 39

One way to support the result in Example 5 is to plot $[f(x + h) - f(x)]/h$ on a graphing calculator with a small value of h. Figure 39 shows a graphing calculator screen of $y = [f(x + 0.1) - f(x)]/0.1$, where f is the function $f(x) = 2x^3 + 4x$, and $y = 6x^2 + 4$, which was just found to be the derivative of f. The two functions, plotted on the window $[-2, 2]$ by $[0, 30]$, appear virtually identical. If $h = 0.01$ had been used, the two functions would be indistinguishable.

EXAMPLE 6 Derivative

Let $f(x) = \dfrac{4}{x}$. Find $f'(x)$.

SOLUTION

Step 1 $f(x + h) = \dfrac{4}{x + h}$

Step 2 $f(x + h) - f(x) = \dfrac{4}{x + h} - \dfrac{4}{x}$

$$= \frac{4x - 4(x + h)}{x(x + h)} \qquad \text{Find a common denominator.}$$

$$= \frac{4x - 4x - 4h}{x(x + h)} \qquad \text{Simplify the numerator.}$$

$$= \frac{-4h}{x(x + h)}$$

Step 3 $\dfrac{f(x + h) - f(x)}{h} = \dfrac{\dfrac{-4h}{x(x + h)}}{h}$

$$= \frac{-4h}{x(x + h)} \cdot \frac{1}{h} \qquad \text{Invert and multiply.}$$

$$= \frac{-4}{x(x + h)}$$

Step 4 $f'(x) = \lim\limits_{h \to 0} \dfrac{f(x + h) - f(x)}{h}$

$$= \lim_{h \to 0} \frac{-4}{x(x + h)}$$

$$= \frac{-4}{x(x + 0)}$$

$$f'(x) = \frac{-4}{x(x)} = \frac{-4}{x^2}$$

TRY YOUR TURN 4

YOUR TURN 4 Let $f(x) = -\dfrac{2}{x}$. Find $f'(x)$.

Notice that in Example 6 neither $f(x)$ nor $f'(x)$ is defined when $x = 0$. Look at a graph of $f(x) = 4/x$ to see why this is true.

| EXAMPLE 7 | **Weight Gain** |

A mathematics professor found that, after introducing his dog Django to a new brand of food, Django's weight began to increase. After x weeks on the new food, Django's weight (in pounds) was approximately given by $w(x) = \sqrt{x} + 40$ for $0 \le x \le 6$. Find the rate of change of Django's weight after x weeks.

SOLUTION

Step 1 $w(x + h) = \sqrt{x + h} + 40$

Step 2 $w(x + h) - w(x) = \sqrt{x + h} + 40 - (\sqrt{x} + 40)$
$$= \sqrt{x + h} - \sqrt{x}$$

Step 3 $\dfrac{w(x + h) - w(x)}{h} = \dfrac{\sqrt{x + h} - \sqrt{x}}{h}$

In order to be able to divide by h, multiply both numerator and denominator by $\sqrt{x + h} + \sqrt{x}$; that is, rationalize the *numerator*.

$$\frac{w(x + h) - w(x)}{h} = \frac{\sqrt{x + h} - \sqrt{x}}{h} \cdot \frac{\sqrt{x + h} + \sqrt{x}}{\sqrt{x + h} + \sqrt{x}} \qquad \text{Rationalize the numerator}$$

$$= \frac{(\sqrt{x + h})^2 - (\sqrt{x})^2}{h(\sqrt{x + h} + \sqrt{x})} \qquad (a - b)(a + b) = a^2 - b^2.$$

$$= \frac{x + h - x}{h(\sqrt{x + h} + \sqrt{x})}$$

$$= \frac{h}{h(\sqrt{x + h} + \sqrt{x})} \qquad \text{Simplify.}$$

$$= \frac{1}{\sqrt{x + h} + \sqrt{x}} \qquad \text{Divide by } h.$$

Step 4 $w'(x) = \lim\limits_{h \to 0} \dfrac{1}{\sqrt{x + h} + \sqrt{x}} = \dfrac{1}{\sqrt{x} + \sqrt{x}} = \dfrac{1}{2\sqrt{x}}$

YOUR TURN 5 Let $f(x) = 2\sqrt{x}$. Find $f'(x)$.

This tells us, for example, that after 4 weeks, when Django's weight is $w(4) = \sqrt{4} + 40 = 42$ lb, her weight is increasing at a rate of $w'(4) = 1/(2\sqrt{4}) = 1/4$ lb per week.

TRY YOUR TURN 5

| EXAMPLE 8 | **Cost Analysis** |

The cost in dollars to manufacture x graphing calculators is given by $C(x) = -0.005x^2 + 20x + 150$ when $0 \le x \le 2000$. Find the rate of change of cost with respect to the number manufactured when 100 calculators are made and when 1000 calculators are made.

SOLUTION The rate of change of cost is given by the derivative of the cost function,

$$C'(x) = \lim_{h \to 0} \frac{C(x + h) - C(x)}{h}.$$

Going through the steps for finding $C'(x)$ gives

$$C'(x) = -0.01x + 20.$$

When $x = 100$,

$$C'(100) = -0.01(100) + 20 = 19.$$

YOUR TURN 6 If cost is given by $C(x) = 10x - 0.002x^2$, find the rate of change when $x = 100$.

This rate of change of cost per calculator gives the marginal cost at $x = 100$, which means the approximate cost of producing the 101st calculator is $19.

When 1000 calculators are made, the marginal cost is

$$C'(1000) = -0.01(1000) + 20 = 10,$$

or $10. **TRY YOUR TURN 6**

We can use the notation for the derivative to write the equation of the tangent line. Using the point-slope form, $y - y_1 = m(x - x_1)$, and letting $y_1 = f(x_1)$ and $m = f'(x_1)$, we have the following formula.

> ### Equation of the Tangent Line
> The tangent line to the graph of $y = f(x)$ at the point $(x_1, f(x_1))$ is given by the equation
>
> $$y - f(x_1) = f'(x_1)(x - x_1),$$
>
> provided $f'(x)$ exists.

■ EXAMPLE 9 Tangent Line

Find the equation of the tangent line to the graph of $f(x) = 4/x$ at $x = 2$.

SOLUTION From the answer to Example 6, we have $f'(x) = -4/x^2$, so $f'(x_1) = f'(2) = -4/2^2 = -1$. Also $f(x_1) = f(2) = 4/2 = 2$. Then the equation of the tangent line is

$$y - 2 = (-1)(x - 2),$$

YOUR TURN 7 Find the equation of the tangent line to the graph $f(x) = 2\sqrt{x}$ at $x = 4$.

or

$$y = -x + 4$$

after simplifying. **TRY YOUR TURN 7**

Existence of the Derivative
The definition of the derivative included the phrase "provided this limit exists." If the limit used to define the derivative does not exist, then of course the derivative does not exist. For example, a derivative cannot exist at a point where the function itself is not defined. If there is no function value for a particular value of x, there can be no tangent line for that value. This was the case in Example 6—there was no tangent line (and no derivative) when $x = 0$.

Derivatives also do not exist at "corners" or "sharp points" on a graph. For example, the function graphed in Figure 40 is the *absolute value function*, defined previously as

$$f(x) = \begin{cases} x & \text{if } x \geq 0 \\ -x & \text{if } x < 0, \end{cases}$$

and written $f(x) = |x|$. By the definition of derivative, the derivative at any value of x is given by

$$f'(x) = \lim_{h \to 0} \frac{f(x+h) - f(x)}{h},$$

provided this limit exists. To find the derivative at 0 for $f(x) = |x|$, replace x with 0 and $f(x)$ with $|0|$ to get

$$f'(0) = \lim_{h \to 0} \frac{|0 + h| - |0|}{h} = \lim_{h \to 0} \frac{|h|}{h}.$$

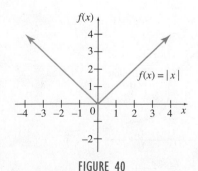

FIGURE 40

In Example 5 in the first section of this chapter, we showed that

$$\lim_{h \to 0} \frac{|h|}{h} \text{ does not exist;}$$

therefore, the derivative does not exist at 0. However, the derivative does exist for all values of x other than 0.

 CAUTION The command `nDeriv(abs(X),X,0)` on a TI-84 Plus calculator gives the answer 0, which is wrong. It does this by investigating a point slightly to the left of 0 and slightly to the right of 0. Since the function has the same value at these two points, it assumes that the function must be flat around 0, which is false in this case because of the sharp corner at 0. Be careful about naively trusting your calculator; think about whether the answer is reasonable.

In Figure 41, we have zoomed in on the origin in Figure 40. Notice that the graph looks essentially the same. The corner is still sharp, and the graph does not resemble a straight line any more than it originally did. As we observed earlier, the derivative only exists at a point when the function more and more resembles a straight line as we zoom in on the point.

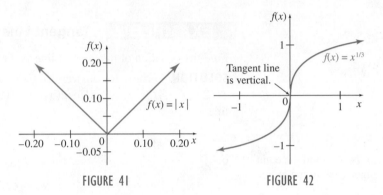

FIGURE 41   FIGURE 42

A graph of the function $f(x) = x^{1/3}$ is shown in Figure 42. As the graph suggests, the tangent line is vertical when $x = 0$. Since a vertical line has an undefined slope, the derivative of $f(x) = x^{1/3}$ cannot exist when $x = 0$. Use the fact that $\lim_{h \to 0} h^{1/3}/h = \lim_{h \to 0} 1/h^{2/3}$ does not exist and the definition of the derivative to verify that $f'(0)$ does not exist for $f(x) = x^{1/3}$.

Figure 43 summarizes the various ways that a derivative can fail to exist. Notice in Figure 43 that at a point where the function is discontinuous, such as x_3, x_4, and x_6, the

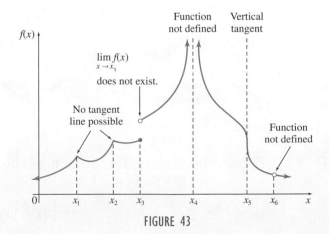

FIGURE 43

derivative does not exist. A function must be continuous at a point for the derivative to exist there. But just because a function is continuous at a point does not mean the derivative necessarily exists. For example, observe that the function in Figure 43 is continuous at x_1 and x_2, but the derivative does not exist at those values of x because of the sharp corners, making a tangent line impossible. This is exactly what happens with the function $f(x) = |x|$ at $x = 0$, as we saw in Figure 40. Also, the function is continuous at x_5, but the derivative doesn't exist there because the tangent line is vertical, and the slope of a vertical line is undefined.

We summarize conditions for the derivative to exist or not below.

Existence of the Derivative

The derivative exists when a function f satisfies *all* of the following conditions at a point.

1. f is continuous,

2. f is smooth, and

3. f does not have a vertical tangent line.

The derivative does *not* exist when *any* of the following conditions are true for a function at a point.

1. f is discontinuous,

2. f has a sharp corner, or

3. f has a vertical tangent line.

EXAMPLE 10 Astronomy

A nova is a star whose brightness suddenly increases and then gradually fades. The cause of the sudden increase in brightness is thought to be an explosion of some kind. The intensity of light emitted by a nova as a function of time is shown in Figure 44. Find where the function is not differentiable. *Source: Astronomy: The Structure of the Universe.*

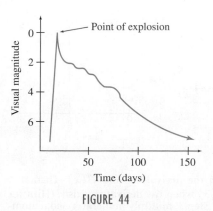

FIGURE 44

SOLUTION Notice that although the graph is a continuous curve, it is not differentiable at the point of explosion.

3.4 EXERCISES

1. By considering, but not calculating, the slope of the tangent line, give the derivative of the following.

 a. $f(x) = 5$ **b.** $f(x) = x$

 c. $f(x) = -x$ **d.** The line $x = 3$

 e. The line $y = mx + b$

2. **a.** Suppose $g(x) = \sqrt[3]{x}$. Use the graph of $g(x)$ to find $g'(0)$.

 b. Explain why the derivative of a function does not exist at a point where the tangent line is vertical.

3. If $f(x) = \dfrac{x^2 - 1}{x + 2}$, where is f not differentiable?

4. If the rate of change of $f(x)$ is zero when $x = a$, what can be said about the tangent line to the graph of $f(x)$ at $x = a$?

Estimate the slope of the tangent line to each curve at the given point (x, y).

5.

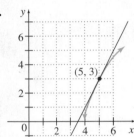

6.

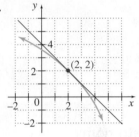

7.

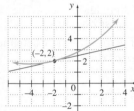

8.

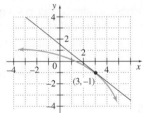

9.

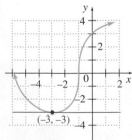

10.

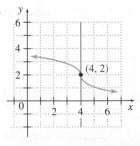

Using the definition of the derivative, find $f'(x)$. Then find $f'(-2)$, $f'(0)$, and $f'(3)$ when the derivative exists. (Hint for Exercises 17 and 18: In Step 3, multiply numerator and denominator by $\sqrt{x + h} + \sqrt{x}$.)

11. $f(x) = 3x - 7$ 12. $f(x) = -2x + 5$

13. $f(x) = -4x^2 + 9x + 2$ 14. $f(x) = 6x^2 - 5x - 1$

15. $f(x) = 12/x$ 16. $f(x) = 3/x$

17. $f(x) = \sqrt{x}$ 18. $f(x) = -3\sqrt{x}$

19. $f(x) = 2x^3 + 5$ 20. $f(x) = 4x^3 - 3$

For each function, find (a) the equation of the secant line through the points where x has the given values, and (b) the equation of the tangent line when x has the first value.

21. $f(x) = x^2 + 2x$; $x = 3$, $x = 5$

22. $f(x) = 6 - x^2$; $x = -1$, $x = 3$

23. $f(x) = 5/x$; $x = 2$, $x = 5$

24. $f(x) = -3/(x + 1)$; $x = 1$, $x = 5$

25. $f(x) = 4\sqrt{x}$; $x = 9$, $x = 16$

26. $f(x) = \sqrt{x}$; $x = 25$, $x = 36$

Use a graphing calculator to find $f'(2)$, $f'(16)$, and $f'(-3)$ for the following when the derivative exists.

27. $f(x) = -4x^2 + 11x$ 28. $f(x) = 6x^2 - 4x$

29. $f(x) = e^x$ 30. $f(x) = \ln|x|$

31. $f(x) = -\dfrac{2}{x}$ 32. $f(x) = \dfrac{6}{x}$

33. $f(x) = \sqrt{x}$ 34. $f(x) = -3\sqrt{x}$

Find the x-values where the following do not have derivatives.

35.

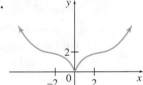

36.

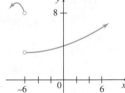

37.

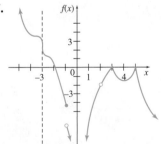

38.

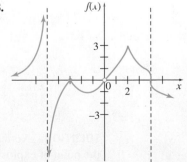

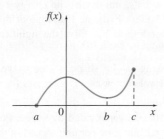

39. For the function shown in the sketch, give the intervals or points on the x-axis where the rate of change of $f(x)$ with respect to x is

   a. positive;   **b.** negative;   **c.** zero.

In Exercises 40 and 41, tell which graph, a or b, represents velocity and which represents distance from a starting point. (Hint: Consider where the derivative is zero, positive, or negative.)

40. a.
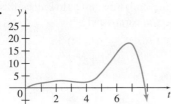

  b.
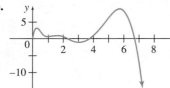

41. a.
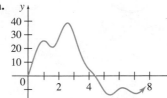

  b.

In Exercises 42–45, find the derivative of the function at the given point.

   a. Approximate the definition of the derivative with small values of h.

   b. Use a graphing calculator to zoom in on the function until it appears to be a straight line, and then find the slope of that line.

42. $f(x) = x^x$; $a = 2$ **43.** $f(x) = x^x$; $a = 3$

44. $f(x) = x^{1/x}$; $a = 2$ **45.** $f(x) = x^{1/x}$; $a = 3$

46. For each function in Column A, graph $[f(x + h) - f(x)]/h$ for a small value of h on the window $[-2, 2]$ by $[-2, 8]$. Then graph each function in Column B on the same window. Compare the first set of graphs with the second set to choose from Column B the derivative of each of the functions in Column A.

Column A	Column B
$\ln |x|$	e^x
e^x	$3x^2$
x^3	$\dfrac{1}{x}$

47. Explain why
$$\frac{f(x + h) - f(x - h)}{2h}$$
should give a reasonable approximation of $f'(x)$ when $f'(x)$ exists and h is small.

48. a. For the function $f(x) = -4x^2 + 11x$, find the value of $f'(3)$, as well as the approximation using
$$\frac{f(x + h) - f(x)}{h}$$
and using the formula in Exercise 47 with $h = 0.1$.

  b. Repeat part a using $h = 0.01$.

  c. Repeat part a using the function $f(x) = -2/x$ and $h = 0.1$.

  d. Repeat part c using $h = 0.01$.

  e. Repeat part a using the function $f(x) = \sqrt{x}$ and $h = 0.1$.

  f. Repeat part e using $h = 0.01$.

  g. Using the results of parts a through f, discuss which approximation formula seems to give better accuracy.

APPLICATIONS

Business and Economics

49. Demand Suppose the demand for a certain item is given by $D(p) = -2p^2 - 4p + 300$, where p represents the price of the item in dollars.

  a. Find the rate of change of demand with respect to price.

  b. Find and interpret the rate of change of demand when the price is $10.

50. Profit The profit (in thousands of dollars) from the expenditure of x thousand dollars on advertising is given by $P(x) = 1000 + 32x - 2x^2$. Find the marginal profit at the following expenditures. In each case, decide whether the firm should increase the expenditure.

  a. $8000   **b.** $6000   **c.** $12,000   **d.** $20,000

51. Revenue The revenue in dollars generated from the sale of x picnic tables is given by $R(x) = 20x - \dfrac{x^2}{500}$.

  a. Find the marginal revenue when 1000 tables are sold.

  b. Estimate the revenue from the sale of the 1001st table by finding $R'(1000)$.

  c. Determine the actual revenue from the sale of the 1001st table.

  d. Compare your answers for parts b and c. What do you find?

52. Cost The cost in dollars of producing x tacos is $C(x) = -0.00375x^2 + 1.5x + 1000$, for $0 \le x \le 180$.

a. Find the marginal cost.

b. Find and interpret the marginal cost at a production level of 100 tacos.

c. Find the exact cost to produce the 101st taco.

d. Compare the answers to parts b and c. How are they related?

e. Show that whenever $C(x) = ax^2 + bx + c$,
$[C(x + 1) - C(x)] - C'(x) = a$. **Source: The College Mathematics Journal.**

f. Show that whenever $C(x) = ax^2 + bx + c$,

$$C(x + 1) - C(x) = C'\left(x + \frac{1}{2}\right).$$

53. Social Security Assets The table gives actual and projected year-end assets in Social Security trust funds, in trillions of current dollars, where Year represents the number of years since 2000. **Source: Social Security Administration.**

The polynomial function defined by

$$f(x) = 0.0000329x^3 - 0.00450x^2 + 0.0613x + 2.34$$

models the data quite well.

a. To verify the fit of the model, find $f(10)$, $f(20)$, and $f(30)$.

b. Use a graphing calculator with a command such as `nDeriv` to find the slope of the tangent line to the graph of f at the following x-values: 0, 10, 20, 30, and 35.

c. Use your results in part b to describe the graph of f and interpret the corresponding changes in Social Security assets.

Year	Trillions of Dollars
10	2.45
20	2.34
30	1.03
40	−0.57
50	−1.86

Life Sciences

54. Flight Speed Many birds, such as cockatiels or the Arctic terns shown below, have flight muscles whose expenditure of power varies with the flight speed in a manner similar to the graph shown in the next column. The horizontal axis of the graph shows flight speed in meters per second, and the vertical axis shows power in watts per kilogram. **Source: Biolog-e: The Undergraduate Bioscience Research Journal.**

a. The speed V_{mp} minimizes energy costs per unit of time. What is the slope of the line tangent to the curve at the point corresponding to V_{mp}? What is the physical significance of the slope at that point?

b. The speed V_{mr} minimizes the energy costs per unit of distance covered. Estimate the slope of the curve at the point corresponding to V_{mr}. Give the significance of the slope at that point.

c. By looking at the shape of the curve, describe how the power level decreases and increases for various speeds.

d. Notice that the slope of the lines found in parts a and b represents the power divided by speed. Power is measured in energy per unit time per unit weight of the bird, and speed is distance per unit time, so the slope represents energy per unit distance per unit weight of the bird. If a line is drawn from the origin to a point on the graph, at which point is the slope of the line (representing energy per unit distance per unit weight of the bird) smallest? How does this compare with your answers to parts a and b?

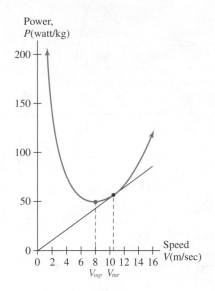

55. Shellfish Population In one research study, the population of a certain shellfish in an area at time t was closely approximated by the following graph. Estimate and interpret the derivative at each of the marked points.

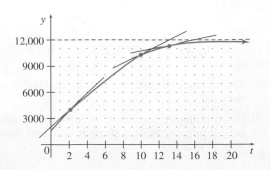

56. Eating Behavior The eating behavior of a typical human during a meal can be described by

$$I(t) = 27 + 72t - 1.5t^2,$$

where t is the number of minutes since the meal began, and $I(t)$ represents the amount (in grams) that the person has eaten at time t. *Source: Appetite.*

a. Find the rate of change of the intake of food for this particular person 5 minutes into a meal and interpret.

b. Verify that the rate in which food is consumed is zero 24 minutes after the meal starts.

c. Comment on the assumptions and usefulness of this function after 24 minutes. Given this fact, determine a logical domain for this function.

57. Quality Control of Cheese It is often difficult to evaluate the quality of products that undergo a ripening or maturation process. Researchers have successfully used ultrasonic velocity to determine the maturation time of Mahon cheese. The age can be determined by

$$M(v) = 0.0312443v^2 - 101.39v + 82,264, \quad v \geq 1620,$$

where $M(v)$ is the estimated age of the cheese (in days) for a velocity v (m per second). *Source: Journal of Food Science.*

a. If Mahon cheese ripens in 150 days, determine the velocity of the ultrasound that one would expect to measure. (*Hint:* Set $M(v) = 150$ and solve for v.)

b. Determine the derivative of this function when $v = 1700$ m per second and interpret.

Physical Sciences

58. Temperature The graph shows the temperature in degrees Celsius as a function of the altitude h in feet when an inversion layer is over Southern California. (See Exercise 41 in the previous section.) Estimate and interpret the derivatives of $T(h)$ at the marked points.

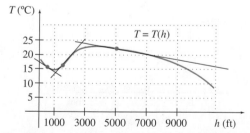

59. Oven Temperature The graph in the next column shows the temperature one Christmas day of a Heat-Kit Bakeoven, a wood-burning oven for baking. *Source: Heatkit.com.* The oven was lit at 8:30 a.m. Let $T(x)$ be the temperature x hours after 8:30 a.m.

a. Find and interpret $T'(0.5)$.

b. Find and interpret $T'(3)$.

c. Find and interpret $T'(4)$.

d. At a certain time a Christmas turkey was put into the oven, causing the oven temperature to drop. Estimate when the turkey was put into the oven.

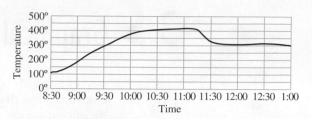

60. Baseball The graph shows how the velocity of a baseball that was traveling at 40 miles per hour when it was hit by a Little League baseball player varies with respect to the weight of the bat. *Source: Biological Cybernetics.*

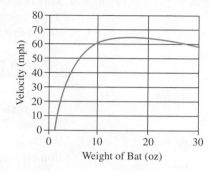

a. Estimate and interpret the derivative for a 16-oz and 25-oz bat.

b. What is the optimal bat weight for this player?

61. Baseball The graph shows how the velocity of a baseball that was traveling at 90 miles per hour when it was hit by a Major League baseball player varies with respect to the weight of the bat. *Source: Biological Cybernetics.*

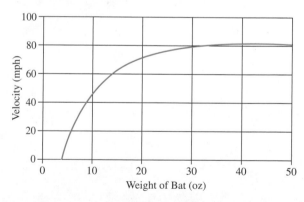

a. Estimate and interpret the derivative for a 40-oz and 30-oz bat.

b. What is the optimal bat weight for this player?

YOUR TURN ANSWERS

1. a. $y = -2x + 2$ **b.** $y = -5x - 4$

2. $f'(x) = 2x - 1$ and $f'(-2) = -5$.

3. $f'(x) = 3x^2$ and $f'(-1) = 3$.

4. $f'(x) = \dfrac{2}{x^2}$.   **5.** $f'(x) = \dfrac{1}{\sqrt{x}}$

6. 9.60 **7.** $y = \frac{1}{2}x + 2$

3.5 Graphical Differentiation

APPLY IT Given a graph of the production function, how can we find the graph of the marginal production function?

We will answer this question in Example 1 using graphical differentiation.

In the previous section, we estimated the derivative at various points of a graph by estimating the slope of the tangent line at those points. We will now extend this process to show how to sketch the graph of the derivative given the graph of the original function. This is important because, in many applications, a graph is all we have, and it is easier to find the derivative graphically than to find a formula that fits the graph and take the derivative of that formula.

EXAMPLE 1 Production of Landscape Mulch

In Figure 45(a), the graph shows total production (*TP*), measured in cubic yards of landscape mulch per week, as a function of labor used, measured in workers hired by a small business. The graph in Figure 45(b) shows the marginal production curve (MP_L), which is the derivative of the total production function. Verify that the graph of the marginal production curve (MP_L) is the graph of the derivative of the total production curve (*TP*).

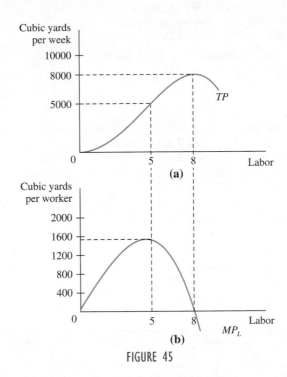

FIGURE 45

APPLY IT **SOLUTION** Let q refer to the quantity of labor. We begin by choosing a point where estimating the derivative of *TP* is simple. Observe that when $q = 8$, *TP* has a horizontal tangent line, so its derivative is 0. This explains why the graph of MP_L equals 0 when $q = 8$.

Now, observe in Figure 45(a) that when $q < 8$, the tangent lines of *TP* have positive slope and the slope is steepest when $q = 5$. This means that the derivative should be positive for $q < 8$ and largest when $q = 5$. Verify that the graph of MP_L has this property.

Finally, as Figure 45(a) shows, the tangent lines of *TP* have negative slope when $q > 8$, so its derivative, represented by the graph of MP_L, should also be negative there. Verify that the graph of MP_L, in Figure 45(b), has this property as well.

In Example 1, we saw how the general shape of the graph of the derivative could be found from the graph of the original function. To get a more accurate graph of the derivative, we need to estimate the slope of the tangent line at various points, as we did in the previous section.

EXAMPLE 2 Temperature

Figure 46 gives the temperature in degrees Celsius as a function, $T(h)$, of the altitude h in feet when an inversion layer is over Southern California. Sketch the graph of the derivative of the function. (This graph appeared in Exercise 58 in the previous section.)

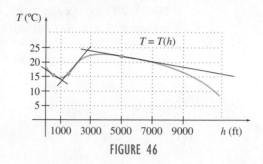

FIGURE 46

SOLUTION First, observe that when $h = 1000$ and $h = 3500$, $T(h)$ has horizontal tangent lines, so $T'(1000) = 0$ and $T'(3500) = 0$.

Notice that the tangent lines would have a negative slope for $0 < h < 1000$. Thus, the graph of the derivative should be negative (below the x-axis) there. Then, for $1000 < h < 3500$, the tangent lines have positive slope, so the graph of the derivative should be positive (above the x-axis) there. Notice from the graph of $T(h)$ that the slope is largest when $h = 1500$. Finally, for $h > 3500$, the tangent lines would be negative again, forcing the graph of the derivative back down below the x-axis to take on negative values.

Now that we have a general shape of the graph, we can estimate the value of the derivative at several points to improve the accuracy of the graph. To estimate the derivative, find two points on each tangent line and compute the slope between the two points. The estimates at selected points are given in the table to the left. (Your answers may be slightly different, since estimation from a picture can be inexact.) Figure 47 shows a graph of these values of $T'(h)$.

Estimates of $T'(h)$

h	$T'(h)$
500	−0.005
1000	0
1500	0.008
3500	0
5000	−0.00125

YOUR TURN 1 Sketch the graph of the derivative of the function $f(x)$.

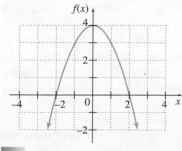

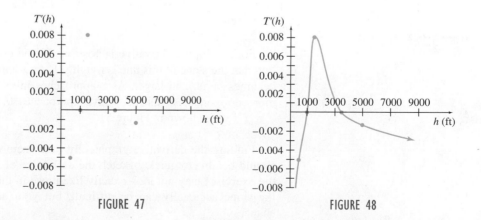

FIGURE 47   **FIGURE 48**

Using all of these facts, we connect the points in the graph $T'(h)$ smoothly, with the result shown in Figure 48.

TRY YOUR TURN 1

| CAUTION | Remember that when you graph the derivative, you are graphing the *slope* of the original function. Do not confuse the slope of the original function with the *y*-value of the original function. In fact, the slope of the original function is equal to the *y*-value of its derivative. |

Sometimes the original function is not smooth or even continuous, so the graph of the derivative may also be discontinuous.

EXAMPLE 3 Graphing a Derivative

Sketch the graph of the derivative of the function shown in Figure 49.

SOLUTION   Notice that when $x < -2$, the slope is 1, and when $-2 < x < 0$, the slope is -1. At $x = -2$, the derivative does not exist due to the sharp corner in the graph. The derivative also does not exist at $x = 0$ because the function is discontinuous there. Using this information, the graph of $f'(x)$ on $x < 0$ is shown in Figure 50.

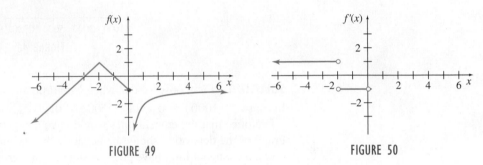

FIGURE 49   FIGURE 50

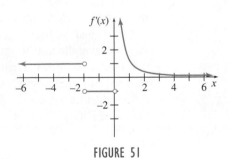

FIGURE 51

YOUR TURN 2 Sketch the graph of the derivative of the function $g(x)$.

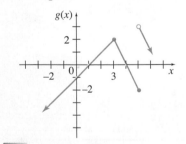

For $x > 0$, the derivative is positive. If you draw a tangent line at $x = 1$, you should find that the slope of this line is roughly 1. As x approaches 0 from the right, the derivative becomes larger and larger. As x approaches infinity, the tangent lines become more and more horizontal, so the derivative approaches 0. The resulting sketch of the graph of $y = f'(x)$ is shown in Figure 51.   **TRY YOUR TURN 2**

Finding the derivative graphically may seem difficult at first, but with practice you should be able to quickly sketch the derivative of any function graphed. Your answers to the exercises may not look exactly like those in the back of the book, because estimating the slope accurately can be difficult, but your answers should have the same general shape.

Figures 52(a), (b), and (c) show the graphs of $y = x^2$, $y = x^4$, and $y = x^{4/3}$ on a graphing calculator. When finding the derivative graphically, all three seem to have the same behavior: negative derivative for $x < 0$, 0 derivative at $x = 0$, and positive derivative for $x > 0$. Beyond these general features, however, the derivatives look quite different, as you can see from Figures 53(a), (b), and (c), which show the graphs of the derivatives. When finding derivatives graphically, detailed information can only be found by very carefully measuring the slope of the tangent line at a large number of points.

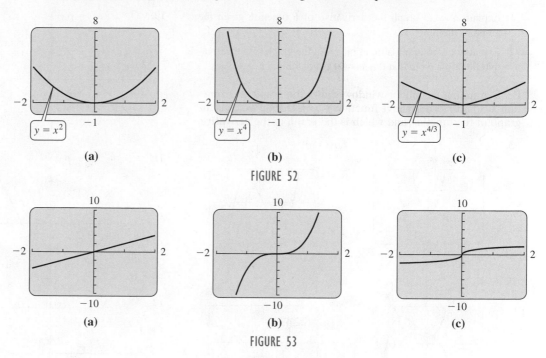

(a) $\qquad$ (b) $\qquad$ (c)

FIGURE 52

(a) $\qquad$ (b) $\qquad$ (c)

FIGURE 53

TECHNOLOGY NOTE

On many calculators, the graph of the derivative can be plotted if a formula for the original function is known. For example, the graphs in Figure 53 were drawn on a TI-84 Plus by using the `nDeriv` command. Define $Y_2 = \frac{d}{dx}(Y_1)\big|_{x\,=\,x}$ after entering the original function into Y_1. You can use this feature to practice finding the derivative graphically. Enter a function into Y_1, sketch the graph on the graphing calculator, and use it to draw by hand the graph of the derivative. Then use `nDeriv` to draw the graph of the derivative, and compare it with your sketch.

EXAMPLE 4   **Graphical Differentiation**

Figure 54 shows the graph of a function f and its derivative function f'. Use slopes to decide which graph is that of f and which is the graph of f'.

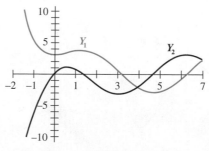

FIGURE 54

SOLUTION   Look at the places where each graph crosses the x-axis; that is, the x-intercepts, since x-intercepts occur on the graph of f' whenever the graph of f has a horizontal tangent line or slope of zero. Also, a decreasing graph corresponds to negative slope or a negative derivative, while an increasing graph corresponds to positive slope or a positive derivative. Y_1 has zero slope near $x = 0$, $x = 1$, and $x = 5$; Y_2 has x-intercepts near these values of x. Y_1 decreases on $(-2, 0)$ and $(1, 5)$; Y_2 is negative on those intervals. Y_1 increases on $(0, 1)$ and $(5, 7)$; Y_2 is positive there. Thus, Y_1 is the graph of f and Y_2 is the graph of f'.

3.5 EXERCISES

 1. Explain how to graph the derivative of a function given the graph of the function.

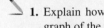 **2.** Explain how to graph a function given the graph of the derivative function. Note that the answer is not unique.

Each graphing calculator window shows the graph of a function $f(x)$ and its derivative function $f'(x)$. Decide which is the graph of the function and which is the graph of the derivative.

3.

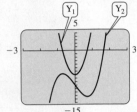

4.

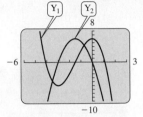

5.

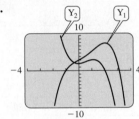

6.

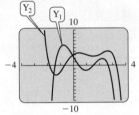

Sketch the graph of the derivative for each function shown.

7.

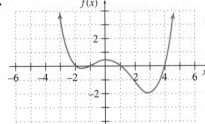

8.

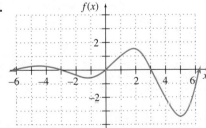

9.

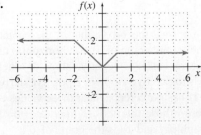

10.

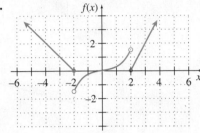

11.

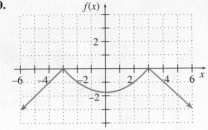

12.

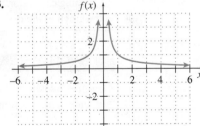

13.

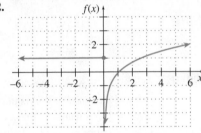

14.

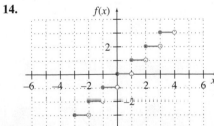

15.

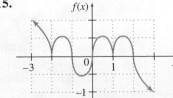

16.

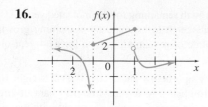

APPLICATIONS

Business and Economics

17. Consumer Demand When the price of an essential commodity rises rapidly, consumption drops slowly at first. If the price continues to rise, however, a "tipping" point may be reached, at which consumption takes a sudden substantial drop. Suppose the accompanying graph shows the consumption of gasoline, $G(t)$, in millions of gallons, in a certain area. We assume that the price is rising rapidly. Here t is the time in months after the price began rising. Sketch a graph of the rate of change in consumption as a function of time.

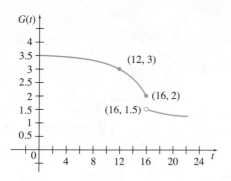

18. Sales The graph shows annual sales (in thousands of dollars) of a Nintendo game at a particular store. Sketch a graph of the rate of change of sales as a function of time.

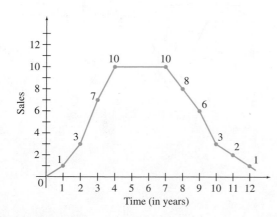

Life Sciences

19. Insecticide The graph in the next column shows how the number of arthropod species resistant to insecticides has varied with time. Sketch a graph of the rate of change of the insecticide–resistant species as a function of time. *Source: Science.*

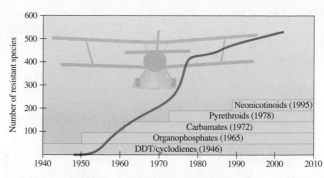

20. Body Mass Index The following graph shows how the body mass index-for-age percentile for boys varies from the age of 2 to 20 years. *Source: Centers for Disease Control.*

a. Sketch a graph of the rate of change of the 95th percentile as a function of age.

b. Sketch a graph of the rate of change of the 50th percentile as a function of age.

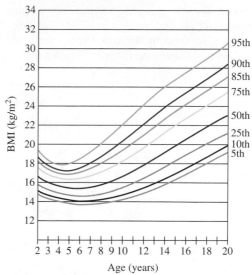

Body Mass Index-for-Age Percentiles: Boys, 2 to 20 years

21. Shellfish Population In one research study, the population of a certain shellfish in an area at time t was closely approximated by the following graph. Sketch a graph of the growth rate of the population.

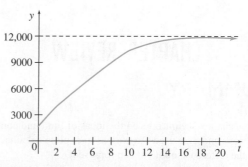

22. Flight Speed The graph on the next page shows the relationship between the speed of a bird in flight and the required power expended by flight muscles. Sketch the graph of the rate of change of the power as a function of the speed. *Source: Biolog-e: The Undergraduate Bioscience Research Journal*.

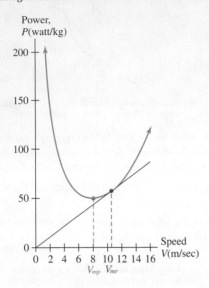

23. Human Growth The growth remaining in sitting height at consecutive skeletal age levels is indicated below for boys. *Source: Standards in Pediatric Orthopedics: Tables, Charts, and Graphs Illustrating Growth*. Sketch a graph showing the rate of change of growth remaining for the indicated years. Use the graph and your sketch to estimate the remaining growth and the rate of change of remaining growth for a 14-year-old boy.

24. Weight Gain The graph below shows the typical weight (in kilograms) of an English boy for his first 18 years of life. Sketch the graph of the rate of change of weight with respect to time. *Source: Human Growth After Birth*.

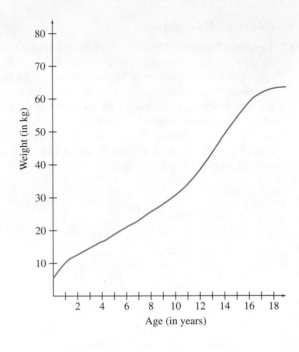

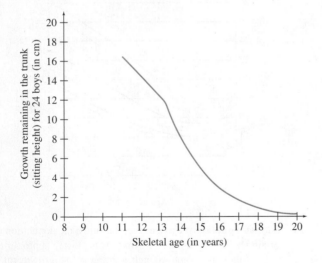

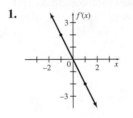 **YOUR TURN ANSWERS**

1.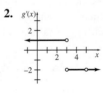

2.

3 CHAPTER REVIEW

SUMMARY

In this chapter we introduced the ideas of limit and continuity of functions and then used these ideas to explore calculus. We saw that the difference quotient can represent

- the average rate of change,
- the slope of the secant line, and
- the average velocity.

We saw that the derivative can represent

- the instantaneous rate of change,
- the slope of the tangent line, and
- the instantaneous velocity.

We also learned how to estimate the value of the derivative using graphical differentiation. In the next chapter, we will take a closer look at the definition of the derivative to develop a set of rules to quickly and easily calculate the derivative of a wide range of functions without the need to directly apply the definition of the derivative each time.

Limit of a Function   Let f be a function and let a and L be real numbers. If

1. as x takes values closer and closer (but not equal) to a on both sides of a, the corresponding values of $f(x)$ get closer and closer (and perhaps equal) to L; and

2. the value of $f(x)$ can be made as close to L as desired by taking values of x close enough to a;

   then L is the limit of $f(x)$ as x approaches a, written

$$\lim_{x \to a} f(x) = L.$$

Existence of Limits   The limit of f as x approaches a may not exist.

1. If $f(x)$ becomes infinitely large in magnitude (positive or negative) as x approaches the number a from either side, we write $\lim_{x \to a} f(x) = \infty$ or $\lim_{x \to a} f(x) = -\infty$. In either case, the limit does not exist.

2. If $f(x)$ becomes infinitely large in magnitude (positive) as x approaches a from one side and infinitely large in magnitude (negative) as x approaches a from the other side, then $\lim_{x \to a} f(x)$ does not exist.

3. If $\lim_{x \to a^-} f(x) = L$ and $\lim_{x \to a^+} f(x) = M$, and $L \neq M$, then $\lim_{x \to a} f(x)$ does not exist.

Limits at Infinity   For any positive real number n,

$$\lim_{x \to \infty} \frac{1}{x^n} = \lim_{x \to -\infty} \frac{1}{x^n} = 0.$$

Finding Limits at Infinity   If $f(x) = p(x)/q(x)$ for polynomials $p(x)$ and $q(x)$, $\lim_{x \to \infty} f(x)$ and $\lim_{x \to -\infty} f(x)$ can be found by dividing $p(x)$ and $q(x)$ by the highest power of x in $q(x)$.

Continuity   A function f is continuous at c if

1. $f(c)$ is defined,

2. $\lim_{x \to c} f(x)$ exists, and

3. $\lim_{x \to c} f(x) = f(c)$.

Average Rate of Change   The average rate of change of $f(x)$ with respect to x as x changes from a to b is

$$\frac{f(b) - f(a)}{b - a}.$$

Difference Quotient   The average rate of change can also be written as

$$\frac{f(x + h) - f(x)}{h}.$$

Derivative   The derivative of $f(x)$ with respect to x is

$$f'(x) = \lim_{h \to 0} \frac{f(x + h) - f(x)}{h}.$$

KEY TERMS

To understand the concepts presented in this chapter, you should know the meaning and use of the following terms.
For easy reference, the section in the chapter where a word (or expression) was first used is provided.

3.1	**3.2**	**3.3**	**3.4**
limit	continuous	average rate of change	secant line
limit from the left/right	discontinuous	difference quotient	tangent line
one-/two-sided limit	removable discontinuity	instantaneous rate of change	slope of the curve
piecewise function	continuous on an open/closed	velocity	derivative
limit at infinity	interval		differentiable
	continuous from the right/left		differentiation
	Intermediate Value Theorem		

REVIEW EXERCISES

CONCEPT CHECK

Determine whether each of the following statements is true or false, and explain why.

1. The limit of a product is the product of the limits when each of the limits exists.

2. The limit of a function may not exist at a point even though the function is defined there.

3. If a rational function has a polynomial in the denominator of higher degree than the polynomial in the numerator, then the limit at infinity must equal zero.

4. If the limit of a function exists at a point, then the function is continuous there.

5. A polynomial function is continuous everywhere.

6. A rational function is continuous everywhere.

7. The derivative gives the average rate of change of a function.

8. The derivative gives the instantaneous rate of change of a function.

9. The instantaneous rate of change is a limit.

10. The derivative is a function.

11. The slope of the tangent line gives the average rate of change.

12. The derivative of a function exists wherever the function is continuous.

PRACTICE AND EXPLORATIONS

13. Is a derivative always a limit? Is a limit always a derivative? Explain.

14. Is every continuous function differentiable? Is every differentiable function continuous? Explain.

15. Describe how to tell when a function is discontinuous at the real number $x = a$.

16. Give two applications of the derivative

$$f'(x) = \lim_{h \to 0} \frac{f(x+h) - f(x)}{h}.$$

Decide whether the limits in Exercises 17–34 exist. If a limit exists, find its value.

17. **a.** $\lim_{x \to -3^-} f(x)$ **b.** $\lim_{x \to -3^+} f(x)$ **c.** $\lim_{x \to -3} f(x)$ **d.** $f(-3)$

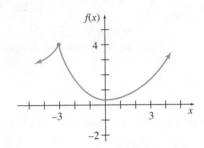

18. **a.** $\lim_{x \to -1^-} g(x)$ **b.** $\lim_{x \to -1^+} g(x)$ **c.** $\lim_{x \to -1} g(x)$ **d.** $g(-1)$

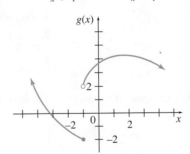

19. **a.** $\lim_{x \to 4^-} f(x)$ **b.** $\lim_{x \to 4^+} f(x)$ **c.** $\lim_{x \to 4} f(x)$ **d.** $f(4)$

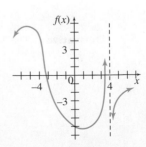

20. a. $\lim\limits_{x \to 2^-} h(x)$ **b.** $\lim\limits_{x \to 2^+} h(x)$ **c.** $\lim\limits_{x \to 2} h(x)$ **d.** $h(2)$

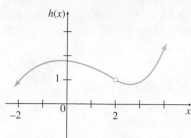

21. $\lim\limits_{x \to -\infty} g(x)$

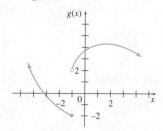

22. $\lim\limits_{x \to \infty} f(x)$

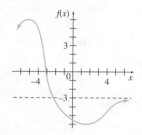

23. $\lim\limits_{x \to 6} \dfrac{2x + 7}{x + 3}$

24. $\lim\limits_{x \to -3} \dfrac{2x + 5}{x + 3}$

25. $\lim\limits_{x \to 4} \dfrac{x^2 - 16}{x - 4}$

26. $\lim\limits_{x \to 2} \dfrac{x^2 + 3x - 10}{x - 2}$

27. $\lim\limits_{x \to -4} \dfrac{2x^2 + 3x - 20}{x + 4}$

28. $\lim\limits_{x \to 3} \dfrac{3x^2 - 2x - 21}{x - 3}$

29. $\lim\limits_{x \to 9} \dfrac{\sqrt{x} - 3}{x - 9}$

30. $\lim\limits_{x \to 16} \dfrac{\sqrt{x} - 4}{x - 16}$

31. $\lim\limits_{x \to \infty} \dfrac{2x^2 + 5}{5x^2 - 1}$

32. $\lim\limits_{x \to \infty} \dfrac{x^2 + 6x + 8}{x^3 + 2x + 1}$

33. $\lim\limits_{x \to -\infty} \left(\dfrac{3}{8} + \dfrac{3}{x} - \dfrac{6}{x^2} \right)$

34. $\lim\limits_{x \to -\infty} \left(\dfrac{9}{x^4} + \dfrac{10}{x^2} - 6 \right)$

Identify the x-values where f is discontinuous.

35.

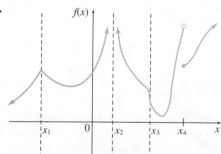

36.

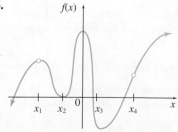

Find all x-values where the function is discontinuous. For each such value, give $f(a)$ and $\lim\limits_{x \to a} f(x)$ or state that it does not exist.

37. $f(x) = \dfrac{-5 + x}{3x(3x + 1)}$

38. $f(x) = \dfrac{7 - 3x}{(1 - x)(3 + x)}$

39. $f(x) = \dfrac{x - 6}{x + 5}$

40. $f(x) = \dfrac{x^2 - 9}{x + 3}$

41. $f(x) = x^2 + 3x - 4$

42. $f(x) = 2x^2 - 5x - 3$

In Exercises 43 and 44, (a) graph the given function, (b) find all values of x where the function is discontinuous, and (c) find the limit from the left and from the right at any values of x found in part b.

43. $f(x) = \begin{cases} 1 - x & \text{if } x < 1 \\ 2 & \text{if } 1 \le x \le 2 \\ 4 - x & \text{if } x > 2 \end{cases}$

44. $f(x) = \begin{cases} 2 & \text{if } x < 0 \\ -x^2 + x + 2 & \text{if } 0 \le x \le 2 \\ 1 & \text{if } x > 2 \end{cases}$

Find each limit (a) by investigating values of the function near the point where the limit is taken and (b) by using a graphing calculator to view the function near the point.

45. $\lim\limits_{x \to 1} \dfrac{x^4 + 2x^3 + 2x^2 - 10x + 5}{x^2 - 1}$

46. $\lim\limits_{x \to -2} \dfrac{x^4 + 3x^3 + 7x^2 + 11x + 2}{x^3 + 2x^2 - 3x - 6}$

Find the average rate of change for the following on the given interval. Then find the instantaneous rate of change at the first x-value.

47. $y = 6x^3 + 2$ from $x = 1$ to $x = 4$

48. $y = -2x^3 - 3x^2 + 8$ from $x = -2$ to $x = 6$

49. $y = \dfrac{-6}{3x - 5}$ from $x = 4$ to $x = 9$

50. $y = \dfrac{x + 4}{x - 1}$ from $x = 2$ to $x = 5$

For each function, find (a) the equation of the secant line through the points where x has the given values, and (b) the equation of the tangent line when x has the first value.

51. $f(x) = 3x^2 - 5x + 7; x = 2, x = 4$

52. $f(x) = \dfrac{1}{x}; x = 1/2, x = 3$

53. $f(x) = \dfrac{12}{x - 1}; x = 3, x = 7$

54. $f(x) = 2\sqrt{x - 1}; x = 5, x = 10$

Use the definition of the derivative to find the derivative of the following.

55. $y = 4x^2 + 3x - 2$

56. $y = 5x^2 - 6x + 7$

In Exercises 57 and 58, find the derivative of the function at the given point (a) by approximating the definition of the derivative with small values of h and (b) by using a graphing calculator to zoom in on the function until it appears to be a straight line, and then finding the slope of that line.

57. $f(x) = (\ln x)^x$; $x_0 = 3$ **58.** $f(x) = x^{\ln x}$; $x_0 = 2$

Sketch the graph of the derivative for each function shown.

59.

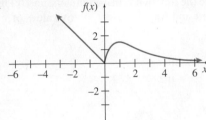

60.

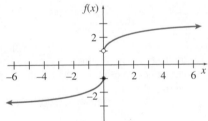

61. Let f and g be differentiable functions such that

$$\lim_{x \to \infty} f(x) = c$$

$$\lim_{x \to \infty} g(x) = d$$

where $c \neq d$. Determine

$$\lim_{x \to \infty} \frac{cf(x) - dg(x)}{f(x) - g(x)}.$$

(Choose one of the following.) *Source: Society of Actuaries.*

a. 0

b. $\dfrac{cf'(0) - dg'(0)}{f'(0) - g'(0)}$

c. $f'(0) - g'(0)$ **d.** $c - d$ **e.** $c + d$

APPLICATIONS

Business and Economics

62. Revenue Waverly Products has found that its revenue is related to advertising expenditures by the function
$$R(x) = 5000 + 16x - 3x^2,$$
where $R(x)$ is the revenue in dollars when x hundred dollars are spent on advertising.

a. Find the marginal revenue function.

b. Find and interpret the marginal revenue when $1000 is spent on advertising.

63. Cost Analysis A company charges $1.50 per lb when a certain chemical is bought in lots of 125 lb or less, with a price per pound of $1.35 if more than 125 lb are purchased. Let $C(x)$ represent the cost of x lb. Find the cost for the following numbers of pounds.

a. 100 **b.** 125 **c.** 140 **d.** Graph $y = C(x)$.

e. Where is C discontinuous?

Find the average cost per pound if the following number of pounds are bought.

f. 100 **g.** 125 **h.** 140

Find and interpret the marginal cost (that is, the instantaneous rate of change of the cost) for the following numbers of pounds.

i. 100 **j.** 140

64. Marginal Analysis Suppose the profit (in cents) from selling x lb of potatoes is given by

$$P(x) = 15x + 25x^2.$$

Find the average rate of change in profit from selling each of the following amounts.

a. 6 lb to 7 lb **b.** 6 lb to 6.5 lb **c.** 6 lb to 6.1 lb

Find the marginal profit (that is, the instantaneous rate of change of the profit) from selling the following amounts.

d. 6 lb **e.** 20 lb **f.** 30 lb

g. What is the domain of x?

h. Is it possible for the marginal profit to be negative here? What does this mean?

i. Find the average profit function. (Recall that average profit is given by total profit divided by the number produced, or $\overline{P}(x) = P(x)/x$.)

j. Find the marginal average profit function (that is, the function giving the instantaneous rate of change of the average profit function).

k. Is it possible for the marginal average profit to vary here? What does this mean?

l. Discuss whether this function describes a realistic situation.

65. Average Cost The graph on the next page shows the total cost $C(x)$ to produce x tons of cement. (Recall that average cost is given by total cost divided by the number produced, or $\overline{C}(x) = C(x)/x$.)

a. Draw a line through $(0, 0)$ and $(5, C(5))$. Explain why the slope of this line represents the average cost per ton when 5 tons of cement are produced.

b. Find the value of x for which the average cost is smallest.

c. What can you say about the marginal cost at the point where the average cost is smallest?

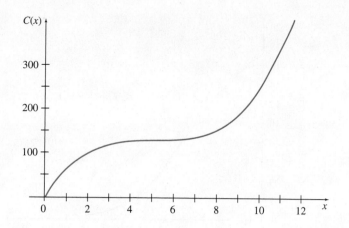

66. Tax Rates A simplified income tax considered in the U.S. Senate in 1986 had two tax brackets. Married couples earning $29,300 or less would pay 15% of their income in taxes. Those earning more than $29,300 would pay $4350 plus 27% of the income over $29,300 in taxes. Let $T(x)$ be the amount of taxes paid by someone earning x dollars in a year. *Source: Wall Street Journal.*

a. Find $\lim_{x \to 29,300^-} T(x)$.　　**b.** Find $\lim_{x \to 29,300^+} T(x)$.

c. Find $\lim_{x \to 29,300} T(x)$.　　**d.** Sketch a graph of $T(x)$.

e. Identify any x-values where T is discontinuous.

f. Let $A(x) = T(x)/x$ be the average tax rate, that is, the amount paid in taxes divided by the income. Find a formula for $A(x)$. (*Note:* The formula will have two parts: one for $x \le 29,300$ and one for $x > 29,300$.)

g. Find $\lim_{x \to 29,300^-} A(x)$.　　**h.** Find $\lim_{x \to 29,300^+} A(x)$.

i. Find $\lim_{x \to 29,300} A(x)$.　　**j.** Find $\lim_{x \to \infty} A(x)$.

k. Sketch the graph of $A(x)$.

67. Unemployment The annual unemployment rates of the U.S. civilian noninstitutional population for 1990–2009 are shown in the graph. Sketch a graph showing the rate of change of the annual unemployment rates for this period. Use the given graph and your sketch to estimate the annual unemployment rate and rate of change of the unemployment rate in 2008. *Source: Bureau of Labor Statistics.*

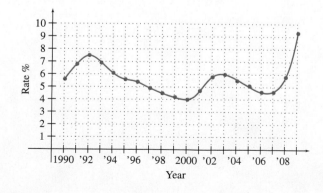

Life Sciences

68. Alzheimer's Disease The graph below shows the projected number of people aged 65 and over in the United States with Alzheimer's Disease. *Source: Alzheimer's Disease Facts and Figures.* Estimate and interpret the derivative in each of the following years

a. 2000　　**b.** 2040

c. Find the average rate of change between 2000 and 2040 in the number of people aged 65 and over in the United States with Alzheimer's disease.

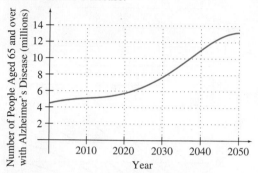

69. Spread of a Virus The spread of a virus is modeled by
$$V(t) = -t^2 + 6t - 4,$$
where $V(t)$ is the number of people (in hundreds) with the virus and t is the number of weeks since the first case was observed.

a. Graph $V(t)$.

b. What is a reasonable domain of t for this problem?

c. When does the number of cases reach a maximum? What is the maximum number of cases?

d. Find the rate of change function.

e. What is the rate of change in the number of cases at the maximum?

f. Give the sign (+ or −) of the rate of change up to the maximum and after the maximum.

70. Whales Diving The following figure, already shown in the section on Properties of Functions, shows the depth of a sperm whale as a function of time, recorded by researchers at the Woods Hole Oceanographic Institution in Massachusetts. *Source: Peter Tyack, Woods Hole Oceanographic Institution.*

a. Find the rate that the whale was descending at the following times.

　i. 17 hours and 37 minutes

　ii. 17 hours and 39 minutes

b. Sketch a graph of the rate the whale was descending as a function of time.

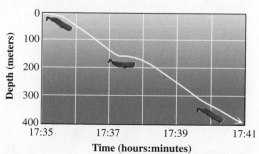

71. Body Mass Index The following graph shows how the body mass index-for-age percentile for girls varies from the age of 2 to 20 years. *Source: Centers for Disease Control.*

a. Sketch a graph of the rate of change of the 95th percentile as a function of age.

b. Sketch a graph of the rate of change of the 50th percentile as a function of age.

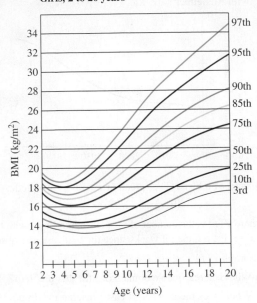

Body Mass Index-for-Age Percentiles: Girls, 2 to 20 years

72. Human Growth The growth remaining in sitting height at consecutive skeletal age levels is indicated below for girls. Sketch a graph showing the rate of change of growth remaining for the indicated years. Use the graph and your sketch to estimate the remaining growth and the rate of change of remaining growth for a 10-year-old girl. *Source: Standards in Pediatric Orthopedics: Tables, Charts, and Graphs Illustrating Growth.*

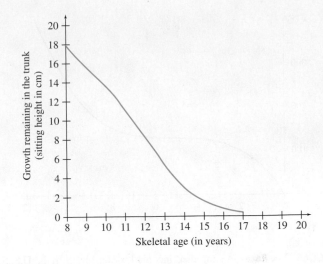

Physical Sciences

73. Temperature Suppose a gram of ice is at a temperature of $-100°C$. The graph shows the temperature of the ice as increasing numbers of calories of heat are applied. It takes 80 calories to melt one gram of ice at $0°C$ into water, and 540 calories to boil one gram of water at $100°C$ into steam.

a. Where is this graph discontinuous?

b. Where is this graph not differentiable?

c. Sketch the graph of the derivative.

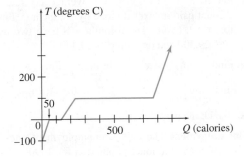

A MODEL FOR DRUGS ADMINISTERED INTRAVENOUSLY

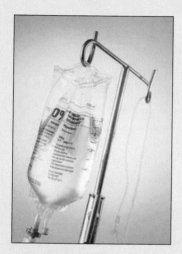

When a drug is administered intravenously it enters the bloodstream immediately, producing an immediate effect for the patient. The drug can be either given as a single rapid injection or given at a constant drip rate. The latter is commonly referred to as an intravenous (IV) infusion. Common drugs administered intravenously include morphine for pain, diazepam (or Valium) to control a seizure, and digoxin for heart failure.

SINGLE RAPID INJECTION

With a single rapid injection, the amount of drug in the bloodstream reaches its peak immediately and then the body eliminates the drug exponentially. The larger the amount of drug there is in the body, the faster the body eliminates it. If a lesser amount of drug is in the body, it is eliminated more slowly.

The amount of drug in the bloodstream t hours after a single rapid injection can be modeled using an exponential decay function, like those found in the chapter on Nonlinear Functions, as follows:

$$A(t) = De^{kt},$$

where D is the size of the dose administered and k is the exponential decay constant for the drug.

EXAMPLE 1 Rapid Injection

The drug labetalol is used for the control of blood pressure in patients with severe hypertension. The half-life of labetalol is 4 hours. Suppose a 35-mg dose of the drug is administered to a patient by rapid injection.

(a) Find a model for the amount of drug in the bloodstream t hours after the drug is administered.

SOLUTION Since $D = 35$ mg, the function has the form

$$A(t) = 35e^{kt}.$$

Recall from the chapter on Nonlinear Functions that the general equation giving the half-life T in terms of the decay constant k was

$$T = -\frac{\ln 2}{k}.$$

Solving this equation for k, we get

$$k = -\frac{\ln 2}{T}.$$

Since the half-life of this drug is 4 hours,

$$k = -\frac{\ln 2}{4} \approx -0.17.$$

Therefore, the model is

$$A(t) = 35e^{-0.17t}.$$

The graph of $A(t)$ is given in Figure 55.

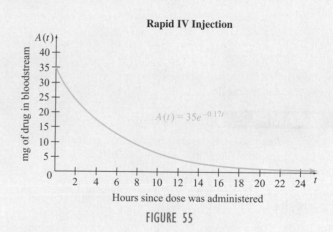

Rapid IV Injection

$A(t) = 35e^{-0.17t}$

Hours since dose was administered

FIGURE 55

(b) Find the average rate of change of drug in the bloodstream between $t = 0$ and $t = 2$. Repeat for $t = 4$ and $t = 6$.

SOLUTION The average rate of change from $t = 0$ to $t = 2$ is

$$\frac{A(2) - A(0)}{2 - 0} \approx \frac{25 - 35}{2} = -5 \text{ mg/hr}.$$

The average rate of change from $t = 4$ to $t = 6$ is

$$\frac{A(6) - A(4)}{6 - 4} \approx \frac{13 - 18}{2} = -2.5 \text{ mg/hr}.$$

Notice that since the half-life of the drug is 4 hours, the average rate of change from $t = 4$ to $t = 6$ is half of the average rate of change from $t = 0$ to $t = 2$. What would the average rate of change be from $t = 8$ to $t = 10$?

(c) What happens to the amount of drug in the bloodstream as t increases? (i.e., What is the limit of the function as t approaches ∞?)

SOLUTION Looking at the graph of $A(t)$, we can see that

$$\lim_{t \to \infty} A(t) = 0.$$

An advantage of an intravenous rapid injection is that the amount of drug in the body reaches a high level immediately. Suppose, however, that the effective level of this drug is between 30 mg and 40 mg. From the graph, we can see that it only takes an hour after the dose is given for the amount of drug in the body to fall below the effective level.

INTRAVENOUS INFUSION

With an IV infusion, the amount of drug in the bloodstream starts at zero and increases until the rate the drug is entering the body equals the rate the drug is being eliminated from the body. At this point, the amount of drug in the bloodstream levels off. This model is a limited growth function, like those from Chapter 2.

The amount of drug in the bloodstream t hours after an IV infusion begins can be modeled using a limited growth function, as follows.

$$A(t) = \frac{r}{-k}(1 - e^{kt}),$$

where r is the rate of infusion per hour and k is the exponential decay constant for the drug.

EXAMPLE 2 IV Infusion

The same drug used in Example 1 is given to a patient by IV infusion at a drip rate of 6 mg/hr. Recall that the half-life of this drug is 4 hours.

(a) Find a model for the amount of drug in the bloodstream t hours after the IV infusion begins.

SOLUTION Since $r = 6$ and $k = -0.17$, the function has the form

$$A(t) = 35(1 - e^{-0.17t}).$$

The graph of $A(t)$ is given in Figure 56.

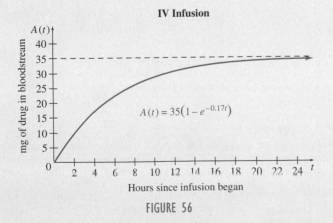

IV Infusion

$$A(t) = 35\left(1 - e^{-0.17t}\right)$$

Hours since infusion began

FIGURE 56

(b) Find the average rate of change of drug in the bloodstream between $t = 0$ and $t = 2$. Repeat for $t = 4$ and $t = 6$.

SOLUTION The average rate of change from $t = 0$ to $t = 2$ is

$$\frac{A(2) - A(0)}{2 - 0} \approx \frac{10 - 0}{2} = 5 \text{ mg/hr.}$$

The average rate of change from $t = 4$ to $t = 6$ is

$$\frac{A(6) - A(4)}{6 - 4} \approx \frac{22 - 17}{2} = 2.5 \text{ mg/hr.}$$

Recall that the average rate of change from $t = 0$ to $t = 2$ for the rapid injection of this drug was -5 mg/hr and the average rate of change from $t = 4$ to $t = 6$ was -2.5 mg/hr. In fact, at any given time, the rapid injection function is decreasing at the same rate the IV infusion function is increasing.

(c) What happens to the amount of drug in the bloodstream as t increases? (i.e., What is the limit of the function as t approaches ∞?)

SOLUTION Looking at the graph of $A(t)$ in Figure 56 and the formula for $A(t)$ in part (a), we can see that

$$\lim_{t \to \infty} A(t) = 35.$$

An advantage of an IV infusion is that a dose can be given such that the limit of $A(t)$ as t approaches ∞ is an effective level. Once the amount of drug has reached this effective level, it will remain there as long as the infusion continues. However, using this method of administration, it may take a while for the amount of drug in the body to reach an effective level. For our example, the effective level is between 30 mg and 40 mg. Looking at the graph, you can see that it takes about 11 hours to reach an effective level. If this patient were experiencing dangerously high blood pressure, you wouldn't want to wait 11 hours for the drug to reach an effective level.

SINGLE RAPID INJECTION FOLLOWED BY AN INTRAVENOUS INFUSION

Giving a patient a single rapid injection immediately followed by an intravenous infusion allows a patient to experience the advantages of both methods. The single rapid injection immediately produces an effective drug level in the patient's bloodstream. While the amount of drug in the bloodstream from the rapid infusion is decreasing, the amount of drug in the system from the IV infusion is increasing.

The amount of drug in the bloodstream t hours after the injection is given and infusion has started can be calculated by finding the sum of the two models.

$$A(t) = De^{kt} + \frac{r}{-k}(1 - e^{kt}).$$

EXAMPLE 3 Combination Model

A 35-mg dose of labetalol is administered to a patient by rapid injection. Immediately thereafter, the patient is given an IV infusion at a drip rate of 6 mg/hr. Find a model for the amount of drug in the bloodstream t hours after the drug is administered.

SOLUTION Recall from Example 1, the amount of drug in the bloodstream t hours after the rapid injection was found to be

$$A(t) = 35e^{-0.17t}.$$

From Example 2, the amount of drug in the bloodstream t hours after the IV infusion began was found to be

$$A(t) = 35(1 - e^{-0.17t}).$$

Therefore, t hours after administering both the rapid injection and the IV infusion, the amount of drug in the bloodstream is

$$A(t) = 35e^{-0.17t} + 35(1 - e^{-0.17t})$$
$$= 35 \text{ mg}.$$

The graph of $A(t)$ is given in Figure 57.

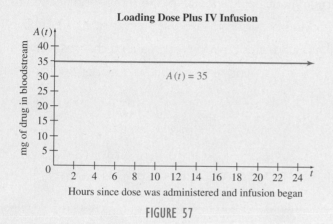

Loading Dose Plus IV Infusion

Hours since dose was administered and infusion began

FIGURE 57

Notice that the constant multiple of the rapid injection function, 35, is equal to the constant multiple of the IV infusion function. When this is the case, the sum of the two functions will be that constant.

EXAMPLE 4   Combination Model

A drug with a half-life of 3 hours is found to be effective when the amount of drug in the bloodstream is 58 mg. A 58-mg loading dose is given by rapid injection followed by an IV infusion. What should the rate of infusion be to maintain this level of drug in the bloodstream?

SOLUTION   Recall that the amount of drug in the bloodstream t hours after both a rapid injection and IV infusion are administered is given by

$$A(t) = De^{kt} + \frac{r}{-k}(1 - e^{kt}).$$

The rapid injection dose, D, is 58 mg. The half-life of the drug is three hours; therefore,

$$k = -\frac{\ln 2}{3} \approx -0.23.$$

It follows that

$$A(t) = 58e^{-0.23t} + \frac{r}{0.23}(1 - e^{-0.23t}).$$

Since we want the sum of the rapid injection function and the IV infusion function to be 58 mg, it follows that

$$\frac{r}{0.23} = 58.$$

Solving for r, we get

$$r = 13.34 \text{ mg/hr.}$$

EXERCISES

1. A 500-mg dose of a drug is administered by rapid injection to a patient. The half-life of the drug is 9 hours.

   a. Find a model for the amount of drug in the bloodstream t hours after the drug is administered.

   b. Find the average rate of change of drug in the bloodstream between $t = 0$ and $t = 2$. Repeat for $t = 9$ and $t = 11$.

2. A drug is given to a patient by IV infusion at a drip rate of 350 mg/hr. The half-life of this drug is 3 hours.

   a. Find a model of the amount of drug in the bloodstream t hours after the IV infusion begins.

   b. Find the average rate of change of drug in the bloodstream between $t = 0$ and $t = 3$. Repeat for $t = 3$ and $t = 6$.

3. A drug with a half-life of 9 hours is found to be effective when the amount of drug in the bloodstream is 250 mg. A 250-mg loading dose is given by rapid injection followed by an IV infusion. What should the rate of infusion be to maintain this level of drug in the bloodstream?

4. Use the table feature on a graphing calculator or a spreadsheet to develop a table that shows how much of the drug is present in a patient's system at the end of each 1/2 hour time interval for 24 hours for the model found in Exercise 1a. A chart such as this provides the health care worker with immediate information about patient drug levels.

5. Use the table feature on a graphing calculator or a spreadsheet to develop a table that shows how much of the drug is present in a patient's system at the end of each 1/2 hour time interval for 10 hours for the model found in Exercise 2a. A chart such as this provides the health care worker with immediate information about patient drug levels.

6. Use the table feature on a graphing calculator or a spreadsheet to develop a table that shows how much of the drug is present in a patient's system at the end of each 1/2 hour time interval for 10 hours for the model found in Exercise 3. Are your results surprising?

DIRECTIONS FOR GROUP PROJECT

Choose a drug that is commonly prescribed by physicians for a common ailment. Develop an analysis for this drug that is similar to the analysis for labetalol in Examples 1 through 3. You can obtain information on the drug from the Internet or from advertisements found in various media. Once you complete the analysis, prepare a professional presentation that can be delivered at a public forum. The presentation should summarize the facts presented in this extended application but at a level that is understandable to a typical layperson.

4

Calculating the Derivative

By differentiating the function defining a mathematical model we can see how the model's output changes with the input. In an exercise in Section 2 we explore a rational-function model for the length of the rest period needed to recover from vigorous exercise such as riding a bike. The derivative indicates how the rest required changes with the work expended in kilocalories per minute.

n the previous chapter, we found the derivative to be a useful tool for describing the rate of change, velocity, and the slope of a curve. Taking the derivative by using the definition, however, can be difficult. To take full advantage of the power of the derivative, we need faster ways of calculating the derivative. That is the goal of this chapter.

4.1 Techniques for Finding Derivatives

APPLY IT

How can a manager determine the best production level if the relationship between profit and production is known? How fast is the number of Americans who are expected to be over 100 years old growing?

These questions can be answered by finding the derivative of an appropriate function. We shall return to them at the end of this section in Examples 8 and 9.

Using the definition to calculate the derivative of a function is a very involved process even for simple functions. In this section we develop rules that make the calculation of derivatives much easier. Keep in mind that even though the process of finding a derivative will be greatly simplified with these rules, *the interpretation of the derivative will not change.* But first, a few words about notation are in order.

In addition to $f'(x)$, there are several other commonly used notations for the derivative.

Notations for the Derivative

The derivative of $y = f(x)$ may be written in any of the following ways:

$$f'(x), \qquad \frac{dy}{dx}, \qquad \frac{d}{dx}[f(x)], \qquad \text{or} \qquad D_x[f(x)].$$

The dy/dx notation for the derivative (read "the derivative of y with respect to x") is sometimes referred to as *Leibniz notation,* named after one of the co-inventors of calculus, Gottfried Wilhelm von Leibniz (1646–1716). (The other was Sir Isaac Newton, 1642–1727.)

With the above notation, the derivative of $y = f(x) = 2x^3 + 4x$, for example, which was found in Example 5 of Section 3.4 to be $f'(x) = 6x^2 + 4$, would be written

$$\frac{dy}{dx} = 6x^2 + 4$$

$$\frac{d}{dx}(2x^3 + 4x) = 6x^2 + 4$$

$$D_x(2x^3 + 4x) = 6x^2 + 4.$$

A variable other than x is often used as the independent variable. For example, if $y = f(t)$ gives population growth as a function of time, then the derivative of y with respect to t could be written

$$f'(t), \qquad \frac{dy}{dt}, \qquad \frac{d}{dt}[f(t)], \qquad \text{or} \qquad D_t[f(t)].$$

Other variables also may be used to name the function, as in $g(x)$ or $h(t)$.

Now we will use the definition

$$f'(x) = \lim_{h \to 0} \frac{f(x+h) - f(x)}{h}$$

to develop some rules for finding derivatives more easily than by the four-step process given in the previous chapter.

The first rule tells how to find the derivative of a constant function defined by $f(x) = k$, where k is a constant real number. Since $f(x+h)$ is also k, by definition $f'(x)$ is

$$f'(x) = \lim_{h \to 0} \frac{f(x+h) - f(x)}{h}$$

$$= \lim_{h \to 0} \frac{k - k}{h} = \lim_{h \to 0} \frac{0}{h} = \lim_{h \to 0} 0 = 0,$$

establishing the following rule.

Constant Rule

If $f(x) = k$, where k is any real number, then

$$f'(x) = 0.$$

(The derivative of a constant is 0.)

This rule is logical because the derivative represents rate of change, and a constant function, by definition, does not change. Figure 1 illustrates this constant rule geometrically; it shows a graph of the horizontal line $y = k$. At any point P on this line, the tangent line at P is the line itself. Since a horizontal line has a slope of 0, the slope of the tangent line is 0. This agrees with the result above: The derivative of a constant is 0.

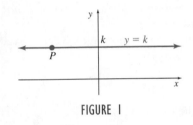

FIGURE 1

EXAMPLE 1 **Derivative of a Constant**

(a) If $f(x) = 9$, then $f'(x) = 0$.

(b) If $h(t) = \pi$, then $D_t[h(t)] = 0$.

(c) If $y = 2^3$, then $dy/dx = 0$.

Functions of the form $y = x^n$, where n is a fixed real number, are very common in applications. To obtain a rule for finding the derivative of such a function, we can use the definition to work out the derivatives for various special values of n. This was done in Section 3.4 in Example 4 to show that for $f(x) = x^2$, $f'(x) = 2x$.

For $f(x) = x^3$, the derivative is found as follows.

$$f'(x) = \lim_{h \to 0} \frac{f(x+h) - f(x)}{h}$$

$$= \lim_{h \to 0} \frac{(x+h)^3 - x^3}{h}$$

$$= \lim_{h \to 0} \frac{(x^3 + 3x^2h + 3xh^2 + h^3) - x^3}{h}.$$

The binomial theorem (discussed in most intermediate and college algebra texts) was used to expand $(x+h)^3$ in the last step. Now, the limit can be determined.

$$f'(x) = \lim_{h \to 0} \frac{3x^2h + 3xh^2 + h^3}{h}$$

$$= \lim_{h \to 0} (3x^2 + 3xh + h^2)$$

$$= 3x^2.$$

The results in the following table were found in a similar way, using the definition of the derivative. (These results are modifications of some of the examples and exercises from the previous chapter.)

Derivative of $f(x) = x^n$		
Function	n	Derivative
$f(x) = x$	1	$f'(x) = 1 = 1x^0$
$f(x) = x^2$	2	$f'(x) = 2x = 2x^1$
$f(x) = x^3$	3	$f'(x) = 3x^2$
$f(x) = x^4$	4	$f'(x) = 4x^3$
$f(x) = x^{-1}$	-1	$f'(x) = -1 \cdot x^{-2} = \dfrac{-1}{x^2}$
$f(x) = x^{1/2}$	$1/2$	$f'(x) = \dfrac{1}{2}x^{-1/2} = \dfrac{1}{2x^{1/2}}$

These results suggest the following rule.

Power Rule

If $f(x) = x^n$ for any real number n, then

$$f'(x) = nx^{n-1}.$$

(The derivative of $f(x) = x^n$ is found by multiplying by the exponent n and decreasing the exponent on x by 1.)

While the power rule is true for every real-number value of n, a proof is given here only for positive integer values of n. This proof follows the steps used above in finding the derivative of $f(x) = x^3$.

For any real numbers p and q, by the binomial theorem,

$$(p + q)^n = p^n + np^{n-1}q + \frac{n(n-1)}{2}p^{n-2}q^2 + \cdots + npq^{n-1} + q^n.$$

Replacing p with x and q with h gives

$$(x + h)^n = x^n + nx^{n-1}h + \frac{n(n-1)}{2}x^{n-2}h^2 + \cdots + nxh^{n-1} + h^n,$$

from which

$$(x + h)^n - x^n = nx^{n-1}h + \frac{n(n-1)}{2}x^{n-2}h^2 + \cdots + nxh^{n-1} + h^n.$$

Dividing each term by h yields

$$\frac{(x + h)^n - x^n}{h} = nx^{n-1} + \frac{n(n-1)}{2}x^{n-2}h + \cdots + nxh^{n-2} + h^{n-1}.$$

Use the definition of derivative, and the fact that each term except the first contains h as a factor and thus approaches 0 as h approaches 0, to get

$$f'(x) = \lim_{h \to 0} \frac{(x + h)^n - x^n}{h}$$

$$= nx^{n-1} + \frac{n(n-1)}{2}x^{n-2}0 + \cdots + nx0^{n-2} + 0^{n-1}$$

$$= nx^{n-1}.$$

This shows that the derivative of $f(x) = x^n$ is $f'(x) = nx^{n-1}$, proving the power rule for positive integer values of n.

—FOR REVIEW——

At this point you may wish to turn back to Sections R.6 and R.7 for a review of negative exponents and rational exponents. The relationship between powers, roots, and rational exponents is explained there.

YOUR TURN 1 If $f(t) = \dfrac{1}{\sqrt{t}}$, find $f'(t)$.

EXAMPLE 2 Power Rule

(a) If $f(x) = x^6$, find $f'(x)$.

SOLUTION $f'(x) = 6x^{6-1} = 6x^5$

(b) If $y = t = t^1$, find $\dfrac{dy}{dt}$.

SOLUTION $\dfrac{dy}{dt} = 1t^{1-1} = t^0 = 1$

(c) If $y = 1/x^3$, find dy/dx.

SOLUTION Use a negative exponent to rewrite this equation as $y = x^{-3}$; then

$$\frac{dy}{dx} = -3x^{-3-1} = -3x^{-4} \quad \text{or} \quad \frac{-3}{x^4}.$$

(d) Find $D_x(x^{4/3})$.

SOLUTION $D_x(x^{4/3}) = \dfrac{4}{3}x^{4/3-1} = \dfrac{4}{3}x^{1/3}$

(e) If $y = \sqrt{z}$, find dy/dz.

SOLUTION Rewrite this as $y = z^{1/2}$; then

$$\frac{dy}{dz} = \frac{1}{2}z^{1/2-1} = \frac{1}{2}z^{-1/2} \quad \text{or} \quad \frac{1}{2z^{1/2}} \quad \text{or} \quad \frac{1}{2\sqrt{z}}. \quad \text{TRY YOUR TURN 1}$$

The next rule shows how to find the derivative of the product of a constant and a function.

Constant Times a Function

Let k be a real number. If $g'(x)$ exists, then the derivative of $f(x) = k \cdot g(x)$ is

$$f'(x) = k \cdot g'(x).$$

(The derivative of a constant times a function is the constant times the derivative of the function.)

This rule is proved with the definition of the derivative and rules for limits.

$$f'(x) = \lim_{h \to 0} \frac{kg(x+h) - kg(x)}{h}$$

$$= \lim_{h \to 0} k\frac{[g(x+h) - g(x)]}{h} \qquad \text{Factor out } k.$$

$$= k \lim_{h \to 0} \frac{g(x+h) - g(x)}{h} \qquad \text{Limit rule 1}$$

$$= k \cdot g'(x) \qquad \text{Definition of derivative}$$

EXAMPLE 3 Derivative of a Constant Times a Function

(a) If $y = 8x^4$, find $\dfrac{dy}{dx}$.

SOLUTION $\dfrac{dy}{dx} = 8(4x^3) = 32x^3$

(b) If $y = -\dfrac{3}{4}x^{12}$, find dy/dx.

SOLUTION $\dfrac{dy}{dx} = -\dfrac{3}{4}(12x^{11}) = -9x^{11}$

(c) Find $D_t(-8t)$.

    SOLUTION $D_t(-8t) = -8(1) = -8$

(d) Find $D_p(10p^{3/2})$.

    SOLUTION $D_p(10p^{3/2}) = 10\left(\dfrac{3}{2}p^{1/2}\right) = 15p^{1/2}$

(e) If $y = \dfrac{6}{x}$, find $\dfrac{dy}{dx}$.

    SOLUTION   Rewrite this as $y = 6x^{-1}$; then

$$\frac{dy}{dx} = 6(-1x^{-2}) = -6x^{-2} \quad \text{or} \quad \frac{-6}{x^2}.$$ **TRY YOUR TURN 2**

YOUR TURN 2 If $y = 3\sqrt{x}$, find dy/dx.

EXAMPLE 4 **Beagles**

Researchers have determined that the daily energy requirements of female beagles who are at least 1 year old change with respect to age according to the function

$$E(t) = 753t^{-0.1321},$$

where $E(t)$ is the daily energy requirements (in $kJ/W^{0.67}$) for a dog that is t years old. *Source: Journal of Nutrition.*

(a) Find $E'(t)$.

    SOLUTION   Using the rules of differentiation we find that

$$E'(t) = 753(-0.1321)t^{-0.1321-1} = -99.4713t^{-1.1321}.$$

(b) Determine the rate of change of the daily energy requirements of a 2-year-old female beagle.

    SOLUTION $E'(2) = -99.4713(2)^{-1.1321} \approx -45.4$

Thus, the daily energy requirements of a 2-year-old female beagle are decreasing at the rate of 45.4 $kJ/W^{0.67}$ per year.

The final rule in this section is for the derivative of a function that is a sum or difference of terms.

Sum or Difference Rule

If $f(x) = u(x) \pm v(x)$, and if $u'(x)$ and $v'(x)$ exist, then

$$f'(x) = u'(x) \pm v'(x).$$

(The derivative of a sum or difference of functions is the sum or difference of the derivatives.)

The proof of the sum part of this rule is as follows: If $f(x) = u(x) + v(x)$, then

$$f'(x) = \lim_{h \to 0} \frac{[u(x+h) + v(x+h)] - [u(x) + v(x)]}{h}$$

$$= \lim_{h \to 0} \frac{[u(x+h) - u(x)] + [v(x+h) - v(x)]}{h}$$

$$= \lim_{h \to 0} \left[\frac{u(x+h) - u(x)}{h} + \frac{v(x+h) - v(x)}{h}\right]$$

$$= \lim_{h \to 0} \frac{u(x+h) - u(x)}{h} + \lim_{h \to 0} \frac{v(x+h) - v(x)}{h}$$

$$= u'(x) + v'(x).$$

A similar proof can be given for the difference of two functions.

EXAMPLE 5 Derivative of a Sum

Find the derivative of each function.

(a) $y = 6x^3 + 15x^2$

SOLUTION Let $u(x) = 6x^3$ and $v(x) = 15x^2$; then $y = u(x) + v(x)$. Since $u'(x) = 18x^2$ and $v'(x) = 30x$,

$$\frac{dy}{dx} = 18x^2 + 30x.$$

(b) $p(t) = 12t^4 - 6\sqrt{t} + \dfrac{5}{t}$

SOLUTION Rewrite $p(t)$ as $p(t) = 12t^4 - 6t^{1/2} + 5t^{-1}$; then

$$p'(t) = 48t^3 - 3t^{-1/2} - 5t^{-2}.$$

Also, $p'(t)$ may be written as $p'(t) = 48t^3 - \dfrac{3}{\sqrt{t}} - \dfrac{5}{t^2}.$

(c) $f(x) = \dfrac{x^3 + 3\sqrt{x}}{x}$

SOLUTION Rewrite $f(x)$ as $f(x) = \dfrac{x^3}{x} + \dfrac{3x^{1/2}}{x} = x^2 + 3x^{-1/2}$. Then

$$D_x[f(x)] = 2x - \frac{3}{2}x^{-3/2},$$

or

$$D_x[f(x)] = 2x - \frac{3}{2\sqrt{x^3}}.$$

(d) $f(x) = (4x^2 - 3x)^2$

SOLUTION Rewrite $f(x)$ as $f(x) = 16x^4 - 24x^3 + 9x^2$ using the fact that $(a - b)^2 = a^2 - 2ab + b^2$; then

$$f'(x) = 64x^3 - 72x^2 + 18x.$$ **TRY YOUR TURN 3**

YOUR TURN 3 If
$$h(t) = -3t^2 + 2\sqrt{t} + \frac{5}{t^4} - 7,$$
find $h'(t)$.

TECHNOLOGY NOTE Some computer programs and calculators have built-in methods for taking derivatives symbolically, which is what we have been doing in this section, as opposed to approximating the derivative numerically by using a small number for h in the definition of the derivative. In the computer program Maple, we would do part (a) of Example 5 by entering

```
> diff(6*x^3+15*x^2,x);
```

where the x after the comma tells what variable the derivative is with respect to. Maple would respond with

```
18*x^2+30*x.
```

Similarly, on the TI-89, we would enter `d(6x^3+15x^2,x)` and the calculator would give "$18 \cdot x^2 + 30 \cdot x$."

Other graphing calculators, such as the TI-84 Plus, do not have built-in methods for taking derivatives symbolically. As we saw in the last chapter, however, they do have the ability to calculate the derivative of a function at a particular point and to simultaneously graph a function and its derivative.

Recall that, on the TI-84 Plus, we could use the `nDeriv` Command, as shown in Figure 2, to approximate the value of the derivative when $x = 1$. Figure 3(a) and Figure 3(b) indicate how to

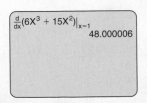

FIGURE 2

input the functions into the calculator and the corresponding graphs of both the function and its derivative. Consult the *Graphing Calculator and Excel Spreadsheet Manual*, available with this book, for assistance.

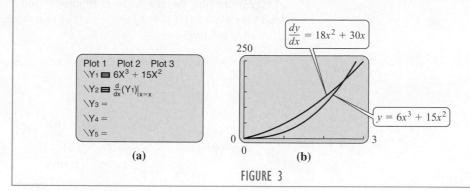

(a)   (b)

FIGURE 3

The rules developed in this section make it possible to find the derivative of a function more directly, so that applications of the derivative can be dealt with more effectively. The following examples illustrate some business applications.

Marginal Analysis

In previous sections we discussed the concepts of marginal cost, marginal revenue, and marginal profit. These concepts of **marginal analysis** are summarized here.

In business and economics the rates of change of such variables as cost, revenue, and profit are important considerations. Economists use the word *marginal* to refer to rates of change. For example, *marginal cost* refers to the rate of change of cost. Since the derivative of a function gives the rate of change of the function, a marginal cost (or revenue, or profit) function is found by taking the derivative of the cost (or revenue, or profit) function. Roughly speaking, the marginal cost at some level of production x is the cost to produce the $(x + 1)$st item. (Similar statements could be made for revenue or profit.)

To see why it is reasonable to say that the marginal cost function is approximately the cost of producing one more unit, look at Figure 4, where $C(x)$ represents the cost of producing x units of some item. Then the cost of producing $x + 1$ units is $C(x + 1)$. The cost of the $(x + 1)$st unit is, therefore, $C(x + 1) - C(x)$. This quantity is shown in the graph in Figure 4.

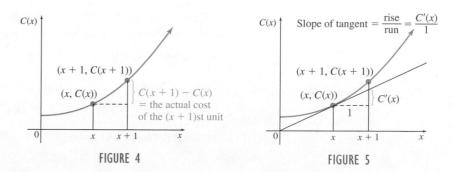

FIGURE 4   FIGURE 5

Now if $C(x)$ is the cost function, then the marginal cost $C'(x)$ represents the slope of the tangent line at any point $(x, C(x))$. The graph in Figure 5 shows the cost function $C(x)$ and the tangent line at a point $(x, C(x))$. Remember what it means for a line to have a given slope. If the slope of the line is $C'(x)$, then

$$\frac{\Delta y}{\Delta x} = C'(x) = \frac{C'(x)}{1},$$

and beginning at any point on the line and moving 1 unit to the right requires moving $C'(x)$ units up to get back to the line again. The vertical distance from the horizontal line to the tangent line shown in Figure 5 is therefore $C'(x)$.

Superimposing the graphs from Figures 4 and 5 as in Figure 6 shows that $C'(x)$ is indeed very close to $C(x + 1) - C(x)$. The two values are closest when x is very large, so that 1 unit is relatively small.

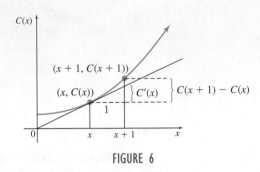

FIGURE 6

EXAMPLE 6 **Marginal Cost**

Suppose that the total cost in hundreds of dollars to produce x thousand barrels of a beverage is given by

$$C(x) = 4x^2 + 100x + 500.$$

Find the marginal cost for the following values of x.

(a) $x = 5$

SOLUTION To find the marginal cost, first find $C'(x)$, the derivative of the total cost function.

$$C'(x) = 8x + 100$$

When $x = 5$,

$$C'(5) = 8(5) + 100 = 140.$$

After 5 thousand barrels of the beverage have been produced, the cost to produce one thousand more barrels will be *approximately* 140 hundred dollars, or $14,000.

The *actual* cost to produce one thousand more barrels is $C(6) - C(5)$:

$$C(6) - C(5) = (4 \cdot 6^2 + 100 \cdot 6 + 500) - (4 \cdot 5^2 + 100 \cdot 5 + 500)$$
$$= 1244 - 1100 = 144,$$

144 hundred dollars, or $14,400.

(b) $x = 30$

SOLUTION After 30 thousand barrels have been produced, the cost to produce one thousand more barrels will be approximately

$$C'(30) = 8(30) + 100 = 340,$$

or $34,000. Notice that the cost to produce an additional thousand barrels of beverage has increased by approximately $20,000 at a production level of 30,000 barrels compared to a production level of 5000 barrels. **TRY YOUR TURN 4**

YOUR TURN 4 If the cost function is given by $C(x) = 5x^3 - 10x^2 + 75$, find the marginal cost when $x = 100$.

Demand Functions

The demand function, defined by $p = D(q)$, relates the number of units q of an item that consumers are willing to purchase to the price p. (Demand functions were also discussed in Chapter 1.) The total revenue $R(q)$ is related to price per unit and the amount demanded (or sold) by the equation

$$R(q) = qp = q \cdot D(q).$$

EXAMPLE 7   Marginal Revenue

The demand function for a certain product is given by

$$p = \frac{50{,}000 - q}{25{,}000}.$$

Find the marginal revenue when $q = 10{,}000$ units and p is in dollars.

SOLUTION   From the given function for p, the revenue function is given by

$$R(q) = qp$$

$$= q\left(\frac{50{,}000 - q}{25{,}000}\right)$$

$$= \frac{50{,}000q - q^2}{25{,}000}$$

$$= 2q - \frac{1}{25{,}000}q^2.$$

The marginal revenue is

$$R'(q) = 2 - \frac{2}{25{,}000}q.$$

When $q = 10{,}000$, the marginal revenue is

$$R'(10{,}000) = 2 - \frac{2}{25{,}000}(10{,}000) = 1.2,$$

YOUR TURN 5 If the demand function is given by $p = 16 - 1.25q$, find the marginal revenue when $q = 5$.

or \$1.20 per unit. Thus, the next item sold (at sales of 10,000) will produce additional revenue of about \$1.20.   **TRY YOUR TURN 5**

Management must be careful to keep track of marginal costs and revenue. If the marginal cost of producing an extra unit exceeds the marginal revenue received from selling it, then the company will lose money on that unit.

EXAMPLE 8   Marginal Profit

Suppose that the cost function for the product in Example 7 is given by

$$C(q) = 2100 + 0.25q, \qquad \text{where } 0 \le q \le 30{,}000.$$

Find the marginal profit from the production of the following numbers of units.

(a) 15,000

APPLY IT

SOLUTION   From Example 7, the revenue from the sale of x units is

$$R(q) = 2q - \frac{1}{25{,}000}q^2.$$

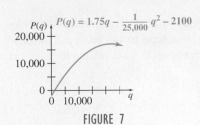

FIGURE 7

Since profit, P, is given by $P = R - C$,

$$P(q) = R(q) - C(q)$$
$$= \left(2q - \frac{1}{25,000}q^2 \right) - (2100 + 0.25q)$$
$$= 2q - \frac{1}{25,000}q^2 - 2100 - 0.25q$$
$$= 1.75q - \frac{1}{25,000}q^2 - 2100. \qquad \text{See Figure 7.}$$

The marginal profit from the sale of q units is

$$P'(q) = 1.75 - \frac{2}{25,000}q = 1.75 - \frac{1}{12,500}q.$$

At $q = 15,000$ the marginal profit is

$$P'(15,000) = 1.75 - \frac{1}{12,500}(15,000) = 0.55,$$

or \$0.55 per unit.

(b) 21,875

SOLUTION When $q = 21,875$, the marginal profit is

$$P'(21,875) = 1.75 - \frac{1}{12,500}(21,875) = 0.$$

(c) 25,000

SOLUTION When $q = 25,000$, the marginal profit is

$$P'(25,000) = 1.75 - \frac{1}{12,500}(25,000) = -0.25,$$

or $-\$0.25$ per unit.

As shown by parts (b) and (c), if more than 21,875 units are sold, the marginal profit is negative. This indicates that increasing production beyond that level will *reduce* profit.

The final example shows an application of the derivative to a problem of demography.

EXAMPLE 9 Centenarians

The number of Americans (in thousands) who are expected to be over 100 years old can be approximated by the function

$$f(t) = 0.00943t^3 - 0.470t^2 + 11.085t + 23.441,$$

where t is the year, with $t = 0$ corresponding to 2000, and $0 \le t \le 50$. ***Source: U.S. Census Bureau.***

(a) Find a formula giving the rate of change of the number of Americans over 100 years old.

 **APPLY IT**   **SOLUTION** Using the techniques for finding the derivative, we have

$$f'(t) = 0.02829t^2 - 0.940t + 11.085.$$

This tells us the rate of change in the number of Americans over 100 years old.

(b) Find the rate of change in the number of Americans who are expected to be over 100 years old in the year 2015.

SOLUTION   The year 2015 corresponds to $t = 15$.

$$f'(15) = 0.02829(15)^2 - 0.940(15) + 11.085 = 3.35025$$

The number of Americans over 100 years old is expected to grow at a rate of about 3.35 thousand, or about 3350, per year in the year 2015.

4.1   EXERCISES

Find the derivative of each function defined as follows.

1. $y = 12x^3 - 8x^2 + 7x + 5$

2. $y = 8x^3 - 5x^2 - \dfrac{x}{12}$

3. $y = 3x^4 - 6x^3 + \dfrac{x^2}{8} + 5$

4. $y = 5x^4 + 9x^3 + 12x^2 - 7x$

5. $f(x) = 6x^{3.5} - 10x^{0.5}$

6. $f(x) = -2x^{1.5} + 12x^{0.5}$

7. $y = 8\sqrt{x} + 6x^{3/4}$

8. $y = -100\sqrt{x} - 11x^{2/3}$

9. $y = 10x^{-3} + 5x^{-4} - 8x$

10. $y = 5x^{-5} - 6x^{-2} + 13x^{-1}$

11. $f(t) = \dfrac{7}{t} - \dfrac{5}{t^3}$

12. $f(t) = \dfrac{14}{t} + \dfrac{12}{t^4} + \sqrt{2}$

13. $y = \dfrac{6}{x^4} - \dfrac{7}{x^3} + \dfrac{3}{x} + \sqrt{5}$

14. $y = \dfrac{3}{x^6} + \dfrac{1}{x^5} - \dfrac{7}{x^2}$

15. $p(x) = -10x^{-1/2} + 8x^{-3/2}$

16. $h(x) = x^{-1/2} - 14x^{-3/2}$

17. $y = \dfrac{6}{\sqrt[4]{x}}$

18. $y = \dfrac{-2}{\sqrt[3]{x}}$

19. $f(x) = \dfrac{x^3 + 5}{x}$

20. $g(x) = \dfrac{x^3 - 4x}{\sqrt{x}}$

21. $g(x) = (8x^2 - 4x)^2$

22. $h(x) = (x^2 - 1)^3$

23. Which of the following describes the derivative function $f'(x)$ of a quadratic function $f(x)$?

  a. Quadratic **b.** Linear **c.** Constant **d.** Cubic (third degree)

24. Which of the following describes the derivative function $f'(x)$ of a cubic (third degree) function $f(x)$?

  a. Quadratic **b.** Linear **c.** Constant **d.** Cubic

25. Explain the relationship between the slope and the derivative of $f(x)$ at $x = a$.

26. Which of the following do *not* equal $\dfrac{d}{dx}(4x^3 - 6x^{-2})$?

  a. $\dfrac{12x^2 + 12}{x^3}$ **b.** $\dfrac{12x^5 + 12}{x^3}$ **c.** $12x^2 + \dfrac{12}{x^3}$

  d. $12x^3 + 12x^{-3}$

Find each derivative.

27. $D_x\left[9x^{-1/2} + \dfrac{2}{x^{3/2}}\right]$

28. $D_x\left[\dfrac{8}{\sqrt[4]{x}} - \dfrac{3}{\sqrt{x^3}}\right]$

29. $f'(-2)$ if $f(x) = \dfrac{x^4}{6} - 3x$

30. $f'(3)$ if $f(x) = \dfrac{x^3}{9} - 7x^2$

In Exercises 31–34, find the slope of the tangent line to the graph of the given function at the given value of x. Find the equation of the tangent line in Exercises 31 and 32.

31. $y = x^4 - 5x^3 + 2$; $x = 2$

32. $y = -3x^5 - 8x^3 + 4x^2$; $x = 1$

33. $y = -2x^{1/2} + x^{3/2}$; $x = 9$

34. $y = -x^{-3} + x^{-2}$; $x = 2$

35. Find all points on the graph of $f(x) = 9x^2 - 8x + 4$ where the slope of the tangent line is 0.

36. Find all points on the graph of $f(x) = x^3 + 9x^2 + 19x - 10$ where the slope of the tangent line is -5.

In Exercises 37–40, for each function find all values of x where the tangent line is horizontal.

37. $f(x) = 2x^3 + 9x^2 - 60x + 4$

38. $f(x) = x^3 + 15x^2 + 63x - 10$

39. $f(x) = x^3 - 4x^2 - 7x + 8$

40. $f(x) = x^3 - 5x^2 + 6x + 3$

41. At what points on the graph of $f(x) = 6x^2 + 4x - 9$ is the slope of the tangent line -2?

42. At what points on the graph of $f(x) = 2x^3 - 9x^2 - 12x + 5$ is the slope of the tangent line 12?

43. At what points on the graph of $f(x) = x^3 + 6x^2 + 21x + 2$ is the slope of the tangent line 9?

44. If $g'(5) = 12$ and $h'(5) = -3$, find $f'(5)$ for $f(x) = 3g(x) - 2h(x) + 3$.

45. If $g'(2) = 7$ and $h'(2) = 14$, find $f'(2)$ for
$$f(x) = \frac{1}{2}g(x) + \frac{1}{4}h(x).$$

46. Use the information given in the figure to find the following values.

   a. $f(1)$ **b.** $f'(1)$ **c.** The domain of f **d.** The range of f

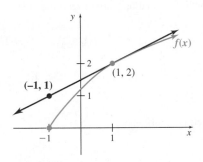

47. Explain the concept of marginal cost. How does it relate to cost? How is it found?

48. In Exercises 43–46 of Section 2.2, the effect of a when graphing $y = af(x)$ was discussed. Now describe how this relates to the fact that $D_x[af(x)] = af'(x)$.

49. Show that, for any constant k,
$$\frac{d}{dx}\left[\frac{f(x)}{k}\right] = \frac{f'(x)}{k}.$$

50. Use the differentiation feature on your graphing calculator to solve the problems (to 2 decimal places) below, where $f(x)$ is defined as follows:
$$f(x) = 1.25x^3 + 0.01x^2 - 2.9x + 1.$$

   a. Find $f'(4)$.

   b. Find all values of x where $f'(x) = 0$.

APPLICATIONS

Business and Economics

51. Revenue Assume that a demand equation is given by $q = 5000 - 100p$. Find the marginal revenue for the following production levels (values of q). (*Hint:* Solve the demand equation for p and use $R(q) = qp$.)

   a. 1000 units   **b.** 2500 units   **c.** 3000 units

52. Profit Suppose that for the situation in Exercise 51 the cost of producing q units is given by $C(q) = 3000 - 20q + 0.03q^2$. Find the marginal profit for the following production levels.

   a. 500 units   **b.** 815 units   **c.** 1000 units

53. Revenue If the price in dollars of a stereo system is given by
$$p(q) = \frac{1000}{q^2} + 1000,$$
where q represents the demand for the product, find the marginal revenue when the demand is 10.

54. Profit Suppose that for the situation in Exercise 53 the cost in dollars of producing q stereo systems is given by $C(q) = 0.2q^2 + 6q + 50$. Find the marginal profit when the demand is 10.

55. Sales Often sales of a new product grow rapidly at first and then level off with time. This is the case with the sales represented by the function
$$S(t) = 100 - 100t^{-1},$$
where t represents time in years. Find the rate of change of sales for the following numbers of years.

   a. 1   **b.** 10

56. Profit An analyst has found that a company's costs and revenues in dollars for one product are given by
$$C(x) = 2x \quad \text{and} \quad R(x) = 6x - \frac{x^2}{1000},$$
respectively, where x is the number of items produced.

   a. Find the marginal cost function.

   b. Find the marginal revenue function.

   c. Using the fact that profit is the difference between revenue and costs, find the marginal profit function.

   d. What value of x makes marginal profit equal 0?

   e. Find the profit when the marginal profit is 0.

(As we shall see in the next chapter, this process is used to find *maximum* profit.)

57. Postal Rates U.S. postal rates have steadily increased since 1932. Using data depicted in the table for the years 1932–2009, the cost in cents to mail a single letter can be modeled using a quadratic formula as follows:
$$C(t) = 0.008446\,t^2 - 0.08924t + 1.254,$$
where t is the number of years since 1932. *Source: U.S. Postal Service.*

Year	Cost	Year	Cost
1932	3	1988	25
1958	4	1991	29
1963	5	1995	32
1968	6	1999	33
1971	8	2001	34
1974	10	2002	37
1975	13	2006	39
1978	15	2007	41
1981	18	2008	42
1981	20	2009	44
1985	22		

a. Find the predicted cost of mailing a letter in 1982 and 2002 and compare these estimates with the actual rates.

b. Find the rate of change of the postage cost for the years 1982 and 2002 and interpret your results.

c. Using the regression feature on a graphing calculator, find a cubic function that models this data, letting $t = 0$ correspond to the year 1932. Then use your answer to find the rate of change of the postage cost for the years 1982 and 2002.

d. Discuss whether the quadratic or cubic function best describes the data. Do the answers from part b or from part c best describe the rate that postage was going up in the years 1982 and 2002?

e. Explore other functions that could be used to model the data, using the various regression features on a graphing calculator, and discuss to what extent any of them are useful descriptions of the data.

58. Money The total amount of money in circulation for the years 1950–2009 can be closely approximated by

$$M(t) = 0.005209t^3 - 0.04159t^2 - 0.3664t + 34.49$$

where t represents the number of years since 1950 and $M(t)$ is in billions of dollars. Find the derivative of $M(t)$ and use it to find the rate of change of money in circulation in the following years. *Source: U.S. Treasury.*

a. 1960   **b.** 1980   **c.** 1990   **d.** 2000

e. What do your answers to parts a–d tell you about the amount of money in circulation in those years?

Life Sciences

59. Cancer Insulation workers who were exposed to asbestos and employed before 1960 experienced an increased likelihood of lung cancer. If a group of insulation workers has a cumulative total of 100,000 years of work experience with their first date of employment t years ago, then the number of lung cancer cases occurring within the group can be modeled using the function

$$N(t) = 0.00437t^{3.2}.$$

Find the rate of growth of the number of workers with lung cancer in a group as described by the following first dates of

employment. *Source: Observation and Inference: An Introduction to the Methods of Epidemiology.*

a. 5 years ago   **b.** 10 years ago

60. Blood Sugar Level Insulin affects the glucose, or blood sugar, level of some diabetics according to the function

$$G(x) = -0.2x^2 + 450,$$

where $G(x)$ is the blood sugar level 1 hour after x units of insulin are injected. (This mathematical model is only approximate, and it is valid only for values of x less than about 40.) Find the blood sugar level after the following numbers of units of insulin are injected.

a. 0   **b.** 25

Find the rate of change of blood sugar level after injection of the following numbers of units of insulin.

c. 10   **d.** 25

61. Bighorn Sheep The cumulative horn volume for certain types of bighorn rams, found in the Rocky Mountains, can be described by the quadratic function

$$V(t) = -2159 + 1313t - 60.82t^2,$$

where $V(t)$ is the horn volume (in cm^3) and t is the year of growth, $2 \le t \le 9$. *Source: Conservation Biology.*

a. Find the horn volume for a 3-year-old ram.

b. Find the rate at which the horn volume of a 3-year-old ram is changing.

62. Brain Mass The brain mass of a human fetus during the last trimester can be accurately estimated from the circumference of the head by

$$m(c) = \frac{c^3}{100} - \frac{1500}{c},$$

where $m(c)$ is the mass of the brain (in grams) and c is the circumference (in centimeters) of the head. *Source: Early Human Development.*

a. Estimate the brain mass of a fetus that has a head circumference of 30 cm.

b. Find the rate of change of the brain mass for a fetus that has a head circumference of 30 cm and interpret your results.

63. Velocity of Marine Organism The typical velocity (in centimeters per second) of a marine organism of length l (in centimeters) is given by $v = 2.69l^{1.86}$. Find the rate of change of the velocity with respect to the length of the organism. *Source: Mathematical Topics in Population Biology Morphogenesis and Neurosciences.*

64. Heart The left ventricular length (viewed from the front of the heart) of a human fetus that is at least 18 weeks old can be estimated by

$$l(x) = -2.318 + 0.2356x - 0.002674x^2,$$

where $l(x)$ is the ventricular length (in centimeters) and x is the age (in weeks) of the fetus. *Source: American Journal of Cardiology.*

a. Determine a meaningful domain for this function.

b. Find $l'(x)$.

c. Find $l'(25)$.

65. Track and Field In 1906 Kennelly developed a simple formula for predicting an upper limit on the fastest time that humans could ever run distances from 100 yards to 10 miles. His formula is given by

$$t = 0.0588s^{1.125},$$

where s is the distance in meters and t is the time to run that distance in seconds. *Source: Proceedings of the American Academy of Arts and Sciences.*

   a. Find Kennelly's estimate for the fastest mile. (*Hint:* 1 mile $\approx$ 1609 meters.)

   b. Find dt/ds when $s = 100$ and interpret your answer.

   c. Compare this and other estimates to the current world records. Have these estimates been surpassed?

66. Human Cough To increase the velocity of the air flowing through the trachea when a human coughs, the body contracts the windpipe, producing a more effective cough. Tuchinsky formulated that the velocity of air that is flowing through the trachea during a cough is

$$V = C(R_0 - R)R^2,$$

where C is a constant based on individual body characteristics, R_0 is the radius of the windpipe before the cough, and R is the radius of the windpipe during the cough. It can be shown that the maximum velocity of the cough occurs when $dV/dR = 0$. Find the value of R that maximizes the velocity.* *Source: COMAP, Inc.*

67. Body Mass Index The body mass index (BMI) is a number that can be calculated for any individual as follows: Multiply weight by 703 and divide by the person's height squared. That is,

$$BMI = \frac{703w}{h^2},$$

where w is in pounds and h is in inches. The National Heart, Lung, and Blood Institute uses the BMI to determine whether a person is "overweight" $(25 \leq BMI < 30)$ or "obese" $(BMI \geq 30)$. *Source: The National Institutes of Health.*

   a. Calculate the BMI for Lebron James, basketball player for the Miami Heat, who is 250 lb. and 6'8" tall.

   b. How much weight would Lebron James have to lose until he reaches a BMI of 24.9 and is no longer "overweight"? Comment on whether BMI cutoffs are appropriate for athletes with considerable muscle mass.

   c. For a 125-lb female, what is the rate of change of BMI with respect to height? (*Hint:* Take the derivative of the function: $f(h) = 703(125)/h^2$.)

   d. Calculate and interpret the meaning of $f'(65)$.

   e. Use the TABLE feature on your graphing calculator to construct a table for BMI for various weights and heights.

Physical Sciences

Velocity We saw in the previous chapter that if a function $s(t)$ gives the position of an object at time t, the derivative gives the velocity, that is, $v(t) = s'(t)$. For each position function in Exercises 68–71, find (a) $v(t)$ and (b) the velocity when $t = 0$, $t = 5$, and $t = 10$.

68. $s(t) = 11t^2 + 4t + 2$

*Interestingly, Tuchinsky also states that X-rays indicate that the body naturally contracts the windpipe to this radius during a cough.

69. $s(t) = 18t^2 - 13t + 8$

70. $s(t) = 4t^3 + 8t^2 + t$

71. $s(t) = -3t^3 + 4t^2 - 10t + 5$

72. Velocity If a rock is dropped from a 144-ft building, its position (in feet above the ground) is given by $s(t) = -16t^2 + 144$, where t is the time in seconds since it was dropped.

   a. What is its velocity 1 second after being dropped? 2 seconds after being dropped?

   b. When will it hit the ground?

   c. What is its velocity upon impact?

73. Velocity A ball is thrown vertically upward from the ground at a velocity of 64 ft per second. Its distance from the ground at t seconds is given by $s(t) = -16t^2 + 64t$.

   a. How fast is the ball moving 2 seconds after being thrown? 3 seconds after being thrown?

   b. How long after the ball is thrown does it reach its maximum height?

   c. How high will it go?

74. Dead Sea Researchers who have been studying the alarming rate in which the level of the Dead Sea has been dropping have shown that the density $d(x)$ (in g per cm^3) of the Dead Sea brine during evaporation can be estimated by the function

$$d(x) = 1.66 - 0.90x + 0.47x^2,$$

where x is the fraction of the remaining brine, $0 \leq x \leq 1$. *Source: Geology.*

   a. Estimate the density of the brine when 50% of the brine remains.

   b. Find and interpret the instantaneous rate of change of the density when 50% of the brine remains.

75. Dog's Human Age From the data printed in the following table from the *Minneapolis Star Tribune* on September 20, 1998, a dog's age when compared to a human's age can be modeled using either a linear formula or a quadratic formula as follows:

$$y_1 = 4.13x + 14.63$$
$$y_2 = -0.033x^2 + 4.647x + 13.347,$$

where y_1 and y_2 represent a dog's human age for each formula and x represents a dog's actual age. *Source: Mathematics Teacher.*

Dog Age	Human Age
1	16
2	24
3	28
5	36
7	44
9	52
11	60
13	68
15	76

a. Find y_1 and y_2 when $x = 5$.

b. Find dy_1/dx and dy_2/dx when $x = 5$ and interpret your answers.

c. If the first two points are eliminated from the table, find the equation of a line that perfectly fits the reduced set of data. Interpret your findings.

d. Of the three formulas, which do you prefer?

YOUR TURN ANSWERS

1. $f'(t) = -\dfrac{1}{2}t^{-\frac{3}{2}}$ or $f'(t) = -\dfrac{1}{2t^{\frac{3}{2}}}$

2. $\dfrac{dy}{dx} = \dfrac{3}{2}x^{-\frac{1}{2}}$ or $\dfrac{dy}{dx} = \dfrac{3}{2\sqrt{x}}$

3. $h'(t) = -6t + t^{-\frac{1}{2}} - 20t^{-5}$ or $h'(t) = -6t + \dfrac{1}{\sqrt{t}} - \dfrac{20}{t^5}$

4. $148,000

5. $3.50

4.2 Derivatives of Products and Quotients

APPLY IT A manufacturer of small motors wants to make the average cost per motor as small as possible. How can this be done?

We show how the derivative is used to solve a problem like this in Example 5, later in this section.

In the previous section we saw that the derivative of a sum of two functions is found from the sum of the derivatives. What about products? Is the derivative of a product equal to the product of the derivatives? For example, if

$$u(x) = 2x + 3 \quad \text{and} \quad v(x) = 3x^2,$$

then

$$u'(x) = 2 \quad \text{and} \quad v'(x) = 6x.$$

Let $f(x)$ be the product of u and v; that is, $f(x) = (2x + 3)(3x^2) = 6x^3 + 9x^2$. By the rules of the preceding section, $f'(x) = 18x^2 + 18x = 18x(x + 1)$. On the other hand, $u'(x) \cdot v'(x) = 2(6x) = 12x \neq f'(x)$. In this example, the derivative of a product is *not* equal to the product of the derivatives, nor is this usually the case.

The rule for finding derivatives of products is as follows.

Product Rule

If $f(x) = u(x) \cdot v(x)$, and if $u'(x)$ and $v'(x)$ both exist, then

$$f'(x) = u(x) \cdot v'(x) + v(x) \cdot u'(x).$$

(The derivative of a product of two functions is the first function times the derivative of the second plus the second function times the derivative of the first.)

FOR REVIEW

This proof uses several of the rules for limits given in the first section of the previous chapter. You may want to review them at this time.

To sketch the method used to prove the product rule, let

$$f(x) = u(x) \cdot v(x).$$

Then $f(x + h) = u(x + h) \cdot v(x + h)$, and, by definition, $f'(x)$ is given by

$$f'(x) = \lim_{h \to 0} \frac{f(x + h) - f(x)}{h}$$

$$= \lim_{h \to 0} \frac{u(x + h) \cdot v(x + h) - u(x) \cdot v(x)}{h}.$$

Now subtract and add $u(x + h) \cdot v(x)$ in the numerator, giving

$$f'(x) = \lim_{h \to 0} \frac{u(x + h) \cdot v(x + h) - u(x + h) \cdot v(x) + u(x + h) \cdot v(x) - u(x) \cdot v(x)}{h}$$

$$= \lim_{h \to 0} \frac{u(x + h)[v(x + h) - v(x)] + v(x)[u(x + h) - u(x)]}{h}$$

$$= \lim_{h \to 0} u(x + h)\left[\frac{v(x + h) - v(x)}{h}\right] + \lim_{h \to 0} v(x)\left[\frac{u(x + h) - u(x)}{h}\right]$$

$$= \lim_{h \to 0} u(x + h) \cdot \lim_{h \to 0} \frac{v(x + h) - v(x)}{h} + \lim_{h \to 0} v(x) \cdot \lim_{h \to 0} \frac{u(x + h) - u(x)}{h}. \quad \text{(1)}$$

If $u'(x)$ and $v'(x)$ both exist, then

$$\lim_{h \to 0} \frac{u(x + h) - u(x)}{h} = u'(x) \qquad \text{and} \qquad \lim_{h \to 0} \frac{v(x + h) - v(x)}{h} = v'(x).$$

The fact that $u'(x)$ exists can be used to prove

$$\lim_{h \to 0} u(x + h) = u(x),$$

and since no h is involved in $v(x)$,

$$\lim_{h \to 0} v(x) = v(x).$$

Substituting these results into Equation (1) gives

$$f'(x) = u(x) \cdot v'(x) + v(x) \cdot u'(x),$$

the desired result.

　　To help see why the product rule is true, consider the special case in which u and v are positive functions. Then $u(x) \cdot v(x)$ represents the area of a rectangle, as shown in Figure 8. If we assume that u and v are increasing, then $u(x + h) \cdot v(x + h)$ represents the area of a slightly larger rectangle when h is a small positive number, as shown in the figure. The change in the area of the rectangle is given by the pink rectangle, with an area of $u(x)$ times the amount v has changed, plus the blue rectangle, with an area of $v(x)$ times the amount u has changed, plus the small green rectangle. As h becomes smaller and smaller, the green rectangle becomes negligibly small, and the change in the area is essentially $u(x)$ times the change in v plus $v(x)$ times the change in u.

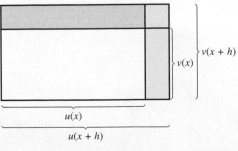

FIGURE 8

EXAMPLE 1 Product Rule

Let $f(x) = (2x + 3)(3x^2)$. Use the product rule to find $f'(x)$.

SOLUTION Here f is given as the product of $u(x) = 2x + 3$ and $v(x) = 3x^2$. By the product rule and the fact that $u'(x) = 2$ and $v'(x) = 6x$,

$$f'(x) = u(x) \cdot v'(x) + v(x) \cdot u'(x)$$
$$= (2x + 3)(6x) + (3x^2)(2)$$
$$= 12x^2 + 18x + 6x^2 = 18x^2 + 18x = 18x(x + 1).$$

This result is the same as that found at the beginning of the section.

EXAMPLE 2 Product Rule

Find the derivative of $y = (\sqrt{x} + 3)(x^2 - 5x)$.

SOLUTION Let $u(x) = \sqrt{x} + 3 = x^{1/2} + 3$, and $v(x) = x^2 - 5x$. Then

$$\frac{dy}{dx} = u(x) \cdot v'(x) + v(x) \cdot u'(x)$$

$$= (x^{1/2} + 3)(2x - 5) + (x^2 - 5x)\left(\frac{1}{2}x^{-1/2}\right).$$

Simplify by multiplying and combining terms.

$$\frac{dy}{dx} = (2x)(x^{1/2}) + 6x - 5x^{1/2} - 15 + (x^2)\left(\frac{1}{2}x^{-1/2}\right) - (5x)\left(\frac{1}{2}x^{-1/2}\right)$$

$$= 2x^{3/2} + 6x - 5x^{1/2} - 15 + \frac{1}{2}x^{3/2} - \frac{5}{2}x^{1/2}$$

$$= \frac{5}{2}x^{3/2} + 6x - \frac{15}{2}x^{1/2} - 15 \qquad \text{TRY YOUR TURN 1} \blacksquare$$

YOUR TURN 1 Find the derivative of $y = (x^3 + 7)(4 - x^2)$.

We could have found the derivatives above by multiplying out the original functions. The product rule then would not have been needed. In the next section, however, we shall see products of functions where the product rule is essential.

What about *quotients* of functions? To find the derivative of the quotient of two functions, use the next rule.

Quotient Rule

If $f(x) = u(x)/v(x)$, if all indicated derivatives exist, and if $v(x) \neq 0$, then

$$f'(x) = \frac{v(x) \cdot u'(x) - u(x) \cdot v'(x)}{[v(x)]^2}.$$

(The derivative of a quotient is the denominator times the derivative of the numerator minus the numerator times the derivative of the denominator, all divided by the square of the denominator.)

The proof of the quotient rule is similar to that of the product rule and is left for the exercises. (See Exercises 37 and 38.)

CAUTION | Just as the derivative of a product is *not* the product of the derivatives, the derivative of a quotient is *not* the quotient of the derivatives. If you are asked to take the derivative of a product or a quotient, it is essential that you recognize that the function contains a product or quotient and then use the appropriate rule.

EXAMPLE 3 Quotient Rule

Find $f'(x)$ if $f(x) = \dfrac{2x - 1}{4x + 3}$.

SOLUTION Let $u(x) = 2x - 1$, with $u'(x) = 2$. Also, let $v(x) = 4x + 3$, with $v'(x) = 4$. Then, by the quotient rule,

$$f'(x) = \frac{v(x) \cdot u'(x) - u(x) \cdot v'(x)}{[v(x)]^2}$$

$$= \frac{(4x + 3)(2) - (2x - 1)(4)}{(4x + 3)^2}$$

$$= \frac{8x + 6 - 8x + 4}{(4x + 3)^2}$$

$$= \frac{10}{(4x + 3)^2}.$$ TRY YOUR TURN 2

YOUR TURN 2 Find $f'(x)$ if $f(x) = \dfrac{3x + 2}{5 - 2x}$.

CAUTION | In the second step of Example 3, we had the expression

$$\frac{(4x + 3)(2) - (2x - 1)(4)}{(4x + 3)^2}.$$

Students often incorrectly "cancel" the $4x + 3$ in the numerator with one factor of the denominator. Because the numerator is a *difference* of two products, however, you must multiply and combine terms *before* looking for common factors in the numerator and denominator.

EXAMPLE 4 Product and Quotient Rules

Find $D_x\left[\dfrac{(3 - 4x)(5x + 1)}{7x - 9}\right]$.

SOLUTION This function has a product within a quotient. Instead of multiplying the factors in the numerator first (which is an option), we can use the quotient rule together with the product rule, as follows. Use the quotient rule first to get

$$D_x\left[\frac{(3 - 4x)(5x + 1)}{7x - 9}\right] = \frac{(7x - 9)D_x[(3 - 4x)(5x + 1)] - [(3 - 4x)(5x + 1)D_x(7x - 9)]}{(7x - 9)^2}.$$

Now use the product rule to find $D_x[(3 - 4x)(5x + 1)]$ in the numerator.

$$= \frac{(7x - 9)[(3 - 4x)5 + (5x + 1)(-4)] - (3 + 11x - 20x^2)(7)}{(7x - 9)^2}$$

$$= \frac{(7x - 9)(15 - 20x - 20x - 4) - (21 + 77x - 140x^2)}{(7x - 9)^2}$$

$$= \frac{(7x - 9)(11 - 40x) - 21 - 77x + 140x^2}{(7x - 9)^2}$$

$$= \frac{-280x^2 + 437x - 99 - 21 - 77x + 140x^2}{(7x - 9)^2}$$

$$= \frac{-140x^2 + 360x - 120}{(7x - 9)^2}$$ TRY YOUR TURN 3

YOUR TURN 3 Find $D_x\left[\dfrac{(5x - 3)(2x + 7)}{3x + 7}\right]$.

Average Cost

Average Cost Suppose $y = C(x)$ gives the total cost to manufacture x items. As mentioned earlier, the average cost per item is found by dividing the total cost by the number of items. The rate of change of average cost, called the *marginal average cost*, is the derivative of the average cost.

> ### Marginal Average Cost
>
> If the total cost to manufacture x items is given by $C(x)$, then the average cost per item is $\overline{C}(x) = C(x)/x$. The **marginal average cost** is the derivative of the average cost function, $\overline{C}'(x)$.

Similarly, the marginal average revenue function, $\overline{R}'(x)$, is defined as the derivative of the average revenue function, $\overline{R}(x) = R(x)/x$, and the marginal average profit function, $\overline{P}'(x)$, is defined as the derivative of the average profit function, $\overline{P}(x) = P(x)/x$.

A company naturally would be interested in making the average cost as small as possible. The next chapter will show that this can be done by using the derivative of $C(x)/x$. This derivative often can be found by means of the quotient rule, as in the next example.

EXAMPLE 5 Minimum Average Cost

Suppose the cost in dollars of manufacturing x hundred small motors is given by

$$C(x) = \frac{3x^2 + 120}{2x + 1}, \quad 10 \le x \le 200.$$

(a) Find the average cost per hundred motors.

SOLUTION The average cost is defined by

$$\overline{C}(x) = \frac{C(x)}{x} = \frac{3x^2 + 120}{2x + 1} \cdot \frac{1}{x} = \frac{3x^2 + 120}{2x^2 + x}.$$

(b) Find the marginal average cost.

SOLUTION The marginal average cost is given by

$$\frac{d}{dx}[\overline{C}(x)] = \frac{(2x^2 + x)(6x) - (3x^2 + 120)(4x + 1)}{(2x^2 + x)^2}$$

$$= \frac{12x^3 + 6x^2 - 12x^3 - 480x - 3x^2 - 120}{(2x^2 + x)^2}$$

$$= \frac{3x^2 - 480x - 120}{(2x^2 + x)^2}.$$

(c) As we shall see in the next chapter, average cost is generally minimized when the marginal average cost is zero. Find the level of production that minimizes average cost.

APPLY IT

SOLUTION Set the derivative $\overline{C}'(x) = 0$ and solve for x.

$$\frac{3x^2 - 480x - 120}{(2x^2 + x)^2} = 0$$

$$3x^2 - 480x - 120 = 0$$

$$3(x^2 - 160x - 40) = 0$$

YOUR TURN 4 Suppose cost is given by $C(x) = \dfrac{4x + 50}{x + 2}$. Find the marginal average cost.

Use the quadratic formula to solve this quadratic equation. Discarding the negative solution leaves $x = (160 + \sqrt{(160)^2 + 160})/2 \approx 160$ as the solution. Since x is in hundreds, production of 160 hundred or 16,000 motors will minimize average cost.

TRY YOUR TURN 4

4.2 EXERCISES

Use the product rule to find the derivative of the following. (Hint for Exercises 3–6: Write the quantity as a product.)

1. $y = (3x^2 + 2)(2x - 1)$

2. $y = (5x^2 - 1)(4x + 3)$

3. $y = (2x - 5)^2$

4. $y = (7x - 6)^2$

5. $k(t) = (t^2 - 1)^2$

6. $g(t) = (3t^2 + 2)^2$

7. $y = (x + 1)(\sqrt{x} + 2)$

8. $y = (2x - 3)(\sqrt{x} - 1)$

9. $p(y) = (y^{-1} + y^{-2})(2y^{-3} - 5y^{-4})$

10. $q(x) = (x^{-2} - x^{-3})(3x^{-1} + 4x^{-4})$

Use the quotient rule to find the derivative of the following.

11. $f(x) = \dfrac{6x + 1}{3x + 10}$

12. $f(x) = \dfrac{8x - 11}{7x + 3}$

13. $y = \dfrac{5 - 3t}{4 + t}$

14. $y = \dfrac{9 - 7t}{1 - t}$

15. $y = \dfrac{x^2 + x}{x - 1}$

16. $y = \dfrac{x^2 - 4x}{x + 3}$

17. $f(t) = \dfrac{4t^2 + 11}{t^2 + 3}$

18. $y = \dfrac{-x^2 + 8x}{4x^2 - 5}$

19. $g(x) = \dfrac{x^2 - 4x + 2}{x^2 + 3}$

20. $k(x) = \dfrac{x^2 + 7x - 2}{x^2 - 2}$

21. $p(t) = \dfrac{\sqrt{t}}{t - 1}$

22. $r(t) = \dfrac{\sqrt{t}}{2t + 3}$

23. $y = \dfrac{5x + 6}{\sqrt{x}}$

24. $y = \dfrac{4x - 3}{\sqrt{x}}$

25. $h(z) = \dfrac{z^{2.2}}{z^{3.2} + 5}$

26. $g(y) = \dfrac{y^{1.4} + 1}{y^{2.5} + 2}$

27. $f(x) = \dfrac{(3x^2 + 1)(2x - 1)}{5x + 4}$

28. $g(x) = \dfrac{(2x^2 + 3)(5x + 2)}{6x - 7}$

29. If $g(3) = 4$, $g'(3) = 5$, $f(3) = 9$, and $f'(3) = 8$, find $h'(3)$ when $h(x) = f(x)g(x)$.

30. If $g(3) = 4$, $g'(3) = 5$, $f(3) = 9$, and $f'(3) = 8$, find $h'(3)$ when $h(x) = f(x)/g(x)$.

31. Find the error in the following work.

$$D_x\left(\frac{2x + 5}{x^2 - 1}\right) = \frac{(2x + 5)(2x) - (x^2 - 1)2}{(x^2 - 1)^2}$$

$$= \frac{4x^2 + 10x - 2x^2 + 2}{(x^2 - 1)^2}$$

$$= \frac{2x^2 + 10x + 2}{(x^2 - 1)^2}$$

32. Find the error in the following work.

$$D_x\left(\frac{x^2 - 4}{x^3}\right) = x^3(2x) - (x^2 - 4)(3x^2) = 2x^4 - 3x^4 + 12x^2$$

$$= -x^4 + 12x^2$$

33. Find an equation of the line tangent to the graph of $f(x) = x/(x - 2)$ at $(3, 3)$.

34. Find an equation of the line tangent to the graph of $f(x) = (2x - 1)(x + 4)$ at $(1, 5)$.

35. Consider the function

$$f(x) = \frac{3x^3 + 6}{x^{2/3}}.$$

   a. Find the derivative using the quotient rule.

   b. Find the derivative by first simplifying the function to

   $$f(x) = \frac{3x^3}{x^{2/3}} + \frac{6}{x^{2/3}} = 3x^{7/3} + 6x^{-2/3}$$

   and using the rules from the previous section.

   c. Compare your answers from parts a and b and explain any discrepancies.

36. What is the result of applying the product rule to the function

$$f(x) = kg(x),$$

where k is a constant? Compare with the rule for differentiating a constant times a function from the previous section.

37. Following the steps used to prove the product rule for derivatives, prove the quotient rule for derivatives.

38. Use the fact that $f(x) = u(x)/v(x)$ can be rewritten as $f(x)v(x) = u(x)$ and the product rule for derivatives to verify the quotient rule for derivatives. (*Hint:* After applying the product rule, substitute $u(x)/v(x)$ for $f(x)$ and simplify.)

For each function, find the value(s) of x in which $f'(x) = 0$, to 3 decimal places.

39. $f(x) = (x^2 - 2)(x^2 - \sqrt{2})$

40. $f(x) = \dfrac{x - 2}{x^2 + 4}$

APPLICATIONS

Business and Economics

41. Average Cost The total cost (in hundreds of dollars) to produce x units of perfume is

$$C(x) = \frac{3x + 2}{x + 4}.$$

Find the average cost for each production level.

a. 10 units   **b.** 20 units   **c.** x units

d. Find the marginal average cost function.

42. Average Profit The total profit (in tens of dollars) from selling x self-help books is

$$P(x) = \frac{5x - 6}{2x + 3}.$$

Find the average profit from each sales level.

a. 8 books   **b.** 15 books   **c.** x books

d. Find the marginal average profit function.

e. Is this a reasonable function for profit? Why or why not?

43. Employee Training A company that manufactures bicycles has determined that a new employee can assemble $M(d)$ bicycles per day after d days of on-the-job training, where

$$M(d) = \frac{100d^2}{3d^2 + 10}.$$

a. Find the rate of change function for the number of bicycles assembled with respect to time.

b. Find and interpret $M'(2)$ and $M'(5)$.

44. Marginal Revenue Suppose that the demand function is given by $p = D(q)$, where q is the quantity that consumers demand when the price is p. Show that the marginal revenue is given by

$$R'(q) = D(q) + qD'(q).$$

45. Marginal Average Cost Suppose that the average cost function is given by $\overline{C}(x) = C(x)/x$, where x is the number of items produced. Show that the marginal average cost function is given by

$$\overline{C}'(x) = \frac{xC'(x) - C(x)}{x^2}.$$

46. Revenue Suppose that at the beginning of the year, a Vermont maple syrup distributor found that the demand for maple syrup, sold at $15 a quart, was 500 quarts each month. At that time, the price was going up at a rate of $0.50 a month, but despite this, the demand was going up at a rate of 30 quarts a month due to increased advertising. How fast was the revenue increasing?

47. Average Cost A gasoline refinery found that the cost to produce 12,500 gallons of gasoline last month was $27,000. At that time, the cost was going up at a rate of $1200 per month, while the number of gallons of gasoline the refinery produced was going up at a rate of 350 gallons per month. At what rate was the average cost to produce a gallon of gasoline increasing or decreasing last month?

Life Sciences

48. Muscle Reaction When a certain drug is injected into a muscle, the muscle responds by contracting. The amount of contraction, s (in millimeters) is related to the concentration of the drug, x (in milliliters) by

$$s(x) = \frac{x}{m + nx},$$

where m and n are constants.

a. Find $s'(x)$.

b. Find the rate of contraction when the concentration of the drug is 50 ml, $m = 10$, and $n = 3$.

49. Growth Models In Exercise 58 of Section 2.3, the formula for the growth rate of a population in the presence of a quantity x of food was given as

$$f(x) = \frac{Kx}{A + x}.$$

This was referred to as Michaelis-Menten kinetics.

a. Find the rate of change of the growth rate with respect to the amount of food.

b. The quantity A in the formula for $f(x)$ represents the quantity of food for which the growth rate is half of its maximum. Using your answer from part a, find the rate of change of the growth rate when $x = A$.

50. Bacteria Population Assume that the total number (in millions) of bacteria present in a culture at a certain time t (in hours) is given by

$$N(t) = 3t(t - 10)^2 + 40.$$

a. Find $N'(t)$.

Find the rate at which the population of bacteria is changing at the following times.

b. 8 hours   **c.** 11 hours

d. The answer in part b is negative, and the answer in part c is positive. What does this mean in terms of the population of bacteria?

51. Work/Rest Cycles Murrell's formula for calculating the total amount of rest, in minutes, required after performing a particular type of work activity for 30 minutes is given by the formula

$$R(w) = \frac{30(w - 4)}{w - 1.5},$$

where w is the work expended in kilocalories per minute, kcal/min. *Source: Human Factors in Engineering and Design.*

a. A value of 5 for w indicates light work, such as riding a bicycle on a flat surface at 10 mph. Find $R(5)$.

b. A value of 7 for w indicates moderate work, such as mowing grass with a pushmower on level ground. Find $R(7)$.

c. Find $R'(5)$ and $R'(7)$ and compare your answers. Explain whether these answers make sense.

52. Optimal Foraging Using data collected by zoologist Reto Zach, the work done by a crow to break open a whelk (large marine snail) can be estimated by the function

$$W = \left(1 + \frac{20}{H - 0.93}\right)H,$$

where H is the height (in meters) of the whelk when it is dropped. *Source: Mathematics Teacher*.

a. Find dW/dH.

b. One can show that the amount of work is minimized when $dW/dH = 0$. Find the value of H that minimizes W.

c. Interestingly, Zach observed the crows dropping the whelks from an average height of 5.23 m. What does this imply?

Social Sciences

53. Memory Retention Some psychologists contend that the number of facts of a certain type that are remembered after t hours is given by

$$f(t) = \frac{90t}{99t - 90}.$$

Find the rate at which the number of facts remembered is changing after the following numbers of hours.

a. 1   **b.** 10

General Interest

54. Vehicle Waiting Time The average number of vehicles waiting in a line to enter a parking ramp can be modeled by the function

$$f(x) = \frac{x^2}{2(1 - x)},$$

where x is a quantity between 0 and 1 known as the traffic intensity. Find the rate of change of the number of vehicles in line with respect to the traffic intensity for the following values of the intensity. *Source: Principles of Highway Engineering and Traffic Control*.

a. $x = 0.1$ **b.** $x = 0.6$

YOUR TURN ANSWERS

1. $\dfrac{dy}{dx} = -5x^4 + 12x^2 - 14x$

2. $f'(x) = \dfrac{19}{(5 - 2x)^2}$

3. $\dfrac{30x^2 + 140x + 266}{(3x + 7)^2}$

4. $\dfrac{-4x^2 - 100x - 100}{(x^2 + 2x)^2}$

4.3 The Chain Rule

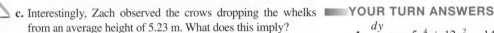

APPLY IT   Suppose we know how fast the radius of a circular oil slick is growing, and we know how much the area of the oil slick is growing per unit of change in the radius. How fast is the area growing?

We will answer this question in Example 4 using the chain rule for derivatives.

Before discussing the chain rule, we consider the composition of functions. Many of the most useful functions for modeling are created by combining simpler functions. Viewing complex functions as combinations of simpler functions often makes them easier to understand and use.

Composition of Functions   Suppose a function f assigns to each element x in set X some element $y = f(x)$ in set Y. Suppose also that a function g takes each element in set Y and assigns to it a value $z = g[f(x)]$ in set Z. By using both f and g, an element x in X is assigned to an element z in Z, as illustrated in Figure 9. The result of this process is a new function called the *composition* of functions g and f and defined as follows.

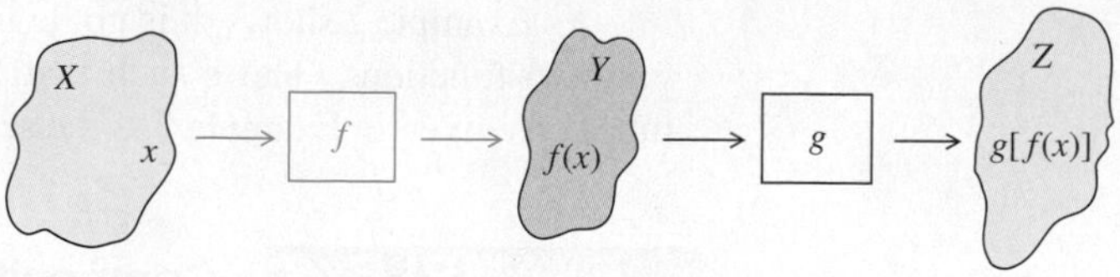

FIGURE 9

─FOR REVIEW─

You may want to review how to find the domain of a function. Domain was discussed in Section 2.1 on Properties of Functions.

Composite Function

Let f and g be functions. The **composite function**, or **composition**, of g and f is the function whose values are given by $g[f(x)]$ for all x in the domain of f such that $f(x)$ is in the domain of g. (Read $g[f(x)]$ as "g of f of x".)

EXAMPLE 1   **Composite Functions**

Let $f(x) = 2x - 1$ and $g(x) = \sqrt{3x + 5}$. Find the following.

(a) $g[f(4)]$

  SOLUTION   Find $f(4)$ first.

$$f(4) = 2 \cdot 4 - 1 = 8 - 1 = 7$$

  Then

$$g[f(4)] = g[7] = \sqrt{3 \cdot 7 + 5} = \sqrt{26}.$$

(b) $f[g(4)]$

  SOLUTION   Since $g(4) = \sqrt{3 \cdot 4 + 5} = \sqrt{17}$,

$$f[g(4)] = 2 \cdot \sqrt{17} - 1 = 2\sqrt{17} - 1.$$

YOUR TURN 1   For the functions in Example 1, find $f[g(0)]$ and $g[f(0)]$.

(c) $f[g(-2)]$

  SOLUTION $f[g(-2)]$ does not exist since -2 is not in the domain of g.

TRY YOUR TURN 1

EXAMPLE 2   **Composition of Functions**

Let $f(x) = 2x^2 + 5x$ and $g(x) = 4x + 1$. Find the following.

(a) $f[g(x)]$

  SOLUTION   Using the given functions, we have

$$\begin{aligned}
f[g(x)] &= f[4x + 1] \\
&= 2(4x + 1)^2 + 5(4x + 1) \\
&= 2(16x^2 + 8x + 1) + 20x + 5 \\
&= 32x^2 + 16x + 2 + 20x + 5 \\
&= 32x^2 + 36x + 7.
\end{aligned}$$

(b) $g[f(x)]$

  SOLUTION   By the definition above, with f and g interchanged,

$$\begin{aligned}
g[f(x)] &= g[2x^2 + 5x] \\
&= 4(2x^2 + 5x) + 1 \\
&= 8x^2 + 20x + 1.
\end{aligned}$$

YOUR TURN 2   Let $f(x) = 2x - 3$ and $g(x) = x^2 + 1$. Find $g[f(x)]$.

TRY YOUR TURN 2

As Example 2 shows, it is not always true that $f[g(x)] = g[f(x)]$. In fact, it is rare to find two functions f and g such that $f[g(x)] = g[f(x)]$. The domain of both composite functions given in Example 2 is the set of all real numbers.

EXAMPLE 3 Composition of Functions

Write each function as the composition of two functions f and g so that $h(x) = f[g(x)]$.

(a) $h(x) = 2(4x + 1)^2 + 5(4x + 1)$

SOLUTION Let $f(x) = 2x^2 + 5x$ and $g(x) = 4x + 1$. Then $f[g(x)] = f(4x + 1) = 2(4x + 1)^2 + 5(4x + 1)$. Notice that $h(x)$ here is the same as $f[g(x)]$ in Example 2(a).

(b) $h(x) = \sqrt{1 - x^2}$

SOLUTION One way to do this is to let $f(x) = \sqrt{x}$ and $g(x) = 1 - x^2$. Another choice is to let $f(x) = \sqrt{1 - x}$ and $g(x) = x^2$. Verify that with either choice, $f[g(x)] = \sqrt{1 - x^2}$. For the purposes of this section, the first choice is better; it is useful to think of f as being the function on the outer layer and g as the function on the inner layer. With this function h, we see a square root on the outer layer, and when we peel that away we see $1 - x^2$ on the inside.   **TRY YOUR TURN 3**

YOUR TURN 3 Write $h(x) = (2x - 3)^3$ as a composition of two functions f and g so that $h(x) = f[g(x)]$.

The Chain Rule
Suppose $f(x) = x^2$ and $g(x) = 5x^3 + 2$. What is the derivative of $h(x) = f[g(x)] = (5x^3 + 2)^2$? At first you might think the answer is just $h'(x) = 2(5x^3 + 2) = 10x^3 + 4$ by using the power rule. You can check this answer by multiplying out $h(x) = (5x^3 + 2)^2 = 25x^6 + 20x^3 + 4$. Now calculate $h'(x) = 150x^5 + 60x^2$. The guess using the power rule was clearly wrong! The error is that the power rule applies to x raised to a power, not to some other function of x raised to a power.

How, then, could we take the derivative of $p(x) = (5x^3 + 2)^{20}$? This seems far too difficult to multiply out. Fortunately, there is a way. Notice from the previous paragraph that $h'(x) = 150x^5 + 60x^2 = 2(5x^3 + 2)15x^2$. So the original guess was almost correct, except it was missing the factor of $15x^2$, which just happens to be $g'(x)$. This is not a coincidence. To see why the derivative of $f[g(x)]$ involves taking the derivative of f and then multiplying by the derivative of g, let us consider a realistic example, the question from the beginning of this section.

EXAMPLE 4 Area of an Oil Slick

A leaking oil well off the Gulf Coast is spreading a circular film of oil over the water surface. At any time t (in minutes) after the beginning of the leak, the radius of the circular oil slick (in feet) is given by

$$r(t) = 4t.$$

Find the rate of change of the area of the oil slick with respect to time.

APPLY IT   **SOLUTION** We first find the rate of change in the radius over time by finding dr/dt:

$$\frac{dr}{dt} = 4.$$

This value indicates that the radius is increasing by 4 ft each minute.

The area of the oil slick is given by

$$A(r) = \pi r^2, \quad \text{with} \quad \frac{dA}{dr} = 2\pi r.$$

The derivative, dA/dr, gives the rate of change in area per unit increase in the radius.

As these derivatives show, the radius is increasing at a rate of 4 ft/min, and for each foot that the radius increases, the area increases by $2\pi r$ ft^2. It seems reasonable, then, that the area is increasing at a rate of

$$2\pi r \text{ ft}^2/\text{ft} \times 4 \text{ ft/min} = 8\pi r \text{ ft}^2/\text{min}.$$

That is,

$$\frac{dA}{dt} = \frac{dA}{dr} \cdot \frac{dr}{dt} = 2\pi r \cdot 4 = 8\pi r.$$

Notice that because area (A) is a function of radius (r), which is a function of time (t), area as a function of time is a composition of two functions, written $A(r(t))$. The last step, then, can also be written as

$$\frac{dA}{dt} = \frac{d}{dt}A[r(t)] = A'[r(t)] \cdot r'(t) = 2\pi r \cdot 4 = 8\pi r.$$

Finally, we can substitute $r(t) = 4t$ to get the derivative in terms of t:

$$\frac{dA}{dt} = 8\pi r = 8\pi(4t) = 32\pi t.$$

The rate of change of the area of the oil slick with respect to time is $32\pi t$ ft^2/min. ▄

To check the result of Example 4, use the fact that $r = 4t$ and $A = \pi r^2$ to get the same result:

$$A = \pi(4t)^2 = 16\pi t^2, \quad \text{with} \quad \frac{dA}{dt} = 32\pi t.$$

The product used in Example 4,

$$\frac{dA}{dt} = \frac{dA}{dr} \cdot \frac{dr}{dt},$$

is an example of the **chain rule**, which is used to find the derivative of a composite function.

> ## Chain Rule
> If y is a function of u, say $y = f(u)$, and if u is a function of x, say $u = g(x)$, then $y = f(u) = f[g(x)]$, and
>
> $$\frac{dy}{dx} = \frac{dy}{du} \cdot \frac{du}{dx}.$$

One way to remember the chain rule is to pretend that dy/du and du/dx are fractions, with du "canceling out." The proof of the chain rule requires advanced concepts and, therefore, is not given here.

EXAMPLE 5   Chain Rule

Find dy/dx if $y = (3x^2 - 5x)^{1/2}$.

SOLUTION   Let $y = u^{1/2}$, and $u = 3x^2 - 5x$. Then

$$\frac{dy}{dx} = \frac{dy}{du} \cdot \frac{du}{dx}$$

$$= \frac{1}{2}u^{-1/2} \cdot (6x - 5).$$

YOUR TURN 4 Find dy/dx
if $y = (5x^2 - 6x)^{-2}$.

Replacing u with $3x^2 - 5x$ gives

$$\frac{dy}{dx} = \frac{1}{2}(3x^2 - 5x)^{-1/2}(6x - 5) = \frac{6x - 5}{2(3x^2 - 5x)^{1/2}}.$$ **TRY YOUR TURN 4**

The following alternative version of the chain rule is stated in terms of composite functions.

> **Chain Rule (Alternate Form)**
>
> If $y = f[g(x)]$, then
>
> $$\frac{dy}{dx} = f'[g(x)] \cdot g'(x).$$
>
> (To find the derivative of $f[g(x)]$, find the derivative of $f(x)$, replace each x with $g(x)$, and then multiply the result by the derivative of $g(x)$.)

In words, the chain rule tells us to first take the derivative of the outer function, then multiply it by the derivative of the inner function.

EXAMPLE 6 Chain Rule

Use the chain rule to find $D_x(x^2 + 5x)^8$.

SOLUTION As in Example 3(b), think of this as a function with layers. The outer layer is something being raised to the 8th power, so let $f(x) = x^8$. Once this layer is peeled away, we see that the inner layer is $x^2 + 5x$, so $g(x) = x^2 + 5x$. Then $(x^2 + 5x)^8 = f[g(x)]$ and

$$D_x(x^2 + 5x)^8 = f'[g(x)]g'(x).$$

Here $f'(x) = 8x^7$, with $f'[g(x)] = 8[g(x)]^7 = 8(x^2 + 5x)^7$ and $g'(x) = 2x + 5$.

$$D_x(x^2 + 5x)^8 = f'[g(x)]g'(x)$$

$$= 8[g(x)]^7 g'(x)$$

$$= 8(x^2 + 5x)^7(2x + 5)$$ **TRY YOUR TURN 5**

YOUR TURN 5
Find $D_x(x^2 - 7)^{10}$.

CAUTION (a) A common error is to forget to multiply by $g'(x)$ when using the chain rule. Remember, the derivative must involve a "chain," or product, of derivatives.

(b) Another common mistake is to write the derivative as $f'[g'(x)]$. Remember to leave $g(x)$ unchanged in $f'[g(x)]$ and then to multiply by $g'(x)$

One way to avoid both of the errors described above is to remember that the chain rule is a two-step process. In Example 6, the first step was taking the derivative of the power, and the second step was multiplying by $g'(x)$. Forgetting to multiply by $g'(x)$ would be an erroneous one-step process. The other erroneous one-step process is to take the derivative inside the power, getting $f'[g'(x)]$, or $8(2x + 5)^7$ in Example 6.

Sometimes both the chain rule and either the product or quotient rule are needed to find a derivative, as the next examples show.

EXAMPLE 7 Derivative Rules

Find the derivative of $y = 4x(3x + 5)^5$.

SOLUTION Write $4x(3x + 5)^5$ as the product

$$(4x) \cdot (3x + 5)^5.$$

To find the derivative of $(3x + 5)^5$, let $g(x) = 3x + 5$, with $g'(x) = 3$. Now use the product rule and the chain rule.

$$\frac{dy}{dx} = 4x\overbrace{[5(3x + 5)^4 \cdot 3]}^{\text{Derivative of } (3x+5)^5} + (3x + 5)^5\overbrace{(4)}^{\text{Derivative of } 4x}$$

$$= 60x(3x + 5)^4 + 4(3x + 5)^5 \qquad \text{Factor out the greatest common factor, } 4(3x + 5)^4.$$

$$= 4(3x + 5)^4[15x + (3x + 5)^1] \qquad \text{Simplify inside brackets.}$$

$$= 4(3x + 5)^4(18x + 5)$$

TRY YOUR TURN 6

YOUR TURN 6 Find the derivative of $y = x^2(5x - 1)^3$.

EXAMPLE 8 Derivative Rules

Find $D_x\left[\dfrac{(3x + 2)^7}{x - 1}\right]$.

SOLUTION Use the quotient rule and the chain rule.

$$D_x\left[\frac{(3x + 2)^7}{x - 1}\right] = \frac{(x - 1)[7(3x + 2)^6 \cdot 3] - (3x + 2)^7(1)}{(x - 1)^2}$$

$$= \frac{21(x - 1)(3x + 2)^6 - (3x + 2)^7}{(x - 1)^2}$$

$$= \frac{(3x + 2)^6[21(x - 1) - (3x + 2)]}{(x - 1)^2} \qquad \text{Factor out the greatest common factor, } (3x + 2)^6.$$

$$= \frac{(3x + 2)^6[21x - 21 - 3x - 2]}{(x - 1)^2} \qquad \text{Simplify inside brackets.}$$

$$= \frac{(3x + 2)^6(18x - 23)}{(x - 1)^2}$$

TRY YOUR TURN 7

YOUR TURN 7

Find $D_x\left[\dfrac{(4x - 1)^3}{x + 3}\right]$.

Some applications requiring the use of the chain rule are illustrated in the next two examples.

EXAMPLE 9 City Revenue

The revenue realized by a small city from the collection of fines from parking tickets is given by

$$R(n) = \frac{8000n}{n + 2},$$

where n is the number of work-hours each day that can be devoted to parking patrol. At the outbreak of a flu epidemic, 30 work-hours are used daily in parking patrol, but during the epidemic that number is decreasing at the rate of 6 work-hours per day. How fast is revenue from parking fines decreasing at the outbreak of the epidemic?

SOLUTION We want to find dR/dt, the change in revenue with respect to time. By the chain rule,

$$\frac{dR}{dt} = \frac{dR}{dn} \cdot \frac{dn}{dt}.$$

First find dR/dn, using the quotient rule, as follows.

$$\frac{dR}{dn} = \frac{(n+2)(8000) - 8000n(1)}{(n+2)^2} = \frac{16,000}{(n+2)^2}$$

Since 30 work-hours were used at the outbreak of the epidemic, $n = 30$, so $dR/dn = 16,000/(30+2)^2 = 15.625$. Also, $dn/dt = -6$. Thus,

$$\frac{dR}{dt} = \frac{dR}{dn} \cdot \frac{dn}{dt} = (15.625)(-6) = -93.75.$$

Revenue is being lost at the rate of about \$94 per day at the outbreak of the epidemic. ▪

EXAMPLE 10 Compound Interest

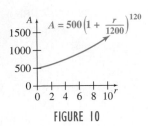

$$A = 500\left(1 + \frac{r}{1200}\right)^{120}$$

FIGURE 10

Suppose a sum of \$500 is deposited in an account with an interest rate of r percent per year compounded monthly. At the end of 10 years, the balance in the account (as illustrated in Figure 10) is given by

$$A = 500\left(1 + \frac{r}{1200}\right)^{120}.$$

Find the rate of change of A with respect to r if $r = 5$ or 7.*

SOLUTION First find dA/dr using the chain rule.

$$\frac{dA}{dr} = (120)(500)\left(1 + \frac{r}{1200}\right)^{119}\left(\frac{1}{1200}\right)$$

$$= 50\left(1 + \frac{r}{1200}\right)^{119}$$

If $r = 5$,

$$\frac{dA}{dr} = 50\left(1 + \frac{5}{1200}\right)^{119}$$

$$\approx 82.01,$$

or \$82.01 per percentage point. If $r = 7$,

$$\frac{dA}{dr} = 50\left(1 + \frac{7}{1200}\right)^{119}$$

$$\approx 99.90,$$

or \$99.90 per percentage point. ▪

NOTE One lesson to learn from this section is that a derivative is always with respect to some variable. In the oil slick example, notice that the derivative of the area with respect to the radius is $2\pi r$, while the derivative of the area with respect to time is $8\pi r$. As another example, consider the velocity of a conductor walking at 2 mph on a train car. Her velocity with respect to the ground may be 50 mph, but the earth on which the train is running is moving about the sun at 1.6 million mph. The derivative of her position function might be 2, 50, or 1.6 million mph, depending on what variable it is with respect to.

*Notice that r is given here as an integer percent, rather than as a decimal, which is why the formula for compound interest has 1200 where you would expect to see 12. This leads to a simpler interpretation of the derivative.

4.3 EXERCISES

Let $f(x) = 5x^2 - 2x$ and $g(x) = 8x + 3$. Find the following.

1. $f[g(2)]$ **2.** $f[g(-5)]$ **3.** $g[f(2)]$

4. $g[f(-5)]$ **5.** $f[g(k)]$ **6.** $g[f(5z)]$

In Exercises 7–14, find $f[g(x)]$ and $g[f(x)]$.

7. $f(x) = \dfrac{x}{8} + 7$; $g(x) = 6x - 1$

8. $f(x) = -8x + 9$; $g(x) = \dfrac{x}{5} + 4$

9. $f(x) = \dfrac{1}{x}$; $g(x) = x^2$

10. $f(x) = \dfrac{2}{x^4}$; $g(x) = 2 - x$

11. $f(x) = \sqrt{x + 2}$; $g(x) = 8x^2 - 6$

12. $f(x) = 9x^2 - 11x$; $g(x) = 2\sqrt{x + 2}$

13. $f(x) = \sqrt{x + 1}$; $g(x) = \dfrac{-1}{x}$

14. $f(x) = \dfrac{8}{x}$; $g(x) = \sqrt{3 - x}$

Write each function as the composition of two functions. (There may be more than one way to do this.)

15. $y = (5 - x^2)^{3/5}$ **16.** $y = (3x^2 - 7)^{2/3}$

17. $y = -\sqrt{13 + 7x}$ **18.** $y = \sqrt{9 - 4x}$

19. $y = (x^2 + 5x)^{1/3} - 2(x^2 + 5x)^{2/3} + 7$

20. $y = (x^{1/2} - 3)^2 + (x^{1/2} - 3) + 5$

Find the derivative of each function defined as follows.

21. $y = (8x^4 - 5x^2 + 1)^4$ **22.** $y = (2x^3 + 9x)^5$

23. $k(x) = -2(12x^2 + 5)^{-6}$ **24.** $f(x) = -7(3x^4 + 2)^{-4}$

25. $s(t) = 45(3t^3 - 8)^{3/2}$ **26.** $s(t) = 12(2t^4 + 5)^{3/2}$

27. $g(t) = -3\sqrt{7t^3 - 1}$ **28.** $f(t) = 8\sqrt{4t^2 + 7}$

29. $m(t) = -6t(5t^4 - 1)^4$ **30.** $r(t) = 4t(2t^5 + 3)^4$

31. $y = (3x^4 + 1)^4(x^3 + 4)$ **32.** $y = (x^3 + 2)(x^2 - 1)^4$

33. $q(y) = 4y^2(y^2 + 1)^{5/4}$ **34.** $p(z) = z(6z + 1)^{4/3}$

35. $y = \dfrac{-5}{(2x^3 + 1)^2}$ **36.** $y = \dfrac{1}{(3x^2 - 4)^5}$

37. $r(t) = \dfrac{(5t - 6)^4}{3t^2 + 4}$ **38.** $p(t) = \dfrac{(2t + 3)^3}{4t^2 - 1}$

39. $y = \dfrac{3x^2 - x}{(2x - 1)^5}$ **40.** $y = \dfrac{x^2 + 4x}{(3x^3 + 2)^4}$

41. In your own words explain how to form the composition of two functions.

42. The generalized power rule says that if $g(x)$ is a function of x and $y = [g(x)]^n$ for any real number n, then

$$\frac{dy}{dx} = n \cdot [g(x)]^{n-1} \cdot g'(x).$$

Explain why the generalized power rule is a consequence of the chain rule and the power rule.

Consider the following table of values of the functions f and g and their derivatives at various points.

x	1	2	3	4
$f(x)$	2	4	1	3
$f'(x)$	-6	-7	-8	-9
$g(x)$	2	3	4	1
$g'(x)$	2/7	3/7	4/7	5/7

Find the following using the table above.

43. a. $D_x(f[g(x)])$ at $x = 1$ **b.** $D_x(f[g(x)])$ at $x = 2$

44. a. $D_x(g[f(x)])$ at $x = 1$ **b.** $D_x(g[f(x)])$ at $x = 2$

In Exercises 45–48, find the equation of the tangent line to the graph of the given function at the given value of x.

45. $f(x) = \sqrt{x^2 + 16}$; $x = 3$

46. $f(x) = (x^3 + 7)^{2/3}$; $x = 1$

47. $f(x) = x(x^2 - 4x + 5)^4$; $x = 2$

48. $f(x) = x^2\sqrt{x^4 - 12}$; $x = 2$

In Exercises 49 and 50, find all values of x for the given function where the tangent line is horizontal.

49. $f(x) = \sqrt{x^3 - 6x^2 + 9x + 1}$

50. $f(x) = \dfrac{x}{(x^2 + 4)^4}$

51. Katie and Sarah are working on taking the derivative of

$$f(x) = \frac{2x}{3x + 4}.$$

Katie uses the quotient rule to get

$$f'(x) = \frac{(3x + 4)2 - 2x(3)}{(3x + 4)^2} = \frac{8}{(3x + 4)^2}.$$

Sarah converts it into a product and uses the product rule and the chain rule:

$$f(x) = 2x(3x + 4)^{-1}$$
$$f'(x) = 2x(-1)(3x + 4)^{-2}(3) + 2(3x + 4)^{-1}$$
$$= 2(3x + 4)^{-1} - 6x(3x + 4)^{-2}.$$

Explain the discrepancies between the two answers. Which procedure do you think is preferable?

52. Margy and Nate are working on taking the derivative of

$$f(x) = \frac{2}{(3x+1)^4}.$$

Margy uses the quotient rule and chain rule as follows:

$$f'(x) = \frac{(3x+1)^4 \cdot 0 - 2 \cdot 4(3x+1)^3 \cdot 3}{(3x+1)^8}$$

$$= \frac{-24(3x+1)^3}{(3x+1)^8} = \frac{-24}{(3x+1)^5}.$$

Nate rewrites the function and uses the power rule and chain rule as follows:

$$f(x) = 2(3x+1)^{-4}$$

$$f'(x) = (-4)2(3x+1)^{-5} \cdot 3 = \frac{-24}{(3x+1)^5}.$$

Compare the two procedures. Which procedure do you think is preferable?

APPLICATIONS

Business and Economics

53. Demand Suppose the demand for a certain brand of vacuum cleaner is given by

$$D(p) = \frac{-p^2}{100} + 500,$$

where p is the price in dollars. If the price, in terms of the cost c, is expressed as

$$p(c) = 2c - 10,$$

find the demand in terms of the cost.

54. Revenue Assume that the total revenue (in dollars) from the sale of x television sets is given by

$$R(x) = 24(x^2 + x)^{2/3}.$$

Find the marginal revenue when the following numbers of sets are sold.

a. 100 **b.** 200 **c.** 300

d. Find the average revenue from the sale of x sets.

e. Find the marginal average revenue.

f. Write a paragraph covering the following questions. How does the revenue change over time? What does the marginal revenue function tell you about the revenue function? What does the average revenue function tell you about the revenue function?

55. Interest A sum of $1500 is deposited in an account with an interest rate of r percent per year, compounded daily. At the end of 5 years, the balance in the account is given by

$$A = 1500\left(1 + \frac{r}{36,500}\right)^{1825}.$$

Find the rate of change of A with respect to r for the following interest rates.

a. 6% **b.** 8% **c.** 9%

56. Demand Suppose a demand function is given by

$$q = D(p) = 30\left(5 - \frac{p}{\sqrt{p^2+1}}\right),$$

where q is the demand for a product and p is the price per item in dollars. Find the rate of change in the demand for the product per unit change in price (i.e., find dq/dp).

57. Depreciation A certain truck depreciates according to the formula

$$V = \frac{60,000}{1 + 0.3t + 0.1t^2},$$

where V is the value of the truck (in dollars), t is time measured in years, and $t = 0$ represents the time of purchase (in years). Find the rate at which the value of the truck is changing at the following times.

a. 2 years **b.** 4 years

58. Cost Suppose the cost in dollars of manufacturing q items is given by

$$C = 2000q + 3500,$$

and the demand equation is given by

$$q = \sqrt{15,000 - 1.5p}.$$

In terms of the demand q,

a. find an expression for the revenue R;

b. find an expression for the profit P;

c. find an expression for the marginal profit.

d. Determine the value of the marginal profit when the price is $5000.

Life Sciences

59. Fish Population Suppose the population P of a certain species of fish depends on the number x (in hundreds) of a smaller fish that serves as its food supply, so that

$$P(x) = 2x^2 + 1.$$

Suppose, also, that the number of the smaller species of fish depends on the amount a (in appropriate units) of its food supply, a kind of plankton. Specifically,

$$x = f(a) = 3a + 2.$$

A biologist wants to find the relationship between the population P of the large fish and the amount a of plankton available, that is, $P[f(a)]$. What is the relationship?

60. Oil Pollution An oil well off the Gulf Coast is leaking, with the leak spreading oil over the surface as a circle. At any time t (in minutes) after the beginning of the leak, the radius of the circular oil slick on the surface is $r(t) = t^2$ feet. Let $A(r) = \pi r^2$ represent the area of a circle of radius r.

a. Find and interpret $A[r(t)]$.

b. Find and interpret $D_t A[r(t)]$ when $t = 100$.

61. Thermal Inversion When there is a thermal inversion layer over a city (as happens often in Los Angeles), pollutants cannot rise vertically but are trapped below the layer and must disperse horizontally. Assume that a factory smokestack begins emitting a pollutant at 8 A.M. Assume that the pollutant disperses horizontally, forming a circle. If t represents the time (in hours) since the factory began emitting pollutants ($t = 0$ represents 8 A.M.), assume that the radius of the circle of pollution is $r(t) = 2t$ miles. Let $A(r) = \pi r^2$ represent the area of a circle of radius r.

a. Find and interpret $A[r(t)]$.

b. Find and interpret $D_t A[r(t)]$ when $t = 4$.

62. Bacteria Population The total number of bacteria (in millions) present in a culture is given by

$$N(t) = 2t(5t + 9)^{1/2} + 12,$$

where t represents time (in hours) after the beginning of an experiment. Find the rate of change of the population of bacteria with respect to time for the following numbers of hours.

a. 0   **b.** 7/5   **c.** 8

63. Calcium Usage To test an individual's use of calcium, a researcher injects a small amount of radioactive calcium into the person's bloodstream. The calcium remaining in the bloodstream is measured each day for several days. Suppose the amount of the calcium remaining in the bloodstream (in milligrams per cubic centimeter) t days after the initial injection is approximated by

$$C(t) = \frac{1}{2}(2t + 1)^{-1/2}.$$

Find the rate of change of the calcium level with respect to time for the following numbers of days.

a. 0   **b.** 4   **c.** 7.5

d. Is C always increasing or always decreasing? How can you tell?

64. Drug Reaction The strength of a person's reaction to a certain drug is given by

$$R(Q) = Q\left(C - \frac{Q}{3}\right)^{1/2},$$

where Q represents the quantity of the drug given to the patient and C is a constant.

a. The derivative $R'(Q)$ is called the *sensitivity* to the drug. Find $R'(Q)$.

b. Find the sensitivity to the drug if $C = 59$ and a patient is given 87 units of the drug.

c. Is the patient's sensitivity to the drug increasing or decreasing when $Q = 87$?

General Interest

65. Candy The volume and surface area of a "jawbreaker" for any radius is given by the formulas

$$V(r) = \frac{4}{3}\pi r^3 \quad \text{and} \quad S(r) = 4\pi r^2,$$

respectively. Roger Guffey estimates the radius of a jawbreaker while in a person's mouth to be

$$r(t) = 6 - \frac{3}{17}t,$$

where $r(t)$ is in millimeters and t is in minutes. *Source: Mathematics Teacher.*

a. What is the life expectancy of a jawbreaker?

b. Find dV/dt and dS/dt when $t = 17$ and interpret your answer.

c. Construct an analogous experiment using some other type of food or verify the results of this experiment.

66. Zenzizenzizenzic Zenzizenzizenzic is an obsolete word with the distinction of containing the most z's of any word found in the Oxford English Dictionary. It was used in mathematics, before powers were written as superscript numbers, to represent the square of the square of the square of a number. In symbols, zenzizenzizenzic is written as $((x^2)^2)^2$. *Source: The Phrontistery.*

a. Use the chain rule twice to find the derivative.

b. Use the properties of exponents to first simplify the expression, and then find the derivative.

67. Zenzizenzicube Zenzizenzicube is another obsolete word (see Exercise 66) that represents the square of the square of a cube. In symbols, zenzizenzicube is written as $((x^3)^2)^2$. *Source: The Phrontistery.*

a. Use the chain rule twice to find the derivative.

b. Use the properties of exponents to first simplify the expression, and then find the derivative.

YOUR TURN ANSWERS

1. $2\sqrt{5} - 1; \sqrt{2}$

2. $4x^2 - 12x + 10$

3. One possible answer is $g(x) = 2x - 3$ and $f(x) = x^3$

4. $\dfrac{dy}{dx} = \dfrac{-2(10x - 6)}{(5x^2 - 6x)^3}$

5. $20x(x^2 - 7)^9$

6. $\dfrac{dy}{dx} = x(5x - 1)^2(25x - 2)$

7. $\dfrac{(4x - 1)^2(8x + 37)}{(x + 3)^2}$

4.4 Derivatives of Exponential Functions

APPLY IT Given a new product whose rate of growth is rapid at first and then slows, how can we find the rate of growth?
We will use a derivative to answer this question in Example 5 at the end of this section.

We can find the derivative of the exponential function by using the definition of the derivative. Thus

$$\frac{d(e^x)}{dx} = \lim_{h \to 0} \frac{e^{x+h} - e^x}{h}$$

$$= \lim_{h \to 0} \frac{e^x e^h - e^x}{h} \qquad \text{Property 1 of exponents}$$

$$= e^x \lim_{h \to 0} \frac{e^h - 1}{h}. \qquad \text{Property 1 of limits}$$

In the last step, since e^x does not involve h, we were able to bring e^x in front of the limit. The result says that the derivative of e^x is e^x times a constant, namely, $\lim_{h \to 0} (e^h - 1)/h$. To investigate this limit, we evaluate the expression for smaller and smaller values of h, as shown in the table in the margin. Based on the table, it appears that $\lim_{h \to 0} (e^h - 1)/h = 1$. This is proved in more advanced courses. We, therefore, have the following formula.

> ### Derivative of e^x
>
> $$\frac{d}{dx}(e^x) = e^x$$

To find the derivative of the exponential function with a base other than e, use the change-of-base theorem for exponentials to rewrite a^x as $e^{(\ln a)x}$. Thus, for any positive constant $a \neq 1$,

$$\frac{d(a^x)}{dx} = \frac{d[e^{(\ln a)x}]}{dx} \qquad \text{Change-of-base theorem for exponentials}$$

$$= e^{(\ln a)x} \ln a \qquad \text{Chain rule}$$

$$= (\ln a)a^x. \qquad \text{Change-of-base theorem again}$$

> ### Derivative of a^x
> For any positive constant $a \neq 1$,
>
> $$\frac{d}{dx}(a^x) = (\ln a)a^x.$$

(The derivative of an exponential function is the original function times the natural logarithm of the base.)

We now see why e is the best base to work with: It has the simplest derivative of all the exponential functions. Even if we choose a different base, e appears in the derivative anyway through the $\ln a$ term. (Recall that $\ln a$ is the logarithm of a to the base e.) In fact,

Approximation of $\lim_{h \to 0} \dfrac{e^h - 1}{h}$	
h	$\dfrac{e^h - 1}{h}$
-0.1	0.9516
-0.01	0.9950
-0.001	0.9995
-0.0001	1.0000
0.00001	1.0000
0.0001	1.0001
0.001	1.0005
0.01	1.0050
0.1	1.0517

of all the functions we have studied, e^x is the simplest to differentiate, because its derivative is just itself.*

The chain rule can be used to find the derivative of the more general exponential function $y = a^{g(x)}$. Let $y = f(u) = a^u$ and $u = g(x)$, so that $f[g(x)] = a^{g(x)}$. Then

$$f'[g(x)] = f'(u) = (\ln a)a^u = (\ln a)a^{g(x)},$$

and by the chain rule,

$$\frac{dy}{dx} = f'[g(x)] \cdot g'(x)$$

$$= (\ln a)a^{g(x)} \cdot g'(x).$$

As before, this formula becomes simpler when we use natural logarithms because $\ln e = 1$. We summarize these results next.

Derivative of $a^{g(x)}$ and $e^{g(x)}$

$$\frac{d}{dx}(a^{g(x)}) = (\ln a)a^{g(x)}g'(x)$$

and

$$\frac{d}{dx}(e^{g(x)}) = e^{g(x)}g'(x)$$

You need not memorize the previous two formulas. They are simply the result of applying the chain rule to the formula for the derivative of a^x.

CAUTION Notice the difference between the derivative of a variable to a constant power, such as $D_x x^3 = 3x^2$, and a constant to a variable power, like $D_x 3^x = (\ln 3)3^x$. Remember, $D_x 3^x \neq x3^{x-1}$.

EXAMPLE 1 **Derivatives of Exponential Functions**

Find the derivative of each function.

(a) $y = e^{5x}$

   SOLUTION Let $g(x) = 5x$, so $g'(x) = 5$. Then

$$\frac{dy}{dx} = 5e^{5x}.$$

(b) $s = 3^t$

   SOLUTION
$$\frac{ds}{dt} = (\ln 3)3^t$$

(c) $y = 10e^{3x^2}$

   SOLUTION
$$\frac{dy}{dx} = 10(e^{3x^2})(6x) = 60xe^{3x^2}$$

(d) $s = 8 \cdot 10^{1/t}$

   SOLUTION
$$\frac{ds}{dt} = 8(\ln 10)10^{1/t}\left(\frac{-1}{t^2}\right)$$

$$= \frac{-8(\ln 10)10^{1/t}}{t^2}$$

YOUR TURN 1 Find dy/dx for
(a) $y = 4^{3x}$,
(b) $y = e^{7x^3+5}$.

TRY YOUR TURN 1

*There is a joke about a deranged mathematician who frightened other inmates at an insane asylum by screaming at them, "I'm going to differentiate you!" But one inmate remained calm and simply responded, "I don't care; I'm e^x."

EXAMPLE 2 Derivative of an Exponential Function

Let $y = e^{x^2+1}\sqrt{5x + 2}$. Find $\dfrac{dy}{dx}$.

SOLUTION   Rewrite y as $y = e^{x^2+1}(5x + 2)^{1/2}$, and then use the product rule and the chain rule.

$$\frac{dy}{dx} = e^{x^2+1} \cdot \frac{1}{2}(5x + 2)^{-1/2} \cdot 5 + (5x + 2)^{1/2}e^{x^2+1} \cdot 2x$$

$$= e^{x^2+1}(5x + 2)^{-1/2}\left[\frac{5}{2} + (5x + 2) \cdot 2x\right] \quad \text{Factor out the greatest common factor,} \quad e^{x^2+1}(5x + 2)^{-1/2}.$$

$$= e^{x^2+1}(5x + 2)^{-1/2}\left[\frac{5 + 4x(5x + 2)}{2}\right] \quad \text{Least common denominator}$$

$$= e^{x^2+1}(5x + 2)^{-1/2}\left[\frac{5 + 20x^2 + 8x}{2}\right]$$

$$= \frac{e^{x^2+1}(20x^2 + 8x + 5)}{2\sqrt{5x + 2}} \quad \text{Simplify.}$$

TRY YOUR TURN 2

YOUR TURN 2 Let $y = (x^2 + 1)^2 e^{2x}$. Find dy/dx.

EXAMPLE 3 Derivative of an Exponential Function

Let $f(x) = \dfrac{100{,}000}{1 + 100e^{-0.3x}}$. Find $f'(x)$.

SOLUTION   Use the quotient rule.

$$f'(x) = \frac{(1 + 100e^{-0.3x})(0) - 100{,}000(-30e^{-0.3x})}{(1 + 100e^{-0.3x})^2}$$

$$= \frac{3{,}000{,}000e^{-0.3x}}{(1 + 100e^{-0.3x})^2}$$

TRY YOUR TURN 3

YOUR TURN 3 Let $f(x) = \dfrac{100}{5 + 2e^{-0.01x}}$. Find $f'(x)$.

In the previous example, we could also have taken the derivative by writing $f(x) = 100{,}000(1 + 100e^{-0.3x})^{-1}$, from which we have

$$f'(x) = -100{,}000(1 + 100e^{-0.3x})^{-2}100e^{-0.3x}(-0.3).$$

This simplifies to the same expression as in Example 3.

EXAMPLE 4 Radioactivity

The amount in grams in a sample of uranium 239 after t years is given by

$$A(t) = 100e^{-0.362t}.$$

Find the rate of change of the amount present after 3 years.

SOLUTION   The rate of change is given by the derivative dA/dt.

$$\frac{dA}{dt} = 100(e^{-0.362t})(-0.362) = -36.2e^{-0.362t}$$

After 3 years ($t = 3$), the rate of change is

$$\frac{dA}{dt} = -36.2e^{-0.362(3)} = -36.2e^{-1.086} \approx -12.2$$

YOUR TURN 4 The quantity (in grams) of a radioactive substance present after t years is $Q(t) = 100e^{-0.421t}$. Find the rate of change of the quantity present after 2 years.

grams per year.

TRY YOUR TURN 4

Frequently a population, or the sales of a certain product, will start growing slowly, then grow more rapidly, and then gradually level off. Such growth can often be approximated by a mathematical model known as the **logistic function**:

$$G(t) = \frac{mG_0}{G_0 + (m - G_0)e^{-kmt}},$$

where t represents time in appropriate units, G_0 is the initial number present, m is the maximum possible size of the population, k is a positive constant, and $G(t)$ is the population at time t. It is sometimes simpler to divide the numerator and denominator of the logistic function by G_0, writing the result as

$$G(t) = \frac{m}{1 + \left(\dfrac{m}{G_0} - 1\right)e^{-kmt}}.$$

Notice that

$$\lim_{t \to \infty} G(t) = \frac{m}{1 + 0} = m$$

because $\lim\limits_{t \to \infty} e^{-kmt} = 0$.

EXAMPLE 5   Product Sales

A company sells 990 units of a new product in the first year and 3213 units in the fourth year. They expect that sales can be approximated by a logistic function, leveling off at around 100,000 in the long run.

(a) Find a formula $S(t)$ for the sales as a function of time.

SOLUTION   We already know that $S_0 = 990$ and $m = 100{,}000$, so

$$S(t) = \frac{100{,}000}{1 + \left(\dfrac{100{,}000}{990} - 1\right)e^{-k100{,}000t}}$$

$$= \frac{100{,}000}{1 + 100.01e^{-k100{,}000t}}.$$

To find k, use the fact that $S(4) = 3213$.

$$3213 = \frac{100{,}000}{1 + 100.01e^{-k100{,}000 \cdot 4}}$$

$$3213 = \frac{100{,}000}{1 + 100.01e^{-k400{,}000}}$$

$$3213(1 + 100.01e^{-k400{,}000}) = 100{,}000 \qquad \text{Cross multiply.}$$

$$3213 + 321{,}332e^{-k400{,}000} = 100{,}000$$

$$321{,}332e^{-k400{,}000} = 96{,}787 \qquad \text{Subtract 3213 from both sides.}$$

$$e^{-k400{,}000} = 0.3012 \qquad \text{Divide both sides by 321,332.}$$

$$-k400{,}000 = \ln 0.3012 \qquad \text{Take the natural logarithm of both sides.}$$

$$k = -\ln 0.3012 / 400{,}000$$

$$k \approx 3 \times 10^{-6}$$

Rounding 100.01 to 100 and simplifying $k100{,}000 = (3 \times 10^{-6})100{,}000 = 0.3,$

$$S(t) = \frac{100{,}000}{1 + 100e^{-k100{,}000t}}$$

$$= \frac{100{,}000}{1 + 100e^{-0.3t}}.$$

(b) Find the rate of change of sales after 4 years.

SOLUTION The derivative of this sales function, which gives the rate of change of sales, was found in Example 3. Using that derivative,

$$S'(4) = \frac{3{,}000{,}000e^{-0.3(4)}}{[1 + 100e^{-0.3(4)}]^2} = \frac{3{,}000{,}000e^{-1.2}}{(1 + 100e^{-1.2})^2}.$$

Using a calculator, $e^{-1.2} \approx 0.3012$, and

$$S'(4) \approx \frac{3{,}000{,}000(0.3012)}{[1 + 100(0.3012)]^2}$$

$$\approx \frac{903{,}600}{(1 + 30.12)^2}$$

$$\approx \frac{903{,}600}{968.5} \approx 933.$$

The rate of change of sales after 4 years is about 933 units per year. The positive number indicates that sales are increasing at this time.

The graph of the function in Example 5 is shown in Figure 11.

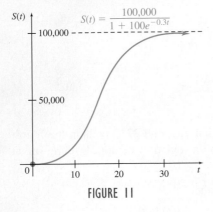

t	S(t)
0	990
5	4300
10	17,000
15	47,000
20	80,000
30	99,000

$$S(t) = \frac{100{,}000}{1 + 100e^{-0.3t}}$$

FIGURE 11

4.4 EXERCISES

Find derivatives of the functions defined as follows.

1. $y = e^{4x}$

2. $y = e^{-2x}$

3. $y = -8e^{3x}$

4. $y = 1.2e^{5x}$

5. $y = -16e^{2x+1}$

6. $y = -4e^{-0.3x}$

7. $y = e^{x^2}$

8. $y = e^{-x^2}$

9. $y = 3e^{2x^2}$

10. $y = -5e^{4x^3}$

11. $y = 4e^{2x^2-4}$

12. $y = -3e^{3x^2+5}$

13. $y = xe^x$

14. $y = x^2e^{-2x}$

15. $y = (x + 3)^2 e^{4x}$

16. $y = (3x^3 - 4x)e^{-5x}$

17. $y = \dfrac{x^2}{e^x}$

18. $y = \dfrac{e^x}{2x + 1}$

19. $y = \dfrac{e^x + e^{-x}}{x}$

20. $y = \dfrac{e^x - e^{-x}}{x}$

21. $p = \dfrac{10{,}000}{9 + 4e^{-0.2t}}$

22. $p = \dfrac{500}{12 + 5e^{-0.5t}}$

23. $f(z) = (2z + e^{-z^2})^2$

24. $f(t) = (e^{t^2} + 5t)^3$

25. $y = 7^{3x+1}$

26. $y = 4^{-5x+2}$

27. $y = 3 \cdot 4^{x^2+2}$

28. $y = -10^{3x^2-4}$

29. $s = 2 \cdot 3^{\sqrt{t}}$

30. $s = 5 \cdot 2^{\sqrt{t-2}}$

31. $y = \dfrac{te^t + 2}{e^{2t} + 1}$

32. $y = \dfrac{t^2 e^{2t}}{t + e^{3t}}$

33. $f(x) = e^{x\sqrt{3x+2}}$

34. $f(x) = e^{x^2/(x^3+2)}$

35. Prove that if $y = y_0 e^{kt}$, where y_0 and k are constants, then $dy/dt = ky$. (This says that for exponential growth and decay, the rate of change of the population is proportional to the size of the population, and the constant of proportionality is the growth or decay constant.)

36. Use a graphing calculator to sketch the graph of $y = [f(x + h) - f(x)]/h$ using $f(x) = e^x$ and $h = 0.0001$. Compare it with the graph of $y = e^x$ and discuss what you observe.

37. Use graphical differentiation to verify that $\dfrac{d}{dx}(e^x) = e^x$.

APPLICATIONS

Business and Economics

38. Sales The sales of a new personal computer (in thousands) are given by

$$S(t) = 100 - 90e^{-0.3t},$$

where t represents time in years. Find the rate of change of sales at each time.

a. After 1 year   **b.** After 5 years

c. What is happening to the rate of change of sales as time goes on?

d. Does the rate of change of sales ever equal zero?

39. Cost The cost in dollars to produce x DVDs can be approximated by

$$C(x) = \sqrt{900 - 800 \cdot 1.1^{-x}}.$$

Find the marginal cost when the following quantities are made.

a. 0   **b.** 20

c. What happens to the marginal cost as the number produced becomes larger and larger?

40. Product Awareness After the introduction of a new product for tanning without sun exposure, the percent of the public that is aware of the product is approximated by

$$A(t) = 10t^2 2^{-t},$$

where t is the time in months. Find the rate of change of the percent of the public that is aware of the product after the following numbers of months.

a. 2   **b.** 4

c. Notice that the answer to part a is positive and the answer to part b is negative. What does this tell you about how public awareness of the product has changed?

41. Product Durability Using data in a car magazine, we constructed the mathematical model

$$y = 100e^{-0.03045t}$$

for the percent of cars of a certain type still on the road after t years. Find the percent of cars on the road after the following numbers of years.

a. 0   **b.** 2   **c.** 4   **d.** 6

Find the rate of change of the percent of cars still on the road after the following numbers of years.

e. 0   **f.** 2

g. Interpret your answers to parts e and f.

42. Investment The value of a particular investment changes over time according to the function

$$S(t) = 5000e^{0.1(e^{0.25t})},$$

where $S(t)$ is the value after t years. Calculate the rate at which the value of the investment is changing after 8 years. (Choose one of the following.) *Source: Society of Actuaries.*

a. 618   **b.** 1934   **c.** 2011   **d.** 7735   **e.** 10,468

43. Internet Users The growth in the number (in millions) of Internet users in the United States between 1990 and 2015 can be approximated by a logistic function with $k = 0.0018$, where t is the number of years since 1990. In 1990 (when $t = 0$), the number of users was about 2 million, and the number is expected to level out around 250 million. *Source: World Bank.*

a. Find the growth function $G(t)$ for the number of Internet users in the United States.

Estimate the number of Internet users in the United States and the rate of growth for the following years.

b. 1995

c. 2000

d. 2010

e. What happens to the rate of growth over time?

Life Sciences

44. Population Growth In Section 2.4, Exercise 47, the growth in world population (in millions) was approximated by the exponential function

$$A(t) = 3100e^{0.0166t},$$

where t is the number of years since 1960. Find the instantaneous rate of change in the world population at the following times. *Source: United Nations.*

a. 2010   **b.** 2015

45. Minority Population In Section 2.4, Exercise 49, we saw that the projected Hispanic population in the United States (in millions) can be approximated by the function

$$h(t) = 37.79(1.021)^t$$

where $t = 0$ corresponds to 2000 and $0 \le t \le 50$. *Source: U.S. Census Bureau.*

a. Estimate the Hispanic population in the United States for the year 2015.

b. What is the instantaneous rate of change of the Hispanic population in the United States when $t = 15$? Interpret your answer.

46. Insect Growth The growth of a population of rare South American beetles is given by the logistic function with $k = 0.00001$ and t in months. Assume that there are 200 beetles initially and that the maximum population size is 10,000.

a. Find the growth function $G(t)$ for these beetles.

Find the population and rate of growth of the population after the following times.

b. 6 months   **c.** 3 years   **d.** 7 years

e. What happens to the rate of growth over time?

47. Clam Population The population of a bed of clams in the Great South Bay off Long Island is described by the logistic function with $k = 0.0001$ and t in years. Assume that there are 400 clams initially and that the maximum population size is 5200.

a. Find the growth function $G(t)$ for the clams.

Find the population and rate of growth of the population after the following times.

b. 1 year   **c.** 4 years   **d.** 10 years

e. What happens to the rate of growth over time?

48. Pollution Concentration The concentration of pollutants (in grams per liter) in the east fork of the Big Weasel River is approximated by

$$P(x) = 0.04e^{-4x},$$

where x is the number of miles downstream from a paper mill that the measurement is taken. Find the following values.

a. The concentration of pollutants 0.5 mile downstream

b. The concentration of pollutants 1 mile downstream

c. The concentration of pollutants 2 miles downstream

Find the rate of change of concentration with respect to distance for the following distances.

d. 0.5 mile   **e.** 1 mile   **f.** 2 miles

49. Breast Cancer It has been observed that the following formula accurately models the relationship between the size of a breast tumor and the amount of time that it has been growing.

$$V(t) = 1100[1023e^{-0.02415t} + 1]^{-4},$$

where t is in months and $V(t)$ is measured in cubic centimeters. *Source: Cancer.*

a. Find the tumor volume at 240 months.

b. Assuming that the shape of a tumor is spherical, find the radius of the tumor from part a. (*Hint:* The volume of a sphere is given by the formula $V = (4/3)\pi r^3$.)

c. If a tumor of size 0.5 cm³ is detected, according to the formula, how long has it been growing? What does this imply?

d. Find $\lim_{t \to \infty} V(t)$ and interpret this value. Explain whether this makes sense.

e. Calculate the rate of change of tumor volume at 240 months and interpret.

50. Mortality The percentage of people of any particular age group that will die in a given year may be approximated by the formula

$$P(t) = 0.00239e^{0.0957t},$$

where t is the age of the person in years. *Source: U.S. Vital Statistics.*

a. Find $P(25)$, $P(50)$, and $P(75)$.

b. Find $P'(25)$, $P'(50)$, and $P'(75)$.

c. Interpret your answers for parts a and b. Are there any limitations of this formula?

51. Medical Literature It has been observed that there has been an increase in the proportion of medical research papers that use the word "novel" in the title or abstract, and that this proportion can be accurately modeled by the function

$$p(x) = 0.001131e^{0.1268x},$$

where x is the number of years since 1970. *Source: Nature.*

a. Find $p(40)$.

b. If this phenomenon continues, estimate the year in which every medical article will contain the word "novel" in its title or abstract.

c. Estimate the rate of increase in the proportion of medical papers using this word in the year 2010.

d. Explain some factors that may be contributing to researchers using this word.

52. Arctic Foxes The age/weight relationship of female Arctic foxes caught in Svalbard, Norway, can be estimated by the function

$$M(t) = 3102e^{-e^{-0.022(t-56)}},$$

where t is the age of the fox in days and $M(t)$ is the weight of the fox in grams. *Source: Journal of Mammalogy.*

a. Estimate the weight of a female fox that is 200 days old.

b. Use $M(t)$ to estimate the largest size that a female fox can attain. (*Hint:* Find $\lim_{t \to \infty} M(t)$.)

c. Estimate the age of a female fox when it has reached 80% of its maximum weight.

d. Estimate the rate of change in weight of an Arctic fox that is 200 days old. (*Hint:* Recall that $D_t[e^{f(t)}] = f'(t)e^{f(t)}$.)

e. Use a graphing calculator to graph $M(t)$ and then describe the growth pattern.

f. Use the table function on a graphing calculator or a spreadsheet to develop a chart that shows the estimated weight and growth rate of female foxes for days 50, 100, 150, 200, 250, and 300.

53. Beef Cattle Researchers have compared two models that are used to predict the weight of beef cattle of various ages,

$$W_1(t) = 509.7(1 - 0.941e^{-0.00181t})$$

and

$$W_2(t) = 498.4(1 - 0.889e^{-0.00219t})^{1.25},$$

where $W_1(t)$ and $W_2(t)$ represent the weight (in kilograms) of a t-day-old beef cow. *Source: Journal of Animal Science.*

a. What is the maximum weight predicted by each function for the average beef cow? Is this difference significant?

b. According to each function, find the age that the average beef cow reaches 90% of its maximum weight.

c. Find $W_1'(750)$ and $W_2'(750)$. Compare your results.

d. Graph the two functions on $[0, 2500]$ by $[0, 525]$ and comment on the differences in growth patterns for each of these functions.

e. Graph the derivative of these two functions on $[0, 2500]$ by $[0, 1]$ and comment on any differences you notice between these functions.

54. Cholesterol Researchers have found that the risk of coronary heart disease rises as blood cholesterol increases. This risk may be approximated by the function

$$R(c) = 3.19(1.006)^c, \quad 100 \le c \le 300,$$

where R is the risk in terms of coronary heart disease incidence per 1000 per year, and c is the cholesterol in mg/dL. Suppose a person's cholesterol is 180 mg/dL and going up at a rate of 15 mg/dL per year. At what rate is the person's risk of coronary heart disease going up? *Source: Circulation*.

Social Sciences

55. Survival of Manuscripts Paleontologist John Cisne has demonstrated that the survival of ancient manuscripts can be modeled by the logistic equation. For example, the number of copies of the Venerable Bede's *De Temporum Ratione* was found to approach a limiting value over the five centuries after its publication in the year 725. Let $G(t)$ represent the proportion of manuscripts known to exist after t centuries out of the limiting value, so that $m = 1$. Cisne found that for Venerable Bede's *De Temporum Ratione*, $k = 3.5$ and $G_0 = 0.00369$. *Source: Science*.

a. Find the growth function $G(t)$ for the proportion of copies of *De Temporum Ratione* found.

Find the proportion of manuscripts and their rate of growth after the following number of centuries.

b. 1   **c.** 2   **d.** 3

e. What happens to the rate of growth over time?

56. Habit Strength According to work by the psychologist C. L. Hull, the strength of a habit is a function of the number of times the habit is repeated. If N is the number of repetitions and $H(N)$ is the strength of the habit, then

$$H(N) = 1000(1 - e^{-kN}),$$

where k is a constant. Find $H'(N)$ if $k = 0.1$ and the number of times the habit is repeated is as follows.

a. 10   **b.** 100   **c.** 1000

d. Show that $H'(N)$ is always positive. What does this mean?

57. Online Learning The growth of the number of students taking at least one online course can be approximated by a logistic function with $k = 0.0440$, where t is the number of years since 2002. In 2002 (when $t = 0$), the number of students enrolled in at least one online course was 1.603 million. Assume that the number will level out at around 6.8 million students. *Source: The Sloan Consortium*.

a. Find the growth function $G(t)$ for students enrolled in at least one online course.

Find the number of students enrolled in at least one online course and the rate of growth in the number in the following years.

b. 2004   **c.** 2006   **d.** 2010

e. What happens to the rate of growth over time?

Physical Sciences

58. Radioactive Decay The amount (in grams) of a sample of lead 214 present after t years is given by

$$A(t) = 500e^{-0.31t}.$$

Find the rate of change of the quantity present after each of the following years.

a. 4   **b.** 6   **c.** 10

d. What is happening to the rate of change of the amount present as the number of years increases?

e. Will the substance ever be gone completely?

59. Electricity In a series resistance-capacitance DC circuit, the instantaneous charge Q on the capacitor as a function of time (where $t = 0$ is the moment the circuit is energized by closing a switch) is given by the equation

$$Q(t) = CV(1 - e^{-t/RC}),$$

where C, V, and R are constants. Further, the instantaneous charging current I_C is the rate of change of charge on the capacitor, or $I_C = dQ/dt$. *Source: Kevin Friedrich*.

a. Find the expression for I_C as a function of time.

b. If $C = 10^{-5}$ farads, $R = 10^7$ ohms, and $V = 10$ volts, what is the charging current after 200 seconds? (*Hint:* When placed into the function in part a the units can be combined into amps.)

60. Heat Index The heat index is a measure of how hot it really feels under different combinations of temperature and humidity. The heat index, in degrees Fahrenheit, can be approximated by

$$H(T) = T - 0.9971e^{0.02086T}\left[1 - e^{0.0445(D-57.2)}\right],$$

where the temperature T and dewpoint D are both expressed in degrees Fahrenheit. *Source: American Meteorological Society*.

a. Assume the dewpoint is $D = 85°$ F. Find the function $H(T)$.

b. Using the function you found in part a, find the heat index when the temperature T is $80°$ F.

c. Find the rate of change of the heat index when $T = 80°$ F.

General Interest

61. Track and Field In 1958, L. Lucy developed a method for predicting the world record for any given year that a human could run a distance of 1 mile. His formula is given as follows:

$$t(n) = 218 + 31(0.933)^n,$$

where $t(n)$ is the world record (in seconds) for the mile run in year $1950 + n$. Thus, $n = 5$ corresponds to the year 1955. *Source: Statistics in Sports*.

a. Find the estimate for the world record in the year 2010.

b. Calculate the instantaneous rate of change for the world record at the end of year 2010 and interpret.

c. Find $\lim\limits_{n \to \infty} t(n)$ and interpret. How does this compare with the current world record?

62. Ballooning Suppose a person is going up in a hot air balloon. The surrounding air temperature in degrees Fahrenheit decreases with height according to the formula

$$T(h) = 80e^{-0.000065h},$$

where h is the height in feet. How fast is the temperature decreasing when the person is at a height of 1000 ft and rising at a height of 800 ft/hr?

63. The Gateway Arch The Gateway Arch in St. Louis, Missouri, is approximately 630 ft wide and 630 ft high. At first glance, the arch resembles a parabola, but its shape is actually known as a modified catenary. The height s (in feet) of the arch, measured from the ground to the middle of the arch, is approximated by the function

$$s(x) = 693.9 - 34.38(e^{0.01003x} + e^{-0.01003x}),$$

where $x = 0$ represents the center of the arch with $-299.2 \leq x \leq 299.2$. *Source: Notices of the AMS.*

a. What is the height of the arch when $x = 0$?

b. What is the slope of the line tangent to the curve when $x = 0$? Does this make sense?

c. Find the rate of change of the height of the arc when $x = 150$ ft.

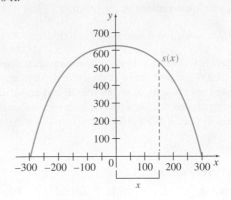

YOUR TURN ANSWERS

1. (a) $\dfrac{dy}{dx} = 3(\ln 4)4^{3x}$ **(b)** $\dfrac{dy}{dx} = 21x^2 e^{7x^3 + 5}$

2. $\dfrac{dy}{dx} = 2e^{2x}(x^2 + 1)(x + 1)^2$

3. $f'(x) = \dfrac{2e^{-0.01x}}{(5 + 2e^{-0.01x})^2}$ **4.** -18.1 grams per year

4.5 Derivatives of Logarithmic Functions

APPLY IT How does the average resale value of an automobile change with the age of the automobile?

We will use the derivative to answer this question in Example 4.

Recall that in the section on Logarithmic Functions, we showed that the logarithmic function and the exponential function are inverses of each other. In the last section we showed that the derivative of a^x is $(\ln a)a^x$. We can use this information and the chain rule to find the derivative of $\log_a x$. We begin by solving the general logarithmic function for x.

$$f(x) = \log_a x$$

$$a^{f(x)} = x \qquad \text{Definition of the logarithm}$$

Now consider the left and right sides of the last equation as functions of x that are equal, so their derivatives with respect to x will also be equal. Notice in the first step that we need to use the chain rule when differentiating $a^{f(x)}$.

$$(\ln a)a^{f(x)}f'(x) = 1 \qquad \text{Derivative of the exponential function}$$

$$(\ln a)xf'(x) = 1 \qquad \text{Substitute } a^{f(x)} = x.$$

Finally, divide both sides of this equation by $(\ln a)x$ to get

$$f'(x) = \frac{1}{(\ln a)x}.$$

Derivative of $\log_a x$

$$\frac{d}{dx}[\log_a x] = \frac{1}{(\ln a)x}$$

(The derivative of a logarithmic function is the reciprocal of the product of the variable and the natural logarithm of the base.)

As with the exponential function, this formula becomes particularly simple when we let $a = e$, because of the fact that $\ln e = 1$.

Derivative of $\ln x$

$$\frac{d}{dx}[\ln x] = \frac{1}{x}$$

This fact can be further justified geometrically. Notice what happens to the slope of the line $y = 2x + 4$ if the x-axis and y-axis are switched. That is, if we replace x with y and y with x, then the resulting line $x = 2y + 4$ or $y = x/2 - 2$ is a reflection of the line $y = 2x + 4$ across the line $y = x$, as seen in Figure 12. Furthermore, the slope of the new line is the reciprocal of the original line. In fact, the reciprocal property holds for all lines.

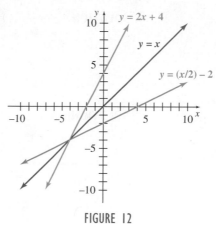

FIGURE 12

In the section on Logarithmic Functions, we showed that switching the x and y variables changes the exponential graph into a logarithmic graph, a defining property of functions that are inverses of each other. We also showed in the previous section that the slope of the tangent line of e^x at any point is e^x—that is, the y-coordinate itself. So, if we switch the x and y variables, the new slope of the tangent line will be $1/y$, except that it is no longer y, it is x. Thus, the slope of the tangent line of $y = \ln x$ must be $1/x$ and hence $D_x \ln x = 1/x$.

EXAMPLE 1 Derivatives of Logarithmic Functions

Find the derivative of each function.

(a) $f(x) = \ln 6x$

SOLUTION Use the properties of logarithms and the rules for derivatives.

$$f'(x) = \frac{d}{dx}(\ln 6x)$$

$$= \frac{d}{dx}(\ln 6 + \ln x)$$

$$= \frac{d}{dx}(\ln 6) + \frac{d}{dx}(\ln x) = 0 + \frac{1}{x} = \frac{1}{x}$$

(b) $y = \log x$

SOLUTION Recall that when the base is not specified, we assume that the logarithm is a common logarithm, which has a base of 10.

$$\frac{dy}{dx} = \frac{1}{(\ln 10)x}$$

TRY YOUR TURN 1

YOUR TURN 1 Find the derivative of $f(x) = \log_3 x$.

Applying the chain rule to the formulas for the derivative of logarithmic functions gives us

$$\frac{d}{dx} \log_a g(x) = \frac{1}{\ln a} \cdot \frac{g'(x)}{g(x)}$$

and

$$\frac{d}{dx} \ln g(x) = \frac{g'(x)}{g(x)}.$$

EXAMPLE 2   Derivatives of Logarithmic Functions

Find the derivative of each function.

(a) $f(x) = \ln(x^2 + 1)$

SOLUTION Here $g(x) = x^2 + 1$ and $g'(x) = 2x$. Thus,

$$f'(x) = \frac{g'(x)}{g(x)} = \frac{2x}{x^2 + 1}.$$

(b) $y = \log_2(3x^2 - 4x)$

YOUR TURN 2 Find the derivative of
(a) $y = \ln(2x^3 - 3)$,
(b) $f(x) = \log_4(5x + 3x^3)$.

SOLUTION

$$\frac{dy}{dx} = \frac{1}{\ln 2} \cdot \frac{6x - 4}{3x^2 - 4x}$$

$$= \frac{6x - 4}{(\ln 2)(3x^2 - 4x)}$$

TRY YOUR TURN 2

If $y = \ln(-x)$, where $x < 0$, the chain rule with $g(x) = -x$ and $g'(x) = -1$ gives

$$\frac{dy}{dx} = \frac{g'(x)}{g(x)} = \frac{-1}{-x} = \frac{1}{x}.$$

The derivative of $y = \ln(-x)$ is the same as the derivative of $y = \ln x$. For this reason, these two results can be combined into one rule using the absolute value of x. A similar situation holds true for $y = \ln[g(x)]$ and $y = \ln[-g(x)]$, as well as for $y = \log_a[g(x)]$ and $y = \log_a[-g(x)]$. These results are summarized as follows.

Derivative of $\log_a|x|$, $\log_a|g(x)|$, $\ln|x|$, and $\ln|g(x)|$

$$\frac{d}{dx}[\log_a|x|] = \frac{1}{(\ln a)x} \qquad \frac{d}{dx}[\log_a|g(x)|] = \frac{1}{\ln a} \cdot \frac{g'(x)}{g(x)}$$

$$\frac{d}{dx}[\ln|x|] = \frac{1}{x} \qquad \frac{d}{dx}[\ln|g(x)|] = \frac{g'(x)}{g(x)}$$

NOTE You need not memorize the previous four formulas. They are simply the result of the chain rule applied to the formula for the derivative of $y = \log_a x$, as well as the fact that when $\log_a x = \ln x$, so that $a = e$, then $\ln a = \ln e = 1$. An absolute value inside of a logarithm has no effect on the derivative, other than making the result valid for more values of x.

> **EXAMPLE 3** **Derivatives of Logarithmic Functions**

Find the derivative of each function.

(a) $y = \ln |5x|$

> **SOLUTION** Let $g(x) = 5x$, so that $g'(x) = 5$. From the previous formula,
>
> $$\frac{dy}{dx} = \frac{g'(x)}{g(x)} = \frac{5}{5x} = \frac{1}{x}.$$

Notice that the derivative of $\ln |5x|$ is the same as the derivative of $\ln |x|$. Also notice that we would have found the exact same answer for the derivative of $y = \ln 5x$ (without the absolute value), but the result would not apply to negative values of x. Also, in Example 1, the derivative of $\ln 6x$ was the same as that for $\ln x$. This suggests that for any constant a,

$$\frac{d}{dx} \ln |ax| = \frac{d}{dx} \ln |x|$$
$$= \frac{1}{x}.$$

Exercise 46 asks for a proof of this result.

(b) $f(x) = 3x \ln x^2$

> **SOLUTION** This function is the product of the two functions $3x$ and $\ln x^2$, so use the product rule.
>
> $$f'(x) = (3x)\left[\frac{d}{dx} \ln x^2\right] + (\ln x^2)\left[\frac{d}{dx} 3x\right]$$
>
> $$= 3x\left(\frac{2x}{x^2}\right) + (\ln x^2)(3)$$
>
> $$= 6 + 3 \ln x^2$$

By the power rule for logarithms,

$$f'(x) = 6 + \ln(x^2)^3$$
$$= 6 + \ln x^6.$$

Alternatively, write the answer as $f'(x) = 6 + 6 \ln x$, except that this last form requires $x > 0$, while negative values of x are acceptable in $6 + \ln x^6$.

Another method would be to use a rule of logarithms to simplify the function to $f(x) = 3x \cdot 2 \ln x = 6x \ln x$ and then to take the derivative.

(c) $s(t) = \dfrac{\log_8(t^{3/2} + 1)}{t}$

> **SOLUTION** Use the quotient rule and the chain rule.
>
> $$s'(t) = \frac{t \cdot \dfrac{1}{(t^{3/2} + 1)\ln 8} \cdot \dfrac{3}{2} t^{1/2} - \log_8(t^{3/2} + 1) \cdot 1}{t^2}$$

YOUR TURN 3 Find the derivative of each function.

(a) $y = \ln|2x + 6|$

(b) $f(x) = x^2 \ln 3x$

(c) $s(t) = \dfrac{\ln(t^2 - 1)}{t + 1}$

This expression can be simplified slightly by multiplying the numerator and the denominator by $2(t^{3/2} + 1) \ln 8$.

$$s'(t) = \frac{t \cdot \dfrac{1}{(t^{3/2} + 1)\ln 8} \cdot \dfrac{3}{2} t^{1/2} - \log_8(t^{3/2} + 1)}{t^2} \cdot \frac{2(t^{3/2} + 1) \ln 8}{2(t^{3/2} + 1) \ln 8}$$

$$= \frac{3t^{3/2} - 2(t^{3/2} + 1)(\ln 8) \log_8(t^{3/2} + 1)}{2t^2(t^{3/2} + 1) \ln 8}$$

TRY YOUR TURN 3

EXAMPLE 4 **Automobile Resale Value**

Based on projections from the *Kelly Blue Book*, the resale value of a 2010 Toyota Corolla 4-door sedan can be approximated by the following function

$$f(x) = 15{,}450 - 13{,}915 \log(t + 1),$$

where t is the number of years since 2010. Find and interpret $f(4)$ and $f'(4)$. *Source: Kelly Blue Book.*

APPLY IT **SOLUTION** Recognizing this function as a common (base 10) logarithm, we have

$$f(4) = 15{,}450 - 13{,}915 \log(4 + 1) \approx 5500.08.$$

The average resale value of a 2010 Toyota Corolla in 2014 would be approximately $5500.08.

The derivative of $f(t)$ is

$$f'(t) = \frac{-13{,}915}{(\ln 10)(t + 1)},$$

so $f'(4) \approx -1208.64$. In 2014, the average resale value of a 2010 Toyota Corolla is decreasing by $1208.64 per year.

4.5 EXERCISES

Find the derivative of each function.

1. $y = \ln(8x)$

2. $y = \ln(-4x)$

3. $y = \ln(8 - 3x)$

4. $y = \ln(1 + x^3)$

5. $y = \ln|4x^2 - 9x|$

6. $y = \ln|-8x^3 + 2x|$

7. $y = \ln\sqrt{x + 5}$

8. $y = \ln\sqrt{2x + 1}$

9. $y = \ln(x^4 + 5x^2)^{3/2}$

10. $y = \ln(5x^3 - 2x)^{3/2}$

11. $y = -5x \ln(3x + 2)$

12. $y = (3x + 7) \ln(2x - 1)$

13. $s = t^2 \ln|t|$

14. $y = x \ln|2 - x^2|$

15. $y = \dfrac{2 \ln(x + 3)}{x^2}$

16. $v = \dfrac{\ln u}{u^3}$

17. $y = \dfrac{\ln x}{4x + 7}$

18. $y = \dfrac{-2 \ln x}{3x - 1}$

19. $y = \dfrac{3x^2}{\ln x}$

20. $y = \dfrac{x^3 - 1}{2 \ln x}$

21. $y = (\ln|x + 1|)^4$

22. $y = \sqrt{\ln|x - 3|}$

23. $y = \ln|\ln x|$

24. $y = (\ln 4)(\ln|3x|)$

25. $y = e^{x^2} \ln x$

26. $y = e^{2x-1} \ln(2x - 1)$

27. $y = \dfrac{e^x}{\ln x}$

28. $p(y) = \dfrac{\ln y}{e^y}$

29. $g(z) = (e^{2z} + \ln z)^3$

30. $s(t) = \sqrt{e^{-t} + \ln 2t}$

31. $y = \log(6x)$

32. $y = \log(4x - 3)$

33. $y = \log|1 - x|$

34. $y = \log|3x|$

35. $y = \log_5\sqrt{5x + 2}$

36. $y = \log_7\sqrt{4x - 3}$

37. $y = \log_3(x^2 + 2x)^{3/2}$

38. $y = \log_2(2x^2 - x)^{5/2}$

39. $w = \log_8(2^p - 1)$

40. $z = 10^y \log y$

41. $f(x) = e^{\sqrt{x}} \ln(\sqrt{x} + 5)$

42. $f(x) = \ln(xe^{\sqrt{x}} + 2)$

43. $f(t) = \dfrac{\ln(t^2 + 1) + t}{\ln(t^2 + 1) + 1}$

44. $f(t) = \dfrac{2t^{3/2}}{\ln(2t^{3/2} + 1)}$

45. Why do we use the absolute value of x or of $g(x)$ in the derivative formulas for the natural logarithm?

46. Prove $\dfrac{d}{dx} \ln|ax| = \dfrac{d}{dx} \ln|x|$ for any constant a.

47. A friend concludes that because $y = \ln 6x$ and $y = \ln x$ have the same derivative, namely $dy/dx = 1/x$, these two functions must be the same. Explain why this is incorrect.

48. Use a graphing calculator to sketch the graph of $y = [f(x + h) - f(x)]/h$ using $f(x) = \ln|x|$ and $h = 0.0001$. Compare it with the graph of $y = 1/x$ and discuss what you observe.

49. Using the fact that

$$\ln[u(x)v(x)] = \ln u(x) + \ln v(x),$$

use the chain rule and the formula for the derivative of ln x to derive the product rule. In other words, find $[u(x)v(x)]'$ without assuming the product rule.

50. Using the fact that

$$\ln \frac{u(x)}{v(x)} = \ln u(x) - \ln v(x),$$

use the chain rule and the formula for the derivative of ln x to derive the quotient rule. In other words, find $[u(x)/v(x)]'$ without assuming the quotient rule.

51. Use graphical differentiation to verify that $\frac{d}{dx} (\ln x) = \frac{1}{x}$.

52. Use the fact that $d \ln x/dx = 1/x$, as well as the change-of-base theorem for logarithms, to prove that

$$\frac{d \log_a x}{dx} = \frac{1}{x \ln a}.$$

53. Let

$$h(x) = u(x)^{v(x)}.$$

a. Using the fact that

$$\ln [u(x)^{v(x)}] = v(x) \ln u(x),$$

use the chain rule, the product rule, and the formula for the derivative of ln x to show that

$$\frac{d}{dx} \ln h(x) = \frac{v(x)u'(x)}{u(x)} + (\ln u(x)) v'(x).$$

b. Use the result from part a and the fact that

$$\frac{d}{dx} \ln h(x) = \frac{h'(x)}{h(x)}$$

to show that

$$\frac{d}{dx} h(x) = u(x)^{v(x)} \left[\frac{v(x)u'(x)}{u(x)} + (\ln u(x)) v'(x) \right].$$

The idea of taking the logarithm of a function before differentiating is known as logarithmic differentiation.

Use the ideas from Exercise 53 to find the derivative of each of the following functions.

54. $h(x) = x^x$　　　　**55.** $h(x) = (x^2 + 1)^{5x}$

APPLICATIONS

Business and Economics

56. Profit If the total revenue received from the sale of x items is given by

$$R(x) = 30 \ln(2x + 1),$$

while the total cost to produce x items is $C(x) = x/2$, find the following.

a. The marginal revenue

b. The profit function $P(x)$

c. The marginal profit when $x = 60$

d. Interpret the results of part c.

57. Revenue Suppose the demand function for q units of a certain item is

$$p = D(q) = 100 + \frac{50}{\ln q}, \quad q > 1,$$

where p is in dollars.

a. Find the marginal revenue.

b. Approximate the revenue from one more unit when 8 units are sold.

c. How might a manager use the information from part b?

58. Profit If the cost function in dollars for q units of the item in Exercise 57 is $C(q) = 100q + 100$, find the following.

a. The marginal cost

b. The profit function $P(q)$

c. The approximate profit from one more unit when 8 units are sold

d. How might a manager use the information from part c?

59. Marginal Average Cost Suppose the cost in dollars to make x oboe reeds is given by

$$C(x) = 5 \log_2 x + 10.$$

Find the marginal average cost when the following numbers of reeds are sold.

a. 10　　　　　　　**b.** 20

Life Sciences

60. Body Surface Area There is a mathematical relationship between an infant's weight and total body surface area (BSA), given by

$$A(w) = 4.688w^{0.8168 - 0.0154 \log_{10} w},$$

where w is the weight (in grams) and $A(w)$ is the BSA in square centimeters. *Source: British Journal of Cancer*.

a. Find the BSA for an infant who weighs 4000 g.

b. Find $A'(4000)$ and interpret your answer.

c. Use a graphing calculator to graph $A(w)$ on $[2000, 10,000]$ by $[0, 6000]$.

61. Bologna Sausage Scientists have developed a model to predict the growth of bacteria in bologna sausage at 32°C. The number of bacteria is given by

$$\ln \left(\frac{N(t)}{N_0} \right) = 9.8901 e^{-e^{2.54197 - 0.2167t}},$$

where N_0 is the number of bacteria present at the beginning of the experiment and $N(t)$ is the number of bacteria present at time t (in hours). *Source: Applied and Environmental Microbiology.*

a. Use the properties of logarithms to find an expression for $N(t)$. Assume that $N_0 = 1000$.

b. Use a graphing calculator to estimate the derivative of $N(t)$ when $t = 20$ and interpret.

c. Let $S(t) = \ln(N(t)/N_0)$. Graph $S(t)$ on $[0, 35]$ by $[0, 12]$.

d. Graph $N(t)$ on $[0, 35]$ by $[0, 20,000,000]$ and compare the graphs from parts c and d.

e. Find $\lim_{t \to \infty} S(t)$ and then use this limit to find $\lim_{t \to \infty} N(t)$.

62. **Pronghorn Fawns** The field metabolic rate (FMR), or the total energy expenditure per day in excess of growth, can be calculated for pronghorn fawns using Nagy's formula,

$$F(x) = 0.774 + 0.727 \log x,$$

where x is the mass (in grams) of the fawn and $F(x)$ is the energy expenditure (in kJ/day). *Source: Animal Behavior.*

a. Determine the total energy expenditure per day in excess of growth for a pronghorn fawn that weighs 25,000 g.

b. Find $F'(25,000)$ and interpret the result.

c. Graph the function on $[5000, 30,000]$ by $[3, 5]$.

63. **Fruit Flies** A study of the relation between the rate of reproduction in *Drosophila* (fruit flies) bred in bottles and the density of the mated population found that the number of imagoes (sexually mature adults) per mated female per day (y) can be approximated by

$$\log y = 1.54 - 0.008x - 0.658 \log x,$$

where x is the mean density of the mated population (measured as flies per bottle) over a 16-day period. *Source: Elements of Mathematical Biology.*

a. Show that the above equation is equivalent to

$$y = 34.7(1.0186)^{-x}x^{-0.658}.$$

b. Using your answer from part a, find the number of imagoes per mated female per day when the density is

i. 20 flies per bottle;

ii. 40 flies per bottle.

c. Using your answer from part a, find the rate of change in the number of imagoes per mated female per day with respect to the density when the density is

i. 20 flies per bottle;

ii. 40 flies per bottle.

64. **Insect Mating** Consider an experiment in which equal numbers of male and female insects of a certain species are permitted to intermingle. Assume that

$$M(t) = (0.1t + 1) \ln\sqrt{t}$$

represents the number of matings observed among the insects in an hour, where t is the temperature in degrees Celsius. (*Note:* The formula is an approximation at best and holds only for specific temperature intervals.)

a. Find the number of matings when the temperature is 15°C.

b. Find the number of matings when the temperature is 25°C.

c. Find the rate of change of the number of matings when the temperature is 15°C.

65. **Population Growth** Suppose that the population of a certain collection of rare Brazilian ants is given by

$$P(t) = (t + 100) \ln(t + 2),$$

where t represents the time in days. Find the rates of change of the population on the second day and on the eighth day.

Social Sciences

66. **Poverty** The passage of the Social Security Amendments of 1965 resulted in the creation of the Medicare and Medicaid programs. Since then, the percent of persons 65 years and over with family income below the poverty level has declined. The percent can be approximated by the following function:

$$P(t) = 30.60 - 5.79 \ln t,$$

where t is the number of years since 1965. Find the percent of persons 65 years and over with family income below the poverty level and the rate of change in the following years. *Source: U.S. Census.*

a. 1970   b. 1990   c. 2010

d. What happens to the rate of change over time?

Physical Sciences

67. **Richter Scale** The Richter scale provides a measure of the magnitude of an earthquake. In fact, the largest Richter number M ever recorded for an earthquake was 8.9 from the 1933 earthquake in Japan. The following formula shows a relationship between the amount of energy released and the Richter number.

$$M = \frac{2}{3} \log \frac{E}{0.007},$$

where E is measured in kilowatt-hours. *Source: Mathematics Teacher.*

a. For the 1933 earthquake in Japan, what value of E gives a Richter number $M = 8.9$?

b. If the average household uses 247 kWh per month, how many months would the energy released by an earthquake of this magnitude power 10 million households?

c. Find the rate of change of the Richter number M with respect to energy when $E = 70,000$ kWh.

d. What happens to dM/dE as E increases?

General Interest

68. Street Crossing Consider a child waiting at a street corner for a gap in traffic that is large enough so that he can safely cross the street. A mathematical model for traffic shows that if the expected waiting time for the child is to be at most 1 minute, then the maximum traffic flow, in cars per hour, is given by

$$f(x) = \frac{29,000(2.322 - \log x)}{x},$$

where x is the width of the street in feet. Find the maximum traffic flow and the rate of change of the maximum traffic flow

with respect to street width for the following values of the street width. *Source: An Introduction to Mathematical Modeling.*

a. 30 ft **b.** 40 ft

■ YOUR TURN ANSWERS

1. $f'(x) = \dfrac{1}{(\ln 3)x}$

2. (a) $\dfrac{dy}{dx} = \dfrac{6x^2}{2x^3 - 3}$

(b) $f'(x) = \dfrac{5 + 9x^2}{(\ln 4)(5x + 3x^3)}$

3. (a) $\dfrac{dy}{dx} = \dfrac{1}{x + 3}$

(b) $f'(x) = x + 2x \ln 3x$

(c) $s'(t) = \dfrac{2t - (t - 1)\ln(t^2 - 1)}{(t - 1)(t + 1)^2}$

4 CHAPTER REVIEW

SUMMARY

In this chapter we used the definition of the derivative to develop techniques for finding derivatives of several types of functions. With the help of the rules that were developed, such as the power rule, product rule, quotient rule, and chain rule, we can now directly compute the derivative of a large variety of functions. In particular, we developed rules for finding derivatives of exponential and

logarithmic functions. We also began to see the wide range of applications that these functions have in business, life sciences, social sciences, and the physical sciences. In the next chapter we will apply these techniques to study the behavior of certain functions, and we will learn that differentiation can be used to find maximum and minimum values of continuous functions.

Assume all indicated derivatives exist.

Constant Function   If $f(x) = k$, where k is any real number, then $f'(x) = 0$.

Power Rule   If $f(x) = x^n$, for any real number n, then $f'(x) = n \cdot x^{n-1}$.

Constant Times a Function   Let k be a real number. Then the derivative of $y = k \cdot f(x)$ is $dy/dx = k \cdot f'(x)$.

Sum or Difference Rule   If $y = u(x) \pm v(x)$, then $\dfrac{dy}{dx} = u'(x) \pm v'(x)$.

Product Rule   If $f(x) = u(x) \cdot v(x)$, then

$$f'(x) = u(x) \cdot v'(x) + v(x) \cdot u'(x).$$

Quotient Rule   If $f(x) = \dfrac{u(x)}{v(x)}$, then

$$f'(x) = \frac{v(x) \cdot u'(x) - u(x) \cdot v'(x)}{[v(x)]^2}.$$

Chain Rule   If y is a function of u, say $y = f(u)$, and if u is a function of x, say $u = g(x)$, then $y = f(u) = f[g(x)]$, and

$$\frac{dy}{dx} = \frac{dy}{du} \cdot \frac{du}{dx}.$$

Chain Rule (Alternative Form)   Let $y = f[g(x)]$. Then $\dfrac{dy}{dx} = f'[g(x)] \cdot g'(x)$.

Exponential Functions $\dfrac{d}{dx}(e^x) = e^x$ $\qquad$ $\dfrac{d}{dx}(a^x) = (\ln a)a^x$

$\dfrac{d}{dx}(e^{g(x)}) = e^{g(x)}g'(x)$ $\qquad$ $\dfrac{d}{dx}(a^{g(x)}) = (\ln a)a^{g(x)}g'(x)$

Logarithmic Functions $\dfrac{d}{dx}(\ln |x|) = \dfrac{1}{x}$ $\qquad$ $\dfrac{d}{dx}(\log_a|x|) = \dfrac{1}{(\ln a)x}$

$\dfrac{d}{dx}(\ln |g(x)|) = \dfrac{g'(x)}{g(x)}$ $\qquad$ $\dfrac{d}{dx}(\log_a|g(x)|) = \dfrac{1}{\ln a}\cdot\dfrac{g'(x)}{g(x)}$

KEY TERMS

4.1
marginal analysis

4.2
marginal average cost

4.3
composite function
composition
chain rule

4.4
logistic function

REVIEW EXERCISES

CONCEPT CHECK

Determine whether each of the following statements is true or false, and explain why.

1. The derivative of π^3 is $3\pi^2$.

2. The derivative of a sum is the sum of the derivatives.

3. The derivative of a product is the product of the derivatives.

4. The marginal cost function is the derivative of the cost function.

5. The chain rule is used to take the derivative of a product of functions.

6. The only function that is its own derivative is e^x.

7. The derivative of 10^x is $x10^{x-1}$.

8. The derivative of $\ln |x|$ is the same as the derivative of $\ln x$.

9. The derivative of $\ln kx$ is the same as the derivative of $\ln x$.

10. The derivative of $\log x$ is the same as the derivative of $\ln x$.

PRACTICE AND EXPLORATIONS

Use the rules for derivatives to find the derivative of each function defined as follows.

11. $y = 5x^3 - 7x^2 - 9x + \sqrt{5}$

12. $y = 7x^3 - 4x^2 - 5x + \sqrt{2}$

13. $y = 9x^{8/3}$

14. $y = -4x^{-3}$

15. $f(x) = 3x^{-4} + 6\sqrt{x}$

16. $f(x) = 19x^{-1} - 8\sqrt{x}$

17. $k(x) = \dfrac{3x}{4x + 7}$

18. $r(x) = \dfrac{-8x}{2x + 1}$

19. $y = \dfrac{x^2 - x + 1}{x - 1}$

20. $y = \dfrac{2x^3 - 5x^2}{x + 2}$

21. $f(x) = (3x^2 - 2)^4$

22. $k(x) = (5x^3 - 1)^6$

23. $y = \sqrt{2t^7 - 5}$

24. $y = -3\sqrt{8t^4 - 1}$

25. $y = 3x(2x + 1)^3$

26. $y = 4x^2(3x - 2)^5$

27. $r(t) = \dfrac{5t^2 - 7t}{(3t + 1)^3}$

28. $s(t) = \dfrac{t^3 - 2t}{(4t - 3)^4}$

29. $p(t) = t^2(t^2 + 1)^{5/2}$

30. $g(t) = t^3(t^4 + 5)^{7/2}$

31. $y = -6e^{2x}$

32. $y = 8e^{0.5x}$

33. $y = e^{-2x^3}$

34. $y = -4e^{x^2}$

35. $y = 5xe^{2x}$

36. $y = -7x^2e^{-3x}$

37. $y = \ln(2 + x^2)$

38. $y = \ln(5x + 3)$

39. $y = \dfrac{\ln |3x|}{x - 3}$

40. $y = \dfrac{\ln |2x - 1|}{x + 3}$

41. $y = \dfrac{xe^x}{\ln(x^2 - 1)}$

42. $y = \dfrac{(x^2 + 1)e^{2x}}{\ln x}$

43. $s = (t^2 + e^t)^2$

44. $q = (e^{2p+1} - 2)^4$

45. $y = 3 \cdot 10^{-x^2}$

46. $y = 10 \cdot 2^{\sqrt{x}}$

47. $g(z) = \log_2(z^3 + z + 1)$

48. $h(z) = \log(1 + e^z)$

49. $f(x) = e^{2x}\ln(xe^x + 1)$

50. $f(x) = \dfrac{e^{\sqrt{x}}}{\ln(\sqrt{x} + 1)}$

Consider the following table of values of the functions f and g and their derivatives at various points.

x	1	2	3	4
$f(x)$	3	4	2	1
$f'(x)$	-5	-6	-7	-11
$g(x)$	4	1	2	3
$g'(x)$	2/9	3/10	4/11	6/13

Find the following using the table.

51. a. $D_x(f[g(x)])$ at $x = 2$ **b.** $D_x(f[g(x)])$ at $x = 3$

52. a. $D_x(g[f(x)])$ at $x = 2$ **b.** $D_x(g[f(x)])$ at $x = 3$

Find the slope of the tangent line to the given curve at the given value of x. Find the equation of each tangent line.

53. $y = x^2 - 6x;$ $x = 2$ **54.** $y = 8 - x^2;$ $x = 1$

55. $y = \dfrac{3}{x - 1};$ $x = -1$ **56.** $y = \dfrac{x}{x^2 - 1};$ $x = 2$

57. $y = \sqrt{6x - 2};$ $x = 3$ **58.** $y = -\sqrt{8x + 1};$ $x = 3$

59. $y = e^x;$ $x = 0$ **60.** $y = xe^x;$ $x = 1$

61. $y = \ln x;$ $x = 1$ **62.** $y = x \ln x;$ $x = e$

63. Consider the graphs of the function $y = \sqrt{2x - 1}$ and the straight line $y = x + k$. Discuss the number of points of intersection versus the change in the value of k. *Source: Japanese University Entrance Examination Problems in Mathematics.*

64. a. Verify that

$$\frac{d \ln f(x)}{dx} = \frac{f'(x)}{f(x)}.$$

This expression is called the *relative rate of change*. It expresses the rate of change of f relative to the size of f. Stephen B. Maurer denotes this expression by $\hat{f}$ and notes that economists commonly work with relative rates of change. *Source: The College Mathematics Journal.*

 b. Verify that

$$\widehat{fg} = \hat{f} + \hat{g}$$

Interpret this equation in terms of relative rates of change.

c. In his article, Maurer uses the result of part b to solve the following problem:

"Last year, the population grew by 1% and the average income per person grew by 2%. By what approximate percent did the national income grow?"

Explain why the result from part b implies that the answer to this question is approximately 3%.

65. Suppose that the student body in your college grows by 2% and the tuition goes up by 3%. Use the result from the previous exercise to calculate the approximate amount that the total tuition collected goes up, and compare this with the actual amount.

 66. Why is e a convenient base for exponential and logarithmic functions?

APPLICATIONS

Business and Economics

Marginal Average Cost Find the marginal average cost function of each function defined as follows.

67. $C(x) = \sqrt{x + 1}$

68. $C(x) = \sqrt{3x + 2}$

69. $C(x) = (x^2 + 3)^3$

70. $C(x) = (4x + 3)^4$

71. $C(x) = 10 - e^{-x}$

72. $C(x) = \ln(x + 5)$

73. Sales. The sales of a company are related to its expenditures on research by

$$S(x) = 1000 + 60\sqrt{x} + 12x,$$

where $S(x)$ gives sales in millions when x thousand dollars is spent on research. Find and interpret dS/dx if the following amounts are spent on research.

a. $9000   **b.** $16,000   **c.** $25,000

d. As the amount spent on research increases, what happens to sales?

74. Profit Suppose that the profit (in hundreds of dollars) from selling x units of a product is given by

$$P(x) = \frac{x^2}{2x + 1}.$$

Find and interpret the marginal profit when the following numbers of units are sold.

a. 4   **b.** 12   **c.** 20

d. What is happening to the marginal profit as the number sold increases?

e. Find and interpret the marginal average profit when 4 units are sold.

75. Costs A company finds that its total costs are related to the amount spent on training programs by

$$T(x) = \frac{1000 + 60x}{4x + 5},$$

where $T(x)$ is costs in thousands of dollars when x hundred dollars are spent on training. Find and interpret $T'(x)$ if the following amounts are spent on training.

a. $900   **b.** $1900

c. Are costs per dollar spent on training always increasing or decreasing?

76. Compound Interest If a sum of $1000 is deposited into an account that pays $r\%$ interest compounded quarterly, the balance after 12 years is given by

$$A = 1000\left(1 + \frac{r}{400}\right)^{48}.$$

Find and interpret $\dfrac{dA}{dr}$ when $r = 5$.

77. Continuous Compounding If a sum of $1000 is deposited into an account that pays $r\%$ interest compounded continuously, the balance after 12 years is given by

$$A = 1000e^{12r/100}.$$

Find and interpret $\dfrac{dA}{dr}$ when $r = 5$.

78. Doubling Time If a sum of money is deposited into an account that pays $r\%$ interest compounded annually, the doubling time (in years) is given by

$$T = \frac{\ln 2}{\ln(1 + r/100)}.$$

Find and interpret dT/dr when $r = 5$.

79. U.S. Post Office The number (in billions) of pieces of mail handled by the U.S. Post Office each year from 1980 through 2009 can be approximated by

$$P(t) = -0.00132t^4 + 0.0665t^3 - 1.127t^2 + 11.581t + 105.655,$$

where t is the number of years since 1980. Find and interpret the rate of change in the volume of mail for the following years. *Source: U.S. Postal Service.*

a. 1995   **b.** 2005

80. Elderly Employment After declining over the last century, the percentage of men aged 65 and older in the workforce has begun to rise in recent years, as shown by the following table. *Source: U.S. Bureau of Labor Statistics.*

Year	Percent of Men 65 and Older in Workforce
1900	63.1
1920	55.6
1930	54.0
1940	41.8
1950	45.8
1960	33.1
1970	26.8
1980	19.0
1990	16.3
2000	17.7
2009	21.9

a. Using the regression feature on a graphing calculator, find a cubic and a quartic function that model this data, letting $t = 0$ correspond to the year 1900.

b. Using each of your answers to part a, find the rate that the percent of men aged 65 and older in the workforce was increasing in 2005.

c. Discuss which model from part a best describes the data, as well as which answer from part b best describes the rate that the percent of men aged 65 and older in the workforce was increasing in 2005.

d. Explore other functions that could be used to model the data, using the various regression features on a graphing calculator, and discuss to what extent any of them are useful descriptions of the data.

81. Value of the Dollar The U.S. dollar has been declining in value over the last century, except during the Great Depression, when it increased in value. The following table shows the number of dollars it took in various years to equal $1 in 1913. *Source: U.S. Bureau of Labor Statistics.*

Year	Number of Dollars It Took to Equal $1 in 1913
1913	1.00
1920	2.02
1930	1.69
1940	1.41
1950	2.53
1960	2.99
1970	3.92
1980	8.32
1990	13.20
2000	17.39
2010	22.02

a. Using the regression feature on a graphing calculator, find a cubic and a quartic function that model this data, letting $t = 0$ correspond to the year 1900.

b. Using each of your answers to part a, find the rate that the number of dollars required to equal $1 in 1913 was increasing in 2005.

c. Discuss which model from part a best describes the data, as well as which answer from part b best describes the rate that the number of dollars required to equal $1 in 1913 was increasing in 2005.

d. Explore other functions that could be used to model the data, using the various regression features on a graphing calculator, and discuss to what extent any of them are useful descriptions of the data.

Life Sciences

82. Exponential Growth Suppose a population is growing exponentially with an annual growth constant $k = 0.05$. How fast is the population growing when it is 1,000,000? Use the derivative to calculate your answer, and then explain how the answer can be obtained without using the derivative.

83. Logistic Growth Suppose a population is growing logistically with $k = 5 \times 10^{-6}, m = 30,000$, and $G_0 = 2000$. Assume time is measured in years.

a. Find the growth function $G(t)$ for this population.

b. Find the population and rate of growth of the population after 6 years.

84. Fish The length of the monkeyface prickleback, a West Coast game fish, can be approximated by

$$L = 71.5(1 - e^{-0.1t})$$

and the weight by

$$W = 0.01289 \cdot L^{2.9},$$

where L is the length in centimeters, t is the age in years, and W is the weight in grams. *Source: California Fish and Game.*

a. Find the approximate length of a 5-year-old monkeyface.

b. Find how fast the length of a 5-year-old monkeyface is growing.

c. Find the approximate weight of a 5-year-old monkeyface. (*Hint:* Use your answer from part a.)

d. Find the rate of change of the weight with respect to length for a 5-year-old monkeyface.

e. Using the chain rule and your answers to parts b and d, find how fast the weight of a 5-year-old monkeyface is growing.

85. Arctic Foxes The age/weight relationship of male Arctic foxes caught in Svalbard, Norway, can be estimated by the function

$$M(t) = 3583e^{-e^{-0.020(t-66)}},$$

where t is the age of the fox in days and $M(t)$ is the weight of the fox in grams. *Source: Journal of Mammology.*

a. Estimate the weight of a male fox that is 250 days old.

b. Use $M(t)$ to estimate the largest size that a male fox can attain. (*Hint:* Find $\lim_{t \to \infty} M(t)$.)

c. Estimate the age of a male fox when it has reached 50% of its maximum weight.

d. Estimate the rate of change in weight of a male Arctic fox that is 250 days old. (*Hint:* Recall that $D_t e^{f(t)} = f'(t)e^{f(t)}$.)

  e. Use a graphing calculator to graph $M(t)$ and then describe the growth pattern.

f. Use the table function on a graphing calculator or a spreadsheet to develop a chart that shows the estimated weight and growth rate of male foxes for days 50, 100, 150, 200, 250, and 300.

86. Hispanic Population In Section 2.4, Exercise 49, we found that the projected Asian population in the United States, in millions, can be approximated by

$$a(t) = 11.14(1.023)^t,$$

where t is the years since 2000. Find the instantaneous rate of change in the projected Asian population in the United States in each of the following years. *Source: U.S. Census.*

a. 2005   **b.** 2025

Physical Sciences

87. Wind Energy In Section 2.4, Exercise 55, we found that the total world wind energy capacity (in megawatts) in recent years could be approximated by the function

$$C(t) = 19,231(1.2647)^t,$$

where t is the number of years since 2000. Find the rate of change in the energy capacity for the following years. *Source: World Wind Energy Association.*

a. 2005   b. 2010   c. 2015

General Interest

88. Cats The distance from Lisa Wunderle's cat, Belmar, to a piece of string he is stalking is given in feet by

$$f(t) = \frac{8}{t+1} + \frac{20}{t^2+1},$$

where t is the time in seconds since he begins.

a. Find Belmar's average velocity between 1 second and 3 seconds.

b. Find Belmar's instantaneous velocity at 3 seconds.

89. Food Surplus In Section 2.4, Example 7, we found that the production of corn (in billions of bushels) in the United States since 1930 could be approximated by

$$p(x) = 1.757(1.0249)^{x-1930}$$

where x is the year. Find and interpret $p'(2000)$.

90. Dating a Language Over time, the number of original basic words in a language tends to decrease as words become obsolete or are replaced with new words. Linguists have used calculus to study this phenomenon and have developed a methodology for dating a language, called *glottochronology*. Experiments have indicated that a good estimate of the number of words that remain in use at a given time is given by

$$N(t) = N_0 e^{-0.217t},$$

where $N(t)$ is the number of words in a particular language, t is measured in the number of millennium, and N_0 is the original number of words in the language. *Source: The UMAP Journal.*

a. In 1950, C. Feng and M. Swadesh established that of the original 210 basic ancient Chinese words from 950 A.D., 167 were still being used. Letting $t = 0$ correspond to 950, with $N_0 = 210$, find the number of words predicted to have been in use in 1950 A.D., and compare it with the actual number in use.

b. Estimate the number of words that will remain in the year 2050 ($t = 1.1$).

c. Find $N'(1.1)$ and interpret your answer.

91. Driving Fatalities A study by the National Highway Traffic Safety Administration found that driver fatalities rates were highest for the youngest and oldest drivers. The rates per 1000 licensed drivers for every 100 million miles may be approximated by the function

$$f(x) = k(x - 49)^6 + 0.8,$$

where x is the driver's age in years and k is the constant 3.8×10^{-9}. Find and interpret the rate of change of the fatality rate when the driver is

a. 20 years old;   **b.** 60 years old.

Source: National Highway Traffic Safety Administration.

ELECTRIC POTENTIAL AND ELECTRIC FIELD

In physics, a major area of study is electricity, including such concepts as electric charge, electric force, and electric current. Two ideas that physicists use a great deal are electric potential and electric field. Electric potential is the same as voltage, such as for a battery. Electric field can be thought of in terms of a force field, which is often referred to in space movies as a deflector shield around a spaceship. An electric field causes an electric force to act on charged objects when they are in the electric field.

Both electric potential and electric field are produced by electric charges. It is important to physicists and electrical engineers to know what the electric field and the electric potential are near a charged object. Usually the problem involves finding the electric potential and taking the (negative) derivative of the electric potential to determine the electric field. More explicitly,

$$E = -\frac{dV}{dz}, \tag{1}$$

where V is the electric potential (voltage) near the charged object, z is the distance from the object, and E is the electric field.

Let's look at an example. Suppose we have a charged disk of radius R and we want to determine the electric potential and electric field at a distance z along the axis of the disk (Figure 13).

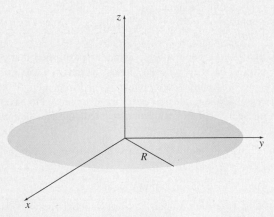

FIGURE 13

*For those interested in the constants k_1 and k_2:

$$k_1 = \frac{Q}{2\pi\varepsilon_0 R^2} \quad \text{and} \quad k_2 = \frac{1}{2}k_1 R^2 = \frac{Q}{4\pi\varepsilon_0},$$

where Q is the charge, ε_0 is called the electric permittivity, and R is the radius of the disk.

Using some basic definitions and integral calculus (that you will learn about later in your calculus book), it can be shown that the electric potential on the axis is

$$V = k_1\left(\sqrt{z^2 + R^2} - z\right) \tag{2}$$

where k_1 is a constant.* To determine the electric field, we apply Equation (1) and find

$$E = -k_1\left(\frac{z}{\sqrt{z^2 + R^2}} - 1\right) \tag{3}$$

You will provide the details for this example by working through Exercise 1 at the end of this section.

Physicists often like to see if complicated expressions such as these can be simplified when certain conditions apply. An example would be to imagine that the location z is very far away from the disk. In that case, the disk doesn't look much like a disk anymore, but more like a point. In Equations (2) and (3), we see that z will be much larger than R. If you reach the topic of Taylor series in calculus, you will learn how these series can be used to approximate square roots, so that for large values of z the voltage is inversely proportional to z and looks like this:

$$V = \frac{k_2}{z}, \tag{4}$$

where k_2 is a constant. To determine the electric field, we apply Equation (1) again to find

$$E = \frac{k_2}{z^2}. \tag{5}$$

You will be asked to prove this result in Exercise 2.

Thus we see that the exact functions for V and E of the disk become much simpler at locations far away from the disk. The voltage is a reciprocal function and the electric field is called an inverse-square law. By the way, these functions are the same ones that would be used for a point charge, which is a charge that takes up very little space.

Now let's look at what happens when we are very close to the surface of the disk. You could imagine that an observer very close to the surface would see the disk as a large flat plane. If we apply the Taylor series once more to the exact function for the potential (Equation (2)), but this time with z much smaller than R, we find

$$V = k_1\left(R - z + \frac{z^2}{2R}\right). \tag{6}$$

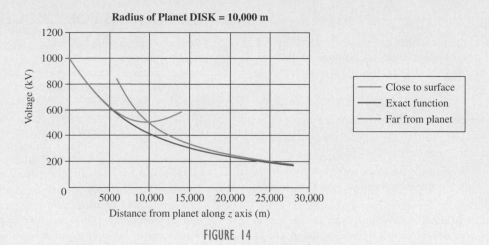

FIGURE 14

Notice that this is a quadratic function, or a parabola. Applying Equation (1) again, we see that the electric field is

$$E = k_1\left(1 - \frac{z}{R}\right). \qquad (7)$$

This is just a linear function that increases as we approach the surface (which would be $z = 0$) and decreases the farther away from the surface we move. Again, the details of this calculation are left for you to do in Exercise 3.

Here is a fun way of making sense of the equations listed above. Suppose you are in a spaceship approaching the planet DISK on a very important mission. Planet DISK is noted for the fact that there is always a sizeable amount of charge on it. You are approaching the planet from very far away along its axis. In Figure 14 the three voltage functions (Equations (2), (4), and (6)) are plotted, and in Figure 15 the three electric field functions (Equations (3), (5), and (7)) are plotted. The graphs were generated using $R = 10,000$, $k_1 = 100$, and $k_2 = 5 \times 10^9$.

Notice that it looks like you can use the reciprocal function for the voltage (Equation (4)) and the inverse-square law function for the electric field (Equation (5)) when you are farther away than about 20,000 m because the exact functions and the approximate functions are almost exactly the same. Also, when you get close to the planet, say less than about 4000 m, you can use the quadratic function for the voltage (Equation (6)) and the linear function for the electric field (Equation (7)) because the exact functions and the approximate functions are nearly the same. This means that you should use the exact functions (Equations (2) and (3)) between about 20,000 m and about 4000 m because the other functions deviate substantially in that region.

There are many other examples that could be studied, but they involve other functions that you haven't covered yet, especially trigonometric functions. But the process is still the same: If you can determine the electric potential in the region around a charged object, then the electric field is found by taking the negative derivative of the electric potential.

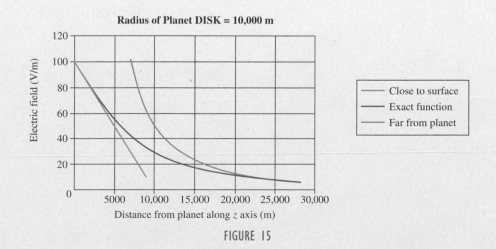

FIGURE 15

EXERCISES

1. Use Equation (1) to prove that the electric field of the disk (Equation (3)) is obtained from the voltage of the disk (Equation (2)). (*Hint:* It may help to write the square root in Equation (2) as a power.)

2. Apply Equation (1) to the voltage of a point charge (Equation (4)) to obtain the electric field of a point charge (Equation (5)).

3. Show that the electric field in Equation (7) results from the electric potential in Equation (6).

4. Sometimes for z very, very close to the disk, the third term in Equation (6) is so small that it can be dismissed. Show that the electric field is constant for this case.

5. Use a graphing calculator or Wolfram|Alpha (which can be found at www.wolframalpha.com) to recreate the graphs of the functions in Figures 14 and 15.

6. Use a spreadsheet to create a table of values for the functions displayed in Figures 14 and 15. Compare the three voltage functions and then compare the three electric field functions.

DIRECTIONS FOR GROUP PROJECT

Determine the electric potential and the electric field at various locations along the axis of a charged compact disc (CD). You will need to measure the radius of a typical CD and use the value for the electric permittivity (sometimes called the permittivity of free space), $\varepsilon_0 = 8.85 \times 10^{-12} C^2/Nm^2$. To estimate the value for the charge on the CD, assume that a typical CD has about one mole of atoms (6.0×10^{23}, which is Avogadro's number) and that one out of every billion of these atoms loses an electron. The charge is found by multiplying 10^{-9} (one billionth) by Avogadro's number and by the charge of one electron (or proton), which is $1.6 \times 10^{-19} C$. With this information you can calculate the constants k_1 and k_2. Use appropriate graphing software, such as Microsoft Excel, to plot all three of the voltage functions (Equations (2), (4), and (6)) on one graph and all three electric field functions (Equations (3), (5), and (7)) on one graph.

5

Graphs and the Derivative

Derivatives provide useful information about the behavior of functions and the shapes of their graphs. The first derivative describes the rate of increase or decrease, while the second derivative indicates how the rate of increase or decrease is changing. In an exercise at the end of this chapter, we will see what changes in the sign of the second derivative tell us about the shape of the graph that shows a weightlifter's performance as a function of age.

The graph in Figure 1 shows the relationship between the number of sleep-related accidents and traffic density during a 24-hour period. The blue line indicates the hourly distribution of sleep-related accidents. The green line indicates the hourly distribution of traffic density. The red line indicates the relative risk of sleep-related accidents. For example, the relative risk graph shows us that a person is nearly seven times as likely to have an accident at 4:00 A.M. than at 10:00 P.M. *Source: Sleep.*

Given a graph like the one in Figure 1, we can often locate maximum and minimum values simply by looking at the graph. It is difficult to get *exact* values or *exact* locations of maxima and minima from a graph, however, and many functions are difficult to graph without the aid of technology. In Chapter 2, we saw how to find exact maximum and minimum values for quadratic functions by identifying the vertex. A more general approach is to use the derivative of a function to determine precise maximum and minimum values of the function. The procedure for doing this is described in this chapter, which begins with a discussion of increasing and decreasing functions.

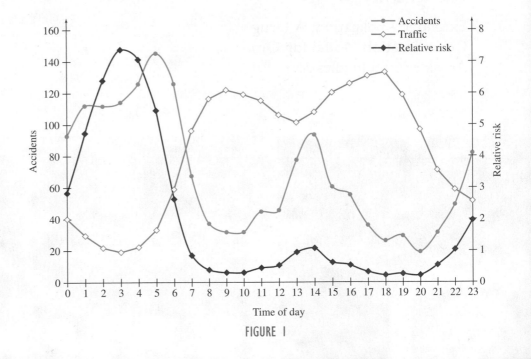

Time of day

FIGURE 1

5.1 Increasing and Decreasing Functions

APPLY IT How long is it profitable to increase production?
We will answer this question in Example 5 after further investigating increasing and decreasing functions.

A function is *increasing* if the graph goes *up* from left to right and *decreasing* if its graph goes *down* from left to right. Examples of increasing functions are shown in Figures 2(a)–(c), and examples of decreasing functions in Figures 2(d)–(f).

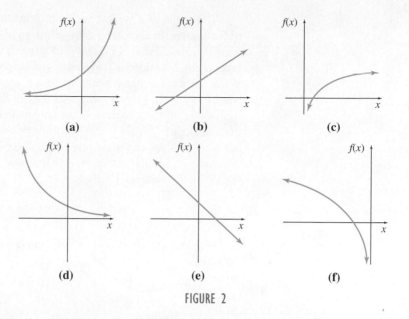

FIGURE 2

Increasing and Decreasing Functions

Let f be a function defined on some interval. Then for any two numbers x_1 and x_2 in the interval, f is **increasing** on the interval if

$$f(x_1) < f(x_2) \quad \text{whenever} \quad x_1 < x_2,$$

and f is **decreasing** on the interval if

$$f(x_1) > f(x_2) \quad \text{whenever} \quad x_1 < x_2.$$

EXAMPLE 1   Increasing and Decreasing

Where is the function graphed in Figure 3 increasing? Where is it decreasing?

YOUR TURN 1 Find where the function is increasing and decreasing.

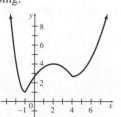

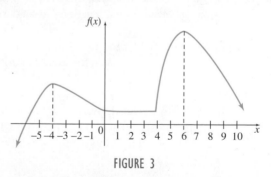

FIGURE 3

SOLUTION Moving from left to right, the function is increasing for x-values up to -4, then decreasing for x-values from -4 to 0, constant (neither increasing nor decreasing) for x-values from 0 to 4, increasing for x-values from 4 to 6, and decreasing for all x-values larger than 6. In interval notation, the function is increasing on $(-\infty, -4)$ and $(4, 6)$, decreasing on $(-4, 0)$ and $(6, \infty)$, and constant on $(0, 4)$.   **TRY YOUR TURN 1**

How can we tell from the equation that defines a function where the graph increases and where it decreases? The derivative can be used to answer this question. Remember that the derivative of a function at a point gives the slope of the line tangent to the function at that point. Recall also that a line with a positive slope rises from left to right and a line with a negative slope falls from left to right.

The graph of a typical function, f, is shown in Figure 4. Think of the graph of f as a roller coaster track moving from left to right along the graph. Now, picture one of the cars on the roller coaster. As shown in Figure 5, when the car is on level ground or parallel to level ground, its floor is horizontal, but as the car moves up the slope, its floor tilts upward. When the car reaches a peak, its floor is again horizontal, but it then begins to tilt downward (very steeply) as the car rolls downhill. The floor of the car as it moves from left to right along the track represents the tangent line at each point. Using this analogy, we can see that the slope of the tangent line will be *positive* when the car travels uphill and f is *increasing*, and the slope of the tangent line will be *negative* when the car travels downhill and f is *decreasing*. (In this case it is also true that the slope of the tangent line will be zero at "peaks" and "valleys.")

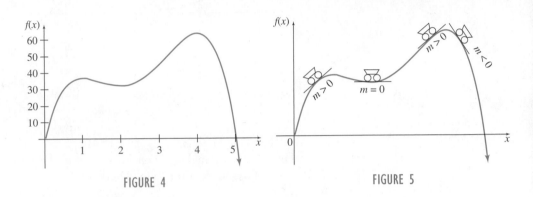

FIGURE 4   FIGURE 5

Thus, on intervals where $f'(x) > 0$, the function $f(x)$ will increase, and on intervals where $f'(x) < 0$, the function $f(x)$ will decrease. We can determine where $f(x)$ peaks by finding the intervals on which it increases and decreases.

Our discussion suggests the following test.

Test for Intervals Where $f(x)$ is Increasing and Decreasing

Suppose a function f has a derivative at each point in an open interval; then

if $f'(x) > 0$ for each x in the interval, f is *increasing* on the interval; ↗

if $f'(x) < 0$ for each x in the interval, f is *decreasing* on the interval; ↘

if $f'(x) = 0$ for each x in the interval, f is *constant* on the interval.   →

NOTE
The third condition must hold for an entire open interval, not a single point. It would not be correct to say that because $f'(x) = 0$ at a point, then $f(x)$ is constant at that point.

The derivative $f'(x)$ can change signs from positive to negative (or negative to positive) at points where $f'(x) = 0$ and at points where $f'(x)$ does not exist. The values of x where this occurs are called *critical numbers*.

Critical Numbers

The **critical numbers** for a function f are those numbers c in the domain of f for which $f'(c) = 0$ or $f'(c)$ does not exist. A **critical point** is a point whose x-coordinate is the critical number c and whose y-coordinate is $f(c)$.

It is shown in more advanced classes that if the critical numbers of a function are used to determine open intervals on a number line, then the sign of the derivative at any point in an interval will be the same as the sign of the derivative at any other point in the interval. This suggests that the test for increasing and decreasing functions be applied as follows (assuming that no open intervals exist where the function is constant).

FOR REVIEW

The method for finding where a function is increasing and decreasing is similar to the method introduced in Section R.5 for solving quadratic inequalities.

Applying the Test

1. Locate the critical numbers for f on a number line, as well as any points where f is undefined. These points determine several open intervals.

2. Choose a value of x in each of the intervals determined in Step 1. Use these values to decide whether $f'(x) > 0$ or $f'(x) < 0$ in that interval.

3. Use the test on the previous page to decide whether f is increasing or decreasing on the interval.

FOR REVIEW

In this chapter you will need all of the rules for derivatives you learned in the previous chapter. If any of these are still unclear, go over the Derivative Summary at the end of that chapter and practice some of the Review Exercises before proceeding.

EXAMPLE 2 Increasing and Decreasing

Find the intervals in which the following function is increasing or decreasing. Locate all points where the tangent line is horizontal. Graph the function.

$$f(x) = x^3 + 3x^2 - 9x + 4$$

SOLUTION Here $f'(x) = 3x^2 + 6x - 9$. To find the critical numbers, set this derivative equal to 0 and solve the resulting equation by factoring.

$$3x^2 + 6x - 9 = 0$$
$$3(x^2 + 2x - 3) = 0$$
$$3(x + 3)(x - 1) = 0$$
$$x = -3 \quad \text{or} \quad x = 1$$

The tangent line is horizontal at $x = -3$ or $x = 1$. Since there are no values of x where $f'(x)$ fails to exist, the only critical numbers are -3 and 1. To determine where the function is increasing or decreasing, locate -3 and 1 on a number line, as in Figure 6. (Be sure to place the values on the number line in numerical order.) These points determine three intervals: $(-\infty, -3)$, $(-3, 1)$, and $(1, \infty)$.

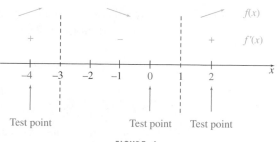

FIGURE 6

Now choose any value of x in the interval $(-\infty, -3)$. Choosing $x = -4$ and evaluating $f'(-4)$ using the factored form of $f'(x)$ gives

$$f'(-4) = 3(-4 + 3)(-4 - 1) = 3(-1)(-5) = 15,$$

which is positive. You could also substitute $x = -4$ in the unfactored form of $f'(x)$, but using the factored form makes it easier to see whether the result is positive or negative, depending upon whether you have an even or an odd number of negative factors. Since one value of x in this interval makes $f'(x) > 0$, all values will do so, and therefore, f is increasing on $(-\infty, -3)$. Selecting 0 from the middle interval gives $f'(0) = -9$, so f is decreasing on $(-3, 1)$. Finally, choosing 2 in the right-hand region gives $f'(2) = 15$, with f increasing on $(1, \infty)$. The arrows in each interval in Figure 6 indicate where f is increasing or decreasing.

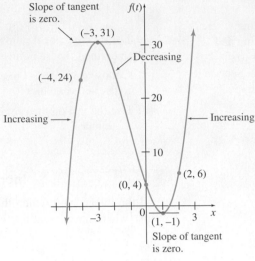

$$f(x) = x^3 + 3x^2 - 9x + 4$$

FIGURE 7

YOUR TURN 2

Find the intervals in which $f(x) = -x^3 - 2x^2 + 15x + 10$ is increasing or decreasing. Graph the function.

We now have an additional tool for graphing functions: the test for determining where a function is increasing or decreasing. (Other tools are discussed in the next few sections.) To graph the function, plot a point at each of the critical numbers by finding $f(-3) = 31$ and $f(1) = -1$. Also plot points for $x = -4$, 0, and 2, the test values of each interval. Use these points along with the information about where the function is increasing and decreasing to get the graph in Figure 7.   **TRY YOUR TURN 2**

CAUTION   Be careful to use $f(x)$, not $f'(x)$, to find the y-value of the points to plot.

Recall critical numbers are numbers c in the domain of f for which $f'(c) = 0$ or $f'(x)$ does not exist. In Example 2, there are no critical values, c, where $f'(c)$ fails to exist. The next example illustrates the case where a function has a critical number at c because the derivative does not exist at c.

EXAMPLE 3   **Increasing and Decreasing**

Find the critical numbers and decide where f is increasing and decreasing if $f(x) = (x - 1)^{2/3}$.

SOLUTION   We find $f'(x)$ first, using the power rule and the chain rule.

$$f'(x) = \frac{2}{3}(x - 1)^{-1/3}(1) = \frac{2}{3(x - 1)^{1/3}}$$

To find the critical numbers, we first find any values of x that make $f'(x) = 0$, but here $f'(x)$ is never 0. Next, we find any values of x where $f'(x)$ fails to exist. This occurs whenever the denominator of $f'(x)$ is 0, so set the denominator equal to 0 and solve.

$$3(x - 1)^{1/3} = 0 \qquad \text{Divide by 3.}$$
$$[(x - 1)^{1/3}]^3 = 0^3 \qquad \text{Raise both sides to the 3rd power.}$$
$$x - 1 = 0$$
$$x = 1$$

Since $f'(1)$ does not exist but $f(1)$ is defined, $x = 1$ is a critical number, the only critical number. This point divides the number line into two intervals: $(-\infty, 1)$ and $(1, \infty)$. Draw a number line for f', and use a test point in each of the intervals to find where f is increasing and decreasing.

$$f'(0) = \frac{2}{3(0-1)^{1/3}} = \frac{2}{-3} = -\frac{2}{3}$$

$$f'(2) = \frac{2}{3(2-1)^{1/3}} = \frac{2}{3}$$

YOUR TURN 3 Find where f is increasing and decreasing if $f(x) = (2x + 4)^{2/5}$. Graph the function.

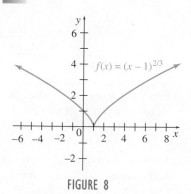

Since f is defined for all x, these results show that f is decreasing on $(-\infty, 1)$ and increasing on $(1, \infty)$. The graph of f is shown in Figure 8. **TRY YOUR TURN 3**

In Example 3, we found a critical number where $f'(x)$ failed to exist. This occurred when the denominator of $f'(x)$ was zero. Be on the alert for such values of x. Also be alert for values of x that would make the expression under a square root, or other even root, negative. For example, if $f(x) = \sqrt{x}$, then $f'(x) = 1/(2\sqrt{x})$. Notice that $f'(x)$ does not exist for $x \leq 0$, but the values of $x < 0$ are not critical numbers because those values of x are not in the domain of f. The function $f(x) = \sqrt{x}$ does have a critical point at $x = 0$, because 0 is in the domain of f.

Sometimes a function may not have any critical numbers, but we are still able to determine where the function is increasing and decreasing, as shown in the next example.

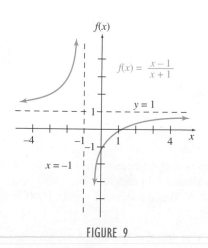

FIGURE 8

EXAMPLE 4 Increasing and Decreasing (No Critical Numbers)

Find the intervals for which the following function increases and decreases. Graph the function.

$$f(x) = \frac{x - 1}{x + 1}$$

SOLUTION Notice that the function f is undefined when $x = -1$, so -1 is not in the domain of f. To determine any critical numbers, first use the quotient rule to find $f'(x)$.

$$f'(x) = \frac{(x + 1)(1) - (x - 1)(1)}{(x + 1)^2}$$

$$= \frac{x + 1 - x + 1}{(x + 1)^2} = \frac{2}{(x + 1)^2}$$

This derivative is never 0, but it fails to exist at $x = -1$, where the function is undefined. Since -1 is not in the domain of f, there are no critical numbers for f.

We can still apply the first derivative test, however, to find where f is increasing and decreasing. The number -1 (where f is undefined) divides the number line into two intervals: $(-\infty, -1)$ and $(-1, \infty)$. Draw a number line for f', and use a test point in each of these intervals to find that $f'(x) > 0$ for all x except -1. (This can also be determined by observing that $f'(x)$ is the quotient of 2, which is positive, and $(x + 1)^2$, which is always positive or 0.) This means that the function f is increasing on both $(-\infty, -1)$ and $(-1, \infty)$.

To graph the function, we find any asymptotes. Since the value $x = -1$ makes the denominator 0 but not the numerator, the line $x = -1$ is a vertical asymptote. To find the horizontal asymptote, we find

YOUR TURN 4 Find where f is increasing and decreasing if

$$f(x) = \frac{-2x}{x + 2}.$$

Graph the function.

$$\lim_{x \to \infty} \frac{x - 1}{x + 1} = \lim_{x \to \infty} \frac{1 - 1/x}{1 + 1/x} \qquad \text{Divide numerator and denominator by } x.$$

$$= 1.$$

We get the same limit as x approaches $-\infty$, so the graph has the line $y = 1$ as a horizontal asymptote. Using this information, as well as the x-intercept $(1, 0)$ and the y-intercept $(0, -1)$, gives the graph in Figure 9. **TRY YOUR TURN 4**

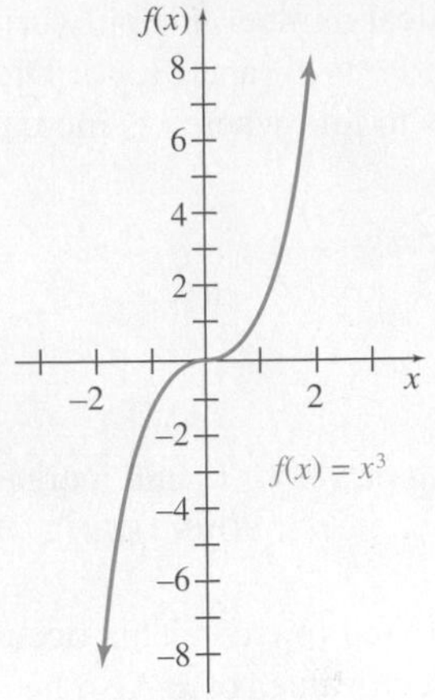

$f(x) = x^3$

FIGURE 10

CAUTION It is important to note that the reverse of the test for increasing and decreasing functions is not true—it is possible for a function to be increasing on an interval even though the derivative is not positive at every point in the interval. A good example is given by $f(x) = x^3$, which is increasing on every interval, even though $f'(x) = 0$ when $x = 0$. See Figure 10.

Similarly, it is incorrect to assume that the sign of the derivative in regions separated by critical numbers must alternate between + and −. If this were always so, it would lead to a simple rule for finding the sign of the derivative: just check one test point, and then make the other regions alternate in sign. But this is not true if one of the factors in the derivative is raised to an even power. In the function $f(x) = x^3$ just considered, $f'(x) = 3x^2$ is positive on both sides of the critical number $x = 0$.

TECHNOLOGY NOTE A graphing calculator can be used to find the derivative of a function at a particular x-value. The screen in Figure 11 supports our results in Example 2 for the test values, −4 and 2. (Notice that the calculator screen does not show the entire command.) The results are not exact because the calculator uses a numerical method to approximate the derivative at the given x-value.

Some graphing calculators can find where a function changes from increasing to decreasing by finding a maximum or minimum. The calculator windows in Figure 12 show this feature for the function in Example 2. Note that these, too, are approximations. This concept will be explored further in the next section.

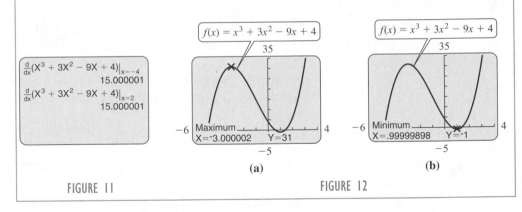

FIGURE 11   FIGURE 12

Knowing the intervals where a function is increasing or decreasing can be important in applications, as shown by the next examples.

EXAMPLE 5   Profit Analysis

A company selling computers finds that the cost per computer decreases linearly with the number sold monthly, decreasing from $1000 when none are sold to $800 when 1000 are sold. Thus, the average cost function has a y-intercept of 1000 and a slope of $-200/1000 = -0.2$, so it is given by the formula

$$\overline{C}(x) = 1000 - 0.2x, \quad 0 \le x \le 1000,$$

where x is the number of computers sold monthly. Since $\overline{C}(x) = C(x)/x$, the cost function is given by

$$C(x) = x\overline{C}(x) = x(1000 - 0.2x)$$
$$= 1000x - 0.2x^2, \quad 0 \le x \le 1000.$$

Suppose the revenue function can be approximated by

$$R(x) = 0.0008x^3 - 2.4x^2 + 2400x, \quad 0 \le x \le 1000.$$

Determine any intervals on which the profit function is increasing.

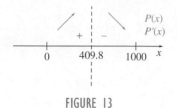

SOLUTION   First find the profit function $P(x)$.

$$P(x) = R(x) - C(x)$$
$$= (0.0008x^3 - 2.4x^2 + 2400x) - (1000x - 0.2x^2)$$
$$= 0.0008x^3 - 2.2x^2 + 1400x$$

To find any intervals where this function is increasing, set $P'(x) = 0$.

$$P'(x) = 0.0024x^2 - 4.4x + 1400 = 0$$

Solving this with the quadratic formula gives the approximate solutions $x = 409.8$ and $x = 1423.6$. The latter number is outside of the domain. Use $x = 409.8$ to determine two intervals on a number line, as shown in Figure 13. Choose $x = 0$ and $x = 1000$ as test points.

$$P'(0) = 0.0024(0^2) - 4.4(0) + 1400 = 1400$$
$$P'(1000) = 0.0024(1000^2) - 4.4(1000) + 1400 = -600$$

This means that when no computers are sold monthly, the profit is going up at a rate of $1400 per computer. When 1000 computers are sold monthly, the profit is going down at a rate of $600 per computer. The test points show that the function increases on $(0, 409.8)$ and decreases on $(409.8, 1000)$. See Figure 13. Thus, the profit is increasing when 409 computers or fewer are sold, and decreasing when 410 or more are sold, as shown in Figure 14.

FIGURE 13

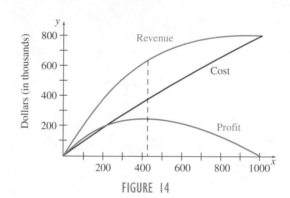

FIGURE 14

As the graph in Figure 14 shows, the profit will increase as long as the revenue function increases faster than the cost function. That is, increasing production will produce more profit as long as the marginal revenue is greater than the marginal cost.

EXAMPLE 6   Recollection of Facts

In the exercises in the previous chapter, the function

$$f(t) = \frac{90t}{99t - 90}$$

gave the number of facts recalled after t hours for $t > 10/11$. Find the intervals in which $f(t)$ is increasing or decreasing.

SOLUTION   First use the quotient rule to find the derivative, $f'(t)$.

$$f'(t) = \frac{(99t - 90)(90) - 90t(99)}{(99t - 90)^2}$$
$$= \frac{8910t - 8100 - 8910t}{(99t - 90)^2} = \frac{-8100}{(99t - 90)^2}$$

Since $(99t - 90)^2$ is positive everywhere in the domain of the function and since the numerator is a negative constant, $f'(t) < 0$ for all t in the domain of $f(t)$. Thus $f(t)$ always decreases and, as expected, the number of words recalled decreases steadily over time.

5.1 EXERCISES

Find the open intervals where the functions graphed as follows are (a) increasing, or (b) decreasing.

1.
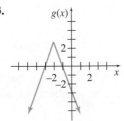

2.
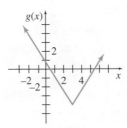

3.
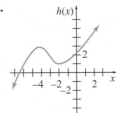

4.
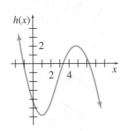

5.
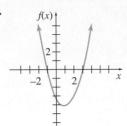

6.
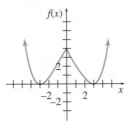

7.
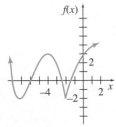

8.
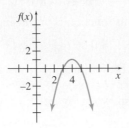

For each of the exercises listed below, suppose that the function that is graphed is not $f(x)$, but $f'(x)$. Find the open intervals where $f(x)$ is (a) increasing or (b) decreasing.

9. Exercise 1

10. Exercise 2

11. Exercise 7

12. Exercise 8

For each function, find (a) the critical numbers; (b) the open intervals where the function is increasing; and (c) the open intervals where it is decreasing.

13. $y = 2.3 + 3.4x - 1.2x^2$

14. $y = 1.1 - 0.3x - 0.3x^2$

15. $f(x) = \frac{2}{3}x^3 - x^2 - 24x - 4$

16. $f(x) = \frac{2}{3}x^3 - x^2 - 4x + 2$

17. $f(x) = 4x^3 - 15x^2 - 72x + 5$

18. $f(x) = 4x^3 - 9x^2 - 30x + 6$

19. $f(x) = x^4 + 4x^3 + 4x^2 + 1$

20. $f(x) = 3x^4 + 8x^3 - 18x^2 + 5$

21. $y = -3x + 6$

22. $y = 6x - 9$

23. $f(x) = \dfrac{x + 2}{x + 1}$

24. $f(x) = \dfrac{x + 3}{x - 4}$

25. $y = \sqrt{x^2 + 1}$

26. $y = x\sqrt{9 - x^2}$

27. $f(x) = x^{2/3}$

28. $f(x) = (x + 1)^{4/5}$

29. $y = x - 4\ln(3x - 9)$

30. $f(x) = \ln\dfrac{5x^2 + 4}{x^2 + 1}$

31. $f(x) = xe^{-3x}$

32. $f(x) = xe^{x^2 - 3x}$

33. $f(x) = x^2 2^{-x}$

34. $f(x) = x2^{-x^2}$

35. $y = x^{2/3} - x^{5/3}$

36. $y = x^{1/3} + x^{4/3}$

37. A friend looks at the graph of $y = x^2$ and observes that if you start at the origin, the graph increases whether you go to the right or the left, so the graph is increasing everywhere. Explain why this reasoning is incorrect.

38. Use the techniques of this chapter to find the vertex and intervals where f is increasing and decreasing, given

$$f(x) = ax^2 + bx + c,$$

where we assume $a > 0$. Verify that this agrees with what we found in Chapter 2.

39. Repeat Exercise 38 under the assumption $a < 0$.

40. Where is the function defined by $f(x) = e^x$ increasing? Decreasing? Where is the tangent line horizontal?

41. Repeat Exercise 40 with the function defined by $f(x) = \ln x$.

42. a. For the function in Exercise 15, find the average of the critical numbers.

b. For the function in Exercise 15, use a graphing calculator to find the roots of the function, and then find the average of those roots.

c. Compare your answers to parts a and b. What do you notice?

d. Repeat part a for the function in Exercise 17.

e. Repeat part b for the function in Exercise 17.

f. Compare your answers to parts d and e. What do you notice?

It can be shown that the average of the roots of a polynomial (including the complex roots, if there are any) and the critical numbers of a polynomial (including complex roots of $f'(x) = 0$, if there are any) are always equal. *Source: The Mathematics Teacher.*

For each of the following functions, use a graphing calculator to find the open intervals where $f(x)$ is (a) increasing, or (b) decreasing.

43. $f(x) = e^{0.001x} - \ln x$

44. $f(x) = \ln(x^2 + 1) - x^{0.3}$

APPLICATIONS

Business and Economics

45. Housing Starts A county realty group estimates that the number of housing starts per year over the next three years will be

$$H(r) = \frac{300}{1 + 0.03r^2},$$

where r is the mortgage rate (in percent).

a. Where is $H(r)$ increasing?

b. Where is $H(r)$ decreasing?

46. Cost Suppose the total cost $C(x)$ (in dollars) to manufacture a quantity x of weed killer (in hundreds of liters) is given by

$$C(x) = x^3 - 2x^2 + 8x + 50.$$

a. Where is $C(x)$ decreasing?

b. Where is $C(x)$ increasing?

47. Profit A manufacturer sells video games with the following cost and revenue functions (in dollars), where x is the number of games sold, for $0 \le x \le 3300$.

$$C(x) = 0.32x^2 - 0.00004x^3$$
$$R(x) = 0.848x^2 - 0.0002x^3$$

Determine the interval(s) on which the profit function is increasing.

48. Profit A manufacturer of CD players has determined that the profit $P(x)$ (in thousands of dollars) is related to the quantity x of CD players produced (in hundreds) per month by

$$P(x) = -(x - 4)e^x - 4, \qquad 0 < x \le 3.9.$$

a. At what production levels is the profit increasing?

b. At what levels is it decreasing?

49. Social Security Assets The projected year-end assets in the Social Security trust funds, in trillions of dollars, where t represents the number of years since 2000, can be approximated by

$$A(t) = 0.0000329t^3 - 0.00450t^2 + 0.0613t + 2.34,$$

where $0 \le t \le 50$. *Source: Social Security Administration.*

a. Where is $A(t)$ increasing?

b. Where is $A(t)$ decreasing?

50. Unemployment The annual unemployment rates of the U.S. civilian noninstitutional population for 1990–2009 are shown in the graph. When is the function increasing? Decreasing? Constant? *Source: Bureau of Labor Statistics.*

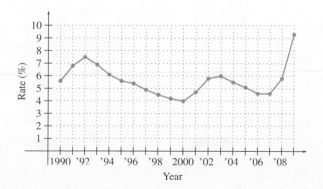

Life Sciences

51. Air Pollution The graph shows the amount of air pollution removed by trees in the Chicago urban region for each month of the year. From the graph we see, for example, that the ozone level starting in May increases up to June, and then abruptly decreases. *Source: National Arbor Day Foundation.*

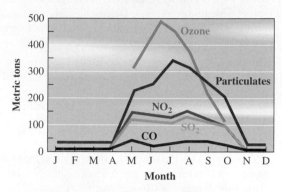

a. Are these curves the graphs of functions?

b. Look at the graph for particulates. Where is the function increasing? Decreasing? Constant?

c. On what intervals do all four lower graphs indicate that the corresponding functions are constant? Why do you think the functions are constant on those intervals?

52. Spread of Infection The number of people $P(t)$ (in hundreds) infected t days after an epidemic begins is approximated by

$$P(t) = \frac{10 \ln(0.19t + 1)}{0.19t + 1}.$$

When will the number of people infected start to decline?

53. Alcohol Concentration In Exercise 55 in the section on Polynomial and Rational Functions, we gave the function defined by

$$A(x) = 0.003631x^3 - 0.03746x^2 + 0.1012x + 0.009$$

as the approximate blood alcohol concentration in a 170-lb woman x hours after drinking 2 oz of alcohol on an empty stomach, for x in the interval [0, 5]. *Source: Medicolegal Aspects of Alcohol Determination in Biological Specimens.*

a. On what time intervals is the alcohol concentration increasing?

b. On what intervals is it decreasing?

54. Drug Concentration The percent of concentration of a drug in the bloodstream x hours after the drug is administered is given by

$$K(x) = \frac{4x}{3x^2 + 27}.$$

a. On what time intervals is the concentration of the drug increasing?

b. On what intervals is it decreasing?

55. Drug Concentration Suppose a certain drug is administered to a patient, with the percent of concentration of the drug in the bloodstream t hours later given by

$$K(t) = \frac{5t}{t^2 + 1}.$$

a. On what time intervals is the concentration of the drug increasing?

b. On what intervals is it decreasing?

56. Cardiology The aortic pressure-diameter relation in a particular patient who underwent cardiac catheterization can be modeled by the polynomial

$$D(p) = 0.000002p^3 - 0.0008p^2 + 0.1141p + 16.683,$$

$$55 \le p \le 130,$$

where $D(p)$ is the aortic diameter (in millimeters) and p is the aortic pressure (in mmHg). Determine where this function is increasing and where it is decreasing within the interval given above. *Source: Circulation*.

57. Thermic Effect of Food The metabolic rate of a person who has just eaten a meal tends to go up and then, after some time has passed, returns to a resting metabolic rate. This phenomenon is known as the thermic effect of food. Researchers have indicated that the thermic effect of food for one particular person is

$$F(t) = -10.28 + 175.9te^{-t/1.3},$$

where $F(t)$ is the thermic effect of food (in kJ/hr) and t is the number of hours that have elapsed since eating a meal. *Source: American Journal of Clinical Nutrition*.

a. Find $F'(t)$.

b. Determine where this function is increasing and where it is decreasing. Interpret your answers.

58. Holstein Dairy Cattle Researchers have developed the following function that can be used to accurately predict the weight of Holstein cows (females) of various ages:

$$W_1(t) = 619(1 - 0.905e^{-0.002t})^{1.2386},$$

where $W_1(t)$ is the weight of the Holstein cow (in kilograms) that is t days old. Where is this function increasing? *Source: Canadian Journal of Animal Science*.

Social Sciences

59. Population The standard normal probability function is used to describe many different populations. Its graph is the well-known normal curve. This function is defined by

$$f(x) = \frac{1}{\sqrt{2\pi}}e^{-x^2/2}.$$

Give the intervals where the function is increasing and decreasing.

60. Nuclear Arsenals The figure shows estimated totals of nuclear weapons inventory for the United States and the former Soviet Union (and its successor states) from 1945 to 2010. *Source: Federation of American Scientists*.

a. On what intervals were the total inventories of both countries increasing?

b. On what intervals were the total inventories of both countries decreasing?

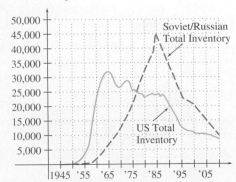

General Interest

61. Sports Cars The following graph shows the horsepower and torque as a function of the engine speed for a 1964 Ford Mustang. *Source: Online with Fuel Line Exhaust*.

a. On what intervals is the power increasing with engine speed?

b. On what intervals is the power decreasing with engine speed?

c. On what intervals is the torque increasing with engine speed?

d. On what intervals is the torque decreasing with engine speed?

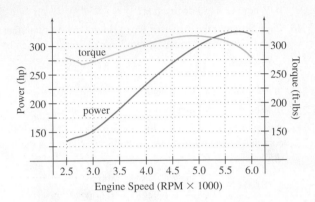

62. Automobile Mileage As a mathematics professor loads more weight in the back of his Subaru, the mileage goes down. Let x be the amount of weight (in pounds) that he adds, and let $y = f(x)$ be the mileage (in mpg).

a. Is $f'(x)$ positive or negative? Explain.

b. What are the units of $f'(x)$?

YOUR TURN ANSWERS

1. Increasing on $(-1, 2)$ and $(4, \infty)$. Decreasing on $(-\infty, -1)$ and $(2, 4)$.

2. Increasing on $(-3, 5/3)$. Decreasing on $(-\infty, -3)$ and $(5/3, \infty)$

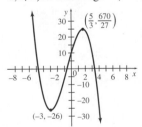

3. Increasing on $(-2, \infty)$ and decreasing on $(-\infty, -2)$.

4. Never increasing. Decreasing on $(-\infty, -2)$ and $(-2, \infty)$.

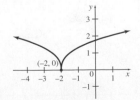

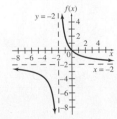

5.2 Relative Extrema

APPLY IT **In a 30-second commercial, when is the best time to present the sales message?**
We will answer this question in Example 1 by investigating the idea of a relative maximum.

As we have seen throughout this text, the graph of a function may have peaks and valleys. It is important in many applications to determine where these points occur. For example, if the function represents the profit of a company, these peaks and valleys indicate maximum profits and losses. When the function is given as an equation, we can use the derivative to determine these points, as shown in the first example.

EXAMPLE 1 Maximizing Viewer's Attention

Suppose that the manufacturer of a diet soft drink is disappointed by sales after airing a new series of 30-second television commercials. The company's market research analysts hypothesize that the problem lies in the timing of the commercial's message, Drink Sparkling Light. Either it comes too early in the commercial, before the viewer has become involved; or it comes too late, after the viewer's attention has faded. After extensive experimentation, the research group finds that the percent of full attention that a viewer devotes to a commercial is a function of time (in seconds) since the commercial began, where

$$\text{Viewer's attention} = f(t) = -\frac{3}{20}t^2 + 6t + 20, \quad 0 \le t \le 30.$$

When is the best time to present the commercial's sales message?

APPLY IT   **SOLUTION**   Clearly, the message should be delivered when the viewer's attention is at a maximum. To find this time, find $f'(t)$.

$$f'(t) = -\frac{3}{10}t + 6 = -0.3t + 6$$

The derivative $f'(t)$ is greater than 0 when

$$-0.3t + 6 > 0,$$
$$-3t > -60, \text{ or}$$
$$t < 20.$$

Similarly, $f'(t) < 0$ when $-0.3t + 6 < 0$, or $t > 20$. Thus, attention increases for the first 20 seconds and decreases for the last 10 seconds. The message should appear about 20 seconds into the commercial. At that time the viewer will devote $f(20) = 80\%$ of his attention to the commercial.

The maximum level of viewer attention (80%) in Example 1 is a *relative maximum*, defined as follows.

Relative Maximum or Minimum

Let c be a number in the domain of a function f. Then $f(c)$ is a **relative** (or **local**) **maximum** for f if there exists an open interval (a, b) containing c such that

$$f(x) \le f(c)$$

for all x in (a, b).

Likewise, $f(c)$ is a **relative** (or **local**) **minimum** for f if there exists an open interval (a, b) containing c such that

$$f(x) \ge f(c)$$

for all x in (a, b).

(continued)

A function has a **relative** (or **local**) **extremum** (plural: **extrema**) at c if it has either a relative maximum or a relative minimum there.

If c is an endpoint of the domain of f, we only consider x in the half-open interval that is in the domain.*

The intuitive idea is that a relative maximum is the greatest value of the function in some region right around the point, although there may be greater values elsewhere. For example, the highest value of the Dow Jones industrial average this week is a relative maximum, although the Dow may have reached a higher value earlier this year. Similarly, a relative minimum is the least value of a function in some region around the point.

A simple way to view these concepts is that a relative maximum is a peak, and a relative minimum is the bottom of a valley, although either a relative minimum or maximum can also occur at the endpoint of the domain.

NOTE
Recall from Section 2.3 on Polynomials and Rational Functions that a relative extremum that is not an endpoint is also referred to as a turning point.

EXAMPLE 2 Relative Extrema

Identify the x-values of all points where the graph in Figure 15 has relative extrema.

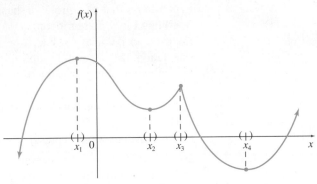

FIGURE 15

YOUR TURN 1 Identify the x-values of all points where the graph has relative extrema.

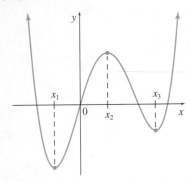

SOLUTION The parentheses around x_1 show an open interval containing x_1 such that $f(x) \leq f(x_1)$, so there is a relative maximum of $f(x_1)$ at $x = x_1$. Notice that many other open intervals would work just as well. Similar intervals around x_2, x_3, and x_4 can be used to find a relative maximum of $f(x_3)$ at $x = x_3$ and relative minima of $f(x_2)$ at $x = x_2$ and $f(x_4)$ at $x = x_4$.   **TRY YOUR TURN 1**

The function graphed in Figure 16 has relative maxima when $x = x_1$ or $x = x_3$ and relative minima when $x = x_2$ or $x = x_4$. The tangent lines at the points having x-values x_1 and

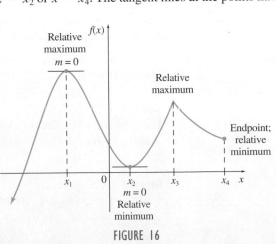

FIGURE 16

*There is disagreement on calling an endpoint a maximum or minimum. We define it this way because this is an applied calculus book, and in an application it would be considered a maximum or minimum value of the function.

x_2 are shown in the figure. Both tangent lines are horizontal and have slope 0. There is no single tangent line at the point where $x = x_3$.

Since the derivative of a function gives the slope of a line tangent to the graph of the function, to find relative extrema we first identify all critical numbers and endpoints. A relative extremum *may* exist at a critical number. (A rough sketch of the graph of the function near a critical number often is enough to tell whether an extremum has been found.) These facts about extrema are summarized below.

If a function f has a relative extremum at c, then c is a critical number or c is an endpoint of the domain.

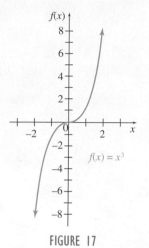

$f(x) = x^3$

FIGURE 17

| **CAUTION** | Be very careful not to get this result backward. It does *not* say that a function has relative extrema at all critical numbers of the function. For example, Figure 17 shows the graph of $f(x) = x^3$. The derivative, $f'(x) = 3x^2$, is 0 when $x = 0$, so that 0 is a critical number for that function. However, as suggested by the graph of Figure 17, $f(x) = x^3$ has neither a relative maximum nor a relative minimum at $x = 0$ (or anywhere else, for that matter). A critical number is a candidate for the location of a relative extremum, but only a candidate.

First Derivative Test

Suppose all critical numbers have been found for some function f. How is it possible to tell from the equation of the function whether these critical numbers produce relative maxima, relative minima, or neither? One way is suggested by the graph in Figure 18.

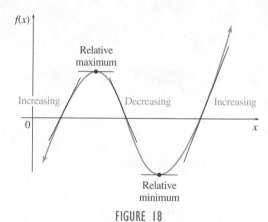

FIGURE 18

As shown in Figure 18, on the left of a relative maximum the tangent lines to the graph of a function have positive slopes, indicating that the function is increasing. At the relative maximum, the tangent line is horizontal. On the right of the relative maximum the tangent lines have negative slopes, indicating that the function is decreasing. Around a relative minimum the opposite occurs. As shown by the tangent lines in Figure 18, the function is decreasing on the left of the relative minimum, has a horizontal tangent at the minimum, and is increasing on the right of the minimum.

Putting this together with the methods from Section 1 for identifying intervals where a function is increasing or decreasing gives the following **first derivative test** for locating relative extrema.

First Derivative Test

Let c be a critical number for a function f. Suppose that f is continuous on (a, b) and differentiable on (a, b) except possibly at c, and that c is the only critical number for f in (a, b).

1. $f(c)$ is a relative maximum of f if the derivative $f'(x)$ is positive in the interval (a, c) and negative in the interval (c, b).

2. $f(c)$ is a relative minimum of f if the derivative $f'(x)$ is negative in the interval (a, c) and positive in the interval (c, b).

The sketches in the following table show how the first derivative test works. Assume the same conditions on a, b, and c for the table as those given for the first derivative test.

Relative Extrema				
$f(x)$ has:	**Sign of f' in (a, c)**	**Sign of f' in (c, b)**	**Sketches**	
Relative maximum	$+$	$-$		
Relative minimum	$-$	$+$		
No relative extrema	$+$	$+$		
No relative extrema	$-$	$-$		

EXAMPLE 3 Relative Extrema

Find all relative extrema for the following functions, as well as where each function is increasing and decreasing.

(a) $f(x) = 2x^3 - 3x^2 - 72x + 15$

Method 1
First Derivative Test

SOLUTION

The derivative is $f'(x) = 6x^2 - 6x - 72$. There are no points where $f'(x)$ fails to exist, so the only critical numbers will be found where the derivative equals 0. Setting the derivative equal to 0 gives

$$6x^2 - 6x - 72 = 0$$
$$6(x^2 - x - 12) = 0$$
$$6(x - 4)(x + 3) = 0$$
$$x - 4 = 0 \quad \text{or} \quad x + 3 = 0$$
$$x = 4 \quad \text{or} \quad x = -3.$$

As in the previous section, the critical numbers 4 and -3 are used to determine the three intervals $(-\infty, -3)$, $(-3, 4)$, and $(4, \infty)$ shown on the number line in Figure 19.

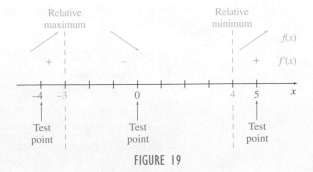

FIGURE 19

Any number from each of the three intervals can be used as a test point to find the sign of f' in each interval. Using $-4, 0,$ and 5 gives the following information.

$$f'(-4) = 6(-8)(-1) > 0$$
$$f'(0) = 6(-4)(3) < 0$$
$$f'(5) = 6(1)(8) > 0$$

YOUR TURN 2 Find all relative extrema of $f(x) = -x^3 - 2x^2 + 15x + 10$.

Thus, the derivative is positive on $(-\infty, -3)$, negative on $(-3, 4)$, and positive on $(4, \infty)$. By Part 1 of the first derivative test, this means that the function has a relative maximum of $f(-3) = 150$ when $x = -3$; by Part 2, f has a relative minimum of $f(4) = -193$ when $x = 4$. The function is increasing on $(-\infty, -3)$ and $(4, \infty)$ and decreasing on $(-3, 4)$. The graph is shown in Figure 20.

TRY YOUR TURN 2

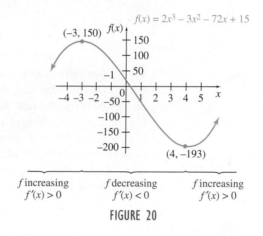

FIGURE 20

Method 2
Graphing Calculator

Many graphing calculators can locate a relative extremum when supplied with an interval containing the extremum. For example, after graphing the function $f(x) = 2x^3 - 3x^2 - 72x + 15$ on a TI-84 Plus, we selected "maximum" from the CALC menu and entered a left bound of -4 and a right bound of 0. The calculator asks for an initial guess, but in this example it doesn't matter what we enter. The result of this process, as well as a similar process for finding the relative minimum, is shown in Figure 21.

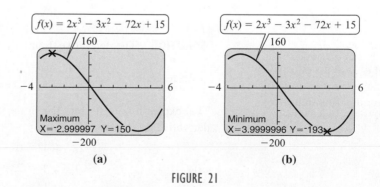

(a)   (b)

FIGURE 21

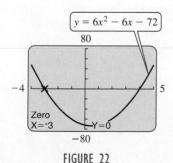

FIGURE 22

Another way to verify the extrema with a graphing calculator is to graph $y = f'(x)$ and find where the graph crosses the x-axis. Figure 22 shows the result of this approach for finding the relative minimum of the previous function.

(b) $f(x) = 6x^{2/3} - 4x$

SOLUTION Find $f'(x)$.

$$f'(x) = 4x^{-1/3} - 4 = \frac{4}{x^{1/3}} - 4$$

The derivative fails to exist when $x = 0$, but the function itself is defined when $x = 0$, making 0 a critical number for f. To find other critical numbers, set $f'(x) = 0$.

$$f'(x) = 0$$

$$\frac{4}{x^{1/3}} - 4 = 0$$

$$\frac{4}{x^{1/3}} = 4$$

$$4 = 4x^{1/3} \qquad \text{Multiply both sides by } x^{1/3}.$$

$$1 = x^{1/3} \qquad \text{Divide both sides by 4.}$$

$$1 = x \qquad \text{Cube both sides.}$$

The critical numbers 0 and 1 are used to locate the intervals $(-\infty, 0)$, $(0, 1)$, and $(1, \infty)$ on a number line as in Figure 23. Evaluating $f'(x)$ at the test points $-1, 1/2$, and 2 and using the first derivative test shows that f has a relative maximum at $x = 1$; the value of this relative maximum is $f(1) = 2$. Also, f has a relative minimum at $x = 0$; this relative minimum is $f(0) = 0$. The function is increasing on $(0, 1)$ and decreasing on $(-\infty, 0)$ and $(1, \infty)$. Notice that the graph, shown in Figure 24, has a sharp point at the critical number where the derivative does not exist. In the last section of this chapter we will show how to verify other features of the graph.   **TRY YOUR TURN 3**

YOUR TURN 3 Find all relative extrema of $f(x) = x^{2/3} - x^{5/3}$.

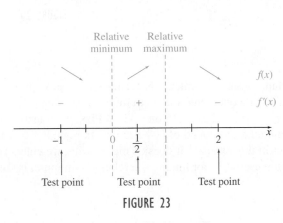

FIGURE 23

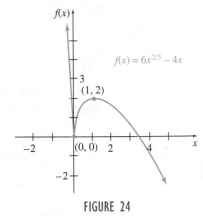

FIGURE 24

(c) $f(x) = xe^{2-x^2}$

SOLUTION The derivative, found by using the product rule and the chain rule, is

$$f'(x) = x(-2x)e^{2-x^2} + e^{2-x^2}$$
$$= e^{2-x^2}(-2x^2 + 1).$$

This expression exists for all x in the domain of f. Since e^{2-x^2} is always positive, the derivative is 0 when

$$-2x^2 + 1 = 0$$

$$1 = 2x^2$$

$$\frac{1}{2} = x^2$$

$$x = \pm\sqrt{1/2}$$

$$x = \pm\frac{1}{\sqrt{2}} \approx \pm 0.707.$$

FOR REVIEW

Recall that $e^x > 0$ for all x, so there can never be a solution to $e^{g(x)} = 0$ for any function $g(x)$.

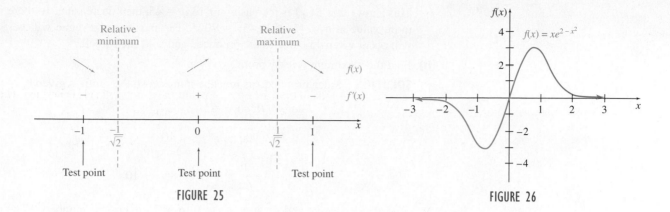

FIGURE 25

FIGURE 26

YOUR TURN 4 Find all relative extrema of $f(x) = x^2 e^x$.

There are two critical points, $-1/\sqrt{2}$ and $1/\sqrt{2}$. Using test points of $-1, 0,$ and 1 gives the results shown in Figure 25.

The function has a relative minimum at $-1/\sqrt{2}$ of $f(-1/\sqrt{2}) \approx -3.17$ and a relative maximum at $1/\sqrt{2}$ of $f(1/\sqrt{2}) \approx 3.17$. It is decreasing on the interval $(-\infty, -1/\sqrt{2})$, increasing on the interval $(-1/\sqrt{2}, 1/\sqrt{2})$, and decreasing on the interval $(1/\sqrt{2}, \infty)$. The graph is shown in Figure 26. **TRY YOUR TURN 4**

> **CAUTION** A critical number must be in the domain of the function. For example, the derivative of $f(x) = x/(x-4)$ is $f'(x) = -4/(x-4)^2$, which fails to exist when $x = 4$. But $f(4)$ does not exist, so 4 is not a critical number, and the function has no relative extrema.

As mentioned at the beginning of this section, finding the maximum or minimum value of a quantity is important in applications of mathematics. The final example gives a further illustration.

EXAMPLE 4 Bicycle Sales

A small company manufactures and sells bicycles. The production manager has determined that the cost and demand functions for q ($q \geq 0$) bicycles per week are

$$C(q) = 10 + 5q + \frac{1}{60}q^3 \qquad \text{and} \qquad p = D(q) = 90 - q,$$

where p is the price per bicycle.

(a) Find the maximum weekly revenue.

SOLUTION The revenue each week is given by

$$R(q) = qp = q(90 - q) = 90q - q^2.$$

To maximize $R(q) = 90q - q^2$, find $R'(q)$. Then find the critical numbers.

$$R'(q) = 90 - 2q = 0$$
$$90 = 2q$$
$$q = 45$$

Since $R'(q)$ exists for all q, 45 is the only critical number. To verify that $q = 45$ will produce a *maximum*, evaluate the derivative on both sides of $q = 45$.

$$R'(40) = 10 \qquad \text{and} \qquad R'(50) = -10$$

This shows that $R(q)$ is increasing up to $q = 45$, then decreasing, so there is a maximum value at $q = 45$ of $R(45) = 2025$. The maximum revenue will be $2025 and will occur when 45 bicycles are produced and sold each week.

(b) Find the maximum weekly profit.

SOLUTION Since profit equals revenue minus cost, the profit is given by

$$P(q) = R(q) - C(q)$$

$$= (90q - q^2) - \left(10 + 5q + \frac{1}{60}q^3\right)$$

$$= -\frac{1}{60}q^3 - q^2 + 85q - 10.$$

Find the derivative and set it equal to 0 to find the critical numbers. (The derivative exists for all q.)

$$P'(q) = -\frac{1}{20}q^2 - 2q + 85 = 0$$

Solving this equation by the quadratic formula gives the solutions $q \approx 25.8$ and $q \approx -65.8$. Since q cannot be negative, the only critical number of concern is 25.8. Determine whether $q = 25.8$ produces a maximum by testing a value on either side of 25.8 in $P'(q)$.

$$P'(0) = 85 \qquad \text{and} \qquad P'(40) = -75$$

These results show that $P(q)$ increases to $q = 25.8$ and then decreases. Since q must be an integer, evaluate $P(q)$ at $q = 25$ and $q = 26$. Since $P(25) = 1229.58$ and $P(26) = 1231.07$, the maximum value occurs when $q = 26$. Thus, the maximum profit of $1231.07 occurs when 26 bicycles are produced and sold each week. Notice that this is not the same as the number that should be produced to yield maximum revenue.

(c) Find the price the company should charge to realize maximum profit.

SOLUTION As shown in part (b), 26 bicycles per week should be produced and sold to get the maximum profit of $1231.07 per week. Since the price is given by

$$p = 90 - q,$$

if $q = 26$, then $p = 64$. The manager should charge $64 per bicycle and produce and sell 26 bicycles per week to get the maximum profit of $1231.07 per week. Figure 27 shows the graphs of the functions used in this example. Notice that the slopes of the revenue and cost functions are the same at the point where the maximum profit occurs. Why is this true? **TRY YOUR TURN 5** ▮▮

YOUR TURN 5 Find the maximum weekly profit and the price a company should charge to realize maximum profit if $C(q) = 100 + 10q$ and $p = D(q) = 50 - 2q$.

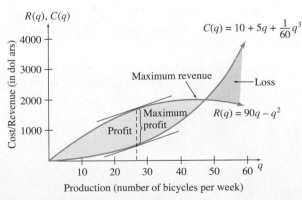

FIGURE 27

> | CAUTION | Be careful to give the y-value of the point where an extremum occurs. Although we solve the equation $f'(x) = 0$ for x to find the extremum, the maximum or minimum value of the function is the corresponding y-value. Thus, in Example 4(a), we found that at $q = 45$, the maximum weekly revenue is $2025 (not $45).

The examples in this section involving the maximization of a quadratic function, such as the advertising example and the bicycle revenue example, could be solved by the methods described in Chapter 2 on Nonlinear Functions. But those involving more complicated functions, such as the bicycle profit example, are difficult to analyze without the tools of calculus.

Finding extrema for realistic problems requires an accurate mathematical model of the problem. In particular, it is important to be aware of restrictions on the values of the variables. For example, if $T(x)$ closely approximates the number of items that can be manufactured daily on a production line when x is the number of employees on the line, x must certainly be restricted to the positive integers or perhaps to a few common fractional values. (We can imagine half-time workers, but not 1/49-time workers.)

On the other hand, to apply the tools of calculus to obtain an extremum for some function, the function must be defined and be meaningful at every real number in some interval. Because of this, the answer obtained from a mathematical model might be a number that is not feasible in the actual problem.

Usually, the requirement that a continuous function be used, rather than one that can take on only certain selected values, is of theoretical interest only. In most cases, the methods of calculus give acceptable results as long as the assumptions of continuity and differentiability are not totally unreasonable. If they lead to the conclusion, say, that $80\sqrt{2}$ workers should be hired, it is usually only necessary to investigate acceptable values close to $80\sqrt{2}$. This was done in Example 4.

5.2 EXERCISES

Find the locations and values of all relative extrema for the functions with graphs as follows. Compare with Exercises 1–8 in the preceding section.

1.
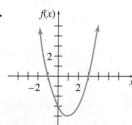

2.
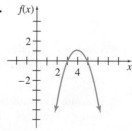

3.
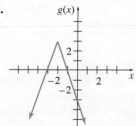

4.
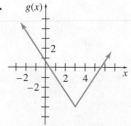

5.
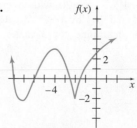

6.
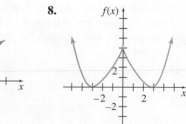

7.
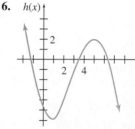

8.

For each of the exercises listed below, suppose that the function that is graphed is not $f(x)$ but $f'(x)$. Find the locations of all relative extrema, and tell whether each extremum is a relative maximum or minimum.

9. Exercise 1   **10.** Exercise 2

11. Exercise 7   **12.** Exercise 8

Find the x-value of all points where the functions defined as follows have any relative extrema. Find the value(s) of any relative extrema.

13. $f(x) = x^2 - 10x + 33$ **14.** $f(x) = x^2 + 8x + 5$

15. $f(x) = x^3 + 6x^2 + 9x - 8$

16. $f(x) = x^3 + 3x^2 - 24x + 2$

17. $f(x) = -\dfrac{4}{3}x^3 - \dfrac{21}{2}x^2 - 5x + 8$

18. $f(x) = -\dfrac{2}{3}x^3 - \dfrac{1}{2}x^2 + 3x - 4$

19. $f(x) = x^4 - 18x^2 - 4$ **20.** $f(x) = x^4 - 8x^2 + 9$

21. $f(x) = 3 - (8 + 3x)^{2/3}$ **22.** $f(x) = \dfrac{(5 - 9x)^{2/3}}{7} + 1$

23. $f(x) = 2x + 3x^{2/3}$ **24.** $f(x) = 3x^{5/3} - 15x^{2/3}$

25. $f(x) = x - \dfrac{1}{x}$ **26.** $f(x) = x^2 + \dfrac{1}{x}$

27. $f(x) = \dfrac{x^2 - 2x + 1}{x - 3}$ **28.** $f(x) = \dfrac{x^2 - 6x + 9}{x + 2}$

29. $f(x) = x^2 e^x - 3$ **30.** $f(x) = 3xe^x + 2$

31. $f(x) = 2x + \ln x$ **32.** $f(x) = \dfrac{x^2}{\ln x}$

33. $f(x) = \dfrac{2^x}{x}$ **34.** $f(x) = x + 8^{-x}$

Use the derivative to find the vertex of each parabola.

35. $y = -2x^2 + 12x - 5$ **36.** $y = ax^2 + bx + c$

Graph each function on a graphing calculator, and then use the graph to find all relative extrema (to three decimal places). Then confirm your answer by finding the derivative and using the calculator to solve the equation $f'(x) = 0$.

37. $f(x) = x^5 - x^4 + 4x^3 - 30x^2 + 5x + 6$

38. $f(x) = -x^5 - x^4 + 2x^3 - 25x^2 + 9x + 12$

39. Graph $f(x) = 2|x + 1| + 4|x - 5| - 20$ with a graphing calculator in the window $[-10, 10]$ by $[-15, 30]$. Use the graph and the function to determine the x-values of all extrema.

40. Consider the function

$$g(x) = \frac{1}{x^{12}} - 2\left(\frac{1000}{x}\right)^6.$$

Source: Mathematics Teacher.

a. Using a graphing calculator, try to find any local minima, or tell why finding a local minimum is difficult for this function.

b. Find any local minima using the techniques of calculus.

c. Based on your results in parts a and b, describe circumstances under which relative extrema are easier to find using the techniques of calculus than using a graphing calculator.

APPLICATIONS

Business and Economics

Profit In Exercises 41–44, find (a) the number, q, of units that produces maximum profit; (b) the price, p, per unit that produces maximum profit; and (c) the maximum profit, P.

41. $C(q) = 80 + 18q; \quad p = 70 - 2q$

42. $C(q) = 25q + 5000; \quad p = 90 - 0.02q$

43. $C(q) = 100 + 20qe^{-0.01q}; \quad p = 40e^{-0.01q}$

44. $C(q) = 21.047q + 3; \quad p = 50 - 5\ln(q + 10)$

45. Power On August 8, 2007, the power used in New York state (in thousands of megawatts) could be approximated by the function

$$P(t) = -0.005846t^3 + 0.1614t^2 - 0.4910t + 20.47,$$

where t is the number of hours since midnight, for $0 \le t \le 24$. Find any relative extrema for power usage, as well as when they occurred. *Source: Current Energy.*

46. Profit The total profit $P(x)$ (in thousands of dollars) from the sale of x units of a certain prescription drug is given by

$$P(x) = \ln(-x^3 + 3x^2 + 72x + 1)$$

for x in $[0, 10]$.

a. Find the number of units that should be sold in order to maximize the total profit.

b. What is the maximum profit?

47. Revenue The demand equation for telephones at one store is

$$p = D(q) = 200e^{-0.1q},$$

where p is the price (in dollars) and q is the quantity of telephones sold per week. Find the values of q and p that maximize revenue.

48. Revenue The demand equation for one type of computer networking system is

$$p = D(q) = 500qe^{-0.0016q^2},$$

where p is the price (in dollars) and q is the quantity of servers sold per month. Find the values of q and p that maximize revenue.

49. Cost Suppose that the cost function for a product is given by $C(x) = 0.002x^3 + 9x + 6912$. Find the production level (i.e., value of x) that will produce the minimum average cost per unit $\overline{C}(x)$.

50. Unemployment The annual unemployment rates of the U.S. civilian noninstitutional population for 1990–2009 are shown in the graph. Identify the years where relative extrema occur, and estimate the unemployment rate at each of these years. *Source: Bureau of Labor Statistics.*

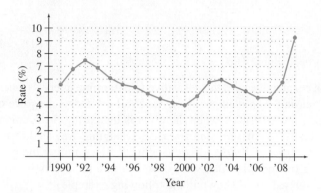

Year

Life Sciences

51. Activity Level In the summer the activity level of a certain type of lizard varies according to the time of day. A biologist has determined that the activity level is given by the function

$$a(t) = 0.008t^3 - 0.288t^2 + 2.304t + 7,$$

where t is the number of hours after 12 noon. When is the activity level highest? When is it lowest?

52. Milk Consumption The average individual daily milk consumption for herds of Charolais, Angus, and Hereford calves can be described by the function

$$M(t) = 6.281t^{0.242}e^{-0.025t}, \qquad 1 \le t \le 26,$$

where $M(t)$ is the milk consumption (in kilograms) and t is the age of the calf (in weeks). *Source: Animal Production.*

a. Find the time in which the maximum daily consumption occurs and the maximum daily consumption.

b. If the general formula for this model is given by

$$M(t) = at^b e^{-ct},$$

find the time where the maximum consumption occurs and the maximum consumption. (*Hint:* Express your answer in terms of a, b, and c.)

53. Alaskan Moose The mathematical relationship between the age of a captive female moose and its mass can be described by the function

$$M(t) = 369(0.93)^t t^{0.36}, \quad t \le 12,$$

where $M(t)$ is the mass of the moose (in kilograms) and t is the age (in years) of the moose. Find the age at which the mass of a female moose is maximized. What is the maximum mass? *Source: Journal of Wildlife Management.*

54. Thermic Effect of Food As we saw in the last section, the metabolic rate after a person eats a meal tends to go up and then, after some time has passed, returns to a resting metabolic rate. This phenomenon is known as the thermic effect of food and can be described for a particular individual as

$$F(t) = -10.28 + 175.9te^{-t/1.3},$$

where $F(t)$ is the thermic effect of food (in kJ/hr), and t is the number of hours that have elapsed since eating a meal. Find the time after the meal when the thermic effect of the food is maximized. *Source: American Journal of Clinical Nutrition.*

Social Sciences

55. Attitude Change Social psychologists have found that as the discrepancy between the views of a speaker and those of an audience increases, the attitude change in the audience also increases to a point but decreases when the discrepancy becomes too large, particularly if the communicator is viewed by the audience as having low credibility. Suppose that the degree of change can be approximated by the function

$$D(x) = -x^4 + 8x^3 + 80x^2,$$

where x is the discrepancy between the views of the speaker and those of the audience, as measured by scores on a questionnaire. Find the amount of discrepancy the speaker should aim for to maximize the attitude change in the audience. *Source: Journal of Personality and Social Psychology.*

56. Film Length A group of researchers found that people prefer training films of moderate length; shorter films contain too little information, while longer films are boring. For a training film on the care of exotic birds, the researchers determined that the ratings people gave for the film could be approximated by

$$R(t) = \frac{20t}{t^2 + 100},$$

where t is the length of the film (in minutes). Find the film length that received the highest rating.

Physical Sciences

57. Height After a great deal of experimentation, two Atlantic Institute of Technology senior physics majors determined that when a bottle of French champagne is shaken several times, held upright, and uncorked, its cork travels according to

$$s(t) = -16t^2 + 40t + 3,$$

where s is its height (in feet) above the ground t seconds after being released.

a. How high will it go?

b. How long is it in the air?

▬▬ YOUR TURN ANSWERS

1. Relative maximum of $f(x_2)$ at $x = x_2$; relative minima of $f(x_1)$ at $x = x_1$ and $f(x_3)$ at $x = x_3$.

2. Relative maximum of $f(5/3) = 670/27 \approx 24.8$ at $x = 5/3$ and relative minimum of $f(-3) = -26$ at $x = -3$.

3. Relative maximum of $f\left(\dfrac{2}{5}\right) = \dfrac{3}{5}\left(\dfrac{2}{5}\right)^{2/3} \approx 0.3257$ at $x = 2/5$ and relative minimum of $f(0) = 0$ at $x = 0$.

4. Relative maximum of $f(-2) = 4e^{-2} \approx 0.5413$ at $x = -2$ and relative minimum of $f(0) = 0$ at $x = 0$.

5. Maximum weekly profit is \$100 when $q = 10$ and the company should charge \$30 per item.

5.3 Higher Derivatives, Concavity, and the Second Derivative Test

APPLY IT Just because the price of a stock is increasing, does that alone make it a good investment?
We will address this question in Example 1.

In the first section of this chapter, we used the derivative to determine intervals where a function is increasing or decreasing. For example, if the function represents the price of a stock, we can use the derivative to determine when the price is increasing. In addition, it would be important for us to know how the *rate of increase* is changing. We can determine how the *rate of increase* (or the *rate of decrease*) is changing by determining the rate of change of the derivative of the function. In other words, we can find the derivative of the derivative, called the **second derivative**, as shown in the following example.

EXAMPLE 1   Stock Prices

Suppose a friend is trying to get you to invest in the stock of a young company. The following function represents the price $P(t)$ of the company's stock since it became available two years ago:

$$P(t) = 17 + t^{1/2},$$

where t is the number of months since the stock became available. He claims that the price of the stock is always increasing and that you will make a fortune on it. Verify his claims. Is the price of the stock increasing? How fast? How much will you make if you invest now?

APPLY IT **SOLUTION**   The derivative of $P(t)$,

$$P'(t) = \frac{1}{2} t^{-1/2} = \frac{1}{2\sqrt{t}},$$

is always positive because $\sqrt{t}$ is positive for $t > 0$. This means that the price function $P(t)$ is always increasing. But *how fast* is it increasing?

The derivative $P'(t) = 1/(2\sqrt{t})$ tells how fast the price is increasing at any number of months, t, since the stock became available. For example, when $t = 1$ month, $P'(t) = 1/2$, and the price is increasing at 1/2 dollar, or 50 cents, per month. When $t = 4$ months, $P'(t) = 1/4$; the stock is increasing at 25 cents per month. By the time you buy in at $t = 24$ months, the price is increasing at 10 cents per month, and the *rate of increase* looks as though it will continue to decrease.

In general, the rate of increase in P' is given by the derivative of $P'(t)$, called the second derivative and denoted by $P''(t)$. Since $P'(t) = (1/2)t^{-1/2}$,

$$P''(t) = -\frac{1}{4} t^{-3/2} = -\frac{1}{4\sqrt{t^3}}.$$

$P''(t)$ is negative for $t > 0$ and, therefore, confirms the suspicion that the *rate* of increase in price does indeed decrease for all $t > 0$. The price of the company's stock will not drop, but the amount of return will certainly not be the fortune your friend predicts.

If you invest now, at $t = 24$ months, the price would be $21.90. A year later, it would be worth $23 a share. If you were rich enough to buy 100 shares for $21.90 each, the total investment would be worth $2300 in a year. The increase of $110 is about 5% of the investment. The only investors to make a lot of money on this stock would be those who bought early, when the rate of increase was much greater.

As mentioned earlier, the second derivative of a function f, written f'', gives the rate of change of the *derivative* of f. Before continuing to discuss applications of the second derivative, we need to introduce some additional terminology and notation.

Higher Derivatives

If a function f has a derivative f', then the derivative of f', if it exists, is the second derivative of f, written f''. The derivative of f'', if it exists, is called the **third derivative** of f, and so on. By continuing this process, we can find **fourth derivatives** and other higher derivatives. For example, if $f(x) = x^4 + 2x^3 + 3x^2 - 5x + 7$, then

$$f'(x) = 4x^3 + 6x^2 + 6x - 5, \qquad \text{First derivative of } f$$
$$f''(x) = 12x^2 + 12x + 6, \qquad \text{Second derivative of } f$$
$$f'''(x) = 24x + 12, \qquad \text{Third derivative of } f$$

and

$$f^{(4)}(x) = 24. \qquad \text{Fourth derivative of } f$$

Notation for Higher Derivatives

The second derivative of $y = f(x)$ can be written using any of the following notations:

$$f''(x), \quad \frac{d^2y}{dx^2}, \quad \text{or} \quad D_x^2[f(x)].$$

The third derivative can be written in a similar way. For $n \geq 4$, the nth derivative is written $f^{(n)}(x)$.

CAUTION Notice the difference in notation between $f^{(4)}(x)$, which indicates the fourth derivative of $f(x)$, and $f^4(x)$, which indicates $f(x)$ raised to the fourth power.

EXAMPLE 2 Second Derivative

Let $f(x) = x^3 + 6x^2 - 9x + 8$.

(a) Find $f''(x)$.

SOLUTION To find the second derivative of $f(x)$, find the first derivative, and then take its derivative.

$$f'(x) = 3x^2 + 12x - 9$$
$$f''(x) = 6x + 12$$

(b) Find $f''(0)$.

SOLUTION Since $f''(x) = 6x + 12$,

$$f''(0) = 6(0) + 12 = 12. \qquad \text{TRY YOUR TURN 1}$$

YOUR TURN 1 Find $f''(1)$ if $f(x) = 5x^4 - 4x^3 + 3x$.

EXAMPLE 3 Second Derivative

Find the second derivative for the functions defined as follows.

(a) $f(x) = (x^2 - 1)^2$

SOLUTION Here, using the chain rule,

$$f'(x) = 2(x^2 - 1)(2x) = 4x(x^2 - 1).$$

Use the product rule to find $f''(x)$.

$$f''(x) = 4x(2x) + (x^2 - 1)(4)$$
$$= 8x^2 + 4x^2 - 4$$
$$= 12x^2 - 4$$

(b) $g(x) = 4x(\ln x)$

SOLUTION Use the product rule.

$$g'(x) = 4x \cdot \frac{1}{x} + (\ln x) \cdot 4 = 4 + 4(\ln x)$$

$$g''(x) = 0 + 4 \cdot \frac{1}{x} = \frac{4}{x}$$

(c) $h(x) = \dfrac{x}{e^x}$

SOLUTION Here, we need the quotient rule.

$$h'(x) = \frac{e^x - xe^x}{(e^x)^2} = \frac{e^x(1-x)}{(e^x)^2} = \frac{1-x}{e^x}$$

$$h''(x) = \frac{e^x(-1) - (1-x)e^x}{(e^x)^2} = \frac{e^x(-1-1+x)}{(e^x)^2} = \frac{-2+x}{e^x}$$

YOUR TURN 2 Find the second derivative for
(a) $f(x) = (x^3 + 1)^2$
(b) $g(x) = xe^x$
(c) $h(x) = \dfrac{\ln x}{x}$.

TRY YOUR TURN 2

Earlier, we saw that the first derivative of a function represents the rate of change of the function. The second derivative, then, represents the rate of change of the first derivative. If a function describes the position of a vehicle (along a straight line) at time t, then the first derivative gives the velocity of the vehicle. That is, if $y = s(t)$ describes the position (along a straight line) of the vehicle at time t, then $v(t) = s'(t)$ gives the velocity at time t.

We also saw that *velocity* is the rate of change of distance with respect to time. Recall, the difference between velocity and speed is that velocity may be positive or negative, whereas speed is always positive. A negative velocity indicates travel in a negative direction (backing up) with regard to the starting point; positive velocity indicates travel in the positive direction (going forward) from the starting point.

The instantaneous rate of change of velocity is called **acceleration**. Since instantaneous rate of change is the same as the derivative, acceleration is the derivative of velocity. Thus if $a(t)$ represents the acceleration at time t, then

$$a(t) = \frac{d}{dt} v(t) = s''(t).$$

If the velocity is positive and the acceleration is positive, the velocity is increasing, so the vehicle is speeding up. If the velocity is positive and the acceleration is negative, the vehicle is slowing down. A negative velocity and a positive acceleration mean the vehicle is backing up and slowing down. If both the velocity and acceleration are negative, the vehicle is speeding up in the negative direction.

EXAMPLE 4 **Velocity and Acceleration**

Suppose a car is moving in a straight line, with its position from a starting point (in feet) at time t (in seconds) given by

$$s(t) = t^3 - 2t^2 - 7t + 9.$$

Find the following.

(a) The velocity at any time t

SOLUTION The velocity is given by

$$v(t) = s'(t) = 3t^2 - 4t - 7$$

feet per second.

(b) The acceleration at any time t

SOLUTION Acceleration is given by

$$a(t) = v'(t) = s''(t) = 6t - 4$$

feet per second per second.

(c) The time intervals (for $t \geq 0$) when the car is going forward or backing up

SOLUTION We first find when the velocity is 0, that is, when the car is stopped.

$$v(t) = 3t^2 - 4t - 7 = 0$$
$$(3t - 7)(t + 1) = 0$$
$$t = 7/3 \quad \text{or} \quad t = -1$$

We are interested in $t \geq 0$. Choose a value of t in each of the intervals $(0, 7/3)$ and $(7/3, \infty)$ to see that the velocity is negative in $(0, 7/3)$ and positive in $(7/3, \infty)$. The car is backing up for the first 7/3 seconds, then going forward.

(d) The time intervals (for $t \geq 0$) when the car is speeding up or slowing down

SOLUTION The car will speed up when the velocity and acceleration are the same sign and slow down when they have opposite signs. Here, the acceleration is positive when $6t - 4 > 0$, that is, $t > 2/3$ seconds, and negative for $t < 2/3$ seconds. Since the velocity is negative in $(0, 7/3)$ and positive in $(7/3, \infty)$, the car is speeding up for $0 < t < 2/3$ seconds, slowing down for $2/3 < t < 7/3$ seconds, and speeding up again for $t > 7/3$ seconds. See the sign graphs.

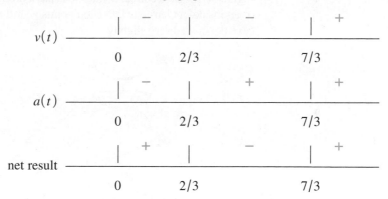

YOUR TURN 3 Find the velocity and acceleration of the car if the distance (in feet) is given by $s(t) = t^3 - 3t^2 - 24t + 10$, at time t (in seconds). When is the car going forward or backing up? When is the car speeding up or slowing down?

TRY YOUR TURN 3

Concavity of a Graph
The first derivative has been used to show where a function is increasing or decreasing and where the extrema occur. The second derivative gives the rate of change of the first derivative; it indicates *how fast* the function is increasing or decreasing. The rate of change of the derivative (the second derivative) affects the *shape* of the graph. Intuitively, we say that a graph is *concave upward* on an interval if it "holds water" and *concave downward* if it "spills water." See Figure 28.

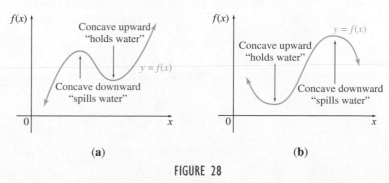

(a)   **(b)**

FIGURE 28

More precisely, a function is **concave upward** on an interval (a, b) if the graph of the function lies above its tangent line at each point of (a, b). A function is **concave downward** on (a, b) if the graph of the function lies below its tangent line at each point of (a, b). A point where a graph changes **concavity** is called an **inflection point**. See Figure 29.

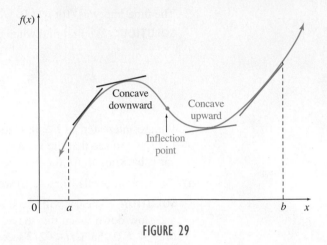

FIGURE 29

Users of soft contact lenses recognize concavity as the way to tell if a lens is inside out. As Figure 30 shows, a correct contact lens has a profile that is entirely concave upward. The profile of an inside-out lens has inflection points near the edges, where the profile begins to turn concave downward very slightly.

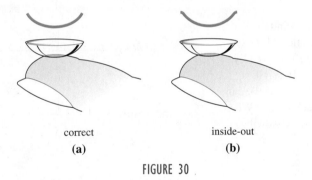

FIGURE 30

Just as a function can be either increasing or decreasing on an interval, it can be either concave upward or concave downward on an interval. Examples of various combinations are shown in Figure 31.

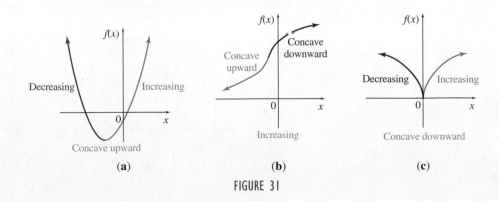

FIGURE 31

Figure 32 shows two functions that are concave upward on an interval (a, b) Several tangent lines are also shown. In Figure 32(a), the slopes of the tangent lines (moving from left to right) are first negative, then 0, and then positive. In Figure 32(b), the slopes are all positive, but they get larger.

In both cases, the slopes are *increasing*. The slope at a point on a curve is given by the derivative. Since a function is increasing if its derivative is positive, its slope is increasing if the derivative of the slope function is positive. Since the derivative of a derivative is the second derivative, a function is concave upward on an interval if its second derivative is positive at each point of the interval.

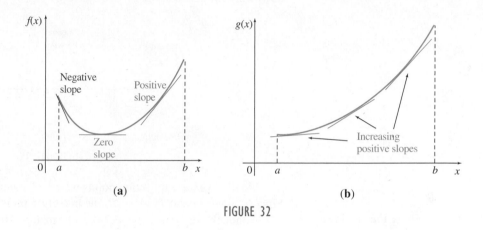

FIGURE 32

A similar result is suggested by Figure 33 for functions whose graphs are concave downward. In both graphs, the slopes of the tangent lines are *decreasing* as we move from left to right. Since a function is decreasing if its derivative is negative, a function is concave downward on an interval if its second derivative is negative at each point of the interval. These observations suggest the following test.

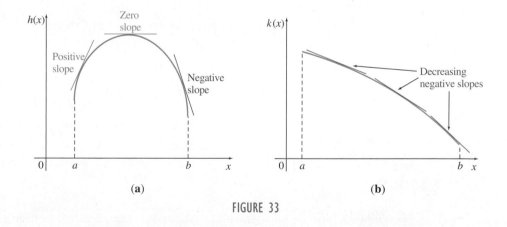

FIGURE 33

Test for Concavity

Let f be a function with derivatives f' and f'' existing at all points in an interval (a, b). Then f is concave upward on (a, b) if $f''(x) > 0$ for all x in (a, b) and concave downward on (a, b) if $f''(x) < 0$ for all x in (a, b).

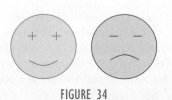

FIGURE 34

An easy way to remember this test is by the faces shown in Figure 34. When the second derivative is positive at a point $(+ +)$, the graph is concave upward $(\smile)$. When the second derivative is negative at a point $(- -)$, the graph is concave downward $(\frown)$.

EXAMPLE 5 Concavity

Find all intervals where $f(x) = x^4 - 8x^3 + 18x^2$ is concave upward or downward, and find all inflection points.

SOLUTION The first derivative is $f'(x) = 4x^3 - 24x^2 + 36x$, and the second derivative is $f''(x) = 12x^2 - 48x + 36$. We factor $f''(x)$ as $12(x - 1)(x - 3)$, and then create a number line for $f''(x)$ as we did in the previous two sections for $f'(x)$.

We see from Figure 35 that $f''(x) > 0$ on the intervals $(-\infty, 1)$ and $(3, \infty)$, so f is concave upward on these intervals. Also, $f''(x) < 0$ on the interval $(1, 3)$, so f is concave downward on this interval.

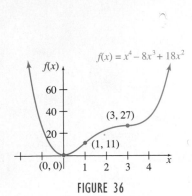

$f(x) = x^4 - 8x^3 + 18x^2$

FIGURE 36

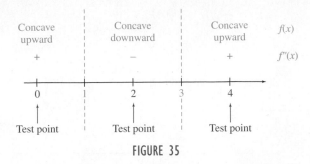

FIGURE 35

Finally, we have inflection points where f'' changes sign, namely, at $x = 1$ and $x = 3$. Since $f(1) = 11$ and $f(3) = 27$, the inflection points are $(1, 11)$ and $(3, 27)$.

Although we were only seeking information about concavity and inflection points in this example, it is also worth noting that $f'(x) = 4x^3 - 24x^2 + 36x = 4x(x - 3)^2$, which has roots at $x = 0$ and $x = 3$. Verify that there is a relative minimum at $(0, 0)$, but that $(3, 27)$ is neither a relative minimum nor a relative maximum. The function is graphed in Figure 36. **TRY YOUR TURN 4**

YOUR TURN 4 Find all intervals where $f(x) = x^5 - 30x^3$ is concave upward or downward, and find all inflection points.

Example 5 suggests the following result.

> At an inflection point for a function f, the second derivative is 0 or does not exist.

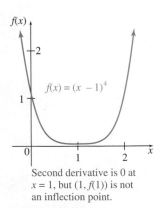

$f(x) = (x - 1)^4$

Second derivative is 0 at $x = 1$, but $(1, f(1))$ is not an inflection point.

FIGURE 37

CAUTION

1. Be careful with the previous statement. Finding a value of x where $f''(x) = 0$ does not mean that an inflection point has been located. For example, if $f(x) = (x - 1)^4$, then $f''(x) = 12(x - 1)^2$, which is 0 at $x = 1$. The graph of $f(x) = (x - 1)^4$ is always concave upward, however, so it has no inflection point. See Figure 37.

2. Note that the concavity of a function might change not only at a point where $f''(x) = 0$ but also where $f''(x)$ does not exist. For example, this happens at $x = 0$ for $f(x) = x^{1/3}$.

TECHNOLOGY NOTE Most graphing calculators do not have a feature for finding inflection points. Nevertheless, a graphing calculator sketch can be useful for verifying that your calculations for finding inflection points and intervals where the function is concave up or down are correct.

Second Derivative Test
The idea of concavity can often be used to decide whether a given critical number produces a relative maximum or a relative minimum. This test, an alternative to the first derivative test, is based on the fact that a curve with a horizontal tangent at a point c and concave downward on an open interval containing c also has a relative maximum at c. A relative minimum occurs when a graph has a horizontal tangent at a point d and is concave upward on an open interval containing d. See Figure 38.

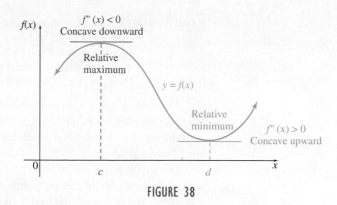

FIGURE 38

A function f is concave upward on an interval if $f''(x) > 0$ for all x in the interval, while f is concave downward on an interval if $f''(x) < 0$ for all x in the interval. These ideas lead to the **second derivative test** for relative extrema.

Second Derivative Test

Let f'' exist on some open interval containing c, (except possibly at c itself) and let $f'(c) = 0$.

1. If $f''(c) > 0$, then $f(c)$ is a relative minimum.

2. If $f''(c) < 0$, then $f(c)$ is a relative maximum.

3. If $f''(c) = 0$ or $f''(c)$ does not exist, then the test gives no information about extrema, so use the first derivative test.

NOTE

In Case 3 of the second derivative test (when $f''(c) = 0$ or does not exist), observe that if $f''(x)$ changes sign at c, there is an inflection point at $x = c$.

EXAMPLE 6 Second Derivative Test

Find all relative extrema for

$$f(x) = 4x^3 + 7x^2 - 10x + 8.$$

SOLUTION   First, find the points where the derivative is 0. Here $f'(x) = 12x^2 + 14x - 10$. Solve the equation $f'(x) = 0$ to get

$$12x^2 + 14x - 10 = 0$$
$$2(6x^2 + 7x - 5) = 0$$
$$2(3x + 5)(2x - 1) = 0$$

$$3x + 5 = 0 \qquad \text{or} \qquad 2x - 1 = 0$$
$$3x = -5 \qquad\qquad\qquad 2x = 1$$
$$x = -\frac{5}{3} \qquad\qquad\qquad x = \frac{1}{2}.$$

Now use the second derivative test. The second derivative is $f''(x) = 24x + 14$. Evaluate $f''(x)$ first at $-5/3$, getting

$$f''\left(-\frac{5}{3}\right) = 24\left(-\frac{5}{3}\right) + 14 = -40 + 14 = -26 < 0,$$

so that by Part 2 of the second derivative test, $-5/3$ leads to a relative maximum of $f(-5/3) = 691/27$. Also, when $x = 1/2$,

YOUR TURN 5

Find all relative extrema of
$f(x) = -2x^3 + 3x^2 + 72x.$

$$f''\left(\frac{1}{2}\right) = 24\left(\frac{1}{2}\right) + 14 = 12 + 14 = 26 > 0,$$

with $1/2$ leading to a relative minimum of $f(1/2) = 21/4$.   **TRY YOUR TURN 5**

| CAUTION | The second derivative test works only for those critical numbers c that make $f'(c) = 0$. This test does not work for critical numbers c for which $f'(c)$ does not exist (since $f''(c)$ would not exist either). Also, the second derivative test does not work for critical numbers c that make $f''(c) = 0$. In both of these cases, use the first derivative test. |

The *law of diminishing returns* in economics is related to the idea of concavity. The function graphed in Figure 39 gives the output y from a given input x. If the input were advertising costs for some product, for example, the output might be the corresponding revenue from sales.

The graph in Figure 39 shows an inflection point at $(c, f(c))$. For $x < c$, the graph is concave upward, so the rate of change of the slope is increasing. This indicates that the output y is increasing at a faster rate with each additional dollar spent. When $x > c$, however, the graph is concave downward, the rate of change of the slope is decreasing, and the increase in y is smaller with each additional dollar spent. Thus, further input beyond c dollars produces diminishing returns. The inflection point at $(c, f(c))$ is called the **point of diminishing returns**. Beyond this point there is a smaller and smaller return for each dollar invested.

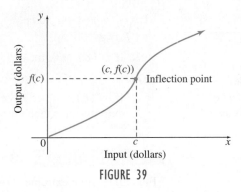

FIGURE 39

As another example of diminishing returns from agriculture, with a fixed amount of land, machinery, fertilizer, and so on, adding workers increases production a lot at first, then less and less with each additional worker.

EXAMPLE 7   Point of Diminishing Returns

The revenue $R(x)$ generated from sales of a certain product is related to the amount x spent on advertising by

$$R(x) = \frac{1}{15{,}000}(600x^2 - x^3), \quad 0 \le x \le 600,$$

where x and $R(x)$ are in thousands of dollars. Is there a point of diminishing returns for this function? If so, what is it?

SOLUTION   Since a point of diminishing returns occurs at an inflection point, look for an x-value that makes $R''(x) = 0$. Write the function as

$$R(x) = \frac{600}{15{,}000}x^2 - \frac{1}{15{,}000}x^3 = \frac{1}{25}x^2 - \frac{1}{15{,}000}x^3.$$

Now find $R'(x)$ and then $R''(x)$.

$$R'(x) = \frac{2x}{25} - \frac{3x^2}{15{,}000} = \frac{2}{25}x - \frac{1}{5000}x^2$$

$$R''(x) = \frac{2}{25} - \frac{1}{2500}x$$

Set $R''(x)$ equal to 0 and solve for x.

$$\frac{2}{25} - \frac{1}{2500}x = 0$$

$$-\frac{1}{2500}x = -\frac{2}{25}$$

$$x = \frac{5000}{25} = 200$$

Test a number in the interval $(0, 200)$ to see that $R''(x)$ is positive there. Then test a number in the interval $(200, 600)$ to find $R''(x)$ negative in that interval. Since the sign of $R''(x)$ changes from positive to negative at $x = 200$, the graph changes from concave upward to concave downward at that point, and there is a point of diminishing returns at the inflection point $(200, 1066\frac{2}{3})$. Investments in advertising beyond \$200,000 return less and less for each dollar invested. Verify that $R'(200) = 8$. This means that when \$200,000 is invested, another \$1000 invested returns approximately \$8000 in additional revenue. Thus it may still be economically sound to invest in advertising beyond the point of diminishing returns. ∎

5.3 EXERCISES

Find $f''(x)$ for each function. Then find $f''(0)$ and $f''(2)$.

1. $f(x) = 5x^3 - 7x^2 + 4x + 3$

2. $f(x) = 4x^3 + 5x^2 + 6x - 7$

3. $f(x) = 4x^4 - 3x^3 - 2x^2 + 6$

4. $f(x) = -x^4 + 7x^3 - \dfrac{x^2}{2}$

5. $f(x) = 3x^2 - 4x + 8$ **6.** $f(x) = 8x^2 + 6x + 5$

7. $f(x) = \dfrac{x^2}{1 + x}$ **8.** $f(x) = \dfrac{-x}{1 - x^2}$

9. $f(x) = \sqrt{x^2 + 4}$ **10.** $f(x) = \sqrt{2x^2 + 9}$

11. $f(x) = 32x^{3/4}$ **12.** $f(x) = -6x^{1/3}$

13. $f(x) = 5e^{-x^2}$ **14.** $f(x) = 0.5e^{x^2}$

15. $f(x) = \dfrac{\ln x}{4x}$ **16.** $f(x) = \ln x + \dfrac{1}{x}$

Find $f'''(x)$, the third derivative of f, and $f^{(4)}(x)$, the fourth derivative of f, for each function.

17. $f(x) = 7x^4 + 6x^3 + 5x^2 + 4x + 3$

18. $f(x) = -2x^4 + 7x^3 + 4x^2 + x$

19. $f(x) = 5x^5 - 3x^4 + 2x^3 + 7x^2 + 4$

20. $f(x) = 2x^5 + 3x^4 - 5x^3 + 9x - 2$

21. $f(x) = \dfrac{x - 1}{x + 2}$ **22.** $f(x) = \dfrac{x + 1}{x}$

23. $f(x) = \dfrac{3x}{x - 2}$ **24.** $f(x) = \dfrac{x}{2x + 1}$

25. Let $f(x) = \ln x$.

    a. Compute $f'(x), f''(x), f'''(x), f^{(4)}(x)$, and $f^{(5)}(x)$.

    b. Guess a formula for $f^{(n)}(x)$, where n is any positive integer.

26. For $f(x) = e^x$, find $f''(x)$ and $f'''(x)$. What is the nth derivative of f with respect to x?

In Exercises 27–48, find the open intervals where the functions are concave upward or concave downward. Find any inflection points.

27.

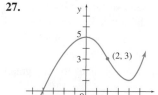

28.

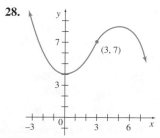

29.

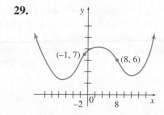

30.

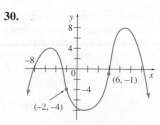

31.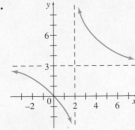

32.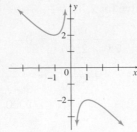

33. $f(x) = x^2 + 10x - 9$

34. $f(x) = 8 - 6x - x^2$

35. $f(x) = -2x^3 + 9x^2 + 168x - 3$

36. $f(x) = -x^3 - 12x^2 - 45x + 2$

37. $f(x) = \dfrac{3}{x-5}$

38. $f(x) = \dfrac{-2}{x+1}$

39. $f(x) = x(x+5)^2$

40. $f(x) = -x(x-3)^2$

41. $f(x) = 18x - 18e^{-x}$

42. $f(x) = 2e^{-x^2}$

43. $f(x) = x^{8/3} - 4x^{5/3}$

44. $f(x) = x^{7/3} + 56x^{4/3}$

45. $f(x) = \ln(x^2 + 1)$

46. $f(x) = x^2 + 8 \ln |x + 1|$

47. $f(x) = x^2 \log |x|$

48. $f(x) = 5^{-x^2}$

For each of the exercises listed below, suppose that the function that is graphed is not $f(x)$, but $f'(x)$. Find the open intervals where the original function is concave upward or concave downward, and find the location of any inflection points.

49. Exercise 27   **50.** Exercise 28

51. Exercise 29   **52.** Exercise 30

53. Give an example of a function $f(x)$ such that $f'(0) = 0$ but $f''(0)$ does not exist. Is there a relative minimum or maximum or an inflection point at $x = 0$?

54. a. Graph the two functions $f(x) = x^{7/3}$ and $g(x) = x^{5/3}$ on the window $[-2, 2]$ by $[-2, 2]$.

    b. Verify that both f and g have an inflection point at $(0, 0)$.

    c. How is the value of $f''(0)$ different from $g''(0)$?

    d. Based on what you have seen so far in this exercise, is it always possible to tell the difference between a point where the second derivative is 0 or undefined based on the graph? Explain.

55. Describe the slope of the tangent line to the graph of $f(x) = e^x$ for the following.

    a. $x \to -\infty$ **b.** $x \to 0$

56. What is true about the slope of the tangent line to the graph of $f(x) = \ln x$ as $x \to \infty$? As $x \to 0$?

Find any critical numbers for f in Exercises 57–64 and then use the second derivative test to decide whether the critical numbers lead to relative maxima or relative minima. If $f''(c) = 0$ or $f''(c)$ does not exist for a critical number c, then the second derivative test gives no information. In this case, use the first derivative test instead.

57. $f(x) = -x^2 - 10x - 25$

58. $f(x) = x^2 - 12x + 36$

59. $f(x) = 3x^3 - 3x^2 + 1$

60. $f(x) = 2x^3 - 4x^2 + 2$

61. $f(x) = (x + 3)^4$

62. $f(x) = x^3$

63. $f(x) = x^{7/3} + x^{4/3}$

64. $f(x) = x^{8/3} + x^{5/3}$

Sometimes the derivative of a function is known, but not the function. We will see more of this later in the book. For each function f' defined in Exercises 65–68, find $f''(x)$, then use a graphing calculator to graph f' and f'' in the indicated window. Use the graph to do the following.

    a. Give the (approximate) x-values where f has a maximum or minimum.

    b. By considering the sign of $f'(x)$, give the (approximate) intervals where $f(x)$ is increasing and decreasing.

    c. Give the (approximate) x-values of any inflection points.

    d. By considering the sign of $f''(x)$, give the intervals where f is concave upward or concave downward.

65. $f'(x) = x^3 - 6x^2 + 7x + 4$; $[-5, 5]$ by $[-5, 15]$

66. $f'(x) = 10x^2(x - 1)(5x - 3)$; $[-1, 1.5]$ by $[-20, 20]$

67. $f'(x) = \dfrac{1 - x^2}{(x^2 + 1)^2}$; $[-3, 3]$ by $[-1.5, 1.5]$

68. $f'(x) = x^2 + x \ln x$; $[0, 1]$ by $[-2, 2]$

69. Suppose a friend makes the following argument. A function f is increasing and concave downward. Therefore, f' is positive and decreasing, so it eventually becomes 0 and then negative, at which point f decreases. Show that your friend is wrong by giving an example of a function that is always increasing and concave downward.

APPLICATIONS

Business and Economics

70. Product Life Cycle The accompanying figure shows the *product life cycle* graph, with typical products marked on it. It illustrates the fact that a new product is often purchased at a faster and faster rate as people become familiar with it. In time, saturation is reached and the purchase rate stays constant until the product is made obsolete by newer products, after which it is purchased less and less. *Source: tutor2u.*

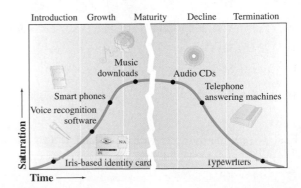

    a. Which products on the left side of the graph are closest to the left-hand inflection point? What does the inflection point mean here?

    b. Which product on the right side of the graph is closest to the right-hand inflection point? What does the inflection point mean here?

    c. Discuss where portable Blu-ray players, iPads, and other new technologies should be placed on the graph.

71. Social Security Assets As seen in the first section of this chapter, the projected year-end assets in the Social Security trust funds, in trillions of dollars, where t represents the number of years since 2000, can be approximated by

$$A(t) = 0.0000329t^3 - 0.00450t^2 + 0.0613t + 2.34,$$

where $0 \le t \le 50$. Find the value of t when Social Security assets will decrease most rapidly. Approximately when does this occur? *Source: Social Security Administration.*

Point of Diminishing Returns In Exercises 72–75, find the point of diminishing returns (x, y) for the given functions, where $R(x)$, represents revenue (in thousands of dollars) and x represents the amount spent on advertising (in thousands of dollars).

72. $R(x) = 10{,}000 - x^3 + 42x^2 + 800x, \quad 0 \le x \le 20$

73. $R(x) = \dfrac{4}{27}(-x^3 + 66x^2 + 1050x - 400), \quad 0 \le x \le 25$

74. $R(x) = -0.3x^3 + x^2 + 11.4x, \quad 0 \le x \le 6$

75. $R(x) = -0.6x^3 + 3.7x^2 + 5x, \quad 0 \le x \le 6$

76. Risk Aversion In economics, an index of *absolute risk aversion* is defined as

$$I(M) = \frac{-U''(M)}{U'(M)},$$

where M measures how much of a commodity is owned and $U(M)$ is a *utility function*, which measures the ability of quantity M of a commodity to satisfy a consumer's wants. Find $I(M)$ for $U(M) = \sqrt{M}$ and for $U(M) = M^{2/3}$, and determine which indicates a greater aversion to risk.

77. Demand Function The authors of an article in an economics journal state that if $D(q)$ is the demand function, then the inequality

$$qD''(q) + D'(q) < 0$$

is equivalent to saying that the marginal revenue declines more quickly than does the price. Prove that this equivalence is true. *Source: Bell Journal of Economics.*

Life Sciences

78. Population Growth When a hardy new species is introduced into an area, the population often increases as shown. Explain the significance of the following function values on the graph.

a. f_0 **b.** $f(a)$ **c.** f_M

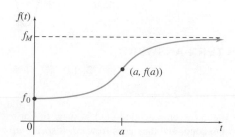

79. Bacteria Population Assume that the number of bacteria $R(t)$ (in millions) present in a certain culture at time t (in hours) is given by

$$R(t) = t^2(t - 18) + 96t + 1000.$$

a. At what time before 8 hours will the population be maximized?

b. Find the maximum population.

80. Ozone Depletion According to an article in *The New York Times*, "Government scientists reported last week that they had detected a slowdown in the rate at which chemicals that deplete the earth's protective ozone layer are accumulating in the atmosphere." Letting $c(t)$ be the amount of ozone-depleting chemicals at time t, what does this statement tell you about $c(t)$, $c'(t)$, and $c''(t)$? *Source: The New York Times.*

81. Drug Concentration The percent of concentration of a certain drug in the bloodstream x hours after the drug is administered is given by

$$K(x) = \frac{3x}{x^2 + 4}.$$

For example, after 1 hour the concentration is given by

$$K(1) = \frac{3(1)}{1^2 + 4} = \frac{3}{5}\% = 0.6\% = 0.006.$$

a. Find the time at which concentration is a maximum.

b. Find the maximum concentration.

82. Drug Concentration The percent of concentration of a drug in the bloodstream x hours after the drug is administered is given by

$$K(x) = \frac{4x}{3x^2 + 27}.$$

a. Find the time at which the concentration is a maximum.

b. Find the maximum concentration.

The next two exercises are a continuation of exercises first given in the section on Derivatives of Exponential Functions. Find the inflection point of the graph of each logistic function. This is the point at which the growth rate begins to decline.

83. Insect Growth The growth function for a population of beetles is given by

$$G(t) = \frac{10{,}000}{1 + 49e^{-0.1t}}.$$

84. Clam Population Growth The population of a bed of clams is described by

$$G(t) = \frac{5200}{1 + 12e^{-0.52t}}.$$

Hints for Exercises 85 and 86: Leave B, c, and k as constants until you are ready to calculate your final answer.

85. Clam Growth Researchers used a version of the Gompertz curve to model the growth of razor clams during the first seven years of the clams' lives with the equation

$$L(t) = Be^{-ce^{-kt}},$$

where $L(t)$ gives the length (in centimeters) after t years, $B = 14.3032$, $c = 7.267963$, and $k - 0.670840$. Find the inflection point and describe what it signifies. *Source: Journal of Experimental Biology.*

86. Breast Cancer Growth Researchers used a version of the Gompertz curve to model the growth of breast cancer tumors with the equation

$$N(t) = e^{c(1-e^{-kt})},$$

where $N(t)$ is the number of cancer cells after t days, $c = 27.3$, and $k = 0.011$. Find the inflection point and describe what it signifies. *Source: Cancer Research*.

87. Popcorn Researchers have determined that the amount of moisture present in a kernel of popcorn affects the volume of the popped corn and can be modeled for certain sizes of kernels by the function

$$v(x) = -35.98 + 12.09x - 0.4450x^2,$$

where x is moisture content (%, wet basis) and $v(x)$ is the expansion volume (in $cm^3/gram$). Describe the concavity of this function. *Source: Cereal Chemistry*.

88. Alligator Teeth Researchers have developed a mathematical model that can be used to estimate the number of teeth $N(t)$ at time t (days of incubation) for *Alligator mississippiensis*, where

$$N(t) = 71.8e^{-8.96e^{-0.0685t}}.$$

Find the inflection point and describe its importance to this research. *Source: Journal of Theoretical Biology*.

Social Sciences

89. Crime In 1995, the rate of violent crimes in New York City continued to decrease, but at a slower rate than in previous years. Letting $f(t)$ be the rate of violent crime as a function of time, what does this tell you about $f(t)$, $f'(t)$, and $f''(t)$? *Source: The New York Times*.

Physical Sciences

90. Chemical Reaction An autocatalytic chemical reaction is one in which the product being formed causes the rate of formation to increase. The rate of a certain autocatalytic reaction is given by

$$V(x) = 12x(100 - x),$$

where x is the quantity of the product present and 100 represents the quantity of chemical present initially. For what value of x is the rate of the reaction a maximum?

91. Velocity and Acceleration When an object is dropped straight down, the distance (in feet) that it travels in t seconds is given by

$$s(t) = -16t^2.$$

Find the velocity at each of the following times.

a. After 3 seconds   **b.** After 5 seconds

c. After 8 seconds

d. Find the acceleration. (The answer here is a constant the acceleration due to the influence of gravity alone near the surface of Earth.)

92. Baseball Roger Clemens, ace pitcher for many major league teams, including the Boston Red Sox, is standing on top of the 37-ft-high "Green Monster" left-field wall in Boston's Fenway Park, to which he has returned for a visit. We have asked him to fire his famous 95 mph (140 ft per second) fastball straight up. The position equation, which gives the height of the ball at any time t, in seconds, is given by $s(t) = -16t^2 + 140t + 37$. Find the following. *Source: Frederick Russell*.

a. The maximum height of the ball

b. The time and velocity when the ball hits the ground

93. Height of a Ball If a cannonball is shot directly upward with a velocity of 256 ft per second, its height above the ground after t seconds is given by $s(t) = 256t - 16t^2$. Find the velocity and the acceleration after t seconds. What is the maximum height the cannonball reaches? When does it hit the ground?

94. Velocity and Acceleration of a Car A car rolls down a hill. Its distance (in feet) from its starting point is given by $s(t) = 1.5t^2 + 4t$, where t is in seconds.

a. How far will the car move in 10 seconds?

b. What is the velocity at 5 seconds? At 10 seconds?

c. How can you tell from $v(t)$ that the car will not stop?

d. What is the acceleration at 5 seconds? At 10 seconds?

e. What is happening to the velocity and the acceleration as t increases?

95. Velocity and Acceleration A car is moving along a straight stretch of road. The acceleration of the car is given by the graph shown. Assume that the velocity of the car is always positive. At what time was the car moving most rapidly? Explain. *Source: Larry Taylor*.

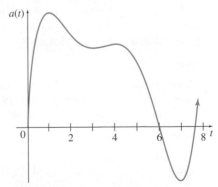

YOUR TURN ANSWERS

1. 36

2. (a) $f''(x) = 30x^4 + 12x$

(b) $g''(x) = 2e^x + xe^x$ **(c)** $h''(x) = \dfrac{-3 + 2\ln x}{x^3}$

3. $v(t) = 3t^2 - 6t - 24$ and $a(t) = 6t - 6$. Car backs up for the first 4 seconds and then goes forward. It speeds up for $0 < t < 1$, slows down for $1 < t < 4$, and then speeds up for $t > 4$.

4. Concave up on $(-3, 0)$ and $(3, \infty)$; concave down on $(-\infty, -3)$ and $(0, 3)$; inflection points are $(-3, 567)$, $(0, 0)$, and $(3, -567)$.

5. Relative maximum of $f(4) = 208$ at $x = 4$ and relative minimum of $f(-3) = -135$ at $x = -3$.

5.4 Curve Sketching

APPLY IT How can we use differentiation to help us sketch the graph of a function, and describe its behavior?

In the following examples, the test for concavity, the test for increasing and decreasing functions, and the concept of limits at infinity will help us sketch the graphs and describe the behavior of a variety of functions. This process, called **curve sketching**, has decreased somewhat in importance in recent years due to the widespread use of graphing calculators. We believe, however, that this topic is worth studying for the following reasons.

For one thing, a graphing calculator picture can be misleading, particularly if important points lie outside the viewing window. Even if all important features are within the viewing windows, there is still the problem that the calculator plots and connects points and misses what goes on between those points. As an example of the difficulty in choosing an appropriate window without a knowledge of calculus, see Exercise 40 in the second section of this chapter.

Furthermore, curve sketching may be the best way to learn the material in the previous three sections. You may feel confident that you understand what increasing and concave upward mean, but using those concepts in a graph will put your understanding to the test.

Curve sketching may be done with the following steps.

Curve Sketching

To sketch the graph of a function f:

1. Consider the domain of the function, and note any restrictions. (That is, avoid dividing by 0, taking a square root of a negative number, or taking the logarithm of 0 or a negative number.)

2. Find the y-intercept (if it exists) by substituting $x = 0$ into $f(x)$. Find any x-intercepts by solving $f(x) = 0$ if this is not too difficult.

3. **a.** If f is a rational function, find any vertical asymptotes by investigating where the denominator is 0, and find any horizontal asymptotes by finding the limits as $x \to \infty$ and $x \to -\infty$.

   b. If f is an exponential function, find any horizontal asymptotes; if f is a logarithmic function, find any vertical asymptotes.

4. Investigate symmetry. If $f(-x) = f(x)$, the function is even, so the graph is symmetric about the y-axis. If $f(-x) = -f(x)$, the function is odd, so the graph is symmetric about the origin.

5. Find $f'(x)$. Locate any critical points by solving the equation $f'(x) = 0$ and determining where $f'(x)$ does not exist, but $f(x)$ does. Find any relative extrema and determine where f is increasing or decreasing.

6. Find $f''(x)$. Locate potential inflection points by solving the equation $f''(x) = 0$ and determining where $f''(x)$ does not exist. Determine where f is concave upward or concave downward.

7. Plot the intercepts, the critical points, the inflection points, the asymptotes, and other points as needed. Take advantage of any symmetry found in Step 4.

8. Connect the points with a smooth curve using the correct concavity, being careful not to connect points where the function is not defined.

9. Check your graph using a graphing calculator. If the picture looks very different from what you've drawn, see in what ways the picture differs and use that information to help find your mistake.

There are four possible combinations for a function to be increasing or decreasing and concave up or concave down, as shown in the following table.

$f''(x)$ $\diagdown$ $f'(x)$	Concavity Summary	
	+ (Function Is Increasing)	− (Function Is Decreasing)
+ (function is concave up)	$\diagup$	$\diagdown$
− (function is concave down)	$\diagup$	$\diagdown$

EXAMPLE 1 Polynomial Function Graph

Graph $f(x) = 2x^3 - 3x^2 - 12x + 1$.

SOLUTION The domain is $(-\infty, \infty)$. The y-intercept is located at $y = f(0) = 1$. Finding the x-intercepts requires solving the equation $f(x) = 0$. But this is a third-degree equation; since we have not covered a procedure for solving such equations, we will skip this step. This is neither a rational nor an exponential function, so we also skip step 3. Observe that $f(-x) = 2(-x)^3 - 3(-x)^2 - 12(-x) + 1 = -2x^3 - 3x^2 + 12x + 1$, which is neither $f(x)$ nor $-f(x)$, so there is no symmetry about the y-axis or origin.

To find the intervals where the function is increasing or decreasing, find the first derivative.

$$f'(x) = 6x^2 - 6x - 12$$

This derivative is 0 when

$$6(x^2 - x - 2) = 0$$
$$6(x - 2)(x + 1) = 0$$
$$x = 2 \quad \text{or} \quad x = -1.$$

These critical numbers divide the number line in Figure 40 into three regions. Testing a number from each region in $f'(x)$ shows that f is increasing on $(-\infty, -1)$ and $(2, \infty)$ and decreasing on $(-1, 2)$. This is shown with the arrows in Figure 41. By the first derivative test, f has a relative maximum when $x = -1$ and a relative minimum when $x = 2$. The relative maximum is $f(-1) = 8$, while the relative minimum is $f(2) = -19$.

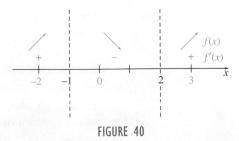

FIGURE 40

Now use the second derivative to find the intervals where the function is concave upward or downward. Here

$$f''(x) = 12x - 6,$$

which is 0 when $x = 1/2$. Testing a point with x less than $1/2$, and one with x greater than $1/2$, shows that f is concave downward on $(-\infty, 1/2)$ and concave upward on $(1/2, \infty)$.

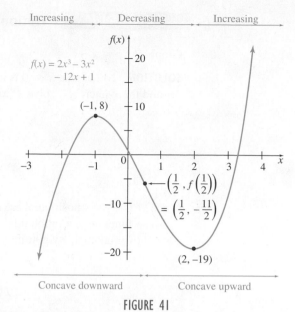

FIGURE 41

The graph has an inflection point at $(1/2, f(1/2))$, or $(1/2, -11/2)$. This information is summarized in the following table.

	Graph Summary			
Interval	$(-\infty, -1)$	$(-1, 1/2)$	$(1/2, 2)$	$(2, \infty)$
Sign of f'	$+$	$-$	$-$	$+$
Sign of f''	$-$	$-$	$+$	$+$
f Increasing or Decreasing	Increasing	Decreasing	Decreasing	Increasing
Concavity of f	Downward	Downward	Upward	Upward
Shape of Graph	⌒	⌒	⌣	⌣

YOUR TURN 1 Graph $f(x) = -x^3 + 3x^2 + 9x - 10$.

Use this information and the critical points to get the graph shown in Figure 41. Notice that the graph appears to be symmetric about its inflection point. It can be shown that is always true for third-degree polynomials. In other words, if you put your pencil point at the inflection point and then spin the book 180° about the pencil point, the graph will appear to be unchanged.

TRY YOUR TURN 1

TECHNOLOGY NOTE

A graphing calculator picture of the function in Figure 41 on the arbitrarily chosen window $[-3, 3]$ by $[-7, 7]$ gives a misleading picture, as Figure 42(a) shows. Knowing where the turning points lie tells us that a better window would be $[-3, 4]$ by $[-20, 20]$, with the results shown in Figure 42(b).

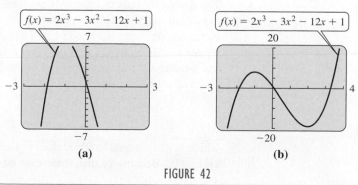

FIGURE 42

EXAMPLE 2 Rational Function Graph

Graph $f(x) = x + \dfrac{1}{x}$.

SOLUTION Notice that $x = 0$ is not in the domain of the function, so there is no y-intercept. To find the x-intercept, solve $f(x) = 0$.

$$x + \frac{1}{x} = 0$$
$$x = -\frac{1}{x}$$
$$x^2 = -1$$

Since x^2 is always positive, there is also no x-intercept.

The function is a rational function, but it is not written in the usual form of one polynomial over another. By getting a common denominator and adding the fractions, it can be rewritten in that form:

$$f(x) = x + \frac{1}{x} = \frac{x^2 + 1}{x}.$$

FOR REVIEW

Asymptotes were discussed in Section 2.3 on Polynomial and Rational Functions. You may wish to refer back to that section to review. To review limits, refer to Section 3.1 in the chapter titled The Derivative.

Because $x = 0$ makes the denominator (but not the numerator) 0, the line $x = 0$ is a vertical asymptote. To find any horizontal asymptotes, we investigate

$$\lim_{x \to \infty} \frac{x^2 + 1}{x} = \lim_{x \to \infty} \left(\frac{x^2}{x} + \frac{1}{x} \right) = \lim_{x \to \infty} \left(x + \frac{1}{x} \right).$$

The second term, $1/x$, approaches 0 as $x \to \infty$, but the first term, x, becomes infinitely large, so the limit does not exist. Verify that $\lim\limits_{x \to -\infty} f(x)$ also does not exist, so there are no horizontal asymptotes.

Observe that as x gets very large, the second term $(1/x)$ in $f(x)$ gets very small, so $f(x) = x + (1/x) \approx x$. The graph gets closer and closer to the straight line $y = x$ as x becomes larger and larger. This is what is known as an **oblique asymptote**.

Observe that

$$f(-x) = (-x) + \frac{1}{-x} = -\left(x + \frac{1}{x} \right) = -f(x),$$

so the graph is symmetric about the origin. This means that the left side of the graph can be found by rotating the right side 180° about the origin.

Here $f'(x) = 1 - (1/x^2)$, which is 0 when

$$\frac{1}{x^2} = 1$$
$$x^2 = 1$$
$$x^2 - 1 = (x - 1)(x + 1) = 0$$
$$x = 1 \quad \text{or} \quad x = -1.$$

The derivative fails to exist at 0, where the vertical asymptote is located. Evaluating $f'(x)$ in each of the regions determined by the critical numbers and the asymptote shows that f is increasing on $(-\infty, -1)$ and $(1, \infty)$ and decreasing on $(-1, 0)$ and $(0, 1)$. See Figure 43(a). By the first derivative test, f has a relative maximum of $y = f(-1) = -2$ when $x = -1$, and a relative minimum of $y = f(1) = 2$ when $x = 1$.

The second derivative is

$$f''(x) = \frac{2}{x^3},$$

which is never equal to 0 and does not exist when $x = 0$. (The function itself also does not exist at 0.) Because of this, there may be a change of concavity, but not an inflection point,

YOUR TURN 2 Graph
$f(x) = 4x + \dfrac{1}{x}$.

when $x = 0$. The second derivative is negative when x is negative, making f concave downward on $(-\infty, 0)$. Also, $f''(x) > 0$ when $x > 0$, making f concave upward on $(0, \infty)$. See Figure 43(b).

Use this information, the asymptotes, and the critical points to get the graph shown in Figure 44.   **TRY YOUR TURN 2**

		Graph Summary		
Interval	$(-\infty, -1)$	$(-1, 0)$	$(0, 1)$	$(1, \infty)$
Sign of f'	$+$	$-$	$-$	$+$
Sign of f"	$-$	$-$	$+$	$+$
f Increasing or Decreasing	Increasing	Decreasing	Decreasing	Increasing
Concavity of f	Downward	Downward	Upward	Upward
Shape of Graph	⌒	⌢	⌣	⌣

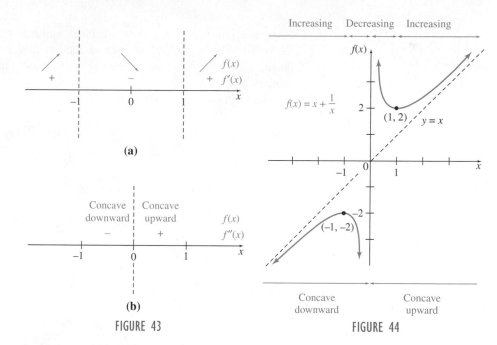

FIGURE 43

FIGURE 44

EXAMPLE 3   Rational Function Graph

Graph $f(x) = \dfrac{3x^2}{x^2 + 5}$.

SOLUTION   The y-intercept is located at $y = f(0) = 0$. Verify that this is also the only x-intercept. There is no vertical asymptote, because $x^2 + 5 \neq 0$ for any value of x. Find any horizontal asymptote by calculating $\lim\limits_{x \to \infty} f(x)$ and $\lim\limits_{x \to -\infty} f(x)$. First, divide both the numerator and the denominator of $f(x)$ by x^2.

$$\lim_{x \to \infty} \frac{3x^2}{x^2 + 5} = \lim_{x \to \infty} \frac{\dfrac{3x^2}{x^2}}{\dfrac{x^2}{x^2} + \dfrac{5}{x^2}} = \frac{3}{1 + 0} = 3$$

Verify that the limit of $f(x)$ as $x \to -\infty$ is also 3. Thus, the horizontal asymptote is $y = 3$.

Observe that

$$f(-x) = \frac{3(-x)^2}{(-x)^2 + 5} = \frac{3x^2}{x^2 + 5} = f(x),$$

so the graph is symmetric about the y-axis. This means that the left side of the graph is the mirror image of the right side.

We now compute $f'(x)$:

$$f'(x) = \frac{(x^2 + 5)(6x) - (3x^2)(2x)}{(x^2 + 5)^2}.$$

Notice that $6x$ can be factored out of each term in the numerator:

$$f'(x) = \frac{(6x)[(x^2 + 5) - x^2]}{(x^2 + 5)^2}$$

$$= \frac{(6x)(5)}{(x^2 + 5)^2} = \frac{30x}{(x^2 + 5)^2}.$$

From the numerator, $x = 0$ is a critical number. The denominator is always positive. (Why?) Evaluating $f'(x)$ in each of the regions determined by $x = 0$ shows that f is decreasing on $(-\infty, 0)$ and increasing on $(0, \infty)$. By the first derivative test, f has a relative minimum when $x = 0$.

The second derivative is

$$f''(x) = \frac{(x^2 + 5)^2(30) - (30x)(2)(x^2 + 5)(2x)}{(x^2 + 5)^4}.$$

Factor $30(x^2 + 5)$ out of the numerator:

$$f''(x) = \frac{30(x^2 + 5)[(x^2 + 5) - (x)(2)(2x)]}{(x^2 + 5)^4}.$$

Divide a factor of $(x^2 + 5)$ out of the numerator and denominator, and simplify the numerator:

$$f''(x) = \frac{30[(x^2 + 5) - (x)(2)(2x)]}{(x^2 + 5)^3}$$

$$= \frac{30[(x^2 + 5) - (4x^2)]}{(x^2 + 5)^3}$$

$$= \frac{30(5 - 3x^2)}{(x^2 + 5)^3}.$$

YOUR TURN 3

Graph $f(x) = \dfrac{4x^2}{x^2 + 4}$.

The numerator of $f''(x)$ is 0 when $x = \pm\sqrt{5/3} \approx \pm 1.29$. Testing a point in each of the three intervals defined by these points shows that f is concave downward on $(-\infty, -1.29)$ and $(1.29, \infty)$, and concave upward on $(-1.29, 1.29)$. The graph has inflection points at $(\pm\sqrt{5/3}, f(\pm\sqrt{5/3})) \approx (\pm 1.29, 0.75)$.

Use this information, the asymptote, the critical point, and the inflection points to get the graph shown in Figure 45. **TRY YOUR TURN 3**

	Graph Summary			
Interval	$(-\infty, -1.29)$	$(-1.29, 0)$	$(0, 1.29)$	$(1.29, \infty)$
Sign of f'	$-$	$-$	$+$	$+$
Sign of f''	$-$	$+$	$+$	$-$
f Increasing or Decreasing	Decreasing	Decreasing	Increasing	Increasing
Concavity of f	Downward	Upward	Upward	Downward
Shape of Graph				

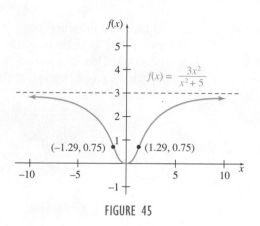

FIGURE 45

EXAMPLE 4   Graph with Logarithm

Graph $f(x) = \dfrac{\ln x}{x^2}$.

SOLUTION   The domain is $x > 0$, so there is no y-intercept. The x-intercept is 1, because $\ln 1 = 0$. We know that $y = \ln x$ has a vertical asymptote at $x = 0$, because $\lim\limits_{x \to 0^+} \ln x = -\infty$. Dividing by x^2 when x is small makes $(\ln x)/x^2$ even more negative than $\ln x$. Therefore, $(\ln x)/x^2$ has a vertical asymptote at $x = 0$ as well. The first derivative is

$$f'(x) = \frac{x^2 \cdot \dfrac{1}{x} - 2x \ln x}{(x^2)^2} = \frac{x(1 - 2\ln x)}{x^4} = \frac{1 - 2\ln x}{x^3}$$

by the quotient rule. Setting the numerator equal to 0 and solving for x gives

$$1 - 2\ln x = 0$$
$$1 = 2\ln x$$
$$\ln x = 0.5$$
$$x = e^{0.5} \approx 1.65.$$

Since $f'(1)$ is positive and $f'(2)$ is negative, f increases on $(0, 1.65)$ then decreases on $(1.65, \infty)$, with a maximum value of $f(1.65) \approx 0.18$.

To find any inflection points, we set $f''(x) = 0$.

$$f''(x) = \frac{x^3\left(-2 \cdot \dfrac{1}{x}\right) - (1 - 2\ln x) \cdot 3x^2}{(x^3)^2}$$

$$= \frac{-2x^2 - 3x^2 + 6x^2\ln x}{x^6} = \frac{-5 + 6\ln x}{x^4}$$

$$\frac{-5 + 6\ln x}{x^4} = 0$$

$-5 + 6\ln x = 0$ Set the numerator equal to 0.

$6\ln x = 5$ Add 5 to both sides.

$\ln x = 5/6$ Divide both sides by 6.

$x = e^{5/6} \approx 2.3$ $e^{\ln x} = x.$

There is an inflection point at $(2.3, f(2.3)) \approx (2.3, 0.16)$. Verify that $f''(1)$ is negative and $f''(3)$ is positive, so the graph is concave downward on $(1, 2.3)$ and upward on $(2.3, \infty)$. This information is summarized in the following table and could be used to sketch the graph. A graph of the function is shown in Figure 46.   **TRY YOUR TURN 4**

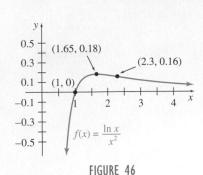

FIGURE 46

YOUR TURN 4   Graph $f(x) = (x + 2)e^{-x}$.
(Recall $\lim\limits_{x \to \infty} x^n e^{-x} = 0$)

	Graph Summary		
Interval	$(0, 1.65)$	$(1.65, 2.3)$	$(2.3, \infty)$
Sign of f'	$+$	$-$	$-$
Sign of f''	$-$	$-$	$+$
f Increasing or Decreasing	Increasing	Decreasing	Decreasing
Concavity of f	Downward	Downward	Upward
Shape of Graph	⌒	⌒	⌣

As we saw earlier, a graphing calculator, when used with care, can be helpful in studying the behavior of functions. This section has illustrated that calculus is also a great help. The techniques of calculus show where the important points of a function, such as the relative extrema and the inflection points, are located. Furthermore, they tell how the function behaves between and beyond the points that are graphed, something a graphing calculator cannot always do.

5.4 EXERCISES

1. By sketching a graph of the function or by investigating values of the function near 0, find $\lim\limits_{x \to 0} x \ln |x|$. (This result will be useful in Exercise 21.)

2. Describe how you would find the equation of the horizontal asymptote for the graph of

$$f(x) = \frac{3x^2 - 2x}{2x^2 + 5}.$$

Graph each function, considering the domain, critical points, symmetry, regions where the function is increasing or decreasing, inflection points, regions where the function is concave upward or concave downward, intercepts where possible, and asymptotes where applicable. (*Hint*: In Exercise 21, use the result of Exercise 1. In Exercises 25–27, recall from Exercise 66 in the section on Limits that $\lim\limits_{x \to \infty} x^n e^{-x} = 0$.)

3. $f(x) = -2x^3 - 9x^2 + 108x - 10$

4. $f(x) = x^3 - \dfrac{15}{2}x^2 - 18x - 1$

5. $f(x) = -3x^3 + 6x^2 - 4x - 1$

6. $f(x) = x^3 - 6x^2 + 12x - 11$

7. $f(x) = x^4 - 24x^2 + 80$ **8.** $f(x) = -x^4 + 6x^2$

9. $f(x) = x^4 - 4x^3$ **10.** $f(x) = x^5 - 15x^3$

11. $f(x) = 2x + \dfrac{10}{x}$ **12.** $f(x) = 16x + \dfrac{1}{x^2}$

13. $f(x) = \dfrac{-x + 4}{x + 2}$ **14.** $f(x) = \dfrac{3x}{x - 2}$

15. $f(x) = \dfrac{1}{x^2 + 4x + 3}$ **16.** $f(x) = \dfrac{-8}{x^2 - 6x - 7}$

17. $f(x) = \dfrac{x}{x^2 + 1}$ **18.** $f(x) = \dfrac{1}{x^2 + 4}$

19. $f(x) = \dfrac{1}{x^2 - 9}$ **20.** $f(x) = \dfrac{-2x}{x^2 - 4}$

21. $f(x) = x \ln |x|$ **22.** $f(x) = x - \ln |x|$

23. $f(x) = \dfrac{\ln x}{x}$ **24.** $f(x) = \dfrac{\ln x^2}{x^2}$

25. $f(x) = xe^{-x}$ **26.** $f(x) = x^2 e^{-x}$

27. $f(x) = (x - 1)e^{-x}$ **28.** $f(x) = e^x + e^{-x}$

29. $f(x) = x^{2/3} - x^{5/3}$ **30.** $f(x) = x^{1/3} + x^{4/3}$

31. The default window on many calculators is $[-10, 10]$ by $[-10, 10]$. For the odd exercises between 3 and 15, tell which would give a poor representation in this window. (*Note:* Your answers may differ from ours, depending on what you consider "poor.")

32. Repeat Exercise 31 for the even exercises between 4 and 16.

33. Repeat Exercise 31 for the odd exercises between 17 and 29.

34. Repeat Exercise 31 for the even exercises between 18 and 30.

In Exercises 35–39, sketch the graph of a single function that has all of the properties listed.

35. a. Continuous and differentiable everywhere except at $x = 1$, where it has a vertical asymptote

   b. $f'(x) < 0$ everywhere it is defined

   c. A horizontal asymptote at $y = 2$

   d. $f''(x) < 0$ on $(-\infty, 1)$ and $(2, 4)$

   e. $f''(x) > 0$ on $(1, 2)$ and $(4, \infty)$

36. a. Continuous for all real numbers

   b. $f'(x) < 0$ on $(-\infty, -6)$ and $(1, 3)$

   c. $f'(x) > 0$ on $(-6, 1)$ and $(3, \infty)$

   d. $f''(x) > 0$ on $(-\infty, -6)$ and $(3, \infty)$

   e. $f''(x) < 0$ on $(-6, 3)$

   f. A y-intercept at $(0, 2)$

37. a. Continuous and differentiable for all real numbers

   b. $f'(x) > 0$ on $(-\infty, -3)$ and $(1, 4)$

   c. $f'(x) < 0$ on $(-3, 1)$ and $(4, \infty)$

   d. $f''(x) < 0$ on $(-\infty, -1)$ and $(2, \infty)$

   e. $f''(x) > 0$ on $(-1, 2)$

   f. $f'(-3) = f'(4) = 0$

   g. $f''(x) = 0$ at $(-1, 3)$ and $(2, 4)$

38. a. Continuous for all real numbers

   b. $f'(x) > 0$ on $(-\infty, -2)$ and $(0, 3)$

   c. $f'(x) < 0$ on $(-2, 0)$ and $(3, \infty)$

   d. $f''(x) < 0$ on $(-\infty, 0)$ and $(0, 5)$

   e. $f''(x) > 0$ on $(5, \infty)$

   f. $f'(-2) = f'(3) = 0$

   g. $f'(0)$ doesn't exist

   h. Differentiable everywhere except at $x = 0$

   i. An inflection point at $(5, 1)$

39. a. Continuous for all real numbers

   b. Differentiable everywhere except at $x = 4$

   c. $f(1) = 5$

   d. $f'(1) = 0$ and $f'(3) = 0$

   e. $f'(x) > 0$ on $(-\infty, 1)$ and $(4, \infty)$

   f. $f'(x) < 0$ on $(1, 3)$ and $(3, 4)$

   g. $\lim\limits_{x \to 4^-} f'(x) = -\infty$ and $\lim\limits_{x \to 4^+} f'(x) = \infty$

   h. $f''(x) > 0$ on $(2, 3)$

   i. $f''(x) < 0$ on $(-\infty, 2)$, $(3, 4)$, and $(4, \infty)$

40. On many calculators, graphs of rational functions produce lines at vertical asymptotes. For example, graphing $y = (x - 1)/(x + 1)$ on the window $[-4.9, 4.9]$ by $[-4.9, 4.9]$ produces such a line at $x = -1$ on the TI-84 Plus and TI-89. But with the window $[-4.7, 4.7]$ by $[-4.7, 4.7]$ on a TI-84 Plus; or $[-7.9, 7.9]$ by $[-7.9, 7.9]$ on a TI-89, the spurious line does not appear. Experiment with this function on your calculator, trying different windows, and try to figure out an explanation for this phenomenon. (*Hint:* Consider the number of pixels on the calculator screen.)

YOUR TURN ANSWERS

1.

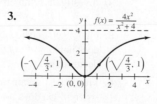

2.

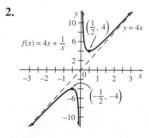

3.

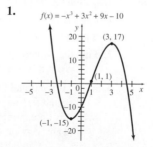

4.

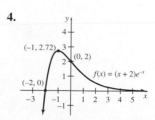

5 — CHAPTER REVIEW

SUMMARY

In this chapter we have explored various concepts related to the graph of a function:

- increasing and decreasing,
- critical numbers (numbers c in the domain of f for which $f'(x) = 0$ or $f'(x)$ does not exist),
- critical points (whose x-coordinate is a critical number c and whose y-coordinate is $f(c)$),
- relative maxima and minima (together known as relative extrema),
- concavity, and
- inflection points (where the concavity changes).

The first and second derivative tests provide ways to locate relative extrema. The last section brings all these concepts together. Also, we investigated two applications of the second derivative:

- acceleration (the second derivative of the position function), and
- the point of diminishing returns (an inflection point on an input/output graph).

Test for Increasing/Decreasing On any open interval,

    if $f'(x) > 0$, then f is increasing;

    if $f'(x) < 0$, then f is decreasing;

    if $f'(x) = 0$, then f is constant.

First Derivative Test If c is a critical number for f on the open interval (a, b), f is continuous on (a, b), and f is differentiable on (a, b) (except possibly at c), then

1. $f(c)$ is a relative maximum if $f'(x) > 0$ on (a, c) and $f'(x) < 0$ on (c, b);
2. $f(c)$ is a relative minimum if $f'(x) < 0$ on (a, c) and $f'(x) > 0$ on (c, b).

Test for Concavity On any open interval,

    if $f''(x) > 0$, then f is concave upward;

    if $f''(x) < 0$, then f is concave downward.

Second Derivative Test Suppose f'' exists on an open interval containing c and $f'(c) = 0$.

1. If $f''(c) > 0$, then $f(c)$ is a relative minimum.
2. If $f''(c) < 0$, then $f(c)$ is a relative maximum.
3. If $f''(c) = 0$ or $f''(c)$ does not exist, then the test gives no information about extrema, so use the first derivative test.

Curve Sketching To sketch the graph of a function f:

1. Consider the domain of the function, and note any restrictions. (That is, avoid dividing by 0, taking a square root of a negative number, or taking the logarithm of 0 or a negative number.)

2. Find the y-intercept (if it exists) by substituting $x = 0$ into $f(x)$. Find any x-intercepts by solving $f(x) = 0$ if this is not too difficult.

3. **a.** If f is a rational function, find any vertical asymptotes by investigating where the denominator is 0, and find any horizontal asymptotes by finding the limits as $x \to \infty$ and $x \to -\infty$.
   b. If f is an exponential function, find any horizontal asymptotes; if f is a logarithmic function, find any vertical asymptotes.

4. Investigate symmetry. If $f(-x) = f(x)$, the function is even, so the graph is symmetric about the y-axis. If $f(-x) = -f(x)$, the function is odd, so the graph is symmetric about the origin.

5. Find $f'(x)$. Locate any critical points by solving the equation $f'(x) = 0$ and determining where $f'(x)$ does not exist, but $f(x)$ does. Find any relative extrema and determine where f is increasing or decreasing.

6. Find $f''(x)$. Locate potential inflection points by solving the equation $f''(x) = 0$ and determining where $f''(x)$ does not exist. Determine where f is concave upward or concave downward.

7. Plot the intercepts, the critical points, the inflection points, the asymptotes, and other points as needed. Take advantage of any symmetry found in Step 4.

8. Connect the points with a smooth curve using the correct concavity, being careful not to connect points where the function is not defined.

9. Check your graph using a graphing calculator. If the picture looks very different from what you've drawn, see in what ways the picture differs and use that information to help find your mistake.

KEY TERMS

5.1
increasing function
decreasing function
critical number
critical point

5.2
relative (or local) maximum

relative (or local) minimum
relative (or local) extremum
first derivative test

5.3
second derivative
third derivative

fourth derivative
acceleration
concave upward and downward
concavity
inflection point

second derivative test
point of diminishing returns

5.4
curve sketching
oblique asymptote

REVIEW EXERCISES

CONCEPT CHECK

For Exercises 1–12 determine whether each of the following statements is true or false, and explain why.

1. A critical number c is a number in the domain of a function f for which $f'(c) = 0$ or $f'(c)$ does not exist.

2. If $f'(x) > 0$ on an interval, the function is positive on that interval.

3. If c is a critical number, then the function must have a relative maximum or minimum at c.

4. If f is continuous on (a, b), $f'(x) < 0$ on (a, c), and $f'(x) > 0$ on (c, b), then f has a relative minimum at c.

5. If $f'(c)$ exists, $f''(c)$ also exists.

6. The acceleration is the second derivative of the position function.

7. If $f''(x) > 0$ on an interval, the function is increasing on that interval.

8. If $f''(c) = 0$, the function has an inflection point at c.

9. If $f''(c) = 0$, the function does not have a relative maximum or minimum at c.

10. Every rational function has either a vertical or a horizontal asymptote.

11. If an odd function has a y-intercept, it must pass through the origin.

12. If $f'(c) = 0$, where c is a value in interval (a, b), then f is a constant on the interval (a, b).

PRACTICE AND EXPLORATIONS

13. When given the equation for a function, how can you determine where it is increasing and where it is decreasing?

14. When given the equation for a function, how can you determine where the relative extrema are located? Give two ways to test whether a relative extremum is a minimum or a maximum.

15. Does a relative maximum of a function always have the largest y-value in the domain of the function? Explain your answer.

16. What information about a graph can be found from the second derivative?

Find the open intervals where f is increasing or decreasing.

17. $f(x) = x^2 + 9x + 8$
18. $f(x) = -2x^2 + 7x + 14$
19. $f(x) = -x^3 + 2x^2 + 15x + 16$
20. $f(x) = 4x^3 + 8x^2 - 16x + 11$
21. $f(x) = \dfrac{16}{9 - 3x}$
22. $f(x) = \dfrac{15}{2x + 7}$
23. $f(x) = \ln|x^2 - 1|$
24. $f(x) = 8xe^{-4x}$

Find the locations and values of all relative maxima and minima.

25. $f(x) = -x^2 + 4x - 8$
26. $f(x) = x^2 - 6x + 4$
27. $f(x) = 2x^2 - 8x + 1$
28. $f(x) = -3x^2 + 2x - 5$
29. $f(x) = 2x^3 + 3x^2 - 36x + 20$
30. $f(x) = 2x^3 + 3x^2 - 12x + 5$
31. $f(x) = \dfrac{xe^x}{x - 1}$
32. $f(x) = \dfrac{\ln(3x)}{2x^2}$

Find the second derivative of each function, and then find $f''(1)$ and $f''(-3)$.

33. $f(x) = 3x^4 - 5x^2 - 11x$
34. $f(x) = 9x^3 + \dfrac{1}{x}$

35. $f(x) = \dfrac{4x + 2}{3x - 6}$

36. $f(x) = \dfrac{1 - 2x}{4x + 5}$

37. $f(t) = \sqrt{t^2 + 1}$

38. $f(t) = -\sqrt{5 - t^2}$

Graph each function, considering the domain, critical points, symmetry, regions where the function is increasing or decreasing, inflection points, regions where the function is concave up or concave down, intercepts where possible, and asymptotes where applicable.

39. $f(x) = -2x^3 - \dfrac{1}{2}x^2 + x - 3$

40. $f(x) = -\dfrac{4}{3}x^3 + x^2 + 30x - 7$

41. $f(x) = x^4 - \dfrac{4}{3}x^3 - 4x^2 + 1$

42. $f(x) = -\dfrac{2}{3}x^3 + \dfrac{9}{2}x^2 + 5x + 1$

43. $f(x) = \dfrac{x - 1}{2x + 1}$

44. $f(x) = \dfrac{2x - 5}{x + 3}$

45. $f(x) = -4x^3 - x^2 + 4x + 5$

46. $f(x) = x^3 + \dfrac{5}{2}x^2 - 2x - 3$

47. $f(x) = x^4 + 2x^2$

48. $f(x) = 6x^3 - x^4$

49. $f(x) = \dfrac{x^2 + 4}{x}$

50. $f(x) = x + \dfrac{8}{x}$

51. $f(x) = \dfrac{2x}{3 - x}$

52. $f(x) = \dfrac{-4x}{1 + 2x}$

53. $f(x) = xe^{2x}$

54. $f(x) = x^2 e^{2x}$

55. $f(x) = \ln(x^2 + 4)$

56. $f(x) = x^2 \ln x$

57. $f(x) = 4x^{1/3} + x^{4/3}$

58. $f(x) = 5x^{2/3} + x^{5/3}$

In Exercises 59 and 60, sketch the graph of a single function that has all of the properties listed.

59. **a.** Continuous everywhere except at $x = -4$, where there is a vertical asymptote

    b. A y-intercept at $y = -2$

    c. x-intercepts at $x = -3, 1$, and 4

    d. $f'(x) < 0$ on $(-\infty, -5), (-4, -1)$, and $(2, \infty)$

    e. $f'(x) > 0$ on $(-5, -4)$ and $(-1, 2)$

    f. $f''(x) > 0$ on $(-\infty, -4)$ and $(-4, -3)$

    g. $f''(x) < 0$ on $(-3, -1)$ and $(-1, \infty)$

    h. Differentiable everywhere except at $x = -4$ and $x = -1$

60. **a.** Continuous and differentiable everywhere except at $x = -3$, where it has a vertical asymptote

    b. A horizontal asymptote at $y = 1$

    c. An x-intercept at $x = -2$

    d. A y-intercept at $y = 4$

    e. $f'(x) > 0$ on the intervals $(-\infty, -3)$ and $(-3, 2)$

    f. $f'(x) < 0$ on the interval $(2, \infty)$

    g. $f''(x) > 0$ on the intervals $(-\infty, -3)$ and $(4, \infty)$

    h. $f''(x) < 0$ on the interval $(-3, 4)$

    i. $f'(2) = 0$

    j. An inflection point at $(4, 3)$

APPLICATIONS

Business and Economics

Stock Prices In Exercises 61 and 62, $P(t)$ is the price of a certain stock at time t during a particular day.

61. **a.** If the price of the stock is falling faster and faster, are $P'(t)$ and $P''(t)$ positive or negative?

    b. Explain your answer.

62. **a.** When the stock reaches its highest price of the day, are $P'(t)$ and $P''(t)$ positive, zero, or negative?

    b. Explain your answer.

63. **Cat Brushes** The cost function to produce q electric cat brushes is given by $C(q) = -10q^2 + 250q$. The demand equation is given by $p = -q^2 - 3q + 299$, where p is the price in dollars.

    a. Find and simplify the profit function.

    b. Find the number of brushes that will produce the maximum profit.

    c. Find the price that produces the maximum profit.

    d. Find the maximum profit.

    e. Find the point of diminishing returns for the profit function.

64. **Gasoline Prices** In 2008, the price of gasoline in the United States spiked and then dropped. The average monthly price (in cents per gallon) of unleaded regular gasoline for 2008 can be approximated by the function

$$p(t) = -0.614t^3 + 6.25t^2 + 1.94t + 297, \quad \text{for } 0 < t < 12,$$

where t is in months and $t = 1$ corresponds to January 2008. *Source: U.S. Energy Information Administration.*

    a. Determine the interval(s) on which the price is increasing.

    b. Determine the interval(s) on which the price is decreasing.

    c. Find any relative extrema for the price of gasoline, as well as when they occurred.

Life Sciences

65. **Weightlifting** An abstract for an article states, "We tentatively conclude that Olympic weightlifting ability in trained subjects undergoes a nonlinear decline with age, in which the second derivative of the performance versus age curve repeatedly changes sign." *Source: Medicine and Science in Sports and Exercise.*

    a. What does this quote tell you about the first derivative of the performance versus age curve?

    b. Describe what you know about the performance versus age curve based on the information in the quote.

66. Scaling Laws Many biological variables depend on body mass, with a functional relationship of the form

$$Y = Y_0 M^b,$$

where M represents body mass, b is a multiple of $1/4$, and Y_0 is a constant. For example, when Y represents metabolic rate, $b = 3/4$. When Y represents heartbeat, $b = -1/4$. When Y represents life span, $b = 1/4$. *Source: Science.*

a. Determine which of metabolic rate, heartbeat, and life span are increasing or decreasing functions of mass. Also determine which have graphs that are concave upward and which have graphs that are concave downward.

b. Verify that all functions of the form given above satisfy the equation

$$\frac{dY}{dM} = \frac{b}{M}Y.$$

This means that the rate of change of Y is proportional to Y and inversely proportional to body mass.

67. Thoroughbred Horses The association between velocity during exercise and blood lactate concentration after submaximal 800-m exercise of thoroughbred racehorses on sand and grass tracks has been studied. The lactate-velocity relationship can be described by the functions.

$$l_1(v) = 0.08e^{0.33v} \quad \text{and}$$
$$l_2(v) = -0.87v^2 + 28.17v - 211.41,$$

where $l_1(v)$ and $l_2(v)$ are the lactate concentrations (in mmol/L) and v is the velocity (in m/sec) of the horse during workout on sand and grass tracks, respectively. Sketch the graph of both functions for $13 \le v \le 17$. *Source: The Veterinary Journal.*

68. Neuron Communications In the FitzHugh-Nagumo model of how neurons communicate, the rate of change of the electric potential v with respect to time is given as a function of v by $f(v) = v(a - v)(v - 1)$, where a is a positive constant. Sketch a graph of this function when $a = 0.25$ and $0 \le v \le 1$. *Source: Mathematical Biology.*

69. Fruit Flies The number of imagoes (sexually mature adult fruit flies) per mated female per day (y) can be approximated by

$$y = 34.7(1.0186)^{-x}x^{-0.658},$$

where x is the mean density of the mated population (measured as flies per bottle) over a 16-day period. Sketch the graph of the function. *Source: Elements of Mathematical Biology.*

70. Blood Volume A formula proposed by Hurley for the red cell volume (RCV) in milliliters for males is

$$RCV = 1486S^2 - 4106S + 4514,$$

where S is the surface area (in square meters). A formula given by Pearson et al., is

$$RCV = 1486S - 825.$$

Source: Journal of Nuclear Medicine and British Journal of Haematology.

a. For the value of S which the RCV values given by the two formulas are closest, find the rate of change of RCV with respect to S for both formulas. What does this number represent?

b. The formula for plasma volume for males given by Hurley is

$$PV = 995e^{0.6085S},$$

while the formula given by Pearson et al., is

$$PV = 1578S,$$

where PV is measured in milliliters and S in square meters. Find the value of S for which the PV values given by the two formulas are the closest. Then find the value of PV that each formula gives for this value of S.

c. For the value of S found in part b, find the rate of change of PV with respect to S for both formulas. What does this number represent?

d. Notice in parts a and c that both formulas give the same instantaneous rate of change at the value of S for which the function values are closest. Prove that if two functions f and g are differentiable and never cross but are closest together when $x = x_0$, then $f'(x_0) = g'(x_0)$.

Social Sciences

71. Learning Researchers used a version of the Gompertz curve to model the rate that children learn with the equation

$$y(t) = A^{c^t},$$

where $y(t)$ is the portion of children of age t years passing a certain mental test, $A = 0.3982 \times 10^{-291}$, and $c = 0.4252$. Find the inflection point and describe what it signifies. (*Hint:* Leave A and c as constants until you are ready to calculate your final answer. If A is too small for your calculator to handle, use common logarithms and properties of logarithms to calculate $(\log A)/(\log e)$.) *Source: School and Society.*

72. Population Under the scenario that the fertility rate in the European Union (EU) remains at 1.8 until 2020, when it rises to replacement level, the predicted population (in millions) of the 15 member countries of the EU can be approximated over the next century by

$$P(t) = 325 + 7.475(t + 10)e^{-(t + 10)/20},$$

where t is the number of years since 2000. *Source: Science.*

a. In what year is the population predicted to be largest? What is the population predicted to be in that year?

b. In what year is the population declining most rapidly?

c. What is the population approaching as time goes on?

73. Nuclear Weapons The graph shows the total inventory of nuclear weapons held by the United States and by the Soviet

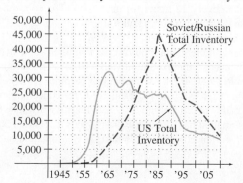

Union and its successor states from 1945 to 2010. (See Exercise 60 in the first section of this chapter.) *Source: Federation of American Scientists*.

a. In what years was the U.S. total inventory of weapons at a relative maximum?

b. When the U.S. total inventory of weapons was at the largest relative maximum, is the graph for the Soviet stockpile concave up or concave down? What does this mean?

Physical Sciences

74. Velocity and Acceleration A projectile is shot straight up with an initial velocity of 512 ft per second. Its height above the ground after t seconds is given by $s(t) = 512t - 16t^2$.

a. Find the velocity and acceleration after t seconds.

b. What is the maximum height attained?

c. When does the projectile hit the ground and with what velocity?

EXTENDED APPLICATION

A DRUG CONCENTRATION MODEL FOR ORALLY ADMINISTERED MEDICATIONS

Finding a range for the concentration of a drug in the bloodstream that is both safe and effective is one of the primary goals in pharmaceutical research and development. This range is called the *therapeutic window*. When determining the proper dosage (both the size of the dose and the frequency of administration), it is important to understand the behavior of the drug once it enters the body. Using data gathered during research we can create a mathematical model that predicts the concentration of the drug in the bloodstream at any given time.

We will look at two examples that explore a mathematical model for the concentration of a particular drug in the bloodstream. We will find the maximum and minimum concentrations of the drug given the size of the dose and the frequency of administration. We will then determine what dose should be administered to maintain concentrations within a given therapeutic window.

The drug tolbutamide is used for the management of mild to moderately severe type 2 diabetes. Suppose a 1000-mg dose of this drug is taken every 12 hours for three days. The concentration of the drug in the bloodstream, t hours after the initial dose is taken, is shown in Figure 47.

Looking at the graph, you can see that after a few doses have been administered, the maximum values of the concentration function begin to level off. The function also becomes periodic, repeating itself between every dose. At this point, the concentration is said to be at steady-state. The time it takes to reach steady-state depends on the elimination half-life of the drug (the time it takes for half the dose to be eliminated from the body). The elimination half-life of

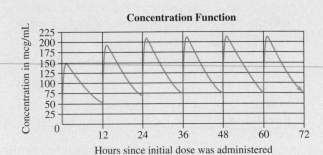

FIGURE 47

the drug used for this function is about 7 hours. Generally speaking, we say that steady-state is reached after about 5 half-lives.

We will define the *steady-state concentration function*, $C_{ss}(t)$, to be the concentration of drug in the bloodstream t hours after a dose has been administered once steady-state has been reached.

The steady-state concentration function can be written as the difference of two exponential decay functions, or

$$C_{ss}(t) = c_1 e^{kt} - c_2 e^{k_a t}.$$

The constants c_1 and c_2 are influenced by several factors, including the size of the dose and how widely the particular drug disperses through the body. The constants k_a and k are decay constants reflecting the rate at which the drug is being absorbed into the bloodstream and eliminated from the bloodstream, respectively.

Consider the following steady-state concentration function:

$$C_{ss}(t) = 0.2473De^{-0.1t} - 0.1728De^{-2.8t} \text{ mcg/mL}$$

where D is the size of the dose (in milligrams) administered every 12 hours. The concentration is given in micrograms per milliliter.

If a single dose is 1000 mg, then the concentration of drug in the bloodstream is

$$C_{ss}(t) = 247.3e^{-0.1t} - 172.8e^{-2.8t} \text{ mcg/mL}.$$

The graph of $C_{ss}(t)$ is given below in Figure 48.

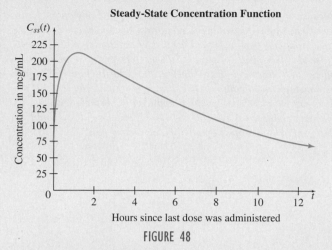

Steady-State Concentration Function

FIGURE 48

EXAMPLE 1 Drug Concentration

Find the maximum and minimum concentrations for the steady-state concentration function

$$C_{ss}(t) = 247.3e^{-0.1t} - 172.8e^{-2.8t} \text{ mcg/mL}.$$

SOLUTION The maximum concentration occurs when $C'_{ss}(t) = 0$. Calculating the derivative, we get:

$$C'_{ss}(t) = 247.3(-0.1)e^{-0.1t} - 172.8(-2.8)e^{-2.8t}$$
$$= -24.73e^{-0.1t} + 483.84e^{-2.8t}.$$

If we factor out $e^{-0.1t}$, we can find where the derivative is equal to zero.

$$C'_{ss}(t) = e^{-0.1t}(-24.73 + 483.84e^{-2.7t}) = 0$$

$C'_{ss}(t) = 0$ when

$$-24.73 + 483.84e^{-2.7t} = 0.$$

Solving this equation for t, we get

$$t = \frac{\ln\left(\dfrac{24.73}{483.84}\right)}{-2.7} \approx 1.1 \text{ hours.}$$

Therefore, the maximum concentration is

$$C_{ss}(1.1) = 247.3e^{-0.1(1.1)} - 172.8e^{-2.8(1.1)} \approx 214 \text{ mcg/mL.}$$

Looking at the graph of $C_{ss}(t)$ in Figure 48, you can see that the minimum concentration occurs at the endpoints (when $t = 0$ and $t = 12$; immediately after a dose is administered and immediately before a next dose is to be administered, respectively).

Therefore, the minimum concentration is

$$C_{ss}(0) = 247.3e^{-0.1(0)} - 172.8e^{-2.8(0)} = 247.3 - 172.8$$
$$= 74.5 \text{ mcg/mL.}$$

Verify that $C_{ss}(12)$ gives the same value.

If the therapeutic window for this drug is 70–240 mcg/mL, then, once steady-state has been reached, the concentration remains safe and effective as long as treatment continues.

Suppose, however, that a new study found that this drug is effective only if the concentration remains between 100 and 400 mcg/mL. How could you adjust the dose so that the maximum and minimum steady-state concentrations fall within this range?

EXAMPLE 2 Therapeutic Window

Find a range for the size of doses such that the steady-state concentration remains within the therapeutic window of 100 to 400 mcg/mL.

SOLUTION Recall that the steady-state concentration function is

$$C_{ss}(t) = 0.2473De^{-0.1t} - 0.1728De^{-2.8t} \text{ mcg/mL},$$

where D is the size of the dose given (in milligrams) every 12 hours.

From Example 1, we found that the minimum concentration occurs when $t = 0$. Therefore, we want the minimum concentration, $C_{ss}(0)$, to be greater than or equal to 100 mcg/mL.

$$C_{ss}(0) = 0.2473De^{-0.1(0)} - 0.1728De^{-2.8(0)} \geq 100$$

or

$$0.2473D - 0.1728D \geq 100$$

Solving for D, we get

$$0.0745D \geq 100$$

$$D \geq 1342 \text{ mg.}$$

In Example 1, we also found that the maximum concentration occurs when $t = 1.1$ hours. If we change the size of the dose, the maximum concentration will change; however, the time it takes to reach the maximum concentration does not change. Can you see why this is true?

Since the maximum concentration occurs when $t = 1.1$, we want $C_{ss}(1.1)$, the maximum concentration, to be less than or equal to 400 mcg/mL.

$$C_{ss}(1.1) = 0.2473De^{-0.1(1.1)} - 0.1728De^{-2.8(1.1)} \leq 400$$

or

$$0.2215D - 0.0079D \leq 400.$$

Solving for D, we get

$$0.2136D \leq 400$$

$$D \leq 1873 \text{ mg.}$$

Therefore, if the dose is between 1342 mg and 1873 mg, the steady-state concentration remains within the new therapeutic window.

EXERCISES

Use the following information to answer Exercises 1–3.

A certain drug is given to a patient every 12 hours. The steady-state concentration function is given by

$$C_{ss}(t) = 1.99De^{-0.14t} - 1.62De^{-2.08t} \text{ mcg/mL},$$

where D is the size of the dose in milligrams.

1. If a 500-mg dose is given every 12 hours, find the maximum and minimum steady-state concentrations.

2. If the dose is increased to 1500 mg every 12 hours, find the maximum and minimum steady-state concentrations.

3. What dose should be given every 12 hours to maintain a steady-state concentration between 80 and 400 mcg/mL?

DIRECTIONS FOR GROUP PROJECT

Because of declining health, many elderly people rely on prescription medications to stabilize or improve their medical condition. Your group has been assigned the task of developing a brochure to be made available at senior citizens' centers and physicians' offices that describes drug concentrations in the body for orally administered medications. The brochure should summarize the facts presented in this extended application but at a level that is understandable to a typical layperson. The brochure should be designed to look professional with a marketing flair.

6

Applications of the Derivative

When several variables are related by a single equation, their rates of change are also related. For example, the height and horizontal distance of a kite are related to the length of the string holding the kite. In an exercise in Section 5 we differentiate this relationship to discover how fast the kite flier must let out the string to maintain the kite at a constant height and constant horizontal speed.

The previous chapter included examples in which we used the derivative to find the maximum or minimum value of a function. This problem is ubiquitous; consider the efforts people expend trying to maximize their income, or to minimize their costs or the time required to complete a task. In this chapter we will treat the topic of optimization in greater depth.

The derivative is applicable in far wider circumstances, however. In roughly 500 B.C., Heraclitus said, "Nothing endures but change," and his observation has relevance here. If change is continuous, rather than in sudden jumps, the derivative can be used to describe the rate of change. This explains why calculus has been applied to so many fields.

6.1 Absolute Extrema

APPLY IT During a 10-year period, when did the U.S. dollar reach a minimum exchange rate with the Canadian dollar, and how much was the U.S. dollar worth then?

We will answer this question in Example 3.

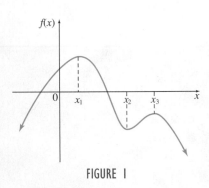

FIGURE 1

A function may have more than one relative maximum. It may be important, however, in some cases to determine if one function value is larger than any other. In other cases, we may want to know whether one function value is smaller than any other. For example, in Figure 1, $f(x_1) \geq f(x)$ for all x in the domain. There is no function value that is smaller than all others, however, because $f(x) \rightarrow -\infty$ as $x \rightarrow \infty$ or as $x \rightarrow -\infty$.

The largest possible value of a function is called the *absolute maximum* and the smallest possible value of a function is called the *absolute minimum*. As Figure 1 shows, one or both of these may not exist on the domain of the function, $(-\infty, \infty)$ here. Absolute extrema often coincide with relative extrema, as with $f(x_1)$ in Figure 1. Although a function may have several relative maxima or relative minima, it never has more than one *absolute maximum* or *absolute minimum,* although the absolute maximum or minimum might occur at more than one value of x.

Absolute Maximum or Minimum

Let f be a function defined on some interval. Let c be a number in the interval. Then $f(c)$ is the **absolute maximum** of f on the interval if

$$f(x) \leq f(c)$$

for every x in the interval, and $f(c)$ is the **absolute minimum** of f on the interval if

$$f(x) \geq f(c)$$

for every x in the interval.

A function has an **absolute extremum** (plural: **extrema**) at c if it has either an absolute maximum or an absolute minimum there.

CAUTION   Notice that, just like a relative extremum, an absolute extremum is a y-value, not an x-value.

Now look at Figure 2, which shows three functions defined on closed intervals. In each case there is an absolute maximum value and an absolute minimum value. These absolute extrema may occur at the endpoints or at relative extrema. As the graphs in Figure 2 show, an absolute extremum is either the largest or the smallest function value occurring on a closed interval, while a relative extremum is the largest or smallest function value in some (perhaps small) open interval.

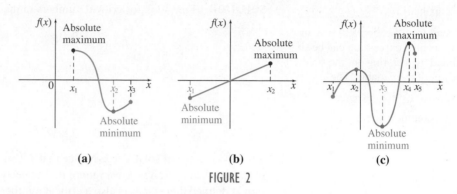

FIGURE 2

Although a function can have only one absolute minimum value and only one absolute maximum value, it can have many points where these values occur. (Note that the absolute maximum value and absolute minimum value are numbers, not points.) As an extreme example, consider the function $f(x) = 2$. The absolute minimum value of this function is clearly 2, as is the absolute maximum value. Both the absolute minimum and the absolute maximum occur at every real number x.

One of the main reasons for the importance of absolute extrema is given by the **extreme value theorem** (which is proved in more advanced courses).

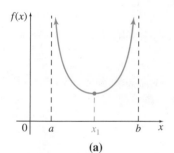

(a)

> ## Extreme Value Theorem
> A function f that is continuous on a closed interval $[a, b]$ will have both an absolute maximum and an absolute minimum on the interval.

A continuous function on an open interval may or may not have an absolute maximum or minimum. For example, the function in Figure 3(a) has an absolute minimum on the interval (a, b) at x_1, but it does not have an absolute maximum. Instead, it becomes arbitrarily large as x approaches a or b. Also, a discontinuous function on a closed interval may or may not have an absolute minimum or maximum. The function in Figure 3(b) has an absolute minimum at $x = a$, yet it has no absolute maximum. It may appear at first to have an absolute maximum at x_1, but notice that $f(x_1)$ has a smaller value than f at values of x less than x_1.

The extreme value theorem guarantees the existence of absolute extrema for a continuous function on a closed interval. To find these extrema, use the following steps.

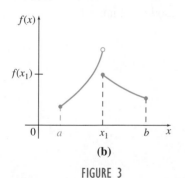

(b)

FIGURE 3

> ## Finding Absolute Extrema
> To find absolute extrema for a function f continuous on a closed interval $[a, b]$:
>
> **1.** Find all critical numbers for f in (a, b).
>
> **2.** Evaluate f for all critical numbers in (a, b).
>
> **3.** Evaluate f for the endpoints a and b of the interval $[a, b]$.
>
> **4.** The largest value found in Step 2 or 3 is the absolute maximum for f on $[a, b]$, and the smallest value found is the absolute minimum for f on $[a, b]$.

EXAMPLE 1 Absolute Extrema

Find the absolute extrema of the function

$$f(x) = x^{8/3} - 16x^{2/3}$$

on the interval $[-1, 8]$.

SOLUTION First look for critical numbers in the interval $(-1, 8)$.

$$f'(x) = \frac{8}{3}x^{5/3} - \frac{32}{3}x^{-1/3}$$

$$= \frac{8}{3}x^{-1/3}(x^2 - 4) \qquad \text{Factor.}$$

$$= \frac{8}{3}\left(\frac{x^2 - 4}{x^{1/3}}\right)$$

Set $f'(x) = 0$ and solve for x. Notice that $f'(x) = 0$ at $x = 2$ and $x = -2$, but -2 is not in the interval $(-1, 8)$, so we ignore it. The derivative is undefined at $x = 0$, but the function is defined there, so 0 is also a critical number.

Evaluate the function at the critical numbers and the endpoints.

	Extrema Candidates Value	
x-Value	of Function	
-1	-15	
0	0	
2	-19.05	← Absolute minimum
8	192	← Absolute maximum

The absolute maximum, 192, occurs when $x = 8$, and the absolute minimum, approximately -19.05, occurs when $x = 2$. A graph of f is shown in Figure 4.

TRY YOUR TURN 1

YOUR TURN 1 Find the absolute extrema of the function $f(x) = 3x^{2/3} - 3x^{5/3}$ on the interval $[0, 8]$.

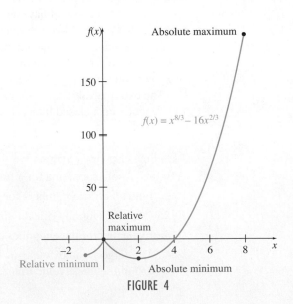

FIGURE 4

⊞ **TECHNOLOGY NOTE**

In Example 1, a graphing calculator that gives the maximum and minimum values of a function on an interval, such as the `fMax` or `fMin` feature of the TI-84 Plus, could replace the table. Alternatively, we could first graph the function on the given interval and then select the feature that gives the maximum or minimum value of the graph of the function instead of completing the table.

EXAMPLE 2 **Absolute Extrema**

Find the locations and values of the absolute extrema, if they exist, for the function

$$f(x) = 3x^4 - 4x^3 - 12x^2 + 2.$$

SOLUTION In this example, the extreme value theorem does not apply since the domain is an open interval, $(-\infty, \infty)$, which has no endpoints. Begin as before by finding any critical numbers.

$$f'(x) = 12x^3 - 12x^2 - 24x = 0$$
$$12x(x^2 - x - 2) = 0$$
$$12x(x + 1)(x - 2) = 0$$
$$x = 0 \quad \text{or} \quad x = -1 \quad \text{or} \quad x = 2$$

There are no values of x where $f'(x)$ does not exist. Evaluate the function at the critical numbers.

Extrema Candidates	
x-Value	Value of Function
-1	-3
0	2
2	-30 ← Absolute minimum

YOUR TURN 2 Find the locations and values of the absolute extrema, if they exist, for the function $f(x) = -x^4 - 4x^3 + 8x^2 + 20$.

For an open interval, rather than evaluating the function at the endpoints, we evaluate the limit of the function when the endpoints are approached. Because the positive x^4-term dominates the other terms as x becomes large,

$$\lim_{x \to \infty}(3x^4 - 4x^3 - 12x^2 + 2) = \infty.$$

The limit is also ∞ as x approaches $-\infty$. Since the function can be made arbitrarily large, it has no absolute maximum. The absolute minimum, -30, occurs at $x = 2$. This result can be confirmed with a graphing calculator, as shown in Figure 5. **TRY YOUR TURN 2**

In many of the applied extrema problems in the next section, a continuous function on an open interval has just one critical number. In that case, we can use the following theorem, which also applies to closed intervals.

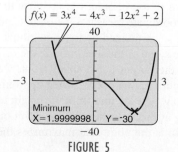

FIGURE 5

Critical Point Theorem

Suppose a function f is continuous on an interval I and that f has exactly one critical number in the interval I, located at $x = c$.

If f has a relative maximum at $x = c$, then this relative maximum is the absolute maximum of f on the interval I.

If f has a relative minimum at $x = c$, then this relative minimum is the absolute minimum of f on the interval I.

The critical point theorem is of no help in the previous two examples because they each had more than one critical point on the interval under consideration. But the theorem could be useful for some of Exercises 31–38 at the end of this section, and we will make good use of it in the next section.

EXAMPLE 3 **U.S. and Canadian Dollar Exchange**

The U.S. and Canadian exchange rate changes daily. The value of the U.S. dollar (in Canadian dollars) between 2000 and 2010 can be approximated by the function

$$f(t) = 0.00316t^3 - 0.0471t^2 + 0.114t + 1.47$$

where t is the number of years since 2000. Based on this approximation, in what year during this period did the value of the U.S. dollar reach its absolute minimum? What is the minimum value of the dollar during this period? *Source: The Federal Reserve.*

APPLY IT

SOLUTION The function is defined on the interval [0, 10]. We first look for critical numbers in this interval. Here $f'(t) = 0.00948t^2 - 0.0942t + 0.114$. We set this derivative equal to 0 and use the quadratic formula to solve for t.

$$0.00948t^2 - 0.0942t + 0.114 = 0$$

$$t = \frac{0.0942 \pm \sqrt{(-0.0942)^2 - 4(0.00948)(0.114)}}{2(0.00948)}$$

$$t = 1.41 \quad \text{or} \quad t = 8.53$$

Both values are in the interval [0,10]. Now evaluate the function at the critical numbers and the endpoints 0 and 10.

t-Value	Extrema Candidates Value of Function	
0	1.47	
1.41	1.55	
8.53	0.977	← Absolute minimum
10	1.06	

About 8.53 years after 2000, that is, around the middle of 2008, the U.S. dollar was worth about $0.98 Canadian, which was an absolute minimum during this period. It is also worth noting the absolute maximum value of the U.S. dollar, which was about $1.55 Canadian, occurred approximately 1.41 years after 2000, or around May of 2001. ∎

Graphical Optimization

Figure 6 shows the production output for a family owned business that produces landscape mulch. As the number of workers (measured in hours of labor) varies, the total production of mulch also varies. In Section 3.5, Example 1, we saw that maximum production occurs with 8 workers (corresponding to 320 hours of labor). A manager, however, may want to know how many hours of labor to use in order to maximize the output per hour of labor. For any point on the curve, the y-coordinate measures the output and the x-coordinate measures the hours of labor, so the y-coordinate divided by the x-coordinate gives the output per hour of labor. This quotient is also the slope of the line through the origin and the point on the curve. Therefore, to maximize the output per hour of labor, we need to find where this slope is greatest. As shown in Figure 6, this occurs when approximately 270 hours of labor are used. Notice that this is also where the line from the origin to the curve is tangent to the curve. Another way of looking at this is to say that the point on the curve where the tangent line passes through the origin is the point that maximizes the output per hour of labor.

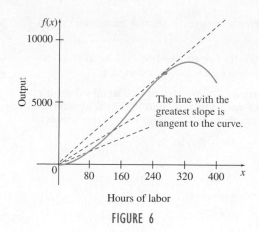

FIGURE 6

We can show that, in general, when $y = f(x)$ represents the output as a function of input, the maximum output per unit input occurs when the line from the origin to a point on the graph of the function is tangent to the function. Our goal is to maximize

$$g(x) = \frac{\text{output}}{\text{input}} = \frac{f(x)}{x}.$$

Taking the derivative and setting it equal to 0 gives

$$g'(x) = \frac{xf'(x) - f(x)}{x^2} = 0$$

$$xf'(x) = f(x)$$

$$f'(x) = \frac{f(x)}{x}.$$

Notice that $f'(x)$ gives the slope of the tangent line at the point, and $f(x)/x$ gives the slope of the line from the origin to the point. When these are equal, as in Figure 6, the output per input is maximized. In other examples, the point on the curve where the tangent line passes through the origin gives a minimum. For a life science example of this, see Exercise 54 in Section 3.4 on the Definition of the Derivative.

6.1 EXERCISES

Find the locations of any absolute extrema for the functions with graphs as follows.

1.

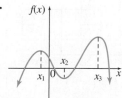

2.

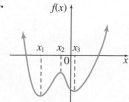

5.

g(x) graph with x_1

6.

g(x) graph with x_1

3.

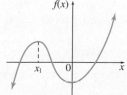

4.

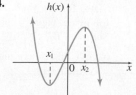

7.

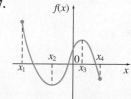

8.

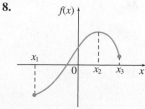

9. What is the difference between a relative extremum and an absolute extremum?

10. Can a relative extremum be an absolute extremum? Is a relative extremum necessarily an absolute extremum?

Find the absolute extrema if they exist, as well as all values of x where they occur, for each function, and specified domain. If you have one, use a graphing calculator to verify your answers.

11. $f(x) = x^3 - 6x^2 + 9x - 8;\quad [0, 5]$

12. $f(x) = x^3 - 3x^2 - 24x + 5;\quad [-3, 6]$

13. $f(x) = \frac{1}{3}x^3 + \frac{3}{2}x^2 - 4x + 1;\quad [-5, 2]$

14. $f(x) = \frac{1}{3}x^3 - \frac{1}{2}x^2 - 6x + 3;\quad [-4, 4]$

15. $f(x) = x^4 - 18x^2 + 1;\quad [-4, 4]$

16. $f(x) = x^4 - 32x^2 - 7;\quad [-5, 6]$

17. $f(x) = \frac{1-x}{3+x};\quad [0, 3]$ **18.** $f(x) = \frac{8+x}{8-x};\quad [4, 6]$

19. $f(x) = \frac{x-1}{x^2+1};\quad [1, 5]$ **20.** $f(x) = \frac{x}{x^2+2};\quad [0, 4]$

21. $f(x) = (x^2 - 4)^{1/3};\quad [-2, 3]$

22. $f(x) = (x^2 - 16)^{2/3};\quad [-5, 8]$

23. $f(x) = 5x^{2/3} + 2x^{5/3};\quad [-2, 1]$

24. $f(x) = x + 3x^{2/3};\quad [-10, 1]$

25. $f(x) = x^2 - 8\ln x;\quad [1, 4]$ **26.** $f(x) = \frac{\ln x}{x^2};\quad [1, 4]$

27. $f(x) = x + e^{-3x};\quad [-1, 3]$ **28.** $f(x) = x^2 e^{-0.5x};\quad [2, 5]$

Graph each function on the indicated domain, and use the capabilities of your calculator to find the location and value of the absolute extrema.

29. $f(x) = \frac{-5x^4 + 2x^3 + 3x^2 + 9}{x^4 - x^3 + x^2 + 7};\quad [-1, 1]$

30. $f(x) = \frac{x^3 + 2x + 5}{x^4 + 3x^3 + 10};\quad [-3, 0]$

Find the absolute extrema if they exist, as well as all values of x where they occur.

31. $f(x) = 2x + \frac{8}{x^2} + 1, \quad x > 0$

32. $f(x) = 12 - x - \frac{9}{x}, \quad x > 0$

33. $f(x) = -3x^4 + 8x^3 + 18x^2 + 2$

34. $f(x) = x^4 - 4x^3 + 4x^2 + 1$

35. $f(x) = \frac{x-1}{x^2 + 2x + 6}$ **36.** $f(x) = \frac{x}{x^2 + 1}$

37. $f(x) = \frac{\ln x}{x^3}$ **38.** $f(x) = x \ln x$

39. Find the absolute maximum and minimum of $f(x) = 2x - 3x^{2/3}$ **(a)** on the interval $[-1,\ 0.5]$; **(b)** on the interval $[0.5, 2]$.

40. Let $f(x) = e^{-2x}$. For $x > 0$, let $P(x)$ be the perimeter of the rectangle with vertices $(0, 0)$, $(x, 0)$, $(x, f(x))$ and $(0, f(x))$. Which of the following statements is true? *Source: Society of Actuaries.*

a. The function P has an absolute minimum but not an absolute maximum on the interval $(0, \infty)$.

b. The function P has an absolute maximum but not an absolute minimum on the interval $(0, \infty)$.

c. The function P has both an absolute minimum and an absolute maximum on the interval $(0, \infty)$.

d. The function P has neither an absolute maximum nor an absolute minimum on the interval $(0, \infty)$, but the graph of the function P does have an inflection point with positive x-coordinate.

e. The function P has neither an absolute maximum nor an absolute minimum on the interval $(0, \infty)$, and the graph of the function P does not have an inflection point with positive x-coordinate.

APPLICATIONS

Business and Economics

41. Bank Robberies The number of bank robberies in the United States for the years 2000–2009 is given in the following figure. Consider the closed interval [2000, 2009]. *Source: FBI.*

a. Give all relative maxima and minima and when they occur on the interval.

b. Give the absolute maxima and minima and when they occur on the interval. Interpret your results.

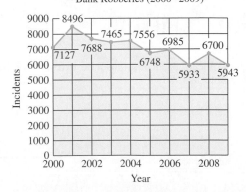

Bank Robberies (2000–2009)

42. Bank Burglaries The number of bank burglaries (entry into or theft from a bank during nonbusiness hours) in the United States for the years 2000–2009 is given in the figure on the following page. Consider the closed interval [2000, 2009]. *Source: FBI.*

a. Give all relative maxima and minima and when they occur on the interval.

b. Give the absolute maxima and minima and when they occur on the interval. Interpret your results.

Bank Burglaries (2000–2009)

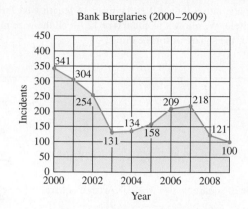

43. Profit The total profit $P(x)$ (in thousands of dollars) from the sale of x hundred thousand automobile tires is approximated by

$$P(x) = -x^3 + 9x^2 + 120x - 400, \quad x \geq 5.$$

Find the number of hundred thousands of tires that must be sold to maximize profit. Find the maximum profit.

44. Profit A company has found that its weekly profit from the sale of x units of an auto part is given by

$$P(x) = -0.02x^3 + 600x - 20,000.$$

Production bottlenecks limit the number of units that can be made per week to no more than 150, while a long-term contract requires that at least 50 units be made each week. Find the maximum possible weekly profit that the firm can make.

Average Cost Find the minimum value of the average cost for the given cost function on the given intervals.

45. $C(x) = x^3 + 37x + 250$ on the following intervals.

    a. $1 \leq x \leq 10$ **b.** $10 \leq x \leq 20$

46. $C(x) = 81x^2 + 17x + 324$ on the following intervals.

    a. $1 \leq x \leq 10$ **b.** $10 \leq x \leq 20$

Cost Each graph gives the cost as a function of production level. Use the method of graphical optimization to estimate the production level that results in the minimum cost per item produced.

47.

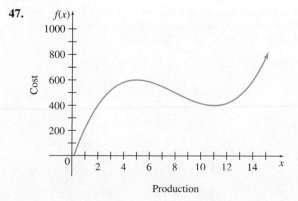

48.

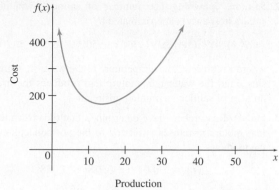

Profit Each graph gives the profit as a function of production level. Use graphical optimization to estimate the production level that gives the maximum profit per item produced.

49.

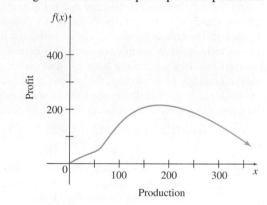

50.

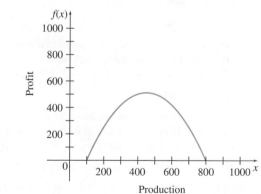

Life Sciences

51. Pollution A marshy region used for agricultural drainage has become contaminated with selenium. It has been determined that flushing the area with clean water will reduce the selenium for a while, but it will then begin to build up again. A biologist has found that the percent of selenium in the soil x months after the flushing begins is given by

$$f(x) = \frac{x^2 + 36}{2x}, \quad 1 \leq x \leq 12.$$

When will the selenium be reduced to a minimum? What is the minimum percent?

52. Salmon Spawning The number of salmon swimming upstream to spawn is approximated by

$$S(x) = -x^3 + 3x^2 + 360x + 5000, \quad 6 \le x \le 20,$$

where x represents the temperature of the water in degrees Celsius. Find the water temperature that produces the maximum number of salmon swimming upstream.

53. Molars Researchers have determined that the crown length of first molars in fetuses is related to the postconception age of the tooth as

$$L(t) = -0.01t^2 + 0.788t - 7.048,$$

where $L(t)$ is the crown length (in millimeters) of the molar t weeks after conception. Find the maximum length of the crown of first molars during weeks 22 through 28. *Source: American Journal of Physical Anthropology*.

54. Fungal Growth Because of the time that many people spend indoors, there is a concern about the health risk of being exposed to harmful fungi that thrive in buildings. The risk appears to increase in damp environments. Researchers have discovered that by controlling both the temperature and the relative humidity in a building, the growth of the fungus *A. versicolor* can be limited. The relationship between temperature and relative humidity, which limits growth, can be described by

$$R(T) = -0.00007T^3 + 0.0401T^2 - 1.6572T + 97.086,$$

$$15 \le T \le 46,$$

where $R(T)$ is the relative humidity (in percent) and T is the temperature (in degrees Celsius). Find the temperature at which the relative humidity is minimized. *Source: Applied and Environmental Microbiology*.

Physical Sciences

55. Gasoline Mileage From information given in a recent business publication, we constructed the mathematical model

$$M(x) = -\frac{1}{45}x^2 + 2x - 20, \quad 30 \le x \le 65,$$

to represent the miles per gallon used by a certain car at a speed of x mph. Find the absolute maximum miles per gallon and the absolute minimum and the speeds at which they occur.

56. Gasoline Mileage For a certain sports utility vehicle,

$$M(x) = -0.015x^2 + 1.31x - 7.3, \quad 30 \le x \le 60,$$

represents the miles per gallon obtained at a speed of x mph. Find the absolute maximum miles per gallon and the absolute minimum and the speeds at which they occur.

General Interest

Area A piece of wire 12 ft long is cut into two pieces. (See the figure.) One piece is made into a circle and the other piece is made into a square. Let the piece of length x be formed into a circle. We allow x to equal 0 or 12, so all the wire may be used for the square or for the circle.

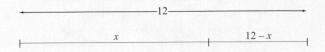

$$\text{Radius of circle} = \frac{x}{2\pi} \qquad \text{Area of circle} = \pi\left(\frac{x}{2\pi}\right)^2$$

$$\text{Side of square} = \frac{12 - x}{4} \qquad \text{Area of square} = \left(\frac{12 - x}{4}\right)^2$$

57. Where should the cut be made in order to minimize the sum of the areas enclosed by both figures?

58. Where should the cut be made in order to make the sum of the areas maximum? (*Hint:* Remember to use the endpoints of a domain when looking for absolute maxima and minima.)

59. For the solution to Exercise 57, show that the side of the square equals the diameter of the circle, that is, that the circle can be inscribed in the square.*

60. Information Content Suppose dots and dashes are transmitted over a telegraph line so that dots occur a fraction p of the time (where $0 < p < 1$) and dashes occur a fraction $1 - p$ of the time. The *information content* of the telegraph line is given by $I(p)$, where

$$I(p) = -p \ln p - (1 - p) \ln(1 - p).$$

a. Show that $I'(p) = -\ln p + \ln(1 - p)$.

b. Set $I'(p) = 0$ and find the value of p that maximizes the information content.

c. How might the result in part b be used?

YOUR TURN ANSWERS

1. Absolute maximum of about 0.977 occurs when $x = 2/5$ and absolute minimum of -84 when $x = 8$.

2. Absolute maximum, 148, occurs at $x = -4$. No absolute minimum.

*For a generalization of this phenomenon, see Cade, Pat and Russell A. Gordon, "An Apothem Apparently Appears," *The College Mathematics Journal*, Vol. 36, No. 1, Jan. 2005, pp. 52–55.

6.2 Applications of Extrema

APPLY IT **How should boxes and cans be designed to minimize the material needed to construct them or to maximize the volume?**
In Examples 3 and 4 we will use the techniques of calculus to find an answer to these questions.

In this section we give several examples showing applications of calculus to maximum and minimum problems. To solve these examples, go through the following steps.

Solving an Applied Extrema Problem

1. Read the problem carefully. Make sure you understand what is given and what is unknown.
2. If possible, sketch a diagram. Label the various parts.
3. Decide on the variable that must be maximized or minimized. Express that variable as a function of *one* other variable.
4. Find the domain of the function.
5. Find the critical points for the function from Step 3.
6. If the domain is a closed interval, evaluate the function at the endpoints and at each critical number to see which yields the absolute maximum or minimum. If the domain is an open interval, apply the critical point theorem when there is only one critical number. If there is more than one critical number, evaluate the function at the critical numbers and find the limit as the endpoints of the interval are approached to determine if an absolute maximum or minimum exists at one of the critical points.

CAUTION Do not skip Step 6 in the preceding box. If a problem asks you to maximize a quantity and you find a critical point at Step 5, do not automatically assume the maximum occurs there, for it may occur at an endpoint, as in Exercise 58 of the previous section, or it may not exist at all.

An infamous case of such an error occurred in a 1945 study of "flying wing" aircraft designs similar to the Stealth bomber. In seeking to maximize the range of the aircraft (how far it can fly on a tank of fuel), the study's authors found that a critical point occurred when almost all of the volume of the plane was in the wing. They claimed that this critical point was a maximum. But another engineer later found that this critical point, in fact, *minimized* the range of the aircraft! *Source: Science*.

EXAMPLE 1 Maximization

Find two nonnegative numbers x and y for which $2x + y = 30$, such that xy^2 is maximized.

SOLUTION Step 1, reading and understanding the problem, is up to you. Step 2 does not apply in this example; there is nothing to draw. We proceed to Step 3, in which we decide what is to be maximized and assign a variable to that quantity. Here, xy^2 is to be maximized, so let

$$M = xy^2.$$

According to Step 3, we must express M in terms of just *one* variable, which can be done using the equation $2x + y = 30$ by solving for either x or y. Solving for y gives

$$2x + y = 30$$
$$y = 30 - 2x.$$

Substitute for y in the expression for M to get

$$M = x(30 - 2x)^2$$
$$= x(900 - 120x + 4x^2)$$
$$= 900x - 120x^2 + 4x^3.$$

We are now ready for Step 4, when we find the domain of the function. Because of the nonnegativity requirement, x must be at least 0. Since y must also be at least 0, we require $30 - 2x \geq 0$, so $x \leq 15$. Thus x is confined to the interval $[0, 15]$.

Moving on to Step 5, we find the critical points for M by finding dM/dx, and then solving the equation $dM/dx = 0$ for x.

Extrema Candidates		
x	M	
0	0	
5	2000	← Maximum
15	0	

$$\frac{dM}{dx} = 900 - 240x + 12x^2 = 0$$

$$12(75 - 20x + x^2) = 0 \quad \text{Factor out the 12.}$$

$$12(5 - x)(15 - x) = 0 \quad \text{Factor the quadratic.}$$

$$x = 5 \quad \text{or} \quad x = 15$$

YOUR TURN 1 Find two nonnegative numbers x and y for which $x + 3y = 30$, such that $x^2 y$ is maximized.

Finally, at Step 6, we find M for the critical numbers $x = 5$ and $x = 15$, as well as for $x = 0$, an endpoint of the domain. The other endpoint, $x = 15$, has already been included as a critical number. We see in the table that the maximum value of the function occurs when $x = 5$. Since $y = 30 - 2x = 30 - 2(5) = 20$, the values that maximize xy^2 are $x = 5$ and $y = 20$.

TRY YOUR TURN 1

NOTE A critical point is only a candidate for an absolute maximum or minimum. The absolute maximum or minimum might occur at a different critical point or at an endpoint.

EXAMPLE 2 Minimizing Time

A math professor participating in the sport of orienteering must get to a specific tree in the woods as fast as possible. He can get there by traveling east along the trail for 300 m and then north through the woods for 800 m. He can run 160 m per minute along the trail but only 70 m per minute through the woods. Running directly through the woods toward the tree minimizes the distance, but he will be going slowly the whole time. He could instead run 300 m along the trail before entering the woods, maximizing the total distance but minimizing the time in the woods. Perhaps the fastest route is a combination, as shown in Figure 7. Find the path that will get him to the tree in the minimum time.

SOLUTION As in Example 1, the first step is to read and understand the problem. If the statement of the problem is not clear to you, go back and reread it until you understand it before moving on.

We have already started Step 2 by providing Figure 7. Let x be the distance shown in Figure 7, so the distance he runs on the trail is $300 - x$. By the Pythagorean theorem, the distance he runs through the woods is $\sqrt{800^2 + x^2}$.

The first part of Step 3 is noting that we are trying to minimize the total amount of time, which is the sum of the time on the trail and the time through the woods. We must express this time as a function of x. Since time = distance/speed, the total time is

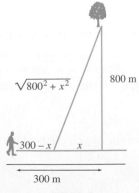

FIGURE 7

$$T(x) = \frac{300 - x}{160} + \frac{\sqrt{800^2 + x^2}}{70}.$$

To complete Step 4, notice in this equation that $0 \leq x \leq 300$.

We now move to Step 5, in which we find the critical points by calculating the derivative and setting it equal to 0. Since $\sqrt{800^2 + x^2} = (800^2 + x^2)^{1/2}$,

$$T'(x) = -\frac{1}{160} + \frac{1}{70}\left(\frac{1}{2}\right)(800^2 + x^2)^{-1/2}(2x) = 0.$$

$$\frac{x}{70\sqrt{800^2 + x^2}} = \frac{1}{160}$$

$16x = 7\sqrt{800^2 + x^2}$	Cross multiply and divide by 10.
$256x^2 = 49(800^2 + x^2) = (49 \cdot 800^2) + 49x^2$	Square both sides.
$207x^2 = 49 \cdot 800^2$	Subtract $49x^2$ from both sides.
$x^2 = \dfrac{49 \cdot 800^2}{207}$	
$x = \dfrac{7 \cdot 800}{\sqrt{207}} \approx 389$	Take the square root of both sides

Since 389 is not in the interval $[0, 300]$, the minimum time must occur at one of the endpoints.

We now complete Step 6 by creating a table with $T(x)$ evaluated at the endpoints. We see from the table that the time is minimized when $x = 300$, that is, when the professor heads straight for the tree. **TRY YOUR TURN 2**

Extrema Candidates

x	$T(x)$	
0	13.30	
300	12.21	← Minimum

YOUR TURN 2 Suppose the professor in Example 2 can only run 40 m per minute through the woods. Find the path that will get him to the tree in the minimum time.

EXAMPLE 3 **Maximizing Volume**

An open box is to be made by cutting a square from each corner of a 12-in. by 12-in. piece of metal and then folding up the sides. What size square should be cut from each corner to produce a box of maximum volume?

APPLY IT

SOLUTION Let x represent the length of a side of the square that is cut from each corner, as shown in Figure 8(a). The width of the box is $12 - 2x$, with the length also $12 - 2x$. As shown in Figure 8(b), the depth of the box will be x inches. The volume of the box is given by the product of the length, width, and height. In this example, the volume, $V(x)$, depends on x:

$$V(x) = x(12 - 2x)(12 - 2x) = 144x - 48x^2 + 4x^3.$$

Clearly, $0 \le x$, and since neither the length nor the width can be negative, $0 \le 12 - 2x$, so $x \le 6$. Thus, the domain of V is the interval $[0, 6]$.

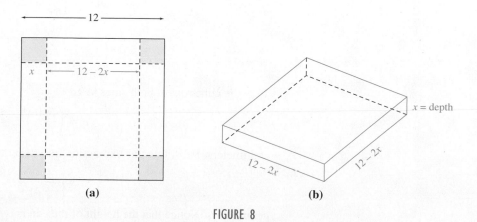

(a)   (b)

FIGURE 8

Extrema Candidates	
x	$V(x)$
0	0
2	128 ← Maximum
6	0

YOUR TURN 3 Repeat Example 3 using an 8-m by 8-m piece of metal.

The derivative is $V'(x) = 144 - 96x + 12x^2$. Set this derivative equal to 0.

$$12x^2 - 96x + 144 = 0$$
$$12(x^2 - 8x + 12) = 0$$
$$12(x - 2)(x - 6) = 0$$
$$x - 2 = 0 \quad \text{or} \quad x - 6 = 0$$
$$x = 2 \quad\quad\quad\quad x = 6$$

Find $V(x)$ for x equal to 0, 2, and 6 to find the depth that will maximize the volume. The table indicates that the box will have maximum volume when $x = 2$ and that the maximum volume will be 128 in³.   **TRY YOUR TURN 3**

EXAMPLE 4   Minimizing Area

A company wants to manufacture cylindrical aluminum cans with a volume of 1000 cm³ (1 liter). What should the radius and height of the can be to minimize the amount of aluminum used?

APPLY IT

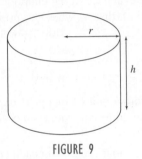

FIGURE 9

SOLUTION   The two variables in this problem are the radius and the height of the can, which we shall label r and h, as in Figure 9. Minimizing the amount of aluminum used requires minimizing the surface area of the can, which we will designate S. The surface area consists of a top and a bottom, each of which is a circle with an area πr^2, plus the side. If the side were sliced vertically and unrolled, it would form a rectangle with height h and width equal to the circumference of the can, which is $2\pi r$. Thus the surface area is given by

$$S = 2\pi r^2 + 2\pi rh.$$

The right side of the equation involves two variables. We need to get a function of a single variable. We can do this by using the information about the volume of the can:

$$V = \pi r^2 h = 1000.$$

(Here we have used the formula for the volume of a cylinder.) Solve this for h:

$$h = \frac{1000}{\pi r^2}.$$

(Solving for r would have involved a square root and a more complicated function.) We now substitute this expression for h into the equation for S to get

$$S = 2\pi r^2 + 2\pi r \frac{1000}{\pi r^2} = 2\pi r^2 + \frac{2000}{r}.$$

There are no restrictions on r other than that it be a positive number, so the domain of S is $(0, \infty)$.

Find the critical points for S by finding dS/dr, then solving the equation $dS/dr = 0$ for r.

$$\frac{dS}{dr} = 4\pi r - \frac{2000}{r^2} = 0$$
$$4\pi r^3 = 2000$$
$$r^3 = \frac{500}{\pi}$$

Take the cube root of both sides to get

$$r = \left(\frac{500}{\pi}\right)^{1/3} \approx 5.419$$

centimeters. Substitute this expression into the equation for h to get

$$h = \frac{1000}{\pi 5.419^2} \approx 10.84$$

centimeters. Notice that the height of the can is twice its radius.

There are several ways to carry out Step 6 to verify that we have found the minimum. Because there is only one critical number, the critical point theorem applies.

Method 1
Critical Point Theorem
with First Derivative Test

Verify that when $r < 5.419$, then $dS/dr < 0$, and when $r > 5.419$, then $dS/dr > 0$. Since the function is decreasing before 5.419 and increasing after 5.419, there must be a relative minimum at $r = 5.419$ cm. By the critical point theorem, there is an absolute minimum there.

Method 2
Critical Point Theorem
with Second Derivative Test

We could also use the critical point theorem with the second derivative test.

$$\frac{d^2S}{dr^2} = 4\pi + \frac{4000}{r^3}$$

Notice that for positive r, the second derivative is always positive, so there is a relative minimum at $r = 5.419$ cm. By the critical point theorem, there is an absolute minimum there.

Method 3
Limits at Endpoints

We could also find the limit as the endpoints are approached.

$$\lim_{r \to 0} S = \lim_{r \to \infty} S = \infty$$

The surface area becomes arbitrarily large as r approaches the endpoints of the domain, so the absolute minimum surface area must be at the critical point.

YOUR TURN 4 Repeat Example 4 if the volume is to be 500 cm³.

The graphing calculator screen in Figure 10 confirms that there is an absolute minimum at $r = 5.419$ cm.

TRY YOUR TURN 4

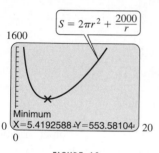

FIGURE 10

Notice that if the previous example had asked for the height and radius that maximize the amount of aluminum used, the problem would have no answer. There is no maximum for a function that can be made arbitrarily large.

Maximum Sustainable Harvest

For most living things, reproduction is *seasonal*— it can take place only at selected times of the year. Large whales, for example, reproduce every two years during a relatively short time span of about two months. Shown on the time axis in Figure 11 are the reproductive periods. Let S = number of adults present during the reproductive period and let R = number of adults that return the next season to reproduce. *Source: Mathematics for the Biosciences*.

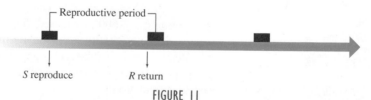

FIGURE 11

If we find a relationship between R and S, $R = f(S)$, then we have formed a **spawner-recruit** function or **parent-progeny** function. These functions are notoriously hard to develop because of the difficulty of obtaining accurate counts and because of the many hypotheses that can be made about the life stages. We will simply suppose that the function f takes various forms.

If $R > S$, we can presumably harvest

$$H = R - S = f(S) - S$$

individuals, leaving S to reproduce. Next season, $R = f(S)$ will return and the harvesting process can be repeated, as shown in Figure 12 on the next page.

Let S_0 be the number of spawners that will allow as large a harvest as possible without threatening the population with extinction. Then $H(S_0)$ is called the **maximum sustainable harvest**.

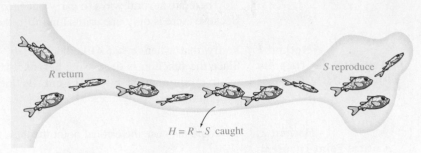

FIGURE 12

EXAMPLE 5 **Maximum Sustainable Harvest**

Suppose the spawner-recruit function for Idaho rabbits is $f(S) = 2.17\sqrt{S}\ln(S + 1)$, where S is measured in thousands of rabbits. Find S_0 and the maximum sustainable harvest, $H(S_0)$.

SOLUTION S_0 is the value of S that maximizes H. Since

$$H(S) = f(S) - S$$
$$= 2.17\sqrt{S}\ln(S + 1) - S,$$
$$H'(S) = 2.17\left(\frac{\ln(S + 1)}{2\sqrt{S}} + \frac{\sqrt{S}}{S + 1}\right) - 1.$$

Now we want to set this derivative equal to 0 and solve for S.

$$0 = 2.17\left(\frac{\ln(S + 1)}{2\sqrt{S}} + \frac{\sqrt{S}}{S + 1}\right) - 1.$$

This equation cannot be solved analytically, so we will graph $H'(S)$ with a graphing calculator and find any S-values where $H'(S)$ is 0. (An alternative approach is to use the equation solver some graphing calculators have.) The graph with the value where $H'(S)$ is 0 is shown in Figure 13.

From the graph we see that $H'(S) = 0$ when $S = 36.557775$, so the number of rabbits needed to sustain the population is about 36,600. A graph of H will show that this is a maximum. From the graph, using the capability of the calculator, we find that the harvest is $H(36.557775) \approx 11.015504$. These results indicate that after one reproductive season, a population of 36,600 rabbits will have increased to 47,600. Of these, 11,000 may be harvested, leaving 36,600 to regenerate the population. Any harvest larger than 11,000 will threaten the future of the rabbit population, while a harvest smaller than 11,000 will allow the population to grow larger each season. Thus 11,000 is the maximum sustainable harvest for this population.

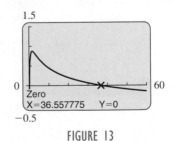

FIGURE 13

6.2 EXERCISES

In Exercises 1–4, use the steps shown in Exercise 1 to find non-negative numbers x and y that satisfy the given requirements. Give the optimum value of the indicated expression.

1. $x + y = 180$ and the product $P = xy$ is as large as possible.

  a. Solve $x + y = 180$ for y.

  b. Substitute the result from part a into $P = xy$, the equation for the variable that is to be maximized.

  c. Find the domain of the function P found in part b.

  d. Find dP/dx. Solve the equation $dP/dx = 0$.

  e. Evaluate P at any solutions found in part d, as well as at the endpoints of the domain found in part c.

  f. Give the maximum value of P, as well as the two numbers x and y whose product is that value.

2. The sum of x and y is 140 and the sum of the squares of x and y is minimized.

3. $x + y = 90$ and x^2y is maximized.

4. $x + y = 105$ and xy^2 is maximized.

APPLICATIONS

Business and Economics

Average Cost In Exercises 5 and 6, determine the average cost function $\overline{C}(x) = C(x)/x$. To find where the average cost is smallest, first calculate $\overline{C}'(x)$, the derivative of the average cost function. Then use a graphing calculator to find where the derivative is 0. Check your work by finding the minimum from the graph of the function $\overline{C}(x)$.

5. $C(x) = \dfrac{1}{2}x^3 + 2x^2 - 3x + 35$

6. $C(x) = 10 + 20x^{1/2} + 16x^{3/2}$

7. Revenue If the price charged for a candy bar is $p(x)$ cents, then x thousand candy bars will be sold in a certain city, where

$$p(x) = 160 - \frac{x}{10}.$$

a. Find an expression for the total revenue from the sale of x thousand candy bars.

b. Find the value of x that leads to maximum revenue.

c. Find the maximum revenue.

8. Revenue The sale of compact disks of "lesser" performers is very sensitive to price. If a CD manufacturer charges $p(x)$ dollars per CD, where

$$p(x) = 12 - \frac{x}{8},$$

then x thousand CDs will be sold.

a. Find an expression for the total revenue from the sale of x thousand CDs.

b. Find the value of x that leads to maximum revenue.

c. Find the maximum revenue.

9. Area A campground owner has 1400 m of fencing. He wants to enclose a rectangular field bordering a river, with no fencing needed along the river. (See the sketch.) Let x represent the width of the field.

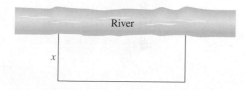

a. Write an expression for the length of the field.

b. Find the area of the field (area = length × width).

c. Find the value of x leading to the maximum area.

d. Find the maximum area.

10. Area Find the dimensions of the rectangular field of maximum area that can be made from 300 m of fencing material. (This fence has four sides.)

11. Area An ecologist is conducting a research project on breeding pheasants in captivity. She first must construct suitable pens. She wants a rectangular area with two additional fences across its width, as shown in the sketch. Find the maximum area she can enclose with 3600 m of fencing.

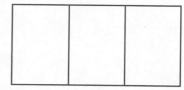

12. Area A farmer is constructing a rectangular pen with one additional fence across its width. Find the maximum area that can be enclosed with 2400 m of fencing.

13. Cost with Fixed Area A fence must be built in a large field to enclose a rectangular area of 25,600 m². One side of the area is bounded by an existing fence; no fence is needed there. Material for the fence costs $3 per meter for the two ends and $1.50 per meter for the side opposite the existing fence. Find the cost of the least expensive fence.

14. Cost with Fixed Area A fence must be built to enclose a rectangular area of 20,000 ft². Fencing material costs $2.50 per foot for the two sides facing north and south and $3.20 per foot for the other two sides. Find the cost of the least expensive fence.

15. Revenue A local club is arranging a charter flight to Hawaii. The cost of the trip is $1600 each for 90 passengers, with a refund of $10 per passenger for each passenger in excess of 90.

a. Find the number of passengers that will maximize the revenue received from the flight.

b. Find the maximum revenue.

16. Profit In planning a restaurant, it is estimated that a profit of $8 per seat will be made if the number of seats is no more than 50, inclusive. On the other hand, the profit on each seat will decrease by 10¢ for each seat above 50.

a. Find the number of seats that will produce the maximum profit.

b. What is the maximum profit?

17. Timing Income A local group of scouts has been collecting aluminum cans for recycling. The group has already collected 12,000 lb of cans, for which they could currently receive $7.50 per hundred pounds. The group can continue to collect cans at the rate of 400 lb per day. However, a glut in the aluminum market has caused the recycling company to announce that it will lower its price, starting immediately, by $0.15 per hundred pounds per day. The scouts can make only one trip to the recycling center. Find the best time for the trip. What total income will be received?

18. Pricing Decide what you would do if your assistant presented the following contract for your signature:

Your firm offers to deliver 250 tables to a dealer, at $160 per table, and to reduce the price per table on the entire order by 50¢ for each additional table over 250.

Find the dollar total involved in the largest possible transaction between the manufacturer and the dealer; then find the smallest possible dollar amount.

19. Packaging Design A television manufacturing firm needs to design an open-topped box with a square base. The box must hold 32 in³. Find the dimensions of the box that can be built with the minimum amount of materials. (See the figure.)

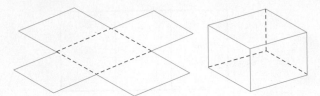

20. Packaging Design A company wishes to manufacture a box with a volume of 36 ft³ that is open on top and is twice as long as it is wide. Find the dimensions of the box produced from the minimum amount of material.

21. Container Design An open box will be made by cutting a square from each corner of a 3-ft by 8-ft piece of cardboard and then folding up the sides. What size square should be cut from each corner in order to produce a box of maximum volume?

22. Container Design Consider the problem of cutting corners out of a rectangle and folding up the sides to make a box. Specific examples of this problem are discussed in Example 3 and Exercise 21.

  a. In the solution to Example 3, compare the area of the base of the box with the area of the walls.

  b. Repeat part a for the solution to Exercise 21.

  c. Make a conjecture about the area of the base compared with the area of the walls for the box with the maximum volume.

23. Packaging Cost A closed box with a square base is to have a volume of 16,000 cm³. The material for the top and bottom of the box costs $3 per square centimeter, while the material for the sides costs $1.50 per square centimeter. Find the dimensions of the box that will lead to the minimum total cost. What is the minimum total cost?

24. Use of Materials A mathematics book is to contain 36 in² of printed matter per page, with margins of 1 in. along the sides and $1\frac{1}{2}$ in. along the top and bottom. Find the dimensions of the page that will require the minimum amount of paper. (See the figure.)

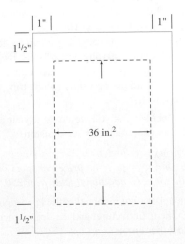

25. Can Design

  a. For the can problem in Example 4, the minimum surface area required that the height be twice the radius. Show that this is true for a can of arbitrary volume V.

  b. Do many cans in grocery stores have a height that is twice the radius? If not, discuss why this may be so.

26. Container Design Your company needs to design cylindrical metal containers with a volume of 16 cubic feet. The top and bottom will be made of a sturdy material that costs $2 per square foot, while the material for the sides costs $1 per square foot. Find the radius, height, and cost of the least expensive container.

In Exercises 27–29, use a graphing calculator to determine where the derivative is equal to zero.

27. Can Design Modify the can problem in Example 4 so the cost must be minimized. Assume that aluminum costs 3¢ per square centimeter, and that there is an additional cost of 2¢ per cm times the perimeter of the top, and a similar cost for the bottom, to seal the top and bottom of the can to the side.

28. Can Design In this modification of the can problem in Example 4, the cost must be minimized. Assume that aluminum costs 3¢ per square centimeter, and that there is an additional cost of 1¢ per cm times the height of the can to make a vertical seam on the side.

29. Can Design This problem is a combination of Exercises 27 and 28. We will again minimize the cost of the can, assuming that aluminum costs 3¢ per square centimeter. In addition, there is a cost of 2¢ per cm to seal the top and bottom of the can to the side, plus 1¢ per cm to make a vertical seam.

30. Packaging Design A cylindrical box will be tied up with ribbon as shown in the figure. The longest piece of ribbon available is 130 cm long, and 10 cm of that are required for the bow. Find the radius and height of the box with the largest possible volume.

31. Cost A company wishes to run a utility cable from point A on the shore (see the figure on the next page) to an installation at point B on the island. The island is 6 miles from the shore. It costs $400 per mile to run the cable on land and $500 per mile underwater. Assume that the cable starts at A and runs along the shoreline, then angles and runs underwater to the island. Find the point at which the line should begin to angle in order to yield the minimum total cost.

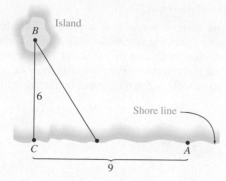

32. Cost Repeat Exercise 31, but make point A 7 miles from point C.

Life Sciences

33. Disease Epidemiologists have found a new communicable disease running rampant in College Station, Texas. They estimate that t days after the disease is first observed in the community, the percent of the population infected by the disease is approximated by

$$p(t) = \frac{20t^3 - t^4}{1000}$$

for $0 \le t \le 20$.

  a. After how many days is the percent of the population infected a maximum?

  b. What is the maximum percent of the population infected?

34. Disease Another disease hits the chronically ill town of College Station, Texas. This time the percent of the population infected by the disease t days after it hits town is approximated by $p(t) = 10te^{-t/8}$ for $0 \le t \le 40$.

  a. After how many days is the percent of the population infected a maximum?

  b. What is the maximum percent of the population infected?

Maximum Sustainable Harvest Find the maximum sustainable harvest in Exercises 35 and 36. See Example 5.

35. $f(S) = 12S^{0.25}$ **36.** $f(S) = \dfrac{25S}{S + 2}$

37. Pollution A lake polluted by bacteria is treated with an antibacterial chemical. After t days, the number N of bacteria per milliliter of water is approximated by

$$N(t) = 20\left(\frac{t}{12} - \ln\left(\frac{t}{12}\right)\right) + 30$$

for $1 \le t \le 15$.

  a. When during this time will the number of bacteria be a minimum?

  b. What is the minimum number of bacteria during this time?

  c. When during this time will the number of bacteria be a maximum?

  d. What is the maximum number of bacteria during this time?

38. Maximum Sustainable Harvest The population of salmon next year is given by $f(S) = Se^{r(1-S/P)}$, where S is this year's salmon population, P is the equilibrium population, and r is a

constant that depends upon how fast the population grows. The number of salmon that can be fished next year while keeping the population the same is $H(S) = f(S) - S$. The maximum value of $H(S)$ is the maximum sustainable harvest. *Source: Journal of the Fisheries Research Board of Canada.*

  a. Show that the maximum sustainable harvest occurs when $f'(S) = 1$. (*Hint:* To maximize, set $H'(S) = 0$.)

  b. Let the value of S found in part a be denoted by S_0. Show that the maximum sustainable harvest is given by

$$S_0\left(\frac{1}{1 - rS_0/P} - 1\right).$$

  (*Hint:* Set $f'(S_0) = 1$ and solve for $e^{r(1 - S_0/P)}$. Then find $H(S_0)$ and substitute the expression for $e^{r(1 - S_0/P)}$.)

Maximum Sustainable Harvest In Exercises 39 and 40, refer to Exercise 38. Find $f'(S_0)$ and solve the equation $f'(S_0) = 1$, using a calculator to find the intersection of the graphs of $f'(S_0)$ and $y = 1$.

39. Find the maximum sustainable harvest if $r = 0.1$ and $P = 100$.

40. Find the maximum sustainable harvest if $r = 0.4$ and $P = 500$.

41. Pigeon Flight Homing pigeons avoid flying over large bodies of water, preferring to fly around them instead. (One possible explanation is the fact that extra energy is required to fly over water because air pressure drops over water in the daytime.) Assume that a pigeon released from a boat 1 mile from the shore of a lake (point B in the figure) flies first to point P on the shore and then along the straight edge of the lake to reach its home at L. If L is 2 miles from point A, the point on the shore closest to the boat, and if a pigeon needs $4/3$ as much energy per mile to fly over water as over land, find the location of point P, which minimizes energy used.

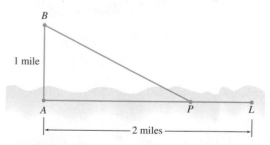

42. Pigeon Flight Repeat Exercise 41, but assume a pigeon needs $10/9$ as much energy to fly over water as over land.

43. Harvesting Cod A recent article described the population $f(S)$ of cod in the North Sea next year as a function of this year's population S (in thousands of tons) by various mathematical models.

    Shepherd: $f(S) = \dfrac{aS}{1 + (S/b)^c}$;

    Ricker: $f(S) = aSe^{-bS}$;

    Beverton–Holt: $f(S) = \dfrac{aS}{1 + (S/b)}$,

where a, b, and c are constants. *Source: Nature.*

a. Find a replacement of variables in the Ricker model above that will make it the same as another form of the Ricker model described in Exercise 38 of this section, $f(S) = Se^{r(1-S/P)}$.

b. Find $f'(S)$ for all three models.

c. Find $f'(0)$ for all three models. From your answer, describe in words the geometric meaning of the constant a.

d. The values of a, b, and c reported in the article for the Shepherd model are 3.026, 248.72, and 3.24, respectively. Find the value of this year's population that maximizes next year's population using the Shepherd model.

e. The values of a and b reported in the article for the Ricker model are 4.151 and 0.0039, respectively. Find the value of this year's population that maximizes next year's population using the Ricker model.

f. Explain why, for the Beverton-Holt model, there is no value of this year's population that maximizes next year's population.

44. Bird Migration Suppose a migrating bird flies at a velocity v, and suppose the amount of time the bird can fly depends on its velocity according to the function $T(v)$. *Source: A Concrete Approach to Mathematical Modelling.*

a. If E is the bird's initial energy, then the bird's effective power is given by kE/T, where k is the fraction of the power that can be converted into mechanical energy. According to principles of aerodynamics,

$$\frac{kE}{T} = aSv^3 + I,$$

where a is a constant, S is the wind speed, and I is the induced power, or rate of working against gravity. Using this result and the fact that distance is velocity multiplied by time, show that the distance that the bird can fly is given by

$$D(v) = \frac{kEv}{aSv^3 + I}.$$

b. Show that the migrating bird can fly a maximum distance by flying at a velocity

$$v = \left(\frac{I}{2aS}\right)^{1/3}.$$

General Interest

45. Travel Time A hunter is at a point along a river bank. He wants to get to his cabin, located 3 miles north and 8 miles west. (See the figure.) He can travel 5 mph along the river but only 2 mph on this very rocky land. How far upriver should he go in order to reach the cabin in minimum time?

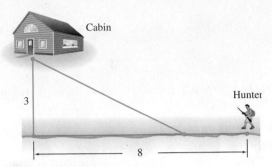

46. Travel Time Repeat Exercise 45, but assume the cabin is 19 miles north and 8 miles west.

47. Postal Regulations The U.S. Postal Service stipulates that any boxes sent through the mail must have a length plus girth totaling no more than 108 in. (See the figure.) Find the dimensions of the box with maximum volume that can be sent through the U.S. mail, assuming that the width and the height of the box are equal. *Source: U.S. Postal Service.*

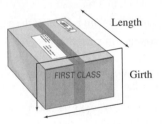

YOUR TURN ANSWERS

1. $x = 20$ and $y = 10/3$
2. Go 93 m along the trail and then head into the woods.
3. Box will have maximum volume when $x = 4/3$ m and the maximum volume is 1024/27 m³.
4. Radius is 4.3 cm and height is 8.6 cm.

6.3 Further Business Applications: Economic Lot Size; Economic Order Quantity; Elasticity of Demand

APPLY IT How many batches of primer should a paint company produce per year to minimize its costs while meeting its customers' demand?

We will answer this question in Example 1 using the concept of economic lot size.

In this section we introduce three common business applications of calculus. The first two, *economic lot size* and *economic order quantity*, are related. A manufacturer must determine the production lot (or batch) size that will result in minimum production and storage costs, while a purchaser must decide what quantity of an item to order in an effort to minimize reordering and storage costs. The third application, *elasticity of demand*, deals with the sensitivity of demand for a product to changes in the price of the product.

Economic Lot Size

Suppose that a company manufactures a constant number of units of a product per year and that the product can be manufactured in several batches of equal size throughout the year. On the one hand, if the company were to manufacture one large batch every year, it would minimize setup costs but incur high warehouse costs. On the other hand, if it were to make many small batches, this would increase setup costs. Calculus can be used to find the number that should be manufactured in each batch in order to minimize the total cost. This number is called the **economic lot size**.

Figure 14 below shows several possibilities for a product having an annual demand of 12,000 units. The top graph shows the results if all 12,000 units are made in one batch per year. In this case an average of 6000 units will be held in a warehouse. If 3000 units are made in each batch, four batches will be made at equal time intervals during the year, and the average number of units in the warehouse falls to only 1500. If 1000 units are made in each of twelve batches, an average of 500 units will be in the warehouse.

The variable in our discussion of economic lot size will be

$$q = \text{number of units in each batch.}$$

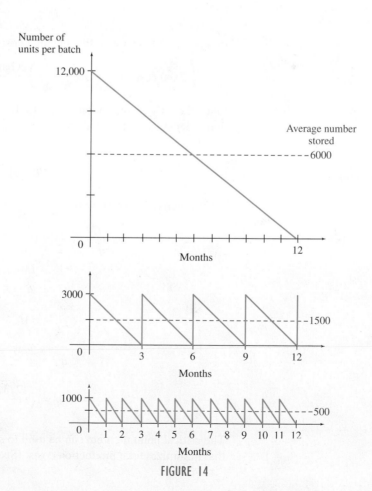

FIGURE 14

In addition, we have the following constants:

$$k = \text{cost of storing one unit of the product for one year;}$$
$$f = \text{fixed setup cost to manufacture the product;}$$
$$g = \text{cost of manufacturing a single unit of the product;}$$
$$M = \text{total number of units produced annually.}$$

The company has two types of costs: a cost associated with manufacturing the item and a cost associated with storing the finished product. Because q units are produced in each batch, and each batch has a fixed cost f and a variable cost g per unit, the manufacturing cost per batch is

$$f + gq.$$

The number of units produced in a year is M, so the number of batches per year must be M/q. Therefore, the total annual manufacturing cost is

$$(f + gq)\frac{M}{q} = \frac{fM}{q} + gM. \tag{1}$$

Since demand is constant, the inventory goes down linearly from q to 0, as in Figure 14, with an average inventory of $q/2$ units per year. The cost of storing one unit of the product for a year is k, so the total storage cost is

$$k\left(\frac{q}{2}\right) = \frac{kq}{2}. \tag{2}$$

The total production cost is the sum of the manufacturing and storage costs, or the sum of Equations (1) and (2). If $T(q)$ is the total cost of producing M units in batches of size q,

$$T(q) = \frac{fM}{q} + gM + \frac{kq}{2}.$$

In words, we have found that the total cost is equal to

$$\left(\text{fixed cost} + \frac{\text{cost}}{\text{unit}} \times \frac{\text{\# units}}{\text{batch}}\right)\frac{\text{\# batches}}{\text{year}} + \text{storage cost} \times \text{\# units in storage.}$$

Since the only constraint on q is that it be a positive number, the domain of T is $(0, \infty)$. To find the value of q that will minimize $T(q)$, remember that f, g, k, and M are constants and find $T'(q)$.

$$T'(q) = \frac{-fM}{q^2} + \frac{k}{2}$$

Set this derivative equal to 0.

$$\frac{-fM}{q^2} + \frac{k}{2} = 0$$

$$\frac{k}{2} = \frac{fM}{q^2} \qquad \text{Add } \frac{fM}{q^2} \text{ to both sides.}$$

$$q^2\frac{k}{2} = fM \qquad \text{Multiply both sides by } q^2.$$

$$q^2 = \frac{2fM}{k} \qquad \text{Multiply both sides by } 2/k.$$

$$q = \sqrt{\frac{2fM}{k}} \qquad \text{Take square root of both sides.} \tag{3}$$

The critical point theorem can be used to show that $\sqrt{(2fM)/k}$ is the economic lot size that minimizes total production costs. (See Exercise 1.)

This application is referred to as the *inventory problem* and is treated in more detail in management science courses. Please note that Equation (3) was derived under very specific assumptions. If the assumptions are changed slightly, a different conclusion might be reached, and it would not necessarily be valid to use Equation (3).

In some examples Equation (3) may not give an integer value, in which case we must investigate the next integer smaller than q and the next integer larger to see which gives the minimum cost.

EXAMPLE 1   Lot Size

APPLY IT

A paint company has a steady annual demand for 24,500 cans of automobile primer. The comptroller for the company says that it costs $2 to store one can of paint for 1 year and $500 to set up the plant for the production of the primer. Find the number of cans of primer that should be produced in each batch, as well as the number of batches per year, in order to minimize total production costs.

SOLUTION   Use Equation (3), with $k = 2$, $M = 24,500$, and $f = 500$.

$$q = \sqrt{\frac{2fM}{k}} = \sqrt{\frac{2(500)(24,500)}{2}} = \sqrt{12,250,000} = 3500$$

The company should make 3500 cans of primer in each batch to minimize production costs. The number of batches per year is $M/q = 24,500/3500 = 7$.

TRY YOUR TURN 1

YOUR TURN 1   Suppose the annual demand in Example 1 is only 18,000 cans, the setup cost is $750, and the storage cost is $3 per can. Find the number of cans that should be produced in each batch and the number of batches per year to minimize production costs.

Economic Order Quantity

We can extend our previous discussion to the problem of reordering an item that is used at a constant rate throughout the year. Here, the company using a product must decide how often to order and how many units to request each time an order is placed; that is, it must identify the **economic order quantity**. In this case, the variable is

$$q = \text{number of units to order each time.}$$

We also have the following constants:

$$k = \text{cost of storing one unit for one year}$$
$$f = \text{fixed cost to place an order}$$
$$M = \text{total units needed per year}$$

The goal is to minimize the total cost of ordering over a year's time, where

$$\text{Total cost} = \text{Storage cost} + \text{Reorder cost.}$$

Again assume an average inventory of $q/2$, so the yearly storage cost is $kq/2$. The number of orders placed annually is M/q. The reorder cost is the product of this quantity and the cost per order, f. Thus, the reorder cost is fM/q, and the total cost is

$$T(q) = \frac{fM}{q} + \frac{kq}{2}.$$

This is almost the same formula we derived for the inventory problem, which also had a constant term gM. Since a constant does not affect the derivative, Equation (3) is also valid for the economic order quantity problem. As before, the number of orders placed annually is M/q. This illustrates how two different applications might have the same mathematical structure, so a solution to one applies to both.

EXAMPLE 2   Order Quantity

A large pharmacy has an annual need for 480 units of a certain antibiotic. It costs $3 to store one unit for one year. The fixed cost of placing an order (clerical time, mailing, and so on) amounts to $31. Find the number of units to order each time, and how many times a year the antibiotic should be ordered.

YOUR TURN 2 Suppose the annual need in Example 2 is 320 units, the fixed cost amounts to $30, and the storage cost is $2 per unit. Find the number of units to order each time and how many times a year to order to minimize cost.

SOLUTION Here $k = 3$, $M = 480$, and $f = 31$. We have

$$q = \sqrt{\frac{2fM}{k}} = \sqrt{\frac{2(31)(480)}{3}} = \sqrt{9920} \approx 99.6$$

$T(99) = 298.803$ and $T(100) = 298.800$, so ordering 100 units of the drug each time minimizes the annual cost. The drug should be ordered $M/q = 480/100 = 4.8$ times a year, or about once every $2\frac{1}{2}$ months. **TRY YOUR TURN 2**

Elasticity of Demand

Anyone who sells a product or service is concerned with how a change in price affects demand. The sensitivity of demand to changes in price varies with different items. Luxury items tend to be more sensitive to price than essentials. For items such as milk, heating fuel, and light bulbs, relatively small percentage changes in price will not change the demand for the item much, so long as the price is not far from its normal range. For cars, home loans, jewelry, and concert tickets, however, small percentage changes in price can have a significant effect on demand.

One way to measure the sensitivity of demand to changes in price is by the relative change—the ratio of percent change in demand to percent change in price. If q represents the quantity demanded and p the price, this ratio can be written as

$$\frac{\Delta q/q}{\Delta p/p},$$

FOR REVIEW

Recall from Chapter 1 that the Greek letter Δ, pronounced *delta*, is used in mathematics to mean "change."

where Δq represents the change in q and Δp represents the change in p. This ratio is always negative, because q and p are positive, while Δq and Δp have opposite signs. (An *increase* in price causes a *decrease* in demand.) If the absolute value of this ratio is large, it suggests that a relatively small increase in price causes a relatively large drop (decrease) in demand.

This ratio can be rewritten as

$$\frac{\Delta q/q}{\Delta p/p} = \frac{\Delta q}{q} \cdot \frac{p}{\Delta p} = \frac{p}{q} \cdot \frac{\Delta q}{\Delta p}.$$

Suppose $q = f(p)$. (Note that this is the inverse of the way our demand functions have been expressed so far; previously we had $p = D(q)$.) Then $\Delta q = f(p + \Delta p) - f(p)$, and

$$\frac{\Delta q}{\Delta p} = \frac{f(p + \Delta p) - f(p)}{\Delta p}.$$

As $\Delta p \to 0$, this quotient becomes

$$\lim_{\Delta p \to 0} \frac{\Delta q}{\Delta p} = \lim_{\Delta p \to 0} \frac{f(p + \Delta p) - f(p)}{\Delta p} = \frac{dq}{dp},$$

and

$$\lim_{\Delta p \to 0} \frac{p}{q} \cdot \frac{\Delta q}{\Delta p} = \frac{p}{q} \cdot \frac{dq}{dp}.$$

The negative of this last quantity is called the *elasticity of demand* (E) and measures the instantaneous responsiveness of demand to price. *

Elasticity of Demand

Let $q = f(p)$, where q is demand at a price p. The **elasticity of demand** is

$$E = -\frac{p}{q} \cdot \frac{dq}{dp}.$$

Demand is inelastic if $E < 1$.

Demand is elastic if $E > 1$.

Demand has unit elasticity if $E = 1$.

---

*Economists often define elasticity as the negative of our definition.

For example, E has been estimated at 0.6 for physician services and at 2.3 for restaurant meals. The demand for medical care is much less responsive to price changes than is the demand for nonessential commodities, such as restaurant meals. *Source: Economics: Private and Public Choice.*

If $E < 1$, the relative change in demand is less than the relative change in price, and the demand is called *inelastic*. If $E > 1$, the relative change in demand is greater than the relative change in price, and the demand is called *elastic*. In other words, inelastic means that a small change in price has little effect on demand, while elastic means that a small change in price has more effect on demand. When $E = 1$, the percentage changes in price and demand are relatively equal and the demand is said to have **unit elasticity**.

EXAMPLE 3 Elasticity

Terrence Wales described the demand for distilled spirits as

$$q = f(p) = -0.00375p + 7.87,$$

where p represents the retail price of a case of liquor in dollars per case. Here q represents the average number of cases purchased per year by a consumer. Calculate and interpret the elasticity of demand when $p = \$118.30$ per case. *Source: The American Economic Review.*

SOLUTION From $q = -0.00375p + 7.87$, we determine $dq/dp = -0.00375$. Now we find E.

$$E = -\frac{p}{q} \cdot \frac{dq}{dp}$$

$$= -\frac{p}{-0.00375p + 7.87}(-0.00375)$$

$$= \frac{0.00375p}{-0.00375p + 7.87}$$

Let $p = 118.30$ to get

$$E = \frac{0.00375(118.30)}{-0.00375(118.30) + 7.87} \approx 0.0597.$$

YOUR TURN 3 Suppose the demand equation for a given commodity is $q = 24,000 - 3p^2$. Calculate and interpret E when $p = \$50$.

Since $0.0597 < 1$, the demand is inelastic, and a percentage change in price will result in a smaller percentage change in demand. Thus an increase in price will increase revenue. For example, a 10% increase in price will cause an approximate decrease in demand of $(0.0597)(0.10) = 0.00597$ or about 0.6%. **TRY YOUR TURN 3**

EXAMPLE 4 Elasticity

The demand for beer was modeled by Hogarty and Elzinga with the function given by $q = f(p) = 1/p$. The price was expressed in dollars per can of beer, and the quantity sold in cans per day per adult. Calculate and interpret the elasticity of demand. *Source: The Review of Economics and Statistics.*

SOLUTION Since $q = 1/p$,

$$\frac{dq}{dp} = \frac{-1}{p^2}, \quad \text{and}$$

YOUR TURN 4 Suppose the demand equation for a given product is $q = 200e^{-0.4p}$. Calculate and interpret E when $p = \$100$.

$$E = -\frac{p}{q} \cdot \frac{dq}{dp} = -\frac{p}{1/p} \cdot \frac{-1}{p^2} = 1.$$

Here, the elasticity is 1, unit elasticity, at every (positive) price. As we will see shortly, this means that revenues remain constant when the price changes. **TRY YOUR TURN 4**

Elasticity can be related to the total revenue, R, by considering the derivative of R. Since revenue is given by price times sales (demand),

$$R = pq.$$

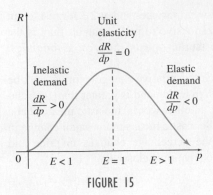

FIGURE 15

Differentiate with respect to p using the product rule.

$$\frac{dR}{dp} = p \cdot \frac{dq}{dp} + q \cdot 1$$

$$= \frac{q}{q} \cdot p \cdot \frac{dq}{dp} + q \qquad \text{Multiply by } \frac{q}{q} \text{ (or 1).}$$

$$= q\left(\frac{p}{q} \cdot \frac{dq}{dp}\right) + q$$

$$= q(-E) + q \qquad -E = \frac{p}{q} \cdot \frac{dq}{dp}.$$

$$= q(-E + 1) \qquad \text{Factor.}$$

$$= q(1 - E)$$

Total revenue R is increasing, optimized, or decreasing depending on whether $dR/dp > 0$, $dR/dp = 0$, or $dR/dp < 0$. These three situations correspond to $E < 1$, $E = 1$, or $E > 1$. See Figure 15.

In summary, total revenue is related to elasticity as follows.

Revenue and Elasticity

1. If the demand is inelastic, total revenue increases as price increases.

2. If the demand is elastic, total revenue decreases as price increases.

3. Total revenue is maximized at the price where demand has unit elasticity.

EXAMPLE 5 **Elasticity**

Assume that the demand for a product is $q = 216 - 2p^2$, where p is the price in dollars.

(a) Find the price intervals where demand is elastic and where demand is inelastic.

SOLUTION Since $q = 216 - 2p^2$, $dq/dp = -4p$, and

$$E = -\frac{p}{q} \cdot \frac{dq}{dp}$$

$$= -\frac{p}{216 - 2p^2}(-4p)$$

$$= \frac{4p^2}{216 - 2p^2}.$$

To decide where $E < 1$ or $E > 1$, solve the corresponding *equation*.

$$E = 1$$

$$\frac{4p^2}{216 - 2p^2} = 1 \qquad \qquad E = \frac{4p^2}{216 - 2p^2}$$

$$4p^2 = 216 - 2p^2 \qquad \text{Multiply both sides by } 216 - 2p^2.$$

$$6p^2 = 216 \qquad \text{Add } 2p^2 \text{ to both sides.}$$

$$p^2 = 36 \qquad \text{Divide both sides by 6.}$$

$$p = 6 \qquad \text{Take square root of both sides.}$$

Substitute a test number on either side of 6 in the expression for E to see which values make $E < 1$ and which make $E > 1$.

$$\text{Let } p = 1: E = \frac{4(1)^2}{216 - 2(1)^2} = \frac{4}{214} < 1.$$

$$\text{Let } p = 10: E = \frac{4(10)^2}{216 - 2(10)^2} = \frac{400}{216 - 200} > 1.$$

YOUR TURN 5 Suppose the demand function for a given product is $q = 3600 - 3p^2$. Find the price intervals where demand is elastic and where demand is inelastic. What price results in maximum revenue? What is the maximum revenue?

Demand is inelastic when $E < 1$. This occurs when $p < 6$. Demand is elastic when $E > 1$; that is, when $p > 6$.

(b) What price results in the maximum revenue? What is the maximum revenue?

SOLUTION Total revenue is maximized at the price where demand has unit elasticity. As we saw in part (a), this occurs when the price is set at $6 per item. The demand at this price is $q = 216 - 2(6)^2 = 144$. The maximum revenue is then $pq = 6 \cdot 144 = \$864$.

TRY YOUR TURN 5

6.3 EXERCISES

1. In the discussion of economic lot size, use the critical point theorem to show that $\sqrt{(2fM)/k}$ is the economic lot size that minimizes total production costs.

2. Why do you think that the cost g does not appear in the equation for q [Equation (3)]?

3. Choose the correct answer. *Source: American Institute of Certified Public Accountants.*

   The economic order quantity formula assumes that

   a. Purchase costs per unit differ due to quantity discounts.

   b. Costs of placing an order vary with quantity ordered.

   c. Periodic demand for the goods is known.

   d. Erratic usage rates are cushioned by safety stocks.

4. Describe elasticity of demand in your own words.

5. A Giffen good is a product for which the demand function is increasing. Economists debate whether such goods actually exist. What is true about the elasticity of a Giffen good? *Source: investopedia.com.*

6. What must be true about the demand function if $E = 0$?

7. Suppose that a demand function is linear—that is, $q = m - np$ for $0 \le p \le m/n$, where m and n are positive constants. Show that $E = 1$ at the midpoint of the demand curve on the interval $0 \le p \le m/n$; that is, at $p = m/(2n)$.

8. Suppose the demand function is of the form $q = Cp^{-k}$, where C and k are positive constants.

   a. Find the elasticity E.

   b. If $0 < k < 1$, what does your answer from part a say about how prices should be set to maximize the revenue?

   c. If $k > 1$, what does your answer from part a say about how prices should be set to maximize the revenue?

   d. If $k = 1$, what does your answer from part a tell you about setting prices to maximize revenue?

   e. Based on your answers above, is a demand function of the form $q = Cp^{-k}$ realistic? Explain your answer.

APPLICATIONS

Business and Economics

9. **Lot Size** Suppose 100,000 lamps are to be manufactured annually. It costs $1 to store a lamp for 1 year, and it costs

$500 to set up the factory to produce a batch of lamps. Find the number of lamps to produce in each batch.

10. **Lot Size** A manufacturer has a steady annual demand for 13,950 cases of sugar. It costs $9 to store 1 case for 1 year, $31 in setup cost to produce each batch, and $16 to produce each case. Find the number of cases per batch that should be produced.

11. **Lot Size** Find the number of batches of lamps that should be manufactured annually in Exercise 9.

12. **Lot Size** Find the number of batches of sugar that should be manufactured annually in Exercise 10.

13. **Order Quantity** A bookstore has an annual demand for 100,000 copies of a best-selling book. It costs $0.50 to store 1 copy for 1 year, and it costs $60 to place an order. Find the optimum number of copies per order.

14. **Order Quantity** A restaurant has an annual demand for 900 bottles of a California wine. It costs $1 to store 1 bottle for 1 year, and it costs $5 to place a reorder. Find the optimum number of bottles per order.

15. **Lot Size** Suppose that in the inventory problem, the storage cost depends on the maximum inventory size, rather than the average. This would be more realistic if, for example, the company had to build a warehouse large enough to hold the maximum inventory, and the cost of storage was the same no matter how full or empty the warehouse was. Show that in this case the number of units that should be ordered or manufactured to minimize the total cost is

$$q = \sqrt{\frac{fM}{k}}.$$

16. **Lot Size** A book publisher wants to know how many times a year a print run should be scheduled. Suppose it costs $1000 to set up the printing process, and the subsequent cost per book is so low it can be ignored. Suppose further that the annual warehouse cost is $6 times the maximum number of books stored. Assuming 5000 copies of the book are needed per year, how many books should be printed in each print run? (See Exercise 15.)

17. **Lot Size** Suppose that in the inventory problem, the storage cost is a combination of the cost described in the text and the cost described in Exercise 15. In other words, suppose there is an annual cost, k_1, for storing a single unit, plus an annual cost per unit, k_2, that must be paid for each unit up to the maximum number of units stored. Show that the number of units that

should be ordered or manufactured to minimize the total cost in this case is

$$q = \sqrt{\frac{2fM}{k_1 + 2k_2}}.$$

18. Lot Size Every year, Corinna Paolucci sells 30,000 cases of her Famous Spaghetti Sauce. It costs her $1 per year in electricity to store a case, plus she must pay annual warehouse fees of $2 per case for the maximum number of cases she will store. If it costs her $750 to set up a production run, plus $8 per case to manufacture a single case, how many production runs should she have each year to minimize her total costs? (See Exercise 17.)

Elasticity For each of the following demand functions, find (a) E, and (b) values of q (if any) at which total revenue is maximized.

19. $q = 50 - \dfrac{p}{4}$

20. $q = 25{,}000 - 50p$

21. $q = 37{,}500 - 5p^2$

22. $q = 48{,}000 - 10p^2$

23. $p = 400e^{-0.2q}$

24. $q = 10 - \ln p$

Elasticity Find the elasticity of demand (E) for the given demand function at the indicated values of p. Is the demand elastic, inelastic, or neither at the indicated values? Interpret your results.

25. $q = 400 - 0.2p^2$

    a. $p = \$20$ **b.** $p = \$40$

26. $q = 300 - 2p$

    a. $p = \$100$ **b.** $p = \$50$

27. Elasticity of Crude Oil The short-term demand for crude oil in the United States in 2008 can be approximated by

$$q = f(p) = 2{,}431{,}129p^{-0.06},$$

where p represents the price of crude oil in dollars per barrel and q represents the per capita consumption of crude oil. Calculate and interpret the elasticity of demand when the price is $40 per barrel. *Source: 2003 OPEC Review.*

28. Elasticity of Rice The demand for rice in Japan for a particular year was estimated by the general function

$$q = f(p) = Ap^{-0.13},$$

where p represents the price of a unit of rice, A represents a constant that can be calculated uniquely for a particular year, and q represents the annual per capita rice demand. Calculate and interpret the elasticity of demand. *Source: Agricultural Economics.*

29. Elasticity of Software In 2008, the Valve Corporation, a software entertainment company, ran a holiday sale on its popular Steam software program. Using data collected from the sale, it is possible to estimate the demand corresponding to various discounts in the price of the software. Assuming that the original price was $40, the demand for the software can be estimated by the function

$$q = 3{,}751{,}000p^{-2.826},$$

where p is the price and q is the demand. Calculate and interpret the elasticity of demand. *Source: codinghorror.com.*

30. Elasticity The price of beef in the United States has been found to depend on the demand (measured by per capita consumption) according to the equation

$$q = \frac{342.5}{p^{0.5314}}.$$

Find the elasticity. Is the demand for beef elastic or inelastic? *Source: SAS Institute Inc.*

31. Elasticity A study of the demand for air travel in Australia found that the demand for discount air travel from Sydney to Melbourne (in revenue passenger kilometre per capita, the product of the number of passengers travelling on a route and the distance of the route, divided by the populations of the host cities) depends on the airfare according to the equation

$$q = 55.2 - 0.022p.$$

Source: International Journal of Transport Economics.

    a. Find the elasticity when the price is $166.10, the average discount airfare deflated by the consumer price index to 1989–1990 prices, according to the authors of the study.

    b. Is the demand for airfare elastic or inelastic at this price?

    c. Find the price that maximizes revenue.

32. Elasticity The price along the West Coast of the United States for Japanese spruce logs (in dollars per cubic meter) based on the demand (in thousands of cubic meters per day) has been approximated by

$$p = 0.604q^2 - 20.16q + 263.067.$$

Source: Kamiak Econometrics.

    a. Find the elasticity when the demand is 11 thousand cubic meters a day. (*Hint:* To find dq/dp when p is expressed in terms of q, you may use the fact that

$$\frac{dq}{dp} = \frac{1}{dp/dq}.$$

Review the explanation of the derivative of the natural logarithm to see why this is true.)

    b. Is the demand for spruce logs elastic or inelastic?

    c. What happens to the elasticity as q approaches 16.6887? Discuss the limitations of this model for the price as a function of the demand.

33. Elasticity A geometric interpretation of elasticity is as follows. Consider the tangent line to the demand curve $q = f(p)$ at the point $P_0 = (p_0, q_0)$. Let the point where the tangent line intersects the p-axis be called A, and the point where it intersects the q-axis be called B. Let P_0A and P_0B be the distances from P_0 to A and to B, respectively. Calculate the ratio P_0B/P_0A in terms of p_0, q_0, and $f'(p_0)$, and show that this ratio equals the elasticity. *Source: The AMATYC Review.*

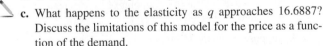

YOUR TURN ANSWERS

1. 6 batches per year with 3000 cans per batch
2. Order 98 units about every 3.675 months.
3. $E = 0.909$, the demand is inelastic.
4. $E = 40$, the demand is elastic.
5. Demand is inelastic when $p < 20$ and demand is elastic when $p > 20$. The maximum revenue is $48,000 when $p = 20$.

6.4 Implicit Differentiation

APPLY IT How does demand for a certain commodity change with respect to price? *We will answer this question in Example 4.*

In almost all of the examples and applications so far, all functions have been defined in the form

$$y = f(x),$$

with y given **explicitly** in terms of x, or as an **explicit function** of x. For example,

$$y = 3x - 2, \qquad y = x^2 + x + 6, \qquad \text{and} \qquad y = -x^3 + 2$$

are all explicit functions of x. The equation $4xy - 3x = 6$ can be expressed as an explicit function of x by solving for y. This gives

$$4xy - 3x = 6$$
$$4xy = 3x + 6$$
$$y = \frac{3x + 6}{4x}.$$

On the other hand, some equations in x and y cannot be readily solved for y, and some equations cannot be solved for y at all. For example, while it would be possible (but tedious) to use the quadratic formula to solve for y in the equation $y^2 + 2yx + 4x^2 = 0$, it is not possible to solve for y in the equation $y^5 + 8y^3 + 6y^2x^2 + 2yx^3 + 6 = 0$. In equations such as these last two, y is said to be given **implicitly** in terms of x.

In such cases, it may still be possible to find the derivative dy/dx by a process called **implicit differentiation**. In doing so, we assume that there exists some function or functions f, which we may or may not be able to find, such that $y = f(x)$ and dy/dx exists. It is useful to use dy/dx here rather than $f'(x)$ to make it clear which variable is independent and which is dependent.

EXAMPLE 1 Implicit Differentiation

Find dy/dx if $3xy + 4y^2 = 10$.

SOLUTION Differentiate with respect to x on both sides of the equation.

$$3xy + 4y^2 = 10$$
$$\frac{d}{dx}(3xy + 4y^2) = \frac{d}{dx}(10) \qquad \text{(1)}$$

Now differentiate each term on the left side of the equation. Think of $3xy$ as the product $(3x)(y)$ and use the product rule and the chain rule. Since

$$\frac{d}{dx}(3x) = 3 \quad \text{and} \quad \frac{d}{dx}(y) = \frac{dy}{dx},$$

the derivative of $(3x)(y)$ is

$$(3x)\frac{dy}{dx} + (y)3 = 3x\frac{dy}{dx} + 3y.$$

To differentiate the second term, $4y^2$, use the chain rule, since y is assumed to be some function of x.

$$\frac{d}{dx}(4y^2) = 4(2y^1)\overbrace{\frac{dy}{dx}}^{\substack{\text{Derivative} \\ \text{of } y^2}} = 8y\frac{dy}{dx}$$

FOR REVIEW

In Chapter 1, we pointed out that when y is given as a function of x, x is the independent variable and y is the dependent variable. We later defined the derivative dy/dx when y is a function of x. In an equation such as $3xy + 4y^2 = 10$, either variable can be considered the independent variable. If a problem asks for dy/dx, consider x the independent variable; if it asks for dx/dy, consider y the independent variable. A similar rule holds when other variables are used.

On the right side of Equation (1), the derivative of 10 is 0. Taking the indicated derivatives in Equation (1) term by term gives

$$3x\frac{dy}{dx} + 3y + 8y\frac{dy}{dx} = 0.$$

Now solve this result for dy/dx.

$$(3x + 8y)\frac{dy}{dx} = -3y$$

YOUR TURN 1 Find dy/dx if $x^2 + y^2 = xy$.

$$\frac{dy}{dx} = \frac{-3y}{3x + 8y}$$

TRY YOUR TURN 1

NOTE Because we are treating y as a function of x, notice that each time an expression has y in it, we use the chain rule.

EXAMPLE 2 Implicit Differentiation

Find dy/dx for $x + \sqrt{x}\,\sqrt{y} = y^2$.

SOLUTION Take the derivative on both sides with respect to x.

$$\frac{d}{dx}(x + \sqrt{x}\,\sqrt{y}) = \frac{d}{dx}(y^2)$$

Since $\sqrt{x} \cdot \sqrt{y} = x^{1/2} \cdot y^{1/2}$, use the product rule and the chain rule as follows.

$$\underbrace{1}_{\substack{\text{Derivative}\\ \text{of } x}} + \overbrace{x^{1/2}\left(\frac{1}{2}y^{-1/2} \cdot \frac{dy}{dx}\right) + y^{1/2}\left(\frac{1}{2}x^{-1/2}\right)}^{\substack{\text{Derivative}\\ \text{of } x^{1/2}y^{1/2}}} = \overbrace{2y\frac{dy}{dx}}^{\substack{\text{Derivative}\\ \text{of } y^2}}$$

$$1 + \frac{x^{1/2}}{2y^{1/2}} \cdot \frac{dy}{dx} + \frac{y^{1/2}}{2x^{1/2}} = 2y\frac{dy}{dx}$$

Multiply both sides by $2x^{1/2} \cdot y^{1/2}$.

$$2x^{1/2} \cdot y^{1/2} + x\frac{dy}{dx} + y = 4x^{1/2} \cdot y^{3/2} \cdot \frac{dy}{dx}$$

Combine terms and solve for dy/dx.

$$2x^{1/2} \cdot y^{1/2} + y = (4x^{1/2} \cdot y^{3/2} - x)\frac{dy}{dx}$$

YOUR TURN 2 Find dy/dx for $xe^y + x^2 = \ln y$.

$$\frac{dy}{dx} = \frac{2x^{1/2} \cdot y^{1/2} + y}{4x^{1/2} \cdot y^{3/2} - x}$$

TRY YOUR TURN 2

EXAMPLE 3 Tangent Line

The graph of $x^3 + y^3 = 9xy$, shown in Figure 16, is a *folium of Descartes*.* Find the equation of the tangent line at the point $(2, 4)$, shown in Figure 16.

SOLUTION Since this is not the graph of a function, y is not a function of x, and dy/dx is not defined. But if we restrict the curve to the vicinity of $(2, 4)$, as shown in Figure 17,

*Information on this curve and others is available on the Famous Curves section of the MacTutor History of Mathematics Archive website at www-history.mcs.st-and.ac.uk/~history. See Exercises 33–36 for more curves.

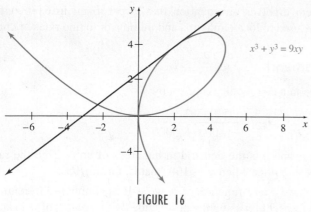

FIGURE 16

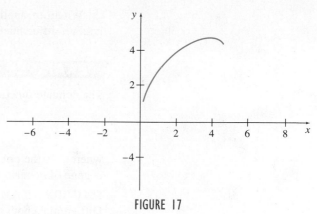

FIGURE 17

the curve does represent the graph of a function, and we can calculate dy/dx by implicit differentiation.

$$3x^2 + 3y^2 \cdot \frac{dy}{dx} = 9x\frac{dy}{dx} + 9y \qquad \text{Chain rule and product rule}$$

$$3y^2 \cdot \frac{dy}{dx} - 9x\frac{dy}{dx} = 9y - 3x^2 \qquad \text{Move all } dy/dx \text{ terms to the same side of the equation.}$$

$$\frac{dy}{dx}(3y^2 - 9x) = 9y - 3x^2 \qquad \text{Factor.}$$

$$\frac{dy}{dx} = \frac{9y - 3x^2}{3y^2 - 9x}$$

$$= \frac{3(3y - x^2)}{3(y^2 - 3x)} = \frac{3y - x^2}{y^2 - 3x}$$

To find the slope of the tangent line at the point $(2, 4)$, let $x = 2$ and $y = 4$. The slope is

$$m = \frac{3y - x^2}{y^2 - 3x} = \frac{3(4) - 2^2}{4^2 - 3(2)} = \frac{8}{10} = \frac{4}{5}.$$

The equation of the tangent line is then found by using the point-slope form of the equation of a line.

$$y - y_1 = m(x - x_1)$$

$$y - 4 = \frac{4}{5}(x - 2)$$

$$y - 4 = \frac{4}{5}x - \frac{8}{5}$$

$$y = \frac{4}{5}x + \frac{12}{5}$$

YOUR TURN 3 The graph of $y^4 - x^4 - y^2 + x^2 = 0$ is called the *devil's curve*. Find the equation of the tangent line at the point $(1, 1)$.

The tangent line is graphed in Figure 16. **TRY YOUR TURN 3**

NOTE In Example 3, we could have substituted $x = 2$ and $y = 4$ immediately after taking the derivative implicitly. You may find that such a substitution makes solving the equation for dy/dx easier.

The steps used in implicit differentiation can be summarized as follows.

Implicit Differentiation

To find dy/dx for an equation containing x and y:

1. Differentiate on both sides of the equation with respect to x, keeping in mind that y is assumed to be a function of x.
2. Using algebra, place all terms with dy/dx on one side of the equals sign and all terms without dy/dx on the other side.
3. Factor out dy/dx, and then solve for dy/dx.

When an applied problem involves an equation that is not given in explicit form, implicit differentiation can be used to locate maxima and minima or to find rates of change.

EXAMPLE 4 Demand

The demand function for a certain commodity is given by

$$p = \frac{500,000}{2q^3 + 400q + 5000},$$

where p is the price in dollars and q is the demand in hundreds of units. Find the rate of change of demand with respect to price when $q = 100$ (that is, find dq/dp).

APPLY IT

SOLUTION Since we don't have q as a function of p, we will use implicit differentiation. Differentiate both sides with respect to p using the power rule (with a power of -1) and the chain rule.

$$1 = \frac{-500,000}{(2q^3 + 400q + 5000)^2}\left(6q^2\frac{dq}{dp} + 400\frac{dq}{dp}\right)$$

$$= \frac{-500,000}{(2q^3 + 400q + 5000)^2}(6q^2 + 400)\frac{dq}{dp}$$

Now substitute $q = 100$.

$$1 = \frac{-500,000}{(2\cdot 100^3 + 400\cdot 100 + 5000)^2}(6\cdot 100^2 + 400)\frac{dq}{dp}$$

$$= \frac{-500,000\cdot 60,400}{2,045,000^2}\cdot\frac{dq}{dp}$$

YOUR TURN 4 Find the rate of change of demand with respect to price when $q = 200$ if the demand function is given by $p = \dfrac{100,000}{q^2 + 100q}$.

Therefore,

$$\frac{dq}{dp} = -\frac{2,045,000^2}{500,000\cdot 60,400} \approx -138.$$

This means that when demand (q) is 100 hundreds, or 10,000, demand is decreasing at the rate of 139 hundred, or 13,900, units per dollar change in price. **TRY YOUR TURN 4**

6.4 EXERCISES

Find dy/dx by implicit differentiation for the following.

1. $6x^2 + 5y^2 = 36$

2. $7x^2 - 4y^2 = 24$

3. $8x^2 - 10xy + 3y^2 = 26$

4. $7x^2 = 5y^2 + 4xy + 1$

5. $5x^3 = 3y^2 + 4y$

6. $3x^3 - 8y^2 = 10y$

7. $3x^2 = \dfrac{2 - y}{2 + y}$

8. $2y^2 = \dfrac{5 + x}{5 - x}$

9. $2\sqrt{x} + 4\sqrt{y} = 5y$

10. $4\sqrt{x} - 8\sqrt{y} = 6y^{3/2}$

11. $x^4y^3 + 4x^{3/2} = 6y^{3/2} + 5$

12. $(xy)^{4/3} + x^{1/3} = y^6 + 1$

13. $e^{x^2y} = 5x + 4y + 2$

14. $x^2e^y + y = x^3$

15. $x + \ln y = x^2y^3$

16. $y\ln x + 2 = x^{3/2}y^{5/2}$

Find the equation of the tangent line at the given point on each curve.

17. $x^2 + y^2 = 25$; $(-3, 4)$

18. $x^2 + y^2 = 100$; $(8, -6)$

19. $x^2y^2 = 1$; $(-1, 1)$

20. $x^2y^3 = 8$; $(-1, 2)$

21. $2y^2 - \sqrt{x} = 4$; $(16, 2)$

22. $y + \dfrac{\sqrt{x}}{y} = 3$; $(4, 2)$

23. $e^{x^2 + y^2} = xe^{5y} - y^2e^{5x/2}$; $(2, 1)$

24. $2xe^{xy} = e^{x^3} + ye^{x^2}$; $(1, 1)$

25. $\ln(x + y) = x^3y^2 + \ln(x^2 + 2) - 4$; $(1, 2)$

26. $\ln(x^2 + y^2) = \ln 5x + \dfrac{y}{x} - 2$; $(1, 2)$

In Exercises 27–32, find the equation of the tangent line at the given value of x on each curve.

27. $y^3 + xy - y = 8x^4$; $x = 1$

28. $y^3 + 2x^2y - 8y = x^3 + 19$; $x = 2$

29. $y^3 + xy^2 + 1 = x + 2y^2;\quad x = 2$

30. $y^4(1 - x) + xy = 2;\quad x = 1$

31. $2y^3(x - 3) + x\sqrt{y} = 3;\quad x = 3$

32. $\dfrac{y}{18}(x^2 - 64) + x^{2/3}y^{1/3} = 12;\quad x = 8$

Information on curves in Exercises 33–36, as well as many other curves, is available on the Famous Curves section of the MacTutor History of Mathematics Archive website at www-history.mcs.st-and.ac.uk/~history.

33. The graph of $x^{2/3} + y^{2/3} = 2$, shown in the figure, is an *astroid*. Find the equation of the tangent line at the point $(1, 1)$.

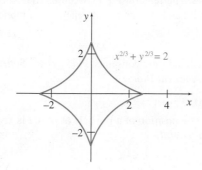

34. The graph of $3(x^2 + y^2)^2 = 25(x^2 - y^2)$, shown in the figure, is a *lemniscate of Bernoulli*. Find the equation of the tangent line at the point $(2, 1)$.

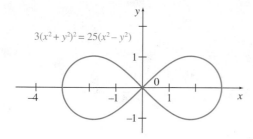

35. The graph of $y^2(x^2 + y^2) = 20x^2$, shown in the figure, is a *kappa curve*. Find the equation of the tangent line at the point $(1, 2)$.

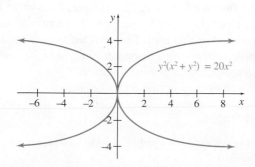

36. The graph of $2(x^2 + y^2)^2 = 25xy^2$, shown in the figure, is a *double folium*. Find the equation of the tangent line at the point $(2, 1)$.

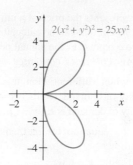

37. The graph of $x^2 + y^2 = 100$ is a circle having center at the origin and radius 10.

  a. Write the equations of the tangent lines at the points where $x = 6$.

  b. Graph the circle and the tangent lines.

38. Much has been written recently about elliptic curves because of their role in Andrew Wiles's 1995 proof of Fermat's Last Theorem. An elliptic curve can be written in the form

$$y^2 = x^3 + ax + b,$$

where a and b are constants, and the cubic function on the right has distinct roots. Find dy/dx for this curve.

39. Let $\sqrt{u} + \sqrt{2v + 1} = 5$. Find each derivative.

  a. $\dfrac{du}{dv}$ **b.** $\dfrac{dv}{du}$

  c. Based on your answers to parts a and b, what do you notice about the relationship between du/dv and dv/du?

40. Let $e^{u^2 - v} - v = 1$. Find each derivative.

  a. $\dfrac{du}{dv}$ **b.** $\dfrac{dv}{du}$

  c. Based on your answers to parts a and b, what do you notice about the relationship between du/dv and dv/du?

41. Suppose $x^2 + y^2 + 1 = 0$. Use implicit differentiation to find dy/dx. Then explain why the result you got is meaningless. (*Hint*: Can $x^2 + y^2 + 1$ equal 0?)

APPLICATIONS

Business and Economics

42. Demand The demand equation for a certain product is $2p^2 + q^2 = 1600$, where p is the price per unit in dollars and q is the number of units demanded.

  a. Find and interpret dq/dp.

  b. Find and interpret dp/dq.

43. Cost and Revenue For a certain product, cost C and revenue R are given as follows, where x is the number of units sold (in hundreds)

$$\text{Cost: } C^2 = x^2 + 100\sqrt{x} + 50$$
$$\text{Revenue: } 900(x - 5)^2 + 25R^2 = 22,500$$

  a. Find and interpret the marginal cost dC/dx at $x = 5$.

  b. Find and interpret the marginal revenue dR/dx at $x = 5$.

44. Elasticity of Demand Researchers found the demand for milk in Mexico for a particular year can be estimated by the implicit equation

$$\ln q = C - 0.678 \ln p,$$

where p represents the price of a unit of fluid milk and C represents a constant that can be calculated uniquely for a particular year. Here q represents the annual per capita fluid milk demand. *Source: Agricultural Economics.*

a. Use implicit differentiation to calculate and interpret the elasticity of demand. (Recall from the previous section that elasticity of demand is $E = -(p/q) \cdot dq/dp$.)

b. Solve the equation for q, then calculate the elasticity of demand.

45. Elasticity of Demand Researchers found the demand for cheese in Mexico for a particular year can be estimated by the implicit equation

$$\ln q = D - 0.44 \ln p,$$

where p represents the price of a unit of cheese and D represents a constant that can be calculated uniquely for a particular year. Here q represents the annual per capita cheese demand. *Source: Agricultural Economics.*

a. Use implicit differentiation to calculate and interpret the elasticity of demand. (Recall from the previous section that elasticity of demand is $E = -(p/q) \cdot dq/dp$.)

b. Solve the equation for q, then calculate the elasticity of demand.

Life Sciences

46. Respiratory Rate Researchers have found a correlation between respiratory rate and body mass in the first three years of life. This correlation can be expressed by the function

$$\log R(w) = 1.83 - 0.43 \log(w),$$

where w is the body weight (in kilograms) and $R(w)$ is the respiratory rate (in breaths per minute). *Source: Archives of Disease in Children.*

a. Find $R'(w)$ using implicit differentiation.

b. Find $R'(w)$ by first solving the equation for $R(w)$.

c. Discuss the two procedures. Is there a situation when you would want to use one method over another?

47. Biochemical Reaction A simple biochemical reaction with three molecules has solutions that oscillate toward a steady state when positive constants a and b are below the curve $b - a = (b + a)^3$. Find the largest possible value of a for which the reaction has solutions that oscillate toward a steady state. (*Hint:* Find where $da/db = 0$. Derive values for $a + b$ and $a - b$, and then solve the equations in two unknowns.) *Source: Mathematical Biology.*

48. Species The relationship between the number of species in a genus (x) and the number of genera (y) comprising x species is given by

$$xy^a = k,$$

where a and k are constants. Find dy/dx. *Source: Elements of Mathematical Biology.*

Physical Sciences

Velocity The position of a particle at time t is given by s. Find the velocity ds/dt.

49. $s^3 - 4st + 2t^3 - 5t = 0$

50. $2s^2 + \sqrt{st} - 4 = 3t$

YOUR TURN ANSWERS

1. $dy/dx = (y - 2x)/(2y - x)$

2. $\dfrac{dy}{dx} = \dfrac{ye^y + 2xy}{1 - xye^y}$

3. $y = x$

4. $dq/dp = -72$

6.5 Related Rates

APPLY IT When a skier's blood vessels contract because of the cold, how fast is the velocity of blood changing?

We use related rates to answer this question in Example 6 of this section.

It is common for variables to be functions of time; for example, sales of an item may depend on the season of the year, or a population of animals may be increasing at a certain rate several months after being introduced into an area. Time is often present implicitly in a mathematical model, meaning that derivatives with respect to time must be found by the method of implicit differentiation discussed in the previous section.

We start with a simple algebraic example.

EXAMPLE 1 Related Rates

Suppose that x and y are both functions of t, which can be considered to represent time, and that x and y are related by the equation

$$xy^2 + y = x^2 + 17.$$

Suppose further that when $x = 2$ and $y = 3$, then $dx/dt = 13$. Find the value of dy/dt at that moment.

SOLUTION We start by taking the derivative of the relationship, using the product and chain rules. Keep in mind that both x and y are functions of t. The result is

$$x\left(2y\frac{dy}{dt}\right) + y^2\frac{dx}{dt} + \frac{dy}{dt} = 2x\frac{dx}{dt}.$$

Now substitute $x = 2$, $y = 3$, and $dx/dt = 13$ to get

$$2\left(6\frac{dy}{dt}\right) + 9(13) + \frac{dy}{dt} = 4(13),$$

$$12\frac{dy}{dt} + 117 + \frac{dy}{dt} = 52.$$

Solve this last equation for dy/dt to get

$$13\frac{dy}{dt} = -65,$$

$$\frac{dy}{dt} = -5.$$

YOUR TURN 1 Suppose x and y are both functions of t and $x^3 + 2xy + y^2 = 1$. If $x = 1$, $y = -2$, and $dx/dt = 6$, then find dy/dt.

TRY YOUR TURN 1

Our next example is typical of the word problems involving related rates.

EXAMPLE 2   Area

A small rock is dropped into a lake. Circular ripples spread over the surface of the water, with the radius of each circle increasing at the rate of $3/2$ ft per second. Find the rate of change of the area inside the circle formed by a ripple at the instant the radius is 4 ft.

SOLUTION As shown in Figure 18, the area A and the radius r are related by

$$A = \pi r^2.$$

Both A and r are functions of the time t in seconds. Take the derivative of both sides with respect to time.

$$\frac{d}{dt}(A) = \frac{d}{dt}(\pi r^2)$$

$$\frac{dA}{dt} = 2\pi r \cdot \frac{dr}{dt} \tag{1}$$

Since the radius is increasing at the rate of $3/2$ ft per second,

$$\frac{dr}{dt} = \frac{3}{2}.$$

The rate of change of area at the instant $r = 4$ is given by dA/dt evaluated at $r = 4$. Substituting into Equation (1) gives

$$\frac{dA}{dt} = 2\pi \cdot 4 \cdot \frac{3}{2}$$

$$\frac{dA}{dt} = 12\pi \approx 37.7 \text{ ft}^2 \text{ per second.}$$

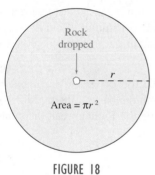

Rock dropped

r

Area $= \pi r^2$

FIGURE 18

In Example 2, the derivatives (or rates of change) dA/dt and dr/dt are related by Equation (1); for this reason they are called **related rates**. As suggested by Example 2, four basic steps are involved in solving problems about related rates.

Solving a Related Rate Problem

1. Identify all given quantities, as well as the quantities to be found. Draw a sketch when possible.

2. Write an equation relating the variables of the problem.

3. Use implicit differentiation to find the derivative of both sides of the equation in Step 2 with respect to time.

4. Solve for the derivative giving the unknown rate of change and substitute the given values.

CAUTION 1. Differentiate *first*, and *then* substitute values for the variables. If the substitutions were performed first, differentiating would not lead to useful results.

2. Some students confuse related rates problems with applied extrema problems, perhaps because they are both word problems. There is an easy way to tell the difference. In applied extrema problems, you are always trying to maximize or minimize something, that is, make it as large or as small as possible. In related rate problems, you are trying to find how fast something is changing; time is always the independent variable.

EXAMPLE 3 Sliding Ladder

A 50-ft ladder is placed against a large building. The base of the ladder is resting on an oil spill, and it slips (to the right in Figure 19) at the rate of 3 ft per minute. Find the rate of change of the height of the top of the ladder above the ground at the instant when the base of the ladder is 30 ft from the base of the building.

SOLUTION Starting with Step 1, let y be the height of the top of the ladder above the ground, and let x be the distance of the base of the ladder from the base of the building. We are trying to find dy/dt when $x = 30$. To perform Step 2, use the Pythagorean theorem to write

$$x^2 + y^2 = 50^2. \qquad (2)$$

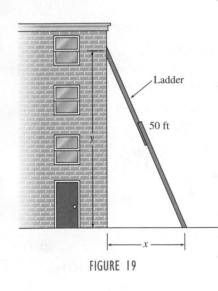

Ladder

50 ft

FIGURE 19

Both x and y are functions of time t (in minutes) after the moment that the ladder starts slipping. According to Step 3, take the derivative of both sides of Equation (2) with respect to time, getting

$$\frac{d}{dt}(x^2 + y^2) = \frac{d}{dt}(50^2)$$

$$2x\frac{dx}{dt} + 2y\frac{dy}{dt} = 0. \qquad (3)$$

To complete Step 4, we need to find the values of x, y, and dx/dt. Once we find these, we can substitute them into Equation (3) to find dy/dt.

Since the base is sliding at the rate of 3 ft per minute,

$$\frac{dx}{dt} = 3.$$

Also, the base of the ladder is 30 ft from the base of the building, so $x = 30$. Use this to find y.

$$50^2 = 30^2 + y^2$$
$$2500 = 900 + y^2$$
$$1600 = y^2$$
$$y = 40$$

In summary, $y = 40$ when $x = 30$. Also, the rate of change of x over time t is $dx/dt = 3$. Substituting these values into Equation (3) to find the rate of change of y over time gives

$$2(30)(3) + 2(40)\frac{dy}{dt} = 0$$

$$180 + 80\frac{dy}{dt} = 0$$

$$80\frac{dy}{dt} = -180$$

$$\frac{dy}{dt} = \frac{-180}{80} = \frac{-9}{4} = -2.25.$$

YOUR TURN 2   A 25-ft ladder is placed against a building. The base of the ladder is slipping away from the building at a rate of 3 ft per minute. Find the rate at which the top of the ladder is sliding down the building when the bottom of the ladder is 7 ft from the base of the building.

At the instant when the base of the ladder is 30 ft from the base of the building, the top of the ladder is sliding down the building at the rate of 2.25 ft per minute. (The minus sign shows that the ladder is sliding *down*, so the distance y is *decreasing*.)*   **TRY YOUR TURN 2**

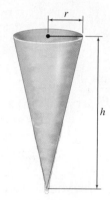

FIGURE 20

EXAMPLE 4   Icicle

A cone-shaped icicle is dripping from the roof. The radius of the icicle is decreasing at a rate of 0.2 cm per hour, while the length is increasing at a rate of 0.8 cm per hour. If the icicle is currently 4 cm in radius and 20 cm long, is the volume of the icicle increasing or decreasing, and at what rate?

SOLUTION   For this problem we need the formula for the volume of a cone:

$$V = \frac{1}{3}\pi r^2 h, \tag{4}$$

where r is the radius of the cone and h is the height of the cone, which in this case is the length of the icicle, as in Figure 20.

In this problem, V, r, and h are functions of the time t in hours. Taking the derivative of both sides of Equation (4) with respect to time yields

$$\frac{dV}{dt} = \frac{1}{3}\pi\left[r^2\frac{dh}{dt} + (h)(2r)\frac{dr}{dt}\right]. \tag{5}$$

Since the radius is decreasing at a rate of 0.2 cm per hour and the length is increasing at a rate of 0.8 cm per hour,

$$\frac{dr}{dt} = -0.2 \quad \text{and} \quad \frac{dh}{dt} = 0.8.$$

YOUR TURN 3   Suppose that in Example 4 the volume of the icicle is decreasing at a rate of 10 cm³ per hour and the radius is decreasing at a rate of 0.4 cm per hour. Find the rate of change of the length of the icicle when the radius is 4 cm and the length is 20 cm.

Substituting these, as well as $r = 4$ and $h = 20$, into Equation (5) yields

$$\frac{dV}{dt} = \frac{1}{3}\pi[4^2(0.8) + (20)(8)(-0.2)]$$

$$= \frac{1}{3}\pi(-19.2) \approx -20.$$

Because the sign of dV/dt is negative, the volume of the icicle is decreasing at a rate of 20 cm³ per hour.   **TRY YOUR TURN 3**

---

*The model in Example 3 breaks down as the top of the ladder nears the ground. As y approaches 0, dy/dt becomes infinitely large. In reality, the ladder loses contact with the wall before y reaches 0.

EXAMPLE 5 Revenue

A company is increasing production of peanuts at the rate of 50 cases per day. All cases produced can be sold. The daily demand function is given by

$$p = 50 - \frac{q}{200},$$

where q is the number of units produced (and sold) and p is price in dollars. Find the rate of change of revenue with respect to time (in days) when the daily production is 200 units.

SOLUTION   The revenue function,

$$R = qp = q\left(50 - \frac{q}{200}\right) = 50q - \frac{q^2}{200},$$

relates R and q. The rate of change of q over time (in days) is $dq/dt = 50$. The rate of change of revenue over time, dR/dt, is to be found when $q = 200$. Differentiate both sides of the equation

$$R = 50q - \frac{q^2}{200}$$

with respect to t.

$$\frac{dR}{dt} = 50\frac{dq}{dt} - \frac{1}{100}q\frac{dq}{dt} = \left(50 - \frac{1}{100}q\right)\frac{dq}{dt}$$

YOUR TURN 4   Repeat Example 5 using the daily demand function given by

$$p = 2000 - \frac{q^2}{100}.$$

Now substitute the known values for q and dq/dt.

$$\frac{dR}{dt} = \left[50 - \frac{1}{100}(200)\right](50) = 2400$$

Thus revenue is increasing at the rate of $2400 per day.   **TRY YOUR TURN 4**

EXAMPLE 6 Blood Flow

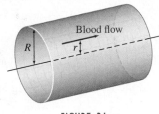

Blood flow

FIGURE 21

Blood flows faster the closer it is to the center of a blood vessel. According to Poiseuille's laws, the velocity V of blood is given by

$$V = k(R^2 - r^2),$$

where R is the radius of the blood vessel, r is the distance of a layer of blood flow from the center of the vessel, and k is a constant, assumed here to equal 375. See Figure 21. Suppose a skier's blood vessel has radius $R = 0.08$ mm and that cold weather is causing the vessel to contract at a rate of $dR/dt = -0.01$ mm per minute. How fast is the velocity of blood changing?

APPLY IT   **SOLUTION**   Find dV/dt. Treat r as a constant. Assume the given units are compatible.

$$V = 375(R^2 - r^2)$$
$$\frac{dV}{dt} = 375\left(2R\frac{dR}{dt} - 0\right) \quad r \text{ is a constant.}$$
$$\frac{dV}{dt} = 750R\frac{dR}{dt}$$

Here $R = 0.08$ and $dR/dt = -0.01$, so

$$\frac{dV}{dt} = 750(0.08)(-0.01) = -0.6.$$

That is, the velocity of the blood is decreasing at a rate of -0.6 mm per minute each minute. The minus sign indicates that this is a deceleration (negative acceleration), since it represents a negative rate of change of velocity.

6.5 EXERCISES

Assume x and y are functions of t. Evaluate dy/dt for each of the following.

1. $y^2 - 8x^3 = -55$; $\dfrac{dx}{dt} = -4, x = 2, y = 3$

2. $8y^3 + x^2 = 1$; $\dfrac{dx}{dt} = 2, x = 3, y = -1$

3. $2xy - 5x + 3y^3 = -51$; $\dfrac{dx}{dt} = -6, x = 3, y = -2$

4. $4x^3 - 6xy^2 + 3y^2 = 228$; $\dfrac{dx}{dt} = 3, x = -3, y = 4$

5. $\dfrac{x^2 + y}{x - y} = 9$; $\dfrac{dx}{dt} = 2, x = 4, y = 2$

6. $\dfrac{y^3 - 4x^2}{x^3 + 2y} = \dfrac{44}{31}$; $\dfrac{dx}{dt} = 5, x = -3, y = -2$

7. $xe^y = 2 - \ln 2 + \ln x$; $\dfrac{dx}{dt} = 6, x = 2, y = 0$

8. $y \ln x + xe^y = 1$; $\dfrac{dx}{dt} = 5, x = 1, y = 0$

APPLICATIONS

Business and Economics

9. Cost A manufacturer of handcrafted wine racks has determined that the cost to produce x units per month is given by $C = 0.2x^2 + 10{,}000$. How fast is cost per month changing when production is changing at the rate of 12 units per month and the production level is 80 units?

10. Cost/Revenue The manufacturer in Exercise 9 has found that the cost C and revenue R (in dollars) in one month are related by the equation

$$C = \dfrac{R^2}{450{,}000} + 12{,}000.$$

Find the rate of change of revenue with respect to time when the cost is changing by $15 per month and the monthly revenue is $25,000.

11. Revenue/Cost/Profit Given the revenue and cost functions $R = 50x - 0.4x^2$ and $C = 5x + 15$ (in dollars), where x is the daily production (and sales), find the following when 40 units are produced daily and the rate of change of production is 10 units per day.

a. The rate of change of revenue with respect to time

b. The rate of change of cost with respect to time

c. The rate of change of profit with respect to time

12. Revenue/Cost/Profit Repeat Exercise 11, given that 80 units are produced daily and the rate of change of production is 12 units per day.

13. Demand The demand function for a certain product is determined by the fact that the product of the price and the quantity demanded equals 8000. The product currently sells for $3.50 per unit. Suppose manufacturing costs are increasing over time at a rate of 15% and the company plans to increase the price p at this rate as well. Find the rate of change of demand over time.

14. Revenue A company is increasing production at the rate of 25 units per day. The daily demand function is determined by the fact that the price (in dollars) is a linear function of q. At a price of $70, the demand is 0, and 100 items will be demanded at a price of $60. Find the rate of change of revenue with respect to time (in days) when the daily production (and sales) is 20 items.

Life Sciences

15. Blood Velocity A cross-country skier has a history of heart problems. She takes nitroglycerin to dilate blood vessels, thus avoiding angina (chest pain) due to blood vessel contraction. Use Poiseuille's law with $k = 555.6$ to find the rate of change of the blood velocity when $R = 0.02$ mm and R is changing at 0.003 mm per minute. Assume r is constant. (See Example 6.)

16. Allometric Growth Suppose x and y are two quantities that vary with time according to the allometric formula $y = nx^m$. (See Exercise 84 in the section on Logarithmic Functions.) Show that the derivatives of x and y are related by the formula

$$\dfrac{1}{y}\dfrac{dy}{dt} = m\dfrac{1}{x}\dfrac{dx}{dt}.$$

(*Hint:* Take natural logarithms of both sides before taking the derivatives.)

17. Brain Mass The brain mass of a fetus can be estimated using the total mass of the fetus by the function

$$b = 0.22m^{0.87},$$

where m is the mass of the fetus (in grams) and b is the brain mass (in grams). Suppose the brain mass of a 25-g fetus is changing at a rate of 0.25 g per day. Use this to estimate the rate of change of the total mass of the fetus, dm/dt. *Source: Archives d'Anatomie, d'Histologie et d'Embryologie.*

18. Birds The energy cost of bird flight as a function of body mass is given by

$$E = 429m^{-0.35},$$

where m is the mass of the bird (in grams) and E is the energy expenditure (in calories per gram per hour). Suppose that the mass of a 10-g bird is increasing at a rate of 0.001 g per hour. Find the rate at which the energy expenditure is changing with respect to time. *Source: Wildlife Feeding and Nutrition.*

19. **Metabolic Rate** The average daily metabolic rate for captive animals from weasels to elk can be expressed as a function of mass by

$$r = 140.2m^{0.75},$$

where m is the mass of the animal (in kilograms) and r is the metabolic rate (in kcal per day). *Source: Wildlife Feeding and Nutrition.*

a. Suppose that the mass of a weasel is changing with respect to time at a rate dm/dt. Find dr/dt.

b. Determine dr/dt for a 250-kg elk that is gaining mass at a rate of 2 kg per day.

20. **Lizards** The energy cost of horizontal locomotion as a function of the body mass of a lizard is given by

$$E = 26.5m^{-0.34},$$

where m is the mass of the lizard (in kilograms) and E is the energy expenditure (in kcal/kg/km). Suppose that the mass of a 5-kg lizard is increasing at a rate of 0.05 kg per day. Find the rate at which the energy expenditure is changing with respect to time. *Source: Wildlife Feeding and Nutrition.*

Social Sciences

21. **Crime Rate** Sociologists have found that crime rates are influenced by temperature. In a midwestern town of 100,000 people, the crime rate has been approximated as

$$C = \frac{1}{10}(T - 60)^2 + 100,$$

where C is the number of crimes per month and T is the average monthly temperature in degrees Fahrenheit. The average temperature for May was 76°, and by the end of May the temperature was rising at the rate of 8° per month. How fast is the crime rate rising at the end of May?

22. **Memorization Skills** Under certain conditions, a person can memorize W words in t minutes, where

$$W(t) = \frac{-0.02t^2 + t}{t + 1}.$$

Find dW/dt when $t = 5$.

Physical Sciences

23. **Sliding Ladder** A 17-ft ladder is placed against a building. The base of the ladder is slipping away from the building at a rate of 9 ft per minute. Find the rate at which the top of the ladder is sliding down the building at the instant when the bottom of the ladder is 8 ft from the base of the building.

24. **Distance**

a. One car leaves a given point and travels north at 30 mph. Another car leaves the same point at the same time and travels west at 40 mph. At what rate is the distance between the two cars changing at the instant when the cars have traveled 2 hours?

b. Suppose that, in part a, the second car left 1 hour later than the first car. At what rate is the distance between the two cars changing at the instant when the second car has traveled 1 hour?

25. **Area** A rock is thrown into a still pond. The circular ripples move outward from the point of impact of the rock so that the radius of the circle formed by a ripple increases at the rate of 2 ft per minute. Find the rate at which the area is changing at the instant the radius is 4 ft.

26. **Volume** A spherical snowball is placed in the sun. The sun melts the snowball so that its radius decreases 1/4 in. per hour. Find the rate of change of the volume with respect to time at the instant the radius is 4 in.

27. **Ice Cube** An ice cube that is 3 cm on each side is melting at a rate of 2 cm³ per min. How fast is the length of the side decreasing?

28. **Volume** A sand storage tank used by the highway department for winter storms is leaking. As the sand leaks out, it forms a conical pile. The radius of the base of the pile increases at the rate of 0.75 in. per minute. The height of the pile is always twice the radius of the base. Find the rate at which the volume of the pile is increasing at the instant the radius of the base is 6 in.

29. **Shadow Length** A man 6 ft tall is walking away from a lamp post at the rate of 50 ft per minute. When the man is 8 ft from the lamp post, his shadow is 10 ft long. Find the rate at which the length of the shadow is increasing when he is 25 ft from the lamp post. (See the figure.)

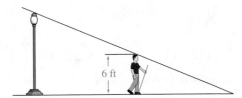

30. **Water Level** A trough has a triangular cross section. The trough is 6 ft across the top, 6 ft deep, and 16 ft long. Water is being pumped into the trough at the rate of 4 ft³ per minute.

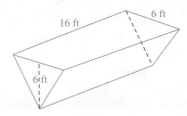

Find the rate at which the height of the water is increasing at the instant that the height is 4 ft.

31. **Velocity** A pulley is on the edge of a dock, 8 ft above the water level. (See the figure on the next page.) A rope is being used to pull in a boat. The rope is attached to the boat at water level. The rope is being pulled in at the rate of 1 ft per second. Find the rate at which the boat is approaching the dock at the instant the boat is 8 ft from the dock.

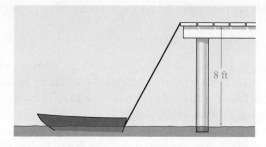

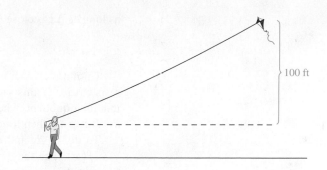

32. Kite Flying Christine O'Brien is flying her kite in a wind that is blowing it east at a rate of 50 ft per minute. She has already let out 200 ft of string, and the kite is flying 100 ft above her hand. How fast must she let out string at this moment to keep the kite flying with the same speed and altitude?

YOUR TURN ANSWERS

1. $dy/dt = -3$.
2. $-7/8$ ft/min
3. Increases 3.4 cm per hour
4. Revenue is increasing at the rate of $40,000 per day.

6.6 Differentials: Linear Approximation

APPLY IT If the estimated sales of cellular telephones turn out to be inaccurate, approximately how much are profits affected?

Using differentials, we will answer this question in Example 4.

As mentioned earlier, the symbol Δx represents a change in the variable x. Similarly, Δy represents a change in y. An important problem that arises in many applications is to determine Δy given specific values of x and Δx. This quantity is often difficult to evaluate. In this section we show a method of approximating Δy that uses the derivative dy/dx. In essence, we use the tangent line at a particular value of x to approximate $f(x)$ for values close to x.

For values x_1 and x_2,

$$\Delta x = x_2 - x_1.$$

Solving for x_2 gives

$$x_2 = x_1 + \Delta x.$$

For a function $y = f(x)$, the symbol Δy represents a change in y:

$$\Delta y = f(x_2) - f(x_1).$$

Replacing x_2 with $x_1 + \Delta x$ gives

$$\Delta y = f(x_1 + \Delta x) - f(x_1).$$

If Δx is used instead of h, the derivative of a function f at x_1 could be defined as

$$\frac{dy}{dx} = \lim_{\Delta x \to 0} \frac{\Delta y}{\Delta x}.$$

If the derivative exists, then

$$\frac{dy}{dx} \approx \frac{\Delta y}{\Delta x}$$

as long as Δx is close to 0. Multiplying both sides by Δx (assume $\Delta x \neq 0$) gives

$$\Delta y \approx \frac{dy}{dx} \cdot \Delta x.$$

Until now, dy/dx has been used as a single symbol representing the derivative of y with respect to x. In this section, separate meanings for dy and dx are introduced in such a way that their quotient, when $dx \neq 0$, is the derivative of y with respect to x. These meanings of dy and dx are then used to find an approximate value of Δy.

To define dy and dx, look at Figure 22 below, which shows the graph of a function $y = f(x)$. The tangent line to the graph has been drawn at the point P. Let Δx be any nonzero real number (in practical problems, Δx is a small number) and locate the point $x + \Delta x$ on the x-axis. Draw a vertical line through $x + \Delta x$. Let this vertical line cut the tangent line at M and the graph of the function at Q.

Define the new symbol dx to be the same as Δx. Define the new symbol dy to equal the length MR. The slope of PM is $f'(x)$. By the definition of slope, the slope of PM is also dy/dx, so that

$$f'(x) = \frac{dy}{dx},$$

or

$$dy = f'(x)\,dx.$$

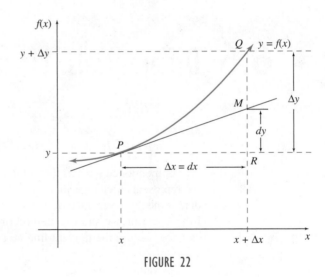

FIGURE 22

In summary, the definitions of the symbols dy and dx are as follows.

Differentials

For a function $y = f(x)$ whose derivative exists, the **differential** of x, written dx, is an arbitrary real number (usually small compared with x); the **differential** of y, written dy, is the product of $f'(x)$ and dx, or

$$dy = f'(x)\,dx.$$

The usefulness of the differential is suggested by Figure 22. As dx approaches 0, the value of dy gets closer and closer to that of Δy, so that for small nonzero values of dx

$$dy \approx \Delta y,$$

or

$$\Delta y \approx f'(x)\,dx.$$

EXAMPLE 1 Differential

Find dy for the following functions.

(a) $y = 6x^2$

> **SOLUTION** The derivative is $dy/dx = 12x$ so
>
> $$dy = 12x\, dx.$$

(b) $y = 800x^{-3/4}$, $x = 16$, $dx = 0.01$

> **SOLUTION**
>
> $$dy = -600x^{-7/4}dx$$
>
> $$= -600(16)^{-7/4}(0.01)$$
>
> $$= -600\left(\frac{1}{2^7}\right)(0.01) = -0.046875$$

YOUR TURN 1 Find dy if $y = 300x^{-2/3}$, $x = 8$, and $dx = 0.05$.

TRY YOUR TURN 1

Differentials can be used to approximate function values for a given x-value (in the absence of a calculator or computer). As discussed above,

$$\Delta y = f(x + \Delta x) - f(x).$$

For small nonzero values of Δx, $\Delta y \approx dy$, so that

$$dy \approx f(x + \Delta x) - f(x),$$

or

$$f(x) + dy \approx f(x + \Delta x).$$

Replacing dy with $f'(x)dx$ gives the following result.

> **Linear Approximation**
>
> Let f be a function whose derivative exists. For small nonzero values of Δx,
>
> $$dy \approx \Delta y,$$
>
> and
>
> $$f(x + \Delta x) \approx f(x) + dy = f(x) + f'(x)dx.$$

EXAMPLE 2 Approximation

Approximate $\sqrt{50}$.

SOLUTION We will use the linear approximation

$$f(x + \Delta x) \approx f(x) + f'(x)dx$$

for a small value of Δx to form this estimate. We first choose a number x that is close to 50 for which we know its square root. Since $\sqrt{49} = 7$, we let $f(x) = \sqrt{x}$, $x = 49$, and $\Delta x = dx = 1$. Using this information, with the fact that when $f(x) = \sqrt{x} = x^{1/2}$,

$$f'(x) = \frac{1}{2}x^{-1/2} = \frac{1}{2x^{1/2}},$$

we have

$$f(x + \Delta x) \approx f(x) + f'(x)dx = \sqrt{x} + \frac{1}{2\sqrt{x}}\, dx.$$

Substituting $x = 49$ and $dx = 1$ into the preceding formula gives

$$f(50) = f(49 + 1) \approx \sqrt{49} + \frac{1}{2\sqrt{49}}(1)$$

$$= 7 + \frac{1}{14}$$

$$= 7\frac{1}{14}.$$

YOUR TURN 2
Approximate $\sqrt{99}$.

A calculator gives $7\frac{1}{14} \approx 7.07143$ and $\sqrt{50} \approx 7.07107$. Our approximation of $7\frac{1}{14}$ is close to the true answer and does not require a calculator. **TRY YOUR TURN 2**

While calculators have made differentials less important, the approximation of functions, including linear approximation, is still important in the branch of mathematics known as numerical analysis.

Marginal Analysis

Differentials are used to find an approximate value of the change in the dependent variable corresponding to a given change in the independent variable. When the concept of marginal cost (or profit or revenue) was used to approximate the change in cost for nonlinear functions, the same idea was developed. Thus the differential dy approximates Δy in much the same way as the marginal quantities approximate changes in functions.

For example, for a cost function $C(x)$,

$$dC = C'(x)dx = C'(x)\Delta x.$$

Since $\Delta C \approx dC$,

$$\Delta C \approx C'(x)\Delta x.$$

If the change in production, Δx, is equal to 1, then

$$C(x + 1) - C(x) = \Delta C$$
$$\approx C'(x)\Delta x$$
$$= C'(x),$$

which shows that marginal cost $C'(x)$ approximates the cost of the next unit produced, as mentioned earlier.

EXAMPLE 3 Cost

Let $C(x) = 2x^3 + 300$.

(a) Use $C'(x)$ to approximate ΔC when $\Delta x = 1$ and $x = 3$. Then compare ΔC with $C'(3)$.

SOLUTION Since $C(x) = 2x^3 + 300$, the derivative is

$$C'(x) = 6x^2$$

and the marginal cost approximation at $x = 3$ is

$$C'(3) = 6(3^2) = 54.$$

Now, the actual cost of the next unit produced is

$$\Delta C = C(4) - C(3) = 428 - 354 = 74.$$

Here, the approximation of $C'(3)$ for ΔC is poor, since $\Delta x = 1$ is large relative to $x = 3$.

(b) Use $C'(x)$ to approximate ΔC when $\Delta x = 1$ and x = 50. Then compare ΔC with $C'(50)$.

SOLUTION The marginal cost approximation at x = 50 is

$$C'(50) = 6(50^2) = 15,000$$

and the actual cost of the next unit produced is

$$\Delta C = C(51) - C(50) = 265,602 - 250,300 = 15,302.$$

This approximation is quite good since $\Delta x = 1$ is small compared to $x = 50$.

EXAMPLE 4 Profit

An analyst for a manufacturer of electronic devices estimates that the profit (in dollars) from the sale of x cellular telephones is given by

$$P(x) = 4000 \ln x.$$

In a report to management, the analyst projected sales for the coming year to be 10,000 phones, for a total profit of about \$36,840. He now realizes that his sales estimate may have been as much as 1000 phones too high. Approximately how far off is his profit estimate?

APPLY IT

SOLUTION Differentials can be used to find the approximate change in P resulting from decreasing x by 1000. This change can be approximated by $dP = P'(x)dx$ where $x = 10,000$ and $dx = -1000$. Since $P'(x) = 4000/x$,

$$\Delta P \approx dP = \frac{4000}{x}dx$$

$$= \frac{4000}{10,000}(-1000)$$

$$= -400.$$

Thus the profit estimate may have been as much as \$400 too high. Computing the actual difference with a calculator gives $4000 \ln 9000 - 4000 \ln 10,000 \approx -421$, which is close to our approximation.

Error Estimation
The final example in this section shows how differentials are used to estimate errors that might enter into measurements of a physical quantity.

EXAMPLE 5 Error Estimation

In a precision manufacturing process, ball bearings must be made with a radius of 0.6 mm, with a maximum error in the radius of ± 0.015 mm. Estimate the maximum error in the volume of the ball bearing.

SOLUTION The formula for the volume of a sphere is

$$V = \frac{4}{3}\pi r^3.$$

If an error of Δr is made in measuring the radius of the sphere, the maximum error in the volume is

$$\Delta V = \frac{4}{3}\pi(r + \Delta r)^3 - \frac{4}{3}\pi r^3.$$

Rather than calculating ΔV, approximate ΔV with dV, where

$$dV = 4\pi r^2 dr.$$

Replacing r with 0.6 and $dr = \Delta r$ with ± 0.015 gives

YOUR TURN 3 Repeat Example 5 for $r = 1.25$ mm with a maximum error in the radius of ± 0.025 mm.

$$dV = 4\pi(0.6)^2(\pm 0.015)$$

$$\approx \pm 0.0679.$$

The maximum error in the volume is about 0.07 mm³. **TRY YOUR TURN 3**

6.6 EXERCISES

For Exercises 1–8, find *dy* for the given values of *x* and Δ*x*.

1. $y = 2x^3 - 5x$; $x = -2, \Delta x = 0.1$

2. $y = 4x^3 - 3x$; $x = 3, \Delta x = 0.2$

3. $y = x^3 - 2x^2 + 3$; $x = 1, \Delta x = -0.1$

4. $y = 2x^3 + x^2 - 4x$; $x = 2, \Delta x = -0.2$

5. $y = \sqrt{3x + 2}$ $x = 4, \Delta x = 0.15$

6. $y = \sqrt{4x - 1}$; $x = 5, \Delta x = 0.08$

7. $y = \dfrac{2x - 5}{x + 1}$; $x = 2, \Delta x = -0.03$

8. $y = \dfrac{6x - 3}{2x + 1}$; $x = 3, \Delta x = -0.04$

Use the differential to approximate each quantity. Then use a calculator to approximate the quantity, and give the absolute value of the difference in the two results to 4 decimal places.

9. $\sqrt{145}$

10. $\sqrt{23}$

11. $\sqrt{0.99}$

12. $\sqrt{17.02}$

13. $e^{0.01}$

14. $e^{-0.002}$

15. $\ln 1.05$

16. $\ln 0.98$

APPLICATIONS

Business and Economics

17. **Demand** The demand for grass seed (in thousands of pounds) at a price of *p* dollars is
$$D(p) = -3p^3 - 2p^2 + 1500.$$
Use the differential to approximate the changes in demand for the following changes in *p*.

   a. \$2 to \$2.10   **b.** \$6 to \$6.15

18. **Average Cost** The average cost (in dollars) to manufacture *x* dozen marking pencils is
$$A(x) = 0.04x^3 + 0.1x^2 + 0.5x + 6.$$
Use the differential to approximate the changes in the average cost for the following changes in *x*.

   a. 3 to 4   **b.** 5 to 6

19. **Revenue** A company estimates that the revenue (in dollars) from the sale of *x* doghouses is given by
$$R(x) = 12{,}000 \ln(0.01x + 1).$$
Use the differential to approximate the change in revenue from the sale of one more doghouse when 100 doghouses are sold.

20. **Profit** The cost function for the company in Exercise 19 is
$$C(x) = 150 + 75x,$$
where *x* represents the demand for the product. Find the approximate change in profit for a 1-unit change in demand when demand is at a level of 100 doghouses. Use the differential.

21. **Material Requirement** A cube 4 in. on an edge is given a protective coating 0.1 in. thick. About how much coating should a production manager order for 1000 such cubes?

22. **Material Requirement** Beach balls 1 ft in diameter have a thickness of 0.03 in. How much material would be needed to make 5000 beach balls?

Life Sciences

23. **Alcohol Concentration** In Exercise 55 in the section on Polynomial and Rational Functions, we gave the function defined by
$$A(x) = 0.003631x^3 - 0.03746x^2 + 0.1012x + 0.009$$
as the approximate blood alcohol concentration in a 170-lb woman *x* hours after drinking 2 oz of alcohol on an empty stomach, for *x* in the interval [0, 5]. *Source: Medicolegal Aspects of Alcohol Determination in Biological Specimens.*

   a. Approximate the change in alcohol level from 1 to 1.2 hours.

   b. Approximate the change in alcohol level from 3 to 3.2 hours.

24. **Drug Concentration** The concentration of a certain drug in the bloodstream *x* hours after being administered is approximately
$$C(x) = \frac{5x}{9 + x^2}.$$
Use the differential to approximate the changes in concentration for the following changes in *x*.

   a. 1 to 1.5   **b.** 2 to 2.25

25. **Bacteria Population** The population of bacteria (in millions) in a certain culture *x* hours after an experimental nutrient is introduced into the culture is
$$P(x) = \frac{25x}{8 + x^2}.$$
Use the differential to approximate the changes in population for the following changes in *x*.

   a. 2 to 2.5   **b.** 3 to 3.25

26. **Area of a Blood Vessel** The radius of a blood vessel is 1.7 mm. A drug causes the radius to change to 1.6 mm. Find the approximate change in the area of a cross section of the vessel.

27. **Volume of a Tumor** A tumor is approximately spherical in shape. If the radius of the tumor changes from 14 mm to 16 mm, find the approximate change in volume.

28. **Area of an Oil Slick** An oil slick is in the shape of a circle. Find the approximate increase in the area of the slick if its radius increases from 1.2 miles to 1.4 miles.

29. **Area of a Bacteria Colony** The shape of a colony of bacteria on a Petri dish is circular. Find the approximate increase in its area if the radius increases from 20 mm to 22 mm.

30. Gray Wolves Accurate methods of estimating the age of gray wolves are important to scientists who study wolf population dynamics. One method of estimating the age of a gray wolf is to measure the percent closure of the pulp cavity of a canine tooth and then estimate age by

$$A(p) = \frac{1.181p}{94.359 - p},$$

where p is the percent closure and $A(p)$ is the age of the wolf (in years). *Source: Journal of Wildlife Management.*

a. What is a sensible domain for this function?

b. Use differentials to estimate how long it will take for a gray wolf that first measures a 60% closure to obtain a 65% closure. Compare this with the actual value of about 0.55 years.

31. Pigs Researchers have observed that the mass of a female (gilt) pig can be estimated by the function

$$M(t) = -3.5 + 197.5e^{-e^{-0.01394(t-108.4)}},$$

where t is the age of the pig (in days) and $M(t)$ is the mass of the pig (in kilograms). *Source: Animal Science.*

a. If a particular gilt is 80 days old, use differentials to estimate how much it will gain before it is 90 days old.

b. What is the actual gain in mass?

Physical Sciences

32. Volume A spherical balloon is being inflated. Find the approximate change in volume if the radius increases from 4 cm to 4.2 cm.

33. Volume A spherical snowball is melting. Find the approximate change in volume if the radius decreases from 3 cm to 2.8 cm.

34. Volume A cubical crystal is growing in size. Find the approximate change in the length of a side when the volume increases from 27 cubic mm to 27.1 cubic mm.

35. Volume An icicle is gradually increasing in length, while maintaining a cone shape with a length 15 times the radius. Find the approximate amount that the volume of the icicle increases when the length increases from 13 cm to 13.2 cm.

General Interest

36. Measurement Error The edge of a square is measured as 3.45 in., with a possible error of ± 0.002 in. Estimate the maximum error in the area of the square.

37. Tolerance A worker is cutting a square from a piece of sheet metal. The specifications call for an area that is 16 cm^2 with an error of no more than 0.01 cm^2. How much error could be tolerated in the length of each side to ensure that the area is within the tolerance?

38. Measurement Error The radius of a circle is measured as 4.87 in., with a possible error of ± 0.040 in. Estimate the maximum error in the area of the circle.

39. Measurement Error A sphere has a radius of 5.81 in., with a possible error of ± 0.003 in. Estimate the maximum error in the volume of the sphere.

40. Tolerance A worker is constructing a cubical box that must contain 125 ft^3, with an error of no more than 0.3 ft^3. How much error could be tolerated in the length of each side to ensure that the volume is within the tolerance?

41. Measurement Error A cone has a known height of 7.284 in. The radius of the base is measured as 1.09 in., with a possible error of ± 0.007 in. Estimate the maximum error in the volume of the cone.

YOUR TURN ANSWERS

1. $-5/16$
2. 9.95
3. About 0.5 mm^3

6   CHAPTER REVIEW

SUMMARY

In this chapter, we began by discussing how to find an absolute maximum or minimum. In contrast to a relative extremum, which is the largest or smallest value of a function on some open interval about the point, an absolute extremum is the largest or smallest value of the function on the entire interval under consideration. We then studied various applications with maximizing or minimizing as the goal. Two more applications, economic lot size and economic order quantity, were covered in a separate section, which also applied the derivative to the economic concept of elasticity of demand. Implicit differentiation is more of a technique than an application, but it underlies related rate problems, in which one or more rates are given and another is to be found. Finally, we studied the differential as a way to find linear approximations of functions.

Finding Absolute Extrema   To find absolute extrema for a function f continuous on a closed interval $[a, b]$:

1. Find all critical numbers for f in (a, b).
2. Evaluate f for all critical numbers in (a, b).
3. Evaluate f for the endpoints a and b of the interval.
4. The largest value found in Step 2 or 3 is the maximum, and the smallest value is the minimum.

Solving an Applied Extrema Problem

1. Read the problem carefully. Make sure you understand what is given and what is unknown.
2. If possible, sketch a diagram. Label the various parts.
3. Decide on the variable that must be maximized or minimized. Express that variable as a function of *one* other variable.
4. Find the domain of the function.
5. Find the critical points for the function from Step 3.
6. If the domain is a closed interval, evaluate the function at the endpoints and at each critical number to see which yields the absolute maximum or minimum. If the domain is an open interval, apply the critical point theorem when there is only one critical number. If there is more than one critical number, evaluate the function at the critical numbers and find the limit as the endpoints of the interval are approached to determine if an absolute maximum or minimum exists at one of the critical points.

Elasticity of Demand

Let $q = f(p)$, where q is demand at a price p.

$$E = -\frac{p}{q} \cdot \frac{dq}{dp}.$$

Demand is inelastic if $E < 1$.
Demand is elastic if $E > 1$.
Demand has unit elasticity if $E = 1$.
Total revenue is maximized at the price where demand has unit elasticity.

Implicit Differentiation

To find dy/dx for an equation containing x and y:

1. Differentiate on both sides of the equation with respect to x, keeping in mind that y is assumed to be a function of x.

2. Place all terms with dy/dx on one side of the equals sign and all terms without dy/dx on the other side.

3. Factor out dy/dx, and then solve for dy/dx.

Solving a Related Rate Problem

1. Identify all given quantities, as well as the quantities to be found. Draw a sketch when possible.
2. Write an equation relating the variables of the problem.
3. Use implicit differentiation to find the derivative of both sides of the equation in Step 2 with respect to time.
4. Solve for the derivative, giving the unknown rate of change, and substitute the given values.

Differentials $dy = f'(x)dx$

Linear Approximation $f(x + \Delta x) \approx f(x) + dy = f(x) + f'(x)dx$

KEY TERMS

6.1
absolute maximum
absolute minimum
absolute extremum
  (or extrema)
extreme value theorem

critical point theorem
graphical optimization

6.2
spawner-recruit function
parent-progeny function
maximum sustainable
  harvest

6.3
economic lot size
economic order quantity
elasticity of demand
unit elasticity

6.4
explicit function

implicit differentiation

6.5
related rates

6.6
differential

REVIEW EXERCISES

CONCEPT CHECK

Determine whether each of the following statements is true or false, and explain why.

1. The absolute maximum of a function always occurs where the derivative has a critical number.

2. A continuous function on a closed interval has an absolute maximum and minimum.

3. A continuous function on an open interval does not have an absolute maximum or minimum.

4. Demand for a product is elastic if the elasticity is greater than 1.

5. Total revenue is maximized at the price where demand has unit elasticity.

6. Implicit differentiation can be used to find dy/dx when x is defined in terms of y.

7. In a related rates problem, all derivatives are with respect to time.

8. In a related rates problem, there can be more than two quantities that vary with time.

9. A differential is a real number.

10. When the change in x is small, the differential of y is approximately the change in y.

PRACTICE AND EXPLORATIONS

Find the absolute extrema if they exist, and all values of x where they occur on the given intervals.

11. $f(x) = -x^3 + 6x^2 + 1;$ $[-1, 6]$

12. $f(x) = 4x^3 - 9x^2 - 3;$ $[-1, 2]$

13. $f(x) = x^3 + 2x^2 - 15x + 3;$ $[-4, 2]$

14. $f(x) = -2x^3 - 2x^2 + 2x - 1;$ $[-3, 1]$

15. When solving applied extrema problems, why is it necessary to check the endpoints of the domain?

16. What is elasticity of demand (in words; no mathematical symbols allowed)? Why is the derivative used to describe elasticity?

17. Find the absolute maximum and minimum of $f(x) = \dfrac{2 \ln x}{x^2}$ on each interval.

   a. $[1, 4]$ **b.** $[2, 5]$

18. Find the absolute maximum and minimum of $f(x) = \dfrac{e^{2x}}{x^2}$ on each interval.

   a. $[1/2, 2]$ **b.** $[1, 3]$

19. When is it necessary to use implicit differentiation?

20. When a term involving y is differentiated in implicit differentiation, it is multiplied by dy/dx. Why? Why aren't terms involving x multiplied by dx/dx?

Find dy/dx.

21. $x^2 - 4y^2 = 3x^3y^4$ **22.** $x^2y^3 + 4xy = 2$

23. $2\sqrt{y-1} = 9x^{2/3} + y$ **24.** $9\sqrt{x} + 4y^3 = 2\sqrt{y}$

25. $\dfrac{6 + 5x}{2 - 3y} = \dfrac{1}{5x}$ **26.** $\dfrac{x + 2y}{x - 3y} = y^{1/2}$

27. $\ln(xy + 1) = 2xy^3 + 4$ **28.** $\ln(x + y) = 1 + x^2 + y^3$

29. Find the equation of the line tangent to the graph of $\sqrt{2y} - 4xy = -22$ at the point $(3, 2)$.

30. Find an equation of the line tangent to the graph of $8y^3 - 4xy^2 = 20$ at the point $(-3, 1)$.

31. What is the difference between a related rate problem and an applied extremum problem?

32. Why is implicit differentiation used in related rate problems?

Find dy/dt.

33. $y = 8x^3 - 7x^2;$ $\dfrac{dx}{dt} = 4, x = 2$

34. $y = \dfrac{9 - 4x}{3 + 2x};$ $\dfrac{dx}{dt} = -1, x = -3$

35. $y = \dfrac{1 + \sqrt{x}}{1 - \sqrt{x}};$ $\dfrac{dx}{dt} = -4, x = 4$

36. $\dfrac{x^2 + 5y}{x - 2y} = 2;$ $\dfrac{dx}{dt} = 1, x = 2, y = 0$

37. $y = xe^{3x};$ $\dfrac{dx}{dt} = -2, x = 1$

38. $y = \dfrac{1}{e^{x^2} + 1};$ $\dfrac{dx}{dt} = 3, x = 1$

39. What is a differential? What is it used for?

40. Describe when linear approximations are most accurate.

Evaluate dy.

41. $y = \dfrac{3x - 7}{2x + 1};$ $x = 2, \Delta x = 0.003$

42. $y = 8 - x^2 + x^3;$ $x = -1, \Delta x = 0.02$

43. Suppose x and y are related by the equation
$$-12x + x^3 + y + y^2 = 4.$$

  a. Find all critical points on the curve.

  b. Determine whether the critical points found in part a are relative maxima or relative minima by taking values of x nearby and solving for the corresponding values of y.

  c. Is there an absolute maximum or minimum for x and y in the relationship given in part a? Why or why not?

44. In Exercise 43, implicit differentiation was used to find the relative extrema. The exercise was contrived to avoid various difficulties that could have arisen. Discuss some of the difficulties that might be encountered in such problems, and how these difficulties might be resolved.

APPLICATIONS

Business and Economics

45. Profit The total profit (in tens of dollars) from the sale of x hundred boxes of candy is given by
$$P(x) = -x^3 + 10x^2 - 12x.$$

  a. Find the number of boxes of candy that should be sold in order to produce maximum profit.

  b. Find the maximum profit.

46. Packaging Design The packaging department of a corporation is designing a box with a square base and no top. The volume is to be 32 m³. To reduce cost, the box is to have minimum surface area. What dimensions (height, length, and width) should the box have?

47. Packaging Design A company plans to package its product in a cylinder that is open at one end. The cylinder is to have a volume of 27π in³. What radius should the circular bottom of the cylinder have to minimize the cost of the material?

48. Packaging Design Fruit juice will be packaged in cylindrical cans with a volume of 40 in³ each. The top and bottom of the can cost 4¢ per in², while the sides cost 3¢ per in². Find the radius and height of the can of minimum cost.

49. Order Quantity A large camera store sells 20,000 batteries annually. It costs 15¢ to store 1 battery for 1 year and $12 to place a reorder. Find the number of batteries that should be ordered each time.

50. Order Quantity A store sells 180,000 cases of a product annually. It costs $12 to store 1 case for 1 year and $20 to place a reorder. Find the number of cases that should be ordered each time.

51. Lot Size A company produces 128,000 cases of soft drink annually. It costs $1 to store 1 case for 1 year and $10 to produce 1 lot. Find the number of lots that should be produced annually.

52. Lot Size In 1 year, a health food manufacturer produces and sells 240,000 cases of vitamins. It costs $2 to store 1 case for 1 year and $15 to produce each batch. Find the number of batches that should be produced annually.

53. Elasticity of Demand The demand for butter in Mexico for a particular year can be estimated by the general function

$$\ln q = D - 0.47 \ln p,$$

where p represents the price of a unit of butter and D represents a constant that can be calculated uniquely for a particular year. Here q represents the annual per capita butter demand. Calculate and interpret the elasticity of demand. *Source: Agricultural Economics.*

54. Elasticity Suppose the demand function for a product is given by $q = A/p^k$, where A and k are positive constants. For what values of k is the demand elastic? Inelastic?

Life Sciences

55. Pollution A circle of pollution is spreading from a broken underwater waste disposal pipe, with the radius increasing at the rate of 4 ft per minute. Find the rate of change of the area of the circle when the radius is 7 ft.

56. Logistic Growth Many populations grow according to the logistic equation

$$\frac{dx}{dt} = rx(N - x),$$

where r is a constant involving the rate of growth and N is the carrying capacity of the environment, beyond which the population decreases. Show that the graph of x has an inflection point where $x = N/2$. (*Hint:* Use implicit differentiation. Then set $d^2x/dt^2 = 0$, and factor.)

57. Dentin Growth The dentinal formation of molars in mice has been studied by researchers in Copenhagen. They determined that the growth curve that best fits dentinal formation for the first molar is

$$M(t) = 1.3386309 - 0.4321173t + 0.0564512t^2$$
$$-0.0020506t^3 + 0.0000315t^4 - 0.0000001785t^5,$$
$$5 \le t \le 51,$$

where t is the age of the mouse (in days), and $M(t)$ is the cumulative dentin volume (in 10^{-1} mm³). *Source: Journal of Craniofacial Genetics and Developmental Biology.*

a. Use a graphing calculator to sketch the graph of this function on $[5, 51]$ by $[0, 7.5]$.

b. Find the time in which the dentin formation is growing most rapidly. (*Hint:* Find the maximum value of the derivative of this function.)

58. Human Skin Surface The surface of the skin is made up of a network of intersecting lines that form polygons. Researchers have discovered a functional relationship between the age of a female and the number of polygons per area of skin according to

$$P(t) = 237.09 - 8.0398t + 0.20813t^2 - 0.0027563t^3$$
$$+ 0.000013016t^4, \quad 0 \le t \le 95,$$

where t is the age of the person (in years), and $P(t)$ is the number of polygons for a particular surface area of skin. *Source: Gerontology.*

a. Use a graphing calculator to sketch a graph of $P(t)$ on $[0, 95]$ by $[0, 300]$.

b. Find the maximum and minimum number of polygons per area predicted by the model.

c. Discuss the accuracy of this model for older people.

Physical Sciences

59. Sliding Ladder A 50-ft ladder is placed against a building. The top of the ladder is sliding down the building at the rate of 2 ft per minute. Find the rate at which the base of the ladder is slipping away from the building at the instant that the base is 30 ft from the building.

60. Spherical Radius A large weather balloon is being inflated with air at the rate of 0.9 ft³ per minute. Find the rate of change of the radius when the radius is 1.7 ft.

61. Water Level A water trough 2 ft across, 4 ft long, and 1 ft deep has ends in the shape of isosceles triangles. (See the figure.) It is being filled with 3.5 ft³ of water per minute. Find the rate at which the depth of water in the tank is changing when the water is 1/3 ft deep.

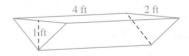

General Interest

62. Volume Approximate the volume of coating on a sphere of radius 4 in. if the coating is 0.02 in. thick.

63. Area A square has an edge of 9.2 in., with a possible error in the measurement of ±0.04 in. Estimate the possible error in the area of the square.

64. Package Dimensions UPS has the following rule regarding package dimensions. The length can be no more than 108 in., and the length plus the girth (twice the sum of the width and the height) can be no more than 130 in. If the width of a package is 4 in. more than its height and it has the maximum length plus girth allowed, find the length that produces maximum volume.

65. Pursuit A boat moves north at a constant speed. A second boat, moving at the same speed, pursues the first boat in such a way that it always points directly at the first boat. When the first boat is at the point $(0, 1)$, the second boat is at the point $(6, 2.5)$, with the positive y-axis pointing north. It can then be shown that the curve traced by the second boat, known as a pursuit curve, is given by

$$y = \frac{x^2}{16} - 2 \ln x + \frac{1}{4} + 2 \ln 6.$$

Find the y-coordinate of the southernmost point of the second boat's path. *Source: Differential Equations: Theory and Applications.*

66. Playground Area The city park department is planning an enclosed play area in a new park. One side of the area will be against an existing building, with no fence needed there. Find the dimensions of the maximum rectangular area that can be made with 900 m of fence.

67. Surfing A mathematician is surfing in Long Beach, New York. He is standing on the shore and wants to paddle out to a spot 40 ft from shore; the closest point on the shore to that spot is 40 ft from where he is now standing. (See the figure.) If he can walk 5 ft per second along the shore and paddle 3 ft per second once he's in the water, how far along the shore

should he walk before paddling toward the desired destination if he wants to complete the trip in the shortest possible time? What is the shortest possible time?

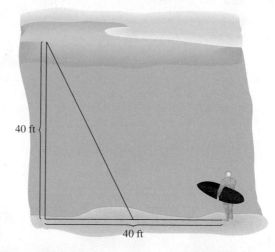

40 ft

40 ft

68. Repeat Exercise 67, but the closest point on the shore to the desired destination is now 25 ft from where he is standing.

EXTENDED APPLICATION

A TOTAL MODEL FOR A TRAINING PROGRAM

In this application, we set up a mathematical model for determining the total costs in setting up a training program. Then we use calculus to find the time interval between training programs that produces the minimum total cost. The model assumes that the demand for trainees is constant and that the fixed cost of training a batch of trainees is known. Also, it is assumed that people who are trained, but for whom no job is readily available, will be paid a fixed amount per month while waiting for a job to open up.

The model uses the following variables.

D = demand for trainees per month

N = number of trainees per batch

C_1 = fixed cost of training a batch of trainees

C_2 = marginal cost of training per trainee per month

C_3 = salary paid monthly to a trainee who has not yet been given a job after training

m = time interval in months between successive batches of trainees

t = length of training program in months

$Z(m)$ = total monthly cost of program

The total cost of training a batch of trainees is given by $C_1 + NtC_2$. However, $N = mD$, so that the total cost per batch is $C_1 + mDtC_2$.

After training, personnel are given jobs at the rate of D per month. Thus, $N - D$ of the trainees will not get a job the first month, $N - 2D$ will not get a job the second month, and so on. The $N - D$ trainees who do not get a job the first month produce total costs of $(N - D)C_3$, those not getting jobs during the second

month produce costs of $(N - 2D)C_3$, and so on. Since $N = mD$, the costs during the first month can be written as

$$(N - D)C_3 = (mD - D)C_3 = (m - 1)DC_3,$$

while the costs during the second month are $(m - 2)DC_3$, and so on. The total cost for keeping the trainees without a job is thus

$$(m - 1)DC_3 + (m - 2)DC_3$$
$$+ (m - 3)DC_3 + \cdots + 2DC_3 + DC_3,$$

which can be factored to give

$$DC_3[(m - 1) + (m - 2) + (m - 3) + \cdots + 2 + 1].$$

The expression in brackets is the sum of the terms of an arithmetic sequence, discussed in most algebra texts. Using formulas for arithmetic sequences, the expression in brackets can be shown to equal $m(m - 1)/2$, so that we have

$$DC_3\left[\frac{m(m - 1)}{2}\right] \qquad \textbf{(1)}$$

as the total cost for keeping jobless trainees.

The total cost per batch is the sum of the training cost per batch, $C_1 + mDtC_2$, and the cost of keeping trainees without a proper job, given by Equation (1). Since we assume that a batch of trainees is trained every m months, the total cost per month, $Z(m)$, is given by

$$Z(m) = \frac{C_1 + mDtC_2}{m} + \frac{DC_3\left[\frac{m(m - 1)}{2}\right]}{m}$$

$$= \frac{C_1}{m} + DtC_2 + DC_3\left(\frac{m - 1}{2}\right).$$

Source: P. L. Goyal and S. K. Goyal.

EXERCISES

1. Find $Z'(m)$.

2. Solve the equation $Z'(m) = 0$.

   As a practical matter, it is usually required that m be a whole number. If m does not come out to be a whole number, then m^+ and m^-, the two whole numbers closest to m, must be chosen. Calculate both $Z(m^+)$ and $Z(m^-)$; the smaller of the two provides the optimum value of Z.

3. Suppose a company finds that its demand for trainees is 3 per month, that a training program requires 12 months, that the fixed cost of training a batch of trainees is $15,000, that the marginal cost per trainee per month is $100, and that trainees are paid $900 per month after training but before going to work. Use your result from Exercise 2 and find m.

4. Since m is not a whole number, find m^+ and m^-.

5. Calculate $Z(m^+)$ and $Z(m^-)$.

6. What is the optimum time interval between successive batches of trainees? How many trainees should be in a batch?

7. The parameters of this model are likely to change over time; it is essential that such changes be incorporated into the model as they change. One way to anticipate this is to create a spreadsheet that gives the manager the total cost of training a batch of trainees for various scenarios. Using the data from Exercise 3 as a starting point, create a spreadsheet that varies these numbers and calculates the total cost of training a group of employees for each scenario. Graph the total cost of training with respect to changes in the various costs associated with training.

DIRECTIONS FOR GROUP PROJECT

Suppose you have read an article in the paper announcing that a new high-tech company is locating in your town. Given that the company is manufacturing very specialized equipment, you realize that it must develop a program to train all new employees. Because you would like to get an internship at this new company, use the information above to develop a hypothetical training program that optimizes the time interval between successive batches of trainees and the number of trainees that should be in each session. Assume that you know the new CEO because you and three of your friends have served her pizza at various times at the local pizza shop (your current jobs) and that she is willing to listen to a proposal that describes your training program. Prepare a presentation for your interview that will describe your training program. Use presentation software such as Microsoft PowerPoint.

7

Integration

If we know the rate at which a quantity is changing, we can find the total change over a period of time by integrating. An exercise in Section 3 illustrates how this process can be used to estimate the number of cars that cross the Tappan Zee Bridge in New York state each day, given information about how the rate of cars per hour varies with time. This same concept allows us to determine how far a car has gone, given its speed as a function of time; how much a culture of bacteria will grow; or how much consumers benefit by buying a product at the price determined by supply and demand.

Up to this point in calculus you have solved problems such as

$$f(x) = x^5; \text{find } f'(x).$$

In this chapter you will be asked to solve problems that are the reverse of these, that is, problems of the form

$$f'(x) = 5x^4; \text{find } f(x).$$

The derivative and its applications, which you studied in previous chapters, are part of what is called *differential calculus*. The next two chapters are devoted to the other main branch of calculus, *integral calculus*. Integrals have many applications: finding areas; determining the lengths of curved paths; solving complicated probability problems; and calculating the location of an object (such as the distance of a space shuttle from Earth) when its velocity and initial position are known. The Fundamental Theorem of Calculus, presented later in this chapter, will reveal a surprisingly close connection between differential and integral calculus.

7.1 Antiderivatives

APPLY IT   If an object is thrown from the top of the Willis Tower in Chicago, how fast is it going when it hits the ground?
Using antiderivatives, we will answer this question in Example 11.

Functions used in applications in previous chapters have provided information about a *total amount* of a quantity, such as cost, revenue, profit, temperature, gallons of oil, or distance. Derivatives of these functions provided information about the rate of change of these quantities and allowed us to answer important questions about the extrema of the functions. It is not always possible to find ready-made functions that provide information about the total amount of a quantity, but it is often possible to collect enough data to come up with a function that gives the *rate of change* of a quantity. We know that derivatives give the rate of change when the total amount is known. The reverse of finding a derivative is known as **antidifferentiation**. The goal is to find an *antiderivative*, defined as follows.

> **Antiderivative**
> If $F'(x) = f(x)$, then $F(x)$ is an **antiderivative** of $f(x)$.

EXAMPLE 1   Antiderivative

(a) If $F(x) = 10x$, then $F'(x) = 10$, so $F(x) = 10x$ is an antiderivative of $f(x) = 10$.
(b) For $F(x) = x^2$, $F'(x) = 2x$, making $F(x) = x^2$ an antiderivative of $f(x) = 2x$.

EXAMPLE 2 Antiderivative

Find an antiderivative of $f(x) = 5x^4$.

SOLUTION To find a function $F(x)$ whose derivative is $5x^4$, work backwards. Recall that the derivative of x^n is nx^{n-1}. If

YOUR TURN 1 Find an antiderivative $f(x) = 8x^7$.

$$nx^{n-1} \text{ is } 5x^4,$$

then $n - 1 = 4$ and $n = 5$, so x^5 is an antiderivative of $5x^4$.   **TRY YOUR TURN 1**

EXAMPLE 3 Population

Suppose a population is growing at a rate given by $f(x) = e^x$, where x is time in years from some initial date. Find a function giving the population at time x.

SOLUTION Let the population function be $F(x)$. Then

$$f(x) = F'(x) = e^x.$$

The derivative of the function defined by $F(x) = e^x$ is $F'(x) = e^x$, so one possible population function with the given growth rate is $F(x) = e^x$.

The function from Example 1(b), defined by $F(x) = x^2$, is not the only function whose derivative is $f(x) = 2x$. For example,

$$F(x) = x^2, \qquad G(x) = x^2 + 2, \qquad \text{and} \qquad H(x) = x^2 - 4$$

are all antiderivatives of $f(x) = 2x$, and any two of them differ only by a constant. These three functions, shown in Figure 1, have the same derivative, $f(x) = 2x$, and the slopes of their tangent lines at any particular value of x are the same. In fact, for any real number C, the function $F(x) = x^2 + C$ has $f(x) = 2x$ as its derivative. This means that there is a *family* or *class* of functions having $2x$ as a derivative. As the next theorem states, if two functions $F(x)$ and $G(x)$ are antiderivatives of $f(x)$, then $F(x)$ and $G(x)$ can differ only by a constant.

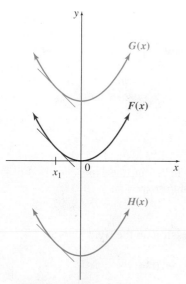

Slopes of the tangent lines
at $x = x_1$ are the same.

FIGURE 1

If $F(x)$ and $G(x)$ are both antiderivatives of a function $f(x)$ on an interval, then there is a constant C such that

$$F(x) - G(x) = C.$$

(Two antiderivatives of a function can differ only by a constant.) The arbitrary real number C is called an integration constant.

The family of all antiderivatives of the function f is indicated by

$$\int f(x)\, dx = F(x) + C.$$

The symbol $\int$ is the **integral sign**, $f(x)$ is the **integrand**, and $\int f(x)\, dx$ is called an **indefinite integral**, the most general antiderivative of f.

Indefinite Integral
If $F'(x) = f(x)$, then

$$\int f(x)\, dx = F(x) + C,$$

for any real number C.

For example, using this notation,

$$\int 2x\, dx = x^2 + C.$$

The dx in the indefinite integral indicates that $\int f(x)\, dx$ is the "integral of $f(x)$ *with respect to x*" just as the symbol dy/dx denotes the "derivative of y with respect to x." For example, in the indefinite integral $\int 2ax\, dx$, dx indicates that a is to be treated as a constant and x as the variable, so that

$$\int 2ax\, dx = \int a(2x)\, dx = ax^2 + C.$$

On the other hand,

$$\int 2ax\, da = a^2x + C = xa^2 + C.$$

A more complete interpretation of dx will be discussed later.

The symbol $\int f(x)\, dx$ was created by G. W. Leibniz (1646–1716) in the latter part of the seventeenth century. The $\int$ is an elongated S from *summa,* the Latin word for *sum.* The word *integral* as a term in the calculus was coined by Jakob Bernoulli (1654–1705), a Swiss mathematician who corresponded frequently with Leibniz. The relationship between sums and integrals will be clarified in a later section.

Finding an antiderivative is the reverse of finding a derivative. Therefore, each rule for derivatives leads to a rule for antiderivatives. For example, the power rule for derivatives tells us that

$$\frac{d}{dx} x^5 = 5x^4.$$

Consequently,

$$\int 5x^4\, dx = x^5 + C,$$

the result found in Example 2. Note that the derivative of x^n is found by multiplying x by n and reducing the exponent on x by 1. To find an indefinite integral—that is, to undo what was done—*increase* the exponent by 1 and *divide* by the new exponent, $n + 1$.

---FOR REVIEW---

Recall that $\dfrac{d}{dx} x^n = nx^{n-1}$.

Power Rule
For any real number $n \neq -1$,

$$\int x^n\, dx = \frac{x^{n+1}}{n+1} + C.$$

(The antiderivative of $f(x) = x^n$ for $n \neq -1$ is found by increasing the exponent n by 1 and dividing x raised to the new power by the new value of the exponent.)

This rule can be verified by differentiating the expression on the right above:

$$\frac{d}{dx}\left(\frac{x^{n+1}}{n+1} + C\right) = \frac{n+1}{n+1} x^{(n+1)-1} + 0 = x^n.$$

(If $n = -1$, the expression in the denominator is 0, and the above rule cannot be used. Finding an antiderivative for this case is discussed later.)

EXAMPLE 4 **Power Rule**

Use the power rule to find each indefinite integral.

(a) $\int t^3 dt$

SOLUTION Use the power rule with $n = 3$.

$$\int t^3\, dt = \frac{t^{3+1}}{3 + 1} + C = \frac{t^4}{4} + C$$

To check the solution, find the derivative of $t^4/4 + C$. The derivative is t^3, the original function.

(b) $\int \frac{1}{t^2}\, dt$

SOLUTION First, write $1/t^2$ as t^{-2}. Then

$$\int \frac{1}{t^2}\, dt = \int t^{-2}\, dt = \frac{t^{-2+1}}{-2 + 1} + C = \frac{t^{-1}}{-1} + C = -\frac{1}{t} + C.$$

Verify the solution by differentiating $-(1/t) + C$ to get $1/t^2$.

(c) $\int \sqrt{u}\, du$

SOLUTION Since $\sqrt{u} = u^{1/2}$,

$$\int \sqrt{u}\, du = \int u^{1/2}\, du = \frac{u^{3/2}}{3/2} + C = \frac{2}{3} u^{3/2} + C.$$

To check this, differentiate $(2/3)u^{3/2} + C$; the derivative is $u^{1/2}$, the original function.

(d) $\int dx$

YOUR TURN 2

Find $\int \dfrac{1}{t^4}\, dt$.

SOLUTION Write dx as $1 \cdot dx$, and use the fact that $x^0 = 1$ for any nonzero number x to get

$$\int dx = \int 1\, dx = \int x^0\, dx = \frac{x^1}{1} + C = x + C. \qquad \textbf{TRY YOUR TURN 2}$$

As shown earlier, the derivative of the product of a constant and a function is the product of the constant and the derivative of the function. A similar rule applies to indefinite integrals. Also, since derivatives of sums or differences are found term by term, indefinite integrals also can be found term by term.

FOR REVIEW

Recall that $\dfrac{d}{dx}[f(x) \pm g(x)] = [f'(x) \pm g'(x)]$ and

$\dfrac{d}{dx}[kf(x)] = kf'(x).$

Constant Multiple Rule and Sum or Difference Rule

If all indicated integrals exist,

$$\int k \cdot f(x)\, dx = k \int f(x)\, dx, \qquad \text{for any real number } k,$$

and

$$\int [f(x) \pm g(x)]\, dx = \int f(x)\, dx \pm \int g(x)\, dx.$$

(The antiderivative of a constant times a function is the constant times the antiderivative of the function. The antiderivative of a sum or difference of functions is the sum or difference of the antiderivatives.)

CAUTION   The constant multiple rule requires k to be a *number*. The rule does not apply to a *variable*. For example,

$$\int x\sqrt{x-1}\,dx \neq x\int\sqrt{x-1}\,dx.$$

EXAMPLE 5   Rules of Integration

Use the rules to find each integral.

(a) $\int 2v^3\,dv$

SOLUTION   By the constant multiple rule and the power rule,

$$\int 2v^3\,dv = 2\int v^3\,dv = 2\left(\frac{v^4}{4}\right) + C = \frac{v^4}{2} + C.$$

Because C represents any real number, it is not necessary to multiply it by 2 in the next-to-last step.

(b) $\int \frac{12}{z^5}\,dz$

SOLUTION   Rewrite $12/z^5$ as $12z^{-5}$, then find the integral.

$$\int \frac{12}{z^5}\,dz = \int 12z^{-5}\,dz = 12\int z^{-5}\,dz = 12\left(\frac{z^{-4}}{-4}\right) + C$$

$$= -3z^{-4} + C = \frac{-3}{z^4} + C$$

(c) $\int (3z^2 - 4z + 5)\,dz$

SOLUTION   By extending the sum and difference rules to more than two terms, we get

$$\int (3z^2 - 4z + 5)\,dz = 3\int z^2\,dz - 4\int z\,dz + 5\int dz$$

$$= 3\left(\frac{z^3}{3}\right) - 4\left(\frac{z^2}{2}\right) + 5z + C$$

$$= z^3 - 2z^2 + 5z + C.$$

YOUR TURN 3

Find $\int (6x^2 + 8x - 9)\,dx$.

Only one constant C is needed in the answer; the three constants from integrating term by term are combined.   TRY YOUR TURN 3

Remember to check your work by taking the derivative of the result. For instance, in Example 5(c) check that $z^3 - 2z^2 + 5z + C$ is the required indefinite integral by taking the derivative

$$\frac{d}{dz}(z^3 - 2z^2 + 5z + C) = 3z^2 - 4z + 5,$$

which agrees with the original information.

EXAMPLE 6 **Rules of Integration**

Use the rules to find each integral.

(a) $\displaystyle\int \frac{x^2 + 1}{\sqrt{x}}\, dx$

SOLUTION First rewrite the integrand as follows.

$$\int \frac{x^2 + 1}{\sqrt{x}}\, dx = \int \left(\frac{x^2}{\sqrt{x}} + \frac{1}{\sqrt{x}}\right) dx \qquad \text{Rewrite as a sum of fractions.}$$

$$= \int \left(\frac{x^2}{x^{1/2}} + \frac{1}{x^{1/2}}\right) dx \qquad \sqrt{a} = a^{1/2}$$

$$= \int (x^{3/2} + x^{-1/2})\, dx \qquad \text{Use } \frac{a^m}{a^n} = a^{m-n}.$$

Now find the antiderivative.

$$\int (x^{3/2} + x^{-1/2})\, dx = \frac{x^{5/2}}{5/2} + \frac{x^{1/2}}{1/2} + C$$

$$= \frac{2}{5} x^{5/2} + 2x^{1/2} + C$$

(b) $\displaystyle\int (x^2 - 1)^2\, dx$

YOUR TURN 4

Find $\displaystyle\int \frac{x^3 - 2}{\sqrt{x}}\, dx$.

SOLUTION Square the binomial first, and then find the antiderivative.

$$\int (x^2 - 1)^2\, dx = \int (x^4 - 2x^2 + 1)\, dx$$

$$= \frac{x^5}{5} - \frac{2x^3}{3} + x + C \qquad \textbf{TRY YOUR TURN 4}$$

It was shown earlier that the derivative of $f(x) = e^x$ is $f'(x) = e^x$, and the derivative of $f(x) = a^x$ is $f'(x) = (\ln a)a^x$. Also, the derivative of $f(x) = e^{kx}$ is $f'(x) = k \cdot e^{kx}$, and the derivative of $f(x) = a^{kx}$ is $f'(x) = k(\ln a)a^{kx}$. These results lead to the following formulas for indefinite integrals of exponential functions.

Indefinite Integrals of Exponential Functions

$$\int e^x\, dx = e^x + C$$

$$\int e^{kx}\, dx = \frac{e^{kx}}{k} + C, \quad k \neq 0$$

For $a > 0,\ a \neq 1$:

$$\int a^x\, dx = \frac{a^x}{\ln a} + C$$

$$\int a^{kx}\, dx = \frac{a^{kx}}{k(\ln a)} + C, \quad k \neq 0$$

(The antiderivative of the exponential function e^x is itself. If x has a coefficient of k, we must divide by k in the antiderivative. If the base is not e, we must divide by the natural logarithm of the base.)

EXAMPLE 7 **Exponential Functions**

(a) $\displaystyle\int 9e^t \, dt = 9\int e^t \, dt = 9e^t + C$

(b) $\displaystyle\int e^{9t} \, dt = \frac{e^{9t}}{9} + C$

(c) $\displaystyle\int 3e^{(5/4)u} \, du = 3\left(\frac{e^{(5/4)u}}{5/4}\right) + C$

$$= 3\left(\frac{4}{5}\right)e^{(5/4)u} + C$$

$$= \frac{12}{5} e^{(5/4)u} + C$$

(d) $\displaystyle\int 2^{-5x} \, dx = \frac{2^{-5x}}{-5(\ln 2)} + C = -\frac{2^{-5x}}{5(\ln 2)} + C$

The restriction $n \neq -1$ was necessary in the formula for $\int x^n \, dx$ since $n = -1$ made the denominator of $1/(n + 1)$ equal to 0. To find $\int x^n \, dx$ when $n = -1$, that is, to find $\int x^{-1} \, dx$, recall the differentiation formula for the logarithmic function: The derivative of $f(x) = \ln |x|$, where $x \neq 0$, is $f'(x) = 1/x = x^{-1}$. This formula for the derivative of $f(x) = \ln |x|$ gives a formula for $\int x^{-1} \, dx$.

Indefinite Integral of x^{-1}

$$\int x^{-1} \, dx = \int \frac{1}{x} \, dx = \ln |x| + C$$

(The antiderivative of $f(x) = x^n$ for $n = -1$ is the natural logarithm of the absolute value of x.)

CAUTION Don't neglect the absolute value sign in the natural logarithm when integrating x^{-1}. If x can take on a negative value, $\ln x$ will be undefined there. Note, however, that the absolute value is redundant (but harmless) in an expression such as $\ln |x^2 + 1|$, since $x^2 + 1$ can never be negative.

EXAMPLE 8 **Integrals**

YOUR TURN 5

Find $\displaystyle\int \left(\frac{3}{x} + e^{-3x}\right) dx.$

(a) $\displaystyle\int \frac{4}{x} \, dx = 4\int \frac{1}{x} \, dx = 4 \ln |x| + C$

(b) $\displaystyle\int \left(-\frac{5}{x} + e^{-2x}\right) dx = -5 \ln |x| - \frac{1}{2}e^{-2x} + C$

TRY YOUR TURN 5

In all these examples, the antiderivative family of functions was found. In many applications, however, the given information allows us to determine the value of the integration constant C. The next examples illustrate this idea.

| **EXAMPLE 9** | **Cost** |

Suppose a publishing company has found that the marginal cost at a level of production of x thousand books is given by

$$C'(x) = \frac{50}{\sqrt{x}}$$

and that the fixed cost (the cost before the first book can be produced) is $25,000. Find the cost function $C(x)$.

SOLUTION Write $50/\sqrt{x}$ as $50/x^{1/2}$ or $50x^{-1/2}$, and then use the indefinite integral rules to integrate the function.

$$C(x) = \int \frac{50}{\sqrt{x}}\, dx = \int 50x^{-1/2}\, dx = 50(2x^{1/2}) + k = 100x^{1/2} + k$$

(Here k is used instead of C to avoid confusion with the cost function $C(x)$.) To find the value of k, use the fact that $C(0)$ is 25,000.

$$C(x) = 100x^{1/2} + k$$
$$25{,}000 = 100 \cdot 0 + k$$
$$k = 25{,}000$$

With this result, the cost function is $C(x) = 100x^{1/2} + 25{,}000$.

| **EXAMPLE 10** | **Demand** |

Suppose the marginal revenue from a product is given by $400e^{-0.1q} + 8$, where q is the number of products produced.

(a) Find the revenue function for the product.

SOLUTION The marginal revenue is the derivative of the revenue function, so

$$R'(q) = 400e^{-0.1q} + 8$$

$$R(q) = \int (400e^{-0.1q} + 8)\, dq$$

$$= 400\,\frac{e^{-0.1q}}{-0.1} + 8q + C$$

$$= -4000e^{-0.1q} + 8q + C.$$

If $q = 0$, then $R = 0$ (no items sold means no revenue), so that

$$0 = -4000e^{-0.1(0)} + 8(0) + C$$
$$0 = -4000 + 0 + C$$
$$4000 = C.$$

Thus, the revenue function is

$$R(q) = -4000e^{-0.1q} + 8q + 4000.$$

(b) Find the demand function for this product.

SOLUTION Recall that $R = qp$, where p is the demand function giving the price p as a function of q. Then

$$-4000e^{-0.1q} + 8q + 4000 = qp$$

$$\frac{-4000e^{-0.1q} + 8q + 4000}{q} = p. \qquad \text{Divide by } q.$$

The demand function is $p = \dfrac{-4000e^{-0.1q} + 8q + 4000}{q}$.

In the next example, integrals are used to find the position of a particle when the acceleration of the particle is given.

EXAMPLE 11 Velocity and Acceleration

Recall that if the function $s(t)$ gives the position of a particle at time t, then its velocity $v(t)$ and its acceleration $a(t)$ are given by

$$v(t) = s'(t) \qquad \text{and} \qquad a(t) = v'(t) = s''(t).$$

(a) Suppose the velocity of an object is $v(t) = 6t^2 - 8t$ and that the object is at 5 when time is 0. Find $s(t)$.

SOLUTION Since $v(t) = s'(t)$, the function $s(t)$ is an antiderivative of $v(t)$:

$$s(t) = \int v(t)\, dt = \int (6t^2 - 8t)\, dt$$
$$= 2t^3 - 4t^2 + C$$

for some constant C. Find C from the given information that $s = 5$ when $t = 0$.

$$s(t) = 2t^3 - 4t^2 + C$$
$$5 = 2(0)^3 - 4(0)^2 + C$$
$$5 = C$$
$$s(t) = 2t^3 - 4t^2 + 5$$

(b) Many experiments have shown that when an object is dropped, its acceleration (ignoring air resistance) is constant. This constant has been found to be approximately 32 ft per second every second; that is,

$$a(t) = -32.$$

The negative sign is used because the object is falling. Suppose an object is thrown down from the top of the 1100-ft-tall Willis Tower (formerly known as the Sears Tower) in Chicago. If the initial velocity of the object is -20 ft per second, find $s(t)$, the distance of the object from the ground at time t.

SOLUTION First find $v(t)$ by integrating $a(t)$:

$$v(t) = \int (-32)\, dt = -32t + k.$$

When $t = 0$, $v(t) = -20$:

$$-20 = -32(0) + k$$
$$-20 = k$$

and

$$v(t) = -32t - 20.$$

Be sure to evaluate the constant of integration k before integrating again to get $s(t)$. Now integrate $v(t)$ to find $s(t)$.

$$s(t) = \int (-32t - 20) \, dt = -16t^2 - 20t + C$$

Since $s(t) = 1100$ when $t = 0$, we can substitute these values into the equation for $s(t)$ to get $C = 1100$ and

$$s(t) = -16t^2 - 20t + 1100$$

as the distance of the object from the ground after t seconds.

(c) Use the equations derived in (b) to find the velocity of the object when it hit the ground and how long it took to strike the ground.

SOLUTION When the object strikes the ground, $s = 0$, so

$$0 = -16t^2 - 20t + 1100.$$

To solve this equation for t, factor out the common factor of -4 and then use the quadratic formula.

$$0 = -4(4t^2 + 5t - 275)$$

$$t = \frac{-5 \pm \sqrt{25 + 4400}}{8} \approx \frac{-5 \pm 66.5}{8}$$

Only the positive value of t is meaningful here: $t \approx 7.69$. It took the object about 7.69 seconds to strike the ground. From the velocity equation, with $t = 7.69$, we find

$$v(t) = -32t - 20$$
$$v(7.69) = -32(7.69) - 20 \approx -266,$$

so the object was falling (as indicated by the negative sign) at about 266 ft per second when it hit the ground. **TRY YOUR TURN 6**

YOUR TURN 6 Repeat Example 11(b) and 11(c) for the Burj Khalifa in Dubai, which is the tallest building in the world, standing 2717 ft. The initial velocity is -20 ft per second.

EXAMPLE 12 Slope

Find a function f whose graph has slope $f'(x) = 6x^2 + 4$ and goes through the point $(1, 1)$.

SOLUTION Since $f'(x) = 6x^2 + 4$,

$$f(x) = \int (6x^2 + 4) \, dx = 2x^3 + 4x + C.$$

The graph of f goes through $(1, 1)$, so C can be found by substituting 1 for x and 1 for $f(x)$.

$$1 = 2(1)^3 + 4(1) + C$$
$$1 = 6 + C$$
$$C = -5$$

YOUR TURN 7 Find an equation of the curve whose tangent line has slope $f'(x) = 3x^{1/2} + 4$ and goes through the point $(1, -2)$.

Finally, $f(x) = 2x^3 + 4x - 5$. **TRY YOUR TURN 7**

APPLY IT

7.1 EXERCISES

1. What must be true of $F(x)$ and $G(x)$ if both are antiderivatives of $f(x)$?

2. How is the antiderivative of a function related to the function?

3. In your own words, describe what is meant by an integrand.

4. Explain why the restriction $n \neq -1$ is necessary in the rule
$$\int x^n \, dx = \frac{x^{n+1}}{n+1} + C.$$

Find the following.

5. $\displaystyle\int 6 \, dk$

6. $\displaystyle\int 9 \, dy$

7. $\displaystyle\int (2z + 3) \, dz$

8. $\displaystyle\int (3x - 5) \, dx$

9. $\displaystyle\int (6t^2 - 8t + 7) \, dt$

10. $\displaystyle\int (5x^2 - 6x + 3) \, dx$

11. $\displaystyle\int (4z^3 + 3z^2 + 2z - 6) \, dz$

12. $\displaystyle\int (16y^3 + 9y^2 - 6y + 3) \, dy$

13. $\displaystyle\int (5\sqrt{z} + \sqrt{2}) \, dz$

14. $\displaystyle\int (t^{1/4} + \pi^{1/4}) \, dt$

15. $\displaystyle\int 5x(x^2 - 8) \, dx$

16. $\displaystyle\int x^2(x^4 + 4x + 3) \, dx$

17. $\displaystyle\int (4\sqrt{v} - 3v^{3/2}) \, dv$

18. $\displaystyle\int (15x\sqrt{x} + 2\sqrt{x}) \, dx$

19. $\displaystyle\int (10u^{3/2} - 14u^{5/2}) \, du$

20. $\displaystyle\int (56t^{5/2} + 18t^{7/2}) \, dt$

21. $\displaystyle\int \left(\frac{7}{z^2}\right) dz$

22. $\displaystyle\int \left(\frac{4}{x^3}\right) dx$

23. $\displaystyle\int \left(\frac{\pi^3}{y^3} - \frac{\sqrt{\pi}}{\sqrt{y}}\right) dy$

24. $\displaystyle\int \left(\sqrt{u} + \frac{1}{u^2}\right) du$

25. $\displaystyle\int (-9t^{-2.5} - 2t^{-1}) \, dt$

26. $\displaystyle\int (10x^{-3.5} + 4x^{-1}) \, dx$

27. $\displaystyle\int \frac{1}{3x^2} \, dx$

28. $\displaystyle\int \frac{2}{3x^4} \, dx$

29. $\displaystyle\int 3e^{-0.2x} \, dx$

30. $\displaystyle\int -4e^{0.2v} \, dv$

31. $\displaystyle\int \left(\frac{-3}{x} + 4e^{-0.4x} + e^{0.1}\right) dx$

32. $\displaystyle\int \left(\frac{9}{x} - 3e^{-0.4x}\right) dx$

33. $\displaystyle\int \frac{1 + 2t^3}{4t} \, dt$

34. $\displaystyle\int \frac{2y^{1/2} - 3y^2}{6y} \, dy$

35. $\displaystyle\int (e^{2u} + 4u) \, du$

36. $\displaystyle\int (v^2 - e^{3v}) \, dv$

37. $\displaystyle\int (x + 1)^2 \, dx$

38. $\displaystyle\int (2y - 1)^2 \, dy$

39. $\displaystyle\int \frac{\sqrt{x} + 1}{\sqrt[3]{x}} \, dx$

40. $\displaystyle\int \frac{1 - 2\sqrt[3]{z}}{\sqrt[3]{z}} \, dz$

41. $\displaystyle\int 10^x \, dx$

42. $\displaystyle\int 3^{2x} \, dx$

43. Find an equation of the curve whose tangent line has a slope of
$$f'(x) = x^{2/3},$$
given that the point $(1, 3/5)$ is on the curve.

44. The slope of the tangent line to a curve is given by
$$f'(x) = 6x^2 - 4x + 3.$$
If the point $(0, 1)$ is on the curve, find an equation of the curve.

APPLICATIONS

Business and Economics

Cost Find the cost function for each marginal cost function.

45. $C'(x) = 4x - 5$; fixed cost is $8

46. $C'(x) = 0.2x^2 + 5x$; fixed cost is $10

47. $C'(x) = 0.03e^{0.01x}$; fixed cost is $8

48. $C'(x) = x^{1/2}$; 16 units cost $45

49. $C'(x) = x^{2/3} + 2$; 8 units cost $58

50. $C'(x) = x + 1/x^2$; 2 units cost $5.50

51. $C'(x) = 5x - 1/x$; 10 units cost $94.20

52. $C'(x) = 1.2^x(\ln 1.2)$; 2 units cost $9.44

Demand Find the demand function for each marginal revenue function. Recall that if no items are sold, the revenue is 0.

53. $R'(x) = 175 - 0.02x - 0.03x^2$

54. $R'(x) = 50 - 5x^{2/3}$

55. $R'(x) = 500 - 0.15\sqrt{x}$

56. $R'(x) = 600 - 5e^{0.0002x}$

57. **Text Messaging** The approximate rate of change in the number (in billions) of monthly text messages is given by
$$f'(t) = 7.50t - 16.8,$$
where t represents the number of years since 2000. In 2005 ($t = 5$) there were approximately 9.8 billion monthly text messages. *Source: Cellular Telecommunication & Internet Association.*

a. Find the function that gives the total number (in billions) of monthly text messages in year t.

b. According to this function, how many monthly text messages were there in 2009? Compare this with the actual number of 152.7 billion.

58. **Profit** The marginal profit of a small fast-food stand is given, in thousands of dollars, by
$$P'(x) = \sqrt{x} + \frac{1}{2},$$
where x is the sales volume in thousands of hamburgers. The "profit" is $-$1000$ when no hamburgers are sold. Find the profit function.

59. Profit The marginal profit in dollars on Brie cheese sold at a cheese store is given by

$$P'(x) = x(50x^2 + 30x),$$

where x is the amount of cheese sold, in hundreds of pounds. The "profit" is $-\$40$ when no cheese is sold.

a. Find the profit function.

b. Find the profit from selling 200 lb of Brie cheese.

Life Sciences

60. Biochemical Excretion If the rate of excretion of a biochemical compound is given by

$$f'(t) = 0.01e^{-0.01t},$$

the total amount excreted by time t (in minutes) is $f(t)$.

a. Find an expression for $f(t)$.

b. If 0 units are excreted at time $t = 0$, how many units are excreted in 10 minutes?

61. Flour Beetles A model for describing the population of adult flour beetles involves evaluating the integral

$$\int \frac{g(x)}{x}\, dx,$$

where $g(x)$ is the per-unit-abundance growth rate for a population of size x. The researchers consider the simple case in which $g(x) = a - bx$ for positive constants a and b. Find the integral in this case. *Source: Ecology.*

62. Concentration of a Solute According to Fick's law, the diffusion of a solute across a cell membrane is given by

$$c'(t) = \frac{kA}{V}[C - c(t)], \tag{1}$$

where A is the area of the cell membrane, V is the volume of the cell, $c(t)$ is the concentration inside the cell at time t, C is the concentration outside the cell, and k is a constant. If c_0 represents the concentration of the solute inside the cell when $t = 0$, then it can be shown that

$$c(t) = (c_0 - C)e^{-kAt/V} + C. \tag{2}$$

a. Use the last result to find $c'(t)$.

b. Substitute back into Equation (1) to show that (2) is indeed the correct antiderivative of (1).

63. Cell Growth Under certain conditions, the number of cancer cells $N(t)$ at time t increases at a rate

$$N'(t) = Ae^{kt},$$

where A is the rate of increase at time 0 (in cells per day) and k is a constant.

a. Suppose $A = 50$, and at 5 days, the cells are growing at a rate of 250 per day. Find a formula for the number of cells after t days, given that 300 cells are present at $t = 0$.

b. Use your answer from part a to find the number of cells present after 12 days.

64. Blood Pressure The rate of change of the volume $V(t)$ of blood in the aorta at time t is given by

$$V'(t) = -kP(t),$$

where $P(t)$ is the pressure in the aorta at time t and k is a constant that depends upon properties of the aorta. The pressure in the aorta is given by

$$P(t) = P_0 e^{-mt},$$

where P_0 is the pressure at time $t = 0$ and m is another constant. Letting V_0 be the volume at time $t = 0$, find a formula for $V(t)$.

Social Sciences

65. Bachelor's Degrees The number of bachelor's degrees conferred in the United States has been increasing steadily in recent decades. Based on data from the National Center for Education Statistics, the rate of change of the number of bachelor's degrees (in thousands) can be approximated by the function

$$B'(t) = 0.06048t^2 - 1.292t + 15.86,$$

where t is the number of years since 1970. *Source: National Center for Education Statistics.*

a. Find $B(t)$, given that about 839,700 degrees were conferred in 1970 ($t = 0$).

b. Use the formula from part a to project the number of bachelor's degrees that will be conferred in 2015 ($t = 45$).

66. Degrees in Dentistry The number of degrees in dentistry (D.D.S. or D.M.D.) conferred to females in the United States has been increasing steadily in recent decades. Based on data from the National Center for Education Statistics, the rate of change of the number of bachelor's degrees can be approximated by the function

$$D'(t) = 29.25e^{0.03572t},$$

where t is the number of years since 1980. *Source: National Center for Education Statistics.*

a. Find $D(t)$, given that about 700 degrees in dentistry were conferred to females in 1980 ($t = 0$).

b. Use the formula from part a to project the number of degrees in dentistry that will be conferred to females in 2015 ($t = 35$).

Physical Sciences

Exercises 67–71 refer to Example 11 in this section.

67. Velocity For a particular object, $a(t) = 5t^2 + 4$ and $v(0) = 6$. Find $v(t)$.

68. Distance Suppose $v(t) = 9t^2 - 3\sqrt{t}$ and $s(1) = 8$. Find $s(t)$.

69. Time An object is dropped from a small plane flying at 6400 ft. Assume that $a(t) = -32$ ft per second and $v(0) = 0$. Find $s(t)$. How long will it take the object to hit the ground?

70. Distance Suppose $a(t) = 18t + 8$, $v(1) = 15$, and $s(1) = 19$. Find $s(t)$.

71. Distance Suppose $a(t) = (15/2)\sqrt{t} + 3e^{-t}$, $v(0) = -3$, and $s(0) = 4$. Find $s(t)$.

72. Motion Under Gravity Show that an object thrown from an initial height h_0 with an initial velocity v_0 has a height at time t given by the function

$$h(t) = \tfrac{1}{2}gt^2 + v_0 t + h_0,$$

where g is the acceleration due to gravity, a constant with value -32 ft/sec^2.

73. Rocket A small rocket was launched straight up from a platform. After 5 seconds, the rocket reached a maximum height of 412 ft. Find the initial velocity and height of the rocket. (*Hint:* See the previous exercise.)

74. Rocket Science In the 1999 movie *October Sky*, Homer Hickum was accused of launching a rocket that started a forest fire. Homer proved his innocence by showing that his rocket could not have flown far enough to reach where the fire started. He used the following reasoning.

a. Using the fact that $a(t) = -32$ (see Example 11(b)), find $v(t)$ and $s(t)$, given $v(0) = v_0$ and $s(0) = 0$.

(The initial velocity was unknown, and the initial height was 0 ft.)

b. Homer estimated that the rocket was in the air for 14 seconds. Use $s(14) = 0$ to find v_0.

c. If the rocket left the ground at a 45° angle, the velocity in the horizontal direction would be equal to v_0, the velocity in the vertical direction, so the distance traveled horizontally would be $v_0 t$. (The rocket left the ground at a steeper angle, so this would overestimate the distance from starting to landing point.) Find the distance the rocket would travel horizontally during its 14-second flight.

■ **YOUR TURN ANSWERS**

1. x^8 or $x^8 + C$
2. $-\dfrac{1}{3t^3} + C$
3. $2x^3 + 4x^2 - 9x + C$
4. $\dfrac{2}{7}x^{7/2} - 4x^{1/2} + C$
5. $3\ln|x| - \dfrac{1}{3}e^{-3x} + C$
6. $s(t) = -16t^2 - 20t + 2717$; 12.42 sec; 417 ft/sec
7. $f(x) = 2x^{3/2} + 4x - 8$

7.2 Substitution

APPLY IT **If a formula for the marginal revenue is known, how can a formula for the total revenue be found?**
Using the method of substitution, this question will be answered in Exercise 39.

In earlier chapters you learned all the rules for finding derivatives of elementary functions. By correctly applying those rules, you can take the derivative of any function involving powers of x, exponential functions, and logarithmic functions, combined in any way using the operations of arithmetic (addition, subtraction, multiplication, division, and exponentiation). By contrast, finding the antiderivative is much more complicated. There are a large number of techniques—more than we can cover in this book. Furthermore, for some functions all possible techniques fail. In the last section we saw how to integrate a few simple functions. In this section we introduce a technique known as *substitution* that will greatly expand the set of functions you can integrate.

The substitution technique depends on the idea of a *differential*, discussed in Chapter 6 on Applications of the Derivative. If $u = f(x)$, the *differential* of u, written du, is defined as

$$du = f'(x)\, dx.$$

For example, if $u = 2x^3 + 1$, then $du = 6x^2\, dx$. In this chapter we will only use differentials as a convenient notational device when finding an antiderivative such as

$$\int (2x^3 + 1)^4 6x^2\, dx.$$

The function $(2x^3 + 1)^4 6x^2$ might remind you of the result when using the chain rule to take the derivative. We will now use differentials and the chain rule in reverse to find the antiderivative. Let $u = 2x^3 + 1$; then $du = 6x^2 dx$. Now substitute u for $2x^3 + 1$ and du for $6x^2 dx$ in the indefinite integral.

$$\int (2x^3 + 1)^4 6x^2 \, dx = \int \overbrace{(2x^3 + 1)}^{u}{}^4 \overbrace{(6x^2 \, dx)}^{du}$$

$$= \int u^4 du$$

With substitution we have changed a complicated integral into a simple one. This last integral can now be found by the power rule.

$$\int u^4 \, du = \frac{u^5}{5} + C$$

Finally, substitute $2x^3 + 1$ for u in the antiderivative to get

$$\int (2x^3 + 1)^4 6x^2 \, dx = \frac{(2x^3 + 1)^5}{5} + C.$$

We can check the accuracy of this result by using the chain rule to take the derivative. We get

$$\frac{d}{dx}\left[\frac{(2x^3 + 1)^5}{5} + C\right] = \frac{1}{5} \cdot 5(2x^3 + 1)^4 (6x^2) + 0$$

$$= (2x^3 + 1)^4 6x^2.$$

This method of integration is called **integration by substitution**. As shown above, it is simply the chain rule for derivatives in reverse. The results can always be verified by differentiation.

EXAMPLE 1   Substitution

Find $\int 6x(3x^2 + 4)^7 \, dx$.

SOLUTION   If we choose $u = 3x^2 + 4$, then $du = 6x \, dx$ and the integrand can be written as the product of $(3x^2 + 4)^7$ and $6x \, dx$. Now substitute.

$$\int 6x(3x^2 + 4)^7 \, dx = \int (3x^2 + 4)^7 (6x \, dx) = \int u^7 \, du$$

Find this last indefinite integral.

$$\int u^7 \, du = \frac{u^8}{8} + C$$

Now replace u with $3x^2 + 4$.

$$\int 6x(3x^2 + 4)^7 \, dx = \frac{u^8}{8} + C = \frac{(3x^2 + 4)^8}{8} + C$$

To verify this result, find the derivative.

YOUR TURN 1

Find $\int 8x(4x^2 + 8)^6 dx$.

$$\frac{d}{dx}\left[\frac{(3x^2 + 4)^8}{8} + C\right] = \frac{8}{8}(3x^2 + 4)^7 (6x) + 0 = (3x^2 + 4)^7 (6x)$$

The derivative is the original function, as required.   **TRY YOUR TURN 1**

EXAMPLE 2 Substitution

Find $\int x^2 \sqrt{x^3 + 1}\, dx$.

SOLUTION

Method 1
Modifying the Integral

An expression raised to a power is usually a good choice for u, so because of the square root or $1/2$ power, let $u = x^3 + 1$; then $du = 3x^2\, dx$. The integrand does not contain the constant 3, which is needed for du. To take care of this, multiply by $3/3$, placing 3 inside the integral sign and $1/3$ outside.

$$\int x^2 \sqrt{x^3 + 1}\, dx = \frac{1}{3}\int 3x^2 \sqrt{x^3 + 1}\, dx = \frac{1}{3}\int \sqrt{x^3 + 1}\,(3x^2\, dx)$$

Now substitute u for $x^3 + 1$ and du for $3x^2\, dx$, and then integrate.

$$\frac{1}{3}\int \sqrt{x^3 + 1}\,(3x^2 dx) = \frac{1}{3}\int \sqrt{u}\, du = \frac{1}{3}\int u^{1/2}\, du$$
$$= \frac{1}{3}\cdot \frac{u^{3/2}}{3/2} + C = \frac{2}{9}u^{3/2} + C$$

Since $u = x^3 + 1$,

$$\int x^2 \sqrt{x^3 + 1}\, dx = \frac{2}{9}(x^3 + 1)^{3/2} + C.$$

Method 2
Eliminating the Constant

As in Method 1, we let $u = x^3 + 1$, so that $du = 3x^2\, dx$. Since there is no 3 in the integral, we divide the equation for du by 3 to get

$$\frac{1}{3}\, du = x^2\, dx.$$

We then substitute u for $x^3 + 1$ and $du/3$ for $x^2\, dx$ to get

$$\int \sqrt{u}\,\frac{1}{3}\, du = \frac{1}{3}\int u^{1/2}\, du$$

and proceed as we did in Method 1. The two methods are just slightly different ways of doing the same thing, but some people prefer one method over the other.

YOUR TURN 2

Find $\int x^3 \sqrt{3x^4 + 10}\, dx$.

TRY YOUR TURN 2

The substitution method given in the examples above *will not always work*. For example, you might try to find

$$\int x^3 \sqrt{x^3 + 1}\, dx$$

by substituting $u = x^3 + 1$, so that $du = 3x^2\, dx$. However, there is no *constant* that can be inserted inside the integral sign to give $3x^2$ alone. This integral, and a great many others, cannot be evaluated by substitution.

With practice, choosing u will become easy if you keep two principles in mind.

1. u should equal some expression in the integral that, when replaced with u, tends to make the integral simpler.

2. u must be an expression whose derivative—disregarding any constant multiplier, such as the 3 in $3x^2$—is also present in the integral.

The substitution should include as much of the integral as possible, as long as its derivative is still present. In Example 1, we could have chosen $u = 3x^2$, but $u = 3x^2 + 4$

is better, because it has the same derivative as $3x^2$ and captures more of the original integral. If we carry this reasoning further, we might try $u = (3x^2 + 4)^4$, but this is a poor choice, for $du = 4(3x^2 + 4)^3(6x)\,dx$, an expression not present in the original integral.

▮ EXAMPLE 3 Substitution

Find $\displaystyle\int \frac{x + 3}{(x^2 + 6x)^2}\,dx$.

SOLUTION Let $u = x^2 + 6x$, so that $du = (2x + 6)\,dx = 2(x + 3)\,dx$. The integral is missing the 2, so multiply by $2 \cdot (1/2)$, putting 2 inside the integral sign and 1/2 outside.

$$\int \frac{x + 3}{(x^2 + 6x)^2}\,dx = \frac{1}{2} \int \frac{2(x + 3)}{(x^2 + 6x)^2}\,dx$$

$$= \frac{1}{2} \int \frac{du}{u^2} = \frac{1}{2} \int u^{-2}\,du = \frac{1}{2} \cdot \frac{u^{-1}}{-1} + C = \frac{-1}{2u} + C$$

Substituting $x^2 + 6x$ for u gives

$$\int \frac{x + 3}{(x^2 + 6x)^2}\,dx = \frac{-1}{2(x^2 + 6x)} + C. \qquad \text{TRY YOUR TURN 3} \ \ \blacksquare$$

YOUR TURN 3

Find $\displaystyle\int \frac{x + 1}{(4x^2 + 8x)^3}\,dx$.

In Example 3, the quantity $x^2 + 6x$ was raised to a power in the denominator. When such an expression is not raised to a power, the function can often be integrated using the fact that

$$\frac{d}{dx} \ln |f(x)| = \frac{1}{f(x)} \cdot f'(x).$$

This suggests that such integrals can be solved by letting u equal the expression in the denominator, as long as the derivative of the denominator is present in the numerator (disregarding any constant multiplier as usual). The next example illustrates this idea.

▮ EXAMPLE 4 Substitution

Find $\displaystyle\int \frac{(2x - 3)\,dx}{x^2 - 3x}$.

SOLUTION Let $u = x^2 - 3x$, so that $du = (2x - 3)\,dx$. Then

$$\int \frac{(2x - 3)\,dx}{x^2 - 3x} = \int \frac{du}{u} = \ln |u| + C = \ln |x^2 - 3x| + C.$$

$$\text{TRY YOUR TURN 4} \ \ \blacksquare$$

YOUR TURN 4

Find $\displaystyle\int \frac{x + 3}{x^2 + 6x}\,dx$.

Recall that if $f(x)$ is a function, then by the chain rule, the derivative of the exponential function $y = e^{f(x)}$ is

$$\frac{d}{dx} e^{f(x)} = e^{f(x)} \cdot f'(x).$$

This suggests that the antiderivative of a function of the form $e^{f(x)}$ can be found by letting u be the exponent, as long as $f'(x)$ is also present in the integral (disregarding any constant multiplier as usual).

EXAMPLE 5 Substitution

Find $\int x^2 e^{x^3} dx$.

SOLUTION Let $u = x^3$, the exponent on e. Then $du = 3x^2 dx$. Multiplying by 3/3 gives

$$\int x^2 e^{x^3} dx = \frac{1}{3} \int e^{x^3}(3x^2 dx)$$

$$= \frac{1}{3} \int e^u du = \frac{1}{3} e^u + C = \frac{1}{3} e^{x^3} + C. \quad \textsf{TRY YOUR TURN 5}$$

YOUR TURN 5

Find $\int x^3 e^{x^4} dx$.

The techniques in the preceding examples can be summarized as follows.

Substitution

Each of the following forms can be integrated using the substitution $u = f(x)$.

Form of the Integral	**Result**
1. $\int [f(x)]^n f'(x) dx, \quad n \neq -1$	$\int u^n du = \dfrac{u^{n+1}}{n+1} + C = \dfrac{[f(x)]^{n+1}}{n+1} + C$
2. $\int \dfrac{f'(x)}{f(x)} dx$	$\int \dfrac{1}{u} du = \ln|u| + C = \ln|f(x)| + C$
3. $\int e^{f(x)} f'(x) dx$	$\int e^u du = e^u + C = e^{f(x)} + C$

The next example shows a more complicated integral in which none of the previous forms apply, but for which substitution still works.

EXAMPLE 6 Substitution

Find $\int x\sqrt{1-x}\, dx$.

SOLUTION Let $u = 1 - x$. To get the x outside the radical in terms of u, solve $u = 1 - x$ for x to get $x = 1 - u$. Then $dx = -du$ and we can substitute as follows.

$$\int x\sqrt{1-x}\, dx = \int (1-u)\sqrt{u}(-du) = \int (u-1)u^{1/2} du$$

$$= \int (u^{3/2} - u^{1/2}) du = \frac{2}{5}u^{5/2} - \frac{2}{3}u^{3/2} + C$$

$$= \frac{2}{5}(1-x)^{5/2} - \frac{2}{3}(1-x)^{3/2} + C$$

TRY YOUR TURN 6

YOUR TURN 6

Find $\int x\sqrt{3+x}\, dx$

The substitution method is useful if the integral can be written in one of the following forms, where u is some function of x.

Substitution Method

In general, for the types of problems we are concerned with, there are three cases. We choose u to be one of the following:

1. the quantity under a root or raised to a power;
2. the quantity in the denominator;
3. the exponent on e.

Remember that some integrands may need to be rearranged to fit one of these cases.

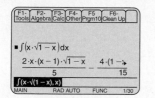

FIGURE 2

Some calculators, such as the TI-89 and TI-Nspire CAS, can find indefinite integrals automatically. Many computer algebra systems, such as Maple, Matlab, and Mathematica, also do this. The website www.wolframalpha.com can also be used to symbolically determine indefinite integrals and derivatives of functions. Figure 2 shows the integral in Example 6 performed on a TI-89. The answer looks different but is algebraically equivalent to the answer found in Example 6.

EXAMPLE 7 Demand

The research department for a hardware chain has determined that at one store the marginal price of x boxes per week of a particular type of nails is

$$p'(x) = \frac{-4000}{(2x + 15)^3}.$$

Find the demand equation if the weekly demand for this type of nails is 10 boxes when the price of a box of nails is $4.

SOLUTION To find the demand function, first integrate $p'(x)$ as follows.

$$p(x) = \int p'(x) \, dx = \int \frac{-4000}{(2x + 15)^3} \, dx$$

Let $u = 2x + 15$. Then $du = 2 \, dx$, and

$$p(x) = \frac{-4000}{2} \int (2x + 15)^{-3} \, 2 \, dx \qquad \text{Multiply by } 2/2.$$

$$= -2000 \int u^{-3} \, du \qquad \text{Substitute.}$$

$$= (-2000)\frac{u^{-2}}{-2} + C$$

$$= \frac{1000}{u^2} + C$$

$$p(x) = \frac{1000}{(2x + 15)^2} + C. \tag{1}$$

Find the value of C by using the given information that $p = 4$ when $x = 10$.

$$4 = \frac{1000}{(2 \cdot 10 + 15)^2} + C$$

$$4 = \frac{1000}{35^2} + C$$

$$4 \approx 0.82 + C$$

$$3.18 \approx C$$

Replacing C with 3.18 in Equation (1) gives the demand function,

$$p(x) = \frac{1000}{(2x + 15)^2} + 3.18.$$

With a little practice, you will find you can skip the substitution step for integrals such as that shown in Example 7, in which the derivative of u is a constant. Recall from the chain rule that when you differentiate a function, such as $p(x) = 1000/(2x + 15)^2 + 3.18$

in the previous example, you multiply by 2, the derivative of $(2x + 15)$. So when taking the antiderivative, simply divide by 2:

$$\int -4000(2x + 15)^{-3} \, dx = \frac{-4000}{2} \cdot \frac{(2x + 15)^{-2}}{-2} + C$$

$$= \frac{1000}{(2x + 15)^2} + C.$$

> **CAUTION** This procedure is valid because of the constant multiple rule presented in the previous section, which says that constant multiples can be brought into or out of integrals, just as they can with derivatives. This procedure is *not* valid with any expression other than a constant.

EXAMPLE 8 Popularity Index

To determine the top 100 popular songs of each year since 1956, Jim Quirin and Barry Cohen developed a function that represents the rate of change on the charts of *Billboard* magazine required for a song to earn a "star" on the *Billboard* "Hot 100" survey. They developed the function

$$f(x) = \frac{A}{B + x},$$

where $f(x)$ represents the rate of change in position on the charts, x is the position on the "Hot 100" survey, and A and B are positive constants. The function

$$F(x) = \int f(x) \, dx$$

is defined as the "Popularity Index." Find $F(x)$. *Source: Chartmasters' Rock 100.*

SOLUTION Integrating $f(x)$ gives

$$F(x) = \int f(x) \, dx$$

$$= \int \frac{A}{B + x} \, dx$$

$$= A \int \frac{1}{B + x} \, dx.$$

Let $u = B + x$, so that $du = dx$. Then

$$F(x) = A \int \frac{1}{u} \, du = A \ln u + C$$

$$= A \ln(B + x) + C.$$

(The absolute value bars are not necessary, since $B + x$ is always positive here.)

7.2 EXERCISES

1. Integration by substitution is related to what differentiation method? What type of integrand suggests using integration by substitution?

2. The following integrals may be solved using substitution. Choose a function u that may be used to solve each problem. Then find du.

a. $\int (3x^2 - 5)^4 \, 2x \, dx$

b. $\int \sqrt{1 - x} \, dx$

c. $\int \frac{x^2}{2x^3 + 1} \, dx$

d. $\int 4x^3 e^{x^4} \, dx$

Use substitution to find each indefinite integral.

3. $\int 4(2x + 3)^4 \, dx$

4. $\int (-4t + 1)^3 \, dt$

5. $\int \dfrac{2 \, dm}{(2m + 1)^3}$

6. $\int \dfrac{3 \, du}{\sqrt{3u - 5}}$

7. $\int \dfrac{2x + 2}{(x^2 + 2x - 4)^4} \, dx$

8. $\int \dfrac{6x^2 \, dx}{(2x^3 + 7)^{3/2}}$

9. $\int z\sqrt{4z^2 - 5} \, dz$

10. $\int r\sqrt{5r^2 + 2} \, dr$

11. $\int 3x^2 e^{2x^3} \, dx$

12. $\int r e^{-r^2} \, dr$

13. $\int (1 - t)e^{2t - t^2} \, dt$

14. $\int (x^2 - 1)e^{x^3 - 3x} \, dx$

15. $\int \dfrac{e^{1/z}}{z^2} \, dz$

16. $\int \dfrac{e^{\sqrt{y}}}{2\sqrt{y}} \, dy$

17. $\int \dfrac{t}{t^2 + 2} \, dt$

18. $\int \dfrac{-4x}{x^2 + 3} \, dx$

19. $\int \dfrac{x^3 + 2x}{x^4 + 4x^2 + 7} \, dx$

20. $\int \dfrac{t^2 + 2}{t^3 + 6t + 3} \, dt$

21. $\int \dfrac{2x + 1}{(x^2 + x)^3} \, dx$

22. $\int \dfrac{y^2 + y}{(2y^3 + 3y^2 + 1)^{2/3}} \, dy$

23. $\int p(p + 1)^5 \, dp$

24. $\int 4r\sqrt{8 - r} \, dr$

25. $\int \dfrac{u}{\sqrt{u - 1}} \, du$

26. $\int \dfrac{2x}{(x + 5)^6} \, dx$

27. $\int (\sqrt{x^2 + 12x})(x + 6) \, dx$ **28.** $\int (\sqrt{x^2 - 6x})(x - 3) \, dx$

29. $\int \dfrac{(1 + 3 \ln x)^2}{x} \, dx$

30. $\int \dfrac{\sqrt{2 + \ln x}}{x} \, dx$

31. $\int \dfrac{e^{2x}}{e^{2x} + 5} \, dx$

32. $\int \dfrac{1}{x(\ln x)} \, dx$

33. $\int \dfrac{\log x}{x} \, dx$

34. $\int \dfrac{(\log_2 (5x + 1))^2}{5x + 1} \, dx$

35. $\int x8^{3x^2 + 1} \, dx$

36. $\int \dfrac{10^{5\sqrt{x + 2}}}{\sqrt{x}} \, dx$

37. Stan and Ollie work on the integral
$$\int 3x^2 e^{x^3} \, dx.$$
Stan lets $u = x^3$ and proceeds to get
$$\int e^u \, du = e^u + C = e^{x^3} + C.$$
Ollie tries $u = e^{x^3}$ and proceeds to get
$$\int du = u + C = e^{x^3} + C.$$
Discuss which procedure you prefer, and why.

38. Stan and Ollie work on the integral
$$\int 2x(x^2 + 2) \, dx.$$

Stan lets $u = x^2 + 2$ and proceeds to get
$$\int u \, du = \dfrac{u^2}{2} + C = \dfrac{(x^2 + 2)^2}{2} + C.$$
Ollie multiplies out the function under the integral and gets
$$\int (2x^3 + 4x) \, dx = \dfrac{x^4}{2} + 2x^2 + C.$$
How can they both be right?

APPLICATIONS

Business and Economics

39. APPLY IT Revenue The marginal revenue (in thousands of dollars) from the sale of x MP3 players is given by
$$R'(x) = 4x(x^2 + 27,000)^{-2/3}.$$

a. Find the total revenue function if the revenue from 125 players is $29,591.

b. How many players must be sold for a revenue of at least $40,000?

40. Debt A company incurs debt at a rate of
$$D'(t) = 90(t + 6)\sqrt{t^2 + 12t}$$
dollars per year, where t is the amount of time (in years) since the company began. By the fourth year the company had accumulated $16,260 in debt.

a. Find the total debt function.

b. How many years must pass before the total debt exceeds $40,000?

41. Cost A company has found that the marginal cost (in thousands of dollars) to produce x central air conditioning units is
$$C'(x) = \dfrac{60x}{5x^2 + e},$$
where x is the number of units produced.

a. Find the cost function, given that the company incurs a fixed cost of $10,000 even if no units are built.

b. The company will seek a new source of investment income if the cost is more than $20,000 to produce 5 units. Should they seek this new source?

42. Profit The rate of growth of the profit (in millions of dollars) from a new technology is approximated by
$$P'(x) = xe^{-x^2},$$
where x represents time measured in years. The total profit in the third year that the new technology is in operation is $10,000.

a. Find the total profit function.

b. What happens to the total amount of profit in the long run?

43. Transportation According to data from the Bureau of Transportation Statistics, the rate of change in the number of local transit vehicles (buses, light rail, etc.), in thousands, in the United States from 1970 to the present can be approximated by

$$f'(t) = 4.0674 \times 10^{-4}t(t - 1970)^{0.4},$$

where t is the year. *Source: National Transportation Statistics 2006.*

a. Using the fact that in 1970 there were 61,298 such vehicles, find a formula giving the approximate number of local transit vehicles as a function of time.

b. Use the answer to part a to forecast the number of local transit vehicles in the year 2015.

Life Sciences

44. **Outpatient Visits** According to data from the American Hospital Association, the rate of change in the number of hospital outpatient visits, in millions, in the United States each year from 1980 to the present can be approximated by

$$f'(t) = 0.001483t(t - 1980)^{0.75},$$

where t is the year. *Source: Hospital Statistics.*

a. Using the fact that in 1980 there were 262,951,000 outpatient visits, find a formula giving the approximate number of outpatient visits as a function of time.

b. Use the answer to part a to forecast the number of outpatient visits in the year 2015.

YOUR TURN ANSWERS

1. $\dfrac{(4x^2 + 8)^7}{7} + C$

2. $\dfrac{(3x^4 + 10)^{3/2}}{18} + C$

3. $-\dfrac{1}{16(4x^2 + 8x)^2} + C$

4. $\frac{1}{2} \ln|x^2 + 6x| + C$

5. $\frac{1}{4}e^{x^4} + C$

6. $\frac{2}{5}(3 + x)^{5/2} - 2(3 + x)^{3/2} + C$

7.3 Area and the Definite Integral

APPLY IT If we know how the rate that oil is leaking from a machine varies with time, how can we estimate the total amount of leakage over a certain period of time?

We will answer this question in Example 3 using a method introduced in this section.

To calculate the areas of geometric figures such as rectangles, squares, triangles, and circles, we use specific formulas. In this section we consider the problem of finding the area of a figure or region that is bounded by curves, such as the shaded region in Figure 3.

The brilliant Greek mathematician Archimedes (about 287 B.C.–212 B.C.) is considered one of the greatest mathematicians of all time. His development of a rigorous method known as *exhaustion* to derive results was a forerunner of the ideas of integral calculus. Archimedes used a method that would later be verified by the theory of integration. His method involved viewing a geometric figure as a sum of other figures. For example, he thought of a plane surface area as a figure consisting of infinitely many parallel line segments. Among the results established by Archimedes' method was the fact that the area of a segment of a parabola (shown in color in Figure 3) is equal to 4/3 the area of a triangle with the same base and the same height.

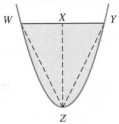

Area of parabolic segment

$= \frac{4}{3}$ (area of triangle *WYZ*)

FIGURE 3

EXAMPLE 1 Approximation of Area

Consider the region bounded by the y-axis, the x-axis, and the graph of $f(x) = \sqrt{4 - x^2}$, shown in Figure 4.

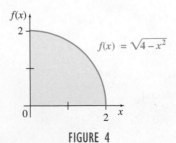

FIGURE 4

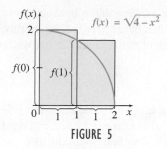

FIGURE 5

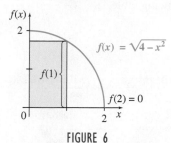

FIGURE 6

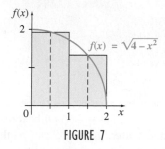

FIGURE 7

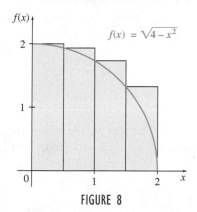

FIGURE 8

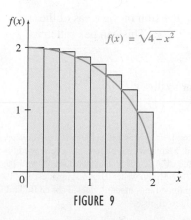

FIGURE 9

(a) Approximate the area of the region using two rectangles. Determine the height of the rectangle by the value of the function at the *left* endpoint.

SOLUTION A very rough approximation of the area of this region can be found by using two rectangles whose heights are determined by the value of the function at the left endpoints, as in Figure 5. The height of the rectangle on the left is $f(0) = 2$ and the height of the rectangle on the right is $f(1) = \sqrt{3}$. The width of each rectangle is 1, making the total area of the two rectangles

$$1 \cdot f(0) + 1 \cdot f(1) = 2 + \sqrt{3} \approx 3.7321 \text{ square units.}$$

Note that $f(x)$ is a decreasing function, and that we will overestimate the area when we evaluate the function at the left endpoint to determine the height of the rectangle in that interval.

(b) Repeat part (a) using the value of the function at the *right* endpoint to determine the height of the rectangle.

SOLUTION Using the right endpoints, as in Figure 6, the area of the two rectangles is

$$1 \cdot f(1) + 1 \cdot f(2) = \sqrt{3} + 0 \approx 1.7321 \text{ square units.}$$

Note that we underestimate the area of this particular region when we use the right endpoints.

If the left endpoint gives an answer too big and the right endpoint an answer too small, it seems reasonable to average the two answers. This produces the method called the *trapezoidal rule*, discussed in more detail later in this chapter. In this example, we get

$$\frac{3.7321 + 1.7321}{2} = 2.7321 \text{ square units.}$$

(c) Repeat part (a) using the value of the function at the *midpoint* of each interval to determine the height of the rectangle.

SOLUTION In Figure 7, the rectangles are drawn with height determined by the midpoint of each interval. This method is called the **midpoint rule**, and gives

$$1 \cdot f(0.5) + 1 \cdot f(1.5) = \sqrt{3.75} + \sqrt{1.75} \approx 3.2594 \text{ square units.}$$

(d) We can improve the accuracy of the previous approximations by increasing the number of rectangles. Repeat part (a) using four rectangles.

SOLUTION Divide the interval from $x = 0$ to $x = 2$ into four equal parts, each of width 1/2. The height of each rectangle is given by the value of f at the left side of the rectangle, as shown in Figure 8. The area of each rectangle is the width, 1/2, multiplied by the height. The total area of the four rectangles is

$$\frac{1}{2} \cdot f(0) + \frac{1}{2} \cdot f\left(\frac{1}{2}\right) + \frac{1}{2} \cdot f(1) + \frac{1}{2} \cdot f\left(1\frac{1}{2}\right)$$

$$= \frac{1}{2}(2) + \frac{1}{2}\left(\frac{\sqrt{15}}{2}\right) + \frac{1}{2}(\sqrt{3}) + \frac{1}{2}\left(\frac{\sqrt{7}}{2}\right)$$

$$= 1 + \frac{\sqrt{15}}{4} + \frac{\sqrt{3}}{2} + \frac{\sqrt{7}}{4} \approx 3.4957 \text{ square units.}$$

This approximation looks better, but it is still greater than the actual area.

(e) Repeat part (a) using eight rectangles.

SOLUTION Divide the interval from $x = 0$ to $x = 2$ into 8 equal parts, each of width 1/4 (see Figure 9). The total area of all of these rectangles is

$$\frac{1}{4} \cdot f(0) + \frac{1}{4} \cdot f\left(\frac{1}{4}\right) + \frac{1}{4} \cdot f\left(\frac{1}{2}\right) + \frac{1}{4} \cdot f\left(\frac{3}{4}\right) + \frac{1}{4} \cdot f(1)$$

$$+ \frac{1}{4} \cdot f\left(\frac{5}{4}\right) + \frac{1}{4} \cdot f\left(\frac{3}{2}\right) + \frac{1}{4} \cdot f\left(\frac{7}{4}\right)$$

$$\approx 3.3398 \text{ square units.}$$

The process used in Example 1 of approximating the area under a curve by using more and more rectangles to get a better and better approximation can be generalized. To do this, divide the interval from $x = 0$ to $x = 2$ into n equal parts. Each of these n intervals has width

$$\frac{2 - 0}{n} = \frac{2}{n},$$

so each rectangle has width $2/n$ and height determined by the function value at the left side of the rectangle, or the right side, or the midpoint. We could also average the left and right side values as before. Using a computer or graphing calculator to find approximations to the area for several values of n gives the results in the following table.

	Approximations to the Area			
n	**Left Sum**	**Right Sum**	**Trapezoidal**	**Midpoint**
2	3.7321	1.7321	2.7321	3.2594
4	3.4957	2.4957	2.9957	3.1839
8	3.3398	2.8398	3.0898	3.1567
10	3.3045	2.9045	3.1045	3.1524
20	3.2285	3.0285	3.1285	3.1454
50	3.1783	3.0983	3.1383	3.1426
100	3.1604	3.1204	3.1404	3.1419
500	3.1455	3.1375	3.1415	3.1416

The numbers in the last four columns of this table represent approximations to the area under the curve, above the x-axis, and between the lines $x = 0$ and $x = 2$. As n becomes larger and larger, all four approximations become better and better, getting closer to the actual area. In this example, the exact area can be found by a formula from plane geometry. Write the given function as

$$y = \sqrt{4 - x^2},$$

then square both sides to get

$$y^2 = 4 - x^2$$
$$x^2 + y^2 = 4,$$

the equation of a circle centered at the origin with radius 2. The region in Figure 4 is the quarter of this circle that lies in the first quadrant. The actual area of this region is one-quarter of the area of the entire circle, or

$$\frac{1}{4}\pi(2)^2 = \pi \approx 3.1416.$$

As the number of rectangles increases without bound, the sum of the areas of these rectangles gets closer and closer to the actual area of the region, π. This can be written as

$$\lim_{n \to \infty} (\text{sum of areas of } n \text{ rectangles}) = \pi.$$

(The value of π was originally found by a process similar to this.)*

---

*The number π is the ratio of the circumference of a circle to its diameter. It is an example of an *irrational number*, and as such it cannot be expressed as a terminating or repeating decimal. Many approximations have been used for π over the years. A passage in the Bible (1 Kings 7:23) indicates a value of 3. The Egyptians used the value 3.16, and Archimedes showed that its value must be between 22/7 and 223/71. A Hindu writer, Brahmagupta, used $\sqrt{10}$ as its value in the seventh century. The search for the digits of π has continued into modern times. Fabrice Bellard, using a desktop computer, recently computed the value to nearly 2.7 trillion digits.

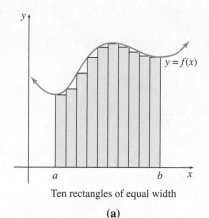

Ten rectangles of equal width

(a)

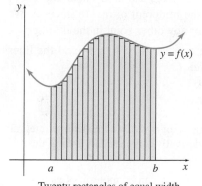

Twenty rectangles of equal width

(b)

FIGURE 10

Notice in the previous table that for a particular value of n, the midpoint rule gave the best answer (the one closest to the true value of $\pi \approx 3.1416$), followed by the trapezoidal rule, followed by the left and right sums. In fact, the midpoint rule with $n = 20$ gives a value (3.1454) that is slightly more accurate than the left sum with $n = 500$ (3.1455). It is usually the case that the midpoint rule gives a more accurate answer than either the left or the right sum.

Now we can generalize to get a method of finding the area bounded by the curve $y = f(x)$, the x-axis, and the vertical lines $x = a$ and $x = b$, as shown in Figure 10. To approximate this area, we could divide the region under the curve first into 10 rectangles (Figure 10(a)) and then into 20 rectangles (Figure 10(b)). The sum of the areas of the rectangles gives an approximation to the area under the curve when $f(x) \geq 0$. In the next section we will consider the case in which $f(x)$ might be negative.

To develop a process that would yield the *exact* area, begin by dividing the interval from a to b into n pieces of equal width, using each of these n pieces as the base of a rectangle (see Figure 11). Let x_1 be an arbitrary point in the first interval, x_2 be an arbitrary point in the second interval, and so on, up to the nth interval. In the graph of Figure 11, the symbol Δx is used to represent the width of each of the intervals. Since the length of the entire interval is $b - a$, each of the n pieces has length

$$\Delta x = \frac{b - a}{n}.$$

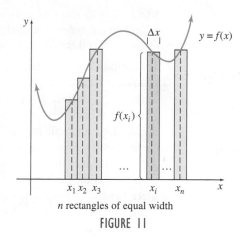

n rectangles of equal width

FIGURE 11

The pink rectangle is an arbitrary rectangle called the ith rectangle. Its area is the product of its length and width. Since the width of the ith rectangle is Δx and the length of the ith rectangle is given by the height $f(x_i)$,

$$\text{Area of the } i\text{th rectangle} = f(x_i) \cdot \Delta x.$$

The total area under the curve is approximated by the sum of the areas of all n of the rectangles. With sigma notation, the approximation to the total area becomes

$$\text{Area of all } n \text{ rectangles} = \sum_{i=1}^{n} f(x_i) \cdot \Delta x.$$

The exact area is defined to be the limit of this sum (if the limit exists) as the number of rectangles increases without bound:

$$\text{Exact area} = \lim_{n \to \infty} \sum_{i=1}^{n} f(x_i) \Delta x.$$

Whenever this limit exists, regardless of whether $f(x)$ is positive or negative, we will call it the *definite integral* of $f(x)$ from a to b. It is written as follows.

FOR REVIEW

Recall from Chapter 1 that the symbol Σ (sigma) indicates "the sum of." Here, we use $\sum_{i=1}^{n} f(x_i)\Delta x$ to indicate the sum $f(x_1)\Delta x + f(x_2)\Delta x + f(x_3)\Delta x + \cdots + f(x_n)\Delta x$, where we replace i with 1 in the first term, 2 in the second term, and so on, ending with n replacing i in the last term.

The Definite Integral

If f is defined on the interval $[a, b]$, the **definite integral** of f from a to b is given by

$$\int_a^b f(x)\, dx = \lim_{n \to \infty} \sum_{i=1}^n f(x_i)\Delta x,$$

provided the limit exists, where $\Delta x = (b - a)/n$ and x_i is *any* value of x in the ith interval.*

The definite integral can be approximated by

$$\sum_{i=1}^n f(x_i)\Delta x.$$

If $f(x) \geq 0$ on the interval $[a, b]$, the definite integral gives the area under the curve between $x = a$ and $x = b$. In the midpoint rule, x_i is the midpoint of the ith interval. We may also let x_i be the left endpoint, the right endpoint, or any other point in the ith interval.

In Example 1, the area bounded by the x-axis, the curve $y = \sqrt{4 - x^2}$, and the lines $x = 0$ and $x = 2$ could be written as the definite integral

$$\int_0^2 \sqrt{4 - x^2}\, dx = \pi.$$

NOTE Notice that unlike the indefinite integral, which is a set of *functions*, the definite integral represents a *number*. The next section will show how antiderivatives are used in finding the definite integral and, thus, the area under a curve.

Keep in mind that finding the definite integral of a function can be thought of as a mathematical process that gives the sum of an infinite number of individual parts (within certain limits). The definite integral represents area only if the function involved is *nonnegative* ($f(x) \geq 0$) at every x-value in the interval $[a, b]$. There are many other interpretations of the definite integral, and all of them involve this idea of approximation by appropriate sums. In the next section we will consider the definite integral when $f(x)$ might be negative.

As indicated in this definition, although the left endpoint of the ith interval has been used to find the height of the ith rectangle, any number in the ith interval can be used. (A more general definition is possible in which the rectangles do not necessarily all have the same width.) The b above the integral sign is called the **upper limit** of integration, and the a is the **lower limit** of integration. This use of the word *limit* has nothing to do with the limit of the sum; it refers to the limits, or boundaries, on x.

TECHNOLOGY NOTE Some calculators have a built-in function for evaluating the definite integral. For example, the TI-84 Plus uses the `fnInt` command, found in the MATH menu, as shown in Figure 12(a). Figure 12(b) shows the command used for Example 1 and gives the answer 3.141593074, with an error of approximately 0.0000004.

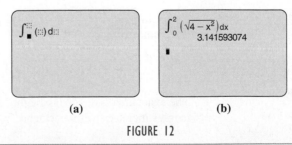

(a)   (b)

FIGURE 12

*The sum in the definition of the definite integral is an example of a Riemann sum, named for the German mathematician Georg Riemann (1826–1866), who at the age of 20 changed his field of study from theology and the classics to mathematics. Twenty years later he died of tuberculosis while traveling in Italy in search of a cure. The concepts of *Riemann sum* and *Riemann integral* are still studied in rigorous calculus textbooks.

EXAMPLE 2 **Approximation of Area**

Approximate $\int_0^4 2x\, dx$, the area of the region under the graph of $f(x) = 2x$, above the x-axis, and between $x = 0$ and $x = 4$, by using four rectangles of equal width whose heights are the values of the function at the midpoint of each subinterval.

Method 1
Calculating by Hand

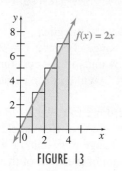

FIGURE 13

SOLUTION

We want to find the area of the shaded region in Figure 13. The heights of the four rectangles given by $f(x_i)$ for $i = 1, 2, 3,$ and 4 are as follows.

	Rectangle Heights	
i	x_i	$f(x_i)$
1	$x_1 = 0.5$	$f(0.5) = 1.0$
2	$x_2 = 1.5$	$f(1.5) = 3.0$
3	$x_3 = 2.5$	$f(2.5) = 5.0$
4	$x_4 = 3.5$	$f(3.5) = 7.0$

The width of each rectangle is $\Delta x = (4 - 0)/4 = 1$. The sum of the areas of the four rectangles is

$$\sum_{i=1}^{4} f(x_i)\Delta x = f(x_1)\Delta x + f(x_2)\Delta x + f(x_3)\Delta x + f(x_4)\Delta x$$
$$= f(0.5)\Delta x + f(1.5)\Delta x + f(2.5)\Delta x + f(3.5)\Delta x$$
$$= 1(1) + 3(1) + 5(1) + 7(1)$$
$$= 16.$$

Using the formula for the area of a triangle, $A = (1/2)bh$, with b, the length of the base, equal to 4 and h, the height, equal to 8, gives

$$A = \frac{1}{2}bh = \frac{1}{2}(4)(8) = 16,$$

the exact value of the area. The approximation equals the exact area in this case because our use of the midpoints of each subinterval distributed the error evenly above and below the graph.

Method 2
Graphing Calculator

A graphing calculator can be used to organize the information in this example. For example, the `seq` feature in the LIST OPS menu of the TI-84 Plus calculator can be used to store the values of i in the list L_1. Using the STAT EDIT menu, the entries for x_i can be generated by entering the formula $-.5 + L_1$ as the heading of L_2. Similarly, entering the formula for $f(x_i)$, $2*L_2$, at the top of list L_3 will generate the values of $f(x_i)$ in L_3. (The entries are listed automatically when the formula is entered.) Then the `sum` feature in the LIST MATH menu can be used to add the values in L_3. The resulting screens are shown in Figure 14.

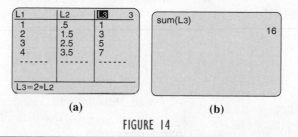

(a)   (b)

FIGURE 14

⊞ **Method 3**
Spreadsheet

The calculations in this example can also be done on a spreadsheet. In Microsoft Excel, for example, store the values of i in column A. Put the command "=A1 − .5" into B1; copying this formula into the rest of column B gives the values of x_i. Similarly, use the formula for $f(x_i)$ to fill column C. Column D is the product of Column C and Δx. Sum column D to get the answer. For more details, see the *Graphing Calculator and Excel Spreadsheet Manual* available with this book.

YOUR TURN 1 Repeat Example 1 to approximate

$$\int_{1}^{5} 4x \, dx.$$

TRY YOUR TURN 1 ▬

Total Change

Suppose the function $f(x) = x^2 + 20$ gives the marginal cost of some item at a particular x-value. Then $f(2) = 24$ gives the rate of change of cost at $x = 2$. That is, a unit change in x (at this point) will produce a change of 24 units in the cost function. Also, $f(3) = 29$ means that each unit of change in x (when $x = 3$) will produce a change of 29 units in the cost function.

To find the *total* change in the cost function as x changes from 2 to 3, we could divide the interval from 2 to 3 into n equal parts, using each part as the base of a rectangle as we did above. The area of each rectangle would approximate the change in cost at the x-value that is the left endpoint of the base of the rectangle. Then the sum of the areas of these rectangles would approximate the net total change in cost from $x = 2$ to $x = 3$. The limit of this sum as $n \to \infty$ would give the exact total change.

This result produces another application of the definite integral: the area of the region under the graph of the marginal cost function $f(x)$ that is above the x-axis and between $x = a$ and $x = b$ gives the *net total change in the cost* as x goes from a to b.

> **Total Change in $F(x)$**
>
> If $f(x)$ gives the rate of change of $F(x)$ for x in $[a, b]$, then the **total change** in $F(x)$ as x goes from a to b is given by
>
> $$\lim_{n \to \infty} \sum_{i=1}^{n} f(x_i)\Delta x = \int_{a}^{b} f(x) \, dx.$$

In other words, the total change in a quantity can be found from the function that gives the rate of change of the quantity, using the same methods used to approximate the area under a curve.

EXAMPLE 3 Oil Leakage

APPLY IT Figure 15 shows the rate that oil is leaking from a machine in a large factory (in cubic centimeters per hour) with specific rates over a 12-hour period given in the table. Approximate the total amount of leakage over a 12-hour shift.

SOLUTION Use approximating rectangles, dividing the interval from 0 to 12 into 12 equal subdivisions. Each subinterval has width 1. Using the left endpoint of each subinterval and the table to determine the height of the rectangle, as shown, the approximation becomes

$$1 \cdot 15.2 + 1 \cdot 18.0 + 1 \cdot 18.8 + 1 \cdot 14.1 + 1 \cdot 9.5 + 1 \cdot 9.6 + 1 \cdot 13.1 + 1 \cdot 17.3$$
$$+ 1 \cdot 20.0 + 1 \cdot 19.2 + 1 \cdot 16.6 + 1 \cdot 16.4 = 187.8.$$

About 187.8 cubic centimeters of oil leak during this time. Mathematically, we could write

$$\int_{0}^{12} f(x) \, dx \approx 187.8,$$

where $f(x)$ is the function shown in Figure 15. ▬

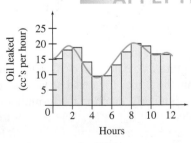

FIGURE 15

Oil Leakage (cc's per hour)			
x	y	x	y
0	15.2	6	13.1
1	18.0	7	17.3
2	18.8	8	20.0
3	14.1	9	19.2
4	9.5	10	16.6
5	9.6	11	16.4

Recall, velocity is the rate of change in distance from time a to time b. Thus the area under the velocity function defined by $v(t)$ from $t = a$ to $t = b$ gives the distance traveled in that time period.

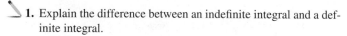

EXAMPLE 4   **Total Distance**

A driver traveling on a business trip checks the speedometer each hour. The table shows the driver's velocity at several times.

Approximate the total distance traveled during the 3-hour period using the left endpoint of each interval, then the right endpoint.

Velocity				
Time (hr)	0	1	2	3
Velocity (mph)	0	52	58	60

YOUR TURN 2 Repeat Example 4 for a driver traveling at the following velocities at various times.

Time (hr)	0	0.5	1	1.5	2
Velocity (mph)	0	50	56	40	48

SOLUTION   Using left endpoints, the total distance is

$$0 \cdot 1 + 52 \cdot 1 + 58 \cdot 1 = 110.$$

With right endpoints, we get

$$52 \cdot 1 + 58 \cdot 1 + 60 \cdot 1 = 170.$$

Again, left endpoints give a total that is too small, while right endpoints give a total that is too large. The average, 140 miles, is a better estimate of the total distance traveled.

TRY YOUR TURN 2

Before discussing further applications of the definite integral, we need a more efficient method for evaluating it. This method will be developed in the next section.

7.3   EXERCISES

1. Explain the difference between an indefinite integral and a definite integral.

2. Complete the following statement.

$$\int_0^4 (x^2 + 3)\, dx = \lim_{n \to \infty} \underline{\quad\quad}, \text{ where } \Delta x = \underline{\quad\quad}, \text{ and } x_i$$

is _____.

3. Let $f(x) = 2x + 5$, $x_1 = 0$, $x_2 = 2$, $x_3 = 4$, $x_4 = 6$, and $\Delta x = 2$.

   a. Find $\sum_{i=1}^{4} f(x_i)\Delta x$.

   b. The sum in part a approximates a definite integral using rectangles. The height of each rectangle is given by the value of the function at the left endpoint. Write the definite integral that the sum approximates.

4. Let $f(x) = 1/x$, $x_1 = 1/2$, $x_2 = 1$, $x_3 = 3/2$, $x_4 = 2$, and $\Delta x = 1/2$.

   a. Find $\sum_{i=1}^{4} f(x_i)\Delta x$.

   b. The sum in part a approximates a definite integral using rectangles. The height of each rectangle is given by the value of the function at the left endpoint. Write the definite integral that the sum approximates.

In Exercises 5–12, approximate the area under the graph of $f(x)$ and above the x-axis using the following methods with $n = 4$. (a) Use left endpoints. (b) Use right endpoints. (c) Average the answers in parts a and b. (d) Use midpoints.

5. $f(x) = 2x + 5$ from $x = 2$ to $x = 4$

6. $f(x) = 3x + 2$ from $x = 1$ to $x = 3$

7. $f(x) = -x^2 + 4$ from $x = -2$ to $x = 2$

8. $f(x) = x^2$ from $x = 1$ to $x = 5$

9. $f(x) = e^x + 1$ from $x = -2$ to $x = 2$

10. $f(x) = e^x - 1$ from $x = 0$ to $x = 4$

11. $f(x) = \dfrac{2}{x}$ from $x = 1$ to $x = 9$

12. $f(x) = \dfrac{1}{x}$ from $x = 1$ to $x = 3$

13. Consider the region below $f(x) = x/2$, above the x-axis, and between $x = 0$ and $x = 4$. Let x_i be the midpoint of the ith subinterval.

    a. Approximate the area of the region using four rectangles.

    b. Find $\int_0^4 f(x)\, dx$ by using the formula for the area of a triangle.

14. Consider the region below $f(x) = 5 - x$, above the x-axis, and between $x = 0$ and $x = 5$. Let x_i be the midpoint of the ith subinterval.

a. Approximate the area of the region using five rectangles.

b. Find $\int_0^5 (5 - x)\,dx$ by using the formula for the area of a triangle.

15. Find $\int_0^4 f(x)\,dx$ for each graph of $y = f(x)$.

a.   **b.**

16. Find $\int_0^6 f(x)\,dx$ for each graph of $y = f(x)$, where $f(x)$ consists of line segments and circular arcs.

a. 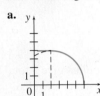   **b.**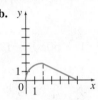

Find the exact value of each integral using formulas from geometry.

17. $\displaystyle\int_{-4}^{0} \sqrt{16 - x^2}\,dx$ **18.** $\displaystyle\int_{-3}^{3} \sqrt{9 - x^2}\,dx$

19. $\displaystyle\int_{2}^{5} (1 + 2x)\,dx$ **20.** $\displaystyle\int_{1}^{3} (5 + x)\,dx$

21. In this exercise, we investigate the value of $\int_0^1 x^2\,dx$ using larger and larger values of n in the definition of the definite integral.

a. First let $n = 10$, so $\Delta x = 0.1$. Fill a list on your calculator with values of x^2 as x goes from 0.1 to 1. (On a TI-84 Plus, use the command `seq(X^2,X,.1,1,.1)→L1`.)

b. Sum the values in the list formed in part a, and multiply by 0.1, to estimate $\int_0^1 x^2\,dx$ with $n = 10$. (On a TI-84 Plus, use the command `.1*sum(L1)`.)

c. Repeat parts a and b with $n = 100$.

d. Repeat parts a and b with $n = 500$.

e. Based on your answers to parts b through d, what do you estimate the value of $\int_0^1 x^2\,dx$ to be?

22. Repeat Exercise 21 for $\int_0^1 x^3\,dx$.

23. The booklet *All About Lawns* published by Ortho Books gives the following instructions for measuring the area of an irregularly shaped region. (See figure in the next column.) *Source: All About Lawns*

Irregular Shapes
(within 5% accuracy)
Measure a long (L) axis of the area. Every 10 feet along the length line, measure the width at right angles to the length line. Total widths and multiply by 10.

Area $= (\overline{A_1 A_2} + \overline{B_1 B_2} + \overline{C_1 C_2}$ etc.$) \times 10$

$A = (40' + 60' + 32') \times 10$

$A = 132' \times 10'$

$A = 1320$ square feet

How does this method relate to the discussion in this section?

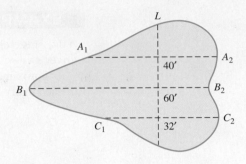

APPLICATIONS

In Exercises 24–28, estimate the area under each curve by summing the area of rectangles. Use the left endpoints, then the right endpoints, then give the average of those answers.

Business and Economics

24. Electricity Consumption The following graph shows the rate of use of electrical energy (in millions of kilowatts) in a certain city on a very hot day. Estimate the total usage of electricity on that day. Let the width of each rectangle be 2 hours.

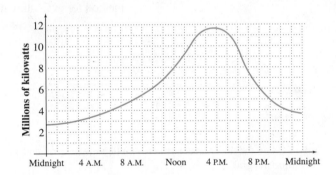

25. Wind Energy Consumption The following graph shows the U.S. wind energy consumption (trillion BTUs) for various years. Estimate the total consumption for the 12-year period from 1997 to 2009 using rectangles of width 3 years. *Source: Annual Energy Review.*

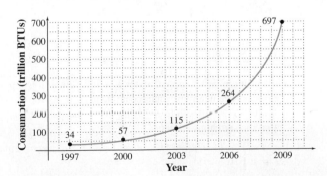

Life Sciences

26. Oxygen Inhalation The graph on the next page shows the rate of inhalation of oxygen (in liters per minute) by a person riding a bicycle very rapidly for 10 minutes. Estimate the total volume of oxygen inhaled in the first 20 minutes after the beginning of the ride. Use rectangles with widths of 1 minute.

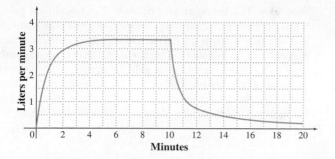

27. Foot-and-Mouth Epidemic In 2001, the United Kingdom suffered an epidemic of foot-and-mouth disease. The graph below shows the reported number of cattle (red) and pigs (blue) that were culled each month from mid-February through mid-October in an effort to stop the spread of the disease. *Source: Department of Environment, Food and Rural Affairs, United Kingdom.*

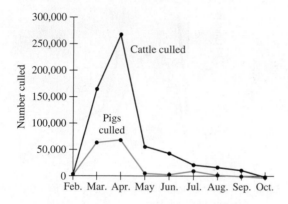

a. Estimate the total number of cattle that were culled from mid-February through mid-October and compare this with 581,801, the actual number of cattle that were culled. Use rectangles that are one month in width, starting with mid-February.

b. Estimate the total number of pigs that were culled from mid-February through mid-October and compare this with 146,145, the actual number of pigs that were culled. Use rectangles that are one month in width starting with mid-February.

Social Sciences

28. Automobile Accidents The graph shows the number of fatal automobile accidents in California for various years. Estimate the total number of accidents in the 8-year period from 2000 to 2008 using rectangles of width 2 years. *Source: California Highway Patrol.*

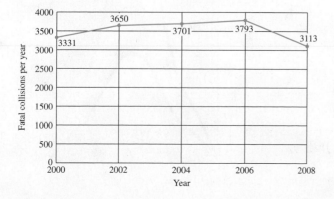

Physical Sciences

Distance The next two graphs are from the Road & Track website. The curves show the velocity at t seconds after the car accelerates from a dead stop. To find the total distance traveled by the car in reaching 130 mph, we must estimate the definite integral

$$\int_0^T v(t)\,dt,$$

where T represents the number of seconds it takes for the car to reach 130 mph.

Use the graphs to estimate this distance by adding the areas of rectangles and using the midpoint rule. To adjust your answer to miles per hour, divide by 3600 (the number of seconds in an hour). You then have the number of miles that the car traveled in reaching 130 mph. Finally, multiply by 5280 ft per mile to convert the answer to feet. *Source: Road & Track.*

29. Estimate the distance traveled by the Lamborghini Gallardo LP560-4 using the graph below. Use rectangles with widths of 3 seconds, except for the last rectangle, which should have a width of 2 seconds. The circle marks the point where the car has gone a quarter mile. Does this seem correct?

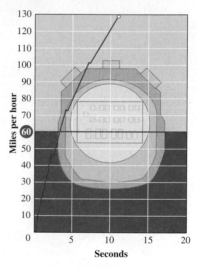

30. Estimate the distance traveled by the Alfa Romeo 8C Competizione using the graph below. Use rectangles with widths of 4 seconds, except for the last rectangle, which should have a width of 3.5 seconds. The circle marks the point where the car has gone a quarter mile. Does this seem correct?

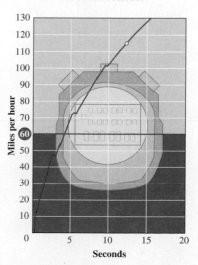

Distance When data are given in tabular form, you may need to vary the size of the interval to calculate the area under the curve. The next two exercises include data from *Car and Driver* magazine. To estimate the total distance traveled by the car (in feet) during the time it took to reach its maximum velocity, estimate the area under the velocity versus time graph, as in the previous two exercises. Use the left endpoint for each time interval (the velocity at the beginning of that interval) and then the right endpoint (the velocity at the end of the interval). Finally, average the two answers together. Calculating and adding up the areas of the rectangles is most easily done on a spreadsheet or graphing calculator. As in the previous two exercises, you will need to multiply by a conversion factor of $5280/3600 = 22/15$, since the velocities are given in miles per hour, but the time is in seconds, and we want the answer in feet. *Source: Car and Driver.*

31. Estimate the distance traveled by the Mercedes-Benz S550, using the table below.

Acceleration	Seconds
Zero to 30 mph	2.0
40 mph	2.9
50 mph	4.1
60 mph	5.3
70 mph	6.9
80 mph	8.7
90 mph	10.7
100 mph	13.2
110 mph	16.1
120 mph	19.3
130 mph	23.4

32. Estimate the distance traveled by the Chevrolet Malibu Maxx SS, using the table below.

Acceleration	Seconds
Zero to 30 mph	2.4
40 mph	3.5
50 mph	5.1
60 mph	6.9
70 mph	8.9
80 mph	11.2
90 mph	14.9
100 mph	19.2
110 mph	24.4

Heat Gain The following graphs show the typical heat gain, in BTUs per hour per square foot, for windows (one with plain glass and one that is triple glazed) in Pittsburgh in June, one facing east and one facing south. The horizontal axis gives the time of the day. Estimate the total heat gain per square foot by summing the areas of rectangles. Use rectangles with widths of 2 hours, and let the function value at the midpoint of the subinterval give the height of the rectangle. *Source: Sustainable by Design.*

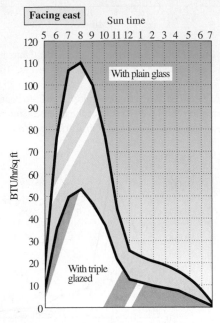

33. a. Estimate the total heat gain per square foot for a plain glass window facing east.

b. Estimate the total heat gain per square foot for a triple glazed window facing east.

34. a. Estimate the total heat gain per square foot for a plain glass window facing south.

b. Estimate the total heat gain per square foot for a triple glazed window facing south.

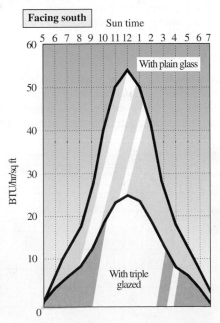

35. Automobile Velocity Two cars start from rest at a traffic light and accelerate for several minutes. The graph shows their velocities (in feet per second) as a function of time (in seconds). Car A is the one that initially has greater velocity. *Source: Stephen Monk.*

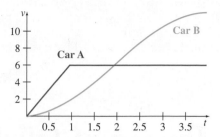

a. How far has car A traveled after 2 seconds? (*Hint*: Use formulas from geometry.)

b. When is car A farthest ahead of car B?

c. Estimate the farthest that car A gets ahead of car B. For car A, use formulas from geometry. For car B, use $n = 4$ and the value of the function at the midpoint of each interval.

d. Give a rough estimate of when car B catches up with car A.

36. Distance Musk the friendly pit bull has escaped again! Here is her velocity during the first 4 seconds of her romp.

t (sec)	0	1	2	3	4
v (ft/sec)	0	8	13	17	18

Give two estimates for the total distance Musk traveled during her 4-second trip, one using the left endpoint of each interval and one using the right endpoint.

37. Distance The speed of a particle in a test laboratory was noted every second for 3 seconds. The results are shown in the following table. Use the left endpoints and then the right endpoints to estimate the total distance the particle moved in the first three seconds.

t (sec)	0	1	2	3
v (ft/sec)	10	6.5	6	5.5

38. Running In 1987, Canadian Ben Johnson set a world record in the 100-m sprint. (The record was later taken away when he was found to have used an anabolic steroid to enhance his performance.) His speed at various times in the race is given in the following table*. *Source: Information Graphics.*

Time (sec)	Speed (mph)
0	0
1.84	12.9
3.80	23.8
6.38	26.3
7.23	26.3
8.96	26.0
9.83	25.7

a. Use the information in the table and left endpoints to estimate the distance that Johnson ran in miles. You will first need to calculate Δt for each interval. At the end, you will need to divide by 3600 (the number of seconds in an hour), since the speed is in miles per hour.

b. Repeat part a, using right endpoints.

c. Wait a minute; we know that the distance Johnson ran is 100 m. Divide this by 1609, the number of meters in a mile, to find how far Johnson ran in miles. Is your answer from part a or part b closer to the true answer? Briefly explain why you think this answer should be more accurate.

39. Traffic The following graph shows the number of vehicles per hour crossing the Tappan Zee Bridge, which spans the Hudson River north of New York City. The graph shows the number of vehicles traveling eastbound (into the city) and westbound (out of the city) as a function of time. *Source: The New York Times.*

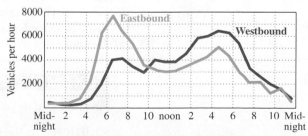

Source: New York Metropolitan Transportation Council

*The world record of 9.58 seconds is currently held by Usain Bolt of Jamaica.

a. Using midpoints on intervals of one hour, estimate the total number of vehicles that cross the bridge going eastbound each day.

b. Repeat the instructions for part a for vehicles going westbound.

c. Discuss whether the answers to parts a and b should be equal, and try to explain any discrepancies.

7.4 The Fundamental Theorem of Calculus

APPLY IT If we know how the rate of consumption of natural gas varies over time, how can we compute the total amount of natural gas used?

We will answer this question in Example 7.

In the first section of this chapter, you learned about antiderivatives. In the previous section, you learned about the definite integral. In this section, we connect these two separate topics and present one of the most powerful theorems of calculus.

We have seen that, if $f(x) \geq 0$,

$$\int_a^b f(x)\, dx$$

gives the area between the graph of $f(x)$ and the x-axis, from $x = a$ to $x = b$. The definite integral was defined and evaluated in the previous section using the limit of a sum. In that section, we also saw that if $f(x)$ gives the rate of change of $F(x)$, the definite integral $\int_a^b f(x)\, dx$ gives the total change of $F(x)$ as x changes from a to b. If $f(x)$ gives the rate of change of $F(x)$, then $F(x)$ is an antiderivative of $f(x)$. Writing the total change in $F(x)$ from $x = a$ to $x = b$ as $F(b) - F(a)$ shows the connection between antiderivatives and definite integrals. This relationship is called the **Fundamental Theorem of Calculus**.

> **Fundamental Theorem of Calculus**
>
> Let f be continuous on the interval $[a, b]$, and let F be *any* antiderivative of f. Then
>
> $$\int_a^b f(x)\, dx = F(b) - F(a) = F(x)\Big|_a^b.$$

The symbol $F(x)\big|_a^b$ is used to represent $F(b) - F(a)$. It is important to note that the Fundamental Theorem does not require $f(x) > 0$. The condition $f(x) > 0$ is necessary only when using the Fundamental Theorem to find area. Also, note that the Fundamental Theorem does not *define* the definite integral; it just provides a method for evaluating it.

EXAMPLE 1 **Fundamental Theorem of Calculus**

First find $\int 4t^3\, dt$ and then find $\int_1^2 4t^3\, dt$.

SOLUTION By the power rule given earlier, the indefinite integral is

$$\int 4t^3\, dt = t^4 + C.$$

By the Fundamental Theorem, the value of the definite integral $\int_1^2 4t^3\,dt$ is found by evaluating $t^4\big|_1^2$, with no constant C required.

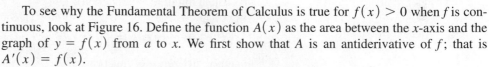

$$\int_1^2 4t^3\,dt = t^4\,\Big|_1^2 = 2^4 - 1^4 = 15$$ **TRY YOUR TURN 1**

YOUR TURN 1 Find $\displaystyle\int_1^3 3x^2\,dx$.

Example 1 illustrates the difference between the definite integral and the indefinite integral. A definite integral is a real number; an indefinite integral is a family of functions in which all the functions are antiderivatives of a function f.

NOTE No constant C is needed, as it is for the indefinite integral, because even if C were added to an antiderivative F, it would be eliminated in the final answer:

$$\int_a^b f(x)\,dx = (F(x) + C)\,\Big|_a^b$$
$$= (F(b) + C) - (F(a) + C)$$
$$= F(b) - F(a).$$

In other words, any antiderivative will give the same answer, so for simplicity, we choose the one with $C = 0$.

To see why the Fundamental Theorem of Calculus is true for $f(x) > 0$ when f is continuous, look at Figure 16. Define the function $A(x)$ as the area between the x-axis and the graph of $y = f(x)$ from a to x. We first show that A is an antiderivative of f; that is $A'(x) = f(x)$.

To do this, let h be a small positive number. Then $A(x + h) - A(x)$ is the shaded area in Figure 16. This area can be approximated with a rectangle having width h and height $f(x)$. The area of the rectangle is $h \cdot f(x)$, and

$$A(x + h) - A(x) \approx h \cdot f(x).$$

Dividing both sides by h gives

$$\frac{A(x + h) - A(x)}{h} \approx f(x).$$

This approximation improves as h gets smaller and smaller. Taking the limit on the left as h approaches 0 gives an exact result.

$$\lim_{h \to 0} \frac{A(x + h) - A(x)}{h} = f(x)$$

This limit is simply $A'(x)$, so

$$A'(x) = f(x).$$

This result means that A is an antiderivative of f, as we set out to show.

$A(b)$ is the area under the curve from a to b, and $A(a) = 0$, so the area under the curve can be written as $A(b) - A(a)$. From the previous section, we know that the area under the curve is also given by $\int_a^b f(x)\,dx$. Putting these two results together gives

$$\int_a^b f(x)\,dx = A(b) - A(a)$$
$$= A(x)\,\Big|_a^b$$

where A is an antiderivative of f. From the note after Example 1, we know that any antiderivative will give the same answer, which proves the Fundamental Theorem of Calculus.

The Fundamental Theorem of Calculus certainly deserves its name, which sets it apart as the most important theorem of calculus. It is the key connection between differential

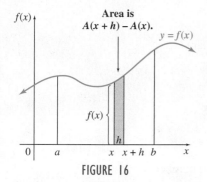

FIGURE 16

calculus and integral calculus, which originally were developed separately without knowledge of this connection between them.

The variable used in the integrand does not matter; each of the following definite integrals represents the number $F(b) - F(a)$.

$$\int_a^b f(x)\,dx = \int_a^b f(t)\,dt = \int_a^b f(u)\,du$$

Key properties of definite integrals are listed below. Some of them are just restatements of properties from Section 1.

Properties of Definite Integrals

If all indicated definite integrals exist,

1. $\displaystyle\int_a^a f(x)\,dx = 0$;

2. $\displaystyle\int_a^b k \cdot f(x)\,dx = k \cdot \int_a^b f(x)\,dx$ for any real constant k
   (constant multiple of a function);

3. $\displaystyle\int_a^b [f(x) \pm g(x)]\,dx = \int_a^b f(x)\,dx \pm \int_a^b g(x)\,dx$
   (sum or difference of functions);

4. $\displaystyle\int_a^b f(x)\,dx = \int_a^c f(x)\,dx + \int_c^b f(x)\,dx$ for any real number c;

5. $\displaystyle\int_a^b f(x)\,dx = -\int_b^a f(x)\,dx$.

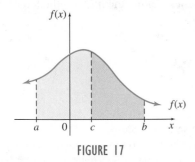

FIGURE 17

For $f(x) \geq 0$, since the distance from a to a is 0, the first property says that the "area" under the graph of f bounded by $x = a$ and $x = a$ is 0. Also, since $\int_a^c f(x)\,dx$ represents the blue region in Figure 17 and $\int_c^b f(x)\,dx$ represents the pink region,

$$\int_a^b f(x)\,dx = \int_a^c f(x)\,dx + \int_c^b f(x)\,dx,$$

as stated in the fourth property. While the figure shows $a < c < b$, the property is true for any value of c where both $f(x)$ and $F(x)$ are defined.

An algebraic proof is given here for the third property; proofs of the other properties are left for the exercises. If $F(x)$ and $G(x)$ are antiderivatives of $f(x)$ and $g(x)$, respectively,

$$\int_a^b [f(x) + g(x)]\,dx = [F(x) + G(x)]\Big|_a^b$$

$$= [F(b) + G(b)] - [F(a) + G(a)]$$

$$= [F(b) - F(a)] + [G(b) - G(a)]$$

$$= \int_a^b f(x)\,dx + \int_a^b g(x)\,dx.$$

EXAMPLE 2   Fundamental Theorem of Calculus

Find $\int_2^5 (6x^2 - 3x + 5)\, dx$.

SOLUTION   Use the properties above and the Fundamental Theorem, along with properties from Section 1.

$$\int_2^5 (6x^2 - 3x + 5)\, dx = 6\int_2^5 x^2\, dx - 3\int_2^5 x\, dx + 5\int_2^5 dx$$

$$= 2x^3 \Big|_2^5 - \frac{3}{2}x^2 \Big|_2^5 + 5x \Big|_2^5$$

$$= 2(5^3 - 2^3) - \frac{3}{2}(5^2 - 2^2) + 5(5 - 2)$$

$$= 2(125 - 8) - \frac{3}{2}(25 - 4) + 5(3)$$

$$= 234 - \frac{63}{2} + 15 = \frac{435}{2}$$

TRY YOUR TURN 2

YOUR TURN 2

Find $\int_3^5 (2x^3 - 3x + 4)\, dx$.

YOUR TURN 3

Find $\int_1^3 \frac{2}{y}\, dy$.

EXAMPLE 3   Fundamental Theorem of Calculus

$$\int_1^2 \frac{dy}{y} = \ln |y| \Big|_1^2 = \ln |2| - \ln |1|$$

$$= \ln 2 - \ln 1 \approx 0.6931 - 0 = 0.6931 \quad \text{TRY YOUR TURN 3}$$

EXAMPLE 4   Substitution

Evaluate $\int_0^5 x\sqrt{25 - x^2}\, dx$.

SOLUTION

Method 1
Changing the Limits   Use substitution. Let $u = 25 - x^2$, so that $du = -2x\, dx$. With a definite integral, the limits should be changed, too. The new limits on u are found as follows.

$$\text{If } x = 5, \text{ then } u = 25 - 5^2 = 0.$$
$$\text{If } x = 0, \text{ then } u = 25 - 0^2 = 25.$$

Then

$$\int_0^5 x\sqrt{25 - x^2}\, dx = -\frac{1}{2}\int_0^5 \sqrt{25 - x^2}(-2x\, dx) \qquad \text{Multiply by } -2/-2.$$

$$= -\frac{1}{2}\int_{25}^0 \sqrt{u}\, du \qquad \text{Substitute and change limits.}$$

$$= -\frac{1}{2}\int_{25}^0 u^{1/2}\, du \qquad \sqrt{u} = u^{1/2}$$

$$= -\frac{1}{2} \cdot \frac{u^{3/2}}{3/2} \Big|_{25}^0 \qquad \text{Use the power rule.}$$

$$= -\frac{1}{2} \cdot \frac{2}{3}[0^{3/2} - 25^{3/2}]$$

$$= -\frac{1}{3}(-125) = \frac{125}{3}.$$

Method 2
Evaluating the Antiderivative

An alternative method that some people prefer is to evaluate the antiderivative first and then calculate the definite integral. To evaluate the antiderivative in this example, ignore the limits on the original integral and use the substitution $u = 25 - x^2$, so that $du = -2x\,dx$. Then

$$\int x\sqrt{25 - x^2}\,dx = -\frac{1}{2}\int \sqrt{25 - x^2}(-2x\,dx)$$

$$= -\frac{1}{2}\int \sqrt{u}\,du$$

$$= -\frac{1}{2}\int u^{1/2}\,du$$

$$= -\frac{1}{2}\frac{u^{3/2}}{3/2} + C$$

$$= -\frac{u^{3/2}}{3} + C$$

$$= -\frac{(25 - x^2)^{3/2}}{3} + C.$$

We will ignore the constant C because it doesn't affect the answer, as we mentioned in the Note following Example 1.

Then, using the Fundamental Theorem of Calculus, we have

$$\int_0^5 x\sqrt{25 - x^2}\,dx = -\frac{(25 - x^2)^{3/2}}{3}\Big|_0^5$$

$$= 0 - \left[-\frac{(25)^{3/2}}{3}\right]$$

$$= \frac{125}{3}.$$

YOUR TURN 4 Evaluate
$$\int_0^4 2x\sqrt{16 - x^2}\,dx.$$

TRY YOUR TURN 4

| CAUTION | Don't confuse these two methods. In Method 1, we never return to the original variable or the original limits of integration. In Method 2, it is essential to return to the original variable and to not change the limits. When using Method 1, we recommend labeling the limits with the appropriate variable to avoid confusion, so the substitution in Example 4 becomes

$$\int_{x=0}^{x=5} x\sqrt{25 - x^2}\,dx = -\frac{1}{2}\int_{u=25}^{u=0} \sqrt{u}\,du.$$

The Fundamental Theorem of Calculus is a powerful tool, but it has a limitation. The problem is that not every function has an antiderivative in terms of the functions and operations you have seen so far. One example of an integral that cannot be evaluated by the Fundamental Theorem of Calculus for this reason is

$$\int_a^b e^{-x^2/2}\,dx,$$

yet this integral is crucial in probability and statistics. Such integrals may be evaluated by numerical integration, which is covered in the last section of this chapter. Fortunately for you, all the integrals in this section can be antidifferentiated using the techniques presented in the first two sections of this chapter.

Area In the previous section we saw that, if $f(x) \geq 0$ in $[a, b]$, the definite integral $\int_a^b f(x)\,dx$ gives the area below the graph of the function $y = f(x)$, above the x-axis, and between the lines $x = a$ and $x = b$.

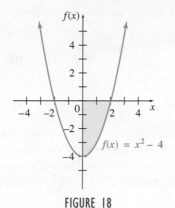

FIGURE 18

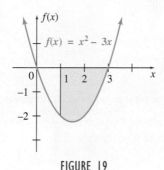

FIGURE 19

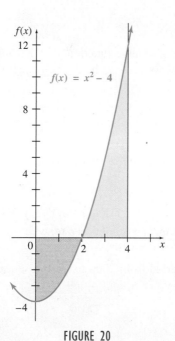

FIGURE 20

To see how to work around the requirement that $f(x) \geq 0$, look at the graph of $f(x) = x^2 - 4$ in Figure 18. The area bounded by the graph of f, the x-axis, and the vertical lines $x = 0$ and $x = 2$ lies below the x-axis. Using the Fundamental Theorem gives

$$\int_0^2 (x^2 - 4)\,dx = \left(\frac{x^3}{3} - 4x\right)\Big|_0^2$$

$$= \left(\frac{8}{3} - 8\right) - (0 - 0) = -\frac{16}{3}.$$

The result is a negative number because $f(x)$ is negative for values of x in the interval $[0, 2]$. Since Δx is always positive, if $f(x) < 0$ the product $f(x) \cdot \Delta x$ is negative, so $\int_0^2 f(x)\,dx$ is negative. Since area is nonnegative, the required area is given by $|-16/3|$ or $16/3$. Using a definite integral, the area could be written as

$$\left|\int_0^2 (x^2 - 4)\,dx\right| = \left|-\frac{16}{3}\right| = \frac{16}{3}.$$

EXAMPLE 5 Area

Find the area of the region between the x-axis and the graph of $f(x) = x^2 - 3x$ from $x = 1$ to $x = 3$.

SOLUTION The region is shown in Figure 19. Since the region lies below the x-axis, the area is given by

$$\left|\int_1^3 (x^2 - 3x)\,dx\right|.$$

By the Fundamental Theorem,

$$\int_1^3 (x^2 - 3x)\,dx = \left(\frac{x^3}{3} - \frac{3x^2}{2}\right)\Big|_1^3 = \left(\frac{27}{3} - \frac{27}{2}\right) - \left(\frac{1}{3} - \frac{3}{2}\right) = -\frac{10}{3}.$$

The required area is $|-10/3| = 10/3$.

EXAMPLE 6 Area

Find the area between the x-axis and the graph of $f(x) = x^2 - 4$ from $x = 0$ to $x = 4$.

SOLUTION Figure 20 shows the required region. Part of the region is below the x-axis. The definite integral over that interval will have a negative value. To find the area, integrate the negative and positive portions separately and take the absolute value of the first result before combining the two results to get the total area. Start by finding the point where the graph crosses the x-axis. This is done by solving the equation

$$x^2 - 4 = 0.$$

The solutions of this equation are 2 and -2. The only solution in the interval $[0, 4]$ is 2. The total area of the region in Figure 20 is

$$\left|\int_0^2 (x^2 - 4)\,dx\right| + \int_2^4 (x^2 - 4)\,dx = \left|\left(\frac{1}{3}x^3 - 4x\right)\Big|_0^2\right| + \left(\frac{1}{3}x^3 - 4x\right)\Big|_2^4$$

$$= \left|\frac{8}{3} - 8\right| + \left(\frac{64}{3} - 16\right) - \left(\frac{8}{3} - 8\right)$$

$$= 16.$$

YOUR TURN 5 Repeat Example 6 for the function $f(x) = x^2 - 9$ from $x = 0$ to $x = 6$.

TRY YOUR TURN 5

Incorrectly using one integral over the entire interval to find the area in Example 6 would have given

$$\int_0^4 (x^2 - 4)\, dx = \left(\frac{x^3}{3} - 4x\right)\Big|_0^4 = \left(\frac{64}{3} - 16\right) - 0 = \frac{16}{3},$$

which is not the correct area. This definite integral does not represent any area but is just a real number.

For instance, if $f(x)$ in Example 6 represents the annual rate of profit of a company, then 16/3 represents the total profit for the company over a 4-year period. The integral between 0 and 2 is $-16/3$; the negative sign indicates a loss for the first two years. The integral between 2 and 4 is 32/3, indicating a profit. The overall profit is $32/3 - 16/3 = 16/3$, although the total shaded area is $32/3 + |-16/3| = 16$.

Finding Area

In summary, to find the area bounded by $f(x)$, $x = a$, $x = b$, and the x-axis, use the following steps.

1. Sketch a graph.

2. Find any x-intercepts of $f(x)$ in $[a, b]$. These divide the total region into subregions.

3. The definite integral will be *positive* for subregions above the x-axis and *negative* for subregions below the x-axis. Use separate integrals to find the (positive) areas of the subregions.

4. The total area is the sum of the areas of all of the subregions.

In the last section, we saw that the area under a rate of change function $f'(x)$ from $x = a$ to $x = b$ gives the total value of $f(x)$ on $[a, b]$. Now we can use the definite integral to solve these problems.

EXAMPLE 7   Natural Gas Consumption

The yearly rate of consumption of natural gas (in trillions of cubic feet) for a certain city is

$$C'(t) = t + e^{0.01t},$$

where t is time in years and $t = 0$ corresponds to 2000. At this consumption rate, what was the total amount the city used in the 10-year period of 2000 to 2010?

APPLY IT   **SOLUTION**   To find the consumption over the 10-year period from 2000 to 2010, use the definite integral.

$$\int_0^{10} (t + e^{0.01t})\, dt = \left(\frac{t^2}{2} + \frac{e^{0.01t}}{0.01}\right)\Big|_0^{10}$$
$$= (50 + 100e^{0.1}) - (0 + 100)$$
$$\approx -50 + 100(1.10517) \approx 60.5$$

Therefore, a total of about 60.5 trillion ft³ of natural gas was used from 2000 to 2010 at this consumption rate.

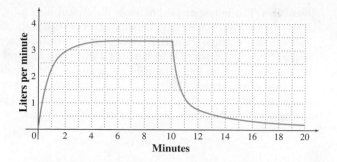

27. Foot-and-Mouth Epidemic In 2001, the United Kingdom suffered an epidemic of foot-and-mouth disease. The graph below shows the reported number of cattle (red) and pigs (blue) that were culled each month from mid-February through mid-October in an effort to stop the spread of the disease. *Source: Department of Environment, Food and Rural Affairs, United Kingdom.*

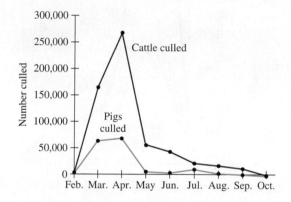

a. Estimate the total number of cattle that were culled from mid-February through mid-October and compare this with 581,801, the actual number of cattle that were culled. Use rectangles that are one month in width, starting with mid-February.

b. Estimate the total number of pigs that were culled from mid-February through mid-October and compare this with 146,145, the actual number of pigs that were culled. Use rectangles that are one month in width starting with mid-February.

Social Sciences

28. Automobile Accidents The graph shows the number of fatal automobile accidents in California for various years. Estimate the total number of accidents in the 8-year period from 2000 to 2008 using rectangles of width 2 years. *Source: California Highway Patrol.*

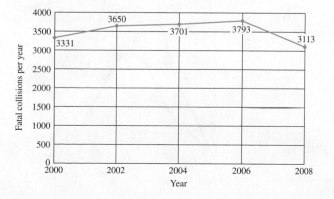

Physical Sciences

Distance The next two graphs are from the Road & Track website. The curves show the velocity at t seconds after the car accelerates from a dead stop. To find the total distance traveled by the car in reaching 130 mph, we must estimate the definite integral

$$\int_0^T v(t)\,dt,$$

where T represents the number of seconds it takes for the car to reach 130 mph.

Use the graphs to estimate this distance by adding the areas of rectangles and using the midpoint rule. To adjust your answer to miles per hour, divide by 3600 (the number of seconds in an hour). You then have the number of miles that the car traveled in reaching 130 mph. Finally, multiply by 5280 ft per mile to convert the answer to feet. *Source: Road & Track.*

29. Estimate the distance traveled by the Lamborghini Gallardo LP560-4 using the graph below. Use rectangles with widths of 3 seconds, except for the last rectangle, which should have a width of 2 seconds. The circle marks thc point where the car has gone a quarter mile. Does this seem correct?

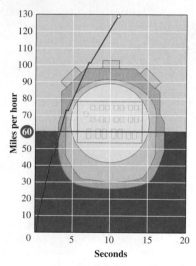

30. Estimate the distance traveled by the Alfa Romeo 8C Competizione using the graph below. Use rectangles with widths of 4 seconds, except for the last rectangle, which should have a width of 3.5 seconds. The circle marks the point where the car has gone a quarter mile. Does this seem correct?

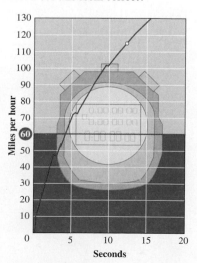

Distance When data are given in tabular form, you may need to vary the size of the interval to calculate the area under the curve. The next two exercises include data from *Car and Driver* magazine. To estimate the total distance traveled by the car (in feet) during the time it took to reach its maximum velocity, estimate the area under the velocity versus time graph, as in the previous two exercises. Use the left endpoint for each time interval (the velocity at the beginning of that interval) and then the right endpoint (the velocity at the end of the interval). Finally, average the two answers together. Calculating and adding up the areas of the rectangles is most easily done on a spreadsheet or graphing calculator. As in the previous two exercises, you will need to multiply by a conversion factor of $5280/3600 = 22/15$, since the velocities are given in miles per hour, but the time is in seconds, and we want the answer in feet. *Source: Car and Driver.*

31. Estimate the distance traveled by the Mercedes-Benz S550, using the table below.

Acceleration	Seconds
Zero to 30 mph	2.0
40 mph	2.9
50 mph	4.1
60 mph	5.3
70 mph	6.9
80 mph	8.7
90 mph	10.7
100 mph	13.2
110 mph	16.1
120 mph	19.3
130 mph	23.4

32. Estimate the distance traveled by the Chevrolet Malibu Maxx SS, using the table below.

Acceleration	Seconds
Zero to 30 mph	2.4
40 mph	3.5
50 mph	5.1
60 mph	6.9
70 mph	8.9
80 mph	11.2
90 mph	14.9
100 mph	19.2
110 mph	24.4

Heat Gain The following graphs show the typical heat gain, in BTUs per hour per square foot, for windows (one with plain glass and one that is triple glazed) in Pittsburgh in June, one facing east and one facing south. The horizontal axis gives the time of the day. Estimate the total heat gain per square foot by summing the areas of rectangles. Use rectangles with widths of 2 hours, and let the function value at the midpoint of the subinterval give the height of the rectangle. *Source: Sustainable by Design.*

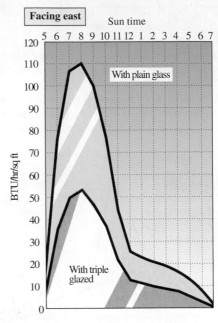

33. a. Estimate the total heat gain per square foot for a plain glass window facing east.

 b. Estimate the total heat gain per square foot for a triple glazed window facing east.

34. a. Estimate the total heat gain per square foot for a plain glass window facing south.

 b. Estimate the total heat gain per square foot for a triple glazed window facing south.

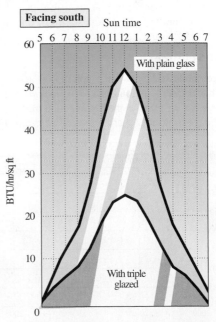

7.4   EXERCISES

Evaluate each definite integral.

1. $\int_{-2}^{4} (-3)\, dp$

2. $\int_{-4}^{1} \sqrt{2}\, dx$

3. $\int_{-1}^{2} (5t - 3)\, dt$

4. $\int_{-2}^{2} (4z + 3)\, dz$

5. $\int_{0}^{2} (5x^2 - 4x + 2)\, dx$

6. $\int_{-2}^{3} (-x^2 - 3x + 5)\, dx$

7. $\int_{0}^{2} 3\sqrt{4u + 1}\, du$

8. $\int_{3}^{9} \sqrt{2r - 2}\, dr$

9. $\int_{0}^{4} 2(t^{1/2} - t)\, dt$

10. $\int_{0}^{4} -(3x^{3/2} + x^{1/2})\, dx$

11. $\int_{1}^{4} (5y\sqrt{y} + 3\sqrt{y})\, dy$

12. $\int_{4}^{9} (4\sqrt{r} - 3r\sqrt{r})\, dr$

13. $\int_{4}^{6} \frac{2}{(2x - 7)^2}\, dx$

14. $\int_{1}^{4} \frac{-3}{(2p + 1)^2}\, dp$

15. $\int_{1}^{5} (6n^{-2} - n^{-3})\, dn$

16. $\int_{2}^{3} (3x^{-3} - 5x^{-4})\, dx$

17. $\int_{-3}^{-2} \left(2e^{-0.1y} + \frac{3}{y}\right) dy$

18. $\int_{-2}^{-1} \left(\frac{-2}{t} + 3e^{0.3t}\right) dt$

19. $\int_{1}^{2} \left(e^{4u} - \frac{1}{(u + 1)^2}\right) du$

20. $\int_{0.5}^{1} (p^3 - e^{4p})\, dp$

21. $\int_{-1}^{0} y(2y^2 - 3)^5\, dy$

22. $\int_{0}^{3} m^2(4m^3 + 2)^3\, dm$

23. $\int_{1}^{64} \frac{\sqrt{z} - 2}{\sqrt[3]{z}}\, dz$

24. $\int_{1}^{8} \frac{3 - y^{1/3}}{y^{2/3}}\, dy$

25. $\int_{1}^{2} \frac{\ln x}{x}\, dx$

26. $\int_{1}^{3} \frac{\sqrt{\ln x}}{x}\, dx$

27. $\int_{0}^{8} x^{1/3}\sqrt{x^{4/3} + 9}\, dx$

28. $\int_{1}^{2} \frac{3}{x(1 + \ln x)}\, dx$

29. $\int_{0}^{1} \frac{e^{2t}}{(3 + e^{2t})^2}\, dt$

30. $\int_{0}^{1} \frac{e^{2z}}{\sqrt{1 + e^{2z}}}\, dz$

In Exercises 31–40, use the definite integral to find the area between the *x*-axis and *f*(*x*) over the indicated interval. Check first to see if the graph crosses the *x*-axis in the given interval.

31. $f(x) = 2x - 14;$ $[6, 10]$

32. $f(x) = 4x - 32;$ $[5, 10]$

33. $f(x) = 2 - 2x^2;$ $[0, 5]$

34. $f(x) = 9 - x^2;$ $[0, 6]$

35. $f(x) = x^3;$ $[-1, 3]$

36. $f(x) = x^3 - 2x;$ $[-2, 4]$

37. $f(x) = e^x - 1;$ $[-1, 2]$

38. $f(x) = 1 - e^{-x};$ $[-1, 2]$

39. $f(x) = \frac{1}{x} - \frac{1}{e};$ $[1, e^2]$

40. $f(x) = 1 - \frac{1}{x};$ $[e^{-1}, e]$

Find the area of each shaded region.

41.

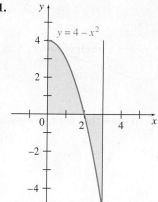

$y = 4 - x^2$

42.

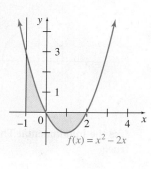

$f(x) = x^2 - 2x$

43.

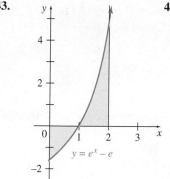

$y = e^x - e$

44.

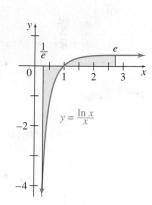

$y = \dfrac{\ln x}{x}$

45. Assume $f(x)$ is continuous for $g \le x \le c$ as shown in the figure. Write an equation relating the three quantities

$$\int_{a}^{b} f(x)\, dx, \qquad \int_{a}^{c} f(x)\, dx, \qquad \int_{b}^{c} f(x)\, dx.$$

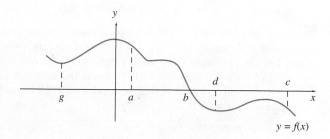

$y = f(x)$

46. Is the equation you wrote for Exercise 45 still true

a. if *b* is replaced by *d*?

b. if *b* is replaced by *g*?

47. The graph of $f(x)$, shown here, consists of two straight line segments and two quarter circles. Find the value of $\int_0^{16} f(x)\,dx$.

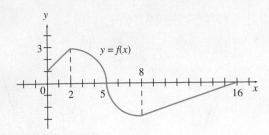

Use the Fundamental Theorem to show that the following are true.

48. $\displaystyle\int_a^b kf(x)\,dx = k\int_a^b f(x)\,dx$

49. $\displaystyle\int_a^b f(x)\,dx = \int_a^c f(x)\,dx + \int_c^b f(x)\,dx$

50. $\displaystyle\int_a^b f(x)\,dx = -\int_b^a f(x)\,dx$

51. Use Exercise 49 to find $\int_{-1}^4 f(x)\,dx$, given

$$f(x) = \begin{cases} 2x + 3 & \text{if } x \le 0 \\ -\dfrac{x}{4} - 3 & \text{if } x > 0. \end{cases}$$

52. You are given $\int_0^1 e^{x^2}\,dx = 1.46265$ and $\int_0^2 e^{x^2}\,dx = 16.45263$. Use this information to find

a. $\displaystyle\int_{-1}^1 e^{x^2}\,dx$; **b.** $\displaystyle\int_1^2 e^{x^2}\,dx$.

53. Let $g(t) = t^4$ and define $f(x) = \int_c^x g(t)\,dt$ with $c = 1$.

a. Find a formula for $f(x)$.

b. Verify that $f'(x) = g(x)$. The fact that

$$\frac{d}{dx}\int_c^x g(t)\,dt = g(x)$$

is true for all continuous functions g is an alternative version of the Fundamental Theorem of Calculus.

c. Let us verify the result in part b for a function whose antiderivative cannot be found. Let $g(t) = e^{t^2}$ and let $c = 0$. Use the integration feature on a graphing calculator to find $f(x)$ for $x = 1$ and $x = 1.01$. Then use the definition of the derivative with $h = 0.01$ to approximate $f'(1)$, and compare it with $g(1)$.

54. Consider the function $f(x) = x(x^2 + 3)^7$.

a. Use the Fundamental Theorem of Calculus to evaluate $\displaystyle\int_{-5}^5 f(x)\,dx$.

b. Use symmetry to describe how the integral from part a could be evaluated without using substitution or finding an antiderivative.

APPLICATIONS

Business and Economics

55. Profit Karla Harby Communications, a small company of science writers, found that its rate of profit (in thousands of dollars) after t years of operation is given by

$$P'(t) = (3t + 3)(t^2 + 2t + 2)^{1/3}.$$

a. Find the total profit in the first three years.

b. Find the profit in the fourth year of operation.

c. What is happening to the annual profit over the long run?

56. Worker Efficiency A worker new to a job will improve his efficiency with time so that it takes him fewer hours to produce an item with each day on the job, up to a certain point. Suppose the rate of change of the number of hours it takes a worker in a certain factory to produce the xth item is given by

$$H'(x) = 20 - 2x.$$

a. What is the total number of hours required to produce the first 5 items?

b. What is the total number of hours required to produce the first 10 items?

Life Sciences

57. Pollution Pollution from a factory is entering a lake. The rate of concentration of the pollutant at time t is given by

$$P'(t) = 140t^{5/2},$$

where t is the number of years since the factory started introducing pollutants into the lake. Ecologists estimate that the lake can accept a total level of pollution of 4850 units before all the fish life in the lake ends. Can the factory operate for 4 years without killing all the fish in the lake?

58. Spread of an Oil Leak An oil tanker is leaking oil at the rate given (in barrels per hour) by

$$L'(t) = \frac{80\ln(t + 1)}{t + 1},$$

where t is the time (in hours) after the tanker hits a hidden rock (when $t = 0$).

a. Find the total number of barrels that the ship will leak on the first day.

b. Find the total number of barrels that the ship will leak on the second day.

c. What is happening over the long run to the amount of oil leaked per day?

59. Tree Growth After long study, tree scientists conclude that a eucalyptus tree will grow at the rate of $0.6 + 4/(t + 1)^3$ ft per year, where t is time (in years).

a. Find the number of feet that the tree will grow in the second year.

b. Find the number of feet the tree will grow in the third year.

60. Growth of a Substance The rate at which a substance grows is given by

$$R'(x) = 150e^{0.2x},$$

where x is the time (in days). What is the total accumulated growth during the first 3.5 days?

61. Drug Reaction For a certain drug, the rate of reaction in appropriate units is given by

$$R'(t) = \frac{5}{t+1} + \frac{2}{\sqrt{t+1}},$$

where t is time (in hours) after the drug is administered. Find the total reaction to the drug over the following time periods.

a. From $t = 1$ to $t = 12$ **b.** From $t = 12$ to $t = 24$

62. Human Mortality If $f(x)$ is the instantaneous death rate for members of a population at time x, then the number of individuals who survive to age T is given by

$$F(T) = \int_0^T f(x)\,dx.$$

In 1825 the biologist Benjamin Gompertz proposed that $f(x) = kb^x$. Find a formula for $F(T)$. *Source: Philosophical Transactions of the Royal Society of London.*

63. Cell Division Let the expected number of cells in a culture that have an x percent probability of undergoing cell division during the next hour be denoted by $n(x)$.

a. Explain why $\int_{20}^{30} n(x)\,dx$ approximates the total number of cells with a 20% to 30% chance of dividing during the next hour.

b. Give an integral representing the number of cells that have less than a 60% chance of dividing during the next hour.

c. Let $n(x) = \sqrt{5x+1}$ give the expected number of cells (in millions) with x percent probability of dividing during the next hour. Find the number of cells with a 5 to 10% chance of dividing.

64. Bacterial Growth A population of *E. coli* bacteria will grow at a rate given by

$$w'(t) = (3t+2)^{1/3},$$

where w is the weight (in milligrams) after t hours. Find the change in weight of the population from $t = 0$ to $t = 3$.

65. Blood Flow In an example from an earlier chapter, the velocity v of the blood in a blood vessel was given as

$$v = k(R^2 - r^2),$$

where R is the (constant) radius of the blood vessel, r is the distance of the flowing blood from the center of the blood vessel, and k is a constant. Total blood flow (in millimeters per minute) is given by

$$Q(R) = \int_0^R 2\pi vr\,dr.$$

a. Find the general formula for Q in terms of R by evaluating the definite integral given above.

b. Evaluate $Q(0.4)$.

66. Rams' Horns The average annual increment in the horn length (in centimeters) of bighorn rams born since 1986 can be approximated by

$$y = 0.1762x^2 - 3.986x + 22.68,$$

where x is the ram's age (in years) for x between 3 and 9. Integrate to find the total increase in the length of a ram's horn during this time. *Source: Journal of Wildlife Management.*

67. Beagles The daily energy requirements of female beagles who are at least 1 year old change with respect to time according to the function

$$E(t) = 753t^{-0.1321},$$

where $E(t)$ is the daily energy requirement (in kJ/$W^{0.67}$), where W is the dog's weight (in kilograms) for a beagle that is t years old. *Source: Journal of Nutrition.*

a. Assuming 365 days in a year, show that the energy requirement for a female beagle that is t days old is given by

$$E(t) = 1642t^{-0.1321}.$$

b. Using the formula from part a, determine the total energy requirements (in kJ/$W^{0.67}$) for a female beagle between her first and third birthday.

68. Sediment The density of sediment (in grams per cubic centimeter) at the bottom of Lake Coeur d'Alene, Idaho, is given by

$$p(x) = p_0 e^{0.0133x},$$

where x is the depth (in centimeters) and p_0 is the density at the surface. The total mass of a square-centimeter column of sediment above a depth of h cm is given by

$$\int_0^h p(x)\,dx.$$

If $p_0 = 0.85$ g per cm^3, find the total mass above a depth of 100 cm. *Source: Mathematics Teacher.*

Social Sciences

69. Age Distribution The U.S. Census Bureau gives an age distribution that is approximately modeled (in millions) by the function

$$f(x) = 40.2 + 3.50x - 0.897x^2$$

where x varies from 0 to 9 decades. The population of a given age group can be found by integrating this function over the interval for that age group. *Source: Ralph DeMarr and the U.S. Census Bureau.*

a. Find the integral of $f(x)$ over the interval [0, 9]. What does this integral represent?

b. Baby boomers are those born between 1945 and 1965, that is, those in the range of 4.5 to 6.5 decades in 2010. Estimate the number of baby boomers.

70. Income Distribution Based on data from the U.S. Census Bureau, an approximate family income distribution for the United States is given by the function

$$f(x) = 0.0353x^3 - 0.541x^2 + 3.78x + 4.29,$$

where x is the annual income in units of $10,000, with $0 \leq x \leq 10$. For example, $x = 0.5$ represents an annual family income of $5000. (*Note*: This function does not give a good representation for family incomes over $100,000.) The percent of the families with an income in a given range can be found by integrating this function over that range. Find the percentage of families with an income between $25,000 and $50,000. *Source: Ralph DeMarr and the U.S. Census Bureau.*

Physical Sciences

71. Oil Consumption Suppose that the rate of consumption of a natural resource is $c'(t)$, where

$$c'(t) = ke^{rt}.$$

Here t is time in years, r is a constant, and k is the consumption in the year when $t = 0$. In 2010, an oil company sold 1.2 billion barrels of oil. Assume that $r = 0.04$.

a. Write $c'(t)$ for the oil company, letting $t = 0$ represent 2010.

b. Set up a definite integral for the amount of oil that the company will sell in the next 10 years.

c. Evaluate the definite integral of part b.

d. The company has about 20 billion barrels of oil in reserve. To find the number of years that this amount will last, solve the equation

$$\int_0^T 1.2e^{0.04t}dt = 20.$$

e. Rework part d, assuming that $r = 0.02$.

72. Oil Consumption The rate of consumption of oil (in billions of barrels) by the company in Exercise 71 was given as

$$1.2e^{0.04t},$$

where $t = 0$ corresponds to 2010. Find the total amount of oil used by the company from 2010 to year T. At this rate, how much will be used in 5 years?

■ YOUR TURN ANSWERS

1. 26
2. 256
3. 2 ln 3 or ln 9
4. 128/3
5. 54

7.5 The Area Between Two Curves

APPLY IT **If an executive knows how the savings from a new manufacturing process decline over time and how the costs of that process will increase, how can she compute when the net savings will cease and what the total savings will be?**
We will answer this question in Example 4.

Many important applications of integrals require finding the area between two graphs. The method used in previous sections to find the area between the graph of a function and the x-axis from $x = a$ to $x = b$ can be generalized to find such an area. For example, the area between the graphs of $f(x)$ and $g(x)$ from $x = a$ to $x = b$ in Figure 21(a) is the same as the area under the graph of $f(x)$, shown in Figure 21(b), minus the area under the graph of $g(x)$ (see Figure 21(c)). That is, the area between the graphs is given by

$$\int_a^b f(x)\, dx - \int_a^b g(x)\, dx,$$

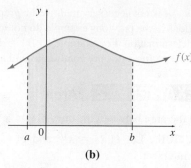

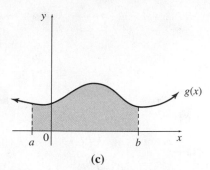

(a)   **(b)**   **(c)**

FIGURE 21

which can be written as

$$\int_a^b [f(x) - g(x)]\, dx.$$

Area Between Two Curves

If f and g are continuous functions and $f(x) \geq g(x)$ on $[a, b]$, then the area between the curves $f(x)$ and $g(x)$ from $x = a$ to $x = b$ is given by

$$\int_a^b [f(x) - g(x)]\, dx.$$

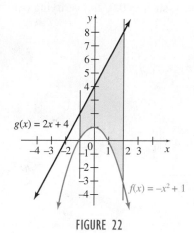

FIGURE 22

EXAMPLE 1   Area

Find the area bounded by $f(x) = -x^2 + 1$, $g(x) = 2x + 4$, $x = -1$, and $x = 2$.

SOLUTION A sketch of the four equations is shown in Figure 22. In general, it is not necessary to spend time drawing a detailed sketch, but only to know whether the two functions intersect, and which function is greater between the intersections. To find out, set the two functions equal.

$$-x^2 + 1 = 2x + 4$$
$$0 = x^2 + 2x + 3$$

Verify by the quadratic formula that this equation has no real roots. Since the graph of f is a parabola opening downward that does not cross the graph of g (a line), the parabola must be entirely under the line, as shown in Figure 22. Therefore $g(x) \geq f(x)$ for x in the interval $[-1, 2]$, and the area is given by

$$\int_{-1}^2 [g(x) - f(x)]\, dx = \int_{-1}^2 [(2x + 4) - (-x^2 + 1)]\, dx$$

$$= \int_{-1}^2 (2x + 4 + x^2 - 1)\, dx$$

$$= \int_{-1}^2 (x^2 + 2x + 3)\, dx$$

$$= \frac{x^3}{3} + x^2 + 3x \Big|_{-1}^2$$

$$= \left(\frac{8}{3} + 4 + 6\right) - \left(\frac{-1}{3} + 1 - 3\right)$$

$$= \frac{8}{3} + 10 + \frac{1}{3} + 2$$

$$= 15.$$

YOUR TURN 1 Repeat Example 1 for $f(x) = 4 - x^2$, $g(x) = x + 2$, $x = -2$, and $x = 1$.

TRY YOUR TURN 1

NOTE It is not necessary to draw the graphs to determine which function is greater. Since the functions in the previous example do not intersect, we can evaluate them at *any* point to make this determination. For example, $f(0) = 1$ and $g(0) = 4$. Because $g(x) > f(x)$ at $x = 4$, and the two functions are continuous and never intersect, $g(x) > f(x)$ for all x.

EXAMPLE 2 Area

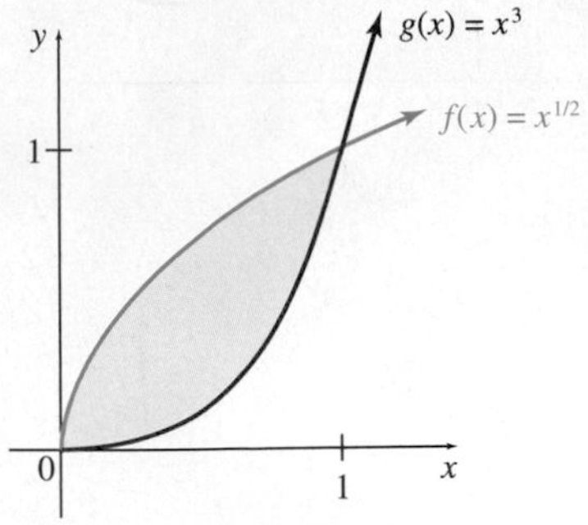

Find the area between the curves $y = x^{1/2}$ and $y = x^3$.

SOLUTION Let $f(x) = x^{1/2}$ and $g(x) = x^3$. As before, set the two equal to find where they intersect.

$$x^{1/2} = x^3$$
$$0 = x^3 - x^{1/2}$$
$$0 = x^{1/2}(x^{5/2} - 1)$$

The only solutions are $x = 0$ and $x = 1$. Verify that the graph of f is concave downward, while the graph of g is concave upward, so the graph of f must be greater between 0 and 1. (This may also be verified by taking a point between 0 and 1, such as 0.5, and verifying that $0.5^{1/2} > 0.5^3$.) The graph is shown in Figure 23.

The area between the two curves is given by

$$\int_a^b [f(x) - g(x)]\, dx = \int_0^1 (x^{1/2} - x^3)\, dx.$$

Using the Fundamental Theorem,

$$\int_0^1 (x^{1/2} - x^3)\, dx = \left(\frac{x^{3/2}}{3/2} - \frac{x^4}{4} \right)\Big|_0^1$$
$$= \left(\frac{2}{3}x^{3/2} - \frac{x^4}{4} \right)\Big|_0^1$$
$$= \frac{2}{3}(1) - \frac{1}{4}$$
$$= \frac{5}{12}.$$

YOUR TURN 2 Repeat Example 2 for $y = x^{1/4}$ and $y = x^2$.

TRY YOUR TURN 2

FIGURE 23

TECHNOLOGY NOTE A graphing calculator is very useful in approximating solutions of problems involving the area between two curves. First, it can be used to graph the functions and identify any intersection points. Then it can be used to approximate the definite integral that represents the area. (A function that gives a numerical approximation to the integral is located in the MATH menu of a TI-84 Plus calculator.) Figure 24 shows the results of using these steps for Example 2. The second window shows that the area closely approximates 5/12.

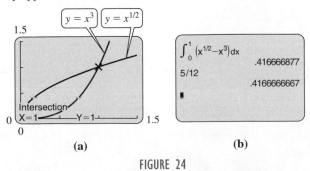

FIGURE 24

The difference between two integrals can be used to find the area between the graphs of two functions even if one graph lies below the *x*-axis. In fact, if $f(x) \geq g(x)$ for all values of x in the interval $[a, b]$, then the area between the two graphs is always given by

$$\int_a^b [f(x) - g(x)]\, dx.$$

To see this, look at the graphs in Figure 25(a), where $f(x) \geq g(x)$ for x in $[a, b]$. Suppose a constant C is added to both functions, with C large enough so that both graphs lie above the x-axis, as in Figure 25(b). The region between the graphs is not changed. By the work above, this area is given by $\int_a^b [f(x) - g(x)]\,dx$ regardless of where the graphs of $f(x)$ and $g(x)$ are located. As long as $f(x) \geq g(x)$ on $[a, b]$, then the area between the graphs from $x = a$ to $x = b$ will equal $\int_a^b [f(x) - g(x)]\,dx$.

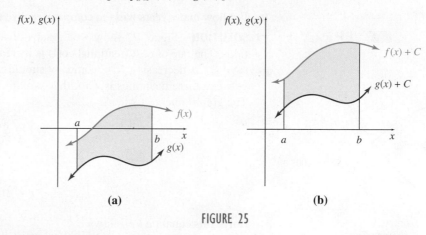

(a)   **(b)**

FIGURE 25

EXAMPLE 3   Area

Find the area of the region enclosed by $y = x^2 - 2x$ and $y = x$ on $[0, 4]$.

SOLUTION   Verify that the two graphs cross at $x = 0$ and $x = 3$. Because the first graph is a parabola opening upward, the parabola must be below the line between 0 and 3 and above the line between 3 and 4. See Figure 26. (The greater function could also be identified by checking a point between 0 and 3, such as 1, and a point between 3 and 4, such as 3.5. For each of these values of x, we could calculate the corresponding value of y for the two functions and see which is greater.) Because the graphs cross at $x = 3$, the area is found by taking the sum of two integrals as follows.

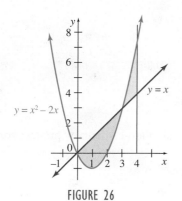

FIGURE 26

$$\text{Area} = \int_0^3 [x - (x^2 - 2x)]\,dx + \int_3^4 [(x^2 - 2x) - x]\,dx$$

$$= \int_0^3 (-x^2 + 3x)\,dx + \int_3^4 (x^2 - 3x)\,dx$$

$$= \left(\frac{-x^3}{3} + \frac{3x^2}{2} \right)\Big|_0^3 + \left(\frac{x^3}{3} - \frac{3x^2}{2} \right)\Big|_3^4$$

$$= \left(-9 + \frac{27}{2} - 0 \right) + \left(\frac{64}{3} - 24 - 9 + \frac{27}{2} \right)$$

$$= \frac{19}{3}$$

YOUR TURN 3   Repeat Example 3 for $y = x^2 - 3x$ and $y = 2x$ on $[0, 6]$.

TRY YOUR TURN 3

In the remainder of this section we will consider some typical applications that require finding the area between two curves.

EXAMPLE 4   Savings

A company is considering a new manufacturing process in one of its plants. The new process provides substantial initial savings, with the savings declining with time t (in years) according to the rate-of-savings function

$$S'(t) = 100 - t^2,$$

where $S'(t)$ is in thousands of dollars per year. At the same time, the cost of operating the new process increases with time t (in years), according to the rate-of-cost function (in thousands of dollars per year)

$$C'(t) = t^2 + \frac{14}{3}t.$$

(a) For how many years will the company realize savings?

APPLY IT

SOLUTION Figure 27 shows the graphs of the rate-of-savings and rate-of-cost functions. The rate of cost (marginal cost) is increasing, while the rate of savings (marginal savings) is decreasing. The company should use this new process until the difference between these quantities is zero; that is, until the time at which these graphs intersect. The graphs intersect when

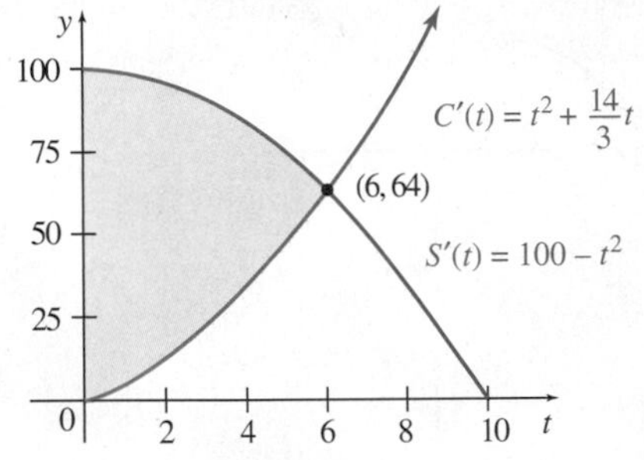

$$C'(t) = S'(t),$$

or

$$t^2 + \frac{14}{3}t = 100 - t^2.$$

Solve this equation as follows.

$$0 = 2t^2 + \frac{14}{3}t - 100$$

$$0 = 3t^2 + 7t - 150 \qquad \text{Multiply by } \tfrac{3}{2}.$$

$$= (t - 6)(3t + 25) \qquad \text{Factor.}$$

Set each factor equal to 0 separately to get

$$t = 6 \qquad \text{or} \qquad t = -25/3.$$

Only 6 is a meaningful solution here. The company should use the new process for 6 years.

(b) What will be the net total savings during this period?

SOLUTION Since the total savings over the 6-year period is given by the area under the rate-of-savings curve and the total additional cost is given by the area under the rate-of-cost curve, the net total savings over the 6-year period is given by the area between the rate-of-cost and the rate-of-savings curves and the lines $t = 0$ and $t = 6$. This area can be evaluated with a definite integral as follows.

$$\text{Total savings} = \int_0^6 \left[(100 - t^2) - \left(t^2 + \frac{14}{3}t \right) \right] dt$$

$$= \int_0^6 \left(100 - \frac{14}{3}t - 2t^2 \right) dt$$

$$= \left(100t - \frac{7}{3}t^2 - \frac{2}{3}t^3 \right) \Big|_0^6$$

$$= 100(6) - \frac{7}{3}(36) - \frac{2}{3}(216) = 372$$

The company will save a total of $372,000 over the 6-year period.

The answer to a problem will not always be an integer. Suppose in solving the quadratic equation in Example 4 we found the solutions to be $t = 6.7$ and $t = -7.3$. It may not be

$C'(t) = t^2 + \frac{14}{3}t$

$(6, 64)$

$S'(t) = 100 - t^2$

FIGURE 27

realistic to use a new process for 6.7 years; it may be necessary to choose between 6 years and 7 years. Since the mathematical model produces a result that is not in the domain of the function in this case, it is necessary to find the total savings after 6 years and after 7 years and then select the best result.

Consumers' Surplus

The market determines the price at which a product is sold. As indicated earlier, the point of intersection of the demand curve and the supply curve for a product gives the equilibrium price. At the equilibrium price, consumers will purchase the same amount of the product that the manufacturers want to sell. Some consumers, however, would be willing to spend more for an item than the equilibrium price. The total of the differences between the equilibrium price of the item and the higher prices that individuals would be willing to pay is thought of as savings realized by those individuals and is called the **consumers' surplus**.

To calculate the total amount that consumers would be willing to pay for q_0 items, first consider the simple case in which everyone is willing to pay exactly p_0, the equilibrium price. Then the total amount everyone would pay would be the price times the quantity, or $p_0 q_0$, which is the green area in Figure 28. In fact, this is the exact total that everyone together pays when the item sells for p_0. Now divide the interval from 0 to q_0 into n intervals, each of width $\Delta q = q_0/n$. Each interval represents Δq people. Specifically, we will assume that the people represented by the ith interval, where i is a number between 1 and n, are those who are willing to pay a price $p_i = D(q_i)$ for the item, where q_i is some quantity on that interval. Then the total amount those people would be willing to pay would be $D(q_i)\Delta q$. The total amount that everyone together would be willing to pay is

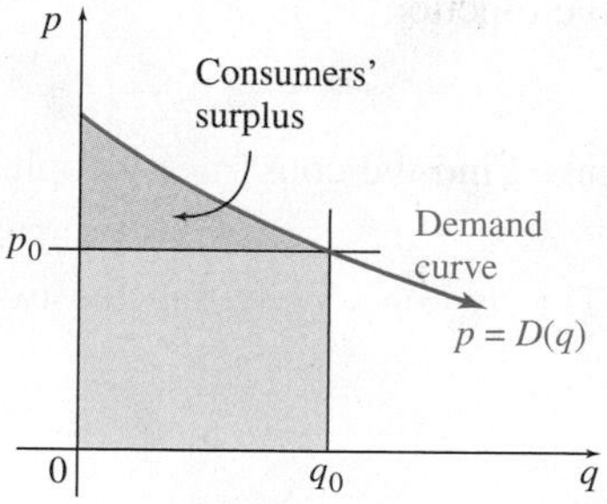

$$\sum_{i=1}^{n} D(q_i)\Delta q.$$

In a more realistic situation, the demand curve changes continuously, so we find the total amount that everyone would be willing to pay by taking the limit as n goes to infinity:

$$\lim_{n \to \infty} \sum_{i=1}^{n} D(q_i)\Delta q = \int_{0}^{q_0} D(q) \, dq.$$

This quantity, which represents the total amount consumers are willing to spend for q_0 items, is the area under the demand curve in Figure 28. The pink shaded area represents the difference between what consumers would be willing to pay and what they actually pay, or the consumers' surplus.

As the figure suggests, the consumers' surplus is given by an area between the two curves $p = D(q)$ and $p = p_0$, so its value can be found with a definite integral as follows.

Consumers' Surplus
If $D(q)$ is a demand function with equilibrium price p_0 and equilibrium demand q_0, then

$$\textbf{Consumers' surplus} = \int_{0}^{q_0} [D(q) - p_0] \, dq.$$

Similarly, if some manufacturers would be willing to supply a product at a price *lower* than the equilibrium price p_0, the total of the differences between the equilibrium price and the lower prices at which the manufacturers would sell the product is considered added income for the manufacturers and is called the **producers' surplus**. Figure 29 shows the

FIGURE 28

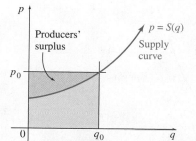

FIGURE 29

(green shaded) total area under the supply curve from $q = 0$ to $q = q_0$, which is the minimum total amount the manufacturers are willing to realize from the sale of q_0 items. The total area under the line $p = p_0$ is the amount actually realized. The difference between these two areas, the producers' surplus, is also given by a definite integral.

Producers' Surplus

If $S(q)$ is a supply function with equilibrium price p_0 and equilibrium supply q_0, then

$$\text{Producers' surplus} = \int_0^{q_0} [p_0 - S(q)] \, dq.$$

EXAMPLE 5   Consumers' and Producers' Surplus

Suppose the price (in dollars per ton) for oat bran is

$$D(q) = 400 - e^{q/2},$$

when the demand for the product is q tons. Also, suppose the function

$$S(q) = e^{q/2} - 1$$

gives the price (in dollars per ton) when the supply is q tons. Find the consumers' surplus and the producers' surplus.

SOLUTION   Begin by finding the equilibrium quantity. This is done by setting the two equations equal.

$$e^{q/2} - 1 = 400 - e^{q/2}$$

$$2e^{q/2} = 401$$

$$e^{q/2} = \frac{401}{2}$$

$$q/2 = \ln\left(\frac{401}{2}\right)$$

$$q = 2\ln\left(\frac{401}{2}\right) \approx 10.60163$$

The result can be further rounded to 10.60 tons as long as this rounded value is not used in future calculations. At the equilibrium point where the supply and demand are both 10.60 tons, the price is

$$S(10.60163) = e^{10.60163/2} - 1 \approx 199.50,$$

or \$199.50. Verify that this same answer is found by computing $D(10.60163)$. The consumers' surplus, represented by the area shown in Figure 30, is

$$\int_0^{10.60163} [(400 - e^{q/2}) - 199.50] \, dq = \int_0^{10.60163} [200.5 - e^{q/2}] \, dq.$$

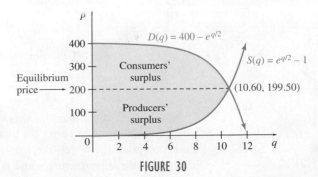

FIGURE 30

Evaluating the definite integral gives

$$(200.5q - 2e^{q/2})\Big|_0^{10.60163} = [200.5(10.60163) - 2e^{10.60163/2}] - (0 - 2)$$
$$\approx 1726.63.$$

Here the consumers' surplus is $1726.63. The producers' surplus, also shown in Figure 30, is given by

YOUR TURN 4 Repeat Example 5 when $D(q) = 600 - e^{q/3}$ and $S(q) = e^{q/3} - 100$.

$$\int_0^{10.60163} [199.50 - (e^{q/2} - 1)]\, dq = \int_0^{10.60163} [200.5 - e^{q/2}]\, dq,$$

which is exactly the same as the expression found for the consumers' surplus, so the producers' surplus is also $1726.63.   **TRY YOUR TURN 4**

NOTE In general, the producers' surplus and consumers' surplus are not the same, as they are in Example 5.

7.5 EXERCISES

Find the area between the curves in Exercises 1–24.

1. $x = -2,\quad x = 1,\quad y = 2x^2 + 5,\quad y = 0$

2. $x = 1,\quad x = 2,\quad y = 3x^3 + 2,\quad y = 0$

3. $x = -3,\quad x = 1,\quad y = x^3 + 1,\quad y = 0$

4. $x = -3,\quad x = 0,\quad y = 1 - x^2,\quad y = 0$

5. $x = -2,\quad x = 1,\quad y = 2x,\quad y = x^2 - 3$

6. $x = 0,\quad x = 6,\quad y = 5x,\quad y = 3x + 10$

7. $y = x^2 - 30,\quad y = 10 - 3x$

8. $y = x^2 - 18,\quad y = x - 6$

9. $y = x^2,\quad y = 2x$

10. $y = x^2,\quad y = x^3$

11. $x = 1,\quad x = 6,\quad y = \dfrac{1}{x},\quad y = \dfrac{1}{2}$

12. $x = 0,\quad x = 4,\quad y = \dfrac{1}{x + 1},\quad y = \dfrac{x - 1}{2}$

13. $x = -1,\quad x = 1,\quad y = e^x,\quad y = 3 - e^x$

14. $x = -1,\quad x = 2,\quad y = e^{-x},\quad y = e^x$

15. $x = -1,\quad x = 2,\quad y = 2e^{2x},\quad y = e^{2x} + 1$

16. $x = 2,\quad x = 4,\quad y = \dfrac{x - 1}{4},\quad y = \dfrac{1}{x - 1}$

17. $y = x^3 - x^2 + x + 1,\quad y = 2x^2 - x + 1$

18. $y = 2x^3 + x^2 + x + 5,\quad y = x^3 + x^2 + 2x + 5$

19. $y = x^4 + \ln(x + 10),\quad y = x^3 + \ln(x + 10)$

20. $y = x^5 - 2\ln(x + 5),\quad y = x^3 - 2\ln(x + 5)$

21. $y = x^{4/3},\quad y = 2x^{1/3}$

22. $y = \sqrt{x},\quad y = x\sqrt{x}$

23. $x = 0,\quad x = 3,\quad y = 2e^{3x},\quad y = e^{3x} + e^6$

24. $x = 0,\quad x = 3,\quad y = e^x,\quad y = e^{4-x}$

In Exercises 25 and 26, use a graphing calculator to find the values of x where the curves intersect and then to find the area between the two curves.

25. $y = e^x,\quad y = -x^2 - 2x$

26. $y = \ln x,\quad y = x^3 - 5x^2 + 6x - 1$

APPLICATIONS

Business and Economics

27. Net Savings Suppose a company wants to introduce a new machine that will produce a rate of annual savings (in dollars) given by

$$S'(x) = 150 - x^2,$$

where x is the number of years of operation of the machine, while producing a rate of annual costs (in dollars) of

$$C'(x) = x^2 + \frac{11}{4}x.$$

a. For how many years will it be profitable to use this new machine?

b. What are the net total savings during the first year of use of the machine?

c. What are the net total savings over the entire period of use of the machine?

28. Net Savings A new smog-control device will reduce the output of sulfur oxides from automobile exhausts. It is estimated that the rate of savings (in millions of dollars per year) to the community from the use of this device will be approximated by

$$S'(x) = -x^2 + 4x + 8,$$

after x years of use of the device. The new device cuts down on the production of sulfur oxides, but it causes an increase in the production of nitrous oxides. The rate of additional costs (in millions of dollars per year) to the community after x years is approximated by

$$C'(x) = \frac{3}{25}x^2.$$

a. For how many years will it pay to use the new device?

b. What will be the net savings over this period of time?

29. Profit Canham Enterprises had an expenditure rate of $E'(x) = e^{0.1x}$ dollars per day and an income rate of $I'(x) = 98.8 - e^{0.1x}$ dollars per day on a particular job, where x was the number of days from the start of the job. The company's profit on that job will equal total income less total expenditures. Profit will be maximized if the job ends at the optimum time, which is the point where the two curves meet. Find the following.

a. The optimum number of days for the job to last

b. The total income for the optimum number of days

c. The total expenditures for the optimum number of days

d. The maximum profit for the job

30. Net Savings A factory of Hollis Sherman Industries has installed a new process that will produce an increased rate of revenue (in thousands of dollars per year) of

$$R'(t) = 104 - 0.4e^{t/2},$$

where t is time measured in years. The new process produces additional costs (in thousands of dollars per year) at the rate of

$$C'(t) = 0.3e^{t/2}.$$

a. When will it no longer be profitable to use this new process?

b. Find the net total savings.

31. Producers' Surplus Find the producers' surplus if the supply function for pork bellies is given by

$$S(q) = q^{5/2} + 2q^{3/2} + 50.$$

Assume supply and demand are in equilibrium at $q = 16$.

32. Producers' Surplus Suppose the supply function for concrete is given by

$$S(q) = 100 + 3q^{3/2} + q^{5/2},$$

and that supply and demand are in equilibrium at $q = 9$. Find the producers' surplus.

33. Consumers' Surplus Find the consumers' surplus if the demand function for grass seed is given by

$$D(q) = \frac{200}{(3q + 1)^2},$$

assuming supply and demand are in equilibrium at $q = 3$.

34. Consumers' Surplus Find the consumers' surplus if the demand function for extra virgin olive oil is given by

$$D(q) = \frac{32,000}{(2q + 8)^3},$$

and if supply and demand are in equilibrium at $q = 6$.

35. Consumers' and Producers' Surplus Suppose the supply function for oil is given (in dollars) by

$$S(q) = q^2 + 10q,$$

and the demand function is given (in dollars) by

$$D(q) = 900 - 20q - q^2.$$

a. Graph the supply and demand curves.

b. Find the point at which supply and demand are in equilibrium.

c. Find the consumers' surplus.

d. Find the producers' surplus.

36. Consumers' and Producers' Surplus Suppose the supply function for a certain item is given by

$$S(q) = (q + 1)^2,$$

and the demand function is given by

$$D(q) = \frac{1000}{q + 1}.$$

a. Graph the supply and demand curves.

b. Find the point at which supply and demand are in equilibrium.

c. Find the consumers' surplus.

d. Find the producers' surplus.

37. Consumers' and Producers' Surplus Suppose that with the supply and demand for oil as in Exercise 35, the government sets the price at $264 per unit.

a. Use the supply function to calculate the quantity that will be produced at the new price.

b. Find the consumers' surplus for the new price, using the quantity found in part a in place of the equilibrium quantity. How much larger is this than the consumers' surplus in Exercise 35?

c. Find the producers' surplus for the new price, using the quantity found in part a in place of the equilibrium quantity. How much smaller is this than the producers' surplus in Exercise 35?

d. Calculate the difference between the total of the consumers' and producers' surplus under the equilibrium price and under the government price. Economists refer to this loss as the *welfare cost* of the government's setting the price.

e. Because of the welfare cost calculated in part d, many economists argue that it is bad economics for the government to set prices. Others point to the increase in the consumers' surplus, calculated in part b, as a justification for such government action. Discuss the pros and cons of this issue.

38. Fuel Economy In an article in the December 1994 *Scientific American* magazine, the authors estimated future gas use. Without a change in U.S. policy, auto fuel use is forecasted to rise along the projection shown at the right in the figure below. The shaded band predicts gas use if the technologies for increased fuel economy are phased in by the year 2010. The moderate estimate (center curve) corresponds to an average of 46 mpg for all cars on the road. *Source: Scientific American.*

 a. Discuss the interpretation of the shaded area and other regions of the graph that pertain to the topic in this section.

 b. According to the Energy Information Administration, the U.S. gasoline consumption in 2010 was 9,030,000 barrels per day. Discuss how this affects the areas considered in part a. *Source: U.S. Energy Information Administration.*

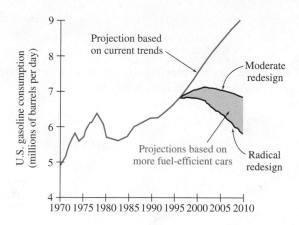

Life Sciences

39. Pollution Pollution begins to enter a lake at time $t = 0$ at a rate (in gallons per hour) given by the formula

$$f(t) = 10(1 - e^{-0.5t}),$$

where t is the time (in hours). At the same time, a pollution filter begins to remove the pollution at a rate

$$g(t) = 0.4t$$

as long as pollution remains in the lake.

a. How much pollution is in the lake after 12 hours?

b. Use a graphing calculator to find the time when the rate that pollution enters the lake equals the rate the pollution is removed.

c. Find the amount of pollution in the lake at the time found in part b.

d. Use a graphing calculator to find the time when all the pollution has been removed from the lake.

40. Pollution Repeat the steps of Exercise 39, using the functions

$$f(t) = 15(1 - e^{-0.05t})$$

and

$$g(t) = 0.3t.$$

Social Sciences

41. Distribution of Income Suppose that all the people in a country are ranked according to their incomes, starting at the bottom. Let x represent the fraction of the community making the lowest income $(0 \le x \le 1)$; $x = 0.4$, therefore, represents the lower 40% of all income producers. Let $I(x)$ represent the proportion of the total income earned by the lowest x of all people. Thus, $I(0.4)$ represents the fraction of total income earned by the lowest 40% of the population. The curve described by this function is known as a *Lorenz curve*. Suppose

$$I(x) = 0.9x^2 + 0.1x.$$

Find and interpret the following.

a. $I(0.1)$ **b.** $I(0.4)$

If income were distributed uniformly, we would have $I(x) = x$. The area under this line of complete equality is $1/2$. As $I(x)$ dips further below $y = x$, there is less equality of income distribution. This inequality can be quantified by the ratio of the area between $I(x)$ and $y = x$ to $1/2$. This ratio is called the *Gini index of income inequality* and equals $2 \int_0^1 [x - I(x)]\, dx$.

c. Graph $I(x) = x$ and $I(x) = 0.9x^2 + 0.1x$, for $0 \le x \le 1$, on the same axes.

d. Find the area between the curves.

 e. For U.S. families, the Gini index was 0.386 in 1968 and 0.466 in 2008. Describe how the distribution of family incomes has changed over this time. *Source: U.S. Census.*

Physical Sciences

42. Metal Plate A worker sketches the curves $y = \sqrt{x}$ and $y = x/2$ on a sheet of metal and cuts out the region between the curves to form a metal plate. Find the area of the plate.

▬ YOUR TURN ANSWERS

1. 9/2   **2.** 7/15
3. 71/3   **4.** $5103.83; $5103.83

7.6 Numerical Integration

APPLY IT   If the velocity of a vehicle is known only at certain points in time, how can the total distance traveled by the vehicle be estimated?
Using numerical integration, we will answer this question in Example 3 of this section.

Some integrals cannot be evaluated by any technique. One solution to this problem was presented in Section 3 of this chapter, in which the area under a curve was approximated by summing the areas of rectangles. This method is seldom used in practice because better methods exist that are more accurate for the same amount of work. These methods are referred to as **numerical integration** methods. We discuss two such methods here: the trapezoidal rule and Simpson's rule.

Trapezoidal Rule
Recall, the trapezoidal rule was mentioned briefly in Section 3, where we found approximations with it by averaging the sums of rectangles found by using left endpoints and then using right endpoints. In this section we derive an explicit formula for the trapezoidal rule in terms of function values.* To illustrate the derivation of the trapezoidal rule, consider the integral

$$\int_1^5 \frac{1}{x}\, dx.$$

The shaded region in Figure 31 shows the area representing that integral, the area under the graph $f(x) = 1/x$, above the x-axis, and between the lines $x = 1$ and $x = 5$.

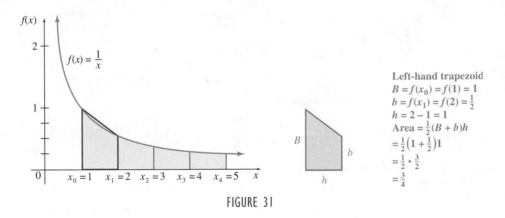

FIGURE 31

Note that this function can be integrated using the Fundamental Theorem of Calculus. Since $\int (1/x)\, dx = \ln |x| + C$,

$$\int_1^5 \frac{1}{x}\, dx = \ln |x| \Big|_1^5 = \ln 5 - \ln 1 = \ln 5 - 0 = \ln 5 \approx 1.609438.$$

We can also approximate the integral using numerical integration. As shown in the figure, if the area under the curve is approximated with trapezoids rather than rectangles, the approximation should be improved.

As in earlier work, to approximate this area we divide the interval $[1, 5]$ into subintervals of equal widths. To get a first approximation to $\ln 5$ by the trapezoidal rule, find the sum of the

---

*In American English a trapezoid is a four-sided figure with two parallel sides, contrasted with a trapezium, which has no parallel sides. In British English, however, it is just the opposite. What Americans call a trapezoid is called a trapezium in Great Britain.

Approximations to $\int_1^5 \frac{1}{x} dx$	
n	Trapezoidal Approximation
6	1.64360
8	1.62897
10	1.62204
20	1.61262
100	1.60957
1000	1.60944

areas of the four trapezoids shown in Figure 31. From geometry, the area of a trapezoid is half the product of the sum of the bases and the altitude. Each of the trapezoids in Figure 31 has altitude 1. (In this case, the bases of the trapezoid are vertical and the altitudes are horizontal.) Adding the areas gives

$$\ln 5 = \int_1^5 \frac{1}{x} dx \approx \frac{1}{2}\left(\frac{1}{1} + \frac{1}{2}\right)(1) + \frac{1}{2}\left(\frac{1}{2} + \frac{1}{3}\right)(1) + \frac{1}{2}\left(\frac{1}{3} + \frac{1}{4}\right)(1) + \frac{1}{2}\left(\frac{1}{4} + \frac{1}{5}\right)(1)$$

$$= \frac{1}{2}\left(\frac{3}{2} + \frac{5}{6} + \frac{7}{12} + \frac{9}{20}\right) \approx 1.68333.$$

To get a better approximation, divide the interval $[1, 5]$ into more subintervals. Generally speaking, the larger the number of subintervals, the better the approximation. The results for selected values of n are shown to 5 decimal places. When $n = 1000$, the approximation agrees with the true value of $\ln 5 \approx 1.609438$ to 5 decimal places.

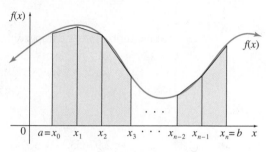

FIGURE 32

Generalizing from this example, let f be a continuous function on an interval $[a, b]$. Divide the interval from a to b into n equal subintervals by the points $a = x_0, x_1, x_2, \ldots, x_n = b$, as shown in Figure 32. Use the subintervals to make trapezoids that approximately fill in the region under the curve. The approximate value of the definite integral $\int_a^b f(x)\,dx$ is given by the sum of the areas of the trapezoids, or

$$\int_a^b f(x)\,dx \approx \frac{1}{2}[f(x_0) + f(x_1)]\left(\frac{b-a}{n}\right) + \frac{1}{2}[f(x_1) + f(x_2)]\left(\frac{b-a}{n}\right)$$

$$+ \cdots + \frac{1}{2}[f(x_{n-1}) + f(x_n)]\left(\frac{b-a}{n}\right)$$

$$= \left(\frac{b-a}{n}\right)\left[\frac{1}{2}f(x_0) + \frac{1}{2}f(x_1) + \frac{1}{2}f(x_1) + \frac{1}{2}f(x_2) + \frac{1}{2}f(x_2) + \cdots + \frac{1}{2}f(x_{n-1}) + \frac{1}{2}f(x_n)\right]$$

$$= \left(\frac{b-a}{n}\right)\left[\frac{1}{2}f(x_0) + f(x_1) + f(x_2) + \cdots + f(x_{n-1}) + \frac{1}{2}f(x_n)\right].$$

This result gives the following rule.

Trapezoidal Rule

Let f be a continuous function on $[a, b]$ and let $[a, b]$ be divided into n equal subintervals by the points $a = x_0, x_1, x_2, \ldots, x_n = b$. Then, by the **trapezoidal rule**,

$$\int_a^b f(x)\,dx \approx \left(\frac{b-a}{n}\right)\left[\frac{1}{2}f(x_0) + f(x_1) + \cdots + f(x_{n-1}) + \frac{1}{2}f(x_n)\right].$$

EXAMPLE 1 **Trapezoidal Rule**

Use the trapezoidal rule with $n = 4$ to approximate

$$\int_0^2 \sqrt{x^2 + 1}\,dx.$$

SOLUTION

Method 1
Calculating by Hand

Here $a = 0$, $b = 2$, and $n = 4$, with $(b - a)/n = (2 - 0)/4 = 1/2$ as the altitude of each trapezoid. Then $x_0 = 0$, $x_1 = 1/2$, $x_2 = 1$, $x_3 = 3/2$, and $x_4 = 2$. Now find the corresponding function values. The work can be organized into a table, as follows.

Calculations for Trapezoidal Rule		
i	x_i	$f(x_i)$
0	0	$\sqrt{0^2 + 1} = 1$
1	1/2	$\sqrt{(1/2)^2 + 1} \approx 1.11803$
2	1	$\sqrt{1^2 + 1} \approx 1.41421$
3	3/2	$\sqrt{(3/2)^2 + 1} \approx 1.80278$
4	2	$\sqrt{2^2 + 1} \approx 2.23607$

Substitution into the trapezoidal rule gives

$$\int_0^2 \sqrt{x^2 + 1}\, dx$$

$$\approx \frac{2 - 0}{4} \left[\frac{1}{2}(1) + 1.11803 + 1.41421 + 1.80278 + \frac{1}{2}(2.23607) \right]$$

$$\approx 2.97653.$$

The approximation 2.97653 found above using the trapezoidal rule with $n = 4$ differs from the true value of 2.95789 by 0.01864. As mentioned above, this error would be reduced if larger values were used for n. For example, if $n = 8$, the trapezoidal rule gives an answer of 2.96254, which differs from the true value by 0.00465. Techniques for estimating such errors are considered in more advanced courses.

Method 2
Graphing Calculator

Just as we used a graphing calculator to approximate area using rectangles, we can also use it for the trapezoidal rule. As before, put the values of i in L_1 and the values of x_i in L_2. In the heading for L_3, put $\sqrt{(L_2^2 + 1)}$. Using the fact that $(b - a)/n = (2 - 0)/4 = 0.5$, the command $.5*(.5*L_3(1)+\text{sum}(L_3,2,4)+.5*L_3(5))$ gives the result 2.976528589. For more details, see the *Graphing Calculator and Excel Spreadsheet Manual* available with this book.

Method 3
Spreadsheet

YOUR TURN 1 Use the trapezoidal rule with $n = 4$ to approximate $\int_1^3 \sqrt{x^2 + 3}\,dx$.

The trapezoidal rule can also be done on a spreadsheet. In Microsoft Excel, for example, store the values of 0 through n in column A. After putting the left endpoint in E1 and Δx in E2, put the command "$=\$E\$1+A1*\$E\$2$" into B1; copying this formula into the rest of column B gives the values of x_i. Similarly, use the formula for $f(x_i)$ to fill column C. Using the fact that $n = 5$ in this example, the command "$\$E\$2*(.5*C1+\text{sum}(C2:C4)+.5*C5)$" gives the result 2.976529. For more details, see the *Graphing Calculator and Excel Spreadsheet Manual* available with this book.

TRY YOUR TURN 1

The trapezoidal rule is not widely used because its results are not very accurate. In fact, the midpoint rule discussed earlier in this chapter is usually more accurate than the trapezoidal rule. We will now consider a method that usually gives more accurate results than either the trapezoidal or midpoint rule.

Simpson's Rule

Another numerical method, *Simpson's rule*, approximates consecutive portions of the curve with portions of parabolas rather than the line segments of the trapezoidal rule. Simpson's rule usually gives a better approximation than the trapezoidal rule for the same number of subintervals. As shown in Figure 33 on the next page, one

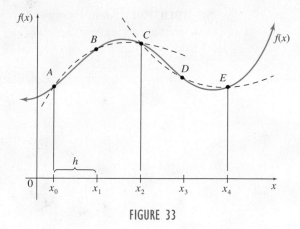

FIGURE 33

parabola is fitted through points A, B, and C, another through C, D, and E, and so on. Then the sum of the areas under these parabolas will approximate the area under the graph of the function. Because of the way the parabolas overlap, it is necessary to have an even number of intervals, and therefore an odd number of points, to apply Simpson's rule.

If h, the length of each subinterval, is $(b - a)/n$, the area under the parabola through points A, B, and C can be found by a definite integral. The details are omitted; the result is

$$\frac{h}{3}[f(x_0) + 4f(x_1) + f(x_2)].$$

Similarly, the area under the parabola through points C, D, and E is

$$\frac{h}{3}[f(x_2) + 4f(x_3) + f(x_4)].$$

When these expressions are added, the last term of one expression equals the first term of the next. For example, the sum of the two areas given above is

$$\frac{h}{3}[f(x_0) + 4f(x_1) + 2f(x_2) + 4f(x_3) + f(x_4)].$$

This illustrates the origin of the pattern of the terms in the following rule.

Simpson's Rule

Let f be a continuous function on $[a, b]$ and let $[a, b]$ be divided into an even number n of equal subintervals by the points $a = x_0, x_1, x_2, \ldots, x_n = b$. Then by **Simpson's rule**,

$$\int_a^b f(x)\, dx \approx \left(\frac{b - a}{3n}\right)[f(x_0) + 4f(x_1) + 2f(x_2) + 4f(x_3) + \cdots$$
$$+ 2f(x_{n-2}) + 4f(x_{n-1}) + f(x_n)].$$

Thomas Simpson (1710–1761), a British mathematician, wrote texts on many branches of mathematics. Some of these texts went through as many as ten editions. His name became attached to this numerical method of approximating definite integrals even though the method preceded his work.

CAUTION   In Simpson's rule, n (the number of subintervals) must be even.

EXAMPLE 2   Simpson's Rule

Use Simpson's rule with $n = 4$ to approximate

$$\int_0^2 \sqrt{x^2 + 1}\, dx,$$

which was approximated by the trapezoidal rule in Example 1.

SOLUTION As in Example 1, $a = 0$, $b = 2$, and $n = 4$, and the endpoints of the four intervals are $x_0 = 0$, $x_1 = 1/2$, $x_2 = 1$, $x_3 = 3/2$, and $x_4 = 2$. The table of values is also the same.

	Calculations for Simpson's Rule	
i	x_i	$f(x_i)$
0	0	1
1	1/2	1.11803
2	1	1.41421
3	3/2	1.80278
4	2	2.23607

YOUR TURN 2 Use Simpson's rule with $n = 4$ to approximate $\int_1^3 \sqrt{x^2 + 3}\,dx$.

Since $(b - a)/(3n) = 2/12 = 1/6$, substituting into Simpson's rule gives

$$\int_0^2 \sqrt{x^2 + 1}\,dx \approx \frac{1}{6}[1 + 4(1.11803) + 2(1.41421) + 4(1.80278) + 2.23607] \approx 2.95796.$$

This differs from the true value by 0.00007, which is less than the trapezoidal rule with $n = 8$. If $n = 8$ for Simpson's rule, the approximation is 2.95788, which differs from the true value by only 0.00001. **TRY YOUR TURN 2**

NOTE

1. Just as we can use a graphing calculator or a spreadsheet for the trapezoidal rule, we can also use such technology for Simpson's rule. For more details, see the *Graphing Calculator and Excel Spreadsheet Manual* available with this book.

2. Let M represent the midpoint rule approximation and T the trapezoidal rule approximation, using n subintervals in each. Then the formula $S = (2M + T)/3$ gives the Simpson's rule approximation with $2n$ subintervals.

Numerical methods make it possible to approximate

$$\int_a^b f(x)\,dx$$

even when $f(x)$ is not known. The next example shows how this is done.

EXAMPLE 3 Total Distance

As mentioned earlier, the velocity $v(t)$ gives the rate of change of distance $s(t)$ with respect to time t. Suppose a vehicle travels an unknown distance. The passengers keep track of the velocity at 10-minute intervals (every $1/6$ of an hour) with the following results.

Velocity of a Vehicle							
Time in Hours, t	1/6	2/6	3/6	4/6	5/6	1	7/6
Velocity in Miles per Hour, $v(t)$	45	55	52	60	64	58	47

What is the total distance traveled in the 60-minute period from $t = 1/6$ to $t = 7/6$?

APPLY IT

SOLUTION The distance traveled in t hours is $s(t)$, with $s'(t) = v(t)$. The total distance traveled between $t = 1/6$ and $t = 7/6$ is given by

$$\int_{1/6}^{7/6} v(t)\,dt.$$

Even though this integral cannot be evaluated since we do not have an expression for $v(t)$, either the trapezoidal rule or Simpson's rule can be used to approximate its value and give the total distance traveled. In either case, let $n = 6$, $a = t_0 = 1/6$, and $b = t_6 = 7/6$. By the trapezoidal rule,

$$\int_{1/6}^{7/6} v(t) \, dt \approx \frac{7/6 - 1/6}{6} \left[\frac{1}{2}(45) + 55 + 52 + 60 + 64 + 58 + \frac{1}{2}(47) \right]$$

$$\approx 55.83.$$

By Simpson's rule,

$$\int_{1/6}^{7/6} v(t) \, dt \approx \frac{7/6 - 1/6}{3(6)} [45 + 4(55) + 2(52) + 4(60) + 2(64) + 4(58) + 47]$$

$$= \frac{1}{18}(45 + 220 + 104 + 240 + 128 + 232 + 47) \approx 56.44.$$

The distance traveled in the 1-hour period was about 56 miles. ▄

As already mentioned, Simpson's rule generally gives a better approximation than the trapezoidal rule. As n increases, the two approximations get closer and closer. For the same accuracy, however, a smaller value of n generally can be used with Simpson's rule so that less computation is necessary. Simpson's rule is the method used by many calculators that have a built-in integration feature.

The branch of mathematics that studies methods of approximating definite integrals (as well as many other topics) is called *numerical* analysis. Numerical integration is useful even with functions whose antiderivatives can be determined if the antidifferentiation is complicated and a computer or calculator programmed with Simpson's rule is handy. You may want to program your calculator for both the trapezoidal rule and Simpson's rule. For some calculators, these programs are in the *Graphing Calculator and Excel Spreadsheet Manual* available with this book.

7.6   EXERCISES

In Exercises 1–10, use $n = 4$ to approximate the value of the given integrals by the following methods: (a) the trapezoidal rule, and (b) Simpson's rule. (c) Find the exact value by integration.

1. $\int_0^2 (3x^2 + 2) \, dx$

2. $\int_0^2 (2x^2 + 1) \, dx$

3. $\int_{-1}^3 \frac{3}{5 - x} \, dx$

4. $\int_1^5 \frac{6}{2x + 1} \, dx$

5. $\int_{-1}^2 (2x^3 + 1) \, dx$

6. $\int_0^3 (2x^3 + 1) \, dx$

7. $\int_1^5 \frac{1}{x^2} \, dx$

8. $\int_2^4 \frac{1}{x^3} \, dx$

9. $\int_0^1 4xe^{-x^2} \, dx$

10. $\int_0^4 x\sqrt{2x^2 + 1} \, dx$

11. Find the area under the semicircle $y = \sqrt{4 - x^2}$ and above the x-axis by using $n = 8$ with the following methods.

    a. The trapezoidal rule   **b.** Simpson's rule

    c. Compare the results with the area found by the formula for the area of a circle. Which of the two approximation techniques was more accurate?

12. Find the area between the x-axis and the upper half of the ellipse $4x^2 + 9y^2 = 36$ by using $n = 12$ with the following methods.

    a. The trapezoidal rule   **b.** Simpson's rule

    (*Hint:* Solve the equation for y and find the area of the semiellipse.)

    c. Compare the results with the actual area, $3\pi \approx 9.4248$ (which can be found by methods not considered in this text). Which approximation technique was more accurate?

13. Suppose that $f(x) > 0$ and $f''(x) > 0$ for all x between a and b, where $a < b$. Which of the following cases is true of a trapezoidal approximation T for the integral $\int_a^b f(x)\,dx$? Explain.

    a. $T < \displaystyle\int_a^b f(x)dx$ **b.** $T > \displaystyle\int_a^b f(x)dx$

    c. Can't say which is larger

14. Refer to Exercise 13. Which of the three cases applies to these functions?

    a. $f(x) = x^2;$ $[0, 3]$ **b.** $f(x) = \sqrt{x};$ $[0, 9]$

    c.

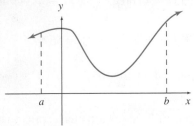

Exercises 15–18 require both the trapezoidal rule and Simpson's rule. They can be worked without calculator programs if such programs are not available, although they require more calculation than the other problems in this exercise set.

Error Analysis The difference between the true value of an integral and the value given by the trapezoidal rule or Simpson's rule is known as the error. In numerical analysis, the error is studied to determine how large n must be for the error to be smaller than some specified amount. For both rules, the error is inversely proportional to a power of n, the number of subdivisions. In other words, the error is roughly k/n^p, where k is a constant that depends on the function and the interval, and p is a power that depends only on the method used. With a little experimentation, you can find out what the power p is for the trapezoidal rule and for Simpson's rule.

15. **a.** Find the exact value of $\int_0^1 x^4\,dx$.

    b. Approximate the integral in part a using the trapezoidal rule with $n = 4, 8, 16,$ and 32. For each of these answers, find the absolute value of the error by subtracting the trapezoidal rule answer from the exact answer found in part a.

    c. If the error is k/n^p, then the error times n^p should be approximately a constant. Multiply the errors in part b times n^p for $p = 1, 2,$ etc., until you find a power p yielding the same answer for all four values of n.

16. Based on the results of Exercise 15, what happens to the error in the trapezoidal rule when the number of intervals is doubled?

17. Repeat Exercise 15 using Simpson's rule.

18. Based on the results of Exercise 17, what happens to the error in Simpson's rule when the number of intervals is doubled?

19. For the integral in Exercise 7, apply the midpoint rule with $n = 4$ and Simpson's rule with $n = 8$ to verify the formula $S = (2M + T)/3$.

20. Repeat the instructions of Exercise 19 using the integral in Exercise 8.

APPLICATIONS

Business and Economics

21. **Total Sales** A sales manager presented the following results at a sales meeting.

Year, x	1	2	3	4	5	6	7
Rate of Sales, $f(x)$	0.4	0.6	0.9	1.1	1.3	1.4	1.6

Find the total sales over the given period as follows.

    a. Plot these points. Connect the points with line segments.

    b. Use the trapezoidal rule to find the area bounded by the broken line of part a, the x-axis, the line $x = 1$, and the line $x = 7$.

    c. Approximate the same area using Simpson's rule.

22. **Total Cost** A company's marginal costs (in hundreds of dollars per year) were as follows over a certain period.

Year, x	1	2	3	4	5	6	7
Marginal Cost, $f(x)$	9.0	9.2	9.5	9.4	9.8	10.1	10.5

Repeat parts a–c of Exercise 21 for these data to find the total cost over the given period.

Life Sciences

23. **Drug Reaction Rate** The reaction rate to a new drug is given by

$$y = e^{-t^2} + \frac{1}{t + 1},$$

where t is time (in hours) after the drug is administered. Find the total reaction to the drug from $t = 1$ to $t = 9$ by letting $n = 8$ and using the following methods.

    a. The trapezoidal rule   **b.** Simpson's rule

24. **Growth Rate** The growth rate of a certain tree (in feet) is given by

$$y = \frac{2}{t + 2} + e^{-t^2/2},$$

where t is time (in years). Find the total growth from $t = 1$ to $t = 7$ by using $n = 12$ with the following methods.

    a. The trapezoidal rule   **b.** Simpson's rule

Blood Level Curves In the study of bioavailability in pharmacy, a drug is given to a patient. The level of concentration of the drug is then measured periodically, producing blood level curves such as the ones shown in the figure.

The areas under the curves give the total amount of the drug available to the patient for each milliliter of blood. Use the trapezoidal rule with $n = 10$ to find the following areas. *Source: Basics of Bioavailability.*

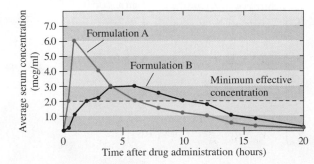

25. Find the total area under the curve for Formulation A. What does this area represent?

26. Find the total area under the curve for Formulation B. What does this area represent?

27. Find the area between the curve for Formulation A and the minimum effective concentration line. What does your answer represent?

28. Find the area between the curve for Formulation B and the minimum effective concentration line. What does this area represent?

29. Calves The daily milk consumption (in kilograms) for calves can be approximated by the function

$$y = b_0 w^{b_1} e^{-b_2 w},$$

where w is the age of the calf (in weeks) and b_0, b_1, and b_2 are constants. *Source: Animal Production.*

a. The age in days is given by $t = 7w$. Use this fact to convert the function above to a function in terms of t.

b. For a group of Angus calves, $b_0 = 5.955$, $b_1 = 0.233$, and $b_2 = 0.027$. Use the trapezoidal rule with $n = 10$, and then Simpson's rule with $n = 10$, to find the total amount of milk consumed by one of these calves over the first 25 weeks of life.

c. For a group of Nelore calves, $b_0 = 8.409$, $b_1 = 0.143$, and $b_2 = 0.037$. Use the trapezoidal rule with $n = 10$, and then Simpson's rule with $n = 10$, to find the total amount of milk consumed by one of these calves over the first 25 weeks of life.

30. Foot-and-Mouth Epidemic In 2001, the United Kingdom suffered an epidemic of foot-and-mouth disease. The graph below shows the reported number of cattle (red) and pigs (blue) that were culled each month from mid-February through mid-October in an effort to stop the spread of the disease. In Section 7.3 on Area and the Definite Integral we estimated the number of cattle and pigs that were culled using rectangles. *Source: Department of Environment, Food and Rural Affairs, United Kingdom.*

a. Estimate the total number of cattle that were culled from mid-February through mid-October and compare this with 581,801, the actual number of cattle that were culled. Use Simpson's rule with interval widths of one month starting with mid-February.

b. Estimate the total number of pigs that were culled from mid-February through mid-October and compare this with 146,145, the actual number of pigs that were culled. Use Simpson's rule with interval widths of one month starting with mid-February.

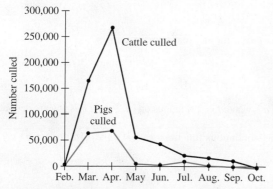

Social Sciences

31. Educational Psychology The results from a research study in psychology were as follows.

Number of Hours of Study, x	1	2	3	4	5	6	7
Rate of Extra Points Earned on a Test, $f(x)$	4	7	11	9	15	16	23

Repeat parts a–c of Exercise 21 for these data.

Physical Sciences

32. Chemical Formation The following table shows the results from a chemical experiment.

Concentration of Chemical A, x	1	2	3	4	5	6	7
Rate of Formation of Chemical B, $f(x)$	12	16	18	21	24	27	32

Repeat parts a–c of Exercise 21 for these data.

If you have a program for Simpson's rule in your graphing calculator, use it with $n = 20$ for Exercises 33–35.

33. Total Revenue An electronics company analyst has determined that the rate per month at which revenue comes in from the calculator division is given by

$$R(x) = 105e^{0.01x} + 32,$$

where x is the number of months the division has been in operation. Find the total revenue between the 12th and 36th months.

34. Milk Consumption As we saw in an earlier chapter, the average individual daily milk consumption for herds of Charolais, Angus, and Hereford calves can be described by a mathematical function. Here we write the consumption in kg/day as a function of the age of the calf in days (t) as

$$M(t) = 3.922t^{0.242}e^{-0.00357t}, \quad 7 \le t \le 182.$$

Find the total amount of milk consumed from 7 to 182 days for a calf. *Source: Animal Production.*

35. Probability The most important function in probability and statistics is the density function for the standard normal distribution, which is the familiar bell-shaped curve. The function is

$$f(x) = \frac{1}{\sqrt{2\pi}} e^{-x^2/2}.$$

a. The area under this curve between $x = -1$ and $x = 1$ represents the probability that a normal random variable is within 1 standard deviation of the mean. Find this probability.

b. Find the area under this curve between $x = -2$ and $x = 2$, which represents the probability that a normal random variable is within 2 standard deviations of the mean.

c. Find the probability that a normal random variable is within 3 standard deviations of the mean.

YOUR TURN ANSWERS

1. 5.3552   **2.** 5.3477

7 CHAPTER REVIEW

SUMMARY

Earlier chapters dealt with the derivative, one of the two main ideas of calculus. This chapter deals with integration, the second main idea. There are two aspects of integration. The first is indefinite integration, or finding an antiderivative; the second is definite integration, which can be used to find the area under a curve. The Fundamental Theorem of Calculus unites these two ideas by showing that the way to find the area under a curve is to use the antiderivative. Substitution is a technique for finding antiderivatives. Numerical integration can be used to find the definite integral when finding an antiderivative is not feasible. The idea of the definite integral can also be applied to finding the area between two curves.

Antidifferentiation Formulas

Power Rule
$$\int x^n \, dx = \frac{x^{n+1}}{n+1} + C, \ n \neq -1$$

Constant Multiple Rule
$$\int k \cdot f(x) \, dx = k \int f(x) \, dx, \quad \text{for any real number } k$$

Sum or Difference Rule
$$\int [f(x) \pm g(x)] \, dx = \int f(x) \, dx \pm \int g(x) \, dx$$

Integration of x^{-1}
$$\int x^{-1} \, dx = \ln |x| + C$$

Integration of Exponential Functions
$$\int e^{kx} \, dx = \frac{e^{kx}}{k} + C, \quad k \neq 0$$

Substitution Method Choose u to be one of the following:

1. the quantity under a root or raised to a power;

2. the quantity in the denominator;

3. the exponent on e.

Definite Integrals

Definition of the Definite Integral
$\int_a^b f(x) \, dx = \lim\limits_{n \to \infty} \sum\limits_{i=1}^{n} f(x_i)\Delta x$, where $\Delta x = (b-a)/n$ and x_i is any value of x in the ith interval. If $f(x)$ gives the rate of change of $F(x)$ for x in $[a, b]$, then this represents the total change in $F(x)$ as x goes from a to b.

Properties of Definite Integrals

1. $\displaystyle\int_a^a f(x) \, dx = 0$

2. $\displaystyle\int_a^b k \cdot f(x) \, dx = k \int_a^b f(x) \, dx, \quad \text{for any real number } k.$

3. $\displaystyle\int_a^b [f(x) \pm g(x)] \, dx = \int_a^b f(x) \, dx \pm \int_a^b g(x) \, dx$

4. $\displaystyle\int_a^b f(x) \, dx = \int_a^c f(x) \, dx + \int_c^b f(x) \, dx, \quad \text{for any real number } c$

5. $\displaystyle\int_a^b f(x) \, dx = -\int_b^a f(x) \, dx$

Fundamental Theorem of Calculus $\int_a^b f(x) \, dx = F(x)|_a^b = F(b) - F(a)$, where f is continuous on $[a, b]$ and F is any antiderivative of f

Area Between Two Curves $\int_a^b [f(x) - g(x)] \, dx$, where f and g are continuous functions and $f(x) \geq g(x)$ on $[a, b]$

Consumers' Surplus $\int_0^{q_0}[D(q) - p_0]\,dq$, where D is the demand function and p_0 and q_0 are the equilibrium price and demand

Producers' Surplus $\int_0^{q_0}[p_0 - S(q)]\,dq$, where S is the supply function and p_0 and q_0 are the equilibrium price and supply

Trapezoidal Rule $\int_a^b f(x)\,dx \approx \left(\dfrac{b - a}{n}\right)\left[\dfrac{1}{2}f(x_0) + f(x_1) + \cdots + f(x_{n-1}) + \dfrac{1}{2}f(x_n)\right]$

Simpson's Rule $\int_a^b f(x)\,dx \approx \left(\dfrac{b - a}{3n}\right)[f(x_0) + 4f(x_1) + 2f(x_2) + \cdots + 4f(x_{n-1}) + f(x_n)]$

KEY TERMS

7.1
antidifferentiation
antiderivative
integral sign
integrand
indefinite integral

7.2
integration by substitution

7.3
midpoint rule
definite integral
limits of integration
total change

7.4
Fundamental Theorem
of Calculus

7.5
consumers' surplus
producers' surplus

7.6
numerical integration
trapezoidal rule
Simpson's rule

REVIEW EXERCISES

CONCEPT CHECK

Determine whether each of the following statements is true or false, and explain why.

1. The indefinite integral is another term for the family of all antiderivatives of a function.

2. The indefinite integral of x^n is $x^{n+1}/(n + 1) + C$ for all real numbers n.

3. The indefinite integral $\int xf(x)\,dx$ is equal to $x\int f(x)\,dx$.

4. The velocity function is an antiderivative of the acceleration function.

5. Substitution can often be used to turn a complicated integral into a simpler one.

6. The definite integral gives the instantaneous rate of change of a function.

7. The definite integral gives an approximation to the area under a curve.

8. The definite integral of a positive function is the limit of the sum of the areas of rectangles.

9. The Fundamental Theorem of Calculus gives a relationship between the definite integral and an antiderivative of a function.

10. The definite integral of a function is always a positive quantity.

11. The area between two distinct curves is always a positive quantity.

12. The consumers' surplus and the producers' surplus equal each other.

13. In the trapezoidal rule, the number of subintervals must be even.

14. Simpson's rule usually gives a better approximation than the trapezoidal rule.

PRACTICE AND EXPLORATION

15. Explain the differences between an indefinite integral and a definite integral.

16. Explain under what circumstances substitution is useful in integration.

17. Explain why the limits of integration are changed when u is substituted for an expression in x in a definite integral.

18. Describe the type of integral for which numerical integration is useful.

In Exercises 19–40, find each indefinite integral.

19. $\displaystyle\int (2x + 3)\,dx$

20. $\displaystyle\int (5x - 1)\,dx$

21. $\displaystyle\int (x^2 - 3x + 2)\,dx$

22. $\displaystyle\int (6 - x^2)\,dx$

23. $\displaystyle\int 3\sqrt{x}\,dx$

24. $\displaystyle\int \frac{\sqrt{x}}{2}\,dx$

25. $\displaystyle\int (x^{1/2} + 3x^{-2/3})\,dx$

26. $\displaystyle\int (2x^{4/3} + x^{-1/2})\,dx$

27. $\displaystyle\int \frac{-4}{x^3}\,dx$

28. $\displaystyle\int \frac{5}{x^4}\,dx$

29. $\displaystyle\int -3e^{2x}\,dx$

30. $\displaystyle\int 5e^{-x}\,dx$

31. $\displaystyle\int xe^{3x^2}\,dx$

32. $\displaystyle\int 2xe^{x^2}\,dx$

33. $\int \dfrac{3x}{x^2 - 1}\, dx$

34. $\int \dfrac{-x}{2 - x^2}\, dx$

35. $\int \dfrac{x^2\, dx}{(x^3 + 5)^4}$

36. $\int (x^2 - 5x)^4(2x - 5)\, dx$

37. $\int \dfrac{x^3}{e^{3x^4}}\, dx$

38. $\int e^{3x^2+4}x\, dx$

39. $\int \dfrac{(3 \ln x + 2)^4}{x}\, dx$

40. $\int \dfrac{\sqrt{5 \ln x + 3}}{x}\, dx$

41. Let $f(x) = 3x + 1$, $x_1 = -1$, $x_2 = 0$, $x_3 = 1$, $x_4 = 2$, and $x_5 = 3$. Find $\sum\limits_{i=1}^{5} f(x_i)$.

42. Find $\int_0^4 f(x)\, dx$ for each graph of $y = f(x)$.

a.
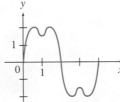

b.
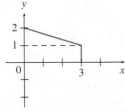

43. Approximate the area under the graph of $f(x) = 2x + 3$ and above the x-axis from $x = 0$ to $x = 4$ using four rectangles. Let the height of each rectangle be the function value on the left side.

44. Find $\int_0^4 (2x + 3)\, dx$ by using the formula for the area of a trapezoid: $A = (1/2)(B + b)h$, where B and b are the lengths of the parallel sides and h is the distance between them. Compare with Exercise 43.

45. In Exercises 29 and 30 of the section on Area and the Definite Integral, you calculated the distance that a car traveled by estimating the integral $\int_0^T v(t)\, dt$.

 a. Let $s(t)$ represent the mileage reading on the odometer. Express the distance traveled between $t = 0$ and $t = T$ using the function $s(t)$.

 b. Since your answer to part a and the original integral both represent the distance traveled by the car, the two can be set equal. Explain why the resulting equation is a statement of the Fundamental Theorem of Calculus.

 46. What does the Fundamental Theorem of Calculus state?

Find each definite integral.

47. $\int_1^2 (3x^2 + 5)\, dx$

48. $\int_1^6 (2x^2 + x)\, dx$

49. $\int_1^5 (3x^{-1} + x^{-3})\, dx$

50. $\int_1^3 (2x^{-1} + x^{-2})\, dx$

51. $\int_0^1 x\sqrt{5x^2 + 4}\, dx$

52. $\int_0^2 x^2(3x^3 + 1)^{1/3}\, dx$

53. $\int_0^2 3e^{-2x}\, dx$

54. $\int_1^5 \dfrac{5}{2}e^{0.4x}\, dx$

55. Use the substitution $u = 4x^2$ and the equation of a semicircle to evaluate

$$\int_0^{1/2} x\sqrt{1 - 16x^4}\, dx.$$

56. Use the substitution $u = x^2$ and the equation of a semicircle to evaluate

$$\int_0^{\sqrt{2}} 4x\sqrt{4 - x^4}\, dx.$$

In Exercises 57 and 58, use substitution to change the integral into one that can be evaluated by a formula from geometry, and then find the value of the integral.

57. $\int_1^{e^5} \dfrac{\sqrt{25 - (\ln x)^2}}{x}\, dx$

58. $\int_1^{\sqrt{7}} 2x\sqrt{36 - (x^2 - 1)^2}\, dx$

In Exercises 59–62, find the area between the x-axis and $f(x)$ over each of the given intervals.

59. $f(x) = \sqrt{4x - 3}$; $[1, 3]$

60. $f(x) = (3x + 2)^6$; $[-2, 0]$

61. $f(x) = xe^{x^2}$; $[0, 2]$

62. $f(x) = 1 + e^{-x}$; $[0, 4]$

Find the area of the region enclosed by each group of curves.

63. $f(x) = 5 - x^2$, $g(x) = x^2 - 3$

64. $f(x) = x^2 - 4x$, $g(x) = x - 6$

65. $f(x) = x^2 - 4x$, $g(x) = x + 6$, $x = -2$, $x = 4$

66. $f(x) = 5 - x^2$, $g(x) = x^2 - 3$, $x = 0$, $x = 4$

Use the trapezoidal rule with $n = 4$ to approximate the value of each integral. Then find the exact value and compare the two answers.

67. $\int_1^3 \dfrac{\ln x}{x}\, dx$

68. $\int_2^{10} \dfrac{x\, dx}{x - 1}$

69. $\int_0^1 e^x\sqrt{e^x + 4}\, dx$

70. $\int_0^2 xe^{-x^2}\, dx$

Use Simpson's rule with $n = 4$ to approximate the value of each integral. Compare your answers with the answers to Exercises 67–70.

71. $\int_1^3 \dfrac{\ln x}{x}\, dx$

72. $\int_2^{10} \dfrac{x\, dx}{x - 1}$

73. $\int_0^1 e^x\sqrt{e^x + 4}\, dx$

74. $\int_0^2 xe^{-x^2}\, dx$

75. Find the area of the region between the graphs of $y = \sqrt{x - 1}$ and $2y = x - 1$ from $x = 1$ to $x = 5$ in three ways.

 a. Use antidifferentiation.

 b. Use the trapezoidal rule with $n = 4$.

 c. Use Simpson's rule with $n = 4$.

76. Find the area of the region between the graphs of $y = \dfrac{1}{x + 1}$ and $y = \dfrac{x + 2}{2}$ from $x = 0$ to $x = 4$ in three ways.

 a. Use antidifferentiation.

 b. Use the trapezoidal rule with $n = 4$.

 c. Use Simpson's rule with $n = 4$.

77. Let $f(x) = [x(x-1)(x+1)(x-2)(x+2)]^2$.

   a. Find $\int_{-2}^{2} f(x)\, dx$ using the trapezoidal rule with $n = 4$.

   b. Find $\int_{-2}^{2} f(x)\, dx$ using Simpson's rule with $n = 4$.

   c. Without evaluating $\int_{-2}^{2} f(x)\, dx$, explain why your answers to parts a and b cannot possibly be correct.

   d. Explain why the trapezoidal rule and Simpson's rule with $n = 4$ give incorrect answers for $\int_{-2}^{2} f(x)\, dx$ with this function.

78. Given $\displaystyle\int_{0}^{2} f(x)\, dx = 3$ and $\displaystyle\int_{2}^{4} f(x)\, dx = 5$, calculate

   $\displaystyle\int_{0}^{2} f(2x)\, dx$. Choose one of the following. *Source: Society of Actuaries*.

   a. 3/2   **b.** 3   **c.** 4   **d.** 6   **e.** 8

APPLICATIONS

Business and Economics

Cost Find the cost function for each of the marginal cost functions in Exercises 79 and 80.

79. $C'(x) = 3\sqrt{2x - 1}$; 13 units cost $270.

80. $C'(x) = \dfrac{8}{2x + 1}$; fixed cost is $18.

81. Investment The curve shown gives the rate that an investment accumulates income (in dollars per year). Use rectangles of width 2 units and height determined by the function value at the midpoint to find the total income accumulated over 10 years.

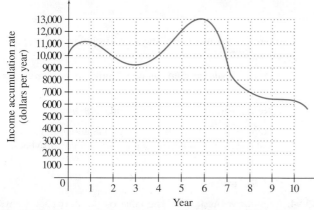

82. Utilization of Reserves A manufacturer of electronic equipment requires a certain rare metal. He has a reserve supply of 4,000,000 units that he will not be able to replace. If the rate at which the metal is used is given by

$$f(t) = 100,000e^{0.03t},$$

where t is time (in years), how long will it be before he uses up the supply? (*Hint*: Find an expression for the total amount used in t years and set it equal to the known reserve supply.)

83. Sales The rate of change of sales of a new brand of tomato soup (in thousands of dollars per month) is given by

$$S'(x) = 3\sqrt{2x + 1} + 3,$$

where x is the time (in months) that the new product has been on the market. Find the total sales after 4 months.

84. Productivity The function defined by

$$f'(x) = -0.1624x + 3.4909$$

approximates marginal U.S. nonfarm productivity from 2000–2009. Productivity is measured as total output per hour compared to a measure of 100 for 2000, and x is the number of years since 2000. *Source: Bureau of Labor Statistics*.

   a. Give the function that describes total productivity in year x.

   b. Use your function from part a to find productivity at the end of 2008. In 2009, productivity actually measured 122.3. How does your value using the function compare with this?

85. Producers' and Consumers' Surplus Suppose that the supply function for some commodity is

$$S(q) = q^2 + 5q + 100$$

and the demand function for the commodity is

$$D(q) = 350 - q^2.$$

   a. Find the producers' surplus.

   b. Find the consumers' surplus.

86. Net Savings A company has installed new machinery that will produce a savings rate (in thousands of dollars per year) of

$$S'(x) = 225 - x^2,$$

where x is the number of years the machinery is to be used. The rate of additional costs (in thousands of dollars per year) to the company due to the new machinery is expected to be

$$C'(x) = x^2 + 25x + 150.$$

For how many years should the company use the new machinery? Find the net savings (in thousands of dollars) over this period.

87. Oil Production The following table shows the amount of crude oil (in billions of barrels) produced in the United States in recent years. *Source: U.S. Energy Information Administration*.

Year	Crude Oil Produced
2000	2.131
2001	2.118
2002	2.097
2003	2.073
2004	1.983
2005	1.890
2006	1.862
2007	1.848
2008	1.812
2009	1.938

In this exercise we are interested in the total amount of crude oil produced over the 9-year period from mid-2000 to mid-2009, using the data for the 10 years above.

a. One approach is to sum up the numbers in the second column, but only count half of the first and last numbers. Give the answer to this calculation.

b. Approximate the amount of crude oil produced over the 9-year period 2000–2009 by taking the average of the left endpoint sum and the right endpoint sum. Explain why this is equivalent to the calculation done in part a.

c. Explain why the answer from part a is the same as using the trapezoidal rule to approximate the amount of crude oil produced over the 9-year period 2000–2009.

d. Find the equation of the least squares line for this data, letting $x = 0$ correspond to 2000. Then integrate this equation over the interval $[0, 9]$ to estimate the amount of crude oil produced over this time period. Compare with your answer to part a.

88. Inventory At time $t = 0$, a store has 19 units of a product in inventory. The cumulative number of units sold is given by $S(t) = e^{3t} - 1$, where t is measured in weeks. The inventory will be replenished when it drops to 1 unit. The cost of carrying inventory until then is 15 per unit per week (prorated for a portion of a week). Calculate the inventory carrying cost that will be incurred before the inventory is replenished. Choose one of the following. *Source: Society of Actuaries.*

a. 90   **b.** 199   **c.** 204   **d.** 210   **e.** 294

Life Sciences

89. Population Growth The rate of change of the population of a rare species of Australian spider for one year is given by

$$f'(t) = 100 - t\sqrt{0.4t^2 + 1},$$

where $f(t)$ is the number of spiders present at time t (in months). Find the total number of additional spiders in the first 10 months.

90. Infection Rate The rate of infection of a disease (in people per month) is given by the function

$$I'(t) = \frac{100t}{t^2 + 1},$$

where t is the time (in months) since the disease broke out. Find the total number of infected people over the first four months of the disease.

91. Insect Cannibalism In certain species of flour beetles, the larvae cannibalize the unhatched eggs. In calculating the population cannibalism rate per egg, researchers needed to evaluate the integral

$$\int_0^A c(x)\, dx,$$

where A is the length of the larval stage and $c(x)$ is the cannibalism rate per egg per larva of age x. The minimum value of A for the flour beetle *Tribolium castaneum* is 17.6 days, which is the value we will use. The function $c(x)$ starts at day 0 with a value of 0, increases linearly to the value 0.024 at day 12, and then stays constant. *Source: Journal of Animal Ecology.* Find the values of the integral using

a. formulas from geometry;

b. the Fundamental Theorem of Calculus.

92. Insulin in Sheep A research group studied the effect of a large injection of glucose in sheep fed a normal diet compared with sheep that were fasting. A graph of the plasma insulin levels (in pM—pico molars, or 10^{-12} of a molar) for both groups is shown below. The red graph designates the fasting sheep and the green graph the sheep fed a normal diet. The researchers compared the area under the curves for the two groups. *Source: Endocrinology.*

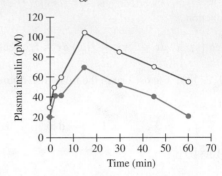

a. For the fasting sheep, estimate the area under the curve using rectangles, first by using the left endpoints, then the right endpoints, and then averaging the two. Note that the width of the rectangles will vary.

b. Repeat part a for the sheep fed a normal diet.

c. How much higher is the area under the curve for the fasting sheep compared with the normal sheep?

93. Milk Production Researchers report that the average amount of milk produced (in kilograms per day) by a 4- to 5-year-old cow weighing 700 kg can be approximated by

$$y = 1.87t^{1.49}e^{-0.189(\ln t)^2},$$

where t is the number of days into lactation. *Source: Journal of Dairy Science.*

a. Approximate the total amount of milk produced from $t = 1$ to $t = 321$ using the trapezoidal rule with $n = 8$.

b. Repeat part a using Simpson's rule with $n = 8$.

c. Repeat part a using the integration feature of a graphing calculator, and compare your answer with the answers to parts a and b.

Social Sciences

94. Automotive Accidents The table on the next page shows the amount of property damage (in dollars) due to automobile accidents in California in recent years. In this exercise we are interested in the total amount of property damage due to automobile accidents over the 8-year period from mid-2000 to mid-2008, using the data for the 9 years. *Source: The California Highway Patrol.*

a. One approach is to sum up the numbers in the second column, but only count half of the first and last numbers. Give the answer to this calculation.

b. Approximate the amount of property damage over the 8-year period 2000–2008 by taking the average of the left endpoint sum and the right endpoint sum. Explain why this is equivalent to the calculation done in part a.

Year	Property Damage ($)
2000	309,569
2001	317,567
2002	335,869
2003	331,055
2004	331,208
2005	330,195
2006	325,453
2007	313,357
2008	278,986

c. Explain why the answer from part a is the same as using the trapezoidal rule to approximate the amount of property damage over the 8-year period 2000–2008.

d. Find the equation of the least squares line for this data, letting $x = 0$ correspond to 2000. Then integrate this equation over the interval $[0, 8]$ to estimate the amount of property damage over this time period. Compare with your answer to part a.

Physical Sciences

95. Linear Motion A particle is moving along a straight line with velocity $v(t) = t^2 - 2t$. Its distance from the starting point after 3 seconds is 8 cm. Find $s(t)$, the distance of the particle from the starting point after t seconds.

EXTENDED APPLICATION

ESTIMATING DEPLETION DATES FOR MINERALS

I t is becoming more and more obvious that the earth contains only a finite quantity of minerals. The "easy and cheap" sources of minerals are being used up, forcing an ever more expensive search for new sources. For example, oil from the North Slope of Alaska would never have been used in the United States during the 1930s because a great deal of Texas and California oil was readily available.

We said in an earlier chapter that population tends to follow an exponential growth curve. Mineral usage also follows such a curve. Thus, if q represents the rate of consumption of a certain mineral at time t, while q_0 represents consumption when $t = 0$, then

$$q = q_0 e^{kt},$$

where k is the growth constant. For example, the world consumption of petroleum in 1970 was 16,900 million barrels. During this period energy use was growing rapidly, and by 1975 annual world consumption had risen to 21,300 million barrels. We can use these two values to make a rough estimate of the constant k, and we find that over this 5-year span the average value of k was about 0.047, representing 4.7% annual growth. If we let $t = 0$ correspond to the base year 1970, then

$$q = 16,900e^{0.047t}$$

is the rate of consumption at time t, assuming that all the trends of the early 1970s have continued. In 1970 a reasonable guess would have put the total amount of oil in provable reserves or likely to be discovered in the future at 1,500,000 million barrels. At the 1970–1975 rate of consumption, in how many years after 1970 would you expect the world's reserves to be depleted? We can use the integral calculus of this chapter to find out. *Source: Energy Information Administration*.

To begin, we need to know the total quantity of petroleum that would be used between time $t = 0$ and some future time $t = T$. Figure 34 on the following page shows a typical graph of the function $q = q_0 e^{kt}$.

Following the work we did in Section 3, divide the time interval from $t = 0$ to $t = T$ into n subintervals. Let each subinterval have width Δt. Let the rate of consumption for the ith subinterval be approximated by q_i^*. Thus, the approximate total consumption for the subinterval is given by

$$q_i^* \cdot \Delta t,$$

and the total consumption over the interval from time $t = 0$ to $t = T$ is approximated by

$$\sum_{i=1}^{n} q_i^* \cdot \Delta t.$$

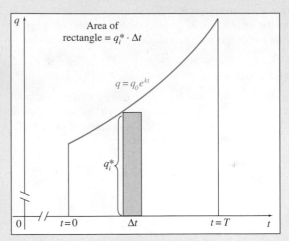

FIGURE 34

The limit of this sum as Δt approaches 0 gives the total consumption from time $t = 0$ to $t = T$. That is,

$$\text{Total consumption} = \lim_{\Delta t \to 0} \sum q_i^* \cdot \Delta t.$$

We have seen, however, that this limit is the definite integral of the function $q = q_0 e^{kt}$ from $t = 0$ to $t = T$, or

$$\text{Total consumption} = \int_0^T q_0 e^{kt} \, dt.$$

We can now evaluate this definite integral.

$$\int_0^T q_0 e^{kt} \, dt = q_0 \int_0^T e^{kt} \, dt = q_0 \left(\frac{e^{kt}}{k} \right) \Big|_0^T$$

$$= \frac{q_0}{k} e^{kt} \Big|_0^T = \frac{q_0}{k} e^{kT} - \frac{q_0}{k} e^0$$

$$= \frac{q_0}{k} e^{kT} - \frac{q_0}{k} (1)$$

$$= \frac{q_0}{k} (e^{kT} - 1) \qquad (1)$$

Now let us return to the numbers we gave for petroleum. We said that $q_0 = 16{,}900$ million barrels, where q_0 represents consumption in the base year of 1970. We have $k = 0.047$ with total petroleum reserves estimated at 1,500,000 million barrels. Thus, using Equation (1) we have

$$1{,}500{,}000 = \frac{16{,}900}{0.047} (e^{0.047T} - 1).$$

Multiply both sides of the equation by 0.047.

$$70{,}500 = 16{,}900 (e^{0.047T} - 1)$$

Divide both sides of the equation by 16,900.

$$4.2 = e^{0.047T} - 1$$

Add 1 to both sides.

$$5.2 = e^{0.047T}$$

Take natural logarithms of both sides.

$$\ln 5.2 = \ln e^{0.047T}$$
$$= 0.047T$$

Finally,

$$T = \frac{\ln 5.2}{0.047} \approx 35.$$

By this result, petroleum reserves would only last 35 years after 1970, that is, until about 2005.

In fact, in the early 1970s some analysts were predicting that reserves would be exhausted before the end of the century, and this was a reasonable guess. But since 1970, more reserves have been discovered. One way to refine our model is to look at the historical data over a longer time span. The following table gives average world annual petroleum consumption in millions of barrels at 5-year intervals from 1970 to 2000. **Source: Energy Information Administration.**

Year	World Consumption (in millions of barrels)
1970	16,900
1975	21,300
1980	22,900
1985	22,200
1990	24,300
1995	25,700
2000	27,900
2005	30,400

The first step in comparing this data with our exponential model is to estimate a value for the growth constant k. One simple way of doing this is to solve the equation

$$30{,}400 = 16{,}900 \cdot e^{k \cdot 35}.$$

Using natural logarithms just as we did in estimating the time to depletion for $k = 0.036$, we find that

$$k = \frac{\ln \left(\dfrac{30{,}400}{16{,}900} \right)}{35} \approx 0.017.$$

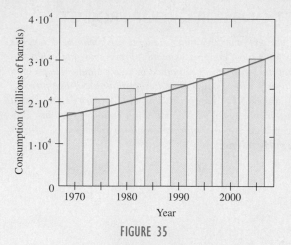

FIGURE 35

So the data from the Bureau of Transportation Statistics suggests a growth constant of about 1.7%. We can check the fit by plotting the function $16{,}900 \cdot e^{0.017t}$ along with a bar graph of the consumption data, shown in Figure 35. The fit looks reasonably good, but over this short range of 35 years, the exponential model is close to a linear model, and the growth in consumption is certainly not smooth.

The exponential model rests on the assumption of a constant growth rate. As already noted, we might expect instead that the growth rate would change as the world comes closer to exhausting its reserves. In particular, scarcity might drive up the price of oil and thus reduce consumption. We can use integration to explore an alternative model in which the factor k changes over time, so that k becomes $k(t)$, a function of time.

As an illustration, we explore a model in which the growth constant k declines toward 0 over time. We'll use 1970 as our base year, so the variable t will count years since 1970. We need a simple positive function $k(t)$ that tends toward 0 as t gets large. To get some numbers to work with, assume that the growth rate was 2% in 1970 and declined to 1% by 1995. There are many possible choices for the function $k(t)$, but a convenient one is

$$k(t) = \frac{0.5}{t + 25}.$$

Using integration to turn the instantaneous rate of consumption into the total consumption up to time T, we can write

$$\text{Total consumption} = 16{,}900 \int_0^T e^{k(t) \cdot t} \, dt$$

$$= 16{,}900 \int_0^T e^{0.5t/(t+25)} \, dt.$$

We'd like to find out when the world will use up its estimated reserves, but as just noted, the estimates have increased since the 1970s. It is estimated that the current global petroleum reserves are 3,000,000 million barrels. *Source: Geotimes.* So we need to solve

$$3{,}000{,}000 = 16{,}900 \int_0^T e^{0.5t/(t+25)} \, dt \qquad (2)$$

But this problem is much harder to solve than the corresponding problem for constant growth, because *there is no formula for evaluating this definite integral!* The function

$$g(t) = e^{0.5t/(t+25)}$$

doesn't have an antiderivative that we can write down in terms of functions that we know how to compute.

Here the numerical integration techniques discussed in Section 6 come to the rescue. We can use one of the integration rules to *approximate* the integral numerically for various values of T, and with some trial and error we can estimate how long the reserves will last. If you have a calculator or computer algebra system that does numerical integration, you can pick some T values and evaluate the right-hand side of Equation (2). Here are the results produced by one computer algebra system:

For $T = 120$ the integral is about 2,797,000.

For $T = 130$ the integral is about 3,053,000.

For $T = 140$ the integral is about 3,311,000.

So using this model we would estimate that starting in 1970 the petroleum reserves would last for about 130 years, that is, until 2100.

Our integration tools are essential in building and exploring models of resource use, but the difference in our two predictions (35 years vs. 130 years) illustrates the difficulty of making accurate predictions. A model that performs well on historical data may not take the changing dynamics of resource use into account, leading to forecasts that are either unduly gloomy or too optimistic.

EXERCISES

1. Find the number of years that the estimated petroleum reserves would last if used at the same rate as in the base year.

2. How long would the estimated petroleum reserves last if the growth constant was only 2% instead of 4.7%?

Estimate the length of time until depletion for each mineral.

3. Bauxite (the ore from which aluminum is obtained): estimated reserves in base year 15,000,000 thousand tons; rate of consumption in base year 63,000 thousand tons; growth constant 6%

4. Bituminous coal: estimated world reserves 2,000,000 million tons; rate of consumption in base year 2200 million tons; growth constant 4%

5. a. Verify that the function $k(t)$ defined on the previous page has the right values at $k = 0$ and $k = 25$.

   b. Find a similar function that has $k(0) = 0.03$ and $k(25) = 0.02$.

6. a. Use the function you defined in Exercise 5 b to write an integral for world petroleum consumption from 1970 until T years after 1970.

   b. If you have access to a numerical integrator, compute some values of your integral and estimate the time required to exhaust the reserve of 3,000,000 million barrels.

7. A reasonable assumption is that over time scarcity might drive up the price of oil and thus reduce consumption. Comment on the fact that the rate of oil consumption actually increased in 2002, connecting current events and economic forecasts to the short-term possibility of a reduction in consumption.

8. Develop a spreadsheet that shows the time to exhaustion for various values of k.

9. Go to the website WolframAlpha.com and enter "integrate." Follow the instructions to find the time to exhaustion for various values of k. Discuss how the solution compares with the solutions provided by a graphing calculator and by Microsoft Excel.

DIRECTIONS FOR GROUP PROJECT

Suppose that you and three other students are spending a summer as interns for a local congresswoman. During your internship you realize that the information contained in your calculus class could be used to help with a new bill under consideration. The primary purpose of the bill is to require, by law, that all cars manufactured after a certain date get at least 60 miles per gallon of gasoline. Prepare a report that uses the information above to make a case for or against a bill of this nature.

8 Further Techniques and Applications of Integration

It might seem that definite integrals with infinite limits have only theoretical interest, but in fact these *improper* integrals provide answers to many practical questions. An example in Section 4 models an environmental cleanup process in which the amount of pollution entering a stream decreases by a constant fraction each year. An improper integral gives the total amount of pollutant that will ever enter the river.

In the previous chapter we discussed indefinite and definite integrals and presented rules for finding the antiderivatives of several types of functions. We showed how numerical methods can be used for functions that cannot be integrated by the techniques presented there. In this chapter we develop additional methods of integrating functions. We also show how to evaluate an integral that has one or both limits at infinity. These new techniques allow us to consider additional applications of integration such as volumes of solids of revolution, the average value of a function, and continuous money flow.

8.1 Integration by Parts

APPLY IT **If we know the rate of growth of a patch of moss, how can we calculate the area the moss covers?**
We will use integration by parts to answer this question in Exercise 42.

The technique of *integration by parts* often makes it possible to reduce a complicated integral to a simpler integral. We know that if u and v are both differentiable functions, then uv is also differentiable and, by the product rule for derivatives,

$$\frac{d(uv)}{dx} = u\frac{dv}{dx} + v\frac{du}{dx}.$$

This expression can be rewritten, using differentials, as

$$d(uv) = u\,dv + v\,du.$$

Integrating both sides of this last equation gives

$$\int d(uv) = \int u\,dv + \int v\,du,$$

or

$$uv = \int u\,dv + \int v\,du.$$

Rearranging terms gives the following formula.

Integration by Parts
If u and v are differentiable functions, then

$$\int u\,dv = uv - \int v\,du.$$

The process of finding integrals by this formula is called **integration by parts**. There are two ways to do integration by parts: the standard method and column integration. Both methods are illustrated in the following example.

EXAMPLE 1 Integration by Parts

Find $\int xe^{5x}\,dx$.

Method 1
Standard Method

SOLUTION

Although this integral cannot be found by using any method studied so far, it can be found with integration by parts. First write the expression $xe^{5x}\,dx$ as a product of two functions u and dv in such a way that $\int dv$ can be found. One way to do this is to choose the two functions x and e^{5x}. Both x and e^{5x} can be integrated, but $\int x\,dx$, which is $x^2/2$, is more complicated than x itself, while the derivative of x is 1, which is simpler than x. Since e^{5x} remains the same (except for the coefficient) whether it is integrated or differentiated, it is best here to choose

$$dv = e^{5x}\,dx \qquad \text{and} \qquad u = x.$$

Then

$$du = dx,$$

and v is found by integrating dv:

$$v = \int dv = \int e^{5x}\,dx = \frac{e^{5x}}{5}.$$

We need not introduce the constant of integration until the last step, because only one constant is needed. Now substitute into the formula for integration by parts and complete the integration.

$$\int u\,dv = uv - \int v\,du$$

$$\int \underbrace{x}_{u}\underbrace{e^{5x}\,dx}_{dv} = \underbrace{x}_{u}\underbrace{\left(\frac{e^{5x}}{5}\right)}_{v} - \int \underbrace{\frac{e^{5x}}{5}}_{v}\underbrace{dx}_{du}$$

$$= \frac{xe^{5x}}{5} - \frac{e^{5x}}{25} + C$$

$$= \frac{e^{5x}}{25}(5x - 1) + C \qquad \text{Factor out } e^{5x}/25.$$

The constant C was added in the last step. As before, check the answer by taking its derivative.

Method 2
Column Integration

A technique called **column integration**, or *tabular integration*, is equivalent to integration by parts but helps in organizing the details.* We begin by creating two columns. The first column, labeled D, contains u, the part to be differentiated in the original integral. The second column, labeled I, contains the rest of the integral: that is, the part to be integrated, but without the dx. To create the remainder of the first column, write the derivative of the function in the first row underneath it in the second row. Now write the derivative of the function in the second row underneath it in the third row. Proceed in this manner down the first column, taking derivatives until you get a 0. Form the second column in a similar manner, except take an antiderivative at each row, until the second column has the same number of rows as the first.

To illustrate this process, consider our goal of finding $\int xe^{5x}\,dx$. Here $u = x$, so e^{5x} is left for the second column. Taking derivatives down the first column and antiderivatives down the second column results in the following table.

D	I
x	e^{5x}
1	$e^{5x}/5$
0	$e^{5x}/25$

*This technique appeared in the 1988 movie *Stand and Deliver*.

Next, draw a diagonal line from each term (except the last) in the left column to the term in the row below it in the right column. Label the first such line with "+", the next with "−", and continue alternating the signs as shown.

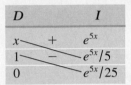

Then multiply the terms on opposite ends of each diagonal line. Finally, sum up the products just formed, adding the "+" terms and subtracting the "−" terms.

$$\int xe^{5x}\, dx = x(e^{5x}/5) - 1(e^{5x}/25) + C$$

$$= \frac{xe^{5x}}{5} - \frac{e^{5x}}{25} + C$$

$$= \frac{e^{5x}}{25}(5x - 1) + C \qquad \text{Factor out } e^{5x}/25.$$

Compare these steps with those of Method 1 and convince yourself that the process is the same.

TRY YOUR TURN 1

YOUR TURN 1
Find $\int xe^{-2x}\, dx$.

Conditions for Integration by Parts

Integration by parts can be used only if the integrand satisfies the following conditions.

1. The integrand can be written as the product of two factors, u and dv.
2. It is possible to integrate dv to get v and to differentiate u to get du.
3. The integral $\int v\, du$ can be found.

EXAMPLE 2 **Integration by Parts**

Find $\int \ln x\, dx$ for $x > 0$.

Method 1
Standard Method

SOLUTION

No rule has been given for integrating $\ln x$, so choose

$$dv = dx \qquad \text{and} \qquad u = \ln x.$$

Then

$$v = x \qquad \text{and} \qquad du = \frac{1}{x}\, dx,$$

and, since $uv = vu$, we have

$$\int \underbrace{\ln x}_{u}\ \underbrace{dx}_{dv} = \underbrace{x \ln x}_{v \cdot u} - \int \underbrace{x \cdot \frac{1}{x}\, dx}_{v \cdot du}$$

$$= x \ln x - \int dx$$

$$= x \ln x - x + C.$$

Method 2
Column Integration

Column integration works a little differently here. As in Method 1, choose ln x as the part to differentiate. The part to be integrated must be 1. (Think of ln x as $1 \cdot \ln x$.) No matter how many times ln x is differentiated, the result is never 0. In this case, stop as soon as the natural logarithm is gone.

D	I
ln x	1
$1/x$	x

Draw diagonal lines with alternating $+$ and $-$ as before. On the last line, because the left column does not contain a 0, draw a horizontal line.

D		I
ln x	$+$	1
$1/x$	$-$	x

The presence of a horizontal line indicates that the product is to be integrated, just as the original integral was represented by the first row of the two columns.

$$\int \ln x \, dx = (\ln x)x - \int \frac{1}{x} \cdot x \, dx$$

$$= x \ln x - \int dx$$

$$= x \ln x - x + C.$$

Note that when setting up the columns, a horizontal line is drawn only when a 0 does not eventually appear in the left column.

YOUR TURN 2
Find $\int \ln 2x \, dx$.

TRY YOUR TURN 2

Sometimes integration by parts must be applied more than once, as in the next example.

EXAMPLE 3 **Integration by Parts**

Find $\int (2x^2 + 5)e^{-3x} \, dx$.

Method 1
Standard Method

SOLUTION

Choose

$$dv = e^{-3x} \, dx \qquad \text{and} \qquad u = 2x^2 + 5.$$

Then

$$v = \frac{-e^{-3x}}{3} \qquad \text{and} \qquad du = 4x \, dx.$$

Substitute these values into the formula for integration by parts.

$$\int u \, dv = uv - \int v \, du$$

$$\int (2x^2 + 5)e^{-3x} \, dx = (2x^2 + 5)\left(\frac{-e^{-3x}}{3}\right) - \int \left(\frac{-e^{-3x}}{3}\right)4x \, dx$$

$$= -(2x^2 + 5)\left(\frac{e^{-3x}}{3}\right) + \frac{4}{3}\int xe^{-3x} \, dx$$

Now apply integration by parts to the last integral, letting

$$dv = e^{-3x} \, dx \qquad \text{and} \qquad u = x,$$

so

$$v = \frac{-e^{-3x}}{3} \quad \text{and} \quad du = dx.$$

$$\int (2x^2 + 5)e^{-3x}\, dx = -(2x^2 + 5)\left(\frac{e^{-3x}}{3}\right) + \frac{4}{3}\int xe^{-3x}\, dx$$

$$= -(2x^2 + 5)\left(\frac{e^{-3x}}{3}\right) + \frac{4}{3}\left[x\left(\frac{-e^{-3x}}{3}\right) - \int \left(\frac{-e^{-3x}}{3}\right) dx\right]$$

$$= -(2x^2 + 5)\left(\frac{e^{-3x}}{3}\right) + \frac{4}{3}\left[-\frac{x}{3}e^{-3x} - \left(\frac{e^{-3x}}{9}\right)\right] + C$$

$$= -(2x^2 + 5)\left(\frac{e^{-3x}}{3}\right) - \frac{4}{9}xe^{-3x} - \frac{4}{27}e^{-3x} + C$$

$$= [-(2x^2 + 5)(9) - 4x(3) - 4]\frac{e^{-3x}}{27} + C \quad \text{Factor out } e^{-3x}/27.$$

$$= (-18x^2 - 12x - 49)\frac{e^{-3x}}{27} + C \quad \text{Simplify.}$$

Method 2
Column Integration

Choose $2x^2 + 5$ as the part to be differentiated, and put e^{-3x} in the integration column.

D		I
$2x^2 + 5$	$+$	e^{-3x}
$4x$	$-$	$-e^{-3x}/3$
4	$+$	$e^{-3x}/9$
0		$-e^{-3x}/27$

Multiplying and adding as before yields

$$\int (2x^2 + 5)e^{-3x}\, dx = (2x^2 + 5)(-e^{-3x}/3) - 4x(e^{-3x}/9) + 4(-e^{-3x}/27) + C$$

$$= (-18x^2 - 12x - 49)\frac{e^{-3x}}{27} + C.$$

YOUR TURN 3

Find $\int (3x^2 + 4)e^{2x}\, dx.$

TRY YOUR TURN 3

With the functions discussed so far in this book, choosing u and dv (or the parts to be differentiated and integrated) is relatively simple. In general, the following strategy should be used.

First see if the integration can be performed using substitution. If substitution does not work:

- See if $\ln x$ is in the integral. If it is, set $u = \ln x$ and dv equal to the rest of the integral. (Equivalently, put $\ln x$ in the D column and the rest of the function in the I column.)

- If $\ln x$ is not present, see if the integral contains x^k, where k is any positive integer, or any other polynomial. If it does, set $u = x^k$ (or the polynomial) and dv equal to the rest of the integral. (Equivalently, put x^k in the D column and the rest of the function in the I column.)

EXAMPLE 4 **Definite Integral**

Find $\displaystyle\int_1^e \frac{\ln x}{x^2}\, dx.$

SOLUTION First find the indefinite integral using integration by parts by the standard method. (You may wish to verify this using column integration.) Whenever $\ln x$ is present, it is selected as u, so let

$$u = \ln x \quad \text{and} \quad dv = \frac{1}{x^2}\, dx.$$

Then

$$du = \frac{1}{x}\, dx \qquad \text{and} \qquad v = -\frac{1}{x}.$$

Substitute these values into the formula for integration by parts, and integrate the second term on the right.

$$\int u\, dv = uv - \int v\, du$$

$$\int \frac{\ln x}{x^2}\, dx = (\ln x)\frac{-1}{x} - \int\left(-\frac{1}{x}\cdot\frac{1}{x}\right) dx$$

$$= -\frac{\ln x}{x} + \int \frac{1}{x^2}\, dx$$

$$= -\frac{\ln x}{x} - \frac{1}{x} + C$$

$$= \frac{-\ln x - 1}{x} + C$$

Now find the definite integral.

$$\int_1^e \frac{\ln x}{x^2}\, dx = \frac{-\ln x - 1}{x}\bigg|_1^e$$

YOUR TURN 4

Find $\displaystyle\int_1^e x^2 \ln x\, dx$.

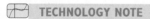

$$= \left(\frac{-1 - 1}{e}\right) - \left(\frac{0 - 1}{1}\right)$$

$$= \frac{-2}{e} + 1 \approx 0.2642411177 \quad \text{TRY YOUR TURN 4}$$

TECHNOLOGY NOTE

Definite integrals can be found with a graphing calculator using the function integral feature or by finding the area under the graph of the function between the limits. For example, using the `fnInt` feature of the TI-84 Plus calculator to find the integral in Example 4 gives 0.2642411177. Using the area under the graph approach gives 0.26424112, the same result rounded.

Many integrals cannot be found by the methods presented so far. For example, consider the integral

$$\int \frac{1}{4 - x^2}\, dx.$$

Substitution of $u = 4 - x^2$ will not help, because $du = -2x\, dx$, and there is no x in the numerator of the integral. We could try integration by parts, using $dv = dx$ and $u = (4 - x^2)^{-1}$. Integration gives $v = x$ and differentiation gives $du = 2x\, dx/(4 - x^2)^2$, with

$$\int \frac{1}{4 - x^2}\, dx = \frac{x}{4 - x^2} - \int \frac{2x^2}{(4 - x^2)^2}\, dx.$$

The integral on the right is more complicated than the original integral, however. A second use of integration by parts on the new integral would only make matters worse. Since we cannot choose $dv = (4 - x^2)^{-1}\, dx$ because it cannot be integrated by the methods studied so far, integration by parts is not possible for this problem.

This integration can be performed using one of the many techniques of integration beyond the scope of this text.* Tables of integrals can also be used, but technology is rapidly

*For example, see Thomas, George B., Maurice D. Weir, and Joel Hass, *Thomas' Calculus*, 12th ed., Pearson, 2010.

making such tables obsolete and even reducing the importance of techniques of integration. The following example shows how the table of integrals given in the appendix of this book may be used.

EXAMPLE 5 Tables of Integrals

Find $\int \dfrac{1}{4 - x^2}\, dx$.

SOLUTION Using formula 7 in the table of integrals in the appendix, with $a = 2$, gives

$$\int \frac{1}{4 - x^2}\, dx = \frac{1}{4} \cdot \ln \left| \frac{2 + x}{2 - x} \right| + C.$$

TRY YOUR TURN 5

YOUR TURN 5

Find $\int \dfrac{1}{x\sqrt{4 + x^2}}\, dx$.

TECHNOLOGY NOTE

We mentioned in the previous chapter how computer algebra systems and some calculators can perform integration. Using a TI-89, the answer to the above integral is

$$\frac{\ln\left(\dfrac{|x + 2|}{|x - 2|} \right)}{4}.$$

(The C is not included.) Verify that this is equivalent to the answer given in Example 5.

If you don't have a calculator or computer program that integrates symbolically, there is a Web site (http://integrals.wolfram.com), as of this writing, that finds indefinite integrals using the computer algebra system Mathematica. It includes instructions on how to enter your function. When the previous integral was entered, it returned the answer

$$\frac{1}{4}\big(\log(-x - 2) - \log(x - 2)\big).$$

Note that Mathematica does not include the C or the absolute value, and that natural logarithms are written as log. Verify that this answer is equivalent to the answer given by the TI-89 and the answer given in Example 5.

Unfortunately, there are integrals that cannot be antidifferentiated by any technique, in which case numerical integration must be used. (See the last section of the previous chapter.) In this book, for simplicity, all integrals to be antidifferentiated can be done with substitution or by parts, except for Exercises 23–28 in this section.

8.1 EXERCISES

Use integration by parts to find the integrals in Exercises 1–10.

1. $\int xe^x\, dx$

2. $\int (x + 6)e^x\, dx$

3. $\int (4x - 12)e^{-8x}\, dx$

4. $\int (6x + 3)e^{-2x}\, dx$

5. $\int x \ln x\, dx$

6. $\int x^3 \ln x\, dx$

7. $\int_0^1 \dfrac{2x + 1}{e^x}\, dx$

8. $\int_0^3 \dfrac{3 - x}{3e^x}\, dx$

9. $\int_1^9 \ln 3x\, dx$

10. $\int_1^2 \ln 5x\, dx$

11. Find the area between $y = (x - 2)e^x$ and the x-axis from $x = 2$ to $x = 4$.

12. Find the area between $y = (x + 1) \ln x$ and the x-axis from $x = 1$ to $x = e$.

Exercises 13–22 are mixed—some require integration by parts, while others can be integrated by using techniques discussed in the chapter on Integration.

13. $\int x^2 e^{2x}\, dx$

14. $\int \dfrac{x^2\, dx}{2x^3 + 1}$

15. $\int x^2\sqrt{x + 4}\, dx$

16. $\int (2x - 1) \ln(3x)\, dx$

17. $\int (8x + 10) \ln(5x)\, dx$

18. $\int x^3 e^{x^4}\, dx$

19. $\int_1^2 (1 - x^2)e^{2x}\, dx$

20. $\int_0^1 \dfrac{x^2\, dx}{2x^3 + 1}$

21. $\int_0^1 \dfrac{x^3\, dx}{\sqrt{3 + x^2}}$

22. $\int_0^5 x\sqrt[3]{x^2 + 2}\, dx$

Use the table of integrals, or a computer or calculator with symbolic integration capabilities, to find each indefinite integral.

23. $\int \dfrac{16}{\sqrt{x^2 + 16}}\, dx$

24. $\int \dfrac{10}{x^2 - 25}\, dx$

25. $\int \dfrac{3}{x\sqrt{121 - x^2}}\, dx$

26. $\int \dfrac{2}{3x(3x - 5)}\, dx$

27. $\int \dfrac{-6}{x(4x + 6)^2}\, dx$

28. $\int \sqrt{x^2 + 15}\, dx$

29. What rule of differentiation is related to integration by parts?

30. Explain why the two methods of solving Example 2 are equivalent.

31. Suppose that u and v are differentiable functions of x with $\int_0^1 v\, du = 4$ and the following functional values.

x	$u(x)$	$v(x)$
0	2	1
1	3	−4

Use this information to determine $\int_0^1 u\, dv$.

32. Suppose that u and v are differentiable functions of x with $\int_1^{20} v\, du = -1$ and the following functional values.

x	$u(x)$	$v(x)$
1	5	−2
20	15	6

Use this information to determine $\int_1^{20} u\, dv$.

33. Suppose we know that the functions r and s are everywhere differentiable and that $r(0) = 0$. Suppose we also know that for $0 \le x \le 2$, the area between the x-axis and the nonnegative function $h(x) = s(x)\dfrac{dr}{dx}$ is 5, and that on the same interval, the area between the x-axis and the nonnegative function $k(x) = r(x)\dfrac{ds}{dx}$ is 10. Determine $r(2)s(2)$.

34. Suppose we know that the functions u and v are everywhere differentiable and that $u(3) = 0$. Suppose we also know that for $1 \le x \le 3$, the area between the x-axis and the nonnegative function $h(x) = u(x)\dfrac{dv}{dx}$ is 15, and that on the same interval, the area between the x-axis and the nonnegative function $k(x) = v(x)\dfrac{du}{dx}$ is 20. Determine $u(1)v(1)$.

35. Use integration by parts to derive the following formula from the table of integrals.

$$\int x^n \cdot \ln |x|\, dx = x^{n+1}\left[\frac{\ln |x|}{n + 1} - \frac{1}{(n + 1)^2}\right] + C, \quad n \ne -1$$

36. Use integration by parts to derive the following formula from the table of integrals.

$$\int x^n e^{ax}\, dx = \frac{x^n e^{ax}}{a} - \frac{n}{a}\int x^{n-1}e^{ax}\, dx + C, \quad a \ne 0$$

37. a. One way to integrate $\int x\sqrt{x + 1}\, dx$ is to use integration by parts. Do so to find the antiderivative.

b. Another way to evaluate the integral in part a is by using the substitution $u = x + 1$. Do so to find the antiderivative.

c. Compare the results from the two methods. If they do not look the same, explain how this can happen. Discuss the advantages and disadvantages of each method.

38. Using integration by parts,

$$\int \frac{1}{x}\, dx = \int \frac{1}{x} \cdot 1\, dx$$
$$= \frac{1}{x} \cdot x - \int \left(-\frac{1}{x^2}\right)x\, dx$$
$$= 1 + \int \frac{1}{x}\, dx.$$

Subtracting $\int \frac{1}{x}\, dx$ from both sides we conclude that $0 = 1$. What is wrong with this logic? *Source: Sam Northshield.*

APPLICATIONS

Business and Economics

39. Rate of Change of Revenue The rate of change of revenue (in dollars per calculator) from the sale of x calculators is

$$R'(x) = (x + 1)\ln(x + 1).$$

Find the total revenue from the sale of the first 12 calculators. (*Hint:* In this exercise, it simplifies matters to write an antiderivative of $x + 1$ as $(x + 1)^2/2$ rather than $x^2/2 + x$.)

Life Sciences

40. Reaction to a Drug The rate of reaction to a drug is given by

$$r'(t) = 2t^2 e^{-t},$$

where t is the number of hours since the drug was administered. Find the total reaction to the drug from $t = 1$ to $t = 6$.

41. Growth of a Population The rate of growth of a microbe population is given by

$$m'(t) = 27te^{3t},$$

where t is time in days. What is the total accumulated growth during the first 2 days?

42. APPLY IT Rate of Growth The area covered by a patch of moss is growing at a rate of

$$A'(t) = \sqrt{t}\,\ln t$$

cm^2 per day, for $t \ge 1$. Find the additional amount of area covered by the moss between 4 and 9 days.

43. Thermic Effect of Food As we saw in an earlier chapter, a person's metabolic rate tends to go up after eating a meal and then, after some time has passed, it returns to a resting metabolic rate. This phenomenon is known as the thermic effect of food, and the effect (in kJ per hour) for one individual is

$$F(t) = -10.28 + 175.9te^{-t/1.3},$$

where t is the number of hours that have elapsed since eating a meal. *Source: American Journal of Clinical Nutrition*. Find the total thermic energy of a meal for the next six hours after a meal by integrating the thermic effect function between $t = 0$ and $t = 6$.

44. Rumen Fermentation The rumen is the first division of the stomach of a ruminant, or cud-chewing animal. An article on the rumen microbial system reports that the fraction of the soluble material passing from the rumen without being fermented during the first hour after its ingestion could be calculated by the integral

$$\int_0^1 ke^{-kt}(1-t)dt,$$

where k measures the rate that the material is fermented. *Source: Annual Review of Ecology and Systematics*.

a. Determine the above integral, and evaluate it for the following values of k used in the article: 1/12, 1/24, and 1/48 hour.

b. The fraction of intermediate material left in the rumen at 1 hour that escapes digestion by passage between 1 and 6 hours is given by

$$\int_1^6 ke^{-kt}(6-t)/5\, dt.$$

Determine this integral, and evaluate it for the values of k given in part a.

◼ **YOUR TURN ANSWERS**

1. $-e^{-2x}(2x+1)/4 + C$ **2.** $x\ln 2x - x + C$

3. $(6x^2 - 6x + 11)e^{2x}/4 + C$ **4.** $(2e^3 + 1)/9$

5. $(-1/2)\ln|(2 + \sqrt{4+x^2})/x| + C$

8.2 Volume and Average Value

APPLY IT **If we have a formula giving the price of a common stock as a function of time, how can we find the average price of the stock over a certain period of time?**

We will answer this question in Example 4 using concepts developed in this section, in which we will discover how to find the average value of a function, as well as how to compute the volume of a solid.

Volume Figure 1 shows the regions below the graph of some function $y = f(x)$, above the x-axis, and between $x = a$ and $x = b$. We have seen how to use integrals to find the area of such a region. Suppose this region is revolved about the x-axis as shown in Figure 2. The resulting figure is called a **solid of revolution**. In many cases, the volume of a solid of revolution can be found by integration.

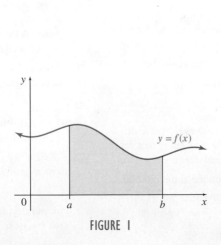

FIGURE 1

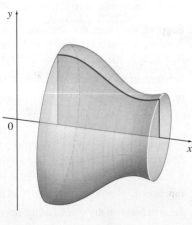

FIGURE 2

To begin, divide the interval $[a, b]$ into n subintervals of equal width Δx by the points $a = x_0, x_1, x_2, \ldots, x_i, \ldots, x_n = b$. Then think of slicing the solid into n slices of equal thickness Δx, as shown in Figure 3(a). If the slices are thin enough, each slice is very close to being a right circular cylinder, as shown in Figure 3(b). The formula for the volume of a right circular cylinder is $\pi r^2 h$, where r is the radius of the circular base and h is the height of the cylinder. As shown in Figure 4, the height of each slice is Δx. (The height is horizontal here, since the cylinder is on its side.) The radius of the circular base of each slice is $f(x_i)$. Thus, the volume of the slice is closely approximated by $\pi[f(x_i)]^2\Delta x$. The volume of the solid of revolution will be approximated by the sum of the volumes of the slices:

$$V \approx \sum_{i=1}^{n} \pi[f(x_i)]^2\Delta x.$$

By definition, the volume of the solid of revolution is the limit of this sum as the thickness of the slices approaches 0, or

$$V = \lim_{\Delta x \to 0} \sum_{i=1}^{n} \pi[f(x_i)]^2\Delta x.$$

This limit, like the one discussed earlier for area, is a definite integral.

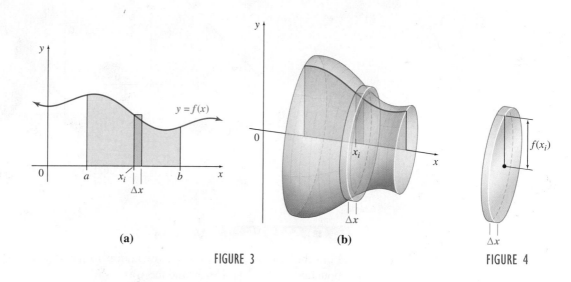

(a)   **(b)**

FIGURE 3   FIGURE 4

Volume of a Solid of Revolution

If $f(x)$ is nonnegative and R is the region between $f(x)$ and the x-axis from $x = a$ to $x = b$, the volume of the solid formed by rotating R about the x-axis is given by

$$V = \lim_{\Delta x \to 0} \sum_{i=1}^{n} \pi[f(x_i)]^2\Delta x = \int_{a}^{b} \pi[f(x)]^2 \, dx.$$

The technique of summing disks to approximate volumes was originated by Johannes Kepler (1571–1630), a famous German astronomer who discovered three laws of planetary motion. He estimated volumes of wine casks used at his wedding by means of solids of revolution.

EXAMPLE 1   **Volume**

Find the volume of the solid of revolution formed by rotating about the x-axis the region bounded by $y = x + 1$, $y = 0$, $x = 1$, and $x = 4$.

SOLUTION The region and the solid are shown in Figure 5. Notice that the orientation of the x-axis is slightly different in Figure 5(b) than in Figure 5(a) to emphasize the three-dimensionality of the figure. Use the formula given above for the volume, with $a = 1$, $b = 4$, and $f(x) = x + 1$.

$$V = \int_1^4 \pi(x + 1)^2 \, dx = \pi \left[\frac{(x + 1)^3}{3} \right]\Big|_1^4$$

$$= \frac{\pi}{3}(5^3 - 2^3)$$

$$= \frac{117\pi}{3} = 39\pi \qquad \text{TRY YOUR TURN 1}$$

YOUR TURN 1 Find the volume of the solid of revolution formed by rotating about the x-axis the region bounded by $y = x^2 + 1$, $y = 0$, $x = -1$, and $x = 1$.

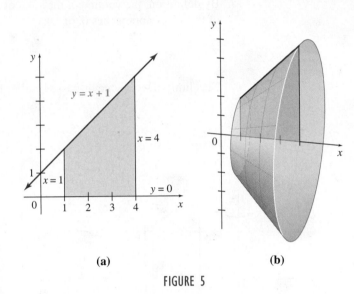

(a)   (b)

FIGURE 5

EXAMPLE 2 Volume

Find the volume of the solid of revolution formed by rotating about the x-axis the area bounded by $f(x) = 4 - x^2$ and the x-axis.

SOLUTION The region and the solid are shown in Figure 6 on the next page. Find a and b from the x-intercepts. If $y = 0$, then $x = 2$ or $x = -2$, so that $a = -2$ and $b = 2$. The volume is

$$V = \int_{-2}^{2} \pi(4 - x^2)^2 \, dx$$

$$= \int_{-2}^{2} \pi(16 - 8x^2 + x^4) \, dx$$

$$= \pi \left(16x - \frac{8x^3}{3} + \frac{x^5}{5} \right)\Big|_{-2}^{2}$$

$$= \frac{512\pi}{15}.$$

TECHNOLOGY NOTE A graphing calculator with the `fnInt` feature gives the value as 107.2330292, which agrees with the approximation of $512\pi/15$ to the 7 decimal places shown.

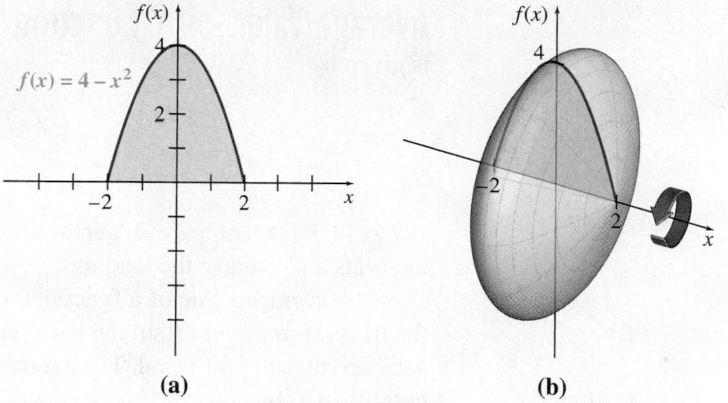

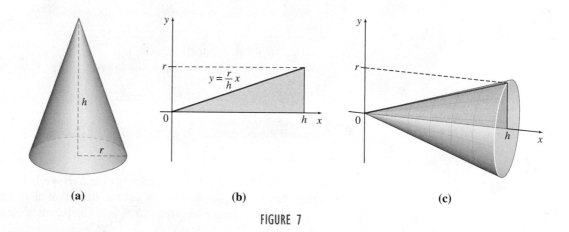

$f(x) = 4 - x^2$

(a)

(b)

FIGURE 6

EXAMPLE 3   Volume

Find the volume of a right circular cone with height h and base radius r.

SOLUTION   Figure 7(a) shows the required cone, while Figure 7(b) shows an area that could be rotated about the x-axis to get such a cone. The cone formed by the rotation is shown in Figure 7(c). Here $y = f(x)$ is the equation of the line through $(0, 0)$ and (h, r). The slope of this line is r/h, and since the y-intercept is 0, the equation of the line is

$$y = \frac{r}{h} x.$$

$y = \frac{r}{h} x$

(a)

(b)

(c)

FIGURE 7

Then the volume is

$$V = \int_0^h \pi \left(\frac{r}{h} x \right)^2 dx = \pi \int_0^h \frac{r^2 x^2}{h^2} dx$$

$$= \pi \frac{r^2 x^3}{3h^2} \bigg|_0^h \qquad \text{Since } r \text{ and } h \text{ are constants}$$

$$= \frac{\pi r^2 h}{3}.$$

This is the familiar formula for the volume of a right circular cone.

Average Value of a Function

The average of the n numbers $v_1, v_2, v_3, \ldots, v_i, \ldots, v_n$ is given by

$$\frac{v_1 + v_2 + v_3 + \cdots + v_n}{n} = \frac{\sum\limits_{i=1}^{n} v_i}{n}.$$

For example, to compute an average temperature, we could take readings at equally spaced intervals and average the readings.

The average value of a function f on $[a, b]$ can be defined in a similar manner; divide the interval $[a, b]$ into n subintervals, each of width Δx. Then choose an x-value, x_i, in each subinterval, and find $f(x_i)$. The average function value for the n subintervals and the given choices of x_i is

$$\frac{f(x_1) + f(x_2) + \cdots + f(x_n)}{n} = \frac{\sum\limits_{i=1}^{n} f(x_i)}{n}.$$

Since $(b - a)/n = \Delta x$, multiply the expression on the right side of the equation by $(b - a)/(b - a)$ and rearrange the expression to get

$$\frac{b - a}{b - a} \cdot \frac{\sum\limits_{i=1}^{n} f(x_i)}{n} = \frac{1}{b - a} \sum\limits_{i=1}^{n} f(x_i) \left(\frac{b - a}{n} \right) = \frac{1}{b - a} \sum\limits_{i=1}^{n} f(x_i) \Delta x.$$

Now, take the limit as $n \to \infty$. If the limit exists, then

$$\lim_{n \to \infty} \frac{1}{b - a} \sum\limits_{i=1}^{n} f(x_i) \Delta x = \frac{1}{b - a} \lim_{n \to \infty} \sum\limits_{i=1}^{n} f(x_i) \Delta x = \frac{1}{b - a} \int_a^b f(x) \, dx.$$

The following definition summarizes this discussion.

Average Value of a Function

The **average value of a function** f on the interval $[a, b]$ is

$$\frac{1}{b - a} \int_a^b f(x) \, dx,$$

provided the indicated definite integral exists.

In Figure 8 the quantity $\bar{y}$ represents the average height of the irregular region. The average height can be thought of as the height of a rectangle with base $b - a$. For $f(x) \geq 0$, this rectangle has area $\bar{y}(b - a)$, which equals the area under the graph of $f(x)$ from $x = a$ to $x = b$, so that

$$\bar{y}(b - a) = \int_a^b f(x) \, dx.$$

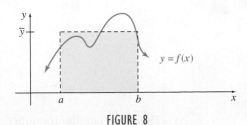

FIGURE 8

| EXAMPLE 4 | **Average Price**

A stock analyst plots the price per share of a certain common stock as a function of time and finds that it can be approximated by the function

$$S(t) = 25 - 5e^{-0.01t},$$

where t is the time (in years) since the stock was purchased. Find the average price of the stock over the first six years.

APPLY IT   **SOLUTION**   Use the formula for average value with $a = 0$ and $b = 6$. The average price is

$$\frac{1}{6-0} \int_0^6 (25 - 5e^{-0.01t})\, dt = \frac{1}{6}\left(25t - \frac{5}{-0.01}e^{-0.01t} \right)\Big|_0^6$$

$$= \frac{1}{6}(25t + 500e^{-0.01t})\Big|_0^6$$

$$= \frac{1}{6}(150 + 500e^{-0.06} - 500)$$

$$\approx 20.147,$$

or approximately \$20.15.

YOUR TURN 2 Find the average value of the function $f(x) = x + \sqrt{x}$ on the interval $[1, 4]$.

TRY YOUR TURN 2

8.2   EXERCISES

Find the volume of the solid of revolution formed by rotating about the *x*-axis each region bounded by the given curves.

1. $f(x) = x, \quad y = 0, \quad x = 0, \quad x = 3$

2. $f(x) = 3x, \quad y = 0, \quad x = 0, \quad x = 2$

3. $f(x) = 2x + 1, \quad y = 0, \quad x = 0, \quad x = 4$

4. $f(x) = x - 4, \quad y = 0, \quad x = 4, \quad x = 10$

5. $f(x) = \frac{1}{3}x + 2, \quad y = 0, \quad x = 1, \quad x = 3$

6. $f(x) = \frac{1}{2}x + 4, \quad y = 0, \quad x = 0, \quad x = 5$

7. $f(x) = \sqrt{x}, \quad y = 0, \quad x = 1, \quad x = 4$

8. $f(x) = \sqrt{x + 5}, \quad y = 0, \quad x = 1, \quad x = 3$

9. $f(x) = \sqrt{2x + 1}, \quad y = 0, \quad x = 1, \quad x = 4$

10. $f(x) = \sqrt{4x + 2}, \quad y = 0, \quad x = 0, \quad x = 2$

11. $f(x) = e^x, \quad y = 0, \quad x = 0, \quad x = 2$

12. $f(x) = 2e^x, \quad y = 0, \quad x = -2, \quad x = 1$

13. $f(x) = \frac{2}{\sqrt{x}}, \quad y = 0, \quad x = 1, \quad x = 3$

14. $f(x) = \frac{2}{\sqrt{x + 2}}, \quad y = 0, \quad x = -1, \quad x = 2$

15. $f(x) = x^2, \quad y = 0, \quad x = 1, \quad x = 5$

16. $f(x) = \frac{x^2}{2}, \quad y = 0, \quad x = 0, \quad x = 4$

17. $f(x) = 1 - x^2, \quad y = 0$

18. $f(x) = 2 - x^2, \quad y = 0$

The function defined by $y = \sqrt{r^2 - x^2}$ has as its graph a semicircle of radius r with center at $(0, 0)$ (see the figure). In Exercises 19–21, find the volume that results when each semicircle is rotated about the *x*-axis. (The result of Exercise 21 gives a formula for the volume of a sphere with radius r.)

19. $f(x) = \sqrt{1 - x^2}$ **20.** $f(x) = \sqrt{36 - x^2}$

21. $f(x) = \sqrt{r^2 - x^2}$

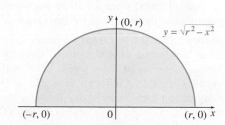

22. Find a formula for the volume of an ellipsoid. See Exercises 19–21 and the following figures.

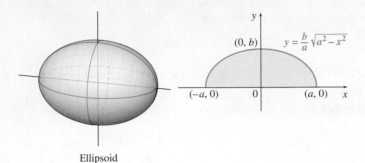

Ellipsoid

23. Use the methods of this section to find the volume of a cylinder with height h and radius r.

Find the average value of each function on the given interval.

24. $f(x) = 2 - 3x^2$; $[1, 3]$ **25.** $f(x) = x^2 - 4$; $[0, 5]$

26. $f(x) = (2x - 1)^{1/2}$; $[1, 13]$

27. $f(x) = \sqrt{x + 1}$; $[3, 8]$ **28.** $f(x) = e^{0.1x}$; $[0, 10]$

29. $f(x) = e^{x/7}$; $[0, 7]$ **30.** $f(x) = x \ln x$; $[1, e]$

31. $f(x) = x^2 e^{2x}$; $[0, 2]$

In Exercises 32 and 33, use the integration feature on a graphing calculator to find the volume of the solid of revolution by rotating about the *x*-axis each region bounded by the given curves.

32. $f(x) = \dfrac{1}{4 + x^2}$, $y = 0$, $x = -2$, $x = 2$

33. $f(x) = e^{-x^2}$, $y = 0$, $x = -1$, $x = 1$

APPLICATIONS

Business and Economics

34. Average Price Otis Taylor plots the price per share of a stock that he owns as a function of time and finds that it can be approximated by the function

$$S(t) = t(25 - 5t) + 18,$$

where t is the time (in years) since the stock was purchased. Find the average price of the stock over the first five years.

35. Average Price A stock analyst plots the price per share of a certain common stock as a function of time and finds that it can be approximated by the function

$$S(t) = 37 + 6e^{-0.03t},$$

where t is the time (in years) since the stock was purchased. Find the average price of the stock over the first six years.

36. Average Inventory The Yasuko Okada Fragrance Company (YOFC) receives a shipment of 400 cases of specialty perfume early Monday morning of every week. YOFC sells the perfume to retail outlets in California at a rate of about 80 cases per day during each business day (Monday through Friday). What is the average daily inventory for YOFC? (*Hint:* Find a function that represents the inventory for any given business day and then integrate.)

37. Average Inventory The DeMarco Pasta Company receives 600 cases of imported San Marzano tomato sauce every 30 days. The number of cases of sauce on inventory t days after the shipment arrives is

$$N(t) = 600 - 20\sqrt{30t}.$$

Find the average daily inventory.

Life Sciences

38. Blood Flow The figure shows the blood flow in a small artery of the body. The flow of blood is *laminar* (in layers), with the velocity very low near the artery walls and highest in the center of the artery. In this model of blood flow, we calculate the total flow in the artery by thinking of the flow as being made up of many layers of concentric tubes sliding one on the other.

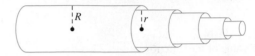

Suppose R is the radius of an artery and r is the distance from a given layer to the center. Then the velocity of blood in a given layer can be shown to equal

$$v(r) = k(R^2 - r^2),$$

where k is a numerical constant.

Since the area of a circle is $A = \pi r^2$, the change in the area of the cross section of one of the layers, corresponding to a small change in the radius, Δr, can be approximated by differentials. For $dr = \Delta r$, the differential of the area A is

$$dA = 2\pi r \, dr = 2\pi r \, \Delta r,$$

where Δr is the thickness of the layer. The total flow in the layer is defined to be the product of velocity and cross-section area, or

$$F(r) = 2\pi r k (R^2 - r^2)\Delta r.$$

a. Set up a definite integral to find the total flow in the artery.

b. Evaluate this definite integral.

39. Drug Reaction The intensity of the reaction to a certain drug, in appropriate units, is given by

$$R(t) = te^{-0.1t},$$

where t is time (in hours) after the drug is administered. Find the average intensity during the following hours.

a. Second hour

b. Twelfth hour

c. Twenty-fourth hour

40. Bird Eggs The average length and width of various bird eggs are given in the following table. *Source: NCTM.*

Bird Name	Length (cm)	Width (cm)
Canada goose	8.6	5.8
Robin	1.9	1.5
Turtledove	3.1	2.3
Hummingbird	1.0	1.0
Raven	5.0	3.3

a. Assume for simplicity that a bird's egg is roughly the shape of an ellipsoid. Use the result of Exercise 22 to estimate the volume of an egg of each bird.

   i. Canada goose

   ii. Robin

   iii. Turtledove

   iv. Hummingbird

   v. Raven

b. In Exercise 12 of Section 1.3, we showed that the average length (in centimeters) of an egg of width w cm is given by

$$l = 1.585w - 0.487.$$

Using this result and the ideas in part a, show that the average volume of an egg of width w centimeters is given by

$$V = \pi(1.585w^3 - 0.487w^2)/6.$$

Use this formula to calculate the average volume for the bird eggs in part a, and compare with your results from part a.

Social Sciences

41. Production Rate Suppose the number of items a new worker on an assembly line produces daily after t days on the job is given by

$$I(t) = 45\ln(t + 1).$$

Find the average number of items produced daily by this employee after the following numbers of days.

a. 5   **b.** 9   **c.** 30

42. Typing Speed The function $W(t) = -3.75t^2 + 30t + 40$ describes a typist's speed (in words per minute) over a time interval $[0, 5]$.

a. Find $W(0)$.

b. Find the maximum W value and the time t when it occurs.

c. Find the average speed over $[0, 5]$.

Physical Sciences

43. Earth's Volume Most people assume that the Earth has a spherical shape. It is actually more of an ellipsoid shape, but not an exact ellipsoid, since there are numerous mountains and valleys. Researchers have found that a *datum*, or a reference ellipsoid, that is offset from the center of the Earth can be used to accurately map different regions. According to one datum, called the Geodetic Reference System 1980, this reference ellipsoid assumes an equatorial radius of 6,378,137 m and a polar radius of 6,356,752.3141 m. *Source: Geodesy Information System.* Use the result of Exercise 22 to estimate the volume of the Earth.

YOUR TURN ANSWERS

1. $56\pi/15$ **2.** $73/18$

8.3 Continuous Money Flow

APPLY IT   Given a changing rate of annual income and a certain rate of interest, how can we find the present value of the income?

We will answer this question in Example 2 using the concept of continuous money flow.

In an earlier chapter we looked at the concepts of present value and future value when a lump sum of money is deposited in an account and allowed to accumulate interest. In some situations, however, money flows into and out of an account almost continuously over a period of time. Examples include income in a store, bank receipts and payments, and highway tolls. Although the flow of money in such cases is not exactly continuous, it can be treated as though it were continuous, with useful results.

EXAMPLE 1   Total Income

The income from a soda machine (in dollars per year) is growing exponentially. When the machine was first installed, it was producing income at a rate of $500 per year. By the end of the first year, it was producing income at a rate of $510.10 per year. Find the total income produced by the machine during its first 3 years of operation.

SOLUTION   Let t be the time (in years) since the installation of the machine. The assumption of exponential growth, coupled with the initial value of 500, implies that the rate of change of income is of the form

$$f(t) = 500e^{kt},$$

where k is some constant. To find k, use the value at the end of the first year.

$$f(1) = 500e^{k(1)} = 510.10$$

$$e^k = 1.0202 \qquad \text{Divide by 500.}$$

$$k = \ln 1.0202 \qquad \text{Take ln of both sides.}$$

$$\approx 0.02$$

Therefore, we have

$$f(t) = 500e^{0.02t}.$$

Since the rate of change of incomes is given, the total income can be determined by using the definite integral.

$$\text{Total income} = \int_0^3 500e^{0.02t}\, dt$$

$$= \frac{500}{0.02}e^{0.02t}\Big|_0^3$$

$$= 25{,}000e^{0.02t}\Big|_0^3 = 25{,}000(e^{0.06} - 1) \approx 1545.91$$

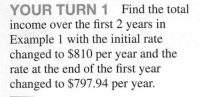

YOUR TURN 1 Find the total income over the first 2 years in Example 1 with the initial rate changed to \$810 per year and the rate at the end of the first year changed to \$797.94 per year.

Thus, the soda machine will produce \$1545.91 total income in its first three years of operation.

TRY YOUR TURN 1

The money in Example 1 is not received as a one-time lump sum payment of \$1545.91. Instead, it comes in on a regular basis, perhaps daily, weekly, or monthly. In discussions of such problems it is usually assumed that the income is received continuously over a period of time.

Total Money Flow

Let the continuous function $f(t)$ represent the rate of flow of money per unit of time. If t is in years and $f(t)$ is in dollars per year, the area under $f(t)$ between two points in time gives the total dollar flow over the given time interval.

The function $f(t) = 2000$, shown in Figure 9, represents a uniform rate of money flow of \$2000 per year. The graph of this money flow is a horizontal line; the *total money flow* over a specified time T is given by the rectangular area below the graph of $f(t)$ and above the t-axis between $t = 0$ and $t = T$. For example, the total money flow over $T = 5$ years would be $2000(5) = 10{,}000$, or \$10,000.

The area in the uniform rate example could be found by using an area formula from geometry. For a variable function like the function in Example 1, however, a definite integral is needed to find the total money flow over a specific time interval. For the function

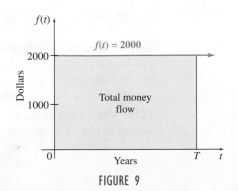

FIGURE 9

$f(t) = 2000e^{0.08t}$, for example, the total money flow over a 5-year period would be given by

$$\int_0^5 2000e^{0.08t}\, dt \approx 12{,}295.62,$$

or \$12,295.62. See Figure 10.

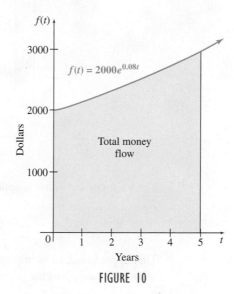

FIGURE 10

Total Money Flow

If $f(t)$ is the rate of money flow, then the **total money flow** over the time interval from $t = 0$ to $t = T$ is given by

$$\int_0^T f(t)\, dt.$$

This "total money flow" does not take into account the interest the money could earn after it is received. It is simply the total income.

Present Value of Money Flow
As mentioned earlier, an amount of money that can be deposited today at a specified interest rate to yield a given sum in the future is called the *present value* of this future sum. The future sum may be called the *future value* or *final amount*. To find the **present value of a continuous money flow** with interest compounded continuously, let $f(t)$ represent the rate of the continuous flow. In Figure 11 on the next page, the time axis from 0 to T is divided into n subintervals, each of width Δt. The amount of money that flows during any interval of time is given by the area between the t-axis and the graph of $f(t)$ over the specified time interval. The area of each subinterval is approximated by the area of a rectangle with height $f(t_i)$, where t_i is the left endpoint of the ith subinterval. The area of each rectangle is $f(t_i)\Delta t$, which (approximately) gives the amount of money flow over that subinterval.

Earlier, we saw that the present value P of an amount A compounded continuously for t years at a rate of interest r is $P = Ae^{-rt}$. Letting t_i represent the time and replacing A with $f(t_i)\Delta t$, the present value of the money flow over the ith subinterval is approximately equal to

$$P_i = [f(t_i)\Delta t]e^{-rt_i}.$$

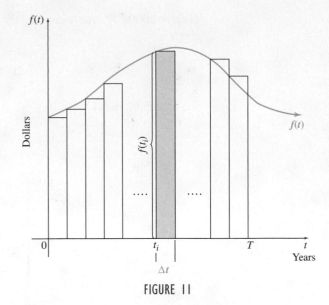

FIGURE 11

The total present value is approximately equal to the sum

$$\sum_{i=1}^{n}[f(t_i)\Delta t]e^{-rt_i}.$$

This approximation is improved as n increases; taking the limit of the sum as n increases without bound gives the present value

$$P = \lim_{n \to \infty}\sum_{i=1}^{n}[f(t_i)\Delta t]e^{-rt_i}.$$

This limit of a summation is given by the following definite integral.

> **Present Value of Money Flow**
> If $f(t)$ is the rate of continuous money flow at an interest rate r for T years, then the present value is
>
> $$P = \int_{0}^{T} f(t)e^{-rt}\,dt.$$

To understand present value of money flow, consider an account that earns interest and has a continuous money flow. The present value of the money flow is the amount that would have to be deposited into a second account that has the same interest rate but does not have a continuous money flow, so the two accounts have the same amount of money after a specified time.

EXAMPLE 2 **Present Value of Income**

A company expects its rate of annual income during the next three years to be given by

$$f(t) = 75{,}000t, \quad 0 \le t \le 3.$$

What is the present value of this income over the 3-year period, assuming an annual interest rate of 8% compounded continuously?

APPLY IT   **SOLUTION**   Use the formula for present value, with $f(t) = 75{,}000t$, $T = 3$, and $r = 0.08$.

$$P = \int_{0}^{3} 75{,}000t e^{-0.08t}\,dt = 75{,}000\int_{0}^{3} t e^{-0.08t}\,dt$$

Using integration by parts, verify that

$$\int te^{-0.08t}\, dt = -12.5te^{-0.08t} - 156.25e^{-0.08t} + C.$$

Therefore,

$$75,000\int_0^3 te^{-0.08t}\, dt = 75,000(-12.5te^{-0.08t} - 156.25e^{-0.08t})\Big|_0^3$$

$$= 75,000\big[-12.5(3)e^{-0.08(3)} - 156.25e^{-0.08(3)} - (0 - 156.25)\big]$$

$$\approx 75,000(-29.498545 - 122.910603 + 156.25)$$

$$\approx 288,064,$$

or about \$288,000. Notice that the actual income over the 3-year period is given by

$$\text{Total money flow} = \int_0^3 75,000t\, dt = \frac{75,000t^2}{2}\Big|_0^3 = 337,500,$$

YOUR TURN 2 Find the present value of an income given by $f(t) = 50,000t$ over the next 5 years if the interest rate is 3.5%.

or \$337,500. This means that it would take a lump-sum deposit of \$288,064 today paying a continuously compounded interest rate of 8% over a 3-year period to equal the total cash flow of \$337,500 with interest. This approach is used as a basis for determining insurance claims involving income considerations. **TRY YOUR TURN 2**

Accumulated Amount of Money Flow at Time T

To find the **accumulated amount of money flow** with interest at any time t, start with the formula $A = Pe^{rt}$, let $t = T$, and in place of P substitute the expression for present value of money flow. The result is the following formula.

> ### Accumulated Amount of Money Flow at Time T
> If $f(t)$ is the rate of money flow at an interest rate r at time t, the accumulated amount of money flow at time T is
> $$A = e^{rT}\int_0^T f(t)e^{-rt}\, dt.$$

Here, the accumulated amount of money A represents the accumulated value or final amount of the money flow *including* interest received on the money after it comes in. (Recall, total money flow *does not* take the interest into account.)

It turns out that most money flows can be expressed as (or at least approximated by) exponential or polynomial functions. When these are multiplied by e^{-rt}, the result is a function that can be integrated. The next example illustrates uniform flow, where $f(t)$ is a constant function. (This is a special case of the polynomial function.)

EXAMPLE 3   Accumulated Amount of Money Flow

If money is flowing continuously at a constant rate of \$2000 per year over 5 years at 6% interest compounded continuously, find the following.

(a) The total money flow over the 5-year period

SOLUTION The total money flow is given by $\int_0^T f(t)\, dt$. Here $f(t) = 2000$ and $T = 5$.

$$\int_0^5 2000\, dt = 2000t\Big|_0^5 = 2000(5) = 10,000$$

The total money flow over the 5-year period is \$10,000.

(b) The accumulated amount of money flow, compounded continuously, at time $T = 5$

SOLUTION At $T = 5$ with $r = 0.06$, the amount is

$$A = e^{rT}\int_0^T f(t)e^{-rt}\,dt = e^{(0.06)5}\int_0^5 (2000)e^{-0.06t}\,dt$$

$$= (e^{0.3})(2000)\int_0^5 e^{-0.06t}\,dt = e^{0.3}(2000)\left(\frac{1}{-0.06}\right)\left(e^{-0.06t}\Big|_0^5\right)$$

$$= \frac{2000e^{0.3}}{-0.06}(e^{-0.3}-1) = \frac{2000}{-0.06}(1-e^{0.3}) \qquad (e^{0.03})(e^{-0.03}) = 1$$

$$\approx 11{,}661.96,$$

or \$11,661.96. The answer to part (a), \$10,000, was the amount of money flow over the 5-year period. The \$11,661.96 gives that amount with interest compounded continuously over the 5-year period.

(c) The total interest earned

SOLUTION This is simply the accumulated amount of money flow minus the total amount of flow, or

$$\$11{,}661.96 - \$10{,}000.00 = \$1661.96.$$

(d) The present value of the amount with interest

SOLUTION Use $P = \int_0^T f(t)e^{-rt}\,dt$ with $f(t) = 2000$, $r = 0.06$, and $T = 5$.

$$P = \int_0^5 2000e^{-0.06t}\,dt = 2000\left(\frac{e^{-0.06t}}{-0.06}\right)\Big|_0^5$$

$$= \frac{2000}{-0.06}(e^{-0.3}-1)$$

$$\approx 8639.39,$$

YOUR TURN 3 Find the accumulated amount of money flow for the income and interest rate in Your Turn 2 in this section.

The present value of the amount with interest in 5 years is \$8639.39, which can be checked by substituting \$11,661.96 for A in $A = Pe^{rt}$. The present value, P, could have been found by dividing the amount found in (b) by $e^{rT} = e^{0.3}$. Check that this would give the same result. **TRY YOUR TURN 3**

If the rate of money flow is increasing or decreasing exponentially, then $f(t) = Ce^{kt}$, where C is a constant that represents the initial amount and k is the (nominal) continuous rate of change, which may be positive or negative.

EXAMPLE 4 Accumulated Amount of Money Flow

A continuous money flow starts at a rate of \$1000 per year and increases exponentially at 2% per year.

(a) Find the accumulated amount of money flow at the end of 5 years at 10% interest compounded continuously.

SOLUTION Here $C = 1000$ and $k = 0.02$, so that $f(t) = 1000e^{0.02t}$. Using $r = 0.10$ and $T = 5$,

$$A = e^{(0.10)5}\int_0^5 1000e^{0.02t}e^{-0.10t}\,dt$$

$$= (e^{0.5})(1000)\int_0^5 e^{-0.08t}\,dt \qquad e^{0.02t}\cdot e^{-0.10t} = e^{-0.08t}$$

$$= 1000e^{0.5}\left(\frac{e^{-0.08t}}{-0.08}\right)\Big|_0^5$$

$$= \frac{1000e^{0.5}}{-0.08}(e^{-0.4}-1) = \frac{1000}{-0.08}(e^{0.1}-e^{0.5}) \approx 6794.38,$$

or \$6794.38.

(b) Find the present value at 5% interest compounded continuously.

SOLUTION Using $f(t) = 1000e^{0.02t}$ with $r = 0.05$ and $T = 5$ in the present value expression,

$$P = \int_0^5 1000e^{0.02t}e^{-0.05t}\,dt$$

$$= 1000 \int_0^5 e^{-0.03t}\,dt = 1000\left(\frac{e^{-0.03t}}{-0.03}\Big|_0^5\right)$$

$$= \frac{1000}{-0.03}(e^{-0.15} - 1) \approx 4643.07,$$

or $4643.07.

If the rate of change of the continuous money flow is given by the polynomial function $f(t) = a_n t^n + a_{n-1} t^{n-1} + \cdots + a_0$, the expressions for present value and accumulated amount can be integrated using integration by parts.

EXAMPLE 5 Present Value of Money Flow

The rate of change of a continuous flow of money is given by

$$f(t) = 1000t^2 + 100t.$$

Find the present value of this money flow at the end of 10 years at 10% compounded continuously.

SOLUTION Evaluate

$$P = \int_0^{10} (1000t^2 + 100t)e^{-0.10t}\,dt.$$

Using integration by parts, verify that

$$\int (1000t^2 + 100t)e^{-0.10t}\,dt =$$

$$(-10,000t^2 - 1000t)e^{-0.1t} - (200,000t + 10,000)e^{-0.1t} - 2,000,000e^{-0.1t} + C.$$

Thus,

$$P = (-10,000t^2 - 1000t)e^{-0.1t} - (200,000t + 10,000)e^{-0.1t}$$

$$- 2,000,000e^{-0.1t}\Big|_0^{10}$$

$$= (-1,000,000 - 10,000)e^{-1} - (2,000,000 + 10,000)e^{-1}$$

$$- 2,000,000e^{-1} - (0 - 10,000 - 2,000,000)$$

$$\approx 163,245.21.$$

TRY YOUR TURN 4

YOUR TURN 4 Find the present value at the end of 8 years of the continuous flow of money given by $f(t) = 200t^2 + 100t + 50$ at 5% compounded continuously.

8.3 EXERCISES

Each of the functions in Exercises 1–14 represents the rate of flow of money in dollars per year. Assume a 10-year period at 8% compounded continuously and find the following: (a) the present value; (b) the accumulated amount of money flow at $t = 10$.

1. $f(t) = 1000$

2. $f(t) = 300$

3. $f(t) = 500$

4. $f(t) = 2000$

5. $f(t) = 400e^{0.03t}$

6. $f(t) = 800e^{0.05t}$

7. $f(t) = 5000e^{-0.01t}$

8. $f(t) = 1000e^{-0.02t}$

9. $f(t) = 25t$

10. $f(t) = 50t$

11. $f(t) = 0.01t + 100$

12. $f(t) = 0.05t + 500$

13. $f(t) = 1000t - 100t^2$

14. $f(t) = 2000t - 150t^2$

APPLICATIONS

Business and Economics

15. Accumulated Amount of Money Flow An investment is expected to yield a uniform continuous rate of money flow of $20,000 per year for 3 years. Find the accumulated amount at an interest rate of 4% compounded continuously.

16. Present Value A real estate investment is expected to produce a uniform continuous rate of money flow of $8000 per year for 6 years. Find the present value at the following rates, compounded continuously.

    a. 2%   **b.** 5%   **c.** 8%

17. Money Flow The rate of a continuous flow of money starts at $5000 and decreases exponentially at 1% per year for 8 years. Find the present value and final amount at an interest rate of 8% compounded continuously.

18. Money Flow The rate of a continuous money flow starts at $1000 and increases exponentially at 5% per year for 4 years. Find the present value and accumulated amount if interest earned is 3.5% compounded continuously.

19. Present Value A money market fund has a continuous flow of money at a rate of $f(t) = 1500 - 60t^2$, reaching 0 in 5 years. Find the present value of this flow if interest is 5% compounded continuously.

20. Accumulated Amount of Money Flow Find the amount of a continuous money flow in 3 years if the rate is given by $f(t) = 1000 - t^2$ and if interest is 5% compounded continuously.

■ **YOUR TURN ANSWERS**

1. $1595.94   2. $556,653
3. $663,111   4. $28,156.02

8.4 Improper Integrals

APPLY IT If we know the rate at which a pollutant is dumped into a stream, how can we compute the total amount released given that the rate of dumping is decreasing over time?

In this section we will learn how to answer questions such as this one, which is answered in Example 3.

Sometimes it is useful to be able to integrate a function over an infinite period of time. For example, we might want to find the total amount of income generated by an apartment building into the indefinite future or the total amount of pollution into a bay from a source that is continuing indefinitely. In this section we define integrals with one or more infinite limits of integration that can be used to solve such problems.

The graph in Figure 12(a) shows the area bounded by the curve $f(x) = x^{-3/2}$, the x-axis, and the vertical line $x = 1$. Think of the shaded region below the curve as extending indefinitely to the right. Does this shaded region have an area?

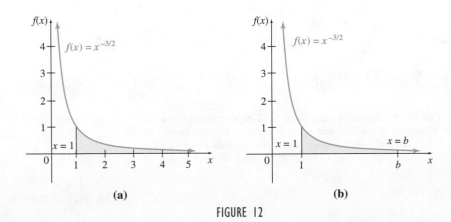

(a)   (b)

FIGURE 12

To see if the area of this region can be defined, introduce a vertical line at $x = b$, as shown in Figure 12(b). This vertical line gives a region with both upper and lower limits of integration. The area of this new region is given by the definite integral

$$\int_1^b x^{-3/2} \, dx.$$

By the Fundamental Theorem of Calculus,

$$\int_1^b x^{-3/2} \, dx = \left. \left(-2x^{-1/2} \right) \right|_1^b$$

$$= -2b^{-1/2} - \left(-2 \cdot 1^{-1/2} \right)$$

$$= -2b^{-1/2} + 2 = 2 - \frac{2}{b^{1/2}}.$$

FOR REVIEW

In Section 3.1 on Limits we saw that for any positive real number n,

$$\lim_{b \to \infty} \frac{1}{b^n} = 0.$$

Suppose we now let the vertical line $x = b$ in Figure 12(b) move farther to the right. That is, suppose $b \to \infty$. The expression $-2/b^{1/2}$ would then approach 0, and

$$\lim_{b \to \infty} \left(2 - \frac{2}{b^{1/2}} \right) = 2 - 0 = 2.$$

This limit is defined to be the *area* of the region shown in Figure 12(a), so that

$$\int_1^\infty x^{-3/2} \, dx = 2.$$

An integral of the form

$$\int_a^\infty f(x) \, dx, \qquad \int_{-\infty}^b f(x) \, dx, \qquad \text{or} \qquad \int_{-\infty}^\infty f(x) \, dx$$

is called an *improper integral*. These **improper integrals** are defined as follows.

Improper Integrals

If f is continuous on the indicated interval and if the indicated limits exist, then

$$\int_a^\infty f(x) \, dx = \lim_{b \to \infty} \int_a^b f(x) \, dx,$$

$$\int_{-\infty}^b f(x) \, dx = \lim_{a \to -\infty} \int_a^b f(x) \, dx,$$

$$\int_{-\infty}^\infty f(x) \, dx = \int_{-\infty}^c f(x) \, dx + \int_c^\infty f(x) \, dx,$$

for real numbers a, b, and c, where c is arbitrarily chosen.

If the expressions on the right side exist, the integrals are **convergent**; otherwise, they are **divergent**. A convergent integral has a value that is a real number. A divergent integral does not, often because the area under the curve is infinitely large.

EXAMPLE 1 Improper Integrals

Evaluate each integral.

(a) $\displaystyle\int_1^\infty \frac{dx}{x}$

SOLUTION A graph of this region is shown in Figure 13. By the definition of an improper integral,

$$\int_1^\infty \frac{dx}{x} = \lim_{b \to \infty} \int_1^b \frac{dx}{x}.$$

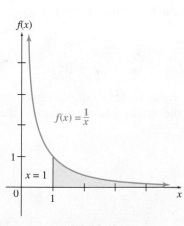

FIGURE 13

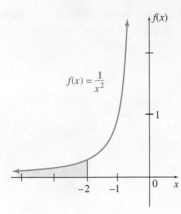

FIGURE 14

YOUR TURN 1 Find each integral.

(a) $\displaystyle\int_8^\infty \frac{1}{x^{1/3}}\,dx$ (b) $\displaystyle\int_8^\infty \frac{1}{x^{4/3}}\,dx$

Find $\displaystyle\int_1^b \frac{dx}{x}$ by the Fundamental Theorem of Calculus.

$$\int_1^b \frac{dx}{x} = \ln |x|\Big|_1^b = \ln |b| - \ln |1| = \ln |b| - 0 = \ln |b|$$

As $b \to \infty$, $\ln |b| \to \infty$, so $\displaystyle\lim_{b\to\infty} \ln |b|$ does not exist. Since the limit does not exist, $\displaystyle\int_1^\infty \frac{dx}{x}$ is divergent.

(b) $\displaystyle\int_{-\infty}^{-2} \frac{1}{x^2}\,dx = \lim_{a\to-\infty} \int_a^{-2} \frac{1}{x^2}\,dx = \lim_{a\to-\infty} \left(\frac{-1}{x}\right)\Big|_a^{-2}$

$$= \lim_{a\to-\infty} \left(\frac{1}{2} + \frac{1}{a}\right) = \frac{1}{2}$$

A graph of this region is shown in Figure 14. Since the limit exists, this integral converges.   **TRY YOUR TURN 1**

It may seem puzzling that the areas under the curves $f(x) = 1/x^{3/2}$ and $f(x) = 1/x^2$ are finite, while $f(x) = 1/x$ has an infinite amount of area. At first glance the graphs of these functions appear similar. The difference is that although all three functions get small as x becomes infinitely large, $f(x) = 1/x$ does not become small enough fast enough. In the graphing calculator screen in Figure 15, notice how much faster $1/x^2$ becomes small compared with $1/x$.

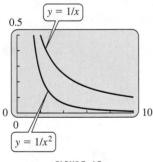

FIGURE 15

CAUTION Since graphing calculators provide only approximations, using them to find improper integrals is tricky and requires skill and care. Although their approximations may be good in some cases, they are wrong in others, and they cannot tell us for certain that an improper integral does not exist. See Exercises 39–41.

EXAMPLE 2   Improper Integral

Find $\int_{-\infty}^\infty 4e^{-3x}\,dx$.

SOLUTION   In the definition of an improper integral with limits of $-\infty$ and ∞, the value of c is arbitrary, so we'll choose the simple value $c = 0$. We can then write the integral as

$$\int_{-\infty}^\infty 4e^{-3x}\,dx = \int_{-\infty}^0 4e^{-3x}\,dx + \int_0^\infty 4e^{-3x}\,dx$$

and evaluate each of the two improper integrals on the right. If they both converge, the original integral will equal their sum. To show you all the details while maintaining the suspense, we will evaluate the second integral first.

By definition,

$$\int_0^\infty 4e^{-3x}\,dx = \lim_{b\to\infty} \int_0^b 4e^{-3x}\,dx = \lim_{b\to\infty} \left(\frac{-4}{3}e^{-3x}\right)\Big|_0^b$$

$$= \lim_{b\to\infty} \left(\frac{-4}{3e^{3b}} + \frac{4}{3}\right) = 0 + \frac{4}{3} = \frac{4}{3}.$$

FOR REVIEW

Recall that

$$\lim_{p\to\infty} \frac{1}{e^p} = 0$$

and

$$\lim_{p\to-\infty} \frac{1}{e^p} = \infty.$$

Similarly, the second integral is evaluated as

$$\int_{-\infty}^{0} 4e^{-3x}\, dx = \lim_{b \to -\infty} \int_{b}^{0} 4e^{-3x}\, dx = \lim_{b \to -\infty} \left(\frac{-4}{3} e^{-3x} \right)\Bigg|_{b}^{0}$$

$$= \lim_{b \to -\infty} \left(-\frac{4}{3} + \frac{4}{3e^{3b}} \right) = \infty.$$

YOUR TURN 2
Find $\int_{0}^{\infty} 5e^{-2x}\, dx$.

Since one of the two improper integrals diverges, the original improper integral diverges.   **TRY YOUR TURN 2**

The following examples describe applications of improper integrals.

EXAMPLE 3   Pollution

The rate at which a pollutant is being dumped into a stream at time t is given by $P_0 e^{-kt}$, where P_0 is the rate that the pollutant is initially released into the stream. Suppose $P_0 = 1000$ and $k = 0.06$. Find the total amount of the pollutant that will be released into the stream into the indefinite future.

APPLY IT   **SOLUTION**   Find

$$\int_{0}^{\infty} P_0 e^{-kt}\, dt = \int_{0}^{\infty} 1000 e^{-0.06t}\, dt.$$

This integral is similar to one of the integrals used to solve Example 2 and may be evaluated by the same method.

$$\int_{0}^{\infty} 1000 e^{-0.06t}\, dt = \lim_{b \to \infty} \int_{0}^{b} 1000 e^{-0.06t}\, dt$$

$$= \lim_{b \to \infty} \left(\frac{1000}{-0.06} e^{-0.06t} \right)\Bigg|_{0}^{b}$$

$$= \lim_{b \to \infty} \left(\frac{1000}{-0.06 e^{0.06b}} - \frac{1000}{-0.06} e^{0} \right) = \frac{-1000}{-0.06} \approx 16{,}667$$

A total of approximately 16,667 units of the pollutant will be released over time.

The *capital value* of an asset is often defined as the present value of all future net earnings of the asset. In other words, suppose an asset provides a continuous money flow that is invested in an account earning a certain rate of interest. A lump sum is invested in a second account earning the same rate of interest, but with no money flow, so that as $t \to \infty$, the amounts in the two accounts approach each other. The lump sum necessary to make this happen is the capital value of the asset. If $R(t)$ gives the annual rate at which earnings are produced by an asset at time t, then the present value formula from Section 3 gives the **capital value** as

$$\int_{0}^{\infty} R(t) e^{-rt}\, dt,$$

where r is the annual rate of interest, compounded continuously.

EXAMPLE 4   Capital Value

Suppose income from a rental property is generated at the annual rate of $4000 per year. Find the capital value of this property at an interest rate of 10% compounded continuously.

SOLUTION This is a continuous income stream with a rate of flow of $4000 per year, so $R(t) = 4000$. Also, $r = 0.10$ or 0.1. The capital value is given by

$$\int_0^\infty 4000e^{-0.1t}\, dt = \lim_{b\to\infty} \int_0^b 4000e^{-0.1t}\, dt$$

$$= \lim_{b\to\infty} \left(\frac{4000}{-0.1} e^{-0.1t} \right) \Big|_0^b$$

$$= \lim_{b\to\infty} (-40{,}000 e^{-0.1b} + 40{,}000) = 40{,}000,$$

or $40,000.

8.4 EXERCISES

Determine whether each improper integral converges or diverges, and find the value of each that converges.

1. $\int_3^\infty \dfrac{1}{x^2}\, dx$

2. $\int_3^\infty \dfrac{1}{(x+1)^3}\, dx$

3. $\int_4^\infty \dfrac{2}{\sqrt{x}}\, dx$

4. $\int_{27}^\infty \dfrac{2}{\sqrt[3]{x}}\, dx$

5. $\int_{-\infty}^{-1} \dfrac{2}{x^3}\, dx$

6. $\int_{-\infty}^{-4} \dfrac{3}{x^4}\, dx$

7. $\int_1^\infty \dfrac{1}{x^{1.0001}}\, dx$

8. $\int_1^\infty \dfrac{1}{x^{0.999}}\, dx$

9. $\int_{-\infty}^{-10} x^{-2}\, dx$

10. $\int_{-\infty}^{-1} (x-2)^{-3}\, dx$

11. $\int_{-\infty}^{-1} x^{-8/3}\, dx$

12. $\int_{-\infty}^{-27} x^{-5/3}\, dx$

13. $\int_0^\infty 8e^{-8x}\, dx$

14. $\int_0^\infty 50e^{-50x}\, dx$

15. $\int_{-\infty}^0 1000e^x\, dx$

16. $\int_{-\infty}^0 5e^{60x}\, dx$

17. $\int_{-\infty}^{-1} \ln|x|\, dx$

18. $\int_1^\infty \ln|x|\, dx$

19. $\int_0^\infty \dfrac{dx}{(x+1)^2}$

20. $\int_0^\infty \dfrac{dx}{(4x+1)^3}$

21. $\int_{-\infty}^{-1} \dfrac{2x-1}{x^2-x}\, dx$

22. $\int_1^\infty \dfrac{4x+6}{x^2+3x}\, dx$

23. $\int_2^\infty \dfrac{1}{x\ln x}\, dx$

24. $\int_2^\infty \dfrac{1}{x(\ln x)^2}\, dx$

25. $\int_0^\infty xe^{4x}\, dx$

26. $\int_{-\infty}^0 xe^{0.2x}\, dx$ (*Hint:* Recall that $\lim\limits_{x\to-\infty} xe^x = 0$.)

27. $\int_{-\infty}^\infty x^3 e^{-x^4}\, dx$ (*Hint:* Recall from Exercise 66 in Section 3.1 on Limits that $\lim\limits_{x\to\infty} x^n e^{-x} = 0$.)

28. $\int_{-\infty}^\infty e^{-|x|}\, dx$ (*Hint:* Recall that when $x < 0$, $|x| = -x$.)

29. $\int_{-\infty}^\infty \dfrac{x}{x^2+1}\, dx$

30. $\int_{-\infty}^\infty \dfrac{2x+4}{x^2+4x+5}\, dx$

Find the area between the graph of the given function and the x-axis over the given interval, if possible.

31. $f(x) = \dfrac{1}{x-1}$, for $(-\infty, 0]$

32. $f(x) = e^{-x}$, for $(-\infty, e]$

33. $f(x) = \dfrac{1}{(x-1)^2}$, for $(-\infty, 0]$

34. $f(x) = \dfrac{3}{(x-1)^3}$, for $(-\infty, 0]$

35. Find $\int_{-\infty}^\infty xe^{-x^2}\, dx$.

36. Find $\int_{-\infty}^\infty \dfrac{x}{(1+x^2)^2}\, dx$.

37. Show that $\int_1^\infty \dfrac{1}{x^p}\, dx$ converges if $p > 1$ and diverges if $p \le 1$.

38. Example 1(b) leads to a paradox. On the one hand, the unbounded region in that example has an area of $1/2$, so theoretically it could be colored with ink. On the other hand, the boundary of that region is infinite, so it cannot be drawn with a finite amount of ink. This seems impossible, because coloring the region automatically colors the boundary. Explain why it is possible to color the region.

39. Consider the functions $f(x) = 1/\sqrt{1+x^2}$ and $g(x) = 1/\sqrt{1+x^4}$.

a. Use your calculator to approximate $\int_1^b f(x)\, dx$ for $b = 20, 50, 100, 1000$, and $10{,}000$.

b. Based on your answers from part a, would you guess that $\int_1^\infty f(x)\, dx$ is convergent or divergent?

c. Use your calculator to approximate $\int_1^b g(x)\, dx$ for $b = 20, 50, 100, 1000,$ and $10,000.$

d. Based on your answers from part c, would you guess that $\int_1^\infty g(x)\, dx$ is convergent or divergent?

e. Show how the answer to parts b and d might be guessed by comparing the integrals with others whose convergence or divergence is known. (*Hint:* For large x, the difference between $1 + x^2$ and x^2 is relatively small.)

Note: The first integral is indeed divergent, and the second convergent, with an approximate value of 0.9270.

40. a. Use your calculator to approximate $\int_0^b e^{-x^2}\, dx$ for $b = 1, 5, 10,$ and $20.$

b. Based on your answers to part a, does $\int_0^\infty e^{-x^2}\, dx$ appear to be convergent or divergent? If convergent, what seems to be its approximate value?

c. Explain why this integral should be convergent by comparing e^{-x^2} with e^{-x} for $x > 1.$

Note: The integral is convergent, with a value of $\sqrt{\pi}/2.$

41. a. Use your calculator to approximate $\int_0^b e^{-0.00001x}\, dx$ for $b = 10, 50, 100,$ and $1000.$

b. Based on your answers to part a, does $\int_0^\infty e^{-0.00001x}\, dx$ appear to be convergent or divergent?

c. To what value does the integral actually converge?

APPLICATIONS

Business and Economics

Capital Value Find the capital values of the properties in Exercises 42–43.

42. A castle for which annual rent of $225,000 will be paid in perpetuity; the interest rate is 6% compounded continuously

43. A fort on a strategic peninsula in the North Sea; the annual rent is $1,000,000, paid in perpetuity; the interest rate is 5% compounded continuously

44. Capital Value Find the capital value of an asset that generates $7200 yearly income if the interest rate is as follows.

a. 5% compounded continuously

b. 10% compounded continuously

45. Capital Value An investment produces a perpetual stream of income with a flow rate of
$$R(t) = 1200e^{0.03t}.$$
Find the capital value at an interest rate of 7% compounded continuously.

46. Capital Value Suppose income from an investment starts (at time 0) at $6000 a year and increases linearly and continuously at a rate of $200 a year. Find the capital value at an interest rate of 5% compounded continuously.

47. Scholarship The Drucker family wants to establish an ongoing scholarship award at a college. Each year in June $3000 will be awarded, starting 1 year from now. What amount must the Druckers provide the college, assuming funds will be invested at 10% compounded continuously?

Social Sciences

48. Drug Reaction The rate of reaction to a drug is given by
$$r'(t) = 2t^2 e^{-t},$$
where t is the number of hours since the drug was administered. Find the total reaction to the drug over all the time since it was administered, assuming this is an infinite time interval. (*Hint:* $\lim_{t \to \infty} t^k e^{-t} = 0$ for all real numbers $k.$)

49. Drug Epidemic In an epidemiological model used to study the spread of drug use, a single drug user is introduced into a population of N non-users. Under certain assumptions, the number of people expected to use drugs as a result of direct influence from each drug user is given by
$$S = N \int_0^\infty \frac{a(1 - e^{-kt})}{k} e^{-bt}\, dt,$$
where a, b, and k are constants. Find the value of S. *Source: Mathematical Biology.*

50. Present Value When harvesting a population, such as fish, the present value of the resource is given by
$$P = \int_0^\infty e^{-rt} n(t) y(t)\, dt,$$
where r is a discount factor, $n(t)$ is the net revenue at time t, and $y(t)$ is the harvesting effort. Suppose $y(t) = K$ and $n(t) = at + b$. Find the present value. *Source: Some Mathematical Questions in Biology.*

Physical Sciences

Radioactive Waste Radioactive waste is entering the atmosphere over an area at a decreasing rate. Use the improper integral
$$\int_0^\infty P e^{-kt}\, dt$$
with $P = 50$ to find the total amount of the waste that will enter the atmosphere for each value of $k.$

51. $k = 0.06$

52. $k = 0.04$

YOUR TURN ANSWERS

1. (a) Divergent (b) 3/2 **2.** 5/2

8 CHAPTER REVIEW

SUMMARY

In this chapter, we introduced another technique of integration and some applications of integration. The technique is known as integration by parts, which is derived from the product rule for derivatives. We also developed definite integral formulas to calculate the volume of a solid of revolution and the average value of a function on some interval. We then used definite integrals to study continuous money flow. Finally, we learned how to evaluate improper integrals that have upper or lower limits of ∞ or $-\infty$.

Integration by Parts If u and v are differentiable functions, then

$$\int u\,dv = uv - \int v\,du.$$

Volume of a Solid of Revolution If $f(x)$ is nonnegative and R is the region between $f(x)$ and the x-axis from $x = a$ to $x = b$, the volume of the solid formed by rotating R about the x-axis is given by

$$V = \int_a^b \pi[f(x)]^2\,dx.$$

Average Value of a Function The average value of a function f on the interval $[a, b]$ is

$$\frac{1}{b-a}\int_a^b f(x)\,dx,$$

provided the indicated definite integral exists.

Total Money Flow If $f(t)$ is the rate of money flow, then the total money flow over the time interval from $t = 0$ to $t = T$ is given by

$$\int_0^T f(t)\,dt.$$

Present Value of Money Flow If $f(t)$ is the rate of continuous money flow at an interest rate r for T years, then the present value is

$$P = \int_0^T f(t)e^{-rt}\,dt.$$

Accumulated Amount of Money Flow at Time T If $f(t)$ is the rate of money flow at an interest rate r at time t, the accumulated amount of money flow at time T is

$$A = e^{rT}\int_0^T f(t)e^{-rt}\,dt.$$

Improper Integrals If f is continuous on the indicated interval and if the indicated limits exist, then

$$\int_a^\infty f(x)\,dx = \lim_{b\to\infty}\int_a^b f(x)\,dx,$$

$$\int_{-\infty}^b f(x)\,dx = \lim_{a\to-\infty}\int_a^b f(x)\,dx,$$

$$\int_{-\infty}^\infty f(x)\,dx = \int_{-\infty}^c f(x)\,dx + \int_c^\infty f(x)\,dx,$$

for real numbers a, b, and c, where c is arbitrarily chosen.

Capital Value If $R(t)$ gives the annual rate at which earnings are produced by an asset at time t, the capital value is given by

$$\int_0^\infty R(t)e^{-rt}\,dt,$$

where r is the annual rate of interest, compounded continuously.

KEY TERMS

8.1
integration by parts
column integration

8.2
solid of revolution
average value of a function

8.3
total money flow
present value of continuous
  money flow
accumulated amount of money
  flow

8.4
improper integral
convergent integral
divergent integral
capital value

REVIEW EXERCISES

CONCEPT CHECK

Determine whether each of the following statements is true or false, and explain why.

1. Integration by parts should be used to evaluate $\int_0^1 \dfrac{x^2}{x^3 + 1}\, dx$.

2. Integration by parts should be used to evaluate $\int_0^1 xe^{10x}\, dx$.

3. We would need to apply the method of integration by parts twice to determine

$$\int x^3 e^{-x^2}\, dx.$$

4. Integration by parts should be used to determine $\int \ln(4x)\, dx$.

5. The average value of the function $f(x) = 2x^2 + 3$ on [1, 4] is given by

$$\frac{1}{3}\int_1^4 \pi(2x^2 + 3)^2\, dx.$$

6. The volume of the solid formed by revolving the function $f(x) = \sqrt{x^2 + 1}$ about the x-axis on the interval [1, 2] is given by

$$\int_1^2 \pi\sqrt{x^2 + 1}\, dx.$$

7. The volume of the solid formed by revolving the function $f(x) = x + 4$ about the x-axis on the interval [−4, 5] is given by

$$\int_{-4}^5 \pi(x + 4)^2\, dx.$$

8. If $f(t) = 1000e^{0.05t}$ represents the rate of flow of money for a vending machine over the first five years of income, then the total money flow for that time period is given by

$$\int_0^5 1000e^{0.05t}\, dt.$$

9. If a company expects an annual flow of money during the next five years to be $f(t) = 1000e^{0.05t}$, the present value of this income, assuming an annual interest rate of 4.5% compounded continuously is given by

$$\int_0^5 1000e^{0.005t}\, dt.$$

10. $\displaystyle\int_{-\infty}^{\infty} xe^{-2x}\, dx = \lim_{c \to \infty}\int_{-c}^c xe^{-2x}\, dx$

PRACTICE AND EXPLORATIONS

11. Describe the type of integral for which integration by parts is useful.

12. Compare finding the average value of a function with finding the average of n numbers.

13. What is an improper integral? Explain why improper integrals must be treated in a special way.

Find each integral, using techniques from this or the previous chapter.

14. $\displaystyle\int x(8 - x)^{3/2}\, dx$

15. $\displaystyle\int \frac{3x}{\sqrt{x - 2}}\, dx$

16. $\displaystyle\int xe^x\, dx$

17. $\displaystyle\int (3x + 6)e^{-3x}\, dx$

18. $\displaystyle\int \ln|4x + 5|\, dx$

19. $\displaystyle\int (x - 1)\ln|x|\, dx$

20. $\displaystyle\int \frac{x}{25 - 9x^2}\, dx$

21. $\displaystyle\int \frac{x}{\sqrt{16 + 8x^2}}\, dx$

22. $\displaystyle\int_1^e x^3 \ln x\, dx$

23. $\displaystyle\int_0^1 x^2 e^{x/2}\, dx$

24. Find the area between $y = (3 + x^2)e^{2x}$ and the x-axis from $x = 0$ to $x = 1$.

25. Find the area between $y = x^3(x^2 - 1)^{1/3}$ and the x-axis from $x = 1$ to $x = 3$.

Find the volume of the solid of revolution formed by rotating each bounded region about the x-axis.

26. $f(x) = 3x - 1$, $y = 0$, $x = 2$

27. $f(x) = \sqrt{x - 4}$, $y = 0$, $x = 13$

28. $f(x) = e^{-x}$, $y = 0$, $x = -2$, $x = 1$

29. $f(x) = \dfrac{1}{\sqrt{x - 1}}$, $y = 0$, $x = 2$, $x = 4$

30. $f(x) = 4 - x^2$, $y = 0$, $x = -1$, $x = 1$

31. $f(x) = \dfrac{x^2}{4}$, $y = 0$, $x = 4$

32. A frustum is what remains of a cone when the top is cut off by a plane parallel to the base. Suppose a right circular frustum (that is, one formed from a right circular cone) has a base with radius r, a top with radius $r/2$, and a height h. (See the figure below.) Find the volume of this frustum by rotating about the x-axis the region below the line segment from $(0, r)$ to $(h, r/2)$.

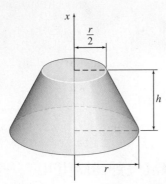

33. How is the average value of a function found?

34. Find the average value of $f(x) = \sqrt{x + 1}$ over the interval $[0, 8]$.

35. Find the average value of $f(x) = 7x^2(x^3 + 1)^6$ over the interval $[0, 2]$.

Find the value of each integral that converges.

36. $\displaystyle\int_{10}^{\infty} x^{-1} \, dx$

37. $\displaystyle\int_{-\infty}^{-5} x^{-2} \, dx$

38. $\displaystyle\int_{0}^{\infty} \frac{dx}{(3x + 1)^2}$

39. $\displaystyle\int_{1}^{\infty} 6e^{-x} \, dx$

40. $\displaystyle\int_{-\infty}^{0} \frac{x}{x^2 + 3} \, dx$

41. $\displaystyle\int_{4}^{\infty} \ln(5x) \, dx$

Find the area between the graph of each function and the x-axis over the given interval, if possible.

42. $f(x) = \dfrac{5}{(x - 2)^2}$, for $(-\infty, 1]$

43. $f(x) = 3e^{-x}$, for $[0, \infty)$

44. How is the present value of money flow found? The accumulated amount of money flow?

APPLICATIONS

Business and Economics

45. Total Revenue The rate of change of revenue from the sale of x toaster ovens is

$$R'(x) = x(x - 50)^{1/2}.$$

Find the total revenue from the sale of the 50th to the 75th ovens.

Present Value of Money Flow Each function in Exercises 46–49 represents the rate of flow of money (in dollars per year) over the given time period, compounded continuously at the given annual interest rate. Find the present value in each case.

46. $f(t) = 5000$, 8 years, 9%

47. $f(t) = 25,000$, 12 years, 10%

48. $f(t) = 150e^{0.04t}$, 5 years, 6%

49. $f(t) = 15t$, 18 months, 8%

Accumulated Amount of Money Flow at Time T Assume that each function gives the rate of flow of money in dollars per year over the given period, with continuous compounding at the given annual interest rate. Find the accumulated amount of money flow at the end of the time period.

50. $f(t) = 1000$, 5 years, 6%

51. $f(t) = 500e^{-0.04t}$, 8 years, 10%

52. $f(t) = 20t$, 6 years, 4%

53. $f(t) = 1000 + 200t$, 10 years, 9%

54. Money Flow An investment scheme is expected to produce a continuous flow of money, starting at $1000 and increasing exponentially at 5% a year for 7 years. Find the present value at an interest rate of 11% compounded continuously.

55. Money Flow The proceeds from the sale of a building will yield a uniform continuous flow of $10,000 a year for 10 years. Find the final amount at an interest rate of 10.5% compounded continuously.

56. Capital Value Find the capital value of an office building for which annual rent of $50,000 will be paid in perpetuity, if the interest rate is 9%.

Life Sciences

57. Drug Reaction The reaction rate to a new drug t hours after the drug is administered is

$$r'(t) = 0.5te^{-t}.$$

Find the total reaction over the first 5 hours.

58. Oil Leak Pollution An oil leak from an uncapped well is polluting a bay at a rate of $f(t) = 125e^{-0.025t}$ gallons per year. Use an improper integral to find the total amount of oil that will enter the bay, assuming the well is never capped.

Physical Sciences

59. Average Temperatures Suppose the temperature (degrees F) in a river at a point x meters downstream from a factory that is discharging hot water into the river is given by

$$T(x) = 160 - 0.05x^2.$$

Find the average temperature over each interval.

a. $[0, 10]$ **b.** $[10, 40]$ **c.** $[0, 40]$

ESTIMATING LEARNING CURVES IN MANUFACTURING WITH INTEGRALS

In the previous chapter you have seen how the trapezoidal rule uses sums of areas of polygons to approximate the area under a smooth curve, that is, a definite integral. In this Extended Application we look at the reverse process, using an integral to estimate a sum, in the context of estimating production costs.

As a manufacturer produces more units of a new product, the individual units generally become cheaper to produce, because with experience, production workers gain skill and speed, and managers spot opportunities for improved efficiency. This decline in unit costs is often called an *experience curve* or *learning curve*. This curve is important when a manufacturer negotiates a contract with a buyer.

Here's an example, based on an actual contract that came before the Armed Services Board of Contract Appeals. *Source: Armed Services Board of Contract Appeals*. The Navy asked the ITT Defense Communications Division to bid on the manufacture of several different kinds of mobile telephone switchboards, including 280 of the model called the SB 3865. ITT figured that the cost of making a single SB 3865 was around $300,000. But they couldn't submit a bid of $300,000 × 280 or $84 million, because multiple units should have a lower unit price. So ITT used a learning curve to estimate an average unit cost of $135,300 for all 280 switchboards and submitted a bid of $135,300 × 280 or $37.9 million.

The contract gave the Navy an option to purchase 280 SB units over three years, but in fact it bought fewer. Suppose the Navy bought 140 SBs. Should it pay half of the original price of $37.9 million? No: ITT's bid was based on the efficiencies of a 280-unit run, so 140 units should be repriced to yield *more* than half the full price. A repricing clause in the contract specified that a learning curve would be used to reprice partial orders, and when the Navy ordered less than the full amount, ITT invoked this clause to reprice the switchboards. The question in dispute at the hearing was which learning curve to use.

There are two common learning curve models. The *unit learning curve* model assumes that each time the number of units doubles, say from n to $2n$, the cost of producing the last unit is some constant fraction r of the cost for the nth unit. Usually the fraction r is given as a percent. If $r = 90\%$ (typical for big pieces of hardware), then the contract would refer to a "90% learning curve." The *cumulative learning curve* model assumes that when the number of units doubles, the *average cost* of producing all $2n$ units is some constant fraction of the *average cost* of the first n units. The Navy's contract with ITT didn't specify which model was to be used—it just referred to "a 90% learning curve." The government used the unit model and ITT used the cumulative average model, and ITT calculated a fair price millions of dollars higher than the government's price!

In practice, ITT used a calculator program to make its estimate, and the government used printed tables, but both the program and the tables were derived using calculus. To see how the computation works, we'll derive the government's unit learning curve.

Each unit has a different cost, with the first unit being the most expensive and the 280th unit the least expensive. So the cost of the nth unit, call it $C(n)$, is a function of n. To find a fair price for n units we'll add up all the unit prices. That is, we will compute $C(1) + C(2) + \cdots + C(n-1) + C(n)$. Before we can do that, we need a formula for $C(n)$ in terms of n, but all we know about $C(n)$ is that

$$C(2n) = r \cdot C(n)$$

for *every* n, with $r = 0.90$. This sort of equation is called a *functional equation*: It relates two different values of the function without giving an explicit formula for the function. In Exercise 3 you'll see how you might discover a solution to this functional equation, but here we'll just give the result:

$$C(n) = C(1) \cdot n^b, \text{ where } b = \frac{\ln r}{\ln 2} \approx -0.152.$$

Thus, to find the price for making 280 units, we need to add up all the values of $C(1) \cdot n^b$ as n ranges from 1 to 280. There's just one problem: We don't know $C(1)$! The only numbers that ITT gave were the *average* cost per unit for 280 units, namely $135,300, and the total price of $37.9 million. But if we write out the formula for the 280-unit price in terms of $C(1)$, we can figure out $C(1)$ by dividing. Here's how it works.

The cost C is a function of an integer variable n, since the contractor can't deliver fractional units of the hardware. But the function $C(x) = C(1) \cdot x^{-0.152}$ is a perfectly good function of the real variable x, and the sum of the first 280 values of $C(n)$ should be close to $\int_1^{280} C(x)\, dx = C(1) \cdot \int_1^{280} x^{-0.152}\, dx$. In Exercise 4 you'll see how to derive the following improved estimate:

$$C(1)\left(\int_1^{281} x^{-0.152}\, dx + \frac{1 - 281^{-0.152}}{2}\right).$$

The integrand is a power function, so you know how to evaluate the integral exactly.

$$\int_1^{281} x^{-0.152}\, dx = \frac{1}{0.848} x^{0.848} \Big|_1^{281} = \frac{1}{0.848}\left(281^{0.848} - 1^{0.848}\right)$$

Thus, the sum is approximately

$$C(1)\left[\frac{1}{0.848}\left(281^{0.848} - 1^{0.848}\right) + \frac{1 - 281^{-0.152}}{2}\right] \approx C(1) \cdot 139.75.$$

Since ITT's price for 280 units was $37.9 million,

$$C(1) = \frac{\$37,900,000}{139.75} \approx \$271,000.$$

Now that we know $C(1)$, we can reprice an order of 140 units by adding up the first 140 values of $C(n)$. An estimate exactly like the one above tells us that according to the government's model, a fair price for 140 units is about $21 million. As we expected, this is more than half the 280-unit price, in fact about $2 million more.

EXERCISES

1. According to the formula for $C(n)$, what is the unit price of the 280th unit, to the nearest thousand dollars?

2. Suppose that instead of using natural logarithms to compute b, we use logarithms with a base of 10 and define $b = (\log r)/(\log 2)$. Does this change the value of b?

3. All power functions satisfy an equation similar to our functional equation: If $f(x) = ax^b$, then $f(2x) = a(2x)^b = a2^b \cdot x^b = 2^b \cdot f(x)$. How can you choose a and b to make $C(x) = ax^b$ a solution to the functional equation $C(2n) = r \cdot C(n)$?

4. Figure 16 indicates how you could use the integral

$$\int_1^5 \frac{1}{x}\, dx + \frac{1 - \dfrac{1}{5}}{2}$$ as an estimate for the sum $1 + \dfrac{1}{2} + \dfrac{1}{3} + \dfrac{1}{4}$.

The graph shows the function $y = \dfrac{1}{x}$.

a. Write a justification for the integral estimate. (Your argument will also justify the integral estimate for the $C(n)$ sum.) Based on your explanation, does the integral expression overestimate the sum?

b. You know how to integrate the function $1/x$. Compute the integral estimate and the actual value. What is the percentage error in the estimate?

5. Go to the website WolframAlpha.com and enter: "integrate." Follow the instructions and use them along with the improved estimate to verify the fair price for 140 units of about $21 million given in the text. Then enter "sum" into Wolfram Alpha and follow the instructions to verify the fair price by summing the first 140 values of $C(n)$. Discuss how the solution compares with the solutions provided by the integration and summation features on a graphing calculator, as well as the solution provided by adding up the values using a spreadsheet.

DIRECTIONS FOR GROUP PROJECT

Suppose that you and three other students have an internship with a manufacturing company that is submitting a bid to make several thousand units of some highly technical equipment. The one problem with the bid is that the number of units that will be purchased is only an estimate and that the actual number needed may greatly vary from the estimate. Using the information given above, prepare a presentation for an internal sales meeting that will describe the case listed above and its applications to the bid at hand. Make your presentation realistic in the sense that the product you are manufacturing should have a name, average price, and so on. Then show how integrals can be used to estimate learning curves in this situation and produce a pricing structure for the bid. Presentation software, such as Microsoft PowerPoint, should be used.

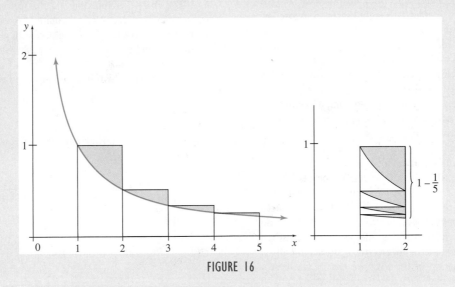

FIGURE 16

9

Multivariable Calculus

Safe diving requires an understanding of how the increased pressure below the surface affects the body's intake of nitrogen. An exercise in Section 2 of this chapter investigates a formula for nitrogen pressure as a function of two variables, depth and dive time. Partial derivatives tell us how this function behaves when one variable is held constant as the other changes. Dive tables based on the formula help divers to choose a safe time for a given depth, or a safe depth for a given time.

W e have thus far limited our study of calculus to functions of one variable. There are other phenomena that require more than one variable to adequately model the situation. For example, the price of an electronics device depends on how long it has been on the market, the number of competing devices, labor costs, demand, and many other factors. In this case, the price is a function of more than one variable. To analyze and better understand situations like this, we will extend the ideas of calculus, including differentiation and integration, to functions of more than one variable.

9.1 Functions of Several Variables

APPLY IT **How are the amounts of labor and capital needed to produce a certain number of items related?**
We will study this question in Example 8 using a production function that depends on the two independent variables of labor and capital.

If a company produces x items at a cost of \$10 per item, then the total cost $C(x)$ of producing the items is given by

$$C(x) = 10x.$$

The cost is a function of one independent variable, the number of items produced. If the company produces two products, with x of one product at a cost of \$10 each, and y of another product at a cost of \$15 each, then the total cost to the firm is a function of *two* independent variables, x and y. By generalizing $f(x)$ notation, the total cost can be written as $C(x, y)$, where

$$C(x, y) = 10x + 15y.$$

When $x = 5$ and $y = 12$ the total cost is written $C(5, 12)$, with

$$C(5, 12) = 10 \cdot 5 + 15 \cdot 12 = 230.$$

A general definition follows.

Function of Two or More Variables
The expression $z = f(x, y)$ is a **function of two variables** if a unique value of z is obtained from each ordered pair of real numbers (x, y). The variables x and y are **independent variables**, and z is the **dependent variable**. The set of all ordered pairs of real numbers (x, y) such that $f(x, y)$ exists is the **domain** of f; the set of all values of $f(x, y)$ is the **range**. Similar definitions could be given for functions of three, four, or more independent variables.

EXAMPLE 1   Evaluating Functions

Let $f(x, y) = 4x^2 + 2xy + 3/y$ and find the following.

(a) $f(-1, 3)$

SOLUTION   Replace x with -1 and y with 3.

$$f(-1, 3) = 4(-1)^2 + 2(-1)(3) + \frac{3}{3} = 4 - 6 + 1 = -1$$

(b) $f(2, 0)$

SOLUTION Because of the quotient $3/y$, it is not possible to replace y with 0, so $f(2, 0)$ is undefined. By inspection, we see that the domain of the function is the set of all (x, y) such that $y \neq 0$.

(c) $\dfrac{f(x + h, y) - f(x, y)}{h}$

SOLUTION Calculate as follows:

$$\frac{f(x + h, y) - f(x, y)}{h} = \frac{4(x + h)^2 + 2(x + h)y + 3/y - [4x^2 + 2xy + 3/y]}{h}$$

$$= \frac{4x^2 + 8xh + 4h^2 + 2xy + 2hy + 3/y - 4x^2 - 2xy - 3/y}{h}$$

$$= \frac{8xh + 4h^2 + 2hy}{h} \qquad \text{Simplify the numerator.}$$

$$= \frac{h(8x + 4h + 2y)}{h} \qquad \text{Factor } h \text{ from the numerator.}$$

$$= 8x + 4h + 2y. \qquad \textbf{TRY YOUR TURN 1}$$

YOUR TURN 1 For the function in Example 1, find $f(2,3)$.

EXAMPLE 2 Volume of a Can

Let r and h represent the radius and height of a can in cm. The volume of the can is then a function of the two variables r and h given by

$$V(r, h) = \pi r^2 h.$$

Find $V(3, 11)$.

SOLUTION Replace r with 3 and h with 11 to get

$$V(3, 11) = \pi \cdot 3^2 \cdot 11 = 99\pi \approx 311 \text{ cm}^3.$$

This says that a can with radius 3 cm and height 11 cm has a volume of approximately 311 cm³.

EXAMPLE 3 Evaluating a Function

Let $f(x, y, z) = 4xz - 3x^2y + 2z^2$. Find $f(2, -3, 1)$.

SOLUTION Replace x with 2, y with -3, and z with 1.

$$f(2, -3, 1) = 4(2)(1) - 3(2)^2(-3) + 2(1)^2 = 8 + 36 + 2 = 46$$

$$\textbf{TRY YOUR TURN 2}$$

YOUR TURN 2 For the function in Example 3, find $f(1, 2, 3)$.

Graphing Functions of Two Independent Variables

Functions of one independent variable are graphed by using an x-axis and a y-axis to locate points in a plane. The plane determined by the x- and y-axes is called the *xy-plane*. A third axis is needed to graph functions of two independent variables—the z-axis, which goes through the origin in the xy-plane and is perpendicular to both the x-axis and the y-axis.

Figure 1 shows one possible way to draw the three axes. In Figure 1, the yz-plane is in the plane of the page, with the x-axis perpendicular to the plane of the page.

Just as we graphed ordered pairs earlier we can now graph **ordered triples** of the form (x, y, z). For example, to locate the point corresponding to the ordered triple $(2, -4, 3)$, start at the origin and go 2 units along the positive x-axis. Then go 4 units in a negative direction (to the left) parallel to the y-axis. Finally, go up 3 units parallel to the z-axis. The point representing $(2, -4, 3)$ is shown in Figure 1, together with several other points. The region of three-dimensional space where all coordinates are positive is called the **first octant**.

In Chapter 1 we saw that the graph of $ax + by = c$ (where a and b are not both 0) is a straight line. This result generalizes to three dimensions.

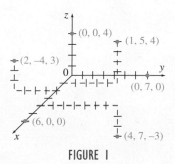

FIGURE 1

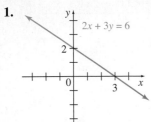
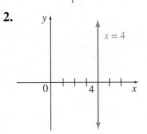
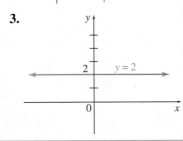

YOUR TURN 3 Graph $x + 2y + 3z = 12$ in the first octant.

Plane

The graph of

$$ax + by + cz = d$$

is a **plane** if a, b, and c are not all 0.

EXAMPLE 4 **Graphing a Plane**

Graph $2x + y + z = 6$.

SOLUTION The graph of this equation is a plane. Earlier, we graphed straight lines by finding x- and y-intercepts. A similar idea helps in graphing a plane. To find the x-intercept, which is the point where the graph crosses the x-axis, let $y = 0$ and $z = 0$.

$$2x + 0 + 0 = 6$$
$$x = 3$$

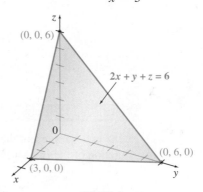

FIGURE 2

The point $(3, 0, 0)$ is on the graph. Letting $x = 0$ and $z = 0$ gives the point $(0, 6, 0)$, while $x = 0$ and $y = 0$ lead to $(0, 0, 6)$. The plane through these three points includes the triangular surface shown in Figure 2. This surface is the first-octant part of the plane that is the graph of $2x + y + z = 6$. The plane does not stop at the axes but extends without bound.

TRY YOUR TURN 3

EXAMPLE 5 **Graphing a Plane**

Graph $x + z = 6$.

SOLUTION To find the x-intercept, let $y = 0$ and $z = 0$, giving $(6, 0, 0)$. If $x = 0$ and $y = 0$, we get the point $(0, 0, 6)$. Because there is no y in the equation $x + z = 6$, there can be no y-intercept. A plane that has no y-intercept is parallel to the y-axis. The first-octant portion of the graph of $x + z = 6$ is shown in Figure 3.

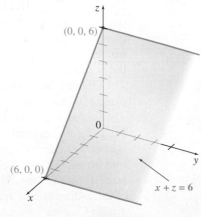

FIGURE 3

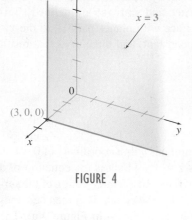

FIGURE 4

EXAMPLE 6   **Graphing Planes**

Graph each equation in the first octant.

(a) $x = 3$

SOLUTION   This graph, which goes through $(3, 0, 0)$, can have no y-intercept and no z-intercept. It is, therefore, a plane parallel to the y-axis and the z-axis and, therefore, to the yz-plane. The first-octant portion of the graph is shown in Figure 4.

(b) $y = 4$

SOLUTION   This graph goes through $(0, 4, 0)$ and is parallel to the xz-plane. The first-octant portion of the graph is shown in Figure 5.

(c) $z = 1$

SOLUTION   The graph is a plane parallel to the xy-plane, passing through $(0, 0, 1)$. Its first-octant portion is shown in Figure 6.

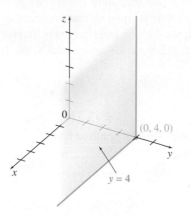

FIGURE 5

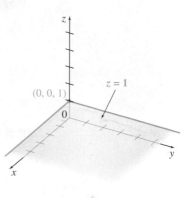

FIGURE 6

The graph of a function of one variable, $y = f(x)$, is a curve in the plane. If x_0 is in the domain of f, the point $(x_0, f(x_0))$ on the graph lies directly above or below the number x_0 on the x-axis, as shown in Figure 7.

The graph of a function of two variables, $z = f(x, y)$, is a **surface** in three-dimensional space. If (x_0, y_0) is in the domain of f, the point $(x_0, y_0, f(x_0, y_0))$ lies directly above or below the point (x_0, y_0) in the xy-plane, as shown in Figure 8.

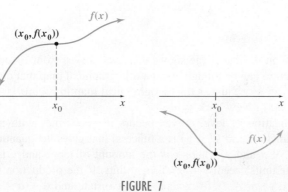

FIGURE 7

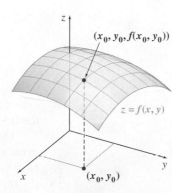

FIGURE 8

Although computer software is available for drawing the graphs of functions of two independent variables, you can often get a good picture of the graph without it by finding various **traces**—the curves that result when a surface is cut by a plane. The **xy-trace** is the intersection of the surface with the xy-plane. The **yz-trace** and **xz-trace** are defined similarly. You can also determine the intersection of the surface with planes parallel to the xy-plane. Such planes are of the form $z = k$, where k is a constant, and the curves that result when they cut the surface are called **level curves**.

EXAMPLE 7 Graphing a Function

Graph $z = x^2 + y^2$.

SOLUTION The yz-plane is the plane in which every point has a first coordinate of 0, so its equation is $x = 0$. When $x = 0$, the equation becomes $z = y^2$, which is the equation of a parabola in the yz-plane, as shown in Figure 9(a). Similarly, to find the intersection of the surface with the xz-plane (whose equation is $y = 0$), let $y = 0$ in the equation. It then becomes $z = x^2$, which is the equation of a parabola in the xz-plane, as shown in Figure 9(a). The xy-trace (the intersection of the surface with the plane $z = 0$) is the single point $(0, 0, 0)$ because $x^2 + y^2$ is equal to 0 only when $x = 0$ and $y = 0$.

Next, we find the level curves by intersecting the surface with the planes $z = 1$, $z = 2$, $z = 3$, etc. (all of which are parallel to the xy-plane). In each case, the result is a circle:

$$x^2 + y^2 = 1, \qquad x^2 + y^2 = 2, \qquad x^2 + y^2 = 3,$$

and so on, as shown in Figure 9(b). Drawing the traces and level curves on the same set of axes suggests that the graph of $z = x^2 + y^2$ is the bowl-shaped figure, called a **paraboloid**, that is shown in Figure 9(c).

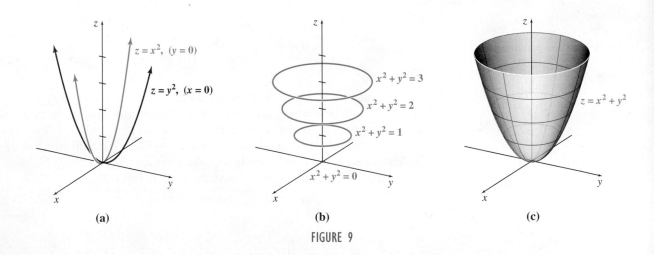

(a)   **(b)**   **(c)**

FIGURE 9

Figure 10 on the next page shows the level curves from Example 7 plotted in the xy-plane. The picture can be thought of as a topographical map that describes the surface generated by $z = x^2 + y^2$, just as the topographical map in Figure 11 describes the surface of the land in a part of New York state.

One application of level curves occurs in economics with production functions. A **production function** $z = f(x, y)$ is a function that gives the quantity z of an item produced as a function of x and y, where x is the amount of labor and y is the amount of capital (in appropriate units) needed to produce z units. If the production function has the special form $z = P(x, y) = Ax^a y^{1-a}$, where A is a constant and $0 < a < 1$, the function is called

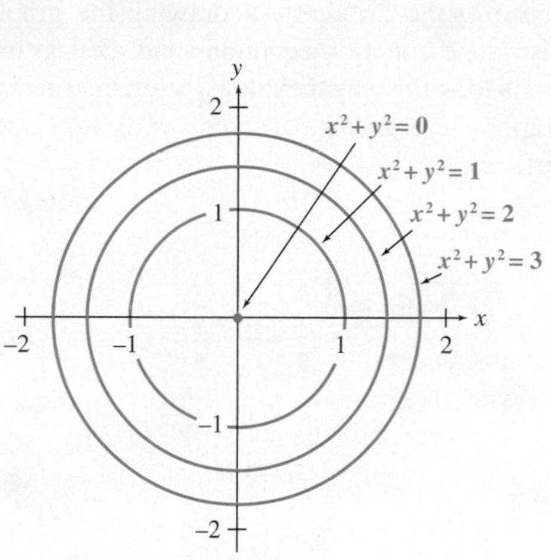

FIGURE 10

FIGURE 11

a **Cobb-Douglas production function**. This function was developed in 1928 by economist Paul H. Douglas (1892–1976), who later became a senator for the state of Illinois, and mathematician Charles W. Cobb. For production functions, level curves are used to indicate combinations of the values of x and y that produce the same value of production z.

EXAMPLE 8   Cobb-Douglas Production Function

Find the level curve at a production of 100 items for the Cobb-Douglas production function $z = x^{2/3}y^{1/3}$.

SOLUTION   Let $z = 100$ to get

$$100 = x^{2/3}y^{1/3}$$
$$\frac{100}{x^{2/3}} = y^{1/3}.$$

Now cube both sides to express y as a function of x.

$$y = \frac{100^3}{x^2} = \frac{1,000,000}{x^2}$$

TRY YOUR TURN 4

The level curve of height 100 found in Example 8 is shown graphed in three dimensions in Figure 12(a) and on the familiar xy-plane in Figure 12(b). The points of the graph correspond to those values of x and y that lead to production of 100 items.

The curve in Figure 12 is called an *isoquant*, for *iso* (equal) and *quant* (amount). In Example 8, the "amounts" all "equal" 100.

APPLY IT

YOUR TURN 4   Find the level curve at a production of 27 items in the form $y = f(x)$ for the Cobb-Douglas production function $z = x^{1/4}y^{3/4}$.

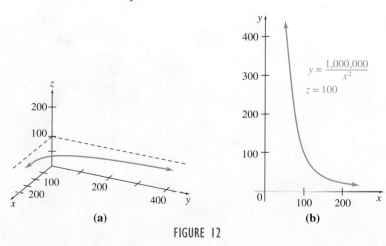

$$y = \frac{1,000,000}{x^2}$$
$$z = 100$$

(a)

(b)

FIGURE 12

Because of the difficulty of drawing the graphs of more complicated functions, we merely list some common equations and their graphs. We encourage you to explore why these graphs look the way they do by studying their traces, level curves, and axis intercepts. These graphs were drawn by computer, a very useful method of depicting three-dimensional surfaces.

Paraboloid, $z = x^2 + y^2$

| xy-trace: point |
| yz-trace: parabola |
| xz-trace: parabola |

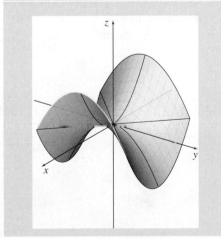

Ellipsoid, $\dfrac{x^2}{a^2} + \dfrac{y^2}{b^2} + \dfrac{z^2}{c^2} = 1$

| xy-trace: ellipse |
| yz-trace: ellipse |
| xz-trace: ellipse |

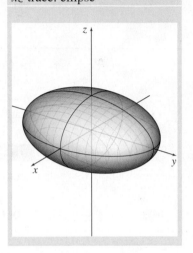

Hyperbolic Paraboloid, $z = x^2 - y^2$
(sometimes called a *saddle*)

| xy-trace: two intersecting lines |
| yz-trace: parabola |
| xz-trace: parabola |

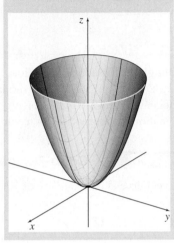

Hyperboloid of Two Sheets,
$-x^2 - y^2 + z^2 = 1$

| xy-trace: none |
| yz-trace: hyperbola |
| xz-trace: hyperbola |

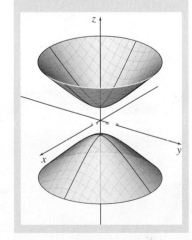

Notice that not all the graphs correspond to functions of two variables. In the ellipsoid, for example, if x and y are both 0, then z can equal c or $-c$, whereas a function can take on only one value. We can, however, interpret the graph as a **level surface** for a function of three variables. Let

$$w(x, y, z) = \frac{x^2}{a^2} + \frac{y^2}{b^2} + \frac{z^2}{c^2}.$$

Then $w = 1$ produces the level surface of the ellipsoid shown, just as $z = c$ gives level curves for the function $z = f(x, y)$.

Another way to draw the graph of a function of two variables is with a graphing calculator. Figure 13 shows the graph of $z = x^2 + y^2$ generated by a TI-89. Figure 14 shows the same graph drawn by the computer program Maple™.

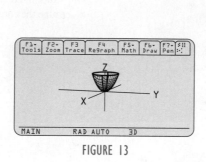

FIGURE 13

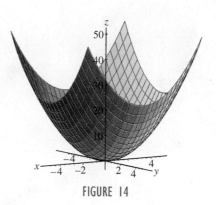

FIGURE 14

9.1   EXERCISES

1. Let $f(x, y) = 2x - 3y + 5$. Find the following.

   a. $f(2, -1)$ **b.** $f(-4, 1)$ **c.** $f(-2, -3)$ **d.** $f(0, 8)$

2. Let $g(x, y) = x^2 - 2xy + y^3$. Find the following.

   a. $g(-2, 4)$ **b.** $g(-1, -2)$ **c.** $g(-2, 3)$ **d.** $g(5, 1)$

3. Let $h(x, y) = \sqrt{x^2 + 2y^2}$. Find the following.

   a. $h(5, 3)$ **b.** $h(2, 4)$ **c.** $h(-1, -3)$ **d.** $h(-3, -1)$

4. Let $f(x, y) = \dfrac{\sqrt{9x + 5y}}{\log x}$. Find the following.

   a. $f(10, 2)$ **b.** $f(100, 1)$

   c. $f(1000, 0)$ **d.** $f\left(\dfrac{1}{10}, 5\right)$

Graph the first-octant portion of each plane.

5. $x + y + z = 9$ **6.** $x + y + z = 15$

7. $2x + 3y + 4z = 12$ **8.** $4x + 2y + 3z = 24$

9. $x + y = 4$ **10.** $y + z = 5$

11. $x = 5$ **12.** $z = 4$

Graph the level curves in the first quadrant of the xy-plane for the following functions at heights of $z = 0$, $z = 2$, and $z = 4$.

13. $3x + 2y + z = 24$ **14.** $3x + y + 2z = 8$

15. $y^2 - x = -z$ **16.** $2y - \dfrac{x^2}{3} = z$

17. Discuss how a function of three variables in the form $w = f(x, y, z)$ might be graphed.

18. Suppose the graph of a plane $ax + by + cz = d$ has a portion in the first octant. What can be said about $a, b, c,$ and d?

19. In the chapter on Nonlinear Functions, the vertical line test was presented, which tells whether a graph is the graph of a function. Does this test apply to functions of two variables? Explain.

20. A graph that was not shown in this section is the *hyperboloid of one sheet*, described by the equation $x^2 + y^2 - z^2 = 1$. Describe it as completely as you can.

Match each equation in Exercises 21–26 with its graph in a–f on the next page

21. $z = x^2 + y^2$ **22.** $z^2 - y^2 - x^2 = 1$

23. $x^2 - y^2 = z$

24. $z = y^2 - x^2$

25. $\dfrac{x^2}{16} + \dfrac{y^2}{25} + \dfrac{z^2}{4} = 1$

26. $z = 5(x^2 + y^2)^{-1/2}$

a.

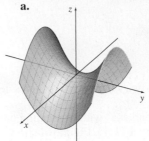

b.

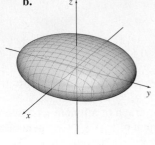

c.

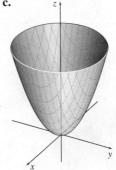

d.

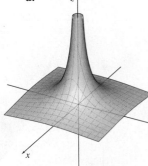

e.

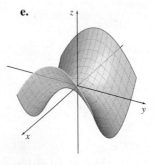

f.

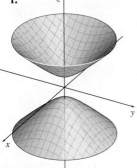

27. Let $f(x, y) = 4x^2 - 2y^2$, and find the following.

a. $\dfrac{f(x + h, y) - f(x, y)}{h}$

b. $\dfrac{f(x, y + h) - f(x, y)}{h}$

c. $\lim\limits_{h \to 0} \dfrac{f(x + h, y) - f(x, y)}{h}$

d. $\lim\limits_{h \to 0} \dfrac{f(x, y + h) - f(x, y)}{h}$

28. Let $f(x, y) = 5x^3 + 3y^2$, and find the following.

a. $\dfrac{f(x + h, y) - f(x, y)}{h}$

b. $\dfrac{f(x, y + h) - f(x, y)}{h}$

c. $\lim\limits_{h \to 0} \dfrac{f(x + h, y) - f(x, y)}{h}$

d. $\lim\limits_{h \to 0} \dfrac{f(x, y + h) - f(x, y)}{h}$

29. Let $f(x, y) = xye^{x^2 + y^2}$. Use a graphing calculator or spreadsheet to find each of the following and give a geometric interpretation of the results. (*Hint:* First factor e^2 from the limit and then evaluate the quotient at smaller and smaller values of h.)

a. $\lim\limits_{h \to 0} \dfrac{f(1 + h, 1) - f(1, 1)}{h}$

b. $\lim\limits_{h \to 0} \dfrac{f(1, 1 + h) - f(1, 1)}{h}$

30. The following table provides values of the function $f(x, y)$. However, because of potential errors in measurement, the functional values may be slightly inaccurate. Using the statistical package included with a graphing calculator or spreadsheet and critical thinking skills, find the function $f(x, y) = a + bx + cy$ that best estimates the table where a, b, and c are integers. (*Hint:* Do a linear regression on each column with the value of y fixed and then use these four regression equations to determine the coefficient c.)

x \ y	0	1	2	3
0	4.02	7.04	9.98	13.00
1	6.01	9.06	11.98	14.96
2	7.99	10.95	14.02	17.09
3	9.99	13.01	16.01	19.02

APPLICATIONS

Business and Economics

31. Production Production of a digital camera is given by

$$P(x, y) = 100\left(\frac{3}{5}x^{-2/5} + \frac{2}{5}y^{-2/5}\right)^{-5},$$

where x is the amount of labor in work-hours and y is the amount of capital. Find the following.

a. What is the production when 32 work-hours and 1 unit of capital are provided?

b. Find the production when 1 work-hour and 32 units of capital are provided.

c. If 32 work-hours and 243 units of capital are used, what is the production output?

Individual Retirement Accounts The multiplier function

$$M = \frac{(1 + i)^n (1 - t) + t}{[1 + (1 - t)i]^n}$$

compares the growth of an Individual Retirement Account (IRA) with the growth of the same deposit in a regular savings account. The function M depends on the three variables n, i, and t, where n represents the number of years an amount is left at interest, i represents the interest rate in both types of accounts, and t represents the income tax rate. Values of $M > 1$ indicate that the IRA grows faster than the savings account. Let $M = f(n, i, t)$ and find the following.

32. Find the multiplier when funds are left for 25 years at 5% interest and the income tax rate is 33%. Which account grows faster?

33. What is the multiplier when money is invested for 40 years at 6% interest and the income tax rate is 28%? Which account grows faster?

Production Find the level curve at a production of 500 for the production functions in Exercises 34 and 35. Graph each level curve in the *xy*-plane.

34. In their original paper, Cobb and Douglas estimated the production function for the United States to be $z = 1.01x^{3/4}y^{1/4}$, where x represents the amount of labor and y the amount of capital. *Source: American Economic Review.*

35. A study of the connection between immigration and the fiscal problems associated with the aging of the baby boom generation considered a production function of the form $z = x^{0.6}y^{0.4}$, where x represents the amount of labor and y the amount of capital. *Source: Journal of Political Economy.*

36. Production For the function in Exercise 34, what is the effect on z of doubling x? Of doubling y? Of doubling both?

37. Cost If labor (x) costs $250 per unit, materials (y) cost $150 per unit, and capital (z) costs $75 per unit, write a function for total cost.

Life Sciences

38. Heat Loss The rate of heat loss (in watts) in harbor seal pups has been approximated by

$$H(m,T,A) = \frac{15.2m^{0.67}(T - A)}{10.23 \ln m - 10.74},$$

where m is the body mass of the pup (in kg), and T and A are the body core temperature and ambient water temperature, respectively (in °C). Find the heat loss for the following data. *Source: Functional Ecology.*

a. Body mass = 21 kg; body core temperature = 36°C; ambient water temperature = 4°C

b. Body mass = 29 kg; body core temperature = 38°C; ambient water temperature = 16°C

39. Body Surface Area The surface area of a human (in square meters) has been approximated by

$$A = 0.024265h^{0.3964}m^{0.5378},$$

where h is the height (in cm) and m is the mass (in kg). Find A for the following data. *Source: The Journal of Pediatrics.*

a. Height, 178 cm; mass, 72 kg

b. Height, 140 cm; mass, 65 kg

c. Height, 160 cm; mass, 70 kg

d. Using your mass and height, find your own surface area.

40. Dinosaur Running An article entitled "How Dinosaurs Ran" explains that the locomotion of different sized animals can be compared when they have the same Froude number, defined as

$$F = \frac{v^2}{gl},$$

where v is the velocity, g is the acceleration of gravity (9.81 m per sec^2), and l is the leg length (in meters). *Source: Scientific American.*

a. One result described in the article is that different animals change from a trot to a gallop at the same Froude number, roughly 2.56. Find the velocity at which this change occurs for a ferret, with a leg length of 0.09 m, and a rhinoceros, with a leg length of 1.2 m.

b. Ancient footprints in Texas of a sauropod, a large herbivorous dinosaur, are roughly 1 m in diameter, corresponding to a leg length of roughly 4 m. By comparing the stride divided by the leg length with that of various modern creatures, it can be determined that the Froude number for these dinosaurs is roughly 0.025. How fast were the sauropods traveling?

41. Pollution Intolerance According to research at the Great Swamp in New York, the percentage of fish that are intolerant to pollution can be estimated by the function

$$P(W, R, A) = 48 - 2.43W - 1.81R - 1.22A,$$

where W is the percentage of wetland, R is the percentage of residential area, and A is the percentage of agricultural area surrounding the swamp. *Source: Northeastern Naturalist.*

a. Use this function to estimate the percentage of fish that will be intolerant to pollution if 5 percent of the land is classified as wetland, 15 percent is classified as residential, and 0 percent is classified as agricultural. (*Note:* The land can also be classified as forest land.)

b. What is the maximum percentage of fish that will be intolerant to pollution?

c. Develop two scenarios that will drive the percentage of fish that are intolerant to pollution to zero.

d. Which variable has the greatest influence on P?

42. Dengue Fever In tropical regions, dengue fever is a significant health problem that affects nearly 100 million people each year. Using data collected from the 2002 dengue epidemic in Colima, Mexico, researchers have estimated that the incidence I (number of new cases in a given year) of dengue can be predicted by the following function.

$$I(p, a, m, n, e) = (25.54 + 0.04p - 7.92a + 2.62m + 4.46n + 0.15e)^2,$$

where p is the precipitation (mm), a is the mean temperature (°C), m is the maximum temperature (°C), n is the minimum temperature (°C), and e is the evaporation (mm). *Source: Journal of Environmental Health.*

a. Estimate the incidence of a dengue fever outbreak for a region with 80 mm of rainfall, average temperature of 23°C, maximum temperature of 34°C, minimum temperature of 16°C, and evaporation of 50 mm.

b. Which variable has a negative influence on the incidence of dengue? Describe this influence and what can be inferred mathematically about the biology of the fever.

43. Deer-Vehicle Accidents Using data collected by the U.S. Forest Service, the annual number of deer-vehicle accidents for any given county in Ohio can be estimated by the function

$$A(L, T, U, C) = 53.02 + 0.383L + 0.0015T + 0.0028U$$
$$- 0.0003C,$$

where A is the estimated number of accidents, L is the road length (in kilometers), T is the total county land area (in hundred-acres (Ha)), U is the urban land area (in hundred-acres), and C is the number of hundred-acres of crop land. *Source: Ohio Journal of Science.*

a. Use this formula to estimate the number of deer-vehicle accidents for Mahoning County, where $L = 266$ km, $T = 107{,}484$ Ha, $U = 31{,}697$ Ha, and $C = 24{,}870$ Ha. The actual value was 396.

b. Given the magnitude and nature of the input numbers, which of the variables have the greatest potential to influence the number of deer-vehicle accidents? Explain your answer.

44. Deer Harvest Using data collected by the U.S. Forest Service, the annual number of deer that are harvested for any given county in Ohio can be estimated by the function

$$N(R, C) = 329.32 + 0.0377R - 0.0171C,$$

where N is the estimated number of harvested deer, R is the rural land area (in hundred-acres), and C is the number of hundred-acres of crop land. *Source: Ohio Journal of Science.*

a. Use this formula to estimate the number of harvested deer for Tuscarawas County, where $R = 141{,}319$ Ha and $C = 37{,}960$ Ha. The actual value was 4925 deer harvested.

b. Sketch the graph of this function in the first octant.

45. Agriculture Pregnant sows tethered in stalls often show high levels of repetitive behavior, such as bar biting and chain chewing, indicating chronic stress. Researchers from Great Britain have developed a function that estimates the relationship between repetitive behavior, the behavior of sows in adjacent stalls, and food allowances such that

$$\ln(T) = 5.49 - 3.00 \ln(F) + 0.18 \ln(C),$$

where T is the percent of time spent in repetitive behavior, F is the amount of food given to the sow (in kilograms per day), and C is the percent of time that neighboring sows spent bar biting and chain chewing. *Source: Applied Animal Behaviour Science.*

a. Solve the above expression for T.

b. Find and interpret T when $F = 2$ and $C = 40$.

General Interest

46. Postage Rates Extra postage is charged for parcels sent by U.S. mail that are more than 84 in. in length and girth combined. (Girth is the distance around the parcel perpendicular to its length. See the figure.) Express the combined length and girth as a function of L, W, and H.

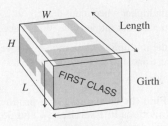

47. Required Material Refer to the figure for Exercise 46. Assume L, W, and H are in feet. Write a function in terms of L, W, and H that gives the total area of the material required to build the box.

48. Elliptical Templates The holes cut in a roof for vent pipes require elliptical templates. A formula for determining the length of the major axis of the ellipse is given by

$$L = f(H, D) = \sqrt{H^2 + D^2},$$

where D is the (outside) diameter of the pipe and H is the "rise" of the roof per D units of "run"; that is, the slope of the roof is H/D. (See the figure below.) The width of the ellipse (minor axis) equals D. Find the length and width of the ellipse required to produce a hole for a vent pipe with a diameter of 3.75 in. in roofs with the following slopes.

a. 3/4   **b.** 2/5

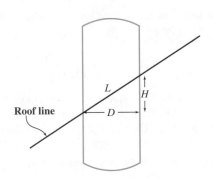

YOUR TURN ANSWERS

1. 29

2. 24

3.

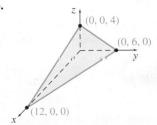

4. $y = 81/x^{1/3}$.

9.2 Partial Derivatives

APPLY IT What is the change in productivity if labor is increased by one work-hour? What if capital is increased by one unit?

We will answer this question in Example 5 using the concept of partial derivatives.

Earlier, we found that the derivative dy/dx gives the rate of change of y with respect to x. In this section, we show how derivatives are found and interpreted for multivariable functions.

A small firm makes only two products, radios and CD players. The profits of the firm are given by

$$P(x, y) = 40x^2 - 10xy + 5y^2 - 80,$$

where x is the number of radios sold and y is the number of CD players sold. How will a change in x or y affect P?

Suppose that sales of radios have been steady at 10 units; only the sales of CD players vary. The management would like to find the marginal profit with respect to y, the number of CD players sold. Recall that marginal profit is given by the derivative of the profit function. Here, x is fixed at 10. Using this information, we begin by finding a new function, $f(y) = P(10, y)$. Let $x = 10$ to get

$$f(y) = P(10, y) = 40(10)^2 - 10(10)y + 5y^2 - 80$$
$$= 3920 - 100y + 5y^2.$$

The function $f(y)$ shows the profit from the sale of y CD players, assuming that x is fixed at 10 units. Find the derivative df/dy to get the marginal profit with respect to y.

$$\frac{df}{dy} = -100 + 10y$$

In this example, the derivative of the function $f(y)$ was taken with respect to y only; we assumed that x was fixed. To generalize, let $z = f(x, y)$. An intuitive definition of the *partial derivatives* of f with respect to x and y follows.

Partial Derivatives (Informal Definition)

The **partial derivative of f with respect to x** is the derivative of f obtained by treating x as a variable and y as a constant.

The **partial derivative of f with respect to y** is the derivative of f obtained by treating y as a variable and x as a constant.

The symbols $f_x(x, y)$ (no prime is used), $\partial z/\partial x$, z_x, and $\partial f/\partial x$ are used to represent the partial derivative of $z = f(x, y)$ with respect to x, with similar symbols used for the partial derivative with respect to y.

Generalizing from the definition of the derivative given earlier, partial derivatives of a function $z = f(x, y)$ are formally defined as follows.

Partial Derivatives (Formal Definition)

Let $z = f(x, y)$ be a function of two independent variables. Let all indicated limits exist. Then the partial derivative of f with respect to x is

$$f_x(x, y) = \frac{\partial f}{\partial x} = \lim_{h \to 0} \frac{f(x + h, y) - f(x, y)}{h},$$

and the partial derivative of f with respect to y is

$$f_y(x, y) = \frac{\partial f}{\partial y} = \lim_{h \to 0} \frac{f(x, y + h) - f(x, y)}{h}.$$

If the indicated limits do not exist, then the partial derivatives do not exist.

Similar definitions could be given for functions of more than two independent variables.

EXAMPLE 1 Partial Derivatives

Let $f(x, y) = 4x^2 - 9xy + 6y^3$. Find $f_x(x, y)$ and $f_y(x, y)$.

SOLUTION To find $f_x(x, y)$, treat y as a constant and x as a variable. The derivative of the first term, $4x^2$, is $8x$. In the second term, $-9xy$, the constant coefficient of x is $-9y$, so the derivative with x as the variable is $-9y$. The derivative of $6y^3$ is zero, since we are treating y as a constant. Thus,

$$f_x(x, y) = 8x - 9y.$$

Now, to find $f_y(x, y)$, treat y as a variable and x as a constant. Since x is a constant, the derivative of $4x^2$ is zero. In the second term, the coefficient of y is $-9x$ and the derivative of $-9xy$ is $-9x$. The derivative of the third term is $18y^2$. Thus,

$$f_y(x, y) = -9x + 18y^2. \qquad \text{TRY YOUR TURN 1}$$

YOUR TURN 1 Let $f(x,y) = 2x^2y^3 + 6x^5y^4$. Find $f_x(x,y)$ and $f_y(x,y)$.

The next example shows how the chain rule can be used to find partial derivatives.

EXAMPLE 2 Partial Derivatives

Let $f(x, y) = \ln|x^2 + 3y|$. Find $f_x(x, y)$ and $f_y(x, y)$.

SOLUTION Recall the formula for the derivative of a natural logarithm function. If $g(x) = \ln|x|$, then $g'(x) = 1/x$. Using this formula and the chain rule,

$$f_x(x, y) = \frac{1}{x^2 + 3y} \cdot \frac{\partial}{\partial x}(x^2 + 3y) = \frac{1}{x^2 + 3y} \cdot 2x = \frac{2x}{x^2 + 3y},$$

and

$$f_y(x, y) = \frac{1}{x^2 + 3y} \cdot \frac{\partial}{\partial y}(x^2 + 3y) = \frac{1}{x^2 + 3y} \cdot 3 = \frac{3}{x^2 + 3y}.$$

YOUR TURN 2 Let $f(x,y) = e^{3x^2y}$. Find $f_x(x,y)$ and $f_y(x,y)$.

$$\text{TRY YOUR TURN 2}$$

The notation

$$f_x(a, b) \qquad \text{or} \qquad \frac{\partial f}{\partial y}(a, b)$$

represents the value of the partial derivative when $x = a$ and $y = b$, as shown in the next example.

EXAMPLE 3 Evaluating Partial Derivatives

Let $f(x, y) = 2x^2 + 9xy^3 + 8y + 5$. Find the following.

(a) $f_x(-1, 2)$

SOLUTION First, find $f_x(x, y)$ by holding y constant.

$$f_x(x, y) = 4x + 9y^3$$

Now let $x = -1$ and $y = 2$.

$$f_x(-1, 2) = 4(-1) + 9(2)^3 = -4 + 72 = 68$$

(b) $\dfrac{\partial f}{\partial y}(-4, -3)$

SOLUTION Since $\partial f/\partial y = 27xy^2 + 8$,

$$\frac{\partial f}{\partial y}(-4, -3) = 27(-4)(-3)^2 + 8 = 27(-36) + 8 = -964.$$

(c) All values of x and y such that both $f_x(x, y) = 0$ and $f_y(x, y) = 0$. (The importance of such points will be shown in the next section.)

SOLUTION From parts (a) and (b),

$$f_x(x, y) = 4x + 9y^3 = 0 \quad \text{and} \quad f_y(x, y) = 27xy^2 + 8 = 0.$$

Solving the first equation for x yields $x = -9y^3/4$. Substituting this into the second equation yields

$$27\left(\frac{-9y^3}{4}\right)y^2 + 8 = 0$$

$$\frac{-243y^5}{4} + 8 = 0$$

$$\frac{-243y^5}{4} = -8$$

$$y^5 = \frac{32}{243}$$

$$y = \frac{2}{3}. \qquad \text{Take the fifth root of both sides.}$$

Substituting $y = 2/3$ yields $x = -9y^3/4 = -9(2/3)^3/4 = -2/3$. Thus, $f_x(x, y) = 0$ and $f_y(x, y) = 0$ when $x = -2/3$ and $y = 2/3$.

(d) $f_x(x, y)$ using the formal definition of the partial derivative.

SOLUTION Calculate as follows:

$$\frac{f(x+h, y) - f(x, y)}{h} = \frac{2(x+h)^2 + 9(x+h)y^3 + 8y + 5 - (2x^2 + 9xy^3 + 8y + 5)}{h}$$

$$= \frac{2x^2 + 4xh + 2h^2 + 9xy^3 + 9hy^3 + 8y + 5 - 2x^2 - 9xy^3 - 8y - 5}{h}$$

$$= \frac{4xh + 2h^2 + 9hy^3}{h} \qquad \text{Simplify the numerator.}$$

$$= \frac{h(4x + 2h + 9y^3)}{h} \qquad \text{Factor } h \text{ from the numerator.}$$

$$= 4x + 2h + 9y^3.$$

Now take the limit as h goes to 0. Thus,

$$f_x(x, y) = \lim_{h \to 0} \frac{f(x + h, y) - f(x, y)}{h} = \lim_{h \to 0}(4x + 2h + 9y^3)$$

YOUR TURN 3 Let $f(x,y) = xye^{x^2+y^3}$. Find $f_x(2,1)$ and $f_y(2,1)$.

$$= 4x + 9y^3,$$

the same answer we found in part (a).

TRY YOUR TURN 3

In some cases, the difference quotient may not simplify as easily as it did in Example 3(d). In such cases, the derivative may be approximated by putting a small value for h into $[f(x + h) - f(x)]/h$. In Example 3(d), with $x = -1$ and $y = 2$, the values $h = 10^{-4}$ and 10^{-5} give approximations for $f_x(-1, 2)$ as 68.0002 and 68.00002, respectively, compared with the exact value of 68 found in Example 3(a).

The derivative of a function of one variable can be interpreted as the slope of the tangent line to the graph at that point. With some modification, the same is true of partial derivatives of functions of two variables. At a point on the graph of a function of two variables, $z = f(x, y)$, there may be many tangent lines, all of which lie in the same tangent plane, as shown in Figure 15.

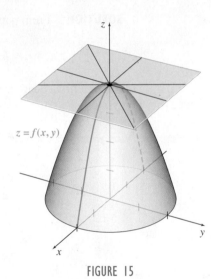

$z = f(x, y)$

FIGURE 15

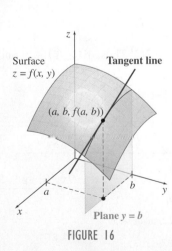

Surface $z = f(x, y)$

Tangent line

$(a, b, f(a, b))$

Plane $y = b$

FIGURE 16

In any particular direction, however, there will be only one tangent line. We use partial derivatives to find the slope of the tangent lines in the x- and y-directions as follows.

Figure 16 shows a surface $z = f(x, y)$ and a plane that is parallel to the xz-plane. The equation of the plane is $y = b$. (This corresponds to holding y fixed.) Since $y = b$ for points on the plane, any point on the curve that represents the intersection of the plane and the surface must have the form $(x, y, z) = (x, b, f(x, b))$. Thus, this curve can be described as $z = f(x, b)$. Since b is constant, $z = f(x, b)$ is a function of one variable. When the derivative of $z = f(x, b)$ is evaluated at $x = a$, it gives the slope of the line tangent to this curve at the point $(a, b, f(a, b))$, as shown in Figure 16. Thus, the partial derivative of f with respect to x, $f_x(a, b)$, gives the rate of change of the surface $z = f(x, y)$ in the x-direction at the point $(a, b, f(a, b))$. In the same way, the partial derivative with respect to y will give the slope of the line tangent to the surface in the y-direction at the point $(a, b, f(a, b))$.

Rate of Change

The derivative of $y = f(x)$ gives the rate of change of y with respect to x. In the same way, if $z = f(x, y)$, then $f_x(x, y)$ gives the rate of change of z with respect to x, if y is held constant.

EXAMPLE 4 Water Temperature

Suppose that the temperature of the water at the point on a river where a nuclear power plant discharges its hot waste water is approximated by

$$T(x, y) = 2x + 5y + xy - 40,$$

where x represents the temperature of the river water (in degrees Celsius) before it reaches the power plant and y is the number of megawatts (in hundreds) of electricity being produced by the plant.

(a) Find and interpret $T_x(9, 5)$.

SOLUTION First, find the partial derivative $T_x(x, y)$.

$$T_x(x, y) = 2 + y$$

This partial derivative gives the rate of change of T with respect to x. Replacing x with 9 and y with 5 gives

$$T_x(9, 5) = 2 + 5 = 7.$$

Just as marginal cost is the approximate cost of one more item, this result, 7, is the approximate change in temperature of the output water if input water temperature changes by 1 degree, from $x = 9$ to $x = 9 + 1 = 10$, while y remains constant at 5 (500 megawatts of electricity produced).

(b) Find and interpret $T_y(9, 5)$.

SOLUTION The partial derivative $T_y(x, y)$ is

$$T_y(x, y) = 5 + x.$$

This partial derivative gives the rate of change of T with respect to y as

$$T_y(9, 5) = 5 + 9 = 14.$$

This result, 14, is the approximate change in temperature resulting from a 1-unit increase in production of electricity from $y = 5$ to $y = 5 + 1 = 6$ (from 500 to 600 megawatts), while the input water temperature x remains constant at 9°C.

As mentioned in the previous section, if $P(x, y)$ gives the output P produced by x units of labor and y units of capital, $P(x, y)$ is a production function. The partial derivatives of this production function have practical implications. For example, $\partial P/\partial x$ gives the marginal productivity of labor. This represents the rate at which the output is changing with respect to labor for a fixed capital investment. That is, if the capital investment is held constant and labor is increased by 1 work-hour, $\partial P/\partial x$ will yield the approximate change in the production level. Likewise, $\partial P/\partial y$ gives the marginal productivity of capital, which represents the rate at which the output is changing with respect to a one-unit change in capital for a fixed labor value. So if the labor force is held constant and the capital investment is increased by 1 unit, $\partial P/\partial y$ will approximate the corresponding change in the production level.

EXAMPLE 5 Production Function

A company that manufactures computers has determined that its production function is given by

$$P(x, y) = 0.1xy^2 \ln(2x + 3y + 2),$$

where x is the size of the labor force (measured in work-hours per week) and y is the amount of capital (measured in units of $1000) invested. Find the marginal productivity of labor and capital when $x = 50$ and $y = 20$, and interpret the results.

APPLY IT

SOLUTION The marginal productivity of labor is found by taking the derivative of P with respect to x.

$$\frac{\partial P}{\partial x} = 0.1\left[\frac{xy^2}{2x + 3y + 2} \cdot 2 + y^2 \ln(2x + 3y + 2)\right] \quad \text{Use the product and chain rules.}$$

$$\frac{\partial P}{\partial x}(50,20) = 0.1\left[\frac{50(20)^2}{2(50) + 3(20) + 2} \cdot 2 + 20^2 \ln(2(50) + 3(20) + 2)\right] \approx 228$$

Thus, if the capital investment is held constant at $20,000 and labor is increased from 50 to 51 work-hours per week, production will increase by about 228 units. In the same way, the marginal productivity of capital is $\partial P/\partial y$.

$$\frac{\partial P}{\partial y} = 0.1\left[\frac{xy^2}{2x + 3y + 2} \cdot 3 + 2xy \ln(2x + 3y + 2)\right] \quad \text{Use the product and chain rules.}$$

$$\frac{\partial P}{\partial y}(50,20) = 0.1\left[\frac{50(20)^2}{2(50) + 3(20) + 2} \cdot 3 + 2(50)(20)\ln(2(50) + 3(20) + 2)\right] \approx 1055$$

If work-hours are held constant at 50 hours per week and the capital investment is increased from $20,000 to $21,000, production will increase by about 1055 units.

Second-Order Partial Derivatives

The second derivative of a function of one variable is very useful in determining relative maxima and minima. **Second-order partial derivatives** (partial derivatives of a partial derivative) are used in a similar way for functions of two or more variables. The situation is somewhat more complicated, however, with more independent variables. For example, $f(x, y) = 4x + x^2y + 2y$ has two first-order partial derivatives,

$$f_x(x, y) = 4 + 2xy \quad \text{and} \quad f_y(x, y) = x^2 + 2.$$

Since each of these has two partial derivatives, one with respect to y and one with respect to x, there are *four* second-order partial derivatives of function f. The notations for these four second-order partial derivatives are given below.

Second-Order Partial Derivatives

For a function $z = f(x, y)$, if the indicated partial derivative exists, then

$$\frac{\partial}{\partial x}\left(\frac{\partial z}{\partial x}\right) = \frac{\partial^2 z}{\partial x^2} = f_{xx}(x, y) = z_{xx} \qquad \frac{\partial}{\partial y}\left(\frac{\partial z}{\partial y}\right) = \frac{\partial^2 z}{\partial y^2} = f_{yy}(x, y) = z_{yy}$$

$$\frac{\partial}{\partial y}\left(\frac{\partial z}{\partial x}\right) = \frac{\partial^2 z}{\partial y \partial x} = f_{xy}(x, y) = z_{xy} \qquad \frac{\partial}{\partial x}\left(\frac{\partial z}{\partial y}\right) = \frac{\partial^2 z}{\partial x \partial y} = f_{yx}(x, y) = z_{yx}$$

NOTE For most functions found in applications and for all of the functions in this book, the second-order partial derivatives $f_{xy}(x, y)$ and $f_{yx}(x, y)$ are equal. This is always true when $f_{xy}(x, y)$ and $f_{yx}(x, y)$ are continuous. Therefore, it is not necessary to be particular about the order in which these derivatives are found.

EXAMPLE 6 **Second-Order Partial Derivatives**

Find all second-order partial derivatives for

$$f(x, y) = -4x^3 - 3x^2y^3 + 2y^2.$$

SOLUTION First find $f_x(x, y)$ and $f_y(x, y)$.

$$f_x(x, y) = -12x^2 - 6xy^3 \quad \text{and} \quad f_y(x, y) = -9x^2y^2 + 4y$$

To find $f_{xx}(x, y)$, take the partial derivative of $f_x(x, y)$ with respect to x.

$$f_{xx}(x, y) = -24x - 6y^3$$

Take the partial derivative of $f_y(x, y)$ with respect to y; this gives f_{yy}.

$$f_{yy}(x, y) = -18x^2y + 4$$

Find $f_{xy}(x, y)$ by starting with $f_x(x, y)$, then taking the partial derivative of $f_x(x, y)$ with respect to y.

YOUR TURN 4 Let $f(x, y) = x^2e^{7y} + x^4y^5$. Find all second partial derivatives.

$$f_{xy}(x, y) = -18xy^2$$

Finally, find $f_{yx}(x, y)$ by starting with $f_y(x, y)$; take its partial derivative with respect to x.

$$f_{yx}(x, y) = -18xy^2 \qquad \text{TRY YOUR TURN 4}$$

EXAMPLE 7 **Second-Order Partial Derivatives**

Let $f(x, y) = 2e^x - 8x^3y^2$. Find all second-order partial derivatives.

SOLUTION Here $f_x(x, y) = 2e^x - 24x^2y^2$ and $f_y(x, y) = -16x^3y$. (Recall: If $g(x) = e^x$, then $g'(x) = e^x$.) Now find the second-order partial derivatives.

$$f_{xx}(x, y) = 2e^x - 48xy^2 \qquad f_{xy}(x, y) = -48x^2y$$
$$f_{yy}(x, y) = -16x^3 \qquad\qquad f_{yx}(x, y) = -48x^2y$$

Partial derivatives of functions with more than two independent variables are found in a similar manner. For instance, to find $f_z(x, y, z)$ for $f(x, y, z)$, hold x and y constant and differentiate with respect to z.

EXAMPLE 8 **Second-Order Partial Derivatives**

Let $f(x, y, z) = 2x^2yz^2 + 3xy^2 - 4yz$. Find $f_x(x, y, z)$, $f_y(x, y, z)$, $f_{xz}(x, y, z)$, and $f_{yz}(x, y, z)$.

SOLUTION
$$f_x(x, y, z) = 4xyz^2 + 3y^2$$
$$f_y(x, y, z) = 2x^2z^2 + 6xy - 4z$$

To find $f_{xz}(x, y, z)$, differentiate $f_x(x, y, z)$ with respect to z.

$$f_{xz}(x, y, z) = 8xyz$$

Differentiate $f_y(x, y, z)$ with respect to z to get $f_{yz}(x, y, z)$.

$$f_{yz}(x, y, z) = 4x^2z - 4$$

9.2 EXERCISES

1. Let $z = f(x, y) = 6x^2 - 4xy + 9y^2$. Find the following using the formal definition of the partial derivative.

   a. $\dfrac{\partial z}{\partial x}$ **b.** $\dfrac{\partial z}{\partial y}$ **c.** $\dfrac{\partial f}{\partial x}(2, 3)$ **d.** $f_y(1, -2)$

2. Let $z = g(x, y) = 8x + 6x^2y + 2y^2$. Find the following using the formal definition of the partial derivative.

   a. $\dfrac{\partial g}{\partial x}$ **b.** $\dfrac{\partial g}{\partial y}$ **c.** $\dfrac{\partial z}{\partial y}(-3, 0)$ **d.** $g_x(2, 1)$

In Exercises 3–20, find $f_x(x, y)$ and $f_y(x, y)$. Then find $f_x(2, -1)$ and $f_y(-4, 3)$. Leave the answers in terms of e in Exercises 7–10, 15–16, and 19–20.

3. $f(x, y) = -4xy + 6y^3 + 5$ 4. $f(x, y) = 9x^2y^2 - 4y^2$

5. $f(x, y) = 5x^2y^3$ 6. $f(x, y) = -3x^4y^3 + 10$

7. $f(x, y) = e^{x+y}$ 8. $f(x, y) = 4e^{3x+2y}$

9. $f(x, y) = -6e^{4x-3y}$ 10. $f(x, y) = 8e^{7x-y}$

11. $f(x, y) = \dfrac{x^2 + y^3}{x^3 - y^2}$ 12. $f(x, y) = \dfrac{3x^2y^3}{x^2 + y^2}$

13. $f(x, y) = \ln|1 + 5x^3y^2|$ 14. $f(x, y) = \ln|4x^4 - 2x^2y^2|$

15. $f(x, y) = xe^{x^2y}$ 16. $f(x, y) = y^2e^{x+3y}$

17. $f(x, y) = \sqrt{x^4 + 3xy + y^4 + 10}$

18. $f(x, y) = (7x^2 + 18xy^2 + y^3)^{1/3}$

19. $f(x, y) = \dfrac{3x^2y}{e^{xy} + 2}$

20. $f(x, y) = (7e^{x+2y} + 4)(e^{x^2} + y^2 + 2)$

Find all second-order partial derivatives for the following.

21. $f(x, y) = 4x^2y^2 - 16x^2 + 4y$

22. $g(x, y) = 5x^4y^2 + 12y^3 - 9x$

23. $R(x, y) = 4x^2 - 5xy^3 + 12y^2x^2$

24. $h(x, y) = 30y + 5x^2y + 12xy^2$

25. $r(x, y) = \dfrac{6y}{x + y}$ 26. $k(x, y) = \dfrac{-7x}{2x + 3y}$

27. $z = 9ye^x$ 28. $z = -6xe^y$

29. $r = \ln|x + y|$ 30. $k = \ln|5x - 7y|$

31. $z = x \ln|xy|$ 32. $z = (y + 1)\ln|x^3y|$

For the functions defined as follows, find all values of x and y such that both $f_x(x, y) = 0$ and $f_y(x, y) = 0$.

33. $f(x, y) = 6x^2 + 6y^2 + 6xy + 36x - 5$

34. $f(x, y) = 50 + 4x - 5y + x^2 + y^2 + xy$

35. $f(x, y) = 9xy - x^3 - y^3 - 6$

36. $f(x, y) = 2200 + 27x^3 + 72xy + 8y^2$

Find $f_x(x, y, z)$, $f_y(x, y, z)$, $f_z(x, y, z)$, and $f_{yz}(x, y, z)$ for the following.

37. $f(x, y, z) = x^4 + 2yz^2 + z^4$

38. $f(x, y, z) = 6x^3 - x^2y^2 + y^5$

39. $f(x, y, z) = \dfrac{6x - 5y}{4z + 5}$

40. $f(x, y, z) = \dfrac{2x^2 + xy}{yz - 2}$

41. $f(x, y, z) = \ln|x^2 - 5xz^2 + y^4|$

42. $f(x, y, z) = \ln|8xy + 5yz - x^3|$

In Exercises 43 and 44, approximate the indicated derivative for each function by using the definition of the derivative with small values of h.

43. $f(x, y) = (x + y/2)^{x+y/2}$

   a. $f_x(1, 2)$ **b.** $f_y(1, 2)$

44. $f(x, y) = (x + y^2)^{2x+y}$

   a. $f_x(2, 1)$ **b.** $f_y(2, 1)$

APPLICATIONS

Business and Economics

45. **Manufacturing Cost** Suppose that the manufacturing cost of a personal digital assistant (PDA) is approximated by

$$M(x, y) = 45x^2 + 40y^2 - 20xy + 50,$$

where x is the cost of electronic chips and y is the cost of labor. Find the following.

   a. $M_y(4, 2)$ **b.** $M_x(3, 6)$ **c.** $(\partial M/\partial x)(2, 5)$
   d. $(\partial M/\partial y)(6, 7)$

46. **Revenue** The revenue from the sale of x units of a sedative and y units of an antibiotic is given by

$$R(x, y) = 5x^2 + 9y^2 - 4xy.$$

Suppose 9 units of sedative and 5 units of antibiotic are sold.

   a. What is the approximate effect on revenue if 10 units of sedative and 5 units of antibiotic are sold?

   b. What is the approximate effect on revenue if the amount of antibiotic sold is increased to 6 units, while sedative sales remain constant?

47. **Sales** A car dealership estimates that the total weekly sales of its most popular model is a function of the car's list price, p, and the interest rate in percent, i, offered by the manufacturer. The approximate weekly sales are given by

$$f(p, i) = 99p - 0.5pi - 0.0025p^2.$$

a. Find the weekly sales if the average list price is $19,400 and the manufacturer is offering an 8% interest rate.

b. Find and interpret $f_p(p, i)$ and $f_i(p, i)$.

c. What would be the effect on weekly sales if the price is $19,400 and interest rates rise from 8% to 9%?

48. Marginal Productivity Suppose the production function of a company is given by

$$P(x, y) = 250\sqrt{x^2 + y^2},$$

where x represents units of labor and y represents units of capital. Find the following when 6 units of labor and 8 units of capital are used.

a. The marginal productivity of labor

b. The marginal productivity of capital

49. Marginal Productivity A manufacturer estimates that production (in hundreds of units) is a function of the amounts x and y of labor and capital used, as follows.

$$f(x, y) = \left(\frac{1}{4}x^{-1/4} + \frac{3}{4}y^{-1/4}\right)^{-4}$$

a. Find the number of units produced when 16 units of labor and 81 units of capital are utilized.

b. Find and interpret $f_x(16, 81)$ and $f_y(16, 81)$.

c. What would be the approximate effect on production of increasing labor by 1 unit from 16 units of labor with 81 units of capital?

50. Marginal Productivity The production function z for the United States was once estimated as

$$z = x^{0.7}y^{0.3},$$

where x stands for the amount of labor and y the amount of capital. Find the marginal productivity of labor and of capital.

51. Marginal Productivity A similar production function for Canada is

$$z = x^{0.4}y^{0.6},$$

with x, y, and z as in Exercise 50. Find the marginal productivity of labor and of capital.

52. Marginal Productivity A manufacturer of automobile batteries estimates that his total production (in thousands of units) is given by

$$f(x, y) = 3x^{1/3}y^{2/3},$$

where x is the number of units of labor and y is the number of units of capital utilized.

a. Find and interpret $f_x(64, 125)$ and $f_y(64, 125)$ if the current level of production uses 64 units of labor and 125 units of capital.

b. Use your answer from part a to calculate the approximate effect on production of increasing labor to 65 units while holding capital at the current level.

c. Suppose that sales have been good and management wants to increase either capital or labor by 1 unit. Which option would result in a larger increase in production?

Life Sciences

53. Calorie Expenditure The average energy expended for an animal to walk or run 1 km can be estimated by the function

$$f(m, v) = 25.92m^{0.68} + \frac{3.62m^{0.75}}{v},$$

where $f(m, v)$ is the energy used (in kcal per hour), m is the mass (in g), and v is the speed of movement (in km per hour) of the animal. *Source: Wildlife Feeding and Nutrition.*

a. Find $f(300, 10)$.

b. Find $f_m(300, 10)$ and interpret.

c. If a mouse could run at the same speed that an elephant walks, which animal would expend more energy? How can partial derivatives be used to explore this question?

54. Heat Loss The rate of heat loss (in watts) in harbor seal pups has been approximated by

$$H(m, T, A) = \frac{15.2m^{0.67}(T - A)}{10.23 \ln m - 10.74},$$

where m is the body mass of the pup (in kg), and T and A are the body core temperature and ambient water temperature, respectively (in °C). Find the approximate change in heat loss under the following conditions. *Source: Functional Ecology.*

a. The body core temperature increases from 37°C to 38°, while the ambient water temperature remains at 8°C and the body mass remains at 24 kg.

b. The ambient water temperature increases from 10°C to 11°, while the body core temperature remains at 37°C and the body mass remains at 26 kg.

55. Body Surface Area The surface area of a human (in square meters) has been approximated by

$$A = 0.024265h^{0.3964}m^{0.5378},$$

where h is the height (in cm) and m is the mass (in kg). *Source: The Journal of Pediatrics.*

a. Find the approximate change in surface area when the mass changes from 72 kg to 73 kg, while the height remains at 180 cm.

b. Find the approximate change in surface area when the height changes from 160 cm to 161 cm, while the mass remains at 70 kg.

56. Blood Flow According to the Fick Principle, the quantity of blood pumped through the lungs depends on the following variables (in milliliters):

b = quantity of oxygen used by the body in one minute

a = quantity of oxygen per liter of blood that has just gone through the lungs

v = quantity of oxygen per liter of blood that is about to enter the lungs

In one minute,

Amount of oxygen used = Amount of oxygen per liter
× Liters of blood pumped.

If C is the number of liters of blood pumped through the lungs in one minute, then

$$b = (a - v) \cdot C \quad \text{or} \quad C = \frac{b}{a - v}.$$

Source: Anaesthesia UK.

a. Find the number of liters of blood pumped through the lungs in one minute if $a = 160$, $b = 200$, and $v = 125$.

b. Find the approximate change in C when a changes from 160 to 161, $b = 200$, and $v = 125$.

c. Find the approximate change in C when $a = 160$, b changes from 200 to 201, and $v = 125$.

d. Find the approximate change in C when $a = 160$, $b = 200$, and v changes from 125 to 126.

e. A change of 1 unit in which quantity of oxygen produces the greatest change in the liters of blood pumped?

57. **Health** A weight-loss counselor has prepared a program of diet and exercise for a client. If the client sticks to the program, the weight loss that can be expected (in pounds per week) is given by

$$\text{Weight loss} = f(n, c) = \frac{1}{8}n^2 - \frac{1}{5}c + \frac{1937}{8},$$

where c is the average daily calorie intake for the week and n is the number of 40-minute aerobic workouts per week.

a. How many pounds can the client expect to lose by eating an average of 1200 cal per day and participating in four 40-minute workouts in a week?

b. Find and interpret $\partial f / \partial n$.

c. The client currently averages 1100 cal per day and does three 40-minute workouts each week. What would be the approximate impact on weekly weight loss of adding a fourth workout per week?

58. **Health** The body mass index is a number that can be calculated for any individual as follows: Multiply a person's weight by 703 and divide by the person's height squared. That is,

$$B = \frac{703w}{h^2},$$

where w is in pounds and h is in inches. The National Heart, Lung and Blood Institute uses the body mass index to determine whether a person is "overweight" ($25 \le B < 30$) or "obese" ($B \ge 30$). *Source: The National Institutes of Health.*

a. Calculate the body mass index for Miami Dolphins offensive tackle Jake Long, who weighs 317 lb and is 6′7″ tall.

b. Calculate $\frac{\partial B}{\partial w}$ and $\frac{\partial B}{\partial h}$ and interpret.

c. Using the fact that 1 in. = 0.0254 m and 1 lb. = 0.4536 kg. transform this formula to handle metric units.

59. **Drug Reaction** The reaction to x units of a drug t hours after it was administered is given by

$$R(x, t) = x^2(a - x)t^2e^{-t},$$

for $0 \le x \le a$ (where a is a constant). Find the following.

a. $\frac{\partial R}{\partial x}$ b. $\frac{\partial R}{\partial t}$ c. $\frac{\partial^2 R}{\partial x^2}$ d. $\frac{\partial^2 R}{\partial x \partial t}$

e. Interpret your answers to parts a and b.

60. **Scuba Diving** In 1908, J. Haldane constructed diving tables that provide a relationship between the water pressure on body tissues for various water depths and dive times. The tables were successfully used by divers to virtually eliminate decompression sickness. The pressure in atmospheres for a no-stop dive is given by the following formula:*

$$p(l, t) = 1 + \frac{l}{33}(1 - 2^{-t/5}),$$

where t is in minutes, l is in feet, and p is in atmospheres (atm). *Source: The UMAP Journal.*

a. Find the pressure at 33 ft for a 10-minute dive.

b. Find $p_l(33, 10)$ and $p_t(33, 10)$ and interpret. (*Hint:* $D_t(a^t) = \ln(a)a^t$.)

c. Haldane estimated that decompression sickness for no-stop dives could be avoided if the diver's tissue pressure did not exceed 2.15 atm. Find the maximum amount of time that a diver could stay down (time includes going down and coming back up) if he or she wants to dive to a depth of 66 ft.

61. **Wind Chill** In 1941, explorers Paul Siple and Charles Passel discovered that the amount of heat lost when an object is exposed to cold air depends on both the temperature of the air and the velocity of the wind. They developed the *Wind Chill Index* as a way to measure the danger of frostbite while doing outdoor activities. The wind chill can be calculated as follows:

$$W(V, T) = 91.4 - \frac{(10.45 + 6.69\sqrt{V} - 0.447V)(91.4 - T)}{22}$$

where V is the wind speed in miles per hour and T is the temperature in Fahrenheit for wind speeds between 4 and 45 mph. *Source: The UMAP Journal.*

a. Find the wind chill for a wind speed of 20 mph and 10°F.

b. If a weather report indicates that the wind chill is −25°F and the actual outdoor temperature is 5°F, use a graphing calculator to find the corresponding wind speed to the nearest mile per hour.

c. Find $W_V(20, 10)$ and $W_T(20, 10)$ and interpret.

d. Using the table command on a graphing calculator or a spreadsheet, develop a wind chill chart for various wind speeds and temperatures.

62. **Heat Index** The chart on the next page shows the heat index, which combines the effects of temperature with humidity to give a measure of the apparent temperature, or how hot it feels to

*These estimates are conservative. Please consult modern dive tables before making a dive.

the body. *Source: The Weather Channel*. For example, when the outside temperature is 90°F and the relative humidity is 40%, then the apparent temperature is approximately 93°F. Let $I = f(T, H)$ give the heat index, I, as a function of the temperature T (in degrees Fahrenheit) and the percent humidity H. Estimate the following.

a. $f(90, 30)$ **b.** $f(90, 75)$ **c.** $f(80, 75)$

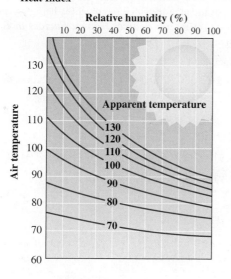

Heat Index

Estimate the following by approximating the partial derivative using a value of $h = 5$ in the difference quotient.

d. $f_T(90, 30)$ **e.** $f_H(90, 30)$ **f.** $f_T(90, 75)$

g. $f_H(90, 75)$

h. Describe in words what your answers in parts d–g mean.

63. Breath Volume The table at the bottom of this page accompanies the Voldyne® 5000 Volumetric Exerciser. The table gives the typical lung capacity (in milliliters) for women of various ages and heights. Based on the chart, it is possible to conclude that the partial derivative of the lung capacity with respect to age and with respect to height has constant values. What are those values?

Social Sciences

64. Education A developmental mathematics instructor at a large university has determined that a student's probability of success in the university's pass/fail remedial algebra course is a function of s, n, and a, where s is the student's score on the departmental placement exam, n is the number of semesters of mathematics passed in high school, and a is the student's mathematics SAT score. She estimates that p, the probability of passing the course (in percent), will be

$$p = f(s, n, a) = 0.05a + 6(sn)^{1/2}$$

for $200 \le a \le 800$, $0 \le s \le 10$, and $0 \le n \le 8$. Assuming that the above model has some merit, find the following.

a. If a student scores 6 on the placement exam, has taken 4 semesters of high school math, and has an SAT score of 460, what is the probability of passing the course?

b. Find p for a student with 5 semesters of high school mathematics, a placement score of 4, and an SAT score of 300.

c. Find and interpret $f_n(4, 5, 480)$ and $f_a(4, 5, 480)$.

Physical Sciences

65. Gravitational Attraction The gravitational attraction F on a body a distance r from the center of Earth, where r is greater than the radius of Earth, is a function of its mass m and the distance r as follows:

$$F = \frac{mgR^2}{r^2},$$

	Height (in.)	58″	60″	62″	64″	66″	68″	70″	72″	74″
	20	1900	2100	2300	2500	2700	2900	3100	3300	3500
A	25	1850	2050	2250	2450	2650	2850	3050	3250	3450
G	30	1800	2000	2200	2400	2600	2800	3000	3200	3400
E	35	1750	1950	2150	2350	2550	2750	2950	3150	3350
	40	1700	1900	2100	2300	2500	2700	2900	3100	3300
I	45	1650	1850	2050	2250	2450	2650	2850	3050	3250
N	50	1600	1800	2000	2200	2400	2600	2800	3000	3200
	55	1550	1750	1950	2150	2350	2550	2750	2950	3150
Y	60	1500	1700	1900	2100	2300	2500	2700	2900	3100
E	65	1450	1650	1850	2050	2250	2450	2650	2850	3050
A	70	1400	1600	1800	2000	2200	2400	2600	2800	3000
R	75	1350	1550	1750	1950	2150	2350	2550	2750	2950
S	80	1300	1500	1700	1900	2100	2300	2500	2700	2900

where R is the radius of Earth and g is the force of gravity—about 32 feet per second per second (ft per sec^2).

a. Find and interpret F_m and F_r.

b. Show that $F_m > 0$ and $F_r < 0$. Why is this reasonable?

66. Velocity In 1931, Albert Einstein developed the following formula for the sum of two velocities, x and y:

$$w(x, y) = \frac{x + y}{1 + \dfrac{xy}{c^2}},$$

where x and y are in miles per second and c represents the speed of light, 186,282 miles per second. *Source: The Mathematics Teacher.*

a. Suppose that, relative to a stationary observer, a new super space shuttle is capable of traveling at 50,000 miles per second and that, while traveling at this speed, it launches a rocket that travels at 150,000 miles per second. How fast is the rocket traveling relative to the stationary observer?

b. What is the instantaneous rate of change of w with respect to the speed of the space shuttle, x, when the space shuttle is traveling at 50,000 miles per second and the rocket is traveling at 150,000 miles per second?

c. Hypothetically, if a person is driving at the speed of light, c, and she turns on the headlights, what is the velocity of the light coming from the headlights relative to a stationary observer?

67. Movement Time Fitts' law is used to estimate the amount of time it takes for a person, using his or her arm, to pick up a light object, move it, and then place it in a designated target area. Mathematically, Fitts' law for a particular individual is given by

$$T(s, w) = 105 + 265 \log_2\left(\frac{2s}{w}\right),$$

where s is the distance (in feet) the object is moved, w is the width of the area in which the object is being placed, and T is the time (in msec). *Source: Human Factors in Engineering Design.*

a. Calculate $T(3, 0.5)$.

b. Find $T_s(3, 0.5)$ and $T_w(3, 0.5)$ and interpret these values. (*Hint:* $\log_2 x = \ln x / \ln 2$.)

■■■YOUR TURN ANSWERS

1. $f_x(x,y) = 4xy^3 + 30x^4y^4$; $f_y(x,y) = 6x^2y^2 + 24x^5y^3$
2. $f_x(x,y) = 6xye^{3x^2y}$; $f_y(x,y) = 3x^2e^{3x^2y}$
3. $9e^5, 8e^5$
4. $f_{xx}(x,y) = 2e^{7y} + 12x^2y^5$;
   $f_{yy}(x,y) = 49x^2e^{7y} + 20x^4y^3$;
   $f_{xy}(x,y) = f_{yx}(x, y) = 14xe^{7y} + 20x^3y^4$

9.3 Maxima and Minima

APPLY IT   What amounts of sugar and flavoring produce the minimum cost per batch of a soft drink? What is the minimum cost?

In this section we will learn how to answer questions such as this one, which is answered in Example 4.

FOR REVIEW

It may be helpful to review Section 5.2 on relative extrema at this point. The concepts presented there are basic to what will be done in this section.

One of the most important applications of calculus is finding maxima and minima of functions. Earlier, we studied this idea extensively for functions of a single independent variable; now we will see that extrema can be found for functions of two variables. In particular, an extension of the second derivative test can be defined and used to identify maxima or minima. We begin with the definitions of relative maxima and minima.

Relative Maxima and Minima

Let (a, b) be the center of a circular region contained in the xy-plane. Then, for a function $z = f(x, y)$ defined for every (x, y) in the region, $f(a, b)$ is a **relative (or local) maximum** if

$$f(a, b) \geq f(x, y)$$

for all points (x, y) in the circular region, and $f(a, b)$ is a **relative (or local) minimum** if

$$f(a, b) \leq f(x, y)$$

for all points (x, y) in the circular region.

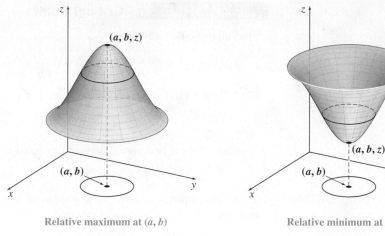

Relative maximum at (a, b)

FIGURE 17

Relative minimum at (a, b)

FIGURE 18

As before, the word *extremum* is used for either a relative maximum or a relative minimum. Examples of a relative maximum and a relative minimum are given in Figures 17 and 18.

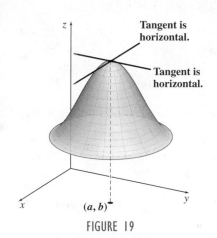

FIGURE 19

NOTE When functions of a single variable were discussed, a distinction was made between relative extrema and absolute extrema. The methods for finding absolute extrema are quite involved for functions of two variables, so we will discuss only relative extrema here. In many practical applications the relative extrema coincide with the absolute extrema. In this brief discussion of extrema for multivariable functions, we omit cases where an extremum occurs on a boundary of the domain.

As suggested by Figure 19, at a relative maximum the tangent line parallel to the *xz*-plane has a slope of 0, as does the tangent line parallel to the *yz*-plane. (Notice the similarity to functions of one variable.) That is, if the function $z = f(x, y)$ has a relative extremum at (a, b), then $f_x(a, b) = 0$ and $f_y(a, b) = 0$, as stated in the next theorem.

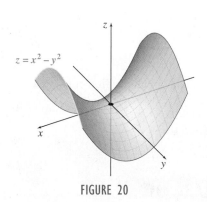

FIGURE 20

Location of Extrema

Let a function $z = f(x, y)$ have a relative maximum or relative minimum at the point (a, b). Let $f_x(a, b)$ and $f_y(a, b)$ both exist. Then

$$f_x(a, b) = 0 \quad \text{and} \quad f_y(a, b) = 0.$$

Just as with functions of one variable, the fact that the slopes of the tangent lines are 0 is no guarantee that a relative extremum has been located. For example, Figure 20 shows the graph of $z = f(x, y) = x^2 - y^2$. Both $f_x(0, 0) = 0$ and $f_y(0, 0) = 0$, and yet $(0, 0)$ leads to neither a relative maximum nor a relative minimum for the function. The point $(0, 0, 0)$ on the graph of this function is called a **saddle point**; it is a minimum when approached from one direction but a maximum when approached from another direction. A saddle point is neither a maximum nor a minimum.

The theorem on location of extrema suggests a useful strategy for finding extrema. First, locate all points (a, b) where $f_x(a, b) = 0$ and $f_y(a, b) = 0$. Then test each of these points separately, using the test given after the next example. For a function $f(x, y)$, the points (a, b) such that $f_x(a, b) = 0$ and $f_y(a, b) = 0$ are called **critical points**.

NOTE When we discussed functions of a single variable, we allowed critical points to include points from the domain where the derivative does not exist. For functions of more than one variable, to avoid complications, we will only consider cases in which the function is differentiable.

> **EXAMPLE 1**   **Critical Points**

Find all critical points for

$$f(x, y) = 6x^2 + 6y^2 + 6xy + 36x - 5.$$

SOLUTION   Find all points (a, b) such that $f_x(a, b) = 0$ and $f_y(a, b) = 0$. Here

$$f_x(x, y) = 12x + 6y + 36 \qquad \text{and} \qquad f_y(x, y) = 12y + 6x.$$

Set each of these two partial derivatives equal to 0.

$$12x + 6y + 36 = 0 \qquad \text{and} \qquad 12y + 6x = 0$$

These two equations make up a system of linear equations. We can use the substitution method to solve this system. First, rewrite $12y + 6x = 0$ as follows:

$$12y + 6x = 0$$
$$6x = -12y$$
$$x = -2y.$$

Now substitute $-2y$ for x in the other equation and solve for y.

$$12x + 6y + 36 = 0$$
$$12(-2y) + 6y + 36 = 0$$
$$-24y + 6y + 36 = 0$$
$$-18y + 36 = 0$$
$$-18y = -36$$
$$y = 2$$

YOUR TURN 1   Find all critical points for $f(x, y) = 4x^3 + 3xy + 4y^3$.

From the equation $x = -2y$, $x = -2(2) = -4$. The solution of the system of equations is $(-4, 2)$. Since this is the only solution of the system, $(-4, 2)$ is the only critical point for the given function. By the theorem above, if the function has a relative extremum, it will occur at $(-4, 2)$.   **TRY YOUR TURN 1**

The results of the next theorem can be used to decide whether $(-4, 2)$ in Example 1 leads to a relative maximum, a relative minimum, or neither.

Test for Relative Extrema

For a function $z = f(x, y)$, let f_{xx}, f_{yy}, and f_{xy} all exist in a circular region contained in the xy-plane with center (a, b). Further, let

$$f_x(a, b) = 0 \qquad \text{and} \qquad f_y(a, b) = 0.$$

Define the number D, known as **the discriminant**, by

$$D = f_{xx}(a, b) \cdot f_{yy}(a, b) - [f_{xy}(a, b)]^2.$$

Then

a. $f(a, b)$ is a relative maximum if $D > 0$ and $f_{xx}(a, b) < 0$;

b. $f(a, b)$ is a relative minimum if $D > 0$ and $f_{xx}(a, b) > 0$;

c. $f(a, b)$ is a saddle point (neither a maximum nor a minimum) if $D < 0$;

d. if $D = 0$, the test gives no information.

This test is comparable to the second derivative test for extrema of functions of one independent variable. The following table summarizes the conclusions of the theorem.

	$f_{xx}(a, b) < 0$	$f_{xx}(a, b) > 0$
$D > 0$	Relative maximum	Relative minimum
$D = 0$	No information	
$D < 0$	Saddle point	

Notice that in parts a and b of the test for relative extrema, it is only necessary to test the second partial $f_{xx}(a, b)$ and not $f_{yy}(a, b)$. This is because if $D > 0$, $f_{xx}(a, b)$ and $f_{yy}(a, b)$ must have the same sign.

EXAMPLE 2 Relative Extrema

The previous example showed that the only critical point for the function

$$f(x, y) = 6x^2 + 6y^2 + 6xy + 36x - 5$$

is $(-4, 2)$. Does $(-4, 2)$ lead to a relative maximum, a relative minimum, or neither?

SOLUTION Find out by using the test above. From Example 1,

$$f_x(-4, 2) = 0 \quad \text{and} \quad f_y(-4, 2) = 0.$$

Now find the various second partial derivatives used in finding D. From $f_x(x, y) = 12x + 6y + 36$ and $f_y(x, y) = 12y + 6x$,

$$f_{xx}(x, y) = 12, \quad f_{yy}(x, y) = 12, \quad \text{and} \quad f_{xy}(x, y) = 6.$$

(If these second-order partial derivatives had not all been constants, they would have had to be evaluated at the point $(-4, 2)$.) Now

$$D = f_{xx}(-4, 2) \cdot f_{yy}(-4, 2) - [f_{xy}(-4, 2)]^2 = 12 \cdot 12 - 6^2 = 108.$$

Since $D > 0$ and $f_{xx}(-4, 2) = 12 > 0$, part b of the theorem applies, showing that $f(x, y) = 6x^2 + 6y^2 + 6xy + 36x - 5$ has a relative minimum at $(-4, 2)$. This relative minimum is $f(-4, 2) = -77$. A graph of this surface drawn by the computer program Maple™ is shown in Figure 21.

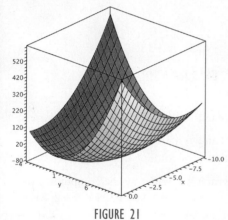

FIGURE 21

EXAMPLE 3 Saddle Point

Find all points where the function

$$f(x, y) = 9xy - x^3 - y^3 - 6$$

has any relative maxima or relative minima.

SOLUTION First find any critical points. Here

$$f_x(x, y) = 9y - 3x^2 \quad \text{and} \quad f_y(x, y) = 9x - 3y^2.$$

Set each of these partial derivatives equal to 0.

$$\begin{aligned} f_x(x, y) &= 0 & f_y(x, y) &= 0 \\ 9y - 3x^2 &= 0 & 9x - 3y^2 &= 0 \\ 9y &= 3x^2 & 9x &= 3y^2 \\ 3y &= x^2 & 3x &= y^2 \end{aligned}$$

The substitution method can be used again to solve the system of equations

$$3y = x^2$$
$$3x = y^2.$$

The first equation, $3y = x^2$, can be rewritten as $y = x^2/3$. Substitute this into the second equation to get

$$3x = y^2 = \left(\frac{x^2}{3}\right)^2 = \frac{x^4}{9}.$$

Solve this equation as follows.

$$27x = x^4 \qquad \text{Multiply both sides by 9.}$$
$$x^4 - 27x = 0$$
$$x(x^3 - 27) = 0 \qquad \text{Factor.}$$
$$x = 0 \quad \text{or} \quad x^3 - 27 = 0 \qquad \text{Set each factor equal to 0.}$$
$$x = 0 \quad \text{or} \quad x^3 = 27$$
$$x = 0 \quad \text{or} \quad x = 3 \qquad \text{Take the cube root on both sides.}$$

Use these values of x, along with the equation $3y = x^2$, rewritten as $y = x^2/3$, to find y. If $x = 0$, $y = 0^2/3 = 0$. If $x = 3$, $y = 3^2/3 = 3$. The critical points are $(0, 0)$ and $(3, 3)$. To identify any extrema, use the test. Here

$$f_{xx}(x, y) = -6x, \qquad f_{yy}(x, y) = -6y, \qquad \text{and} \qquad f_{xy}(x, y) = 9.$$

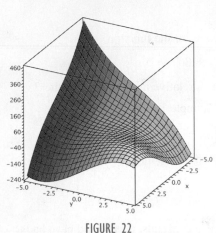

FIGURE 22

Test each of the possible critical points.

For $(0, 0)$:	For $(3, 3)$:
$f_{xx}(0, 0) = -6(0) = 0$	$f_{xx}(3, 3) = -6(3) = -18$
$f_{yy}(0, 0) = -6(0) = 0$	$f_{yy}(3, 3) = -6(3) = -18$
$f_{xy}(0, 0) = 9$	$f_{xy}(3, 3) = 9$
$D = 0 \cdot 0 - 9^2 = -81.$	$D = -18(-18) - 9^2 = 243.$
Since $D < 0$, there is a saddle point at $(0, 0)$.	Here $D > 0$ and $f_{xx}(3, 3) = -18 < 0$; there is a relative maximum at $(3, 3)$.

YOUR TURN 2 Identify each of the critical points in YOUR TURN 1 as a relative maximum, relative minimum, or saddle point.

Notice that these values are in accordance with the graph generated by the computer program Maple™ shown in Figure 22.

TRY YOUR TURN 2

EXAMPLE 4 Production Costs

A company is developing a new soft drink. The cost in dollars to produce a batch of the drink is approximated by

$$C(x, y) = 2200 + 27x^3 - 72xy + 8y^2,$$

where x is the number of kilograms of sugar per batch and y is the number of grams of flavoring per batch. Find the amounts of sugar and flavoring that result in the minimum cost per batch. What is the minimum cost?

APPLY IT

Method 1
Calculating by Hand

SOLUTION

Start with the following partial derivatives.

$$C_x(x, y) = 81x^2 - 72y \qquad \text{and} \qquad C_y(x, y) = -72x + 16y$$

Set each of these equal to 0 and solve for y.

$$81x^2 - 72y = 0 \qquad\qquad -72x + 16y = 0$$
$$-72y = -81x^2 \qquad\qquad 16y = 72x$$
$$y = \frac{9}{8}x^2 \qquad\qquad\qquad y = \frac{9}{2}x$$

Since $(9/8)x^2$ and $(9/2)x$ both equal y, they are equal to each other. Set them equal, and solve the resulting equation for x.

$$\frac{9}{8}x^2 = \frac{9}{2}x$$

$$9x^2 = 36x$$

$$9x^2 - 36x = 0 \qquad \text{Subtract } 36x \text{ from both sides.}$$

$$9x(x - 4) = 0 \qquad \text{Factor.}$$

$$9x = 0 \quad \text{or} \quad x - 4 = 0 \qquad \text{Set each factor equal to 0.}$$

The equation $9x = 0$ leads to $x = 0$ and $y = 0$, which cannot be a minimizer of $C(x, y)$ since, for example, $C(1, 1) < C(0, 0)$. This fact can also be verified by the test for relative extrema. Substitute $x = 4$, the solution of $x - 4 = 0$, into $y = (9/2)x$ to find y.

$$y = \frac{9}{2}x = \frac{9}{2}(4) = 18$$

Now check to see whether the critical point $(4, 18)$ leads to a relative minimum. Here

$$C_{xx}(x, y) = 162x, \quad C_{yy}(x, y) = 16, \quad \text{and} \quad C_{xy}(x, y) = -72.$$

For $(4, 18)$,

$$C_{xx}(4, 18) = 162(4) = 648, \quad C_{yy}(4, 18) = 16, \quad \text{and} \quad C_{xy}(4, 18) = -72,$$

so that

$$D = (648)(16) - (-72)^2 = 5184.$$

Since $D > 0$ and $C_{xx}(4, 18) > 0$, the cost at $(4, 18)$ is a minimum.

To find the minimum cost, go back to the cost function and evaluate $C(4, 18)$.

$$C(x, y) = 2200 + 27x^3 - 72xy + 8y^2$$

$$C(4, 18) = 2200 + 27(4)^3 - 72(4)(18) + 8(18)^2 = 1336$$

The minimum cost for a batch of soft drink is $1336.00. A graph of this surface drawn by the computer program Maple™ is shown in Figure 23.

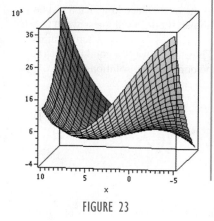

FIGURE 23

⊞ **Method 2 Spreadsheets**

Finding the maximum or minimum of a function of one or more variables can be done using a spreadsheet. The Solver included with Excel is located in the Tools menu and requires that cells be identified ahead of time for each variable in the problem. (On some versions of Excel, the Solver must be installed from an outside source. For details, see the *Graphing Calculator and Excel Spreadsheet Manual* available with this text.) It also requires that another cell be identified where the function, in terms of the variable cells, is placed. For example, to solve the above problem, we could identify cells A1 and B1 to represent the variables x and y, respectively. The Solver requires that we place a guess for the answer in these cells. Thus, our initial value or guess will be to place the number 5 in each of these cells. An expression for the function must be placed in another cell, with x and y replaced by A1 and B1. If we choose cell A3 to represent the function, in cell A3 we would type "= 2200 + 27*A1^3 - 72*A1*B1 + 8*B1^2."

We now click on the Tools menu and choose Solver. This solver will attempt to find a solution that either maximizes or minimizes the value of cell A3. Figure 24 illustrates the Solver box and the items placed in it.

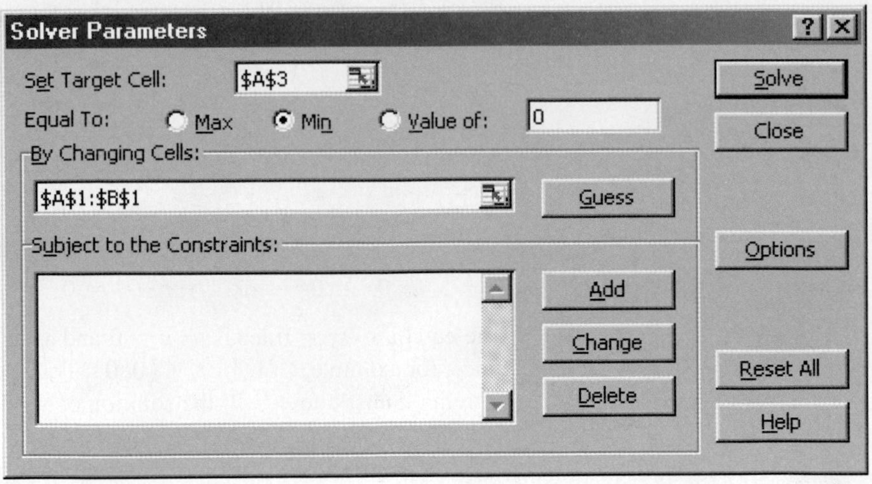

Solver Parameters [?][X]

Set Target Cell: [$A$3] [⬚]

Equal To: ○ Max ● Min ○ Value of: [0]

By Changing Cells:

[$A$1:$B$1] [⬚] [Guess]

Subject to the Constraints:

[ ] [Add]
[ ] [Change]
[ ] [Delete]

[Solve] [Close] [Options] [Reset All] [Help]

FIGURE 24

To obtain a solution, click on Solve. The rounded solution $x = 4$ and $y = 18$ is located in cells A1 and B1. The minimum cost $C(4, 18) = 1336$ is located in cell A3.

CAUTION One must be careful when using Solver because it will not find a maximizer or minimizer of a function if the initial guess is the exact place in which a saddle point occurs. For example, in the problem above, if our initial guess was $(0, 0)$, the Solver would have returned the value of $(0, 0)$ as the place where a minimum occurs. But $(0, 0)$ is a saddle point. Thus, it is always a good idea to run the Solver for two different initial values and compare the solutions.

9.3 EXERCISES

Find all points where the functions have any relative extrema. Identify any saddle points.

1. $f(x, y) = xy + y - 2x$

2. $f(x, y) = 3xy + 6y - 5x$

3. $f(x, y) = 3x^2 - 4xy + 2y^2 + 6x - 10$

4. $f(x, y) = x^2 + xy + y^2 - 6x - 3$

5. $f(x, y) = x^2 - xy + y^2 + 2x + 2y + 6$

6. $f(x, y) = 2x^2 + 3xy + 2y^2 - 5x + 5y$

7. $f(x, y) = x^2 + 3xy + 3y^2 - 6x + 3y$

8. $f(x, y) = 5xy - 7x^2 - y^2 + 3x - 6y - 4$

9. $f(x, y) = 4xy - 10x^2 - 4y^2 + 8x + 8y + 9$

10. $f(x, y) = 4y^2 + 2xy + 6x + 4y - 8$

11. $f(x, y) = x^2 + xy - 2x - 2y + 2$

12. $f(x, y) = x^2 + xy + y^2 - 3x - 5$

13. $f(x, y) = 3x^2 + 2y^3 - 18xy + 42$

14. $f(x, y) = 7x^3 + 3y^2 - 126xy - 63$

15. $f(x, y) = x^2 + 4y^3 - 6xy - 1$

16. $f(x, y) = 3x^2 + 7y^3 - 42xy + 5$

17. $f(x, y) = e^{x(y+1)}$

18. $f(x, y) = y^2 + 2e^x$

19. Describe the procedure for finding critical points of a function in two independent variables.

20. How are second-order partial derivatives used in finding extrema?

Figures a–f show the graphs of the functions defined in Exercises 21–26. Find all relative extrema for each function, and then match the equation to its graph.

21. $z = -3xy + x^3 - y^3 + \dfrac{1}{8}$

22. $z = \dfrac{3}{2}y - \dfrac{1}{2}y^3 - x^2y + \dfrac{1}{16}$

23. $z = y^4 - 2y^2 + x^2 - \dfrac{17}{16}$

24. $z = -2x^3 - 3y^4 + 6xy^2 + \dfrac{1}{16}$

25. $z = -x^4 + y^4 + 2x^2 - 2y^2 + \dfrac{1}{16}$

26. $z = -y^4 + 4xy - 2x^2 + \dfrac{1}{16}$

a.

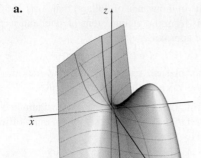

b.

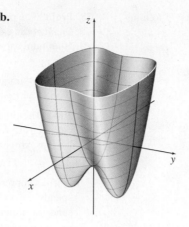

c.

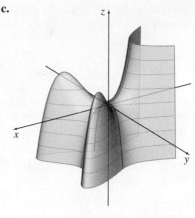

d.

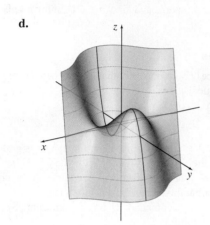

e.

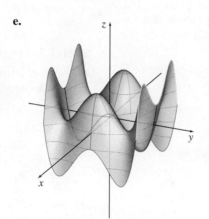

f.

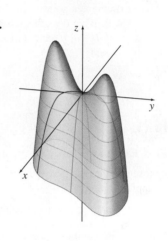

27. Show that $f(x, y) = 1 - x^4 - y^4$ has a relative maximum, even though D in the theorem is 0.

28. Show that $D = 0$ for $f(x, y) = x^3 + (x - y)^2$ and that the function has no relative extrema.

29. A friend taking calculus is puzzled. She remembers that for a function of one variable, if the first derivative is zero at a point and the second derivative is positive, then there must be a relative minimum at the point. She doesn't understand why that isn't true for a function of two variables—that is, why $f_x(x, y) = 0$ and $f_{xx}(x, y) > 0$ doesn't guarantee a relative minimum. Provide an explanation.

30. Let $f(x, y) = y^2 - 2x^2y + 4x^3 + 20x^2$. The only critical points are $(-2, 4)$, $(0, 0)$, and $(5, 25)$. Which of the following correctly describes the behavior of f at these points? *Source: Society of Actuaries*.

a. $(-2, 4)$: local (relative) minimum
$(0, 0)$: local (relative) minimum
$(5, 25)$: local (relative) maximum

b. $(-2, 4)$: local (relative) minimum
$(0, 0)$: local (relative) maximum
$(5, 25)$: local (relative) maximum

c. $(-2, 4)$: neither a local (relative) minimum nor a local (relative) maximum
$(0, 0)$: local (relative) maximum
$(5, 25)$: local (relative) minimum

d. $(-2, 4)$: local (relative) maximum
$(0, 0)$: neither a local (relative) minimum nor a local (relative) maximum
$(5, 25)$: local (relative) minimum

e. $(-2, 4)$: neither a local (relative) minimum nor a local (relative) maximum
$(0, 0)$: local (relative) minimum
$(5, 25)$: neither a local (relative) minimum nor a local (relative) maximum

31. Consider the function $f(x, y) = x^2(y + 1)^2 + k(x + 1)^2y^2$.

a. For what values of k is the point $(x, y) = (0, 0)$ a critical point?

b. For what values of k is the point $(x, y) = (0, 0)$ a relative minimum of the function?

32. In Exercise 9 of Section 1.3, we found the least squares line through a set of n points $(x_1, y_1), (x_2, y_2), \ldots, (x_n, y_n)$ by choosing the slope of the line m and the y-intercept b to minimize the quantity

$$S(m, b) = \sum (mx + b - y)^2,$$

where the summation symbol Σ means that we sum over all the data points. Minimize S by setting $S_m(m, b) = 0$ and $S_b(m, b) = 0$, and then rearrange the results to derive the equations from Section 1.3

$$\left(\sum x\right)b + \left(\sum x^2\right)m = \sum xy$$
$$nb + \left(\sum x\right)m = \sum y.$$

33. Suppose a function $z = f(x, y)$ satisfies the criteria for the test for relative extrema at a point (a, b), and $f_{xx}(a, b) > 0$, while $f_{yy}(a, b) < 0$. What does this tell you about $f(a, b)$? Based on the sign of $f_{xx}(a, b)$ and $f_{yy}(a, b)$, why does this seem intuitively plausible?

APPLICATIONS

Business and Economics

34. **Profit** Suppose that the profit (in hundreds of dollars) of a certain firm is approximated by

$$P(x, y) = 1500 + 36x - 1.5x^2 + 120y - 2y^2,$$

where x is the cost of a unit of labor and y is the cost of a unit of goods. Find values of x and y that maximize profit. Find the maximum profit.

35. **Labor Costs** Suppose the labor cost (in dollars) for manufacturing a precision camera can be approximated by

$$L(x, y) = \frac{3}{2}x^2 + y^2 - 2x - 2y - 2xy + 68,$$

where x is the number of hours required by a skilled craftsperson and y is the number of hours required by a semiskilled person. Find values of x and y that minimize the labor cost. Find the minimum labor cost.

36. **Cost** The total cost (in dollars) to produce x units of electrical tape and y units of packing tape is given by

$$C(x, y) = 2x^2 + 2y^2 - 3xy + 4x - 94y + 4200.$$

Find the number of units of each kind of tape that should be produced so that the total cost is a minimum. Find the minimum total cost.

37. **Revenue** The total revenue (in hundreds of dollars) from the sale of x spas and y solar heaters is approximated by

$$R(x, y) = 15 + 169x + 182y - 5x^2 - 7y^2 - 7xy.$$

Find the number of each that should be sold to produce maximum revenue. Find the maximum revenue.

38. **Profit** The profit (in thousands of dollars) that Aunt Mildred's Metalworks earns from producing x tons of steel and y tons of aluminum can be approximated by

$$P(x, y) = 36xy - x^3 - 8y^3.$$

Find the amounts of steel and aluminum that maximize the profit, and find the value of the maximum profit.

39. **Time** The time (in hours) that a branch of Amalgamated Entities needs to spend to meet the quota set by the main office can be approximated by

$$T(x, y) = x^4 + 16y^4 - 32xy + 40,$$

where x represents how many thousands of dollars the factory spends on quality control and y represents how many thousands of dollars they spend on consulting. Find the amount of money they should spend on quality control and on consulting to minimize the time spent, and find the minimum number of hours.

Social Sciences

40. **Political Science** The probability that a three-person jury will make a correct decision is given by

$$P(\alpha, r, s) = \alpha[3r^2(1 - r) + r^3]$$
$$+ (1 - \alpha)[3s^2(1 - s) + s^3],$$

where $0 < \alpha < 1$ is the probability that the person is guilty of the crime, r is the probability that a given jury member will vote "guilty" when the defendant is indeed guilty of the crime, and s is the probability that a given jury member will vote "innocent" when the defendant is indeed innocent. *Source: Frontiers of Economics.*

a. Calculate $P(0.9, 0.5, 0.6)$ and $P(0.1, 0.8, 0.4)$ and interpret your answers.

b. Using common sense and without using calculus, what value of r and s would maximize the jury's probability of making the correct verdict? Do these values depend on α in this problem? Should they? What is the maximum probability?

c. Verify your answer for part b using calculus. (*Hint:* There are two critical points. Argue that the maximum value occurs at one of these points.)

Physical Sciences

41. **Computer Chips** The table on the following page, which illustrates the dramatic increase in the number of transistors in personal computers since 1985, was given in the chapter on Nonlinear Functions, Section 2.4, Exercise 54.

a. To fit the data to a function of the form $y = ab^t$, where t is the number of years since 1985 and y is the number of transistors (in millions), we could take natural logarithms of both sides of the equation to get $\ln y = \ln a + t \ln b$. We could then let $w = \ln y$, $r = \ln a$, and $s = \ln b$ to form $w = r + st$. Using linear regression, find values for r and s that will fit the data. Then find the function $y = ab^t$. (*Hint:* Take the natural logarithm of the values in the transistors column and then use linear regression to find values of r and s that fit the data. Once you know r and s, you can determine the values of a and b by calculating $a = e^r$ and $b = e^s$.)

Year (since 1985)	Chip	Transistors (in millions)
0	386	0.275
4	486	1.2
8	Pentium	3.1
12	Pentium II	7.5
14	Pentium III	9.5
15	Pentium 4	42
20	Pentium D	291
22	Penryn	820
24	Nehalem	1900

b. Use the solver capability of a spreadsheet to find a function of the form $y = ab^t$ that fits the data above. (*Hint:* Using the ideas from part a, find values for a and b that minimize the function

$$f(a, b) = [\ln(0.275) - 0 \ln b - \ln a]^2$$
$$+ [\ln(1.2) - 4 \ln b - \ln a]^2$$
$$+ [\ln(3.1) - 8 \ln b - \ln a]^2$$
$$+ [\ln(7.5) - 12 \ln b - \ln a]^2$$
$$+ [\ln(9.5) - 14 \ln b - \ln a]^2$$
$$+ [\ln(42) - 15 \ln b - \ln a]^2$$
$$+ [\ln(291) - 20 \ln b - \ln a]^2$$
$$+ [\ln(820) - 22 \ln b - \ln a]^2$$
$$+ [\ln(1900) - 24 \ln b - \ln a]^2 .)$$

c. Compare your answer to this problem with the one found with a graphing calculator in the chapter on Nonlinear Functions, Section 2.4, Exercise 54.

General Interest

42. Food Frying The process of frying food changes its quality, texture, and color. According to research done at the University of Saskatchewan, the total change in color E (which is measured in the form of energy as kJ/mol) of blanched potato strips can be estimated by the function

$$E(t, T) = 436.16 - 10.57t - 5.46T - 0.02t^2 + 0.02T^2 + 0.08Tt,$$

where T is the temperature (in °C) and t is the frying time (in min). *Source: Critical Reviews in Food Science and Nutrition.*

a. What is the value of E prior to cooking? (Assume that $T = 0$.)

b. Use this function to estimate the total change in color of a potato strip that has been cooked for 10 minutes at 180°C.

c. Determine the critical point of this function and determine if a maximum, minimum, or saddle point occurs at that point. Describe what may be happening at this point.

YOUR TURN ANSWERS

1. $(0, 0)$ and $(-1/4, -1/4)$
2. Saddle point at $(0, 0)$; relative maximum at $(-1/4, -1/4)$

9.4 Lagrange Multipliers

APPLY IT What dimensions for a new building will maximize the floor space at a fixed cost?

Using Lagrange multipliers, we will answer this question in Example 2.

In Section 6.2 on Applications of Extrema, it was possible to express problems involving two variables as equivalent problems requiring only a single variable. This method works well, provided that it is possible to use algebra to express the one variable in terms of the other. It is not always possible to do this, however, and most real applications require more than two variables and one or more additional restrictions, called **constraints**.

An approach that works well when there is a constraint in the problem uses an additional variable, called the **Lagrange multiplier**. For example, in the opening question, suppose a builder wants to maximize the floor space in a new building while keeping the costs fixed at $500,000. The building will be 40 ft high, with a rectangular floor plan and three stories. The costs, which depend on the dimensions of the rectangular floor plan, are given by

$$\text{Costs} = xy + 20y + 20x + 474,000,$$

where x is the width and y the length of the rectangle. Thus, the builder wishes to maximize the area $A(x, y) = xy$ and satisfy the condition

$$xy + 20y + 20x + 474,000 = 500,000.$$

In addition to maximizing area, then, the builder must keep costs at (or below) $500,000. We will see how to solve this problem in Example 2 of this section.

Problems with constraints are often solved by the method of Lagrange multipliers, named for the French mathematician Joseph Louis Lagrange (1736–1813). The method of Lagrange multipliers is used for problems of the form:

Find the relative extrema for $z = f(x, y)$,

subject to $g(x, y) = 0$.

We state the method only for functions of two independent variables, but it is valid for any number of variables.

Lagrange Multipliers

All relative extrema of the function $z = f(x, y)$, subject to a constraint $g(x, y) = 0$, will be found among those points (x, y) for which there exists a value of λ such that

$$F_x(x, y, \lambda) = 0, \qquad F_y(x, y, \lambda) = 0, \qquad F_\lambda(x, y, \lambda) = 0,$$

where

$$F(x, y, \lambda) = f(x, y) - \lambda \cdot g(x, y),$$

and all indicated partial derivatives exist.

In the theorem, the function $F(x, y, \lambda) = f(x, y) - \lambda \cdot g(x, y)$ is called the Lagrange function; λ, the Greek letter *lambda*, is the *Lagrange multiplier*.

CAUTION If the constraint is not of the form $g(x, y) = 0$, it must be put in that form before using the method of Lagrange multipliers. For example, if the constraint is $x^2 + y^3 = 5$, subtract 5 from both sides to get $g(x, y) = x^2 + y^3 - 5 = 0$.

EXAMPLE 1   Lagrange Multipliers

Find the minimum value of

$$f(x, y) = 5x^2 + 6y^2 - xy,$$

subject to the constraint $x + 2y = 24$.

SOLUTION   Go through the following steps.

Step 1 Rewrite the constraint in the form $g(x, y) = 0$.

In this example, the constraint $x + 2y = 24$ becomes

$$x + 2y - 24 = 0,$$

with

$$g(x, y) = x + 2y - 24.$$

Step 2 Form the Lagrange function $F(x, y, \lambda)$, the difference of the function $f(x, y)$ and the product of λ and $g(x, y)$.

Here,

$$\begin{aligned} F(x, y, \lambda) &= f(x, y) - \lambda \cdot g(x, y) \\ &= 5x^2 + 6y^2 - xy - \lambda(x + 2y - 24) \\ &= 5x^2 + 6y^2 - xy - \lambda x - 2\lambda y + 24\lambda. \end{aligned}$$

Step 3 Find $F_x(x, y, \lambda)$, $F_y(x, y, \lambda)$, and $F_\lambda(x, y, \lambda)$.

$$F_x(x, y, \lambda) = 10x - y - \lambda$$
$$F_y(x, y, \lambda) = 12y - x - 2\lambda$$
$$F_\lambda(x, y, \lambda) = -x - 2y + 24$$

Step 4 Form the system of equations $F_x(x, y, \lambda) = 0$, $F_y(x, y, \lambda) = 0$, and $F_\lambda(x, y, \lambda) = 0$.

$$10x - y - \lambda = 0 \tag{1}$$
$$12y - x - 2\lambda = 0 \tag{2}$$
$$-x - 2y + 24 = 0 \tag{3}$$

Step 5 Solve the system of equations from Step 4 for x, y, and λ.

One way to solve this system is to begin by solving each of the first two equations for λ, then set the two results equal and simplify, as follows.

$$10x - y - \lambda = 0 \quad \text{becomes} \quad \lambda = 10x - y$$

$$12y - x - 2\lambda = 0 \quad \text{becomes} \quad \lambda = \frac{-x + 12y}{2}$$

$$10x - y = \frac{-x + 12y}{2} \qquad \text{Set the expressions for } \lambda \text{ equal.}$$
$$20x - 2y = -x + 12y$$
$$21x = 14y$$
$$x = \frac{14y}{21} = \frac{2y}{3}$$

Now substitute $2y/3$ for x in Equation (3).

$$-x - 2y + 24 = 0$$
$$-\frac{2y}{3} - 2y + 24 = 0 \qquad \text{Let } x = \frac{2y}{3}.$$
$$2y + 6y - 72 = 0 \qquad \text{Multiply by } -3.$$
$$8y = 72$$
$$y = \frac{72}{8} = 9$$

Since $x = 2y/3$ and $y = 9$, $x = 6$. It is not necessary to find the value of λ.

Thus, if $f(x, y) = 5x^2 + 6y^2 - xy$ has a minimum value subject to the constraint $x + 2y = 24$, it is at the point $(6, 9)$. The value of $f(6, 9)$ is 612.

We need to convince ourselves that $f(6, 9) = 612$ is indeed a minimum for the function. How can we tell that it is not a maximum? The second derivative test from the previous section does not apply to the solutions found by Lagrange multipliers. (See Exercise 21.) We could gain some insight by trying a point very close to $(6, 9)$ that also satisfies the constraint $x + 2y = 24$. For example, let $y = 9.1$, so $x = 24 - 2y = 24 - 2(9.1) = 5.8$. Then $f(5.8, 9.1) = 5(5.8)^2 + 6(9.1)^2 - (5.8)(9.1) = 612.28$, which is greater than 612. Because a nearby point has a value larger than 612, the value 612 is probably not a maximum. Another method would be to use a computer to sketch the graph of the function and see that it has a minimum but not a maximum. In practical problems, such as Example 2, it is often obvious whether a function has a minimum or a maximum. **TRY YOUR TURN 1**

YOUR TURN 1 Find the minimum value of $f(x, y) = x^2 + 2x + 9y^2 + 3y + 6xy$ subject to the constraint $2x + 3y = 12$.

NOTE In Example 1, we solved the system of equations by solving each equation with λ in it for λ. We then set these expressions for λ equal and solved for one of the original variables. This is a good general approach to use in solving these systems of equations, since we are usually not interested in the value of λ.

CAUTION Lagrange multipliers give only the relative extrema, not the absolute extrema. In many applications, the relative extrema will be the absolute extrema, but this is not guaranteed. In some cases in which the method of Lagrange multipliers finds a solution, there may not even be any absolute extrema. For example, see Exercises 18 and 19 at the end of this section.

Before looking at applications of Lagrange multipliers, let us summarize the steps involved in solving a problem by this method.

Using Lagrange Multipliers

1. Write the constraint in the form $g(x, y) = 0$.

2. Form the Lagrange function

$$F(x, y, \lambda) = f(x, y) - \lambda \cdot g(x, y).$$

3. Find $F_x(x, y, \lambda)$, $F_y(x, y, \lambda)$, and $F_\lambda(x, y, \lambda)$.

4. Form the system of equations

$$F_x(x, y, \lambda) = 0, \qquad F_y(x, y, \lambda) = 0, \qquad F_\lambda(x, y, \lambda) = 0.$$

5. Solve the system in Step 4; the relative extrema for f are among the solutions of the system.

The proof of this method is complicated and is not given here, but we can explain why the method is plausible. Consider the curve formed by points in the xy-plane that satisfy $F_\lambda(x, y, \lambda) = -g(x, y) = 0$ (or just $g(x, y) = 0$). Figure 25 shows how such a curve might look. Crossing this region are curves $f(x, y) = k$ for various values of k. Notice that at the points where the curve $f(x, y) = k$ is tangent to the curve $g(x, y) = 0$, the largest and smallest meaningful values of f occur. It can be shown that this is equivalent to $f_x(x, y) = \lambda g_x(x, y)$ and $f_y(x, y) = \lambda g_y(x, y)$ for some constant λ. In Exercise 20, you are asked to show that this is equivalent to the system of equations found in Step 4 above.

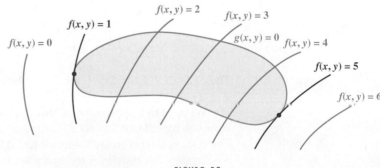

FIGURE 25

Lagrange multipliers are widely used in economics, where a frequent goal is to maximize a utility function, which measures how well consumption satisfies the consumers' desires, subject to constraints on income or time.

EXAMPLE 2 **Lagrange Multipliers**

Complete the solution of the problem given in the introduction to this section. Maximize the area, $A(x, y) = xy$, subject to the cost constraint

$$xy + 20y + 20x + 474,000 = 500,000.$$

APPLY IT

SOLUTION   Go through the five steps presented earlier.

> *Step 1* $g(x, y) = xy + 20y + 20x - 26,000 = 0$
>
> *Step 2* $F(x, y, \lambda) = xy - \lambda(xy + 20y + 20x - 26,000)$
>
> *Step 3* $F_x(x, y, \lambda) = y - \lambda y - 20\lambda$
>
> $F_y(x, y, \lambda) = x - \lambda x - 20\lambda$
>
> $F_\lambda(x, y, \lambda) = -xy - 20y - 20x + 26,000$
>
> *Step 4* $y - \lambda y - 20\lambda = 0$ (4)
>
> $x - \lambda x - 20\lambda = 0$ (5)
>
> $-xy - 20y - 20x + 26,000 = 0$ (6)
>
> *Step 5* Solving Equations (4) and (5) for λ gives

$$\lambda = \frac{y}{y + 20} \quad \text{and} \quad \lambda = \frac{x}{x + 20}$$

$$\frac{y}{y + 20} = \frac{x}{x + 20}$$

$$y(x + 20) = x(y + 20)$$

$$xy + 20y = xy + 20x$$

$$x = y.$$

Now substitute y for x in Equation (6) to get

$$-y^2 - 20y - 20y + 26,000 = 0$$

$$-y^2 - 40y + 26,000 = 0.$$

Use the quadratic formula to find $y \approx -182.5$ or $y \approx 142.5$. We eliminate the negative value because length cannot be negative. Since $x = y$, we know that $x \approx 142.5$.

The maximum area of $(142.5)^2 \approx 20,306 \text{ ft}^2$ will be achieved if the floor plan is a square with a side of 142.5 ft. You can verify that this answer is a maximum using the method at the end of Example 1.

As mentioned earlier, the method of Lagrange multipliers works for more than two independent variables. The next example shows how to find extrema for a function of three independent variables.

EXAMPLE 3 **Volume of a Box**

Find the dimensions of the closed rectangular box of maximum volume that can be produced from 6 ft^2 of material.

Method 1
Lagrange Multipliers

SOLUTION   In Chapter 6 on Applications of the Derivative, we were able to solve problems such as this by adding an extra constraint, such as requiring the bottom of the box to be square. Here we have no such constraint. Let x, y, and z represent the dimensions of the box, as shown in Figure 26 on the next page. The volume of the box is given by

$$f(x, y, z) = xyz.$$

FIGURE 26

As shown in Figure 26, the total amount of material required for the two ends of the box is $2xy$, the total needed for the sides is $2xz$, and the total needed for the top and bottom is $2yz$. Since 6 ft^2 of material is available,

$$2xy + 2xz + 2yz = 6 \qquad \text{or} \qquad xy + xz + yz = 3.$$

In summary, $f(x, y, z) = xyz$ is to be maximized subject to the constraint $xy + xz + yz = 3$. Go through the steps that were given.

Step 1 $g(x, y, z) = xy + xz + yz - 3 = 0$

Step 2 $F(x, y, z, \lambda) = xyz - \lambda(xy + xz + yz - 3)$

Step 3 $F_x(x, y, z, \lambda) = yz - \lambda y - \lambda z$
$F_y(x, y, z, \lambda) = xz - \lambda x - \lambda z$
$F_z(x, y, z, \lambda) = xy - \lambda x - \lambda y$
$F_\lambda(x, y, z, \lambda) = -xy - xz - yz + 3$

Step 4 $yz - \lambda y - \lambda z = 0$
$xz - \lambda x - \lambda z = 0$
$xy - \lambda x - \lambda y = 0$
$-xy - xz - yz + 3 = 0$

Step 5 Solve each of the first three equations for λ. You should get

$$\lambda = \frac{yz}{y + z}, \qquad \lambda = \frac{xz}{x + z}, \qquad \text{and} \qquad \lambda = \frac{xy}{x + y}.$$

Set these expressions for λ equal, and simplify as follows. Notice in the second and last steps that since none of the dimensions of the box can be 0, we can divide both sides of each equation by x or z.

$$\frac{yz}{y + z} = \frac{xz}{x + z} \qquad \text{and} \qquad \frac{xz}{x + z} = \frac{xy}{x + y}$$

$$\frac{y}{y + z} = \frac{x}{x + z} \qquad\qquad \frac{z}{x + z} = \frac{y}{x + y}$$

$$xy + yz = xy + xz \qquad\qquad zx + zy = yx + yz$$

$$yz = xz \qquad\qquad\qquad zx = yx$$

$$y = x \qquad\qquad\qquad\quad z = y$$

(Setting the first and third expressions equal gives no additional information.) Thus $x = y = z$. From the fourth equation in Step 4, with $x = y$ and $z = y$,

$$-xy - xz - yz + 3 = 0$$
$$-y^2 - y^2 - y^2 + 3 = 0$$
$$-3y^2 = -3$$
$$y^2 = 1$$
$$y = \pm 1.$$

The negative solution is not applicable, so the solution of the system of equations is $x = 1$, $y = 1$, $z = 1$. In other words, the box with maximum volume under the constraint is a cube that measures 1 ft on each side. As in the previous examples, verify that this is a maximum.

Method 2
Spreadsheets

Finding extrema of a constrained function of one or more variables can be done using a spreadsheet. In addition to the requirements stated in the last section, the constraint must also be input into the Excel Solver. To do this, we need to input the left-hand or variable part of the constraint into a designated cell. If A5 is the designated cell, then in cell A5 we would type "=A1*B1 + A1*C1 + B1*C1."

We now click on the Tools menu and choose Solver. This solver will attempt to find a solution that either maximizes or minimizes the value of cell A3, depending on which option we choose. Figure 27 illustrates the Solver box and the items placed in it.

To obtain a solution, click on Solve. The solution $x = 1$ and $y = 1$ and $z = 1$ is located in cells A1, B1, and C1, respectively. The maximum volume $f(1, 1, 1) = 1$ is located in cell A3.

YOUR TURN 2 Solve
Example 3 with the box changed so
that the front and the top are missing.

TRY YOUR TURN 2

| CAUTION | One must be careful when using Solver because the solution may depend on the initial value. Thus, it is always a good idea to run the Solver for two different initial values and compare the solutions. |

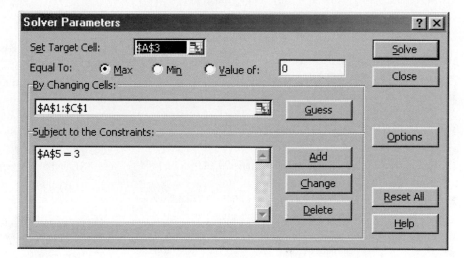

FIGURE 27

Utility Functions

A **utility function** of two variables is a function $z = f(x, y)$ in which x and y give the quantity of items that a consumer might value, such as cereal and milk, and z is a measure of the value that the consumer places on the combination of items represented by the point (x, y). (Naturally, this definition can be extended to any number of variables, but for simplicity we will restrict this discussion to two variables.) For example, if $z = f(x, y) = x^2 y^4$, where x represents the number of quarts of milk and y represents the number of pounds of cereal, then 4 quarts of milk and 3 pounds of cereal has a utility of $4^2 \cdot 3^4 = 1296$. The consumer would value this combination as much as 36 quarts of milk and 1 pound of cereal, since this combination also has a utility of $36^2 \cdot 1^4 = 1296$, but less so than 3 quarts of milk and 4 pounds of cereal, which has a utility of $3^2 \cdot 4^4 = 2304$. The set of points $c = f(x, y)$ form an **indifference curve**, since the consumer considers all points on this curve to be equally desirable.

Now suppose a quart of milk costs \$2 and a pound of cereal costs \$3, so that the combination of milk and cereal represented by the point (x, y) costs $2x + 3y$. Suppose further that the consumer has \$90 to spend. A natural question would be how much of each quantity to buy to maximize the consumer's utility. In other words, the consumer wishes to maximize $f(x, y) = x^2 y^4$ subject to the constraint that $2x + 3y = 90$, or $g(x, y) = 2x + 3y - 90 = 0$. This is exactly the type of problem that Lagrange multipliers were designed to solve. Use Lagrange multipliers to verify that the function $f(x, y) = x^2 y^4$ subject to the constraint $g(x, y) = 2x + 3y - 90 = 0$ has a maximum value of 36,000,000 when $x = 15$ and $y = 20$, so the consumer should purchase 15 quarts of milk and 20 pounds of cereal.

9.4 EXERCISES

Find the relative maxima or minima in Exercises 1–10.

1. Maximum of $f(x, y) = 4xy$, subject to $x + y = 16$

2. Maximum of $f(x, y) = 2xy + 4$, subject to $x + y = 20$

3. Maximum of $f(x, y) = xy^2$, subject to $x + 2y = 15$

4. Maximum of $f(x, y) = 8x^2y$, subject to $3x - y = 9$

5. Minimum of $f(x, y) = x^2 + 2y^2 - xy$, subject to $x + y = 8$

6. Minimum of $f(x, y) = 3x^2 + 4y^2 - xy - 2$, subject to $2x + y = 21$

7. Maximum of $f(x, y) = x^2 - 10y^2$, subject to $x - y = 18$

8. Maximum of $f(x,y) = 12xy - x^2 - 3y^2$, subject to $x + y = 16$

9. Maximum of $f(x, y, z) = xyz^2$, subject to $x + y + z = 6$

10. Maximum of $f(x, y, z) = xy + 2xz + 2yz$, subject to $xyz = 32$

11. Find positive numbers x and y such that $x + y = 24$ and $3xy^2$ is maximized.

12. Find positive numbers x and y such that $x + y = 48$ and $5x^2y + 10$ is maximized.

13. Find three positive numbers whose sum is 90 and whose product is a maximum.

14. Find three positive numbers whose sum is 240 and whose product is a maximum.

15. Find the maximum and minimum values of $f(x,y) = x^3 + 2xy + 4y^2$ subject to $x + 2y = 12$. Be sure to use the method at the end of Example 1 to determine whether each solution is a maximum or a minimum.

16. Explain the difference between the two methods we used in Sections 3 and 4 to solve extrema problems.

17. Why is it unnecessary to find the value of λ when using the method explained in this section?

18. Show that the function $f(x, y) = xy^2$ in Exercise 3, subject to $x + 2y = 15$, does not have an absolute minimum or maximum. (*Hint*: Solve the constraint for x and substitute into f.)

19. Show that the function $f(x, y) = 8x^2y$ in Exercise 4, subject to $3x - y = 9$, does not have an absolute minimum or maximum. (*Hint*: Solve the constraint for y and substitute into f.)

20. Show that the three equations in Step 4 of the box "Using Lagrange Multipliers" are equivalent to the three equations

$$f_x(\lambda, y) = \lambda g_x(\lambda, y), \quad f_y(x, y) = \lambda g_y(x, y), \quad g(x, y) = 0.$$

21. Consider the problem of minimizing $f(x,y) = x^2 + 2x + 9y^2 + 4y + 8xy$ subject to $x + y = 1$.

    a. Find the solution using the method of Lagrange multipliers.

    b. Experiment with points very near the point from part a to convince yourself that the point from part a actually gives a minimum. (*Hint:* See the last paragraph of Example 1.)

    c. Solve $x + y = 1$ for y and substitute the expression for y into $f(x,y)$. Then explain why the resulting expression in x has a minimum but no maximum.

    d. Suppose you erroneously applied the method of finding the discriminant D from the previous section to determine whether the point found in part a is a minimum. What does the test erroneously tell you about the point?

22. Discuss the advantages and disadvantages of the method of Lagrange multipliers compared with solving the equation $g(x, y) = 0$ for y (or x), substituting that expression into f and then minimizing or maximizing f as a function of one variable. You might want to try some examples both ways and consider what happens when there are more than two variables.

APPLICATIONS

Business and Economics

Utility Maximize each of the following utility functions, with the cost of each commodity and total amount available to spend given.

23. $f(x,y) = xy^2$, cost of a unit of x is \$1, cost of a unit of y is \$2, and \$60 is available.

24. $f(x,y) = x^2y^3$, cost of a unit of x is \$2, cost of a unit of y is \$1, and \$80 is available.

25. $f(x,y) = x^4y^2$, cost of a unit of x is \$2, cost of a unit of y is \$4, and \$60 is available.

26. $f(x,y) = x^3y^4$, cost of a unit of x is \$3, cost of a unit of y is \$3, and \$42 is available.

27. **Maximum Area for Fixed Expenditure** Because of terrain difficulties, two sides of a fence can be built for \$6 per ft, while the other two sides cost \$4 per ft. (See the sketch.) Find the field of maximum area that can be enclosed for \$1200.

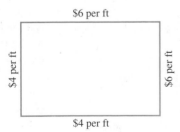

28. **Maximum Area for Fixed Expenditure** To enclose a yard, a fence is built against a large building, so that fencing material is used only on three sides. Material for the ends costs \$15 per ft; material for the side opposite the building costs \$25 per ft. Find the dimensions of the yard of maximum area that can be enclosed for \$2400.

29. **Cost** The total cost to produce x large jewelry-making kits and y small ones is given by

$$C(x, y) = 2x^2 + 6y^2 + 4xy + 10.$$

If a total of ten kits must be made, how should production be allocated so that total cost is minimized?

30. Profit The profit from the sale of x units of radiators for automobiles and y units of radiators for generators is given by

$$P(x, y) = -x^2 - y^2 + 4x + 8y.$$

Find values of x and y that lead to a maximum profit if the firm must produce a total of 6 units of radiators.

31. Production A manufacturing firm estimates that its total production of automobile batteries in thousands of units is

$$f(x, y) = 3x^{1/3}y^{2/3},$$

where x is the number of units of labor and y is the number of units of capital utilized. Labor costs are \$80 per unit, and capital costs are \$150 per unit. How many units each of labor and capital will maximize production, if the firm can spend \$40,000 for these costs?

32. Production For another product, the manufacturing firm in Exercise 31 estimates that production is a function of labor x and capital y as follows:

$$f(x, y) = 12x^{3/4}y^{1/4}.$$

If \$25,200 is available for labor and capital, and if the firm's costs are \$100 and \$180 per unit, respectively, how many units of labor and capital will give maximum production?

33. Area A farmer has 500 m of fencing. Find the dimensions of the rectangular field of maximum area that can be enclosed by this amount of fencing.

34. Area Find the area of the largest rectangular field that can be enclosed with 600 m of fencing. Assume that no fencing is needed along one side of the field.

35. Surface Area A cylindrical can is to be made that will hold 250π in^3 of candy. Find the dimensions of the can with minimum surface area.

36. Surface Area An ordinary 12-oz beer or soda pop can holds about 25 in^3. Find the dimensions of a can with minimum surface area. Measure a can and see how close its dimensions are to the results you found.

37. Volume A rectangular box with no top is to be built from 500 m^2 of material. Find the dimensions of such a box that will enclose the maximum volume.

38. Surface Area A 1-lb soda cracker box has a volume of 185 in^3. The end of the box is square. Find the dimensions of such a box that has minimum surface area.

39. Cost A rectangular closed box is to be built at minimum cost to hold 125 m^3. Since the cost will depend on the surface area, find the dimensions that will minimize the surface area of the box.

40. Cost Find the dimensions that will minimize the surface area (and hence the cost) of a rectangular fish aquarium, open on top, with a volume of 32 ft^3.

41. Container Construction A company needs to construct a box with an open top that will be used to transport 400 yd^3 of material, in several trips, from one place to another. Two of the sides and bottom of the box can be made of a free, lightweight material, but only 4 yd^2 of the material is available. Because of the nature of the material to be transported, the two ends of the box must be made from a heavyweight material that costs \$20 per yd^2. Each trip costs 10 cents. *Source: Geometric Programming*.

a. Let x, y, and z denote the length, width, and height of the box, respectively. If we want to use all of the free material, show that the total cost in dollars is given by the function

$$f(x, y, z) = \frac{40}{xyz} + 40yz,$$

subject to the constraint $2xz + xy = 4$.

b. Use the Solver feature on a spreadsheet to find the dimensions of the box that minimize the transportation cost, subject to the constraint.

Social Sciences

42. Political Science The probability that the majority of a three-person jury will convict a guilty person is given by the formula:

$$P(r, s, t) = rs(1 - t) + (1 - r)st + r(1 - s)t + rst$$

subject to the constraint that

$$r + s + t = \alpha,$$

where r, s, and t represent each of the three jury members' probability of reaching a guilty verdict and α is some fixed constant that is generally less than or equal to the number of jurors. *Source: Mathematical Social Sciences*.

a. Form the Lagrange function.

b. Find the values of r, s, and t that maximize the probability of convicting a guilty person when $\alpha = 0.75$.

c. Find the values of r, s, and t that maximize the probability of convicting a guilty person when $\alpha = 3$.

YOUR TURN ANSWERS

1. $f(12, -4) = 12$

2. The box should be 2 ft wide and 1 ft high and long.

9.5 Total Differentials and Approximations

APPLY IT How do errors in measuring the length and radius of a blood vessel affect the calculation of its volume?

In Example 3 in this section, we will see how to answer this question using a total differential.

In the second section of this chapter we used partial derivatives to find the marginal productivity of labor and of capital for a production function. The marginal productivity approximates the change of production for a 1-unit change in labor or capital. To estimate the change in productivity for a small change in both labor and capital, we can extend the concept of differential, introduced in an earlier chapter for functions of one variable, to the concept of *total differential*.

Total Differential for Two Variables

Let $z = f(x, y)$ be a function of x and y. Let dx and dy be real numbers. Then the **total differential** of z is

$$dz = f_x(x, y) \cdot dx + f_y(x, y) \cdot dy.$$

(Sometimes dz is written df.)

Recall that the differential for a function of one variable $y = f(x)$ is used to approximate the function by its tangent line. This works because a differentiable function appears very much like a line when viewed closely. Similarly, the differential for a function of two variables $z = f(x, y)$ is used to approximate a function by its tangent plane. A differentiable function of two variables looks like a plane when viewed closely, which is why the earth looks flat when you are standing on it.

FOR REVIEW

In Chapter 6 on Applications of the Derivative, we introduced the differential. Recall that the differential of a function defined by $y = f(x)$ is

$$dy = f'(x) \cdot dx,$$

where dx, the differential of x, is any real number (usually small). We saw that the differential dy is often a good approximation of Δy, where $\Delta y = f(x + \Delta x) - f(x)$ and $\Delta x = dx$.

YOUR TURN 1 For the function $f(x,y) = 3x^2y^4 + 6\sqrt{x^2 - 7y^2}$, find **(a)** dz, and **(b)** the value of dz when $x = 4, y = 1, dx = 0.02$, and $dy = -0.03$.

EXAMPLE 1 Total Differentials

Consider the function $z = f(x, y) = 9x^3 - 8x^2y + 4y^3$.

(a) Find dz.

SOLUTION First find $f_x(x, y)$ and $f_y(x, y)$.

$$f_x(x, y) = 27x^2 - 16xy \quad \text{and} \quad f_y(x, y) = -8x^2 + 12y^2$$

By the definition,

$$dz = (27x^2 - 16xy)\, dx + (-8x^2 + 12y^2)\, dy.$$

(b) Evaluate dz when $x = 1, y = 3, dx = 0.01$, and $dy = -0.02$.

SOLUTION Putting these values into the result from part (a) gives

$$dz = [27(1)^2 - 16(1)(3)](0.01) + [-8(1)^2 + 12(3)^2](-0.02)$$
$$= (-21)(0.01) + (100)(-0.02)$$
$$= -2.21.$$

This result indicates that an increase of 0.01 in x and a decrease of 0.02 in y, when $x = 1$ and $y = 3$, will produce an approximate *decrease* of 2.21 in $f(x, y)$.

TRY YOUR TURN 1

Approximations Recall that with a function of one variable, $y = f(x)$, the differential dy approximates the change in y, Δy, corresponding to a change in x, Δx or dx. The approximation for a function of two variables is similar.

Approximations

For small values of dx and dy,

$$dz \approx \Delta z,$$

where $\Delta z = f(x + dx, y + dy) - f(x, y)$.

EXAMPLE 2   Approximations

Approximate $\sqrt{2.98^2 + 4.01^2}$.

SOLUTION   Notice that $2.98 \approx 3$ and $4.01 \approx 4$, and we know that $\sqrt{3^2 + 4^2} = \sqrt{25} = 5$. We, therefore, let $f(x, y) = \sqrt{x^2 + y^2}$, $x = 3$, $dx = -0.02$, $y = 4$, and $dy = 0.01$. We then use dz to approximate $\Delta z = \sqrt{2.98^2 + 4.01^2} - \sqrt{3^2 + 4^2}$.

$$
\begin{aligned}
dz &= f_x(x, y) \cdot dx + f_y(x, y) \cdot dy \\
&= \left(\frac{1}{2\sqrt{x^2 + y^2}} \cdot 2x \right) dx + \left(\frac{1}{2\sqrt{x^2 + y^2}} \cdot 2y \right) dy \\
&= \left(\frac{x}{\sqrt{x^2 + y^2}} \right) dx + \left(\frac{y}{\sqrt{x^2 + y^2}} \right) dy \\
&= \frac{3}{5}(-0.02) + \frac{4}{5}(0.01) \\
&= -0.004
\end{aligned}
$$

YOUR TURN 2

Approximate $\sqrt{5.03^2 + 11.99^2}$.

Thus, $\sqrt{2.98^2 + 4.01^2} \approx 5 + (-0.004) = 4.996$. A calculator gives $\sqrt{2.98^2 + 4.01^2} \approx 4.996048$. The error is approximately 0.000048.   **TRY YOUR TURN 2**

For small values of dx and dy, the values of Δz and dz are approximately equal. Since $\Delta z = f(x + dx, y + dy) - f(x, y)$,

$$f(x + dx, y + dy) = f(x, y) + \Delta z$$

or

$$f(x + dx, y + dy) \approx f(x, y) + dz.$$

Replacing dz with the expression for the total differential gives the following result.

Approximations by Differentials

For a function f having all indicated partial derivatives, and for small values of dx and dy,

$$f(x + dx, y + dy) \approx f(x, y) + dz,$$

or

$$f(x + dx, y + dy) \approx f(x, y) + f_x(x, y) \cdot dx + f_y(x, y) \cdot dy.$$

The idea of a total differential can be extended to include functions of three or more independent variables.

Total Differential for Three Variables

If $w = f(x, y, z)$, then the total differential dw is

$$dw = f_x(x, y, z)\, dx + f_y(x, y, z)\, dy + f_z(x, y, z)\, dz,$$

provided all indicated partial derivatives exist.

EXAMPLE 3   Blood Vessels

A short length of blood vessel is in the shape of a right circular cylinder (see Figure 28).

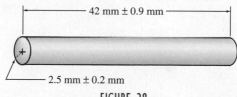

FIGURE 28

(a) The length of the vessel is measured as 42 mm, and the radius is measured as 2.5 mm. Suppose the maximum error in the measurement of the length is 0.9 mm, with an error of no more than 0.2 mm in the measurement of the radius. Find the maximum possible error in calculating the volume of the blood vessel.

APPLY IT

SOLUTION The volume of a right circular cylinder is given by $V = \pi r^2 h$. To approximate the error in the volume, find the total differential, dV.

$$dV = (2\pi rh) \cdot dr + (\pi r^2) \cdot dh$$

Here, $r = 2.5$, $h = 42$, $dr = 0.2$, and $dh = 0.9$. Substitution gives

$$dV = [(2\pi)(2.5)(42)](0.2) + [\pi(2.5)^2](0.9) \approx 149.6.$$

The maximum possible error in calculating the volume is approximately 149.6 mm^3.

(b) Suppose that the errors in measuring the radius and length of the vessel are at most 1% and 3%, respectively. Estimate the maximum percent error in calculating the volume.

SOLUTION To find the percent error, calculate dV/V.

$$\frac{dV}{V} = \frac{(2\pi rh)dr + (\pi r^2)dh}{\pi r^2 h} = 2\frac{dr}{r} + \frac{dh}{h}$$

YOUR TURN 3 In Example 3, estimate the maximum percent error in calculating the volume if the errors in measuring the radius and length of the vessel are at most 4% and 2%, respectively.

Because $dr/r = 0.01$ and $dh/h = 0.03$,

$$\frac{dV}{V} = 2(0.01) + 0.03 = 0.05.$$

The maximum percent error in calculating the volume is approximately 5%.

TRY YOUR TURN 3

EXAMPLE 4 **Volume of a Can of Beer**

The formula for the volume of a cylinder given in Example 3 also applies to cans of beer, for which $r \approx 1.5$ in. and $h \approx 5$ in. How sensitive is the volume to changes in the radius compared with changes in the height?

SOLUTION Using the formula for dV from the previous example with $r = 1.5$ and $h = 5$ gives

$$dV = (2\pi)(1.5)(5)dr + \pi(1.5)^2 dh = \pi(15dr + 2.25dh).$$

The factor of 15 in front of dr in this equation, compared with the factor of 2.25 in front of dh, shows that a small change in the radius has almost 7 times the effect on the volume as a small change in the height. One author argues that this is the reason that beer cans are so tall and thin. *Source: The College Mathematics Journal.* The brewers can reduce the radius by a tiny amount and compensate by making the can taller. The resulting can appears larger in volume than the shorter, wider can. (Others have argued that a shorter, wider can does not fit as easily in the hand.)

9.5 EXERCISES

Evaluate dz using the given information.

1. $z = 2x^2 + 4xy + y^2$; $x = 5$, $y = -1$, $dx = 0.03$, $dy = -0.02$

2. $z = 5x^3 + 2xy^2 - 4y$; $x = 1$, $y = 3$, $dx = 0.01$, $dy = 0.02$

3. $z = \dfrac{y^2 + 3x}{y^2 - x}$; $x = 4$, $y = -4$, $dx = 0.01$, $dy = 0.03$

4. $z = \ln(x^2 + y^2)$; $x = 2$, $y = 3$, $dx = 0.02$, $dy = -0.03$

Evaluate dw using the given information.

5. $w = \dfrac{5x^2 + y^2}{z + 1}$; $x = -2$, $y = 1$, $z = 1$, $dx = 0.02$, $dy = -0.03$, $dz = 0.02$

6. $w = x \ln(yz) - y \ln \dfrac{x}{z}$; $x = 2$, $y = 1$, $z = 4$, $dx = 0.03$, $dy = 0.02$, $dz = -0.01$

Use the total differential to approximate each quantity. Then use a calculator to approximate the quantity, and give the absolute value of the difference in the two results to 4 decimal places.

7. $\sqrt{8.05^2 + 5.97^2}$

8. $\sqrt{4.96^2 + 12.06^2}$

9. $(1.92^2 + 2.1^2)^{1/3}$

10. $(2.93^2 - 0.94^2)^{1/3}$

11. $1.03e^{0.04}$

12. $0.98e^{-0.04}$

13. $0.99 \ln 0.98$

14. $2.03 \ln 1.02$

APPLICATIONS

Business and Economics

15. Manufacturing Approximate the volume of aluminum needed for a beverage can of radius 2.5 cm and height 14 cm. Assume the walls of the can are 0.08 cm thick.

16. Manufacturing Approximate the volume of material needed to make a water tumbler of diameter 3 cm and height 9 cm. Assume the walls of the tumbler are 0.2 cm thick.

17. Volume of a Coating An industrial coating 0.1 in. thick is applied to all sides of a box of dimensions 10 in. by 9 in. by 18 in. Estimate the volume of the coating used.

18. Manufacturing Cost The manufacturing cost of a smartphone is approximated by

$$M(x, y) = 45x^2 + 40y^2 - 20xy + 50,$$

where x is the cost of the parts and y is the cost of labor. Right now, the company spends $8 on parts and $14 on labor. Use differentials to approximate the change in cost if the company spends $8.25 on parts and $13.75 on labor.

19. Production The production function for one country is

$$z = x^{0.65}y^{0.35},$$

where x stands for units of labor and y for units of capital. At present, 50 units of labor and 29 units of capital are available. Use differentials to estimate the change in production if the number of units of labor is increased to 52 and capital is decreased to 27 units.

20. Production The production function for another country is

$$z = x^{0.8}y^{0.2},$$

where x stands for units of labor and y for units of capital. At present, 20 units of labor and 18 units of capital are being provided. Use differentials to estimate the change in production if an additional unit of labor is provided and if capital is decreased to 16 units.

Life Sciences

21. Bone Preservative Volume A piece of bone in the shape of a right circular cylinder is 7 cm long and has a radius of 1.4 cm. It is coated with a layer of preservative 0.09 cm thick. Estimate the volume of preservative used.

22. Blood Vessel Volume A portion of a blood vessel is measured as having length 7.9 cm and radius 0.8 cm. If each measurement could be off by as much as 0.15 cm, estimate the maximum possible error in calculating the volume of the vessel.

23. Blood Volume In Exercise 56 of Section 2 in this chapter, we found that the number of liters of blood pumped through the lungs in one minute is given by

$$C = \frac{b}{a - v}.$$

Suppose $a = 160$, $b = 200$, and $v = 125$. Estimate the change in C if a becomes 145, b becomes 190, and v changes to 130.

24. Heat Loss In Exercise 54 of Section 2 of this chapter, we found that the rate of heat loss (in watts) in harbor seal pups could be approximated by

$$H(m, T, A) = \frac{15.2m^{0.67}(T - A)}{10.23 \ln m - 10.74},$$

where m is the body mass of the pup (in kg), and T and A are the body core temperature and ambient water temperature, respectively (in °C). Suppose m is 25 kg, T is 36.0°, and A is 12.0°C. Approximate the change in H if m changes to 26 kg, T to 36.5°, and A to 10.0°C.

25. Dialysis A model that estimates the concentration of urea in the body for a particular dialysis patient, following a dialysis session, is given by

$$C(t, g) = 0.6(0.96)^{(210t/1500) - 1}$$
$$+ \frac{gt}{126t - 900}[1 - (0.96)^{(210t/1500) - 1}],$$

where t represents the number of minutes of the dialysis session and g represents the rate at which the body generates urea in mg per minute. *Source: Clinical Dialysis.*

a. Find $C(180, 8)$.

b. Using the total differential, estimate the urea concentration if the dialysis session of part a was cut short by 10 minutes and the urea generation rate was 9 mg per minute. Compare this with the actual concentration. (*Hint:* First, replace the variable g with the number 8, thus reducing the function to one variable. Then use your graphing calculator to calculate the partial derivative $C_t(180, 8)$. A similar procedure can be done for $C_g(180, 8)$.)

26. Horn Volume The volume of the horns from bighorn sheep was estimated by researchers using the equation

$$V = \frac{h\pi}{3}(r_1^2 + r_1r_2 + r_2^2),$$

where h is the length of a horn segment (in centimeters) and r_1 and r_2 are the radii of the two ends of the horn segment (in centimeters). *Source: Conservation Biology.*

a. Determine the volume of a segment of horn that is 40 cm long with radii of 5 cm and 3 cm, respectively.

b. Use the total differential to estimate the volume of the segment of horn if the horn segment from part a was actually 42 cm long with radii of 5.1 cm and 2.9 cm, respectively. Compare this with the actual volume.

27. Eastern Hemlock Ring shake, which is the separation of the wood between growth rings, is a serious problem in hemlock trees. Researchers have developed the following function that estimates the probability P that a given hemlock tree has ring shake.

$$P(A, B, D) = \frac{1}{1 + e^{3.68 - 0.016A - 0.77B - 0.12D}},$$

where A is the age of the tree (yr), B is 1 if bird pecking is present and 0 otherwise, and D is the diameter (in.) of the tree at breast height. *Source: Forest Products Journal.*

a. Estimate the probability that a 150-year-old tree, with bird pecking present and a breast height diameter of 20 in., will have ring shake.

b. Estimate the probability that a 150-year-old tree, with no presence of bird pecking and a breast height diameter of 20 in., will have ring shake.

c. Develop a statement about what can be said about the influence that the three variables have on the probability of ring shake.

d. Using the total differential, estimate the probability if the actual age of the tree was 160 years and the diameter at breast height was 25 in. Assume that no bird pecking was present. Compare your answer to the actual value. (*Hint:* Assume that $B = 0$ and exclude that variable from your calculations.)

e. Comment on the practicality of using differentials in part d.

Physical Sciences

28. **Swimming** The amount of time in seconds it takes for a swimmer to hear a single, hand-held, starting signal is given by the formula

$$t(x, y, p, C) = \frac{\sqrt{x^2 + (y - p)^2}}{331.45 + 0.6C},$$

where (x, y) is the location of the starter (in meters), $(0, p)$ is the location of the swimmer (in meters), and C is the air temperature (in degrees Celsius). *Source: COMAP.* Assume that the starter is located at the point $(x, y) = (5, -2)$. See the diagram.

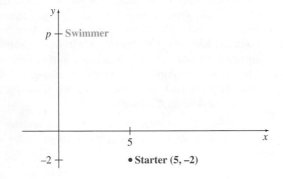

a. Calculate $t(5, -2, 20, 20)$ and $t(5, -2, 10, 20)$. Could the difference in time change the outcome of a race?

b. Calculate the total differential for t if the starter remains stationary, the swimmer moves from 20 m to 20.5 m away from the starter in the y direction, and the temperature decreases from 20°C to 15°C. Interpret your answer.

General Interest

29. **Estimating Area** The height of a triangle is measured as 37.5 cm, with the base measured as 15.8 cm. The measurement of the height can be off by as much as 0.8 cm and that of the base by no more than 1.1 cm. Estimate the maximum possible error in calculating the area of the triangle.

30. **Estimating Volume** The height of a cone is measured as 9.3 cm and the radius as 3.2 cm. Each measurement could be off by as much as 0.1 cm. Estimate the maximum possible error in calculating the volume of the cone.

31. **Estimating Volume** Suppose that in measuring the length, width, and height of a box, there is a maximum 1% error in each measurement. Estimate the maximum error in calculating the volume of the box.

32. **Estimating Volume** Suppose there is a maximum error of $a\%$ in measuring the radius of a cone and a maximum error of $b\%$ in measuring the height. Estimate the maximum percent error in calculating the volume of the cone, and compare this value with the maximum percent error in calculating the volume of a cylinder.

33. **Ice Cream Cone** An ice cream cone has a radius of approximately 1 in. and a height of approximately 4 in. By what factor does a change in the radius affect the volume compared with a change in the height?

34. **Hose** A hose has a radius of approximately 0.5 in. and a length of approximately 20 ft. By what factor does a change in the radius affect the volume compared with a change in the length?

YOUR TURN ANSWERS

1. (a) $dz = (6xy^4 + 6x/\sqrt{x^2 - 7y^2})dx +$
   $(12x^2y^3 - 42y/\sqrt{x^2 - 7y^2})dy$ (b) -4.7

2. 13.0023   3. 10%

9.6 Double Integrals

APPLY IT How can we find the volume of a bottle with curved sides?
We will answer this question in Example 6 using a double integral, the key idea in this section.

In an earlier chapter, we saw how integrals of functions with one variable may be used to find area. In this section, this idea is extended and used to find volume. We found partial derivatives of functions of two or more variables at the beginning of this chapter by holding constant all variables except one. A similar process is used in this section to find antiderivatives of functions of two or more variables. For example, in

$$\int (5x^3y^4 - 6x^2y + 2)\, dy$$

the notation dy indicates integration with respect to y, so we treat y as the variable and x as a constant. Using the rules for antiderivatives gives

$$\int (5x^3y^4 - 6x^2y + 2)\, dy = x^3y^5 - 3x^2y^2 + 2y + C(x).$$

The constant C used earlier must be replaced with $C(x)$ to show that the "constant of integration" here can be any function involving only the variable x. Just as before, check this work by taking the derivative (actually the partial derivative) of the answer:

$$\frac{\partial}{\partial y}[x^3y^5 - 3x^2y^2 + 2y + C(x)] = 5x^3y^4 - 6x^2y + 2 + 0,$$

which shows that the antiderivative is correct.

We can use this antiderivative to evaluate a definite integral.

EXAMPLE 1 Definite Integral

Evaluate $\displaystyle\int_1^2 (5x^3y^4 - 6x^2y + 2)\, dy$.

SOLUTION

$$\int_1^2 (5x^3y^4 - 6x^2y + 2)\, dy = \left[x^3y^5 - 3x^2y^2 + 2y + C(x)\right]\Big|_1^2$$

Use the indefinite integral previously found.

$$= x^3 2^5 - 3x^2 2^2 + 2\cdot 2 + C(x)$$
$$\quad - [x^3 1^5 - 3x^2 1^2 + 2\cdot 1 + C(x)]$$
$$= 32x^3 - 12x^2 + 4 + C(x)$$
$$\quad - [x^3 - 3x^2 + 2 + C(x)]$$
$$= 31x^3 - 9x^2 + 2 \qquad \textit{Simplify.}$$

YOUR TURN 1 Evaluate
$$\int_1^3 (6x^2y^2 + 4xy + 8x^3 + 10y^4 + 3)\, dy.$$

In the second step, we substituted $y = 2$ and $y = 1$ and subtracted, according to the Fundamental Theorem of Calculus. Notice that $C(x)$ does not appear in the final answer, just as the constant does not appear in a regular definite integral. Therefore, from now on we will not include $C(x)$ when we find the antiderivative for a definite integral with respect to y.

TRY YOUR TURN 1

By integrating the result from Example 1 with respect to x, we can evaluate a double integral.

EXAMPLE 2 Definite Integral

Evaluate $\displaystyle\int_0^3 \left[\int_1^2 (5x^3y^4 - 6x^2y + 2)\, dy\right] dx$.

SOLUTION

$$\int_0^3 \left[\int_1^2 (5x^3y^4 - 6x^2y + 2)\, dy\right] dx = \int_0^3 [31x^3 - 9x^2 + 2]\, dx$$

Use the result from Example 1.

$$= \frac{31}{4}x^4 - 3x^3 + 2x \Big|_0^3$$

Use the Fundamental Theorem of Calculus.

$$= \frac{31}{4}\cdot 3^4 - 3\cdot 3^3 + 2\cdot 3 - \left(\frac{31}{4}\cdot 0^4 - 3\cdot 0^3 + 2\cdot 0\right)$$

$$= \frac{2211}{4}$$

We can integrate the inner integral with respect to y and the outer integral with respect to x, as in Example 2, or in the reverse order. The next example shows the same integral done both ways.

EXAMPLE 3 Definite Integrals

Evaluate each integral.

(a) $\displaystyle\int_1^2 \left[\int_3^5 (6xy^2 + 12x^2y + 4y)\, dx \right] dy$

SOLUTION

$$\int_1^2 \left[\int_3^5 (6xy^2 + 12x^2y + 4y)\,dx \right] dy = \int_1^2 \left[(3x^2y^2 + 4x^3y + 4xy) \Big|_3^5 \right] dy \qquad \text{Integrate with respect to } x.$$

$$= \int_1^2 [(3\cdot 5^2 \cdot y^2 + 4\cdot 5^3 \cdot y + 4\cdot 5\cdot y)$$

$$\qquad - (3\cdot 3^2 \cdot y^2 + 4\cdot 3^3 \cdot y + 4\cdot 3\cdot y)]\,dy$$

$$= \int_1^2 [(75y^2 + 500y + 20y)$$

$$\qquad - (27y^2 + 108y + 12y)]\,dy$$

$$= \int_1^2 (48y^2 + 400y)\,dy$$

$$= (16y^3 + 200y^2) \Big|_1^2 \qquad \text{Integrate with respect to } y.$$

$$= 16\cdot 2^3 + 200\cdot 2^2 - (16\cdot 1^3 + 200\cdot 1^2)$$

$$= 128 + 800 - (16 + 200)$$

$$= 712.$$

(b) $\displaystyle\int_3^5 \left[\int_1^2 (6xy^2 + 12x^2y + 4y)\, dy \right] dx$

SOLUTION   (This is the same integrand with the same limits of integration as in part (a), but the order of integration is reversed.)

$$\int_3^5 \left[\int_1^2 (6xy^2 + 12x^2y + 4y)\,dy \right] dx = \int_3^5 \left[(2xy^3 + 6x^2y^2 + 2y^2) \Big|_1^2 \right] dx \qquad \text{Integrate with respect to } y.$$

$$= \int_3^5 [(2x\cdot 2^3 + 6x^2 \cdot 2^2 + 2\cdot 2^2)$$

$$\qquad - (2x\cdot 1^3 \cdot\ + 6x^2 \cdot 1^2 + 2\cdot 1^2)]\,dx$$

$$= \int_3^5 [(16x + 24x^2 + 8)$$

$$\qquad - (2x + 6x^2 + 2)]\,dx$$

$$= \int_3^5 (14x + 18x^2 + 6)\,dx \qquad \text{Integrate with respect to } x.$$

$$= (7x^2 + 6x^3 + 6x) \Big|_3^5$$

$$= 7\cdot 5^2 + 6\cdot 5^3 + 6\cdot 5 - (7\cdot 3^2 + 6\cdot 3^3 + 6\cdot 3)$$

$$= 175 + 750 + 30 - (63 + 162 + 18) = 712$$

TRY YOUR TURN 2

YOUR TURN 2 Evaluate $\displaystyle\int_0^2 \left[\int_1^3 (6x^2y^2 + 4xy + 8x^3 + 10y^4 + 3)\, dy \right] dx$, and then integrate with the order of integration changed.

NOTE In the second step of Example 3 (a), it might help you avoid confusion as to whether to put the limits of 3 and 5 into x or y by writing the integral as

$$\int_1^2 \left[(3x^2y^2 + 4x^3y + 4xy) \Big|_{x=3}^{x=5} \right] dy$$

The brackets we have used for the inner integral in Example 3 are not essential because the order of integration is indicated by the order of $dx\, dy$ or $dy\, dx$. For example, if the integral is written as

$$\int_1^2 \int_3^5 (6xy^2 + 12x^2y + 4y) \, dx\, dy,$$

we first integrate with respect to x, letting x vary from 3 to 5, and then with respect to y, letting y vary from 1 to 2, as in Example 3(a).

The answers in the two parts of Example 3 are equal. It can be proved that for a large class of functions, including most functions that occur in applications, the following equation holds true.

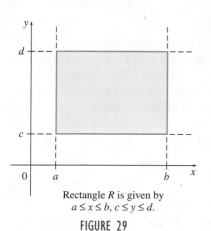

Rectangle R is given by
$a \le x \le b, c \le y \le d.$

FIGURE 29

Fubini's Theorem

$$\int_a^b \int_c^d f(x, y) \, dy\, dx = \int_c^d \int_a^b f(x, y) \, dx\, dy$$

Either of these integrals is called an **iterated integral** since it is evaluated by integrating twice, first using one variable and then using the other. The fact that the iterated integrals above are equal makes it possible to define a *double integral*. First, the set of points (x, y), with $a \le x \le b$ and $c \le y \le d$, defines a rectangular region R in the plane, as shown in Figure 29. Then, the *double integral over R* is defined as follows.

Double Integral

The **double integral** of $f(x, y)$ over a rectangular region R is written

$$\iint_R f(x, y) \, dy\, dx \qquad \text{or} \qquad \iint_R f(x, y) \, dx\, dy,$$

and equals either

$$\int_a^b \int_c^d f(x, y) \, dy\, dx \qquad \text{or} \qquad \int_c^d \int_a^b f(x, y) \, dx\, dy.$$

Extending earlier definitions, $f(x, y)$ is the **integrand** and R is the **region of integration**.

EXAMPLE 4   Double Integrals

Find $\displaystyle \iint_R \frac{3\sqrt{x} \cdot y}{y^2 + 1} \, dx\, dy$ over the rectangular region R defined by $0 \le x \le 4$ and $0 \le y \le 2$.

SOLUTION Integrate first with respect to x; then integrate the result with respect to y.

$$\iint\limits_R \frac{3\sqrt{x} \cdot y}{y^2 + 1} \, dx \, dy = \int_0^2 \int_0^4 \frac{3\sqrt{x} \cdot y}{y^2 + 1} \, dx \, dy$$

$$= \int_0^2 \left. \frac{2x^{3/2} \cdot y}{y^2 + 1} \right|_0^4 dy \qquad \text{Use the power rule with } x^{1/2}.$$

$$= \int_0^2 \left(\frac{2(4)^{3/2} \cdot y}{y^2 + 1} - \frac{2(0)^{3/2} \cdot y}{y^2 + 1} \right) dy$$

$$= 8 \int_0^2 \frac{2y}{y^2 + 1} \, dy \qquad \text{Factor out } 4^{3/2} = 8.$$

$$= 8 \int_1^5 \frac{du}{u} \qquad \begin{array}{l}\text{Let } u = y^2 + 1. \text{ Change limits}\\ \text{of integration.}\end{array}$$

$$= 8 \ln u \big|_1^5$$

$$= 8 \ln 5 - 8 \ln 1 = 8 \ln 5$$

As a check, integrate with respect to y first. The answer should be the same.

TRY YOUR TURN 3

YOUR TURN 3 Find

$$\iint\limits_R \frac{1}{\sqrt{x + y + 3}} \, dx \, dy$$

over the rectangular region R defined by $0 \le x \le 5$ and $1 \le y \le 6$.

Volume As shown earlier, the definite integral $\int_a^b f(x) \, dx$ can be used to find the area under a curve. In a similar manner, double integrals are used to find the *volume under a surface*. Figure 30 shows that portion of a surface $f(x, y)$ directly over a rectangle R in the xy-plane. Just as areas were approximated by a large number of small rectangles, volume could be approximated by adding the volumes of a large number of properly drawn small boxes. The height of a typical box would be $f(x, y)$ with the length and width given by dx and dy. The formula for the volume of a box would then suggest the following result.

Volume

Let $z = f(x, y)$ be a function that is never negative on the rectangular region R defined by $a \le x \le b$, $c \le y \le d$. The volume of the solid under the graph of f and over the region R is

$$\iint\limits_R f(x, y) \, dx \, dy.$$

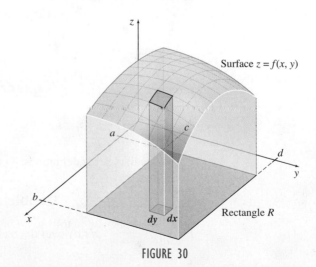

FIGURE 30

EXAMPLE 5 Volume

Find the volume under the surface $z = x^2 + y^2$ shown in Figure 31.

SOLUTION By the equation just given, the volume is

$$\iint\limits_R f(x, y)\, dx\, dy,$$

where $f(x, y) = x^2 + y^2$ and R is the region $0 \le x \le 4,\ 0 \le y \le 4$. By definition,

$$\iint\limits_R f(x, y)\, dx\, dy = \int_0^4 \int_0^4 (x^2 + y^2)\, dx\, dy$$

$$= \int_0^4 \left(\frac{1}{3}x^3 + xy^2\right)\bigg|_0^4 dy$$

$$= \int_0^4 \left(\frac{64}{3} + 4y^2\right) dy = \left(\frac{64}{3}y + \frac{4}{3}y^3\right)\bigg|_0^4$$

$$= \frac{64}{3}\cdot 4 + \frac{4}{3}\cdot 4^3 - 0 = \frac{512}{3}.$$

YOUR TURN 4 Find the volume under the surface $z = 4 - x^3 - y^3$ over the rectangular region $0 \le x \le 1, 0 \le y \le 1$.

TRY YOUR TURN 4

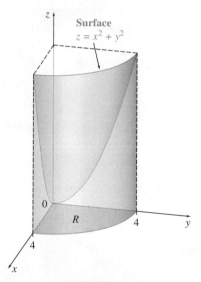

FIGURE 31

EXAMPLE 6 Perfume Bottle

A product design consultant for a cosmetics company has been asked to design a bottle for the company's newest perfume. The thickness of the glass is to vary so that the outside of the bottle has straight sides and the inside has curved sides, with flat ends shaped like parabolas on the 4-cm sides, as shown in Figure 32. Before presenting the design to management, the consultant needs to make a reasonably accurate estimate of the amount each bottle will hold. If the base of the bottle is to be 4 cm by 3 cm, and if a cross section of its interior is to be a parabola of the form $z = -y^2 + 4y$, what is its internal volume?

4 cm   3 cm

FIGURE 32

APPLY IT

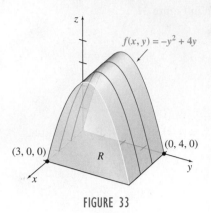

$f(x, y) = -y^2 + 4y$

(3, 0, 0)   R   (0, 4, 0)

x   *y*

FIGURE 33

SOLUTION The interior of the bottle can be graphed in three-dimensional space, as shown in Figure 33, where $z = 0$ corresponds to the base of the bottle. Its volume is simply the volume above the region R in the xy-plane and below the graph of $f(x, y) = -y^2 + 4y$. This volume is given by the double integral

$$\int_0^3 \int_0^4 (-y^2 + 4y)\, dy\, dx = \int_0^3 \left(\frac{-y^3}{3} + \frac{4y^2}{2} \right)\Big|_0^4 dx$$

$$= \int_0^3 \left(\frac{-64}{3} + 32 - 0 \right) dx$$

$$= \frac{32}{3} x \Big|_0^3$$

$$= 32 - 0 = 32.$$

The bottle holds 32 cm³.

Double Integrals Over Other Regions
In this section, we found double integrals over rectangular regions by evaluating iterated integrals with constant limits of integration. We can also evaluate iterated integrals with *variable* limits of integration. (Notice in the following examples that the variable limits always go on the *inner* integral sign.)

The use of variable limits of integration permits evaluation of double integrals over the types of regions shown in Figure 34. Double integrals over more complicated regions are discussed in more advanced books. Integration over regions such as those in Figure 34 is done with the results of the following theorem.

Double Integrals Over Variable Regions
Let $z = f(x, y)$ be a function of two variables. If R is the region (in Figure 34(a)) defined by $a \leq x \leq b$ and $g(x) \leq y \leq h(x)$, then

$$\iint_R f(x, y)\, dy\, dx = \int_a^b \left[\int_{g(x)}^{h(x)} f(x, y)\, dy \right] dx.$$

If R is the region (in Figure 34(b)) defined by $g(y) \leq x \leq h(y)$ and $c \leq y \leq d$, then

$$\iint_R f(x, y)\, dx\, dy = \int_c^d \left[\int_{g(y)}^{h(y)} f(x, y)\, dx \right] dy.$$

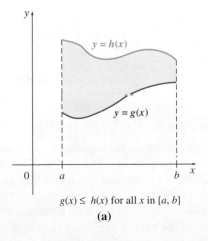

$y = h(x)$

$y = g(x)$

0   *a*   *b*   *x*

$g(x) \leq h(x)$ for all x in $[a, b]$

(a)

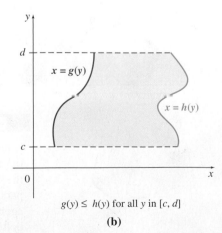

d $x = g(y)$

$x = h(y)$

c

0   *x*

$g(y) \leq h(y)$ for all y in $[c, d]$

(b)

FIGURE 34

> [!NOTE] EXAMPLE 7 Double Integrals

Evaluate $\int_1^2 \int_y^{y^2} xy \, dx \, dy$.

SOLUTION The region of integration is shown in Figure 35. Integrate first with respect to x, then with respect to y.

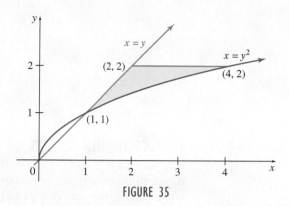

FIGURE 35

$$\int_1^2 \int_y^{y^2} xy \, dx \, dy = \int_1^2 \left(\int_y^{y^2} xy \, dx \right) dy = \int_1^2 \left(\frac{1}{2}x^2 y \right)\Big|_y^{y^2} dy$$

Replace x first with y^2 and then with y, and subtract.

$$\int_1^2 \int_y^{y^2} xy \, dx \, dy = \int_1^2 \left[\frac{1}{2}(y^2)^2 y - \frac{1}{2}(y)^2 y \right] dy$$

$$= \int_1^2 \left(\frac{1}{2}y^5 - \frac{1}{2}y^3 \right) dy = \left(\frac{1}{12}y^6 - \frac{1}{8}y^4 \right)\Big|_1^2$$

$$= \left(\frac{1}{12} \cdot 2^6 - \frac{1}{8} \cdot 2^4 \right) - \left(\frac{1}{12} \cdot 1^6 - \frac{1}{8} \cdot 1^4 \right)$$

$$= \frac{64}{12} - \frac{16}{8} - \frac{1}{12} + \frac{1}{8} = \frac{27}{8}$$

YOUR TURN 5 Find $\iint\limits_R (x^3 + 4y) \, dy \, dx$ over the region bounded by $y = 4x$ and $y = x^3$ for $0 \le x \le 2$.

TRY YOUR TURN 5

> [!NOTE] EXAMPLE 8 Double Integrals

Let R be the shaded region in Figure 36, and evaluate

$$\iint\limits_R (x + 2y) \, dy \, dx.$$

SOLUTION Region R is bounded by $h(x) = 2x$ and $g(x) = x^2$, with $0 \le x \le 2$. By the first result in the previous theorem,

$$\iint\limits_R (x + 2y) \, dy \, dx = \int_0^2 \int_{x^2}^{2x} (x + 2y) \, dy \, dx$$

$$= \int_0^2 (xy + y^2)\Big|_{x^2}^{2x} dx$$

$$= \int_0^2 [x(2x) + (2x)^2 - [x \cdot x^2 + (x^2)^2]] \, dx$$

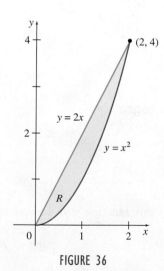

FIGURE 36

$$= \int_0^2 \left[2x^2 + 4x^2 - (x^3 + x^4)\right] dx$$

$$= \int_0^2 (6x^2 - x^3 - x^4) \, dx$$

$$= \left(2x^3 - \frac{1}{4}x^4 - \frac{1}{5}x^5\right)\Big|_0^2$$

$$= 2 \cdot 2^3 - \frac{1}{4} \cdot 2^4 - \frac{1}{5} \cdot 2^5 - 0$$

$$= 16 - 4 - \frac{32}{5} = \frac{28}{5}.$$

Interchanging Limits of Integration

Sometimes it is easier to integrate first with respect to x and then y, while with other integrals the reverse process is easier. The limits of integration can be reversed whenever the region R is like the region in Figure 36, which has the property that it can be viewed as either type of region shown in Figure 34. In practice, this means that all boundaries can be written in terms of y as a function of x, or by solving for x as a function of y.

For instance, in Example 8, the same result would be found if we evaluated the double integral first with respect to x and then with respect to y. In that case, we would need to define the equations of the boundaries in terms of y rather than x, so R would be defined by $y/2 \le x \le \sqrt{y}, 0 \le y \le 4$. The resulting integral is

$$\int_0^4 \int_{y/2}^{\sqrt{y}} (x + 2y) \, dx \, dy = \int_0^4 \left(\frac{x^2}{2} + 2xy\right)\Big|_{y/2}^{\sqrt{y}} dy$$

$$= \int_0^4 \left[\left(\frac{y}{2} + 2y\sqrt{y}\right) - \left(\frac{y^2}{8} + 2\left(\frac{y}{2}\right)y\right)\right] dy$$

$$= \int_0^4 \left(\frac{y}{2} + 2y^{3/2} - \frac{9}{8}y^2\right) dy$$

$$= \left(\frac{y^2}{4} + \frac{4}{5}y^{5/2} - \frac{3}{8}y^3\right)\Big|_0^4$$

$$= 4 + \frac{4}{5} \cdot 4^{5/2} - 24$$

$$= \frac{28}{5}.$$

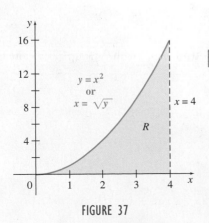

FIGURE 37

EXAMPLE 9 Interchanging Limits of Integration

Evaluate

$$\int_0^{16} \int_{\sqrt{y}}^4 \sqrt{x^3 + 4} \, dx \, dy.$$

SOLUTION Notice that it is impossible to first integrate this function with respect to x. Thus, we attempt to interchange the limits of integration.

For this integral, region R is given by $\sqrt{y} \le x \le 4$, $0 \le y \le 16$. A graph of R is shown in Figure 37.

The same region R can be written in an alternate way. As Figure 37 shows, one boundary of R is $x = \sqrt{y}$. Solving for y gives $y = x^2$. Also, Figure 37 shows that $0 \le x \le 4$. Since R can be written as $0 \le y \le x^2, 0 \le x \le 4$, the double integral above can be written

$$\int_0^4 \int_0^{x^2} \sqrt{x^3 + 4}\, dy\, dx = \int_0^4 y\sqrt{x^3 + 4}\Big|_0^{x^2} dx$$

$$= \int_0^4 x^2\sqrt{x^3 + 4}\, dx$$

$$= \frac{1}{3}\int_0^4 3x^2\sqrt{x^3 + 4}\, dx \quad \text{Let } u = x^3 + 4. \text{ Change limits of integration.}$$

$$= \frac{1}{3}\int_4^{68} u^{1/2}du$$

$$= \frac{2}{9}u^{3/2}\Big|_4^{68}$$

$$= \frac{2}{9}[68^{3/2} - 4^{3/2}]$$

$$\approx 122.83.$$

CAUTION Fubini's Theorem cannot be used to interchange the order of integration when the limits contain variables, as in Example 9. Notice in Example 9 that after the order of integration was changed, the new limits were completely different. It would be a serious error to rewrite the integral in Example 9 as

$$\int_{\sqrt{y}}^4 \int_0^{16} \sqrt{x^3 + 4}\, dy\, dx.$$

9.6 EXERCISES

Evaluate each integral.

1. $\int_0^5 (x^4 y + y)\, dx$

2. $\int_1^2 (xy^3 - x)\, dy$

3. $\int_4^5 x\sqrt{x^2 + 3y}\, dy$

4. $\int_3^6 x\sqrt{x^2 + 3y}\, dx$

5. $\int_4^9 \frac{3 + 5y}{\sqrt{x}}\, dx$

6. $\int_2^7 \frac{3 + 5y}{\sqrt{x}}\, dy$

7. $\int_2^6 e^{2x+3y}\, dx$

8. $\int_{-1}^1 e^{2x+3y}\, dy$

9. $\int_0^3 ye^{4x+y^2}\, dy$

10. $\int_1^5 ye^{4x+y^2}\, dx$

Evaluate each iterated integral. (Many of these use results from Exercises 1–10.)

11. $\int_1^2 \int_0^5 (x^4 y + y)\, dx\, dy$

12. $\int_0^3 \int_1^2 (xy^3 - x)\, dy\, dx$

13. $\int_0^1 \int_3^6 x\sqrt{x^2 + 3y}\, dx\, dy$

14. $\int_0^3 \int_4^5 x\sqrt{x^2 + 3y}\, dy\, dx$

15. $\int_1^2 \int_4^9 \frac{3 + 5y}{\sqrt{x}}\, dx\, dy$

16. $\int_{16}^{25} \int_2^7 \frac{3 + 5y}{\sqrt{x}}\, dy\, dx$

17. $\int_1^3 \int_1^3 \frac{1}{xy}\, dy\, dx$

18. $\int_1^5 \int_2^4 \frac{1}{y}\, dx\, dy$

19. $\int_2^4 \int_3^5 \left(\frac{x}{y} + \frac{y}{3}\right) dx\, dy$

20. $\int_3^4 \int_1^2 \left(\frac{6x}{5} + \frac{y}{x}\right) dx\, dy$

Find each double integral over the rectangular region R with the given boundaries.

21. $\iint\limits_R (3x^2 + 4y)\, dx\, dy; \quad 0 \le x \le 3, 1 \le y \le 4$

22. $\iint\limits_R (x^2 + 4y^3)\, dy\, dx; \quad 1 \le x \le 2, 0 \le y \le 3$

23. $\iint\limits_R \sqrt{x + y}\, dy\, dx; \quad 1 \le x \le 3, 0 \le y \le 1$

24. $\iint\limits_R x^2\sqrt{x^3 + 2y}\, dx\, dy; \quad 0 \le x \le 2, 0 \le y \le 3$

25. $\iint\limits_R \frac{3}{(x + y)^2}\, dy\, dx; \quad 2 \le x \le 4, 1 \le y \le 6$

26. $\displaystyle\iint_R \frac{y}{\sqrt{2x + 5y^2}}\, dx\, dy;\quad 0 \le x \le 2, 1 \le y \le 3$

27. $\displaystyle\iint_R y e^{x + y^2}\, dx\, dy;\quad 2 \le x \le 3, 0 \le y \le 2$

28. $\displaystyle\iint_R x^2 e^{x^3 + 2y}\, dx\, dy;\quad 1 \le x \le 2, 1 \le y \le 3$

Find the volume under the given surface $z = f(x, y)$ and above the rectangle with the given boundaries.

29. $z = 8x + 4y + 10;\quad -1 \le x \le 1, 0 \le y \le 3$

30. $z = 3x + 10y + 20;\quad 0 \le x \le 3, -2 \le y \le 1$

31. $z = x^2;\quad 0 \le x \le 2, 0 \le y \le 5$

32. $z = \sqrt{y};\quad 0 \le x \le 4, 0 \le y \le 9$

33. $z = x\sqrt{x^2 + y};\quad 0 \le x \le 1, 0 \le y \le 1$

34. $z = yx\sqrt{x^2 + y^2};\quad 0 \le x \le 4, 0 \le y \le 1$

35. $z = \dfrac{xy}{(x^2 + y^2)^2};\quad 1 \le x \le 2, 1 \le y \le 4$

36. $z = e^{x + y};\quad 0 \le x \le 1, 0 \le y \le 1$

Although it is often true that a double integral can be evaluated by using either dx or dy first, sometimes one choice over the other makes the work easier. Evaluate the double integrals in Exercises 37 and 38 in the easiest way possible.

37. $\displaystyle\iint_R x e^{xy}\, dx\, dy;\quad 0 \le x \le 2, 0 \le y \le 1$

38. $\displaystyle\iint_R 2x^3 e^{x^2 y}\, dx\, dy;\quad 0 \le x \le 1, 0 \le y \le 1$

Evaluate each double integral.

39. $\displaystyle\int_2^4 \int_2^{x^2} (x^2 + y^2)\, dy\, dx$

40. $\displaystyle\int_0^2 \int_0^{3y} (x^2 + y)\, dx\, dy$

41. $\displaystyle\int_0^4 \int_0^x \sqrt{xy}\, dy\, dx$

42. $\displaystyle\int_1^4 \int_0^x \sqrt{x + y}\, dy\, dx$

43. $\displaystyle\int_2^6 \int_{2y}^{4y} \frac{1}{x}\, dx\, dy$

44. $\displaystyle\int_1^4 \int_x^{x^2} \frac{1}{y}\, dy\, dx$

45. $\displaystyle\int_0^4 \int_1^{e^x} \frac{x}{y}\, dy\, dx$

46. $\displaystyle\int_0^1 \int_{2x}^{4x} e^{x + y}\, dy\, dx$

Use the region R with the indicated boundaries to evaluate each double integral.

47. $\displaystyle\iint_R (5x + 8y)\, dy\, dx;\quad 1 \le x \le 3, 0 \le y \le x - 1$

48. $\displaystyle\iint_R (2x + 6y)\, dy\, dx;\quad 2 \le x \le 4, 2 \le y \le 3x$

49. $\displaystyle\iint_R (4 - 4x^2)\, dy\, dx;\quad 0 \le x \le 1, 0 \le y \le 2 - 2x$

50. $\displaystyle\iint_R \frac{1}{x}\, dy\, dx;\quad 1 \le x \le 2, 0 \le y \le x - 1$

51. $\displaystyle\iint_R e^{x/y^2}\, dx\, dy;\quad 1 \le y \le 2, 0 \le x \le y^2$

52. $\displaystyle\iint_R (x^2 - y)\, dy\, dx;\quad -1 \le x \le 1, -x^2 \le y \le x^2$

53. $\displaystyle\iint_R x^3 y\, dy\, dx;\quad R$ bounded by $y = x^2, y = 2x$

54. $\displaystyle\iint_R x^2 y^2\, dx\, dy;\quad R$ bounded by $y = x, y = 2x, x = 1$

55. $\displaystyle\iint_R \frac{1}{y}\, dy\, dx;\quad R$ bounded by $y = x, y = \dfrac{1}{x}, x = 2$

56. $\displaystyle\iint_R e^{2y/x}\, dy\, dx;\quad R$ bounded by $y = x^2, y = 0, x = 2$

Evaluate each double integral. If the function seems too difficult to integrate, try interchanging the limits of integration, as in Exercises 37 and 38.

57. $\displaystyle\int_0^{\ln 2} \int_{e^y}^2 \frac{1}{\ln x}\, dx\, dy$

58. $\displaystyle\int_0^2 \int_{y/2}^1 e^{x^2}\, dx\, dy$

59. Recall from the Volume and Average Value section in the previous chapter that volume could be found with a single integral. In this section volume is found using a double integral. Explain when volume can be found with a single integral and when a double integral is needed.

60. Give an example of a region that cannot be expressed by either of the forms shown in Figure 34. (One example is the disk with a hole in the middle between the graphs of $x^2 + y^2 = 1$ and $x^2 + y^2 = 2$ in Figure 10.)

The idea of the average value of a function, discussed earlier for functions of the form $y = f(x)$, can be extended to functions of more than one independent variable. For a function $z = f(x, y)$, the average value of f over a region R is defined as

$$\frac{1}{A} \iint_R f(x, y)\, dx\, dy,$$

where A is the area of the region R. Find the average value for each function over the regions R having the given boundaries.

61. $f(x, y) = 6xy + 2x;\quad 2 \le x \le 5, 1 \le y \le 3$

62. $f(x, y) = x^2 + y^2;\quad 0 \le x \le 2, 0 \le y \le 3$

63. $f(x, y) = e^{-5y + 3x};\quad 0 \le x \le 2, 0 \le y \le 2$

64. $f(x, y) = e^{2x + y};\quad 1 \le x \le 2, 2 \le y \le 3$

APPLICATIONS

Business and Economics

65. Packaging The manufacturer of a fruit juice drink has decided to try innovative packaging in order to revitalize sagging sales. The fruit juice drink is to be packaged in containers in the shape of tetrahedra in which three edges are perpendicular, as shown in the figure on the next page. Two of the perpendicular edges will be 3 in. long, and the third edge will be 6 in. long. Find the volume of the container. (*Hint:* The equation of the plane shown in the figure is $z = f(x, y) = 6 - 2x - 2y$.)

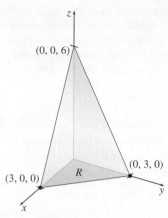

(0, 0, 6)

(3, 0, 0)

(0, 3, 0)

R

z

x

y

66. **Average Cost** A company's total cost for operating its two warehouses is

$$C(x, y) = \frac{1}{9}x^2 + 2x + y^2 + 5y + 100$$

dollars, where x represents the number of units stored at the first warehouse and y represents the number of units stored at the second. Find the average cost to store a unit if the first warehouse has between 40 and 80 units, and the second has between 30 and 70 units. (*Hint:* Refer to Exercises 61–64.)

67. **Average Production** A production function is given by

$$P(x, y) = 500x^{0.2}y^{0.8},$$

where x is the number of units of labor and y is the number of units of capital. Find the average production level if x varies from 10 to 50 and y from 20 to 40. (*Hint:* Refer to Exercises 61–64.)

68. **Average Profit** The profit (in dollars) from selling x units of one product and y units of a second product is

$$P = -(x - 100)^2 - (y - 50)^2 + 2000.$$

The weekly sales for the first product vary from 100 units to 150 units, and the weekly sales for the second product vary from 40 units to 80 units. Estimate average weekly profit for these two products. (*Hint:* Refer to Exercises 61–64.)

69. **Average Revenue** A company sells two products. The demand functions of the products are given by

$$q_1 = 300 - 2p_1 \quad \text{and} \quad q_2 = 500 - 1.2p_2,$$

where q_1 units of the first product are demanded at price p_1 and q_2 units of the second product are demanded at price p_2. The total revenue will be given by

$$R = q_1p_1 + q_2p_2.$$

Find the average revenue if the price p_1 varies from \$25 to \$50 and the price p_2 varies from \$50 to \$75. (*Hint:* Refer to Exercises 61–64.)

70. **Time** In an exercise earlier in this chapter, we saw that the time (in hours) that a branch of Amalgamated Entities needs to spend to meet the quota set by the main office can be approximated by

$$T(x, y) = x^4 + 16y^4 - 32xy + 40,$$

where x represents how many thousands of dollars the factory spends on quality control and y represents how many thousands of dollars they spend on consulting. Find the average time if the amount spent on quality control varies from \$0 to \$4000 and the amount spent on consulting varies from \$0 to \$2000. (*Hint:* Refer to Exercises 61–64.)

71. **Profit** In an exercise earlier in this chapter, we saw that the profit (in thousands of dollars) that Aunt Mildred's Metalworks earns from producing x tons of steel and y tons of aluminum can be approximated by

$$P(x, y) = 36xy - x^3 - 8y^3.$$

Find the average profit if the amount of steel produced varies from 0 to 8 tons, and the amount of aluminum produced varies from 0 to 4 tons. (*Hint:* Refer to Exercises 61–64.)

YOUR TURN ANSWERS

1. $52x^2 + 16x + 16x^3 + 490$ 2. $3644/3$
3. $(56\sqrt{14} - 184)/3$ 4. $7/2$ 5. $5888/105$

9 CHAPTER REVIEW

SUMMARY

In this chapter, we extended our study of calculus to include functions of several variables. We saw that it is possible to produce three-dimensional graphs of functions of two variables and that the process is greatly enhanced using level curves. Level curves are formed by determining the values of x and y that produce a particular functional value. We also saw the graphs of several surfaces, including the

- paraboloid, whose equation is $z = x^2 + y^2$,
- ellipsoid, whose general equation is $\dfrac{x^2}{a^2} + \dfrac{y^2}{b^2} + \dfrac{z^2}{c^2} = 1$,
- hyperbolic paraboloid, whose equation is $z = x^2 - y^2$, and
- hyperboloid of two sheets, whose equation is $-x^2 - y^2 + z^2 = 1$.

Level curves are also important in economics and are used to indicate combinations of the values of x and y that produce the same value of production z. This procedure was used to analyze the Cobb-Douglas production function, which has the general form

$$z = P(x, y) = Ax^ay^{1-a}, \text{ where } A \text{ is constant and } 0 < a < 1.$$

Partial derivatives are the extension of the concept of differentiation with respect to one of the variables while the other variables are held constant. Partial derivatives were used to identify extrema of a function of several variables. In particular, we identified all points where the partial with respect to x and the partial with respect to y are both zero, which we called critical points. We then classified each critical point as a relative maximum, a relative minimum, or a saddle point. Recall that a saddle point is a minimum when approached from one direction but a maximum when approached from another direction. We introduced the method of Lagrange multipliers to determine extrema in problems with constraints. Differentials, introduced earlier for functions of one variable, were generalized to define the total differential. We saw that total differentials can be used to approximate the value of a function using its tangent plane. We concluded the chapter by introducing double integrals, which are simply two iterated integrals, one for each variable. Double integrals were then used to find volume.

Function of Two Variables   The expression $z = f(x, y)$ is a function of two variables if a unique value of z is obtained from each ordered pair of real numbers (x, y). The variables x and y are independent variables, and z is the dependent variable. The set of all ordered pairs of real numbers (x, y) such that $f(x, y)$ exists is the domain of f; the set of all values of $f(x, y)$ is the range.

Plane   The graph of $ax + by + cz = d$ is a plane if a, b, and c are not all 0.

Partial Derivatives (Informal Definition)   The partial derivative of f with respect to x is the derivative of f obtained by treating x as a variable and y as a constant.

The partial derivative of f with respect to y is the derivative of f obtained by treating y as a variable and x as a constant.

Partial Derivatives (Formal Definition)   Let $z = f(x, y)$ be a function of two independent variables. Let all indicated limits exist. Then the partial derivative of f with respect to x is

$$f_x(x, y) = \frac{\partial f}{\partial x} = \lim_{h \to 0} \frac{f(x + h, y) - f(x, y)}{h},$$

and the partial derivative of f with respect to y is

$$f_y(x, y) = \frac{\partial f}{\partial y} = \lim_{h \to 0} \frac{f(x, y + h) - f(x, y)}{h}.$$

If the indicated limits do not exist, then the partial derivatives do not exist.

Second-Order Partial Derivatives   For a function $z = f(x, y)$, if the partial derivative exists, then

$$\frac{\partial}{\partial x}\left(\frac{\partial z}{\partial x}\right) = \frac{\partial^2 z}{\partial x^2} = f_{xx}(x, y) = z_{xx} \qquad \frac{\partial}{\partial y}\left(\frac{\partial z}{\partial y}\right) = \frac{\partial^2 z}{\partial y^2} = f_{yy}(x, y) = z_{yy}$$

$$\frac{\partial}{\partial y}\left(\frac{\partial z}{\partial x}\right) = \frac{\partial^2 z}{\partial y \partial x} = f_{xy}(x, y) = z_{xy} \qquad \frac{\partial}{\partial x}\left(\frac{\partial z}{\partial y}\right) = \frac{\partial^2 z}{\partial x \partial y} = f_{yx}(x, y) = z_{yx}$$

Relative Extrema   Let (a, b) be the center of a circular region contained in the xy-plane. Then, for a function $z = f(x, y)$ defined for every (x, y) in the region, $f(a, b)$ is a relative maximum if

$$f(a, b) \geq f(x, y)$$

for all points (x, y) in the circular region, and $f(a, b)$ is a relative minimum if

$$f(a, b) \leq f(x, y)$$

for all points (x, y) in the circular region.

Location of Extrema   Let a function $z = f(x, y)$ have a relative maximum or relative minimum at the point (a, b). Let $f_x(a, b)$ and $f_y(a, b)$ both exist. Then

$$f_x(a, b) = 0 \text{ and } f_y(a, b) = 0.$$

Test for Relative Extrema   For a function $z = f(x, y)$, let f_{xx}, f_{yy}, and f_{xy} all exist in a circular region contained in the xy-plane with center (a, b). Further, let

$$f_x(a, b) = 0 \text{ and } f_y(a, b) = 0.$$

Define D, known as the discriminant, by

$$D = f_{xx}(a, b) \cdot f_{yy}(a, b) - [f_{xy}(a, b)]^2.$$

Then

a. $f(a, b)$ is a relative maximum if $D > 0$ and $f_{xx}(a, b) < 0$;
b. $f(a, b)$ is a relative minimum if $D > 0$ and $f_{xx}(a, b) > 0$;
c. $f(a, b)$ is a saddle point (neither a maximum nor a minimum) if $D < 0$;
d. if $D = 0$, the test gives no information.

Lagrange Multipliers   All relative extrema of the function $z = f(x, y)$, subject to the constraint $g(x, y) = 0$, will be found among those points (x, y) for which there exists a value of λ such that

$$F_x(x, y, \lambda) = 0, F_y(x, y, \lambda) = 0, \text{ and } F_\lambda(x, y, \lambda) = 0,$$

where

$$F(x, y, \lambda) = f(x, y) - \lambda \cdot g(x, y),$$

and all indicated partial derivatives exist.

Using Lagrange Multipliers   **1.** Write the constraint in the form $g(x, y) = 0$.
2. Form the Lagrange function

$$F(x, y, \lambda) = f(x, y) - \lambda \cdot g(x, y).$$

3. Find $F_x(x, y, \lambda)$, $F_y(x, y, \lambda)$, and $F_\lambda(x, y, \lambda)$.
4. Form the system of equations

$$F_x(x, y, \lambda) = 0, F_y(x, y, \lambda) = 0, \text{ and } F_\lambda(x, y, \lambda) = 0.$$

5. Solve the system in Step 4; the relative extrema for f are among the solutions of the system.

Total Differential for Two Variables   Let $z = f(x, y)$ be a function of x and y. Let dx and dy be real numbers. Then the total differential of z is

$$dz = f_x(x, y) \cdot dx + f_y(x, y) \cdot dy.$$

(Sometimes dz is written df.)

Approximations   For small values of dx and dy,

$$dz \approx \Delta z$$

where $\Delta z = f(x + dx, y + dy) - f(x, y)$.

Approximations by Differentials   For a function f having all indicated partial derivatives, and for small values of dx and dy,

$$f(x + dx, y + dy) \approx f(x, y) + dz,$$

or

$$f(x + dx, y + dy) \approx f(x, y) + f_x(x, y) \cdot dx + f_y(x, y) \cdot dy.$$

Total Differential for Three Variables   If $w = f(x, y, z)$, then the total differential dw is

$$dw = f_x(x, y, z) \cdot dx + f_y(x, y, z) \cdot dy + f_z(x, y, z) \cdot dz,$$

provided all indicated partial derivatives exist.

Double Integral   The double integral of $f(x, y)$ over a rectangular region R defined by $a \le x \le b, c \le y \le d$ is written

$$\iint\limits_R f(x, y)\, dy\, dx \quad \text{or} \quad \iint\limits_R f(x, y)\, dx\, dy,$$

and equals either

$$\int_a^b \int_c^d f(x, y)\, dy\, dx \quad \text{or} \quad \int_c^d \int_a^b f(x, y)\, dx\, dy.$$

Volume   Let $z = f(x, y)$ be a function that is never negative on the rectangular region R defined by $a \le x \le b, c \le y \le d$. The volume of the solid under the graph of f and over the region R is

$$\iint\limits_R f(x, y)\, dy\, dx.$$

Double Integrals over Variable Regions   Let $z = f(x, y)$ be a function of two variables. If R is the region defined by $a \le x \le b$ and $g(x) \le y \le h(x)$, then

$$\iint\limits_R f(x, y)\, dy\, dx = \int_a^b \left[\int_{g(x)}^{h(x)} f(x, y)\, dy \right] dx.$$

If R is the region defined by $g(y) \le x \le h(y)$ and $c \le y \le d$, then

$$\iint\limits_R f(x, y)\, dx\, dy = \int_c^d \left[\int_{g(y)}^{h(y)} f(x, y)\, dx \right] dy.$$

KEY TERMS

9.1
function of two variables
independent variable
dependent variable
domain
range
ordered triple
first octant
plane
surface
trace
level curves

paraboloid
production function
Cobb-Douglas production
    function
level surface
ellipsoid
hyperbolic paraboloid
hyperboloid of two sheets

9.2
partial derivative
second-order partial derivative

9.3
relative maximum
relative minimum
saddle point
critical point
discriminant

9.4
constraints
Lagrange multiplier
utility function
indifference curve

9.5
total differential
9.6
Fubini's Theorem
iterated integral
double integral
integrand
region of integration

REVIEW EXERCISES

CONCEPT CHECK

Determine whether each of the following statements is true or false, and explain why.

1. The graph of $6x - 2y + 7z = 14$ is a plane.

2. The graph of $2x + 4y = 10$ is a plane that is parallel to the z-axis.

3. A level curve for a paraboloid could be a single point.

4. If the partial derivatives with respect to x and y at some point are both 0, the tangent plane to the function at that point is horizontal.

5. If $f(x, y) = 3x^2 + 2xy + y^2$, then $f(x + h, y) = 3(x + h)^2 + 2xy + h + y^2$.

6. For a function $z = f(x, y)$, suppose that the point (a, b) has been identified such that $f_x(a, b) = f_y(a, b) = 0$. We can conclude that a relative maximum or a relative minimum must exist at (a, b).

7. A saddle point can be a relative maximum or a relative minimum.

8. A function of two variables may have both a relative maximum and an absolute maximum at the same point.

9. The method of Lagrange multipliers tells us whether a point identified by the method is a maximum or minimum.

10. $\int_2^4 \int_1^5 (3x + 4y)\, dy\, dx = \int_2^4 \int_1^5 (3x + 4y)\, dx\, dy$

11. $\int_0^1 \int_{-2}^2 xe^y\, dy\, dx = \int_{-2}^2 \int_0^1 xe^y\, dx\, dy$

12. $\int_0^4 \int_1^x (x + xy^2)\, dy\, dx = \int_1^x \int_0^4 (x + xy^2)\, dx\, dy$

PRACTICE AND EXPLORATIONS

13. Describe in words how to take a partial derivative.

14. Describe what a partial derivative means geometrically.

15. Describe what a total differential is and how it is useful.

16. Suppose you are walking through the region of New York state shown in the topographical map in Figure 11 in the first section of this chapter. Assume you are heading north, toward the top of the map, over the western side of the mountain at the left, but not directly over the peak. Explain why you reach your highest point when you are going in the same direction as a contour line. Explain how this relates to Lagrange multipliers. (*Hint:* See Figure 25.)

Find $f(-1, 2)$ and $f(6, -3)$ for the following.

17. $f(x, y) = -4x^2 + 6xy - 3$

18. $f(x, y) = 2x^2y^2 - 7x + 4y$

19. $f(x, y) = \dfrac{x - 2y}{x + 5y}$

20. $f(x, y) = \dfrac{\sqrt{x^2 + y^2}}{x - y}$

Graph the first-octant portion of each plane.

21. $x + y + z = 4$

22. $x + 2y + 6z = 6$

23. $5x + 2y = 10$

24. $4x + 3z = 12$

25. $x = 3$

26. $y = 4$

27. Let $z = f(x, y) = 3x^3 + 4x^2y - 2y^2$. Find the following.

  a. $\dfrac{\partial z}{\partial x}$ **b.** $\dfrac{\partial z}{\partial y}(-1, 4)$ **c.** $f_{xy}(2, -1)$

28. Let $z = f(x, y) = \dfrac{x + y^2}{x - y^2}$. Find the following.

  a. $\dfrac{\partial z}{\partial y}$ **b.** $\dfrac{\partial z}{\partial x}(0, 2)$ **c.** $f_{xx}(-1, 0)$

Find $f_x(x, y)$ and $f_y(x, y)$.

29. $f(x, y) = 6x^2y^3 - 4y$

30. $f(x, y) = 5x^4y^3 - 6x^5y$

31. $f(x, y) = \sqrt{4x^2 + y^2}$

32. $f(x, y) = \dfrac{2x + 5y^2}{3x^2 + y^2}$

33. $f(x, y) = x^3e^{3y}$

34. $f(x, y) = (y - 2)^2 e^{x+2y}$

35. $f(x, y) = \ln|2x^2 + y^2|$

36. $f(x, y) = \ln|2 - x^2y^3|$

Find $f_{xx}(x, y)$ and $f_{xy}(x, y)$.

37. $f(x, y) = 5x^3y - 6xy^2$

38. $f(x, y) = -3x^2y^3 + x^3y$

39. $f(x, y) = \dfrac{3x}{2x - y}$

40. $f(x, y) = \dfrac{3x + y}{x - 1}$

41. $f(x, y) = 4x^2e^{2y}$

42. $f(x, y) = ye^{x^2}$

43. $f(x, y) = \ln|2 - x^2y|$

44. $f(x, y) = \ln|1 + 3xy^2|$

Find all points where the functions defined below have any relative extrema. Find any saddle points.

45. $z = 2x^2 - 3y^2 + 12y$

46. $z = x^2 + y^2 + 9x - 8y + 1$

47. $f(x, y) = x^2 + 3xy - 7x + 5y^2 - 16y$

48. $z = x^3 - 8y^2 + 6xy + 4$

49. $z = \dfrac{1}{2}x^2 + \dfrac{1}{2}y^2 + 2xy - 5x - 7y + 10$

50. $f(x, y) = 2x^2 + 4xy + 4y^2 - 3x + 5y - 15$

51. $z = x^3 + y^2 + 2xy - 4x - 3y - 2$

52. $f(x, y) = 7x^2 + y^2 - 3x + 6y - 5xy$

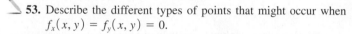 **53.** Describe the different types of points that might occur when $f_x(x, y) = f_y(x, y) = 0$.

Use Lagrange multipliers to find the extrema of the functions defined in Exercises 54 and 55.

54. $f(x, y) = x^2y; \quad x + y = 4$

55. $f(x, y) = x^2 + y^2; \quad x = y - 6$

56. Find positive numbers x and y, whose sum is 80, such that x^2y is maximized.

57. Find positive numbers x and y, whose sum is 75, such that xy^2 is maximized.

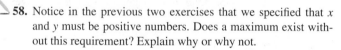 **58.** Notice in the previous two exercises that we specified that x and y must be positive numbers. Does a maximum exist without this requirement? Explain why or why not.

Evaluate dz using the given information.

59. $z = 6x^2 - 7y^2 + 4xy; \quad x = 3, y = -1, dx = 0.03, dy = 0.01$

60. $z = \dfrac{x + 5y}{x - 2y}; \quad x = 1, y = -2, dx = -0.04, dy = 0.02$

Use the total differential to approximate each quantity. Then use a calculator to approximate the quantity, and give the absolute value of the difference in the two results to 4 decimal places.

61. $\sqrt{5.1^2 + 12.05^2}$

62. $\sqrt{4.06}\, e^{0.04}$

Evaluate the following.

63. $\displaystyle\int_1^4 \dfrac{4y - 3}{\sqrt{x}}\, dx$

64. $\displaystyle\int_1^5 e^{3x + 5y}\, dx$

65. $\displaystyle\int_0^5 \dfrac{6x}{\sqrt{4x^2 + 2y^2}}\, dx$

66. $\displaystyle\int_1^3 \dfrac{y^2}{\sqrt{7x + 11y^3}}\, dy$

Evaluate each iterated integral.

67. $\displaystyle\int_0^2 \int_0^4 (x^2y^2 + 5x)\, dx\, dy$

68. $\displaystyle\int_0^3 \int_0^5 (2x + 6y + y^2)\, dy\, dx$

69. $\displaystyle\int_3^4 \int_2^5 \sqrt{6x + 3y}\, dx\, dy$

70. $\displaystyle\int_1^2 \int_3^5 e^{2x - 7y}\, dx\, dy$

71. $\displaystyle\int_2^4 \int_2^4 \dfrac{1}{y}\, dx\, dy$

72. $\displaystyle\int_1^2 \int_1^2 \dfrac{1}{x}\, dx\, dy$

Find each double integral over the region R with boundaries as indicated.

73. $\displaystyle\iint_R (x^2 + 2y^2)\, dx\, dy; \quad 0 \le x \le 5, 0 \le y \le 2$

74. $\displaystyle\iint_R \sqrt{2x + y}\, dx\, dy; \quad 1 \le x \le 3, 2 \le y \le 5$

75. $\displaystyle\iint_R \sqrt{y + x}\, dx\, dy; \quad 0 \le x \le 7, 1 \le y \le 9$

76. $\displaystyle\iint_R ye^{y^2 + x}\, dx\, dy; \quad 0 \le x \le 1, 0 \le y \le 1$

Find the volume under the given surface $z = f(x, y)$ and above the given rectangle.

77. $z = x + 8y + 4; \quad 0 \le x \le 3, 1 \le y \le 2$

78. $z = x^2 + y^2; \quad 3 \le x \le 5, 2 \le y \le 4$

Evaluate each double integral. If the function seems too difficult to integrate, try interchanging the limits of integration.

79. $\displaystyle\int_0^1 \int_0^{2x} xy\, dy\, dx$

80. $\displaystyle\int_1^2 \int_2^{2x^2} y\, dy\, dx$

81. $\displaystyle\int_0^1 \int_{x^2}^x x^3y\, dy\, dx$

82. $\displaystyle\int_0^1 \int_y^{\sqrt{y}} x\, dx\, dy$

83. $\displaystyle\int_0^2 \int_{x/2}^1 \dfrac{1}{y^2 + 1}\, dy\, dx$

84. $\displaystyle\int_0^8 \int_{x/2}^4 \sqrt{y^2 + 4}\, dy\, dx$

Use the region R, with boundaries as indicated, to evaluate the given double integral.

85. $\displaystyle\iint_R (2x + 3y)\, dx\, dy; \quad 0 \le y \le 1, y \le x \le 2 - y$

86. $\displaystyle\iint_R (2 - x^2 - y^2)\, dy\, dx; \quad 0 \le x \le 1, x^2 \le y \le x$

APPLICATIONS

Business and Economics

87. Charge for Auto Painting The charge (in dollars) for painting a sports car is given by

$$C(x, y) = 4x^2 + 5y^2 - 4xy + \sqrt{x},$$

where x is the number of hours of labor needed and y is the number of gallons of paint and sealant used. Find the following.

a. The charge for 10 hours and 5 gal of paint and sealant

b. The charge for 15 hours and 10 gal of paint and sealant

c. The charge for 20 hours and 20 gal of paint and sealant

88. Manufacturing Costs The manufacturing cost (in dollars) for a certain computer is given by

$$c(x, y) = 2x + y^2 + 4xy + 25,$$

where x is the memory capacity of the computer in gigabytes (GB) and y is the number of hours of labor required. For 640 GB and 6 hours of labor, find the following.

a. The approximate change in cost for an additional 1 GB of memory

b. The approximate change in cost for an additional hour of labor

89. Productivity The production function z for one country is

$$z = x^{0.7}y^{0.3},$$

where x represents the amount of labor and y the amount of capital. Find the marginal productivity of the following.

a. Labor   **b.** Capital

90. Cost The cost (in dollars) to manufacture x solar cells and y solar collectors is

$$c(x, y) = x^2 + 5y^2 + 4xy - 70x - 164y + 1800.$$

a. Find values of x and y that produce minimum total cost.

b. Find the minimum total cost.

Utility **Maximize each of the following utility functions, with the cost of each commodity and total amount available to spend given.**

91. $f(x,y) = xy^3$, cost of a unit of x is $2, cost of a unit of y is $4, and $80 is available.

92. $f(x,y) = x^5y^2$, cost of a unit of x is $10, cost of a unit of y is $6, and $42 is available.

93. Cost The cost (in dollars) to produce x satellite receiving dishes and y transmitters is given by

$$C(x, y) = 100 \ln(x^2 + y) + e^{xy/20}.$$

Production schedules now call for 15 receiving dishes and 9 transmitters. Use differentials to approximate the change in costs if 1 more dish and 1 fewer transmitter are made.

94. Production Materials Approximate the volume of material needed to manufacture a cone of radius 2 cm, height 8 cm, and wall thickness 0.21 cm.

95. Production Materials A sphere of radius 2 ft is to receive an insulating coating 1 in. thick. Approximate the volume of the coating needed.

96. Production Error The height of a sample cone from a production line is measured as 11.4 cm, while the radius is measured as 2.9 cm. Each of these measurements could be off by 0.2 cm. Approximate the maximum possible error in the volume of the cone.

97. Profit The total profit from 1 acre of a certain crop depends on the amount spent on fertilizer, x, and on hybrid seed, y, according to the model

$$P(x, y) = 0.01(-x^2 + 3xy + 160x - 5y^2 + 200y + 2600).$$

The budget for fertilizer and seed is limited to $280.

a. Use the budget constraint to express one variable in terms of the other. Then substitute into the profit function to get a function with one independent variable. Use the method shown in Chapter 6 on Applications of the Derivative to find the amounts spent on fertilizer and seed that will maximize profit. What is the maximum profit per acre? (*Hint:* Throughout this exercise you may ignore the coefficient of 0.01 until you need to find the maximum profit.)

b. Find the amounts spent on fertilizer and seed that will maximize profit using the first method shown in this chapter. (*Hint:* You will not need to use the budget constraint.)

c. Use the Lagrange multiplier method to solve the original problem.

d. Look for the relationships among these methods.

Life Sciences

98. Blood Vessel Volume A length of blood vessel is measured as 2.7 cm, with the radius measured as 0.7 cm. If each of these measurements could be off by 0.1 cm, estimate the maximum possible error in the volume of the vessel.

99. Total Body Water Accurate prediction of total body water is critical in determining adequate dialysis doses for patients with renal disease. For African American males, total body water can be estimated by the function

$$T(A, M, S) = -18.37 - 0.09A + 0.34M + 0.25S,$$

where T is the total body water (in liters), A is age (in years), M is mass (in kilograms), and S is height (in centimeters). *Source: Kidney International.*

a. Find $T(65, 85, 180)$.

b. Find and interpret $T_A(A, M, S)$, $T_M(A, M, S)$, and $T_S(A, M, S)$.

100. Brown Trout Researchers from New Zealand have determined that the length of a brown trout depends on both its mass and age and that the length can be estimated by

$$L(m, t) = (0.00082t + 0.0955)e^{(\ln m + 10.49)/2.842},$$

where $L(m, t)$ is the length of the trout (in centimeters), m is the mass of the trout (in grams), and t is the age of the trout (in years). *Source: Transactions of the American Fisheries Society.*

a. Find $L(450, 4)$.

b. Find $L_m(450, 7)$ and $L_t(450, 7)$ and interpret.

101. Survival Curves The following figure shows survival curves (percent surviving as a function of age) for people in the United States in 1900 and 2000. *Source: National Vital Statistics Report.* Let $f(x, y)$ give the proportion surviving at age x in year y. Use the graph to estimate the following. Interpret each answer in words.

a. $f(60, 1900)$ **b.** $f(70, 2000)$

c. $f_x(60, 1900)$ **d.** $f_x(70, 2000)$

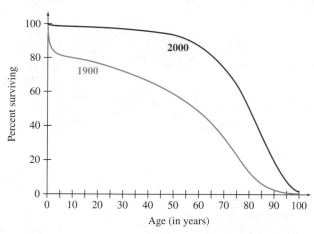

General Interest

102. Area The bottom of a planter is to be made in the shape of an isosceles triangle, with the two equal sides 3 ft long and the third side 2 ft long. The area of an isosceles triangle with two equal sides of length a and third side of length b is

$$f(a, b) = \frac{1}{4} b \sqrt{4a^2 - b^2}.$$

a. Find the area of the bottom of the planter.

b. The manufacturer is considering changing the shape so that the third side is 2.5 ft long. What would be the approximate effect on the area?

103. Surface Area A closed box with square ends must have a volume of 125 in^3. Use Lagrange multipliers to find the dimensions of such a box that has minimum surface area.

104. Area Use Lagrange multipliers to find the maximum rectangular area that can be enclosed with 400 ft of fencing, if no fencing is needed along one side.

EXTENDED APPLICATION

USING MULTIVARIABLE FITTING TO CREATE A RESPONSE SURFACE DESIGN

Suppose you are designing a flavored drink with orange and banana flavors. You want to find the ideal concentrations of orange and banana flavoring agents, but since the concentrations could range from 0% to 100%, you can't try every possibility. A common design technique in the food industry is to make up several test drinks using different combinations of flavorings and have them rated for taste appeal by a panel of tasters. Such ratings are called *hedonic responses* and are often recorded on a 10-point scale from 0 (worst) to 9 (best). One combination will most likely get the highest average score, but since you have only tried a few of the infinite number of flavor combinations, the winning combination on the taste test might be far from the mix that would be the most popular in the market. How can you use the information from your test to locate the best point on the *flavor plane*?

One approach to this problem uses *response surfaces*, three-dimensional surfaces that approximate the data points from your flavor test.* For your test, you might choose mixtures that are spread out over the flavor plane. For example, you could combine low, medium, and high orange with low, medium, and high banana to get 9 different flavors. If you had 15 tasters and used intensities of 20, 50, and 80 for each fruit, the test data might look like the table.

Average Hedonic Scores ($n = 15$)			
Banana Intensity (0 to 100)			
		20	50	80
Orange	20	3.2	4.9	2.8
Intensity	50	6.0	7.2	5.1
0 to 100	80	4.5	5.5	4.8

For example, the table shows that the drink with orange intensity 20 and banana intensity 80 got an average flavor rating of 2.8 from the test panel (they didn't like it).

Your test results are points in space, where you can think of the x-axis as the orange axis, the y-axis as banana, and the z-axis as taste score. A three-dimensional bar chart is a common way of displaying data of this kind. Figure 38 on the next page is a bar chart of the flavor test results.

Looking at the bar chart, we can guess that the best flavor mix will be somewhere near the middle. We'd like to "drape" a smooth surface over the bars and see where that surface has a maximum. But as with any sample, our tasters are not perfectly representative

*For a brief introduction to response surfaces, see Devore, Jay L. and Nicholas R. Farnum, *Applied Statistics for Engineers and Scientists*, Duxbury, 2004.

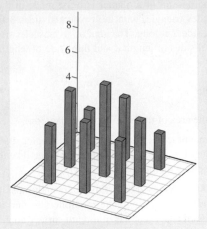

FIGURE 38

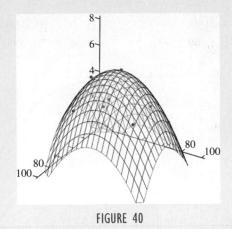

FIGURE 40

of the whole population: Our test results give the general shape of the true population response, but each bar includes an error that results from our small sample size. The solution is to fit a *smooth* surface to the data points.

A simple type of function for modeling such data sets is a quadratic function. You've seen many quadratic functions of two variables in the examples and exercises for this chapter, and you know that they can have maxima, minima, and saddle points. We don't know in advance which quadratic shape will give us the best fit, so we'll use the most general quadratic,

$$G(x, y) = Ax^2 + By^2 + Cxy + Dx + Ey + F.$$

Our job is to find the six coefficients, A through F, that give the best fit to our nine data points. As with the least squares line formula you used in Section 1.3, there are formulas for these six coefficients. Most statistical software packages will generate them directly from your data set, and here is the best-fitting quadratic found by one such program:

$$G(x, y) = -0.00202x^2 - 0.00163y^2 + 0.000194xy$$
$$+ 0.21380x + 0.14768y - 2.36204.$$

In this case the response surface shows how the dependent variable, taste rating, *responds* to the two independent variables, orange and banana intensity. Figures 39 and 40 are two views of

the surface together with the data: a surface superimposed on the bar chart, and the same surface with the data shown as points in space.

In research papers, response surface models are often reported using level curves. A contour map for the surface we have found looks like Figure 41, with orange increasing from left to right and banana from bottom to top.

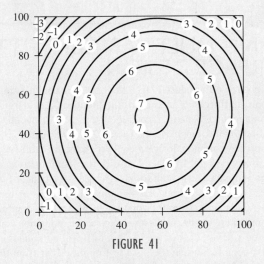

FIGURE 41

It's quite easy to estimate the location of the maximum by marking a point in the middle of the central ellipse and finding its coordinates on the two axes (try it!). You can also use the techniques you learned in the section on Maxima and Minima, computing partial derivatives of G with respect to x and y and solving the resulting linear system. The numbers are awkward, but with some help from a calculator you'll find that the maximum occurs at approximately $(55.3, 48.6)$. So the quadratic model predicts that the most popular drink will have an orange concentration of 55.3 and a banana concentration of 48.6. The model also predicts the public's flavor rating for this drink: We would expect it to be $G(55.3, 48.6)$, which turns

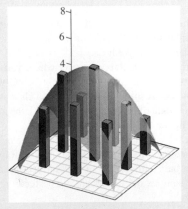

FIGURE 39

out to be about 7.14. When food technologists design a new food, this kind of modeling is often a first step. The next step might be to make a new set of test drinks with concentrations clustered around the point $(55.3, 48.6)$, and use further tests to explore this region of the "flavor plane" in greater detail.

Response surfaces are also helpful for constrained optimization. In the section on Lagrange Multipliers, you saw how Lagrange multipliers could solve problems of the form:

> Find the relative extrema for $z = f(x, y)$,
> subject to the constraint $g(x, y) = 0$.

Sometimes the constraints have a different form: You may have *several* dependent variables that respond to the same inputs, and the design goal is to keep each variable *within a given range*. Here's an example based on the data in U.S. Patent No. 4,276,316, which is titled *Process for Treating Nuts*. **Source: United States Patent and Trademark Office.** The patent granted to researcher Shri C. Sharma and assigned to CPC International, Inc. covers a method for preparing nuts for blanching (that is, having their skins removed). The patent summary reads in part:

> *The nuts are heated with a gas at a temperature of 125° to 175°C for 30 to 180 seconds and then immediately cooled to below 35°C within 5 minutes prior to blanching. This provides improved blanching, sorting and other steps in a process for producing products ranging from nuts per se to peanut butters or spreads.*

In support of the effectiveness of the method, the patent offers data that describe the effects of nine different combinations of air temperature and treatment time on three variables of interest for blanched peanuts: blanching efficiency, roasted peanut flavor, and overall flavor. Efficiency is given in percent, and the two hedonic variables were rated by tasters on a scale of 0 to 9.

The time variable has been converted into a natural logarithm because treatment time effects typically scale with the log of the time. The problem is now to pick a temperature and time range that give the optimum combination of efficiency, roasted flavor, and overall flavor. Each of these three dependent variables responds to the inputs in a different way, and the patent documentation includes quadratic response surfaces for each variable. The lighter shading in Figures 42–44 indicates higher values, which are more

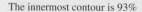

The innermost contour is 93%

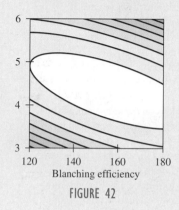

Blanching efficiency

FIGURE 42

The innermost contours are 5.2

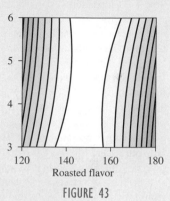

Roasted flavor

FIGURE 43

The "pointed" contours are 5.5

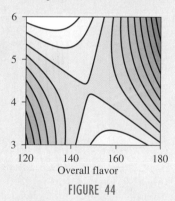

Overall flavor

FIGURE 44

Air Temperature, °C	Treatment Time, Seconds	Log of Treatment Time	Blanching Efficiency	Roasted Peanut Flavor	Overall Flavor
138	45	3.807	93.18	4.94	5.51
160	120	4.787	94.99	5.24	5.37
149	75	4.317	98.43	5.27	5.10
138	120	4.787	96.42	5.05	5.71
160	45	3.807	96.48	5.17	5.62
127	75	4.317	93.56	4.64	5.04
149	180	5.193	94.99	5.24	5.37
149	30	3.401	87.30	5.43	5.44
171	45	3.807	94.40	4.37	5.18

523

desirable. Temperature is plotted across the bottom, and the log of treatment time increases from bottom to top.

Sometimes process designers faced with this kind of problem will combine the dependent variables into a single function by taking a weighted average of their values, and then use a single response surface to optimize this function. Here we look at a different scenario. Suppose we set the following process goals: We want blanching efficiency of at least 93%, a roasted flavor rating of at least 5.2, and an overall flavor rating of at least 5.5. Is there a combination of time and temperature that meets these criteria? If so, what is it?

The first step is to identify the "successful" area on each response surface, which we can do by shading the corresponding region in the contour plot, shown in Figures 45–47.

Now the strategy is clear: We want to stack the three plots on top of each other and see if the shaded regions overlap. Figure 48 is the result.

So we can see that there are two regions on the temperature–time plane that will work. For example, the upper area of overlap suggests a processing temperature of 140°C to 150°C, with a processing time between 90 and 150 seconds (remember that the numbers on the vertical axis are *natural logarithms* of the time in seconds).

Response surfaces are a standard tool in designing everything from food to machine parts, and we have touched on only a small part of the theory here. Frequently a process depends on more than two independent variables. For example, a soft-drink formula

might include three flavorings, an acidifying agent, and a sweetener. The response "surface" now lives in six dimensions and we can no longer draw nice pictures, but the same multivariable mathematics that generated our quadratic response surfaces will lead us to the optimal combination of variables.

EXERCISES

1. The general quadratic function of two variables has six terms. How many terms are in the general cubic function of two variables?

2. Use the contour plot of orange-banana flavor to estimate the "flavor coordinates" of the best-tasting drink.

3. Find the maximum on the flavor response surface by finding the critical point of the function $G(x, y)$.

4. Without shading or numbers on the contours, how would you know that the point you found in Exercises 2 and 3 represents the best flavor rather than the worst flavor? (*Hint:* Compute the discriminant D as described in the section on Maxima and Minima.)

5. Our best drink has a predicted flavor rating of 7.14, but one of our test drinks got a *higher* rating, 7.2. What's going on?

6. Blanching efficiency has a maximum near the center of the temperature–time plane. What is going on near the center of the plane for the roasted flavor and overall flavor response surfaces? Within the domain plotted, where does overall flavor reach a maximum?

7. In the overall flavor contour plot, if we move one contour toward higher flavor from the "pointed" 5.5 contours, we find curved contours that represent an overall flavor rating of about 5.6. If instead of requiring an overall flavor rating of 5.5 we decided to require a rating of 5.6, what would happen to our process design?

8. Use the last figure to describe the other region in the temperature–time plane that delivers a successful process for preparing nuts for blanching.

9. At the website WolframAlpha.com, you can enter "maximize" followed by an expression $f(x, y)$, such as $4 - x^2 - y^2$, and you will be told the maximum value of the expression, as well as the point where the maximum occurs. You will also be shown a graph of the surface $z = f(x, y)$ and a contour plot. Try this for the function $G(x, y)$ given in this Extended Application, and compare with your answers to Exercises 2 and 3.

DIRECTIONS FOR GROUP PROJECT

Perform an experiment that is similar to the flavored drink example from the text on some other product. For example, you could perform an experiment where you develop hedonic responses for various levels of salt and butter on popcorn. Using technology, to the extent that it is available to you, carry out the analysis of your experiment to determine an optimal mixture of each ingredient.

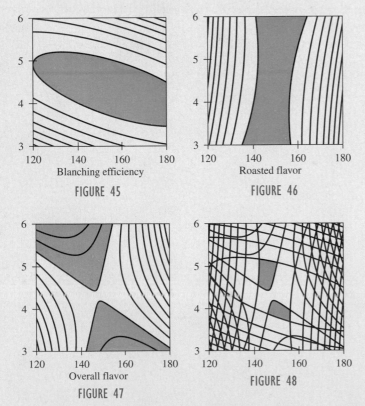

FIGURE 45
Blanching efficiency

FIGURE 46
Roasted flavor

FIGURE 47
Overall flavor

FIGURE 48

10

Differential Equations

When these sky divers open their parachutes, their speed will decrease until air resistance exactly balances the force of gravity. An exercise at the end of this chapter explores solutions to the differential equation that describes free fall with air resistance. The limiting speed with an open parachute is on the order of 10 miles per hour, slow enough for a safe landing.

S uppose that an economist wants to develop an equation that will forecast interest rates. By studying data on previous changes in interest rates, she hopes to find a relationship between the level of interest rates and their rate of change. A function giving the rate of change of interest rates would be the derivative of the function describing the level of interest rates. A **differential equation** is an equation that involves an unknown function $y = f(x)$ and a finite number of its derivatives. Solving the differential equation for y would give the unknown function to be used for forecasting interest rates.

Differential equations have been important in the study of physical science and engineering since the eighteenth century. More recently, differential equations have become useful in the social sciences, life sciences, and economics for solving problems about population growth, ecological balance, and interest rates. In this chapter, we will introduce some methods for solving differential equations and give examples of their applications.

10.1 Solutions of Elementary and Separable Differential Equations

APPLY IT How can we predict the future population of a flock of mountain goats? *Using differential equations, we will answer this question in Example 6.*

Usually a solution of an equation is a *number*. A solution of a differential equation, however, is a *function* that satisfies the equation.

> **EXAMPLE 1** Solving a Differential Equation

Find all solutions of the differential equation

$$\frac{dy}{dx} = 3x^2 - 2x. \tag{1}$$

SOLUTION To say that a function $y(x)$ is a solution of Equation (1) simply means that the derivative of the function y is $3x^2 - 2x$. This is the same as saying that y is an antiderivative of $3x^2 - 2x$, or

$$y = \int (3x^2 - 2x)\, dx = x^3 - x^2 + C. \tag{2}$$

YOUR TURN 1 Find all solutions of the differential equation $\frac{dy}{dx} = 12x^5 + \sqrt{x} + e^{5x}.$

We can verify that the function given by Equation (2) is a solution by taking its derivative. The result is the differential equation (1).   **TRY YOUR TURN 1**

─FOR REVIEW─
For review on finding antiderivatives, see the first two sections in Chapter 7 on Integration.

Each different value of C in Equation (2) leads to a different solution of Equation (1), showing that a differential equation can have an infinite number of solutions. Equation (2) is the **general solution** of the differential equation (1). Some of the solutions of Equation (1) are graphed in Figure 1.

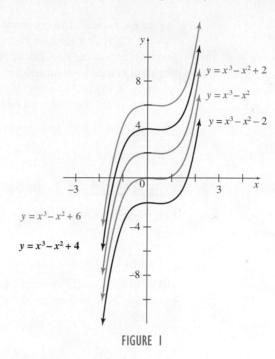

$y = x^3 - x^2 + 2$

$y = x^3 - x^2$

$y = x^3 - x^2 - 2$

$y = x^3 - x^2 + 6$

$y = x^3 - x^2 + 4$

FIGURE 1

The simplest kind of differential equation has the form

$$\frac{dy}{dx} = g(x).$$

Since Equation (1) has this form, the solution of Equation (1) suggests the following generalization.

General Solution of $\dfrac{dy}{dx} = g(x)$

The general solution of the differential equation $\dfrac{dy}{dx} = g(x)$ is

$$y = \int g(x)\, dx.$$

EXAMPLE 2   Population

The population P of a flock of birds is growing exponentially so that

$$\frac{dP}{dt} = 20e^{0.05t},$$

where t is time in years. Find P in terms of t if there were 20 birds in the flock initially.

SOLUTION   To solve the differential equation, first determine the antiderivative of $20e^{0.05t}$, that is,

$$P = \int 20e^{0.05t}\, dt = \frac{20}{0.05}e^{0.05t} + C = 400e^{0.05t} + C.$$

Initially, there were 20 birds, so $P = 20$ at time $t = 0$. We can substitute this information into the equation to determine the value of C that satisfies this condition.

$$20 = 400e^0 + C$$

$$-380 = C$$

Therefore, $P = 400e^{0.05t} - 380$. Verify that P is a solution to the differential equation by taking its derivative. The result should be the original differential equation.

NOTE
We can write the condition $P = 20$ when $t = 0$ as $P(0) = 20$. We will use this notation from now on.

In Example 2, the given information was used to produce a solution with a specific value of C. Such a solution is called a **particular solution** of the differential equation. The given information, $P = 20$ when $t = 0$, is called an **initial condition**. An **initial value problem** is a differential equation with a value of y given at $x = x_0$ (or $t = t_0$ in this case), where x_0 is any real number.

Sometimes a differential equation must be rewritten in the form

$$\frac{dy}{dx} = g(x)$$

before it can be solved.

EXAMPLE 3   Initial Value Problem

Find the particular solution of

$$\frac{dy}{dx} - 2x = \frac{2x}{\sqrt{x^2 + 3}},$$

given that $y(1) = 8$.

SOLUTION   This differential equation is not in the proper form, but we can easily fix this by adding $2x$ to both sides of the equation. That is,

$$\frac{dy}{dx} = 2x + \frac{2x}{\sqrt{x^2 + 3}}.$$

To find the general solution, integrate this expression, using the substitution $u = x^2 + 3$ in the second term.

$$y = \int \left(2x + \frac{2x}{\sqrt{x^2 + 3}} \right) dx$$

$$= x^2 + \int \frac{1}{\sqrt{u}} \, du \qquad\qquad du = 2x \, dx$$

$$= x^2 + 2\sqrt{u} + C \qquad\qquad \text{Use the power rule with } n = -1/2.$$

$$= x^2 + 2\sqrt{x^2 + 3} + C$$

Now use the initial condition to find the value of C. Substituting 8 for y and 1 for x gives

$$8 = 1^2 + 2\sqrt{1^2 + 3} + C$$
$$8 = 5 + C$$
$$C = 3.$$

YOUR TURN 2   Find the particular solution of $\frac{dy}{dx} - 12x^3 = 6x^2$, given that $y(2) = 60$.

The particular solution is $y = x^2 + 2\sqrt{x^2 + 3} + 3$. Verify that $y(1) = 8$ and that differentiating y leads to the original differential equation.   **TRY YOUR TURN 2**

So far in this section, we have used a method that is essentially the same as that used in the section on antiderivatives, when we first started the topic of integration. But not all differential equations can be solved so easily. For example, if interest on an investment is compounded continuously, then the investment grows at a rate proportional to the amount of money present. If A is the amount in an account at time t, then for some constant k, the differential equation

$$\frac{dA}{dt} = kA \qquad\qquad\qquad (3)$$

gives the rate of growth of A with respect to t. This differential equation is different from those discussed previously, which had the form

$$\frac{dy}{dx} = g(x).$$

CAUTION   Since the right-hand side of Equation (3) is a function of A, rather than a function of t, it would be completely invalid to simply integrate both sides as we did before. The previous method only works when the side opposite the derivative is simply a function of the independent variable.

Equation (3) is an example of a more general differential equation we will now learn to solve; namely, those that can be written in the form

$$\frac{dy}{dx} = \frac{p(x)}{q(y)}.$$

Suppose we think of dy/dx as a fraction dy over dx. This is incorrect, of course; the derivative is actually the limit of a small change in y over a small change in x, but the notation is chosen so that this interpretation gives a correct answer, as we shall see. Multiply on both sides by $q(y)\,dx$ to get

$$q(y)\,dy = p(x)\,dx.$$

In this form all terms involving y (including dy) are on one side of the equation and all terms involving x (and dx) are on the other side. A differential equation that can be put into this form is said to be *separable*, since the variables x and y can be separated. After separation, a **separable differential equation** may be solved by integrating each side with respect to the variable given. This method is known as **separation of variables**.

$$\int q(y)\,dy = \int p(x)\,dx$$

$$Q(y) = P(x) + C,$$

where P and Q are antiderivatives of p and q. (We don't need a constant of integration on the left side of the equation; it can be combined with the constant of integration on the right side as C.) To show that this answer is correct, differentiate implicitly with respect to x.

$$Q'(y)\frac{dy}{dx} = P'(x) \qquad \text{Use the chain rule on the left side.}$$

$$q(y)\frac{dy}{dx} = p(x) \qquad \begin{array}{l} q(y) \text{ is the derivative of } Q(y) \text{ and} \\ p(x) \text{ is the derivative of } P(x). \end{array}$$

$$\frac{dy}{dx} = \frac{p(x)}{q(y)}$$

This last equation is the one we set out to solve.

EXAMPLE 4   **Separation of Variables**

Find the general solution of

$$y\frac{dy}{dx} = x^2.$$

SOLUTION   Begin by separating the variables to get

$$y\,dy = x^2\,dx.$$

The general solution is found by determining the antiderivatives of each side.

$$\int y\,dy = \int x^2\,dx$$

$$\frac{y^2}{2} = \frac{x^3}{3} + C$$

$$y^2 = \frac{2}{3}x^3 + 2C$$

$$y^2 = \frac{2}{3}x^3 + K$$

YOUR TURN 3 Find the general solution of $\dfrac{dy}{dx} = \dfrac{x^2 + 1}{xy^2}$.

The constant K was substituted for $2C$ in the last step. The solution is left in implicit form, not solved explicitly for y. In general, we will use C for the arbitrary constant, so that our final answer would be written as $y^2 = (2/3)x^3 + C$. It would also be nice to solve for y by taking the square root of both sides, but since we don't know anything about the sign of y, we don't know whether the solution is $y = \sqrt{(2/3)x^3 + C}$ or $y = -\sqrt{(2/3)x^3 + C}$. If y were raised to the third power, we could solve for y by taking the cube root of both sides, since the cube root is unique. **TRY YOUR TURN 3**

EXAMPLE 5   Separation of Variables

Find the general solution of the differential equation for interest compounded continuously,

$$\frac{dA}{dt} = kA.$$

SOLUTION   Separating variables leads to

$$\frac{1}{A} dA = k\,dt.$$

To solve this equation, determine the antiderivative of each side.

$$\int \frac{1}{A}\,dA = \int k\,dt$$
$$\ln|A| = kt + C$$

(Here A represents a nonnegative quantity, so the absolute value is unnecessary, but we wish to show how to solve equations for which this may not be true.) Use the definition of logarithm to write the equation in exponential form as

$$|A| = e^{kt+C} = e^{kt}e^{C}. \qquad \text{Use the property } e^{m+n} = e^m e^n.$$

Finally, use the definition of absolute value to get

$$A = e^{kt}e^{C} \quad \text{or} \quad A = -e^{kt}e^{C}.$$

Because e^C and $-e^C$ are constants, replace them with the constant M, which may be any nonzero real number. (We use M rather than K because we already have a constant k in this example. We could also relabel M as C, as we did in Example 4.) The resulting equation,

$$A = Me^{kt},$$

not only describes interest compounded continuously but also defines the exponential growth or decay function that was discussed in Chapter 2 on Nonlinear Functions.

─FOR REVIEW─
In Chapter 2 on Nonlinear Functions, we saw that the amount of money in an account with interest compounded continuously is given by

$$A = Pe^{rt},$$

where P is the initial amount in the account, r is the annual interest rate, and t is the time in years. Observe that this is the same as the equation for the amount of money in an account derived here, where P and r have been replaced with M and k, respectively. For more applications of exponential growth and decay, see Chapter 2 on Nonlinear Functions.

CAUTION   Notice that $y = 0$ is also a solution to the differential equation in Example 5, but after we divide by A (which is not possible if $A = 0$) and integrate, the resulting equation $|A| = e^{kt+C}$ does not allow y to equal 0. In this example, the lost solution can be recovered in the final answer if we allow M to equal 0, a value that was previously excluded. When dividing by an expression in separation of variables, look for solutions that would make this expression 0 and may be lost.

Recall that equations of the form $A = Me^{kt}$ arise in situations where the rate of change of a quantity is proportional to the amount present at time t, which is precisely what the differential equation (3) describes. The constant k is called the **growth constant**, while M represents the amount present at time $t = 0$. A positive value of k indicates growth, while a negative value of k indicates decay. The equation was often written in the form $y = y_0 e^{kt}$ in Chapter 2 on Nonlinear Functions, where we discussed other applications of this equation, such as radioactive decay.

As a model of population growth, the equation $y = Me^{kt}$ is not realistic over the long run for most populations. As shown by graphs of functions of the form $y = Me^{kt}$, with both M and k positive, growth would be unbounded. Additional factors, such as space restrictions or a limited amount of food, tend to inhibit growth of populations as time goes on. In an alternative model that assumes a maximum population of size N, the rate of growth of a population is proportional to how close the population is to that maximum, that is, to the difference between N and y. These assumptions lead to the differential equation

$$\frac{dy}{dt} = k(N - y),$$

whose solution is the limited growth function mentioned in the last section of Chapter 2 on Nonlinear Functions.

EXAMPLE 6 Population

A certain nature reserve can support no more than 4000 mountain goats. Assume that the rate of growth is proportional to how close the population is to this maximum, with a growth rate of 20 percent. There are currently 1000 goats in the area.

(a) Solve the general limited growth differential equation, and then write a function describing the goat population at time t.

SOLUTION   To solve the equation

$$\frac{dy}{dt} = k(N - y),$$

first separate the variables.

$$\frac{dy}{N - y} = k \, dt$$

$$\int \frac{dy}{N - y} = \int k \, dt \qquad \text{Integrate both sides.}$$

$$-\ln(N - y) = kt + C \qquad \text{We assume the population is less than } N, \text{ so } N - y > 0.$$

$$\ln(N - y) = -kt - C$$

$$N - y = e^{-kt-C} \qquad \text{Apply the function } e^x \text{ to both sides.}$$

$$N - y = e^{-kt}e^{-C} \qquad \text{Use the property } e^{m+n} = e^m e^n.$$

$$y = N - e^{-kt}e^{-C} \qquad \text{Solve for } y.$$

$$y = N - Me^{-kt} \qquad \text{Relabel the constant } e^{-C} \text{ as } M.$$

Now apply the initial condition that $y(0) = y_0$.

$$y_0 = N - Me^{-k \cdot 0} = N - M \qquad e^0 = 1$$

$$M = N - y_0 \qquad \text{Solve for } M.$$

Substituting this value of M into the previous solution gives

$$y = N - (N - y_0)e^{-kt}.$$

The graph of this function is shown in Figure 2.

For the goat problem, the maximum population is $N = 4000$, the initial population is $y_0 = 1000$, and the growth rate constant is 20%, or $k = 0.2$. Therefore,

$$y = 4000 - (4000 - 1000)e^{-0.2t} = 4000 - 3000e^{-0.2t}.$$

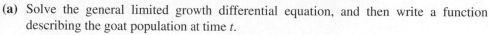

FIGURE 2

(b) What will the goat population be in 5 years?

SOLUTION In 5 years, the population will be

$$y = 4000 - 3000e^{-(0.2)(5)} = 4000 - 3000e^{-1}$$

$$\approx 4000 - 1103.6 = 2896.4,$$

or about 2900 goats.

YOUR TURN 4 In Example 6, find the goat population in 5 years if the reserve can support 6000 goats, the growth rate is 15%, and there are currently 1200 goats in the area.

TRY YOUR TURN 4

Logistic Growth

Let y be the size of a certain population at time t. In the standard model for unlimited growth given by Equation (3), the rate of growth is proportional to the current population size. The constant k, the growth rate constant, is the difference between the birth and death rates of the population. The unlimited growth model predicts that the population's growth rate is a constant, k.

Growth usually is not unlimited, however, and the population's growth rate is usually not constant because the population is limited by environmental factors to a maximum size N, called the **carrying capacity** of the environment for the species. In the limited growth model already given,

$$\frac{dy}{dt} = k(N - y),$$

the rate of growth is proportional to the remaining room for growth, $N - y$.

In the **logistic growth model**

$$\frac{dy}{dt} = k\left(1 - \frac{y}{N}\right)y \tag{4}$$

the rate of growth is proportional to both the current population size y and a factor $(1 - y/N)$ that is equal to the remaining room for growth, $N - y$, divided by N. Equation (4) is called the **logistic equation**. Notice that $(1 - y/N) \to 1$ as $y \to 0$, and the differential equation can be approximated as

$$\frac{dy}{dt} = k\left(1 - \frac{y}{N}\right)y \approx k(1)y = ky.$$

In other words, when y is small, the growth of the population behaves as if it were unlimited. On the other hand, $(1 - y/N) \to 0$ as $y \to N$, so

$$\frac{dy}{dt} = k\left(1 - \frac{y}{N}\right)y \approx k(0)y = 0.$$

That is, population growth levels off as y nears the maximum population N. Thus, the logistic Equation (4) is the unlimited growth Equation (3) with a damping factor $(1 - y/N)$ to account for limiting environmental factors when y nears N. Let y_0 denote the initial population size. Under the assumption $0 < y < N$, the general solution of Equation (4) is

$$y = \frac{N}{1 + be^{-kt}}, \tag{5}$$

where $b = (N - y_0)/y_0$ (see Exercise 31). This solution, called a **logistic curve**, is shown in Figure 3. This function was introduced in Section 4.4 on Derivatives of Exponential Functions in the form

$$G(t) = \frac{m}{1 + \left(\frac{m}{G_0}\right)e^{-kmt}},$$

where m is the limiting value of the population, G_0 is the initial number present, and k is a positive constant.

As expected, the logistic curve begins exponentially and subsequently levels off. Another important feature is the point of inflection $((\ln b)/k, N/2)$, where dy/dx is a maximum (see Exercise 33). Notice that the point of inflection is when the population is half of the carrying capacity and that at this point, the population is increasing most rapidly.

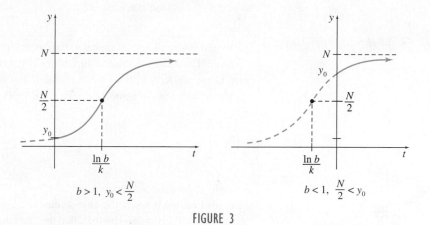

$$b > 1, \quad y_0 < \frac{N}{2} \qquad\qquad b < 1, \quad \frac{N}{2} < y_0$$

FIGURE 3

Logistic equations arise frequently in the study of populations. In about 1840 the Belgian sociologist P. F. Verhulst fitted a logistic curve to U.S. census figures and made predictions about the population that were subsequently proved to be quite accurate. American biologist Raymond Pearl (circa 1920) found that the growth of a population of fruit flies in a limited space could be modeled by the logistic equation

$$\frac{dy}{dt} = 0.2y - \frac{0.2}{1035}y^2.$$

TECHNOLOGY NOTE Some calculators can fit a logistic curve to a set of data points. For example, the TI-84 Plus has this capability, listed as `Logistic` in the STAT CALC menu, along with other types of regression. See Exercises 40, 47, 48, and 52.

Logistic growth is an example of how a model is modified over time as new insights occur. The model for population growth changed from the early exponential curve $y = Me^{kt}$ to the logistic curve

$$y = \frac{N}{1 + be^{-kt}}.$$

Many other quantities besides population grow logistically. That is, their initial rate of growth is slow, but as time progresses, their rate of growth increases to a maximum value and subsequently begins to decline and to approach zero.

EXAMPLE 7 Logistic Curve

Rapid technological advancements in the last 20 years have made many products obsolete practically overnight. J. C. Fisher and R. H. Pry successfully described the phenomenon of a technically superior new product replacing another product by the logistic equation

$$\frac{dz}{dt} = k(1 - z)z, \tag{6}$$

where z is the market share of the new product and $1 - z$ is the market share of the other product. *Source: Technological Forecasting and Social Change*. The new product will initially have little or no market share; that is, $z_0 \approx 0$. Thus, the constant b in Equation (5) will have to be determined in a different way. Let t_0 be the time at which $z = 1/2$. Under the assumption $0 < z < 1$, the general solution of Equation (6) is

$$z = \frac{1}{1 + be^{-kt}}, \tag{7}$$

where $b = e^{kt_0}$ (see Exercise 32).

Fisher and Pry applied their model to the fraction of fabric consumed in the United States that was synthetic. Their data is shown in the table below. At the time of the study, natural fabrics in clothing were being replaced with synthetic fabric. Using the logistic regression function on a TI-84 Plus calculator, with t as the number of years since 1930, the best logistic function to fit this data can be shown to be

$$z = \frac{1.293}{1 + 21.80e^{-0.06751t}}.$$

A graph of the data and this function is shown in Figure 4(a). Although the data fits the function well, this function is not of the form studied by Fisher and Pry because it has a numerator of 1.293, rather than 1. This function predicts that the percentage of fabric that is synthetic approaches 129%! A more appropriate function can be found by rewriting Equation (7) as

$$be^{-kt} = \frac{1}{z} - 1,$$

and then finding the best fit exponential function to the points $(t, 1/z - 1)$. This leads to the equation

$$z = \frac{1}{1 + 18.93(0.92703)^t} = \frac{1}{1 + 18.93e^{(\ln 0.92703)t}} = \frac{1}{1 + 18.93e^{-0.07577t}}.$$

As Figure 4(b) shows, this more realistic model also fits the data well. The dangers of extrapolating beyond the data are illustrated by this equation's prediction that 95.8% of fabrics in the United States would be synthetic by 2010. In fact, cotton is still more popular than synthetic fabrics. *Source: Fabrics Manufacturers*.

Synthetic Fabric as Percent of U.S. Consumption									
Year	1930	1935	1940	1945	1950	1955	1960	1965	1967
Fraction synthetic	0.044	0.079	0.10	0.14	0.22	0.28	0.29	0.43	0.47

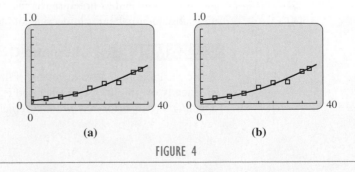

(a)   (b)

FIGURE 4

10.1 EXERCISES

Find the general solution for each differential equation. Verify that each solution satisfies the original differential equation.

1. $\dfrac{dy}{dx} = -4x + 6x^2$

2. $\dfrac{dy}{dx} = 4e^{-3x}$

3. $4x^3 - 2\dfrac{dy}{dx} = 0$

4. $3x^2 - 3\dfrac{dy}{dx} = 2$

5. $y\dfrac{dy}{dx} = x^2$

6. $y\dfrac{dy}{dx} = x^2 - x$

7. $\dfrac{dy}{dx} = 2xy$

8. $\dfrac{dy}{dx} = x^2 y$

9. $\dfrac{dy}{dx} = 3x^2 y - 2xy$

10. $(y^2 - y)\dfrac{dy}{dx} = x$

11. $\dfrac{dy}{dx} = \dfrac{y}{x}, \; x > 0$

12. $\dfrac{dy}{dx} = \dfrac{y}{x^2}$

13. $\dfrac{dy}{dx} = \dfrac{y^2 + 6}{2y}$

14. $\dfrac{dy}{dx} = \dfrac{e^{y^2}}{y}$

15. $\dfrac{dy}{dx} = y^2 e^{2x}$

16. $\dfrac{dy}{dx} = \dfrac{e^x}{e^y}$

Find the particular solution for each initial value problem.

17. $\dfrac{dy}{dx} + 3x^2 = 2x; \quad y(0) = 5$

18. $\dfrac{dy}{dx} = 4x^3 - 3x^2 + x; \quad y(1) = 0$

19. $2\dfrac{dy}{dx} = 4xe^{-x}; \quad y(0) = 42$

20. $x\dfrac{dy}{dx} = x^2 e^{3x}; \quad y(0) = \dfrac{8}{9}$

21. $\dfrac{dy}{dx} = \dfrac{x^3}{y}; \quad y(0) = 5$

22. $x^2\dfrac{dy}{dx} - y\sqrt{x} = 0; \quad y(1) = e^{-2}$

23. $(2x + 3)y = \dfrac{dy}{dx}; \quad y(0) = 1$

24. $\dfrac{dy}{dx} = \dfrac{x^2 + 5}{2y - 1}; \quad y(0) = 11$

25. $\dfrac{dy}{dx} = \dfrac{2x + 1}{y - 3}; \quad y(0) = 4$

26. $x^2\dfrac{dy}{dx} = y; \quad y(1) = -1$

27. $\dfrac{dy}{dx} = \dfrac{y^2}{x}; \quad y(e) = 3$

28. $\dfrac{dy}{dx} = x^{1/2}y^2; \quad y(4) = 9$

29. $\dfrac{dy}{dx} = (y - 1)^2 e^{x-1}; \quad y(1) = 2$

30. $\dfrac{dy}{dx} = (x + 2)^2 e^y; \quad y(1) = 0$

31. a. Solve the logistic Equation (4) in this section by observing that

$$\frac{1}{y} + \frac{1}{N - y} = \frac{N}{(N - y)y}.$$

b. Assume $0 < y < N$. Verify that $b = (N - y_0)/y_0$ in Equation (5), where y_0 is the initial population size.

c. Assume $0 < N < y$ for all y. Verify that $b = (y_0 - N)/y_0$.

32. Suppose that $0 < z < 1$ for all z. Solve the logistic Equation (6) as in Exercise 31. Verify that $b = e^{kx_0}$, where x_0 is the time at which $z = 1/2$.

33. Suppose that $0 < y_0 < N$. Let $b = (N - y_0)/y_0$, and let $y(x) = N/(1 + be^{-kx})$ for all x. Show the following.

a. $0 < y(x) < N$ for all x.

b. The lines $y = 0$ and $y = N$ are horizontal asymptotes of the graph.

c. $y(x)$ is an increasing function.

d. $((\ln b)/k, N/2)$ is a point of inflection of the graph.

e. dy/dx is a maximum at $x_0 = (\ln b)/k$.

34. Suppose that $0 < N < y_0$. Let $b = (y_0 - N)/y_0$ and let

$$y(x) = \frac{N}{1 - be^{-kx}} \quad \text{for all } x \neq \frac{\ln b}{k}.$$

See the figure. Show the following.

a. $0 < b < 1$

b. The lines $y = 0$ and $y = N$ are horizontal asymptotes of the graph.

c. The line $x = (\ln b)/k$ is a vertical asymptote of the graph.

d. $y(x)$ is decreasing on $((\ln b)/k, \infty)$ and on $(-\infty, (\ln b)/k)$.

e. $y(x)$ is concave upward on $((\ln b)/k, \infty)$ and concave downward on $(-\infty, (\ln b)/k)$.

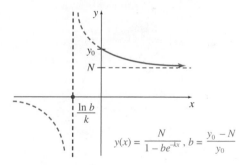

$$y(x) = \frac{N}{1 - be^{-kx}}, \; b = \frac{y_0 - N}{y_0}$$

APPLICATIONS

Business and Economics

35. Profit The marginal profit of a certain company is given by

$$\frac{dy}{dx} = \frac{100}{32 - 4x},$$

where x represents the amount of money (in thousands of dollars) that the company spends on advertising. Find the profit for each advertising expenditure if the profit is $1000 when nothing is spent on advertising.

a. $3000

b. $5000

c. Can advertising expenditures ever reach $8000 according to this model? Explain why or why not.

36. Sales Decline Sales (in thousands) of a certain product are declining at a rate proportional to the amount of sales, with a decay constant of 15% per year.

a. Write a differential equation to express the rate of sales decline.

b. Find a general solution to the equation in part a.

c. How much time will pass before sales become 25% of their original value?

37. Inflation If inflation grows continuously at a rate of 5% per year, how long will it take for $1 to lose half its value?

Elasticity of Demand Elasticity of demand was discussed in Chapter 6 on Applications of the Derivative, where it was defined as

$$E = -\frac{p}{q} \cdot \frac{dq}{dp},$$

for demand q and price p. Find the general demand equation $q = f(p)$ for each elasticity function. (*Hint*: Set each elasticity function equal to $-\frac{p}{q} \cdot \frac{dq}{dp}$, then solve for q. Write the constant of integration as $\ln C$ in Exercise 39.)

38. $E = \dfrac{4p^2}{q^2}$

39. $E = 2$

40. Internet Usage During the early days of the Internet, growth in the number of users worldwide could be approximated by an exponential function. The following table gives the number of worldwide users of the Internet. *Source: Internet World Stats.*

Year	Number of Users (in millions)
1995	16
1996	36
1997	70
1998	147
1999	248
2000	361
2001	513
2002	587
2003	719
2004	817
2005	1018
2006	1093
2007	1319
2008	1574
2009	1802

Use a calculator with exponential and logistic regression capabilities to complete the following.

a. Letting t represent the years since 1990, plot the number of worldwide users of the Internet on the y-axis against the year on the t-axis. Discuss the shape of the graph.

b. Use the exponential regression function on your calculator to determine the exponential equation that best fits the data. Plot the exponential equation on the same graph as the data points. Discuss the appropriateness of fitting an exponential function to these data.

c. Use the logistic regression function on your calculator to determine the logistic equation that best fits the data. Plot the logistic equation on the same graph. Discuss the appropriateness of fitting a logistic function to these data. Which graph better fits the data?

d. Assuming that the logistic function found in part c continues to be accurate, what seems to be the limiting size of the number of worldwide Internet users?

41. Life Insurance A life insurance company invests $5000 in a bank account in order to fund a death benefit of $20,000. Growth in the investment over time can be modeled by the differential equation

$$\frac{dA}{dt} = Ai$$

where i is the interest rate and $A(t)$ is the amount invested at time t (in years). Calculate the interest rate that the investment must earn in order for the company to fund the death benefit in 24 years. Choose one of the following. *Source: Society of Actuaries.*

a. $\dfrac{-\ln 2}{12}$ b. $\dfrac{-\ln 2}{24}$ c. $\dfrac{\ln 2}{24}$ d. $\dfrac{\ln 2}{12}$ e. $\dfrac{\ln 2}{6}$

Life Sciences

42. Tracer Dye The amount of a tracer dye injected into the bloodstream decreases exponentially, with a decay constant of 3% per minute. If 6 cc are present initially, how many cubic centimeters are present after 10 minutes? (Here k will be negative.)

43. Soil Moisture The evapotranspiration index I is a measure of soil moisture. An article on 10- to 14-year-old heath vegetation described the rate of change of I with respect to W, the amount of water available, by the equation

$$\frac{dI}{dW} = 0.088(2.4 - I).$$

Source: Australian Journal of Botany.

a. According to the article, I has a value of 1 when $W = 0$. Solve the initial value problem.

b. What happens to I as W becomes larger and larger?

44. Fish Population An isolated fish population is limited to 4000 by the amount of food available. If there are now 320 fish and the population is growing with a growth constant of 2% a year, find the expected population at the end of 10 years.

Dieting A person's weight depends both on the daily rate of energy intake, say C calories per day, and on the daily rate of energy consumption, typically between 15 and 20 calories per pound per day. Using an average value of 17.5 calories per pound per day, a person weighing w pounds uses $17.5w$ calories per day. If $C = 17.5w$, then weight remains constant, and weight gain or loss occurs according to whether C is greater or less than $17.5w$. *Source: The College Mathematics Journal.*

45. To determine how fast a change in weight will occur, the most plausible assumption is that dw/dt is proportional to the net excess (or deficit) $C - 17.5w$ in the number of calories per day.

   a. Assume C is constant and write a differential equation to express this relationship. Use k to represent the constant of proportionality. What does C being constant imply?

   b. The units of dw/dt are pounds per day, and the units of $C - 17.5w$ are calories per day. What units must k have?

   c. Use the fact that 3500 calories is equivalent to 1 lb to rewrite the differential equation in part a.

   d. Solve the differential equation.

   e. Let w_0 represent the initial weight and use it to express the coefficient of $e^{-0.005t}$ in terms of w_0 and C.

46. (Refer to Exercise 45.) Suppose someone initially weighing 180 lb adopts a diet of 2500 calories per day.

   a. Write the weight function for this individual.

   b. Graph the weight function on the window $[0, 300]$ by $[120, 200]$. What is the asymptote? This value of w is the equilibrium weight w_{eq}. According to the model, can a person ever achieve this weight?

   c. How long will it take a dieter to reach a weight just 2 lb more than w_{eq}?

47. H1N1 Virus The cumulative number of deaths worldwide due to the H1N1 virus, or swine flu, at various days into the epidemic are listed below, where April 21, 2009 was day 1. *Source: BBC.*

Day	Deaths	Day	Deaths
14	27	148	3696
28	74	163	4334
43	114	182	4804
57	164	206	6704
71	315	221	8450
84	580	234	9797
99	1049	266	14,024
117	2074	274	14,378
132	2967		

Use a calculator with logistic regression capability to complete the following.

   a. Plot the number of deaths y against the number of days t. Discuss the appropriateness of fitting a logistic function to this data.

   b. Use the logistic regression function on your calculator to determine the logistic equation that best fits the data.

   c. Plot the logistic regression function from part b on the same graph as the data points. Discuss how well the logistic equation fits the data.

   d. Assuming the logistic equation found in part b continues to be accurate, what seems to be the limiting size of

the number of deaths due to this outbreak of the H1N1 virus?

   e. Discuss whether a logistic model is more appropriate than an exponential model for estimating the number of deaths due to the H1N1 virus.

48. Population Growth The following table gives the historic and projected populations (in millions) of China and India. *Source: United Nations.*

Year	China	India
1950	545	372
1960	646	448
1970	816	553
1980	981	693
1990	1142	862
2000	1267	1043
2010	1354	1214
2020	1431	1367
2030	1462	1485
2040	1455	1565
2050	1417	1614

Use a calculator with logistic regression capability to complete the following.

   a. Letting t represent the years since 1950, plot the Chinese population on the y-axis against the year on the t-axis. Discuss the appropriateness of fitting a logistic function to these data.

   b. Use the logistic regression function on your calculator to determine the logistic equation that best fits the data. Plot the logistic function on the same graph as the data points. Discuss how well the logistic function fits the data.

   c. Assuming the logistic equation found in part b continues to be accurate, what seems to be the limiting size of the Chinese population?

   d. Repeat parts a–c using the population for India.

49. U.S. Hispanic Population A recent report by the U.S. Census Bureau predicts that the U.S. Hispanic population will increase from 35.6 million in 2000 to 102.6 million in 2050. *Source: U.S. Census Bureau.* Assuming the unlimited growth model $dy/dt = ky$ fits this population growth, express the population y as a function of the year t. Let 2000 correspond to $t = 0$.

50. U.S. Asian Population (Refer to Exercise 49.) The report also predicted that the U.S. Asian population would increase from 10.7 million in 2000 to 33.4 million in 2050. *Source: U.S. Census Bureau.* Repeat Exercise 49 using this data.

51. Spread of a Rumor Suppose the rate at which a rumor spreads—that is, the number of people who have heard the rumor over a period of time—increases with the number of people who have heard it. If y is the number of people who have heard the rumor, then

$$\frac{dy}{dt} = ky,$$

where t is the time in days.

a. If y is 1 when $t = 0$, and y is 5 when $t = 2$, find k. Using the value of k from part a, find y for each time.

b. $t = 3$ **c.** $t = 5$ **d.** $t = 10$

52. World Population The following table gives the population of the world at various times over the last two centuries, plus projections for this century. *Source: The New York Times.*

Year	Population (billions)
1804	1
1927	2
1960	3
1974	4
1987	5
1999	6
2011	7
2025	8
2041	9
2071	10

Use a calculator with logistic regression capability to complete the following.

a. Use the logistic regression function on your calculator to determine the logistic equation that best fits the data.

b. Plot the logistic function found in part a and the original data in the same window. Does the logistic function seem to fit the data from 1927 on? Before 1927?

c. To get a better fit, subtract 0.99 from each value of the population in the table. (This makes the population in 1804 small, but not 0 or negative.) Find a logistic function that fits the new data.

d. Plot the logistic function found in part c and the modified data in the same window. Does the logistic function now seem to be a better fit than in part b?

e. Based on the results from parts c and d, predict the limiting value of the world's population as time increases. For comparison, the *New York Times* article predicts a value of 10.73 billion. (*Hint*: After taking the limit, remember to add the 0.99 that was removed earlier.)

f. Based on the results from parts c and d, predict the limiting value of the world population as you go further and further back in time. Does that seem reasonable? Explain.

53. Worker Productivity A company has found that the rate at which a person new to the assembly line produces items is

$$\frac{dy}{dx} = 7.5e^{-0.3y},$$

where x is the number of days the person has worked on the line. How many items can a new worker be expected to produce on the eighth day if he produces none when $x = 0$?

Physical Sciences

54. Radioactive Decay The amount of a radioactive substance decreases exponentially, with a decay constant of 3% per month.

a. Write a differential equation to express the rate of change.

b. Find a general solution to the differential equation from part a.

c. If there are 75 g at the start of the decay process, find a particular solution for the differential equation from part a.

d. Find the amount left after 10 months.

55. Snowplow One morning snow began to fall at a heavy and constant rate. A snowplow started out at 8:00 A.M. At 9:00 A.M. it had traveled 2 miles. By 10:00 A.M. it had traveled 3 miles. Assuming that the snowplow removes a constant volume of snow per hour, determine the time at which it started snowing. (*Hint:* Let t denote the time since the snow started to fall, and let T be the time when the snowplow started out. Let x, the distance the snowplow has traveled, and h, the height of the snow, be functions of t. The assumption that a constant volume of snow per hour is removed implies that the speed of the snowplow times the height of the snow is a constant. Set up and solve differential equations involving dx/dt and dh/dt.) *Source: The American Mathematical Monthly.*

Newton's Law of Cooling Newton's law of cooling states that the rate of change of temperature of an object is proportional to the difference in temperature between the object and the surrounding medium. Thus, if T is the temperature of the object after t hours and T_M is the (constant) temperature of the surrounding medium, then

$$\frac{dT}{dt} = -k(T - T_M),$$

where k is a constant. Use this equation in Exercises 56–59.

56. Show that the solution of this differential equation is

$$T = Ce^{-kt} + T_M,$$

where C is a constant.

57. According to the solution of the differential equation for Newton's law of cooling, what happens to the temperature of an object after it has been in a surrounding medium with constant temperature for a long period of time? How well does this agree with reality?

Newton's Law of Cooling When a dead body is discovered, one of the first steps in the ensuing investigation is for a medical examiner to determine the time of death as closely as possible. Have you ever wondered how this is done? If the temperature of the medium (air, water, or whatever) has been fairly constant and less than 48 hours have passed since the death, Newton's law of cooling can be used. The medical examiner does not actually solve the equation for each case. Instead, a table based on the formula is consulted. Use Newton's law of cooling to work the following exercises. *Source: The College Mathematics Journal.*

58. Assume the temperature of a body at death is 98.6°F, the temperature of the surrounding air is 68°F, and at the end of one hour the body temperature is 90°F.

a. What is the temperature of the body after 2 hours?

b. When will the temperature of the body be 75°F?

c. Approximately when will the temperature of the body be within 0.01° of the surrounding air?

59. Suppose the air temperature surrounding a body remains at a constant 10°F, $C = 88.6$, and $k = 0.24$.

a. Determine a formula for the temperature at any time t.

 b. Use a graphing calculator to graph the temperature T as a function of time t on the window $[0, 30]$ by $[0, 100]$.

c. When does the temperature of the body decrease more rapidly: just after death, or much later? How do you know?

d. What will the temperature of the body be after 4 hours?

e. How long will it take for the body temperature to reach 40°F? Use your calculator graph to verify your answer.

YOUR TURN ANSWERS

1. $y = 2x^6 + (2/3)x^{3/2} + e^{5x}/5 + C$ **2.** $y = 3x^4 + 2x^3 - 4$

3. $y = ((3/2)x^2 + 3\ln|x| + C)^{1/3}$ **4.** 3733

10.2 Linear First-Order Differential Equations

APPLY IT   What happens over time to the glucose level in a patient's bloodstream? *The solution to a linear differential equation gives us an answer in Example 4.*

Recall that $f^{(n)}(x)$ represents the nth derivative of $f(x)$, and that $f^{(n)}(x)$ is called an nth-order derivative. By this definition, the derivative $f'(x)$ is first-order, $f''(x)$ is second-order, and so on. The *order of a differential equation* is that of the highest-order derivative in the equation. In this section only first-order differential equations are discussed.

A **linear first-order differential equation** is an equation of the form

$$\frac{dy}{dx} + P(x)y = Q(x).$$

Notice that a linear differential equation has a dy/dx term, a y term, and a term that is just a function of x. It does not have terms involving nonlinear expressions such as y^2 or e^y, nor does it have terms involving the product or quotient of y and dy/dx. Many useful models produce such equations. In this section we develop a general method for solving first-order linear differential equations.

EXAMPLE 1   **Linear Differential Equation**

Solve the equation

$$x\frac{dy}{dx} + 6y + 2x^4 = 0. \tag{1}$$

SOLUTION   We need to first get the equation in the form of a linear first-order differential equation. Thus, dy/dx should have a coefficient of 1. To accomplish this, we divide both sides of the equation by x and rearrange the terms to get the linear differential equation

$$\frac{dy}{dx} + \frac{6}{x}y = -2x^3.$$

This equation is not separable and cannot be solved by the methods discussed so far. (Verify this.) Instead, multiply both sides of the equation by x^6 (the reason will be explained shortly) to get

$$x^6\frac{dy}{dx} + 6x^5y = -2x^9. \tag{2}$$

On the left, $6x^5$, the coefficient of y, is the derivative of x^6, the coefficient of dy/dx. Recall the product rule for derivatives:

$$D_x(uv) = u\frac{dv}{dx} + \frac{du}{dx}v.$$

If $u = x^6$ and $v = y$, the product rule gives

$$D_x(x^6 y) = x^6 \frac{dy}{dx} + 6x^5 y,$$

which is the left side of Equation (2). Substituting $D_x(x^6 y)$ for the left side of Equation (2) gives

$$D_x(x^6 y) = -2x^9.$$

Assuming $y = f(x)$, as usual, both sides of this equation can be integrated with respect to x and the result solved for y to get

$$x^6 y = \int -2x^9 \, dx = -2\left(\frac{x^{10}}{10}\right) + C = -\frac{x^{10}}{5} + C$$

$$y = -\frac{x^4}{5} + \frac{C}{x^6}. \tag{3}$$

Equation (3) is the general solution of Equation (2) and, therefore, of Equation (1). ▬

This procedure has given us a solution, but what motivated our choice of the multiplier x^6? To see where x^6 came from, let $I(x)$ represent the multiplier, and multiply both sides of the general equation

$$\frac{dy}{dx} + P(x)y = Q(x)$$

by $I(x)$:

$$I(x)\frac{dy}{dx} + I(x)P(x)y = I(x)Q(x). \tag{4}$$

The method illustrated above will work only if the left side of the equation is the derivative of the product function $I(x) \cdot y$, which is

$$I(x)\frac{dy}{dx} + I'(x)y. \tag{5}$$

Comparing the coefficients of y in Equations (4) and (5) shows that $I(x)$ must satisfy

$$I'(x) = I(x)P(x),$$

or

$$\frac{I'(x)}{I(x)} = P(x).$$

Integrating both sides of this last equation gives

$$\ln |I(x)| = \int P(x) \, dx + C$$

$$|I(x)| = e^{\int P(x) \, dx + C}$$

or

$$I(x) = \pm e^C e^{\int P(x) \, dx}.$$

Only one value of $I(x)$ is needed, so let $C = 0$, so that $e^C = 1$, and use the positive result, giving

$$I(x) = e^{\int P(x) \, dx}.$$

In summary, choosing $I(x)$ as $e^{\int P(x) \, dx}$ and multiplying both sides of a linear first-order differential equation by $I(x)$ puts the equation in a form that can be solved by integration.

Integrating Factor

The function $I(x) = e^{\int P(x)\,dx}$ is called an **integrating factor** for the differential equation

$$\frac{dy}{dx} + P(x)y = Q(x).$$

For Equation (1), written as the linear differential equation

$$\frac{dy}{dx} + \frac{6}{x}y = -2x^3,$$

$P(x) = 6/x$, and the integrating factor is

$$I(x) = e^{\int (6/x)\,dx} = e^{6\ln|x|} = e^{\ln|x|^6} = e^{\ln x^6} = x^6.$$

This last step used the fact that $e^{\ln a} = a$ for all positive a.

In summary, we solve a linear first-order differential equation with the following steps.

Solving a Linear First-Order Differential Equation

1. Put the equation in the linear form $\dfrac{dy}{dx} + P(x)y = Q(x)$.
2. Find the integrating factor $I(x) = e^{\int P(x)\,dx}$.
3. Multiply each term of the equation from Step 1 by $I(x)$.
4. Replace the sum of terms on the left with $D_x[I(x)y]$.
5. Integrate both sides of the equation.
6. Solve for y.

EXAMPLE 2   Linear Differential Equation

Give the general solution of $\dfrac{dy}{dx} + 2xy = x$.

SOLUTION

Step 1 This equation is already in the required form.

Step 2 The integrating factor is

$$I(x) = e^{\int 2x\,dx} = e^{x^2}.$$

Step 3 Multiplying each term by e^{x^2} gives

$$e^{x^2}\frac{dy}{dx} + 2xe^{x^2}y = xe^{x^2}.$$

Step 4 The sum of terms on the left can now be replaced with $D_x(e^{x^2}y)$, to get

$$D_x(e^{x^2}y) = xe^{x^2}.$$

Step 5 Integrating on both sides gives

$$e^{x^2}y = \int xe^{x^2}\,dx,$$

or

$$e^{x^2}y = \frac{1}{2}e^{x^2} + C.$$

YOUR TURN 1
Give the general solution of
$$x\frac{dy}{dx} - y - x^2 e^x = 0, x > 0.$$

Step 6 Divide both sides by e^{x^2} to get the general solution

$$y = \frac{1}{2} + Ce^{-x^2}.$$

TRY YOUR TURN 1

EXAMPLE 3 **Linear Differential Equation**

Solve the initial value problem $2\left(\frac{dy}{dx}\right) - 6y - e^x = 0$ with $y(0) = 5$.

SOLUTION Write the equation in the required form by adding e^x to both sides and dividing both sides by 2:

$$\frac{dy}{dx} - 3y = \frac{1}{2}e^x.$$

The integrating factor is

$$I(x) = e^{\int(-3)\,dx} = e^{-3x}.$$

Multiplying each term by $I(x)$ gives

$$e^{-3x}\frac{dy}{dx} - 3e^{-3x}y = \frac{1}{2}e^x e^{-3x},$$

or

$$e^{-3x}\frac{dy}{dx} - 3e^{-3x}y = \frac{1}{2}e^{-2x}.$$

The left side can now be replaced by $D_x(e^{-3x}y)$ to get

$$D_x(e^{-3x}y) = \frac{1}{2}e^{-2x}.$$

Integrating on both sides gives

$$e^{-3x}y = \int \frac{1}{2}e^{-2x}\,dx,$$

$$e^{-3x}y = \frac{1}{2}\left(\frac{e^{-2x}}{-2}\right) + C.$$

Now, multiply both sides by e^{3x} to get

$$y = -\frac{e^x}{4} + Ce^{3x},$$

the general solution. Find the particular solution by substituting 0 for x and 5 for y:

$$5 = -\frac{e^0}{4} + Ce^0 = -\frac{1}{4} + C$$

or

$$\frac{21}{4} = C,$$

which leads to the particular solution

$$y = -\frac{e^x}{4} + \frac{21}{4}e^{3x}.$$

TRY YOUR TURN 2

YOUR TURN 2
Solve the initial value problem
$$\frac{dy}{dx} + 2xy - xe^{-x^2} = 0 \text{ with } y(0) = 3.$$

EXAMPLE 4   Glucose

Suppose glucose is infused into a patient's bloodstream at a constant rate of a grams per minute. At the same time, glucose is removed from the bloodstream at a rate proportional to the amount of glucose present. Then the amount of glucose, $G(t)$, present at time t satisfies

$$\frac{dG}{dt} = a - KG$$

for some constant K. Solve this equation for G. Does the glucose concentration eventually reach a constant? That is, what happens to G as $t \to \infty$? *Source: Ordinary Differential Equations with Applications.*

APPLY IT

SOLUTION   The equation can be written in the form of the linear first-order differential equation

$$\frac{dG}{dt} + KG = a. \tag{6}$$

The integrating factor is

$$I(t) = e^{\int K dt} = e^{Kt}.$$

Multiply both sides of Equation (6) by $I(t) = e^{Kt}$.

$$e^{Kt}\frac{dG}{dt} + Ke^{Kt}G = ae^{Kt}$$

Write the left side as $D_t(e^{Kt}G)$ and solve for G by integrating on both sides.

$$D_t(e^{Kt}G) = ae^{Kt}$$

$$e^{Kt}G = \int ae^{Kt}dt$$

$$e^{Kt}G = \frac{a}{K}e^{Kt} + C$$

Multiply both sides by e^{-Kt} to get

$$G = \frac{a}{K} + Ce^{-Kt}.$$

NOTE
The equation in Example 4 can also be solved by separation of variables. You are asked to do this in Exercise 22.

As $t \to \infty$,

$$\lim_{t\to\infty} G = \lim_{t\to\infty}\left(\frac{a}{K} + Ce^{-Kt}\right) = \lim_{t\to\infty}\left(\frac{a}{K} + \frac{C}{e^{Kt}}\right) = \frac{a}{K}.$$

Thus, the glucose concentration stabilizes at a/K.

10.2 EXERCISES

Find the general solution for each differential equation.

1. $\dfrac{dy}{dx} + 3y = 6$

2. $\dfrac{dy}{dx} + 5y = 12$

3. $\dfrac{dy}{dx} + 2xy = 4x$

4. $\dfrac{dy}{dx} + 4xy = 4x$

5. $x\dfrac{dy}{dx} - y - x = 0, \quad x > 0$

6. $x\dfrac{dy}{dx} + 2xy - x^2 = 0$

7. $2\dfrac{dy}{dx} - 2xy - x = 0$

8. $3\dfrac{dy}{dx} + 6xy + x = 0$

9. $x\dfrac{dy}{dx} + 2y = x^2 + 6x, \quad x > 0$

10. $x^3\dfrac{dy}{dx} - x^2y = x^4 - 4x^3, \quad x > 0$

11. $y - x\dfrac{dy}{dx} = x^3, \quad x > 0$

12. $2xy + x^3 = x\dfrac{dy}{dx}$

Solve each differential equation, subject to the given initial condition.

13. $\dfrac{dy}{dx} + y = 4e^x; \quad y(0) = 50$

14. $\dfrac{dy}{dx} + 4y = 9e^{5x}; \quad y(0) = 25$

15. $\dfrac{dy}{dx} - 2xy - 4x = 0; \quad y(1) = 20$

16. $x\dfrac{dy}{dx} - 3y + 2 = 0; \quad y(1) = 8$

17. $x\dfrac{dy}{dx} + 5y = x^2; \quad y(2) = 12$

18. $2\dfrac{dy}{dx} - 4xy = 5x; \quad y(1) = 10$

19. $x\dfrac{dy}{dx} + (1 + x)y = 3; \quad y(4) = 50$

20. $\dfrac{dy}{dx} + 3x^2 y - 2xe^{-x^3} = 0; \quad y(0) = 1000$

APPLICATIONS

Life Sciences

21. Population Growth The logistic equation introduced in Section 1,

$$\frac{dy}{dx} = k\left(1 - \frac{y}{N}\right)y \qquad (7)$$

can be written as

$$\frac{dy}{dx} = cy - py^2, \qquad (8)$$

where c and p are positive constants. Although this is a non-linear differential equation, it can be reduced to a linear equation by a suitable substitution for the variable y.

a. Letting $y = 1/z$ and $dy/dx = (-1/z^2)dz/dx$, rewrite Equation (8) in terms of z. Solve for z and then for y.

b. Let $z(0) = 1/y_0$ in part a and find a particular solution for y.

c. Find the limit of y as $x \to \infty$. This is the saturation level of the population.

22. Glucose Level Solve the glucose level example (Example 4) using separation of variables.

23. Drug Use The rate of change in the concentration of a drug with respect to time in a user's blood is given by

$$\frac{dC}{dt} = -kC + D(t),$$

where $D(t)$ is dosage at time t and k is the rate that the drug leaves the bloodstream. *Source: Mathematical Biosciences.*

a. Solve this linear equation to show that, if $C(0) = 0$, then

$$C(t) = e^{-kt}\int_0^t e^{ky}D(y)\, dy.$$

(*Hint:* To integrate both sides of the equation in Step 5 of "Solving a Linear First-Order Differential Equation," integrate from 0 to t, and change the variable of integration to y.)

b. Show that if $D(y)$ is a constant D, then

$$C(t) = \frac{D(1 - e^{-kt})}{k}.$$

24. Mouse Infection A model for the spread of an infectious disease among mice is

$$\frac{dN}{dt} = rN - \frac{\alpha r(\alpha + b + v)}{\beta\left[\alpha - r\left(1 + \dfrac{v}{b + \gamma}\right)\right]},$$

where N is the size of the population of mice, α is the mortality rate due to infection, b is the mortality rate due to natural causes for infected mice, β is a transmission coefficient for the rate that infected mice infect susceptible mice, v is the rate the mice recover from infection, and γ is the rate that mice lose immunity. *Source: Lectures on Mathematics in the Life Sciences.* Show that the solution to this equation, with the initial condition $N(0) = (\alpha + b + v)/\beta$, can be written as

$$N(t) = \frac{(\alpha + b + v)}{\beta R}[(R - \alpha)e^{rt} + \alpha],$$

where

$$R = \alpha - r\left(1 + \frac{v}{b + \gamma}\right).$$

Social Sciences

Immigration and Emigration If population is changed either by immigration or emigration, the exponential growth model discussed in Section 1 is modified to

$$\frac{dy}{dt} = ky + f(t),$$

where y is the population at time t and $f(t)$ is some (other) function of t that describes the net effect of the emigration/immigration. Assume $k = 0.02$ and $y(0) = 10,000$. Solve this differential equation for y, given the following functions $f(t)$.

25. $f(t) = e^t$

26. $f(t) = e^{-t}$

27. $f(t) = -t$

28. $f(t) = t$

Physical Sciences

29. Newton's Law of Cooling In Exercises 56–59 in the previous section, we saw that Newton's Law of Cooling states that the rate of change of the temperature of an object is proportional to the difference in temperature between the object and the surrounding medium. This leads to the differential equation

$$\frac{dT}{dt} = -k(T - T_M),$$

where T is the temperature of the object after time t, T_M is the temperature of the surrounding medium, and k is a constant. In the previous section, we solved this equation by separation of variables. Show that this equation is also linear, and find the solution by the method of this section.

10.3 Euler's Method

APPLY IT How many people have heard a rumor 3 hours after it is started?
This question will be answered in Exercise 35 at the end of this section.

Applications sometimes involve differential equations such as

$$\frac{dy}{dx} = \frac{x + y}{y}$$

that cannot be solved by the methods discussed so far, but approximate solutions to these equations often can be found by numerical methods. For many applications, these approximations are quite adequate. In this section we introduce Euler's method, which is only one of numerous mathematical contributions made by Leonhard Euler (1707–1783) of Switzerland. (His name is pronounced "oiler.") He also introduced the $f(x)$ notation used throughout this text. Despite becoming blind during his later years, he was the most prolific mathematician of his era. In fact it took nearly 50 years after his death to publish the works he created in all mathematical fields during his lifetime.

Euler's method of solving differential equations gives approximate solutions to differential equations involving $y = f(x)$ where the initial values of x and y are known; that is, equations of the form

$$\frac{dy}{dx} = g(x, y), \quad \text{with} \quad y(x_0) = y_0.$$

Geometrically, Euler's method approximates the graph of the solution $y = f(x)$ with a polygonal line whose first segment is tangent to the curve at the point (x_0, y_0), as shown in Figure 5.

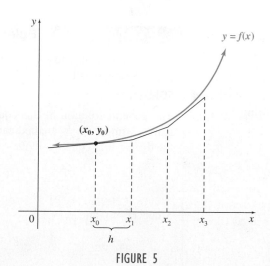

FIGURE 5

FOR REVIEW

In Section 6.6 on Differentials: Linear Approximation, we defined Δy to be the actual change in y as x changed by an amount Δx:

$$\Delta y = f(x + \Delta x) - f(x).$$

The differential dy is an approximation to Δy. We find dy by following the tangent line from the point $(x, f(x))$, rather than by following the actual function. Then dy is found by using the formula $dy = (dy/dx)\, dx$ where $dx = \Delta x$. For example, let $f(x) = x^3$, $x = 1$, and $dx = \Delta x = 0.2$. Then $dy = f'(x)\, dx = 3x^2\, dx = 3(1^2)(0.2) = 0.6$. The actual change in y as x changes from 1 to 1.2 is

$$f(x + \Delta x) - f(x)$$
$$= f(1.2) - f(1)$$
$$= 1.2^3 - 1$$
$$= 0.728.$$

To use Euler's method, divide the interval from x_0 to another point x_n into n subintervals of equal width (see Figure 5.) The width of each subinterval is $h = (x_n - x_0)/n$.

Recall from Section 6.6 on Differentials: Linear Approximation that if Δx is a small change in x, then the corresponding small change in y, Δy, is approximated by

$$\Delta y \approx dy = \frac{dy}{dx} \cdot \Delta x.$$

The differential dy is the change in y along the tangent line. On the interval from x_i to x_{i+1}, note that dy is just $y_{i+1} - y_i$, where y_i is the approximate solution at x_i. We also have $dy/dx = g(x_i, y_i)$ and $\Delta x = h$. Putting these into the previous equation yields

$$y_{i+1} - y_i = g(x_i, y_i)h$$
$$y_{i+1} = y_i + g(x_i, y_i)h.$$

Because y_0 is given, we can use the equation just derived with $i = 0$ to get y_1. We can then use y_1 and the same equation with $i = 1$ to get y_2 and continue in this manner until we get y_n. A summary of Euler's method follows.

Euler's Method

Let $y = f(x)$ be the solution of the differential equation

$$\frac{dy}{dx} = g(x, y), \quad \text{with} \quad y(x_0) = y_0,$$

for $x_0 \le x \le x_n$. Let $x_{i+1} = x_i + h$, where $h = (x_n - x_0)/n$ and

$$y_{i+1} = y_i + g(x_i, y_i)h,$$

for $0 \le i \le n - 1$. Then

$$f(x_{i+1}) \approx y_{i+1}.$$

As the following examples will show, the accuracy of the approximation varies for different functions. As h gets smaller, however, the approximation improves, although making h too small can make things worse. (See the discussion at the end of this section.) Euler's method is not difficult to program; it then becomes possible to try smaller and smaller values of h to get better and better approximations. Graphing calculator programs for Euler's method are included in the *Graphing Calculator and Excel Spreadsheet Manual* available with this book.

EXAMPLE 1 Euler's Method

Use Euler's method to approximate the solution of $dy/dx + 2xy = x$, with $y(0) = 1.5$, for $[0, 1]$. Use $h = 0.1$.

Method 1
Calculating by Hand

SOLUTION

The general solution of this equation was found in Example 2 of the last section, so the results using Euler's method can be compared with the actual solution. Begin by writing the differential equation in the required form as

$$\frac{dy}{dx} = x - 2xy, \quad \text{so that} \quad g(x, y) = x - 2xy.$$

Since $x_0 = 0$ and $y_0 = 1.5$,

$$g(x_0, y_0) = 0 - 2(0)(1.5) = 0,$$

and

$$y_1 = y_0 + g(x_0, y_0)h = 1.5 + 0(0.1) = 1.5.$$

Now $x_1 = 0.1$, $y_1 = 1.5$, and $g(x_1, y_1) = 0.1 - 2(0.1)(1.5) = -0.2$. Then

$$y_2 = 1.5 + (-0.2)(0.1) = 1.48.$$

The 11 values for x_i and y_i for $0 \le i \le 10$ are shown in the table below, together with the actual values using the result from Example 2 in the last section. (Since the result was only a general solution, replace x with 0 and y with 1.5 to get the particular solution $y = 1/2 + e^{-x^2}$.)

The results in the table look quite good. The graphs in Figure 6 show that the polygonal line follows the actual graph of $f(x)$ quite closely.

	Approximate Solution Using $h = 0.1$		
	Euler's Method	Actual Solution	Difference
x_i	y_i	$f(x_i)$	$y_i - f(x_i)$
0	1.5	1.5	0
0.1	1.5	1.49004983	0.0099502
0.2	1.48	1.46078944	0.0192106
0.3	1.4408	1.41393119	0.0268688
0.4	1.384352	1.35214379	0.0322082
0.5	1.31360384	1.27880078	0.0348031
0.6	1.23224346	1.19767633	0.0345671
0.7	1.14437424	1.11262639	0.0317478
0.8	1.05416185	1.02729242	0.0268694
0.9	0.96549595	0.94485807	0.0206379
1.0	0.88170668	0.86787944	0.0138272

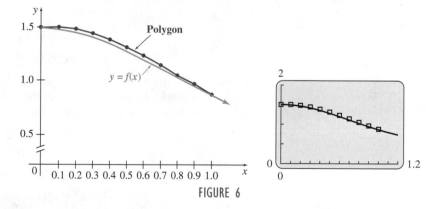

FIGURE 6

Method 2
Graphing Calculator

A graphing calculator can readily implement Euler's method. To do this example on a TI-84 Plus, start x with the value of $x_0 - h$ by storing -0.1 in X, store y_0, or 1.5, in Y, and put $X - 2X*Y$ into the function Y_1. Then the command $X + .1 \to X : Y + Y_1 * .1 \to Y$ gives the next value of y, which is still 1.5. Continue to press the ENTER key to get subsequent values of y. For more details, see the *Graphing Calculator and Excel Spreadsheet Manual* available with this book.

Method 3
Spreadsheet

Euler's method can also be performed on a spreadsheet. In Microsoft Excel, for example, store the values of x in column A and the initial value of y in B1. Then put the command "=B1+(A1−2*A1*B1)*.1" into B2 to get the next value of y, using the formula for $g(x, y)$ in this example. Copy this formula into the rest of column B to get subsequent values of y. For more details, see the *Graphing Calculator and Excel Spreadsheet Manual* available with this book.

YOUR TURN 1 Use Euler's method to approximate the solution of $dy/dx - x^2 y^2 = 1$, with $y(0) = 2$, for [0,1]. Use $h = 0.2$.

TRY YOUR TURN 1

Euler's method produces a very good approximation for this differential equation because the slope of the solution $f(x)$ is not steep in the interval under investigation. The next example shows that such good results cannot always be expected.

EXAMPLE 2 Euler's Method

Use Euler's method to solve $dy/dx = 3y + (1/2)e^x$, with $y(0) = 5$, for $[0, 1]$, using 10 subintervals.

SOLUTION This is the differential equation of Example 3 in the last section. The general solution found there, with the initial condition given above, leads to the particular solution

$$y = -\frac{1}{4}e^x + \frac{21}{4}e^{3x}.$$

To solve by Euler's method, start with $g(x, y) = 3y + (1/2)e^x$, $x_0 = 0$, and $y_0 = 5$. For $n = 10$, $h = (1 - 0)/10 = 0.1$ again, and

$$y_{i+1} = y_i + g(x_i, y_i)h = y_i + \left(3y_i + \frac{1}{2}e^{x_i}\right)h.$$

For y_1, this gives

$$y_1 = y_0 + \left(3y_0 + \frac{1}{2}e^{x_0}\right)h$$

$$= 5 + \left[3(5) + \frac{1}{2}(e^0)\right](0.1)$$

$$= 5 + (15.5)(0.1) = 6.55.$$

Similarly,

$$y_2 = 6.55 + \left[3(6.55) + \frac{1}{2}e^{0.1}\right](0.1)$$

$$= 6.55 + (19.65 + 0.55258546)(0.1)$$

$$= 8.57025855.$$

These and the remaining values for the interval $[0, 1]$ are shown in the table below.

In this example the absolute value of the differences grows very rapidly as x_i gets farther from x_0. See Figure 7 on the next page. These large differences come from the term e^{3x} in the solution; this term grows very quickly as x increases.

	Approximate Solution Using $h = 0.1$		
x_i	Euler's Method y_i	Actual Solution $f(x_i)$	Difference $y_i - f(x_i)$
0	5	5	0
0.1	6.55	6.810466	−0.260466
0.2	8.570259	9.260773	−0.690514
0.3	11.202406	12.575452	−1.373045
0.4	14.630621	17.057658	−2.427037
0.5	19.094399	23.116687	−4.022289
0.6	24.905154	31.305119	−6.399965
0.7	32.467806	42.368954	−9.901147
0.8	42.308836	57.315291	−15.006455
0.9	55.112764	77.503691	−22.390927
1	71.769573	104.769498	−32.999925

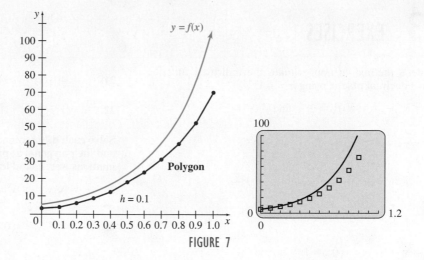

FIGURE 7

As these examples show, numerical methods may produce large errors. The error often can be reduced by using more subintervals of smaller width—letting $n = 100$ or 1000, for example. Approximations for the function in Example 3 with $n = 100$ and $h = (1 - 0)/100 = 0.01$ are shown in the table below. The approximations are considerably improved.

	Approximate Solution Using $h = 0.01$		
x_i	**Euler's Method** y_i	**Actual Solution** $f(x_i)$	**Difference** $y_i - f(x_i)$
0	5	5	0
0.1	6.779418	6.810466	−0.031048
0.2	9.177101	9.260773	−0.083672
0.3	12.406341	12.575452	−0.169111
0.4	16.753855	17.057658	−0.303803
0.5	22.605046	23.116687	−0.511642
0.6	30.477945	31.305119	−0.827175
0.7	41.068839	42.368954	−1.300115
0.8	55.313581	57.315291	−2.001710
0.9	74.469995	77.503691	−3.033695
1	100.228621	104.769498	−4.540878

We could improve the accuracy of Euler's method by using a smaller h, but there are two difficulties. First, this requires more calculations and, consequently, more time. Such calculations are usually done by computer, so the increased time may not matter. But this introduces a second difficulty: The increased number of calculations causes more round-off error, so there is a limit to how small we can make h and still get improvement. The preferred way to get greater accuracy is to use a more sophisticated procedure, such as the Runge-Kutta method. Such methods are beyond the scope of this book but are discussed in numerical analysis and differential equations courses.*

*For example, see Nagle, R. K., E. B. Saff, and A. D. Snider, *Fundamentals of Differential Equations*, 8th ed., Pearson, 2012.

10.3 EXERCISES

Use Euler's method to approximate the indicated function value to 3 decimal places, using $h = 0.1$.

1. $\dfrac{dy}{dx} = x^2 + y^2$; $y(0) = 2$; find $y(0.5)$

2. $\dfrac{dy}{dx} = xy + 4$; $y(0) = 0$; find $y(0.5)$

3. $\dfrac{dy}{dx} = 1 + y$; $y(0) = 2$; find $y(0.6)$

4. $\dfrac{dy}{dx} = x + y^2$; $y(0) = 0$; find $y(0.6)$

5. $\dfrac{dy}{dx} = x + \sqrt{y}$; $y(0) = 1$; find $y(0.4)$

6. $\dfrac{dy}{dx} = 1 + \dfrac{y}{x}$; $y(1) = 0$; find $y(1.4)$

7. $\dfrac{dy}{dx} = 2x\sqrt{1 + y^2}$; $y(1) = 2$; find $y(1.5)$

8. $\dfrac{dy}{dx} = e^{-y} + e^x$; $y(1) = 1$; find $y(1.5)$

Use Euler's method to approximate the indicated function value to 3 decimal places, using $h = 0.1$. Next, solve the differential equation and find the indicated function value to 3 decimal places. Compare the result with the approximation.

9. $\dfrac{dy}{dx} = -4 + x$; $y(0) = 1$; find $y(0.4)$

10. $\dfrac{dy}{dx} = 4x + 3$; $y(1) = 0$; find $y(1.5)$

11. $\dfrac{dy}{dx} = x^3$; $y(0) = 4$; find $y(0.5)$

12. $\dfrac{dy}{dx} = \dfrac{3}{x}$; $y(1) = 2$; find $y(1.4)$

13. $\dfrac{dy}{dx} = 2xy$; $y(1) = 1$; find $y(1.6)$

14. $\dfrac{dy}{dx} = x^2y$; $y(0) = 1$; find $y(0.6)$

15. $\dfrac{dy}{dx} = ye^x$; $y(0) = 2$; find $y(0.4)$

16. $\dfrac{dy}{dx} = \dfrac{2x}{y}$; $y(0) = 3$; find $y(0.6)$

17. $\dfrac{dy}{dx} + y = 2e^x$; $y(0) = 100$; find $y(0.3)$

18. $\dfrac{dy}{dx} - 2y = e^{2x}$; $y(0) = 10$; find $y(0.4)$

Use Method 2 or 3 in Example 1 to construct a table like the ones in the examples for $0 \le x \le 1$, with $h = 0.2$.

19. $\dfrac{dy}{dx} = \sqrt[3]{x}$; $y(0) = 0$ 20. $\dfrac{dy}{dx} = y$; $y(0) = 1$

21. $\dfrac{dy}{dx} = 4 - y$; $y(0) = 0$

22. $\dfrac{dy}{dx} = x - 2xy$; $y(0) = 1$

Solve each differential equation and graph the function $y = f(x)$ and the polygonal approximation on the same axes. (The approximations were found in Exercises 19–22.)

23. $\dfrac{dy}{dx} = \sqrt[3]{x}$; $y(0) = 0$ 24. $\dfrac{dy}{dx} = y$; $y(0) = 1$

25. $\dfrac{dy}{dx} = 4 - y$; $y(0) = 0$

26. $\dfrac{dy}{dx} = x - 2xy$; $y(0) = 1$

27. **a.** Use Euler's method with $h = 0.2$ to approximate $f(1)$, where $f(x)$ is the solution to the differential equation

$$\frac{dy}{dx} = y^2; \quad y(0) = 1.$$

  b. Solve the differential equation in part a using separation of variables, and discuss what happens to $f(x)$ as x approaches 1.

  c. Based on what you learned from parts a and b, discuss what might go wrong when using Euler's method. (More advanced courses on differential equations discuss the question of whether a differential equation has a solution for a given interval in x.)

APPLICATIONS

Solve Exercises 28–35 using Euler's method.

Business and Economics

28. **Bankruptcy** Suppose 125 small business firms are threatened by bankruptcy. If y is the number bankrupt by time t, then $125 - y$ is the number not yet bankrupt by time t. The rate of change of y is proportional to the product of y and $125 - y$. Let 2010 correspond to $t = 0$. Assume 20 firms are bankrupt at $t = 0$.

  a. Write a differential equation using the given information. Use 0.002 for the constant of proportionality.

  b. Approximate the number of firms that are bankrupt in 2015, using $h = 1$.

Life Sciences

29. **Growth of Algae** The phosphate compounds found in many detergents are highly water soluble and are excellent fertilizers for algae. Assume that there are 5000 algae present at time $t = 0$ and conditions will support at most 500,000 algae. Assume that the rate of growth of algae, in the presence of sufficient phosphates, is proportional both to the number present (in thousands) and to the difference between 500,000 and the number present (in thousands).

  a. Write a differential equation using the given information. Use 0.01 for the constant of proportionality.

  b. Approximate the number present when $t = 2$, using $h = 0.5$.

30. Immigration An island is colonized by immigration from the mainland, where there are 100 species. Let the number of species on the island at time t (in years) equal y, where $y = f(t)$. Suppose the rate at which new species immigrate to the island is

$$\frac{dy}{dt} = 0.02(100 - y^{1/2}).$$

Use Euler's method with $h = 0.5$ to approximate y when $t = 5$ if there were 10 species initially.

31. Insect Population A population of insects, y, living in a circular colony grows at a rate

$$\frac{dy}{dt} = 0.05y - 0.1y^{1/2},$$

where t is time in weeks. If there were 60 insects initially, use Euler's method with $h = 1$ to approximate the number of insects after 6 weeks.

32. Whale Population Under certain conditions a population may exhibit a polynomial growth rate function. A population of blue whales is growing according to the function

$$\frac{dy}{dt} = -y + 0.02y^2 + 0.003y^3.$$

Here y is the population in thousands and t is measured in years. Use Euler's method with $h = 1$ to approximate the population in 4 years if the initial population is 15,000.

33. Goat Growth The growth of male Saanen goats can be approximated by the equation

$$\frac{dW}{dt} = -0.01189W + 0.92389W^{0.016},$$

where W is the weight (in kilograms) after t weeks. *Source: Annales de Zootechnie*. Find the weight of a goat at 5 weeks, given that the weight at birth is 3.65 kg. Use Euler's method with $h = 1$.

Social Sciences

34. Learning In an early article describing how people learn, the rate of change of the probability that a person performs a task correctly (p) with respect to time (t) is given by

$$\frac{dp}{dt} = \frac{2k}{\sqrt{m}}(p - p^2)^{3/2},$$

where k and m are constants related to the rate that the person learns the task. *Source: The Journal of General Psychology*. For this exercise, let $m = 4$ and $k = 0.5$.

a. Letting $p = 0.1$ when $t = 0$, use Method 2 or 3 in Example 1 to construct a table for t_i and p_i like the ones in the examples for $0 \le x \le 30$, with $h = 5$.

b. Based on your answer to part a, what does p seem to approach as t increases? Explain why this answer makes sense.

35. APPLY IT Spread of Rumors A rumor spreads through a community of 500 people at the rate

$$\frac{dN}{dt} = 0.02(500 - N)N^{1/2},$$

where N is the number of people who have heard the rumor at time t (in hours). Use Euler's method with $h = 0.5$ to find the number who have heard the rumor after 3 hours, if only 2 people heard it initially.

YOUR TURN ANSWER

1.

x_i	y_i
0	2
0.2	2.2
0.4	2.43872
0.6	2.82903537
0.8	3.60528313
1.0	5.46903563

10.4 Applications of Differential Equations

APPLY IT How do the populations of a predator and its prey change over time?
We will answer this question in Example 2 using a pair of differential equations.

Continuous Deposits An amount of money A invested at an annual interest rate r, compounded continuously, grows according to the differential equation

$$\frac{dA}{dt} = rA,$$

where t is the time in years. Suppose money is deposited into this account at a rate of D dollars per year and that the money is deposited at a rate that is essentially constant and continuous. The differential equation for the growth of the account then becomes

$$\frac{dA}{dt} = rA + D.$$

EXAMPLE 1   Continuous Deposits

When Michael was born, his grandfather arranged to deposit $5000 in an account for him at 8% annual interest compounded continuously. Grandfather plans to add to the account "continuously" at the rate of $1000 a year. How much will be in the account when Michael is 18?

SOLUTION   Here $r = 0.08$ and $D = 1000$, so the differential equation is

$$\frac{dA}{dt} = 0.08A + 1000.$$

Separate the variables and integrate on both sides.

$$\frac{1}{0.08A + 1000} dA = dt$$

$$\frac{1}{0.08} \ln(0.08A + 1000) = t + C$$

$$\ln(0.08A + 1000) = 0.08t + K \qquad\qquad K = 0.08C$$

$$0.08A + 1000 = e^{0.08t + K} \qquad\qquad \text{Apply the function } e^x \text{ to both sides.}$$

$$0.08A = -1000 + e^{0.08t}e^K$$

$$0.08A = -1000 + Me^{0.08t} \qquad\qquad M = e^K$$

$$A = -12{,}500 + \frac{M}{0.08}e^{0.08t}$$

Use the fact that the initial amount deposited was $5000 to find M.

$$5000 = -12{,}500 + \frac{M}{0.08}e^{(0.08)0}$$

$$5000 = -12{,}500 + \frac{M}{0.08}(1)$$

$$1400 = M$$

$$A = -12{,}500 + \frac{1400}{0.08}e^{0.08t}$$

$$A = -12{,}500 + 17{,}500e^{0.08t}$$

YOUR TURN 1   Modify Example 1 so that the initial amount is $6000, the interest rate is 5%, $1200 a year is added continuously, and Michael must wait until he is 21 to collect.

When Michael is 18, $t = 18$, the amount in the account will be

$$A = -12{,}500 + 17{,}500e^{(0.08)18}$$

$$= -12{,}500 + 17{,}500e^{1.44}$$

$$\approx 61{,}362.18,$$

or about $61,400.

TRY YOUR TURN 1

A Predator-Prey Model
The Austrian mathematician A. J. Lotka (1880–1949) and the Italian mathematician Vito Volterra (1860–1940) proposed the following simple model for the way in which the fluctuations of populations of a predator and its prey affect each other. *Source: Elements of Mathematical Biology.* Let $x = f(t)$ denote the population of

the predator and $y = g(t)$ denote the population of the prey at time t. The predator might be a wolf and its prey a moose, or the predator might be a ladybug and the prey an aphid.

Assume that if there were no predators present, the population of prey would increase at a rate py proportional to their number, but that the predators consume the prey at a rate qxy proportional to the product of the number of prey and the number of predators. The net rate of change dy/dt of y is the rate of increase of the prey minus the rate at which the prey are eaten, that is,

$$\frac{dy}{dt} = py - qxy, \tag{1}$$

with positive constants p and q.

Assume that if there were no prey, the predators would starve and their population would *decrease* at a rate rx proportional to their number, but that in the presence of prey the rate of growth of the population of predators is increased by an amount sxy. These assumptions give a second differential equation,

$$\frac{dx}{dt} = -rx + sxy, \tag{2}$$

with additional positive constants r and s.

Equations (1) and (2) form a system of differential equations known as the **Lotka-Volterra equations**. They cannot be solved for x and y as functions of t, but an equation relating the variables x and y can be found. Dividing Equation (1) by Equation (2) gives the separable differential equation

$$\frac{dy}{dx} = \frac{dy/dt}{dx/dt} = \frac{py - qxy}{-rx + sxy}$$

or

$$\frac{dy}{dx} = \frac{y(p - qx)}{x(sy - r)}. \tag{3}$$

Equation (3) is solved for specific values of the constants p, q, r, and s in the next example.

EXAMPLE 2 Predator-Prey

Suppose that $x = f(t)$ (hundreds of predators) and $y = g(t)$ (thousands of prey) satisfy the Lotka-Volterra Equations (1) and (2) with $p = 3$, $q = 1$, $r = 4$, and $s = 2$. Suppose that at a time when there are 100 predators ($x = 1$), there are 1000 prey ($y = 1$). Find an equation relating x and y.

APPLY IT **SOLUTION** With the given values of the constants, p, q, r, and s, Equation (3) reads

$$\frac{dy}{dx} = \frac{y(3 - x)}{x(2y - 4)}.$$

Separating the variables yields

$$\frac{2y - 4}{y} dy = \frac{3 - x}{x} dx,$$

or

$$\int \left(2 - \frac{4}{y}\right) dy = \int \left(\frac{3}{x} - 1\right) dx.$$

Evaluating the integrals gives

$$2y - 4 \ln y = 3 \ln x - x + C.$$

(It is not necessary to use absolute value for the logarithms since x and y are positive.) Use the initial conditions $x = 1$ and $y = 1$ to find C.

$$2 - 4(0) = 3(0) - 1 + C$$
$$C = 3$$

The desired equation is

$$x + 2y - 3 \ln x - 4 \ln y = 3. \tag{4}$$

A graph of Equation (4), in Figure 8, shows that the solution is located on a closed curve. By looking at the original differential equation for y,

$$\frac{dy}{dt} = y(3 - x),$$

we can see that when $x < 3$ (the left side of the curve), then $dy/dt > 0$, so y increases. Similarly, when $x > 3$ (the right side of the curve), y decreases. This means that when there are few predators, the population of prey increases, but when there are many predators, the population of prey decreases, as we would expect. Similarly, by looking at the original differential equation for x,

$$\frac{dx}{dt} = x(2y - 4),$$

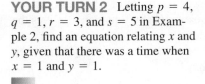

YOUR TURN 2 Letting $p = 4$, $q = 1, r = 3$, and $s = 5$ in Example 2, find an equation relating x and y, given that there was a time when $x = 1$ and $y = 1$.

confirm that when there are few prey ($y < 2$), the population of predators decreases (as we would expect, because the predators don't have enough to eat), and when there are many prey ($y > 2$), the population of predators increases. Convince yourself that, for these reasons, the solution (x, y) must move clockwise on the curve in Figure 8. The pattern repeats indefinitely. **TRY YOUR TURN 2**

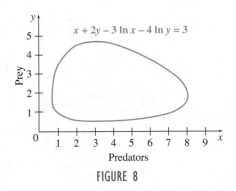

FIGURE 8

Epidemics Under certain conditions, the spread of a contagious disease can be described with the logistic growth model, as in the next example.

EXAMPLE 3 Spread of an Epidemic

Consider a population of size N that satisfies the following conditions.

1. Initially there is only one infected individual.
2. All uninfected individuals are susceptible, and infection occurs when an uninfected individual contacts an infected individual.
3. Contact between any two individuals is just as likely as contact between any other two individuals.
4. Infected individuals remain infectious.

Let $t =$ the time (in days) and $y =$ the number of individuals infected at time t. At any moment there are $(N - y)y$ possible contacts between an uninfected individual and an infected individual. Thus, it is reasonable to assume that the rate of spread of the disease satisfies the following logistic equation (discussed in Section 10.1):

$$\frac{dy}{dt} = k\left(1 - \frac{y}{N}\right)y. \tag{5}$$

As shown in Section 10.1, the general solution of Equation (5) is

$$y = \frac{N}{1 + be^{-kt}}, \tag{6}$$

where $b = (N - y_0)/y_0$. Since just one individual is infected initially, $y_0 = 1$. Substituting these values into Equation (6) gives

$$y = \frac{N}{1 + (N - 1)e^{-kt}} \tag{7}$$

as the specific solution of Equation (5).

The infection rate dy/dt will be a maximum when its derivative is 0, that is, when $d^2y/dt^2 = 0$. Since

$$\frac{dy}{dt} = k\left(1 - \frac{y}{N}\right)y,$$

we have

$$\frac{d^2y}{dt^2} = k\left[\left(1 - \frac{y}{N}\right)\left(\frac{dy}{dt}\right) + y\left(-\frac{1}{N}\right)\left(\frac{dy}{dt}\right)\right]$$

$$= k\left(1 - \frac{2y}{N}\right)\left(\frac{dy}{dt}\right)$$

$$= k\left(1 - \frac{2y}{N}\right)k\left(1 - \frac{y}{N}\right)y$$

$$= k^2 y\left(1 - \frac{y}{N}\right)\left(1 - \frac{2y}{N}\right).$$

Set $\dfrac{d^2y}{dt^2} = 0$ to get

$$y = 0, \quad 1 - \frac{y}{N} = 0, \quad \text{or} \quad 1 - \frac{2y}{N} = 0.$$

That is, $\dfrac{d^2y}{dt^2} = 0$ when $y = 0$, $y = N$, or $y = \dfrac{N}{2}$.

Notice that since the infection rate $\dfrac{dy}{dt} = 0$ when $y = 0$ or $y = N$, the maximum infection rate does not occur there. Also observe that $\dfrac{d^2y}{dt^2} > 0$ when $0 < y < \dfrac{N}{2}$ and $\dfrac{d^2y}{dt^2} < 0$ when $\dfrac{N}{2} < y < N$. Thus, the maximum infection rate occurs when exactly half the total population is still uninfected and equals

$$\frac{dy}{dt} = k\left(1 - \frac{N/2}{N}\right)\frac{N}{2} = \frac{kN}{4}.$$

Letting $y = N/2$ in Equation (7) and solving for t shows that the maximum infection rate occurs at time

$$t_m = \frac{\ln(N-1)}{k}.$$

YOUR TURN 3 Suppose that an epidemic in a community of 50,000 starts with 80 people infected, and that 15 days later, 640 are infected. How many are infected 25 days into the epidemic?

Because y is a function of t, the infection rate dy/dt is also a function of t. Its graph, shown in Figure 9, is called the *epidemic curve*. It is symmetric about the line $t = t_m$.

TRY YOUR TURN 3

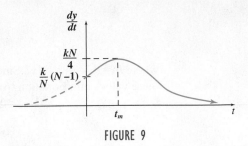

FIGURE 9

Mixing Problems
The mixing of two solutions can lead to a first-order differential equation, as the next example shows.

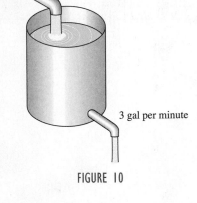

4 gal per minute

3 gal per minute

FIGURE 10

EXAMPLE 4   Salt Concentration

Suppose a tank contains 100 gal of a solution of dissolved salt and water, which is kept uniform by stirring. If pure water is allowed to flow into the tank at the rate of 4 gal per minute, and the mixture flows out at the rate of 3 gal per minute (see Figure 10), how much salt will remain in the tank after t minutes if 15 lb of salt are in the mixture initially? *Source: Ordinary Differential Equations with Boundary Value Problems.*

SOLUTION   Let the amount of salt present in the tank at any specific time be $y = f(t)$. The net rate at which y changes is given by

$$\frac{dy}{dt} = (\text{Rate of salt in}) - (\text{Rate of salt out}).$$

Since pure water is coming in, the rate of salt entering the tank is zero. The rate at which salt is leaving the tank is the product of the amount of salt per gallon (in V gallons) and the number of gallons per minute leaving the tank:

$$\text{Rate of salt out} = \left(\frac{y}{V} \text{ lb per gal}\right)(3 \text{ gal per minute}).$$

The differential equation, therefore, can be written as

$$\frac{dy}{dt} = -\frac{3y}{V}; \quad y(0) = 15,$$

where $y(0)$ is the initial amount of salt in the solution. We must take into account the fact that the volume, V, of the mixture is not constant but is determined by

$$\frac{dV}{dt} = (\text{Rate of liquid in}) - (\text{Rate of liquid out}) = 4 - 3 = 1,$$

or

$$\frac{dV}{dt} = 1,$$

from which

$$V(t) = t + C_1.$$

Because the volume is known to be 100 at time $t = 0$, we have $C_1 = 100$, and

$$\frac{dy}{dt} = \frac{-3y}{t + 100}; \quad y(0) = 15,$$

a separable equation with solution

$$\frac{dy}{y} = \frac{-3}{t + 100} dt$$

$$\ln y = -3 \ln(t + 100) + C.$$

Since $y = 15$ when $t = 0$,

$$\ln 15 = -3 \ln 100 + C$$

$$\ln 15 + 3 \ln 10^2 = C$$

$$\ln(15 \times 10^6) = C.$$

YOUR TURN 4 Suppose that a tank initially contains 500 liters of a solution of water and 5 kg of salt. Suppose that pure water flows in at a rate of 6 L/min, and the solution flows out at a rate of 4 L/min. How many kg of salt remain after 20 minutes?

Finally,

$$\ln y = \ln(t + 100)^{-3} + \ln(15 \times 10^6)$$
$$= \ln[(t + 100)^{-3}(15 \times 10^6)]$$

$$y = \frac{15 \times 10^6}{(t + 100)^3}.$$

TRY YOUR TURN 4

10.4 EXERCISES

APPLICATIONS

Business and Economics

1. **Continuous Deposits** Kimberly Austin deposits $5000 in an IRA at 6% interest compounded continuously for her retirement in 10 years. She intends to make continuous deposits at the rate of $3000 a year until she retires. How much will she have accumulated at that time?

2. **Continuous Deposits** In Exercise 1, how long will it take Kimberly to accumulate $30,000 in her retirement account?

3. **Continuous Deposits** To provide for a future expansion, a company plans to make continuous deposits to a savings account at the rate of $50,000 per year, with no initial deposit. The managers want to accumulate $500,000. How long will it take if the account earns 10% interest compounded continuously?

4. **Continuous Deposits** Suppose the company in Exercise 3 wants to accumulate $500,000 in 3 years. Find the approximate yearly deposit that will be required.

5. **Continuous Deposits** An investor deposits $8000 into an account that pays 6% compounded continuously and then begins to *withdraw* from the account continuously at a rate of $1200 per year.

   a. Write a differential equation to describe the situation.

   b. How much will be left in the account after 2 years?

   c. When will the account be completely depleted?

Life Sciences

6. **Predator-Prey** Explain in your own words why the solution (x, y) must move clockwise on the curve in Figure 8.

7. **Competing Species** The system of equations

$$\frac{dy}{dt} = 4y - 2xy$$

$$\frac{dx}{dt} = -3x + 2xy$$

describes the influence of the populations (in thousands) of two competing species on their growth rates.

a. Following Example 2, find an equation relating x and y, assuming $y = 1$ when $x = 1$.

b. Find values of x and y so that both populations are constant. (*Hint:* Set both differential equations equal to 0.)

8. Symbiotic Species When two species, such as the rhinoceros and birds pictured below, coexist in a symbiotic (dependent) relationship, they either increase together or decrease together. Typical equations for the growth rates of two such species might be

$$\frac{dx}{dt} = -4x + 4xy$$
$$\frac{dy}{dt} = -3y + 2xy.$$

a. Find an equation relating x and y if $x = 5$ when $y = 1$.

b. Find values of x and y so that both populations are constant. (See Exercise 7.)

c. A graph of the relationship found in part a is shown in the figure. Based on the differential equations for the growth rate and this graph, what happens to both populations when $y > 1$? When $y < 1$?

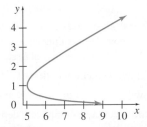

9. Spread of an Epidemic The native Hawaiians lived for centuries in isolation from other peoples. When foreigners finally came to the islands they brought with them diseases such as measles, whooping cough, and smallpox, which decimated the population. Suppose such an island has a native population of 5000, and a sailor from a visiting ship introduces measles, which has an infection rate of 0.00005. Also suppose that the model for spread of an epidemic described in Example 3 applies.

a. Write an equation for the number of natives who remain uninfected. Let t represent time in days.

b. How many are uninfected after 30 days?

c. How many are uninfected after 50 days?

d. When will the maximum infection rate occur?

10. Spread of an Epidemic In Example 3, the number of infected individuals is given by Equation (7).

a. Show that the number of uninfected individuals is given by

$$N - y = \frac{N(N-1)}{N - 1 + e^{kt}}.$$

b. Graph the equation in part a and Equation (7) on the same axes when $N = 100$ and $k = 1$.

c. Find the common inflection point of the two graphs.

d. What is the significance of the common inflection point?

e. What are the limiting values of y and $N - y$?

11. Spread of an Epidemic An influenza epidemic spreads at a rate proportional to the product of the number of people infected and the number not yet infected. Assume that 100 people are infected at the beginning of the epidemic in a community of 20,000 people, and 400 are infected 10 days later.

a. Write an equation for the number of people infected, y, after t days.

b. When will half the community be infected?

12. Spread of an Epidemic The Gompertz growth law,

$$\frac{dy}{dt} = kye^{-at},$$

for constants k and a, is another model used to describe the growth of an epidemic. Repeat Exercise 11, using this differential equation with $a = 0.02$.

13. Spread of Gonorrhea Gonorrhea is spread by sexual contact, takes 3 to 7 days to incubate, and can be treated with antibiotics. There is no evidence that a person ever develops immunity. One model proposed for the rate of change in the number of men infected by this disease is

$$\frac{dy}{dt} = -ay + b(f - y)Y,$$

where y is the fraction of men infected, f is the fraction of men who are promiscuous, Y is the fraction of women infected, and a and b are appropriate constants. *Source: An Introduction to Mathematical Modeling*.

a. Assume $a = 1$, $b = 1$, and $f = 0.5$. Choose $Y = 0.01$, and solve for y using $y = 0.02$ when t is 0 as an initial condition. Round your answer to 3 decimal places.

b. A comparable model for women is

$$\frac{dY}{dt} = -AY + B(F - Y)y,$$

where F is the fraction of women who are promiscuous and A and B are constants. Assume $A = 1$, $B = 1$, and $F = 0.03$. Choose $y = 0.1$ and solve for Y, using $Y = 0.01$ as an initial condition.

Social Sciences

Spread of a Rumor The equation developed in the text for the spread of an epidemic also can be used to describe diffusion of information. In a population of size N, let y be the number who have heard a particular piece of information. Then

$$\frac{dy}{dt} = k\left(1 - \frac{y}{N}\right)y$$

for a positive constant k. Use this model in Exercises 14–16.

14. Suppose a rumor starts among 3 people in a certain office building. That is, $y_0 = 3$. Suppose 500 people work in the building and 50 people have heard the rumor in 2 days. Using Equation (6), write an equation for the number who have heard the rumor in t days. How many people will have heard the rumor in 5 days?

15. A rumor spreads at a rate proportional to the product of the number of people who have heard it and the number who have not heard it. Assume that 3 people in an office with 45 employees heard the rumor initially, and 12 people have heard it 3 days later.

    a. Write an equation for the number, y, of people who have heard the rumor in t days.

    b. When will 30 employees have heard the rumor?

16. A news item is heard on the late news by 5 of the 100 people in a small community. By the end of the next day 20 people have heard the news. Using Equation (6), write an equation for the number of people who have heard the news in t days. How many have heard the news after 3 days?

17. Repeat Exercise 15 using the Gompertz growth law,

$$\frac{dy}{dt} = kye^{-at},$$

   for constants k and a, with $a = 0.1$.

Physical Sciences

18. **Salt Concentration** A tank holds 100 gal of water that contains 20 lb of dissolved salt. A brine (salt) solution is flowing

into the tank at the rate of 2 gal per minute while the solution flows out of the tank at the same rate. The brine solution entering the tank has a salt concentration of 2 lb per gal.

   a. Find an expression for the amount of salt in the tank at any time.

   b. How much salt is present after 1 hour?

   c. As time increases, what happens to the salt concentration?

19. Solve Exercise 18 if the brine solution is introduced at the rate of 3 gal per minute while the rate of outflow remains the same.

20. Solve Exercise 18 if the brine solution is introduced at the rate of 1 gal per minute while the rate of outflow stays the same.

21. Solve Exercise 18 if pure water is added instead of brine.

22. **Chemical in a Solution** Five grams of a chemical is dissolved in 100 liters of alcohol. Pure alcohol is added at the rate of 2 liters per minute and at the same time the solution is being drained at the rate of 1 liter per minute.

   a. Find an expression for the amount of the chemical in the mixture at any time.

   b. How much of the chemical is present after 30 minutes?

23. Solve Exercise 22 if a 25% solution of the same mixture is added instead of pure alcohol.

24. **Soap Concentration** A prankster puts 4 lb of soap in a fountain that contains 200 gal of water. To clean up the mess a city crew runs clear water into the fountain at the rate of 8 gal per minute allowing the excess solution to drain off at the same rate. How long will it be before the amount of soap in the mixture is reduced to 1 lb?

YOUR TURN ANSWERS

1. $61,729.53
2. $x + 5y - 4\ln x - 3\ln y = 6$
3. 2483
4. 4.29 kg

10 CHAPTER REVIEW

SUMMARY

In this chapter, we studied differential equations, which are equations involving derivatives. Our goal has been to find a function that satisfies the equation. We learned to solve two different types of equations:

  • separable equations, using separation of variables, and
  • linear equations, using an integrating factor.

For equations that cannot be solved by either of the previous two methods, we introduced a numerical method known as Euler's method.

Differential equations have a large number of applications; some of those we studied in this chapter include:

  • continuous deposits;
  • the logistic equation for populations;
  • a predator-prey model;
  • and mixing problems.

General Solution of $\dfrac{dy}{dx} = g(x)$ The general solution of the differential equation $dy/dx = g(x)$ is

$$y = \int g(x)\,dx.$$

Separable Differential Equation An equation is separable if it can be written in the form

$$\frac{dy}{dx} = \frac{p(x)}{q(y)}.$$

By separating the variables, it is transformed into the equation

$$\int q(y)\,dy = \int p(x)\,dx.$$

Solving a Linear First-Order Differential Equation

1. Put the equation in the linear form

$$\frac{dy}{dx} + P(x)y = Q(x).$$

2. Find the integrating factor $I(x) = e^{\int P(x)\,dx}$.

3. Multiply each term of the equation from Step 1 by $I(x)$.

4. Replace the sum of terms on the left with $D_x[I(x)y]$.

5. Integrate both sides of the equation.

6. Solve for y.

Euler's Method Let $y = f(x)$ be the solution of the differential equation

$$\frac{dy}{dx} = g(x, y), \text{ with } y(x_0) = y_0,$$

for $x_0 \le x \le x_n$. Let $x_{i+1} = x_i + h$, where $h = (x_n - x_0)/n$ and

$$y_{i+1} = y_i + g(x_i, y_i)h,$$

for $0 \le i \le n - 1$. Then

$$f(x_{i+1}) \approx y_{i+1}.$$

KEY TERMS

differential equation

10.1

general solution
particular solution
initial condition
initial value problem

separable differential
 equation
separation of variables
growth constant
carrying capacity
logistic growth model
logistic equation
logistic curve

10.2

linear first-order
 differential equation
integrating factor

10.3

Euler's method

10.4

Lotka-Volterra equations

REVIEW EXERCISES

CONCEPT CHECK

Determine whether each of the following statements is true or false, and explain why.

1. To determine a particular solution to a differential equation, you first must find a general solution to the differential equation.

2. The function $y = e^{2x} + 5$ satisfies the differential equation $\dfrac{dy}{dx} = 2y$.

3. The function $y = \dfrac{100}{1 + 99e^{-5t}}$ satisfies the differential equation $\dfrac{dy}{dt} = 5\left(1 - \dfrac{y}{100}\right)y$.

4. The differential equation $y\dfrac{dy}{dx} + xy = 2500e^y$ is a first-order linear differential equation.

5. Every differential equation is either separable or linear.

6. It is possible to solve the following differential equation using the method of separation of variables.
$$x\frac{dy}{dx} = (x + 1)(y + 1)$$

7. It is possible to solve the following differential equation using the method of separation of variables.
$$\frac{dy}{dx} = x^2 + 4y^2$$

8. The function $I(x) = x^3$ can be used as an integrating factor for the differential equation
$$\frac{dy}{dx} + 3\frac{y}{x} = \frac{1}{x^2}.$$

9. The function $I(x) = e^{5x}$ can be used as an integrating factor for the differential equation
$$x\frac{dy}{dx} + 5y = e^{2x}.$$

10. Euler's method can be used to find the general solution to a differential equation.

11. If Euler's method is being used to solve the differential equation $\dfrac{dy}{dx} = x + \sqrt{y + 4}$ with $h = 0.1$, then
$$y_{i+1} = y_i + 0.1(x_i + \sqrt{y_i + 4}).$$

12. The differential equation describing continuous deposits is separable.

PRACTICE AND EXPLORATIONS

13. What is a differential equation? What is it used for?

14. What is the difference between a particular solution and a general solution to a differential equation?

15. How can you tell that a differential equation is separable? That it is linear?

16. Can a differential equation be both separable and linear? Explain why not, or give an example of an equation that is both.

Classify each equation as separable, linear, both, or neither.

17. $y\dfrac{dy}{dx} = 2x + y$

18. $\dfrac{dy}{dx} + y^2 = xy^2$

19. $\sqrt{x}\dfrac{dy}{dx} = \dfrac{1 + \ln x}{y}$

20. $\dfrac{dy}{dx} = xy + e^x$

21. $\dfrac{dy}{dx} + x = xy$

22. $\dfrac{x}{y}\dfrac{dy}{dx} = 4 + x^{3/2}$

23. $x\dfrac{dy}{dx} + y = e^x(1 + y)$

24. $\dfrac{dy}{dx} = x^2 + y^2$

Find the general solution for each differential equation.

25. $\dfrac{dy}{dx} = 3x^2 + 6x$

26. $\dfrac{dy}{dx} = 4x^3 + 6x^5$

27. $\dfrac{dy}{dx} = 4e^{2x}$

28. $\dfrac{dy}{dx} = \dfrac{1}{3x + 2}$

29. $\dfrac{dy}{dx} = \dfrac{3x + 1}{y}$

30. $\dfrac{dy}{dx} = \dfrac{e^x + x}{y - 1}$

31. $\dfrac{dy}{dx} = \dfrac{2y + 1}{x}$

32. $\dfrac{dy}{dx} = \dfrac{3 - y}{e^x}$

33. $\dfrac{dy}{dx} + y = x$

34. $x^4\dfrac{dy}{dx} + 3x^3y = 1$

35. $x \ln x \dfrac{dy}{dx} + y = 2x^2$

36. $x\dfrac{dy}{dx} + 2y - e^{2x} = 0$

Find the particular solution for each initial value problem. (Some solutions may give y implicitly.)

37. $\dfrac{dy}{dx} = x^2 - 6x;\quad y(0) = 3$

38. $\dfrac{dy}{dx} = 5(e^{-x} - 1);\quad y(0) = 17$

39. $\dfrac{dy}{dx} = (x + 2)^3 e^y;\quad y(0) = 0$

40. $\dfrac{dy}{dx} = (3 - 2x)y;\quad y(0) = 5$

41. $\dfrac{dy}{dx} = \dfrac{1 - 2x}{y + 3};\quad y(0) = 16$

42. $\sqrt{x}\dfrac{dy}{dx} = xy;\quad y(1) = 4$

43. $e^x\dfrac{dy}{dx} - e^x y = x^2 - 1;\quad y(0) = 42$

44. $\dfrac{dy}{dx} + 3x^2 y = x^2;\quad y(0) = 2$

45. $x\dfrac{dy}{dx} - 2x^2y + 3x^2 = 0; \quad y(0) = 15$

46. $x^2\dfrac{dy}{dx} + 4xy - e^{2x^3} = 0; \quad y(1) = e^2$

47. When is Euler's method useful?

Use Euler's method to approximate the indicated function value for $y = f(x)$ to 3 decimal places, using $h = 0.2$.

48. $\dfrac{dy}{dx} = x + y^{-1}; \quad y(0) = 1; \quad$ find $y(1)$

49. $\dfrac{dy}{dx} = e^x + y; \quad y(0) = 1; \quad$ find $y(0.6)$

50. Let $y = f(x)$ and $dy/dx = (x/2) + 4$, with $y(0) = 0$. Use Euler's method with $h = 0.1$ to approximate $y(0.3)$ to 3 decimal places. Then solve the differential equation and find $f(0.3)$ to 3 decimal places. Also, find $y_3 - f(x_3)$.

51. Let $y = f(x)$ and $dy/dx = 3 + \sqrt{y}$, with $y(0) = 0$. Construct a table for x_i and y_i like the one in Section 10.3, Example 2, for $[0, 1]$, with $h = 0.2$. Then graph the polygonal approximation of the graph of $y = f(x)$.

52. What is the logistic equation? Why is it useful?

APPLICATIONS

Business and Economics

53. Marginal Sales The marginal sales (in hundreds of dollars) of a computer software company are given by

$$\frac{dy}{dx} = 6e^{0.3x},$$

where x is the number of months the company has been in business. Assume that sales were 0 initially.

a. Find the sales after 6 months.

b. Find the sales after 12 months.

54. Production Rate The rate at which a new worker in a certain factory produces items is given by

$$\frac{dy}{dx} = 0.1(150 - y),$$

where y is the number of items produced by the worker per day, x is the number of days worked, and the maximum production per day is 150 items. Assume that the worker produces 15 items at the beginning of the first day on the job $(x = 0)$.

a. Find the number of items the new worker will produce in 10 days.

b. Determine the number of days for a new worker to produce 100 items per day.

55. Continuous Withdrawals A retirement savings account contains $300,000 and earns 5% interest compounded continuously. The retiree makes *continuous* withdrawals of $20,000 per year.

a. Write a differential equation to describe the situation.

b. How much will be left in the account after 10 years?

56. In Exercise 55, approximately how long will it take to use up the account?

Life Sciences

57. Effect of Insecticide After use of an experimental insecticide, the rate of decline of an insect population is

$$\frac{dy}{dt} = \frac{-10}{1 + 5t},$$

where t is the number of hours after the insecticide is applied. Assume that there were 50 insects initially.

a. How many are left after 24 hours?

b. How long will it take for the entire population to die?

58. Growth of a Mite Population A population of mites grows at a rate proportional to the number present, y. If the growth constant is 10% and 120 mites are present at time $t = 0$ (in weeks), find the number present after 6 weeks.

59. Competing Species Find an equation relating x to y given the following equations, which describe the interaction of two competing species and their growth rates.

$$\frac{dx}{dt} = 0.2x - 0.5xy$$
$$\frac{dy}{dt} = -0.3y + 0.4xy$$

Find the values of x and y for which both growth rates are 0.

60. Smoke Content in a Room The air in a meeting room of 15,000 ft^3 has a smoke content of 20 parts per million (ppm). An air conditioner is turned on, which brings fresh air (with no smoke) into the room at a rate of 1200 ft^3 per minute and forces the smoky air out at the same rate. How long will it take to reduce the smoke content to 5 ppm?

61. In Exercise 60, how long will it take to reduce the smoke content to 10 ppm if smokers in the room are adding smoke at the rate of 5 ppm per minute?

62. Spread of Influenza A small, isolated mountain community with a population of 700 is visited by an outsider who carries influenza. After 6 weeks, 300 people are uninfected.

a. Write an equation for the number of people who remain uninfected at time t (in weeks).

b. Find the number still uninfected after 7 weeks.

c. When will the maximum infection rate occur?

63. Population Growth Let

$$y = \frac{N}{1 + be^{-kt}}.$$

If y is y_1, y_2, and y_3 at times t_1, t_2, and $t_3 = 2t_2 - t_1$ (that is, at three equally spaced times), then prove that

$$N = \frac{1/y_1 + 1/y_3 - 2/y_2}{1/(y_1y_3) - 1/y_2^2}.$$

Population Growth In the following table of U.S. Census figures, y is the population in millions. *Source: U.S. Census Bureau.*

Year	y	Year	y
1790	3.9	1910	92.0
1800	5.3	1920	105.7
1810	7.2	1930	122.8
1820	9.6	1940	131.7
1830	12.9	1950	150.7
1840 ·	17.1	1960	179.3
1850	23.2	1970	203.3
1860	31.4	1980	226.5
1870	39.8	1990	248.7
1880	50.2	2000	281.4
1890	62.9	2010	308.7
1900	76.0		

64. Use Exercise 63 and the table to find the following.

   a. Find N using the years 1800, 1850, and 1900.

   b. Find N using the years 1850, 1900, and 1950.

   c. Find N using the years 1870, 1920, and 1970.

   d. Explain why different values of N were obtained in parts a–c. What does this suggest about the validity of this model and others?

65. Let $t = 0$ correspond to 1870, and let every decade correspond to an increase in t of 1. Use the table from Exercise 63.

   a. Use 1870, 1920, and 1970 to find N, 1870 to find b, and 1920 to find k in the equation

$$y = \frac{N}{1 + be^{-kt}}.$$

   b. Estimate the population of the United States in 2010 and compare your estimate to the actual population in 2010.

   c. Predict the populations of the United States in 2030 and 2050.

66. Let $t = 0$ correspond to 1790, and let every decade correspond to an increase in t of 1. Use a calculator with logistic regression capability to complete the following. Use the table from Exercise 63.

   a. Plot the data points. Do the points suggest that a logistic function is appropriate here?

   b. Use the logistic regression function on your calculator to determine the logistic equation that best fits the data.

   c. Plot the logistic equation from part a on the same graph as the data points. How well does the logistic equation seem to fit the data?

   d. What seems to be the limiting size of the U.S. population?

Social Sciences

67. Education Researchers have proposed that the amount a full-time student is educated (x) changes with respect to the student's age t according to the differential equation

$$\frac{dx}{dt} = 1 - kx,$$

where k is a constant measuring the rate that education depreciates due to forgetting or technological obsolescence. *Source: Operations Research.*

   a. Solve the equation using the method of separation of variables.

   b. Solve the equation using an integrating factor.

   c. What does x approach over time?

68. Spread of a Rumor A rumor spreads through the offices of a company with 200 employees, starting in a meeting with 10 people. After 3 days, 35 people have heard the rumor.

   a. Write an equation for the number of people who have heard the rumor in t days. (*Hint:* Refer to Exercises 14–16 in Section 10.4.)

   b. How many people have heard the rumor in 5 days?

Physical Sciences

69. Newton's Law of Cooling A roast at a temperature of 40°F is put in a 300°F oven. After 1 hour the roast has reached a temperature of 150°F. Newton's law of cooling states that

$$\frac{dT}{dt} = k(T - T_M),$$

where T is the temperature of an object, the surrounding medium has temperature T_M at time t, and k is a constant. Use Newton's law to find the temperature of the roast after 2 hours.

70. In Exercise 69, how long does it take for the roast to reach a temperature of 250°F?

71. Air Resistance In Section 7.1 on Antiderivatives, we saw that the acceleration of gravity is a constant if air resistance is ignored. But air resistance cannot always be ignored, or parachutes would be of little use. In the presence of air resistance, the equation for acceleration also contains a term roughly proportional to the velocity squared. Since acceleration forces a falling object downward and air resistance pushes it upward, the air resistance term is opposite in sign to the acceleration of gravity. Thus,

$$a(t) = \frac{dv}{dt} = g - kv^2,$$

where g and k are positive constants. Future calculations will be simpler if we replace g and k by the squared constants G^2 and K^2, giving

$$\frac{dv}{dt} = G^2 - K^2v^2.$$

   a. Use separation of variables and the fact that

$$\frac{1}{G^2 - K^2v^2} = \frac{1}{2G}\left(\frac{1}{G - Kv} + \frac{1}{G + Kv}\right)$$

to solve the differential equation above. Assume $v < G/K$, which is certainly true when the object starts falling (with $v = 0$). Write your solution in the form of v as a function of t.

b. Find $\lim\limits_{t \to \infty} v(t)$, where $v(t)$ is the solution you found in part a. What does this tell you about a falling object in the presence of air resistance?

c. According to *Harper's Index*, the terminal velocity of a cat falling from a tall building is 60 mph. *Source: Harper's*. Use your answers from part b, plus the fact that 60 mph = 88 ft per second and g, the acceleration of gravity, is 32 ft per second^2, to find a formula for the velocity of a falling cat (in ft per second) as a function of time (in seconds). (*Hint:* Find K in terms of G. Then substitute into the answer from part a.)

EXTENDED APPLICATION

POLLUTION OF THE GREAT LAKES

Industrial nations are beginning to face the problems of water pollution. Lakes present a problem, because a polluted lake contains a considerable amount of water that must somehow be cleaned. The main cleanup mechanism is the natural process of gradually replacing the water in the lake. This application deals with pollution in the Great Lakes. The basic idea is to regard the flow in the Great Lakes as a mixing problem.

We make the following assumptions.

1. Rainfall and evaporation balance each other, so the average rates of inflow and outflow are equal.

2. The average rates of inflow and outflow do not vary much seasonally.

3. When water enters the lake, perfect mixing occurs, so that the pollutants are uniformly distributed.

4. Pollutants are not removed from the lake by decay, sedimentation, or in any other way except outflow.

5. Pollutants flow freely out of the lake; they are not retained (as DDT is).

(The first two are valid assumptions; however, the last three are questionable.)

We will use the following variables in the discussion to follow.

V = volume of the lake

P_L = pollution concentration in the lake at time t

P_i = pollution concentration in the inflow to the lake at time t

r = rate of flow

t = time in years

Source: An Introduction to Mathematical Modeling.

By the assumptions stated above, the net change in total pollutants during the time interval Δt is (approximately)

$$V \cdot \Delta P_L = (P_i - P_L)(r \cdot \Delta t),$$

where ΔP_L is the change in the pollution concentration. Dividing this equation by Δt and by V and taking the limit as $\Delta t \to 0$, we get the differential equation

$$\frac{dP_L}{dt} = \frac{(P_i - P_L)r}{V}.$$

Since we are treating V and r as constants, we replace r/V with k, so the equation can be written as the first-order linear equation

$$\frac{dP_L}{dt} + kP_L = kP_i.$$

The solution is

$$P_L(t) = e^{-kt}\left[P_L(0) + k\int_0^t P_i(x)e^{kx}\,dx \right]. \tag{1}$$

Figure 11 shows values of $1/k$ for each lake (except Huron) measured in years. *If the model is reasonable*, the numbers in the figure can be used in Equation (1) to determine the effect of various pollution abatement schemes. Lake Ontario is excluded from the discussion because about 84% of its inflow comes from Erie and can be controlled only indirectly.

FIGURE 11

The fastest possible cleanup will occur if all pollution inflow ceases. This means that $P_i = 0$. In this case, Equation (1) leads to

$$t = \frac{1}{k} \ln\left(\frac{P_L(0)}{P_L(t)}\right).$$

From this we can tell the length of time necessary to reduce pollution to a given percentage of its present level. For example, from the figure, for Lake Superior $1/k = 189$. Thus, to reduce pollution to 50% of its present level, $P_L(0)$, we want

$$\frac{P_L(t)}{P_L(0)} = 0.5 \qquad \text{or} \qquad \frac{P_L(0)}{P_L(t)} = 2,$$

from which

$$t = 189 \ln 2 \approx 131.$$

The following figures, representing years, were found in this way. Fortunately, the pollution in Lake Superior is quite low at present.

As mentioned before, assumptions 3, 4, and 5 are questionable. For persistent pollutants like DDT, the estimated cleanup times may be too low. For other pollutants, how assumptions 4 and 5 affect cleanup times is unclear. However, the values of $1/k$ given in the figure probably provide rough lower bounds for the cleanup times of persistent pollutants.

Lake	50%	20%	10%	5%
Erie	2	4	6	8
Michigan	21	50	71	92
Superior	131	304	435	566

EXERCISES

1. Calculate the number of years to reduce pollution in Lake Erie to each level.
   a. 40%
   b. 30%

2. Repeat Exercise 1 for Lake Michigan.

3. Repeat Exercise 1 for Lake Superior.

4. We claim that Equation (1) is a solution of the differential equation

$$\frac{dP_L}{dt} + kP_L(t) = kP_i(t),$$

where t measures time from the present. The constant $k = r/V$ measures how quickly the water in the lake is replaced through inflow and corresponding outflow. The constant $P_L(0)$ is the current pollution level.

a. To show that Equation (1) does define a solution of the differential equation, multiply both sides of Equation (1) by e^{kt} and then differentiate both sides with respect to t. Remember from the section on the Fundamental Theorem of Calculus that you can differentiate an integral by using the version of the Fundamental Theorem that says

$$\frac{d}{dt} \int_a^t f(x)\, dx = f(t).$$

b. When you substitute $t = 0$ into the right-hand side of Equation (1), you should get $P_L(0)$. Do you? What happens to the integral? What happens to the factor of e^{-kt}?

c. The map indicates a value of 30.8 for Lake Michigan. What value of k does this correspond to? What percent of the water in Lake Michigan is replaced each year by inflow? Which lake has the biggest annual water turnover?

5. Suppose that instead of assuming that all pollution inflow immediately ceases, we model $P_i(t)$ by a decaying exponential of the form $a \cdot e^{-pt}$, where p is a constant that tells us how fast the inflow is being cleaned of pollution. To simplify things, we'll also assume that initially the inflow and the lake have the same pollution concentration, so $a = P_L(0)$. Now substitute $P_L(0)e^{-px}$ for $P_i(x)$ in Equation (1), and evaluate the integral as a function of t.

6. When you simplify the right-hand side of Equation (1) using your new expression for the integral, and then factor out and divide by $P_L(0)$, you'll get the following nice expression for the ratio $P_L(t)/P_L(0)$:

$$\frac{P_L(t)}{P_L(0)} = \frac{1}{k - p}\left(ke^{-pt} - pe^{-kt}\right).$$

a. Suppose that for Lake Michigan the constant p is equal to 0.02. Use a graph of the ratio $P_L(t)/P_L(0)$ to estimate how long it will take to reduce pollution to 50% of its current value. How does this compare with the time, assuming pollution-free inflow?

b. If the constant p has the value 0 for Lake Michigan, what does that tell you about the pollution level in the inflow? In this case, what happens to the ratio $P_L(t)/P_L(0)$ over time?

7. At the website Wolfram|Alpha.com, you can enter "$y'(t) = (f(t) - y) * k, y(0) = a$" to solve the initial value problem in this Extended Application, where we have used $y(t)$ to represent $P_L(t)$, $f(t)$ to represent $P_i(t)$, and a to represent $P_L(0)$. Try this, and verify that the solution is equivalent to Equation (1).

8. Repeat Exercise 7, but in place of $f(t)$, put $a * e \wedge (-p * t)$, the form of $P_L(t)$ used in Exercises 5 and 6. Verify that the solution is equivalent to the solution given in Exercise 6.

9. Repeat Exercise 8, trying other functions of t in place of $f(t)$, such as t^3. Find which functions give a recognizable answer, and verify that answer using Equation (1).

DIRECTIONS FOR GROUP PROJECT

Suppose you and three others are employed by an agency that is concerned about the environmental health of one of the Great Lakes. Choose one of the lakes, and collect information about levels of pollution in it. Then, using the information you collected along with the information given in this application, prepare a public presentation for a local community organization that describes the lake and gives possible timelines for reducing pollution in the lake. Use presentation software such as Microsoft PowerPoint.

Probability and Calculus

Though earthquakes may appear to strike at random, the times between quakes can be modeled with an exponential density function. Such *continuous probability models* have many applications in science, engineering, and medicine. In an exercise in Section 1 of this chapter we'll use an exponential density function to describe the times between major earthquakes in Southern California, and in Section 3 we will compute the mean and standard deviation for this distribution.

In recent years, probability has become increasingly useful in fields ranging from manufacturing to medicine, as well as in all types of research. The foundations of probability were laid in the seventeenth century by Blaise Pascal (1623–1662) and Pierre de Fermat (1601–1665), who investigated *the problem of the points*. This problem dealt with the fair distribution of winnings in an interrupted game of chance between two equally matched players whose scores were known at the time of the interruption.

Probability has advanced from a study of gambling to a well-developed, deductive mathematical system. In this chapter we give a brief introduction to the use of calculus in probability.

11.1 Continuous Probability Models

APPLY IT   **What is the probability that there is a bird's nest within 0.5 kilometers of a given point?**

In Example 3, we will answer the question posed above.

In this section, we show how calculus is used to find the probability of certain events. Before discussing probability, however, we need to introduce some new terminology.

Suppose that a bank is studying the transaction times of its tellers. The lengths of time spent on observed transactions, rounded to the nearest minute, are shown in the following table.

Frequency of Transaction Times										
Time	1	2	3	4	5	6	7	8	9	10	
Frequency	3	5	9	12	15	11	10	6	3	1	(Total: 75)

The table shows, for example, that 9 of the 75 transactions in the study took 3 minutes, 15 transactions took 5 minutes, and 1 transaction took 10 minutes. Because the time for any particular transaction is a random event, the number of minutes for a transaction is called a **random variable**. The frequencies can be converted to probabilities by dividing each frequency by the total number of transactions (75) to get the results shown in the next table.*

Probability of Transaction Times										
Time	1	2	3	4	5	6	7	8	9	10
Probability	0.04	0.07	0.12	0.16	0.20	0.15	0.13	0.08	0.04	0.01

Because each value of the random variable is associated with just one probability, this table defines a function. Such a function is called a **probability function**, and it has the following special properties.

*One definition of the *probability of an event* is the number of outcomes that favor the event divided by the total number of equally likely outcomes in an experiment.

Probability Function of a Random Variable

If the function f is a probability function with domain $\{x_1, x_2, \ldots, x_n\}$, and $f(x_i)$ is the probability that event x_i occurs, then for $1 \le i \le n$,

$$0 \le f(x_i) \le 1,$$

and

$$f(x_1) + f(x_2) + \cdots + f(x_n) = 1.$$

Note that $f(x_i) = 0$ implies that event x_i will not occur and $f(x_i) = 1$ implies that event x_i will occur.

The information in the second table can be displayed graphically with a special kind of bar graph called a **histogram**. The bars of a histogram have the same width, and their heights are determined by the probabilities of the various values of the random variable. See Figure 1.

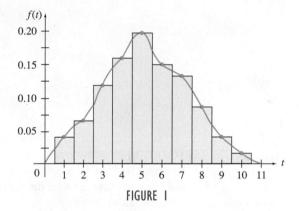

FIGURE 1

The probability function in the second table is a **discrete probability function** because it has a finite domain—the integers from 1 to 10, inclusive. A discrete probability function has a finite domain or an infinite domain that can be listed. For example, if we flip a coin until we get heads, and let the random variable be the number of flips, then the domain is 1, 2, 3, 4, ... On the other hand, the distribution of heights (in inches) of college women includes infinitely many possible measurements, such as 53, 54.2, 66.5, 72.$\overline{3}$, and so on, *within some real number interval*. Probability functions with such domains are called *continuous probability distributions*.

Continuous Probability Distribution

A **continuous random variable** can take on any value in some interval of real numbers. The distribution of this random variable is called a **continuous probability distribution**.

Some probability functions are inherently discrete. For example, the number of houses that a real estate agent sells in a year must be an integer, such as 0, 1, or 2, and could never take on any value in between. But the bank example discussed earlier is different, because you could think of it as a simplification of a continuous distribution. It would be possible to time the teller transactions with greater precision—to the nearest tenth of a minute, or even to the nearest 1/60 of a minute if desired. Theoretically, at least, t could take on any positive real-number value between, say, 0 and 11 minutes. The graph of the probabilities $f(t)$ of these transaction times can be thought of as the continuous curve shown in Figure 1. As indicated in Figure 1, the curve was derived from our table by connecting the points at the tops of the bars in the corresponding histogram and smoothing the resulting polygon into a curve.

To clarify some concepts in probability, we will follow the common convention of using capital letters to indicate random variables and lower case letters to indicate the values that the random variables take on. For example, to indicate the probability that a random variable takes on the value 2, we will write $P(X = 2)$. To indicate the probability that a random variable takes on the arbitrary value x, we will write $P(X = x)$.

For a discrete probability function, the area of each bar (or rectangle) gives the probability of a particular transaction time. Thus, by considering the possible transaction times T as all the real numbers between 0 and 11, the area under the curve of Figure 2 between any two values of T can be interpreted as the probability that a transaction time will be between those two numbers. For example, the shaded region in Figure 2 corresponds to the probability that T is between a and b, written $P(a \leq T \leq b)$.

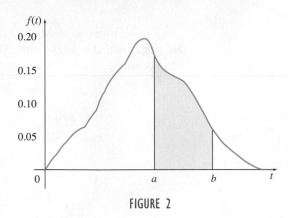

FIGURE 2

FOR REVIEW

The connection between area and the definite integral is discussed in Chapter 7 on Integration. For example, in that chapter we solved such problems as the following:

Find the area between the x-axis and the graph of $f(x) = x^2$ from $x = 1$ to $x = 4$.

Answer: $\int_1^4 x^2 \, dx = 21$

It was shown earlier that the definite integral of a continuous function f, where $f(x) \geq 0$, gives the area under the graph of $f(x)$ from $x = a$ to $x = b$. If a function f can be found to describe a continuous probability distribution, then the definite integral can be used to find the area under the curve from a to b that represents the probability that x will be between a and b.

If X is a continuous random variable whose distribution is described by the function f on $[a, b]$, then

$$P(a \leq X \leq b) = \int_a^b f(x) \, dx.$$

Probability Density Functions

A function f that describes a continuous probability distribution is called a *probability density function*. Such a function must satisfy the following conditions.

Probability Density Function

The function f is a **probability density function** of a random variable X in the interval $[a, b]$ if

1. $f(x) \geq 0$ for all x in the interval $[a, b]$; and

2. $\int_a^b f(x) \, dx = 1$.

Intuitively, Condition 1 says that the probability of a particular event can never be negative. Condition 2 says that the total probability for the interval must be 1; *something must happen.*

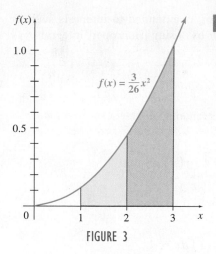

FIGURE 3

YOUR TURN 1 Repeat Example 1(a) for the function $f(x) = 2/x^2$ on $[1, 2]$. What is the probability that X will be between $3/2$ and 2?

EXAMPLE 1 Probability Density Function

(a) Show that the function defined by $f(x) = (3/26)x^2$ is a probability density function for the interval $[1, 3]$.

SOLUTION First, note that Condition 1 holds; that is, $f(x) \geq 0$ for the interval $[1, 3]$. Next show that Condition 2 holds.

$$\int_1^3 \frac{3}{26}x^2 dx = \frac{3}{26}\left(\frac{x^3}{3}\right)\Big|_1^3 = \frac{3}{26}\left(9 - \frac{1}{3}\right) = 1$$

Since both conditions hold, $f(x)$ is a probability density function.

(b) Find the probability that X will be between 1 and 2.

SOLUTION The desired probability is given by the area under the graph of $f(x)$ between $x = 1$ and $x = 2$, as shown in blue in Figure 3. The area is found by using a definite integral.

$$P(1 \leq X \leq 2) = \int_1^2 \frac{3}{26}x^2\, dx = \frac{3}{26}\left(\frac{x^3}{3}\right)\Big|_1^2 = \frac{7}{26} \quad \text{TRY YOUR TURN 1}$$

Earlier, we noted that determining a suitable function is the most difficult part of applying mathematics to actual situations. Sometimes a function appears to model an application well but does not satisfy the requirements for a probability density function. In such cases, we may be able to change the function into a probability density function by multiplying it by a suitable constant, as shown in the next example.

EXAMPLE 2 Probability Density Function

Is there a constant k such that $f(x) = kx^2$ is a probability density function for the interval $[0, 4]$?

SOLUTION First,

$$\int_0^4 kx^2\, dx = \frac{kx^3}{3}\Big|_0^4 = \frac{64k}{3}.$$

YOUR TURN 2 Repeat Example 2 for the function $f(x) = kx^3$ on the interval $[0, 4]$.

The integral must be equal to 1 for the function to be a probability density function. To convert it to one, let $k = 3/64$. The function defined by $(3/64)x^2$ for $[0, 4]$ will be a probability density function, since $(3/64)x^2 \geq 0$ for all x in $[0, 4]$ and

$$\int_0^4 \frac{3}{64}x^2 = 1. \quad \text{TRY YOUR TURN 2}$$

An important distinction is made between a discrete probability function and a probability density function (which is continuous). In a discrete distribution, the probability that the random variable, X, will assume a specific value is given in the distribution for every possible value of X. In a probability density function, however, the probability that X equals a specific value, say, c, is

$$P(X = c) = \int_c^c f(x)\, dx = 0.$$

For a probability density function, only probabilities of *intervals* can be found. For example, suppose the random variable is the annual rainfall for a given region. The amount of rainfall in one year can take on any value within some continuous interval that depends on the region; however, the probability that the rainfall in a given year will be some specific amount, say 33.25 in., is actually zero.

The definition of a probability density function is extended to intervals such as $(-\infty, b]$, $(-\infty, b)$, $[a, \infty)$, (a, ∞), or $(-\infty, \infty)$ by using improper integrals, as follows.

Probability Density Functions on $(-\infty, \infty)$

If f is a probability density function for a continuous random variable X on $(-\infty, \infty)$, then

$$P(X \le b) = P(X < b) = \int_{-\infty}^{b} f(x)\,dx,$$

$$P(X \ge a) = P(X > a) = \int_{a}^{\infty} f(x)\,dx,$$

$$P(-\infty < X < \infty) = \int_{-\infty}^{\infty} f(x)\,dx = 1.$$

The total area under the graph of a probability density function of this type must still equal 1.

EXAMPLE 3 **Location of a Bird's Nest**

Suppose the random variable X is the distance (in kilometers) from a given point to the nearest bird's nest, with the probability density function of the distribution given by $f(x) = 2xe^{-x^2}$ for $x \ge 0$.

(a) Show that $f(x)$ is a probability density function.

SOLUTION Since $e^{-x^2} = 1/e^{x^2}$ is always positive, and $x \ge 0$,

$$f(x) = 2xe^{-x^2} \ge 0,$$

and Condition 1 holds.

Use substitution to evaluate the definite integral $\int_0^{\infty} 2xe^{-x^2}\,dx$. Let $u = -x^2$, so that $du = -2x\,dx$, and

$$\int 2xe^{-x^2}\,dx = -\int e^{-x^2}(-2x\,dx)$$

$$= -\int e^u\,du = -e^u = -e^{-x^2}.$$

Then

$$\int_0^{\infty} 2xe^{-x^2}\,dx = \lim_{b\to\infty} \int_0^{b} 2xe^{-x^2}\,dx = \lim_{b\to\infty} \left(-e^{-x^2}\right)\Big|_0^b$$

$$= \lim_{b\to\infty} \left(-\frac{1}{e^{b^2}} + e^0\right) = 0 + 1 = 1,$$

and Condition 2 holds.

The function defined by $f(x) = 2xe^{-x^2}$ satisfies the two conditions required of a probability density function.

(b) Find the probability that there is a bird's nest within 0.5 km of the given point.

SOLUTION Find $P(X \le 0.5)$ where $X \ge 0$. This probability is given by

$$P(0 \le X \le 0.5) = \int_0^{0.5} 2xe^{-x^2}\,dx.$$

FOR REVIEW

Improper integrals, those with one or two infinite limits, were discussed in Chapter 8 on Further Techniques and Applications of Integration. The type of improper integral we shall need was defined as

$$\int_a^{\infty} f(x)\,dx = \lim_{b\to\infty} \int_a^b f(x)\,dx.$$

For example,

$$\int_1^{\infty} x^{-2}\,dx = \lim_{b\to\infty} \int_1^b x^{-2}\,dx$$

$$= \lim_{b\to\infty} \left(-\frac{1}{x}\Big|_1^b\right)$$

$$= \lim_{b\to\infty} \left(-\frac{1}{b} + \frac{1}{1}\right)$$

$$= 0 + 1 = 1.$$

APPLY IT

Now evaluate the integral. The indefinite integral was found in part (a).

$$P(0 \leq X \leq 0.5) = \int_0^{0.5} 2xe^{-x^2}\, dx = \left(-e^{-x^2}\right)\Big|_0^{0.5}$$

$$= -e^{-(0.5)^2} - \left(-e^0\right) = -e^{-0.25} + 1$$

$$\approx -0.7788 + 1 = 0.2212$$

YOUR TURN 3 Using the probability density function of Example 3, find the probability that there is a bird's nest within 1 km of the given point.

The probability that a bird's nest will be found within 0.5 km of the given point is about 0.22. **TRY YOUR TURN 3**

EXAMPLE 4   Computing Mortality

According to the National Center for Health Statistics, if we start with 100,000 people who are 50 years old, we can expect a certain number of them to die within each 5-year interval, as indicated by the following table.* *Source: National Vital Statistics Reports.*

Years from Age 50	Life Table Midpoint of Interval	Number Dying in Each Interval
0–5	2.5	2565
5–10	7.5	3659
10–15	12.5	5441
15–20	17.5	7622
20–25	22.5	10,498
25–30	27.5	13,858
30–35	32.5	16,833
35–40	37.5	16,720
40–45	42.5	13,211
45–50	47.5	7068
50–55	52.5	2525

(a) Plot the data.

SOLUTION   Figure 4 shows that the plot appears to have the shape of a polynomial.

(b) Find a polynomial equation that models the number of deaths, $N(t)$, as a function of the number of years, t, since age 50. Use the midpoints and the number of deaths in each interval from the table above.

SOLUTION   The highest degree polynomial that the regression feature on a TI-84 Plus calculator can find is fourth degree. As Figure 5(a) shows, this roughly captures the behavior of the data, but it has two drawbacks. For one, it doesn't reach the highest data points. Also, it's decreasing in the beginning when it should be increasing. Higher degree polynomials can be fit using Excel or using the Multiple Regression tool on the Statistics with List Editor application for the TI-89. We were thus able to find that the function

$$N(t) = 5.03958 \times 10^{-5} t^6 - 0.006603 t^5 + 0.2992 t^4 - 6.0507 t^3 +$$
$$67.867 t^2 - 110.3 t + 2485.1$$

fits the data quite well, as shown in Figure 5(b).

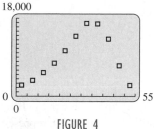

FIGURE 4

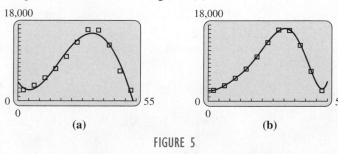

(a)   (b)

FIGURE 5

*For simplicity, we have placed all those who lived past 100 in the class of those who lived from 100 to 105.

(c) Use the answer from part (b) to find a probability density function for the random variable T representing the number of additional years that a 50-year-old person lives.

SOLUTION We will construct a density function $S(t) = kN(t)$ by finding a suitable constant k, as we did in Example 2. The graph of the function turns up after $t = 52.5$, which is unlikely for the actual mortality function, so we will restrict the domain of the density function to the interval $[0, 52.5]$, even though this ignores those who live more than 102.5 years. Using the integration feature on our calculator, we find that

$$\int_0^{52.5} S(t)\,dt = k\int_0^{52.5} N(t)\,dt = 497{,}703k.$$

Notice that this number is close to the product of 5 years (interval length) and 100,000 (the total number of people). This is not a coincidence! We set the above integral equal to 1 to get $k = 1/497{,}703$. The function defined by

$$S(t) = \frac{1}{497{,}703}N(t)$$

$$= \frac{1}{497{,}703}(5.03958 \times 10^{-5}t^6 - 0.006603t^5 + 0.2992t^4$$

$$- 6.0507t^3 + 67.867t^2 - 110.3t + 2485.1)$$

is a probability density function for $[0, 52.5]$ because

$$\int_0^{52.5} S(t)\,dt = 1, \quad \text{and } S(t) \geq 0 \text{ for all } t \text{ in } [0, 52.5].$$

(d) Find the probability that a randomly chosen 50-year-old person will live at least until age 70.

SOLUTION Again using the integration feature on our calculator,

$$P(T \geq 20) = \int_{20}^{52.5} S(t)\,dt \approx 0.8054.$$

Thus a 50-year-old person has a 80.54% chance of living at least until age 70.

Notice that this value could also be estimated from the table by finding the number of people who have not died by age 70 and then dividing this number by 100,000. Thus, according to our table, there are 80,713 people still alive at age 70, representing 80.7% of the original population. As you can see, our estimate agrees quite well with the actual number.

Another important concept in probability is the *cumulative distribution function*, which gives the probability that a random variable X is less than or equal to an arbitrary value x.

Cumulative Distribution Function

If f is a probability density function of a random variable in the interval $[a, b]$, then the **cumulative distribution function** is defined as

$$F(x) = P(X \leq x) = \int_a^x f(t)\,dt$$

for $x \geq a$. Also, $F(x) = 0$ for $x < a$.

NOTE
1. We integrate with respect to the variable t in the integral, rather than x, because we are already using x for the upper limit on the integral. It doesn't matter what variable of integration is used in a definite integral, since that variable doesn't appear in the final answer. We just need to use a variable that's not being used for another purpose.
2. If the random variable is defined on the interval $(-\infty, \infty)$, simply replace a with $-\infty$ in the above definition.

EXAMPLE 5 Cumulative Distribution Function

Consider the random variable X defined in Example 3, giving the distance (in kilometers) from a given point to the nearest bird's nest, with probability density function $f(x) = 2xe^{-x^2}$ for $x \geq 0$.

(a) Find the cumulative distribution function for this random variable.

SOLUTION The cumulative distribution function is given by

$$F(x) = P(X \leq x) = \int_0^x 2te^{-t^2}\, dt \qquad \text{Use the density function with } t \text{ as the variable}$$

$$= -e^{-t^2}\Big|_0^x \qquad \text{Use the antiderivative found in Example 3.}$$

$$= -e^{-x^2} + 1$$

for $x \geq 0$. The cumulative distribution function can be written as $F(x) = 1 - e^{-x^2}$ for $x \geq 0$. Note that for $x < 0$, $F(x) = 0$.

(b) Use the solution to part (a) to calculate the probability that there is a bird's nest within 0.5 km of the given point.

SOLUTION To find $P(X \leq 0.5)$, calculate $F(0.5) = 1 - e^{-0.5^2} \approx 0.2212$. Notice that this is the same answer that we found in Example 3(b). **TRY YOUR TURN 4**

YOUR TURN 4 Use part (a) of Example 5 to calculate the probability that there is a bird's nest within 1 km of the given point.

11.1 EXERCISES

Decide whether the functions defined as follows are probability density functions on the indicated intervals. If not, tell why.

1. $f(x) = \dfrac{1}{9}x - \dfrac{1}{18};\ [2, 5]$

2. $f(x) = \dfrac{1}{3}x - \dfrac{1}{6};\ [3, 4]$

3. $f(x) = \dfrac{x^2}{21};\ [1, 4]$

4. $f(x) = \dfrac{3}{98}x^2;\ [3, 5]$

5. $f(x) = 4x^3;\ [0, 3]$

6. $f(x) = \dfrac{x^3}{81};\ [0, 3]$

7. $f(x) = \dfrac{x^2}{16};\ [-2, 2]$

8. $f(x) = 2x^2;\ [-1, 1]$

9. $f(x) = \dfrac{5}{3}x^2 - \dfrac{5}{90};\ [-1, 1]$

10. $f(x) = \dfrac{3}{13}x^2 - \dfrac{12}{13}x + \dfrac{45}{52};\ [0, 4]$

Find a value of k that will make f a probability density function on the indicated interval.

11. $f(x) = kx^{1/2};\ [1, 4]$

12. $f(x) = kx^{3/2};\ [4, 9]$

13. $f(x) = kx^2;\ [0, 5]$

14. $f(x) = kx^2;\ [-1, 2]$

15. $f(x) = kx;\ [0, 3]$

16. $f(x) = kx;\ [2, 3]$

17. $f(x) = kx;\ [1, 5]$

18. $f(x) = kx^3;\ [2, 4]$

Find the cumulative distribution function for the probability density function in each of the following exercises.

19. Exercise 1

20. Exercise 2

21. Exercise 3

22. Exercise 4

23. Exercise 11

24. Exercise 12

25. The total area under the graph of a probability density function always equals _____.

26. In your own words, define a random variable.

27. What is the difference between a discrete probability function and a probability density function?

28. Why is $P(X = c) = 0$ for any number c in the domain of a probability density function?

Show that each function defined as follows is a probability density function on the given interval; then find the indicated probabilities.

29. $f(x) = \frac{1}{2}(1 + x)^{-3/2}$; $[0, \infty)$

   a. $P(0 \le X \le 2)$ **b.** $P(1 \le X \le 3)$

   c. $P(X \ge 5)$

30. $f(x) = e^{-x}$; $[0, \infty)$

   a. $P(0 \le X \le 1)$ **b.** $P(1 \le X \le 2)$

   c. $P(X \le 2)$

31. $f(x) = (1/2)e^{-x/2}$; $[0, \infty)$

   a. $P(0 \le X \le 1)$ **b.** $P(1 \le X \le 3)$

   c. $P(X \ge 2)$

32. $f(x) = \dfrac{20}{(x + 20)^2}$; $[0, \infty)$

   a. $P(0 \le X \le 1)$ **b.** $P(1 \le X \le 5)$

   c. $P(X \ge 5)$

33. $f(x) = \begin{cases} \dfrac{x^3}{12} & \text{if } 0 \le x \le 2 \\ \dfrac{16}{3x^3} & \text{if } x > 2 \end{cases}$

   a. $P(0 \le X \le 2)$ **b.** $P(X \ge 2)$

   c. $P(1 \le X \le 3)$

34. $f(x) = \begin{cases} \dfrac{20x^4}{9} & \text{if } 0 \le x \le 1 \\ \dfrac{20}{9x^5} & \text{if } x > 1 \end{cases}$

   a. $P(0 \le X \le 1)$ **b.** $P(X \ge 1)$

   c. $P(0 \le X \le 2)$

APPLICATIONS

Business and Economics

35. Life Span of a Computer Part The life (in months) of a certain electronic computer part has a probability density function defined by

$$f(t) = \frac{1}{2}e^{-t/2} \quad \text{for } t \text{ in } [0, \infty).$$

Find the probability that a randomly selected component will last the following lengths of time.

   a. At most 12 months

   b. Between 12 and 20 months

   c. Find the cumulative distribution function for this random variable.

   d. Use the answer to part c to find the probability that a randomly selected component will last at most 6 months.

36. Machine Life A machine has a useful life of 4 to 9 years, and its life (in years) has a probability density function defined by

$$f(t) = \frac{1}{11}\left(1 + \frac{3}{\sqrt{t}}\right).$$

Find the probabilities that the useful life of such a machine selected at random will be the following.

   a. Longer than 6 years

   b. Less than 5 years

   c. Between 4 and 7 years

   d. Find the cumulative distribution function for this random variable.

   e. Use the answer to part d to find the probability that a randomly selected machine has a useful life of at most 8 years.

37. Machine Part The lifetime of a machine part has a continuous distribution on the interval (0, 40) with probability density function f, where $f(x)$ is proportional to $(10 + x)^{-2}$. Calculate the probability that the lifetime of the machine part is less than 6. Choose one of the following. *Source: Society of Actuaries.*

   a. 0.04   **b.** 0.15   **c.** 0.47   **d.** 0.53   **e.** 0.94

38. Insurance An insurance policy pays for a random loss X subject to a deductible of C, where $0 < C < 1$. The loss amount is modeled as a continuous random variable with density function

$$f(x) = \begin{cases} 2x & \text{for } 0 < x < 1 \\ 0 & \text{otherwise.} \end{cases}$$

Given a random loss X, the probability that the insurance payment is less than 0.5 is equal to 0.64. Calculate C. Choose one of the following. (*Hint:* The payment is 0 unless the loss is greater than the deductible, in which case the payment is the loss minus the deductible.) *Source: Society of Actuaries.*

   a. 0.1   **b.** 0.3   **c.** 0.4   **d.** 0.6   **e.** 0.8

Life Sciences

39. Petal Length The length of a petal on a certain flower varies from 1 cm to 4 cm and has a probability density function defined by

$$f(x) = \frac{1}{2\sqrt{x}}.$$

Find the probabilities that the length of a randomly selected petal will be as follows.

   a. Greater than or equal to 3 cm

   b. Less than or equal to 2 cm

   c. Between 2 cm and 3 cm

40. Clotting Time of Blood The clotting time of blood is a random variable t with values from 1 second to 20 seconds and probability density function defined by

$$f(t) = \frac{1}{(\ln 20)t}.$$

Find the following probabilities for a person selected at random.

a. The probability that the clotting time is between 1 and 5 seconds

b. The probability that the clotting time is greater than 10 seconds

41. Flour Beetles Researchers who study the abundance of the flour beetle, *Tribolium castaneum*, have developed a probability density function that can be used to estimate the abundance of the beetle in a population. The density function, which is a member of the gamma distribution, is

$$f(x) = 1.185 \times 10^{-9} x^{4.5222} e^{-0.049846x},$$

where x is the size of the population. *Source: Ecology.*

a. Estimate the probability that a randomly selected flour beetle population is between 0 and 150.

b. Estimate the probability that a randomly selected flour beetle population is between 100 and 200.

42. Flea Beetles The mobility of an insect is an important part of its survival. Researchers have determined that the probability that a marked flea beetle, *Phyllotreta cruciferae* and *Phyllotreta striolata*, will be recaptured within a certain distance and time after release can be calculated from the probability density function

$$p(x, t) = \frac{e^{-x^2/(4Dt)}}{\displaystyle\int_0^L e^{-u^2/(4Dt)} \, du},$$

where t is the time after release (in hours), x is the distance (in meters) from the release point that recaptures occur, L is the maximum distance from the release point that recaptures can occur, and D is the diffusion coefficient. *Source: Ecology Monographs.*

a. If $t = 12$, $L = 6$, and $D = 38.3$, find the probability that a flea beetle will be recaptured within 3 m of the release point.

b. Using the same values for t, L, and D, find the probability that a flea beetle will be recaptured between 1 and 5 m of the release point.

Social Sciences

43. Social Network The number of U.S. users (in millions) on Facebook, a computer social network, in 2009 is given in the table below. *Source: Inside Facebook.*

Age Interval (years)	Midpoint of Interval (year)	Number of Users in Each Interval (millions)
13–17	15	6.049
18–25	21.5	19.461
26–34	30	13.423
35–44	39.5	9.701
45–54	49.5	4.582
55–65	60	2.849
Total		56.065

a. Plot the data. What type of function appears to best match this data?

b. Use the regression feature on your graphing calculator to find a quartic equation that models the number of years, t, since birth and the number of Facebook users, $N(t)$. Use the midpoint value to estimate the point in each interval for the age of the Facebook user. Graph the function with the plot of the data. Does the function resemble the data?

c. By finding an appropriate constant k, find a function $S(t) = kN(t)$ that is a probability density function describing the probability of the age of a Facebook user. (*Hint:* Because the function in part b is negative for values less than 13.4 and greater than 62.0, restrict the domain of the density function to the interval [13.4, 62.0]. That is, integrate the function you found in part b from 13.4 to 62.0.)

d. For a randomly chosen person who uses Facebook, find the probabilities that the person was at least 35 but less than 45 years old, at least 18 but less than 35 years old, and at least 45 years old. Compare these with the actual probabilities.

44. Time to Learn a Task The time required for a person to learn a certain task is a random variable with probability density function defined by

$$f(t) = \frac{8}{7(t-2)^2}.$$

The time required to learn the task is between 3 and 10 minutes. Find the probabilities that a randomly selected person will learn the task in the following lengths of time.

a. Less than 4 minutes

b. More than 5 minutes

Physical Sciences

45. Annual Rainfall The annual rainfall in a remote Middle Eastern country varies from 0 to 5 in. and is a random variable with probability density function defined by

$$f(x) = \frac{5.5 - x}{15}.$$

Find the following probabilities for the annual rainfall in a randomly selected year.

a. The probability that the annual rainfall is greater than 3 in.

b. The probability that the annual rainfall is less than 2 in.

c. The probability that the annual rainfall is between 1 in. and 4 in.

46. Earthquakes The time between major earthquakes in the Southern California region is a random variable with probability density function

$$f(t) = \frac{1}{960} e^{-t/960},$$

where t is measured in days. *Source: Journal of Seismology*.

a. Find the probability that the time between a major earth-quake and the next one is less than 365 days.

b. Find the probability that the time between a major earth-quake and the next one is more than 960 days.

47. Earthquakes The time between major earthquakes in the Taiwan region is a random variable with probability density function

$$f(t) = \frac{1}{3650.1} e^{-t/3650.1},$$

where t is measured in days. *Source: Journal of Seismology*.

a. Find the probability that the time between a major earth-quake and the next one is more than 1 year but less than 3 years.

b. Find the probability that the time between a major earth-quake and the next one is more than 7300 days.

General Interest

48. Drunk Drivers The frequency of alcohol-related traffic fatali-ties has dropped in recent years but is still high among young people. Based on data from the National Highway Traffic Safety Administration, the age of a randomly selected, alcohol-impaired driver in a fatal car crash is a random variable with probability density function given by

$$f(t) = \frac{4.045}{t^{1.532}} \quad \text{for } t \text{ in } [16, 80].$$

Find the following probabilities of the age of such a driver. *Source: Traffic Safety Facts*.

a. Less than or equal to 25

b. Greater than or equal to 35

c. Between 21 and 30

d. Find the cumulative distribution function for this random variable.

e. Use the answer to part d to find the probability that a ran-domly selected alcohol-impaired driver in a fatal car crash is at most 21 years old.

49. Driving Fatalities We saw in a review exercise in Chapter 4 on Calculating the Derivative that driver fatality rates were highest for the youngest and oldest drivers. When adjusted for the number of miles driven by people in each age group, the number of drivers in fatal crashes goes down with age, and the age of a randomly selected driver in a fatal car crash is a ran-dom variable with probability density function given by

$$f(t) = 0.06049 e^{-0.03211t} \quad \text{for } t \text{ in } [16, 84].$$

Find the following probabilities of the age of such a driver. *Source: National Highway Traffic Safety Administration*.

a. Less than or equal to 25

b. Greater than or equal to 35

c. Between 21 and 30

d. Find the cumulative distribution function for this random variable.

e. Use the answer to part d to find the probability that a ran-domly selected driver in a fatal crash is at most 21 years old.

50. Length of a Telephone Call The length of a telephone call (in minutes), t, for a certain town is a continuous random variable with probability density function defined by

$$f(t) = 3t^{-4}, \quad \text{for } t \text{ in } [1, \infty).$$

Find the probabilities for the following situations.

a. The call lasts between 1 and 2 minutes.

b. The call lasts between 3 and 5 minutes.

c. The call lasts longer than 3 minutes.

51. Time of Traffic Fatality The National Highway Traffic Safety Administration records the time of day of fatal crashes. The following table gives the time of day (in hours since midnight) and the frequency of fatal crashes. *Source: The National Highway Traffic Safety Administration*.

Time of Day	Midpoint of Interval (hours)	Frequency
0–3	1.5	4486
3–6	4.5	2774
6–9	7.5	3236
9–12	10.5	3285
12–15	13.5	4356
15–18	16.5	5325
18–21	19.5	5342
21–24	22.5	4952
Total		33,756

a. Plot the data. What type of function appears to best match this data?

b. Use the regression feature on your graphing calculator to find a cubic equation that models the time of day, t, and the number of traffic fatalities, $T(t)$. Use the midpoint value to estimate the time in each interval. Graph the function with the plot of the data. Does the graph fit the data?

c. By finding an appropriate constant k, find a function $S(t) = kT(t)$ that is a probability density function describing the probability of a traffic fatality at a particular time of day.

d. For a randomly chosen traffic fatality, find the probabilities that the accident occurred between 12 am and 2 am ($t = 0$ to $t = 2$) and between 4 pm and 5:30 pm ($t = 16$ to $t = 17.5$).

YOUR TURN ANSWERS

1. $P\left(\frac{3}{2} \le X \le 2\right) = \frac{1}{3}$ **2.** $k = 1/64$

3. 0.6321 **4.** 0.6321

11.2 Expected Value and Variance of Continuous Random Variables

APPLY IT **What is the average age of a drunk driver in a fatal car crash?**
You will be asked to answer this question in Exercise 40.

It often is useful to have a single number, a typical or "average" number, that represents a random variable. The *mean* or *expected value* for a discrete random variable is found by multiplying each value of the random variable by its corresponding probability, as follows.

Expected Value

Suppose the random variable X can take on the n values, $x_1, x_2, x_3, \ldots, x_n$. Also, suppose the probabilities that each of these values occurs are, respectively, $p_1, p_2, p_3, \ldots, p_n$. Then the **mean**, or **expected value**, of the random variable is

$$\mu = x_1 p_1 + x_2 p_2 + x_3 p_3 + \cdots + x_n p_n = \sum_{i=1}^{n} x_i p_i.$$

For the banking example in the previous section, the expected value is given by

$$\mu = 1(0.04) + 2(0.07) + 3(0.12) + 4(0.16) + 5(0.20) + 6(0.15)$$
$$+ 7(0.13) + 8(0.08) + 9(0.04) + 10(0.01)$$
$$= 5.09.$$

Thus, the average time a person can expect to spend with the bank teller is 5.09 minutes.

This definition can be extended to continuous random variables by using definite integrals. Suppose a continuous random variable has probability density function f on $[a, b]$. We can divide the interval from a to b into n subintervals of length Δx, where $\Delta x = (b - a)/n$. In the ith subinterval, the probability that the random variable takes a value close to x_i is approximately $f(x_i) \Delta x$, and so

$$\mu \approx \sum_{i=1}^{n} x_i f(x_i) \Delta x.$$

As $n \to \infty$, the limit of this sum gives the expected value

$$\mu = \int_{a}^{b} x f(x) \, dx.$$

The **variance** of a probability distribution is a measure of the *spread* of the values of the distribution. For a discrete distribution, the variance is found by taking the expected value of the squares of the differences of the values of the random variable and the mean. If the random variable X takes the values $x_1, x_2, x_3, \ldots, x_n$, with respective probabilities $p_1, p_2, p_3, \ldots, p_n$ and mean μ, then the variance of X is

$$\text{Var}(X) = \sum_{i=1}^{n} (x_i - \mu)^2 p_i.$$

Think of the variance as the expected value of $(X - \mu)^2$, which measures how far X is from the mean μ. The **standard deviation** of X is defined as

$$\sigma = \sqrt{\text{Var}(X)}.$$

For the banking example in the previous section, the variance and standard deviation are

$$\begin{aligned}
\text{Var}(X) = &(1 - 5.09)^2(0.04) + (2 - 5.09)^2(0.07) + (3 - 5.09)^2(0.12)\\
&+ (4 - 5.09)^2(0.16) + (5 - 5.09)^2(0.20) + (6 - 5.09)^2(0.15)\\
&+ (7 - 5.09)^2(0.13) + (8 - 5.09)^2(0.08) + (9 - 5.09)^2(0.04)\\
&+ (10 - 5.09)^2(0.01)\\
= &\ 4.1819
\end{aligned}$$

and

$$\sigma = \sqrt{\text{Var}(X)} \approx 2.0450.$$

Like the mean or expected value, the variance of a continuous random variable is an integral.

$$\text{Var}(X) = \int_a^b (x - \mu)^2 f(x)\, dx$$

To find the standard deviation of a continuous probability distribution, like that of a discrete distribution, we find the square root of the variance. The formulas for the expected value, variance, and standard deviation of a continuous probability distribution are summarized here.

> ## Expected Value, Variance, and Standard Deviation
> If X is a continuous random variable with probability density function f on $[a, b]$, then the expected value of X is
>
> $$E(X) = \mu = \int_a^b xf(x)\, dx.$$
>
> The variance of X is
>
> $$\text{Var}(X) = \int_a^b (x - \mu)^2 f(x)\, dx,$$
>
> and the standard deviation of X is
>
> $$\sigma = \sqrt{\text{Var}(X)}.$$

NOTE   In the formulas for expected value, variance, and standard deviation, and all other formulas in this section, it is possible that $a = -\infty$ or $b = \infty$, in which case the density function f is defined on $[a, \infty)$, $(-\infty, b]$, or $(-\infty, \infty)$. In this case, the integrals in these formulas become improper integrals, which are handled according to the procedure described in Section 8.4 on Improper Integrals. Example 2 will illustrate this procedure.

Geometrically, the expected value (or mean) of a probability distribution represents the balancing point of the distribution. If a fulcrum were placed at μ on the x-axis, the figure would be in balance. See Figure 6.

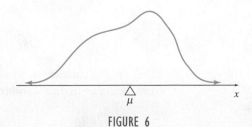

FIGURE 6

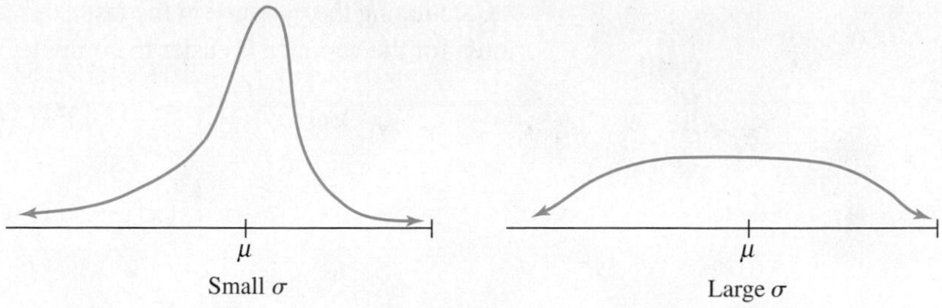

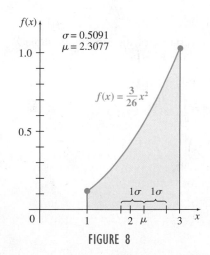

FIGURE 7

The variance and standard deviation of a probability distribution indicate how closely the values of the distribution cluster about the mean. These measures are most useful for comparing different distributions, as in Figure 7.

EXAMPLE 1   Expected Value and Variance

Find the expected value and variance of the random variable X with probability density function defined by $f(x) = (3/26)x^2$ on $[1, 3]$.

SOLUTION   By the definition of expected value just given,

$$\mu = \int_1^3 xf(x)dx$$

$$= \int_1^3 x\left(\frac{3}{26}x^2\right)dx$$

$$= \frac{3}{26}\int_1^3 x^3\,dx$$

$$= \frac{3}{26}\left(\frac{x^4}{4}\right)\Big|_1^3 = \frac{3}{104}(81 - 1) = \frac{30}{13},$$

or about 2.3077.

The variance is

$$\text{Var}(X) = \int_1^3 \left(x - \frac{30}{13}\right)^2\left(\frac{3}{26}x^2\right)dx$$

$$= \int_1^3 \left(x^2 - \frac{60}{13}x + \frac{900}{169}\right)\left(\frac{3}{26}x^2\right)dx \qquad \text{Square } \left(x - \tfrac{30}{13}\right).$$

$$= \frac{3}{26}\int_1^3 \left(x^4 - \frac{60}{13}x^3 + \frac{900}{169}x^2\right)dx \qquad \text{Multiply.}$$

$$= \frac{3}{26}\left(\frac{x^5}{5} - \frac{60}{13}\cdot\frac{x^4}{4} + \frac{900}{169}\cdot\frac{x^3}{3}\right)\Big|_1^3 \qquad \text{Integrate.}$$

$$= \frac{3}{26}\left[\left(\frac{243}{5} - \frac{60(81)}{52} + \frac{900(27)}{169(3)}\right) - \left(\frac{1}{5} - \frac{60}{52} + \frac{300}{169}\right)\right]$$

$$\approx 0.2592.$$

From the variance, the standard deviation is $\sigma \approx \sqrt{0.2592} \approx 0.5091$. The expected value and standard deviation are shown on the graph of the probability density function in Figure 8.   **TRY YOUR TURN 1**

YOUR TURN 1   Repeat Example 1 for the probability density function $f(x) = \dfrac{8}{3x^3}$ on $[1, 2]$.

$f(x)$

$\sigma = 0.5091$
$\mu = 2.3077$

$f(x) = \dfrac{3}{26}x^2$

1σ 1σ

FIGURE 8

Calculating the variance in the last example was a messy job. An alternative version of the formula for the variance is easier to compute. This alternative formula is derived as follows.

$$\text{Var}(X) = \int_a^b (x - \mu)^2 f(x)dx$$

$$= \int_a^b (x^2 - 2\mu x + \mu^2) f(x)dx$$

$$= \int_a^b x^2 f(x)\, dx - 2\mu \int_a^b xf(x)\, dx + \mu^2 \int_a^b f(x)dx \qquad \textbf{(1)}$$

By definition,

$$\int_a^b xf(x)\, dx = \mu,$$

and, since $f(x)$ is a probability density function,

$$\int_a^b f(x)\, dx = 1.$$

Substitute back into Equation (1) to get the alternative formula,

$$\text{Var}(X) = \int_a^b x^2 f(x)\, dx - 2\mu^2 + \mu^2 = \int_a^b x^2 f(x)\, dx - \mu^2.$$

Alternative Formula for Variance

If X is a random variable with probability density function f on $[a, b]$, and if $E(X) = \mu$, then

$$\textbf{Var}(X) = \int_a^b x^2 f(x)\, dx - \mu^2.$$

CAUTION Notice that the term μ^2 comes *after* the dx and so is *not* integrated.

EXAMPLE 2   Variance

Use the alternative formula for variance to compute the variance of the random variable X with probability density function defined by $f(x) = 3/x^4$ for $x \geq 1$.

SOLUTION   To find the variance, first find the expected value:

$$\mu = \int_1^\infty xf(x)dx = \int_1^\infty x \cdot \frac{3}{x^4}\, dx = \int_1^\infty \frac{3}{x^3}\, dx$$

$$= \lim_{b \to \infty} \int_1^b \frac{3}{x^3}\, dx = \lim_{b \to \infty} \left(\frac{3}{-2x^2} \right)\Big|_1^b = \frac{3}{2},$$

or 1.5. Now find the variance by the alternative formula for variance:

$$\text{Var}(X) = \int_1^\infty x^2 \left(\frac{3}{x^4} \right) dx - \left(\frac{3}{2} \right)^2$$

$$= \int_1^\infty \frac{3}{x^2}\, dx - \frac{9}{4}$$

$$= \lim_{b \to \infty} \int_1^b \frac{3}{x^2}\, dx - \frac{9}{4}$$

$$= \lim_{b \to \infty} \left(\frac{-3}{x} \right)\Big|_1^b - \frac{9}{4}$$

$$= 3 - \frac{9}{4} = \frac{3}{4}, \quad \text{or } 0.75. \qquad \textbf{TRY YOUR TURN 2}$$

YOUR TURN 2 Repeat Example 2 for the probability density function $f(x) = 4/x^5$ for $x \geq 1$.

EXAMPLE 3   Passenger Arrival

A recent study has shown that airline passengers arrive at the gate with the amount of time (in hours) before the scheduled flight time given by the probability density function $f(t) = (3/4)(2t - t^2)$ for $0 \leq t \leq 2$.

(a) Find and interpret the expected value for this distribution.

> **SOLUTION**   The expected value is

$$\mu = \int_0^2 t\left(\frac{3}{4}\right)(2t - t^2)\,dt = \int_0^2 \left(\frac{3}{4}\right)(2t^2 - t^3)\,dt$$

$$= \left(\frac{3}{4}\right)\left(\frac{2t^3}{3} - \frac{t^4}{4}\right)\Big|_0^2 = \left(\frac{3}{4}\right)\left(\frac{16}{3} - 4\right) = 1.$$

> This result indicates that passengers arrive at the gate an average of 1 hour before the scheduled flight time.

(b) Compute the standard deviation.

> **SOLUTION**   First compute the variance. We use the alternative formula.

$$\mathrm{Var}(T) = \int_0^2 t^2\left(\frac{3}{4}\right)(2t - t^2)\,dt - 1^2$$

$$= \int_0^2 \left(\frac{3}{4}\right)(2t^3 - t^4)\,dt - 1$$

$$= \left(\frac{3}{4}\right)\left(\frac{t^4}{2} - \frac{t^5}{5}\right)\Big|_0^2 - 1$$

$$= \left(\frac{3}{4}\right)\left(8 - \frac{32}{5}\right) - 1 = \frac{6}{5} - 1 = \frac{1}{5}$$

> The standard deviation is $\sigma = \sqrt{1/5} \approx 0.45$.

(c) Calculate the probability that passengers will arrive at the gate within one standard deviation of the mean.

> **SOLUTION**   Since the mean, or expected value, is 1 and the standard deviation is approximately 0.45, we are calculating the probability that passengers will arrive between

$$\mu - \sigma = 1 - 0.45 = 0.55 \text{ hours}$$

and

$$\mu + \sigma = 1 + 0.45 = 1.45 \text{ hours}$$

before the scheduled flight time. The probability is given by

$$P(0.55 \leq T \leq 1.45) = \int_{0.55}^{1.45} \frac{3}{4}(2t - t^2)\,dt.$$

Evaluating the integral gives

$$\int_{0.55}^{1.45} \frac{3}{4}(2t - t^2)\,dt = \frac{3}{4}\left(t^2 - \frac{t^3}{3}\right)\Big|_{0.55}^{1.45} \approx 0.6294.$$

The probability that passengers will arrive within one standard deviation of the mean is about 0.63.

EXAMPLE 4 Life Expectancy

In the previous section of this chapter we used statistics compiled by the National Center for Health Statistics to determine a probability density function that can be used to study the proportion of all 50-year-olds who will be alive in t years. The function is given by

$$S(t) = \frac{1}{497,703}(5.03958 \times 10^{-5}t^6 - 0.006603t^5 + 0.2992t^4$$
$$- 6.0507t^3 + 67.867t^2 - 110.3t + 2485.1)$$

for $0 \leq t \leq 52.5$.

(a) Find the life expectancy of a 50-year-old person.

SOLUTION Since this is a complicated function that is tedious to integrate analytically, we will employ the integration feature on a TI-84 Plus calculator to calculate

$$\mu = \int_0^{52.5} tS(t)dt \approx 30.38 \text{ years.}$$

According to life tables, the life expectancy of a person between the ages of 50 and 55 is 30.6 years. Our estimate is remarkably accurate given the limited number of data points and the function used in our original analysis. Life expectancy is generally calculated with techniques from life table analysis. *Source: National Center for Health Statistics.*

(b) Find the standard deviation of this probability function.

SOLUTION Using the alternate formula, we first calculate the variance.

$$\text{Var}(T) = \int_0^{52.5} t^2 S(t)dt - \mu^2 = 1057.7195 - (30.38)^2 \approx 134.775$$

Thus, $\sigma = \sqrt{\text{Var}(T)} \approx 11.61$ years.

As we mentioned earlier, the expected value is also referred to as the mean of the random variable. It is a type of average. There is another type of average, known as the *median*, that is often used. It is the value of the random variable for which there is a 50% probability of being larger and a 50% probability of being smaller. The precise definition is as follows.

Median

If X is a random variable with probability density function f on $[a, b]$, then the **median** of X is the number m such that

$$\int_a^m f(x)\,dx = \frac{1}{2}.$$

The median is particularly useful when the random variable is not distributed symmetrically about the mean. An example of this would be a random variable representing the price of homes in a city. There is a small probability that a home will be much more expensive than most of the homes in the city, and this tends to make the mean abnormally high. The median price is a better representation of the average price of a home.

EXAMPLE 5 Median

Find the median for the random variable described in Example 2, with density function defined by $f(x) = 3/x^4$ for $x \geq 1$.

SOLUTION According to the formula,

$$\int_1^m \frac{3}{x^4}\,dx = \frac{1}{2}.$$

Evaluating the integral on the left, we have

$$-\frac{1}{x^3}\Big|_1^m = -\frac{1}{m^3} + 1.$$

Set this equal to 1/2.

$$-\frac{1}{m^3} + 1 = \frac{1}{2}$$

$$\frac{1}{2} = \frac{1}{m^3} \qquad \text{Subtract 1/2 from both sides, and add } 1/m^3.$$

$$m^3 = 2 \qquad \text{Cross multiply.}$$

$$m = \sqrt[3]{2}$$

YOUR TURN 3 Repeat Example 5 for the probability density function $f(x) = 4/x^5$ for $x \geq 1$.

The median value is, therefore, $\sqrt[3]{2} \approx 1.2599$. Notice that this is smaller than the mean of 1.5 found in Example 2. This is because the random variable can take on arbitrarily large values, which pulls up the mean but doesn't affect the median.   **TRY YOUR TURN 3**

Using the notion of cumulative distribution function from the previous section, we can say that the median m is the value for which the cumulative distribution function is 0.5; that is, $F(m) = 0.5$.

11.2 EXERCISES

In Exercises 1–8, a probability density function of a random variable is defined. Find the expected value, the variance, and the standard deviation. Round answers to the nearest hundredth.

1. $f(x) = \frac{1}{4}$; $[3, 7]$

2. $f(x) = \frac{1}{10}$; $[0, 10]$

3. $f(x) = \frac{x}{8} - \frac{1}{4}$; $[2, 6]$

4. $f(x) = 2(1 - x)$; $[0, 1]$

5. $f(x) = 1 - \frac{1}{\sqrt{x}}$; $[1, 4]$

6. $f(x) = \frac{1}{11}\left(1 + \frac{3}{\sqrt{x}}\right)$; $[4, 9]$

7. $f(x) = 4x^{-5}$; $[1, \infty)$

8. $f(x) = 3x^{-4}$; $[1, \infty)$

9. What information does the mean (expected value) of a continuous random variable give?

10. Suppose two random variables have standard deviations of 0.10 and 0.23, respectively. What does this tell you about their distributions?

In Exercises 11–14, the probability density function of a random variable is defined.

a. Find the expected value to the nearest hundredth.

b. Find the variance to the nearest hundredth.

c. Find the standard deviation. Round to the nearest hundredth.

d. Find the probability that the random variable has a value greater than the mean.

e. Find the probability that the value of the random variable is within 1 standard deviation of the mean. Use the value of the standard deviation to the accuracy of your calculator.

11. $f(x) = \frac{\sqrt{x}}{18}$; $[0, 9]$

12. $f(x) = \frac{x^{-1/3}}{6}$; $[0, 8]$

13. $f(x) = \frac{1}{4}x^3$, $[0, 2]$

14. $f(x) = \frac{3}{16}(4 - x^2)$; $[0, 2]$

For Exercises 15–20, **(a)** find the median of the random variable with the probability density function given, and **(b)** find the probability that the random variable is between the expected value (mean) and the median. The expected value for each of these functions was found in Exercises 1–8.

15. $f(x) = \frac{1}{4}$; $[3, 7]$

16. $f(x) = \frac{1}{10}$; $[0, 10]$

17. $f(x) = \frac{x}{8} - \frac{1}{4}$; $[2, 6]$

18. $f(x) = 2(1 - x)$; $[0, 1]$

19. $f(x) = 4x^{-5}$; $[1, \infty)$

20. $f(x) = 3x^{-4}$; $[1, \infty)$

Find the expected value, the variance, and the standard deviation, when they exist, for each probability density function.

21. $f(x) = \begin{cases} \dfrac{x^3}{12} & \text{if } 0 \le x \le 2 \\ \dfrac{16}{3x^3} & \text{if } x > 2 \end{cases}$

22. $f(x) = \begin{cases} \dfrac{20x^4}{9} & \text{if } 0 \le x \le 1 \\ \dfrac{20}{9x^5} & \text{if } x > 1 \end{cases}$

23. Let X be a continuous random variable with density function

$$f(x) = \begin{cases} \dfrac{|x|}{10} & \text{for } -2 \le x \le 4 \\ 0 & \text{otherwise.} \end{cases}$$

Calculate the expected value of X. *Source: Society of Actuaries.* Choose one of the following.

a. 1/5   **b.** 3/5   **c.** 1   **d.** 28/15   **e.** 12/5

APPLICATIONS

Business and Economics

24. Life of a Light Bulb The life (in hours) of a certain kind of light bulb is a random variable with probability density function defined by

$$f(t) = \frac{1}{58\sqrt{t}} \quad \text{for } t \text{ in } [1,900].$$

a. What is the expected life of such a bulb?

b. Find σ.

c. Find the probability that one of these bulbs lasts longer than 1 standard deviation above the mean.

d. Find the median life of these bulbs.

25. Machine Life The life (in years) of a certain machine is a random variable with probability density function defined by

$$f(t) = \frac{1}{11}\left(1 + \frac{3}{\sqrt{t}}\right) \quad \text{for } t \text{ in } [4,9].$$

a. Find the mean life of this machine.

b. Find the standard deviation of the distribution.

c. Find the probability that a particular machine of this kind will last longer than the mean number of years.

26. Life of an Automobile Part The life span of a certain automobile part (in months) is a random variable with probability density function defined by

$$f(t) = \frac{1}{2}e^{-t/2} \quad \text{for } t \text{ in } [0, \infty).$$

a. Find the expected life of this part.

b. Find the standard deviation of the distribution.

c. Find the probability that one of these parts lasts less than the mean number of months.

d. Find the median life of these parts.

27. Losses After Deductible A manufacturer's annual losses follow a distribution with density function

$$f(x) = \begin{cases} \dfrac{2.5(0.6)^{2.5}}{x^{3.5}} & \text{for } x > 0.6 \\ 0 & \text{otherwise.} \end{cases}$$

To cover its losses, the manufacturer purchases an insurance policy with an annual deductible of 2. What is the mean of the manufacturer's annual losses not paid by the insurance policy? Choose one of the following. (*Hint:* The loss not paid by the insurance policy will equal the actual loss if the actual loss is less than the deductible. Otherwise it will equal the deductible.) *Source: Society of Actuaries.*

a. 0.84   **b.** 0.88   **c.** 0.93   **d.** 0.95   **e.** 1.00

28. Insurance Reimbursement An insurance policy reimburses a loss up to a benefit limit of 10. The policyholder's loss, Y, follows a distribution with density function:

$$f(y) = \begin{cases} \dfrac{2}{y^3} & \text{for } y > 1 \\ 0 & \text{otherwise.} \end{cases}$$

What is the expected value of the benefit paid under the insurance policy? Choose one of the following. (*Hint:* The benefit paid will be equal to the actual loss if the actual loss is less than the limit. Otherwise it will equal the limit.) *Source: Society of Actuaries.*

a. 1.0   **b.** 1.3   **c.** 1.8   **d.** 1.9   **e.** 2.0

29. Insurance Claims An insurance company's monthly claims are modeled by a continuous, positive random variable X, whose probability density function is proportional to $(1 + x)^{-4}$, where $0 < x < \infty$. Determine the company's expected monthly claims. Choose one of the following. *Source: Society of Actuaries.*

a. 1/6   **b.** 1/3   **c.** 1/2   **d.** 1   **e.** 3

30. Dental Insurance An insurance policy reimburses dental expense, X, up to a maximum benefit of 250. The probability density function for X is

$$f(x) = \begin{cases} ce^{-0.004x} & \text{for } x \ge 0 \\ 0 & \text{otherwise,} \end{cases}$$

where c is a constant. Calculate the median benefit for this policy. Choose one of the following. (*Hint:* As long as the expenses are less than 250, the expenses and the benefit are equal.) *Source: Society of Actuaries.*

a. 161   **b.** 165   **c.** 173   **d.** 182   **e.** 250

Life Sciences

31. Blood Clotting Time The clotting time of blood (in seconds) is a random variable with probability density function defined by

$$f(t) = \frac{1}{(\ln 20)t} \quad \text{for } t \text{ in } [1, 20].$$

a. Find the mean clotting time.

b. Find the standard deviation.

c. Find the probability that a person's blood clotting time is within 1 standard deviation of the mean.

d. Find the median clotting time.

32. **Length of a Leaf** The length of a leaf on a tree is a random variable with probability density function defined by

$$f(x) = \frac{3}{32}(4x - x^2) \quad \text{for } x \text{ in } [0, 4].$$

a. What is the expected leaf length?

b. Find σ for this distribution.

c. Find the probability that the length of a given leaf is within 1 standard deviation of the expected value.

33. **Petal Length** The length (in centimeters) of a petal on a certain flower is a random variable with probability density function defined by

$$f(x) = \frac{1}{2\sqrt{x}} \quad \text{for } x \text{ in } [1, 4].$$

a. Find the expected petal length.

b. Find the standard deviation.

c. Find the probability that a petal selected at random has a length more than 2 standard deviations above the mean.

d. Find the median petal length.

34. **Flea Beetles** As we saw in Exercise 42 of the previous section, the probability that a marked flea beetle, *Phyllotreta cruciferae* and *Phyllotreta striolata*, will be recaptured within a certain distance and time after release can be calculated from the probability density function

$$p(x, t) = \frac{e^{-x^2/(4Dt)}}{\int_0^L e^{-u^2/(4Dt)} \, du},$$

where t is the time (in hours) after release, x is the distance (in meters) from the release point that recaptures occur, L is the maximum distance from the release point that recaptures can occur, and D is the diffusion coefficient. If $t = 12$, $L = 6$, and $D = 38.3$, find the expected recapture distance. *Source: Ecology Monographs.*

35. **Flour Beetles** As we saw in Exercise 41 of the previous section, a probability density function has been developed to estimate the abundance of the flour beetle, *Tribolium castaneum*. The density function, which is a member of the gamma distribution, is

$$f(x) = 1.185 \times 10^{-9} x^{4.5222} e^{-0.049846x},$$

where x is the size of the population. Calculate the expected size of a flour beetle population. (*Hint:* Use 1000 as the upper limit of integration.) *Source: Ecology.*

Social Sciences

36. **Time to Learn a Task** In Exercise 44 of the previous section, the probability density function for the time required for a person to learn a certain task was given by

$$f(t) = \frac{8}{7(t - 2)^2},$$

for $3 \leq t \leq 10$ minutes. Find the median time for a person to learn the task.

37. **Social Network** In Exercise 43 of the previous section, the probability density function for the number of U.S. users of Facebook, a computer social network, was found to be

$$S(t) = \frac{1}{466.26}(-0.00007445t^4 + 0.01243t^3 - 0.7419t^2 + 18.18t - 137.5)$$

where t was the number of years since birth on $[13.4, 62.0]$. Calculate the expected age of a Facebook user, as well as the standard deviation. *Source: Inside Facebook.*

Physical Sciences

38. **Earthquakes** The time between major earthquakes in the Southern California region is a random variable with probability density function defined by

$$f(t) = \frac{1}{960} e^{-t/960},$$

where t is measured in days. *Source: Journal of Seismology.* Find the expected value and the standard deviation of this probability density function.

39. **Annual Rainfall** The annual rainfall in a remote Middle Eastern country is a random variable with probability density function defined by

$$f(x) = \frac{5.5 - x}{15}, \quad \text{for } x \text{ in } [0, 5].$$

a. Find the mean annual rainfall.

b. Find the standard deviation.

c. Find the probability of a year with rainfall less than 1 standard deviation below the mean.

General Interest

40. **Drunk Drivers** In the last section, we saw that the age of a randomly selected, alcohol-impaired driver in a fatal car crash is a random variable with probability density function given by

$$f(t) = \frac{4.045}{t^{1.532}} \quad \text{for } t \text{ in } [16, 80].$$

Source: Traffic Safety Facts.

a. **APPLY IT** Find the expected age of a drunk driver in a fatal car crash.

b. Find the standard deviation of the distribution.

c. Find the probability that such a driver will be younger than 1 standard deviation below the mean.

d. Find the median age of a drunk driver in a fatal car crash.

41. **Driving Fatalities** In the last section, we saw that the age of a randomly selected driver in a fatal car crash is a random variable with probability density function given by

$$f(t) = 0.06049e^{-0.03211t} \quad \text{for } t \text{ in } [16, 84].$$

Source: National Highway Traffic Safety Administration.

a. Find the expected age of a driver in a fatal car crash.

b. Find the standard deviation of the distribution.

c. Find the probability that such a driver will be younger than 1 standard deviation below the mean.

d. Find the median age of a driver in a fatal car crash.

42. **Length of a Telephone Call** The length of a telephone call (in minutes), t, for a certain town is a continuous random variable with probability density function defined by

$$f(t) = 3t^{-4}$$

for t in $[1, \infty)$. Find the expected length of a phone call.

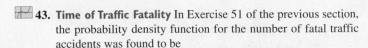 43. **Time of Traffic Fatality** In Exercise 51 of the previous section, the probability density function for the number of fatal traffic accidents was found to be

$$S(t) = \frac{1}{101,370}(-2.564t^3 + 99.11t^2 - 964.6t + 5631)$$

where t is the number of hours since midnight on $[0, 24]$. Calculate the expected time of day at which a fatal accident will occur. *Source: The National Highway Traffic Safety Administration.*

■■■YOUR TURN ANSWERS

1. 4/3, 0.0706 **2.** 2/9 **3.** $\sqrt[4]{2} \approx 1.1892$

11.3 Special Probability Density Functions

APPLY IT What is the probability that the maximum outdoor temperature will be higher than 24°C? What is the probability that a flashlight battery will last longer than 40 hours?

These questions, presented in Examples 1 and 2, can be answered if the probability density function for the maximum temperature and for the life of the battery are known.

In practice, it is not feasible to construct a probability density function for every experiment. Instead, a researcher uses one of several probability density functions that are well known, matching the shape of the experimental distribution to one of the known distributions. In this section we discuss some of the most commonly used probability distributions.

Uniform Distribution
The simplest probability distribution occurs when the probability density function of a random variable remains constant over the sample space. In this case, the random variable is said to be *uniformly distributed* over the sample space. The probability density function for the **uniform distribution** is defined by

$$f(x) = \frac{1}{b - a} \quad \text{for } x \text{ in } [a, b],$$

where a and b are constant real numbers. The graph of $f(x)$ is shown in Figure 9.

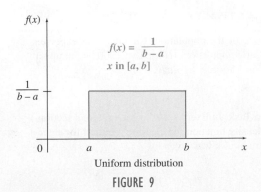

Uniform distribution

FIGURE 9

Since $b - a$ is positive, $f(x) \geq 0$, and

$$\int_a^b \frac{1}{b-a}\,dx = \frac{1}{b-a}x\Big|_a^b = \frac{1}{b-a}(b-a) = 1.$$

Therefore, the function is a probability density function.

The expected value for the uniform distribution is

$$\mu = \int_a^b \left(\frac{1}{b-a}\right)x\,dx = \left(\frac{1}{b-a}\right)\frac{x^2}{2}\Big|_a^b$$

$$= \frac{1}{2(b-a)}(b^2 - a^2) = \frac{1}{2}(b+a). \quad b^2 - a^2 = (b-a)(b+a)$$

The variance is given by

$$\text{Var}(X) = \int_a^b \left(\frac{1}{b-a}\right)x^2\,dx - \left(\frac{b+a}{2}\right)^2$$

$$= \left(\frac{1}{b-a}\right)\frac{x^3}{3}\Big|_a^b - \frac{(b+a)^2}{4}$$

$$= \frac{1}{3(b-a)}(b^3 - a^3) - \frac{1}{4}(b+a)^2$$

$$= \frac{b^2 + ab + a^2}{3} - \frac{b^2 + 2ab + a^2}{4} \qquad b^3 - a^3 = (b-a)(b^2 + ab + a^2)$$

$$= \frac{b^2 - 2ab + a^2}{12}. \qquad \text{Get a common denominator; subtract.}$$

Thus

$$\text{Var}(X) = \frac{1}{12}(b-a)^2, \qquad \text{Factor.}$$

and

$$\sigma = \frac{1}{\sqrt{12}}(b-a).$$

These properties of the uniform distribution are summarized below.

Uniform Distribution

If X is a random variable with probability density function

$$f(x) = \frac{1}{b-a} \quad \text{for } x \text{ in } [a, b],$$

then

$$\mu = \frac{1}{2}(b+a) \qquad \text{and} \qquad \sigma = \frac{1}{\sqrt{12}}(b-a).$$

EXAMPLE 1 **Daily Temperature**

A couple is planning to vacation in San Francisco. They have been told that the maximum daily temperature during the time they plan to be there ranges from 15°C to 27°C. Assume that the probability of any temperature between 15°C and 27°C is equally likely for any given day during the specified time period.

(a) What is the probability that the maximum temperature on the day they arrive will be higher than 24°C?

SOLUTION If the random variable T represents the maximum temperature on a given day, then the uniform probability density function for T is defined by $f(t) = 1/12$ for the interval $[15, 27]$. By definition,

$$P(T > 24) = \int_{24}^{27} \frac{1}{12} \, dt = \frac{1}{12}t \Big|_{24}^{27} = \frac{1}{4}.$$

(b) What average maximum temperature can they expect?

SOLUTION The expected maximum temperature is

$$\mu = \frac{1}{2}(27 + 15) = 21,$$

or 21°C.

(c) What is the probability that the maximum temperature on a given day will be one standard deviation or more below the mean?

SOLUTION First find σ.

YOUR TURN 1 The next vacation for the couple in Example 1 is to a desert with a maximum daily temperature that uniformly ranges from 27°C to 42°C. Find the expected maximum temperature and the probability that the maximum temperature will be within one standard deviation of the mean.

$$\sigma = \frac{1}{\sqrt{12}}(27 - 15) = \frac{12}{\sqrt{12}} = \sqrt{12} = 2\sqrt{3} \approx 3.464.$$

One standard deviation below the mean indicates a temperature of $21 - 3.464 = 17.536°C$.

$$P(T \leq 17.536) = \int_{15}^{17.536} \frac{1}{12} \, dt = \frac{1}{12}t \Big|_{15}^{17.536} \approx 0.2113$$

The probability is about 0.21 that the temperature will not exceed 17.5°C.

TRY YOUR TURN 1

Exponential Distribution
The next distribution is very important in reliability and survival analysis. When manufactured items and living things have a constant failure rate over a period of time, the exponential distribution is used to describe their probability of failure. In this case, the random variable is said to be *exponentially distributed* over the sample space. The probability density function for the **exponential distribution** is defined by

$$f(x) = ae^{-ax} \qquad \text{for } x \text{ in } [0, \infty),$$

where a is a positive constant. The graph of $f(x)$ is shown in Figure 10.

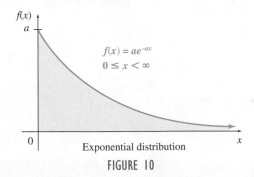

FIGURE 10

Here $f(x) \geq 0$, since e^{-ax} and a are both positive for all values of x. Also,

$$\int_0^{\infty} ae^{-ax} \, dx = \lim_{b \to \infty} \int_0^b ae^{-ax} \, dx$$

$$= \lim_{b \to \infty} (-e^{-ax}) \Big|_0^b = \lim_{b \to \infty} \left(\frac{-1}{e^{ab}} + \frac{1}{e^0} \right) = 1,$$

so the function is a probability density function.

The expected value and the standard deviation of the exponential distribution can be found using integration by parts. The results are given below. (See Exercise 20 at the end of this section.)

> **Exponential Distribution**
> If X is a random variable with probability density function
>
> $$f(x) = ae^{-ax} \quad \text{for } x \text{ in } [0, \infty),$$
>
> then
>
> $$\mu = \frac{1}{a} \quad \text{and} \quad \sigma = \frac{1}{a},$$

EXAMPLE 2 Flashlight Battery

Suppose the useful life (in hours) of a flashlight battery is the random variable T, with probability density function given by the exponential distribution

$$f(t) = \frac{1}{20} e^{-t/20} \quad \text{for } t \geq 0.$$

(a) Find the probability that a particular battery, selected at random, has a useful life of less than 100 hours.

SOLUTION The probability is given by

$$P(T \leq 100) = \int_0^{100} \frac{1}{20} e^{-t/20} \, dt = \frac{1}{20} \left(-20 e^{-t/20} \right) \Big|_0^{100}$$
$$= -\left(e^{-100/20} - e^0 \right) = -\left(e^{-5} - 1 \right)$$
$$\approx 1 - 0.0067 = 0.9933.$$

(b) Find the expected value and standard deviation of the distribution.

SOLUTION Use the formulas given above. Both μ and σ equal $1/a$, and since $a = 1/20$ here,

$$\mu = 20 \quad \text{and} \quad \sigma = 20.$$

This means that the average life of a battery is 20 hours, and no battery lasts less than 1 standard deviation below the mean.

APPLY IT

YOUR TURN 2 Repeat Example 2 for a flashlight battery with a useful life given by the probability density function $f(t) = \frac{1}{25} e^{-t/25}$ for $t \geq 0$.

(c) What is the probability that a battery will last longer than 40 hours?

SOLUTION The probability is given by

$$P(T > 40) = \int_{40}^{\infty} \frac{1}{20} e^{-t/20} \, dt = \lim_{b \to \infty} \left(-e^{-t/20} \right) \Big|_{40}^{b} = \frac{1}{e^2} \approx 0.1353,$$

or about 14%.

TRY YOUR TURN 2

Normal Distribution

The **normal distribution**, with its well-known bell-shaped graph, is undoubtedly the most important probability density function. It is widely used in various applications of statistics. The random variables associated with these applications are said to be normally distributed. The probability density function for the normal distribution has the following characteristics.

Normal Distribution

If μ and σ are real numbers, $\sigma > 0$, and if X is a random variable with probability density function defined by

$$f(x) = \frac{1}{\sigma\sqrt{2\pi}}e^{-(x-\mu)^2/(2\sigma^2)} \quad \text{for } x \text{ in } (-\infty, \infty),$$

then

$$E(X) = \mu \quad \text{and} \quad \text{Var}(X) = \sigma^2, \quad \text{with standard deviation } \sigma.$$

Notice that the definition of the probability density function includes σ, which is the standard deviation of the distribution.

Advanced techniques can be used to show that

$$\int_{-\infty}^{\infty} \frac{1}{\sigma\sqrt{2\pi}}e^{-(x-\mu)^2/(2\sigma^2)}\,dx = 1.$$

Deriving the expected value and standard deviation for the normal distribution also requires techniques beyond the scope of this text.

Each normal probability distribution has associated with it a bell-shaped curve, called a **normal curve**, such as the one in Figure 11. Each normal curve is symmetric about a vertical line through the mean, μ. Vertical lines at points $+1\sigma$ and -1σ from the mean show the inflection points of the graph. (See Exercise 22 at the end of this section.) A normal curve never touches the x-axis; it extends indefinitely in both directions.

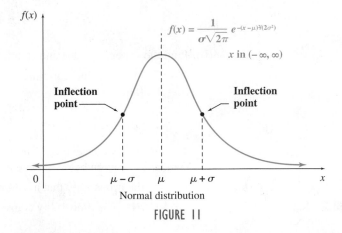

Normal distribution

FIGURE 11

The development of the normal curve is credited to the Frenchman Abraham De Moivre (1667–1754). Three of his publications dealt with probability and associated topics: *Annuities upon Lives* (which contributed to the development of actuarial studies), *Doctrine of Chances*, and *Miscellanea Analytica*.

Many different normal curves have the same mean. In such cases, a larger value of σ produces a "flatter" normal curve, while smaller values of σ produce more values near the mean, resulting in a "taller" normal curve. See Figure 12 on the next page.

It would be far too much work to calculate values for the normal probability distribution for various values of μ and σ. Instead, values are calculated for the **standard normal**

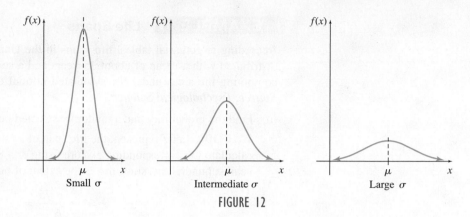

FIGURE 12

distribution, which has $\mu = 0$ and $\sigma = 1$. The graph of the standard normal distribution is shown in Figure 13.

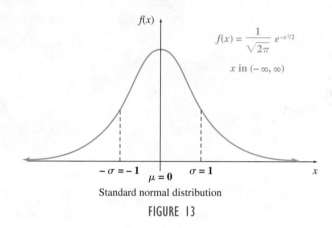

Standard normal distribution

FIGURE 13

Probabilities for the standard normal distribution come from the definite integral

$$\int_a^b \frac{1}{\sqrt{2\pi}} e^{-x^2/2}\, dx.$$

Since $f(x) = e^{-x^2/2}$ does not have an antiderivative that can be expressed in terms of functions used in this course, numerical methods are used to find values of this definite integral. A table in the appendix of this book gives areas under the standard normal curve, along with a sketch of the curve. Each value in this table is the total area under the standard normal curve to the left of the number z.

If a normal distribution does not have $\mu = 0$ and $\sigma = 1$, we use the following theorem, which is proved in Exercise 21.

z-Scores Theorem

Suppose a normal distribution has mean μ and standard deviation σ. The area under the associated normal curve that is to the left of the value x is exactly the same as the area to the left of

$$z = \frac{x - \mu}{\sigma}$$

for the standard normal curve.

Using this result, the table can be used for *any* normal distribution, regardless of the values of μ and σ. The number z in the theorem is called a **z-score**.

EXAMPLE 3 Life Spans

According to actuarial tables, life spans in the United States are approximately normally distributed with a mean of about 75 years and a standard deviation of about 16 years. By computing the areas under the associated normal curve, find the following probabilities. *Source: Psychological Science.*

(a) Find the probability that a randomly selected person lives less than 88 years.

SOLUTION Let T represent the life span of a random individual. To find $P(T < 88)$, we calculate the corresponding z-score using $t = 88$, $\mu = 75$, and $\sigma = 16$. Round to the nearest hundredths, since this is the extent of our normal curve table.

$$z = \frac{88 - 75}{16} = \frac{13}{16} \approx 0.81$$

Look up 0.81 in the normal curve table in the Appendix. The corresponding area is 0.7910. Thus, the shaded area shown in Figure 14 is 0.7910. This means that the probability of a randomly selected person living less than 88 years is $P(T < 88) = P(Z < 0.81) = 0.7910$, or about 79%.

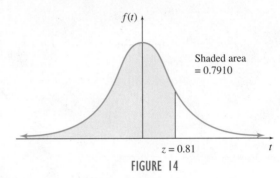

FIGURE 14

(b) Find the probability that a randomly selected person lives more than 67 years.

SOLUTION To calculate $P(T > 67)$, first find the corresponding z-score.

$$z = \frac{67 - 75}{16} = -0.5$$

From the normal curve table, the area to the *left* of $z = -0.5$ is 0.3085. Therefore, the area to the *right* is $P(T > 67) = P(Z > -0.5) = 1 - 0.3085 = 0.6915$. See Figure 15. Thus, the probability of a randomly selected person living more than 67 years is 0.6915, or about 69%.

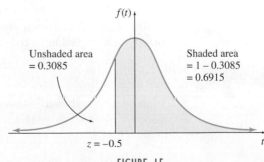

FIGURE 15

(c) Find the probability that a randomly selected person lives between 61 and 70 years.

SOLUTION Find z-scores for both values.

$$z = \frac{61 - 75}{16} = -0.88 \qquad \text{and} \qquad z = \frac{70 - 75}{16} = -0.31$$

YOUR TURN 3 Using the information provided in Example 3, find the probability that a randomly selected person lives (a) more than 79 years and (b) between 67 and 83 years.

Start with the area to the left of $z = -0.31$ and subtract the area to the left of $z = -0.88$. Thus,

$$P(61 \leq T \leq 70) = P(-0.88 \leq Z \leq -0.31) = 0.3783 - 0.1894 = 0.1889.$$

The required area is shaded in Figure 16. The probability of a randomly selected person living between 61 and 70 years is about 19%. **TRY YOUR TURN 3**

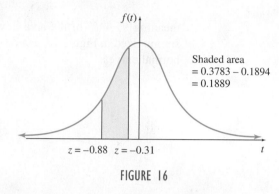

FIGURE 16

It's worth noting that there is always some error in approximating a discrete distribution with a continuous distribution. For example, when T is the life span of a randomly selected person, then $P(T = 65)$ is clearly positive, but if T is a normal random variable, then $P(T = 65) = 0$, since it represents no area. The problem is that a person's age jumps from 65 to 66, but a continuous random variable takes on all real numbers in between. If we were to measure a person's age to the nearest nanosecond, the probability that someone's age is exactly 65 years and 0 nanoseconds would be virtually 0.

Furthermore, a bit of thought shows us that the approximation of life spans by a normal distribution can't be perfect. After all, three standard deviations to the left and right of 75 give $75 - 3 \times 16 = 27$ and $75 + 3 \times 16 = 123$. Because of the symmetry of the normal distribution, $P(T < 27)$ and $P(T > 123)$ should be equal. Yet there are people who die before the age of 27, and no human has been verified to live beyond the age of 123.

TECHNOLOGY NOTE

As an alternative to using the normal curve table, we can use a graphing calculator. Enter the formula for the normal distribution into the calculator, using $\mu = 75$ and $\sigma = 16$. Plot the function on a window that contains at least four standard deviations to the left and right of μ; for Example 3(a), we will let $0 \leq t \leq 140$. Then use the integration feature (under CALC on a TI-84 Plus) to find the area under the curve to the left of 88.

The result is shown in Figure 17. In place of $-\infty$, we have used $t = 0$ as the left endpoint. This is far enough to the left of $\mu = 75$ that it can be considered as $-\infty$ for all practical purposes. It also makes sense in this application, since life span can't be a negative number. You can verify that choosing a slightly different lower limit makes little difference in the answer. In fact, the answer of 0.79174622 is more accurate than the answer of 0.7910 that we found in Example 3(a), where we needed to round $13/16 = 0.8125$ to 0.81 in order to use the table.

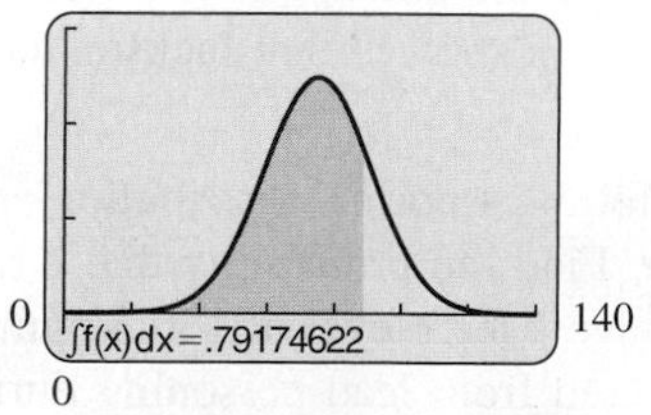

We could get the answer on a TI-84 Plus without generating a graph using the command `fnInt` and entering $\int_0^{88} (Y_1)dx$, where Y_1 is the formula for the normal distribution with $\mu = 75$ and $\sigma = 16$.

The numerical integration method works with any probability density function. In addition, many graphing calculators are programmed with information about specific density functions, such as the normal. We can solve the first part of Example 3 on the TI-84 Plus by entering `normalcdf` $(-1E99, 88, 75, 16)$. The calculator responds with 0.7917476687. ($-1E99$ stands for -1×10^{99}, which the calculator uses for $-\infty$.) If you use this method in the exercises, your answers will differ slightly from those in the back of the book, which were generated using the normal curve table in the Appendix.

0.03

0   ∫f(x)dx=.79174622   140

0

FIGURE 17

The z-scores are actually standard deviation multiples; that is, a z-score of 2.5 corresponds to a value 2.5 standard deviations above the mean. For example, looking up $z = 1.00$ and $z = -1.00$ in the table shows that

$$0.8413 - 0.1587 = 0.6826,$$

so that 68.26% of the area under a normal curve is within 1 standard deviation of the mean. Also, using $z = 2.00$ and $z = -2.00$,

$$0.9772 - 0.0228 = 0.9544,$$

meaning 95.44% of the area is within 2 standard deviations of the mean. These results, summarized in Figure 18, can be used to get a quick estimate of results when working with normal curves.

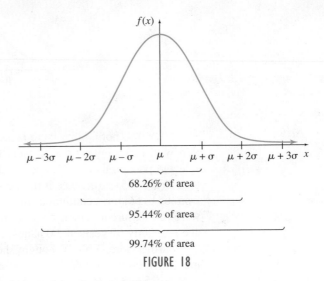

FIGURE 18

Manufacturers make use of the fact that a normal random variable is almost always within 3 standard deviations of the mean to design control charts. When a sample of items produced by a machine has a mean farther than 3 standard deviations from the desired specification, the machine is assumed to be out of control, and adjustments are made to ensure that the items produced meet the tolerance required.

EXAMPLE 4   Lead Poisoning

Historians and biographers have collected evidence suggesting that President Andrew Jackson suffered from lead poisoning. Recently, researchers measured the amount of lead in samples of Jackson's hair from 1815. The results of this experiment showed that Jackson had a mean lead level of 130.5 ppm. *Source: JAMA.*

(a) Levels of lead in hair samples from that time period follow a normal distribution with mean 93 and standard deviation 16. *Source: Science.* Find the probability that a randomly selected person from this time period would have a lead level of 130.5 ppm or higher. Does this provide evidence that Jackson suffered from lead poisoning during this time period?

SOLUTION $P(X \geq 130.5) = P\left(Z \geq \dfrac{130.5 - 93}{16}\right) = P(Z \geq 2.34) = 0.0096$

Since this probability is so low, it is likely that Jackson suffered from lead poisoning during this time period.*

*Although this provides evidence that Andrew Jackson had elevated lead levels, the authors of the paper concluded that Andrew Jackson did not die from lead poisoning.

(b) Today's normal lead levels follow a normal distribution with approximate mean of 10 ppm and standard deviation of 5 ppm. *Source: Clinical Chemistry.* By today's standards, calculate the probability that a randomly selected person from today would have a lead level of 130.5 ppm or higher. From this, can we conclude that Andrew Jackson had lead poisoning?

SOLUTION $P(X \geq 130.5) = P\left(Z \geq \dfrac{130.5 - 10}{5}\right) = P(Z \geq 24.1) \approx 0$

By today's standards, which may not be valid for this experiment, Jackson certainly suffered from lead poisoning.

11.3 EXERCISES

Find **(a)** the mean of the distribution, **(b)** the standard deviation of the distribution, and **(c)** the probability that the random variable is between the mean and 1 standard deviation above the mean.

1. The length (in centimeters) of the leaf of a certain plant is a continuous random variable with probability density function defined by

$$f(x) = \frac{5}{7} \quad \text{for } x \text{ in } [3, 4.4].$$

2. The price of an item (in hundreds of dollars) is a continuous random variable with probability density function defined by

$$f(x) = 4 \quad \text{for } x \text{ in } [2.75, 3].$$

3. The length of time (in years) until a particular radioactive particle decays is a random variable t with probability density function defined by

$$f(t) = 4e^{-4t} \quad \text{for } t \text{ in } [0, \infty).$$

4. The length of time (in years) that a seedling tree survives is a random variable t with probability density function defined by

$$f(t) = 0.05e^{-0.05t} \quad \text{for } t \text{ in } [0, \infty).$$

5. The length of time (in days) required to learn a certain task is a random variable t with probability density function defined by

$$f(t) = \frac{e^{-t/3}}{3} \quad \text{for } t \text{ in } [0, \infty).$$

6. The distance (in meters) that seeds are dispersed from a certain kind of plant is a random variable x with probability density function defined by

$$f(x) = 0.1e^{-0.1x} \quad \text{for } x \text{ in } [0, \infty).$$

Find the proportion of observations of a standard normal distribution that are between the mean and the given number of standard deviations above the mean.

7. 3.50

8. 1.68

Find the proportion of observations of a standard normal distribution that are between the given z-scores.

9. 1.28 and 2.05

10. −2.13 and −0.04

Find a z-score satisfying the conditions given in Exercises 11–14. (*Hint:* Use the table backwards.)

11. 10% of the total area is to the left of z.

12. 2% of the total area is to the left of z.

13. 18% of the total area is to the right of z.

14. 22% of the total area is to the right of z.

15. Describe the standard normal distribution. What are its characteristics?

16. What is a z-score? How is it used?

17. Describe the shape of the graph of each probability distribution.

   a. Uniform   **b.** Exponential   **c.** Normal

In the second section of this chapter, we defined the median of a probability distribution as an integral. The median also can be defined as the number m such that $P(X \leq m) = P(X \geq m)$.

18. Find an expression for the median of the uniform distribution.

19. Find an expression for the median of the exponential distribution.

20. Verify the expected value and standard deviation of the exponential distribution given in the text.

21. Prove the z-scores theorem. (*Hint:* Write the formula for the normal distribution with mean μ and standard deviation σ, using t instead of x as the variable. Then write the integral representing the area to the left of the value x, and make the substitution $u = (t - \mu)/\sigma$.)

22. Show that a normal random variable has inflection points at $x = \mu - \sigma$ and $x = \mu + \sigma$.

23. Use Simpson's rule with $n = 140$, or use the integration feature on a graphing calculator, to approximate the following integrals.

   a. $\displaystyle\int_0^{35} 0.5e^{-0.5x}\,dx$ **b.** $\displaystyle\int_0^{35} 0.5xe^{-0.5x}\,dx$

   c. $\displaystyle\int_0^{35} 0.5x^2e^{-0.5x}\,dx$

24. Use your results from Exercise 23 to verify that, for the exponential distribution with $a = 0.5$, the total probability is 1, and both the mean and the standard deviation are equal to $1/a$.

25. Use Simpson's rule with $n = 40$, or the integration feature on a graphing calculator, to approximate the following for the standard normal probability distribution. Use limits of -6 and 6 in place of $-\infty$ and ∞.

a. The mean   **b.** The standard deviation

26. A very important distribution for analyzing the reliability of manufactured goods is the Weibull distribution, whose probability density function is defined by

$$f(x) = abx^{b-1}e^{-ax^b} \quad \text{for } x \text{ in } [0, \infty),$$

where a and b are constants. Notice that when $b = 1$, this reduces to the exponential distribution. The Weibull distribution is more general than the exponential, because it applies even when the failure rate is not constant. Use Simpson's rule with $n = 100$, or the integration feature on a graphing calculator, to approximate the following for the Weibull distribution with $a = 4$ and $b = 1.5$. Use a limit of 3 in place of ∞.

a. The mean   **b.** The standard deviation

27. Determine the cumulative distribution function for the uniform distribution.

28. Determine the cumulative distribution function for the exponential distribution.

APPLICATIONS

Business and Economics

29. Insurance Sales The amount of insurance (in thousands of dollars) sold in a day by a particular agent is uniformly distributed over the interval $[10, 85]$.

   a. What amount of insurance does the agent sell on an average day?

   b. Find the probability that the agent sells more than $50,000 of insurance on a particular day.

30. Fast-Food Outlets The number of new fast-food outlets opening during June in a certain city is exponentially distributed, with a mean of 5.

   a. Give the probability density function for this distribution.

   b. What is the probability that the number of outlets opening is between 2 and 6?

31. Sales Expense A salesperson's monthly expenses (in thousands of dollars) are exponentially distributed, with an average of 4.25 (thousand dollars).

   a. Give the probability density function for the expenses.

   b. Find the probability that the expenses are more than $10,000.

In Exercises 32–34, assume a normal distribution.

32. Machine Accuracy A machine that fills quart bottles with apple juice averages 32.8 oz per bottle, with a standard deviation of 1.1 oz. What are the probabilities that the amount of juice in a bottle is as follows?

   a. Less than 1 qt

   b. At least 1 oz more than 1 qt

33. Machine Accuracy A machine produces screws with a mean length of 2.5 cm and a standard deviation of 0.2 cm. Find the probabilities that a screw produced by this machine has lengths as follows.

   a. Greater than 2.7 cm

   b. Within 1.2 standard deviations of the mean

34. Customer Expenditures Customers at a certain pharmacy spend an average of $54.40, with a standard deviation of $13.50. What are the largest and smallest amounts spent by the middle 50% of these customers?

35. Insured Loss An insurance policy is written to cover a loss, X, where X has a uniform distribution on $[0, 1000]$. At what level must a deductible be set in order for the expected payment to be 25% of what it would be with no deductible? Choose one of the following. (*Hint:* Use a variable, such as D, for the deductible. The payment is 0 if the loss is less than D, and the loss minus D if the loss is greater than D.) *Source: Society of Actuaries.*

   a. 250   **b.** 375   **c.** 500   **d.** 625   **e.** 750

36. High-Risk Drivers The number of days that elapse between the beginning of a calendar year and the moment a high-risk driver is involved in an accident is exponentially distributed. An insurance company expects that 30% of high-risk drivers will be involved in an accident during the first 50 days of a calendar year. What portion of high-risk drivers are expected to be involved in an accident during the first 80 days of a calendar year? Choose one of the following. *Source: Society of Actuaries.*

   a. 0.15   **b.** 0.34   **c.** 0.43   **d.** 0.57   **e.** 0.66

37. Printer Failure The lifetime of a printer costing $200 is exponentially distributed with mean 2 years. The manufacturer agrees to pay a full refund to a buyer if the printer fails during the first year following its purchase, and a one-half refund if it fails during the second year. If the manufacturer sells 100 printers, how much should it expect to pay in refunds? Choose one of the following. *Source: Society of Actuaries.*

   a. 6321   **b.** 7358   **c.** 7869   **d.** 10,256   **e.** 12,642

38. Electronic Device The time to failure of a component in an electronic device has an exponential distribution with a median of four hours. Calculate the probability that the component will work without failing for at least five hours. Choose one of the following. *Source: Society of Actuaries.*

   a. 0.07   **b.** 0.29   **c.** 0.38   **d.** 0.42   **e.** 0.57

Life Sciences

39. Insect Life Span The life span of a certain insect (in days) is uniformly distributed over the interval $[20, 36]$.

   a. What is the expected life of this insect?

   b. Find the probability that one of these insects, randomly selected, lives longer than 30 days.

40. Location of a Bee Swarm A swarm of bees is released from a certain point. The proportion of the swarm located at least 2 m from the point of release after 1 hour is a random variable that is exponentially distributed with $a = 2$.

   a. Find the expected proportion under the given conditions.

   b. Find the probability that fewer than $1/3$ of the bees are located at least 2 m from the release point after 1 hour.

41. Digestion Time The digestion time (in hours) of a fixed amount of food is exponentially distributed with $a = 1$.

  a. Find the mean digestion time.

  b. Find the probability that the digestion time is less than 30 minutes.

42. Pygmy Heights The average height of a member of a certain tribe of pygmies is 3.2 ft, with a standard deviation of 0.2 ft. If the heights are normally distributed, what are the largest and smallest heights of the middle 50% of this population?

43. Finding Prey H. R. Pulliam found that the time (in minutes) required by a predator to find a prey is a random variable that is exponentially distributed, with $\mu = 25$. *Source: American Naturalist.*

  a. According to this distribution, what is the longest time within which the predator will be 90% certain of finding a prey?

  b. What is the probability that the predator will have to spend more than 1 hour looking for a prey?

44. Life Expectancy According to the National Center for Health Statistics, the life expectancy for a 55-year-old African American female is 26.1 years. Assuming that from age 55, the survival of African American females follows an exponential distribution, determine the following probabilities. *Source: National Vital Statistics Report.*

  a. The probability that a randomly selected 55-year-old African American female will live beyond 80 years of age (at least 25 more years)

  b. The probability that a randomly selected 55-year-old African American female will live less than 20 more years

45. Life Expectancy According to the National Center for Health Statistics, life expectancy for a 70-year-old African American male is 12.3 years. Assuming that from age 70, the survival of African American males follows an exponential distribution, determine the following probabilities. *Source: National Vital Statistics Report.*

  a. The probability that a randomly selected 70-year-old African American male will live beyond 90 years of age

  b. The probability that a randomly selected 70-year-old African American male will live between 10 and 20 more years

46. Mercury Poisoning Historians and biographers have collected evidence that suggests that President Andrew Jackson suffered from mercury poisoning. Recently, researchers measured the amount of mercury in samples of Jackson's hair from 1815. The results of this experiment showed that Jackson had a mean mercury level of 6.0 ppm. *Source: JAMA.*

  a. Levels of mercury in hair samples from that time period followed a normal distribution with mean 6.9 and standard deviation 4.6. *Source: Science of the Total Environment.* Find the probability that a randomly selected person from that time period would have a mercury level of 6.0 ppm or higher. Discuss whether this provides evidence that Jackson suffered from mercury poisoning during this time period.

  b. Today's accepted normal mercury levels follow a normal distribution with approximate mean 0.6 ppm and standard deviation 0.3 ppm. *Source: Clinical Chemistry.* By today's standards, how likely is it that a randomly selected person from today would have a mercury level of 6.0 ppm or higher? Discuss whether we can conclude from this that Andrew Jackson suffered from mercury poisoning.

Social Sciences

47. Dating a Language Over time, the number of original basic words in a language tends to decrease as words become obsolete or are replaced with new words. In 1950, C. Feng and M. Swadesh established that of the original 210 basic ancient Chinese words from 950 A.D., 167 were still being used. The proportion of words that remain after t millennia is a random variable that is exponentially distributed with $a = 0.229$. *Source: The UMAP Journal.*

  a. Find the life expectancy and standard deviation of a Chinese word.

  b. Calculate the probability that a randomly chosen Chinese word will remain after 2000 years.

Physical Sciences

48. Rainfall The rainfall (in inches) in a certain region is uniformly distributed over the interval $[32, 44]$.

  a. What is the expected number of inches of rainfall?

  b. What is the probability that the rainfall will be between 38 and 40 in.?

49. Dry Length Days Researchers have shown that the number of successive dry days that occur after a rainstorm for particular regions of Catalonia, Spain, is a random variable that is distributed exponentially with a mean of 8 days. *Source: International Journal of Climatology.*

  a. Find the probability that 10 or more successive dry days occur after a rainstorm.

  b. Find the probability that fewer than 2 dry days occur after a rainstorm.

50. Earthquakes The proportion of the times (in days) between major earthquakes in the north-south seismic belt of China is a random variable that is exponentially distributed, with $a = 1/609.5$. *Source: Journal of Seismology.*

  a. Find the expected number of days and the standard deviation between major earthquakes for this region.

  b. Find the probability that the time between a major earthquake and the next one is more than 1 year.

General Interest

51. Soccer The time between goals (in minutes) for the Wolves soccer team in the English Premier League during a recent season can be approximated by an exponential distribution with $a = 1/90$. *Source: The Mathematical Spectrum.*

  a. The Wolves scored their first goal of the season 71 minutes into their first game. Find the probability that the time for a goal is no more than 71 minutes.

b. It was 499 minutes later (in game time) before the Wolves scored their next goal. Find the probability that the time for a goal is 499 minutes or more.

52. **Football** The margin of victory over the point spread (defined as the number of points scored by the favored team minus the number of points scored by the underdog minus the point spread, which is the difference between the previous two, as predicted by oddsmakers) in National Football League games has been found to be normally distributed with mean 0 and standard deviation 13.861. Suppose New England is favored over Miami by 3 points. What is the probability that New England wins? (*Hint:* Calculate the probability that the margin of victory over the point spread is greater than −3.) *Source: The American Statistician.*

CHAPTER REVIEW

SUMMARY

In this chapter, we gave a brief introduction to the use of calculus in the study of probability. In particular, the idea of a random variable and its connection to a probability density function and a cumulative distribution function were given. We explored four important concepts:

- expected value (the average value of a random variable that we would expect in the long run),
- variance (a measure of the spread of the values of a distribution),
- standard deviation (the square root of the variance), and
- median (the value of a random variable for which there is a 50% probability of being larger and a 50% probability of being smaller).

Integration techniques were used to determine probabilities, expected value, and variance of continuous random variables. Three probability density functions that have a wide range of applications were studied in detail:

- uniform (when the probability density function remains constant over the sample space),
- exponential (for items that have a constant failure rate over time), and
- normal (for random variables with a bell-shaped distribution).

Probability Density Function on [a, b]	**1.** $f(x) \geq 0$ for all x in the interval $[a, b]$.
	2. $\int_a^b f(x)\,dx = 1$.
	3. $P(c \leq X \leq d) = \int_c^d f(x)\,dx$ for c, d in $[a, b]$.
Cumulative Distribution Function	$F(x) = P(X \leq x) = \int_a^x f(t)\,dt$
Expected Value for a Density Function on [a, b]	$E(X) = \mu = \int_a^b x f(x)\,dx$
Variance for a Density Function on [a, b]	$\mathrm{Var}(X) = \int_a^b (x - \mu)^2 f(x)\,dx$
Alternative Formula for Variance	$\mathrm{Var}(X) = \int_a^b x^2 f(x)\,dx - \mu^2$
Standard Deviation	$\sigma = \sqrt{\mathrm{Var}(X)}$
Median	The value m such that $\int_a^m f(x)\,dx = \frac{1}{2}$.

Uniform Distribution $f(x) = \dfrac{1}{b-a}$ on $[a, b]$

$$\mu = \frac{1}{2}(b+a) \quad \text{and} \quad \sigma = \frac{1}{\sqrt{12}}(b-a)$$

Exponential Distribution $f(x) = ae^{-ax}$ on $[0, \infty)$

$$\mu = \frac{1}{a} \quad \text{and} \quad \sigma = \frac{1}{a}$$

Normal Distribution $f(x) = \dfrac{1}{\sigma\sqrt{2\pi}}\, e^{-(x-\mu)^2/(2\sigma^2)}$ on $(-\infty, \infty)$

$$E(X) = \mu \quad \text{and} \quad \text{Var}(X) = \sigma^2$$

z-Scores Theorem   For a normal curve with mean μ and standard deviation σ, the area to the left of x is the same as the area to the left of

$$z = \frac{x - \mu}{\sigma}$$

for the standard normal curve.

KEY TERMS

11.1
random variable
probability function
histogram
discrete probability function
continuous random variable

continuous probability
   distribution
probability density
   function
cumulative distribution
   function

11.2
mean
expected value
variance
standard deviation
median

11.3
uniform distribution
exponential distribution
normal distribution
normal curve
standard normal distribution
z-score

REVIEW EXERCISES

CONCEPT CHECK

Determine whether each of the following statements is true or false, and explain why.

1. A continuous random variable can take on values greater than 1.

2. A probability density function can take on values greater than 1.

3. A continuous random variable can take on values less than 0.

4. A probability density function can take on values less than 0.

5. The expected value of a random variable must always be at least 0.

6. The variance of a random variable must always be at least 0.

7. The expected value of a uniform random variable is the average of the endpoints of the interval over which the density function is positive.

8. For an exponential random variable, the expected value and standard deviation are always equal.

9. The normal distribution and the exponential distribution have approximately the same shape.

10. In the standard normal distribution, the expected value is 1 and the standard deviation is 0.

PRACTICE AND EXPLORATIONS

11. In a probability function, the y-values (or function values) represent _____.

12. Define a continuous random variable.

13. Give the two conditions that a probability density function for $[a, b]$ must satisfy.

14. In a probability density function, the probability that X equals a specific value, $P(X = c)$, is _____.

Decide whether each function defined as follows is a probability density function for the given interval.

15. $f(x) = \sqrt{x};\ [4, 9]$

16. $f(x) = \dfrac{1}{27}(2x + 4);\ [1, 4]$

17. $f(x) = 0.7e^{-0.7x};\ [0, \infty)$

18. $f(x) = 0.4;\ [4, 6.5]$

In Exercises 19 and 20, find a value of k that will make $f(x)$ define a probability density function for the indicated interval.

19. $f(x) = kx^2$; $[1, 4]$ **20.** $f(x) = k\sqrt{x}$; $[4, 9]$

21. The probability density function of a random variable X is defined by

$$f(x) = \frac{1}{10} \quad \text{for } x \text{ in } [10, 20].$$

Find the following probabilities.

a. $P(X \le 12)$ **b.** $P(X \ge 31/2)$ **c.** $P(10.8 \le X \le 16.2)$

22. The probability density function of a random variable X is defined by

$$f(x) = 1 - \frac{1}{\sqrt{x-1}} \quad \text{for } x \text{ in } [2, 5].$$

Find the following probabilities.

a. $P(X \ge 3)$ **b.** $P(X \le 4)$ **c.** $P(3 \le X \le 4)$

 23. Describe what the expected value or mean of a probability distribution represents geometrically.

24. The probability density functions shown in the graphs have the same mean. Which has the smallest standard deviation?

a.

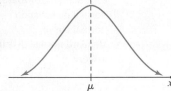

b.

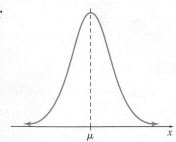

c.

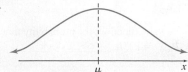

For the probability density functions defined in Exercises 25–28, find (a) the expected value, (b) the variance, (c) the standard deviation, (d) the median, and (e) the cumulative distribution function.

25. $f(x) = \frac{2}{9}(x - 2)$; $[2, 5]$ **26.** $f(x) = \frac{1}{5}$; $[4, 9]$

27. $f(x) = 5x^{-6}$; $[1, \infty)$

28. $f(x) = \frac{1}{20}\left(1 + \frac{3}{\sqrt{x}}\right)$; $[1, 9]$

29. The probability density function of a random variable is defined by $f(x) = 4x - 3x^2$ for x in $[0, 1]$. Find the following for the distribution.

a. The mean   **b.** The standard deviation

c. The probability that the value of the random variable will be less than the mean

d. The probability that the value of the random variable will be within 1 standard deviation of the mean

30. Find the median of the random variable of Exercise 29. Then find the probability that the value of the random variable will lie between the median and the mean of the distribution.

For Exercises 31 and 32, find (a) the mean of the distribution, (b) the standard deviation of the distribution, and (c) the probability that the value of the random variable is within 1 standard deviation of the mean.

31. $f(x) = 0.01e^{-0.01x}$ for x in $[0, \infty)$

32. $f(x) = \frac{5}{112}(1 - x^{-3/2})$ for x in $[1, 25]$

In Exercises 33–40, find the proportion of observations of a standard normal distribution for each region.

33. The region to the left of $z = -0.43$

34. The region to the right of $z = 1.62$

35. The region between $z = -1.17$ and $z = -0.09$

36. The region between $z = -1.39$ and $z = 1.28$

37. The region that is 1.2 standard deviations or more below the mean

38. The region that is up to 2.5 standard deviations above the mean

39. Find a z-score so that 52% of the area under the normal curve is to the right of z.

40. Find a z-score so that 21% of the area under the normal curve is to the left of z.

∞ The topics in this short chapter involved much of the material studied earlier in this book, including functions, domain and range, exponential functions, area and integration, improper integrals, integration by parts, and numerical integration. For the following special probability density functions, give

a. the type of distribution;

b. the domain and range;

c. the graph;

d. the mean and standard deviation;

e. $P(\mu - \sigma \le X \le \mu + \sigma)$.

41. $f(x) = 0.05$ for x in $[10, 30]$

42. $f(x) = e^{-x}$ for x in $[0, \infty)$

43. $f(x) = \frac{e^{-x^2}}{\sqrt{\pi}}$ for x in $(-\infty, \infty)$ (*Hint:* $\sigma = 1/\sqrt{2}$.)

∞ **44.** The chi-square distribution is important in statistics for testing whether data comes from a specified distribution and for testing the independence of two characteristics of a set of data.

When a quantity called the *degrees of freedom* is equal to 4, the probability density function is given by

$$f(x) = \frac{xe^{-x/2}}{4} \text{ for } x \text{ in } [0, \infty).$$

  a. Verify that this is a probability density function by noting that $f(x) \geq 0$ and by finding $P(0 \leq X < \infty)$.

  b. Find $P(0 \leq X \leq 3)$.

45. When the degrees of freedom in the chi-square distribution (see the previous exercise) is 1, the probability density function is given by

$$f(x) = \frac{x^{-1/2}e^{-x/2}}{\sqrt{2\pi}} \text{ for } x \text{ in } (0, \infty).$$

Calculating probabilities is now complicated by the fact that the density function cannot be antidifferentiated. Numerical integration is complicated because the density function becomes unbounded as x approaches 0.

  a. Show that one application of integration by parts (or column integration with just two rows, similar to Example 2 in Section 8.1 on Integration by Parts) allows $P(0 < X \leq b)$ to be rewritten as

$$\frac{1}{\sqrt{2\pi}} \left[2x^{1/2}e^{-x/2} \Big|_0^b + \int_0^b x^{1/2}e^{-x/2} \, dx \right].$$

  b. Using Simpson's rule with $n = 12$ in the result from part a, approximate $P(0 < X \leq 1)$.

  c. Using Simpson's rule with $n = 12$ in the result from part a, approximate $P(0 < X \leq 10)$.

  d. What should be the limit as $b \to \infty$ of the expression in part a? Do the results from parts b and c support this?

APPLICATIONS

Business and Economics

46. Mutual Funds The price per share (in dollars) of a particular mutual fund is a random variable x with probability density function defined by

$$f(x) = \frac{3}{4}(x^2 - 16x + 65) \quad \text{for } x \text{ in } [8, 9].$$

  a. Find the probability that the price will be less than $8.50.

  b. Find the expected value of the price.

  c. Find the standard deviation.

47. Machine Repairs The time (in years) until a certain machine requires repairs is a random variable t with probability density function defined by

$$f(t) = \frac{5}{112}(1 - t^{-3/2}) \quad \text{for } t \text{ in } [1, 25].$$

  a. Find the probability that no repairs are required in the first three years by finding the probability that a repair will be needed in years 4 through 25.

  b. Find the expected value for the number of years before the machine requires repairs.

  c. Find the standard deviation.

48. Retail Outlets The number of new outlets for a clothing manufacturer is an exponential distribution with probability density function defined by

$$f(x) = \frac{1}{6}e^{-x/6} \quad \text{for } x \text{ in } [0, \infty).$$

Find the following for this distribution.

  a. The mean

  b. The standard deviation

  c. The probability that the number of new outlets will be greater than the mean

49. Product Repairs The number of repairs required by a new product each month is exponentially distributed with an average of 8.

  a. What is the probability density function for this distribution?

  b. Find the expected number of repairs per month.

  c. Find the standard deviation.

  d. What is the probability that the number of repairs per month will be between 5 and 10?

50. Useful Life of an Appliance Part The useful life of a certain appliance part (in hundreds of hours) is 46.2, with a standard deviation of 15.8. Find the probability that one such part would last for at least 6000 (60 hundred) hours. Assume a normal distribution.

51. Equipment Insurance A piece of equipment is being insured against early failure. The time from purchase until failure of the equipment is exponentially distributed with mean 10 years. The insurance will pay an amount x if the equipment fails during the first year, and it will pay $0.5x$ if failure occurs during the second or third year. If failure occurs after the first three years, no payment will be made. At what level must x be set if the expected payment made under this insurance is to be 1000? *Source: Society of Actuaries.* Choose one of the following.

  a. 3858   **b.** 4449   **c.** 5382   **d.** 5644   **e.** 7235

Life Sciences

52. Weight Gain of Rats The weight gain (in grams) of rats fed a certain vitamin supplement is a continuous random variable with probability density function defined by

$$f(x) = \frac{8}{7}x^{-2} \quad \text{for } x \text{ in } [1, 8].$$

  a. Find the mean of the distribution.

  b. Find the standard deviation of the distribution.

  c. Find the probability that the value of the random variable is within 1 standard deviation of the mean.

53. Movement of a Released Animal The distance (in meters) that a certain animal moves away from a release point is exponentially distributed, with a mean of 100 m. Find the probability that the animal will move no farther than 100 m away.

54. Snowfall The snowfall (in inches) in a certain area is uniformly distributed over the interval $[2, 30]$.

  a. What is the expected snowfall?

  b. What is the probability of getting more than 20 inches of snow?

55. **Body Temperature of a Bird** The body temperature (in degrees Celsius) of a particular species of bird is a continuous random variable with probability density function defined by

$$f(x) = \frac{3}{19,696}(x^2 + x) \quad \text{for } x \text{ in } [38, 42].$$

a. What is the expected body temperature of this species?

b. Find the probability of a body temperature below the mean.

56. **Average Birth Weight** The average birth weight of infants in the United States is 7.8 lb, with a standard deviation of 1.1 lb. Assuming a normal distribution, what is the probability that a newborn will weigh more than 9 lb?

57. **Heart Muscle Tension** In a pilot study on tension of the heart muscle in dogs, the mean tension was 2.2 g, with a standard deviation of 0.4 g. Find the probability of a tension of less than 1.9 g. Assume a normal distribution.

58. **Life Expectancy** According to the National Center for Health Statistics, the life expectancy for a 65-year-old American male is 17.0 years. Assuming that from age 65, the survival of American males follows an exponential distribution, determine the following probabilities. *Source: National Vital Statistics Report.*

a. The probability that a randomly selected 65-year-old American male will live beyond 80 years of age (at least 15 more years)

b. The probability that a randomly selected 65-year-old American male will live less than 10 more years

59. **Life Expectancy** According to the National Center for Health Statistics, the life expectancy for a 50-year-old American female is 32.5 years. Assuming that from age 50, the survival of American females follows an exponential distribution, determine the following probabilities. *Source: National Vital Statistics Report.*

a. The probability that a randomly selected 50-year-old American female will live beyond 90 years of age (at least 40 more years)

b. The probability that a randomly selected 50-year-old American female will live between 30 and 50 more years

Social Sciences

60. **Assaults** The number of deaths in the United States caused by assault (murder) for each age group is given in the following table. *Source: National Vital Statistics.*

a. Plot the data. What type of function appears to best match this data?

Age Interval (years)	Midpoint of Interval (year)	Number Dying in Each Interval
0–14	7	1096
15–24	19.5	5729
25–34	29.5	4729
35–44	39.5	3013
45–54	49.5	2207
55–64	59.5	1011
65–74	69.5	397
75–84	79.5	274
85 +	89.5 (est)	32
Total		18,488

b. Use the regression feature on your graphing calculator to find a quartic equation that models the number of years, t, since birth and the number of deaths caused by assault, $N(t)$. Use the midpoint value to estimate the point in each interval when the person died. Graph the function with the plot of the data. Does the function resemble the data?

c. By finding an appropriate constant k, find a function $S(t) = kN(t)$ that is a probability density function describing the probability of death by assault. (*Hint:* Because the function in part b is negative for values less than 5.2 and greater than 88.9, restrict the domain of the density function to the interval $[5.2, 88.9]$. That is, integrate the function you found in part b from 5.2 to 88.9.)

d. For a randomly chosen person who was killed by assault, find the probabilities that the person killed was less than 25 years old, at least 45 but less than 65 years old, and at least 75 years old, and compare these with the actual probabilities.

e. Estimate the expected age at which a person will die by assault.

f. Find the standard deviation of this distribution.

Physical Sciences

61. **Earthquakes** The time between major earthquakes in the Taiwan region is a random variable with probability density function defined by

$$f(t) = \frac{1}{3650.1} e^{-t/3650.1},$$

where t is measured in days. Find the expected value and standard deviation of this probability density function. *Source: Journal of Seismology.*

General Interest

62. **State-Run Lotteries** The average state "take" on lotteries is 40%, with a standard deviation of 13%. Assuming a normal distribution, what is the probability that a state-run lottery will have a "take" of more than 50%?

EXPONENTIAL WAITING TIMES

We have seen in this chapter how probabilities that are spread out over continuous time intervals can be modeled by continuous probability density functions. The exponential distribution you met in the last section of this chapter is often used to model *waiting times,* the gaps between events that are randomly distributed in time, such as decays of a radioactive nucleus or arrivals of customers in the waiting line at a bank. In this application we investigate some properties of the exponential family of distributions.

Suppose that in a badly run subway system, the times between arrivals of subway trains at your station are exponentially distributed with a mean of 10 minutes. Sometimes trains arrive very close together, sometimes far apart, but if you keep track over many days, you'll find that the *average* time between trains is 10 minutes. According to the last section of this chapter, the exponential distribution with density function $f(t) = ae^{-at}$ has mean $1/a$, so the probability density function for our interarrival times is

$$f(t) = \frac{1}{10}e^{-t/10}.$$

First let's see what these waiting times look like. We have used a random-number generator from a statistical software package to draw 25 waiting times from this distribution. Figure 19 shows cumulative arrival times, which is what you would observe if you recorded the arrival time of each train measured in minutes from an arbitrary 0 point.

You can see that 25 trains arrive in a span of about 260 minutes, so the average interarrival time was indeed close to 10 minutes. You may also notice that there are some large gaps and some cases where trains arrived very close together.

To get a better feeling for the distribution of long and short interarrival times, look at the following list, which gives the 25 interarrival times in minutes, sorted from smallest to largest.

0.016	4.398	15.659
0.226	4.573	15.954
0.457	5.415	16.403
0.989	9.570	18.978
1.576	10.413	20.736
1.988	10.916	33.013
2.738	13.109	39.073
3.133	13.317	
3.895	14.622	

You can see that there were some very short waits. (In fact, the shortest time between trains is only 1 second, which means our model needs to be adjusted somehow to allow for the time trains spend stopped in the station.) The longest time between trains was 39 minutes, almost four times as long as the average! Although the exponential model exaggerates the irregularities of typical subway service, the problem of pile-ups and long gaps is very real for public transportation, especially for bus routes that are subject to unpredictable traffic delays. Anyone who works at a customer service job is also familiar with this behavior: The waiting line at a bank may be empty for minutes at a stretch, and then several customers walk in at nearly the same time. In this case, the customer interarrival times are exponentially distributed.

Planners who are involved with scheduling need to understand this "clumping" behavior. One way to explore it is to find probabilities for ranges of interarrival times. Here integrals are the natural tool. For example, if we want to estimate the fraction of interarrival times that will be less than 2 minutes, we compute

$$\frac{1}{10}\int_0^2 e^{-t/10}\,dt = 1 - e^{-1/5} \approx 0.1813.$$

So on average, 18% of the interarrival times will be less than 2 minutes, which indicates that clustering of trains will be a problem in our system. (If you have ridden a system like the one in New York City, you may have boarded a train that was ordered to "stand by" for several minutes to spread out a cluster of trains.) We can also compute the probability of a gap of 30 minutes or longer. It will be

$$\frac{1}{10}\int_{30}^{\infty} e^{-t/10}\,dt = e^{-3} \approx 0.0498.$$

So in a random sample of 25 interarrival times we might expect one or two long waits, and our simulation, which includes times of 33 and 39 minutes, is not a fluke. Of course, the rider's experience depends on when she arrives at the station, which is another random input to our model. If she arrives in the middle of a cluster, she'll get a train right away, but if she arrives at the beginning of a long gap she may have a half-hour wait. So we would also like to model the rider's *waiting time,* the time between the rider's arrival at the station and the arrival of the next train.

A remarkable fact about the exponential distribution is that if our passenger arrives at the station at a random time, the distribution of the rider's waiting times is *the same* as the distribution of

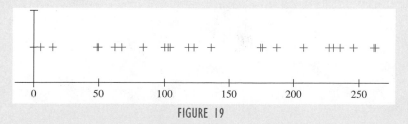

FIGURE 19

interarrival times (that is, exponential with mean 10 minutes). At first this seems paradoxical; since she usually arrives between trains, she should wait less, on average, than the average time between trains. But remember that she's more likely to arrive at the station in one of those long gaps. In our simulation, 72 out of 260 minutes is taken up with long gaps, and even if the rider arrives at the middle of such a gap she'll still wait longer than 15 minutes. Because of this feature the exponential distribution is often called *memoryless:* If you dip into the process at random, it is as if you were starting all over. If you arrive at the station just as a train leaves, your waiting time for the next one still has an exponential distribution with mean 10 minutes. The next train doesn't "know" anything about the one that just left.*

Because the riders' waiting times are exponential, the calculations we have already made tell us what riders will experience: A wait of less than 2 minutes has probability about 0.18. The average wait is 10 minutes, but long waits of more than 30 minutes are not all that rare (probability about 0.05).

Customers waiting for service care about the average wait, but they may care even more about the *predictability* of the wait. In this chapter we stated that the standard deviation for an exponential distribution is the same as the mean, so in our model the standard deviation of riders' waiting times will be 10 minutes. This indicates that a wait of twice the average length is not a rare event. (See Exercise 3.)

Let's compare the experience of riders on our exponential subway with the experience of riders of a perfectly regulated service in which trains arrive *exactly* 10 minutes apart. We'll still assume that the passenger arrives at random. But now the waiting time is uniformly distributed on the time interval $[0 \text{ minutes}, 10 \text{ minutes}]$. This uniform distribution has density function

$$f(t) = \begin{cases} \dfrac{1}{10} & \text{for } 0 \leq t \leq 10 \\ 0 & \text{otherwise} \end{cases}$$

The mean waiting time is

$$\int_0^{10} \frac{1}{10} \cdot t \, dt = 5 \text{ minutes}$$

and the standard deviation of the waiting times is

$$\sqrt{\int_0^{10} \frac{(t-5)^2}{10} \, dt} = \sqrt{\frac{25}{3}} \approx 2.89 \text{ minutes.}$$

Clearly the rider has a better experience on this system. Even though the same average number of trains is running per hour as in the exponential subway, the average wait for the uniform subway is only 5 minutes with a standard deviation of 2.89 minutes, and no one ever waits longer than 10 minutes!

*See Chapter 1 in Volume 2 of Feller, William, *An Introduction to Probability Theory and Its Applications,* 2nd ed., New York: Wiley, 1971.

Any subway run is subject to unpredictable accidents and variations, and this random input is always pushing the riders' waiting times toward the exponential model. Indeed, even with uniform scheduling of trains, there will be service bottlenecks because the exponential distribution is also a reasonable model (over a short time period) for interarrival times of *passengers* entering the station. The goal of schedulers is to move passengers efficiently in spite of random train delays and random input of passengers. One proposed solution, the PRT or personal rapid transit system, uses small vehicles holding just a few passengers that can be scheduled to match a fluctuating demand.

The subway scheduling problem is part of a branch of statistics called *queueing theory*, the study of any process in which inputs arrive at a service point and wait in a line or queue to be served. Examples include telephone calls arriving at a customer service center, our passengers entering the subway station, packets of information traveling through the Internet, and even pieces of code waiting for a processor in a multiprocessor computer. The following Web sites provide a small sampling of work in this very active research area.

- *http://web2.uwindsor.ca/math/hlynka/queue.html* (A collection of information on queueing theory)

- *http://faculty.washington.edu/jbs/itrans/ingsim.htm* (an article on scheduling a PRT)

EXERCISES

1. If X is a continuous random variable, $P(a \leq X \leq b)$ is the same as $P(a < X < b)$. Since these are different events, how can they have the same probability?

2. Someone who rides the subway back and forth to work each weekday makes about 40 trips a month. On the exponential subway, how many times a month can this commuter expect a wait longer than half an hour?

3. Find the probability that a rider of the exponential subway waits more than 20 minutes for a train; that is, find the probability of a wait more than twice as long as the average.

4. On the exponential subway, what is the probability that a randomly arriving passenger has a wait of between 9 and 10 minutes? What is the corresponding probability on the uniform subway?

5. If our system is aiming for an average interarrival time of 10 minutes, we might set a tolerance of plus or minus 2 minutes and try to keep the interarrival times between 8 and 12 minutes. Under the exponential model, what fraction of interarrival times fall in this range? How about under the uniform model?

6. Most mathematical software includes routines for generating "pseudo-random" numbers (that is, numbers that behave randomly even though they are generated by arithmetic). That's what we used to simulate the exponential waiting times for our subway system. But a source on the Internet (http://www.fourmilab.ch/hotbits/) delivers random numbers based on the times between decay events in a sample of Krypton-85. As noted above, the waiting times between decay events have an exponential distribution, so we can see what nature's random numbers look like. Here's a short sample:

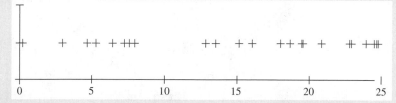

Actually, this source builds its random numbers from random bits, that is, 0's and 1's that occur with equal probability. See if you can think of a way of turning a sequence of exponential waiting times into a random sequence of 0's and 1's.

DIRECTIONS FOR GROUP PROJECT

Find a situation in which you and your group can gather actual wait times, such as a bus stop, doctor's office, teller line at a bank, or check-out line at a grocery store. Collect data on interarrival/service times and determine the mean service time. Using this average, determine whether the data appears to follow an exponential distribution. Develop a table that lists the percentage of the time that particular waiting times occur using both the data and the exponential function. Construct a poster that could be placed near the location where people wait that estimates the waiting time for service.

12 Sequences and Series

Extended Application:
Living Assistance and Subsidized Housing

In sports that place high stress on the body, such as basketball, professional athletes often have relatively short playing careers. In Section 2 of this chapter we look at annuities, a kind of investment that allows a player to use her current high earnings to purchase a sequence of guaranteed annual payments that will begin when she retires.

A function whose domain is the set of natural numbers, such as

$$a(n) = 2n, \quad \text{for} \quad n = 1, 2, 3, 4, \ldots$$

is a **sequence**. The sequence $a(n) = 2n$ can be written by listing its *terms*, 2, 4, 6, 8, ..., $2n$, ... The letter n is used instead of x as a variable to emphasize the fact that the domain includes only natural numbers. For the same reason, a is used instead of f to name the function.

Sequences have many different applications; one example is the sequence of payments used to pay off a car loan or a home mortgage. (For most practical problems, the domain is a *subset* of the set of natural numbers.) This use of sequences is discussed in Section 12.2. The remaining sections of this chapter cover topics related to sequences.

12.1 Geometric Sequences

APPLY IT If a person saved 1¢ on June 1, 2¢ on June 2, 4¢ on June 3, and so forth, continuing the pattern of saving twice as much each day as the previous day, how much would she have saved by the last day of June?
We will answer this question in Example 6 of this section.

In our definition of sequence we used the example $a(n) = 2n$. The range values of this sequence function,

$$a(1) = 2, \quad a(2) = 4, \quad a(3) = 6, \ldots,$$

are called the **elements** or **terms** of the sequence. Instead of writing $a(5)$ for the fifth term of a sequence, it is customary to write a_5; for the sequence above

$$a_5 = 10.$$

In the same way, for the sequence above, $a_1 = 2$, $a_2 = 4$, $a_8 = 16$, $a_{20} = 40$, and $a_{51} = 102$.

The symbol a_n is used for the **general** or **nth term** of a sequence. For example, for the sequence 4, 7, 10, 13, 16, ... the general term might be given by $a_n = 1 + 3n$. This formula for a_n can be used to find any term of the sequence that might be needed. For example, the first three terms of the sequence are

$$a_1 = 1 + 3(1) = 4, \qquad a_2 = 1 + 3(2) = 7, \qquad \text{and} \qquad a_3 = 1 + 3(3) = 10.$$

Also, $a_8 = 25$ and $a_{12} = 37$.

EXAMPLE 1 Sequence

Find the first four terms of the sequence having general term $a_n = -4n + 2$.

SOLUTION Replace n, in turn, with 1, 2, 3, and 4.

$$\text{If } n = 1, \quad a_1 = -4(1) + 2 = -4 + 2 = -2.$$
$$\text{If } n = 2, \quad a_2 = -4(2) + 2 = -8 + 2 = -6.$$
$$\text{If } n = 3, \quad a_3 = -4(3) + 2 = -12 + 2 = -10.$$
$$\text{If } n = 4, \quad a_4 = -4(4) + 2 = -16 + 2 = -14.$$

The first four terms of this sequence are -2, -6, -10, and -14. **TRY YOUR TURN 1**

YOUR TURN 1 Find the first four terms of the sequence having general term $a_n = 3n - 6$.

A sequence in which each term after the first is found by *multiplying* the preceding term by the same number is called a **geometric sequence**. The ratio of any two consecutive terms is a constant r,

$$r = \frac{a_{n+1}}{a_n}, \quad \text{where } n \geq 1,$$

called the **common ratio**. For example, to find r in the following sequence:

$$3, -6, 12, -24, 48, -96, \ldots$$

take $-6/3$ or $12/(-6)$ or $-24/12$, etc. and get $r = -2$. Thus, it is a geometric sequence in which each term after the first is found by multiplying the preceding term by the number -2, the common ratio.

If a is the first term of a geometric sequence and r is the common ratio, then the second term is given by $a_2 = ar$ and the third term by $a_3 = a_2r = ar^2$. Also, $a_4 = ar^3$ and $a_5 = ar^4$. These results are generalized below.

General Term of a Geometric Sequence

If a geometric sequence has first term a and common ratio r, then

$$a_n = ar^{n-1}.$$

EXAMPLE 2 Geometric Sequences

Find the indicated term for each geometric sequence.

(a) Find a_7 for $6, 24, 96, 384, \ldots$.

SOLUTION Here $a = a_1 = 6$. To verify that the sequence is geometric, divide each term except the first by the preceding term.

$$\frac{24}{6} = \frac{96}{24} = \frac{384}{96} = 4$$

Since the ratio is constant, the sequence is geometric with $r = 4$. To find a_7, use the formula for a_n with $n = 7, a = 6$, and $r = 4$.

$$a_7 = 6(4)^{7-1} = 6(4)^6 = 6(4096) = 24{,}576$$

(b) Find a_6 for $8, -16, 32, -64, 128, \ldots$.

SOLUTION As before, verify that $r = -16/8 = 32/(-16) = -64/32 = 128/(-64) = -2$. Here $a = a_1 = 8$, so

$$a_6 = 8(-2)^{6-1} = 8(-2)^5 = 8(-32) = -256. \quad \text{TRY YOUR TURN 2} \blacksquare$$

YOUR TURN 2 Find a_7 for $2, -6, 18, -54, \ldots$.

EXAMPLE 3 Depreciation

A new machine is purchased for $150,000. Each year the machine loses 25% of its value. Find its value at the end of the sixth year.

SOLUTION Since the machine loses 25% of its value each year, it retains 75% of its value. At the end of its first year, its value is 75% of $150,000. Its value at the end of each of the following years is 75% of the previous year's value. These values form a geometric sequence, with $r = 0.75$. If we let $a = a_1 = 150{,}000$ then the value at the end of the first year is a_2, at the end of the second year is a_3, and so on. The value at the end of the sixth year is

$$a_7 = 150{,}000(0.75)^{7-1} = 150{,}000(0.75)^6 = 26{,}696.77734.$$

The value of the machine is $26,696.78 at the end of the sixth year. **TRY YOUR TURN 3** ■

YOUR TURN 3 A machine purchased for $10,000 loses 10% of its value each year. Find its value at the end of its tenth year.

In the next section we will need to know how to find the sum of the first n terms of a geometric sequence. To get a general rule for finding such a sum, begin by writing the sum S_n of the first n terms of a geometric sequence with first term $a = a_1$ and common ratio r as

$$S_n = a_1 + a_2 + a_3 + \cdots + a_n.$$

Since $a_n = ar^{n-1}$, this sum can be written as

$$S_n = a + ar + ar^2 + \cdots + ar^{n-1}. \tag{1}$$

If $r = 1$, all the terms are equal to a, and $S_n = n \cdot a$, the correct result for this case. If $r \neq 1$, multiply both sides of Equation (1) by r, obtaining

$$rS_n = ar + ar^2 + ar^3 + \cdots + ar^n. \tag{2}$$

Now subtract corresponding sides of Equation (1) from Equation (2):

$$rS_n - S_n = ar^n - a.$$

Factoring yields

$$S_n(r - 1) = a(r^n - 1),$$

and dividing by $r - 1$ on both sides gives

$$S_n = \frac{a(r^n - 1)}{r - 1}.$$

This result is summarized below.

Sum of The First n Terms of a Geometric Sequence

If a geometric sequence has first term a and common ratio r, then the sum of the first n terms, S_n, is given by

$$S_n = \frac{a(r^n - 1)}{r - 1}, \quad \textbf{where } r \neq \textbf{1.}$$

EXAMPLE 4 Summing Terms of a Geometric Sequence

Find the sum of the first six terms of the geometric sequence $3, 12, 48, \ldots$.

Method 1
Using the Formula

SOLUTION Here $a = a_1 = 3$ and $r = 4$. Find S_6, the sum of the first six terms, by the formula above.

$$S_6 = \frac{3(4^6 - 1)}{4 - 1} \qquad \text{Let } n = 6, a = 3, r = 4.$$

$$= \frac{3(4096 - 1)}{3}$$

$$= 4095$$

Method 2
Graphing Calculator

The sum of a sequence can be conveniently calculated on a TI-84 Plus calculator using the command `sum(seq(3*4^N, N, 0, 5))` as shown in Figure 1. Notice that to find the sum of the first six terms of the sequence we begin with $N = 0$ and end with $N = 5$ for a total of 4095.

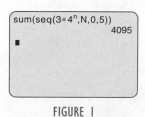

```
sum(seq(3*4ⁿ,N,0,5))
                  4095
■
```

FIGURE 1

YOUR TURN 4 Find the sum of the first seven terms of the geometric sequence $2, -8, 32, \ldots$.

TRY YOUR TURN 4

Using summation notation, we can write $S_n = a + ar + ar^2 + \cdots + ar^{n-1}$ as

$$S_n = \sum_{i=0}^{n-1} ar^i.$$

EXAMPLE 5 Summing Terms of a Geometric Sequence

Use the formula for the sum of the first n terms of a geometric sequence to evaluate the following sum.

$$\sum_{i=0}^{8} \frac{1}{2}(-2)^i$$

YOUR TURN 5

Evaluate the sum $\sum_{i=0}^{4} 81\left(\frac{1}{3}\right)^i$.

SOLUTION Here $a = 1/2$ and $r = -2$. The summation is from $i = 0$ to $n-1 = 8$, so $n = 9$. Using the formula gives

$$S_9 = \frac{\frac{1}{2}[(-2)^9 - 1]}{-2 - 1} = \frac{\frac{1}{2}(-512 - 1)}{-3} = \frac{\frac{1}{2}(-513)}{-3} = 85.5.$$

TRY YOUR TURN 5

EXAMPLE 6 Savings

A person saved 1¢ on June 1, 2¢ on June 2, 4¢ on June 3, and so forth, continuing the pattern of saving twice as much each day as the previous day. How much would she have saved by the end of June?

APPLY IT **SOLUTION** This is a geometric sequence with $a = 1$ and $r = 2$. We want to find S_{30}.

$$S_{30} = \frac{1(2^{30} - 1)}{2 - 1} = 1{,}073{,}741{,}823$$

By the end of June, the person will have saved 1,073,741,823 cents, or $10,737,418.23!

12.1 EXERCISES

List the first n terms of the geometric sequence satisfying the following conditions.

1. $a_1 = 2, r = 3, n = 4$ **2.** $a_1 = 4, r = 2, n = 5$

3. $a_1 = 1/2, r = 4, n = 4$ **4.** $a_1 = 2/3, r = 6, n = 3$

5. $a_3 = 6, a_4 = 12, n = 5$ **6.** $a_2 = 9, a_3 = 3, n = 4$

Find a_5 and a_n for the following geometric sequences.

7. $a_1 = 4, r = 3$ **8.** $a_1 = 8, r = 4$

9. $a_1 = -3, r = -5$ **10.** $a_1 = -4, r = -2$

11. $a_2 = 12, r = 1/2$ **12.** $a_3 = 2, r = 1/3$

13. $a_4 = 64, r = -4$ **14.** $a_4 = 81, r = -3$

For each sequence that is geometric, find r and a_n.

15. $6, 12, 24, 48, \ldots$ **16.** $4, 16, 64, 256, \ldots$

17. $3/4, 3/2, 3, 6, 12, \ldots$ **18.** $-7, -5, -3, -1, 1, 3, \ldots$

19. $4, 8, -16, 32, 64, -128, \ldots$ **20.** $6, 8, 10, 12, 14, \ldots$

21. $-5/8, 5/12, -5/18, 5/27, \ldots$

22. $7/4, -7/12, 7/36, -7/108, \ldots$

Find the sum of the first five terms of each geometric sequence.

23. $3, 6, 12, 24, \ldots$ **24.** $5, 20, 80, 320, \ldots$

25. $12, -6, 3, -3/2, \ldots$ **26.** $18, -3, 1/2, -1/12, \ldots$

27. $a_1 = 3, r = -2$ **28.** $a_1 = -5, r = 4$

29. $a_1 = 6.324, r = 2.598$ **30.** $a_1 = -2.772, r = 1.335$

Use the formula for the sum of the first n terms of a geometric sequence to evaluate the following sums.

31. $\sum_{i=0}^{7} 8(2)^i$ **32.** $\sum_{i=0}^{6} 4(3)^i$

33. $\sum_{i=0}^{8} \frac{3}{2}(4)^i$ **34.** $\sum_{i=0}^{9} \frac{3}{4}(2)^i$

35. $\sum_{i=0}^{4} \frac{3}{4}(-3)^i$

36. $\sum_{i=0}^{5} \frac{3}{2}(-2)^i$

37. $\sum_{i=0}^{8} 64\left(\frac{1}{2}\right)^i$

38. $\sum_{i=0}^{6} 81\left(\frac{2}{3}\right)^i$

APPLICATIONS

Business and Economics

39. Depreciation A certain machine annually loses 20% of the value it had at the beginning of that year. If its initial value is $12,000, find its value at the following times.

a. The end of the fifth year

b. The end of the eighth year

40. Income An oil well produced $4,000,000 of income its first year. Each year thereafter, the well produced 3/4 as much income as the previous year. What is the total amount of income produced by the well in 8 years?

41. Savings Suppose you could save $1 on January 1, $2 on January 2, $4 on January 3, and so on. What amount would you save on January 31? What would be the total amount of your savings during January?

42. Depreciation Each year a machine loses 30% of the value it had at the beginning of the year. Find the value of the machine at the end of 6 years if it cost $200,000 new.

Life Sciences

43. Population The population of a certain colony of bacteria increases by 5% each hour. After 7 hours, what is the percent increase in the population over the initial population?

Physical Sciences

44. Radioactive Decay The half-life of a radioactive substance is the time it takes for half the substance to decay. Suppose the half-life of a substance is 3 years and that 10^{15} molecules of the substance are present initially. How many molecules will be unchanged after 15 years?

45. Rotation of a Wheel A bicycle wheel rotates 400 times per minute. If the rider removes his or her feet from the pedals, the wheel will start to slow down. Each minute, it will rotate only 3/4 as many times as in the preceding minute. How many times will the wheel rotate in the fifth minute after the rider's feet are removed from the pedals?

General Interest

46. Thickness of a Paper Stack A piece of paper is 0.008 in. thick.

a. Suppose the paper is folded in half, so that its thickness doubles, for 12 times in a row. How thick is the final stack of paper?

b. Suppose it were physically possible to fold the paper 50 times in a row. How thick would the final stack of paper be?

47. Sports In the NCAA Men's Basketball Tournament, 64 teams are initially paired off. By playing a series of single-elimination games, a champion is crowned. *Source: Mathematics Teacher*.

a. Write a geometric sequence whose sum determines the number of games that must be played to determine the champion team.

b. How many games must be played to produce the champion?

c. Generalize parts a and b to a tournament where 2^n teams are initially present.

d. Discuss a quick way to determine the answers to parts b and c, based on the fact that each game produces one loser, and all teams except the champion lose one game.

48. Game Shows Some game shows sponsor tournaments where in each game, three individuals play against each other, yielding one winner and two losers. The winners of three such games then play each other, until the final game of three players produces a tournament winner. Suppose 81 people begin such a tournament. *Source: Mathematics Teacher*.

a. Write a geometric sequence whose sum determines the number of games that must be played to determine the tournament champion.

b. How many games must be played to produce the champion?

c. Generalize parts a and b to a tournament where 3^n players are initially present.

d. Further generalize parts a and b to a tournament where t^n players are initially present.

▄▄ **YOUR TURN ANSWERS**

1. −3, 0, 3, 6   **2.** 1458   **3.** $3486.78   **4.** 6554   **5.** 121

12.2 Annuities: An Application of Sequences

APPLY IT Suppose $1500 is deposited at the end of each year for the next 6 years in an account paying 8% per year, compounded annually. How much will be in the account after 6 years?

Such a sequence of equal payments made at equal periods of time is called an **annuity**. If each payment is made at the end of a period, and if the frequency of payments is the same as the frequency of compounding, the annuity is called an **ordinary annuity**. The time between payments is the **payment period**, and the time from the beginning of the first period to the end of the last period is called the **term of the annuity**. The **amount of the**

APPLY IT

annuity, the final sum on deposit in the account, is defined as the sum of the compound amounts of all the payments, compounded to the end of the term.

Figure 2 shows the annuity described above. To find the amount of this annuity, look at each of the $1500 payments separately. The first of these payments will produce a compound amount of

$$1500(1 + 0.08)^5 = 1500(1.08)^5$$

at the end of 6 years. Use 5 as the exponent instead of 6 because the money is deposited at the *end* of the first year and thus earns interest for only 5 years.

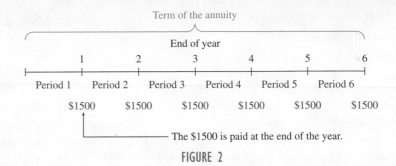

FIGURE 2

The second payment of $1500 will produce a compound amount (at the end of 5 years) of $1500(1.08)^4$. As shown in Figure 3, the total amount of the annuity is

$$1500(1.08)^5 + 1500(1.08)^4 + 1500(1.08)^3 + 1500(1.08)^2$$
$$+ 1500(1.08)^1 + 1500. \tag{1}$$

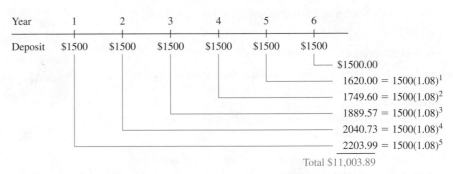

FIGURE 3

(The last payment earns no interest.) Reversing the order of the terms, so the last term is first, shows that Equation (1) is the sum of the terms of a geometric sequence with $a = 1500, r = 1.08$, and $n = 6$. Using the formula for the sum of the first n terms of a geometric sequence gives

$$1500 + 1500(1.08)^1 + 1500(1.08)^2 + 1500(1.08)^3$$
$$+ 1500(1.08)^4 + 1500(1.08)^5 = \frac{1500(1.08^6 - 1)}{1.08 - 1}$$
$$\approx \$11{,}003.89.$$

To generalize this result, suppose that R dollars are paid into an account at the end of each period for n periods, at a rate of interest i per period. The first payment of R dollars will produce a compound amount of $R(1 + i)^{n-1}$ dollars, the second payment will produce

$R(1 + i)^{n-2}$ dollars, and so on; the final payment earns no interest and contributes just R dollars to the total. If S represents the future value (or sum) of the annuity, then

$$S = R(1 + i)^{n-1} + R(1 + i)^{n-2} + R(1 + i)^{n-3} + \cdots + R(1 + i) + R,$$

or, written in reverse order,

$$S = R + R(1 + i)^1 + R(1 + i)^2 + \cdots + R(1 + i)^{n-1}.$$

This is the sum of the first n terms of the geometric sequence having first term R and common ratio $1 + i$. Using the formula for the sum of the first n terms of a geometric sequence gives

$$S = \frac{R[(1 + i)^n - 1]}{(1 + i) - 1} = \frac{R[(1 + i)^n - 1]}{i} = R\left[\frac{(1 + i)^n - 1}{i}\right].$$

The quantity in brackets is commonly written $s_{\overline{n}|i}$ (read "s-angle-n at i"), so that

$$S = R \cdot s_{\overline{n}|i}.$$

 **TECHNOLOGY NOTE**

> Values of $s_{\overline{n}|i}$ can be found by a calculator. The TI-84 Plus has a special `Finance` menu with this formula built in. For more details, see the *Graphing Calculator and Excel Spreadsheet Manual* available with this book.

Our work with annuities can be summarized as follows.

Amount of Annuity

The amount S of an annuity of payments of R dollars each, made at the end of each period for n consecutive interest periods at a rate of interest i per period, is given by

$$S = R\left[\frac{(1 + i)^n - 1}{i}\right] \qquad \text{or} \qquad S = R \cdot s_{\overline{n}|i}.$$

NOTE In this section we are assuming that each payment is made at the end of a period, as is the case in an ordinary annuity. There are situations in which each payment is made at the beginning of a period; this is known as an *annuity due*. The details are slightly different, and we will not go into them here. For more information, see Chapter 5 of our other textbook, *Finite Mathematics,* by Lial, Greenwell, and Ritchey.

EXAMPLE 1 Annuity

Erin D'Aquanni is an athlete who feels that her playing career will last 7 more years. To prepare for her future, she deposits $22,000 at the end of each year for 7 years in an account paying 8% compounded annually. How much will she have on deposit after 7 years?

SOLUTION Her deposits form an ordinary annuity with $R = 22{,}000, n = 7$, and $i = 0.08$. The amount of this annuity is (by the formula above)

$$S = 22{,}000\left[\frac{(1.08)^7 - 1}{0.08}\right].$$

YOUR TURN 1 In Example 1, how much will Erin have on deposit after 7 years if she deposits $22,000 annually and earns only 5% interest?

The number in brackets, $s_{\overline{7}|0.08}$, is 8.92280336, so that

$$S = 22{,}000(8.92280336) = 196{,}301.67,$$

or $196,301.67.

TRY YOUR TURN 1

EXAMPLE 2 **Annuity**

Suppose $1000 is deposited at the end of each 6-month period for 5 years in an account paying 6% per year compounded semiannually. Find the amount of the annuity.

SOLUTION Interest of $i = 0.06/2 = 0.03$ is earned semiannually. In 5 years there are $5 \times 2 = 10$ semiannual periods. Since $s_{\overline{10}|0.03} = [(1.03)^{10} - 1]/0.03 = 11.46388$, the $1000 deposits will produce a total of

$$S = 1000(11.46388) = 11{,}463.88,$$

or $11,463.88.

YOUR TURN 2 Suppose $125 is deposited monthly for 5 years into an account paying 2% per year compounded monthly. Find the amount of the annuity.

TRY YOUR TURN 2

The formula for S involves the variables R, i, and n. The next example shows how to solve for one of these other variables.

EXAMPLE 3 **Annuity**

Melissa Abruzese wants to buy an expensive video camera three years from now. She plans to deposit an equal amount at the end of each quarter for three years in order to accumulate enough money to pay for the camera. Melissa expects the camera to cost $2400 at that time. The bank pays 6% interest per year compounded quarterly. Find the amount of each of the 12 equal deposits she must make.

SOLUTION This example describes an ordinary annuity with $S = 2400$. Since interest is compounded quarterly, $i = 0.06/4 = 0.015$ and $n = 3 \cdot 4 = 12$ periods. The unknown here is the amount of each payment, R. By the formula for the amount of an annuity given above,

$$2400 = R \cdot s_{\overline{12}|0.015}$$

$$2400 = R\left[\frac{1.015^{12} - 1}{0.015}\right]$$

$$2400 = R(13.04121)$$

$$R = 184.03, \qquad \text{Divide both sides by 13.04121.}$$

YOUR TURN 3 Repeat Example 3 if the interest rate is only 2.5% compounded quarterly.

or $184.03.

TRY YOUR TURN 3

Sinking Fund

A **sinking fund** is a fund set up to receive periodic payments; these periodic payments plus the interest on them are designed to produce a given total at some time in the future. As an example, a corporation might set up a sinking fund to receive money that will be needed to pay off a loan in the future. The deposits in Examples 1 and 3 form sinking funds.

EXAMPLE 4 **Sinking Fund**

The Toussaints are close to retirement. They agree to sell an antique urn to a local museum for $17,000. Their tax adviser suggests that they defer receipt of this money until they retire, 5 years in the future. (At that time, they might well be in a lower tax bracket.) The museum agrees to pay them the $17,000 in a lump sum in 5 years. Find the amount of each payment the museum must make into a sinking fund so that it will have the necessary $17,000 in 5 years. Assume that the museum can earn 8% compounded annually on its money and that the payments are made annually.

SOLUTION These payments make up an ordinary annuity. The annuity will amount to $17,000 in 5 years at 8% compounded annually, so

$$17{,}000 = R \cdot s_{\overline{5}|0.08}$$

$$R = \frac{17{,}000}{s_{\overline{5}|0.08}} = \frac{17{,}000}{5.86660} \approx 2897.76,$$

or $2897.76. If the museum deposits $2897.76 at the end of each year for 5 years in an account paying 8% compounded annually, it will have the needed $17,000. This result is shown in the following table. In other cases, the last payment might differ slightly from the others due to rounding R to the nearest penny.

	Sinking Fund Amounts		
Payment Number	Amount of Deposit	Interest Earned	Total in Account
1	$2897.76	$0	$2897.76
2	$2897.76	$231.82	$6027.34
3	$2897.76	$482.19	$9407.29
4	$2897.76	$752.58	$13,057.63
5	$2897.76	$1044.61	$17,000.00

Present Value of an Annuity

As shown above, if a deposit of R dollars is made at the end of each period for n periods, at a rate of interest i per period, then the account will contain

$$S = R \cdot s_{\overline{n}|i} = R\left[\frac{(1 + i)^n - 1}{i}\right]$$

dollars after n periods. Now suppose we want to find the *lump sum P* that must be deposited today at a rate of interest i per period in order to produce the same amount S after n periods.

First recall that P dollars deposited today will amount to $P(1 + i)^n$ dollars after n periods at a rate of interest i per period. This amount, $P(1 + i)^n$, should be the same as S, the amount of the annuity. Substituting $P(1 + i)^n$ for S in the formula above gives

$$P(1 + i)^n = R\left[\frac{(1 + i)^n - 1}{i}\right].$$

To solve this equation for P, multiply both sides of the equation by $(1 + i)^{-n}$.

$$P = R(1 + i)^{-n}\left[\frac{(1 + i)^n - 1}{i}\right]$$

Use the distributive property and the fact that $(1 + i)^{-n}(1 + i)^n = 1$.

$$P = R\left[\frac{(1 + i)^{-n}(1 + i)^n - (1 + i)^{-n}}{i}\right] = R\left[\frac{1 - (1 + i)^{-n}}{i}\right]$$

The amount P is called the **present value of the annuity**. The quantity in brackets is abbreviated as $a_{\overline{n}|i}$.

Present Value of an Annuity

The present value P of an annuity of payments of R dollars each, made at the end of each period for n consecutive interest periods at a rate of interest i per period is given by

$$P = R\left[\frac{1 - (1 + i)^{-n}}{i}\right] \qquad \text{or} \qquad P = R \cdot a_{\overline{n}|i}.$$

EXAMPLE 5 Present Value

What lump sum deposited today at 6% interest compounded annually will yield the same total amount as payments of $1500 at the end of each year for 12 years, also at 6% compounded annually?

SOLUTION Find the present value of an annuity of $1500 per year for 12 years at 6% compounded annually. From the present value formula, $a_{\overline{12}|0.06} = [1 - (1.06)^{-12}]/0.06$ = 8.383844, so

$$P = 1500(8.383844) = 12,575.77,$$

or $12,575.77. A lump sum deposit of $12,575.77 today at 6% compounded annually will yield the same total after 12 years as deposits of $1500 at the end of each year for 12 years at 6% compounded annually.

Check this result as follows. The compound amount in 12 years of a deposit of $12,575.77 today at 6% compounded annually can be found by the formula $A = P(1 + i)^n$:

$$12,575.77(1.06)^{12} = (12,575.77)(2.012196) \approx 25,304.91,$$

or $25,304.91. On the other hand, a deposit of $1500 into an annuity at the end of each year for 12 years, at 6% compounded annually, gives an amount of

$$1500[(1.06)^{12} - 1]/0.06 = 1500(16.86994) = 25,304.91,$$

YOUR TURN 4 Repeat
Example 5 if the annual interest rate
is 4.5% and the payments are $2500
at the end of each year for 12 years.

or $25,304.91.

In summary, there are two ways to have $25,304.91 in 12 years at 6% compounded annually—a single deposit of $12,575.77 today, or payments of $1500 at the end of each year for 12 years. **TRY YOUR TURN 4**

The formula above can be used if the lump sum is known and the periodic payment of the annuity must be found. The next example shows how to do this.

EXAMPLE 6 Payments

A used car costs $6000. After a down payment of $1000, the balance will be paid off in 36 monthly payments with interest of 12% per year, compounded monthly. Find the amount of each payment.

SOLUTION A single lump sum payment of $5000 today would pay off the loan, so $5000 is the present value of an annuity of 36 monthly payments with interest of 12%/12 = 1% per month. We can find R, the amount of each payment, by using the formula

$$P = R \cdot a_{\overline{n}|i}$$

YOUR TURN 5 A used car
costs $11,000, and there is no down
payment. The car will be paid off in
48 monthly payments with interest
of 6% per year, compounded
monthly. Find the amount of each
payment.

and replacing P with 5000, n with 36, and i with 0.01. From the present value formula, $a_{\overline{36}|0.01} = 30.10751$, so

$$5000 = R(30.10751)$$
$$R \approx 166.07$$

or $166.07. Monthly payments of $166.07 each will be needed.   **TRY YOUR TURN 5**

Amortization   A loan is **amortized** if both the principal and interest are paid by a sequence of equal periodic payments. In Example 6 above, a loan of $5000 at 12% interest compounded monthly could be amortized by paying $166.07 per month for 36 months, or (it turns out) $131.67 per month for 48 months.

EXAMPLE 7 Amortization

A speculator agrees to pay $15,000 for a parcel of land. Payments will be made twice each year for 4 years at an interest rate of 12% compounded semiannually.

(a) Find the amount of each payment.

SOLUTION If the speculator immediately paid $15,000, there would be no need for any payments at all. Thus, $15,000 is the present value of an annuity of R dollars, with $2 \cdot 4 = 8$ periods, and $i = 0.12/2 = 0.06$ per period. If P is the present value of an annuity,

$$P = R \cdot a_{\overline{n}|i}.$$

In this example, $P = 15,000$, with

$$15,000 = R \cdot a_{\overline{8}|0.06}$$

or

$$R = \frac{15,000}{a_{\overline{8}|0.06}}.$$

$$= \frac{15,000}{6.20979} \approx 2415.54,$$

or $2415.54. Each payment is $2415.54.

(b) Find the portion of the first payment that is applied to the reduction of the debt.

SOLUTION Interest is 12% per year, compounded semiannually. During the first period, the entire $15,000 is owed. Interest on this amount for 6 months (1/2 year) is found by the formula for simple interest, $I = Prt$, so that

$$I = 15,000(0.12)\left(\frac{1}{2}\right) = 900,$$

or $900. At the end of 6 months, the speculator makes a payment of $2415.54; since $900 of this represents interest, a total of

$$\$2415.54 - \$900 = \$1515.54$$

is applied to the reduction of the original debt.

(c) Find the balance due after 6 months.

SOLUTION The original balance due is $15,000. After 6 months, $1515.54 is applied to reduction of the debt. The debt owed after 6 months is

$$\$15,000 - \$1515.54 = \$13,484.46.$$

(d) How much interest is owed for the second 6-month period? How much will be applied to the debt?

SOLUTION A total of $13,484.46 is owed for the second 6 months. Interest on this amount is

$$I = 13,484.46(0.12)\left(\frac{1}{2}\right) \approx 809.07,$$

or $809.07. A payment of $2415.54 is made at the end of this period; a total of

$$\$2415.54 - \$809.07 = \$1606.47$$

is applied to the reduction of the debt.

Continuing this process gives the *amortization schedule* shown below. As the schedule shows, each payment is the same, except perhaps for a small adjustment in the final payment. Payment 0 represents the original amount of the loan.

		Amortization Schedule		
Payment Number	Amount of Payment	Interest for Period	Portion to Principal	Principal at End of Period
0	—	—	—	$15,000.00
1	$2415.54	$900	$1515.54	$13,484.46
2	$2415.54	$809.07	$1606.47	$11,877.99
3	$2415.54	$712.68	$1702.86	$10,175.13
4	$2415.54	$610.51	$1805.03	$8370.10
5	$2415.54	$502.21	$1913.33	$6456.77
6	$2415.54	$387.41	$2028.13	$4428.64
7	$2415.54	$265.72	$2149.82	$2278.82
8	$2415.54	$136.73	$2278.82	$0

The unpaid balance of a loan after x payments is equivalent to the present value of an annuity after $n - x$ consecutive payments and is given by the function

$$y = R\left[\frac{1 - (1 + i)^{-(n-x)}}{i}\right]$$

For Example 7, the unpaid balance after two payments is

$$y = 2415.54\left[\frac{1 - (1 + 0.12/2)^{-(8-2)}}{0.12/2}\right] \approx 11{,}877.99,$$

or $11,877.99.

This formula can also be used to produce a graph of the unpaid balance. For Example 7, the graph of

$$y = 2415.54\left[\frac{1 - (1 + 0.12/2)^{-(8-x)}}{0.12/2}\right]$$

is shown in Figure 4.

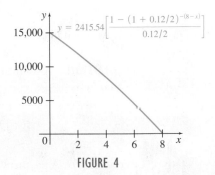

FIGURE 4

EXAMPLE 8 **Amortization**

The Millers buy a house for $174,000, with a down payment of $26,000. Interest is charged at 7.25% per year for 30 years compounded monthly. Find the amount of each monthly payment to amortize the loan.

Method 1
Calculation by Hand

SOLUTION Here, the present value, P, is 148,000 (or 174,000 − 26,000). Also, $i = 0.0725/12 \approx 0.0060416667$, and $n = 12 \cdot 30 = 360$. The monthly payment R must be found. From the formula for the present value of an annuity,

$$148,000 = R \cdot a_{\overline{360}|0.0060416667}$$

$$= R\left[\frac{1 - (1 + 0.0060416667)^{-360}}{0.0060416667}\right]$$

$$= R\left(\frac{1 - 0.1143540397}{0.0060416667}\right)$$

$$= R\left(\frac{0.8856459603}{0.0060416667}\right),$$

or

$$R \approx 1009.62.$$

Monthly payments of $1009.62 will be required to amortize the loan.

Method 2
Graphing Calculator

We can find the monthly payments to amortize this loan using the Finance Application of a TI-84 Plus calculator. To solve this problem, press the APPS button on the calculator and then select the `Finance` option. To input the particular information into the application, select the `TVM Solver` as shown in Figure 5 and then press ENTER. Then input the relevant values needed for the Solver, as shown in Figure 6. Note that the value of PMT is zero in `TVM Solver`. At this point, the particular value of PMT does not matter since we are going to calculate that value. Once the information is input into the solver, you must press 2nd QUIT to leave `TVM Solver`. To find the payment, press APPS, then `Finance`, and then select the `tvm_Pmt` button and press ENTER twice. The result shown in Figure 7 agrees with our work above.

YOUR TURN 6 In Example 8, suppose the house costs $256,000, and the Millers make a down payment of $32,000. If interest is charged at 4.9% per year for 30 years compounded monthly, find the amount of each monthly payment.

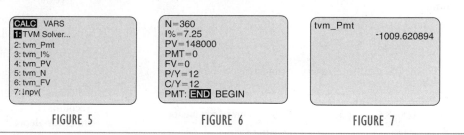

FIGURE 5   FIGURE 6   FIGURE 7

TRY YOUR TURN 6

12.2 EXERCISES

Find the amount of each ordinary annuity. (Interest is compounded annually.)

1. $R = \$120, i = 0.05, n = 10$

2. $R = \$1500, i = 0.04, n = 12$

3. $R = \$9000, i = 0.06, n = 18$

4. $R = \$80,000, i = 0.07, n = 24$

5. $R = \$11,500, i = 0.055, n - 30$

6. $R = \$13,400, i = 0.045, n = 25$

Find the amount of each ordinary annuity based on the information given.

7. $R = \$10,500$, 10% interest compounded semiannually for 7 years

8. $R = \$4200$, 6% interest compounded semiannually for 11 years

9. $R = \$1800$, 8% interest compounded quarterly for 12 years

10. $R = \$5300$, 4% interest compounded quarterly for 9 years

Find the periodic payments that will amount to the given sums under the given conditions.

11. $S = \$10,000$; interest is 8% compounded annually; payments are made at the end of each year for 12 years.

12. $S = \$80,000$; interest is 6% compounded semiannually; payments are made at the end of each semiannual period for 9 years.

13. $S = \$50,000$; interest is 12% compounded quarterly; payments are made at the end of each quarter for 8 years.

14. $S = \$8000$; interest is 4% compounded monthly; payments are made at the end of each month for 5 years.

Find the present value of each ordinary annuity.

15. Payments of $5000 are made annually for 11 years at 6% compounded annually.

16. Payments of $1280 are made annually for 9 years at 7% compounded annually.

17. Payments of $1400 are made semiannually for 8 years at 6% compounded semiannually.

18. Payments of $960 are made semiannually for 16 years at 5% compounded semiannually.

19. Payments of $50,000 are made quarterly for 10 years at 8% compounded quarterly.

20. Payments of $9800 are made quarterly for 15 years at 4% compounded quarterly.

Find the lump sum deposited today that will yield the same total amount as payments of $10,000 at the end of each year for 15 years, at the following interest rates. Interest is compounded annually.

21. 4%   **22.** 5%

23. 6%   **24.** 8%

Find the payments necessary to amortize each loan.

25. $2500, 16% compounded quarterly, 6 quarterly payments

26. $1000, 8% compounded annually, 9 annual payments

27. $90,000, 8% compounded annually, 12 annual payments

28. $41,000, 12% compounded semiannually, 10 semiannual payments

29. $55,000, 6% compounded monthly, 36 monthly payments

30. $6800, 12% compounded monthly, 24 monthly payments

APPLICATIONS

Business and Economics

31. Amount of an Annuity Sara Swangard wants to deposit $12,000 at the end of each year for 9 years into an annuity.

a. Sara's local bank offers an account paying 5% interest compounded annually. Find the final amount she will have on deposit.

b. Sara's brother-in-law works in a bank that pays 3% compounded annually. If she deposits her money in this bank instead, how much money will she have in her account?

c. How much would Sara lose over 9 years by using her brother-in-law's bank instead of her local bank?

32. Amount of an Annuity For 8 years, Tobi Casper deposits $100 at the end of each month into an annuity paying 6% annual interest compounded monthly.

a. Find the total amount Tobi deposits into the account over the 8 years.

b. Find the final amount Tobi will have on deposit at the end of the 8 years.

c. How much interest did Tobi earn?

33. Sinking Fund Steve Day wants $20,000 in 8 years.

a. What amount should he deposit at the end of each quarter at 6% annual interest compounded quarterly to accumulate the $20,000?

b. Find his quarterly deposit if the money is deposited at 4% compounded quarterly.

34. Sinking Fund Megan Donnelly wants to buy a car that she estimates will cost $24,000 in 5 years. How much money must she deposit at the end of each quarter at 5% interest compounded quarterly in order to have enough in 5 years to pay for her car?

35. Sinking Fund Harv's Meats will need to buy a new deboner machine in 4 years. At that time Harv expects the machine to cost $12,000. To accumulate enough money to pay for the machine, Harv decides to deposit a sum of money at the end of each 6-month period in an account paying 4% compounded semiannually. How much should each payment be?

36. Sinking Fund A firm must pay off $40,000 worth of bonds in 7 years.

a. Find the amount of each annual payment to be made into a sinking fund, if the money earns 6% compounded annually.

b. What annual payment should be made if the firm can get interest of 8% compounded annually?

Individual Retirement Accounts With Individual Retirement Accounts (IRAs), a worker whose income does not exceed certain limits can deposit up to a certain amount annually, with taxes deferred on the principal and interest. To attract depositors, banks have been advertising the amount that would accumulate by retirement. Suppose a 40-year-old person deposits $2000 per year until age 65. Find the total in the account with the interest rates stated in Exercises 37–40. Assume semiannual compounding with payments of $1000 made at the end of each semiannual period.

37. 6%   **38.** 8%   **39.** 10%   **40.** 12%

41. Sinking Fund Rebecca Nasman sells some land in Nevada. She will be paid a lump sum of $60,000 in 7 years. Until then, the buyer pays 8% interest, compounded quarterly.

a. Find the amount of each quarterly interest payment.

b. The buyer sets up a sinking fund so that enough money will be present to pay off the $60,000. The buyer wants to make semiannual payments into the sinking fund; the account pays 6% compounded semiannually. Find the amount of each payment into the fund.

c. Prepare a table showing the amount in the sinking fund after each deposit.

42. Sinking Fund Jerry Higgins bought a rare stamp for his collection. He agreed to pay a lump sum of $4000 after 5 years. Until then, he pays 6% interest, compounded semiannually.

a. Find the amount of each semiannual interest payment.

b. Jerry sets up a sinking fund so that enough money will be present to pay off the $4000. He wants to make annual payments into the fund. The account pays 8% compounded annually. Find the amount of each payment into the fund.

c. Prepare a table showing the amount in the sinking fund after each deposit.

43. Investment In 1995, Oseola McCarty donated $150,000 to the University of Southern Mississippi to establish a scholarship fund. What is unusual about her is that the entire amount came from what she was able to save each month from her work as a washer woman, a job she began in 1916 at the age of 8, when she dropped out of school. *Source: The New York Times*.

a. How much would Ms. McCarty have to put into her savings account at the end of every 3 months to accumulate $150,000 over 79 years? Assume she received an interest rate of 5.25% compounded quarterly.

b. Answer part a using a 2% and a 7% interest rate.

44. Present Value of an Annuity What lump sum deposited today at 5% compounded annually for 8 years will provide the same amount as $1000 deposited at the end of each year for 8 years at 6% compounded annually?

45. Present Value of an Annuity In his will the late Mr. Hudspeth said that each child in his family could have an annuity of $2000 at the end of each year for 9 years or the equivalent present value. If money can be deposited at 8% compounded annually, what is the present value?

46. Lottery Winnings In the "Million Dollar Lottery," a winner is paid a million dollars at the rate of $50,000 per year for 20 years. Assume that these payments form an ordinary annuity and that the lottery managers can invest money at 6% compounded annually. Find the lump sum that the management must put away to pay off the "million dollar" winner.

47. Lottery Winnings In most states, the winnings of million-dollar lottery jackpots are divided into equal payments given annually for 20 years. (In Colorado, the results are distributed over 25 years.) This means that the present value of the jackpot is worth less than the stated prize, with the actual value determined by the interest rate at which the money could be invested. *Source: The New York Times Magazine*.

a. Find the present value of a $1 million lottery jackpot distributed in equal annual payments over 20 years, using an interest rate of 5%.

b. Find the present value of a $1 million lottery jackpot distributed in equal annual payments over 20 years, using an interest rate of 9%.

c. Calculate the answer for part a using the 25-year distribution time in Colorado.

d. Calculate the answer for part b using the 25-year distribution time in Colorado.

48. Car Payments Kristina Walters buys a new car costing $22,000. She agrees to make payments at the end of each month for 4 years. If she pays 9% interest, compounded monthly, what is the amount of each payment? Find the total amount of interest Kristina will pay.

House Payments Find the monthly house payment necessary to amortize each of the loans in Exercises 49–52. Then find the unpaid balance after 5 years for each loan. Assume that interest is compounded monthly.

49. $249,560 at 7.75% for 25 years

50. $170,892 at 8.11% for 30 years

51. $353,700 at 7.95% for 30 years

52. $196,511 at 7.57% for 25 years

53. Annuity When Ms. Thompson died, she left $25,000 to her husband, which he deposited at 6% compounded annually. He wants to make annual withdrawals from the account so that the money (principal and interest) is gone in exactly 8 years.

a. Find the amount of each withdrawal.

b. Find the amount of each withdrawal if the money must last 12 years.

54. Annuity The trustees of a college have accepted a gift of $150,000. The donor has directed the trustees to deposit the money in an account paying 6% per year, compounded semiannually. The trustees may withdraw an equal amount of money at the end of each 6-month period; the money must last 5 years.

a. Find the amount of each withdrawal.

b. Find the amount of each withdrawal if the money must last 7 years.

55. Amortization An insurance firm pays $4000 for a new printer for its computer. It amortizes the loan for the printer in 4 annual payments at 8% compounded annually. Prepare an amortization schedule for this machine.

56. Amortization Certain large semitrailer trucks cost $72,000 each. Ace Trucking buys such a truck and agrees to pay for it with a loan that will be amortized with 9 semiannual payments at 9.5% compounded semiannually. Prepare an amortization schedule for this truck.

57. Amortization A printer manufacturer charges $1048 for a high-speed printer. A firm of tax accountants buys 8 of these machines. They make a down payment of $1200 and agree to amortize the balance with monthly payments at 10.5% compounded monthly for 4 years. Prepare an amortization schedule showing the first six payments.

58. Amortization When Barbara Essenmacher opened her law office, she bought $14,000 worth of law books and $7200 worth of office furniture. She paid $1200 down and agreed to amortize the balance with semiannual payments for 5 years at 8% compounded semiannually. Prepare an amortization schedule for this purchase.

YOUR TURN ANSWERS

1. $179,124.19   2. $7880.92

3. $193.22   4. $22,796.45

5. $258.34   6. $1188.83

12.3 Taylor Polynomials at 0

APPLY IT

How can we determine the length of time before a machine part must be replaced?

We shall see in Exercise 39 that a Taylor polynomial can be used to approximate the answer to this question.

Although exponential and logarithmic functions are quite different from polynomials, they can be closely approximated by polynomials. These approximating polynomials are called **Taylor polynomials** after British mathematician Brook Taylor (1685–1731), who published his work on them in 1715.

One of the most useful nonpolynomial functions is the exponential function $f(x) = e^x$. Let us begin our discussion of Taylor polynomials by finding polynomials that approximate e^x for values of x close to 0. As a first approximation to e^x, choose the straight line that is tangent to the graph of $f(x) = e^x$ at the point $(0, 1)$. (See Figure 8.) Since the slope of a tangent line is given by the derivative of the function, and the derivative of $f(x) = e^x$ is $f'(x) = e^x$, the slope of the tangent line at $x = 0$ is $f'(0) = e^0 = 1$.

The tangent line goes through $(0, f(0)) = (0, 1)$ and has slope $f'(0) = 1$. By the point-slope form of the equation of a line, the equation of the tangent line is

$$y - y_1 = m(x - x_1)$$
$$y - f(0) = f'(0)(x - 0)$$
$$y = f(0) + f'(0) \cdot x,$$

or, after substituting 1 for $f(0)$ and 1 for $f'(0)$,

$$y = 1 + x.$$

If $P_1(x)$ is used to represent $1 + x$, then

$$P_1(x) = 1 + x$$

is called the *Taylor polynomial of degree 1* for $f(x) = e^x$ at $x = 0$.

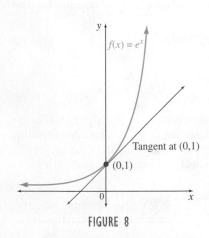

FIGURE 8

To be useful, $P_1(x)$ should approximate e^x for values of x close to 0. To check the accuracy of this approximation, compare values of $P_1(x)$ and values of e^x, for x close to 0, in the following table.

\multicolumn{3}{c}{Approximations and Exact Values of e^x}		
x	$P_1(x) = 1 + x$	$f(x) = e^x$
-1	0	0.3678794412
-0.1	0.9	0.904837418
-0.01	0.99	0.9900498337
-0.001	0.999	0.9990004998
0	1	1
0.001	1.001	1.0010005
0.01	1.01	1.010050167
0.1	1.1	1.105170918
1	2	2.718281828

This table agrees with the graph in Figure 8: the polynomial $P_1(x)$ is a good approximation for $f(x) = e^x$ only when x is close to 0.

For the polynomial $P_1(x)$,

$$P_1(0) = f(0) \quad \text{and} \quad P_1'(0) = f'(0);$$

that is, $P_1(x)$ and $f(x)$ are equal at 0 and their derivatives are equal at 0. A better approximation could be found with a curve. Since $P_1(x)$ is first-degree, we use a second-degree polynomial and require the second derivatives to be equal at 0. If $P_1(x)$ is written as

$$P_1(x) = a_0 + a_1 x,$$

with $a_0 = f(0)$ and $a_1 = f'(0)$, then a second-degree polynomial can be written as

$$P_2(x) = a_0 + a_1 x + a_2 x^2.$$

Just as above, we want

$$P_2(0) = f(0) \quad \text{and} \quad P_2'(0) = f'(0),$$

but we also want

$$P_2''(0) = f''(0).$$

Since $P_2(0) = a_0$, and $f(0) = 1$, then $a_0 = 1$. Also, $P_2'(x) = a_1 + 2a_2 x$, so $P_2'(0) = a_1$. Since $f'(0) = e^0 = 1$, we also must have $a_1 = 1$. Finally, $P_2''(x) = 2a_2$. Since $f''(x) = e^x$, $P_2''(0) = 2a_2$ and $f''(0) = 1$, so that

$$2a_2 = 1$$

$$a_2 = \frac{1}{2}.$$

Our second approximation, the *Taylor polynomial of degree 2* for $f(x) = e^x$ at $x = 0$, is thus

$$P_2(x) = 1 + x + \frac{1}{2}x^2.$$

A graph of this polynomial, along with the graph of $f(x) = e^x$, is shown in Figure 9.

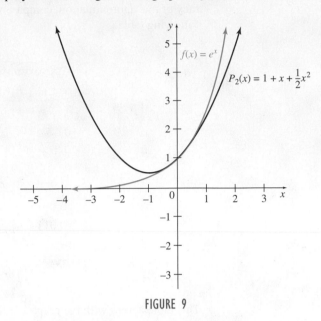

FIGURE 9

As above, the accuracy of this approximation can be checked with a table comparing values of $P_2(x)$ with those of $P_1(x)$ and $f(x)$, as shown.

	Approximations and Exact Values of e^x		
x	$P_1(x) = 1 + x$	$P_2(x) = 1 + x + \dfrac{1}{2}x^2$	$f(x) = e^x$
-1	0	0.5	0.3678794412
-0.1	0.9	0.905	0.9048374180
-0.01	0.99	0.99005	0.9900498337
-0.001	0.999	0.9990005	0.9990004998
0	1	1	1
0.001	1.001	1.0010005	1.001000500
0.01	1.01	1.01005	1.010050167
0.1	1.1	1.105	1.105170918
1	2	2.5	2.718281828

Although the approximations provided by $P_2(x)$ are better than those provided by $P_1(x)$, they are still accurate only for values of x close to 0. A better approximation would be given by a polynomial $P_3(x)$ that equals $f(x)$ when $x = 0$ and has the first, second, and *third* derivatives of $P_3(x)$ and $f(x) = e^x$ equal when $x = 0$. If we let

$$P_3(x) = a_0 + a_1x + a_2x^2 + a_3x^3,$$

we can find the first three derivatives:

$$P_3^{(1)}(x) = a_1 + 2a_2x + 3a_3x^2$$
$$P_3^{(2)}(x) = 2a_2 + 6a_3x$$
$$P_3^{(3)}(x) = 6a_3.$$

Letting $x = 0$ in $P_3(x)$ and in each derivative, in turn, yields

$$P_3(0) = a_0, \qquad P_3^{(1)}(0) = a_1, \qquad P_3^{(2)}(0) = 2a_2, \qquad P_3^{(3)}(0) = 6a_3.$$

Since $f(0) = 1$, $f^{(1)}(0) = 1$, $f^{(2)}(0) = 1$, and $f^{(3)}(0) = 1$ for $f(x) = e^x$,

$$a_0 = 1, \qquad a_1 = 1, \qquad a_2 = \frac{1}{2}, \qquad \text{and} \qquad a_3 = \frac{1}{6},$$

with

$$P_3(x) = 1 + x + \frac{1}{2}x^2 + \frac{1}{6}x^3.$$

A graph of $f(x) = e^x$ and $P_3(x)$ is shown in Figure 10.

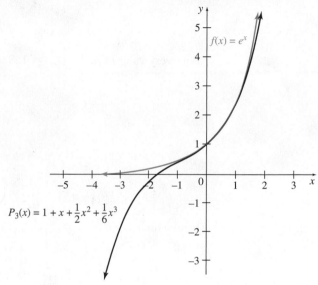

FIGURE 10

To generalize the work above, let $f(x) = e^x$ be approximated by

$$P_n(x) = a_0 + a_1x + a_2x^2 + \cdots + a_nx^n,$$

where

$$P_n(0) = f(0),$$
$$P_n^{(1)}(0) = f^{(1)}(0)$$
$$\vdots$$
$$P_n^{(n)}(0) = f^{(n)}(0).$$

Taking derivatives of $P_n(x)$ gives

$$P_n^{(1)}(x) = a_1 + 2a_2x + 3a_3x^2 + \cdots + n \cdot a_nx^{n-1}$$
$$P_n^{(2)}(x) = 2a_2 + 2 \cdot 3a_3x + \cdots + n(n-1)a_nx^{n-2}$$
$$P_n^{(3)}(x) = 2 \cdot 3a_3 + \cdots + n(n-1)(n-2)a_nx^{n-3}$$
$$\vdots$$
$$P_n^{(n)}(x) = n(n-1)(n-2)(n-3)\cdots 3 \cdot 2 \cdot 1 \cdot a_n = n!a_n.$$

The symbol $n!$ (read "n-factorial") is used for the product

$$n(n-1)(n-2)(n-3)\cdots 3\cdot 2\cdot 1.$$

For example, $3! = 3\cdot 2\cdot 1 = 6$, while $5! = 120$. By convention, $0! = 1.$* If we use factorials and replace x with 0, the various derivatives of $P_n(x)$ become

$$P_n^{(1)}(0) = 1!a_1$$
$$P_n^{(2)}(0) = 2!a_2$$
$$P_n^{(3)}(0) = 3!a_3$$
$$\cdot$$
$$\cdot$$
$$\cdot$$
$$P_n^{(n)}(0) = n!a_n.$$

For every value of n, $f^{(n)}(0) = 1$. Setting corresponding derivatives equal gives

$$1!a_1 = 1$$
$$2!a_2 = 1$$
$$3!a_3 = 1$$
$$\cdot$$
$$\cdot$$
$$\cdot$$
$$n!a_n = 1,$$

from which

$$a_1 = \frac{1}{1!}, \quad a_2 = \frac{1}{2!}, \quad a_3 = \frac{1}{3!}, \quad \cdots, \quad a_n = \frac{1}{n!}.$$

Finally, the *Taylor polynomial of degree n* for $f(x) = e^x$ at $x = 0$ is

$$P_n(x) = 1 + \frac{1}{1!}x + \frac{1}{2!}x^2 + \frac{1}{3!}x^3 + \cdots + \frac{1}{n!}x^n.$$

Using the convention that the *zeroth derivative* of $y = f(x)$ is just f itself, and using Σ to represent a sum, we can write this result in the following way.

Taylor Polynomial for $f(x) = e^x$
The **Taylor polynomial of degree n** for $f(x) = e^x$ at $x = 0$ is

$$P_n(x) - \sum_{i=0}^{n} \frac{f^{(i)}(0)}{i!}x^i = \sum_{i=0}^{n} \frac{1}{i!}r^i$$

---

*The symbol $n!$ for the product

$$n(n-1)(n-2)(n-3)\cdots 3\cdot 2\cdot 1$$

came into use during the late 19th century, although it was by no means the only symbol for n-factorial. Another popular symbol was $\lfloor n$. The exclamation point notation has won out, probably because it is more convenient for printers of textbooks.

Taylor polynomials of degree up to 10 for $f(x) = e^x$ are shown in Figure 11.

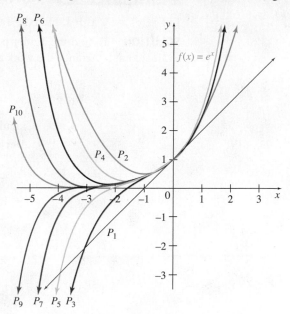

FIGURE 11

Graphing calculators simplify the creation of a sequence of Taylor polynomials. For example, to create Taylor polynomials of degree 1, 2, and 3 for e^x on a TI-84 Plus, let $Y_1 = 1 + X$, $Y_2 = Y_1 + X^2/2$, and $Y_3 = Y_2 + X^3/6$.

EXAMPLE 1   Taylor Polynomial

Use a Taylor polynomial of degree 5 to approximate $e^{-0.2}$.

SOLUTION   In the work above, we found Taylor polynomials for e^x at $x = 0$. As the graphs in Figure 11 suggest, these polynomials can be used to find approximate values of e^x for values of x near 0. The Taylor polynomial of degree 5 for $f(x) = e^x$ is

$$P_5(x) = 1 + \frac{1}{1!}x + \frac{1}{2!}x^2 + \frac{1}{3!}x^3 + \frac{1}{4!}x^4 + \frac{1}{5!}x^5.$$

Replacing x with -0.2 gives

$$1 + \frac{1}{1!}(-0.2) + \frac{1}{2!}(-0.2)^2 + \frac{1}{3!}(-0.2)^3 + \frac{1}{4!}(-0.2)^4$$

$$+ \frac{1}{5!}(-0.2)^5 \approx 0.8187307.$$

YOUR TURN 1   Use a Taylor polynomial of degree 5 to approximate $e^{-0.15}$.

Using a calculator to evaluate $e^{-0.2}$ directly gives 0.8187308, which agrees with our approximation to 6 decimal places.   **TRY YOUR TURN 1**

Generalizing our work in finding the Taylor polynomials for $f(x) = e^x$ leads to the following definition of Taylor polynomials for any appropriate function f.

Taylor Polynomial of Degree n
Let f be a function that can be differentiated n times at 0. The Taylor polynomial of degree n for f at 0 is

$$P_n(x) = f(0) + \frac{f^{(1)}(0)}{1!}x + \frac{f^{(2)}(0)}{2!}x^2 + \frac{f^{(3)}(0)}{3!}x^3 + \cdots + \frac{f^{(n)}(0)}{n!}x^n = \sum_{i=0}^{n} \frac{f^{(n)}(0)}{i!}x^i.$$

NOTE
Because of the $f(0)$ term, a Taylor polynomial of degree n has $n + 1$ terms.

EXAMPLE 2 Taylor Polynomial

Let $f(x) = \sqrt{x + 1}$. Find the Taylor polynomial of degree 4 at $x = 0$.

SOLUTION To find the Taylor polynomial of degree 4, use the first four derivatives of f, evaluated at 0. Arrange the work as follows.

Calculations for Taylor Polynomial	
Derivative	**Value at 0**
$f(x) = \sqrt{x + 1} = (x + 1)^{1/2}$	$f(0) = 1$
$f^{(1)}(x) = \dfrac{1}{2}(x + 1)^{-1/2} = \dfrac{1}{2(x + 1)^{1/2}}$	$f^{(1)}(0) = \dfrac{1}{2}$
$f^{(2)}(x) = -\dfrac{1}{4}(x + 1)^{-3/2} = \dfrac{-1}{4(x + 1)^{3/2}}$	$f^{(2)}(0) = -\dfrac{1}{4}$
$f^{(3)}(x) = \dfrac{3}{8}(x + 1)^{-5/2} = \dfrac{3}{8(x + 1)^{5/2}}$	$f^{(3)}(0) = \dfrac{3}{8}$
$f^{(4)}(x) = -\dfrac{15}{16}(x + 1)^{-7/2} = \dfrac{-15}{16(x + 1)^{7/2}}$	$f^{(4)}(0) = -\dfrac{15}{16}$

Now use the definition of a Taylor polynomial.

YOUR TURN 2 Let $f(x) = \sqrt{x + 4}$. Find the Taylor polynomial of degree 4 at $x = 0$.

$$P_4(x) = f(0) + \frac{f^{(1)}(0)}{1!}x + \frac{f^{(2)}(0)}{2!}x^2 + \frac{f^{(3)}(0)}{3!}x^3 + \frac{f^{(4)}(0)}{4!}x^4$$

$$= 1 + \frac{1/2}{1!}x + \frac{-1/4}{2!}x^2 + \frac{3/8}{3!}x^3 + \frac{-15/16}{4!}x^4$$

$$= 1 + \frac{1}{2}x - \frac{1}{8}x^2 + \frac{1}{16}x^3 - \frac{5}{128}x^4 \qquad \text{TRY YOUR TURN 2}$$

EXAMPLE 3 Approximation

Use the result of Example 2 to approximate $\sqrt{0.9}$.

SOLUTION To approximate $\sqrt{0.9}$, we must evaluate $f(-0.1) = \sqrt{-0.1 + 1} = \sqrt{0.9}$. Using $P_4(x)$ from Example 2, with $x = -0.1$, gives

YOUR TURN 3 Use the result of Your Turn 2 to approximate $\sqrt{4.05}$.

$$P_4(-0.1) = 1 + \frac{1}{2}(-0.1) - \frac{1}{8}(-0.1)^2 + \frac{1}{16}(-0.1)^3 - \frac{5}{128}(-0.1)^4$$

$$= 1 - 0.05 - 0.00125 - 0.0000625 - 0.000003906 = 0.948683594$$

Thus, $\sqrt{0.9} \approx 0.948683594$. A calculator gives a value of 0.9486832981 for the square root of 0.9. **TRY YOUR TURN 3**

EXAMPLE 4 Taylor Polynomial

Find the Taylor polynomial of degree n at $x = 0$ for

$$f(x) = \frac{1}{1 - x}.$$

SOLUTION   As above, find the first n derivatives, and evaluate each at 0.

Calculations for Taylor Polynomial	
Derivative	**Value at 0**
$f(x) = \dfrac{1}{1-x} = (1-x)^{-1}$	$f(0) = 1$
$f^{(1)}(x) = -1(1-x)^{-2}(-1) = (1-x)^{-2}$	$f^{(1)}(0) = 1 = 1!$
$f^{(2)}(x) = 2(1-x)^{-3}$	$f^{(2)}(0) = 2 = 2!$
$f^{(3)}(x) = 3!(1-x)^{-4}$	$f^{(3)}(0) = 3!$
$f^{(4)}(x) = 4!(1-x)^{-5}$	$f^{(4)}(0) = 4!$

Continuing this process,

$$f^{(n)}(x) = n!(1-x)^{-1-n} \quad \text{and} \quad f^{(n)}(0) = n!.$$

By the definition of Taylor polynomials,

$$P_n(x) = 1 + \frac{1!}{1!}x + \frac{2!}{2!}x^2 + \frac{3!}{3!}x^3 + \frac{4!}{4!}x^4 + \cdots + \frac{n!}{n!}x^n$$

$$= 1 + x + x^2 + x^3 + x^4 + \cdots + x^n.$$

EXAMPLE 5   Approximation

Use a Taylor polynomial of degree 4 to approximate $1/0.98$.

SOLUTION   Use the function f from Example 4, with $x = 0.02$, to get

$$f(0.02) = \frac{1}{1 - 0.02} = \frac{1}{0.98}.$$

Based on the result obtained in Example 4,

$$P_4(x) = 1 + x + x^2 + x^3 + x^4,$$

with

$$P_4(0.02) = 1 + (0.02) + (0.02)^2 + (0.02)^3 + (0.02)^4$$

$$= 1 + 0.02 + 0.0004 + 0.000008 + 0.00000016$$

$$= 1.02040816.$$

A calculator gives $1/0.98 = 1.020408163$.

12.3   EXERCISES

For the functions defined as follows, find the Taylor polynomials of degree 4 at 0.

1. $f(x) = e^{-2x}$

2. $f(x) = e^{3x}$

3. $f(x) = e^{x+1}$

4. $f(x) = e^{-x}$

5. $f(x) = \sqrt{x+9}$

6. $f(x) = \sqrt{x+16}$

7. $f(x) = \sqrt[3]{x-1}$

8. $f(x) = \sqrt[3]{x+8}$

9. $f(x) = \sqrt[4]{x+1}$

10. $f(x) = \sqrt[4]{x+16}$

11. $f(x) = \ln(1-x)$

12. $f(x) = \ln(1+2x)$

13. $f(x) = \ln(1+2x^2)$

14. $f(x) = \ln(1-x^3)$

15. $f(x) = xe^{-x}$

16. $f(x) = x^2e^x$

17. $f(x) = (9-x)^{3/2}$

18. $f(x) = (1+x)^{3/2}$

19. $f(x) = \dfrac{1}{1+x}$

20. $f(x) = \dfrac{1}{x-1}$

Use Taylor polynomials of degree 4 at $x = 0$, found in Exercises 1–14 above, to approximate the quantities in Exercises 21–34. Round answers to 4 decimal places.

21. $e^{-0.04}$

22. $e^{0.06}$

23. $e^{1.02}$

24. $e^{-0.07}$

25. $\sqrt{8.92}$

26. $\sqrt{16.3}$

27. $\sqrt[3]{-1.05}$

28. $\sqrt[3]{7.91}$

29. $\sqrt[4]{1.06}$

30. $\sqrt[4]{15.88}$

31. $\ln 0.97$

32. $\ln 1.06$

33. $\ln 1.008$

34. $\ln 0.992$

35. Find a polynomial of degree 3 such that $f(0) = 3$, $f'(0) = 6$, $f''(0) = 12$, and $f'''(0) = 24$.

36. Find a polynomial of degree 4 such that $f(0) = 1$, $f'(0) = 1$, $f''(0) = 2$, $f'''(0) = 6$, and $f^{(4)}(0) = 24$. Generalize this result to a polynomial of degree n, assuming that $f^{(n)}(0) = n!$

37. a. Generalize the result of Example 2 to show that if x is small compared with a,

$$(a + x)^{1/n} \approx a^{1/n} + \frac{xa^{1/n}}{na}.$$

b. Use the result of part a to approximate $\sqrt[3]{66}$ to 5 decimal places, and compare with the actual value.

APPLICATIONS

Business and Economics

38. Duration Let D represent *duration*, a term in finance that measures the length of time an investor must wait to receive half of the value of a cash flow stream totaling S dollars. Let r be the rate of interest and V the value of the investment. The value of S can be calculated by two formulas that are approximately equal:

$$S \approx V(1 + rD)$$

and $\qquad S \approx V(1 + r)^D.$

a. Show that the first approximation follows from the second by using the Taylor polynomial of degree 1 for the function $f(r) = (1 + r)^D$. *Source: Robert D. Campbell.*

b. For $V = \$1000$, $r = 0.1$, and $D = 3.2$, calculate and compare the two expressions for S.

39. APPLY IT Replacement Time for a Part A book on management science gives the equation

$$\frac{e^{\lambda N}}{\lambda} - N = \frac{1}{\lambda} + k$$

to determine N, the time until a particular part can be expected to need replacing. (λ and k are constants for a particular machine.) To find a useful approximate value for N when λN is near 0, go through the following steps.

a. Find a Taylor polynomial of degree 2 at $N = 0$ for $e^{\lambda N}$.

b. Substitute this polynomial into the given equation and solve for N.

In Exercises 40–44, use a Taylor polynomial of degree 2 at $x = 0$ to approximate the desired value. Compare your answers with the results obtained by direct substitution.

40. Profit The profit (in thousands of dollars) when x thousand tons of apples are sold is

$$P(x) = \frac{20 + x^2}{50 + x}.$$

Find $P(0.3)$.

41. Profit The profit (in thousands of dollars) from the sale of x thousand packages of note paper is

$$P(x) = \ln(100 + 3x).$$

Find $P(0.6)$ if $\ln 100$ is given as 4.605.

42. Cost For a certain electronic part, the cost to make the part declines as more parts are made. Suppose that the cost (in dollars) to manufacture the xth part is

$$C(x) = e^{-x/50}.$$

Find $C(5)$.

43. Revenue Revenue from selling agricultural products often increases at a slower and slower rate as more of the products are sold. Suppose the revenue from the xth unit of a product is

$$R(x) = 500 \ln\left(4 + \frac{x}{50}\right).$$

Find $R(10)$ if $\ln 4$ is given as 1.386.

Life Sciences

44. Amount of a Drug in the Bloodstream The amount (in milliliters) of a certain drug in the bloodstream x minutes after being administered is

$$A(x) = \frac{6x}{1 + 10x}.$$

Find $A(0.05)$.

45. Species Survival According to a text on species survival, the probability P that a certain species survives is given by

$$P = 1 - e^{-2k},$$

where k is a constant. Use a Taylor polynomial to show that if k is small, P is approximately $2k$.

Physical Sciences

46. Electric Potential In the Extended Application for Chapter 4, the electric potential at a distance z from an electrically charged disk of radius R was given as

$$V = k_1(\sqrt{z^2 + R^2} - z),$$

where k_1 is a constant.

a. Suppose z is much larger than R. By writing $\sqrt{z^2 + R^2}$ as $z\sqrt{1 + R^2/z^2}$ and using the Taylor polynomial of degree 1 for $\sqrt{1 + x}$, show that the potential can be approximated by

$$V \approx \frac{k_2}{z},$$

the result given in the Extended Application, where k_2 is a constant.

b. Suppose z is much smaller than R. By writing $\sqrt{z^2 + R^2}$ as $R\sqrt{1 + z^2/R^2}$ and using the Taylor polynomial of degree 1 for $\sqrt{1 + x}$, show that the potential can be approximated by

$$V \approx k_1\left(R - z + \frac{z^2}{2R}\right),$$

the result given in the Extended Application.

YOUR TURN ANSWERS

1. 0.860708

2. $P_4(x) = 2 + \frac{1}{4}x - \frac{1}{64}x^2 + \frac{1}{512}x^3 - \frac{5}{16,384}x^4$

3. 2.0124612

12.4 Infinite Series

APPLY IT If some fraction of a particular gene in a population experiences a mutation each generation, can we expect that the entire population will have this mutation over time?

The answer to this question is found in Example 5 by considering the sum of an infinite series.

A repeating decimal such as $0.66666\ldots$ is really the sum of an infinite number of terms:

$$0.66666\ldots = 0.6 + 0.06 + 0.006 + 0.0006 + \cdots$$

$$= \frac{6}{10} + \frac{6}{10^2} + \frac{6}{10^3} + \frac{6}{10^4} + \cdots.$$

In this section we will show how an infinite number of terms can sometimes be added to get a finite sum by a limit process. To do this, we need the following definition.

Infinite Series

An **infinite series** is an expression of the form

$$a_1 + a_2 + a_3 + a_4 + \cdots + a_n + \cdots = \sum_{i=1}^{\infty} a_i.$$

To find the sum $a_1 + a_2 + a_3 + a_4 + \cdots + a_n + \cdots$, first find the sum S_n of the first n terms, called the **nth partial sum**. For example,

$$S_1 = a_1$$
$$S_2 = a_1 + a_2$$
$$S_3 = a_1 + a_2 + a_3$$
$$\vdots$$
$$S_n = a_1 + a_2 + a_3 + \cdots + a_n = \sum_{i=1}^{n} a_i.$$

EXAMPLE 1 Partial Sums

Find the first five partial sums for the sequence

$$1, \frac{1}{2}, \frac{1}{4}, \frac{1}{8}, \frac{1}{16}, \ldots.$$

SOLUTION By the definition of partial sum,

$$S_1 = 1$$
$$S_2 = 1 + \frac{1}{2} = \frac{3}{2}$$
$$S_3 = 1 + \frac{1}{2} + \frac{1}{4} = \frac{7}{4}$$
$$S_4 = 1 + \frac{1}{2} + \frac{1}{4} + \frac{1}{8} = \frac{15}{8}$$
$$S_5 = 1 + \frac{1}{2} + \frac{1}{4} + \frac{1}{8} + \frac{1}{16} = \frac{31}{16}.$$

YOUR TURN 1 Find the first five partial sums for the sequence $1, 1/4, 1/9, 1/16, 1/25, \ldots$

TRY YOUR TURN 1

As n gets larger, the partial sum $S_n = a_1 + a_2 + \cdots + a_n$ includes more and more terms from the infinite series. It is thus reasonable to define the *sum of the infinite series* as $\lim\limits_{n \to \infty} S_n$, if it exists.

> ## Sum of the Infinite Series
>
> Let $S_n = a_1 + a_2 + a_3 + \cdots + a_n$ be the nth partial sum for the series $a_1 + a_2 + a_3 + \cdots + a_n + \cdots$. Suppose
>
> $$\lim_{n \to \infty} S_n = L$$
>
> for some real number L. Then L is called the **sum of the infinite series** $a_1 + a_2 + a_3 + \cdots + a_n + \cdots$, and the infinite series **converges**. If no such limit exists, then the infinite series has no sum and **diverges**.

Infinite Geometric Series

Some good examples of convergent and divergent series come from the study of infinite geometric series, which are the sums of the terms of geometric sequences, discussed in this chapter's first section. For example,

$$1, \frac{1}{2}, \frac{1}{4}, \frac{1}{8}, \frac{1}{16}, \cdots, \frac{1}{2^n}, \cdots$$

is a geometric sequence with first term $a_1 = 1$ and common ratio $r = 1/2$. The first five partial sums for this sequence were found in Example 1. To find S_n, the nth partial sum, use the formula given in the first section: The sum of the first n terms of a geometric sequence having first term $a = a_1$ and common ratio r is

$$S_n = \frac{a(r^n - 1)}{r - 1}.$$

For any value of n, S_n can be found for the geometric sequence by using the formula with $a = 1$ and $r = 1/2$.

$$S_n = \frac{a(r^n - 1)}{r - 1} = \frac{1\left[\left(\dfrac{1}{2}\right)^n - 1\right]}{\dfrac{1}{2} - 1}$$

$$= \frac{\left(\dfrac{1}{2}\right)^n - 1}{-\dfrac{1}{2}} = -2\left[\left(\frac{1}{2}\right)^n - 1\right] = 2\left[1 - \left(\frac{1}{2}\right)^n\right]$$

As n gets larger and larger, that is, as $n \to \infty$, the value of $(1/2)^n$ gets closer and closer to 0, so that

$$\lim_{n \to \infty}\left(\frac{1}{2}\right)^n = 0.$$

Using properties of limits from Chapter 3,

$$\lim_{n \to \infty} S_n = \lim_{n \to \infty} 2\left[1 - \left(\frac{1}{2}\right)^n\right] = 2(1 - 0) = 2.$$

By the definition of the sum of an infinite series,

$$1 + \frac{1}{2} + \frac{1}{4} + \frac{1}{8} + \cdots = 2,$$

and the series converges.

To generalize from this example, start with the formula for the sum of the first n terms of a geometric sequence.

$$S_n = \frac{a(r^n - 1)}{r - 1}.$$

If r is in the interval $(-1, 1)$, then $\lim_{n \to \infty} r^n = 0$. (Consider what happens to a small number as you raise it to a larger and larger power.) In that case,

$$\lim_{n \to \infty} S_n = \lim_{n \to \infty} \frac{a(r^n - 1)}{r - 1} = \frac{a(0 - 1)}{r - 1}$$

$$= \frac{a}{1 - r}.$$

On the other hand, if $r > 1$, then $\lim_{n \to \infty} r^n = \infty$. (Consider what happens to a large number as you raise it to a larger and larger power.) In that case,

$$\lim_{n \to \infty} S_n = \infty,$$

and the series diverges because the terms of the series are getting larger and larger. If $r < -1$, then $\lim_{n \to \infty} r^n$ does not exist, because r^n becomes larger and larger in magnitude while alternating in sign, and the same thing happens to the partial sums, so the series diverges. If $r = 1$, all the terms of the series equal a, so the series diverges (except in the trivial case when $a = 0$.) Finally, if $r = -1$, the terms of the series alternate between a and $-a$, and the partial sums alternate between a and 0, so the series diverges.

Sum of a Geometric Series

The infinite geometric series

$$a + ar + ar^2 + ar^3 + \cdots$$

converges, if r is in $(-1, 1)$, to the sum

$$\frac{a}{1 - r}.$$

The series diverges if r is not in $(-1, 1)$.

EXAMPLE 2   **Geometric Series**

Determine if the following geometric series converge. Give the sum of each convergent series.

(a) $3 + \frac{3}{8} + \frac{3}{64} + \frac{3}{512} + \cdots$

SOLUTION   This is a geometric series, with $a = a_1 = 3$ and $r = 1/8$. Since r is in $(-1, 1)$, the series converges and has sum

$$\frac{a}{1 - r} = \frac{3}{1 - 1/8} = \frac{3}{7/8} = 3 \cdot \frac{8}{7} = \frac{24}{7}.$$

YOUR TURN 2 Determine if the following geometric series converge. Give the sum of each convergent series.

(a) $2 + 2/3 + 2/9 + 2/27 + \cdots$

(b) $5 - 5/4 + 5/16 - 5/64 + \cdots$

(c) $2 - 2(1.01) + 2(1.01)^2 - 2(1.01)^3 + \cdots$

(b) $\dfrac{3}{4} - \dfrac{9}{16} + \dfrac{27}{64} - \dfrac{81}{256} + \cdots$

SOLUTION This geometric series has $a = a_1 = 3/4$ and $r = -3/4$. Since r is in $(-1, 1)$, the series converges. The sum of the series is

$$\frac{3/4}{1 - (-3/4)} = \frac{3/4}{1 + 3/4} = \frac{3/4}{7/4} = \frac{3}{7}.$$

(c) $1 + 1.1 + (1.1)^2 + (1.1)^3 + (1.1)^4 + \cdots + (1.1)^{n-1} + \cdots$

SOLUTION This is a geometric series with common ratio $r = 1.1$. Since $r > 1$, the series diverges. (The partial sum S_n will eventually exceed any preassigned number, no matter how large.) **TRY YOUR TURN 2**

EXAMPLE 3 Trains

Suppose a train leaves a station at noon travelling at 50 mph. Two hours later, on an adjacent track, a second train leaves the station heading in the same direction with a velocity of 60 mph. Determine the time at which the trains are both the same distance from the station.

(a) Solve this problem using algebra.

SOLUTION Let t be the number of hours since 2:00 pm. Since the first train left 2 hours earlier and has already traveled 100 miles, the total distance that the first train has traveled is $d_1 = 100 + 50t$. The second train has traveled $d_2 = 60t$. The two trains will be the same distance from the station when $d_1 = d_2$. Setting the two equations equal to each other and solving for t gives

$$d_1 = d_2$$
$$100 + 50t = 60t$$
$$100 = 10t$$
$$t = 10$$

The trains are equal distances from the station 10 hours after 2:00 pm, or at midnight.

(b) Solve this problem using a geometric series.

SOLUTION At 2:00 pm, the first train is 100 miles from the station. Since the second train is traveling at 60 mph, it will take $100/60 = 5/3$ hours to make up the 100 miles. But, during that $5/3$ hours, the first train will travel another $50(5/3) = 250/3$ miles. So the second train will have to travel another $(250/3)/60 = 50(5/3)/60 = (5/6)(5/3)$ hours to travel this distance. In the meantime, the first train has now traveled another $50(5/3)(5/6)$ miles. It will take the second train $(5/3)(5/6)(5/6)$ hours to make up this time, and so on. The total time that it takes for the trains to be an equal distance apart is found by summing the sequence of times it will take the second train to make up the distance. That is,

$$t = 5/3 + (5/3)(5/6) + (5/3)(5/6)^2 + (5/3)(5/6)^3 + \cdots$$

This is a geometric series, with $a = a_1 = 5/3$ and $r = 5/6$. Since r is in $(-1, 1)$, the series converges and the sum is

$$\frac{5/3}{1 - 5/6} = \frac{5/3}{1/6} = 10.$$

YOUR TURN 3 Suppose Turtle starts a race at 8:00 am and travels at 15 feet per minute. Rabbit, who is much faster than Turtle, starts the race 6 hours later, traveling at 45 feet per minute. Determine the time when Rabbit catches up with Turtle.

Thus, the trains will be an equal distance from the station 10 hours after 2:00, or at midnight. **TRY YOUR TURN 3**

EXAMPLE 4 Multiplier Effect

Suppose a company spends $1,000,000 on payroll in a certain city. Suppose also that the employees of the company reside in the city. Assume that on the average the inhabitants of this city spend 80% of their income in the same city. Then 80% of the original $1,000,000, or $(0.80)(\$1,000,000) = \$800,000$ will be spent in that city. An additional 80% of this $800,000, or $640,000, will in turn be spent in the city, as will 80% of the $640,000, and so on. Find the total expenditure in the city initiated by the original $1,000,000 payroll.

SOLUTION   These amounts, $1,000,000, $800,000, $640,000, $512,000, and so on, form an infinite series with $a = a_1 = \$1,000,000$ and $r = 0.80$. The sum of these amounts is

$$\frac{a}{1-r} = \frac{\$1,000,000}{1 - 0.80} = \$5,000,000.$$

The original $1,000,000 payroll leads to a total expenditure of $5,000,000 in the city. In economics, the quotient of these numbers, $\$5,000,000/\$1,000,000 = 5$, is called the *multiplier*.

EXAMPLE 5   Mutation

Retinoblastoma is a kind of cancer of the eye in children. Medical researchers believe that the disease depends on a single dominant gene, say A. Let a be the normal gene. It is believed that a fraction m of the population, $m = 2 \times 10^{-5}$, per generation will experience *mutation*, a sudden unaccountable change, of a into A. (We exclude the possibility of back mutations of A into a.) With medical care, approximately 70% of those affected with the disease survive. According to past data, the survivors reproduce at half the normal rate. The net fraction of affected persons who produce offspring is thus $r = 35\% = 0.35$. Since gene A is extremely rare, practically all the affected persons are of genotype Aa, so that we may neglect the few individuals of genotype AA. Find the total fraction of the population having the disease.

APPLY IT   **SOLUTION**   We start by defining the following variables.

m = fraction of population with disease due to mutation in this generation

mr = fraction of population with disease due to mutation in the previous generation

mr^2 = fraction of population with disease due to mutation two generations ago

mr^n = fraction of population with disease due to mutation n generations ago

The total fraction p of the population having the disease in this generation is thus

$$p = m + mr + mr^2 + \cdots + mr^n + \cdots$$

Use the formula for the sum of an infinite geometric series to find

$$p = \frac{m}{1-r} = \frac{2 \times 10^{-5}}{1 - 0.35} \approx 3.1 \times 10^{-5}.$$

The fraction of the population having retinoblastoma is about 3×10^{-5}, or about 50% more than the fraction of each generation that experiences mutation.

12.4   EXERCISES

Identify which geometric series converge. Give the sum of each convergent series.

1. $20 + 10 + 5 + \dfrac{5}{2} + \cdots$

2. $1 + 0.8 + 0.64 + 0.512 + \cdots$

3. $2 + 6 + 18 + 54 + \cdots$

4. $3 + 6 + 12 + 24 + \cdots$

5. $27 + 9 + 3 + 1 + \cdots$

6. $64 + 16 + 4 + 1 + \cdots$

7. $100 + 10 + 1 + \cdots$

8. $44 + 22 + 11 + \cdots$

9. $\dfrac{5}{4} + \dfrac{5}{8} + \dfrac{5}{16} + \cdots$

10. $\dfrac{4}{5} + \dfrac{2}{5} + \dfrac{1}{5} + \cdots$

11. $\dfrac{1}{3} - \dfrac{2}{9} + \dfrac{4}{27} - \dfrac{8}{81} + \cdots$

12. $1 + \dfrac{1}{1.01} + \dfrac{1}{(1.01)^2} + \cdots$

13. $e - 1 + \dfrac{1}{e} - \dfrac{1}{e^2} + \cdots$

14. $e + e^2 + e^3 + e^4 + \cdots$

The nth term of a sequence is given. Calculate the first five partial sums.

15. $a_n = \dfrac{1}{n}$

16. $a_n = \dfrac{1}{n + 1}$

17. $a_n = \dfrac{1}{2n + 5}$

18. $a_n = \dfrac{1}{3n - 1}$

19. $a_n = \dfrac{1}{(n + 1)(n + 2)}$

20. $a_n = \dfrac{1}{(n + 3)(2n + 1)}$

21. The repeating decimal 0.222222 . . . can be expressed as infinite geometric series

$$0.2 + 0.2\left(\frac{1}{10}\right) + 0.2\left(\frac{1}{10}\right)^2 + 0.2\left(\frac{1}{10}\right)^3 + \cdots.$$

By finding the sum of the series, determine the rational number whose decimal expansion is 0.222222. . . .

22. The repeating decimal 0.18181818 . . . can be expressed as the infinite geometric series

$$0.18 + 0.18\left(\frac{1}{100}\right) + 0.18\left(\frac{1}{100}\right)^2 + 0.18\left(\frac{1}{100}\right)^3 + \cdots.$$

Determine the rational number whose decimal expression is 0.18181818 . . . .

23. The following classical formulas for computing the value of π were developed by François Viète (1540–1603) and Gottfried von Leibniz (1646–1716), respectively:

$$\frac{2}{\pi} = \frac{\sqrt{2}}{2} \cdot \frac{\sqrt{2 + \sqrt{2}}}{2} \cdot \frac{\sqrt{2 + \sqrt{2 + \sqrt{2}}}}{2} \cdots$$

and

$$\frac{\pi}{4} = 1 - \frac{1}{3} + \frac{1}{5} - \frac{1}{7} + \cdots.$$

Sources: Mathematics Teacher and A History of Mathematics.

a. Multiply the first three terms of Viète's formula together, and compare this with the sum of the first four terms of Leibniz's formula. Which formula is more accurate?

b. Use the table function on a graphing calculator or a spreadsheet to determine how many terms of the second formula must be added together to produce the same accuracy as the product of the first three terms of the first formula. [*Hint:* On a TI-84 Plus, use the command `Y1=4*sum(seq ((−1)^(N−1)/(2N−1),N,1,X))`.]

APPLICATIONS

Business and Economics

24. **Production Orders** A sugar factory receives an order for 1000 units of sugar. The production manager thus orders production of 1000 units of sugar. He forgets, however, that the production of sugar requires some sugar (to prime the machines, for example), and so he ends up with only 900 units of sugar. He then orders an additional 100 units, and receives only 90 units. A further order for 10 units produces 9 units. Finally seeing he is wrong, the manager decides to try mathematics. He views the production process as an infinite geometric series with $a_1 = 1000$ and $r = 0.1$.

a. Using this, find the number of units of sugar that he should have ordered originally.

b. Afterwards, the manager realizes a much simpler solution to his problem. If x is the amount of sugar he orders, and he only gets 90% of what he orders, he should solve $0.9x = 1000$. What is the solution?

c. Explain why the answers to parts a and b are the same.

25. **Tax Rebate** The government claims to be able to stimulate the economy substantially by giving each taxpayer a $200 tax rebate. They reason that 90% of this amount, or $(0.90)(\$200) = \$180$, will be spent. An additional 90% of this $180 will then be spent, and so on.

a. If the government claim is true, how much total expenditure will result from this $200 rebate?

b. Calculate the value of the multiplier. (See Example 4.)

26. **Present Value** In Section 8.3, we computed the present value of a continuous flow of money. Suppose that instead of a continuous flow, an amount C is deposited each year, and the annual interest rate is r. Then the present value of the cash flow over n years is

$$P = C(1 + r)^{-1} + C(1 + r)^{-2} + C(1 + r)^{-3} + \cdots + C(1 + r)^{-n}.$$

a. Show that the present value can be simplified to

$$P = C\frac{(1 + r)^n - 1}{r(1 + r)^n}.$$

b. Show that the present value, taken over an infinite amount of time, is given by $P = C/r$.

27. **Malpractice Insurance** An insurance company determines it cannot write medical malpractice insurance profitably and stops selling the coverage. In spite of this action, the company will have to pay claims for many years on existing medical malpractice policies. The company pays 60 for medical malpractice claims the year after it stops selling the coverage. Each subsequent year's payments are 20% less than those of the previous year. Calculate the total medical malpractice payments that the company pays in all years after it stops selling the coverage. Choose one of the following. (*Hint:* When the problem says "pays 60," you can think of it as paying $60,000, but the units do not actually matter.) *Source: Society of Actuaries.*

a. 94

b. 150

c. 240

d. 300

e. 360

28. Automobile Insurance In modeling the number of claims filed by an individual under an automobile policy during a three-year period, an actuary makes the simplifying assumption that for all integers $n \geq 0$, $p_{n+1} = \frac{1}{5} p_n$, where p_n represents the probability that the policyholder files n claims during the period. Under this assumption, what is the probability that a policyholder files more than one claim during the period? Choose one of the following. (*Hint:* The total probability must equal 1.) *Source: Society of Actuaries.*

a. 0.04

b. 0.16

c. 0.20

d. 0.80

e. 0.96

Physical Sciences

29. Distance Mitzi drops a ball from a height of 10 m and notices that on each bounce the ball returns to about 3/4 of its previous height. About how far will the ball travel before it comes to rest?

30. Rotation of a Wheel After a person pedaling a bicycle removes his or her feet from the pedals, the wheel rotates 400 times the first minute. As it continues to slow down, in each minute it rotates only 3/4 as many times as in the previous minute. How many times will the wheel rotate before coming to a complete stop?

31. Pendulum Arc Length A pendulum bob swings through an arc 40 cm long on its first swing. Each swing thereafter, it swings only 80% as far as on the previous swing. How far will it swing altogether before coming to a complete stop?

General Interest

32. Perimeter A sequence of equilateral triangles is constructed as follows: The first triangle has sides 2 m in length. To get the next triangle, midpoints of the sides of the previous triangle are connected. If this process could be continued indefinitely, what would be the total perimeter of all the triangles?

33. Area What would be the total area of all the triangles of Exercise 32, disregarding the overlaps?

34. Trains Suppose a train leaves a station at noon traveling 100 mph. Two hours later, on an adjacent track, a second train leaves the station heading in the same direction traveling 125 mph. Determine when both trains are the same distance from the station.

a. Solve this problem using algebra.

b. Solve this problem using a geometric series. (See Example 3.)

35. Zeno's Paradox In the fifth century B.C., the Greek philosopher Zeno posed a paradox involving a race between Achilles (the fastest runner at the time) and a tortoise. The tortoise was given a head start, but once the race began, Achilles quickly reached the point where the tortoise had started. By then the tortoise had moved on to a new point. Achilles quickly reached that second point, but the tortoise had now moved to another point. Zeno concluded that Achilles could never reach the tortoise because every time he reached the point where the tortoise had been, the tortoise had moved on to a new point. This conclusion was absurd, yet people had trouble finding an error in Zeno's logic.

Suppose Achilles runs 10 m per second, the tortoise runs 1 m per second, and the tortoise has a 10-m head start.

a. Solve this problem using a geometric series. (See Example 3.)

b. Solve this problem using algebra.

c. Explain the error in Zeno's reasoning.

36. Bikers A famous story about the outstanding mathematician John von Neumann (1903–1957) concerns the following problem: Two bicyclists start 20 miles apart and head toward each other, each going 10 miles per hour. At the same time, a fly traveling 15 miles per hour leaves the front wheel of one bicycle, flies to the front wheel of the other bicycle, turns around and flies back to the wheel of the first bicycle, and so on, continuing in this manner until trapped between the two wheels. What total distance did the fly fly? There is a quick way to solve this problem. However, von Neumann allegedly solved this problem instantly by summing an infinite series. Solve this problem using both methods. *Source: American Mathematical Monthly.*

YOUR TURN ANSWERS

1. 1, 5/4, 49/36, 205/144, 5269/3600
2. (a) Converges to 3 (b) Converges to 4 (c) Diverges
3. 5 pm

12.5 Taylor Series

APPLY IT How many years will it take to double an amount invested at 9% annual interest?

Using Taylor series, we derive the rule of 70 and the rule of 72 to answer this question in Example 5.

As we saw in the previous section, the sum of the infinite geometric series having first term a and common ratio r is

$$\frac{a}{1-r} \quad \text{for } r \text{ in } (-1, 1).$$

If the first term of an infinite geometric series is $a = 1$ and the common ratio is x, then the series is written

$$1 + x + x^2 + x^3 + x^4 + \cdots + x^{n-1} + \cdots.$$

If x is in $(-1, 1)$, then by the formula for the sum of an infinite geometric series, the sum of this series is

$$\frac{1}{1-x}.$$

That is,

$$\frac{1}{1-x} = 1 + x + x^2 + x^3 + x^4 + \cdots \quad \text{for } x \text{ in } (-1, 1).$$

The interval $(-1, 1)$ is called the **interval of convergence** for the series. This series is not an approximation for $1/(1 - x)$; the sum of the series is actually *equal to* $1/(1 - x)$ for any x in $(-1, 1)$.

Earlier in this chapter, we found that the Taylor polynomial of degree n at $x = 0$ for $1/(1 - x)$ is

$$P_n(x) = 1 + x + x^2 + x^3 + x^4 + \cdots + x^n.$$

Since the series given above for $1/(1 - x)$ is just an extension of this Taylor polynomial, it seems natural to call the series a *Taylor series*.

Taylor Series

If all derivatives of a function f exist at 0, then the **Taylor series** for f at 0 is defined to be

$$f(0) + f^{(1)}(0)x + \frac{f^{(2)}(0)}{2!}x^2 + \frac{f^{(3)}(0)}{3!}x^3 + \cdots.$$

The particular Taylor series at 0 is also called a **Maclaurin series**. Scotsman Colin Maclaurin (1698–1746) used this series in his work *Treatise of Fluxions*, published in 1742. In this text, we will only consider Taylor series at 0. Taylor series at other points, as well as methods for finding the interval of convergence, are beyond the scope of this text. For more information, see *Thomas' Calculus*, 12th ed., by Maurice D. Weir and Joel R. Hass, Addison-Wesley, 2010.

Calculations for Taylor Series	
Derivative	**Value at 0**
$f(x) = e^x$	$f(0) = 1$
$f^{(1)}(x) = e^x$	$f^{(1)}(0) = 1$
$f^{(2)}(x) = e^x$	$f^{(2)}(0) = 1$
$f^{(3)}(x) = e^x$	$f^{(3)}(0) = 1$
.	.
.	.
.	.
$f^{(n)}(x) = e^x$	$f^{(n)}(0) = 1$

EXAMPLE 1 **Taylor Series**

Find the Taylor series for $f(x) = e^x$ at 0.

SOLUTION Work as in Section 12.3. The result is in the table to the left. Using the definition given in this section, the Taylor series for $f(x) = e^x$ is

$$1 + x + \frac{1}{2!}x^2 + \frac{1}{3!}x^3 + \frac{1}{4!}x^4 + \cdots + \frac{1}{n!}x^n + \cdots.$$

While we cannot prove it here, the interval of convergence is $(-\infty, \infty)$.

The process for finding the interval of convergence for a given Taylor series is discussed in more advanced calculus courses. Three of the most common Taylor series are listed below, along with their intervals of convergence. Note that these three functions are equal to their respective Taylor series expansion for all values of x contained in the given interval of convergence. (As is customary, these series are written so that the initial term is a *zeroth* term.)

Common Taylor Series

$f(x)$	Taylor Series	Interval of Convergence
e^x	$1 + x + \dfrac{1}{2!}x^2 + \dfrac{1}{3!}x^3 + \cdots + \dfrac{1}{n!}x^n + \cdots$	$(-\infty, \infty)$
$\ln(1 + x)$	$x - \dfrac{x^2}{2} + \dfrac{x^3}{3} - \dfrac{x^4}{4} + \cdots + \dfrac{(-1)^n x^{n+1}}{n+1} + \cdots$	$(-1, 1]$
$\dfrac{1}{1 - x}$	$1 + x + x^2 + x^3 + \cdots + x^n + \cdots$	$(-1, 1)$

Operations on Taylor Series

The first n terms of a Taylor series form a polynomial. Because of this, we would expect many of the operations on polynomials to generalize to Taylor series; some properties of series concerning these operations are given in the following theorems.

Operations on Taylor Series

Let f and g be functions having Taylor series with

$$f(x) = a_0 + a_1 x + a_2 x^2 + a_3 x^3 + \cdots + a_n x^n + \cdots$$

and

$$g(x) = b_0 + b_1 x + b_2 x^2 + b_3 x^3 + \cdots + b_n x^n + \cdots.$$

1. The Taylor series for $f + g$ is

$$(a_0 + b_0) + (a_1 + b_1)x + (a_2 + b_2)x^2 + \cdots$$
$$+ (a_n + b_n)x^n + \cdots,$$

for all x in the interval of convergence of both f and g. (Convergent series may be added term by term.)

2. For a real number c, the Taylor series for $c \cdot f(x)$ is

$$c \cdot a_0 + c \cdot a_1 x + c \cdot a_2 x^2 + \cdots + c \cdot a_n x^n + \cdots,$$

for all x in the interval of convergence of f.

3. For any positive integer k, the Taylor series for $x^k \cdot f(x)$ is

$$a_0 x^k + a_1 x^k \cdot x + a_2 x^k \cdot x^2 + \cdots + a_n x^k \cdot x^n + \cdots$$
$$= a_0 x^k + a_1 x^{k+1} + a_2 x^{k+2} + \cdots + a_n x^{k+n} + \cdots,$$

for all x in the interval of convergence of f.

These properties follow from the properties of derivatives and from the definition of a Taylor series.

EXAMPLE 2 **Taylor Series**

Find Taylor series for the following functions.

(a) $f(x) = 5e^x$

SOLUTION Use property 2, with $c = 5$, along with the Taylor series for $f(x) = e^x$ given earlier. The Taylor series for $5e^x$ is

$$5 \cdot 1 + 5 \cdot x + 5 \cdot \frac{1}{2!}x^2 + 5 \cdot \frac{1}{3!}x^3 + \cdots + 5 \cdot \frac{1}{n!}x^n + \cdots$$

$$= 5 + 5x + \frac{5}{2!}x^2 + \frac{5}{3!}x^3 + \cdots + \frac{5}{n!}x^n + \cdots$$

for all x in $(-\infty, \infty)$.

(b) $f(x) = x^3 \ln(1 + x)$

SOLUTION Use the Taylor series for $\ln(1 + x)$ with Property 3. With $k = 3$, this gives the Taylor series for $x^3 \ln(1 + x)$.

> **YOUR TURN 1** Find Taylor series for the following functions.
> **(a)** $f(x) = -7 \ln(1 + x)$
> **(b)** $g(x) = x^2/(1 - x)$

$$x^3 \cdot x - x^3 \cdot \frac{1}{2}x^2 + x^3 \cdot \frac{1}{3}x^3 - x^3 \cdot \frac{1}{4}x^4 + \cdots + \frac{x^3(-1)^n \cdot x^{n+1}}{n + 1} + \cdots$$

$$= x^4 - \frac{1}{2}x^5 + \frac{1}{3}x^6 - \frac{1}{4}x^7 + \cdots + \frac{(-1)^n x^{4+n}}{n + 1} + \cdots \quad \text{TRY YOUR TURN 1}$$

To see why the properties are so useful, try writing the Taylor series for $x^3 \ln(1 + x)$ directly from the definition of a Taylor series.

The final property of Taylor series is perhaps the most useful of all.

Composition with Taylor Series

Let a function f have a Taylor series such that

$$f(x) = a_0 + a_1x + a_2x^2 + a_3x^3 + \cdots + a_nx^n + \cdots .$$

Then replacing each x with $g(x) = cx^k$ for some constant c and positive integer k gives the Taylor series for $f[g(x)]$:

$$a_0 + a_1g(x) + a_2[g(x)]^2 + a_3[g(x)]^3 + \cdots + a_n[g(x)]^n + \cdots .$$

The interval of convergence of this new series may be different from that of the first series.

EXAMPLE 3 **Composition with Taylor Series**

Find the Taylor series for each function.

(a) $f(x) = e^{-x^2/2}$

SOLUTION We know that the Taylor series for e^x is

$$1 + x + \frac{1}{2!}x^2 + \frac{1}{3!}x^3 + \cdots + \frac{1}{n!}x^n + \cdots$$

for all x in $(-\infty, \infty)$. Use the composition property, and replace each x with $-x^2/2$ to get the Taylor series for $e^{-x^2/2}$.

$$1 + \left(-\frac{x^2}{2}\right) + \frac{1}{2!}\left(-\frac{x^2}{2}\right)^2 + \frac{1}{3!}\left(-\frac{x^2}{2}\right)^3 + \cdots + \frac{1}{n!}\left(-\frac{x^2}{2}\right)^n + \cdots$$

$$= 1 - \frac{1}{2}x^2 + \frac{1}{2!2^2}x^4 - \frac{1}{3!2^3}x^6 + \cdots + \frac{(-1)^n}{n!2^n}x^{2n} + \cdots$$

The Taylor series for $e^{-x^2/2}$ has the same interval of convergence, $(-\infty, \infty)$, as the Taylor series for e^x.

(b) $f(x) = \dfrac{1}{1 + 4x}$

SOLUTION   Write $1/(1 + 4x)$ as

$$\frac{1}{1 + 4x} = \frac{1}{1 - (-4x)},$$

which is $1/(1 - x)$ with x replaced with $-4x$. Start with

$$\frac{1}{1 - x} = 1 + x + x^2 + x^3 + \cdots + x^n + \cdots,$$

which converges for x in $(-1, 1)$, and replace each x with $-4x$ to get

$$\frac{1}{1 + 4x} = \frac{1}{1 - (-4x)}$$

$$= 1 + (-4x) + (-4x)^2 + (-4x)^3 + \cdots + (-4x)^n + \cdots$$

$$= 1 - 4x + 16x^2 - 64x^3 + \cdots + (-1)^n 4^n x^n + \cdots.$$

The interval of convergence of the original series is $(-1, 1)$, or $-1 < x < 1$. Replacing x with $-4x$ gives

$$-1 < -4x < 1 \qquad \text{or} \qquad -\frac{1}{4} < x < \frac{1}{4},$$

so that the interval of convergence of the new series is $(-1/4, 1/4)$.

(c) $f(x) = \dfrac{1}{2 - x^2}$

SOLUTION   This function most nearly matches $1/(1 - x)$. To get 1 in the denominator, instead of 2, divide the numerator and denominator by 2.

$$\frac{1}{2 - x^2} = \frac{1/2}{1 - x^2/2}$$

Thus, we can find the Taylor series for $1/(2 - x^2)$ by starting with the Taylor series for $1/(1 - x)$, multiplying each term by $1/2$, and replacing each x with $x^2/2$.

$$\frac{1}{2 - x^2} = \frac{1/2}{1 - x^2/2}$$

$$= \frac{1}{2} \cdot 1 + \frac{1}{2}\left(\frac{x^2}{2}\right) + \frac{1}{2}\left(\frac{x^2}{2}\right)^2 + \frac{1}{2}\left(\frac{x^2}{2}\right)^3 + \cdots + \frac{1}{2}\left(\frac{x^2}{2}\right)^n + \cdots$$

$$= \frac{1}{2} + \frac{x^2}{4} + \frac{x^4}{8} + \frac{x^6}{16} + \cdots + \frac{x^{2n}}{2^{n+1}} + \cdots$$

The Taylor series for $1/(1 - x)$ is valid when $-1 < x < 1$. Replacing x with $x^2/2$ gives

$$-1 < \frac{x^2}{2} < 1 \quad \text{or} \quad -2 < x^2 < 2.$$

This inequality is satisfied by any x in the interval $(-\sqrt{2}, \sqrt{2})$.

YOUR TURN 2   Find the Taylor series for each function.
(a) $f(x) = \ln(2x^2 + 1)$

(b) $g(x) = \dfrac{3}{4 - x^2}$

TRY YOUR TURN 2

Although we do not go into detail in this book, the Taylor series we discuss may be differentiated and integrated term by term. This result is used in the next example.

EXAMPLE 4   Integrating a Taylor Series

The standard normal curve of statistics is given by

$$f(x) = \frac{1}{\sqrt{2\pi}}e^{-x^2/2}.$$

Find the area bounded by this curve and the lines $x = 0$, $x = 1$, and the x-axis.

SOLUTION   The desired area is shown in Figure 12 below. By earlier methods, this area is given by the definite integral

$$\int_0^1 \frac{1}{\sqrt{2\pi}}e^{-x^2/2}dx = \frac{1}{\sqrt{2\pi}}\int_0^1 e^{-x^2/2}dx.$$

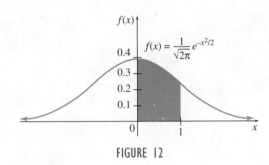

FIGURE 12

This integral cannot be evaluated by any method we have used, but recall that Example 3(a) gave the Taylor series for $f(x) = e^{-x^2/2}$:

$$e^{-x^2/2} = 1 - \frac{1}{2}x^2 + \frac{1}{2!2^2}x^4 - \frac{1}{3!2^3}x^6 + \cdots + \frac{(-1)^n}{n!2^n}x^{2n} + \cdots.$$

An approximation to $\int_0^1 e^{-x^2/2}dx$ can be found by integrating this series term by term. Using, say, the first six terms of this series gives

$$\frac{1}{\sqrt{2\pi}}\int_0^1 e^{-x^2/2}\,dx \approx \frac{1}{\sqrt{2\pi}}\int_0^1 \left(1 - \frac{1}{2}x^2 + \frac{1}{2!2^2}x^4 - \frac{1}{3!2^3}x^6 + \frac{1}{4!2^4}x^8 - \frac{1}{5!2^5}x^{10}\right)dx$$

$$= \frac{1}{\sqrt{2\pi}}\int_0^1 \left(1 - \frac{1}{2}x^2 + \frac{1}{8}x^4 - \frac{1}{48}x^6 + \frac{1}{384}x^8 - \frac{1}{3840}x^{10}\right)dx$$

$$= \frac{1}{\sqrt{2\pi}}\left(x - \frac{1}{6}x^3 + \frac{1}{40}x^5 - \frac{1}{336}x^7 + \frac{1}{3456}x^9 - \frac{1}{42{,}240}x^{11}\right)\Bigg|_0^1$$

$$= \frac{1}{\sqrt{2\pi}}\left(1 - \frac{1}{6} + \frac{1}{40} - \frac{1}{336} + \frac{1}{3456} - \frac{1}{42{,}240} - 0\right)$$

$$\approx \frac{1}{\sqrt{2(3.1416)}}(0.855623)$$

$$\approx 0.3413.$$

This result agrees with the value 0.3413 obtained from the normal curve table in the Appendix. We could have obtained a more accurate result by using more terms of the Taylor series.

In Example 4, we used terms of the Taylor series for $f(x) = e^{-x^2/2}$ up to the term containing x^{10}. This is exactly the same as finding the Taylor polynomial of degree 10 for the function. In general, taking terms up to degree n of a Taylor series is the same as finding the Taylor polynomial of degree n.

In Section 2.5, we saw that the **doubling time** (in years) for a quantity that increases at an annual rate r is given by

$$n = \frac{\ln 2}{\ln(1 + r)},$$

and we approximated n using the rule of 70 and the rule of 72. Now we can derive these rules by using a Taylor series. As shown in the list of common Taylor series,

$$\ln(1 + x) = x - \frac{x^2}{2} + \frac{x^3}{3} - \frac{x^4}{4} + \cdots$$

for x in $(-1, 1]$. Further,

$$\ln(1 + x) = x - \left(\frac{x^2}{2} - \frac{x^3}{3}\right) - \left(\frac{x^4}{4} - \frac{x^5}{5}\right) - \cdots < x$$

because each term in parentheses is positive for x in $(-1, 1]$. Therefore, for $0 < r < 1$, the doubling time,

$$n = \frac{\ln 2}{\ln(1 + r)} = \frac{\ln 2}{r - \dfrac{r^2}{2} + \dfrac{r^3}{3} - \dfrac{r^4}{4} + \cdots},$$

is just slightly larger than the quotient

$$\frac{\ln 2}{r} \approx \frac{0.693}{r} = \frac{69.3}{100r}.$$

Since the actual value of

$$\ln(1 + r) = r - \frac{r^2}{2} + \frac{r^3}{3} - \frac{r^4}{4} + \cdots$$

is slightly smaller than r for $0 < r < 1$, the quotients

$$\frac{70}{100r} \quad \text{and} \quad \frac{72}{100r}$$

give good approximations for the doubling time, the *rule of 70* and the *rule of 72*, discussed in Section 2.5.

> ## Rule of 70 and Rule of 72
>
> **Rule of 70** If a quantity is increasing at a constant rate r compounded annually, where $0.001 \le r \le 0.05$,
>
> $$\text{Doubling time} \approx \frac{70}{100r} \text{ years.}$$
>
> **Rule of 72** If a quantity is increasing at a constant rate r compounded annually, where $0.05 < r \le 0.12$, then
>
> $$\text{Doubling time} \approx \frac{72}{100r} \text{ years.}$$

The rule of 70 is used by demographers because populations usually grow at rates of less than 5 percent. The rule of 72 is preferred by economists and investors, since money frequently grows at a rate of between 5 percent and 12 percent. Because the difference between compounding continuously and compounding several times a year is small, both the rule of 70 and the rule of 72 can be used to approximate the doubling time in any interval.

EXAMPLE 5 Doubling Time

Find the doubling time for an investment at each interest rate.

(a) 9%

APPLY IT

SOLUTION By the formula for doubling time, at an interest rate of 9%, money will double in

$$\frac{\ln 2}{\ln(1 + 0.09)} \approx 8 \text{ years.}$$

(b) 1%

SOLUTION At an interest rate of 1%, money will double in

$$\frac{\ln 2}{\ln(1 + 0.01)} \approx 70 \text{ years.}$$

(c) Use the rule of 70 and the rule of 72 to verify the results in parts (a) and (b).

YOUR TURN 3 Repeat Example 5(c) for interest rates of 3.5% and 8%.

SOLUTION The rule of 70 predicts that at a growth rate of 1%, a population will double in 70 years, in agreement with part (b). The rule of 72 predicts that at an interest rate of 9%, money will double in 8 years, in agreement with part (a). **TRY YOUR TURN 3**

The following table gives the actual doubling time n in years for various growth rates r, together with the approximate doubling times given by the rules of 70 and 72.

Doubling Times													
r	0.001	0.005	0.01	0.02	0.03	0.04	0.05	0.06	0.07	0.08	0.09	0.10	0.11	0.12
n	693	139	69.7	35.0	23.4	17.7	14.2	11.9	10.2	9.0	8.0	7.3	6.6	6.1
$\frac{70}{100r}$	700	140	70	35	23.3	17.5	14	11.7	10	8.8	7.8	7.0	6.4	5.8
$\frac{72}{100r}$	720	144	72	36	24	18	14.4	12	10.3	9	8	7.2	6.5	6

The last row in the table is particularly easy to compute because 72 has so many integral divisors. Therefore, the rule of 72 is frequently used by economists and investors for any interest rate r.

It can be shown that the rule of 70 will give the doubling time with an error of 2% or less if $0.001 \leq n \leq 0.05$, and the rule of 72 will give the doubling time with a 2% error or less if $0.05 < r \leq 0.12$. The above table shows the accuracy of the approximations, and the graph in Figure 13 shows that the graphs of

$$\frac{\ln 2}{\ln(1 + r)}, \qquad \frac{70}{100r}, \qquad \text{and} \qquad \frac{72}{100r}$$

are virtually indistinguishable over the domains just indicated.

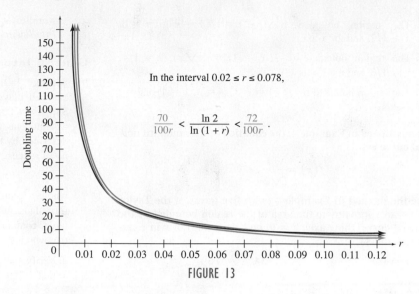

In the interval $0.02 \le r \le 0.078$,

$$\frac{70}{100r} < \frac{\ln 2}{\ln(1+r)} < \frac{72}{100r} \, .$$

FIGURE 13

12.5 EXERCISES

Find the Taylor series for the functions defined as follows. Give the interval of convergence for each series.

1. $f(x) = \dfrac{6}{1-x}$

2. $f(x) = \dfrac{-3}{1-x}$

3. $f(x) = x^2 e^x$

4. $f(x) = x^5 e^x$

5. $f(x) = \dfrac{5}{2-x}$

6. $f(x) = \dfrac{-3}{4-x}$

7. $f(x) = \dfrac{8x}{1+3x}$

8. $f(x) = \dfrac{7x}{1+2x}$

9. $f(x) = \dfrac{x^2}{4-x}$

10. $f(x) = \dfrac{9x^4}{1-x}$

11. $f(x) = \ln(1+4x)$

12. $f(x) = \ln\left(1 - \dfrac{x}{2}\right)$

13. $f(x) = e^{4x^2}$

14. $f(x) = e^{-3x^2}$

15. $f(x) = x^3 e^{-x}$

16. $f(x) = x^4 e^{2x}$

17. $f(x) = \dfrac{2}{1+x^2}$

18. $f(x) = \dfrac{6}{3+x^2}$

19. $f(x) = \dfrac{e^x + e^{-x}}{2}$

20. $f(x) = \dfrac{e^x - e^{-x}}{2}$

21. $f(x) = \ln(1 + 2x^4)$

22. $f(x) = \ln(1 - 5x^2)$

23. Use the fact that

$$\frac{1+x}{1-x} = \frac{1}{1-x} + \frac{x}{1-x}$$

to find a Taylor series for $(1+x)/(1-x)$.

24. By properties of logarithms,

$$\ln\left(\frac{1+x}{1-x}\right) = \ln(1+x) - \ln(1-x).$$

Use this to find a Taylor series for $\ln[(1+x)/(1-x)]$.

25. Use the Taylor series for e^x to suggest that

$$e^x \approx 1 + x + \frac{x^2}{2}$$

for all x close to zero.

26. Use the Taylor series for e^{-x} to suggest that

$$e^{-x} \approx 1 - x + \frac{x^2}{2}$$

for all x close to zero.

27. Use the Taylor series for e^x to show that

$$e^x \ge 1 + x$$

for all x.

28. Use the Taylor series for e^{-x} to show that

$$e^{-x} \ge 1 - x$$

for all x.

Use the method in Example 4 (with five terms of the appropriate Taylor series) to approximate the areas of the following regions.

29. The region bounded by $f(x) = e^{x^2}$, $x = 0$, $x = 1/3$, and the x-axis

30. The region bounded by $f(x) = 1/(1 - x^3)$, $x = 0$, $x = 1/2$, and the x-axis

31. The region bounded by $f(x) = 1/(1 - \sqrt{x})$, $x = 1/4$, $x = 1/3$, and the x-axis

32. The region bounded by $f(x) = e^{\sqrt{x}}$, $x = 0$, $x = 1$, and the x-axis

As mentioned in Example 4, the equation of the standard normal curve is

$$f(x) = \frac{1}{\sqrt{2\pi}} e^{-x^2/2}.$$

Use the method in Example 4 (with five terms of the Taylor series) to approximate the area of the region bounded by the normal curve, the x-axis, $x = 0$, and the values of x in Exercise 33 and 34.

33. $x = 0.4$ **34.** $x = 0.6$

APPLICATIONS

Business and Economics

35. Investment Ray Mesing has invested \$12,000 in a certificate of deposit that has a 4.75% annual interest rate. Determine the doubling time for this investment using the doubling-time formula. How does this compare with the estimate given by the rule of 70?

36. Investment It is anticipated that a bank stock that Katie Vales has invested \$15,000 in will achieve an annual interest rate of 6%. Determine the doubling time for this investment using the doubling-time formula. How does this compare with the estimate given by the rule of 72?

Life Sciences

37. Infant Mortality Infant mortality is an example of a relatively rare event that can be described by the *Poisson distribution*, for which the probability of x occurrences is given by

$$f(x) = \frac{\lambda^x e^{-\lambda}}{x!}, \qquad x = 0, 1, 2, \ldots$$

a. Verify that f describes a probability distribution by showing that

$$\sum_{x=0}^{\infty} f(x) = 1.$$

b. Calculate the expected value for f, given by

$$\sum_{x=0}^{\infty} x f(x).$$

c. In 2010, the U.S. infant mortality rate was estimated at 6.14 per 1000 live births. Assuming that this is the expected value for a Poisson distribution, find the probability that in a random sample of 1000 live births, there were fewer than 4 cases of infant mortality. *Source: Central Intelligence Agency*.

General Interest

38. Baseball In the year 2010, the proportion of U.S. major league baseball players who were foreign born was 231 out of 833. Suppose we begin to randomly select major league players until we find one who is foreign born. Such an experiment can be described by the *geometric distribution*, for which the probability of success after x tries is given by

$$f(x) = (1 - p)^{x-1} p, \quad x = 1, 2, 3, \ldots$$

where p is the probability of success on a given try. (*Note:* This formula is only accurate if the number of baseball players is very large, compared with the number that we select before meeting one who is foreign born.) *Source: Major League Baseball*.

a. Verify that f describes a probability distribution by showing that

$$\sum_{x=1}^{\infty} f(x) = 1.$$

b. Calculate the expected value for f, given by

$$\sum_{x=1}^{\infty} x f(x).$$

(*Hint:* Let $g(z) = p \sum_{x=1}^{\infty} z^x$, and evaluate $g'(1 - p)$.)

c. On average, how many major league baseball players would you expect to meet before meeting one who is foreign born?

d. What is the probability that you meet a foreign-born player within the first three major league players that you meet?

39. Trouble In the Milton Bradley game Trouble™, each player takes turns pressing a "popper" that contains a single die. To begin moving a game piece around the board a player must first pop a 6 on the die. The number of tries required to get a 6 can be described by the geometric distribution. (See Exercise 38.)

a. Using the result of Exercise 38b, what is the expected number of times a popper must be pressed before a success occurs?

b. What is the probability that you will have to press the popper four or more times before a 6 pops up?

YOUR TURN ANSWERS

1. (a) $-7x + 7x^2/2 - 7x^3/3 + 7x^4/4 + \cdots + \dfrac{7(-1)^n x^n}{n} + \cdots$, for all x in $(-1, 1]$

(b) $x^2 + x^3 + x^4 + \cdots + x^n + \cdots$, for all x in $(-1, 1)$

2. (a) $2x^2 - 2x^4 + (8/3)x^6 - 4x^8 + \cdots + \dfrac{(-1)^{n+1} 2^n x^{2n}}{n} + \cdots$, for all x in $\left[-\dfrac{1}{\sqrt{2}}, \dfrac{1}{\sqrt{2}} \right]$

(b) $3/4 + 3x^2/16 + 3x^4/64 + 3x^6/256 + \cdots + \dfrac{3x^{2n}}{4^{n+1}} + \cdots$, for all x in $(-2, 2)$

3. 20 yr; 9 yr

12.6 Newton's Method

How can the true interest rate be found, given the amount loaned, the number of payments, and the amount of each payment?

We will answer this question in Exercise 34 using a technique developed in this section.

Given a function f, a number r such that $f(r) = 0$ is called a zero of f. For example, if $f(x) = x^2 - 4x + 3$, then $f(3) = 0$ and $f(1) = 0$, so that both 3 and 1 are zeros of f. The zeros of linear and quadratic functions can be found with the methods of algebra. More complicated methods exist for finding zeros of third-degree or fourth-degree polynomial functions, but there is no general method for finding zeros of higher-degree polynomials.

In practical applications of mathematics, it is seldom necessary to find *exact* zeros of a function; usually a decimal approximation is all that is needed. We have seen earlier how a graphing calculator may be used to find approximate values of zeros. In this section, we will explore a calculus-based method to do the same. The method provides a sequence of values, c_1, $c_2, \ldots$, whose limit is the true value in a wide variety of applications. Of course, you may simply prefer to use the `zero` feature on your graphing calculator, but Newton's method is the basis for some of the techniques used by mathematicians to solve more complex problems.

The zeros of a differentiable function f can be approximated as follows. Find a closed interval $[a, b]$ so that $f(a)$ and $f(b)$ are of opposite sign, one positive and one negative. As suggested by Figure 14, this means there must exist at least one value c in the interval (a, b) such that $f(c) = 0$. This number c is a zero of the function f.

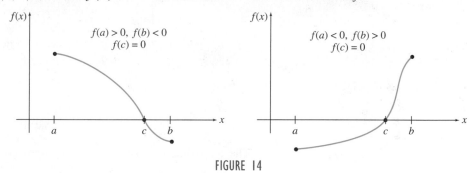

FIGURE 14

To find an approximate value for c, first make a guess for c. Let c_1 be the initial guess. (See Figure 15.) Then locate the point $(c_1, f(c_1))$ on the graph of $y = f(x)$ and identify the tangent line at this point. This tangent line will cut the x-axis at a point c_2. The number c_2 is often a better approximation of c than was c_1.

To locate c_2, first find the equation of the tangent line through $(c_1, f(c_1))$. The slope of this tangent line is $f'(c_1)$. The point-slope form of the equation of the tangent line is

$$y - f(c_1) = f'(c_1)(x - c_1).$$

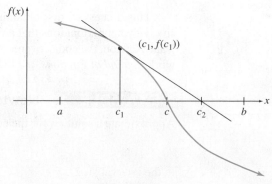

FIGURE 15

When $x = c_2$, we know that $y = 0$. Substituting into the equation of the tangent line gives

$$0 - f(c_1) = f'(c_1)(c_2 - c_1)$$

or

$$-\frac{f(c_1)}{f'(c_1)} = c_2 - c_1,$$

from which

$$c_2 = c_1 - \frac{f(c_1)}{f'(c_1)}.$$

If $f'(c_1)$ should be 0, the tangent line would be horizontal and not cut the x-axis. For this reason, assume $f'(c_1) \neq 0$. This new value, c_2, is usually a better approximation to c than was c_1. To improve the approximation further, locate the tangent line to the curve at $(c_2, f(c_2))$. Let this tangent cut the x-axis at c_3. (See Figure 16.) Find c_3 by a process similar to that used above: if $f'(c_2) \neq 0$,

$$c_3 = c_2 - \frac{f(c_2)}{f'(c_2)}.$$

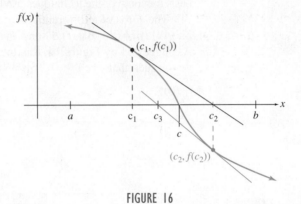

FIGURE 16

The approximation to c often can be improved by repeating this process as many times as desired. In general, if c_n is an approximation to c, a better approximation, c_{n+1}, frequently can be found by the following formula.

Newton's Method
If $f'(c_n) \neq 0$, then

$$c_{n+1} = c_n - \frac{f(c_n)}{f'(c_n)}.$$

This process of first obtaining a rough approximation for c, then replacing it successively by approximations that are often better, is called **Newton's method**, named after Sir Isaac Newton, the codiscoverer of calculus. An early version of this method appeared in his work *Method of Fluxions*, published in 1736.

EXAMPLE 1 Newton's Method

Approximate a solution for the equation

$$3x^3 - x^2 + 5x - 12 = 0$$

in the interval $[1, 2]$.

SOLUTION Let $f(x) = 3x^3 - x^2 + 5x - 12$, so that $f'(x) = 9x^2 - 2x + 5$. Check that $f(1) < 0$ with $f(2) > 0$. Since $f(1)$ and $f(2)$ have opposite signs, there is a solution for the equation in the interval $(1, 2)$. As an initial guess, let $c_1 = 1$. A better guess, c_2, can be found as follows.

$$c_2 = c_1 - \frac{f(c_1)}{f'(c_1)} = 1 - \frac{-5}{12} = 1.4167$$

A third approximation, c_3, can now be found.

$$c_3 = c_2 - \frac{f(c_2)}{f'(c_2)} = 1.4167 - \frac{1.6066}{20.230} = 1.3373$$

In the same way,

$$c_4 = 1.3373 - \frac{0.072895}{18.421} = 1.3333 \quad \text{and} \quad c_5 = 1.3333 - \frac{-6.111 \times 10^{-4}}{18.333} = 1.3333.$$

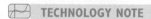

YOUR TURN 1 Approximate a solution for the equation $2x^3 - 5x^2 + 6x - 10 = 0$ on $[1, 3]$.

Subsequent approximations yield no further accuracy, either to the 4 decimal places to which we have rounded or to the digits displayed in a TI-84 Plus calculator. Thus $x = 1.3333$ is a reasonably accurate solution of $3x^3 - x^2 + 5x - 12 = 0$. (The exact solution is $4/3$.)

TRY YOUR TURN 1

TECHNOLOGY NOTE

Newton's method is easily implemented on a graphing calculator. For the previous example on a TI-84 Plus, start by storing 1 in X, the function $f(x)$ in Y_1, and the function $f'(x)$ in Y_2. The command $X - Y_1/Y_2 \rightarrow X$ gives the next value of x. Continue to press the ENTER key for subsequent calculations.

In Example 1 we had to go through five steps to get the degree of accuracy that we wanted. The solutions of similar polynomial equations usually can be found in about as many steps, although other types of equations might require more steps, particularly if the initial guess is far from the true solution.

In any case, if a solution can be found by Newton's method, it usually can be found by a computer in a small fraction of a second. But in some cases, the method will not find the solution, or will only do so for a good initial guess. Figure 17 shows an example in which Newton's method does not give a solution. Because of the symmetry of the graph in Figure 17, all the odd steps (c_3, c_5, and so forth) give c_1, while all the even steps (c_4, c_6, and so forth) give c_2, so the approximations never approach the true solution. Such cases are rare in practice. If you find that Newton's method is not producing a solution, verify that there is a solution, and then try a better initial guess.

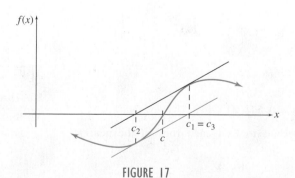

FIGURE 17

Newton's method also can be used to approximate the values of radicals, as shown by the next example.

EXAMPLE 2 **Approximation**

Approximate $\sqrt{12}$ to the nearest thousandth.

SOLUTION First, note that $\sqrt{12}$ is a solution of the equation $x^2 - 12 = 0$. Therefore, let $f(x) = x^2 - 12$, so that $f'(x) = 2x$. Since $3 < \sqrt{12} < 4$, use $c_1 = 3$ as the first approximation to $\sqrt{12}$. A better approximation is given by c_2:

$$c_2 = 3 - \frac{-3}{6} = 3.5.$$

Now find c_3 and c_4:

$$c_3 = 3.5 - \frac{0.25}{7} = 3.464,$$

YOUR TURN 2 Approximate $\sqrt[3]{15}$ to the nearest thousandth.

$$c_4 = 3.464 - \frac{-0.0007}{6.928} = 3.464.$$

Since $c_3 = c_4 = 3.464$, to the nearest thousandth, $\sqrt{12} = 3.464$.   **TRY YOUR TURN 2**

12.6 EXERCISES

Use Newton's method to find a solution for each equation in the given intervals. Find all solutions to the nearest hundredth.

1. $5x^2 - 3x - 3 = 0$; $[1, 2]$

2. $2x^2 - 8x + 3 = 0$; $[3, 4]$

3. $2x^3 - 6x^2 - x + 2 = 0$; $[3, 4]$

4. $-x^3 + 4x^2 - 5x + 4 = 0$; $[2, 3]$

5. $-3x^3 + 5x^2 + 3x + 2 = 0$; $[2, 3]$

6. $4x^3 - 5x^2 - 6x + 6 = 0$; $[1, 2]$

7. $2x^4 - 2x^3 - 3x^2 - 5x - 8 = 0$; $[-2, -1], [2, 3]$

8. $3x^4 + 4x^3 - 6x^2 - 2x - 12 = 0$; $[-3, -2], [1, 2]$

9. $4x^{1/3} - 2x^2 + 4 = 0$; $[-3, 0]$

10. $4x^{1/3} - 2x^2 + 4 = 0$; $[0, 3]$

11. $e^x + x - 2 = 0$; $[0, 3]$

12. $e^{2x} + 3x - 4 = 0$; $[0, 3]$

13. $x^2 e^{-x} + x^2 - 2 = 0$; $[0, 3]$

14. $x^2 e^{-x} + x^2 - 2 = 0$; $[-3, 0]$

15. $\ln x + x - 2 = 0$; $[1, 4]$

16. $2 \ln x + x - 3 = 0$; $[1, 4]$

Use Newton's method to find each root to the nearest thousandth.

17. $\sqrt{2}$ 18. $\sqrt{3}$

19. $\sqrt{11}$ 20. $\sqrt{15}$

21. $\sqrt{250}$ 22. $\sqrt{300}$

23. $\sqrt[3]{9}$ 24. $\sqrt[3]{15}$

25. $\sqrt[3]{100}$ 26. $\sqrt[3]{121}$

Use Newton's method to find the critical points for the functions defined as follows. Approximate them to the nearest hundredth. Decide whether each critical point leads to a relative maximum or a relative minimum.

27. $f(x) = x^3 - 3x^2 - 18x + 4$

28. $f(x) = x^3 + 9x^2 - 6x + 4$

29. $f(x) = x^4 - 3x^3 + 6x - 1$

30. $f(x) = x^4 + 2x^3 - 5x + 2$

31. Use Newton's method to attempt to find a solution for the equation

$$f(x) = (x - 1)^{1/3} = 0$$

by starting with a value very close to 1, which is obviously the true solution. Verify that the approximations get worse with each iteration of Newton's method. This is one of those rare cases in which Newton's method doesn't work at all. Discuss why this is so by considering what happens to the tangent line at $x = 1$.

APPLICATIONS

Business and Economics

32. **Break-Even Point** For a particular product, the revenue and cost functions are

$$R(x) = 10x^{2/3} \quad \text{and} \quad C(x) = 2x - 9$$

Approximate the break-even point to the nearest hundredth.

33. **Manufacturing** A new manufacturing process produces savings of

$$S(x) = x^2 + 40x + 20$$

dollars after x years, with increased costs of

$$C(x) = x^3 + 5x^2 + 9$$

dollars. For how many years, to the nearest hundredth, should the process be used?

34. APPLY IT **True Annual Interest Rate** Federal government regulations require that people loaning money to consumers disclose the true annual interest rate of the loan. The formulas for calculating this interest rate are very complex. For example, suppose P dollars is loaned, with the money to be repaid in n monthly payments of M dollars each. Then the true annual interest rate is found by solving the equation

$$\frac{1 - (1 + i)^{-n}}{i} - \frac{P}{M} = 0$$

for i, the monthly interest rate, and then multiplying i by 12 to get the true annual rate. This equation can best be solved

by Newton's method. (This is how the financial function IRR (Internal Rate of Return) is computed in Microsoft Excel.)

a. Let $f(i) = \dfrac{1 - (1 + i)^{-n}}{i} - \dfrac{P}{M}$. Find $f'(i)$.

b. Form the quotient $f(i)/f'(i)$.

c. Suppose that $P = \$4000$, $n = 24$, and $M = \$197$. Let the initial guess for i be $i_1 = 0.01$. Use Newton's method and find i_2.

d. Find i_3. (*Note:* For the accuracy required by federal law, it is usually sufficient to stop after two successive values of i differ by no more than 10^{-7}.)

Find i_2 and i_3.

35. $P = \$600$, $M = \$57$, $n = 12$, $i_1 = 0.02$

36. $P = \$15,000$, $M = \$337$, $n = 60$, $i_1 = 0.01$

YOUR TURN ANSWERS

1. 2.177   **2.** 2.466

12.7 L'Hospital's Rule

We began our study of calculus with a discussion of *limits*. For example,

$$\lim_{x \to 1} \frac{x^2 + 1}{x + 4} = \frac{2}{5},$$

which can be found by direct substitution using the limit rules for a rational function from Section 3.1. In this section we will use derivatives to find limits of quotients of functions that could not easily be found using the techniques of Chapter 3.

If we try to find

$$\lim_{x \to 1} \frac{x^2 - 1}{x - 1}$$

by evaluating the numerator and denominator at $x = 1$, we get

$$\frac{1^2 - 1}{1 - 1} = \frac{0}{0},$$

an **indeterminate form**. Any attempt to assign a value to $0/0$ leads to a meaningless result. The limit exists, however; as shown earlier, it is found by factoring.

$$\lim_{x \to 1} \frac{x^2 - 1}{x - 1} = \lim_{x \to 1} \frac{(x + 1)(x - 1)}{x - 1} = \lim_{x \to 1} (x + 1) = 2$$

As a second example,

$$\lim_{x \to 1} \frac{\ln x}{(x - 1)^2}$$

also leads to the indeterminate form $0/0$. Selecting values of x close to 1 and using a calculator gives the following table.

x	0.99	0.999	0.9999	1.0001	1.001	1.01
$\dfrac{\ln x}{(x - 1)^2}$	-100.5	-1000.5	$-10{,}000.5$	9999.5	999.5	99.5

As this table suggests, $\lim_{x \to 1} (\ln x)/(x - 1)^2$ does not exist.

In the first example, trying to find $\lim_{x \to 1}(x^2 - 1)/(x - 1)$ by evaluating the expression at $x = 1$ led to the indeterminate form $0/0$, but factoring the expression led to the actual limit, 2. Evaluating $\lim_{x \to 1}(\ln x)/(x - 1)^2$ in the second example led to the indeterminate form $0/0$, but using a table of values showed that this limit did not exist. **L'Hospital's rule** gives a quicker way to decide whether a quotient with the indeterminate form $0/0$ has a limit.

L'Hospital's Rule

Let f and g be functions and let a be a real number such that

$$\lim_{x \to a} f(x) = 0 \qquad \text{and} \qquad \lim_{x \to a} g(x) = 0,$$

or

$$\lim_{x \to a} f(x) = \pm\infty \qquad \text{and} \qquad \lim_{x \to a} g(x) = \pm\infty.$$

Let f and g have derivatives that exist at each point in some open interval containing a.

If $\lim_{x \to a} \dfrac{f'(x)}{g'(x)} = L$, then $\lim_{x \to a} \dfrac{f(x)}{g(x)} = L.$

If $\lim_{x \to a} \dfrac{f'(x)}{g'(x)}$ does not exist because $\left|\dfrac{f'(x)}{g'(x)}\right|$ becomes large without bound

for values of x near a, then $\lim_{x \to a} \dfrac{f(x)}{g(x)}$ also does not exist.

A partial proof of this rule is given at the end of this section. L'Hospital's rule is another example of a mathematical misnomer. Although named after the Marquis de l'Hospital (1661–1704), it was actually developed by Johann Bernoulli (1667–1748) in a textbook published in 1696. (Johann Bernoulli was the brother of Jakob Bernoulli, mentioned in the section on Antiderivatives.) L'Hospital was a student of Bernoulli and published, with a financial arrangement, the works of his teacher under his own name.

EXAMPLE 1 L'Hospital's Rule

Find $\lim_{x \to 2} \dfrac{3x - 6}{\sqrt{2 + x} - 2}$.

SOLUTION It is very important to first make sure that the conditions of l'Hospital's rule are satisfied. Here

$$\lim_{x \to 2}(3x - 6) = 0 \qquad \text{and} \qquad \lim_{x \to 2}(\sqrt{2 + x} - 2) = 0.$$

Since the limits of both numerator and denominator are 0, l'Hospital's rule applies. Now take the derivatives of both numerator and denominator separately. (Do *not* use the quotient rule for derivatives.)

For $f(x) = 3x - 6$, we have $f'(x) = 3$.

For $g(x) = \sqrt{2 + x} - 2$, we have $g'(x) = \dfrac{1}{2\sqrt{2 + x}}.$

Find the limit of the quotient of the derivatives.

$$\lim_{x \to 2} \frac{f'(x)}{g'(x)} = \lim_{x \to 2} \frac{3}{1/(2\sqrt{2 + x})} = \lim_{x \to 2} 6\sqrt{2 + x} = 12.$$

By l'Hospital's rule, this result is the desired limit:

$$\lim_{x \to 2} \frac{3x - 6}{\sqrt{2 + x} - 2} = 12. \qquad \textbf{TRY YOUR TURN 1}$$

YOUR TURN 1

Find $\lim\limits_{x \to 4} \dfrac{3x - 12}{\sqrt[3]{x + 4} - 2}$.

EXAMPLE 2   L'Hospital's Rule

Find $\lim\limits_{x \to 1} \dfrac{\ln x}{(x - 1)^2}$.

SOLUTION   Make sure that l'Hospital's rule applies.

$$\lim_{x \to 1} \ln x = \ln 1 = 0 \qquad \text{and} \qquad \lim_{x \to 1}(x - 1)^2 = 0$$

Since the conditions of l'Hospital's rule are satisfied, we can now take the derivatives of the numerator and denominator separately.

$$D_x(\ln x) = \frac{1}{x} \qquad \text{and} \qquad D_x[(x - 1)^2] = 2(x - 1)$$

Next, we find the limit of the quotient of these derivatives:

$$\lim_{x \to 1} \frac{1/x}{2(x - 1)} = \lim_{x \to 1} \frac{1}{2x(x - 1)} \text{ does not exist.}$$

By l'Hospital's rule, this means that

$$\lim_{x \to 1} \frac{\ln x}{(x - 1)^2} \text{ does not exist.} \qquad \textbf{TRY YOUR TURN 2}$$

YOUR TURN 2

Find $\lim\limits_{x \to 1} \dfrac{e^{x-1} - 1}{x^2 - 2x + 1}$.

Before looking at more examples of l'Hospital's rule, consider the following summary.

Using L'Hospital's Rule

1. Be sure that $\lim\limits_{x \to a} \dfrac{f(x)}{g(x)}$ leads to the indeterminate form $0/0$ or $\pm \infty/\pm \infty$.
2. Take the derivatives of f and g separately.
3. Find $\lim\limits_{x \to a} \dfrac{f'(x)}{g'(x)}$; this limit, if it exists, equals $\lim\limits_{x \to a} \dfrac{f(x)}{g(x)}$.
4. If necessary, apply l'Hospital's rule more than once.

EXAMPLE 3   L'Hospital's Rule

Find $\lim\limits_{x \to 0} \dfrac{x^3}{e^x - 1}$.

SOLUTION   The limit in the numerator is 0, as is the limit in the denominator, so that l'Hospital's rule applies. Taking derivatives separately in the numerator and denominator gives

$$\lim_{x \to 0} \frac{3x^2}{e^x} = \frac{0}{e^0} = \frac{0}{1} = 0.$$

By l'Hospital's rule,

$$\lim_{x \to 0} \frac{x^3}{e^x - 1} = 0. \qquad \textbf{TRY YOUR TURN 3}$$

YOUR TURN 3

Find $\lim\limits_{x \to 0} \dfrac{x^3}{\ln(x + 1)}$.

EXAMPLE 4 L'Hospital's Rule

Find $\lim\limits_{x \to 0} \dfrac{e^x - x - 1}{x^2}$.

SOLUTION Find the limit in both the numerator and denominator to verify that l'Hospital's rule applies. Then take derivatives of both the numerator and denominator separately.

$$\lim_{x \to 0} \frac{e^x - 1}{2x} = \frac{e^0 - 1}{2 \cdot 0} = \frac{1 - 1}{0} = \frac{0}{0}$$

The result is still the indeterminate form $0/0$; use l'Hospital's rule a second time. Taking derivatives of $e^x - 1$ and $2x$ gives

$$\lim_{x \to 0} \frac{e^x}{2} = \frac{e^0}{2} = \frac{1}{2}.$$

YOUR TURN 4

Find $\lim\limits_{x \to 0} \dfrac{e^{3x} - \frac{9}{2}x^2 - 3x - 1}{x^3}$.

Finally, by l'Hospital's rule,

$$\lim_{x \to 0} \frac{e^x - x - 1}{x^2} = \frac{1}{2}.$$ **TRY YOUR TURN 4**

EXAMPLE 5 L'Hospital's Rule

Find $\lim\limits_{x \to 1} \dfrac{x^2 - 1}{\sqrt{x}}$.

SOLUTION Taking derivatives of the numerator and denominator separately gives

$$\lim_{x \to 1} \frac{2x}{(1/2)x^{-1/2}} = \lim_{x \to 1} 4x^{3/2} = 4 \cdot 1^{3/2} = 4 \cdot 1 = 4. \quad \text{Incorrect}$$

Unfortunately, 4 is the wrong answer. What happened? We did not verify that the conditions of l'Hospital's rule were satisfied. In fact,

$$\lim_{x \to 1} (x^2 - 1) = 0, \quad \text{but} \quad \lim_{x \to 1} \sqrt{x} = 1 \neq 0.$$

Since l'Hospital's rule does not apply, we must use another method to find the limit. By substitution,

$$\lim_{x \to 1} \frac{x^2 - 1}{\sqrt{x}} = \frac{1^2 - 1}{\sqrt{1}} = \frac{0}{1} = 0.$$

L'Hospital's rule also applies when

$$\lim_{x \to a} f(x) = \infty \quad \text{and} \quad \lim_{x \to a} g(x) = \infty.$$

EXAMPLE 6 Limit of $0 \cdot \infty$ or ∞/∞

Find each of the following limits.

(a) $\lim\limits_{x \to 0^+} x \ln x$

SOLUTION It is not immediately clear what the limit is. The factor x is getting smaller and smaller as x approaches 0, but the factor $\ln x$ is approaching $-\infty$. We have a limit of the form $0 \cdot \infty$. To evaluate this limit, use the fact that

$$x = \frac{1}{1/x}$$

to rewrite the expression as

$$x \ln x = \frac{\ln x}{1/x}.$$

Now both the numerator and the denominator become infinite in magnitude, and l'Hospital's rule applies to limits of the form ∞/∞.

$$\lim_{x \to 0^+} x \ln x = \lim_{x \to 0^+} \frac{\ln x}{1/x} \qquad \text{Rewrite as a quotient of the form } \infty/\infty.$$

$$= \lim_{x \to 0^+} \frac{1/x}{-1/x^2} \qquad \text{Differentiate the numerator and denominator.}$$

$$= \lim_{x \to 0^+} -x \qquad \text{Simplify.}$$

$$= 0$$

Therefore, by l'Hospital's rule,

$$\lim_{x \to 0^+} x \ln x = 0.$$

(b) $\displaystyle \lim_{x \to 0^+} x \, (\ln x)^2.$

SOLUTION This limit has the form $0 \cdot \infty$ and is similar to the limit in part (a), so we will handle it in the same manner.

$$\lim_{x \to 0^+} x \, (\ln x)^2 = \lim_{x \to 0^+} \frac{(\ln x)^2}{1/x} \qquad \text{Rewrite as a quotient of the form } \infty/\infty.$$

$$= \lim_{x \to 0^+} \frac{2(\ln x)(1/x)}{-1/x^2} \qquad \text{Differentiate the numerator and denominator.}$$

$$= \lim_{x \to 0^+} -2x(\ln x) \qquad \text{Simplify.}$$

This problem is similar to what we started with, but with $\ln x$ raised to the first power, rather than the second power. It seems that we have made progress, so let's try the same idea again.

$$\lim_{x \to 0^+} -2x(\ln x) = \lim_{x \to 0^+} \frac{-2 \ln x}{1/x} \qquad \text{Rewrite as a quotient of the form } \infty/\infty.$$

$$= \lim_{x \to 0^+} \frac{-2/x}{-1/x^2} \qquad \text{Differentiate the numerator and denominator.}$$

$$= \lim_{x \to 0^+} 2x \qquad \text{Simplify.}$$

$$= 0$$

YOUR TURN 5 Find $\displaystyle \lim_{x \to 0^+} x^2 \ln(3x).$

(We could have avoided this second step by noticing that the limit at the end of the first step is just -2 times the limit in part (a).) Therefore, by l'Hospital's rule,

$$\lim_{x \to 0^+} x(\ln x)^2 = 0. \qquad \text{TRY YOUR TURN 5} \ \blacksquare$$

We could use the same idea as in Example 6 repeatedly to show that

$$\lim_{x \to 0^+} x(\ln x)^n = 0$$

for any positive integer n. This limit was investigated graphically in an exercise in Chapter 3, and used again in the section on Curve Sketching in Chapter 5. We finally have a way to demonstrate this result. The intuitive reason for this result is that although $\ln x$ approaches $-\infty$ as x approaches 0 from the right, the logarithm is a very slowly changing function, so it doesn't get large very quickly.

Limits at Infinity

L'Hospital's rule also applies to limits at infinity. The next example illustrates this idea.

EXAMPLE 7 Limit at Infinity

Find each of the following limits.

(a) $\lim_{x \to \infty} xe^{-x}$

SOLUTION As in the previous example, it's not obvious what the limit is. The factor x is getting larger and larger, but the factor e^{-x} is getting smaller and smaller. This is another example of a limit of the form $0 \cdot \infty$. To find out what happens to the product, we will rewrite the product as a quotient. This converts the problem to a limit of the form ∞/∞, as in the previous example.

$$\lim_{x \to \infty} xe^{-x} = \lim_{x \to \infty} \frac{x}{e^x} \qquad \text{Rewrite as a quotient of the form } \infty/\infty.$$

$$= \lim_{x \to \infty} \frac{1}{e^x} \qquad \text{Differentiate the numerator and denominator.}$$

$$= 0$$

Therefore, by l'Hospital's rule.

$$\lim_{x \to \infty} xe^{-x} = 0.$$

(b) $\lim_{x \to \infty} x^n e^{-x}$, where n is a positive integer

SOLUTION Like the limit in part (a), this limit is of the form $0 \cdot \infty$, and it can be evaluated by rewriting the product as a quotient.

$$\lim_{x \to \infty} x^n e^{-x} = \lim_{x \to \infty} \frac{x^n}{e^x} \qquad \text{Rewrite as a quotient of the form } \infty/\infty.$$

$$= \lim_{x \to \infty} \frac{nx^{n-1}}{e^x} \qquad \text{Differentiate the numerator and denominator.}$$

This leaves us with a new problem similar to the original, but with the numerator of degree one less. We could continue to apply l'Hospital's rule until the numerator becomes $n!$, which happens to x^n when it is differentiated n times. Then

$$\lim_{x \to \infty} \frac{n!}{e^x} = 0.$$

YOUR TURN 6

Find $\lim_{x \to \infty} \dfrac{\ln x}{e^x}$ and $\lim_{x \to \infty} \dfrac{\ln x}{x^2}$.

Therefore, by l'Hospital's rule,

$$\lim_{x \to \infty} x^n e^{-x} = 0. \qquad \text{TRY YOUR TURN 6}$$

The limit in Example 7(b) was investigated graphically in an exercise in Chapter 3 and used again in the section on Curve Sketching in Chapter 5. We finally have a way to demonstrate this result. The intuitive reason for this result is that e^{-x} approaches 0 very rapidly as x goes to infinity. Alternatively, we could say that e^x gets large much faster than any power of x as x goes to infinity.

Proof of l'Hospital's Rule
Because the proof of l'Hospital's rule is too advanced for this text we will not prove it here. We will, however, prove the theorem for the special case where f, g, f', and g' are continuous on some open interval containing a, and $g'(a) \neq 0$. We will only consider the case in which

$$\lim_{x \to a} f(x) = 0 \qquad \text{and} \qquad \lim_{x \to a} g(x) = 0.$$

The assumption that f and g are continuous means that both $f(a) = 0$ and $g(a) = 0$. Thus,

$$\lim_{x \to a} \frac{f(x)}{g(x)} = \lim_{x \to a} \frac{f(x) - f(a)}{g(x) - g(a)}, \qquad f(a) = 0 \text{ and } g(a) = 0$$

where we subtracted 0 in both the numerator and denominator. Multiplying the numerator and denominator by $1/(x - a)$ gives

$$\lim_{x \to a} \frac{f(x)}{g(x)} = \lim_{x \to a} \frac{\dfrac{f(x) - f(a)}{x - a}}{\dfrac{g(x) - g(a)}{x - a}}.$$

By a property of limits, this becomes

$$\lim_{x \to a} \frac{f(x)}{g(x)} = \frac{\displaystyle\lim_{x \to a} \frac{f(x) - f(a)}{x - a}}{\displaystyle\lim_{x \to a} \frac{g(x) - g(a)}{x - a}}.$$

By the definition of the derivative, the limit in the numerator is $f'(a)$, and the limit in the denominator is $g'(a)$. From our assumption that both f' and g' are continuous, and if $g'(a) \neq 0$, the quotient on the right above becomes

$$\frac{f'(a)}{g'(a)} = \frac{\displaystyle\lim_{x \to a} f'(x)}{\displaystyle\lim_{x \to a} g'(x)} = \lim_{x \to a} \frac{f'(x)}{g'(x)}.$$

Thus,

$$\lim_{x \to a} \frac{f(x)}{g(x)} = \lim_{x \to a} \frac{f'(x)}{g'(x)},$$

which is what we wanted to show.

12.7 EXERCISES

Use l'Hospital's rule where applicable to find each limit.

1. $\displaystyle\lim_{x \to 1} \frac{x^3 + x^2 - x - 1}{x^2 - x}$

2. $\displaystyle\lim_{x \to 3} \frac{x^3 + x^2 - 11x - 3}{x^2 - 3x}$

3. $\displaystyle\lim_{x \to 0} \frac{x^5 - 2x^3 + 4x^2}{8x^5 - 2x^2 + 5x}$

4. $\displaystyle\lim_{x \to 0} \frac{8x^6 + 3x^4 - 9x}{9x^7 - 2x^4 + x^3}$

5. $\displaystyle\lim_{x \to 2} \frac{\ln(x - 1)}{x - 2}$

6. $\displaystyle\lim_{x \to 0} \frac{\ln(x + 1)}{x}$

7. $\displaystyle\lim_{x \to 0} \frac{e^x - 1}{x^4}$

8. $\displaystyle\lim_{x \to 0} \frac{e^{2x} - 1}{5x^2 - x}$

9. $\displaystyle\lim_{x \to 0} \frac{xe^x}{e^x - 1}$

10. $\displaystyle\lim_{x \to 0} \frac{xe^{-x}}{2e^{2x} - 2}$

11. $\displaystyle\lim_{x \to 0} \frac{e^x}{2x^3 + 9x^2 - 11x}$

12. $\displaystyle\lim_{x \to 0} \frac{e^x}{8x^5 - 3x^4}$

13. $\displaystyle\lim_{x \to 0} \frac{\sqrt{2 + x} - \sqrt{2}}{x}$

14. $\displaystyle\lim_{x \to 0} \frac{\sqrt{9 + x} - 3}{x}$

15. $\displaystyle\lim_{x \to 4} \frac{\sqrt{x} - 2}{x - 4}$

16. $\displaystyle\lim_{x \to 9} \frac{\sqrt{x} - 3}{x - 9}$

17. $\displaystyle\lim_{x \to 8} \frac{\sqrt[3]{x} - 2}{x - 8}$

18. $\displaystyle\lim_{x \to 27} \frac{\sqrt[3]{x} - 3}{x - 27}$

19. $\displaystyle\lim_{x \to 1} \frac{x^9 + 3x^8 + 4x^5 - 8}{x - 1}$

20. $\displaystyle\lim_{x \to 2} \frac{x^7 - 5x^6 + 5x^5 + 32}{x - 2}$

21. $\displaystyle\lim_{x \to 0} \frac{e^x + e^{-x} - 2}{x}$

22. $\displaystyle\lim_{x \to 0} \frac{e^x - 1 + x}{e^{-x} - 1 - x}$

23. $\displaystyle\lim_{x \to 3} \frac{\sqrt{x^2 + 7} - 4}{x^2 - 9}$

24. $\displaystyle\lim_{x \to 5} \frac{\sqrt{x^2 + 11} - 6}{x^2 - 25}$

25. $\displaystyle\lim_{x \to 0} \frac{1 + \dfrac{1}{3}x - (1 + x)^{1/3}}{x^2}$

26. $\displaystyle\lim_{x \to 0} \frac{2e^{5x} - 25x^2 - 10x - 2}{5x^3}$

27. $\displaystyle\lim_{x \to 0} \frac{\sqrt{1 + x} - \sqrt{1 - x}}{x}$

28. $\displaystyle\lim_{x \to 0} \frac{\sqrt{3 - x} - \sqrt{3 + x}}{x}$

29. $\displaystyle\lim_{x \to 0} \frac{\sqrt{x^2 - 5x + 4}}{x}$

30. $\displaystyle\lim_{x \to 1} \frac{\sqrt{x^2 + 5x + 9}}{x - 1}$

31. $\displaystyle\lim_{x \to 0} \frac{(5 + x)\ln(x + 1)}{e^x - 1}$

32. $\displaystyle\lim_{x \to 0} \frac{(7 - x)\ln(1 - x)}{e^{-x} - 1}$

33. $\lim\limits_{x \to 0^+} x^2 (\ln x)^2$

34. $\lim\limits_{x \to 0^+} xe^{1/x}$

35. $\lim\limits_{x \to 0^+} x \ln(e^x - 1)$

36. $\lim\limits_{x \to \infty} \dfrac{(\ln x)^2}{x}$

37. $\lim\limits_{x \to \infty} \dfrac{\sqrt{x}}{\ln(\ln x)}$

38. $\lim\limits_{x \to \infty} \dfrac{e^{\sqrt{x}}}{x^3}$

39. $\lim\limits_{x \to \infty} \dfrac{\ln(e^x + 1)}{5x}$

40. $\lim\limits_{x \to \infty} \dfrac{\ln(4e^{\sqrt{x}} - 1)}{3\sqrt{x}}$

41. $\lim\limits_{x \to \infty} x^5 e^{-0.001x}$

42. $\lim\limits_{x \to \infty} \dfrac{x^3 + 1}{x^2 \ln x}$

In Exercises 43–46, first get a common denominator; then find the limits that exist.

43. $\lim\limits_{x \to 0} \left(\dfrac{e^x}{x^2} - \dfrac{1}{x^2} - \dfrac{1}{x} \right)$

44. $\lim\limits_{x \to 0} \left(\dfrac{12e^x}{x^3} - \dfrac{12}{x^3} - \dfrac{12}{x^2} - \dfrac{6}{x} \right)$

45. $\lim\limits_{x \to 1} \left(\dfrac{x}{x - 1} - \dfrac{1}{\ln x} \right)$

46. $\lim\limits_{x \to 0} \left(\dfrac{2}{x} - \dfrac{\ln(1 + 2x)}{x^2} \right)$

47. Explain what is wrong with the following calculation using l'Hospital's rule.

$$\lim_{x \to 0} \frac{x^2}{x^2 + 3} = \lim_{x \to 0} \frac{2x}{2x} = 1$$

48. Find the following limit, which is the first one given by l'Hospital in his calculus text *Analysis of Infinitely Small Quantities for the Understanding of Curves*, published in 1696. *Source: A History of Mathematics: An Introduction.*

$$\lim_{x \to a} \frac{\sqrt{2a^3 x - x^4} - a\sqrt[3]{a^2 x}}{a - \sqrt[4]{ax^3}}$$

YOUR TURN ANSWERS

1. 36
2. Does not exist
3. 0
4. 9/2
5. 0
6. 0; 0

12 CHAPTER REVIEW

SUMMARY

We have provided a brief introduction to the topics of sequences, series, and l'Hospital's rule. Geometric sequences are comparatively simple to analyze and arise in various applications, including annuities. We next investigated infinite series, as well as a particular form known as Taylor series. Because Taylor series have an infinite number of terms, it is often more practical to take a small number of terms, creating Taylor polynomials. We then discussed Newton's method, which produces a sequence that approaches a zero of a function. Finally, l'Hospital's rule provides a method for evaluating certain limits.

General Term of a Geometric Series	$a_n = ar^{n-1}$
Sum of the First n Terms of a Geometric Series	$S_n = \dfrac{a(r^n - 1)}{r - 1}$ for $r \neq 1$
Sum of an Infinite Geometric Series	For $-1 < r < 1$, $$\sum_{n=1}^{\infty} ar^{n-1} = \frac{a}{1 - r}$$
Amount of an Annuity	$S = R\left[\dfrac{(1 + i)^n - 1}{i}\right]$ or $S = R \cdot s_{\overline{n}|i}$
Present Value of an Annuity	$P = R\left[\dfrac{1 - (1 + i)^{-n}}{i}\right]$ or $P = R \cdot a_{\overline{n}|i}$
Rule of 70	For a rate of increase r compounded annually, where $0.001 \leq r \leq 0.05$, $$\text{Doubling Time} \approx \frac{70}{100r} \text{ years.}$$

Rule of 72 For a rate of increase r compounded annually, where $0.05 < r \leq 0.12$,

$$\text{Doubling Time} \approx \frac{72}{100r} \text{ years.}$$

Taylor Polynomial at 0 $\displaystyle P_n(x) = \sum_{i=0}^{n} \frac{f^{(i)}(0)}{i!} x^i$

Taylor Series at 0 $\displaystyle f(x) = \sum_{i=0}^{\infty} \frac{f^{(i)}(0)}{i!} x^i$

Taylor Series for e^x $\displaystyle e^x = \sum_{i=0}^{\infty} \frac{1}{i!} x^i \text{ for } -\infty < x < \infty$

Taylor Series for ln $(1 + x)$ $\displaystyle \ln(1 + x) = \sum_{i=1}^{\infty} \frac{(-1)^{i+1}}{i} x^i \text{ for } -1 < x \leq 1$

Taylor Series for $\dfrac{1}{1 - x}$ $\displaystyle \frac{1}{1 - x} = \sum_{i=0}^{\infty} x^i \text{ for } -1 < x < 1$

Newton's Method If $f'(c_n) \neq 0$, then

$$c_{n+1} = c_n - \frac{f(c_n)}{f'(c_n)}.$$

L'Hospital's Rule If $\displaystyle \lim_{x \to a} f(x) = \lim_{x \to a} g(x) = 0$ or $\displaystyle \lim_{x \to a} f(x) = \pm\infty$ and $\displaystyle \lim_{x \to a} g(x) = \pm\infty$, then

$$\lim_{x \to a} \frac{f(x)}{g(x)} = \lim_{x \to a} \frac{f'(x)}{g'(x)}.$$

KEY TERMS

12.1
sequence
element
term
general term
nth term
geometric sequence
common ratio

12.2
annuity
ordinary annuity

payment period
term of an annuity
amount of an annuity
sinking fund
present value of an
  annuity
amortization

12.3
Taylor polynomial
Taylor polynomial of
  degree n

12.4
infinite series
nth partial sum
sum of an infinite
  series
convergence
divergence

12.5
interval of convergence
Taylor series

Maclaurin series
doubling time
rule of 70
rule of 72

12.6
Newton's method

12.7
indeterminate form
l'Hospital's rule

REVIEW EXERCISES

CONCEPT CHECK

Determine whether each of the following statements is true or false, and explain why.

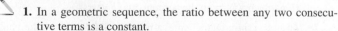

1. In a geometric sequence, the ratio between any two consecutive terms is a constant.

2. The amounts paid into an annuity form a geometric sequence.

3. A loan is amortized if both the principal and interest are paid by a sequence of equal periodic payments.

4. The Taylor polynomial of degree 4 for f at 0 has the same second derivative as f at 0.

5. The Taylor polynomial of degree 4 for f at 0 has the same fifth derivative as f at 0.

6. The Taylor polynomial of a discontinuous function is continuous.

7. An infinite geometric series converges as long as $-1 \leq r \leq 1$.

8. If an infinite series doesn't converge, then it diverges.

9. The Taylor series for e^x at 0 converges for all x.

10. The Taylor series for ln $(1 + x)$ at 0 converges for all x.

11. Newton's method converges as long as there is a real root and the function is differentiable.

12. L'Hospital's rule says that to take the derivative of a quotient, divide the derivative of the numerator by the derivative of the denominator.

PRACTICE AND EXPLORATIONS

Find a_4 and a_n for the following geometric sequences. Then find the sum of the first five terms.

13. $a_1 = 5, r = -2$

14. $a_1 = 128, r = 1/2$

15. $a_1 = 27, r = 1/3$

16. $a_1 = 2, r = -5$

Find Taylor polynomials of degree 4 at 0 for the functions defined as follows.

17. $f(x) = e^{2-x}$

18. $f(x) = 5e^{2x}$

19. $f(x) = \sqrt{x+1}$

20. $f(x) = \sqrt[3]{x+27}$

21. $f(x) = \ln(2-x)$

22. $f(x) = \ln(3+2x)$

23. $f(x) = (1+x)^{2/3}$

24. $f(x) = (4+x)^{3/2}$

Use Taylor polynomials of degree 4 at $x = 0$, found in Exercises 17–24 above, to approximate the quantities in Exercises 25–32. Round to 4 decimal places.

25. $e^{1.93}$

26. $5e^{0.04}$

27. $\sqrt{1.03}$

28. $\sqrt[3]{26.94}$

29. $\ln 2.05$

30. $\ln 3.06$

31. $0.92^{2/3}$

32. $4.02^{3/2}$

Identify the geometric series that converge. Give the sum of each convergent series.

33. $9 - 6 + 4 - 8/3 + \cdots$

34. $2 + 1.4 + 0.98 + 0.686 + \cdots$

35. $3 + 9 + 27 + 81 + \cdots$

36. $4 + 4.8 + 5.76 + 6.912 + \cdots$

37. $\dfrac{2}{5} - \dfrac{2}{25} + \dfrac{2}{125} - \dfrac{2}{625} + \cdots$

38. $36 + 3 + \dfrac{1}{4} + \dfrac{1}{48} + \cdots$

In Exercises 39–40, the nth term of a sequence is given. Calculate the first five partial sums.

39. $a_n = \dfrac{1}{2n-1}$

40. $a_n = \dfrac{1}{(n+2)(n+3)}$

Use the Taylor series given in the text to find the Taylor series for the functions defined as follows. Give the interval of convergence of each series.

41. $f(x) = \dfrac{4}{3-x}$

42. $f(x) = \dfrac{2x}{1+3x}$

43. $f(x) = \dfrac{x^2}{x+1}$

44. $f(x) = \dfrac{3x^3}{2-x}$

45. $f(x) = \ln(1-2x)$

46. $f(x) = \ln\left(1 + \dfrac{1}{3}x\right)$

47. $f(x) = e^{-2x^2}$

48. $f(x) = e^{-5x}$

49. $f(x) = 2x^3 e^{-3x}$

50. $f(x) = x^6 e^{-x}$

Use l'Hospital's rule, where applicable, to find each limit.

51. $\lim\limits_{x \to 2} \dfrac{x^3 - x^2 - x - 2}{x^2 - 4}$

52. $\lim\limits_{x \to 0} \dfrac{x^3 - 4x^2 + 6x}{3x}$

53. $\lim\limits_{x \to -5} \dfrac{x^3 - 3x^2 + 4x - 1}{x^2 - 25}$

54. $\lim\limits_{x \to 0} \dfrac{\ln(3x+1)}{x}$

55. $\lim\limits_{x \to 0} \dfrac{5e^x - 5}{x^3 - 8x^2 + 7x}$

56. $\lim\limits_{x \to 0} \dfrac{\sqrt{5+x} - \sqrt{5}}{x}$

57. $\lim\limits_{x \to 0} \dfrac{-xe^{2x}}{e^{2x} - 1}$

58. $\lim\limits_{x \to 16} \dfrac{\sqrt{x} - 4}{x - 16}$

59. $\lim\limits_{x \to 0} \dfrac{1 + 2x - (1+x)^{1/2}}{x^3}$

60. $\lim\limits_{x \to 0} \dfrac{\sqrt{5+x} - \sqrt{5-x}}{2x}$

61. $\lim\limits_{x \to \infty} x^2 e^{-\sqrt{x}}$

62. $\lim\limits_{x \to \infty} \dfrac{\sqrt{x}}{\ln(x^3+1)}$

In Exercises 63–66, first get a common denominator; then find the limits that exist.

63. $\lim\limits_{x \to 0} \left(\dfrac{e^{3x}}{x^2} - \dfrac{1}{x^2} - \dfrac{3}{x}\right)$

64. $\lim\limits_{x \to 0} \left(\dfrac{2}{x^3} + \dfrac{2}{x^2} + \dfrac{1}{x} - \dfrac{2e^x}{x^3}\right)$

65. $\lim\limits_{x \to 0} \left(\dfrac{\ln(1-4x)}{x^2} + \dfrac{4}{x}\right)$

66. $\lim\limits_{x \to 0} \left(\dfrac{1}{x^3} + \dfrac{1}{x^2}\right)$

Use Newton's method to find a solution to the nearest hundredth for each equation in the given interval.

67. $x^3 - 8x^2 + 18x - 12 = 0$; $[4, 5]$

68. $3x^3 - 4x^2 - 4x - 7 = 0$; $[2, 3]$

69. $x^4 + 3x^3 - 4x^2 - 21x - 21 = 0$; $[2, 3]$

70. $x^4 + x^3 - 14x^2 - 15x - 15 = 0$; $[3, 4]$

Use Newton's method to approximate each radical to the nearest thousandth.

71. $\sqrt{37.6}$

72. $\sqrt{51.7}$

73. $\sqrt[3]{94.7}$

74. $\sqrt[4]{102.6}$

APPLICATIONS

Business and Economics

75. Total Income A mine produced $750,000 of income during its first year. Each year thereafter, income increased by 18%. Find the total income produced in the first 8 years of the mine's life.

76. Sinking Fund In 4 years, Jack McCanna must pay a pledge of $5000 to his church's building fund. He wants to set up a sinking fund to accumulate that amount. What should each semiannual payment into the fund be at 8% compounded semiannually?

77. Annuity Cathy Schneider deposits $491 at the end of each quarter for 9 years. If the account pays 9.4% compounded quarterly, find the final amount in the account.

78. Annuity J. Euclid deposits $1526.38 at the end of each 6-month period in an account paying 7.6% compounded semiannually. How much will be in the account after 5 years?

79. Amortization Diane Antaya borrows $20,000 from the bank to help her expand her business. She agrees to repay the money in equal payments at the end of each year for 9 years. Interest is at 8.9% compounded annually. Find the amount of each payment.

80. Amortization Ross Craycraft wants to expand his pharmacy. To do this, he takes out a bank loan of $49,275 and agrees to repay it at 12.2% compounded monthly over 48 months. Find the amount of each payment necessary to amortize this loan.

House Payments Find the monthly house payments for the following mortgages.

81. $156,890 at 7.74% for 25 years

82. $177,110 at 8.45% for 30 years

83. Investment Michael Dew has invested $14,000 in a certificate of deposit that has a 3.25% annual interest rate. Determine the doubling time for this investment using the doubling-time formula. How does this compare with the estimate given by the rule of 70?

84. Investment It is anticipated that a bank stock that Jeff Marsalis has invested $16,000 in will achieve an annual interest rate of 9%. Determine the doubling time for this investment using the doubling-time formula. How does this compare with the estimate given by the rule of 72?

Life Sciences

85. Bacteria At a summer picnic, the number of bacteria in a bowl of potato salad doubles every 20 minutes. Assume that there are 1000 bacteria at the beginning of the picnic. How many bacteria are present after 2 hours, assuming that no one has eaten any of the potato salad?

Social Sciences

86. Crime The number of reported crimes in a city was about 22,700 in a recent year. Due to the creation of a neighborhood crime program, the city hopes the number of crimes decreases each year by 8%. Let x_n denote the number of crimes in the city n years after the neighborhood crime program began. Find a formula for x_n in terms of n. Determine the number of crimes in the city at the end of five years.

EXTENDED APPLICATION

LIVING ASSISTANCE AND SUBSIDIZED HOUSING

Mr. Jones receives living assistance, in the form of a monthly stipend from the State of New York. He is also living in subsidized housing. This means that the amount he pays in rent depends on his income. He has entered into contracts with the State of New York and his landlord specifying how his stipend and rent are computed. The unusual aspect of these contracts is that, to a degree, each depends on the other. Thus a single change in one contract leads to a potentially infinite sequence of changes in both contracts.

The relevant portion of the contract between the State of New York and Mr. Jones is:

The State of New York agrees to pay Mr. Jones a monthly stipend of $1000. This figure is arrived at by considering his living expenses. The stipend will be increased or decreased by 30% of any increase or decrease in rent.

Mr. Jones is also living in subsidized housing and has worked out a contract with his landlord that specifies: *The monthly rent is $300. However, if Mr. Jones's income increases during the period of the contract, the monthly rent will be increased by 20% of the change.*

The situation gets complicated shortly after Mr. Jones receives the good news that his stipend from the government is being increased by $100/month to $1100/month. As required, he reports to his landlord that his income has increased and, as specified in his contract, his rent increases by 20% of $100. Thus his new rent is $300 + $20 = $320.

Since the contract with the State of New York has a housing allowance built in to it, he reports his $20 rent increase to the state, and his monthly stipend of $1100 is increased by 30% of $20 to $1100 + $6 = $1106.

At this point it becomes clear that Mr. Jones is facing a never ending sequence of stipend and rent adjustments. Although it looks like the adjustments are eventually going to be quite small, he knows he must honor both contracts, and this is going to require a lot of round trips and paperwork. On his way back to his landlord with the news that his state stipend had been raised from $1000 to $1100 to $1106, Mr. Jones decided to consult a lawyer.

The lawyer took a look at the contracts and decided to consult a mathematician to see if it is mathematically possible to make sense of an unending sequence of stipend and rent hikes. As we shall see, this is exactly what infinite series are made for.

To help recognize the pattern, notice that the next term in the infinite series for the state stipend is 30% of the last rent increase, that is, $0.3 \times 20 = \$6$. In other words, every time the state decides to increase the stipend by x, the landlord increases the rent by $0.2x$, and the state is obligated to increase the stipend by $0.3(0.2x) = 0.06x$.

The infinite series for the state stipend (in dollars) is

$$\text{Stipend} = 1000 + 100 + 0.06(100) + (0.06)^2(100) + \cdots.$$

After the first term, this is a geometric series with $r = 0.06$; thus, the sum is $\$1000 + \$100/(1 - 0.06) = \$1,106.38$.

The analysis of the rent is similar. Each rent hike of y dollars is followed by a stipend increase of $0.3y$ and a subsequent rent increase $0.2(0.3y) = 0.06y$. Thus, the infinite series for rent (in dollars) is

$$\text{Rent} = 300 + 20 + 0.06(20) + (0.06)^2(20) + \cdots,$$

which converges to $\$300 + \$20/(1 - 0.06) = \$321.28$.

There is a surprising aspect to this problem. The interrelated nature of the contracts seems to demand an infinite series solution;

yet, it can also be solved without using infinite series. How can this be possible?

The key is to anticipate that the original \$100 increase in stipend is going to necessitate subsequent increases. So we can express the ultimate stipend as \$1000 + \$100 + S, where S is yet to be determined. Similarly, the rent will ultimately be \$300 + R, where R also needs to be determined. The question becomes, can we find values of S and R such that neither contract is violated?

Mr. Jones's contract with the state requires that his stipend be increased by 30% of the change in his rent, that is $S = 0.3R$. On the other hand, his contract with his landlord requires that his 20% stipend increase of $100 + S$ be included in his rent, or in terms of equations, $0.2(100 + S) = R$. In Exercise 1, you will show that solving these simultaneous equations leads to the same stipend and rent as found with infinite series.

EXERCISES

1. Find values for S and R that satisfy $S = 0.3R$ and $0.2(100 + S) = R$. Show that these solutions give the same stipend and rent as found by summing the infinite series.

2. Suppose that instead of a stipend increase of \$100, the state cuts Mr. Jones's stipend by \$50. Assuming that Mr. Jones is able to convince his landlord that he should have his rent decreased by 20% of the change, this also leads to an infinite cycle of stipend and rent changes. Express his stipend and rent as infinite series, and find the sum of each series.

3. Eastville is located 12 miles from Westville. The town councils decide to pool resources and build a single fire station to serve the needs of both towns. The negotiations on where to build the fire station start with both towns proposing the fire station be built in their town. The impasse is broken when Eastville proposes to move the site halfway to Westville, i.e., 6 miles to the west. Westville in turn proposes to move the site halfway to the Eastville proposed site, i.e., 3 miles to the east. This sets off an infinite round of negotiations in which each party proposes moving the site halfway towards the other's previous proposal. Give an infinite series expressing the changes in location proposed by Eastville, and give a similar series for the changes proposed by Westville. Where is the fire station eventually located? (*Hint:* The surest way to recognize a pattern is to work out a few terms, and this calls for simple, but careful, record keeping. Initial separation is 12 miles. Eastville moves 6 miles. Now the separation is 6 miles. Westville moves 3 miles. Separation is 3 miles. Eastville moves 3/2 miles. Separation is . . .)

4. There was enough money leftover after building the fire station in Exercise 3 for a swimming pool. This time, Eastville and Westville approach the negotiations more warily. Eastville starts by suggesting the pool be located just 1/3 of the way towards Westville. From that point on, Westville agrees to split the difference, while at every stage, Eastville proposes moving the pool just 1/3 of the way towards Westville's last proposal. Are the towns able to reach an agreement on the final location of the pool?

5. The sum of the series for the stipend paid to Mr. Jones is approximately \$1,106.3829787. Understandably, an accountant for the State of New York would view this as needless precision. To gain an appreciation of how quickly geometric series converge, particularly with a small value of R, like 0.06, use a calculator to answer the following questions. How many terms of the series do you need to add up so that the sum is within one dollar of the final answer? How many terms do you need to add up to be within a dime or a penny of the final answer?

6. Not all series converge as quickly as geometric series. We know from Section 12.5

$$\ln(2) = 1 - 1/2 + 1/3 - 1/4 + 1/5 \ldots,$$

so the nth term of this series is $(-1)^{n+1}/n$. Use the website WolframAlpha.com to decide how many terms you need to add up so the sum is within 0.01 of $\ln(2)$. To sum a series on WolframAlpha.com, enter the following:

sum <formula for nth term of your series> from n = <first value of n> to <final value of n, and this can even be infinity>.

13

The Trigonometric Functions

The time when the sun sets depends both on your location on the globe and on the time of year. Because Earth's motion around the Sun is periodic, the sunset time for a particular location is a periodic function of time measured in days. We explore this trigonometric model in the exercises for Section 1 in this chapter.

Throughout this book we have discussed many different types of functions, including linear, quadratic, exponential, and logarithmic functions. In this chapter we introduce the *trigonometric functions*, which differ in a fundamental way from those previously studied: the trigonometric functions describe periodic or repetitive relationships.

An example of a periodic relationship is given by an electrocardiogram (EKG), a graph of a human heartbeat. The EKG in Figure 1 shows electrical impulses from a heart. *Source: Nancy Schiller.* Each small square represents 0.04 second. How often does this heart beat?

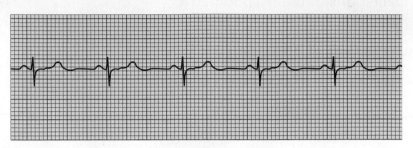

FIGURE 1

Trigonometric functions describe many natural phenomena and are important in the study of optics, heat, electronics, acoustics, and seismology. Also, many algebraic functions have integrals involving trigonometric functions.

13.1 Definitions of the Trigonometric Functions

APPLY IT How far from a camera should an object be to put it in focus?
In Exercise 90 in this section, we will use trigonometry to answer this question.

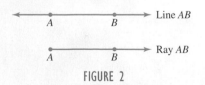

FIGURE 2

The angle is one of the basic concepts of trigonometry. The definition of an angle depends on that of a ray: a **ray** is the portion of a line that starts at a given point and continues indefinitely in one direction. Figure 2 shows a line through the two points A and B. The portion of the line AB that starts at A and continues through and past B is called ray AB. Point A is the **endpoint** of the ray.

An **angle** is formed by rotating a ray about its endpoint. The initial position of the ray is called the **initial side** of the angle, and the endpoint of the ray is called the **vertex** of the angle. The location of the ray at the end of its rotation is called the **terminal side** of the angle. See Figure 3.

An angle can be named by its vertex. For example, the angle in Figure 3 can be called angle A. An angle also can be named by using three letters, with the vertex letter in the middle. For example, the angle in Figure 3 could be named angle BAC or angle CAB.

An angle is in **standard position** if its vertex is at the origin of a coordinate system and if its initial side is along the positive x-axis. The angles in Figures 4 and 5 on the next page are in standard position. An angle in standard position is said to be in the quadrant of its terminal side. For example, the angle in Figure 4(a) is in quadrant I, while the angle in Figure 4(b) is in quadrant II.

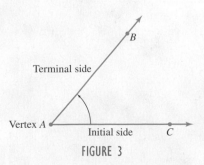

FIGURE 3

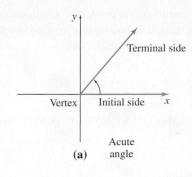

(a) Acute angle

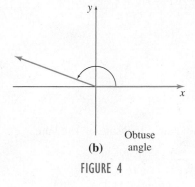

(b) Obtuse angle

FIGURE 4

Notice that the angles in Figures 3 and 4 are formed with a counterclockwise rotation from the positive x-axis. This is true for any positive angle. A negative angle is measured clockwise from the positive x-axis, as we shall see in Example 5.

Degree Measure
The sizes of angles are often indicated in *degrees*. **Degree measure** has remained unchanged since the Babylonians developed it over 4000 years ago. In degree measure, 360 degrees represents a complete rotation of a ray. *One degree*, written $1°$, is $1/360$ of a rotation. Also, $90°$ is $90/360$ or $1/4$ of a rotation, and $180°$ is $180/360$ or $1/2$ of a rotation. See Figure 5.

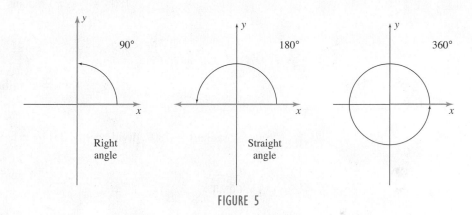

Right angle   Straight angle

FIGURE 5

An angle having a degree measure between $0°$ and $90°$ is called an **acute angle**. An angle of $90°$ is a **right angle**. An angle having measure more than $90°$ but less than $180°$ is an **obtuse angle**, while an angle of $180°$ is a **straight angle**. See Figures 4 and 5.

A complete rotation of a ray results in an angle of measure $360°$. But there is no reason why the rotation need stop at $360°$. By continuing the rotation, angles of measure larger than $360°$ can be produced. The angles in Figure 6 have measures $60°$ and $420°$. These two angles have the same initial side and the same terminal side, but different amounts of rotation.

Radian Measure
While degree measure works well for some applications, using degree measurement in calculus is complicated. Fortunately, there is an alternative system, called **radian measure**, that helps to keep the formulas for derivatives and antiderivatives as simple as possible. To see how this system for measuring angles is obtained, look at angle θ (the Greek letter *theta*) in Figure 7(a). The angle θ is in standard position; Figure 7(a) also shows a circle of radius 1, known as the **unit circle**, centered at the origin.

The vertex of θ is at the center of the circle in Figure 7(a). Angle θ cuts a piece of the circle called an **arc**. The length of this arc is the measure of the angle in radians. In other words, an angle in radians is the length of arc formed by the angle on a unit circle. The term *radian* comes from the phrase *radial angle*. Two nineteenth century scientists, mathematician Thomas Muir and physicist James Thomson, are credited with the development of the radian as a unit of angular measure, although the concept originated over 100 years earlier.

On a circle, the length of an arc is proportional to the radius of the circle. Thus, for a specific angle θ, as shown in Figure 7(b), the ratio of the length of arc s to the radius r of the circle is the same, regardless of the radius of the circle. This allows us to define radian measure on a circle of arbitrary radius as follows:

$$\text{Radian measure of } \theta = \frac{\text{Length of arc}}{\text{Radius}} = \frac{s}{r}.$$

Note that the formula gives the same radian measure of an angle, regardless of the size of the circle, and that one radian results when the angle cuts an arc on the circle equal in length to the radius of the circle.

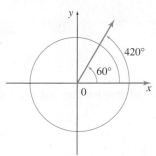

FIGURE 6

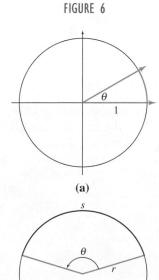

(a)

(b)

FIGURE 7

Since the circumference of a circle is 2π times the radius of the circle, the radius could be marked off 2π times around the circle. Therefore, an angle of 360°—that is, a complete circle—cuts off an arc equal in length to 2π times the radius of the circle, or

$$360° = 2\pi \text{ radians.}$$

This result gives a basis for comparing degree and radian measure.

Since an angle of 180° is half the size of an angle of 360°, an angle of 180° would have half the radian measure of an angle of 360°, or

$$180° = \frac{1}{2}(2\pi) \text{ radians} = \pi \text{ radians.}$$

Degree and Radians

$$180° = \pi \text{ radians}$$

Since π radians = 180°, divide both sides by π to find the degree measure of 1 radian.

1 Radian

$$1 \text{ radian} = \left(\frac{180°}{\pi}\right)$$

This quotient is approximately 57.29578°. Since 180° = π radians, we can find the radian measure of 1° by dividing by 180° on both sides.

1 Degree

$$1° = \frac{\pi}{180} \text{ radians}$$

One degree is approximately equal to 0.0174533 radians.

Graphing calculators and many scientific calculators have the capability of changing from degree to radian measure or from radian to degree measure. If your calculator has this capability, you can practice using it with the angle measures in Example 1. *The most important thing to remember when using a calculator to work with angle measures is to be sure the calculator mode is set for degrees or radians, as appropriate.*

EXAMPLE 1 Equivalent Angles

Convert degree measures to radians and radian measure to degrees.

(a) 45°

SOLUTION Since $1° = \pi/180$ radians,

$$45° = 45\left(\frac{\pi}{180}\right) \text{ radians} = \frac{45\pi}{180} \text{ radians} = \frac{\pi}{4} \text{ radians.}$$

The word *radian* is often omitted, so the answer could be written as just $45° = \pi/4$.

YOUR TURN 1 (a) Convert $210°$ to radians. (b) Convert $3\pi/4$ radians to degrees.

(b) $\dfrac{9\pi}{4}$

SOLUTION Since 1 radian $= 180°/\pi$,

$$\frac{9\pi}{4} \text{ radians} = \frac{9\pi}{4}\left(\frac{180°}{\pi}\right) = 405°. \qquad \textbf{TRY YOUR TURN 1}$$

The following table shows the equivalent radian and degree measure for several angles that we will encounter frequently.

Degrees and Radians of Common Angles								
Degrees	$0°$	$30°$	$45°$	$60°$	$90°$	$180°$	$270°$	$360°$
Radians	0	$\pi/6$	$\pi/4$	$\pi/3$	$\pi/2$	π	$3\pi/2$	2π

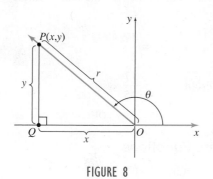

FIGURE 8

The Trigonometric Functions

To define the six basic trigonometric functions, we start with an angle θ in standard position, as shown in Figure 8. Next, we choose an arbitrary point P having coordinates (x, y), located on the terminal side of angle θ. (The point P must not be the vertex of θ.)

Drawing a line segment perpendicular to the x-axis from P to point Q sets up a right triangle having vertices at O (the origin), P, and Q. The distance from P to O is r. Since the distance from P to O can never be negative, $r > 0$. The six **trigonometric functions** of angle θ are defined as follows.

Trigonometric Functions

Let (x, y) be a point other than the origin on the terminal side of an angle θ in standard position. Let r be the distance from the origin to (x, y). Then

$$\text{sine } \theta = \sin \theta = \frac{y}{r} \qquad\qquad \text{cosecant } \theta = \csc \theta = \frac{r}{y} \quad (y \neq 0)$$

$$\text{cosine } \theta = \cos \theta = \frac{x}{r} \qquad\qquad \text{secant } \theta = \sec \theta = \frac{r}{x} \quad (x \neq 0)$$

$$\text{tangent } \theta = \tan \theta = \frac{y}{x} \quad (x \neq 0) \qquad \text{cotangent } \theta = \cot \theta = \frac{x}{y} \quad (y \neq 0).$$

From these definitions, it is easy to prove the following elementary trigonometric identities.

Elementary Trigonometric Identities

$$\csc \theta = \frac{1}{\sin \theta} \qquad \sec \theta = \frac{1}{\cos \theta} \qquad \cot \theta = \frac{1}{\tan \theta}$$

$$\tan \theta = \frac{\sin \theta}{\cos \theta} \qquad \cot \theta = \frac{\cos \theta}{\sin \theta} \qquad \sin^2 \theta + \cos^2 \theta = 1$$

These identities are meaningless when the denominator is zero.

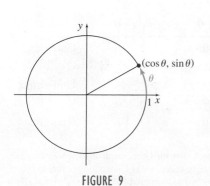

FIGURE 9

If we let $r = 1$ in the definitions of the trigonometric functions, then we can think of θ as the length of the arc, and $\cos \theta$ and $\sin \theta$ as the x- and y-coordinates, respectively, of a point on the unit circle, as shown in Figure 9.

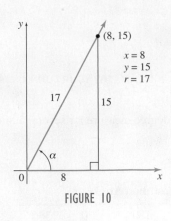

FIGURE 10

EXAMPLE 2 Values of Trigonometric Functions

The terminal side of an angle α (the Greek letter *alpha*) goes through the point $(8, 15)$. Find the values of the six trigonometric functions of angle α.

SOLUTION Figure 10 shows angle α with terminal side through point $(8, 15)$ and the triangle formed by dropping a perpendicular from the point $(8, 15)$. To find the distance r, use the Pythagorean theorem:* In a triangle with a right angle, if the longest side of the triangle (called the hypotenuse) is r and the shorter sides are x and y, then

$$r^2 = x^2 + y^2,$$

or

$$r = \sqrt{x^2 + y^2}.$$

(Recall that $\sqrt{b}$ represents the *positive* square root of b.)

Substituting the known values $x = 8$ and $y = 15$ in the equation gives

$$r = \sqrt{8^2 + 15^2} = \sqrt{64 + 225} = \sqrt{289} = 17.$$

We have $x = 8$, $y = 15$, and $r = 17$. The values of the six trigonometric functions of angle α are found by using the definitions.

$$\sin \alpha = \frac{y}{r} = \frac{15}{17} \qquad \tan \alpha = \frac{y}{x} = \frac{15}{8} \qquad \sec \alpha = \frac{r}{x} = \frac{17}{8}$$

$$\cos \alpha = \frac{x}{r} = \frac{8}{17} \qquad \cot \alpha = \frac{x}{y} = \frac{8}{15} \qquad \csc \alpha = \frac{r}{y} = \frac{17}{15}$$

YOUR TURN 2 Find the values of the six trigonometric functions of an angle α whose terminal side goes through the point $(9, 40)$.

TRY YOUR TURN 2

EXAMPLE 3 Values of Trigonometric Functions

Find the values of the six trigonometric functions for an angle of $\pi/2$.

SOLUTION Select any point on the terminal side of an angle of measure $\pi/2$ radians (or $90°$). See Figure 11. Selecting the point $(0, 1)$ gives $x = 0$ and $y = 1$. Check that $r = 1$ also. Then

$$\sin \frac{\pi}{2} = \frac{1}{1} = 1 \qquad \cot \frac{\pi}{2} = \frac{0}{1} = 0$$

$$\cos \frac{\pi}{2} = \frac{0}{1} = 0 \qquad \csc \frac{\pi}{2} = \frac{1}{1} = 1.$$

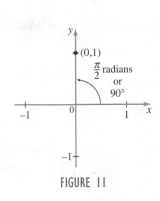

FIGURE 11

The values of $\tan(\pi/2)$ and $\sec(\pi/2)$ are undefined because the denominator is 0.

Methods similar to the procedure in Example 3 can be used to find the values of the six trigonometric functions for the angles with measure 0, π, and $3\pi/2$. These results are summarized in the following table. The table shows that the results for 2π are the same as those for 0.

Trigonometric Functions at Multiples of $\pi/2$							
θ (in radians)	θ (in degrees)	$\sin \theta$	$\cos \theta$	$\tan \theta$	$\cot \theta$	$\sec \theta$	$\csc \theta$
0	0°	0	1	0	Undefined	1	Undefined
$\pi/2$	90°	1	0	Undefined	0	Undefined	1
π	180°	0	-1	0	Undefined	-1	Undefined
$3\pi/2$	270°	-1	0	Undefined	0	Undefined	-1
2π	360°	0	1	0	Undefined	1	Undefined

*Although one of the most famous theorems in mathematics is named after the Greek mathematician Pythagoras, there is much evidence that the relationship between the sides of a right triangle was known long before his time. The Babylonian mathematical tablet identified as *Plimpton 322* has been interpreted by many to be essentially a list of *Pythagorean triples*—sets of three numbers a, b, and c that satisfy the equation $a^2 + b^2 = c^2$.

NOTE When considering the trigonometric functions, it is customary to use x (rather than θ) for the domain elements, as we did with earlier functions, and to write $y = \sin x$ instead of $y = \sin \theta$.

Special Angles
The values of the trigonometric functions for most angles must be found by using a calculator with trigonometric keys. For a few angles called **special angles**, however, the function values can be found exactly. These values are found with the aid of two kinds of right triangles that will be described in this section.

30°–60°–90° Triangle
In a right triangle having angles of 30°, 60°, and 90°, the hypotenuse is always twice as long as the shortest side, and the middle side has a length that is $\sqrt{3}$ times as long as that of the shortest side. Also, the shortest side is opposite the 30° angle.

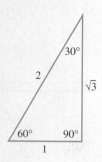

EXAMPLE 4 Values of Trigonometric Functions

Find the values of the trigonometric functions for an angle of $\pi/6$ radians.

SOLUTION Since $\pi/6$ radians $= 30°$, find the necessary values by placing a 30° angle in standard position, as in Figure 12. Choose a point P on the terminal side of the angle so that $r = 2$. From the description of 30°$-$60°$-$90° triangles, P will have coordinates $(\sqrt{3}, 1)$, with $x = \sqrt{3}$, $y = 1$, and $r = 2$. Using the definitions of the trigonometric functions gives the following results.

$$\sin\frac{\pi}{6} = \frac{1}{2} \qquad \tan\frac{\pi}{6} = \frac{1}{\sqrt{3}} = \frac{\sqrt{3}}{3} \qquad \sec\frac{\pi}{6} = \frac{2}{\sqrt{3}} = \frac{2\sqrt{3}}{3}$$

$$\cos\frac{\pi}{6} = \frac{\sqrt{3}}{2} \qquad \cot\frac{\pi}{6} = \sqrt{3} \qquad \csc\frac{\pi}{6} = 2$$

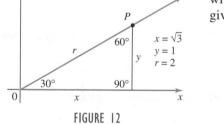

FIGURE 12

We can find the trigonometric function values for 45° angles by using the properties of a right triangle having two sides of equal length.

45°–45°–90° Triangle
In a right triangle having angles of 45°, 45°, and 90°, the hypotenuse has a length that is $\sqrt{2}$ times as long as the length of either of the shorter (equal) sides.

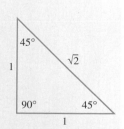

For a derivation of the properties of the 30°–60°–90° and 45°–45°–90° triangles, see Exercises 74 and 75.

EXAMPLE 5 Values of Trigonometric Functions

Find the trigonometric function values for an angle of $-\pi/4$.

SOLUTION Place an angle of $-\pi/4$ radians, or $-45°$, in standard position, as in Figure 13. Choose point P on the terminal side so that $r = \sqrt{2}$. By the description of $45°-45°-90°$ triangles, P has coordinates $(1, -1)$, with $x = 1$, $y = -1$, and $r = \sqrt{2}$.

$$\sin\left(-\frac{\pi}{4}\right) = -\frac{1}{\sqrt{2}} = -\frac{\sqrt{2}}{2} \quad \tan\left(-\frac{\pi}{4}\right) = -1 \quad \sec\left(-\frac{\pi}{4}\right) = \sqrt{2}$$

$$\cos\left(-\frac{\pi}{4}\right) = \frac{1}{\sqrt{2}} = \frac{\sqrt{2}}{2} \quad \cot\left(-\frac{\pi}{4}\right) = -1 \quad \csc\left(-\frac{\pi}{4}\right) = -\sqrt{2}$$

YOUR TURN 3 Find the trigonometric function values for an angle of $7\pi/6$.

TRY YOUR TURN 3

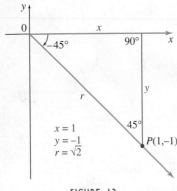

FIGURE 13

For angles other than the special angles of $30°$, $45°$, $60°$, and their multiples, a calculator should be used. Many calculators have keys labeled sin, cos, and tan. To get the other trigonometric functions, use the fact that $\sec x = 1/\cos x$, $\csc x = 1/\sin x$, and $\cot x = 1/\tan x$. (The x^{-1} key is also useful here.)

CAUTION Whenever you use a calculator to compute trigonometric functions, check whether the calculator is set on radians or degrees. If you want one and your calculator is set on the other, you will get erroneous answers. Most calculators have a way of switching back and forth; check the calculator manual for details. On the TI-84 Plus, press the MODE button and then select Radian or Degree.

EXAMPLE 6 Values of Trigonometric Functions

Use a calculator to verify the following results.

(a) $\sin 10° \approx 0.1736$
(b) $\cos 48° \approx 0.6691$

YOUR TURN 4 Use a calculator to find each of the following.
(a) cos 6° **(b)** sec 4

(c) $\tan 82° \approx 7.1154$
(d) $\sin 0.2618 \approx 0.2588$

(e) $\cot 1.2043 = 1/\tan 1.2043 \approx 1/2.6053 \approx 0.3838$

(f) $\sec 0.7679 = 1/\cos 0.7679 \approx 1/0.71937 \approx 1.3901$

TRY YOUR TURN 4

EXAMPLE 7 Values of Trigonometric Functions

Find all values of x between 0 and 2π that satisfy each of the following equations.

(a) $\sin x = 1/2$

SOLUTION The sine function is positive in quadrants I and II. In quadrant I, we draw a triangle with an angle whose sine is $1/2$, as shown in Figure 14 (a). We recognize this triangle as the $30°-60°-90°$ triangle, with angle $x = \pi/6$. In Figure 14 (b), we show the same triangle in quadrant II. The angle is now $x = \pi - \pi/6 = 5\pi/6$. There are two solutions between 0 and 2π, namely, $x = \pi/6$ and $5\pi/6$.

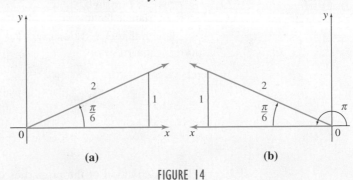

(a)   (b)

FIGURE 14

(b) $\sec x = -2$

SOLUTION The secant function is negative in quadrants II and III. In quadrant II, we draw a triangle with an angle whose secant is -2, as shown in Figure 15 (a). We recognize the 30°–60°–90° triangle once again, with angle $x = \pi - \pi/3 = 2\pi/3$. In Figure 15 (b), we show the same triangle in quadrant III. The angle is now $x = \pi + \pi/3 = 4\pi/3$. There are two solutions between 0 and 2π, namely, $x = 2\pi/3$ and $4\pi/3$.

TRY YOUR TURN 5

YOUR TURN 5 Find all values of x between 0 and 2π that satisfy the equation $\cos x = -\sqrt{2}/2$.

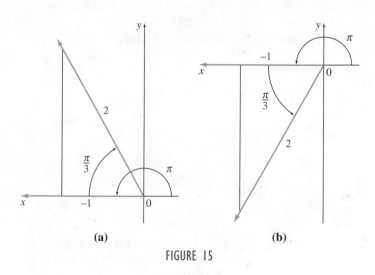

(a)   (b)

FIGURE 15

Graphs of the Trigonometric Functions
Because of the way the trigonometric functions are defined (using a circle), the same function values will be obtained for any two angles that differ by 2π radians (or 360°). For example,

$$\sin(x + 2\pi) = \sin x \quad \text{and} \quad \cos(x + 2\pi) = \cos x$$

for any value of x. Because of this property, the trigonometric functions are *periodic functions.*

Periodic Function

A function $y = f(x)$ is **periodic** if there exists a positive real number a such that

$$f(x) = f(x + a)$$

for all values of x in the domain of the function. The smallest positive value of a is called the **period** of the function.*

Intuitively, a function with period a repeats itself over intervals of length a. Once we know what the graph looks like over one period of length a, we know what the entire graph looks like by simply repeating. Because sine is periodic with period 2π, the graph is found by first finding the graph on the interval between 0 and 2π and then repeating as many times as necessary.

To find values of $y = \sin x$ for values of x between 0 and 2π, think of a point moving counterclockwise around a circle, tracing out an arc for angle x. The value of $\sin x$ gradually increases from 0 to 1 as x increases from 0 to $\pi/2$. The value of $\sin x$ then decreases back to 0 as x goes from $\pi/2$ to π. For $\pi < x < 2\pi$, $\sin x$ is negative. A few typical values from

*Some authors define the period of the function as any value of a that satisfies $f(x) = f(x + a)$.

these intervals are given in the following table, where decimals have been rounded to the nearest tenth.

				Values of the Sine Function					
x	0	$\pi/4$	$\pi/2$	$3\pi/4$	π	$5\pi/4$	$3\pi/2$	$7\pi/4$	2π
$\sin x$	0	0.7	1	0.7	0	-0.7	-1	-0.7	0

Plotting the points from the table of values and connecting them with a smooth curve gives the solid portion of the graph in Figure 16. Since $y = \sin x$ is periodic, the graph continues in both directions indefinitely, as suggested by the dashed lines. The solid portion of the graph in Figure 16 gives the graph over one period.

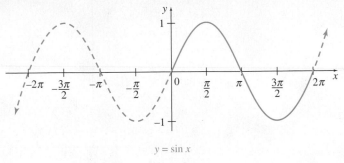

$y = \sin x$

FIGURE 16

The graph of $y = \cos x$ in Figure 17 below can be found in much the same way. Again, the period is 2π. (These graphs could also be drawn using a graphing calculator or a computer.)

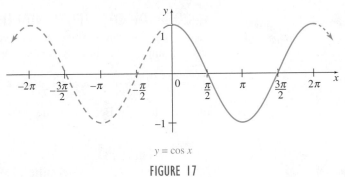

$y = \cos x$

FIGURE 17

Finally, Figure 18 shows the graph of $y = \tan x$. Since $\tan x$ is undefined (because of zero denominators) for $x = \pi/2, 3\pi/2, -\pi/2$, and so on, the graph has vertical asymptotes at these values. As the graph suggests, the tangent function is periodic, with a period of π.

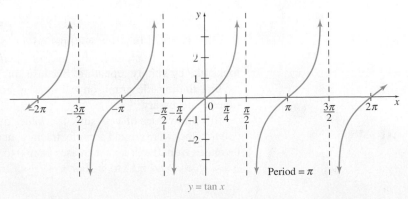

$y = \tan x$

FIGURE 18

The graphs of the secant, cosecant, and cotangent functions are not used as often as these three, so they are not given here.

Translating Graphs of Sine and Cosine Functions

In an earlier section we saw that the graph of the function $y = f(x + c) + d$ was simply the graph of $y = f(x)$ translated c units horizontally and d units vertically. The same facts hold true with trigonometric functions. The constants a, b, c, and d affect the graphs of the functions $y = a \sin(bx + c) + d$ and $y = a \cos(bx + c) + d$ in a similar manner.

In addition, the constants a, b, and c have particular properties. Since the sine and cosine functions range between -1 and 1, the value of a, whose absolute value is called the **amplitude**, can be interpreted as half the difference between the maximum and minimum values of the function.

The period of the function is determined by the constant b, which we will assume to be greater than 0. Recall that the period of both $\sin x$ and $\cos x$ is 2π. The value of $b > 0$ will increase or decrease the period, depending on its value. A similar phenomenon occurs when $b < 0$, but it is not covered in this textbook. Thus, the graph of $y = \sin(bx)$ will look like that of $y = \sin x$, but with a period of $T = 2\pi/b$. The results are similar for $y = \cos(bx)$.

The quantity c/b is called the **phase shift** and corresponds to the number of units that the graph of $\sin bx$ or $\cos bx$ is shifted horizontally. The constant d determines the vertical shift of $\sin x$ or $\cos x$.

EXAMPLE 8 **Graphing Trigonometric Functions**

Graph each function.

(a) $y = \sin 3x$

SOLUTION The graph of this function has amplitude $a = 1$ and no vertical or horizontal shifts. The period of this function is $T = 2\pi/b = 2\pi/3$. Hence, the graph of $y = \sin 3x$ is the same as $y = \sin x$ except that the period is different. See Figure 19.

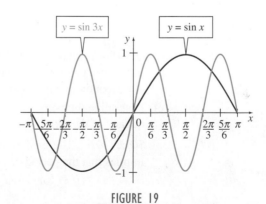

FIGURE 19

(b) $f(x) = 4 \cos\left(\dfrac{1}{2}x + \dfrac{\pi}{4}\right) - 1$

SOLUTION The amplitude is $a = 4$. The graph of $f(x)$ is shifted down 1 unit vertically. The phase shift is $c/b = (\pi/4)/(1/2) = \pi/2$. This shifts the graph $\pi/2$ units to the left, relative to the graph $g(x) = \cos((1/2)x)$. The period of $f(x)$ is $2\pi/(1/2) = 4\pi$. Making these translations on $y = \cos x$ leads to Figure 20.

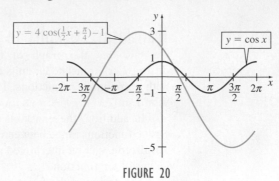

FIGURE 20

EXAMPLE 9 Music

A change in pressure on the eardrum occurs when a pure musical tone is played. For some tones, the pressure on the eardrum follows the sine curve

$$P(t) = 0.004 \sin\left(2\pi ft + \frac{\pi}{7}\right),$$

where P is the pressure in pounds per square foot at time t seconds and f is the frequency on the sound wave in cycles per second. *Source: The Physics and Psychophysics of Music: An Introduction.* When $P(t)$ is positive there is an increase in pressure and the eardrum is pushed inward; when $P(t)$ is negative there is a decrease in pressure and the eardrum is pushed outward.

(a) Graph the pressure on the eardrum for Middle C, which has a frequency of $f = 261.63$ cycles per second, on $[0, 0.005]$.

SOLUTION A graphing calculator graph of the function $P(t) = 0.004 \sin(2\pi ft + \pi/7) = 0.004 \sin(523.26\pi t + \pi/7)$ is given in Figure 21.

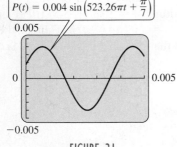

FIGURE 21

(b) Determine analytically the values of t for which $P = 0$ on $[0, 0.005]$.

SOLUTION Since the sine function is zero for multiples of π, we can determine the value(s) of t where $P = 0$ by setting $523.26\pi t + \pi/7 = n\pi$, where n is an integer, and solving for t. After some algebraic manipulations,

$$t = \frac{n - \dfrac{1}{7}}{523.26}$$

and $P = 0$ when $n = 0, \pm 1, \pm 2, \ldots$. However, only values of $n = 1$ or $n = 2$ produce values of t that lie in the interval $[0, 0.005]$. Thus, $P = 0$ when $t \approx 0.0016$ and 0.0035, corresponding to $n = 1$ and $n = 2$, respectively.

(c) Determine the period T of $P(t)$. What is the relationship between the period and frequency of the tone?

SOLUTION The period is $T = 2\pi/b = 2\pi/(523.26\pi) = 1/261.63 \approx 0.004$. This implies that the period of the pressure equation is the reciprocal of the frequency. That is, $T = 2\pi/b = 2\pi/(2\pi f) = 1/f$.

TECHNOLOGY

EXAMPLE 10 Sunrise

The table on the next page lists the approximate number of minutes after midnight, Eastern Standard Time, that the sun rises in Boston for specific days of the year. *Source: The Old Farmer's Almanac.*

(a) Plot the data. Is it reasonable to assume that the times of sunrise are periodic?

SOLUTION Figure 22 shows a graphing calculator plot of the data. Because of the cyclical nature of the days of the year, it is reasonable to assume that the data are periodic.

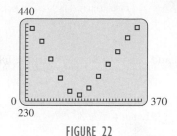

440

230 — 370

FIGURE 22

| Time of Boston Sunrise | |
Day of the Year	Sunrise (minutes after midnight)
21	428
52	393
81	345
112	293
142	257
173	247
203	266
234	298
265	331
295	365
326	403
356	431

$s(t) = 92.1414 \sin(0.016297t + 1.80979) + 342.934$

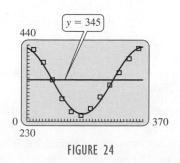

440

230 — 370

FIGURE 23

(b) Find a trigonometric function of the form $s(t) = a \sin(bt + c) + d$ that models this data when t is the day of the year and $s(t)$ is the number of minutes past midnight, Eastern Standard Time, that the sun rises. Use the data from the table.

SOLUTION The function $s(t)$, derived by a TI-84 Plus using the sine regression function under the STAT-CALC menu, is given by

$$s(t) = 92.1414 \sin(0.016297t + 1.80979) + 342.934.$$

Figure 23 shows that this function fits the data well.

(c) Estimate the time of sunrise for days 30, 90, and 240. Round answers to the nearest minute.

SOLUTION

$$s(30) = 92.1414 \sin(0.016297(30) + 1.80979) + 342.934$$
$$\approx 412 \text{ minutes} = 412/60 \text{ hours} \approx 6.867 \text{ hours}$$
$$= 6 \text{ hours} + 0.867(60) \text{ minutes} = 6:52 \text{ A.M.}$$

Similarly,

$$s(90) \approx 331 \text{ minutes} = 5:31 \text{ A.M.}$$
$$s(240) \approx 294 \text{ minutes} + 60 \text{ minutes (daylight savings)}$$
$$= 5:54 \text{ A.M.}$$

(d) Estimate the days of the year that the sun rises at 5:45 A.M.

SOLUTION Figure 24 shows the graphs of $s(t)$ and $y = 345$ (corresponding to a sunrise of 5:45 A.M.). These graphs first intersect on day 80. However, because of daylight savings time, to find the second value we find where the graphs of $s(t)$ and $y = 345 - 60 = 285$ intersect. These graphs intersect on day 233. Thus, the sun rises at approximately 5:45 A.M. on the 80th and 233rd days of the year.

(e) What is the period of the function found in part (b)?

SOLUTION The period of the function given above is $T = 2\pi/0.016297 \approx 385.5$ days. This is close to the true period of about 365 days. The discrepancy could be due to many factors. For example, the underlying function may be more complex than a simple sine function.

$y = 345$

440

230 — 370

FIGURE 24

13.1 EXERCISES

Convert the following degree measures to radians. Leave answers as multiples of π.

1. $60°$
2. $90°$
3. $150°$
4. $135°$

5. $270°$
6. $320°$
7. $495°$
8. $510°$

Convert the following radian measures to degrees.

9. $\dfrac{5\pi}{4}$
10. $\dfrac{2\pi}{3}$
11. $-\dfrac{13\pi}{6}$
12. $-\dfrac{\pi}{4}$

13. $\dfrac{8\pi}{5}$
14. $\dfrac{5\pi}{9}$
15. $\dfrac{7\pi}{12}$
16. 5π

Find the values of the six trigonometric functions for the angles in standard position having the points in Exercises 17–20 on their terminal sides.

17. $(-3, 4)$
18. $(-12, -5)$

19. $(7, -24)$
20. $(20, 15)$

In quadrant *I*, *x*, *y*, and *r* are all positive, so that all six trigonometric functions have positive values. In quadrant II, *x* is negative and *y* is positive (*r* is always positive). Thus, in quadrant II, sine is positive, cosine is negative, and so on. For Exercises 21–24, complete the following table of values for the signs of the trigonometric functions.

Quadrant of θ	$\sin \theta$	$\cos \theta$	$\tan \theta$	$\cot \theta$	$\sec \theta$	$\csc \theta$
21. I	+					
22. II						
23. III						
24. IV						

For Exercises 25–32, complete the following table. Use the 30°–60°–90° and 45°–45°–90° triangles. Do not use a calculator.

θ	$\sin \theta$	$\cos \theta$	$\tan \theta$	$\cot \theta$	$\sec \theta$	$\csc \theta$
25. 30°	$1/2$	$\sqrt{3}/2$			$2\sqrt{3}/3$	
26. 45°			1	1		
27. 60°		$1/2$	$\sqrt{3}$		2	
28. 120°	$\sqrt{3}/2$		$-\sqrt{3}$			$2\sqrt{3}/3$
29. 135°	$\sqrt{2}/2$	$-\sqrt{2}/2$			$-\sqrt{2}$	$\sqrt{2}$
30. 150°		$-\sqrt{3}/2$	$-\sqrt{3}/3$			2
31. 210°	$-1/2$		$\sqrt{3}/3$	$\sqrt{3}$		-2
32. 240°	$-\sqrt{3}/2$	$-1/2$			-2	$-2\sqrt{3}/3$

Find the following function values without using a calculator.

33. $\sin \dfrac{\pi}{3}$
34. $\cos \dfrac{\pi}{6}$
35. $\tan \dfrac{\pi}{4}$
36. $\cot \dfrac{\pi}{3}$

37. $\csc \dfrac{\pi}{6}$
38. $\sin \dfrac{3\pi}{2}$
39. $\cos 3\pi$
40. $\sec \pi$

41. $\sin \dfrac{7\pi}{4}$
42. $\tan \dfrac{5\pi}{2}$
43. $\sec \dfrac{5\pi}{4}$
44. $\cos 5\pi$

45. $\cot -\dfrac{3\pi}{4}$
46. $\tan -\dfrac{5\pi}{6}$
47. $\sin -\dfrac{7\pi}{6}$
48. $\cos -\dfrac{\pi}{6}$

Find all values of *x* between 0 and 2π that satisfy each of the following equations.

49. $\cos x = 1/2$
50. $\sin x = -1/2$

51. $\tan x = -1$
52. $\tan x = \sqrt{3}$

53. $\sec x = -2/\sqrt{3}$
54. $\sec x = \sqrt{2}$

Use a calculator to find the following function values.

55. $\sin 39°$
56. $\cos 67°$

57. $\tan 123°$
58. $\tan 54°$

59. $\sin 0.3638$
60. $\tan 1.0123$

61. $\cos 1.2353$
62. $\sin 1.5359$

Find the amplitude (*a*) and period (*T*) of each function.

63. $f(x) = \cos(3x)$

64. $g(t) = 2 \sin\left(\dfrac{\pi}{4}t + 2\right)$

65. $s(t) = 3 \sin(880\pi t - 7)$

Graph each function defined as follows over a two-period interval.

66. $y = 2 \sin x$
67. $y = 2 \cos x$

68. $y = -\sin x$
69. $y = -\dfrac{1}{2}\cos x$

70. $y = 2 \cos\left(3x - \dfrac{\pi}{4}\right) + 1$
71. $y = 4 \sin\left(\dfrac{1}{2}x + \pi\right) + 2$

72. $y = \dfrac{1}{2}\tan x$
73. $y = -3 \tan x$

74. Consider the triangle shown on the next page, in which the three angles θ are equal and all sides have length 2.

   a. Using the fact that the sum of the angles in a triangle is 180°, what are the measures of the three equal angles θ?

   b. Suppose the triangle is cut in half as shown by a vertical line. What are the measures of the angles in the blue triangle on the left?

   c. What are the measures of the sides of the blue triangle on the left? (*Hint*: Once you've found the length of the base, use the Pythagorean Theorem to find the height.)

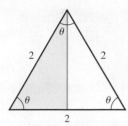

75. Consider the right triangle shown, in which the two sides have length 1.

    a. Using the Pythagorean Theorem, what is the length of the hypotenuse?

    b. Using the fact that the sum of the angles in a triangle is 180°, what are the measures of the three angles?

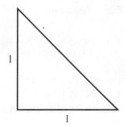

APPLICATIONS

Business and Economics

76. Sales Sales of snowblowers are seasonal. Suppose the sales of snowblowers in one region of the country are approximated by

$$S(t) = 500 + 500 \cos\left(\frac{\pi}{6}t\right),$$

where t is time in months, with $t = 0$ corresponding to November. Find the sales for a–e.

    a. November   **b.** January   **c.** February

    d. May   **e.** August   **f.** Graph $y = S(t)$.

77. Electricity Consumption The amount of electricity (in trillion BTUs) consumed by U.S. residential customers in 2009 is given in the following table. *Source: Energy Information Administration*.

Month	Electricity (trillion BTUs)
January	464
February	394
March	362
April	312
May	321
June	390
July	469
August	472
September	393
October	336
November	316
December	421

a. Plot the data, letting $t = 1$ correspond to January, $t = 2$ to February, and so on. Is it reasonable to assume that electrical consumption is periodic?

b. Use a calculator with trigonometric regression to find a trigonometric function of the form

$$C(t) = a \sin(bt + c) + d$$

that models this data when t is the month and $C(t)$ is the amount of electricity consumed (in trillion BTUs). Graph the function on the same calculator window as the data.

c. Determine the period, T, of the function found in part b. Discuss the reasonableness of this period.

d. Use the function from part b to estimate the consumption for the month of September, and compare it to the actual value.

Life Sciences

78. Monkey Eyes In a study of how monkeys' eyes pursue a moving object, an image was moved sinusoidally through a monkey's field of vision with an amplitude of $2°$ and a period of 0.350 seconds. *Source: Journal of Neurophysiology*.

    a. Find an equation giving the position of the image in degrees as a function of time in seconds.

    b. After how many seconds does the image reach its maximum amplitude?

    c. What is the position of the object after 2 seconds?

79. Transylvania Hypothesis The "Transylvania hypothesis" claims that the full moon has an effect on health-related behavior. A study investigating this effect found a significant relationship between the phase of the moon and the number of general practice consultations nationwide, given by

$$y = 100 + 1.8 \cos\left[\frac{(t - 6)\pi}{14.77}\right],$$

where y is the number of consultations as a percentage of the daily mean and t is the days since the last full moon. *Source: Family Practice*.

    a. What is the period of this function? What is the significance of this period?

    b. There was a full moon on October 11, 2011. On what day in October 2011 does this formula predict the maximum number of consultations? What percent increase would be predicted for that day?

    c. What does the formula predict for October 28, 2011?

80. Air Pollution The amount of pollution in the air fluctuates with the seasons. It is lower after heavy spring rains and higher after periods of little rain. In addition to this seasonal fluctuation, the long-term trend in many areas is upward. An idealized graph of this situation is shown in the figure on the next page. Trigonometric functions can be used to describe the fluctuating part of the pollution levels. Powers of the number e can be used to show the long-term growth. In fact, the pollution level in a certain area might be given by

$$P(t) = 7(1 - \cos 2\pi t)(t + 10) + 100e^{0.2t},$$

where t is time in years, with $t = 0$ representing January 1 of the base year. Thus, July 1 of the same year would be represented by $t = 0.5$, while October 1 of the following year would be

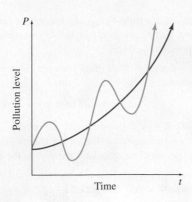

represented by $t = 1.75$. Find the pollution levels on the following dates.

a. January 1, base year

b. July 1, base year

c. January 1, following year

d. July 1, following year

81. Air Pollution Using a computer or a graphing calculator, sketch the function for air pollution given in Exercise 80 over the interval $[0, 6]$.

Physical Sciences

Light Rays When a light ray travels from one medium, such as air, to another medium, such as water or glass, the speed of the light changes, and the direction that the ray is traveling changes. (This is why a fish under water is in a different position from the place at which it appears to be.) These changes are given by Snell's law,

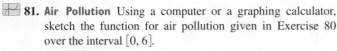

$$\frac{c_1}{c_2} = \frac{\sin \theta_1}{\sin \theta_2},$$

where c_1 is the speed in the first medium, c_2 is the speed in the second medium, and θ_1 and θ_2 are the angles shown in the figure.

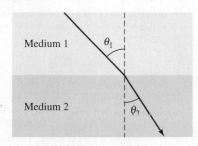

Medium 1 — If this medium is less dense, light travels at a faster speed, c_1.

Medium 2 — If this medium is more dense, light travels at a slower speed, c_2.

In Exercises 82 and 83, assume that $c_1 = 3 \times 10^8$ m per second, and find the speed of light in the second medium.

82. $\theta_1 = 39°, \theta_2 = 28°$

83. $\theta_1 = 46°, \theta_2 = 31°$

Sound Pure sounds produce single sine waves on an oscilloscope. Find the period of each sine wave in the photographs in Exercises 84 and 85. On the vertical scale each square

represents 0.5, and on the horizontal scale each square represents 30°.

84.

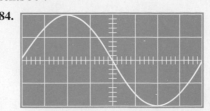

85.

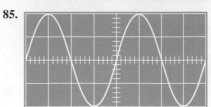

86. Sound Suppose the A key above Middle C is played as a pure tone. For this tone,

$$P(t) = 0.002 \sin(880\pi t),$$

where $P(t)$ is the change of pressure (in pounds per square foot) on a person's eardrum at time t (in seconds). *Source: The Physics and Psychophysics of Music: An Introduction.*

a. Graph this function on $[0, 0.003]$.

b. Determine analytically the values of t for which $P = 0$ on $[0, 0.003]$ and check graphically.

c. Determine the period T of $P(t)$ and the frequency of the A note.

87. Temperature The maximum afternoon temperature (in degrees Fahrenheit) in a given city is approximated by

$$T(t) = 60 - 30 \cos(t/2),$$

where t represents the month, with $t = 0$ representing January, $t = 1$ representing February, and so on. Use a calculator to find the maximum afternoon temperature for the following months.

a. February   **b.** April   **c.** September

d. July   **e.** December

88. Temperature A mathematical model for the temperature in Fairbanks is

$$T(t) = 37 \sin\left[\frac{2\pi}{365}(t - 101)\right] + 25,$$

where $T(t)$ is the temperature (in degrees Fahrenheit) on day t, with $t = 0$ corresponding to January 1 and $t = 364$ corresponding to December 31. *Source: Mathematics Teacher.* Use a calculator to estimate the temperature for a–d.

a. March 16 (Day 74)   **b.** May 2 (Day 121)

c. Day 250   **d.** Day 325

e. Find maximum and minimum values of T.

f. Find the period, T.

89. Sunset The number of minutes after noon, Eastern Standard Time, that the sun sets in Boston for specific days of the year is approximated in the following table. *Source: The Old Farmer's Almanac.*

Day of the Year	Sunset (minutes after noon)
21	283
52	323
81	358
112	393
142	425
173	445
203	434
234	396
265	343
295	292
326	257
356	255

a. Plot the data. Is it reasonable to assume that the times of sunset are periodic?

b. Use a calculator with trigonometric regression to find a trigonometric function of the form $s(t) = a \sin(bt + c) + d$ that models this data when t is the day of the year and $s(t)$ is the number of minutes past noon, Eastern Standard Time, that the sun sets.

c. Estimate the time of sunset for days 60, 120, 240. Round answers to the nearest minute. (*Hint:* Don't forget about daylight savings time.)

d. Use part b to estimate the days of the year that the sun sets at 6:00 P.M. In reality, the days are close to 82 and 290.

90. APPLY IT Cameras In the Kodak Customer Service Pamphlet AA-26, *Optical Formulas and Their Applications,* the near and far limits of the depth of field (how close or how far away an object can be placed and still be in focus) are given by

$$w_1 = \frac{u^2(\tan\theta)}{L + u(\tan\theta)} \quad \text{and} \quad w_2 = \frac{u^2(\tan\theta)}{L - u(\tan\theta)}.$$

In these equations, θ represents the angle between the lens and the "circle of confusion," which is the circular image on the film of a point that is not exactly in focus. (The pamphlet suggests letting $\theta = 1/30°$.) L is the diameter of the lens opening, which is found by dividing the focal length by the f-stop. (This is camera jargon you need not worry about here.) For this problem, let the focal length be 50 mm, or 0.05 m; if the lens is set at f/8, then $L = 0.05/8 = 0.00625$ m. Finally, u is the distance to the object being photographed. Find the near and far limits of the depth of field when the object being photographed is 6 m from the camera.

91. Measurement A surveyor standing 65 m from the base of a building measures the angle to the top of the building and finds it to be 42.8°. (See the figure.) Use trigonometry to find the height of the building.

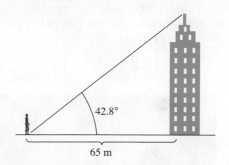

42.8°

65 m

92. Measurement Jenny Crum stands on a cliff at the edge of a canyon. On the opposite side of the canyon is another cliff equal in height to the one she is on. (See the figure.) By dropping a rock and timing its fall, she determines that it is 105 ft to the bottom of the canyon. She also determines that the angle to the base of the opposite cliff is 27°. How far is it to the opposite side of the canyon?

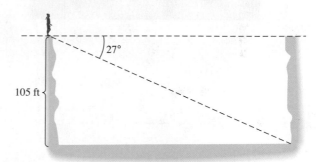

27°

105 ft

93. Whitewater Rafting A mathematics textbook author rafting down the Colorado River was told by a guide that the river dropped an average of 26 ft per mile as it ran through Cataract Canyon. Find the average angle of the river with the horizontal in degrees. (*Hint:* Find the tangent of the angle, and then use a calculator to find the angle where the tangent has that value. There are 5280 ft in a mile. Be sure your calculator is set on degrees.)

94. Computer Drawing A mathematics professor wanted to use a computer drawing program to draw a picture of a regular pentagon (a five-sided figure with sides of equal length and with equal angles). He first made a 1-in. base by drawing a line from $(0, 0)$ to $(1, 0)$. (See the figure.) He then needed to find the coordinates of the other three vertex points. Use trigonometry to find them. (*Hint:* The sum of the exterior angles of any polygon is 360°.)

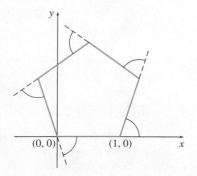

y

$(0, 0)$ $(1, 0)$ x

General Interest

95. Amusement Rides A proud father is attempting to take a picture of his daughters while they are riding on a merry-go-round. Horses on this particular ride move up and down as the ride progresses according to the function

$$h(t) = \sin\left(\frac{t}{\pi} - 2\right) + 4,$$

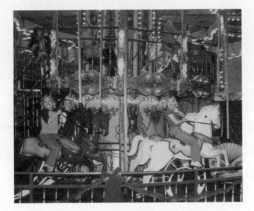

where $h(t)$ represents the height (in feet) of the horse's nose at time t, relative to the merry-go-round platform. However, because of safety fencing surrounding the ride, it is only possible to get a good picture when the height of the horse's nose is between 3.5 and 4 ft off the merry-go-round platform. Find the first time interval that the father has to take the picture.

YOUR TURN ANSWERS

1. (a) $7\pi/6$ (b) $135°$
2. $\sin\alpha = 40/41, \cos\alpha = 9/41, \tan\alpha = 40/9,$
   $\cot\alpha = 9/40, \sec\alpha = 41/9, \csc\alpha = 41/40$
3. $\sin(7\pi/6) = -1/2, \cos(7\pi/6) = -\sqrt{3}/2,$
   $\tan(7\pi/6) = 1/\sqrt{3} = \sqrt{3}/3, \cot(7\pi/6) = \sqrt{3},$
   $\sec(7\pi/6) = -2/\sqrt{3} = -2\sqrt{3}/3,$
   $\csc(7\pi/6) = -2.$
4. (a) 0.9945 (b) -1.5299
5. $3\pi/4, 5\pi/4$

13.2 Derivatives of Trigonometric Functions

APPLY IT How long must a ladder be to reach over a 9-foot-high fence and lean against a nearby building?
In Exercise 50 in this section, we will use trigonometry to answer this question.

In this section, we derive formulas for the derivatives of some of the trigonometric functions. All these derivatives can be found from the formula for the derivative of $y = \sin x$.

We will need to use the following identities, which are listed without proof, to find the derivatives of the trigonometric functions.

Basic Identities

$$\sin^2 x + \cos^2 x = 1$$

$$\tan x = \frac{\sin x}{\cos x}$$

$$\sin(x + y) = \sin x \cos y + \cos x \sin y$$

$$\sin(x - y) = \sin x \cos y - \cos x \sin y$$

$$\cos(x + y) = \cos x \cos y - \sin x \sin y$$

$$\cos(x - y) = \cos x \cos y + \sin x \sin y$$

The derivative of $y = \sin x$ also depends on the value of

$$\lim_{x \to 0} \frac{\sin x}{x}.$$

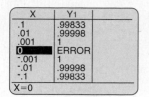

FIGURE 25

To estimate this limit, find the quotient $(\sin x)/x$ for various values of x close to 0. (Be sure that your calculator is set for radian measure.) For example, we used the TABLE feature of the TI-84 Plus calculator to get the values of this quotient shown in Figure 25 as x approaches 0 from either side. Note that, although the calculator shows the quotient equal to 1 for $x = \pm 0.001$, it is an approximation—the value is not exactly 1. Why does the calculator show ERROR when $x = 0$?

These results suggest, and it can be proved, that

$$\lim_{x \to 0} \frac{\sin x}{x} = 1.$$

In Example 1, this limit is used to obtain another limit. Then the derivative of $y = \sin x$ can be found.

EXAMPLE 1 Trigonometric Limit

Find $\lim\limits_{h \to 0} \dfrac{\cos h - 1}{h}$.

SOLUTION Use the limit above and some trigonometric identities.

$$\lim_{h \to 0} \frac{\cos h - 1}{h} = \lim_{h \to 0} \frac{(\cos h - 1)}{h} \cdot \frac{(\cos h + 1)}{(\cos h + 1)} \qquad \text{Multiply by } 1 = \frac{\cos h + 1}{\cos h + 1}.$$

$$= \lim_{h \to 0} \frac{\cos^2 h - 1}{h(\cos h + 1)}$$

$$= \lim_{h \to 0} \frac{-\sin^2 h}{h(\cos h + 1)} \qquad \cos^2 h - 1 = -\sin^2 h$$

$$= \lim_{h \to 0} (-\sin h)\left(\frac{\sin h}{h}\right)\left(\frac{1}{\cos h + 1}\right)$$

$$= (0)(1)\left(\frac{1}{1 + 1}\right) = 0$$

Therefore,

$$\lim_{h \to 0} \frac{\cos h - 1}{h} = 0.$$

FOR REVIEW
Recall from Section 3.1 on Limits: When taking the limit of a product, if the limit of each factor exists, the limit of the product is simply the product of the limits.

We can now find the derivative of $y = \sin x$ by using the general definition for the derivative of a function f given in Chapter 3:

$$f'(x) = \lim_{h \to 0} \frac{f(x + h) - f(x)}{h},$$

provided this limit exists. By this definition, the derivative of $f(x) = \sin x$ is

$$f'(x) = \lim_{h \to 0} \frac{\sin(x + h) - \sin x}{h}$$

$$= \lim_{h \to 0} \frac{\sin x \cdot \cos h + \cos x \cdot \sin h - \sin x}{h} \qquad \text{Identity for } \sin(x + h)$$

$$= \lim_{h \to 0} \frac{(\sin x \cdot \cos h - \sin x) + \cos x \cdot \sin h}{h} \qquad \text{Rearrange terms.}$$

$$= \lim_{h \to 0} \frac{\sin x(\cos h - 1) + \cos x \cdot \sin h}{h} \qquad \text{Factor.}$$

$$f'(x) = \lim_{h \to 0}\left(\sin x \frac{\cos h - 1}{h}\right) + \lim_{h \to 0}\left(\cos x \frac{\sin h}{h}\right) \qquad \text{Limit rule for sums}$$

$$= (\sin x)(0) + (\cos x)(1)$$

$$= \cos x.$$

This result is summarized below.

Derivative of $\sin x$

$$D_x(\sin x) = \cos x$$

We can use the chain rule to find derivatives of other sine functions, as shown in the following examples.

EXAMPLE 2   Derivatives of $\sin x$

Find the derivative of each function.

(a) $y = \sin 6x$

SOLUTION   By the chain rule,

$$\frac{dy}{dx} = (\cos 6x) \cdot D_x(6x)$$

$$= (\cos 6x) \cdot 6$$

$$\frac{dy}{dx} = 6 \cos 6x.$$

(b) $y = 5 \sin(9x^2 + 2) + \cos\left(\dfrac{\pi}{7}\right)$

SOLUTION   By the chain rule,

$$\frac{dy}{dx} = [5 \cos(9x^2 + 2)] \cdot D_x(9x^2 + 2) + 0 \qquad \cos\left(\frac{\pi}{7}\right) \text{ is a constant.}$$

$$= [5 \cos(9x^2 + 2)]18x$$

$$\frac{dy}{dx} = 90x \cos(9x^2 + 2). \qquad\qquad \text{TRY YOUR TURN 1} \blacksquare$$

YOUR TURN 1 Find the derivative of $y = 5 \sin(3x^4)$.

EXAMPLE 3   Chain Rule

Find $D_x(\sin^4 x)$.

SOLUTION   The expression $\sin^4 x$ means $(\sin x)^4$. By the chain rule,

$$D_x(\sin^4 x) = 4 \cdot \sin^3 x \cdot D_x(\sin x)$$

$$= 4 \sin^3 x \cos x. \qquad\qquad \text{TRY YOUR TURN 2} \blacksquare$$

YOUR TURN 2 Find the derivative of $y = 2 \sin^3(\sqrt{x})$.

The derivative of $y = \cos x$ is found from trigonometric identities and from the fact that $D_x(\sin x) = \cos x$. First, use the identity for $\sin(x - y)$ to get

$$\sin\left(\frac{\pi}{2} - x\right) = \sin\frac{\pi}{2} \cdot \cos x - \cos\frac{\pi}{2} \cdot \sin x$$

$$= 1 \cdot \cos x - 0 \cdot \sin x$$

$$= \cos x.$$

In the same way, $\cos\left(\dfrac{\pi}{2} - x\right) = \sin x$. Therefore,

$$D_x(\cos x) = D_x\left[\sin\left(\frac{\pi}{2} - x\right)\right].$$

By the chain rule,

$$D_x\left[\sin\left(\frac{\pi}{2} - x\right)\right] = \cos\left(\frac{\pi}{2} - x\right) \cdot D_x\left(\frac{\pi}{2} - x\right)$$

$$= \cos\left(\frac{\pi}{2} - x\right) \cdot (-1) \qquad \pi/2 \text{ is constant.}$$

$$= -\cos\left(\frac{\pi}{2} - x\right)$$

$$= -\sin x.$$

Derivative of $\cos x$

$$D_x(\cos x) = -\sin x$$

EXAMPLE 4 Derivatives of $\cos x$

Find each derivative.

(a) $D_x[\cos(3x)] = -\sin(3x) \cdot D_x(3x) = -3 \sin 3x$

(b) $D_x(\cos^4 x) = 4 \cos^3 x \cdot D_x(\cos x) = 4 \cos^3 x(-\sin x)$
$$= -4 \sin x \cos^3 x$$

(c) $D_x(3x \cos x)$

SOLUTION Use the product rule.

$$D_x(3x \cos x) = 3x(-\sin x) + (\cos x)(3)$$

$$= -3x \sin x + 3 \cos x \qquad \text{TRY YOUR TURN 3}$$

YOUR TURN 3 Find the derivative of $y = x \cos(x^2)$.

As mentioned in the list of basic identities at the beginning of this section, $\tan x = (\sin x)/\cos x$. The derivative of $y = \tan x$ can be found by using the quotient rule to find the derivative of $y = (\sin x)/\cos x$.

$$D_x(\tan x) = D_x\left(\frac{\sin x}{\cos x}\right) = \frac{\cos x \cdot D_x(\sin x) - \sin x \cdot D_x(\cos x)}{\cos^2 x}$$

$$= \frac{\cos x(\cos x) - \sin x(-\sin x)}{\cos^2 x}$$

$$= \frac{\cos^2 x + \sin^2 x}{\cos^2 x}$$

$$= \frac{1}{\cos^2 x} = \sec^2 x \qquad \cos^2 x + \sin^2 x = 1$$

The last step follows from the definitions of the trigonometric functions, which could be used to show that $1/\cos x = \sec x$. A similar calculation leads to the derivative of $\cot x$.

Derivatives of $\tan x$ and $\cot x$

$$D_x(\tan x) = \sec^2 x$$
$$D_x(\cot x) = -\csc^2 x$$

EXAMPLE 5 Derivatives of $\tan x$ and $\cot x$

Find each derivative.

(a) $D_x(\tan 9x) = \sec^2 9x \cdot D_x(9x) = 9 \sec^2 9x$

(b) $D_x(\cot^6 x) = 6 \cot^5 x \cdot D_x(\cot x) = -6 \cot^5 x \csc^2 x$

YOUR TURN 4 Find the derivative of $y = x \tan^2 x$.

(c) $D_x(\ln |6 \tan x|) = \dfrac{D_x(6 \tan x)}{6 \tan x} = \dfrac{6 \sec^2 x}{6 \tan x} = \dfrac{\sec^2 x}{\tan x}$ 　　TRY YOUR TURN 4

Using the facts that $\sec x = 1/\cos x$ and $\csc x = 1/\sin x$, it is possible to use the quotient rule to find the derivative of each of these functions. In Exercises 34 and 35 at the end of this section, you will be asked to verify the following.

Derivatives of $\sec x$ and $\csc x$

$$D_x \sec x = \sec x \tan x$$

$$D_x \csc x = -\csc x \cot x$$

EXAMPLE 6 Derivatives of $\sec x$ and $\csc x$

Find each derivative.

(a) $D_x(x^2 \sec x) = x^2 \sec x \tan x + 2x \sec x$

(b) $D_x(\csc e^{2x}) = -\csc e^{2x} \cot e^{2x} \cdot D_x(e^{2x})$

$$= -\csc e^{2x} \cot e^{2x} \cdot (2e^{2x})$$

$$= -2e^{2x} \csc e^{2x} \cot e^{2x}$$

YOUR TURN 5 Find the derivative of $y = \sec^2(\sqrt{x})$.

TRY YOUR TURN 5 ▮

EXAMPLE 7 Derivatives of Trigonometric Functions

Find the derivative of each function at the specified value of x.

(a) $f(x) = \sin(\pi e^x)$, when $x = 0$

SOLUTION Using the chain rule, the derivative of $f(x)$ is

$$f'(x) = \cos(\pi e^x) \cdot \pi e^x.$$

Thus,

$$f'(0) = \cos(\pi e^0) \cdot \pi e^0 = (-1)\pi(1) = -\pi.$$

(b) $g(x) = e^x \sin(\pi x)$, when $x = 0$

SOLUTION Using the product rule, the derivative of $g(x)$ is

$$g'(x) = e^x \cos(\pi x)\pi + \sin(\pi x)e^x.$$

Thus,

$$g'(0) = e^0 \cos(\pi 0)\pi + \sin(\pi 0)e^0$$

$$= 1 \cdot 1 \cdot \pi + 0 \cdot 1 = \pi.$$

(c) $h(x) = \tan(\cot x)$, when $x = \dfrac{\pi}{4}$

SOLUTION Using the chain rule, the derivative of $h(x)$ is

$$h'(x) = \sec^2[\cot(x)] \cdot [-\csc^2(x)].$$

Thus,

$$h'\left(\frac{\pi}{4}\right) = \sec^2\left[\cot\left(\frac{\pi}{4}\right)\right] \cdot \left[-\csc^2\left(\frac{\pi}{4}\right)\right] = \sec^2(1) \cdot (-2) \approx -6.851.$$

(d) $k(x) = \tan x \cot x$, when $x = \dfrac{\pi}{4}$

YOUR TURN 6 Find the derivative of $f(x) = \sin(\cos x)$ when $x = \pi/2$.

SOLUTION Since $k(x) = \tan x \cot x = \dfrac{\sin x}{\cos x} \cdot \dfrac{\cos x}{\sin x} = 1$, $k'(x) = 0$. In particular,

$$k'\left(\frac{\pi}{4}\right) = 0.$$

TRY YOUR TURN 6 ▮

EXAMPLE 8   Carbon Dioxide Levels

At Mauna Loa, Hawaii, atmospheric carbon dioxide levels in parts per million (ppm) have been measured regularly since 1958. The function defined by

$$L(t) = 0.022t^2 + 0.55t + 316 + 3.5 \sin(2\pi t)$$

can be used to model these levels, where t is in years and $t = 0$ corresponds to 1960. *Source: Greenhouse Earth.*

(a) Graph $L(t)$ on $[0, 30]$.

SOLUTION   Figure 26 shows a graphing calculator plot of the data. Notice that the overall trend is upwards. The annual fluctuations are modeled by the sine term in the function.

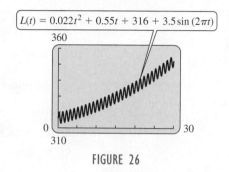

FIGURE 26

(b) Find $L(25)$, $L(35.5)$, and $L(50.2)$.

SOLUTION
$$L(25) = 0.022(25)^2 + 0.55(25) + 316 + 3.5 \sin(50\pi) = 343.5 \text{ ppm},$$
$$L(35.5) = 0.022(35.5)^2 + 0.55(35.5) + 316 + 3.5 \sin(71\pi) \approx 363.25 \text{ ppm},$$
$$L(50.2) = 0.022(50.2)^2 + 0.55(50.2) + 316 + 3.5 \sin(100.4\pi) \approx 402.38 \text{ ppm}$$

(c) Find $L'(50.2)$.

SOLUTION   Since $L'(t) = 0.044t + 0.55 + 7\pi \cos(2\pi t)$, $L'(50.2)$ is given by $L'(50.2) = 0.044(50.2) + 0.55 + 7\pi \cos(100.4\pi) \approx 9.55 \text{ ppm per year.}$

EXAMPLE 9   Volume

The owners of a boarding stable wish to construct a watering trough from which the horses can drink. They have 9 ft by 9 ft pieces of metal, which they can bend into three parts to make the bottom and sides of the trough, as shown in Figure 27(a). They can then weld pieces of scrap metal to the ends to form a trough. At what angle θ should they bend the metal to create the largest possible volume? What is the largest possible volume?

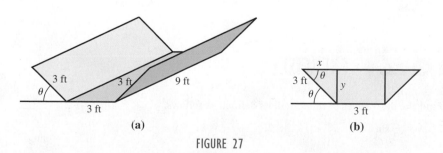

FIGURE 27

SOLUTION   The volume of the trough is its length, 9 ft, times the cross-sectional area. (This is true for any shape with parallel ends and straight sides. For example, the volume of a cylinder is the height times the area of the circular ends, or $h\pi r^2$.) Notice from Figure 27(b) that the cross-sectional area can be broken up into a rectangle with base 3 and height y, and two triangles, each with base x, height y, and angle θ. Since $x/3 = \cos \theta$, we have $x = 3 \cos \theta$. Similarly, since $y/3 = \sin \theta$, we have $y = 3 \sin \theta$. Therefore,

$$A = 3y + 2 \cdot \frac{1}{2}xy$$

$$= 3(3 \sin \theta) + (3 \cos \theta)(3 \sin \theta)$$

$$= 9(\sin \theta + \cos \theta \sin \theta).$$

Because the volume is the length times the area,

$$V = 9A = 9 \cdot 9(\sin \theta + \cos \theta \sin \theta)$$

$$= 81(\sin \theta + \cos \theta \sin \theta),$$

where $0 \le \theta \le \pi/2$. To find the maximum volume, set the derivative equal to 0.

$$\frac{dV}{d\theta} = 81[\cos \theta + \cos \theta \cos \theta + \sin \theta(-\sin \theta)] \quad \text{Product rule}$$

$$= 81(\cos \theta + \cos^2 \theta - \sin^2 \theta)$$

$$= 81[\cos \theta + \cos^2 \theta - (1 - \cos^2 \theta)] \quad \text{Use } \sin^2 x + \cos^2 x = 1.$$

$$= 81(2 \cos^2 \theta + \cos \theta - 1) \quad \text{Rearrange terms.}$$

$$= 81(2 \cos \theta - 1)(\cos \theta + 1) \quad \text{Factor.}$$

Notice in the third line of the above derivation that we used a trigonometric identity to put the expression entirely in terms of $\cos \theta$. To make $dV/d\theta = 0$, set either factor equal to 0.

$$2 \cos \theta - 1 = 0 \qquad\qquad \cos \theta + 1 = 0$$

$$\cos \theta = \frac{1}{2} \qquad\qquad \cos \theta = -1$$

$$\theta = \frac{\pi}{3} \qquad\qquad \text{No solution on } [0, \pi/2]$$

The only value of θ for which $dV/d\theta = 0$ is $\theta = \pi/3$, where $\cos \theta = 1/2$, $\sin \theta = \sqrt{3}/2$, and

$$V = 81\left(\frac{\sqrt{3}}{2} + \frac{1}{2} \cdot \frac{\sqrt{3}}{2}\right)$$

$$= \frac{243\sqrt{3}}{4} \approx 105.2 \text{ ft}^3$$

We must also check the endpoints. At $\theta = 0$, we have $V = 0$, while at $\theta = \pi/2$, we have $V = 81$. Thus the maximum volume of about 105.2 ft³ is achieved with $\theta = \pi/3$. ▬

13.2 EXERCISES

Find the derivatives of the functions defined as follows.

1. $y = \dfrac{1}{2} \sin 8x$

2. $y = -\cos 2x + \cos \dfrac{\pi}{6}$

3. $y = 12 \tan(9x + 1)$

4. $y = -4 \cos(7x^2 - 4)$

5. $y = \cos^4 x$

6. $y = -9 \sin^5 x$

7. $y = \tan^8 x$

8. $y = 3 \cot^5 x$

9. $y = -6x \sin 2x$

10. $y = 2x \sec 4x$

11. $y = \dfrac{\csc x}{x}$

12. $y = \dfrac{\tan x}{x - 1}$

13. $y = \sin e^{4x}$

14. $y = \cos(4e^{2x})$

15. $y = e^{\cos x}$

16. $y = -2e^{\cot x}$

17. $y = \sin(\ln 3x^4)$

18. $y = \cos(\ln |2x^3|)$

19. $y = \ln |\sin x^2|$

20. $y = \ln |\tan^2 x|$

21. $y = \dfrac{2 \sin x}{3 - 2 \sin x}$

22. $y = \dfrac{3 \cos x}{5 - \cos x}$

23. $y = \sqrt{\dfrac{\sin x}{\sin 3x}}$

24. $y = \sqrt{\dfrac{\cos 4x}{\cos x}}$

25. $y = 3 \tan\left(\dfrac{1}{4}x\right) + 4 \cot 2x - 5 \csc x + e^{-2x}$

26. $y = [\sin 3x + \cot(x^3)]^8$

In Exercises 27–32, recall that the slope of the tangent line to a graph is given by the derivative of the function. Find the slope of the tangent line to the graph of each equation at the given point. You may wish to use a graphing calculator to support your answers.

27. $y = \sin x; \quad x = 0$

28. $y = \sin x; \quad x = \pi/4$

29. $y = \cos x; \quad x = -5\pi/6$

30. $y = \cos x; \quad x = -\pi/4$

31. $y = \tan x; \quad x = 0$

32. $y = \cot x; \quad x = \pi/4$

33. Find the derivative of $\cot x$ by using the quotient rule and the fact that $\cot x = \cos x/\sin x$.

34. Verify that the derivative of $\sec x$ is $\sec x \tan x$. (*Hint:* Use the fact that $\sec x = 1/\cos x$.)

35. Verify that the derivative of $\csc x$ is $-\csc x \cot x$. (*Hint:* Use the fact that $\csc x = 1/\sin x$.)

36. In the discussion of the limit of the quotient $(\sin x)/x$, explain why the calculator gave ERROR for the value of $(\sin x)/x$ when $x = 0$.

APPLICATIONS

Business and Economics

37. Revenue from Seasonal Merchandise The revenue received from the sale of electric fans is seasonal, with maximum revenue in the summer. Let the revenue (in dollars) received from the sale of fans be approximated by

$$R(t) = 120 \cos 2\pi t + 150,$$

where t is time in years, measured from July 1.

a. Find $R'(t)$.

b. Find $R'(t)$ for August 1. (*Hint:* August 1 is 1/12 of a year from July 1.)

c. Find $R'(t)$ for January 1.

d. Find $R'(t)$ for June 1.

e. Discuss whether the answers in parts b–d are reasonable for this model.

Life Sciences

38. Swing of a Runner's Arm A runner's arm swings rhythmically according to the equation

$$y = \frac{\pi}{8} \cos\left[3\pi\left(t - \frac{1}{3}\right)\right],$$

where y denotes the angle between the actual position of the upper arm and the downward vertical position (as shown in the

figure) and where t denotes time (in seconds). *Source: Calculus for the Life Sciences.*

a. Graph y as a function of t.

b. Calculate the velocity and the acceleration of the arm.

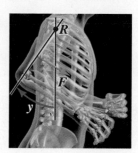

c. Verify that the angle y and the acceleration d^2y/dt^2 are related by the differential equation

$$\frac{d^2y}{dt^2} + 9\pi^2 y = 0.$$

d. Apply the fact that the force exerted by the muscle as the arm swings is proportional to the acceleration of y, with a positive constant of proportionality, to find the direction of the force (counterclockwise or clockwise) at $t = 1$ second, $t = 4/3$ seconds, and $t = 5/3$ seconds. What is the position of the arm at each of these times?

39. Swing of a Jogger's Arm A jogger's arm swings according to the equation

$$y = \frac{1}{5} \sin[\pi(t - 1)].$$

Proceed as directed in parts a–d of the preceding exercise, with the following exceptions: in part c, replace the differential equation with

$$\frac{d^2y}{dt^2} + \pi^2 y = 0,$$

and in part d, consider the times $t = 1.5$ seconds, $t = 2.5$ seconds, and $t = 3.5$ seconds.

40. Carbon Dioxide Levels At Barrow, Alaska, atmospheric carbon dioxide levels (in parts per million) can be modeled using the function defined by

$$C(t) = 0.04t^2 + 0.6t + 330 + 7.5\sin(2\pi t),$$

where t is in years and $t = 0$ corresponds to 1960. *Source: Introductory Astronomy and Astrophysics.*

a. Graph C on $[0, 25]$.

b. Find $C(25)$, $C(35.5)$, and $C(50.2)$.

c. Find $C'(50.2)$ and interpret.

d. C is the sum of a quadratic function and a sine function. What is the significance of each of these functions? Discuss what physical phenomena may be responsible for each function.

41. Population Growth Many biological populations, both plant and animal, experience seasonal growth. For example, an animal population might flourish during the spring and summer and die back in the fall. The population, $f(t)$, at time t, is often modeled by

$$f(t) = f(0)e^{c\,\sin(t)},$$

where $f(0)$ is the size of the population when $t = 0$. Suppose that $f(0) = 1000$ and $c = 2$. Find the functional values in parts a–d.

a. $f(0.2)$ **b.** $f(1)$ **c.** $f'(0)$ **d.** $f'(0.2)$

e. Use a graphing calculator to graph $f(t)$ on $[0, 11]$.

f. Find the maximum and minimum values of $f(t)$ and the values of t where they occur.

Physical Sciences

42. Piston Velocity The distance s of a piston from the center of the crankshaft as it rotates in a 1937 John Deere B engine with respect to the angle θ of the connecting rod, as indicated by the figure, is given by the formula

$$s(\theta) = 2.625 \cos \theta + 2.625(15 + \cos^2 \theta)^{1/2},$$

where s is measured in inches and θ in radians. *Source: The AMATYC Review.*

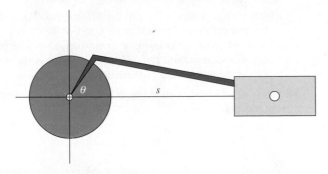

a. Find $s\left(\dfrac{\pi}{2}\right)$.

b. Find $\dfrac{ds}{d\theta}$.

c. Find the value(s) of θ where the maximum velocity of the piston occurs.

43. Sound If a string with a fundamental frequency of 110 hertz is plucked in the middle, it will vibrate at the odd harmonics of $110, 330, 550, \ldots$ hertz but not at the even harmonics of $220, 440, 660, \ldots$ hertz. The resulting pressure P on the eardrum caused by the string can be approximated using the equation

$$P(t) = 0.003 \sin(220\pi t) + \frac{0.003}{3} \sin(660\pi t)$$

$$+ \frac{0.003}{5} \sin(1100\pi t) + \frac{0.003}{7} \sin(1540\pi t),$$

where P is in pounds per square foot at a time of t seconds after the string is plucked. *Source: Fundamentals of Musical Acoustics and The Physics and Psychophysics of Music: An Introduction.*

a. Graph $P(t)$ on $[0, 0.01]$.

b. Find $P'(0.002)$ and interpret.

44. Ground Temperature Mathematical models of ground temperature variation usually involve Fourier series or other sophisticated methods. However, the elementary model

$$u(x, t) = T_0 + A_0 e^{-ax} \cos\left(\frac{\pi}{6}t - ax\right)$$

has been developed for temperature $u(x, t)$ at a given location at a variable time t (in months) and a variable depth x (in centimeters) beneath Earth's surface, T_0 is the annual average surface temperature, and A_0 is the amplitude of the seasonal surface temperature variation. *Source: Applications in School Mathematics 1979 Yearbook.*

Assume that $T_0 = 16°C$ and $A_0 = 11°C$ at a certain location. Also assume that $a = 0.00706$ in cgs (centimeter-gram-second) units.

a. At what minimum depth x is the amplitude of $u(x, t)$ at most $1°C$?

b. Suppose we wish to construct a cellar to keep wine at a temperature between $14°C$ and $18°C$. What minimum depth will accomplish this?

c. At what minimum depth x does the ground temperature model predict that it will be winter when it is summer at the surface, and vice versa? That is, when will the phase shift correspond to $1/2$ year?

d. Show that the ground temperature model satisfies the *heat equation*

$$\frac{\partial u}{\partial t} = k \frac{\partial^2 u}{\partial x^2},$$

where k is a constant.

45. Flying Gravel The grooves or tread in a tire occasionally pick up small pieces of gravel, which then are often thrown into the air as they work loose from the tire. When following behind a vehicle on a highway with loose gravel, it is possible to determine a safe distance to travel behind the vehicle so that your automobile is not hit with flying debris by analyzing the function

$$y = x \tan \alpha - \frac{16x^2}{V^2} \sec^2 \alpha,$$

where y is the height (in feet) of a piece of gravel that leaves the bottom of a tire at an angle α relative to the roadway, x is the horizontal distance (in feet) of the gravel, and V is the velocity of the automobile (in feet per second). *Source: UMAP Journal.*

a. If a car is traveling 30 mph (44 ft per second), find the height of a piece of gravel thrown from a car tire at an angle of $\pi/4$, when the stone is 40 ft from the car.

b. Putting $y = 0$ and solving for x gives the distance that the gravel will fly. Show that the function that gives a relationship between x and the angle α for $y = 0$ is given by

$$x = \frac{V^2}{32} \sin(2\alpha).$$

(*Hint:* $2 \sin \alpha \cos \alpha = \sin(2\alpha)$.)

c. Using part b, if the gravel is thrown from the car at an angle of $\pi/3$ and initial velocity of 44 ft per second, determine how far the gravel will travel.

d. Find $dx/d\alpha$ and use it to determine the value of α that gives the maximum distance that a stone could fly.

e. Find the maximum distance that a stone can fly from a car that is traveling 60 mph.

46. Engine Velocity As shown in Exercise 42, a formula that can be used to determine the distance of a piston with respect to the crankshaft for a 1937 John Deere B engine is

$$s(\theta) = 2.625 \cos(\theta) + 2.625(15 + \cos^2 \theta)^{1/2},$$

where s is measured in inches and θ in radians. *Source: The AMATYC Review.*

a. Given that the angle θ is changing with respect to time, that is, it is a function of t, use the chain rule to find the derivative of s with respect to t, ds/dt.

b. Use part a, with $\theta = 4.944$ and $d\theta/dt = 1340$ rev per minute, to find the maximum velocity of the engine. Express your answer in miles per hour. (*Hint:* 1340 rev per minute = 505,168.1 rad per hour. Use this value and then convert your answer from inches to miles, where 1 mile = 5280 ft.)

47. Motion of a Particle A particle moves along a straight line. The distance of the particle from the origin at time t is given by

$$s(t) = \sin t + 2 \cos t.$$

Find the velocity at the following times.

a. $t = 0$ **b.** $t = \pi/4$ **c.** $t = 3\pi/2$

Find the acceleration at the following times.

d. $t = 0$ **e.** $t = \pi/4$ **f.** $t = \pi$

General Interest

⊕ 48. Rotating Lighthouse The beacon on a lighthouse 50 m from a straight shoreline rotates twice per minute. (See the figure.)

a. How fast is the beam moving along the shoreline at the moment when the light beam and the shoreline are at right angles? (*Hint:* This is a related rate exercise. Find an equation relating θ, the angle between the beam of light and the line from the lighthouse to the shoreline, and x, the distance along the shoreline from the point on the shoreline closest to the lighthouse and the point where the beam hits the shoreline. You need to express $d\theta/dt$ in radians per minute.)

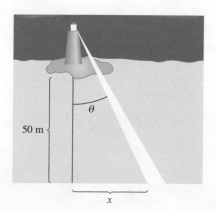

b. In part a, how fast is the beam moving along the shoreline when the beam hits the shoreline 50 m from the point on the shoreline closest to the lighthouse?

⊕ 49. Rotating Camera A television camera on a tripod 60 ft from a road is filming a car carrying the president of the United States. (See the figure.) The car is moving along the road at 600 ft per minute.

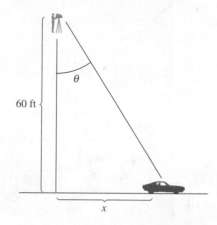

a. How fast is the camera rotating (in revolutions per minute) when the car is at the point on the road closest to the camera? (See the hint for Exercise 48.)

b. How fast is the camera rotating 6 seconds after the moment in part a?

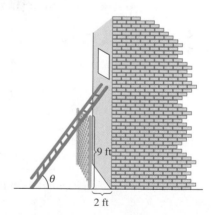

⊕ 50. APPLY IT Ladder A thief tries to enter a building by placing a ladder over a 9-ft-high fence so it rests against the building, which is 2 ft back from the fence. (See the figure above.) What is the length of the shortest ladder that can be used? (*Hint:* Let θ be the angle between the ladder and the ground. Express the length of the ladder in terms of θ, and then find the value of θ that minimizes the length of the ladder.)

⊕ 51. Ladder A janitor in a hospital needs to carry a ladder around a corner connecting a 10-ft-wide corridor and a 5-ft-wide corridor. (See the figure on the next page.) What is the longest such ladder that can make it around the corner? (*Hint:* Find the narrowest point in the corridor by minimizing the length of the ladder as a function of θ, the angle the ladder makes with the 5-ft-wide corridor.)

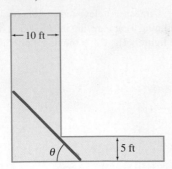

13.3 Integrals of Trigonometric Functions

APPLY IT Given a sales equation, how many snowblowers are sold in a year?

In Exercise 39 in this section, we will use trigonometry and integration to answer this question.

Any differentiation formula leads to a corresponding formula for integration. In particular, the formulas of the last section lead to the following indefinite integrals.

Basic Trigonometric Integrals

$$\int \sin x \, dx = -\cos x + C \qquad \int \cos x \, dx = \sin x + C$$

$$\int \sec^2 x \, dx = \tan x + C \qquad \int \csc^2 x \, dx = -\cot x + C$$

$$\int \sec x \tan x \, dx = \sec x + C \qquad \int \csc x \cot x \, dx = -\csc x + C$$

EXAMPLE 1 Integrals of Trigonometric Functions

Find each integral.

(a) $\displaystyle\int \sin 7x \, dx$

SOLUTION Use substitution. Let $u = 7x$, so that $du = 7\, dx$. Then

$$\int \sin 7x \, dx = \frac{1}{7}\int \sin 7x \, (7\, dx)$$

$$= \frac{1}{7}\int \sin u \, du$$

$$= -\frac{1}{7}\cos u + C$$

$$= -\frac{1}{7}\cos 7x + C.$$

(b) $\displaystyle\int \cos \frac{2}{3} x \, dx$

SOLUTION Let $u = (2/3)x$, then $du = (2/3) \, dx$. This gives

$$\int \cos \frac{2}{3} x \, dx = \frac{3}{2} \int \cos \frac{2}{3} x \left(\frac{2}{3} \, dx \right)$$

$$= \frac{3}{2} \int \cos u \, du$$

$$= \frac{3}{2} \sin u + C$$

$$= \frac{3}{2} \sin \frac{2}{3} x + C.$$

(c) $\displaystyle\int \sin^2 x \cos x \, dx$

SOLUTION Let $u = \sin x$, with $du = \cos x \, dx$. This gives

$$\int \sin^2 x \cos x \, dx = \int u^2 \, du = \frac{1}{3} u^3 + C.$$

Replacing u with $\sin x$ gives

$$\int \sin^2 x \cos x \, dx = \frac{1}{3} \sin^3 x + C.$$

(d) $\displaystyle\int \frac{\sin x}{\sqrt{\cos x}} \, dx$

SOLUTION Rewrite the integrand as

$$\int (\cos x)^{-1/2} \sin x \, dx.$$

If $u = \cos x$, then $du = -\sin x \, dx$, with

$$\int (\cos x)^{-1/2} \sin x \, dx = -\int (\cos x)^{-1/2} (-\sin x \, dx)$$

$$= -\int u^{-1/2} \, du$$

$$= -2u^{1/2} + C$$

$$= -2 \cos^{1/2} x + C.$$

(e) $\displaystyle\int \sec^2 12x \, dx$

SOLUTION If $u = 12x$, then $du = 12 \, dx$, with

$$\int \sec^2 12x \, dx = \frac{1}{12} \int \sec^2 12x \, (12 \, dx)$$

$$= \frac{1}{12} \int \sec^2 u \, du$$

$$= \frac{1}{12} \tan 12x + C.$$

(f) $\int e^{3x} \sec e^{3x} \tan e^{3x} \, dx$

SOLUTION Use substitution. Let $u = e^{3x}$, so that $du = 3e^{3x} \, dx$. Then,

$$\int e^{3x} \sec e^{3x} \tan e^{3x} \, dx = \frac{1}{3} \int \sec e^{3x} \tan e^{3x} \, (3e^{3x} \, dx)$$

$$= \frac{1}{3} \int \sec u \tan u \, du$$

$$= \frac{1}{3} \sec u + C$$

$$= \frac{1}{3} \sec e^{3x} + C.$$

YOUR TURN 1 Find each integral. **(a)** $\int \sin(x/2) \, dx$

(b) $\int 6 \sec^2 x \sqrt{\tan x} \, dx$

TRY YOUR TURN 1 ▬

As mentioned earlier, $\tan x = (\sin x)/\cos x$, so that

$$\int \tan x \, dx = \int \frac{\sin x}{\cos x} \, dx.$$

To find $\int \tan x \, dx$, let $u = \cos x$, with $du = -\sin x \, dx$. Then

$$\int \tan x \, dx = \int \frac{\sin x}{\cos x} \, dx = -\int \frac{du}{u} = -\ln |u| + C.$$

Replacing u with $\cos x$ gives the formula for integrating $\tan x$. The integral for $\cot x$ is found in a similar way.

> **Integrals of $\tan x$ and $\cot x$**
>
> $$\int \tan x \, dx = -\ln |\cos x| + C$$
>
> $$\int \cot x \, dx = \ln |\sin x| + C$$

EXAMPLE 2 **Integrals of $\tan x$ and $\cot x$**

Find each integral.

(a) $\int \tan 6x \, dx$

SOLUTION Let $u = 6x$, so that $du = 6 \, dx$. Then

$$\int \tan 6x \, dx = \frac{1}{6} \int \tan 6x \, (6 \, dx)$$

$$= \frac{1}{6} \int \tan u \, du$$

$$= -\frac{1}{6} \ln |\cos u| + C$$

$$= -\frac{1}{6} \ln |\cos 6x| + C.$$

(b) $\int x \cot x^2 \, dx$

SOLUTION   Let $u = x^2$, so that $du = 2x \, dx$. Then

$$\int x \cot x^2 \, dx = \frac{1}{2} \int (\cot x^2)(2x \, dx)$$

$$= \frac{1}{2} \int \cot u \, du$$

$$= \frac{1}{2} \ln |\sin u| + C$$

$$= \frac{1}{2} \ln |\sin x^2| + C. \qquad \textbf{TRY YOUR TURN 2}$$

YOUR TURN 2
Find $\int \tan(\sqrt{x})/\sqrt{x} \, dx$.

The method of integration by parts discussed in Chapter 8 is often useful for finding certain integrals involving trigonometric functions.

EXAMPLE 3   Integration by Parts

Find $\int 2x \sin x \, dx$.

SOLUTION   Let $u = 2x$ and $dv = \sin x \, dx$. Then $du = 2 \, dx$ and $v = -\cos x$. Use the formula for integration by parts,

$$\int u \, dv = uv - \int v \, du,$$

to get

$$\int 2x \sin x \, dx = -2x \cos x - \int (-\cos x)(2 \, dx)$$

$$= -2x \cos x + 2 \int \cos x \, dx$$

$$= -2x \cos x + 2 \sin x + C.$$

YOUR TURN 3
Find $\int x \cos(3x) \, dx$.

Check the result by differentiating. (This integral could also have been found by using column integration.)   **TRY YOUR TURN 3**

As in Chapter 7, we can find the area under a curve by setting up an appropriate definite integral.

EXAMPLE 4   Area Under the Curve

Find the shaded area in Figure 28.

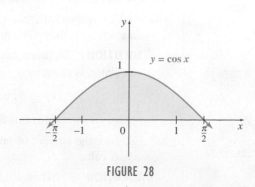

FIGURE 28

SOLUTION The shaded area in Figure 28 is bounded by $y = \cos x$, $y = 0$, $x = -\pi/2$, and $x = \pi/2$. By the Fundamental Theorem of Calculus, this area is given by

$$\int_{-\pi/2}^{\pi/2} \cos x \, dx = \sin x \Big|_{-\pi/2}^{\pi/2}$$

$$= \sin \frac{\pi}{2} - \sin\left(-\frac{\pi}{2}\right)$$

$$= 1 - (-1)$$

$$= 2.$$

YOUR TURN 4 Find the area under the curve $y = \sec^2(x/3)$ between $x = -\pi$ and $x = \pi$.

By symmetry, the same area could be found by evaluating

$$2\int_0^{\pi/2} \cos x \, dx.$$

TRY YOUR TURN 4

TECHNOLOGY NOTE The area in Example 4 could also be found using the definite integral feature of a graphing calculator, entering the expression $\cos x$, the variable x, and the limits of integration.

TECHNOLOGY

EXAMPLE 5 Precipitation in Vancouver, Canada

The average monthly precipitation (in inches) for Vancouver, Canada, is found in the following table. *Source: Weather.com.*

Month	Precipitation (inches)
January	5.9
February	4.9
March	4.3
April	3.0
May	2.4
June	1.8
July	1.4
August	1.5
September	2.5
October	4.5
November	6.7
December	7.0

(a) Plot the data, letting $t = 1$ correspond to January, $t = 2$ to February, and so on. Is it reasonable to assume that average monthly precipitation is periodic?

SOLUTION Figure 29 shows a graphing calculator plot of the data. Because of the cyclical nature of the four seasons, it is reasonable to assume that the data are periodic.

(b) Find a trigonometric function of the form $C(t) = a \sin(bt + c) + d$ that models this data when t is the month and $C(t)$ is the amount of precipitation. Use the table.

SOLUTION The function $C(t)$, derived by the sine regression feature on a TI-84 Plus calculator, is given by

$$C(t) = 2.56349 \sin(0.528143t + 1.30663) + 3.79128.$$

Figure 30 shows that this function fits the data fairly well.

(c) Estimate the amount of precipitation for the month of October, and compare it to the actual value.

SOLUTION $C(10) \approx 4.6$ in. The actual value is 4.5 inches.

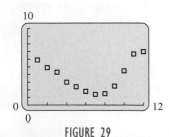

FIGURE 29

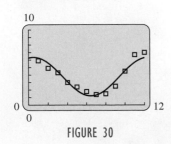

FIGURE 30

(d) Estimate the rate at which the amount of precipitation is changing in October.

SOLUTION The derivative of $C(t)$ is

$$C'(t) = (0.528143)2.56349 \cos(0.528143t + 1.30663)$$
$$= 1.35389 \cos(0.528143t + 1.30663).$$

In October, $t = 10$ and

$$C'(10) \approx 1.29 \text{ inches per month.}$$

(e) Estimate the total precipitation for the year and compare it to the actual value.

SOLUTION To estimate the total precipitation, we use integration as follows.

$$\int_0^{12} C(t)\, dt = \int_0^{12} (2.56349 \sin(0.528143t + 1.30663) + 3.79128)\, dt$$

$$= \left[-\frac{2.56349}{0.528143} \cos(0.528143t + 1.30663) + 3.79128t \right]\Big|_0^{12}$$

$$\approx 45.75 \text{ inches}$$

The actual value is 45.9 inches.

(f) What would you expect the period of a function that models annual precipitation to be? What is the period of the function found in part (b)?

SOLUTION If we assume that the annual rainfall in Vancouver is periodic, we would expect the period to be 12 so that it repeats itself every 12 months. The period for the function given above is

$$T = 2\pi/b = 2\pi/0.528143 \approx 11.90,$$

or about 12 months.

13.3 EXERCISES

Find each integral.

1. $\displaystyle\int \cos 3x\, dx$

2. $\displaystyle\int \sin 5x\, dx$

3. $\displaystyle\int (3 \cos x - 4 \sin x)\, dx$

4. $\displaystyle\int (9 \sin x + 8 \cos x)\, dx$

5. $\displaystyle\int x \sin x^2\, dx$

6. $\displaystyle\int 2x \cos x^2\, dx$

7. $\displaystyle -\int 3 \sec^2 3x\, dx$

8. $\displaystyle -\int 2 \csc^2 8x\, dx$

9. $\displaystyle\int \sin^7 x \cos x\, dx$

10. $\displaystyle\int \sin^4 x \cos x\, dx$

11. $\displaystyle\int 3\sqrt{\cos x}\,(\sin x)\, dx$

12. $\displaystyle\int \frac{\cos x}{\sqrt{\sin x}}\, dx$

13. $\displaystyle\int \frac{\sin x}{1 + \cos x}\, dx$

14. $\displaystyle\int \frac{\cos x}{1 - \sin x}\, dx$

15. $\displaystyle\int 2x^7 \cos x^8\, dx$

16. $\displaystyle\int (x + 2)^4 \sin(x + 2)^5\, dx$

17. $\displaystyle\int \tan \frac{1}{3} x\, dx$

18. $\displaystyle\int \cot\left(-\frac{3}{8}x\right)\, dx$

19. $\displaystyle\int x^5 \cot x^6\, dx$

20. $\displaystyle\int \frac{x}{4} \tan\left(\frac{x}{4}\right)^2\, dx$

21. $\displaystyle\int e^x \sin e^x\, dx$

22. $\displaystyle\int e^{-x} \tan e^{-x}\, dx$

23. $\displaystyle\int e^x \csc e^x \cot e^x\, dx$

24. $\displaystyle\int x^4 \sec x^5 \tan x^5\, dx$

25. $\displaystyle\int -6x \cos 5x\, dx$

26. $\displaystyle\int 7x \sin 5x\, dx$

27. $\int 4x \sin x\, dx$

28. $\int -11x \cos x\, dx$

29. $\int -6x^2 \cos 8x\, dx$

30. $\int 10x^2 \sin \frac{x}{2}\, dx$

Evaluate each definite integral. Use the integration feature of a graphing calculator, if you wish, to support your answers.

31. $\int_0^{\pi/4} \sin x\, dx$

32. $\int_{-\pi/2}^0 \cos x\, dx$

33. $\int_0^{\pi/6} \tan x\, dx$

34. $\int_{\pi/4}^{\pi/2} \cot x\, dx$

35. $\int_{\pi/2}^{2\pi/3} \cos x\, dx$

36. $\int_{\pi/6}^{\pi/4} \sin x\, dx$

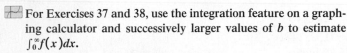 **For Exercises 37 and 38, use the integration feature on a graphing calculator and successively larger values of b to estimate $\int_0^\infty f(x)\,dx$.**

37. $\int_0^b e^{-x} \sin x\, dx$

38. $\int_0^b e^{-x} \cos x\, dx$

APPLICATIONS

Business and Economics

39. APPLY IT Sales Sales of snowblowers are seasonal. Suppose the sales of snowblowers in one region of the country are approximated by

$$S(t) = 500 + 500 \cos\left(\frac{\pi}{6}t\right),$$

where t is time (in months), with $t = 0$ corresponding to November. The figure below shows a graph of S. Use a definite integral to find total sales over a year.

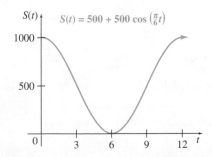

40. Petroleum Consumption The monthly residential consumption of petroleum (in trillions of BTUs) in the United States for 2009 is found in the following table. *Source: Energy Information Administration.*

a. Plot the data, letting $t = 1$ correspond to January, $t = 2$ correspond to February, and so on. Is it reasonable to assume that petroleum consumption is periodic?

Month	Petroleum (trillion BTUs)
January	133.4
February	115.6
March	112.9
April	92.8
May	76.1
June	70.9
July	77.8
August	83.3
September	86.7
October	92.4
November	97.0
December	132.1

b. Use a calculator with trigonometric regression to find a trigonometric function of the form

$$C(t) = a \sin(bt + c) + d$$

that models this data when t is the month and $C(t)$ is the amount of petroleum consumed (in trillions of BTUs). Graph this function on the same calculator window as the data.

c. Estimate the consumption for the month of September and compare it to the actual value.

d. Estimate the rate at which the consumption is changing in September.

e. Estimate the total petroleum consumption for the year for residential customers and compare it to the actual value.

f. What would you expect the period of a function that models annual petroleum consumption to be? What is the period of the function found in part b? Discuss possible reasons for the discrepancy in the values.

Life Sciences

41. Migratory Animals The number of migratory animals (in hundreds) counted at a certain checkpoint is given by

$$T(t) = 50 + 50 \cos\left(\frac{\pi}{6}t\right),$$

where t is time in months, with $t = 0$ corresponding to July. The figure below shows a graph of T. Use a definite integral to find the number of animals passing the checkpoint in a year.

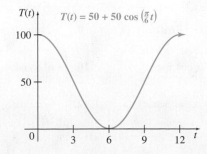

Physical Sciences

42. Voltage The electrical voltage from a standard wall outlet is given as a function of time t by

$$V(t) = 170 \sin(120 \pi t).$$

This is an example of alternating current, which is electricity that reverses direction at regular intervals. The common method for measuring the level of voltage from an alternating current is the *root mean square,* which is given by

$$\text{Root mean square} = \sqrt{\frac{\int_0^T V^2(t)\, dt}{T}},$$

where T is one period of the current.

a. Verify that $T = 1/60$ second for $V(t)$ given above.

b. You may have seen that the voltage from a standard wall outlet is 120 volts. Verify that this is the root mean square value for $V(t)$ given above. (*Hint:* Use the trigonometric identity $\sin^2 x = (1 - \cos 2x)/2$. This identity can be derived by letting $y = x$ in the basic identity for $\cos(x + y)$, and then eliminating $\cos^2 x$ by using the identity $\cos^2 x = 1 - \sin^2 x$.)

43. Length of Day The following function can be used to estimate the number of minutes of daylight in Boston for any given day of the year.

$$N(t) = 183.549 \sin(0.0172t - 1.329) + 728.124,$$

where t is the day of the year. *Source: The Old Farmer's Almanac.* Use this function to estimate the total amount of daylight in a year and compare it to the total amount of daylight reported to be 4467.57 hours.

General Interest

Self-Answering Problems The problems in Exercises 44–46 are called self-answering problems because the answers are embedded in the question. For example, how many ways can you arrange the letters in the word "six"? The answer is six. *Source: Math Horizons.*

44. At time $t = 0$, water begins pouring into an empty sink so that the volume of water is changing at a rate $V'(t) = \cos t$. For time $t = k$, where $0 \le k \le \pi/2$, determine the amount of water in the sink.

45. At time $t = 0$, water begins pouring into an empty tank so that the volume of water is changing at a rate $V'(t) = \sec^2 t$. For time $t = k$, where $0 \le k \le \pi/2$, determine the amount of water in the tank.

46. The cost of a widget varies according to the formula $C'(t) = -\sin t$. At time $t = 0$, the cost is \$1. For arbitrary time t, determine a formula for the cost.

YOUR TURN ANSWERS

1. (a) $-2\cos(x/2) + C$ **(b)** $4(\tan x)^{3/2} + C$

2. $-2\ln|\cos(\sqrt{x})| + C$

3. $x\sin(3x)/3 + \cos(3x)/9 + C$

4. $6\sqrt{3}$

13 CHAPTER REVIEW

SUMMARY

In this chapter, we introduced the trigonometric functions and we studied some of their properties, including their periodic or repetitive nature. To develop the trigonometric functions, it was necessary to make the following definitions.

- A ray is the portion of a line that starts at a given point, called the endpoint, and continues indefinitely in one direction.
- An angle is formed by rotating a ray about its endpoint.
  - The initial position of the ray is called the initial side of the angle, and the endpoint of the ray is called the vertex of the angle.
  - The location of the ray at the end of its rotation is called the terminal side of the angle.

- An angle that
  - measures between 0° and 90° is an acute angle;
  - measures 90° is a right angle;
  - measures more than 90° but less than 180° is an obtuse angle;
  - measures 180° is a straight angle.
- An angle measured in radians is the arc length formed by the angle, on a unit circle.

We saw that trigonometric functions describe many natural phenomena and have many applications in business, economics, and science. We extended the techniques of earlier chapters to find derivatives and integrals involving trigonometric functions. We then used differentiation and integration of trigonometric functions to analyze a variety of applications.

1 Radian 1 radian $= \left(\dfrac{180°}{\pi}\right)$

1 Degree $1° = \dfrac{\pi}{180}$ radians

Degrees/Radians

Degrees	0°	30°	45°	60°	90°	180°	270°	360°
Radians	0	$\pi/6$	$\pi/4$	$\pi/3$	$\pi/2$	π	$3\pi/2$	2π

Trigonometric Functions Let (x, y) be a point other than the origin on the terminal side of an angle θ in standard position. Let r be the distance from the origin to (x, y). Then

$$\sin \theta = \frac{y}{r} \qquad\qquad \csc \theta = \frac{r}{y} \quad (y \neq 0)$$

$$\cos \theta = \frac{x}{r} \qquad\qquad \sec \theta = \frac{r}{x} \quad (x \neq 0)$$

$$\tan \theta = \frac{y}{x} \quad (x \neq 0) \qquad \cot \theta = \frac{x}{y} \quad (y \neq 0).$$

Elementary Trigonometric Identities

$$\csc \theta = \frac{1}{\sin \theta} \qquad \sec \theta = \frac{1}{\cos \theta} \qquad \cot \theta = \frac{1}{\tan \theta}$$

$$\tan \theta = \frac{\sin \theta}{\cos \theta} \qquad \cot \theta = \frac{\cos \theta}{\sin \theta} \qquad \sin^2 \theta + \cos^2 \theta = 1$$

Values of Trigonometric Functions for Common Angles

θ (in radians)	θ (in degrees)	$\sin \theta$	$\cos \theta$	$\tan \theta$	$\cot \theta$	$\sec \theta$	$\csc \theta$
0	0°	0	1	0	Undefined	1	Undefined
$\pi/2$	90°	1	0	Undefined	0	Undefined	1
π	180°	0	-1	0	Undefined	-1	Undefined
$3\pi/2$	270°	-1	0	Undefined	0	Undefined	-1
2π	360°	0	1	0	Undefined	1	Undefined

Periodic Function A function $y = f(x)$ is periodic if there exists a positive real number a such that
$$f(x) = f(x + a)$$
for all values of x in the domain. The smallest positive value of a is called the period of the function.

Basic Identities

$$\sin(x + y) = \sin x \cos y + \cos x \sin y$$
$$\sin(x - y) = \sin x \cos y - \cos x \sin y$$
$$\cos(x + y) = \cos x \cos y - \sin x \sin y$$
$$\cos(x - y) = \cos x \cos y + \sin x \sin y$$

Important Limit $\displaystyle\lim_{x \to 0} \frac{\sin x}{x} = 1$

Basic Trigonometric Derivatives

$$D_x (\sin x) = \cos x \qquad\qquad D_x (\cos x) = -\sin x$$
$$D_x (\tan x) = \sec^2 x \qquad\qquad D_x (\cot x) = -\csc^2 x$$
$$D_x (\sec x) = \sec x \tan x \qquad\quad D_x (\csc x) = -\csc x \cot x$$

Basic Trigonometric Integrals

$$\int \sin x\, dx = -\cos x + C \qquad\qquad \int \cos x\, dx = \sin x + C$$

$$\int \sec^2 x\, dx = \tan x + C \qquad\qquad \int \csc^2 x\, dx = -\cot x + C$$

$$\int \sec x \tan x\, dx = \sec x + C \qquad\quad \int \csc x \cot x\, dx = -\csc x + C$$

$$\int \tan x\, dx = -\ln|\cos x| + C \qquad\quad \int \cot x\, dx = \ln|\sin x| + C$$

KEY TERMS

13.1

ray	standard position	unit circle	secant
endpoint	degree measure	arc	cosecant
angle	acute angle	trigonometric functions	special angles
initial side	right angle	sine	periodic functions
vertex	obtuse angle	cosine	period
terminal side	straight angle	tangent	amplitude
	radian measure	cotangent	phase shift

REVIEW EXERCISES

CONCEPT CHECK

Determine whether each of the following statements is true or false, and explain why.

1. The function $f(x) = \cos x$ is periodic with a period of π.

2. All six of the basic trigonometric functions are periodic.

3. It is reasonable to expect that the Dow Jones Industrial Average is periodic and can be modeled using a sine function.

4. $\cos(a + b) = \cos a + \cos b$

5. $\sin^2\left(\dfrac{\pi}{7}\right) + \cos^2\left(2\pi + \dfrac{\pi}{7}\right) = 1$

6. $D_x \tan(x^2) = \sec^2(x^2)$

7. The cosine function has an infinite number of critical points where an absolute minimum occurs.

8. The secant function has an infinite number of critical points where an absolute maximum occurs.

9. The area between the x-axis and the curve $y = \sin x$ on the interval $[0, 2\pi]$ is given by the definite integral $\displaystyle\int_0^{2\pi} \sin x \, dx$.

10. The method of integration by parts should be used to determine $\displaystyle\int \dfrac{\cos x}{5 + \sin x}\, dx$.

PRACTICE AND EXPLORATIONS

11. What is the relationship between the degree measure and the radian measure of an angle?

12. Under what circumstances should radian measure be used instead of degree measure? Degree measure instead of radian measure?

13. Describe in words how each of the six trigonometric functions is defined.

14. At what angles (given as rational multiples of π) can you determine the exact values for the trigonometric functions?

Convert the following degree measures to radians. Leave answers as multiples of π.

15. $90°$
16. $160°$
17. $225°$
18. $270°$
19. $360°$
20. $405°$

Convert the following radian measures to degrees.

21. 5π
22. $\dfrac{3\pi}{4}$
23. $\dfrac{9\pi}{20}$
24. $\dfrac{3\pi}{10}$
25. $\dfrac{13\pi}{20}$
26. $\dfrac{13\pi}{15}$

Find each function value *without* using a calculator.

27. $\sin 60°$
28. $\tan 120°$
29. $\cos(-45°)$
30. $\sec 150°$
31. $\csc 120°$
32. $\cot 300°$
33. $\sin \dfrac{\pi}{6}$
34. $\cos \dfrac{7\pi}{3}$
35. $\sec \dfrac{5\pi}{3}$
36. $\csc \dfrac{7\pi}{3}$

Find each function value.

37. $\sin 47°$
38. $\cos 72°$
39. $\tan 115°$
40. $\sin(-123°)$
41. $\sin 2.3581$
42. $\cos 0.8215$
43. $\cos 0.5934$
44. $\tan 1.2915$

Graph one period of each function.

45. $y = 4 \cos x$
46. $y = \dfrac{1}{2} \tan x$
47. $y = -\tan x$
48. $y = -\dfrac{2}{3} \sin x$

49. Because the derivative of $y = \sin x$ is $dy/dx = \cos x$, the slope of $y = \sin x$ varies from _____ to _____.

Find the derivative of each function.

50. $y = -4 \sin 7x$
51. $y = 2 \tan 5x$
52. $y = \tan(4x^2 + 3)$
53. $y = \cot(6 - 3x^2)$
54. $y = 2 \cos^5 x$
55. $y = 2 \sin^4(4x^2)$
56. $y = \cot\left(\dfrac{1}{2} x^4\right)$
57. $y = \cos(1 + x^2)$
58. $y = x^2 \csc x$
59. $y = e^{-2x} \sin x$
60. $y = \dfrac{\sin x - 1}{\sin x + 1}$
61. $y = \dfrac{\cos^2 x}{1 - \cos x}$
62. $y = \dfrac{6 - x}{\sec x}$
63. $y = \dfrac{\tan x}{1 + x}$
64. $y = \ln|\cos x|$
65. $y = \ln|5 \sin x|$

Find each integral.

66. $\int \sin 2x \, dx$

67. $\int \cos 5x \, dx$

68. $\int \tan 7x \, dx$

69. $\int \sec^2 5x \, dx$

70. $\int 8 \sec^2 x \, dx$

71. $\int 4 \csc^2 x \, dx$

72. $\int x^2 \sin 4x^3 \, dx$

73. $\int 5x \sec 2x^2 \tan 2x^2 \, dx$

74. $\int \sqrt{\cos x} \sin x \, dx$

75. $\int \cos^8 x \sin x \, dx$

76. $\int x \tan 11x^2 \, dx$

77. $\int x^2 \cot 8x^3 \, dx$

78. $\int (\sin x)^{3/2} \cos x \, dx$

79. $\int (\cos x)^{-4/3} \sin x \, dx$

80. $\int \sec^2 5x \tan 5x \, dx$

Find each definite integral.

81. $\int_0^{\pi/2} \cos x \, dx$

82. $\int_{-\pi}^{2\pi/3} -\sin x \, dx$

83. $\int_0^{2\pi} (10 + 10 \cos x) \, dx$

84. $\int_0^{\pi/3} (3 - 3 \sin x) \, dx$

APPLICATIONS

Business

85. Energy Consumption The monthly residential consumption of natural gas in Pennsylvania for 2009 is found in the following table. *Source: Energy Information Administration.*

Month	Consumption (million cubic feet)
January	47,599
February	40,659
March	30,877
April	19,169
May	8726
June	5317
July	4576
August	4084
September	4711
October	11,175
November	16,944
December	33,740

a. Plot the data, letting $t = 1$ correspond to January, $t = 2$ to February, and so on. Is it reasonable to assume that the monthly consumption of energy is periodic?

b. Find the trigonometric function of the form

$$C(t) = a \sin(bt + c) + d$$

that models this data when t is the month of the year and $C(t)$ is the natural gas consumption. Graph the function on the same calculator window as the data.

c. Estimate the total natural gas consumption for the year for residential customers in Pennsylvania and compare it to the actual value.

d. Calculate the period of the function found in part b. Is this period reasonable?

Life Sciences

86. Blood Pressure A person's blood pressure at time t (in seconds) is given by

$$P(t) = 90 + 15 \sin 144\pi t.$$

Find the maximum and minimum values of P on the interval $[0, 1/72]$. Graph one period of $y = P(t)$.

Blood Vessel System The body's system of blood vessels is made up of arteries, arterioles, capillaries, and veins. The transport of blood from the heart through all organs of the body and back to the heart should be as efficient as possible. One way this can be done is by having large enough blood vessels to avoid turbulence, with blood cells small enough to minimize viscosity.

In Exercises 87–100, we will find the value of angle θ (see the figure) such that total resistance to the flow of blood is minimized. Assume that a main vessel of radius r_1 runs along the horizontal line from A to B. A side artery, of radius r_2, heads for a point C. Choose point B so that CB is perpendicular to AB. Let $CB = s$ and let D be the point where the axis of the branching vessel cuts the axis of the main vessel.

According to Poiseuille's law, the resistance R in the system is proportional to the length L of the vessel and inversely proportional to the fourth power of the radius r. That is,

$$R = k \cdot \frac{L}{r^4}, \tag{1}$$

where k is a constant determined by the viscosity of the blood. Let $AB = L_0$, $AD = L_1$, and $DC = L_2$. *Source: Introduction to Mathematics for Life Scientists.*

87. Use right triangle BDC to find $\sin \theta$.

88. Solve the result of Exercise 87 for L_2.

89. Find $\cot \theta$ in terms of s and $L_0 - L_1$.

90. Solve the result of Exercise 89 for L_1.

91. Write an expression similar to Equation (1) for the resistance R_1 along AD.

92. Write a formula for the resistance along DC.

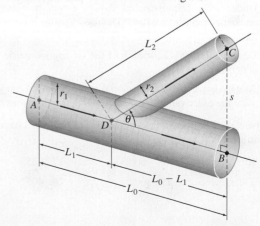

93. The total resistance R is given by the sum of the resistances along AD and DC. Use your answers to Exercises 91 and 92 to write an expression for R.

94. In your formula for R, replace L_1 with the result of Exercise 90 and L_2 with the result of Exercise 88. Simplify your answer.

95. Find $dR/d\theta$. Simplify your answer. (Remember that k, L_1, L_0, s, r_1, and r_2 are constants.)

96. Set $dR/d\theta$ equal to 0.

97. Multiply through by $(\sin^2\theta)/s$.

98. Solve for $\cos\theta$.

99. Suppose $r_1 = 1$ cm and $r_2 = 1/4$ cm. Find $\cos\theta$ and then find θ.

100. Find θ if $r_1 = 1.4$ cm and $r_2 = 0.8$ cm.

Physical Sciences

101. Simple Harmonic Motion The differential equation $s''(t) = -B^2 s(t)$ approximately describes the motion of a pendulum, known as *simple harmonic motion*. Verify that

$$s(t) = A\cos(Bt + C)$$

satisfies this differential equation.

102. Temperature The table lists the average monthly temperatures in Vancouver, Canada. *Source: Weather.com.*

Month	Jan	Feb	Mar	Apr	May	June
Temperature	37	41	43	48	54	59

Month	July	Aug	Sep	Oct	Nov	Dec
Temperature	63	63	58	50	43	38

These average temperatures cycle yearly and change only slightly over many years. Because of the repetitive nature of temperatures from year to year, they can be modeled with a sine function. Some graphing calculators have a sine regression feature. If the table is entered into a calculator, the points can be plotted automatically, as shown in the early chapters of this book with other types of functions.

a. Use a graphing calculator to plot the ordered pairs (month, temperature) in the interval $[0, 12]$ by $[30, 70]$.

b. Use a graphing calculator with a sine regression feature to find an equation of the sine function that models this data.

c. Graph the equation from part b.

d. Calculate the period for the function found in part b. Is this period reasonable?

103. Tennis It is possible to model the flight of a tennis ball that has just been served down the center of the court by the equation

$$y = x\tan\alpha - \frac{16x^2}{V^2}\sec^2\alpha + h,$$

where y is the height (in feet) of a tennis ball that is being served at an angle α relative to the horizontal axis, x is the horizontal distance (in feet) that the ball has traveled, h is the height of the ball when it leaves the server's racket and V is the velocity of the tennis ball when it leaves the server's racket. Source: *UMAP Journal.*

a. If a tennis ball is served from a height of 9 ft and the net is 3 ft high and 39 ft away from the server, does thc tennis ball that is hit with a velocity of 50 mph (approximately 73 ft per second) make it over the net if it is served at an angle of $\pi/24$?

b. When $y = 0$, the corresponding value of x gives the total distance that the tennis ball has traveled while in flight (provided that it cleared the net). For a serving height of 9 ft, the equation for calculating the distance traveled is given by

$$x = \frac{V^2\sin\alpha\cos\alpha \pm V^2\cos^2\alpha\sqrt{\tan^2\alpha + \dfrac{576}{V^2}\sec^2\alpha}}{32}.$$

Use the TABLE function on a graphing calculator or a spreadsheet to determine a range of angles for which the tennis ball will clear the net and travel between 39 and 60 ft when it is hit with an initial velocity of 44 ft per second.

c. Because calculating $dx/d\alpha$ is so complicated analytically, use a graphing calculator to estimate this derivative when the initial velocity is 44 ft per second and $\alpha = \pi/8$. Interpret your answer.

104. Energy Usage A mathematics textbook author has determined that her monthly gas usage y approximately follows the sine curve

$$y = 12.5\sin\left(\frac{\pi}{6}(t + 1.2)\right) + 14.7,$$

where y is measured in thousands of cubic feet (MCF) and t is the month of the year ranging from 1 to 12.

a. Graph this function on a graphing calculator.

b. Find the approximate gas usage for the months of February and July.

c. Find dy/dt, when $t = 7$. Interpret your answer.

General Interest

105. Area A 6-ft board is placed against a wall as shown in the figure below, forming a triangular-shaped area beneath it. At what angle θ should the board be placed to make the triangular area as large as possible?

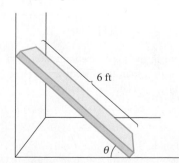

6 ft

θ

106. Mercator's World Map Before Gerardus Mercator designed his map of the world in 1569, sailors who traveled in a fixed compass direction could follow a straight line on a map only over short distances. Over long distances, such a course would be a curve on existing maps, which tried to make area on the map proportional to the actual area. Mercator's map greatly simplified navigation: even over long distances, straight lines on the map corresponded to fixed compass bearings. This was accomplished by distorting distances. On Mercator's map, the distance of an object from the equator to a parallel at latitude θ is given by

$$D(\theta) = k\int_0^\theta \sec x \, dx,$$

where k is a constant of proportionality. Calculus had not yet been discovered when Mercator designed his map; he approximated the distance between parallels of latitude by hand. *Source: Mathematics Magazine*.

a. Verify that

$$\frac{d}{dx}\ln|\sec x + \tan x| = \sec x.$$

b. Verify that

$$\frac{d}{dx}(-\ln|\sec x - \tan x|) = \sec x.$$

c. Using parts a and b, give two different formulas for $\int \sec x \, dx$. Explain how they can both be correct.

d. Los Angeles has a latitude of 34°03′N. (The 03′ represents 3 minutes of latitude. Each minute of latitude is 1/60 of a degree.) If Los Angeles is to be 7 in. from the equator on a Mercator map, how far from the equator should we place New York City, which has a latitude of 40°45′N?

e. Repeat part d for Miami, which has a latitude of 25°46′N.

f. If you do not live in Los Angeles, New York City, or Miami, repeat part d for your town or city.

EXTENDED APPLICATION

THE SHORTEST TIME AND THE CHEAPEST PATH

In an application at the end of Section 13.1 we stated Snell's law relating the angle of refraction when light passes from one medium to another to the speed of light in each medium. Figure 31 represents the relationship graphically.

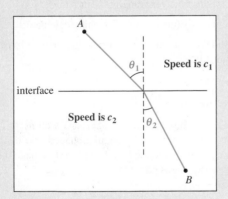

FIGURE 31

If the speed of light in the upper medium is c_1, and in the lower medium the speed is c_2, then the speeds are related to the angles (called *angles of refraction*) by the equation

$$\frac{c_1}{c_2} = \frac{\sin \theta_1}{\sin \theta_2}.$$

You might think this law is of interest only to physicists, but the same minimization problem shows up in other contexts, such as planning the path of a pipeline or road. First we'll use some calculus to derive Snell's law, and then look at some applications.

Let's see why the law, or something like it, should be true. Snell's law is based on the fact that in traveling from A to B, a light ray will follow the path that takes the *minimum time*. If the speeds c_1 and c_2 are equal, the shortest path will also be the fastest, so the best route is a straight line from A to B. In this case the angles θ_1 and θ_2 are equal, since they are vertical angles. But if $c_1 > c_2$, the light ray will "do better" by spending more time in the upper medium, where it travels faster, and less time in the lower medium. Therefore the point where it crosses the interface will move to the right, which will make θ_1 greater than θ_2, as shown in Figure 32.

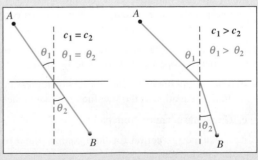

FIGURE 32

The sine is increasing on the interval $(0, \pi/2)$, so $\theta_1 > \theta_2$ implies $\sin \theta_1 > \sin \theta_2$, and Snell's law at least agrees with our intuitive reasoning about how the angles and speeds should be related.

Let's get a more complete picture of the geometric setup. Figure 33 illustrates the relationship.

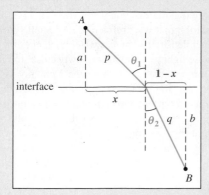

FIGURE 33

Since time is distance divided by speed, the total transit time from A to B is

$$\frac{p}{c_1} + \frac{q}{c_2}.$$

This is the expression we want to minimize, but before we can use the techniques we learned in Chapter 6 we need to choose a variable. We might try expressing both distances in terms of x and come up with an expression for the total time T as a function of x:

$$T(x) = \frac{\sqrt{a^2 + x^2}}{c_1} + \frac{\sqrt{b^2 + (1 - x)^2}}{c_2}.$$

The prospect of differentiating this expression with respect to x—setting the derivative equal to 0—and solving for x is not attractive. We'll zoom in on the point where the light ray crosses the interface, and look at what happens if we move this crossing point a little bit to the right. Using the delta notation we introduced back in Chapter 1, the picture looks like Figure 34.

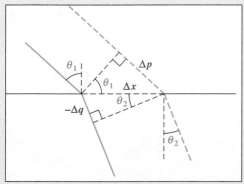

FIGURE 34

If we increase x by a small amount, Δx, the upper path gets *longer* by Δp and the lower path gets *shorter* by Δq. (Since Δq is negative, we label the triangle side with $-\Delta q$ so that it will be a positive length.) When we have found the best crossing point, the amount

we add to the time by making the upper path longer must be exactly balanced by the amount of time we save by making the lower path shorter, which means that

$$\frac{\Delta p}{c_1} = \frac{-\Delta q}{c_2} \quad \text{or} \quad \frac{c_1}{c_2} = \frac{\Delta p}{-\Delta q}.$$

Using the two small right triangles, we find that $\Delta p = \Delta x \sin \theta_1$ and $-\Delta q = \Delta x \sin \theta_2$, so

$$\frac{c_1}{c_2} = \frac{\Delta p}{-\Delta q} = \frac{\Delta x \sin \theta_1}{\Delta x \sin \theta_2} = \frac{\sin \theta_1}{\sin \theta_2},$$

which is Snell's law.

Some of the equations in this derivation are only approximate, and we will look at these approximations more closely in the exercises. But engineers use this kind of reasoning with small increments all the time, and we could make it precise and rigorous using the language of differentials introduced in Section 6.6. The argument with increments shows why the sines of the angles appear in Snell's law: They measure the rate at which moving along the interface changes the lengths of the upper and lower path segments.

Let's use Snell's law to solve an optimization problem that involves *cost* rather than *time*. A road is to be built from town A to town B. Part of the region between the towns is swampy land, over which the road will have to be elevated on a causeway. The rest of the territory is dry land on which a conventional road can be built. The territory looks like Figure 35.

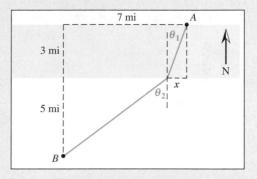

FIGURE 35

Suppose building a road over a swamp is three times more expensive per mile than building it over land. We can reduce the cost by making the road a bit longer than a straight connection and shortening the portion built over the swamp, but what is the right tradeoff? Where should the road emerge from the swamp? How long is the distance x?

This is a classic calculus problem, often solved as an exercise in simplifying complicated derivative expressions, but now that we have Snell's law, the minimization is already done. The preferred medium is the cheaper one, so our equation will look like this:

$$\frac{1}{3} = \frac{\sin \theta_1}{\sin \theta_2} \quad \text{or} \quad \sin \theta_2 = 3 \sin \theta_1.$$

Of course we don't know θ_1 and θ_2, but we have enough information to figure out x. Using the basic identities and writing everything in terms of x, we get

$$\frac{7 - x}{\sqrt{(7 - x)^2 + 5^2}} = 3 \frac{x}{\sqrt{x^2 + 3^2}}$$

The solver on your calculator will find the root easily; it's $x \approx 0.806$ miles. As we expected, the cheapest route goes almost perpendicularly across the swamp. The angle θ_1 has tangent equal to $0.806/3$, which means that θ_1 is about $15°$.

EXERCISES

1. In Figure 34, we drew the two rays coming from point A as if they were nearly parallel. Why can't they be parallel? If they *were* parallel, how would you prove that the two angles labeled θ_1 are equal? How could you make the rays more nearly parallel?

2. In Figure 34, we claim that the segment labeled Δp is the change in the length of the ray from point A. Is it? How could you improve the approximation?

3. If you wear glasses, you've probably been offered the choice between glass and "high-index plastic" for your lenses. The typical high-index plastic lens has an *index of refraction* of 1.6, which means that the ratio (speed of light in air)/(speed of light in plastic) is equal to 1.6. What percent of the speed of light in air is the speed of light in 1.6-index plastic?

4. Without doing the calculation, describe the cheapest route for the road between A and B in the case where swamp construction is 100 times as expensive as construction over land.

5. In the road construction example, what would change in the equation for x if construction over land was actually *more expensive* than construction over the swamp? Find the best x for the case where dry-land construction is twice as expensive as swamp construction, and show that the corresponding route lies to the west of the straight-line route.

6. A light ray that reaches you when you look at a sunset is bent by the same process of refraction that Snell's law describes. The higher density and higher water content of the air close to Earth cause light to travel more slowly closer to Earth, so as it moves through the atmosphere the light ray is bent down toward Earth. Rather than happening all at once at a sharp interface between one medium and another, this *atmospheric refraction* happens gradually, so the light follows a curved path. Light rays coming at an angle of $90°$ to the vertical (that is, directly from the horizon) are bent by an angle of about $0.57°$. The diameter of the sun's disk as we see it is about $0.53°$. When you see the sun begin to set, where is it actually located?*

7. At the website WolframAlpha.com, enter "Snell's law." The page allows you to calculate the angle of refraction when the angle of incidence and the index of refraction of the two media are given, or to calculate the index of refraction when the angles are given. Experiment with this page, and compare the results with those of the previous exercises.

DIRECTIONS FOR GROUP PROJECT

Prepare a demonstration of Snell's law that illustrates the phenomenon of light refraction. Your demonstration should include an explanation of what Snell's law is, why it is true, and some of its uses. Assume that the audience for this presentation has a conceptual understanding of angles but no formal studies in either trigonometry or calculus. Be sure to use Exercises 1–6 in making your presentation. Presentation software such as Microsoft PowerPoint should be used.

*See the U.S. Naval Observatory's Web site at http://aa.usno.navy.mil/AA/faq/docs/RST_defs.html.

Appendix A

Solutions to Prerequisite Skills Diagnostic Test   *(with references to Ch. R)*

For more practice on the material in questions 1–4, see *Beginning and Intermediate Algebra* (5th ed.) by Margaret L. Lial, John Hornsby, and Terry McGinnis, Pearson, 2012.

1. $10/50 = 0.20 = 20\%$

2.
$$\frac{13}{7} - \frac{2}{5} = \frac{13}{7} \cdot \frac{5}{5} - \frac{2}{5} \cdot \frac{7}{7} \qquad \text{Get a common denominator.}$$
$$= \frac{65}{35} - \frac{14}{35}$$
$$= \frac{51}{35}$$

3. The total number of apples and oranges is $x + y$, so $x + y = 75$.

4. The sentence can be rephrased as "The number of students is at least four times the number of professors," or $s \geq 4p$.

5.
$$7k + 8 = -4(3 - k)$$
$$7k + 8 = -12 + 4k \qquad\qquad\qquad \text{Multiply out.}$$
$$7k - 4k + 8 - 8 = -12 - 8 + 4k - 4k \qquad \text{Subtract 8 and } 4k \text{ from both sides.}$$
$$3k = -20 \qquad\qquad\qquad\qquad \text{Simplify.}$$
$$k = -20/3 \qquad\qquad\qquad\qquad \text{Divide both sides by 3.}$$

For more practice, see Sec. R.4.

6.
$$\frac{5}{8}x + \frac{1}{16}x = \frac{11}{16} + x$$

$$\frac{5}{8}x + \frac{1}{16}x - x = \frac{11}{16} \qquad\qquad \text{Subtract } x \text{ from both sides.}$$

$$\frac{5x}{8} \cdot \frac{2}{2} + \frac{x}{16} - x \cdot \frac{16}{16} = \frac{11}{16} \qquad\qquad \text{Get a common denominator.}$$

$$\frac{10x}{16} + \frac{x}{16} - \frac{16x}{16} = \frac{11}{16} \qquad\qquad \text{Simplify.}$$

$$\frac{-5x}{16} = \frac{11}{16} \qquad\qquad \text{Simplify.}$$

$$x = \frac{11}{16} \cdot \frac{16}{-5} = -\frac{11}{5} \qquad\qquad \text{Multiply both sides by the reciprocal of } -5/16.$$

For more practice, see Sec. R.4.

7. The interval $-2 < x \leq 5$ is written as $(-2, 5]$. For more practice, see Sec. R.5.

8. The interval $(-\infty, -3]$ is written as $x \leq -3$. For more practice, see Sec. R.5.

9.
$$5(y - 2) + 1 \leq 7y + 8$$
$$5y - 9 \leq 7y + 8 \qquad\qquad\qquad \text{Multiply out and simplify.}$$
$$5y - 7y - 9 + 9 \leq 7y - 7y + 8 + 9 \qquad \text{Subtract } 7y \text{ from both sides and add 9.}$$
$$-2y \leq 17 \qquad\qquad\qquad\qquad \text{Simplify.}$$
$$y \geq -17/2 \qquad\qquad\qquad\qquad \text{Divide both sides by 3.}$$

10.
$$\frac{2}{3}(5p - 3) > \frac{3}{4}(2p + 1)$$

$$\frac{10p}{3} - 2 > \frac{3p}{2} + \frac{3}{4} \qquad\qquad \text{Multiply out and simplify.}$$

$$\frac{10p}{3} - \frac{3p}{2} - 2 + 2 > \frac{3p}{2} - \frac{3p}{2} + \frac{3}{4} + 2 \qquad \text{Subtract } 3p/2 \text{ from both sides, and add 2.}$$

$$\frac{10p}{3} \cdot \frac{2}{2} - \frac{3p}{2} \cdot \frac{3}{3} > \frac{3}{4} + 2 \cdot \frac{4}{4} \qquad\qquad \text{Simplify and get a common denominator.}$$

$$\frac{11p}{6} > \frac{11}{4} \qquad\qquad \text{Simplify.}$$

$$p > \frac{11}{4} \cdot \frac{6}{11} \qquad\qquad \text{Multiply both sides by the reciprocal of } 11/6.$$

$$p > \frac{3}{2} \qquad\qquad \text{Simplify.}$$

For more practice, see Sec. R.5.

11. $(5y^2 - 6y - 4) - 2(3y^2 - 5y + 1) = 5y^2 - 6y - 4 - 6y^2 + 10y - 2 = -y^2 + 4y - 6$. For more practice, see Sec. R.1.

12. $(x^2 - 2x + 3)(x + 1) = (x^2 - 2x + 3)x + (x^2 - 2x + 3) = x^3 - 2x^2 + 3x + x^2 - 2x + 3 = x^3 - x^2 + x + 3$. For more practice, see Sec. R.1.

13. $(a - 2b)^2 = a^2 - 4ab + 4b^2$. For more practice, see Sec. R.1.

14. $3pq + 6p^2q + 9pq^2 = 3pq(1 + 2p + 3q)$. For more practice, see Sec. R.2.

15. $3x^2 - x - 10 = (3x + 5)(x - 2)$. For more practice, see Sec. R.2.

16.
$$\frac{a^2 - 6a}{a^2 - 4} \cdot \frac{a - 2}{a} = \frac{a(a - 6)}{(a - 2)(a + 2)} \cdot \frac{a - 2}{a} \qquad \text{Factor.}$$

$$= \frac{a - 6}{a + 2} \qquad \text{Simplify.}$$

For more practice, see Sec. R.3.

17.
$$\frac{x + 3}{x^2 - 1} + \frac{2}{x^2 + x} = \frac{x + 3}{(x - 1)(x + 1)} + \frac{2}{x(x + 1)} \qquad \text{Factor.}$$

$$= \frac{x + 3}{(x - 1)(x + 1)} \cdot \frac{x}{x} + \frac{2}{x(x + 1)} \cdot \frac{x - 1}{x - 1} \qquad \text{Get a common denominator.}$$

$$= \frac{x^2 + 3x}{x(x - 1)(x + 1)} + \frac{2x - 2}{x(x + 1)(x - 1)} \qquad \text{Multiply out.}$$

$$= \frac{x^2 + 5x - 2}{x(x - 1)(x + 1)} \qquad \text{Add fractions.}$$

For more practice, see Sec. R.3.

18.
$$3x^2 + 4x = 1$$
$$3x^2 + 4x - 1 = 0 \qquad \text{Subtract 1 from both sides.}$$

$$x = \frac{-4 \pm \sqrt{4^2 - 4 \cdot 3(-1)}}{2 \cdot 3} \qquad \text{Use the quadratic formula.}$$

$$= \frac{-4 \pm \sqrt{28}}{6} \qquad \text{Simplify.}$$

$$= \frac{-2 \pm \sqrt{7}}{2} \qquad \text{Simplify.}$$

For more practice, see Sec. R.4.

19. First solve the corresponding equation

$$\frac{8x}{z + 3} = 2$$
$$8z = 2(z + 3) \qquad \text{Multiply both sides by } (z + 3).$$
$$8z = 2z + 6 \qquad \text{Multiply out.}$$
$$6z = 6 \qquad \text{Subtract } 2z \text{ from both sides.}$$
$$z = 1 \qquad \text{Divide both sides by 6.}$$

The fraction may also change from being less than 2 to being greater than 2 when the denominator equals 0, namely, at $z = -3$. Testing each of the intervals determined by the numabers -3 and 0 shows that the fraction on the left side of the inequality is less than or equal to 2 on $(-\infty, -3) \cup [1, \infty)$. We do not include $x = -3$ in the solution because that would make the denominator 0. For more practice, see Sec. R.5.

20.
$$\frac{4^{-1}(x^2y^3)^2}{x^{-2}y^5} = \frac{x^2(x^4y^6)}{4y^5} \qquad \text{Simplify.}$$

$$= \frac{x^6y}{4} \qquad \text{Simplify.}$$

For more practice, see Sec. R.6.

21.

$$\frac{4^{1/4}(p^{2/3}q^{-1/3})^{-1}}{4^{-1/4}p^{4/3}q^{4/3}} = \frac{4^{1/2}p^{-2/3}q^{1/3}}{p^{4/3}q^{4/3}}$$ Simplify.

$$= \frac{2}{p^2 q}$$ Simplify.

For more practice, see Sec. R.6.

22.

$$k^{-1} - m^{-1} = \frac{1}{k} \cdot \frac{m}{m} - \frac{1}{m} \cdot \frac{k}{k}$$ Get a common denominator.

$$= \frac{m - k}{km}$$ Simplify.

For more practice, see Sec. R.6.

23. $(x^2 + 1)^{-1/2}(x + 2) + 3(x^2 + 1)^{1/2} = (x^2 + 1)^{-1/2}[x + 2 + 3(x^2 + 1)] = (x^2 + 1)^{-1/2}(3x^2 + x + 5)$. For more practice, see Sec. R.6.

24. $\sqrt[3]{64b^6} = 4b^2$. For more practice, see Sec. R.7.

25. $\dfrac{2}{4 - \sqrt{10}} \cdot \dfrac{4 + \sqrt{10}}{4 + \sqrt{10}} = \dfrac{2(4 + \sqrt{10})}{16 - 10} = \dfrac{4 + \sqrt{10}}{3}$. For more practice, see Sec. R.7.

26. $\sqrt{y^2 - 10y + 25} = \sqrt{(y - 5)^2} = |y - 5|$. For more practice, see Sec. R.7.

Appendix B

Learning Objectives

CHAPTER R: Algebra Reference

R.1: Polynomials

1. Simplify polynomials

R.2: Factoring

1. Factor polynomials

R.3: Rational Expressions

1. Simplify rational expression using properties of rational expressions and order of operations

R.4: Equations

1. Solve linear, quadratic, and rational equations

R.5: Inequalities

1. Solve linear, quadratic, and rational inequalities
2. Graph the solution of linear, quadratic, and rational inequalities
3. Write the solutions of linear, quadratic, and rational inequalities using interval notation

R.6: Exponents

1. Evaluate exponential expressions
2. Simplify exponential expressions

R.7: Radicals

1. Simplify radical expressions
2. Rationalize the denominator in radical expressions
3. Rationalize the numerator in radical expressions

CHAPTER 1:   Linear Functions

1.1: Slopes and Equations of Lines

1. Find the slope of a line
2. Find the equation of a line using a point and the slope
3. Find the equation of parallel and perpendicular lines
4. Graph the equation of a line
5. Solve application problems using linear functions

1.2: Linear Functions and Applications

1. Evaluate linear functions
2. Write equations for linear models
3. Solve application problems

1.3: The Least Squares Line

1. Interpret the different value meanings for linear correlation
2. Draw a scatterplot

3. Calculate the correlation coefficient
4. Calculate the least squares line
5. Compute the response variable in a linear model

CHAPTER 2:   Nonlinear Functions

2.1: Properties of Functions

1. Evaluate nonlinear functions
2. Identify the domain and range of functions
3. Classify functions as odd or even
4. Solve application problems

2.2: Quadratic Functions; Translation and Reflection

1. Complete the square for a given quadratic function
2. Obtain the vertex, y-intercepts and x-intercepts
3. Perform translation and reflection rules to graphs of functions
4. Solve application problems

2.3: Polynomial and Rational Functions

1. Perform translation and reflection rules
2. Identify the degree of a polynomial
3. Find horizontal and vertical asymptotes
4. Solve application problems

2.4: Exponential Functions

1. Evaluate exponential functions
2. Solve exponential equations
3. Graph exponential functions
4. Solve application problems

2.5: Logarithmic Functions

1. Convert expressions between exponential and logarithmic forms
2. Evaluate logarithmic expressions
3. Simplify expressions using logarithmic properties
4. Solve logarithmic equations
5. Identify the domain of a logarithmic function
6. Solve application problems

2.6: Applications: Growth and Decay; Mathematics of Finance

1. Obtain the effective rate
2. Compute future values
3. Solve growth and decay application problems
4. Solve finance application problems

CHAPTER 3:   The Derivative

3.1: Limits

1. Determine if the limit exists
2. Evaluate limits by completing tables and graphically
3. Use limit rules to evaluate limits
4. Solve application problems

3.2: Continuity

1. Identify and apply the conditions for continuity
2. Find points of discontinuity
3. Solve application problems

3.3: Rates of Change

1. Find the average rate of change for functions
2. Find the instantaneous rate of change
3. Solve application problems

3.4: Definition of the Derivative

1. Obtain the slope of the tangent line
2. Obtain the equation of the secant line
3. Find the derivative of a function (using the definition of the derivative)
4. Evaluate the derivative
5. Solve application problems

3.5: Graphical Differentiation

1. Interpret graph of a function and its derivative
2. Sketch the graph of the derivative
3. Solve application problems

CHAPTER 4:   Calculating the Derivative

4.1: Techniques for Finding Derivative

1. Determine the derivative of given functions
2. Find the slope of the tangent line
3. Obtain horizontal tangent points
4. Solve application problems

4.2: Derivative of Products and Quotients

1. Apply the product rule to find the derivative
2. Apply the quotient rule to find the derivative
3. Solve application problems

4.3: The Chain Rule

1. Find the composition of two functions
2. Find the derivative using the chain rule
3. Solve application problems

4.4: Derivatives of Exponential Functions

1. Find the derivative of exponential functions
2. Solve application problems

4.5: Derivatives of Logarithmic Functions

1. Find the derivative of logarithmic functions
2. Solve application problems

CHAPTER 5:   Graphs and the Derivative

5.1: Increasing and Decreasing Functions

1. Identify increasing and decreasing intervals for functions
2. Solve application problems

5.2: Relative Extrema

1. Identify relative extrema points graphically
2. Apply the first derivative test
3. Solve application problems

5.3: Higher Derivatives, Concavity and the Second Derivative Test

1. Find higher order derivatives
2. Identify intervals of upward and downward concavity
3. Identify inflection points
4. Apply the second derivative test
5. Solve application problems

5.4: Curve Sketching

1. Graph a function based on derivative (first and second) information

CHAPTER 6:   Applications of the Derivative

6.1: Absolute Extrema

1. Identify absolute extrema locations graphically
2. Find absolute extrema points
3. Solve application problems

6.2: Applications of Extrema

1. Solve maxima/minima application problems

6.3: Further Business Applications: Economic Lot Size; Economic Order Quantity; Elasticity of Demand

1. Solve business and economics application problems

6.4: Implicit Differentiation

1. Differentiate functions of two variables implicitly
2. Find the equation of the tangent line at a given point
3. Solve application problems

CHAPTER 10:   Differential Equations

10.1: Solution of Elementary and Separable Differential Equations

1. Find the general solution of a separable differential equations
2. Find the particular solution of a separable differential equations
3. Solve application problems

10.2: Linear First-Order Differential Equations

1. Find the general solution of a linear differential equations
2. Find the particular solution of a linear differential equations
3. Solve application problems

10.3: Euler's Method

1. Find approximate solutions to differential equations
2. Find approximate values of functions
3. Solve application problems

10.4: Applications of Differential Equations

1. Solve application problems involving differential equations

CHAPTER 11:   Probability and Calculus

11.1: Continuous Probability Models

1. Identify probability density functions
2. Find the cumulative distribution function of probability density functions
3. Solve application problems

11.2: Expected Value and Variance of Continuous Random Variables

1. Find the expected value and variance of probability density functions
2. Find the median of a random variable with a given probability density function
3. Solve application problems

11.3: Special Probability Density Functions

1. Find mean and standard deviation of a probability distribution
2. Compute the z-score
3. Solve application problems

CHAPTER 12:   Sequences and Series

12.1: Geometric Sequences

1. Identify the terms of a geometric sequence
2. Find the sum of terms in a geometric sequences
3. Solve application problems

12.2: Annuities: An Application of Sequences

1. Find future value of an annuity
2. Find the payment required for an amortized loan
3. Find the present value of an annuity
4. Solve application problems

12.3: Taylor Polynomials at 0

1. Identify the Taylor polynomials of different degrees for given functions
2. Use the Taylor polynomials to approximate values of given functions
3. Solve application problems

12.4: Infinite Series

1. Determine the convergence or divergence of a series
2. Solve application problems

12.5: Taylor Series

1. Find the Taylor series for a given function
2. Approximate areas of regions using Taylor series approximations
3. Solve application problems

12.6: Newton's Method

1. Solve equations using Newton's method
2. Solve application problems

12.7: L'Hospital's Rule

1. Find the limit of rational expressions using L'Hospitals rule

CHAPTER 13: The Trigonometric Functions

13.1: Definitions of the Trigonometric Function

1. Convert between degrees and radian measurements
2. Evaluate trigonometric functions
3. Find the amplitude and period of trigonometric functions
4. Graph trigonometric functions
5. Solve application problems

13.2: Derivatives of Trigonometric Functions

1. Find the derivative of trigonometric functions
2. Solve application problems

13.3: Integrals of Trigonometric Functions

1. Find the integral of trigonometric functions
2. Solve application problems

Appendix C

MathPrint Operating System for TI-84 and TI-84 Plus Silver Edition

The graphing calculator screens in this text display math in the format of the TI MathPrint operating system. With MathPrint, the math looks more like that seen in a printed book. You can obtain MathPrint and install it by following the instructions in the *Graphing Calculator and Spreadsheet Manual*. Only the TI-84 family of graphing calculators can be updated with the MathPrint operating system. If you own a TI-83 graphing calculator, you can use this brief appendix to help you "translate" what you see in the Classic mode shown on your calculator.

Translating between MathPrint Mode and Classic Mode
The following table compares displays of several types in MathPrint mode and Classic mode (on a calculator without MathPrint installed).

Feature	MathPrint	Classic (MathPrint not installed)
Improper fractions	$\frac{2}{5} - \frac{1}{3}$ $\frac{1}{15}$ Press **ALPHA** and F1 and select 1: n/d.	2/5 − 1/3 ► Frac 1/15 Enter an expression and press **MATH** and select 1:► Frac.
Mixed fractions	$2\frac{1}{5}*(3\frac{2}{3})$ $\frac{121}{15}$ Press **ALPHA** and F1 and select 2: Un/d.	Not Supported
Absolute values	$\lvert 10 - 15 \rvert$ 5 Press **ALPHA** and F2 and select 1:abs(.	abs(10 − 15) 5 Press **MATH** and ► and select 1:abs(.
Summation	$\sum_{I=1}^{10}(I^2)$ 385 Press **ALPHA** and F2 and select 2:Σ(.	sum(seq(I²,I,1,10)) 385 Press **2ND** and LIST and then ► twice and select 5:sum(for sum, and press **2ND** and LIST and then ► and select 5:seq(for sequence.
Numerical derivatives	$\frac{d}{dX}(X^2)\big\vert_{X=3}$ 6 Press **ALPHA** and F2 and select 3:nDeriv(.	nDeriv(X²,X,3) 6 Press **MATH** and select 8:nDeriv(.
Numerical values of integrals	$\int_1^5 (X^2)dX$ 41.33333333 Press **ALPHA** and F2 and select 4:fnInt(.	fnInt(X²,X,1,5) 41.33333333 Press **MATH** and select 9:fnInt(.

(continued)

Feature	MathPrint	Classic (MathPrint not installed)
Logarithms	$\log_2(32)$ $\qquad\qquad\qquad 5$ Press **ALPHA** and F2 and select 5:logBASE(.	Evaluating logs with bases other than 10 or e cannot be done on a graphing calculator if the MathPrint operating system is not installed. To evaluate $\log_2 32$, use the change-of-base formula: $\log(32)/\log(2)$ $\qquad\qquad\qquad 5$

The $\boxed{Y=}$ Editor

MathPrint features can be accessed from the $\boxed{Y=}$ editor as well as from the home screen. The following table shows examples that illustrate differences between MathPrint in the $\boxed{Y=}$ editor and Classic mode.

Feature	MathPrint	Classic (MathPrint not installed)
Graphing the derivative of $y = x^2$	Plot 1 Plot 2 Plot 3 $\setminus Y_1 \blacksquare \left.\dfrac{d}{dX}(X^2)\right\vert_{X=X}$ $\setminus Y_2 =$	Plot 1 Plot 2 Plot 3 $\setminus Y_1 \blacksquare$ nDeriv(X^2,X,X) $\setminus Y_2 =$
Graphing an antiderivative of $y = x^2$	Plot 1 Plot 2 Plot 3 $\setminus Y_1 \blacksquare \displaystyle\int_0^X (X^2)dX$ $\setminus Y_2 =$	Plot 1 Plot 2 Plot 3 $\setminus Y_1 \blacksquare$ fnInt(X^2,X,0,X) $\setminus Y_2 =$

Appendix D

Table 1 — Formulas from Geometry

PYTHAGOREAN THEOREM

For a right triangle with legs of lengths a and b and hypotenuse of length c, $a^2 + b^2 = c^2$.

CIRCLE

Area: $A = \pi r^2$

Circumference: $C = 2\pi r$

RECTANGLE

Area: $A = lw$

Perimeter: $P = 2l + 2w$

TRIANGLE

Area: $A = \dfrac{1}{2}bh$

SPHERE

Volume: $V = \dfrac{4}{3}\pi r^3$

Surface area: $A = 4\pi r^2$

CONE

Volume: $V = \dfrac{1}{3}\pi r^2 h$

RECTANGULAR BOX

Volume: $V = lwh$

Surface area: $A = 2lh + 2wh + 2lw$

CIRCULAR CYLINDER

Volume: $V = \pi r^2 h$

Surface area: $A = 2\pi r^2 + 2\pi rh$

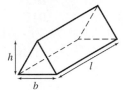

TRIANGULAR PRISM

Volume: $V = \dfrac{1}{2}bhl$

GENERAL INFORMATION ON SURFACE AREA

To find the surface area of a figure, break down the total surface area into the individual components and add up the areas of the components. For example, a rectangular box has six sides, each of which is a rectangle. A circular cylinder has two ends, each of which is a circle, plus the side, which forms a rectangle when opened up.

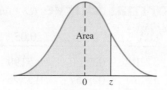

Table 2 — Area Under a Normal Curve to the Left of z, where $z = \dfrac{x - \mu}{\sigma}$

z	0.00	0.01	0.02	0.03	0.04	0.05	0.06	0.07	0.08	0.09
−3.4	0.0003	0.0003	0.0003	0.0003	0.0003	0.0003	0.0003	0.0003	0.0003	0.0002
−3.3	0.0005	0.0005	0.0005	0.0004	0.0004	0.0004	0.0004	0.0004	0.0004	0.0003
−3.2	0.0007	0.0007	0.0006	0.0006	0.0006	0.0006	0.0006	0.0005	0.0005	0.0005
−3.1	0.0010	0.0009	0.0009	0.0009	0.0008	0.0008	0.0008	0.0008	0.0007	0.0007
−3.0	0.0013	0.0013	0.0013	0.0012	0.0012	0.0011	0.0011	0.0011	0.0010	0.0010
−2.9	0.0019	0.0018	0.0017	0.0017	0.0016	0.0016	0.0015	0.0015	0.0014	0.0014
−2.8	0.0026	0.0025	0.0024	0.0023	0.0023	0.0022	0.0021	0.0021	0.0020	0.0019
−2.7	0.0035	0.0034	0.0033	0.0032	0.0031	0.0030	0.0029	0.0028	0.0027	0.0026
−2.6	0.0047	0.0045	0.0044	0.0043	0.0041	0.0040	0.0039	0.0038	0.0037	0.0036
−2.5	0.0062	0.0060	0.0059	0.0057	0.0055	0.0054	0.0052	0.0051	0.0049	0.0048
−2.4	0.0082	0.0080	0.0078	0.0075	0.0073	0.0071	0.0069	0.0068	0.0066	0.0064
−2.3	0.0107	0.0104	0.0102	0.0099	0.0096	0.0094	0.0091	0.0089	0.0087	0.0084
−2.2	0.0139	0.0136	0.0132	0.0129	0.0125	0.0122	0.0119	0.0116	0.0113	0.0110
−2.1	0.0179	0.0174	0.0170	0.0166	0.0162	0.0158	0.0154	0.0150	0.0146	0.0143
−2.0	0.0228	0.0222	0.0217	0.0212	0.0207	0.0202	0.0197	0.0192	0.0188	0.0183
−1.9	0.0287	0.0281	0.0274	0.0268	0.0262	0.0256	0.0250	0.0244	0.0239	0.0233
−1.8	0.0359	0.0352	0.0344	0.0336	0.0329	0.0322	0.0314	0.0307	0.0301	0.0294
−1.7	0.0446	0.0436	0.0427	0.0418	0.0409	0.0401	0.0392	0.0384	0.0375	0.0367
−1.6	0.0548	0.0537	0.0526	0.0516	0.0505	0.0495	0.0485	0.0475	0.0465	0.0455
−1.5	0.0668	0.0655	0.0643	0.0630	0.0618	0.0606	0.0594	0.0582	0.0571	0.0559
−1.4	0.0808	0.0793	0.0778	0.0764	0.0749	0.0735	0.0722	0.0708	0.0694	0.0681
−1.3	0.0968	0.0951	0.0934	0.0918	0.0901	0.0885	0.0869	0.0853	0.0838	0.0823
−1.2	0.1151	0.1131	0.1112	0.1093	0.1075	0.1056	0.1038	0.1020	0.1003	0.0985
−1.1	0.1357	0.1335	0.1314	0.1292	0.1271	0.1251	0.1230	0.1210	0.1190	0.1170
−1.0	0.1587	0.1562	0.1539	0.1515	0.1492	0.1469	0.1446	0.1423	0.1401	0.1379
−0.9	0.1841	0.1814	0.1788	0.1762	0.1736	0.1711	0.1685	0.1660	0.1635	0.1611
−0.8	0.2119	0.2090	0.2061	0.2033	0.2005	0.1977	0.1949	0.1922	0.1894	0.1867
−0.7	0.2420	0.2389	0.2358	0.2327	0.2296	0.2266	0.2236	0.2206	0.2177	0.2148
−0.6	0.2743	0.2709	0.2676	0.2643	0.2611	0.2578	0.2546	0.2514	0.2483	0.2451
−0.5	0.3085	0.3050	0.3015	0.2981	0.2946	0.2912	0.2877	0.2843	0.2810	0.2776
−0.4	0.3446	0.3409	0.3372	0.3336	0.3300	0.3264	0.3228	0.3192	0.3156	0.3121
−0.3	0.3821	0.3783	0.3745	0.3707	0.3669	0.3632	0.3594	0.3557	0.3520	0.3483
−0.2	0.4207	0.4168	0.4129	0.4090	0.4052	0.4013	0.3974	0.3936	0.3897	0.3859
−0.1	0.4602	0.4562	0.4522	0.4483	0.4443	0.4404	0.4364	0.4325	0.4286	0.4247
−0.0	0.5000	0.4960	0.4920	0.4880	0.4840	0.4801	0.4761	0.4721	0.4681	0.4641

(continued)

Table 2 — Area Under a Normal Curve (continued)

z	0.00	0.01	0.02	0.03	0.04	0.05	0.06	0.07	0.08	0.09
0.0	0.5000	0.5040	0.5080	0.5120	0.5160	0.5199	0.5239	0.5279	0.5319	0.5359
0.1	0.5398	0.5438	0.5478	0.5517	0.5557	0.5596	0.5636	0.5675	0.5714	0.5753
0.2	0.5793	0.5832	0.5871	0.5910	0.5948	0.5987	0.6026	0.6064	0.6103	0.6141
0.3	0.6179	0.6217	0.6255	0.6293	0.6331	0.6368	0.6406	0.6443	0.6480	0.6517
0.4	0.6554	0.6591	0.6628	0.6664	0.6700	0.6736	0.6772	0.6808	0.6844	0.6879
0.5	0.6915	0.6950	0.6985	0.7019	0.7054	0.7088	0.7123	0.7157	0.7190	0.7224
0.6	0.7257	0.7291	0.7324	0.7357	0.7389	0.7422	0.7454	0.7486	0.7517	0.7549
0.7	0.7580	0.7611	0.7642	0.7673	0.7704	0.7734	0.7764	0.7794	0.7823	0.7852
0.8	0.7881	0.7910	0.7939	0.7967	0.7995	0.8023	0.8051	0.8078	0.8106	0.8133
0.9	0.8159	0.8186	0.8212	0.8238	0.8264	0.8289	0.8315	0.8340	0.8365	0.8389
1.0	0.8413	0.8438	0.8461	0.8485	0.8508	0.8531	0.8554	0.8577	0.8599	0.8621
1.1	0.8643	0.8665	0.8686	0.8708	0.8729	0.8749	0.8770	0.8790	0.8810	0.8830
1.2	0.8849	0.8869	0.8888	0.8907	0.8925	0.8944	0.8962	0.8980	0.8997	0.9015
1.3	0.9032	0.9049	0.9066	0.9082	0.9099	0.9115	0.9131	0.9147	0.9162	0.9177
1.4	0.9192	0.9207	0.9222	0.9236	0.9251	0.9265	0.9278	0.9292	0.9306	0.9319
1.5	0.9332	0.9345	0.9357	0.9370	0.9382	0.9394	0.9406	0.9418	0.9429	0.9441
1.6	0.9452	0.9463	0.9474	0.9484	0.9495	0.9505	0.9515	0.9525	0.9535	0.9545
1.7	0.9554	0.9564	0.9573	0.9582	0.9591	0.9599	0.9608	0.9616	0.9625	0.9633
1.8	0.9641	0.9649	0.9656	0.9664	0.9671	0.9678	0.9686	0.9693	0.9699	0.9706
1.9	0.9713	0.9719	0.9726	0.9732	0.9738	0.9744	0.9750	0.9756	0.9761	0.9767
2.0	0.9772	0.9778	0.9783	0.9788	0.9793	0.9798	0.9803	0.9808	0.9812	0.9817
2.1	0.9821	0.9826	0.9830	0.9834	0.9838	0.9842	0.9846	0.9850	0.9854	0.9857
2.2	0.9861	0.9864	0.9868	0.9871	0.9875	0.9878	0.9881	0.9884	0.9887	0.9890
2.3	0.9893	0.9896	0.9898	0.9901	0.9904	0.9906	0.9909	0.9911	0.9913	0.9916
2.4	0.9918	0.9920	0.9922	0.9925	0.9927	0.9929	0.9931	0.9932	0.9934	0.9936
2.5	0.9938	0.9940	0.9941	0.9943	0.9945	0.9946	0.9948	0.9949	0.9951	0.9952
2.6	0.9953	0.9955	0.9956	0.9957	0.9959	0.9960	0.9961	0.9962	0.9963	0.9964
2.7	0.9965	0.9966	0.9967	0.9968	0.9969	0.9970	0.9971	0.9972	0.9973	0.9974
2.8	0.9974	0.9975	0.9976	0.9977	0.9977	0.9978	0.9979	0.9979	0.9980	0.9981
2.9	0.9981	0.9982	0.9982	0.9983	0.9984	0.9984	0.9985	0.9985	0.9986	0.9986
3.0	0.9987	0.9987	0.9987	0.9988	0.9988	0.9989	0.9989	0.9989	0.9990	0.9990
3.1	0.9990	0.9991	0.9991	0.9991	0.9992	0.9992	0.9992	0.9992	0.9993	0.9993
3.2	0.9993	0.9993	0.9994	0.9994	0.9994	0.9994	0.9994	0.9995	0.9995	0.9995
3.3	0.9995	0.9995	0.9995	0.9996	0.9996	0.9996	0.9996	0.9996	0.9996	0.9997
3.4	0.9997	0.9997	0.9997	0.9997	0.9997	0.9997	0.9997	0.9997	0.9997	0.9998

Table 3 — Integrals

(C is an arbitrary constant.)

---

1. $\displaystyle \int x^n\,dx = \frac{x^{n+1}}{n+1} + C \quad (\text{if } n \neq -1)$

2. $\displaystyle \int e^{kx}\,dx = \frac{e^{kx}}{k} + C$

3. $\displaystyle \int \frac{a}{x}\,dx = a\ln|x| + C$

4. $\displaystyle \int \ln|ax|\,dx = x(\ln|ax| - 1) + C$

5. $\displaystyle \int \frac{1}{\sqrt{x^2 + a^2}}\,dx = \ln\left|x + \sqrt{x^2 + a^2}\right| + C$

6. $\displaystyle \int \frac{1}{\sqrt{x^2 - a^2}}\,dx = \ln\left|x + \sqrt{x^2 - a^2}\right| + C$

7. $\displaystyle \int \frac{1}{a^2 - x^2}\,dx = \frac{1}{2a}\cdot\ln\left|\frac{a+x}{a-x}\right| + C \quad (a \neq 0)$

8. $\displaystyle \int \frac{1}{x^2 - a^2}\,dx = \frac{1}{2a}\cdot\ln\left|\frac{x-a}{x+a}\right| + C \quad (a \neq 0)$

9. $\displaystyle \int \frac{1}{x\sqrt{a^2 - x^2}}\,dx = -\frac{1}{a}\cdot\ln\left|\frac{a + \sqrt{a^2 - x^2}}{x}\right| + C \quad (a \neq 0)$

10. $\displaystyle \int \frac{1}{x\sqrt{a^2 + x^2}}\,dx = -\frac{1}{a}\cdot\ln\left|\frac{a + \sqrt{a^2 + x^2}}{x}\right| + C \quad (a \neq 0)$

11. $\displaystyle \int \frac{x}{ax + b}\,dx = \frac{x}{a} - \frac{b}{a^2}\cdot\ln|ax + b| + C \quad (a \neq 0)$

12. $\displaystyle \int \frac{x}{(ax + b)^2}\,dx = \frac{b}{a^2(ax + b)} + \frac{1}{a^2}\cdot\ln|ax + b| + C \quad (a \neq 0)$

13. $\displaystyle \int \frac{1}{x(ax + b)}\,dx = \frac{1}{b}\cdot\ln\left|\frac{x}{ax + b}\right| + C \quad (b \neq 0)$

14. $\displaystyle \int \frac{1}{x(ax + b)^2}\,dx = \frac{1}{b(ax + b)} + \frac{1}{b^2}\cdot\ln\left|\frac{x}{ax + b}\right| + C \quad (b \neq 0)$

15. $\displaystyle \int \sqrt{x^2 + a^2}\,dx = \frac{x}{2}\sqrt{x^2 + a^2} + \frac{a^2}{2}\cdot\ln\left|x + \sqrt{x^2 + a^2}\right| + C$

16. $\displaystyle \int x^n\cdot\ln|x|\,dx = x^{n+1}\left[\frac{\ln|x|}{n+1} - \frac{1}{(n+1)^2}\right] + C \quad (n \neq -1)$

17. $\displaystyle \int x^n e^{ax}\,dx = \frac{x^n e^{ax}}{a} - \frac{n}{a}\cdot\int x^{n-1}e^{ax}\,dx + C \quad (a \neq 0)$

Table 4 — Integrals Involving Trigonometric Functions

18. $\int \sin u\,du = -\cos u + C$

19. $\int \cos u\,du = \sin u + C$

20. $\int \sec^2 u\,du = \tan u + C$

21. $\int \csc^2 u\,du = -\cot u + C$

22. $\int \sec u \tan u\,du = \sec u + C$

23. $\int \csc u \cot u\,du = -\csc u + C$

24. $\int \tan u\,du = \ln|\sec u| + C$

25. $\int \cot u\,du = \ln|\sin u| + C$

26. $\int \sec u\,du = \ln|\sec u + \tan u| + C$

27. $\int \csc u\,du = \ln|\csc u - \cot u| + C$

28. $\int \sin^n u\,du = -\dfrac{1}{n}\sin^{n-1} u \cos u + \dfrac{n-1}{n}\int \sin^{n-2} u\,du \quad (n \neq 0)$

29. $\int \cos^n u\,du = \dfrac{1}{n}\cos^{n-1} u \sin u + \dfrac{n-1}{n}\int \cos^{n-2} u\,du \quad (n \neq 0)$

30. $\int \tan^n u\,du = \dfrac{1}{n-1}\tan^{n-1} u - \int \tan^{n-2} u\,du \quad (n \neq 1)$

31. $\int \sec^n u\,du = \dfrac{1}{n-1}\tan u \sec^{n-2} u + \dfrac{n-2}{n-1}\int \sec^{n-2} u\,du \quad (n \neq 1)$

32. $\int \sin au \sin bu\,du = \dfrac{\sin(a-b)u}{2(a-b)} - \dfrac{\sin(a+b)u}{2(a+b)} + C, \quad |a| \neq |b|$

33. $\int \cos au \cos bu\,du = \dfrac{\sin(a-b)u}{2(a-b)} + \dfrac{\sin(a+b)u}{2(a+b)} + C, \quad |a| \neq |b|$

34. $\int \sin au \cos bu\,du = -\dfrac{\cos(a-b)u}{2(a-b)} - \dfrac{\cos(a+b)u}{2(a+b)} + C, \quad |a| \neq |b|$

35. $\int u \sin u\,du = \sin u - u \cos u + C$

36. $\int u^n \sin u\,du = -u^n \cos u + n\int u^{n-1} \cos u\,du$

37. $\int e^{au} \sin bu\,du = \dfrac{e^{au}}{a^2 + b^2}(a \sin bu - b \cos bu) + C$

38. $\int e^{au} \cos bu\,du = \dfrac{e^{au}}{a^2 + b^2}(a \cos bu + b \sin bu) + C$

Answers to Selected Exercises

Answers to selected writing exercises are provided.

Answers to Prerequisite Skills Test

1. 20%   **2.** 51/35   **3.** $x + y = 75$ **4.** $s \geq 4p$ **5.** $-20/3$ (Sec. R.4)   **6.** $-11/5$ (Sec. R.4)   **7.** $(-2, 5]$ (Sec. R.5)
8. $x \leq -3$ (Sec. R.5)   **9.** $y \geq -17/2$ (Sec. R.5)   **10.** $p > 3/2$ (Sec. R.5)   **11.** $-y^2 + 4y - 6$ (Sec. R.1)
12. $x^3 - x^2 + x + 3$ (Sec. R.1)   **13.** $a^2 - 4ab + 4b^2$ (Sec. R.1)   **14.** $3pq(1 + 2p + 3q)$ (Sec. R.2)   **15.** $(3x + 5)(x - 2)$
(Sec. R.2)   **16.** $(a - 6)/(a + 2)$ (Sec. R.3)   **17.** $(x^2 + 5x - 2)/[x(x - 1)(x + 1)]$ (Sec. R.3)   **18.** $(-2 \pm \sqrt{7})/3$ (Sec. R.4)
19. $[-\infty, -3) \cup [1, \infty)$ (Sec. R.5)   **20.** $x^6 y/4$ (Sec. R.6)   **21.** $2/(p^2 q)$ (Sec. R.6)   **22.** $(m - k)/(km)$ (Sec. R.6)
23. $(x^2 + 1)^{-1/2}(3x^2 + x + 5)$ (Sec. R.6)   **24.** $4b^2$ (Sec. R.7)   **25.** $(4 + \sqrt{10})/3$ (Sec. R.7)   **26.** $|y - 5|$ (Sec. R.7)

Chapter R Algebra Reference

Exercises R.1 (page R-5)

For exercises . . .	1–6	7–8,15–22	9–14	23–26
Refer to example . . .	2	3	4	5

1. $-x^2 + x + 9$ **2.** $-6y^2 + 3y + 10$ **3.** $-16q^2 + 4q + 6$
4. $9r^2 - 4r + 19$ **5.** $-0.327x^2 - 2.805x - 1.458$ **6.** $0.8r^2 + 3.6r - 1.5$ **7.** $-18m^3 - 27m^2 + 9m$ **8.** $-12x^4 + 30x^2 + 36x$
9. $9t^2 + 9ty - 10y^2$ **10.** $18k^2 - 7kq - q^2$ **11.** $4 - 9x^2$ **12.** $36m^2 - 25$ **13.** $(6/25)y^2 + (11/40)yz + (1/16)z^2$
14. $(15/16)r^2 - (7/12)rs - (2/9)s^2$ **15.** $27p^3 - 1$ **16.** $15p^3 + 13p^2 - 10p - 8$ **17.** $8m^3 + 1$ **18.** $12k^4 + 21k^3 - 5k^2 + 3k + 2$
19. $3x^2 + xy + 2xz - 2y^2 - 3yz - z^2$ **20.** $2r^2 + 2rs - 5rt - 4s^2 + 8st - 3t^2$ **21.** $x^3 + 6x^2 + 11x + 6$ **22.** $x^3 - 2x^2 - 5x + 6$
23. $x^2 + 4x + 4$ **24.** $4a^2 - 16ab + 16b^2$ **25.** $x^3 - 6x^2 y + 12xy^2 - 8y^3$ **26.** $27x^3 + 27x^2 y + 9xy^2 + y^3$

Exercises R.2 (page R-7)

For exercises . . .	1–4	5–15	16–20	21–32
Refer to example . . .	1	3	2nd CAUTION	4

1. $7a^2(a + 2)$ **2.** $3y(y^2 + 8y + 3)$ **3.** $13p^2 q(p^2 q - 3p + 2q)$
4. $10m^2(6m^2 - 12mn + 5n^2)$ **5.** $(m + 2)(m - 7)$ **6.** $(x + 5)(x - 1)$ **7.** $(z + 4)(z + 5)$ **8.** $(b - 7)(b - 1)$
9. $(a - 5b)(a - b)$ **10.** $(s - 5t)(s + 7t)$ **11.** $(y - 7z)(y + 3z)$ **12.** $(3x + 7)(x - 1)$ **13.** $(3a + 7)(a + 1)$
14. $(5y + 2)(3y - 1)$ **15.** $(7m + 2n)(3m + n)$ **16.** $6(a - 10)(a + 2)$ **17.** $3m(m + 3)(m + 1)$ **18.** $2(2a + 3)(a + 1)$
19. $2a^2(4a - b)(3a + 2b)$ **20.** $12x^2(x - y)(2x + 5y)$ **21.** $(x + 8)(x - 8)$ **22.** $(3m + 5)(3m - 5)$ **23.** $10(x + 4)(x - 4)$
24. Prime   **25.** $(z + 7y)^2$ **26.** $(s - 5t)^2$ **27.** $(3p - 4)^2$ **28.** $(a - 6)(a^2 + 6a + 36)$ **29.** $(3r - 4s)(9r^2 + 12rs + 16s^2)$
30. $3(m + 5)(m^2 - 5m + 25)$ **31.** $(x - y)(x + y)(x^2 + y^2)$ **32.** $(2a - 3b)(2a + 3b)(4a^2 + 9b^2)$

Exercises R.3 (page R-10)

For exercises . . .	1–12	13–38
Refer to example . . .	1	2

1. $v/7$ **2.** $5p/2$ **3.** $8/9$ **4.** $2/(t + 2)$ **5.** $x - 2$ **6.** $4(y + 2)$ **7.** $(m - 2)/(m + 3)$
8. $(r + 2)/(r + 4)$ **9.** $3(x - 1)/(x - 2)$ **10.** $(z - 3)/(z + 2)$ **11.** $(m^2 + 4)/4$ **12.** $(2y + 1)/(y + 1)$ **13.** $3k/5$
14. $25p^2/9$ **15.** $9/(5c)$ **16.** 2   **17.** $1/4$ **18.** $3/10$ **19.** $2(a + 4)/(a - 3)$ **20.** $2/(r + 2)$ **21.** $(k - 2)/(k + 3)$
22. $(m + 6)/(m + 3)$ **23.** $(m - 3)/(2m - 3)$ **24.** $2(2n - 1)/(3n - 5)$ **25.** 1   **26.** $(6 + p)/(2p)$ **27.** $(12 - 15y)/(10y)$
28. $137/(30m)$ **29.** $(3m - 2)/[m(m - 1)]$ **30.** $(r - 6)/[r(2r + 3)]$ **31.** $14/[3(a - 1)]$ **32.** $23/[20(k - 2)]$
33. $(7x + 1)/[(x - 2)(x + 3)(x + 1)]$ **34.** $(y^2 + 1)/[(y + 3)(y + 1)(y - 1)]$ **35.** $k(k - 13)/[(2k - 1)(k + 2)(k - 3)]$
36. $m(3m - 19)/[(3m - 2)(m + 3)(m - 4)]$ **37.** $(4a + 1)/[a(a + 2)]$ **38.** $(5x^2 + 4x - 4)/[x(x - 1)(x + 1)]$

Exercises R.4 (page R-16)

For exercises . . .	1–8	9–26	27–37
Refer to example . . .	2	3–5	6,7

1. -12 **2.** $3/4$ **3.** 12   **4.** $-3/8$ **5.** $-7/8$ **6.** $-6/11$ **7.** 4   **8.** $-10/19$ **9.** $-3, -2$
10. $-1, 3$ **11.** 7   **12.** $-2, 5/2$ **13.** $-1/4, 2/3$ **14.** $2, 5$ **15.** $-3, 3$ **16.** $-4, 1/2$ **17.** $0, 4$ **18.** $(5 + \sqrt{13})/6 \approx 1.434$,
$(5 - \sqrt{13})/6 \approx 0.232$ **19.** $(2 + \sqrt{10})/2 \approx 2.581, (2 - \sqrt{10})/2 \approx -0.581$ **20.** $(-1 + \sqrt{5})/2 \approx 0.618, (-1 - \sqrt{5})/2 \approx$
-1.618 **21.** $5 + \sqrt{5} \approx 7.236, 5 - \sqrt{5} \approx 2.764$ **22.** $(4 + \sqrt{6})/5 \approx 1.290, (4 - \sqrt{6})/5 \approx 0.310$ **23.** $1, 5/2$ **24.** No real
number solutions   **25.** $(-1 + \sqrt{73})/6 \approx 1.257, (-1 - \sqrt{73})/6 \approx -1.591$ **26.** $-1, 0$ **27.** 3   **28.** 12   **29.** $-59/6$ **30.** 6
31. 3   **32.** $-5/2$ **33.** $2/3$ **34.** 1   **35.** 2   **36.** No solution   **37.** No solution

Exercises R.5 (page R-21)

For exercises . . .	1–14	15–26	27–38	39–42	43–54
Refer to example . . .	Figure 1, Example 2	2	3	4	5–7

1. $(-\infty, 4)$

2. $[-3, \infty)$

3. $[1, 2)$

4. $[-2, 3]$

5. $(-\infty, -9)$

6. $[6, \infty)$

7. $-7 \leq x \leq -3$ **8.** $4 \leq x < 10$

9. $x \leq -1$ **10.** $x > 3$ **11.** $-2 \leq x < 6$ **12.** $0 < x < 8$ **13.** $x \leq -4$ or $x \geq 4$ **14.** $x < 0$ or $x \geq 3$

15. $(-\infty, 2]$

16. $(-\infty, 1)$

17. $(3, \infty)$

A-15

18. $(-\infty, 1]$

19. $(1/5, \infty)$

20. $(1/3, \infty)$

21. $(-4, 6)$

22. $[7/3, 4]$

23. $[-5, 3)$

24. $[-1, 2]$

25. $[-17/7, \infty)$

26. $(-\infty, 50/9]$

27. $(-5, 3)$

28. $(-\infty, -6] \cup [1, \infty)$

29. $(1, 2)$

30. $(-\infty, -4) \cup (1/2, \infty)$

31. $(-\infty, -4) \cup (4, \infty)$

32. $[-3/2, 5]$

33. $(-\infty, -1] \cup [5, \infty)$

34. $[-1/2, 2/5]$

35. $(-\infty, -1) \cup (1/3, \infty)$

36. $(-\infty, -2) \cup (5/3, \infty)$

37. $(-\infty, -3] \cup [3, \infty)$

38. $(-\infty, 0) \cup (16, \infty)$

39. $[-2, 0] \cup [2, \infty)$

40. $(-\infty, -4] \cup [-3, 0]$

41. $(-\infty, 0) \cup (1, 6)$

42. $(-1, 0) \cup (4, \infty)$ **43.** $(-5, 3]$ **44.** $(-\infty, -1) \cup (1, \infty)$ **45.** $(-\infty, -2)$

46. $(-2, 3/2)$ **47.** $[-8, 5)$ **48.** $(-\infty, -3/2) \cup [-13/9, \infty)$ **49.** $[2, 3)$ **50.** $(-\infty, -1)$ **51.** $(-2, 0] \cup (3, \infty)$

52. $(-4, -2) \cup (0, 2)$ **53.** $(1, 3/2]$ **54.** $(-\infty, -2) \cup (-2, 2) \cup [4, \infty)$

Exercises R.6 (page R-25)

For exercises . . .	1–8	9–26	27–36	37–50	51–56
Refer to example . . .	1	2	3,4	5	6

1. $1/64$ **2.** $1/81$ **3.** 1 **4.** 1 **5.** $-1/9$ **6.** $1/9$ **7.** 36 **8.** $27/64$
9. $1/64$ **10.** 8^5 **11.** $1/10^8$ **12.** 7 **13.** x^2 **14.** 1 **15.** 2^3k^3 **16.** $1/(3z^7)$ **17.** $x^5/(3y^3)$ **18.** $m^3/5^4$ **19.** a^3b^6
20. $49/(c^6d^4)$ **21.** $(a + b)/(ab)$ **22.** $(1 - ab^2)/b^2$ **23.** $2(m - n)/[mn(m + n^2)]$ **24.** $(3n^2 + 4m)/(mn^2)$
25. $xy/(y - x)$ **26.** $y^4/(xy - 1)^2$ **27.** 11 **28.** 3 **29.** 4 **30.** -25 **31.** $1/2$ **32.** $4/3$ **33.** $1/16$ **34.** $1/5$ **35.** $4/3$
36. $1000/1331$ **37.** 9 **38.** 3 **39.** 64 **40.** 1 **41.** x^4/y^4 **42.** b/a^3 **43.** r **44.** $12^3/y^8$ **45.** $3k^{3/2}/8$ **46.** $1/(2p^2)$
47. $a^{2/3}b^2$ **48.** $y^2/(x^{1/6}z^{5/4})$ **49.** $h^{1/3}t^{1/5}/k^{2/5}$ **50.** m^3p/n **51.** $3x(x^2 + 3x)^2(x^2 - 5)$ **52.** $6x(x^3 + 7)(-2x^3 - 5x + 7)$
53. $5x(x^2 - 1)^{-1/2}(x^2 + 1)$ **54.** $3(6x + 2)^{-1/2}(27x + 5)$ **55.** $(2x + 5)(x^2 - 4)^{-1/2}(4x^2 + 5x - 8)$
56. $(4x^2 + 1)(2x - 1)^{-1/2}(36x^2 - 16x + 1)$

Exercises R.7 (page R-28)

For exercises . . .	1–22	23–26	27–40	41–44
Refer to example . . .	1,2	3	4	5

1. 5 **2.** 6 **3.** -5 **4.** $5\sqrt{2}$ **5.** $20\sqrt{5}$ **6.** $4y^2\sqrt{2y}$ **7.** 9 **8.** 8
9. $7\sqrt{2}$ **10.** $9\sqrt{3}$ **11.** $9\sqrt{7}$ **12.** $-2\sqrt{7}$ **13.** $5\sqrt[3]{2}$ **14.** $3\sqrt[3]{5}$ **15.** $xyz^2\sqrt{2x}$ **16.** $4r^3s^4t^6\sqrt{10rs}$ **17.** $4xy^2z^3\sqrt[3]{2y^2}$
18. $x^2yz^2\sqrt[4]{y^3z^3}$ **19.** $ab\sqrt{ab}(b - 2a^2 + b^3)$ **20.** $p^2\sqrt{pq}(pq - q^4 + p^2)$ **21.** $\sqrt[6]{a^5}$ **22.** $b^2\sqrt[4]{b}$ **23.** $|4 - x|$
24. $|3y + 5|$ **25.** Cannot be simplified **26.** Cannot be simplified **27.** $5\sqrt{7}/7$ **28.** $\sqrt{10}/2$ **29.** $-\sqrt{3}/2$ **30.** $\sqrt{2}$
31. $-3(1 + \sqrt{2})$ **32.** $-5(2 + \sqrt{6})/2$ **33.** $3(2 - \sqrt{2})$ **34.** $(5 - \sqrt{10})/3$ **35.** $(\sqrt{r} + \sqrt{3})/(r - 3)$
36. $5(\sqrt{m} + \sqrt{5})/(m - 5)$ **37.** $\sqrt{y} + \sqrt{5}$ **38.** $(z + \sqrt{5z} - \sqrt{z} - \sqrt{5})/(z - 5)$ **39.** $-2x - 2\sqrt{x(x + 1)} - 1$
40. $[p^2 + p + 2\sqrt{p(p^2 - 1)} - 1]/(-p^2 + p + 1)$ **41.** $-1/[2(1 - \sqrt{2})]$ **42.** $1/(3 + \sqrt{3})$
43. $-1/[2x - 2\sqrt{x(x + 1)} + 1]$ **44.** $2/[p + \sqrt{p(p - 2)}]$

Chapter 1 Linear Functions

Exercises 1.1 (page 13)

For exercises . . .	1–4	5–8	13,29,30	14,31–34	15–17	18,27	19–24	25,26	28	45–60	61–75
Refer to example . . .	1	3	8	9	4	7	5	2	6	11–13	10,14

1. $3/5$ **3.** Not defined **5.** 1
7. $5/9$ **9.** Not defined **11.** 0 **13.** 2 **15.** $y = -2x + 5$ **17.** $y = -7$ **19.** $y = -(1/3)x + 10/3$ **21.** $y = 6x - 7/2$

23. $x = -8$ **25.** $x + 2y = -6$ **27.** $x = -6$ **29.** $3x + 2y = 0$ **31.** $x - y = 7$ **33.** $5x - y = -4$ **35.** No **39.** a **41.** -4

45. 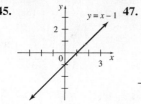 **47.** 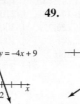 **49.** 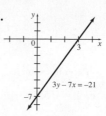 **51.**

53. 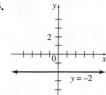 **55.**   **57.** 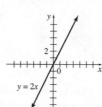 **59.**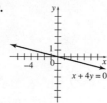

61. a. 12,000 $y = 12,000x + 3000$ **b.** 8 years 1 month **63. a.** $y = 4.612t + 86.164$ **b.** 178.4, which is slightly more than the actual CPI. **c.** It is increasing at a rate of approximately 4.6 per year. **65. a.** $u = 0.85(220 - x) = 187 - 0.85x$, $l = 0.7(220 - x) = 154 - 0.7x$ **b.** 140 to 170 beats per minute. **c.** 126 to 153 beats per minute. **d.** The women are 16 and 52. Their pulse is 143 beats per minute. **67.** Approximately 86 yr **69. a.** $y = 0.3444t + 14.1$ **b.** 2022
71. a. $y = 14,792.05t - 490,416$ **b.** 1,210,670 **73. a.** There appears to be a linear relationship. **b.** $y = 76.9x$ **c.** About 780 megaparsecs (about 1.5×10^{22} mi) **d.** About 12.4 billion yr
75. a.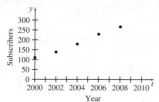

Yes, the data are approximately linear.
b. $y = 1133.4t + 16,072$; the slope 1133.4 indicates that tuition and fees have increased approximately \$1133 per year. **c.** The year 2025 is too far in the future to rely on this equation to predict costs; too many other factors may influence these costs by then.

Exercises 1.2 (page 23)

1. -3 **3.** 22 **5.** 0 **7.** -4 **9.** $7 - 5t$ **11.** True

For exercises …	1–10	19–22	23–26	27–32	33–36,48	37–44	45–47
Refer to example …	1	4	5	2,3	5	6	7

13. True **19.** If $R(x)$ is the cost of renting a snowboard for x hours, then $R(x) = 2.25x + 10$. **21.** If $C(x)$ is the cost of parking a car for x half-hours, then $C(x) = 0.75x + 2$. **23.** $C(x) = 30x + 100$ **25.** $C(x) = 75x + 550$ **27. a.** \$16 **b.** \$11 **c.** \$6
d. 640 watches **e.** 480 watches **f.** 320 watches
g. 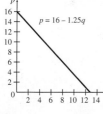 **h.** 0 watches **i.** About 1333 watches **j.** About 2667 watches
k. 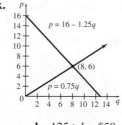 **l.** 800 watches, \$6

29. a. 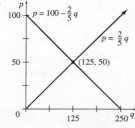 **b.** 125 tubs, \$50 **31.** $D(q) = 6.9 - 0.4q$ **33. a.** $C(x) = 3.50x + 90$ **b.** 17 shirts
c. 108 shirts **35. a.** $C(x) = 0.097x + 1.32$ **b.** \$1.32 **c.** \$98.32
d. \$98.42 **e.** 9.7¢ **f.** 9.7¢, the cost of producing one additional cup of coffee would be 9.7¢. **37. a.** 2 units **b.** \$980 **c.** 52 units
39. Break-even quantity is 45 units; don't produce; $P(x) = 20x - 900$
41. Break-even quantity is -50 units; impossible to make a profit when $C(x) > R(x)$ for all positive x; $P(x) = -10x - 500$ (always a loss)
43. 5 **45. a.** 14.4°C **b.** -28.9°C **c.** 122°C **47.** -40°

Exercises 1.3 (page 32)

For exercises . . .	3b,4,10d, 11d,12e,13e, 15b,16d,17b, 18d,19c,20b, 22d,24c,25d, 26a	3c,10a,11a, 12a,13a,15c, 16b,17c,18a, 19a,21c,22a, 23a,24b, 25a,26b	3d,10b,11b, 12c,13c,18b, 21d,22bc,23b, 26c	5ab,6ab, 7a,8a, 14ab,21ab	5c,6c, 15e	10c,11c, 12d,13d, 18c,23c
Refer to example . . .	4	1	2	1,4	5	3

3. a. 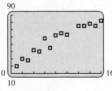   **b.** 0.993   **c.** $Y = 0.555x - 0.5$ 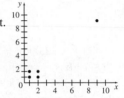   **d.** 5.6

5. a. $Y = 0.9783x + 0.0652, 0.9783$ **b.** $Y = 1.5, 0$ **c.** The point (9,9) is an outlier that has a strong effect on the least squares line and the correlation coefficient.

7. a. 0.7746   **b.**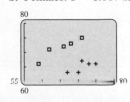

11. a. $Y = -0.1534x + 11.36$ **b.** About 6760   **c.** 2025   **d.** -0.9890 **13. a.** $Y = 97.73x + 1833.3$ **b.** About \$97.73 billion per year   **c.** \$3299 billion   **d.** 2023   **e.** 0.9909

15. a. 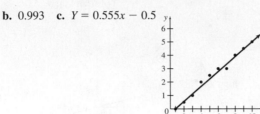   They lie in a linear pattern.   **b.** $r = 0.693$; there is a positive correlation between the price and the distance.   **c.** $Y = 0.0738x + 111.83$; the marginal cost is 7.38 cents per mile.
d. In 2000 marginal cost was 2.43 cents per mile; it has increased to 7.38 cents per mile.
e. Phoenix

17. a. 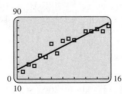   **b.** 0.959, yes   **c.** $Y = 3.98x + 22.7$ **19. a.** $Y = -0.1175x + 17.74$ **b.** 14.2   **c.** -0.9326

21. a. $Y = -0.08915x + 74.28, r = -0.1035$. The taller the student, the shorter is the ideal partner's height.
b. Females: $Y = 0.6674x + 27.89, r = 0.9459$; males: $Y = 0.4348x + 34.04, r = 0.7049$

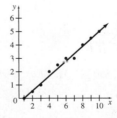

23. a. (graph)   **b.** $Y = 0.366x + 0.803$; the line seems to fit the data.

c. $r = 0.995$ indicates a good fit, which confirms the conclusion in part b.   **25. a.** -0.995; yes   **b.** $Y = -0.0769x + 5.91$ **c.** 2.07 points
27. a. 4.298 miles per hour   **b.** 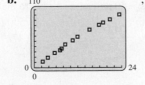 ; yes   **c.** $Y = 4.317x + 3.419$ **d.** 0.9971; yes   **e.** 4.317 miles per hour

Chapter 1 Review Exercises (page 39)

For exercises ...	1–6,15–44	7–10,13,45–59b,62,63	11,12,14,59c–61,64,65
Refer to section ...	1	2	3

1. False **2.** False **3.** True **4.** False **5.** True **6.** False **7.** True **8.** False **9.** False **10.** False **11.** False **12.** True
15. 1 **17.** $-2/11$ **19.** $-4/3$ **21.** 0 **23.** 5 **25.** $y = (2/3)x - 13/3$ **27.** $y = -x - 3$ **29.** $y = -10$ **31.** $2x - y = 10$
33. $x = -1$ **35.** $y = -5$

37. 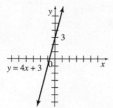 **39.** 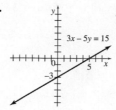 **41.** 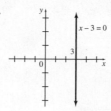 **43.**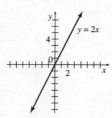

45. a. $E = 352 + 42x$ (where x is in thousands) **b.** $R = 130x$ (where x is in thousands) **c.** More than 4000 chips
47. $S(q) = 0.5q + 10$ **49.** $41.25, 62.5$ diet pills **51.** $C(x) = 180x + 2000$ **53.** $C(x) = 46x + 120$ **55. a.** 40 pounds
b. $280 **57.** $y = 7.23t + 11.9$ **59. a.** $y = 836x + 7500$ **b.** $y = 795x + 8525$ **c.** $Y = 843.7x + 7662$
d.

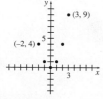

f. 0.9995 **61. a.** $Y = 0.9724x + 31.43$ **b.** About 216 **c.** 0.9338 **63.** $y = 1.22t + 48.9$
65. a. 0.6998; yes, but the fit is not very good. **b.**
c. $Y = 3.396x + 117.2$ **d.** $3396

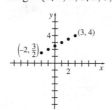

Chapter 2 Nonlinear Functions

Exercises 2.1 (page 53)

For exercises ...	1–8	9–16	17–32	33–40	41–56,76,77	57–62	63–70	71–74	75,78	79,80
Refer to example ...	1,2	3 (b)	3(a), (c)–(e)	2 (d)	4	5	6	7	Dow Jones	8

1. Not a function **3.** Function **5.** Function **7.** Not a function
9. $(-2, -1), (-1, 1), (0, 3), (1, 5), (2, 7), (3, 9)$; **11.** $(-2, 3/2), (-1, 2), (0, 5/2),$
range: $\{-1, 1, 3, 5, 7, 9\}$ $(1, 3), (2, 7/2), (3, 4);$ **13.** $(-2, 0), (-1, -1), (0, 0), (1, 3),$
range: $\{3/2, 2, 5/2, 3, 7/2, 4\}$ $(2, 8), (3, 15)$; range: $\{-1, 0, 3, 8, 15\}$

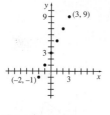 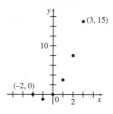

15. $(-2, 4), (-1, 1), (0, 0),$
$(1, 1), (2, 4), (3, 9)$; range: $\{0, 1, 4, 9\}$

17. $(-\infty, \infty)$ **19.** $(-\infty, \infty)$ **21.** $[-2, 2]$ **23.** $[3, \infty)$ **25.** $(-\infty, -1) \cup (-1, 1) \cup (1, \infty)$
27. $(-\infty, -4) \cup (4, \infty)$ **29.** $(-\infty, -1] \cup [5, \infty)$ **31.** $(-\infty, -1) \cup (1/3, \infty)$ **33.** Domain:
$[-5, 4)$; range: $[-2, 6]$ **35.** Domain: $(-\infty, \infty)$; range: $(-\infty, 12]$ **37.** Domain: $[-2, 4]$; range: $[0, 4]$
a. 0 **b.** 4 **c.** 3 **d.** $-1.5, 1.5, 2.5$ **39.** Domain: $[-2, 4]$; range: $[-3, 2]$ **a.** -3 **b.** -2
c. -1 **d.** 2.5 **41. a.** 33 **b.** 15/4 **c.** $3a^2 - 4a + 1$ **d.** $12/m^2 - 8/m + 1$ or
$(12 - 8m + m^2)/m^2$ **e.** $0, 4/3$ **43. a.** 7 **b.** 0 **c.** $(2a + 1)/(a - 4)$ if $a \neq 4, 7$ if $a = 4$
d. $(4 + m)/(2 - 4m)$ if $m \neq 1/2, 7$ if $m = 1/2$ **e.** -5 **45.** $6t^2 + 12t + 4$ **47.** $r^2 + 2rh + h^2 - 2r - 2h + 5$
49. $9/q^2 - 6/q + 5$ or $(9 - 6q + 5q^2)/q^2$ **51. a.** $2x + 2h + 1$ **b.** $2h$ **c.** 2 **53. a.** $2x^2 + 4xh + 2h^2 - 4x - 4h - 5$
b. $4xh + 2h^2 - 4h$ **c.** $4x + 2h - 4$ **55. a.** $1/(x + h)$ **b.** $-h/[x(x + h)]$ **c.** $-1/[x(x + h)]$ **57.** Function

59. Not a Function **61.** Function **63.** Odd **65.** Even **67.** Even **69.** Odd **71. a.** $36 **b.** $36 **c.** $64 **d.** $120 **e.** $120 **f.** $148 **g.** $148 **i.** x, the number of full and partial days **j.** S, the cost of renting a saw **73. a.** $93,300. Attorneys can receive a maximum of $93,300 on a jury award of $250,000. **b.** $124,950. Attorneys can receive a maximum of $124,950 on a jury award of $350,000. **c.** $181,950. Attorneys can receive a maximum of $181,950 on a jury award of $550,000.

d.

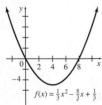

75. a. About 140 m **b.** About 250 m **77. a. i.** 3.6 kcal/km **ii.** 61 kcal/km **b.** $x = g(z) = 1000z$ **c.** $y = 4.4z^{0.88}$ **79. a.** $P(w) = 1000/w + 2w$ **b.** $(0, \infty)$

c.

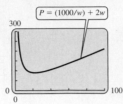

Exercises 2.2 (page 64)

For exercises . . .	3–8	9–24,58–62,64–67	25–38,47,48	39–46	49–53	54–57,68,69
Refer to example . . .	1–3	4	8	Before Example 8	7	6

3. D **5.** A **7.** C

9. $y = 3(x + 3/2)^2 - 7/4, (-3/2, -7/4)$, x-intercepts are -3 and -2; y-intercept is 6. **11.** $y = -2(x - 2)^2 - 1, (2, -1)$ **13.** Vertex is $(-5/2, -1/4)$; axis is $x = -5/2$; x-intercepts are -3 and -2; y-intercept is 6.

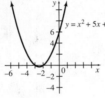

15. Vertex is $(-3, 2)$; axis is $x = -3$; x-intercepts are -4 and -2; y-intercept is -16.

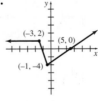

17. Vertex is $(-2, -16)$; axis is $x = -2$; x-intercepts are $-2 \pm 2\sqrt{2} \approx 0.83$ or -4.83; y-intercept is -8.

19. Vertex is $(1, 3)$; axis is $x = 1$; no x-intercepts; y-intercept is 5.

21. Vertex is $(4, 11)$; axis is $x = 4$; x-intercepts are $4 \pm \sqrt{22}/2 \approx 6.35$ or 1.65; y-intercept is -21.

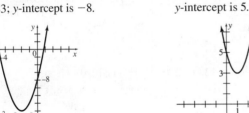

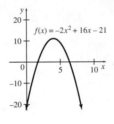

23. Vertex is $(4, -5)$; axis is $x = 4$; x-intercepts are $4 \pm \sqrt{15} \approx 7.87$ or 0.13; y-intercept is 1/3.

25. D **27.** C **29.** E **31.**

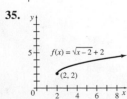

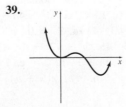

33.

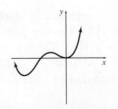

35. **37.** **39.** **41.**

43.
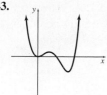

45.
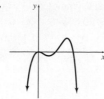

47. a. r **b.** $-r$ **c.** $-r$ **49. a.**
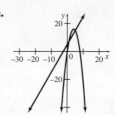
b. 1 batch
c. $16,000
d. $4000

51. a.

b. 2.5 batches **c.** $31,250 **d.** $5000 **53.** Maximum revenue is $9225; 35 seats are unsold.

55. a. $R(x) = x(500 - x) = 500x - x^2$ **b.**
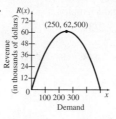
c. $250 **d.** $62,500

57. a. $800 + 25x$ **b.** $80 - x$ **c.** $R(x) = 64,000 + 1200x - 25x^2$ **d.** 24 **e.** $78,400 **59. a.** 87 yr **b.** 98 yr

61. a. 28.5 weeks **b.** 0.81 **c.** 0 weeks or 57 weeks of gestation; no

63. a.
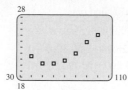
b. Quadratic **c.** $y = 0.002726x^2 - 0.3113x + 29.33$;
d. $f(x) = 0.003(x - 60)^2 + 20.3$
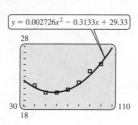

e.
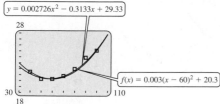

65. 49 yr; 3.98 **67. a.** 61.70 ft **b.** 43.08 mph **69.** 9025 ft^2

71. $y = (4/27)x^2$; $6\sqrt{3}$ ft $\approx$ 10.39 ft
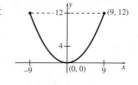

Exercises 2.3 (page 73)

3.
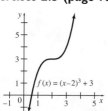

5.
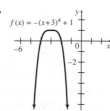

For exercises . . .	3–6	7–15,21–26	16–20,27–42,57,58	46,47,50–52	54–56,59
Refer to example . . .	1	4	6	7	2,3

7. D **9.** E **11.** I **13.** G **15.** A **17.** D **19.** E **21.** 4, 6, etc.
(true degree $= 4$); $+$ **23.** 5, 7, etc. (true degree $= 5$); $+$
25. 7, 9, etc. (true degree $= 7$); $-$

27. Horizontal asymptote: $y = 0$; vertical asymptote: $x = -2$; no x-intercept; y-intercept $= -2$
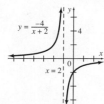

29. Horizontal asymptote: $y = 0$; vertical asymptote: $x = -3/2$; no x-intercept; y-intercept $= 2/3$
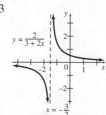

31. Horizontal asymptote: $y = 2$; vertical asymptote: $x = 3$; x-intercept $= 0$; y-intercept $= 0$
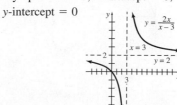

33. Horizontal asymptote: $y = 1$; vertical asymptote: $x = 4$; x-intercept $= -1$; y-intercept $= -1/4$

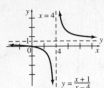

35. Horizontal asymptote: $y = -1/2$; vertical asymptote: $x = -5$; x-intercept $= 3/2$; y-intercept $= 3/20$

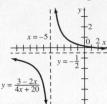

37. Horizontal asymptote: $y = -1/3$; vertical asymptote: $x = -2$; x-intercept $= -4$; y-intercept $= -2/3$

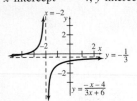

39. No asymptotes; hole at $x = -4$; x-intercept $= -3$; y-intercept $= 3$

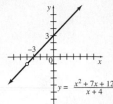

41. One possible answer is $y = 2x/(x - 1)$.   **43. a.** 0   **b.** $2, -3$ **d.** $(x + 1)(x - 1)(x + 2)$ **e.** $3(x + 1)(x - 1)(x + 2)$
f. $(x - a)$ **45. a.** Two; one at $x = -1.4$ and one at $x = 1.4$ **b.** Three; one at $x = -1.414$, one at $x = 1.414$, and one at $x = 1.442$ **47. a.** \$440; \$419; \$383; \$326; \$284; \$251   **b.** Vertical asymptote at $x = -475$; horizontal asymptote at $y = 0$

c. $y = 463.2$ **d.**

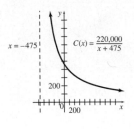

49. $f_1(x) = x(100 - x)/25$, $f_2(x) = x(100 - x)/10$, $f(x) = x^2(100 - x)^2/250$

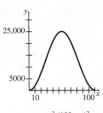

$f(x) = \dfrac{x^2 (100 - x)^2}{250}$

51. a. \$6700; \$15,600; \$26,800; \$60,300; \$127,300; \$328,300; \$663,300   **b.** No

c.

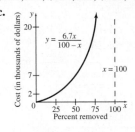

53. a. $a = 337/d$ **b.** 8.32 (using $k = 337$)   **55. a.**

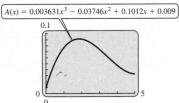

$A(x) = 0.003631x^3 - 0.03746x^2 + 0.1012x + 0.009$

b. Close to 2 hours   **c.** About 1.1 to 2.7 hours

57. a. $[0, \infty)$ **b.**

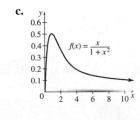

c.

$f(x) = \dfrac{x}{1 + x^2}$

d. The population of the next generation, $f(x)$, gets smaller when the current generation, x, is larger.
59. a. 220 g; 602.5 g; 1220 g   **b.** $c < 19.68$
c.

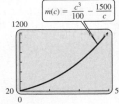

d. 41.9 cm

61. a. 1.80, 0.812, 0.366 **b.**

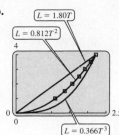

c. 2.48 sec **d.** The period increases by a factor of $\sqrt{2}$.
e. $L = 0.822T^2$, which is very close to the function found in part b.

Exercises 2.4 (page 86)

For exercises . . .	3–11,29–32	13–28	37–45,46	47–49	50–55
Refer to example . . .	1,2	3	4,5	6	7

1. 2, 4, 8, 16, 32, . . . , 1024; $1.125899907 \times 10^{15}$ **3.** E **5.** C **7.** F
9. A **11.** C **13.** 5 **15.** -4 **17.** -3 **19.** $21/4$ **21.** $-12/5$ **23.** 2, -2 **25.** 0, -1 **27.** 0, $1/2$
29.

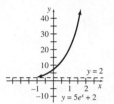

31.

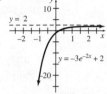

37. a. \$2166.53 **b.** \$2189.94 **c.** \$2201.90 **d.** \$2209.97 **e.** \$2214.03
39. He should choose the 5.9% investment, which would yield \$23.74 additional
interest. **41. a.** \$10.94 **b.** \$11.27 **c.** \$11.62 **43.** 6.30%
45. a. 1, 0.92, 0.85, 0.78, 0.72, 0.66, 0.61, 0.56, 0.51, 0.47, 0.43
b.

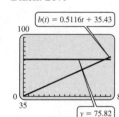

c. About \$384,000 **d.** About \$98
47. a. The function gives a population of about
3660 million, which is close to the actual population.
b. 6022 million **c.** 7725 million

49. a. 41.93 million **b.** 12.48 million **c.** 2.1%, 2.3%; Asian **d.** 37.99 million **51. a.**

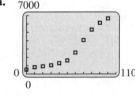

exponentially

e. Hispanic: 2039; Asian: 2036; Black: 2079

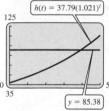

b. $f(x) = 534(1.026)^x$ **c.** 2.6%

d.

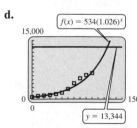

2026 **53. a.** $P = 1013e^{-1.34 \times 10^{-4}x}$; $P = -0.0748x + 1013$; $P = 1/(2.79 \times 10^{-7}x + 9.87 \times 10^{-4})$
b. $P = 1013e^{-1.34 \times 10^{-4}x}$ is the best fit.
c. 829 millibars, 232 millibars
d. $P = 1038(0.99998661)^x$. This is
slightly different from the function
found in part b, which can be rewritten as $P = 1013(0.99998660)^x$.

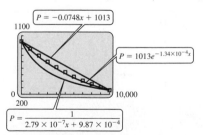

55. a. $C = 16,861t + 7461, C = 1686t^2 + 22,636, C = 19,231(1.2647)^t$ **b.**
c. $C = 19,250(1.2579)^t$; this is close to the function found in part b.
d. 176,100, 191,200, 201,300, 190,900

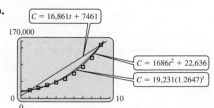

$C = 19,231(1.2647)^t$
is the best fit.

Exercises 2.5 (page 98)

For exercises . . .	1–12	13–24	27–36	37–40	41–56	57–64	65–68	69–70	75–77,79,87	75d,76c,78	81–83
Refer to example . . .	1	2	3	4	5	6	7	first CAUTION	8	9	10

1. $\log_5 125 = 3$ **3.** $\log_3 81 = 4$ **5.** $\log_3(1/9) = -2$ **7.** $2^5 = 32$ **9.** $e^{-1} = 1/e$ **11.** $10^5 = 100{,}000$ **13.** 2 **15.** 3
17. -4 **19.** $-2/3$ **21.** 1 **23.** 5/3 **25.** $\log_3 4$ **27.** $\log_5 3 + \log_5 k$ **29.** $1 + \log_3 p - \log_3 5 - \log_3 k$
31. $\ln 3 + (1/2)\ln 5 - (1/3)\ln 6$ **33.** $5a$ **35.** $2c + 3a + 1$ **37.** 2.113 **39.** -0.281 **41.** $x = 1/6$ **43.** $z = 4/3$
45. $r = 25$ **47.** $x = 1$ **49.** No solution **51.** $x = 3$ **53.** $x = 5$ **55.** $x = 1/\sqrt{3e} \approx 0.3502$ **57.** $x = (\ln 6)/(\ln 2) \approx 2.5850$
59. $k = 1 + \ln 6 \approx 2.7918$ **61.** $x = (\ln 3)/(\ln (5/3)) \approx 2.1507$ **63.** $x = (\ln 1.25)/(\ln 1.2) \approx 1.224$ **65.** $e^{(\ln 10)(x+1)}$
67. 20.09^x **69.** $x < 5$ **75. a.** 23.4 yr **b.** 11.9 yr **c.** 9.0 yr **d.** 23.3 yr; 12 yr; 9 yr **77.** 6.25% **79.** 2035
81. a. About 0.693 **b.** $\ln 2$ **c.** Yes **83. a.** About 1.099 **b.** About 1.386 **85.** About every 7 hr, $T = (3\ln 5)/\ln 2$
87. a. 2039 **b.** 2036 **89.** $s/n = 2^{C/B} - 1$ **91.** No; 1/10 **93. a.** 1000 times greater **b.** 1,000,000 times greater

Exercises 2.6 (page 107)

For exercises . . .	6,28b,29b	7–10,15–17,21c	11–14,18–20	19d,20d	23,24	25–28,30,31,42	21d,e,22,29,41	32–40
Refer to example . . .	3	4	6	7	8	1	5	2

7. 4.06% **9.** 8.33% **11.** $6209.93 **13.** $6283.17 **15.** 9.20% **17.** 6.17% **19. a.** $257,107.67 **b.** $49,892.33
c. $68,189.54 **d.** 14.28% **21. a.** The 8% investment compounded quarterly. **b.** $759.26 **c.** 8.24% and 8.06% **d.** 3.71 yr
e. 3.75 yr **23. a.** 200 **b.** About 1/2 year **c.** No **d.** Yes; 1000 **25. a.** $0.003285e^{0.007232t}$ **b.** 3300 **c.** No; it is too small.
Exponential growth does not accurately describe population growth for the world over a long period of time.
27. a. $y = 25{,}000e^{0.047t}$ **b.** $y = 25{,}000(1.048)^t$ **c.** About 18.6 hours **29. a.** 17.9 days **b.** January 17 **31. a.** 2, 5, 24, and
125 **b.** 0.061 **c.** 0.24 **d.** No, the values of k are different. **e.** Between 3 and 4 **33.** About 13 years
35. a. 0.0193 gram **b.** 69 years **37. a.** $y = 500e^{-0.0863t}$ **b.** $y = 500(386/500)^{t/3}$ **c.** 8.0 days **39. a.** 19.5 watts
b. 173 days **41. a.** 67% **b.** 37% **c.** 23 days **d.** 46 days **43.** 18.02° **45.** 1 hour

Chapter 2 Review Exercises (page 113)

1. True **2.** False **3.** True
4. True **5.** False **6.** False
7. True **8.** False **9.** False
10. False **11.** False **12.** False
13. False **14.** True **15.** False
16. False **17.** True **18.** True

For exercises . . .	1,2,6,20,21,35 –42,87,110, 115,122	3,4,31–34, 105,109	5,9,43–46, 51–54,83, 88–93,101,108, 111,112,114, 116–118,120	7,10–17,22, 29,30,47–50, 55–82,84, 94–95,121	8,23–28,86	18,96–100. 102–104
Refer to section . . .	3	2	4	5	1	6

23. $(-3, 14), (-2, 5), (-1, 0), (0, -1), (1, 2), (2, 9), (3, 20)$; **25. a.** 17 **b.** 4 **c.** $5k^2 - 3$ **d.** $-9m^2 + 12m + 1$
range: $\{-1, 0, 2, 5, 9, 14, 20\}$
e. $5x^2 + 10xh + 5h^2 - 3$ **f.** $-x^2 - 2xh - h^2 + 4x + 4h + 1$
g. $10x + 5h$ **h.** $-2x - h + 4$ **27.** $(-\infty, 0) \cup (0, \infty)$
29. $(-7, \infty)$ **31.**
33.

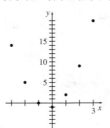

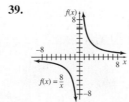

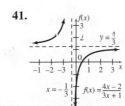

35. 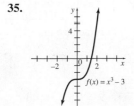 **37.** 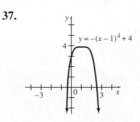 **39.** 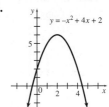 **41.**

43. 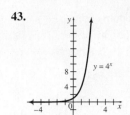 **45.** 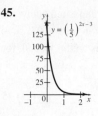 **47.** 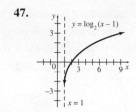 **49.**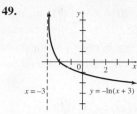

51. -5 **53.** -6 **55.** $\log_3 243 = 5$ **57.** $\ln 2.22554 = 0.8$ **59.** $2^5 = 32$ **61.** $e^{4.41763} = 82.9$ **63.** 4 **65.** 3/2
67. $\log_5(21k^4)$ **69.** $\log_3(y^4/x^2)$ **71.** $p = 1.581$ **73.** $m = -1.807$ **75.** $x = -3.305$ **77.** $m = 2.156$ **79.** $k = 2$
81. $p = 3/4$ **83. a.** $(-\infty, \infty)$ **b.** $(0, \infty)$ **c.** 1 **d.** $y = 0$ **e.** Greater than 1 **f.** Between 0 and 1 **87. a.** $28,000
b. $7000 **c.** $63,000 **d.**

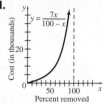

e. No. **89.** $921.95 **91.** $15,510.79 **93.** $17,901.90 **95.** 70 quarters or 17.5
years; 111 quarters or 27.75 years **97.** 6.17% **99.** $1494.52 **101.** $17,339.86
103. About 9.59% **105. a.** $n = 1500 - 10p$ **b.** $R = p(1500 - 10p)$
c. $50 \le p \le 150$ **d.** $R = (1500n - n^2)/10$ **e.** $0 \le n \le 1000$ **f.** $75
g. 750 **h.** $56,250 **i.**

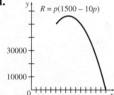

j. The revenue starts at $50,000
when the price is $50, rises to a
maximum of $56,250 when the
price is $75, and falls to 0 when
the price is $150.

107. a.

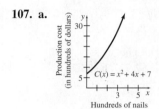

b. $2x + 5$ **c.** $A(x) = x + 4 + 7/x$
d. $1 - 7/[x(x + 1)]$ **109.** The third
day; 104.2°F

111. a.

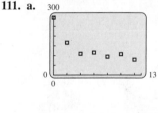

b. $y = 2.384t^2 - 42.55t + 269.2$, $y = -0.4931t^3 + 11.26t^2 - 82.00t + 292.9$, $y = 213.8(0.9149)^t$
c.

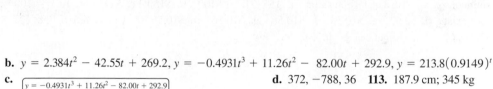

d. 372, -788, 36 **113.** 187.9 cm; 345 kg
115. a. 10.915 billion; this is about
4.006 billion more than the estimate of
6.909 billion. **b.** 26.56 billion; 96.32 billion
117. 0.25; 0.69 minutes

119. a. $S = -3.404 + 103.2 \ln A$ **b.** $S = 81.26 A^{0.3011}$ **c.**

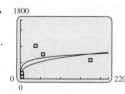

d. 234, 163

121. a. 0 yr **b.** 1.85×10^9 yr
c. As r increases, t increases, but at a slower and slower rate.
As r decreases, t decreases at a faster and faster rate.

Chapter 3 The Derivative

Exercises 3.1 (page 135)

For exercises . . .	1,7b, 8b,9a, 82–84	2,7a, 10a, 53–56	3,5a, 6a, 17–20	4	5a,6b, 8a,9b, 10b,15,16	11, 12	21–30	31–38, 41,42	39,40	43–52, 75–81, 84–93	57–63, 65, 71–74
Refer to example . . .	5	3	2	4	1	11	6	8	9	12	10

1. c **3.** b **5. a.** 3 **b.** 1 **7. a.** 0 **b.** Does not exist **9. a. i.** -1; **ii.** $-1/2$ **iii.** Does not exist; **iv.** Does not exist
b. i. $-1/2$ **ii.** $-1/2$ **iii.** $-1/2$ **iv.** $-1/2$ **11.** 3 **15.** 4 **17.** 10 **19.** Does not exist **21.** -18 **23.** 1/3 **25.** 3
27. 512 **29.** 2/3 **31.** 6 **33.** 3/2 **35.** -5 **37.** $-1/9$ **39.** 1/10 **41.** $2x$ **43.** 3/7 **45.** 3/2 **47.** 0 **49.** ∞ (does not
exist) **51.** $-\infty$ (does not exist) **53.** 1 **55. a.** 2 **b.** Does not exist **57.** 6 **59.** 1.5 **61. a.** Does not exist **b.** $x = -2$
c. If $x = a$ is a vertical asymptote for the graph of $f(x)$, then $\lim_{x \to a} f(x)$ does not exist. **65. a.** 0 **b.** $y = 0$ **67. a.** $-\infty$ (does
not exist) **b.** $x = 0$ **71.** 5 **73.** 0.3333 or 1/3 **75. a.** 1.5 **77. a.** -2 **79. a.** 8 **83. a.** 3 cents **b.** 7.25 cents
c. 8.25 cents **d.** Does not exist **e.** 8.25 cents **85.** $6; the average cost approaches $6 as the number of DVDs produced
becomes very large. **87.** 63 items; the number of items a new employee produces gets closer and closer to 63 as the number of

days of training increases. **89.** $R/(i - g)$ **91. a.** 36.2 cm; the depth of the sediment layer deposited below the bottom of the lake in 1970 is 36.2 cm. **b.** 155 cm; the depth of the sediment approaches 155 cm going back in time. **93. a.** 0.572 **b.** 0.526 **c.** 0.503 **d.** 0.5; the numbers in a, b, and c give the probability that the legislator will vote yes on the second, fourth, and eighth votes. In d, as the number of roll calls increases, the probability of a yes vote approaches 0.5 but is never less than 0.5.

Exercises 3.2 (page 146)

For exercises . . .	1–6,34,35	7–18,31–33	19–28	36–41
Refer to example . . .	1	2	3	4

1. $a = -1$: **a.** $f(-1)$ does not exist. **b.** 1/2 **c.** 1/2 **d.** 1/2
e. $f(-1)$ does not exist. **3.** $a = 1$: **a.** 2 **b.** -2 **c.** -2 **d.** -2
e. $f(1)$ does not equal the limit. **5.** $a = -5$: **a.** $f(-5)$ does not exist **b.** ∞ (does not exist) **c.** $-\infty$ (does not exist)
d. Limit does not exist. **e.** $f(-5)$ does not exist and the limit does not exist; $a = 0$: **a.** $f(0)$ does not exist. **b.** 0 **c.** 0
d. 0 **e.** $f(0)$ does not exist. **7.** $a = 0$, limit does not exist; $a = 2$, limit does not exist. **9.** $a = 2$, limit is 4 **11.** Nowhere
13. $a = -2$, limit does not exist. **15.** $a < 1$, limit does not exist. **17.** $a = 0$, $-\infty$ (limit does not exist); $a = 1$, ∞ (limit does not exist).
19. a. 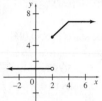 **b.** 2 **c.** 1, 5 **21. a.** 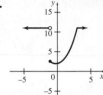 **b.** -1 **c.** 11, 3 **23. a.**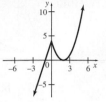

b. None **25.** 2/3 **27.** 4 **31. a.** Discontinuous at $x = 1.2$ **33.** a **35. a.** $500 **b.** $1500 **c.** $1000 **d.** Does not exist
e. Discontinuous at $x = 10$; a change in shifts **f.** 15 **37. a.** $100 **b.** $150 **c.** $125 **d.** At $x = 100$ **39. a.** $2.92 **b.** $3.76
c. Does not exist **d.** $2.92 **e.** $10.56 **f.** $10.56 **g.** $10.56 **h.** $10.56 **i.** 1, 2, 3, 4, 5, 6, 7, 8, 12, 16, 20, 24, 28, 32, 36, 40,
44, 48, 52, 56, 60 **41.** About 687 g **b.** No **c.**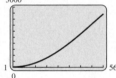

Exercises 3.3 (page 158)

For exercises . . .	1–8	9–18,33, 36,38,44	19–22, 28,29,37	25–27	30–32, 34,35, 39–41	42,43
Refer to example . . .	1	3	6	4 and 5	2	7

1. 6 **3.** -15 **5.** 1/3 **7.** 0.4323 **9.** 17
11. 18 **13.** 5 **15.** 2 **17.** 2 **19.** 6.773
21. 1.121 **25. a.** $700 per item **b.** $500 per item
c. $300 per item **d.** $1100 per item **27. a.** -25 boxes per dollar **b.** -20 boxes per dollar **c.** -30 boxes per dollar
d. Demand is decreasing. Yes, a higher price usually reduces demand. **29. a.** $56.81 per year **b.** $72.94 per year **c.** $64.20
per year **31. a.** 17.7 cents per gallon per month **b.** -48.0 cents per gallon per month **c.** -11.2 cents per gallon per month
33. a. 6% per day **b.** 7% per day **35. a.** 2; from 1 min to 2 min, the population of bacteria increases, on the average, 2 million
per min. **b.** -0.8; from 2 min to 3 min, the population of bacteria decreases, on the average, 0.8 million or 800,000 per min.
c. -2.2; from 3 min to 4 min, the population of bacteria decreases, on the average, 2.2 million per min. **d.** -1; from 4 min to 5 min,
the population decreases, on the average, 1 million per min. **e.** 2 min **f.** 3 min **37. a.**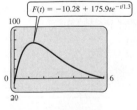
b. 81.51 kilojoules per hour per hour **c.** 18.81 kilojoules per hour per hour
d. 1.3 hours **39. a.** $-15,760$ immigrants per year **b.** 17,680 immigrants per year
c. 960 immigrants per year **d.** They are equal; no **e.** 1,126,000 immigrants
41. a. 4° per 1000 ft **b.** 1.75° per 1000 ft **c.** $-4/3$° per 1000 ft **d.** 0° per 1000 ft
e. 3000 ft; 1000 ft; if 7000 ft is changed to 10,000 ft, the lowest temperature would be at
10,000 ft. **f.** 9000 ft **43. a.** 50 mph **b.** 44 mph

Exercises 3.4 (page 176)

For exercises . . .	5–10, 54,55, 58–61	11–14	15, 16	17, 18	19, 20	21–26	27–34, 42–45, 53	35–38	49–52, 56–57
Refer to example . . .	3	4	6	7	5	1 and 9	2	10	8

1. a. 0 **b.** 1 **c.** -1 **d.** Does not exist
e. m **3.** At $x = -2$ **5.** 2 **7.** 1/4 **9.** 0
11. 3; 3; 3; 3 **13.** $-8x + 9$; 25; 9; -15

15. $-12/x^2$; -3; does not exist; $-4/3$ **17.** $1/(2\sqrt{x})$; does not exist; does not exist; $1/(2\sqrt{3})$ **19.** $6x^2$; 24; 0; 54
21. a. $y = 10x - 15$ **b.** $y = 8x - 9$ **23. a.** $y = -(1/2)x + 7/2$ **b.** $y = -(5/4)x + 5$ **25. a.** $y = (4/7)x + 48/7$
b. $y = (2/3)x + 6$ **27.** -5; -117; 35 **29.** 7.389; 8,886,112; 0.0498 **31.** 1/2; 1/128; 2/9 **33.** $1/(2\sqrt{2})$; 1/8; does not exist

35. 0 **37.** $-3; -1; 0; 2; 3; 5$ **39. a.** $(a, 0)$ and (b, c) **b.** $(0, b)$ **c.** $x = 0$ and $x = b$ **41. a.** Distance **b.** Velocity
43. 56.66 **45.** -0.0158 **49. a.** $-4p - 4$ **b.** -44; demand is decreasing at a rate of about 44 items for each increase in price
of \$1. **51. a.** \$16 per table **b.** \$15.996 (or \$16) **c.** \$15.998 (or \$16) **d.** The marginal revenue gives a good approximation of
the actual revenue from the sale of the 1001st table. **53.** Answers are in trillion of dollars. **a.** 2.54; 2.03; 1.02 **b.** 0.061;
$-0.019; -0.079; -0.120; -0.133$ **55.** 1000; the population is increasing at a rate of 1000 shellfish per time unit. 570; the population is increasing more slowly at 570 shellfish per time unit. 250; the population is increasing at a much slower rate of 250 shellfish
per time unit. **57. a.** 1690 m per sec **b.** 4.84 days per m per sec; an increase in velocity from 1700 m per sec to 1701 m per sec
indicates an approximate increase in the age of the cheese of 4.84 days. **59. a.** About 270; the temperature was increasing at a rate
of about 270° per hour at 9:00 a.m. **b.** About ±150; the temperature was decreasing at a rate of about 150° per hour at 11:30 a.m.
c. About 0; the temperature staying constant at 12:30 p.m. **d.** About 11:15 a.m. **61. a.** 0; About 0.5 mph/oz

Exercises 3.5 (page 184)

3. $f:Y_2; f':Y_1$ **5.** $f:Y_1; f':Y_2$

For exercises . . .	2–6	7–24
Refer to example . . .	4	1,2, and 3

7.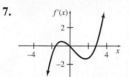

9.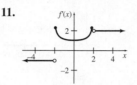

11.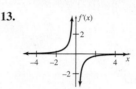

13.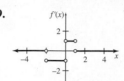

15.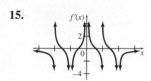

17.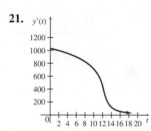

19.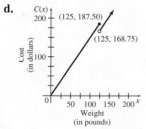

21.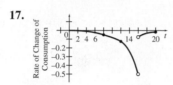

23. 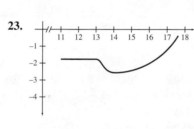 About 9 cm; about 2.6 cm less per year

Chapter 3 Review Exercises (page 188)

1. True **2.** True **3.** True **4.** False **5.** True **6.** False
7. False **8.** True **9.** True **10.** True **11.** False **12.** False
17. a. 4 **b.** 4 **c.** 4 **d.** 4 **19. a.** ∞ **b.** $-\infty$
c. Does not exist **d.** Does not exist **21.** ∞ **23.** 19/9

For exercises . . .	1–3, 17–34, 45–46, 61,66	4–6,15, 35–44, 63,74	7–9, 47–50, 63–65	10–12,16, 51–58,62, 68–69,73	59–60,67, 70–72,74
Refer to section . . .	1	2	3	4	5

25. 8 **27.** -13 **29.** 1/6 **31.** 2/5 **33.** 3/8 **35.** Discontinuous at x_2 and x_4 **37.** 0, does not exist, does not exist; $-1/3$, does
not exist, does not exist **39.** -5, does not exist, does not exist **41.** Continuous everywhere **43. a.**
b. 1 **c.** 0, 2 **45.** 2 **47.** 126; 18 **49.** 9/77; 18/49 **51. a.** $y = 13x - 17$ **b.** $y = 7x - 5$
53. a. $y = -x + 9$ **b.** $y = -3x + 15$ **55.** $8x + 3$ **57.** 1.332 **59.**
61. e **63. a.** \$150 **b.** \$187.50 **c.** \$189

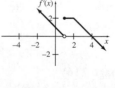

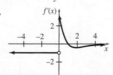

d.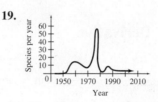

e. Discontinuous at $x = \$125$
f. \$1.50 **g.** \$1.50 **h.** \$1.35
i. 1.5; when 100 lb are purchased, an
additional pound will cost about \$1.50 more. **j.** 1.35; when 140 lb are purchased, an additional
pound will cost about \$1.35 more. **65. b.** $x = 7.5$ **c.** The marginal cost equals the average cost
at the point where the average cost is smallest.

67.

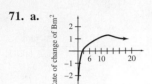 ; about 5.8; about 2

69. a.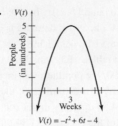

$V(t) = -t^2 + 6t - 4$

b. $[0.8, 5.2]$ **c.** 3 weeks; 500 cases
d. $V'(t) = -2t + 6$ **e.** 0
f. $+; -$

71. a.   **b.** 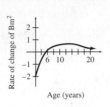   **73. a.** Nowhere   **b.** 50, 130, 230, 770

c.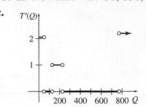

Chapter 4 Calculating the Derivative

Exercises 4.1 (page 207)

For exercises . . .	1–22,27–30	47,55,56	51,53	52,54	57,58,60–62, 64,68–75	59,63, 65,67	31–43
Refer to example . . .	1,2,3,5	6	7	8	9	4	1 (in 4th section of previous chapter)

1. $dy/dx = 36x^2 - 16x + 7$ **3.** $dy/dx = 12x^3 - 18x^2 + (1/4)x$ **5.** $f'(x) = 21x^{2.5} - 5x^{-0.5}$ or $21x^{2.5} - 5/x^{0.5}$
7. $dy/dx = 4x^{-1/2} + (9/2)x^{-1/4}$ or $4/x^{1/2} + 9/(2x^{1/4})$ **9.** $dy/dx = -30x^{-4} - 20x^{-5} - 8$ or $-30/x^4 - 20/x^5 - 8$
11. $f'(t) = -7t^{-2} + 15t^{-4}$ or $-7/t^2 + 15/t^4$ **13.** $dy/dx = -24x^{-5} + 21x^{-4} - 3x^{-2}$ or $-24/x^5 + 21/x^4 - 3/x^2$
15. $p'(x) = 5x^{-3/2} - 12x^{-5/2}$ or $5/x^{3/2} - 12/x^{5/2}$ **17.** $dy/dx = (-3/2)x^{-5/4}$ or $-3/(2x^{5/4})$
19. $f'(x) = 2x - 5x^{-2}$ or $2x - 5/x^2$ **21.** $g'(x) = 256x^3 - 192x^2 + 32x$ **23.** b   **27.** $-(9/2)x^{-3/2} - 3x^{-5/2}$ or
$-9/(2x^{3/2}) - 3/x^{5/2}$ **29.** $-25/3$ **31.** $-28; y = -28x + 34$ **33.** $25/6$ **35.** $(4/9, 20/9)$ **37.** $-5, 2$ **39.** $(4 \pm \sqrt{37})/3$
41. $(-1/2, -19/2)$ **43.** $(-2, -24)$ **45.** 7   **51. a.** 30   **b.** 0   **c.** -10 **53.** \$990   **55. a.** 100   **b.** 1   **57. a.** 18¢; 36¢
b. 0.755¢ per year; 1.09¢ per year   **c.** $C(t) = -0.0001549t^3 + 0.02699t^2 - 0.6484t + 3.212$; 0.889¢ per year; 0.853¢ per year
59. a. 0.4824   **b.** 2.216   **61. a.** 1232.62 cm³   **b.** 948.08 cm³/yr   **63.** $5.00l^{0.86}$ **65. a.** 3 minutes, 58.1 seconds   **b.** 0.118
sec/m; at 100 meters, the fastest possible time increases by 0.118 seconds for each additional meter.   **c.** Yes   **67. a.** 27.5
b. 23 pounds   **c.** $-175,750/h^3$ **d.** -0.64; for a 125-lb female with a height of 65 in. (5'5"), the BMI decreases by 0.64 for each
additional inch of height.   **69. a.** $v(t) = 36t - 13$ **b.** $-13; 167; 347$ **71. a.** $v(t) = -9t^2 + 8t - 10$ **b.** $-10; -195; -830$
73. 0 ft/sec; -32 ft/sec   **b.** 2 seconds   **c.** 64 ft   **75. a.** 35, 36   **b.** When $x = 5$, $dy_1/dx = 4.13$ and $dy_2/dx \approx 4.32$. These
values are fairly close and represent the rate of change of four years for a dog for one year of a human, for a dog that is actually
5 years old.   **c.** $y = 4x + 16$

Exercises 4.2 (page 216)

1. $dy/dx = 18x^2 - 6x + 4$ **3.** $dy/dx = 8x - 20$
5. $k'(t) = 4t^3 - 4t$

For exercises . . .	1–10,29,34,39, 42,44–47,50	11–26,31–33,35,40, 43,45,48,49,51–54	27,28,30	41,42
Refer to example . . .	1,2	3	4	5

7. $dy/dx = (3/2)x^{1/2} + (1/2)x^{-1/2} + 2$ or $3x^{1/2}/2 + 1/(2x^{1/2}) + 2$ **9.** $p'(y) = -8y^{-5} + 15y^{-6} + 30y^{-7}$
11. $f'(x) = 57/(3x + 10)^2$ **13.** $dy/dt = -17/(4 + t)^2$ **15.** $dy/dx = (x^2 - 2x - 1)/(x - 1)^2$
17. $f'(t) = 2t/(t^2 + 3)^2$ **19.** $g'(x) = (4x^2 + 2x - 12)/(x^2 + 3)^2$ **21.** $p'(t) = [-\sqrt{t}/2 - 1/(2\sqrt{t})]/(t - 1)^2$ or
$(-t - 1)/[2\sqrt{t}(t - 1)^2]$ **23.** $dy/dx = (5/2)x^{-1/2} - 3x^{-3/2}$ or $(5x - 6)/(2x\sqrt{x})$
25. $h'(z) = (-z^{4.4} + 11z^{1.2})/(z^{3.2} + 5)^2$ **27.** $f'(x) = (60x^3 + 57x^2 - 24x + 13)/(5x + 4)^2$ **29.** 77
31. In the first step, the numerator should be $(x^2 - 1)2 - (2x + 5)(2x)$.   **33.** $y = -2x + 9$

35. a. $f'(x) = (7x^3 - 4)/x^{5/3}$ **b.** $f'(x) = 7x^{4/3} - 4x^{-5/3}$ **39.** 0, -1.307, and 1.307 **41. a.** \$22.86 per unit
b. \$12.92 per unit **c.** $(3x + 2)/(x^2 + 4x)$ per unit **d.** $\overline{C}'(x) = (-3x^2 - 4x - 8)/(x^2 + 4x)^2$
43. a. $M'(d) = 2000d/(3d^2 + 10)^2$ **b.** 8.3; the new employee can assemble about 8.3 additional bicycles per day after 2 days
of training. 1.4; the new employee can assemble about 1.4 additional bicycles per day after 5 days of training.
47. Increasing at a rate of \$0.03552 per gallon per month **49. a.** $AK/(A + x)^2$ **b.** $K/(4A)$ **51. a.** 8.57 min
b. 16.36 min **c.** 6.12 min²/kcal; 2.48 min²/kcal **53. a.** -100 facts/hr **b.** -0.01 facts/hr

Exercises 4.3 (page 225)

For exercises . . .	1–6	7–14,53,59	15–20	21–28,54,57, 62,63,66,67	29–34,64	35–40,56	58,65	55	60,61	45–50
Refer to example . . .	1	2	3	5,6	7	8	9	10	4	1 (in 3rd section of previous chapter)

1. 1767 **3.** 131 **5.** $320k^2 + 224k + 39$ **7.** $(6x + 55)/8; (3x + 164)/4$ **9.** $1/x^2; 1/x^2$ **11.** $\sqrt{8x^2 - 4}; 8x + 10$
13. $\sqrt{(x - 1)/x}; -1/\sqrt{x + 1}$ **15.** If $f(x) = x^{3/5}$ and $g(x) = 5 - x^2$, then $y = f[g(x)]$.
17. If $f(x) = -\sqrt{x}$ and $g(x) = 13 + 7x$, then $y = f[g(x)]$. **19.** If $f(x) = x^{1/3} - 2x^{2/3} + 7$ and $g(x) = x^2 + 5x$, then
$y = f[g(x)]$. **21.** $dy/dx = 4(8x^4 - 5x^2 + 1)^3(32x^3 - 10x)$ **23.** $k'(x) = 288x(12x^2 + 5)^{-7}$
25. $s'(t) = (1215/2)t^2(3t^3 - 8)^{1/2}$ **27.** $g'(t) = -63t^2/(2\sqrt{7t^3 - 1})$ **29.** $m'(t) = -6(5t^4 - 1)^3(85t^4 - 1)$
31. $dy/dx = 3x^2(3x^4 + 1)^3(19x^4 + 64x + 1)$ **33.** $q'(y) = 2y(y^2 + 1)^{1/4}(9y^2 + 4)$ **35.** $dy/dx = 60x^2/(2x^3 + 1)^3$
37. $r'(t) = 2(5t - 6)^3(15t^2 + 18t + 40)/(3t^2 + 4)^2$ **39.** $dy/dx = (-18x^2 + 2x + 1)/(2x - 1)^6$ **43. a.** -2 **b.** $-24/7$
45. $y = (3/5)x + 16/5$ **47.** $y = x$ **49.** 1, 3 **53.** $D(c) = (-c^2 + 10c + 12,475)/25$ **55. a.** \$101.22 **b.** \$111.86
c. \$117.59 **57. a.** $-\$10,500$ **b.** $-\$4570.64$ **59.** $P[f(a)] = 18a^2 + 24a + 9$ **61. a.** $A[r(t)] = A(t) = 4\pi t^2$; this function
gives the area of the pollution in terms of the time since the pollutants were first emitted. **b.** 32π; at 12 P.M. the area of pollution is
changing at the rate of 32π mi² per hour **63. a.** -0.5 **b.** $-1/54 \approx -0.02$ **c.** $-1/128 \approx -0.008$ **d.** Always decreasing; the
derivative is negative for all $t \geq 0$. **65. a.** 34 minutes **b.** $-(108/17)\pi$ mm³ per minute, $-(72/17)\pi$ mm² per minute
67. a. $12x^{11}$ **b.** $x^{12}; 12x^{11}$

Exercises 4.4 (page 232)

For exercises . . .	1–12,25–30,33,34, 38,41,44–45,48,50, 51–54,56,59–63	13–16,23,24, 39,40,42, 49	17–22,31,32,51	43,46,47,55,57	58
Refer to example . . .	1	2	3	5	4

1. $dy/dx = 4e^{4x}$ **3.** $dy/dx = -24e^{3x}$ **5.** $dy/dx = -32e^{2x+1}$ **7.** $dy/dx = 2xe^{x^2}$ **9.** $dy/dx = 12xe^{2x^2}$
11. $dy/dx = 16xe^{2x^2-4}$ **13.** $dy/dx = xe^x + e^x = e^x(x + 1)$ **15.** $dy/dx = 2(x + 3)(2x + 7)e^{4x}$
17. $dy/dx = (2xe^x - x^2e^x)/e^{2x} = x(2 - x)/e^x$ **19.** $dy/dx = [x(e^x - e^{-x}) - (e^x + e^{-x})]/x^2$
21. $dp/dt = 8000e^{-0.2t}/(9 + 4e^{-0.2t})^2$ **23.** $f'(z) = 4(2z + e^{-z^2})(1 - ze^{-z^2})$ **25.** $dy/dx = 3(\ln 7)7^{3x+1}$
27. $dy/dx = 6x(\ln 4)4^{x^2+2}$ **29.** $ds/dt = (\ln 3)3^{\sqrt{t}}/\sqrt{t}$ **31.** $dy/dt = [(1 - t)e^{3t} - 4e^{2t} + (1 + t)e^t]/(e^{2t} + 1)^2$
33. $f'(x) = (9x + 4)e^{x\sqrt{3x+2}}/[2\sqrt{3x + 2}]$ **39. a.** \$3.81 **b.** \$0.20 **c.** $C'(x)$ approaches zero. **41. a.** 100% **b.** 94%
c. 89% **d.** 83% **e.** -3.045 **f.** -2.865 **g.** The percent of these cars on the road is decreasing but at a slower rate as they age.
43. a. $G(t) = 250/(1 + 124e^{-0.45t})$ **b.** 17.8 million; 7.4 million/yr **c.** 105.2 million; 27.4 million/yr **d.** 246.2 million; 1.7
million/yr **e.** It increases for a while and then gradually decreases to 0. **45. a.** 51,600,000 **b.** 1,070,000 people/year
47. a. $5200/(1 + 12e^{-0.52t})$ **b.** 639, 292 **c.** 2081, 649 **d.** 4877, 157 **e.** It increases for a while and then gradually
decreases to 0. **49. a.** 3.857 cm³ **b.** 0.973 cm **c.** 18 years **d.** 1100 cm³ **e.** 0.282; at 240 months old, the tumor is
increasing in volume at the instantaneous rate of 0.282 cm³/month. **51. a.** 0.180 **b.** 2024 **c.** The marginal increase in the
proportion per year in 2010 is approximately 0.023. **53. a.** 509.7 kg, 498.4 kg **b.** 1239 days, 1095 days **c.** 0.22 kg/day,
0.22 kg/day **d.** The growth patterns of the two functions are **e.** The graphs of the rates of change of the two functions are
very similar.

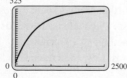

also very similar.

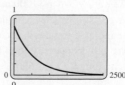

55. a. $G(t) = 1/(1 + 270e^{-3.5t})$ **b.** 0.109, 0.341 per century **c.** 0.802, 0.555 per century **d.** 0.993, 0.0256 per century
e. It increases for a while and then gradually decreases to 0. **57. a.** $G(t) = 6.8/(1 + 3.242e^{-0.2992t})$ **b.** 2.444 million
students; 0.468 million students/yr **c.** 3.435 million students; 0.509 million students/yr **d.** 5.247 million students; 0.359 million
students/yr **e.** The rate increases at first and then decreases toward 0. **59. a.** $(V/R)e^{-t/RC}$ **b.** 1.35×10^{-7} amps
61. a. 218.5 seconds **b.** The record is decreasing by 0.034 seconds per year at the end of 2010. **c.** 218 seconds. If the
estimate is correct, then this is the least amount of time that it will ever take for a human to run a mile. **63. a.** 625.14 ft
b. 0; yes **c.** −1.476 ft per ft from center

Exercises 4.5 (page 240)

For exercises . . .	1,2,59,62,66	3,4,7–10,31,32,56,67	5,6,11–30,33–44, 57,58,60,61,64,65	68
Refer to example . . .	1	2	3	4

1. $dy/dx = 1/x$ **3.** $dy/dx = -3/(8 - 3x)$
or $3/(3x - 8)$ **5.** $dy/dx = (8x - 9)/(4x^2 - 9x)$
7. $dy/dx = 1/[2(x + 5)]$ **9.** $dy/dx = 3(2x^2 + 5)/[x(x^2 + 5)]$ **11.** $dy/dx = -15x/(3x + 2) - 5\ln(3x + 2)$
13. $ds/dt = t + 2t\ln|t|$ **15.** $dy/dx = [2x - 4(x + 3)\ln(x + 3)]/[x^3(x + 3)]$
17. $dy/dx = (4x + 7 - 4x\ln x)/[x(4x + 7)^2]$ **19.** $dy/dx = (6x\ln x - 3x)/(\ln x)^2$ **21.** $dy/dx = 4(\ln|x + 1|)^3/(x + 1)$
23. $dy/dx = 1/(x\ln x)$ **25.** $dy/dx = e^{x^2}/x + 2xe^{x^2}\ln x$ **27.** $dy/dx = (xe^x\ln x - e^x)/[x(\ln x)^2]$
29. $g'(z) = 3(e^{2z} + \ln z)^2(2ze^{2z} + 1)/z$ **31.** $dy/dx = 1/(x\ln 10)$ **33.** $dy/dx = -1/[(\ln 10)(1 - x)]$ or $1/[(\ln 10)(x - 1)]$
35. $dy/dx = 5/[(2\ln 5)(5x + 2)]$ **37.** $dy/dx = 3(x + 1)/[(\ln 3)(x^2 + 2x)]$ **39.** $dw/dp = (\ln 2)2^p/[(\ln 8)(2^p - 1)]$
41. $f'(x) = e^{\sqrt{x}}\{1/[2\sqrt{x}(\sqrt{x} + 5)] + \ln(\sqrt{x} + 5)/[2\sqrt{x}]\}$
43. $f'(t) = [(t^2 + 1)\ln(t^2 + 1) - t^2 + 2t + 1]/\{(t^2 + 1)[\ln(t^2 + 1) + 1]^2\}$
55. $h(x) = (x^2 + 1)^{5x}\left[\dfrac{10x^2}{x^2 + 1} + 5\ln(x^2 + 1)\right]$ **57. a.** $dR/dq = 100 + 50(\ln q - 1)/(\ln q)^2$ **b.** $112.48
c. To decide whether it is reasonable to sell additional items. **59. a.** −0.19396 **b.** −0.06099 **61. a.** $N(t) = 1000e^{9.8901e^{-e^{2.54197 - 0.2167t}}}$
b. 1,307,416 bacteria per hour; the number of bacteria is increasing at a rate of 1,307,416 per hour, 20 hours after the experiment began.

c.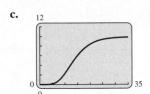

d.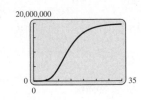

e. 9.8901; $1000e^{9.8901} \approx 19,734,033$
63. b. i. 3.343 **ii.** 1.466 **c. i.** −0.172 **ii.** −0.0511 **65.** 26.9; 13.1
67. a. 1.567×10^{11} kWh **b.** 63.4 months **c.** 4.14×10^{-6}
d. dM/dE decreases and approac.hes zero.

Chapter 4 Review Exercises (page 244)

For exercises . . .	1,2,4,11–16, 53,54,73,79–81	3,17–20,55,56, 74,75,88	5,21–30,51,52, 57,58,67–70,76,91	6–7,31–36,43–46,59,60, 71,77,82–87,89,90	8–10,37–42,47–50, 61,62,72,78
Refer to section . . .	1	2	3	4	5

1. False **2.** True **3.** False **4.** True **5.** False **6.** False **7.** False **8.** True **9.** True **10.** False
11. $dy/dx = 15x^2 - 14x - 9$ **13.** $dy/dx = 24x^{5/3}$ **15.** $f'(x) = -12x^{-5} + 3x^{-1/2}$ or $-12/x^5 + 3/x^{1/2}$
17. $k'(x) = 21/(4x + 7)^2$ **19.** $dy/dx = (x^2 - 2x)/(x - 1)^2$ **21.** $f'(x) = 24x(3x^2 - 2)^3$ **23.** $dy/dt = 7t^6/(2t^7 - 5)^{1/2}$
25. $dy/dx = 3(2x + 1)^2(8x + 1)$ **27.** $r'(t) = (-15t^2 + 52t - 7)/(3t + 1)^4$ **29.** $p'(t) = t(t^2 + 1)^{3/2}(7t^2 + 2)$
31. $dy/dx = -12e^{2x}$ **33.** $dy/dx = -6x^2e^{-2x^3}$ **35.** $dy/dx = 10xe^{2x} + 5e^{2x} = 5e^{2x}(2x + 1)$ **37.** $dy/dx = 2x/(2 + x^2)$
39. $dy/dx = (x - 3 - x\ln|3x|)/[x(x - 3)^2]$ **41.** $dy/dx = [e^x(x + 1)(x^2 - 1)\ln(x^2 - 1) - 2x^2e^x]/[(x^2 - 1)[\ln(x^2 - 1)]^2]$
43. $ds/dt = 2(t^2 + e^t)(2t + e^t)$ **45.** $dy/dx = -6x(\ln 10) \cdot 10^{-x^2}$ **47.** $g'(z) = (3z^2 + 1)/[(\ln 2)(z^3 + z + 1)]$
49. $f'(x) = (x + 1)e^{3x}/(xe^x + 1) + 2e^{2x}\ln(xe^x + 1)$ **51. a.** −3/2 **b.** −24/11 **53.** −2; $y = -2x - 4$
55. −3/4; $y = -(3/4)x - 9/4$ **57.** 3/4; $y = (3/4)x + 7/4$ **59.** 1; $y = x + 1$ **61.** 1; $y = x - 1$ **63.** No points if $k > 0$;
exactly one point if $k = 0$ or if $k < -1/2$; exactly two points if $-1/2 \le k < 0$. **65.** 5%; 5.06%
67. $\overline{C}'(x) = (-x - 2)/[2x^2(x + 1)^{1/2}]$ **69.** $\overline{C}'(x) = (x^2 + 3)^2(5x^2 - 3)/x^2$ **71.** $\overline{C}'(x) = [e^{-x}(x + 1) - 10]/x^2$
73. a. 22; sales will increase by $22 million when $1000 more is spent on research. **b.** 19.5; sales will increase by $19.5 million
when $1000 more is spent on research. **c.** 18; sales will increase by $18 million when $1000 more is spent on research.
d. As more is spent on research, the increase in sales is decreasing. **75. a.** −2.201; costs will decrease by $2201 for the next $100
spent on training. **b.** −0.564; costs will decrease by $564 for the next $100 spent on training. **c.** Decreasing
77. $218.65. The balance increases by roughly $218.65 for every 1% increase in the interest rate when the rate is 5%.

79. a. 4.839 billion/yr; the volume of mail is increasing by about 4,839,000,000 pieces per year **b.** –2.582 billion/yr; the volume of mail is decreasing by about 2,582,000,000 pieces per year **81. a.** $y = 1.799 \times 10^{-5}t^3 + 3.177 \times 10^{-4}t^2 - 0.06866t + 2.504$, $y = -1.112 \times 10^{-6}t^4 + 2.920 \times 10^{-4}t^3 - 0.02238t^2 + 0.6483t - 4.278$ **b.** \$0.59/yr, \$0.46/yr

83. a. $G(t) = 30{,}000/(1 + 14e^{-0.15t})$ **b.** 4483; 572 **85. a.** 3493.76 grams **b.** 3583 grams **c.** 84 days **d.** 1.76 g/day

e.

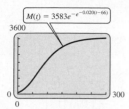

Growth is initially rapid, then tapers off. **f.**

Day	Weight	Rate
50	904	24.90
100	2159	21.87
150	2974	11.08
200	3346	4.59
250	3494	1.76
300	3550	0.66

87. a. 14,612 megawatts/yr **b.** 47,276 megawatts/yr **c.** 152,960 megawatts/yr **89.** 0.242; the production of corn is increasing at a rate of 0.242 billion bushels a year in 2000. **91. a.** -0.4677 fatalities per 1000 licensed drivers per 100 million miles per year; at the age of 20, each extra year results in a decrease of 0.4677 fatalities per 1000 licensed drivers per 100 million miles.
b. 0.003672 fatalities per 1000 licensed drivers per 100 million miles per year; at the age of 60, each extra year results in an increase of 0.003672 fatalities per 1000 licensed drivers per 100 million miles.

Chapter 5 Graphs and the Derivative

Exercises 5.1 (page 260)

1. a. $(1, \infty)$ **b.** $(-\infty, 1)$ **3. a.** $(-\infty, -2)$
b. $(-2, \infty)$ **5. a.** $(-\infty, -4), (-2, \infty)$
b. $(-4, -2)$ **7. a.** $(-7, -4), (-2, \infty)$
b. $(-\infty, -7), (-4, -2)$

For exercises . . .	1–8, 43,44 50,51, 60,61	13–22,25, 29–34,38, 39,42,49, 52,53,56–59	21–24	26–28, 35,36	45,54,55	46–48
Refer to example . . .	1	2	4	3	6	5

9. a. $(-\infty, -1), (3, \infty)$ **b.** $(-1, 3)$ **11. a.** $(-\infty, -8), (-6, -2.5), (-1.5, \infty)$ **b.** $(-8, -6), (-2.5, -1.5)$ **13. a.** 17/12
b. $(-\infty, 17/12)$ **c.** $(17/12, \infty)$ **15. a.** $-3, 4$ **b.** $(-\infty, -3), (4, \infty)$ **c.** $(-3, 4)$ **17. a.** $-3/2, 4$
b. $(-\infty, -3/2), (4, \infty)$ **c.** $(-3/2, 4)$ **19. a.** $-2, -1, 0$ **b.** $(-2, -1), (0, \infty)$ **c.** $(-\infty, -2), (-1, 0)$ **21. a.** None
b. None **c.** $(-\infty, \infty)$ **23. a.** None **b.** None **c.** $(-\infty, -1), (-1, \infty)$ **25. a.** 0 **b.** $(0, \infty)$ **c.** $(-\infty, 0)$ **27. a.** 0
b. $(0, \infty)$ **c.** $(-\infty, 0)$ **29. a.** 7 **b.** $(7, \infty)$ **c.** $(3, 7)$ **31. a.** 1/3 **b.** $(-\infty, 1/3)$ **c.** $(1/3, \infty)$ **33. a.** $0, 2/\ln 2$
b. $(0, 2/\ln 2)$ **c.** $(-\infty, 0), (2/\ln 2, \infty)$ **35. a.** $0, 2/5$ **b.** $(0, 2/5)$ **c.** $(-\infty, 0), (2/5, \infty)$ **39.** Vertex: $(-b/(2a),$
$(4ac - b^2)/(4a))$; increasing on $(-\infty, -b/(2a))$, decreasing on $(-b/(2a), \infty)$ **41.** On $(0, \infty)$; nowhere; nowhere
43. a. About $(567, \infty)$ **b.** About $(0, 567)$ **45. a.** Nowhere **b.** $(0, \infty)$ **47.** $(0, 2200)$ **49. a.** Increasing on $(0, 7.4)$ or from
2000 to about the middle of 2007 **b.** Decreasing on $(7.4, 50)$ or from about the middle of 2007 to 2050 **51. a.** Yes **b.** April to
July; July to November; January to April and November to December **c.** January to April and November to December
53. a. $(0, 1.85)$ **b.** $(1.85, 5)$ **55. a.** $(0, 1)$ **b.** $(1, \infty)$ **57. a.** $F'(t) = 175.9e^{-t/1.3}(1 - 0.769t)$ **b.** $(0, 1.3); (1.3, \infty)$
59. $(-\infty, 0); (0, \infty)$ **61. a.** $(2500, 5750)$ **b.** $(5750, 6000)$ **c.** $(2800, 4800)$ **d.** $(2500, 2800)$ and $(4800, 6000)$

Exercises 5.2 (page 271)

1. Relative minimum of -4 at 1 **3.** Relative maximum
of 3 at -2 **5.** Relative maximum of 3 at -4;
relative minimum of 1 at -2 **7.** Relative maximum

For exercises . . .	1–8, 37–39, 50	13–20, 35,36, 51	21–24	25–34, 52–54, 56	41–44, 46–49	45,55,57
Refer to example . . .	2	3a	3b	3c	4	1

of 3 at -4; relative minimum of -2 at -7 and -2 **9.** Relative maximum at -1; relative minimum at 3 **11.** Relative maxima at
-8 and -2.5; relative minimum at -6 and -1.5 **13.** Relative minimum of 8 at 5 **15.** Relative maximum of -8 at -3; relative
minimum of -12 at -1 **17.** Relative maximum of 827/96 at $-1/4$; relative minimum of $-377/6$ at -5 **19.** Relative maximum
of -4 at 0; relative minimum of -85 at 3 and -3 **21.** Relative maximum of 3 at $-8/3$ **23.** Relative maximum of 1 at -1; relative minimum of 0 at 0 **25.** No relative extrema **27.** Relative maximum of 0 at 1; relative minimum of 8 at 5 **29.** Relative maximum of -2.46 at -2; relative minimum of -3 at 0 **31.** No relative extrema **33.** Relative minimum of $e \ln 2$ at $1/\ln 2$
35. $(3, 13)$ **37.** Relative maximum of 6.211 at 0.085; relative minimum of -57.607 at 2.161
39. Relative minimum at $x = 5$

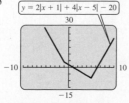

41. a. 13 **b.** \$44 **c.** \$258 **43. a.** 100 **b.** \$14.72
c. \$635.76 **45.** Relative maximum of 20,470 megawatts at midnight
($t = 0$); relative minimum of 20,070 megawatts at 1:40 A.M.; relative
maximum of 30,060 megawatts at 4:44 P.M.; relative minimum of
20,840 megawatts at midnight ($t = 24$). **47.** $q = 10$; $p \approx \$73.58$
49. 120 units **51.** 5:04 P.M.; 6:56 A.M. **53.** 4.96 years; 458.22 kg
55. 10 **57. a.** 28 ft **b.** 2.57 sec

Exercises 5.3 (page 283)

For exercises . . .	1–16	17–26, 76,77	27–48,54, 65–68, 70,80, 85–88	57–64,79, 81,82, 90,92,93	72–75 78, 83,84	91,92, 94,95
Refer to example . . .	2	3	5	6	7	4

1. $f''(x) = 30x - 14$; -14; 46

3. $f''(x) = 48x^2 - 18x - 4$; -4; 152

5. $f''(x) = 6$; 6; 6

7. $f''(x) = 2/(1 + x)^3$; 2; $2/27$

9. $f''(x) = 4/(x^2 + 4)^{3/2}$; $1/2$; $1/(4\sqrt{2})$ **11.** $f''(x) = -6x^{-5/4}$ or $-6/x^{5/4}$; $f''(0)$ does not exist; $-3/2^{1/4}$

13. $f''(x) = 20x^2 e^{-x^2} - 10e^{-x^2}$; -10; $70e^{-4} \approx 1.282$ **15.** $f''(x) = (-3 + 2\ln x)/(4x^3)$; does not exist; -0.050

17. $f'''(x) = 168x + 36$; $f^{(4)}(x) = 168$ **19.** $f'''(x) = 300x^2 - 72x + 12$; $f^{(4)}(x) = 600x - 72$ **21.** $f'''(x) = 18(x + 2)^{-4}$ or $18/(x + 2)^4$; $f^{(4)}(x) = -72(x + 2)^{-5}$ or $-72/(x + 2)^5$ **23.** $f'''(x) = -36(x - 2)^{-4}$ or $-36/(x - 2)^4$; $f^{(4)}(x) = 144(x - 2)^{-5}$ or $144/(x - 2)^5$ **25. a.** $f'(x) = 1/x$; $f''(x) = -1/x^2$; $f'''(x) = 2/x^3$; $f^{(4)}(x) = -6/x^4$; $f^{(5)}(x) = 24/x^5$ **b.** $f^{(n)}(x) = (-1)^{n-1}[1 \cdot 2 \cdot 3 \cdots (n - 1)]/x^n$ or, using factorial notation, $f^{(n)}(x) = (-1)^{n-1}(n - 1)!/x^n$

27. Concave upward on $(2, \infty)$; concave downward on $(-\infty, 2)$; inflection point at $(2, 3)$ **29.** Concave upward on $(-\infty, -1)$ and $(8, \infty)$; concave downward on $(-1, 8)$; inflection points at $(-1, 7)$ and $(8, 6)$ **31.** Concave upward on $(2, \infty)$; concave downward on $(-\infty, 2)$; no inflection points **33.** Always concave upward; no inflection points **35.** Concave upward on $(-\infty, 3/2)$; concave downward on $(3/2, \infty)$; inflection point at $(3/2, 525/2)$ **37.** Concave upward on $(5, \infty)$; concave downward on $(-\infty, 5)$; no inflection points **39.** Concave upward on $(-10/3, \infty)$; concave downward on $(-\infty, -10/3)$; inflection point at $(-10/3, -250/27)$ **41.** Never concave upward; always concave downward; no inflection points **43.** Concave upward on $(-\infty, 0)$ and $(1, \infty)$; concave downward on $(0, 1)$; inflection points at $(0, 0)$ and $(1, -3)$ **45.** Concave upward on $(-1, 1)$; concave downward on $(-\infty, -1)$ and $(1, \infty)$; inflection points at $(-1, \ln 2)$ and $(1, \ln 2)$ **47.** Concave upward on $(-\infty, -e^{-3/2})$ and $(e^{-3/2}, \infty)$; concave downward on $(-e^{-3/2}, 0)$ and $(0, e^{-3/2})$; inflection points at $(-e^{-3/2}, -3e^{-3}/(2\ln 10))$ and $(e^{-3/2}, -3e^{-3}/(2\ln 10))$ **49.** Concave upward on $(-\infty, 0)$ and $(4, \infty)$; concave downward on $(0, 4)$; inflection points at 0 and 4 **51.** Concave upward on $(-7, 3)$ and $(12, \infty)$; concave downward on $(-\infty, -7)$ and $(3, 12)$; inflection points at $-7, 3$, and 12 **53.** Choose $f(x) = x^k$ where $1 < k < 2$. For example, $f(x) = x^{4/3}$ has a relative minimum at $x = 0$, and $f(x) = x^{5/3}$ has an inflection point at $x = 0$. **55. a.** Close to 0 **b.** Close to 1 **57.** Relative maximum at -5. **59.** Relative maximum at 0; relative minimum at 2/3. **61.** Relative minimum at -3 **63.** Relative maximum at $-4/7$; relative minimum at 0 **65. a.** Minimum at about -0.4 and 4.0; maximum at about 2.4 **b.** Increasing on about $(-0.4, 2.4)$ and about $(4.0, \infty)$; decreasing on about $(-\infty, -0.4)$ and $(2.4, 4.0)$ **c.** About 0.7 and 3.3 **d.** Concave upward on about $(-\infty, 0.7)$ and $(3.3, \infty)$; concave downward on about $(0.7, 3.3)$ **67. a.** Maximum at 1; minimum at -1 **b.** Increasing on $(-1, 1)$; decreasing on $(-\infty, -1)$ and $(1, \infty)$ **c.** About $-1.7, 0$, and about 1.7 **d.** Concave upward on about $(-1.7, 0)$ and $(1.7, \infty)$; concave downward on about $(-\infty, -1.7)$ and $(0, 1.7)$ **71.** 45.6; mid 2045 **73.** $(22, 6517.9)$ **75.** $(2.06, 20.8)$ **79. a.** 4 hours **b.** 1160 million **81. a.** After 2 hours **b.** 3/4% **83.** $(38.92, 5000)$ **85.** Inflection point at $t = (\ln c)/k \approx 2.96$ years; this signifies the time when the rate of growth begins to slow down, since L changes from concave up to concave down at this inflection point. **87.** Always concave down **89.** $f(t)$ is decreasing and concave up; $f'(t) < 0$, $f''(t) > 0$. **91. a.** -96 ft/sec **b.** -160 ft/sec **c.** -256 ft/sec **d.** -32 ft/sec^2 **93.** $v(t) = 256 - 32t$; $a(t) = -32$; 1024 ft; 16 seconds after being thrown **95.** $t = 6$

Exercises 5.4 (page 294)

For exercises . . .	3–10,29–30	11,12	13–20	21–28
Refer to example . . .	1	2	3	4

1. 0 **3.**

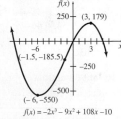

$f(x) = -2x^3 - 9x^2 + 108x - 10$

5.

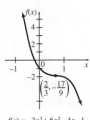

$f(x) = -3x^3 + 6x^2 - 4x - 1$

7.

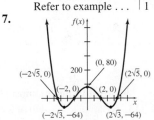

$f(x) = x^4 - 24x^2 + 80$

9.

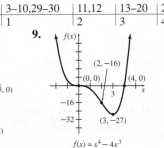

$f(x) = x^4 - 4x^3$

11.

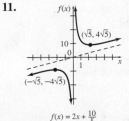

$f(x) = 2x + \frac{10}{x}$

13.

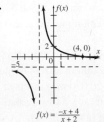

$f(x) = \frac{-x + 4}{x + 2}$

15.

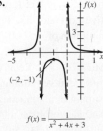

$f(x) = \frac{1}{x^2 + 4x + 3}$

17.

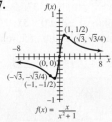

$f(x) = \frac{x}{x^2 + 1}$

19.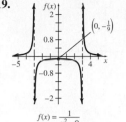

$$f(x) = \frac{1}{x^2 - 9}$$

21.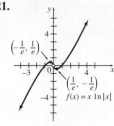

$$f(x) = x \ln|x|$$

23.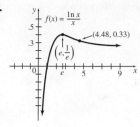

$$f(x) = \frac{\ln x}{x}$$

25.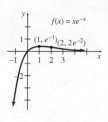

$$f(x) = xe^{-x}$$

27.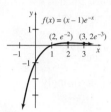

$$f(x) = (x - 1)e^{-x}$$

29.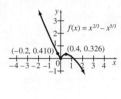

$$f(x) = x^{2/3} - x^{5/3}$$

31. 3, 7, 9, 11, 15   **33.** 17, 19, 23, 25, 27

In Exercises 35–39, other answers are possible.   **35.** 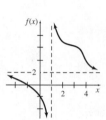   **37.** 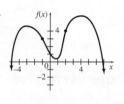   **39.**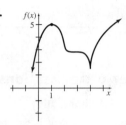

Chapter 5 Review Exercises (page 297)

1. True   **2.** False   **3.** False   **4.** True   **5.** False
6. True   **7.** False   **8.** False   **9.** False
10. False   **11.** True   **12.** False

For exercises …	1,2,12,13, 17–24,70	3,4,14,15, 25–32,63,64, 72,73a,74b,c	5–9,16,33–38, 61,62,65,66,71, 73b,74a	10,11, 39–60,63e, 67–69
Refer to section …	1	2	3	4

17. Increasing on $(-9/2, \infty)$; decreasing on $(-\infty, -9/2)$ **19.** Increasing on $(-5/3, 3)$; decreasing on $(-\infty, -5/3)$ and $(3, \infty)$
21. Never decreasing; increasing on $(-\infty, 3)$ and $(3, \infty)$ **23.** Decreasing on $(-\infty, -1)$ and $(0, 1)$; increasing on $(-1, 0)$ and
$(1, \infty)$ **25.** Relative maximum of -4 at 2   **27.** Relative minimum of -7 at 2   **29.** Relative maximum of 101 at -3; relative
minimum of -24 at 2   **31.** Relative maximum at $(-0.618, 0.206)$; relative minimum at $(1.618, 13.203)$
33. $f''(x) = 36x^2 - 10$; 26; 314   **35.** $f''(x) = 180(3x - 6)^{-3}$ or $180/(3x - 6)^3$; $-20/3$; $-4/75$
37. $f''(t) = (t^2 + 1)^{-3/2}$ or $1/(t^2 + 1)^{3/2}$; $1/2^{3/2} \approx 0.354$; $1/10^{3/2} \approx 0.032$

39.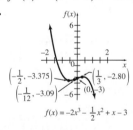

$$f(x) = -2x^3 - \frac{1}{2}x^2 + x - 3$$

41.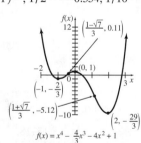

$$f(x) = x^4 - \frac{4}{3}x^3 - 4x^2 + 1$$

43.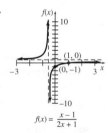

$$f(x) = \frac{x - 1}{2x + 1}$$

45.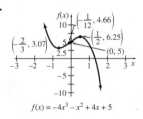

$$f(x) = -4x^3 - x^2 + 4x + 5$$

47.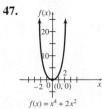

$$f(x) = x^4 + 2x^2$$

49.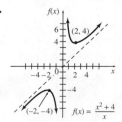

$$f(x) = \frac{x^2 + 4}{x}$$

51.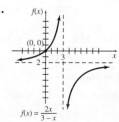

$$f(x) = \frac{2x}{3 - x}$$

53.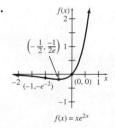

$$f(x) = xe^{2x}$$

55.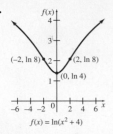
$f(x) = \ln(x^2 + 4)$

57.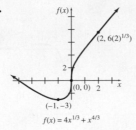
$f(x) = 4x^{1/3} + x^{4/3}$

In Exercise 59, other answers are possible. **59.**

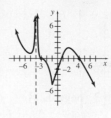

61. a. Both are negative. **63. a.** $P(q) = -q^3 + 7q^2 + 49q$ **b.** 7 brushes **c.** \$229 **d.** \$343 **e.** $q = 7/3$; between 2 and 3 brushes **65. a.** The first derivative has many critical numbers **b.** The curve is always decreasing except at frequent inflection points. **67.**

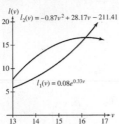

69.

$y = 34.7(1.0186)^{-x} x^{-0.658}$

71. 7.405 yr; the age at which the rate of learning to pass the test begins to slow down **73. a.** At 1965, 1973, 1976, 1983, 1986, and 1988 **b.** Concave upward; this means that the stockpile was increasing at an increasingly rapid rate.

Chapter 6 Applications of the Derivative

Exercises 6.1 (page 309)

For exercises . . .	1–6,31–38	7,8,11–30,39	41–46,51–60	47–50
Refer to example . . .	2	1	3	On Graphical Optimization

1. Absolute maximum at x_3; no absolute minimum
3. No absolute extrema **5.** Absolute minimum at x_1; no absolute maximum **7.** Absolute maximum at x_1; absolute minimum at x_2 **11.** Absolute maximum of 12 at $x = 5$; absolute minimum of -8 at $x = 0$ and $x = 3$ **13.** Absolute maximum of 19.67 at $x = -4$; absolute minimum of -1.17 at $x = 1$ **15.** Absolute maximum of 1 at $x = 0$; absolute minimum of -80 at $x = -3$ and $x = 3$ **17.** Absolute maximum of 1/3 at $x = 0$; absolute minimum of $-1/3$ at $x = 3$ **19.** Absolute maximum of 0.21 at $x = 1 + \sqrt{2} \approx 2.4$; absolute minimum of 0 at $x = 1$ **21.** Absolute maximum of 1.710 at $x = 3$; absolute minimum of -1.587 at $x = 0$ **23.** Absolute maximum of 7 at $x = 1$; absolute minimum of 0 at $x = 0$ **25.** Absolute maximum of 4.910 at $x = 4$; absolute minimum of -1.545 at $x = 2$ **27.** Absolute maximum of 19.09 at $x = -1$; absolute minimum of 0.6995 at $x = (\ln 3)/3$ **29.** Absolute maximum of 1.356 at $x = 0.6085$; absolute minimum of 0.5 at $x = -1$ **31.** Absolute minimum of 7 at $x = 2$; no absolute maximum **33.** Absolute maximum of 137 at $x = 3$; no absolute minimum **35.** Absolute maximum of 0.1 at $x = 4$; absolute minimum of -0.5 at $x = -2$ **37.** Absolute maximum of 0.1226 at $x = e^{1/3}$; no absolute minimum
39. a. Absolute minimum of -5 at $x = -1$; absolute maximum of 0 at $x = 0$ **b.** Absolute maximum of about -0.76 at $x = 2$; absolute minimum of -1 at $x = 1$ **41. a.** Relative maxima of 8496 in 2001, 7556 in 2004, 6985 in 2006, and 6700 in 2008; relative minima of 7127 in 2000, 7465 in 2003, 6748 in 2005, 5933 in 2007, and 5943 in 2009. **b.** Absolute maximum of 8496 in 2001 and absolute minimum of 5933 in 2007. **43.** The maximum profit is \$700,000 when 1,000,000 tires are sold. **45. a.** 112 **b.** 162 **47.** About 11.5 units **49.** 100 units **51.** 6 mo; 6% **53.** About 7.2 mm **55.** Maximum of 25 mpg at 45 mph; minimum of 16.1 mpg at 65 mph **57.** The piece formed into a circle should have length $12\pi/(4 + \pi)$ ft, or about 5.28 ft.

Exercises 6.2 (page 318)

For exercises . . .	1–8,15–18, 33,34,37,44	9–14, 19–24,47	25–30	31,32,41, 42,45,46	35,36, 38–40,43
Refer to example . . .	1	3	4	2	5

1. a. $y = 180 - x$ **b.** $P = x(180 - x)$
c. $[0, 180]$ **d.** $dP/dx = 180 - 2x$; $x = 90$
e. $P(0) = 0$; $P(180) = 0$; $P(90) = 8100$ **f.** 8100; 90 and 90 **3. a.** $y = 90 - x$ **b.** $P = x^2(90 - x)$ **c.** $[0, 90]$
d. $dP/dx = 180x - 3x^2$; $x = 0$, $x = 60$ **e.** $P(0) = 0$, $P(60) = 108,000$, $P(90) = 0$ **f.** 108,000; 60 and 30
5. $\overline{C}(x) = x^2/2 + 2x - 3 + 35/x$; $x = 2.722$ **7. a.** $R(x) = 160,000x - 100x^2$ **b.** 800 **c.** \$640,000 **9. a.** $1400 - 2x$
b. $A(x) = 1400x - 2x^2$ **c.** 350 m **d.** 245,000 m² **11.** 405,000 m² **13.** \$960 **15. a.** 125 passengers **b.** \$156,250
17. In 10 days; \$960 **19.** 4 in. by 4 in. by 2 in. **21.** 2/3 ft (or 8 in.) **23.** 20 cm by 20 cm by 40 cm; \$7200
27. Radius = 5.206 cm, height = 11.75 cm **29.** Radius = 5.242 cm; height = 11.58 cm **31.** 1 mile from point A
33. a. 15 days **b.** 16.875% **35.** 12.98 thousand **37. a.** 12 days **b.** 50 per ml **c.** 1 day **d.** 81.365 per ml

39. 49.37 **41.** Point P is $3\sqrt{7}/7 \approx 1.134$ mi from Point A **43. a.** Replace a with e^r and b with r/P.
b. Shepherd: $f'(S) = a[1 + (1 - c)(S/b)^c]/[1 + (S/b)^c]^2$; Ricker: $f'(S) = ae^{-bS}(1 - bS)$; Beverton-Holt:
$f'(S) = a/[1 + (S/b)]^2$ **c.** Shepherd: a; Ricker: a; Beverton-Holt: a; the constant a represents the slope of the graph of
$f(S)$ at $S = 0$. **d.** 194,000 tons **e.** 256,000 tons **45.** $(56 - 2\sqrt{21})/7 \approx 6.7$ mi **47.** 36 in. by 18 in. by 18 in.

Exercises 6.3 (page 329)

3. c **5.** It is negative. **9.** 10,000 **11.** 10

For exercises . . .	1,2,9–12, 15–18	3,13,14	4,27,32	5,7,8,25,26, 28–30	19–24,31
Refer to example . . .	1	2	3	4	5

13. 4899 **19. a.** $E = p/(200 - p)$ **b.** 25
21. a. $E = 2p^2/(7500 - p^2)$ **b.** 25,000 **23. a.** $E = 5/q$ **b.** 5 **25. a.** $E = 0.5$; inelastic; total revenue increases as price
increases. **b.** $E = 8$; elastic; total revenue decreases as price increases. **27.** 0.06; the demand is inelastic. **29.** 2.826; the
demand is elastic. **31. a.** 0.071 **b.** Inelastic **c.** \$1255

Exercises 6.4 (page 334)

For exercises . . .	1–16,38–40	17–37	41–50
Refer to example . . .	1,2	3	4

1. $dy/dx = -6x/(5y)$ **3.** $dy/dx = (8x - 5y)/(5x - 3y)$
5. $dy/dx = 15x^2/(6y + 4)$ **7.** $dy/dx = -3x(2 + y)^2/2$
9. $dy/dx = \sqrt{y}/[\sqrt{x}(5\sqrt{y} - 2)]$ **11.** $dy/dx = (4x^3y^3 + 6x^{1/2})/(9y^{1/2} - 3x^4y^2)$ **13.** $dy/dx = (5 - 2xye^{x^2y})/(x^2e^{x^2y} - 4)$
15. $dy/dx = y(2xy^3 - 1)/(1 - 3x^2y^3)$ **17.** $y = (3/4)x + 25/4$ **19.** $y = x + 2$ **21.** $y = x/64 + 7/4$
23. $y = (11/12)x - 5/6$ **25.** $y = -(37/11)x + 59/11$ **27.** $y = (5/2)x - 1/2$ **29.** $y = 1$ **31.** $y = -2x + 7$
33. $y = -x + 2$ **35.** $y = (8/9)x + 10/9$ **37. a.** $y = -(3/4)x + 25/2$; $y = (3/4)x - 25/2$
b.

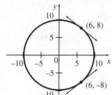

39. a. $du/dv = -2u^{1/2}/(2v + 1)^{1/2}$ **b.** $dv/du = -(2v + 1)^{1/2}/(2u^{1/2})$ **c.** They are reciprocals.
41. $dy/dx = -x/y$; there is no function $y = f(x)$ that satisfies $x^2 + y^2 + 1 = 0$.
43. a. \$0.94; the approximate increase in cost of an additional unit
b. \$0; the approximate change in revenue for a unit increase in sales **45 a.** 0.44; inelastic
b. 0.44 **47.** $1/(3\sqrt{3})$ **49.** $ds/dt = (4s - 6t^2 + 5)/(3s^2 - 4t)$

Exercises 6.5 (page 341)

For exercises . . .	1–8, 16–22	9–14	15	23,24, 29,31,32	25	26–28,30
Refer to example . . .	1	5	6	3	2	4

1. -64 **3.** $-9/7$ **5.** $1/5$ **7.** $-3/2$ **9.** \$384 per month
11. a. Revenue is increasing at a rate of \$180 per day.
b. Cost is increasing at a rate of \$50 per day. **c.** Profit is
increasing at a rate of \$130 per day. **13.** Demand is decreasing at a rate of approximately 98 units per unit time.
15. 0.067 mm per min **17.** About 1.9849 g per day **19. a.** 105.15 m$^{-0.25}$ dm/dt **b.** About 52.89 kcal per day^2
21. 25.6 crimes per month **23.** 24/5 ft/min **25.** 16π ft^2/min **27.** 2/27 cm/min **29.** 62.5 ft per min **31.** $\sqrt{2} \approx 1.41$ ft per sec

Exercises 6.6 (page 348)

For exercises . . .	1–8	9–16	17–20	21–35	36–41
Refer to example . . .	1	2	3	4	5

1. 1.9 **3.** 0.1 **5.** 0.060 **7.** -0.023 **9.** 12.0417; 12.0416; 0.0001
11. 0.995; 0.9950; 0 **13.** 1.01; 1.0101; 0.0001 **15.** 0.05; 0.0488; 0.0012 **17. a.** -4.4 thousand lb **b.** -52.2 thousand lb
19. \$60 **21.** About 9600 in^3 **23. a.** 0.007435 **b.** -0.005105 **25. a.** 0.347 million **b.** -0.022 million **27.** 1568π mm^3
29. 80π mm^2 **31. a.** About 9.3 kg **b.** About 9.5 kg **33.** -7.2π cm^3 **35.** 0.472 cm^3 **37.** 0.00125 cm **39.** ± 1.273 in^3
41. ± 0.116 in^3

Chapter 6 Review Exercises (page 349)

For exercises . . .	1–3, 11–18	4,5, 49–54	6, 19–30, 43,44,56	7,8, 31–38, 55,59–61	9,10, 39–42, 62,63	45–48, 57,58, 64–68
Refer to section . . .	1	3	4	5	6	2

1. False **2.** True **3.** False **4.** True **5.** True
6. True **7.** True **8.** True **9.** True **10.** True
11. Absolute maximum of 33 at 4; absolute minimum of 1 at 0
and 6 **13.** Absolute maximum of 39 at -3; absolute minimum of $-319/27$ at 5/3 **17. a.** Maximum $= 0.37$; minimum $= 0$
b. Maximum $= 0.35$; minimum $= 0.13$ **21.** $dy/dx = (2x - 9x^2y^4)/(8y + 12x^3y^3)$ **23.** $dy/dx = 6\sqrt{y - 1}/[x^{1/3}(1 - \sqrt{y - 1})]$
25. $dy/dx = -(30 + 50x)/3$ **27.** $dy/dx = (2xy^4 + 2y^3 - y)/(x - 6x^2y^3 - 6xy^2)$ **29.** $y = (-16/23)x + 94/23$ **33.** 272
35. -2 **37.** $-8e^3$ **41.** 0.00204 **43. a.** $(2, -5)$ and $(2, 4)$ **b.** $(2, -5)$ is a relative minimum; $(2, 4)$ is a relative maximum.
c. No **45. a.** 600 boxes **b.** \$720 **47.** 3 in. **49.** 1789 **51.** 80 **53.** 0.47; inelastic **55.** 56π ft^2 per min
57. a.

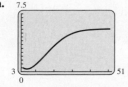

b. About the 15th day **59.** 8/3 ft per min **61.** $21/16 = 1.3125$ ft per min **63.** ± 0.736 in^2
65. $1.25 + 2 \ln 1.5$ **67.** 10 ft; 18.67 sec

Chapter 7 Integration

Exercises 7.1 (page 366)

1. They differ only by a constant.

5. $6k + C$ **7.** $z^2 + 3z + C$

9. $2t^3 - 4t^2 + 7t + C$

11. $z^4 + z^3 + z^2 - 6z + C$

13. $10z^{3/2}/3 + \sqrt{2}z + C$

15. $5x^4/4 - 20x^2 + C$

17. $8v^{3/2}/3 - 6v^{5/2}/5 + C$

For exercises . . .	5–24, 27,28,	25,26, 29–32, 35,36, 41,42, 60,63, 66	33,34, 37–40, 61	43,44	45–52, 57–59, 65	53–56	67–74
Refer to example . . .	4,5	7,8	6	12	9	10	11

19. $4u^{5/2} - 4u^{7/2} + C$ **21.** $-7/z + C$ **23.** $-\pi^3/(2y^2) - 2\sqrt{\pi y} + C$ **25.** $6t^{-1.5} - 2\ln|t| + C$ **27.** $-1/(3x) + C$

29. $-15e^{-0.2x} + C$ **31.** $-3\ln|x| - 10e^{-0.4x} + e^{0.1}x + C$ **33.** $(1/4)\ln|t| + t^3/6 + C$ **35.** $e^{2u}/2 + 2u^2 + C$

37. $x^3/3 + x^2 + x + C$ **39.** $6x^{7/6}/7 + 3x^{2/3}/2 + C$ **41.** $10^x/(\ln 10) + C$ **43.** $f(x) = 3x^{5/3}/5$

45. $C(x) = 2x^2 - 5x + 8$ **47.** $C(x) = 3e^{0.01x} + 5$ **49.** $C(x) = 3x^{5/3}/5 + 2x + 114/5$

51. $C(x) = 5x^2/2 - \ln|x| - 153.50$ **53.** $p = 175 - 0.01x - 0.01x^2$ **55.** $p = 500 - 0.1\sqrt{x}$

57. a. $f(t) = 3.75t^2 - 16.8t + 0.05$ **b.** Approximately 152.6 billion monthly text messages **59. a.** $P(x) = 25x^4/2 + 10x^3 - 40$

b. \$240 **61.** $a\ln x - bx + C$ **63. a.** $N(t) = 155.3e^{0.3219t} + 144.7$ **b.** 7537

65. a. $B(t) = 0.02016t^3 - 0.6460t^2 + 15.86t + 839.7$ **b.** About 2,082,000 **67.** $v(t) = 5t^3/3 + 4t + 6$

69. $s(t) = -16t^2 + 6400$; 20 sec **71.** $s(t) = 2t^{5/2} + 3e^{-t} + 1$ **73.** 160 ft/sec, 12 ft

Exercises 7.2 (page 374)

3. $2(2x + 3)^5/5 + C$

5. $-(2m + 1)^{-2}/2 + C$

7. $-(x^2 + 2x - 4)^{-3}/3 + C$

For exercises . . .	3,4,29, 33,34	5–8, 21,22	9,10,27, 28,30	11–16, 31,35, 36	17–20, 32	23–26	39–44
Refer to example . . .	1	3	2	5	4	6	7

9. $(4z^2 - 5)^{3/2}/12 + C$ **11.** $e^{2x^3}/2 + C$ **13.** $e^{2t-t^2}/2 + C$ **15.** $-e^{1/z} + C$ **17.** $[\ln(t^2 + 2)]/2 + C$

19. $[\ln(x^4 + 4x^2 + 7)]/4 + C$ **21.** $-1/[2(x^2 + x)^2] + C$ **23.** $(p + 1)^7/7 - (p + 1)^6/6 + C$

25. $2(u - 1)^{3/2}/3 + 2(u - 1)^{1/2} + C$ **27.** $(x^2 + 12x)^{3/2}/3 + C$ **29.** $(1 + 3\ln x)^3/9 + C$ **31.** $(1/2)\ln(e^{2x} + 5) + C$

33. $(\ln 10)(\log x)^2/2 + C$ **35.** $8^{3x^2+1}/(6\ln 8) + C$ **39. a.** $R(x) = 6(x^2 + 27{,}000)^{1/3} - 180$ **b.** 150 players

41. a. $C(x) = 6\ln(5x^2 + e) + 4$ **b.** Yes **43. a.** $f(t) = 4.0674 \times 10^{-4}[(t - 1970)^{2.4}/2.4 + 1970(t - 1970)^{1.4}/1.4] + 61.298$ **b.** About 181,000

Exercises 7.3 (page 383)

3. a. 88 **b.** $\int_0^8 (2x + 5)\,dx$

For exercises . . .	5–14,17–22	15,16,24–30, 33–35,39	31,32,36–38
Refer to example . . .	1,2	3	4

5. a. 21 **b.** 23 **c.** 22 **d.** 22 **7. a.** 10 **b.** 10 **c.** 10 **d.** 11 **9. a.** 8.22 **b.** 15.48 **c.** 11.85 **d.** 10.96 **11. a.** 6.70

b. 3.15 **c.** 4.93 **d.** 4.17 **13. a.** 4 **b.** 4 **15. a.** 4 **b.** 5 **17.** 4π **19.** 24 **21. b.** 0.385 **c.** 0.33835 **d.** 0.334334

e. 0.333333 **25.** Left: 1410 trillion BTUs; right: 3399 trillion BTUs; average: 2404.5 trillion BTUs **27. a.** Left: about 582,000

cases; right: about 580,000 cases; average: about 581,000 cases **b.** Left: about 146,000 cases; right: about 144,000 cases; average:

about 145,000 cases **29.** About 1300 ft; yes **31.** 2751 ft, 3153 ft, 2952 ft **33. a.** About 660 BTU/ft^2 **b.** About 320 BTU/ft^2

35. a. 9 ft **b.** 2 sec **c.** 4.6 ft **d.** Between 3 and 3.5 sec **37.** 22.5 and 18 ft **39. a.** About 75,600 **b.** About 77,300

Exercises 7.4 (page 395)

1. -18 **3.** $-3/2$ **5.** $28/3$ **7.** 13

9. $16/3$ **11.** 76 **13.** $4/5$ **15.** $108/25$

17. $20e^{0.3} - 20e^{0.2} + 3\ln 2 - 3\ln 3 \approx 1.353$

For exercises . . .	1–6,9–12, 15–20, 23,24	7,8,13,14, 21,22, 27–30,54	31–44,47	55–61, 63–72
Refer to example . . .	1,2,3	4	5, 6	7

19. $e^8/4 - e^4/4 - 1/6 \approx 731.4$ **21.** $91/3$

23. $447/7 \approx 63.86$ **25.** $(\ln 2)^2/2 \approx 0.2402$ **27.** 49 **29.** $1/8 - 1/[2(3 + e^2)] \approx 0.07687$ **31.** 10 **33.** 76 **35.** $41/2$

37. $e^2 - 3 + 1/e \approx 4.757$ **39.** $e - 2 + 1/e \approx 1.086$ **41.** $23/3$ **43.** $e^2 - 2e + 1 \approx 2.952$

45. $\int_a^c f(x)\,dx = \int_a^b f(x)\,dx + \int_b^c f(x)\,dx$ **47.** -8 **51.** -12 **53. a.** $x^5/5 - 1/5$ **c.** $f'(1) \approx 2.746$, and $g(1) = e \approx 2.718$

55. a. $(9000/8)(17^{4/3} - 2^{4/3}) \approx \$46{,}341$ **b.** $(9000/8)(26^{4/3} - 17^{4/3}) \approx \$37{,}477$ **c.** It is slowly increasing without bound.

57. No **59. a.** 0.8778 ft **b.** 0.6972 ft **61. a.** 18.12 **b.** 8.847 **63. b.** $\int_0^{60} n(x)\,dx$ **c.** $2(51^{3/2} - 26^{3/2})/15 \approx 30.89$ million

65. a. $Q(R) = \pi k R^4/2$ **b.** $0.04k$ mm per min **67. b.** About 505,000 kJ/W$^{0.67}$ **69. a.** About 286 million; the total population
aged 0 to 90 **b.** About 64 million **71. a.** $c'(t) = 1.2e^{0.04t}$ **b.** $\int_0^{10} 1.2e^{0.04t}\, dt$ **c.** $30e^{0.4} - 30 \approx 14.75$ billion **d.** About 12.8 yr
e. About 14.4 yr

Exercises 7.5 (page 405)
1. 21 **3.** 20 **5.** 23/3 **7.** 366.2
9. 4/3 **11.** $2\ln 2 - \ln 6 + 3/2 \approx 1.095$
13. $6\ln(3/2) - 6 + 2e^{-1} + 2e \approx 2.605$

For exercises . . .	1,2	3–6, 11–18,20, 23,24	7–10, 19,21,22, 25,26,41,42	27–30, 39,40	31–37
Refer to example . . .	1	3	2	4	5

15. $(e^{-2} + e^4)/2 - 2 \approx 25.37$ **17.** 1/2 **19.** 1/20 **21.** $3(2^{4/3})/2 - 3(2^{7/3})/7 \approx 1.620$ **23.** $(e^9 + e^6 + 1)/3 \approx 2836$
25. $-1.9241, -0.4164, 0.6650$ **27. a.** 8 yr **b.** About \$148 **c.** About \$771 **29. a.** 39 days **b.** \$3369.18 **c.** \$484.02
d. \$2885.16 **31.** 12,931.66 **33.** 54
35. a.

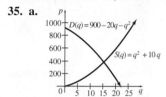

b. $(15, 375)$ **c.** \$4500 **d.** \$3375 **37. a.** 12 **b.** \$5616, \$1116 **c.** \$1872, \$1503
d. \$387 **39. a.** About 71.25 gal **b.** About 25 hr **c.** About 105 gal **d.** About 47.91 hr
41. a. 0.019; the lower 10% of the income producers earn 1.9% of the total income of the
population. **b.** 0.184; the lower 40% of the income producers earn 18.4% of the total income of
the population. **c.**

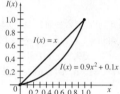

d. 0.15 **e.** Income is distributed less equally in
2008 than in 1968.

Exercises 7.6 (page 413)
1. a. 12.25 **b.** 12 **c.** 12 **3. a.** 3.35 **b.** 3.3
c. $3\ln 3 \approx 3.296$ **5. a.** 11.34 **b.** 10.5 **c.** 10.5
7. a. 0.9436 **b.** 0.8374 **c.** $4/5 = 0.8$ **9. a.** 1.236

For exercises . . .	1a–12a,15, 23a–24a,29	1b–12b,17,19,20, 23b–24b,29,30, 33–35	21,22,25–28, 31,32
Refer to example . . .	1	2	3

b. 1.265 **c.** $2 - 2e^{-1} \approx 1.264$ **11. a.** 5.991 **b.** 6.167
c. 6.283; Simpson's rule **13.** b is true. **15. a.** 0.2 **b.** 0.220703, 0.205200, 0.201302, 0.200325, 0.020703, 0.005200, 0.001302,
0.000325 **c.** $p = 2$ **17. a.** 0.2 **b.** 0.2005208, 0.2000326, 0.2000020, 0.2000001, 0.0005208, 0.0000326, 0.0000020, 0.0000001
c. $p = 4$ **19.** $M = 0.7355; S = 0.8048$
21. a.

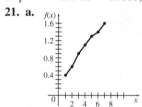

b. 6.3 **c.** 6.27 **23. a.** 1.831 **b.** 1.758 **25.** About 30 mcg(h)/ml; this represents the total
amount of drug available to the patient for each ml of blood. **27.** About 9 mcg(h)/ml; this represents the total effective amount of the drug available to the patient for each ml of blood.
29. a. $y = b_0(t/7)^{b_1}e^{-b_2 t/7}$ **b.** About 1212 kg; about 1231 kg **c.** About 1224 kg; about 1250 kg
31. a.
b. 71.5 **c.** 69.0 **33.** 3979 **35. a.** 0.6827 **b.** 0.9545 **c.** 0.9973

Chapter 7 Review Exercises (page 417)
1. True **2.** False **3.** False **4.** True
5. True **6.** False **7.** False **8.** True
9. True **10.** False **11.** True **12.** False
13. False **14.** True **19.** $x^2 + 3x + C$

For exercises . . .	1–4, 19–30, 79,80, 84	5,16,17, 31–40, 87,94	6–8,15, 41–43, 81,92,96	9,10, 45–62, 82,83, 89–91, 95	11,12, 63–66, 85,86	13,14, 18, 44, 67–77, 93
Refer to section . . .	1	2	3	4	5	6

21. $x^3/3 - 3x^2/2 + 2x + C$ **23.** $2x^{3/2} + C$ **25.** $2x^{3/2}/3 + 9x^{1/3} + C$ **27.** $2x^{-2} + C$ **29.** $-3e^{2x}/2 + C$ **31.** $e^{3x}/6 + C$
33. $(3\ln|x^2 - 1|)/2 + C$ **35.** $-(x^3 + 5)^{-3}/9 + C$ **37.** $-e^{-3x^4}/12 + C$ **39.** $(3\ln x + 2)^5/15 + C$ **41.** 20 **43.** 24
45. a. $s(T) - s(0)$ **b.** $\int_0^T v(t)\, dt = s(T) - s(0)$ is equivalent to the Fundamental Theorem with $a = 0$ and $b = T$ because $s(t)$
is an antiderivative of $v(t)$. **47.** 12 **49.** $3\ln 5 + 12/25 \approx 5.308$ **51.** 19/15 **53.** $3(1 - e^{-4})/2 \approx 1.473$ **55.** $\pi/32$
57. $25\pi/4$ **59.** 13/3 **61.** $(e^4 - 1)/2 \approx 26.80$ **63.** 64/3 **65.** 149/3 **67.** 0.5833; 0.6035 **69.** 4.187; 4.155 **71.** 0.6011
73. 4.156 **75. a.** 4/3 **b.** 1.146 **c.** 1.252 **77. a.** 0 **b.** 0 **79.** $C(x) = (2x - 1)^{3/2} + 145$ **81.** \$96,000 **83.** \$38,000
85. a. \$916.67 **b.** \$666.67 **87. a.** 17.718 billion barrels **b.** 17.718 billion barrels **d.** $y = -0.03545x + 2.1347$; 17.777
billion barrels **89.** 782 **91. a.** 0.2784 **b.** 0.2784 **93. a.** About 8208 kg **b.** About 8430 kg **c.** About 8558 kg
95. $s(t) = t^3/3 - t^2 + 8$.

Chapter 8 Further Techniques and Applications of Integration

Exercises 8.1 (page 432)

For exercises . . .	1–4,36,40,41,43,44	5,6,35,39,42	7–10,20–22	13–19	23–28
Refer to example . . .	1,3	2	4	1–3	5

1. $xe^x - e^x + C$ **3.** $(-x/2 + 23/16)\,e^{-8x} + C$ **5.** $(x^2 \ln x)/2 - x^2/4 + C$ **7.** $-5e^{-1} + 3 \approx 1.161$
9. $26 \ln 3 - 8 \approx 20.56$ **11.** $e^4 + e^2 \approx 61.99$ **13.** $x^2 e^{2x}/2 - xe^{2x}/2 + e^{2x}/4 + C$
15. $(2/7)(x + 4)^{7/2} - (16/5)(x + 4)^{5/2} + (32/3)(x + 4)^{3/2} + C$ or $(2/3)x^2(x + 4)^{3/2} - (8/15)x(x + 4)^{5/2} + (16/105)(x + 4)^{7/2} + C$
17. $(4x^2 + 10x) \ln 5x - 2x^2 - 10x + C$ **19.** $(-e^2/4)(3e^2 + 1) \approx -42.80$ **21.** $2\sqrt{3} - 10/3 \approx 0.1308$
23. $16 \ln |x + \sqrt{x^2 + 16}| + C$ **25.** $-(3/11) \ln |(11 + \sqrt{121 - x^2})/x| + C$ **27.** $-1/(4x + 6) - (1/6) \ln |x/(4x + 6)| + C$
31. -18 **33.** 15   **37. a.** $(2/3)x(x + 1)^{3/2} - (4/15)(x + 1)^{5/2} + C$ **b.** $(2/5)(x + 1)^{5/2} - (2/3)(x + 1)^{3/2} + C$
39. $(169/2) \ln 13 - 42 \approx \$174.74$ **41.** $15e^6 + 3 \approx 6054$ **43.** About 219 kJ

Exercises 8.2 (page 439)

For exercises . . .	1–23,32,33,40,43	24–31,34–37,39,41,42	38
Refer to example . . .	1–3	4	Derivation of volume formula

1. 9π **3.** $364\pi/3$ **5.** $386\pi/27$ **7.** $15\pi/2$ **9.** 18π **11.** $\pi(e^4 - 1)/2 \approx 84.19$ **13.** $4\pi \ln 3 \approx 13.81$ **15.** $3124\pi/5$
17. $16\pi/15$ **19.** $4\pi/3$ **21.** $4\pi r^3/3$ **23.** $\pi r^2 h$ **25.** $13/3 \approx 4.333$ **27.** $38/15 \approx 2.533$ **29.** $e - 1 \approx 1.718$
31. $(5e^4 - 1)/8 \approx 34.00$ **33.** 3.758   **35.** \$42.49   **37.** 200 cases   **39. a.** $110e^{-0.1} - 120e^{-0.2} \approx 1.284$
b. $210e^{-1.1} - 220e^{-1.2} \approx 3.640$ **c.** $330e^{-2.3} - 340e^{-2.4} \approx 2.241$ **41. a.** $9(6 \ln 6 - 5) \approx 51.76$ **b.** $5(10 \ln 10 - 9) \approx 70.13$
c. $3(31 \ln 31 - 30)/2 \approx 114.7$ **43.** $1.083 \times 10^{21}\,\text{m}^3$

Exercises 8.3 (page 447)

For exercises . . .	1(a)–8(a),16	1(b)–8(b),15,17,18	9–14,19,20
Refer to example . . .	2	3,4	5

1. a. \$6883.39   **b.** \$15,319.26   **3. a.** \$3441.69   **b.** \$7659.63   **5. a.** \$3147.75   **b.** \$7005.46   **7. a.** \$32,968.35
b. \$73,372.42   **9. a.** \$746.91   **b.** \$1662.27   **11. a.** \$688.64   **b.** \$1532.59   **13. a.** \$11,351.78   **b.** \$25,263.84
15. \$63,748.43   **17.** \$28,513.76, \$54,075.81   **19.** \$4560.94

Exercises 8.4 (page 452)

For exercises . . .	1–26,31–34,37	27–30,35,36	42–47,50	48,49,51,52
Refer to example . . .	1	2	4	3

1. 1/3   **3.** Divergent   **5.** -1 **7.** 10,000   **9.** 1/10
11. 3/5   **13.** 1   **15.** 1000   **17.** Divergent   **19.** 1   **21.** Divergent
23. Divergent   **25.** Divergent   **27.** 0   **29.** Divergent   **31.** Divergent   **33.** 1   **35.** 0   **39. a.** 2.808, 3.724, 4.417, 6.720, 9.022
b. Divergent   **c.** 0.8770, 0.9070, 0.9170, 0.9260, 0.9269   **d.** Convergent   **41. a.** 9.9995, 49.9875, 99.9500, 995.0166
b. Divergent   **c.** 100,000   **43.** \$20,000,000   **45.** \$30,000   **47.** \$30,000   **49.** $Na/[b(b + k)]$ **51.** About 833.3

Chapter 8 Review Exercises (page 455)

For exercises . . .	1–4,14–25	5–7,26–35,59	8,9,44–55,57	10,36–43,56,58
Refer to section . . .	1	2	3	4

1. False   **2.** True   **3.** False   **4.** True
5. False   **6.** False   **7.** True   **8.** True
9. True   **10.** False
15. $6x(x - 2)^{1/2} - 4(x - 2)^{3/2} + C$ **17.** $-(x + 2)e^{-3x} - (1/3)e^{-3x} + C$ **19.** $(x^2/2 - x)\ln |x| - x^2/4 + x + C$
21. $(1/8)\sqrt{16 + 8x^2} + C$ **23.** $10e^{1/2} - 16 \approx 0.4872$ **25.** $234/7 \approx 33.43$ **27.** $81\pi/2 \approx 127.2$ **29.** $\pi \ln 3 \approx 3.451$
31. $64\pi/5 \approx 40.21$ **35.** 2,391,484/3   **37.** 1/5   **39.** $6/e \approx 2.207$ **41.** Divergent   **43.** 3   **45.** $16,250/3 \approx \$5416.67$
47. \$174,701.45   **49.** \$15.58   **51.** \$5354.97   **53.** \$30,035.17   **55.** \$176,919.15   **57.** 0.4798
59. a. $158.3°$ **b.** $125°$ **c.** $133.3°$

Chapter 9 Multivariable Calculus

Exercises 9.1 (page 467)

For exercises . . .	1–4,27,28,32, 33,38–45,48	5–12	13–16	21–26	23–28	31,34–36
Refer to example . . .	1–3	4–6	7,8	material after Example 8	5	8

1. a. 12 **b.** -6 **c.** 10 **d.** -19 **3. a.** $\sqrt{43}$ **b.** 6 **c.** $\sqrt{19}$ **d.** $\sqrt{11}$

5.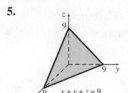

7.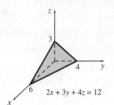

9.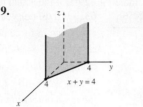

11.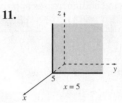

13.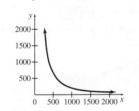

15.

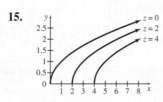

21. c **23.** e **25.** b **27. a.** $8x + 4h$
b. $-4y - 2h$ **c.** $8x$ **d.** $-4y$ **29. a.** $3e^2$;
slope of tangent line in the direction of x at
$(1, 1)$ **b.** $3e^2$; slope of tangent line in the
direction of y at $(1, 1)$ **31. a.** 1987
(rounded) **b.** 595 (rounded) **c.** 359,768
(rounded) **33.** 1.416; the IRA account
grows faster.

35. $y = 500^{5/2}/x^{3/2} \approx 5{,}590{,}170/x^{3/2}$

37. $C(x, y, z) = 250x + 150y + 75z$ **39. a.** 1.89 m² **b.** 1.62 m² **c.** 1.78 m²
41. a. 8.7% **b.** 48% **c.** Multiple solutions: $W = 19.75, R = 0, A = 0$ or
$W = 10, R = 10, A = 4.59$ **d.** Wetland percentage **43. a.** 397 accidents
45. a. $T = 242.257\, C^{0.18}/F^3$ **b.** 58.82; a tethered sow spends nearly 59% of the time doing
repetitive behavior when she is fed 2 kg of food a day and neighboring sows spend 40% of the
time doing repetitive behavior. **47.** $g(L, W, H) = 2LW + 2WH + 2LH$ ft²

Exercises 9.2 (page 478)

For exercises . . .	1,2,33–36,43,44	3–20	21–32	37–42	45–47,53–67	48–52
Refer to example . . .	3	1–3	6,7	8	4	5

1. a. $12x - 4y$ **b.** $-4x + 18y$ **c.** 12 **d.** -40 **3.** $f_x(x, y) = -4y$; $f_y(x, y) = -4x + 18y^2$; 4; 178 **5.** $f_x(x, y) = 10xy^3$;
$f_y(x, y) = 15x^2y^2$; -20; 2160 **7.** $f_x(x, y) = e^{x+y}$; $f_y(x, y) = e^{x+y}$; e^1 or e; e^{-1} or $1/e$ **9.** $f_x(x, y) = -24e^{4x-3y}$; $f_y(x, y) = 18e^{4x-3y}$;
$-24e^{11}$; $18e^{-25}$ **11.** $f_x(x, y) = (-x^4 - 2xy^2 - 3x^2y^3)/(x^3 - y^2)^2$; $f_y(x, y) = (3x^3y^2 - y^4 + 2x^2y)/(x^3 - y^2)^2$; $-8/49$; $-1713/5329$
13. $f_x(x, y) = 15x^2y^2/(1 + 5x^3y^2)$; $f_y(x, y) = 10x^3y/(1 + 5x^3y^2)$; 60/41; 1920/2879 **15.** $f_x(x, y) = e^{x^2y}(2x^2y + 1)$;
$f_y(x, y) = x^3e^{x^2y}$; $-7e^{-4}$; $-64e^{48}$ **17.** $f_x(x, y) = (1/2)(4x^3 + 3y)/(x^4 + 3xy + y^4 + 10)^{1/2}$;
$f_y(x, y) = (1/2)(3x + 4y^3)/(x^4 + 3xy + y^4 + 10)^{1/2}$; $29/(2\sqrt{21})$; $48/\sqrt{311}$
19. $f_x(x, y) = [6xy(e^{xy} + 2) - 3x^2y^2e^{xy}]/(e^{xy} + 2)^2$; $f_y(x, y) = [3x^2(e^{xy} + 2) - 3x^3ye^{xy}]/(e^{xy} + 2)^2$; $-24(e^{-2} + 1)/(e^{-2} + 2)^2$;
$(624e^{-12} + 96)/(e^{-12} + 2)^2$ **21.** $f_{xx}(x, y) = 8y^2 - 32$; $f_{yy}(x, y) = 8x^2$; $f_{xy}(x, y) = f_{yx}(x, y) = 16xy$
23. $R_{xx}(x, y) = 8 + 24y^2$; $R_{yy}(x, y) = -30xy + 24x^2$; $R_{xy}(x, y) = R_{yx}(x, y) = -15y^2 + 48xy$
25. $r_{xx}(x, y) = 12y/(x + y)^3$; $r_{yy}(x, y) = -12x/(x + y)^3$; $r_{xy}(x, y) = r_{yx}(x, y) = (6y - 6x)/(x + y)^3$
27. $z_{xx} = 9ye^x$; $z_{yy} = 0$; $z_{xy} = z_{yx} = 9e^x$ **29.** $r_{xx} = -1/(x + y)^2$; $r_{yy} = -1/(x + y)^2$; $r_{xy} = r_{yx} = -1/(x + y)^2$
31. $z_{xx} = 1/x$; $z_{yy} = -x/y^2$; $z_{xy} = z_{yx} = 1/y$ **33.** $x = -4, y = 2$ **35.** $x = 0, y = 0$; or $x = 3, y = 3$ **37.** $f_x(x, y, z) = 4x^3$;
$f_y(x, y, z) = 2z^2$; $f_z(x, y, z) = 4yz + 4z^3$; $f_{yz}(x, y, z) = 4z$ **39.** $f_x(x, y, z) = 6/(4z + 5)$; $f_y(x, y, z) = -5/(4z + 5)$;
$f_z(x, y, z) = -4(6x - 5y)/(4z + 5)^2$; $f_{yz}(x, y, z) = 20/(4z + 5)^2$ **41.** $f_x(x, y, z) = (2x - 5z^2)/(x^2 - 5xz^2 + y^4)$;
$f_y(x, y, z) = 4y^3/(x^2 - 5xz^2 + y^4)$; $f_z(x, y, z) = -10xz/(x^2 - 5xz^2 + y^4)$; $f_{yz}(x, y, z) = 40xy^3z/(x^2 - 5xz^2 + y^4)^2$
43. a. 6.773 **b.** 3.386 **45. a.** 80 **b.** 150 **c.** 80 **d.** 440 **47. a.** $902,100 **b.** $f_p(p, i) = 99 - 0.5i - 0.005p$; $f_i(p, i) = -0.5p$;
the rate at which weekly sales are changing per unit of change in price when the interest rate remains constant $(f_p(p, i))$ or per unit change in
interest rate when the price remains constant $(f_i(p, i))$ **c.** A weekly sales decrease of $9700 **49. a.** 50.57 hundred units
b. $f_x(16, 81) = 1.053$ hundred units and is the rate at which production is changing when labor changes by 1 unit (from 16 to 17) and capital
remains constant; $f_y(16, 81) = 0.4162$ hundred units and is the rate at which production is changing when capital changes by 1 unit (from 81
to 82) and labor remains constant. **c.** Production would increase by approximately 105 units. **51.** $0.4x^{-0.6}y^{0.6}$; $0.6x^{0.4}y^{-0.4}$

53. a. 1279 kcal per hr   **b.** 2.906 kcal per hr per g; the instantaneous rate of change of energy usage for a 300-kg animal traveling at 10 km per hr is about 2.9 kcal per hr per g.   **55. a.** 0.0142 m^2 **b.** 0.00442 m^2 **57. a.** 4.125 lb   **b.** $\partial f/\partial n = n/4$; the rate of change of weight loss per unit change in workouts   **c.** An additional loss of 3/4 lb   **59. a.** $(2ax - 3x^2)t^2e^{-t}$
b. $x^2(a - x)(2t - t^2)e^{-t}$ **c.** $(2a - 6x)t^2e^{-t}$ **d.** $(2ax - 3x^2)(2t - t^2)e^{-t}$ **e.** $\partial R/\partial x$ gives the rate of change of the reaction per unit of change in the amount of drug administered. $\partial R/\partial t$ gives the rate of change of the reaction for a 1-hour change in the time after the drug is administered.   **61. a.** $-24.9°$F   **b.** 15 mph   **c.** $W_V(20, 10) = -1.114$; while holding the temperature fixed at 10°F, the wind chill decreases approximately 1.1°F when the wind velocity increases by 1 mph; $W_T(20, 10) = 1.429$; while holding the wind velocity fixed at 20 mph, the wind chill increases approximately 1.429°F if the actual temperature increases from 10°F to 11°F.
d. Sample table

T\V	5	10	15	20
30	27	16	9	4
20	16	3	−5	−11
10	6	−9	−18	−25
0	−5	−21	−32	−39

63. −10 ml per year, 100 ml per in.   **65. a.** $F_m = gR^2/r^2$; the rate of change in force per unit change in mass while the distance is held constant; $F_r = -2mgR^2/r^3$; the rate of change in force per unit change in distance while the mass is held constant
67. a. 1055   **b.** $T_s(3, 0.5) = 127.4$ msec per ft. If the distance to move an object increases from 3 ft to 4 ft, while keeping w fixed at 0.5, the approximate increase in movement time is 127.4 msec.
$T_w(3, 0.5) = -764.6$ msec per ft. If the width of the target area increases by 1 ft, while keeping s fixed at 3 ft, the approximate decrease in movement time is 764.6 msec.

Exercises 9.3 (page 488)

For exercises . . .	1–18,21–28	34–40,42
Refer to example . . .	1–3	4

1. Saddle point at $(-1, 2)$ **3.** Relative minimum at $(-3, -3)$
5. Relative minimum at $(-2, -2)$ **7.** Relative minimum at $(15, -8)$ **9.** Relative maximum at $(2/3, 4/3)$ **11.** Saddle point at $(2, -2)$ **13.** Saddle point at $(0, 0)$; relative minimum at $(27, 9)$ **15.** Saddle point at $(0, 0)$; relative minimum at $(9/2, 3/2)$
17. Saddle point at $(0, -1)$ **21.** Relative maximum of 9/8 at $(-1, 1)$; saddle point at $(0, 0)$; a   **23.** Relative minima of $-33/16$ at $(0, 1)$ and at $(0, -1)$; saddle point at $(0, 0)$; b   **25.** Relative maxima of 17/16 at $(1, 0)$ and $(-1, 0)$; relative minima of $-15/16$ at $(0, 1)$ and $(0, -1)$; saddle points at $(0, 0)$, $(-1, 1)$, $(1, -1)$, $(1, 1)$, and $(-1, -1)$; e   **31. a.** all values of k **b.** $k \geq 0$
35. Minimum cost of $59 when $x = 4$, $y = 5$ **37.** Sell 12 spas and 7 solar heaters for a maximum revenue of $166,600.
39. $2000 on quality control and $1000 on consulting, for a minimum time of 8 hours
41. a. $r = -1.722$, $s = 0.3652$, $y = 0.1787(1.441)^t$ **b.** Same as a   **c.** Same as a

Exercises 9.4 (page 498)

For exercises . . .	1–8,11,12,15,21	9,10,13,14,37–42	23–26	27–36
Refer to example . . .	1	3	material after Example 3	2

1. $f(8, 8) = 256$ **3.** $f(5, 5) = 125$ **5.** $f(5, 3) = 28$ **7.** $f(20, 2) = 360$ **9.** $f(3/2, 3/2, 3) = 81/4 = 20.25$
11. $x = 8$, $y = 16$ **13.** 30, 30, 30   **15.** Minimum value of 128 at $(2,5)$, maximum value of 160 at $(-2,7)$
21. a. Minimum value of -5 at $(3,-2)$.   **d.** $(3,-2)$ is a saddle point.   **23.** Purchase 20 units of x and 20 units of y for a maximum utility of 8000.   **25.** Purchase 20 units of x and 5 units of y for a maximum utility of 4,000,000.
27. 60 feet by 60 feet   **29.** 10 large kits and no small kits   **31.** 167 units of labor and 178 units of capital   **33.** 125 m by 125 m   **35.** Radius = 5 in.; height = 10 in.   **37.** 12.91 m by 12.91 m by 6.455 m   **39.** 5 m by 5 m by 5 m
41. b. 2 yd by 1 yd by 1/2 yd

Exercises 9.5 (page 502)

For exercises . . .	1–6,18–20,23–28	7–14	15–17,21,22,29–32	33,34
Refer to example . . .	1	2	3	4

1. 0.12   **3.** 0.0311   **5.** −0.335
7. 10.022; 10.0221; 0.0001   **9.** 2.0067; 2.0080; 0.0013   **11.** 1.07; 1.0720; 0.0020   **13.** −0.02; −0.0200; 0   **15.** 20.73 cm^3
17. 86.4 in^3 **19.** 0.07694 unit   **21.** 6.65 cm^3 **23.** 2.98 liters   **25. a.** 0.2649   **b.** Actual 0.2817; approximation 0.2816
27. a. 87%   **b.** 75%   **d.** 89%; 87%   **29.** 26.945 cm^2 **31.** 3%   **33.** 8

Exercises 9.6 (page 513)

For exercises . . .	1–10	11–20	21–28,61–64,66–71	29–38,65	39–46	47–56	57,58
Refer to example . . .	1	2,3	4	5,6	7	8	9

1. $630y$ **3.** $(2x/9)[(x^2 + 15)^{3/2} - (x^2 + 12)^{3/2}]$ **5.** $6 + 10y$ **7.** $(1/2)(e^{12+3y} - e^{4+3y})$ **9.** $(1/2)(e^{4x+9} - e^{4x})$ **11.** 945
13. $(2/45)(39^{5/2} - 12^{5/2} - 7533)$ **15.** 21   **17.** $(\ln 3)^2$ **19.** $8 \ln 2 + 4$ **21.** 171   **23.** $(4/15)(33 - 2^{5/2} - 3^{5/2})$
25. $-3 \ln(3/4)$ or $3 \ln(4/3)$ **27.** $(1/2)(e^7 - e^6 - e^3 + e^2)$ **29.** 96   **31.** 40/3   **33.** $(2/15)(2^{5/2} - 2)$
35. $(1/4) \ln(17/8)$ **37.** $e^2 - 3$ **39.** 97,632/105   **41.** 128/9   **43.** ln 16 or 4 ln 2   **45.** 64/3   **47.** 34   **49.** 10/3
51. $7(e - 1)/3$ **53.** 16/3   **55.** $4 \ln 2 - 2$ **57.** 1   **61.** 49   **63.** $(e^6 + e^{-10} - e^{-4} - 1)/60$ **65.** 9 in^3 **67.** 14,753 units
69. $34,833   **71.** $32,000

Chapter 9 Review Exercises (page 518)

For exercises . . .	1–5,17–26,87	6,13,14,27–44, 89,98–101	7,8,45–53,90, 103,104	9,54–57, 91,92	10–12,67–86, 100	15,59–66,88, 93–96,98
Refer to section . . .	1	2	3	4	6	5

1. True **2.** True **3.** True **4.** True **5.** False **6.** False **7.** False **8.** True **9.** False **10.** False **11.** True **12.** False
17. $-19; -255$ **19.** $-5/9; -4/3$ **21.**

23.

25.

27. a. $9x^2 + 8xy$ **b.** -12 **c.** 16 **29.**
$f_x(x, y) = 12xy^3; f_y(x, y) = 18x^2y^2 - 4$ **31.** $f_x(x, y) = 4x/(4x^2 + y^2)^{1/2}$;
$f_y(x, y) = y/(4x^2 + y^2)^{1/2}$ **33.** $f_x(x, y) = 3x^2e^{3y}; f_y(x, y) = 3x^3e^{3y}$ **35.** $f_x(x, y) = 4x/(2x^2 + y^2); f_y(x, y) = 2y/(2x^2 + y^2)$
37. $f_{xx}(x, y) = 30xy; f_{xy}(x, y) = 15x^2 - 12y$ **39.** $f_{xx}(x, y) = 12y/(2x - y)^3; f_{xy}(x, y) = (-6x - 3y)/(2x - y)^3$
41. $f_{xx}(x, y) = 8e^{2y}; f_{xy}(x, y) = 16xe^{2y}$ **43.** $f_{xx}(x, y) = (-2x^2y^2 - 4y)/(2 - x^2y)^2; f_{xy}(x, y) = -4x/(2 - x^2y)^2$
45. Saddle point at $(0, 2)$ **47.** Relative minimum at $(2, 1)$ **49.** Saddle point at $(3, 1)$ **51.** Saddle point at $(-1/3, 11/6)$;
relative minimum at $(1, 1/2)$ **55.** Minimum of 18 at $(-3, 3)$ **57.** $x = 25, y = 50$ **59.** 1.22 **61.** 13.0846; 13.0848; 0.0002
63. $8y - 6$ **65.** $(3/2)[(100 + 2y^2)^{1/2} - (2y^2)^{1/2}]$ **67.** 1232/9 **69.** $(2/135)[(42)^{5/2} - (24)^{5/2} - (39)^{5/2} + (21)^{5/2}]$
71. $2 \ln 2$ or $\ln 4$ **73.** 110 **75.** $(4/15)(782 - 8^{5/2})$ **77.** 105/2 **79.** 1/2 **81.** 1/48 **83.** $\ln 2$ **85.** 3
87. a. $\$(325 + \sqrt{10}) \approx \$328.16$ **b.** $\$(800 + \sqrt{15}) \approx \$803.87$ **c.** $\$(2000 + \sqrt{20}) \approx \$2004.47$ **89. a.** $0.7y^{0.3}/x^{0.3}$
b. $0.3x^{0.7}/y^{0.7}$ **91.** Purchase 10 units of x and 15 units of y for a maximum utility of 33,750. **93.** Decrease by $243.82
95. 4.19 ft^3 **97. a.** $200 spent on fertilizer and $80 spent on seed will produce a maximum profit of $266 per acre.
b. Same as a. **c.** Same as a **99. a.** 49.68 liters **b.** -0.09, the approximate change in total body water if age is increased by 1
yr and mass and height are held constant is -0.09 liters; 0.34, the approximate change in total body water if mass is increased by 1
kg and age and height are held constant is 0.34 liters; 0.25, the approximate change in total body water if height is increased by 1
cm and age and mass are held constant is 0.25 liters. **101. a.** 50; in 1900, 50% of those born 60 years earlier are still alive.
b. 75; in 2000, 75% of those born 70 years earlier are still alive. **c.** -1.25; in 1900, the percent of those born 60 years earlier
who are still alive was dropping at a rate of 1.25 percent per additional year of life. **d.** -2; in 2000, the percent of those
born 70 years earlier who are still alive was dropping at a rate of 2 percent per additional year of life. **103.** 5 in. by 5 in. by 5 in.

Chapter 10 Differential Equations

Exercises 10.1 (page 535) For exercises . . .

	1,2	3,4,17–20	5–16,21–32, 38,39,53	35	36,37,41,42, 49–51,54	40,47,48,52	43–46,56–59
Refer to example . . .	1	3	4,5,6	2	5	7	6

1. $y = -2x^2 + 2x^3 + C$ **3.** $y = x^4/2 + C$ **5.** $y^2 = 2x^3/3 + C$ **7.** $y = ke^{x^2}$ **9.** $y = ke^{x^3 - x^2}$ **11.** $y = Cx$
13. $\ln(y^2 + 6) = x + C$ **15.** $y = -1/(e^{2x}/2 + C)$ **17.** $y = x^2 - x^3 + 5$ **19.** $y = -2xe^{-x} - 2e^{-x} + 44$ **21.** $y^2 = x^4/2 + 25$
23. $y = e^{x^2 + 3x}$ **25.** $y^2/2 - 3y = x^2 + x - 4$ **27.** $y = -3/(3 \ln |x| - 4)$ **29.** $y = (e^{x-1} - 3)/(e^{x-1} - 2)$ **35. a.** $1011.75
b. $1024.52 **c.** No **37.** About 13.9 yr **39.** $q = C/p^2$ **41.** d **43. a.** $I = 2.4 - 1.4e^{-0.088W}$ **b.** I approaches 2.4.
45. a. $dw/dt = k(C - 17.5w)$; the calorie intake per day is constant. **b.** lb/calorie **c.** $dw/dt = (C - 17.5w)/3500$
d. $w = C/17.5 - e^{-0.005M}e^{-0.005t}/17.5$ **e.** $w = C/17.5 + (w_0 - C/17.5)e^{-0.005t}$

47. a.

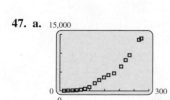

b. $y = \dfrac{25{,}538}{1 + 110.28e^{-0.01819t}}$ **c.**

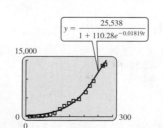

d. 25,538

49. $y = 35.6e^{0.02117t}$ **51. a.** $k \approx 0.8$ **b.** 11 **c.** 55 **d.** About 3000 **53.** About 10 **55.** 7:22:55 A.M. **57.** The temperature approaches T_M, the temperature of the surrounding medium. **59. a.** $T = 88.6e^{-0.24t} + 10$ **b.**

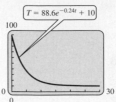

c. Just after death—the graph shows that the most rapid decrease occurs in the first few hours. **d.** About 43.9°F **e.** About 4.5 hours

Exercises 10.2 (page 543)

For exercises . . .	1–20,25–29
Refer to example . . .	2–4

1. $y = 2 + Ce^{-3x}$ **3.** $y = 2 + Ce^{-x^2}$ **5.** $y = x \ln x + Cx$ **7.** $y = -1/2 + Ce^{x^2/2}$
9. $y = x^2/4 + 2x + C/x^2$ **11.** $y = -x^3/2 + Cx$ **13.** $y = 2e^x + 48e^{-x}$ **15.** $y = -2 + 22e^{x^2-1}$ **17.** $y = x^2/7 + 2560/(7x^5)$
19. $y = (3 + 197e^{4-x})/x$ **21. a.** $y = c/(p + Kce^{-cx})$ **b.** $y = cy_0/[py_0 + (c - py_0)e^{-cx}]$ **c.** c/p
25. $y = 1.02e^t + 9999e^{0.02t}$ (rounded) **27.** $y = 50t + 2500 + 7500e^{0.02t}$ **29.** $T = Ce^{-kt} + T_M$

Exercises 10.3 (page 550)

For exercises . . .	1–35
Refer to example . . .	1,2

1. 8.273 **3.** 4.315 **5.** 1.491 **7.** 6.191 **9.** $-0.540; -0.520$ **11.** 4.010; 4.016 **13.** 3.806; 4.759
15. 3.112; 3.271 **17.** 73.505; 74.691

19.

x_i	y_i	$y(x_i)$	$y_i - y(x_i)$
0	0	0	0
0.2	0	0.08772053	-0.08772053
0.4	0.11696071	0.22104189	-0.10408118
0.6	0.26432197	0.37954470	-0.11522273
0.8	0.43300850	0.55699066	-0.12398216
1.0	0.61867206	0.75000000	-0.13132794

21.

x_i	y_i	$y(x_i)$	$y_i - y(x_i)$
0	0	0	0
0.2	0.8	0.725077	0.07492
0.4	1.44	1.3187198	0.12128
0.6	1.952	1.8047535	0.14725
0.8	2.3616	2.2026841	0.15892
1.0	2.68928	2.5284822	0.16080

23.

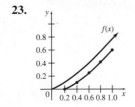

25.

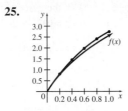

27. a. 4.109 **b.** $y = 1/(1 - x)$; y approaches ∞.
29. a. $dy/dt = 0.01y(500 - y) = 5y - 0.01y^2$
b. About 484 thousand
31. About 75 **33.** About 8.07 kg
35. About 157 people

Exercises 10.4 (page 557)

For exercises . . .	1–5	6–8	9–17	18–24
Refer to example . . .	1	2	3	4

1. $50,216.53 **3.** About 6.9 years **5. a.** $dA/dt = 0.06A - 1200$ **b.** $6470.04
c. 8.51 years **7. a.** $2y - 3 \ln y - 4 \ln x + 2x = 4$ **b.** $x = 2, y = 3/2$, or $x = 0, y = 0$ **9. a.** $y = 24,995,000/(4999 + e^{0.25t})$
b. 3672 **c.** 91 **d.** 34th day **11. a.** $y = 20,000/(1 + 199e^{-0.14t})$ or $20,000e^{0.14t}/(e^{0.14t} + 199)$ **b.** About 38 days
13. a. $y = 0.005 + 0.015e^{-1.010t}$ **b.** $Y = 0.00727e^{-1.1t} + 0.00273$ **15. a.** $y = 45/(1 + 14e^{-0.54t})$ **b.** About 6 days
17. a. $y = 347e^{-4.24e^{-0.1t}}$ **b.** About 5.5 days **19. a.** $y = [2(t + 100)^3 - 1,800,000]/(t + 100)^2$ **b.** About 250 lb of salt
c. Increases **21. a.** $y = 20e^{-0.02t}$ **b.** About 6 lb of salt **c.** Decreases **23. a.** $y = [0.25(t + 100)^2 - 2000]/(t + 100)$
b. About 17.1 g

Chapter 10 Review Exercises (page 561)

For exercises . . .	1–3,6,7,13,14,25–32, 37–42,52–54,57,58, 63–67,69–71	5,15–24	4,8,9,33–36,43–46	10,11,47–51	12,55,56,59–62,68
Refer to section . . .	1	1, 2	2	3	4

1. True **2.** False **3.** True **4.** False **5.** False **6.** True **7.** False **8.** True **9.** False **10.** False **11.** True **12.** True
17. Neither **19.** Separable **21.** Both **23.** Linear **25.** $y = x^3 + 3x^2 + C$ **27.** $y = 2e^{2x} + C$ **29.** $y^2 = 3x^2 + 2x + C$

31. $y = (Cx^2 - 1)/2$ **33.** $y = x - 1 + Ce^{-x}$ **35.** $y = (x^2 + C)/\ln x$ **37.** $y = x^3/3 - 3x^2 + 3$ **39.** $y = -\ln[5 - (x + 2)^4/4]$
41. $y^2 + 6y = 2x - 2x^2 + 352$ **43.** $y = -x^2e^{-x}/2 - xe^{-x}/2 + e^{-x}/4 + 41.75e^x$ **45.** $y = 3/2 + 27e^{x^2}/2$ **49.** 2.608
51.

x_i	y_i
0	0
0.2	0.6
0.4	1.355
0.6	2.188
0.8	3.084
1.0	4.035

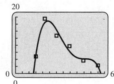

53. a. \$10,099   **b.** \$71,196   **55. a.** $dA/dt = 0.05A - 20,000$
b. \$235,127.87   **57. a.** About 40   **b.** About 1.44×10^{10} hours
59. $0.2 \ln y - 0.5y + 0.3 \ln x - 0.4x = C$; $x = 3/4$ units, $y = 2/5$ units
61. It is not possible (t is negative).   **65. a.** $N = 329$, $b = 7.23$; $k = 0.247$
b. $y \approx 268$ million, which is less than the table value of 308.7 million.
c. About 289 million for 2030, about 303 million for 2050
67. a. and b. $x = 1/k + Ce^{-kt}$ **c.** $1/k$ **69.** $213°$
71. a. $v = (G/K)(e^{2GKt} - 1)/(e^{2GKt} + 1)$ **b.** G/K
c. $v = 88(e^{0.727t} - 1)/(e^{0.727t} + 1)$

Chapter 11 Probability and Calculus

Exercises 11.1 (page 575)

For exercises . . .	1–10	11–18,29–34	19–24,35c,d,36c,d, 48d,e,49d,e	35a,b,36a,b,39–42,44–47, 48a,b,c,49a,b,c,50	43,51
Refer to example . . .	1	2	5	3	4

1. Yes   **3.** Yes   **5.** No; $\int_0^3 4x^3\, dx \neq 1$
7. No; $\int_{-2}^2 x^2/16\, dx \neq 1$
9. No; $f(x) < 0$ for some x values in $[-1, 1]$.   **11.** $k = 3/14$ **13.** $k = 3/125$ **15.** $k = 2/9$ **17.** $k = 1/12$
19. $F(x) = (x^2 - x - 2)/18, 2 \leq x \leq 5$ **21.** $F(x) = (x^3 - 1)/63, 1 \leq x \leq 4$ **23.** $F(x) = (x^{3/2} - 1)/7, 1 \leq x \leq 4$
25. 1   **29. a.** 0.4226   **b.** 0.2071   **c.** 0.4082   **31. a.** 0.3935   **b.** 0.3834   **c.** 0.3679   **33. a.** 1/3   **b.** 2/3   **c.** 295/432
35. a. 0.9975   **b.** 0.0024   **c.** $F(t) = 1 - e^{-t/2}, t \geq 0$ **d.** 0.9502   **37.** c   **39. a.** 0.2679   **b.** 0.4142   **c.** 0.3178
41. a. 0.8131   **b.** 0.4901
43. a.

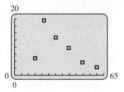

polynomial function   **b.** $N(t) = -0.00007445t^4 + 0.01243t^3 - 0.7419t^2 + 18.18t - 137.5$

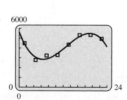

Yes

c. $S(t) = \dfrac{1}{466.26}(-0.00007445t^4 + 0.01243t^3 - 0.7419t^2 + 18.18t - 137.5)$ **d.** Estimates: 0.1688, 0.5896, 0.1610; actual:
0.1730, 0.5865, 0.1325   **45. a.** 0.2   **b.** 0.6   **c.** 0.6   **47. a.** 0.1640   **b.** 0.1353   **49. a.** 0.2829   **b.** 0.4853   **c.** 0.2409
d. $F(t) = 1.8838(0.5982 - e^{-0.03211t}), 16 \leq t \leq 84$ **e.** 0.1671

51. a.

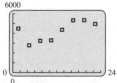

A polynomial function
b. $T(t) = -2.564t^3 + 99.11t^2 - 964.6t + 5631$

c. $S(t) = \dfrac{1}{101,370}(-2.564t^3 + 99.11t^2 - 964.6t + 5631)$ **d.** 0.09457; 0.07732

Exercises 11.2 (page 585)

For exercises . . .	1–6,11–14	7,8	11d,e–14d,e,24–26,31–33, 34,35,38–42	15a–20a,24d,26d,31d, 33d,36,40d,41d	37,43
Refer to example . . .	1	2	3	5	4

1. $\mu = 5$; $\mathrm{Var}(X) \approx 1.33$; $\sigma \approx 1.15$
3. $\mu = 14/3 \approx 4.67$; $\mathrm{Var}(X) \approx 0.89$; $\sigma \approx 0.94$ **5.** $\mu = 2.83$; $\mathrm{Var}(X) \approx 0.57$; $\sigma \approx 0.76$
7. $\mu = 4/3 \approx 1.33$; $\mathrm{Var}(X) = 2/9 \approx 0.22$; $\sigma \approx 0.47$ **11. a.** 5.40   **b.** 5.55   **c.** 2.36   **d.** 0.5352   **e.** 0.6043   **13. a.** 1.6
b. 0.11   **c.** 0.33   **d.** 0.5904   **e.** 0.6967   **15. a.** 5   **b.** 0   **17. a.** 4.828   **b.** 0.0553   **19. a.** $\sqrt[4]{2} \approx 1.189$ **b.** 0.1836
21. 16/5; does not exist; does not exist   **23.** d   **25. a.** 6.409 yr   **b.** 1.447 yr   **c.** 0.4910   **27.** c   **29.** c   **31. a.** 6.342 seconds
b. 5.135 sec   **c.** 0.7518   **d.** 4.472 sec   **33. a.** 2.333 cm   **b.** 0.8692 cm   **c.** 0   **d.** 2.25 cm   **35.** 111   **37.** 31.75 years; 11.55
years   **39. a.** 1.806   **b.** 1.265   **c.** 0.1886   **41. a.** 38.51 years   **b.** 17.56 years   **c.** 0.1656   **d.** 34.26 years   **43.** About 1 pm

Exercises 11.3 (page 597)

For exercises . . .	1,2,29,35,39,48	3–6,30,31,36–38,40, 41,43–45,47,49–51	7–14,32–34,42,52	46
Refer to example . . .	1	2	3	4

1. a. 3.7 cm **b.** 0.4041 cm **c.** 0.2886
3. a. 0.25 years **b.** 0.25 years **c.** 0.2325
5. a. 3 days **b.** 3 days **c.** 0.2325 **7.** 49.98% **9.** 8.01% **11.** −1.28 **13.** 0.92 **19.** $m = (−\ln 0.5)/a$ or $(\ln 2)/a$
23. a. 1.00000 **b.** 1.99999 **c.** 8.00000 **25. a.** $\mu \approx 0$ **b.** $\sigma = 0.9999999251 \approx 1$ **27.** $F(x) = (x − a)/(b − a), a \le x \le b$
29. a. \$47,500 **b.** 0.4667 **31. a.** $f(x) = 0.235e^{−0.235x}$ on $[0, \infty)$ **b.** 0.0954 **33. a.** 0.1587 **b.** 0.7698 **35.** c **37.** d
39. a. 28 days **b.** 0.375 **41. a.** 1 hour **b.** 0.3935 **43. a.** 58 minutes **b.** 0.0907 **45. a.** 0.1967 **b.** 0.2468
47. a. 4.37 millennia; 4.37 millennia **b.** 0.6325 **49. a.** 0.2865 **b.** 0.2212 **51. a.** 0.5457 **b.** 0.0039

Chapter 11 Review Exercises (page 601)

For exercises . . .	1–4,11–22,41, 46a–47a,60a-d	5,6,23–31,42, 46b,c–47b,c,52, 55,60e,f,61	7–10,33–40,43, 48–51,53,54,56, 57–59,62
Refer to section . . .	1	2	3

1. True **2.** True **3.** True **4.** False **5.** False **6.** True
7. True **8.** True **9.** False **10.** False **11.** probabilities
13. 1. $f(x) \ge 0$ for all x in $[a, b]$; 2. $\int_a^b f(x)dx = 1$
15. Not a probability density function **17.** Probability density function **19.** $k = 1/21$ **21. a.** $1/5 = 0.2$ **b.** $9/20 = 0.45$
c. 0.54 **25. a.** 4 **b.** 0.5 **c.** 0.7071 **d.** 4.121 **e.** $F(x) = (x − 2)^2/9, 2 \le x \le 5$ **27. a.** 5/4 **b.** $5/48 \approx 0.1042$
c. 0.3227 **d.** 1.149 **e.** $F(x) = 1 − 1/x^5, x \ge 1$ **29. a.** 0.5833 **b.** 0.2444 **c.** 0.4821 **d.** 0.6114 **31. a.** 100
b. 100 **c.** 0.8647 **33.** 33.36% **35.** 34.31% **37.** 11.51% **39.** −0.05 **41. a.** Uniform **b.** Domain: [10, 30], range: {0.05}
c. 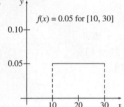 **d.** $\mu = 20; \sigma \approx 5.77$ **e.** 0.577 **43. a.** Normal **b.** Domain: $(−\infty, \infty)$, range: $(0, 1/\sqrt{\pi}]$
c. 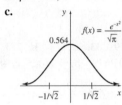 **d.** $\mu = 0; \sigma = 1/\sqrt{2}$ **e.** 0.6826 **45. b.** 0.6819 **c.** 0.9716
d. 1; yes **47. a.** 0.9107 **b.** 13.57 years **c.** 6.68 years
49. a. $f(x) = e^{−x/8}/8; [0, \infty)$ **b.** 8 **c.** 8 **d.** 0.2488 **51.** d
53. 0.6321 **55. a.** 40.07°C **b.** 0.4928 **57.** 0.2266
59. a. 0.2921 **b.** 0.1826 **61.** 3650.1 days; 3650.1 days

Chapter 12 Sequences and Series

Exercises 12.1 (page 612)

For exercises . . .	1–6	7–22	23–30	31–38	39,42,43, 44–46	40,41, 47,48
Refer to example . . .	1	2	4	5	3	6

1. 2, 6, 18, 54 **3.** 1/2, 2, 8, 32 **5.** 3/2, 3, 6, 12, 24
7. $a_5 = 324; a_n = 4(3)^{n−1}$ **9.** $a_5 = −1875; a_n = −3(−5)^{n−1}$
11. $a_5 = 3/2; a_n = 24/2^{n−1}$ **13.** $a_5 = −256; a_n = −(−4)^{n−1}$ **15.** $r = 2; a_n = 6(2)^{n−1}$ **17.** $r = 2; a_n = (3/4)(2)^{n−1}$
19. Not geometric **21.** $r = −2/3; a_n = (−5/8)(−2/3)^{n−1}$ **23.** 93 **25.** 33/4 **27.** 33 **29.** 464.4 **31.** 2040
33. 262,143/2 **35.** 183/4 **37.** 511/4 **39. a.** \$3932 **b.** \$2013 **41.** 2^{30} or \$1,073,741,824; $2^{31} − 1$ or \$2,147,483,647
43. About 41% **45.** About 95 times **47. a.** $1 + 2 + 2^2 + 2^3 + 2^4 + 2^5$ **b.** 63 **c.** $1 + 2 + 2^2 + \cdots + 2^{n−1} = 2^n − 1$

Exercises 12.2 (page 621)

(*Note:* Answers in this section may differ by a few cents, depending on how calculators are used.)

For exercises . . .	1–6,31, 37–40	7–10,32	11–14, 33–36, 41–43	15–24, 44–47	25–30,48, 53,54	49–52	55–58
Refer to example . . .	1	2	3,4	5	6	8	7

1. \$1509.35 **3.** \$278,150.87 **5.** \$833,008.00
7. \$205,785.64 **9.** \$142,836.33 **11.** \$526.95 **13.** \$952.33 **15.** \$39,434.37 **17.** \$17,585.54 **19.** \$1,367,773.96
21. \$111,183.87 **23.** \$97,122.49 **25.** \$476.90 **27.** \$11,942.55 **29.** \$1673.21 **31. a.** \$132,318.77 **b.** \$121,909.27
c. \$10,409.50 **33. a.** \$491.54 **b.** \$533.42 **35.** \$1398.12 **37.** \$112,796.87 **39.** \$209,348.00

41. a. $1200 **b.** $3511.58 **c.**

Payment Number	Amount of Deposit	Interest Earned	Total
1	$3511.58	$0	$3511.58
2	$3511.58	$105.35	$7128.51
3	$3511.58	$213.86	$10,853.95
4	$3511.58	$325.62	$14,691.15
5	$3511.58	$440.73	$18,643.46
6	$3511.58	$559.30	$22,714.34
7	$3511.58	$681.43	$26,907.35
8	$3511.58	$807.22	$31,226.15
9	$3511.58	$936.78	$35,674.51
10	$3511.58	$1070.24	$40,256.33
11	$3511.58	$1207.69	$44,975.60
12	$3511.58	$1349.27	$49,836.45
13	$3511.58	$1495.09	$54,843.12
14	$3511.59	$1645.29	$60,000.00

43. a. $32.49 **b.** $195.52; $10.97 **45.** $12,493.78
47. a. $623,110.52
b. $456,427.28 **c.** $563,757.78
d. $392,903.18 **49.** $1885.00;
$229,612.44 **51.** $2583.01;
$336,107.59 **53. a.** $4025.90
b. $2981.93

55.

Payment Number	Amount of Payment	Interest for Period	Portion to Principal	Principal at End of Period
0	—	—	—	$4000
1	$1207.68	$320.00	$887.68	$3112.32
2	$1207.68	$248.99	$958.69	$2153.63
3	$1207.68	$172.29	$1035.39	$1118.24
4	$1207.70	$89.46	$1118.24	$0

57.

Payment Number	Amount of Payment	Interest for Period	Portion to Principal	Principal at End of Period
0	—	—	—	—
1	$183.93	$62.86	$121.07	$7062.93
2	$183.93	$61.80	$122.13	$6940.80
3	$183.93	$60.73	$123.20	$6817.60
4	$183.93	$59.65	$124.28	$6693.32
5	$183.93	$58.57	$125.36	$6567.96
6	$183.93	$57.47	$126.46	$6441.50

Exercises 12.3 (page 631)

For exercises . . .	1–20,35–39, 40–44,45,46	21–34,40–44
Refer to example . . .	1,2,4	3,5

1. $1 - 2x + 2^2x^2/2! - 2^3x^3/3! + 2^4x^4/4!$ or $1 - 2x + 2x^2 - (4/3)x^3 + (2/3)x^4$
3. $e + ex + ex^2/2! + ex^3/3! + ex^4/4!$ or $e + ex + ex^2/2 + ex^3/6 + ex^4/24$
5. $3 + x/6 - x^2/216 + x^3/3888 - (5/279,936)x^4$
7. $-1 + x/3 + x^2/9 + (5/81)x^3 + (10/243)x^4$ **9.** $1 + x/4 - (3/32)x^2 + (7/128)x^3 - (77/2048)x^4$
11. $-x - x^2/2 - x^3/3 - x^4/4$ **13.** $2x^2 - 2x^4$ **15.** $x - x^2 + x^3/2 - x^4/6$ **17.** $27 - (9/2)x + x^2/8 + x^3/432 + x^4/10,368$
19. $1 - x + x^2 - x^3 + x^4$ **21.** 0.9608 **23.** 2.7732 **25.** 2.9866 **27.** -1.0164 **29.** 1.0147 **31.** -0.0305 **33.** 0.0080
35. $P_3(x) = 3 + 6x + 6x^2 + 4x^3$ **37. b.** 4.04167; actual value is 4.04124. **39. a.** $1 + \lambda N + \lambda^2 N^2/2$ **b.** $N = \sqrt{2k/\lambda}$
41. $4623; $4623 **43.** $718; $718

Exercises 12.4 (page 637)

For exercises . . .	1–14, 21,22, 24,26–33	15–20	25	34,35,36
Refer to example . . .	2	1	4	3

1. Converges to 40 **3.** Diverges **5.** Converges to 81/2 **7.** Converges to 1000/9
9. Converges to 5/2 **11.** Converges to 1/5 **13.** Converges to $e^2/(e + 1)$
15. $S_1 = 1; S_2 = 3/2; S_3 = 11/6; S_4 = 25/12; S_5 = 137/60$
17. $S_1 = 1/7; S_2 = 16/63; S_3 = 239/693; S_4 = 3800/9009; S_5 = 22,003/45,045$
19. $S_1 = 1/6; S_2 = 1/4; S_3 = 3/10; S_4 = 1/3; S_5 = 5/14$ **21.** 2/9 **23. a.** First 3.12; second 2.90 **b.** 38
25. a. $2000 **b.** 10 **27.** d **29.** 70 meters **31.** 200 centimeters **33.** $4\sqrt{3}/3$ square meters **35. a.** 10/9 sec **b.** 10/9 sec

Exercises 12.5 (page 647)

For exercises . . .	1–4,23	5–22,24, 37–39	29–34	35,36
Refer to example . . .	2	3	4	5

1. $6 + 6x + 6x^2 + 6x^3 + \cdots + 6x^n + \cdots; (-1, 1)$

3. $x^2 + x^3 + x^4/2! + x^5/3! + \cdots + x^{n+2}/n! + \cdots; (-\infty, \infty)$

5. $5/2 + (5/2^2)x + (5/2^3)x^2 + (5/2^4)x^3 + \cdots + (5/2^{n+1})x^n + \cdots; (-2, 2)$

7. $8x - 8 \cdot 3x^2 + 8 \cdot 3^2 x^3 - 8 \cdot 3^3 x^4 + \cdots + (-1)^n \cdot 8 \cdot 3^n x^{n+1} + \cdots; (-1/3, 1/3)$

9. $x^2/4 + x^3/4^2 + x^4/4^3 + x^5/4^4 + \cdots + x^{n+2}/4^{n+1} + \cdots; (-4, 4)$

11. $4x - (4^2/2)x^2 + (4^3/3)x^3 - (4^4/4)x^4 + \cdots + (-1)^n 4^{n+1} x^{n+1}/(n + 1) + \cdots; (-1/4, 1/4]$

13. $1 + 4x^2 + (4^2/2!)x^4 + (4^3/3!)x^6 + \cdots + (4^n/n!)x^{2n} + \cdots; (-\infty, \infty)$

15. $x^3 - x^4 + x^5/2! - x^6/3! + \cdots + (-1)^n x^{n+3}/n! + \cdots; (-\infty, \infty)$

17. $2 - 2x^2 + 2x^4 - 2x^6 + \cdots + (-1)^n 2 x^{2n} + \cdots; (-1, 1)$

19. $1 + x^2/2! + x^4/4! + x^6/6! + \cdots + x^{2n}/(2n)! + \cdots; (-\infty, \infty)$

21. $2x^4 - (2^2/2)x^8 + (2^3/3)x^{12} - (2^4/4)x^{16} + \cdots + (-1)^n 2^{n+1} x^{4n+4}/(n + 1) + \cdots; [-1/\sqrt[4]{2}, 1/\sqrt[4]{2}]$

23. $1 + 2x + 2x^2 + 2x^3 + \cdots + 2x^n + \cdots$ **29.** 0.3461 **31.** 0.1729 **33.** 0.1554 **35.** About 14.94 years; about 14.74 years; a difference of 0.2 years, or about 10 weeks **37. b.** λ **c.** 0.1391 **39. a.** 6 **b.** 0.5787

Exercises 12.6 (page 652)

For exercises . . .	1–16,27–30,32–36	17–26
Refer to example . . .	1	2

1. 1.13 **3.** 3.06 **5.** 2.24 **7.** $-1.13, 2.37$ **9.** -0.58 **11.** 0.44 **13.** 1.25 **15.** 1.56
17. 1.414 **19.** 3.317 **21.** 15.811 **23.** 2.080 **25.** 4.642 **27.** Relative maximum at -1.65; relative minimum at 3.65
29. Relative minima at -0.71 and 1.77; relative maximum at 1.19 **33.** 4.80 years **35.** $i_2 = 0.02075485$; $i_3 = 0.02075742$

Exercises 12.7 (page 659)

For exercises . . .	1–3,5,6, 8–10,13–24, 27,28,31,32, 43–46	4,7	11,12, 29,30, 47	25,26	33–36	37–42
Refer to example . . .	1,3	2	5	4	6	7

1. 4 **3.** 0 **5.** 1 **7.** Does not exist **9.** 1
11. Does not exist **13.** $1/(2\sqrt{2})$ or $\sqrt{2}/4$ **15.** 1/4
17. 1/12 **19.** 53 **21.** 0 **23.** 1/8 **25.** 1/9 **27.** 1
29. Does not exist **31.** 5 **33.** 0 **35.** 0
37. ∞ (does not exist) **39.** 1/5 **41.** 0
43. 1/2 **45.** 1/2 **47.** $\lim_{x \to 0}(x^2 + 3) \neq 0$, so l'Hospital's rule does not apply.

Chapter 12 Review Exercises (page 661)

For exercises . . .	1, 13–16, 75, 85,86	2,3, 76–82	4–6, 17–32	7,8, 33–40	9,10, 41–50, 83,84	11, 67–74	12, 51–66
Refer to section . . .	1	2	3	4	5	6	7

1. True **2.** False **3.** True **4.** True **5.** False
6. True **7.** False **8.** True **9.** True **10.** False
11. False **12.** False **13.** -40; $a_n = 5(-2)^{n-1}$; 55
15. 1; $a_n = 27(1/3)^{n-1}$; 121/3 **17.** $e^2 - e^2 x + (e^2/2!)x^2 - (e^2/3!)x^3 + (e^2/4!)x^4$
19. $1 + x/2 - x^2/8 + x^3/16 - (5/128)x^4$ **21.** $\ln 2 - x/2 - x^2/8 - x^3/24 - x^4/64$
23. $1 + (2/3)x - x^2/9 + (4/81)x^3 - (7/243)x^4$ **25.** 6.8895 **27.** 1.0149 **29.** 0.7178 **31.** 0.9459 **33.** Converges to 27/5
35. Diverges **37.** Converges to 1/3 **39.** $S_1 = 1$; $S_2 = 4/3$; $S_3 = 23/15$; $S_4 = 176/105$; $S_5 = 563/315$
41. $4/3 + (4/3^2)x + (4/3^3)x^2 + (4/3^4)x^3 + \cdots + (4/3^{n+1})x^n + \cdots; (-3, 3)$
43. $x^2 - x^3 + x^4 - x^5 + \cdots + (-1)^n x^{n+2} + \cdots; (-1, 1)$
45. $-2x - (2^2/2)x^2 - (2^3/3)x^3 - (2^4/4)x^4 - \cdots - 2^{n+1} x^{n+1}/(n + 1) - \cdots; [-1/2, 1/2)$
47. $1 - 2x^2 + (2^2/2!)x^4 - (2^3/3!)x^6 + \cdots + (-1)^n (2^n/n!)x^{2n} + \cdots; (-\infty, \infty)$
49. $2x^3 - 2 \cdot 3x^4 + (2 \cdot 3^2/2!)x^5 - (2 \cdot 3^3/3!)x^6 + \cdots + (-1)^n (2 \cdot 3^n/n!)x^{n+3} + \cdots; (-\infty, \infty)$ **51.** 7/4 **53.** Does not exist
55. 5/7 **57.** $-1/2$ **59.** Does not exist **61.** 0 **63.** 9/2 **65.** -8 **67.** 4.73 **69.** 2.65 **71.** 6.132
73. 4.558 **75.** \$11,495,247 **77.** \$27,320.71 **79.** \$3322.43 **81.** \$1184.01 **83.** About 21.67 years; about 21.54 years; differ by 0.13 years, or about 7 weeks **85.** 64,000 bacteria

Chapter 13 The Trigonometric Functions

Exercises 13.1 (page 678)

For exercises . . .	1–16	17–20, 89,90	21–24	25–48, 74,86	49–54	55–62,78, 80,81, 86,88	63–73,76	75,87	77,79, 82–84
Refer to example . . .	1	2	5	4,5	6	7	8	10	9

1. $\pi/3$ **3.** $5\pi/6$ **5.** $3\pi/2$
7. $11\pi/4$ **9.** $225°$ **11.** $-390°$
13. $288°$ **15.** $105°$

Note: In Exercises 17–23 we give the answers in the following order: sine, cosine, tangent, cotangent, secant, and cosecant.

17. $4/5; -3/5; -4/3; -3/4; -5/3; 5/4$ **19.** $-24/25; 7/25; -24/7; -7/24; 25/7; -25/24$ **21.** $+ + + + + +$
23. $- - + + - -$ **25.** $\sqrt{3}/3; \sqrt{3}; 2$ **27.** $\sqrt{3}/2; \sqrt{3}/3; 2\sqrt{3}/3$ **29.** $-1; -1$ **31.** $-\sqrt{3}/2; -2\sqrt{3}/3$ **33.** $\sqrt{3}/2$
35. 1 **37.** 2 **39.** -1 **41.** $-\sqrt{2}/2$ **43.** $-\sqrt{2}$ **45.** 1 **47.** 1/2 **49.** $\pi/3, 5\pi/3$ **51.** $3\pi/4, 7\pi/4$ **53.** $5\pi/6, 7\pi/6$

55. 0.6293   **57.** −1.5399   **59.** 0.3558   **61.** 0.3292   **63.** $a = 1; T = 2\pi/3$ **65.** $a = 3; T = 1/440$

67. 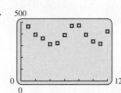   **69.** 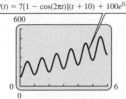   **71.** 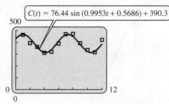   **73.**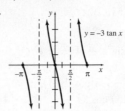

75. a. $\sqrt{2}$ **b.** 45°, 45°, 90°   **77. a.** 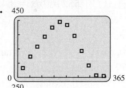   ; yes

b. $C(t) = 76.44 \sin(0.9953t + 0.5686) + 390.3$;   **c.** About 6.3 months   **d.** About 383 BTUs

79. a. 29.54; there is a lunar cycle every 29.54 days.   **b.** October 17; 1.8%   **c.** 98.75%

81. 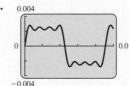   **83.** 2.1×10^8 m per second   **85.** 120°   **87. a.** 34°F   **b.** 58°F   **c.** 80°F   **d.** 90°F   **e.** 39°F   **89. a.** ; yes

b. $s(t) = 94.0872 \sin(0.0166t − 1.2213) + 347.4158$ **c.** 5:26 P.M.; 7:53 P.M.; 7:22 P.M.   **d.** 82nd and 295th days   **91.** 60.2 m
93. 0.28°   **95.** [4.6, 6.3]

Exercises 13.2 (page 688)

For exercises . . .	1–26	27–32,37,47	33–35	36	38–46	50,51
Refer to example . . .	2–6	7	2–4	1	8	9

1. $dy/dx = 4 \cos 8x$ **3.** $dy/dx = 108 \sec^2(9x + 1)$
5. $dy/dx = −4 \cos^3 x \sin x$ **7.** $dy/dx = 8 \tan^7 x \sec^2 x$ **9.** $dy/dx = −12x \cos 2x − 6 \sin 2x$
11. $dy/dx = −(x \csc x \cot x + \csc x)/x^2$ **13.** $dy/dx = 4e^{4x} \cos e^{4x}$ **15.** $dy/dx = (−\sin x)e^{\cos x}$
17. $dy/dx = (4/x) \cos(\ln 3x^4)$ **19.** $dy/dx = (2x \cos x^2)/\sin x^2$ or $2x \cot x^2$ **21.** $dy/dx = (6 \cos x)/(3 − 2 \sin x)^2$
23. $dy/dx = \sqrt{\sin 3x} (\sin 3x \cos x − 3 \sin x \cos 3x)/(2\sqrt{\sin x} (\sin^2 3x))$
25. $dy/dx = (3/4) \sec^2 (x/4) − 8 \csc^2 2x + 5 \csc x \cot x − 2e^{−2x}$ **27.** 1   **29.** 1/2   **31.** 1   **33.** $−\csc^2 x$
37. a. $R'(t) = −240\pi \sin 2\pi t$ **b.** −120π per year   **c.** $0 per year   **d.** 120π per year
39. a.   **b.** $v = dy/dt = (\pi/5) \cos [\pi(t − 1)]; a = d^2y/dt^2 = (−\pi^2/5) \sin [\pi(t − 1)]$ **d.** At $t = 1.5$, acceleration is negative, arm is moving clockwise and is at an angle of 1/5 radian from vertical; at $t = 2.5$, acceleration is positive, arm is moving counterclockwise and is at an angle of −1/5 radian from vertical; at $t = 3.5$, acceleration is negative, arm is moving clockwise and is at an angle of 1/5 radian from vertical.

41. a. About 1488   **b.** About 5381   **c.** 2000   **d.** About 2916
e.   **f.** Maximum is 7389 when $t = \pi/2 + 2\pi n$, where n is any integer; minimum is 135 when $t = 3\pi/2 + 2\pi n$.   **43. a.**   **b.** The pressure is decreasing at a rate of 1.05 lb per ft^2 per sec when $t = 0.002$.
45. a. 13.55 ft   **c.** 52.39 ft   **d.** $dx/d\alpha = (V^2/16) \cos(2\alpha)$ and x is maximized when $\alpha = \pi/4$.   **e.** 242 ft

47. a. 1   **b.** $-\sqrt{2}/2 \approx -0.7071$ **c.** 2   **d.** -2 **e.** $-3\sqrt{2}/2 \approx -2.1213$ **f.** 2   **49. a.** $5/\pi$ rev per minute
b. $5/(2\pi)$ rev per minute   **51.** 20.81 ft

Exercises 13.3 (page 697)

For exercises . . .	1–30	31–36,39,41,42,44	40,43
Refer to example . . .	1–3	4	5

1. $(1/3)\sin 3x + C$ **3.** $3\sin x + 4\cos x + C$
5. $(-\cos x^2)/2 + C$ **7.** $-\tan 3x + C$ **9.** $(1/8)\sin^8 x + C$ **11.** $-2(\cos x)^{3/2} + C$ **13.** $-\ln|1 + \cos x| + C$
15. $(1/4)\sin x^8 + C$ **17.** $-3\ln|\cos(x/3)| + C$ **19.** $(1/6)\ln|\sin x^6| + C$ **21.** $-\cos e^x + C$ **23.** $-\csc e^x + C$
25. $(-6/5)x\sin 5x - (6/25)\cos 5x + C$ **27.** $-4x\cos x + 4\sin x + C$
29. $(-3/4)x^2\sin 8x - (3/16)x\cos 8x + (3/128)\sin 8x + C$ **31.** $1 - \sqrt{2}/2$ **33.** $-\ln(\sqrt{3}/2)$ **35.** $\sqrt{3}/2 - 1$ **37.** 0.5
39. 6000   **41.** 60,000   **43.** 4430 hours; this result is relatively close to the actual value.   **45.** $\tan k$ ("tank")

Chapter 13 Review Exercises (page 699)

For exercises . . .	1–5,11–48,87–94,102	6–8,49–65,86,95–101,103–105	9,10,66–85,106
Refer to section . . .	1	2	3

1. False   **2.** True   **3.** False   **4.** False
5. True   **6.** False   **7.** True   **8.** False   **9.** False   **10.** False
15. $\pi/2$ **17.** $5\pi/4$ **19.** 2π **21.** $900°$ **23.** $81°$ **25.** $117°$ **27.** $\sqrt{3}/2$ **29.** $\sqrt{2}/2$ **31.** $2\sqrt{3}/3$ **33.** $1/2$ **35.** 2
37. 0.7314   **39.** -2.1445 **41.** 0.7058   **43.** 0.8290   **45.**   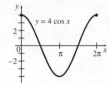   **47.**   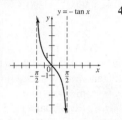   **49.** $-1, 1$

51. $dy/dx = 10\sec^2 5x$ **53.** $dy/dx = 6x\csc^2(6 - 3x^2)$ **55.** $dy/dx = 64x\sin^3(4x^2)\cos(4x^2)$
57. $dy/dx = -2x\sin(1 + x^2)$ **59.** $dy/dx = e^{-2x}(\cos x - 2\sin x)$
61. $dy/dx = (-2\cos x\sin x + \cos^2 x\sin x)/(1 - \cos x)^2$ **63.** $dy/dx = (\sec^2 x + x\sec^2 x - \tan x)/(1 + x)^2$
65. $dy/dx = (\cos x)/(\sin x)$ or $\cot x$ **67.** $(1/5)\sin 5x + C$ **69.** $(1/5)\tan 5x + C$ **71.** $-4\cot x + C$
73. $(5/4)\sec 2x^2 + C$ **75.** $(-1/9)\cos^9 x + C$ **77.** $(1/24)\ln|\sin 8x^3| + C$ **79.** $3(\cos x)^{-1/3} + C$ **81.** 1   **83.** 20π
85. a.    ; yes   **b.** $C(t) = 31,188\sin(0.32644t + 2.2787) + 33,057$;
c. 239,389 million cubic feet; 227,577 million cubic feet   **d.** 19.2 months

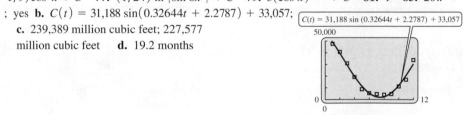

87. s/L_2 **89.** $(L_0 - L_1)/s$ **91.** $R_1 = k(L_1/r_1^4)$ **93.** $R = k(L_1/r_1^4 + L_2/r_2^4)$ **95.** $dR/d\theta = ks\csc^2\theta/r_1^4 - ks\cos\theta/r_2^4\sin^2\theta$
97. $k/r_1^4 - k\cos\theta/r_2^4 = 0$ **99.** $0.0039; \theta \approx 90°$ **103. a.** Yes   **b.** $0.18 \le \alpha \le 0.41$ in radians or $10.3 \le \alpha \le 23.5$ in degrees
c. 0.995 feet/degree; the distance the tennis ball travels will increase by approximately 1 foot by increasing the angle of the tennis
racket by one degree.   **105.** $\theta = \pi/4$ or $45°$

Credits

Text Credits

page 33 Exercise 4: Copyright © Society of Actuaries. Used by permission.   **page 45** Figure 1: Data from finfacts.ie   **page 56** Exercise 75: Peter Tyack, © Woods Hole Oceanographic Institution. Reprinted by permission.   **page 66** Exercise 59: Ralph DeMarr   **page 67** Exercise 65: Ralph DeMarr   **page 87** Exercise 46: Copyright © Society of Actuaries. Used by permission.   **page 100** Exercise 86: Data from *The Quarterly Review of Biology*   **page 109** Exercise 31: Data from downsyndrome.about.com   **page 139** Exercises 88 and 89: Professor Robert D. Campbell   **page 147** Exercise 33: Copyright © Society of Actuaries. Used by permission.   **page 175** Example 10: William J. Kaufmann III   **page 178** Exercise 54: Data from *Biolog-e: The Undergraduate Bioscience Research Journal*   **page 179** Exercise 59: From http://heatkit.com/html/bakeov03.htm. Reprinted by permission; Exercises 60 and 61: Data from *Biological Cybernetics*   **page 185** Exercise 22: Data from *Biolog-e: The Undergraduate Bioscience Research Journal*   **page 190** Exercise 61: Copyright © Society of Actuaries. Used by permission.   **page 191** Exercise 68: Data from Alzheimer's Association; Exercise 70: Peter Tyack, © Woods Hole Oceanographic Institution. Reprinted by permission.   **page 233** Exercise 42: Copyright © Society of Actuaries. Used by permission.   **page 235** Exercise 59: Kevin Friedrich   **page 252** Figure 1: Reprinted with permission. Associated Professional Sleep Societies, LLC, 2011   **page 261** Exercise 51: Recreated from a National Arbor Day Foundation ad, Feb. 4, 1996   **page 286** Exercise 92: Frederick Russell; Exercise 95: Professor Larry Taylor   **page 310** Exercise 40: Copyright © Society of Actuaries. Used by permission.   **page 329** Exercise 3: Copyright 1973-1991, American Institute of Certified Public Accountants, Inc. All rights reserved. Used with permission.   **page 374** Example 8: "Popularity Index" from Quirin, Jim and Barry Cohen, *Chartmasters Rock 100*, 5/e. Copyright 1992 by Chartmasters. Reprinted by permission.
page 385 Exercise 27: Department of Environment, Food and Rural Affairs, Foot and Mouth Disease, http://footandmouth.csl.gov.uk/; Exercises 29 and 30: From http://www.roadandtrack.com/tests/date-panel-archive. Reprinted by permission of *Road & Track*, July 2011.
page 386 Exercises 33 and 34: Data from www.susdesign.com/windowheatgain/   **page 387** Exercise 35: Stephen Monk
page 414 Exercises 25–28: Data from Chodos, D. J., and A. R. DeSantos, *Basics of Bioavailability*, Upjohn Company, 1978
page 415 Exercise 30: Department of Environment, Food and Rural Affairs, Foot and Mouth Disease, http://footandmouth.csl.gov.uk/
page 419 Exercise 78: Copyright © Society of Actuaries. Used by permission.   **page 420** Exercise 88: Copyright © Society of Actuaries. Used by permission; Exercise 92: Data from Oliver, M. H., et al., "Material Undernutrition During the Periconceptual Period Increases Plasma Taurine Levels and Insulin Response to Glucose but not Arginine in the Late Gestation Fetal Sheep," *Endocrinology*, Vol. 14, No. 10
page 433 Exercise 38: Professor Sam Northshield   **page 489** Exercise 30: Copyright © Society of Actuaries. Used by permission.
page 533 Example 7: Data from Fisher, J.C., and R. H. Pry, "A Simple Substitution Model of Technological Change," *Technological Forecasting and Social Change*," Vol. 3, 1971-1972, © 1972   **page 536** Exercise 41: Copyright © Society of Actuaries. Used by permission.
page 556 Example 4: Larry C. Andrews   **page 558** Exercise 13: From Bender, Edward A., *An Introduction to Mathematical Modeling*, 1978. Published by Dover Publications. Reprinted by permission of the author.   **page 564** Extended Application: From Bender, Edward A., *An Introduction to Mathematical Modeling*, 1978, 2000. Reprinted by permission of Dover Publications.   **page 576** Exercises 37 and 38: Copyright © Society of Actuaries. Used by permission.   **page 586** Exercises 23, 27–30: Copyright © Society of Actuaries. Used by permission.   **page 598** Exercises 35-38: Copyright © Society of Actuaries. Used by permission.   **page 603** Exercise 51: Copyright © Society of Actuaries. Used by permission.   **page 632** Exercise 38: Professor Robert D. Campbell   **page 638** Exercise 27: Copyright © Society of Actuaries. Used by permission.   **page 639** Exercise 28: Copyright © Society of Actuaries. Used by permission.   **page 666** Figure 1: Nancy Schiller

Photo Credits

page R-1 Ng Yin Chern/Shutterstock   **page 1** Stockbyte/Getty Images   **page 37** Comstock/Thinkstock   **page 42** Forster Forest/Shutterstock   **page 44** Photos/Thinkstock   **page 77** Olga Miltsova/Shutterstock   **page 100** Cuson/Shutterstock   **page 117** Andrea Danti/ Shutterstock   **page 118** John Good/National Park Service   **page 121** Digital Vision/Thinkstock   **page 157** Courtesy of Raymond N. Greenwell   **page 172** Courtesy of Raymond N. Greenwell   **page 178** Tony Brindley/ Shutterstock   **page 193** BananaStock/Thinkstock   **page 196** Comstock/Thinkstock   **page 218** Pack-Shot/ Shutterstock   **page 241** AardLumens/Fotolia   **page 251** Wallenrock/Shutterstock   **page 286** Theroff97/Dreamstime   **page 300** Stockbyte/Thinkstock   **page 303** John R. Smith/Shutterstock   **page 354** Auremar/ Shutterstock   **page 355** Maureen Plainfield/ Shutterstock   **page 364** Rudy Balasko/ Shutterstock   **page 387** (left) Courtesy of Raymond N. Greenwell; (right) Gary Kemper/AP Images   **page 425** B.W. Folsom/ Shutterstock   **page 459** Specta/ Shutterstock   **page 525** Joggie Botma/ Shutterstock   **page 531** Randal Sedler/ Shutterstock   **page 558** Thinkstock   **page 567** U.S. Geological Survey Photographic Library   **page 596** Library of Congress Prints and Photographs Division [LC-USZC4-2109]   **page 608** Image Source/SuperStock   **page 623** AP Images   **page 665** Berna Namoglu/ Shutterstock   **page 682** Courtesy of Nathan P. Ritchey   **page 687** Gilles Baechler/ Shutterstock   **page 689** Pargeter/Shutterstock

Chapter opener photos are repeated at a smaller size on table of contents and preface pp. v–ix.

Index of Applications

Index

Note: A complete Index of Applications can be found on p. I-1

Sources

This is a sample of the comprehensive source list available for the tenth edition of Calculus with Applications. The complete list is available at the Downloadable Student Resources site, www.pearsonhighered.com/mathstatsresources, as well as to qualified instructors within MyMathLab or through the Pearson Instructor Resource Center, www.pearsonhighered.com/irc.

Chapter 1

Section 1.1

1. Example 10 from *Morbidity and Mortality Weekly Report*, Centers for Disease Control and Prevention, Vol. 58, No. 44, Nov. 13, 2009, p. 1227.
2. Example 14 from http://www.trends-collegeboard.com/college_ pricing/1_3_over_time_current_dollars.html.
3. Exercise 62 from *Time Almanac 2010*, p. 150.
4. Exercise 63 from *Time Almanac 2010*, pp. 637–638.
5. Exercise 64 from Alcabes, P., A. Munoz, D. Vlahov, and G. Friedland, "Incubation Period of Human Immunodeficiency Virus," *Epidemiologic Review*, Vol. 15, No. 2, The Johns Hopkins University School of Hygiene and Public Health, 1993, pp. 303–318.
6. Exercise 65 from Hockey, Robert V., *Physical Fitness: The Pathway to Healthful Living*, Times Mirror/ Mosby College Publishing, 1989, pp. 85–87.
7. Exercise 66 from *Science*, Vol. 253, No. 5017, July 19, 1991, pp. 306–308.
8. Exercise 67 from *Science*, Vol. 254, No. 5034, Nov. 15, 1991, pp. 936–938, and http://www.cdc.gov/nchs/data.
9. Exercise 68 from *World Health Statistics 2010*, World Health Organization, pp. 56–57.
10. Exercise 69 from *The New York Times*, Sept. 11, 2009, p. A12.
11. Exercise 70 from U. S. Census Bureau, http://www.census.gov/ population/socdemo/hh-fam/ms2.pdf.
12. Exercise 71 from *2008 Yearbook of Immigration Statistics*, Office of Immigration Statistics, Aug. 2009, p. 5.
13. Exercise 72 from *Science News*, June 23, 1990, p. 391.
14. Exercise 73 from Acker, A. and C. Jaschek, *Astronomical Methods and Calculations*, John Wiley & Sons, 1986; Karttunen, H. (editor), *Fundamental Astronomy*, Springer-Verlag, 1994.
15. Exercise 74 from http://www.stateofthemedia.org/2009/narrative_audio_audience.php?media=10&cat=2#1listeningtoradio.
16. Exercise 75 from http://www.trends-collegeboard.com/college_ pricing/1_3_over_time_current_dollars.html.

Section 1.2

1. Page 18 from http://www.agmrc.org/media/cms/oceanspray_4BB99D38246C8.pdf.
2. Exercise 46 from *Science News*, Sept. 26, 1992, p. 195, *Science News*, Nov. 7, 1992, p. 399.
3. Exercise 48 from http://www.calstate.edu/budget/fybudget/2009-2010/supportbook2/challenges-off-campus-costs.shtml.

Section 1.3

1. Page 25 from U.S. Dept. of Health and Human Services, National Center for Health Statistics, found in *New York Times 2010 Almanac*, p. 394.
2. Example 5 from *Public Education Finances 2007*, U.S. Census Bureau, July 2009, Table 8. http://www2.census.gov/govs/school/07f33pub.pdf; *The Nation's Report Card: Reading 2007*, National Center for Education Statistics, U.S. Department of Education, Sept. 2007, Table 11. http://nces.ed.gov/nationsreportcard/pdf/main2007/2007496.pdf.
3. Exercise 4 from "November 1989 Course 120 Examination Applied Statistical Methods" of the *Education and Examination Committee of The Society of Actuaries*. Reprinted by permission of The Society of Actuaries.
4. Exercise 10 from http://www.bea.gov/national/FA2004/SelectTable.asp.
5. Exercise 11 from http://www2.fdic.gov/hsob/hsobRpt.asp.
6. Exercise 12 from http://www.ncta.com/Stats/CableAvailableHomes.aspx.
7. Exercise 13 from http://www.federalreserve.gov/releases/g19/Current/.
8. Exercise 14 from http://www.nada.org/NR/rdonlyres/0FE75B2C-69F0-4039-89FE-1366B5B86C97/0/NADAData08_no.pdf.
9. Exercise 15 from American Airlines, http://www.aa.com.
10. Exercise 15 from *The New York Times*, Jan. 7, 2000.
11. Exercise 16 from www.nctm.org/wlme/wlme6/five.htm.
12. Exercise 17 from Stanford, Craig B., "Chimpanzee Hunting Behavior and Human Evolution," *American Scientist*, Vol. 83, May–June 1995, pp. 256–261, and Goetz, Albert, "Using Open-Ended Problems for Assessment," *Mathematics Teacher*, Vol. 99, No. 1, August 2005, pp. 12–17.
13. Exercise 18 from Pierce, George W., *The Songs of Insects*, Cambridge, Mass., Harvard University Press, Copyright © 1948 by the President and Fellows of Harvard College.
14. Exercise 19 from *Digest of Education Statistics 2006*, National Center for Education Statistics, Table 63.
15. Exercise 20 from *Historical Poverty Tables*, U.S. Census Bureau.
16. Exercise 21 from Lee, Grace, Paul Velleman, and Howard Wainer, "Giving the Finger to Dating Services," *Chance*, Vol. 21, No. 3, 2008, pp. 59–61.
17. Exercise 23 from data provided by Gary Rockswold, Mankato State University, Minnesota.
18. Exercise 25 from Carter, Virgil and Robert E. Machol, *Operations Research*, Vol. 19, 1971, pp. 541–545.
19. Exercise 26 from Whipp, Brian J. and Susan Ward, "Will Women Soon Outrun Men?" *Nature*, Vol. 355, Jan. 2, 1992, p. 25. The data are from Peter Matthews, *Track and Field Athletics: The Records*, Guinness, 1986, pp. 11, 44; from Robert W. Schultz and Yuanlong Liu, in *Statistics in Sports*, edited by Bennett, Jay and Jim Arnold, 1998, p. 189; and from *The World Almanac and Book of Facts 2006*, p. 880.
20. Exercise 27 from http://www.run100s.com/HR/.

Review Exercises

1. Exercises 56 and 57 from TradeStats Express™, http://tse.export.gov.
2. Exercise 58 from U.S. Census Bureau, Historical Income Tables–Households, Table H-6, 2008.
3. Exercise 59 from *Chicago Tribune*, Feb. 4, 1996, Sec. 5, p. 4, and NADA Industry Analysis Division, 2006.
4. Exercise 60 from Food and Agriculture Organization Statistical Yearbook, Table D1, Table G5, http://www.fao.org/economic/ess/publications-studies/statistical-yearbook/fao-statistical-yearbook-2009/en/.
5. Exercise 62 from http://www.ers.usda.gov/Data/FoodConsumption/spreadsheets/mtpcc.xls.
6. Exercise 63 from http://www.census.gov/population/socdemo/hh-fam/ms1.xls.
7. Exercise 64 from http://www.census.gov/hhes/www/poverty/histpov/famindex.html.
8. Exercise 65 from http://www.census.gov/popest/states/NST-ann-est.html; http://doa.alaska.gov/dop/fileadmin/socc/pdf/bkgrnd_socc23.pdf.
9. Exercise 66 from Moore, Thomas L., "Paradoxes in Film Rating," *Journal of Statistics Education*, Vol. 14, 2006, http://www.amstat.org/publications/jse/v14n1/datasets.moore.htm1

Extended Application

1. Page 43 from *Health, United States, 2009*, National Center for Health Statistics, U.S. Department of Health and Human Services, Table 24, http://www.cdc.gov/nchs/data/hus/hus09.pdf.

Chapter 2

Section 2.1

1. Page 45 from http://www.finfacts.ie/Private/curency/goldmarketprice.htm.
2. Page 46 using data from Yahoo! Finance: www.yahoo.com.
3. Exercise 73 from Gawande, Atul, "The Malpractice Mess," *The New Yorker*, Nov. 14, 2005, p. 65.
4. Exercise 74 from http://www.tax.state.ny.us/pdf/2009/inc/it150_01i_2009.pdf].

5. Exercise 75 from Peter Tyack, © Woods Hole Oceanographic Institution.
6. Exercise 76 from Robbins, Charles T., *Wildlife Feeding and Nutrition*, 2nd ed., Academic Press, 1993, p. 125.
7. Exercise 77 from Robbins, Charles T., *Wildlife Feeding and Nutrition*, 2nd ed., Academic Press, 1993, p. 142.
8. Exercise 78 from *The New York Times*, Oct. 31, 1999, p. 38.

Section 2.2

1. Exercise 59 from Ralph DeMarr, University of New Mexico.
2. Exercise 60 from Harris, Edward F., Joseph D. Hicks, and Betsy D. Barcroft, "Tissue Contributions to Sex and Race: Differences in Tooth Crown Size of Deciduous Molars," *American Journal of Physical Anthropology*, Vol. 115, 2001, pp. 223–237.
3. Exercise 61 from Abuhamad, A. Z. et al. "Doppler Flow Velocimetry of the Splenic Artery in the Human Fetus: Is It a Marker of Chronic Hypoxia?" *American Journal of Obsterics and Gynecology*, Vol. 172, No. 3, March 1995, pp. 820–825.
4. Exercise 62 from http://seer.cancer.gov/csr/1975_2007/results_merged/sect_15_lung_bronchus.pdf
5. Exercise 63 from http://www.census.gov/population/socdemo/hh-fam/ms2.pdf
6. Exercise 64 from *The New York Times 2010 Almanac*, p. 294.
7. Exercise 65 from Ralph DeMarr, University of New Mexico.
8. Exercise 67 from *National Traffic Safety Institute Student Workbook*, 1993, p. 7.

Section 2.3

1. Exercise 44 from Donley, Edward and Elizabeth Ann George, "Hidden Behavior in Graphs," *Mathematics Teacher*, Vol. 86, No. 6, Sept. 1993.
2. Exercise 45 from Donley, Edward and Elizabeth Ann George, "Hidden Behavior in Graphs," *Mathematics Teacher*, Vol. 86, No. 6, Sept. 1993.
3. Exercise 50 from Dana Lee Ling, College of Micronesia-FSM.
4. Exercise 53 data from Bausch & Lomb. The original chart gave all data to 2 decimal places.
5. Exercise 55 from Garriott, James C. (ed.), *Medical Aspects of Alcohol Determination in Biological Specimens*, PSG Publishing Company, 1988, p. 57.
6. Exercise 56 from http://www.aamc.org/data/facts/applicantmatriculant/table3-fact2009sl-web.pdf
7. Exercise 57 from Smith, J. Maynard, *Models in Ecology*, Oxford: Cambridge University Press, 1974.
8. Exercise 58 from Edelstein-Keshet, Leah, *Mathematical Models in Biology*, Random House, 1988.

9. Exercise 59 from Dobbing, John and Jean Sands, "Head Circumference, Biparietal Diameter and Brain Growth in Fetal and Postnatal Life," *Early Human Development*, Vol. 2, No. 1, April 1978, pp. 81–87.
10. Exercise 60 from http://www.acf.hhs.gov/programs/ohs/about/fy2008.html
11. Exercise 61 from Gary Rockswold, Mankato State University, Mankato, Minnesota.
12. Exercise 62 from *Annual Energy Review*, U.S. Department of Energy, 2005.

Section 2.4

1. Page 84 from Pollan, Michael, "The (Agri)Cultural Contradictions of Obesity," *The New York Times Magazine*, Oct. 12, 2003, p. 41; and USDA–National Agriculture Statistics Service, 2006.
2. Exercise 1 from Thomas, Jamie, "Exponential Functions," *The AMATYC Review*, Vol. 18, No. 2, Spring 1997.
3. Exercise 46 from Problem 5 from "November 1989 Course 140 Examination, Mathematics of Compound Interest" of the Education and Examination Committee of The Society of Actuaries. Reprinted by permission of The Society of Actuaries.
4. Exercise 47 from http://esa.un.org/unpp/index.asp
5. Exercise 49 from U.S. Census Bureau, U.S. Interim Projections by Age, Sex, Race, and Hispanic Origin, http://www.census.gov/ipc/www/usinterimproj/.
6. Exercise 50 from *The Complexities of Physician Supply and Demand: Projections Through 2025*, Association of American Medical Colleges, Nov. 2008, p. 20.
7. Exercise 51 from Marland, G., T.A. Boden, and R.J. Andres. 2006. Global, Regional, and National CO_2 Emissions. In *Trends: A Compendium of Data on Global Change. Carbon Dioxide Information Analysis Center*, Oak Ridge National Laboratory, U.S. Department of Energy, Oak Ridge, TN.
8. Exercise 53 from Miller, A. and J. Thompson, *Elements of Meteorology*, Charles Merrill, 1975.
9. Exercise 54 from http://www.intel.com/technology/mooreslaw/
10. Exercise 55 from http://www.wwindea.org/

Section 2.5

1. Page 98 from Ludwig, John and James Reynolds, *Statistical Ecology: A Primer on Methods and Computing*, New York: Wiley, 1988, p. 92.
2. Exercise 71 from Lucky Larry #16 by Joan Page, *The AMATYC Review*, Vol. 16, No. 1, Fall 1994, p. 67.
3. Exercise 80 from *Science*, Vol. 284, June 18, 1999, p. 1937.
4. Exercise 84 from Huxley, J. S., *Problems of Relative Growth*, Dover, 1968.
5. Exercise 85 from Horelick, Brindell and Sinan Koont, "Applications of

Calculus to Medicine: Prescribing Safe and Effective Dosage," *UMAP Module 202*, 1977.
6. Exercise 86 from McNab, Brian K., "Complications Inherent in Scaling the Basal Rate of Metabolism in Mammals," *The Quarterly Review of Biology*, Vol. 63, No.1, Mar. 1988, pp. 25–54.
7. Exercise 87 from U.S. Census Bureau, U.S. Interim Projections by Age, Sex, Race, and Hispanic Origin, http://www.census.gov/ipc/www/usinterimproj/.
8. Exercise 89 from *Scientific American*, Oct. 1999, p. 103.
9. Exercise 90 from *The New York Times*, June 6, 1999, p. 41.
10. Exercise 91 from http://www.npr.org/templates/rundowns/rundown.php?prgId=2&prgDate=5-7-2002
11. Exercise 92c from http://www.west.net/rperry/PueblaTlaxcala/puebla.html
12. Exercise 92d from http://www.history.com/this-day-in-history/earthquake-shakes-mexico-city
13. Exercise 92g from *The New York Times*, Jan. 13, 1995.
14. Exercise 94 from Tymoczko, Dmitri, "The Geometry of Music Chords," *Science*, Vol. 313, July 7, 2006, p. 72–74.

Section 2.6

1. Exercise 25 from http://www.census.gov/ipc/www/idb/worldpopinfo.php
2. Exercise 26 from Brody, Jane, "Sly Parasite Menaces Pets and Their Owners," *The New York Times*, Dec. 21, 1999, p. F7.
3. Exercise 30 from Speroff, Theodore et al., "A Risk-Benefit Analysis of Elective Bilateral Oophorectomy: Effect of Changes in Compliance with Estrogen Therapy on Outcome," *American Journal of Obstetrics and Gynecology*, Vol. 164, Jan. 1991, pp. 165–174.
4. Exercise 31 from http://downsyndrome.about.com/od/diagnosingdownsyndrome/a/Matagechart.htm.
5. Exercise 40 from *Science*, Vol. 277, July 25, 1997, p. 483.

Review Exercises

1. Exercise 108 from ftp://ftp.bls.gov/pub/special.requests/cpi/cpiai.txt.
2. Exercise 110 from *Family Practice*, May 17, 1993, p. 55.
3. Exercise 111 from http://www.cdc.gov/hiv/topics/surveillance/resources/reports/2007report/table25.htm
4. Exercise 112 from Rusconi, Franca et al., "Reference Values for Respiratory Rate in the First 3 Years of Life," *Pediatrics*, Vol. 94, No. 3, Sept. 1994, pp. 350–355.
5. Exercise 113 from Cattet, Marc R. L. et al., "Predicting Body Mass in Polar Bears: Is Morphometry Useful?" *Journal of Wildlife Management*, Vol. 61. No. 4, 1997, pp. 1083–1090.
6. Exercise 115 from Von Foerster, Heinz, Patricia M. Mora, and Lawrence W. Amiot,

"Doomsday: Friday, 13 November, A.D. 2026," *Science,* Vol. 132, Nov. 4, 1960, pp. 1291–1295.

7. Exercise 115a from http://esa.un.org/unpp/index.asp

8. Exercise 119 from Johnson, Michael P., and Daniel S. Simberloff, "Environmental Determinants of Island Species Numbers in the British Isles," *Journal of Biogeography,* Vol. 1, 1974, pp. 149–154.]

9. Exercise 122 from Ronan, C., *The Natural History of the Universe*, Macmillan, 1991.

Extended Application

1. Page 118 from Bornstein, Marc H. and Helen G. Bornstein, "The Pace of Life," *Nature,* Vol. 259, Feb. 19, 1976, pp. 557–559.

2. Page 119 from Johnson, Michael P., and Daniel S. Simberloff, "Environmental Determinants of Island Species Numbers in the British Isles," *Journal of Biogeography,* Vol. 1, 1974, pp. 149–154.

3. Page 119 from Juan Camilo Bohorquez, "Common ecology quantifies human insurgency," *Nature,* Vol. 462, Dec. 17, 2009, pp. 911–914.

4. Exercise 1 from Gwartney, James D., Richard L. Stroup, and Russell S. Sobel, *Economics: Private and Public Choice*, 9th ed., The Dryden Press, 2000, p. 59.

5. Exercise 2 from White, Craig R. et al., "Phylogenetically informed analysis of the allometry of Mammalian Basal metabolic rate supports neither geometric nor quarter-power scaling," *Evolution,* Vol. 63, Oct. 2009, pp. 2658–2667.

Chapter 3

Section 3.1

1. Exercise 83 from ca.gov.

2. Exercise 84 from United States Postal Service.

3. Exercise 86 from American Airlines, http://www.aa.com.

4. Exercise 88 contributed by Robert D. Campbell of the Frank G. Zarb School of Business at Hofstra University.

5. Exercise 89 contributed by Robert D. Campbell of the Frank G. Zarb School of Business at Hofstra University.

6. Exercise 90 from Kulesa, P., G. Cruywagen et al. "On a Model Mechanism for the Spatial Patterning of Teeth Primordia in the Alligator," *Journal of Theoretical Biology,* Vol. 180, 1996, pp. 287–296.

7. Exercise 91 from Nord, Gail and John Nord, "Sediment in Lake Coeur d' Alene, Idaho," *Mathematics Teacher,* Vol. 91, No. 4, April 1998, pp. 292–295.

8. Exercise 93 from Bishir, John W. and Donald W. Drewes, *Mathematics in the Behavioral and Social Sciences*, New York: Harcourt Brace Jovanovich, 1970, p. 538.

Section 3.2

1. Page 140 from U.S. Department of Labor.

2. Exercise 33 from Problem 26 from May 2003 Course 1 Examination of the *Education and Examination Committee of the Society of Actuaries*. Reprinted by permission of the Society of Actuaries.

3. Exercise 39 from U.S. Postal Service.

4. Exercise 41 from Xin, H., I. Berry, T. Barton, and G. Tabler, "Feed and Water Consumption, Growth, and Mortality of Male Broiler," *Poultry Science,* Vol. 73, No. 5, May 1994, pp. 610–616.

Section 3.3

1. Page 150 from U.S. Census Bureau.

2. Page 151 from Blumberg, S. J. and J. V. Luke, "Wireless Substitution: Early Release of Estimates From the National Health Interview Survey, July–December 2009," National Center for Health Statistics, May, 2010.

3. Exercise 31 from Monthly Energy Review April 2010, Table 9.4 Motor Gasoline Retail Prices, U.S. City Average, www.eia.doe.gov/mer/pdf/mer.pdf.

4. Exercise 32 from Geithner, Timothy F., et al "2009 Annual Report of the Boards of Trustees of the Federal Hospital Insurance and Federal Supplementary Medical Insurance Trust Funds," Table V.E4, May 12, 2009, p. 207.

5. Exercise 34 from Carl Haub, Population Reference Bureau, 2000.

6. Exercise 36 from Harris, E. F., J. D. Hicks, and B. D. Barcroft, "Tissue Contributions to Sex and Race: Differences in Tooth Crown Size of Deciduous Molars," *American Journal of Physical Anthropology,* Vol. 115, 2001, pp. 223–237.

7. Exercise 37 from Reed, G. and J. Hill, "Measuring the Thermic Effect of Food," *American Journal of Clinical Nutrition,* Vol. 63, 1996, pp. 164–169.

8. Exercise 38 from Jorgenson, J., M. Festa-Bianchet, M. Lucherini, and W. Wishart, "Effects of Body Size, Population Density, and Maternal Characteristics on Age at First Reproduction of Bighorn Ewes," *Canadian Journal of Zoology,* Vol. 71, No. 12, Dec. 1993, pp. 2509–2517.

9. Exercise 39 from Homeland Security.

10. Exercise 40 from www.drugabuse.gov.

Section 3.4

1. Page 166 from downsyndrome.about.com.

2. Page 175 from Kaufmann III, William J., "The Light Curve of a Nova," *Astronomy: The Structure of the Universe*, New York: Macmillan, 1977. Reprinted by permission of William J. Kaufmann III.

3. Exercise 52 from Michael W. Ecker in "Controlling the Discrepancy in Marginal Analysis Calculations," *The College Mathematics Journal,* Vol. 37, No. 4, Sept. 2006, pp. 299–300.

4. Exercise 53 from 2009 OASDI Trustees Report, Social Security Administration, http://www.ssa.gov/OACT/TR/2009/LD_figI VB1.html.

5. Exercise 54 from Brighton, Caroline H., "Mechanical Power Output of Cockatiel Flight in Relation to Flight Speed" *Biolog-e: The Undergraduate Bioscience Research Journal,* 2007, http://www.fbs.leeds.ac.uk/students/ejournal/Biolog-e/uploads/Caroline%20Brighton_synopsis.pdf.

6. Exercise 56 from Kissileff, H. R. and J. L. Guss, "Microstructure of Eating Behavior in Humans," *Appetite,* Vol. 36, No. 1, Feb. 2001, pp. 70–78.

7. Exercise 57 from Benedito, J., J. Carcel, M. Gisbert, and A. Mulet, "Quality Control of Cheese Maturation and Defects Using Ultrasonics," *Journal of Food Science,* Vol. 66, No. 1, 2001, pp. 100–104.

8. Exercise 59 from http://heatkit.com/html/bakeov03.htm.

9. Exercises 60 and 61 from Bahill, A. T. and W. J. Karnavas, "Determining Ideal Baseball Bat Weights Using Muscle Force-Velocity Relationships," *Biological Cybernetics,* Vol. 62, 1989, pp.89–97.

Section 3.5

1. Exercise 19 from *Science,* Vol. 297, Sept. 27, 2002, p. 2222.

2. Exercise 20 from Centers for Disease Control, http://www.cdc.gov/nchs/about/major/nhanes/growthcharts/charts.htm.

3. Exercise 22 from Brighton, Caroline H., "Mechanical Power Output of Cockatiel Flight in Relation to Flight Speed" *Biolog-e: The Undergraduate Bioscience Research Journal,* 2007, http://www.fbs.leeds.ac.uk/students/ejournal/Biolog-e/uploads/Caroline%20Brighton_synopsis.pdf.

4. Exercise 23 from Hensinger, Robert, *Standards in Pediatric Orthopedics: Tables, Charts, and Graphs Illustrating Growth*, New York: Raven Press, 1986, p. 192.

5. Exercise 24 from Sinclair, David, *Human Growth After Birth*, New York: Oxford University Press, 1985.

Review Exercises

1. Exercise 61 from Problem 3 from May 2003 Course 1 Examination of the *Education and Examination Committee of the Society of Actuaries*. Reprinted by permission of the Society of Actuaries.

2. Exercise 66 from Murray, Alan, "Winners? Losers? Estimates Show How Impact of Tax Proposal Varies," *Wall Street Journal*, May 9, 1986, p. 29.

3. Exercise 67 from Household Data Annual Averages, Table 1: Employment Status of the Civilian Noninstitutional Population, 1940 to Date, www.bls.gov/cps/cpsaat1.pdf.

4. Exercise 68 from *2011 Alzheimer's Disease Facts and Figures*, Alzheimer's Association, 2011, p. 17.

5. Exercise 70 from Peter Tyack, © Woods Hole Oceanographic Institution.

6. Exercise 71 from Centers for Disease Control, http://www.cdc.gov/nchs/about/major/nhanes/growthcharts/charts.htm.

7. Exercise 72 from Hensinger, Robert, *Standards in Pediatric Orthopedics: Tables, Charts, and Graphs Illustrating Growth*, New York: Raven Press, 1986, p. 193.

Chapter 4

Section 4.1

1. Page 201 from Finke, M., "Energy Requirements of Adult Female Beagles," *Journal of Nutrition*, Vol. 124, 1994, pp. 2604s–2608s.

2. Page 206 from Table 2. Projections of the Population by Selected Age Groups and Sex for the United States: 2010 to 2050 (NP2008-T2), Population Division, *U.S. Census Bureau*, August 14, 2008.

3. Exercise 57 from "Rates for Domestic Letters, 1863–2009," Historian, U.S. Postal Service, May 2009. www.usps.com/postalhistory.

4. Exercise 58 from "Introduction: U.S. Currency and Coin Outstanding and in Circulation," Financial Management Service, U.S. Department of the Treasury, Dec. 31, 2009. www.fms.treas.gov/bulletin/index.html

5. Exercise 59 from Walker, A., *Observation and Inference: An Introduction to the Methods of Epidemiology*, Epidemology Resources, Inc., 1991.

6. Exercise 61 from Fitzsimmons, N. S. Buskirk and M. Smith. "Population History, Genetic Variability, and Horn Growth in Bighorn Sheep," *Conservation Biology*. Vol. 9, No. 2, April 1995, pp. 314–323.

7. Exercise 62 from Dobbing, John and Jean Sands, "Head Circumference, Biparietal Diameter and Brain Growth in Fetal and Postnatal Life," *Early Human Development*, Vol. 2, No. 1, April 1978, pp. 81–87.

8. Exercise 63 from Okubo, Akira, "Fantastic Voyage into the Deep: Marine Biofluid Mechanics," in *Mathematical Topics in Population Biology Morphogenesis and Neurosciences*, edited by E. Teramoto and M. Yamaguti, Springer-Verlag, 1987, pp. 32–47.

9. Exercise 64 from Tan J., N. Silverman, J. Hoffman, M. Villegas, and K. Schmidt, "Cardiac Dimensions Determined by Cross-Sectional Echocardiography in the Normal Human Fetus From 18 Weeks to Term," *American Journal of Cardiology*, Vol. 70, No. 18, Dec. 1, 1992, pp. 1459–1467.

10. Exercise 65 from Kennelly, A., "An Approximate Law of Fatigue in Speeds of Racing Animals," *Proceedings of the American Academy of Arts and Sciences*, Vol. 42, 1906, pp. 275–331.

11. Exercise 66 from Tuchinsky, Philip, "The Human Cough," *UMAP Module 211*, Lexington, MA, COMAP, Inc., 1979, pp. 1–9.

12. Exercise 67 from http://win.niddk.nih.gov/publications/tools.htm

13. Exercise 74 from Yechieli, Yoseph, Ittai Gavrieli, Brian Berkowitz, and Daniel Ronen, "Will the Dead Sea Die?" *Geology*, Vol. 26, No. 8, Aug. 1998, pp. 755–758. These researchers have predicted that the Dead Sea will not die but reach an equilibrium level.

14. Exercise 75 from Vennebush, Patrick, "Media Clips: A Dog's Human Age," *Mathematics Teacher*, Vol. 92, 1999, pp. 710–712.

Section 4.2

1. Exercise 51 from Sanders, Mark and Ernest McCormick, *Human Factors in Engineering and Design*, 7th ed., New York: McGraw-Hill, 1993, pp. 243–246.

2. Exercise 52 from Kellar, Brian and Heather Thompson, "Whelk-come to Mathematics," *Mathematics Teacher*, Vol. 92, No. 6, September 1999, pp. 475–481.

3. Exercise 54 from Mannering, F. and W. Kilareski, *Principles of Highway Engineering and Traffic Control*, New York: Wiley, 1990.

Section 4.3

1. Exercise 65 from Guffey, Roger, "The Life Expectancy of a Jawbreaker: The Application of the Composition of Functions," *Mathematics Teacher*, Vol. 92, No. 2, Feb. 1999, pp. 125–127.

2. Exercise 66 from Chrisomalis, Stephen, "Numerical Prefixes," 2007. http://phrontistery.info/numbers.html

3. Exercise 67 from Chrisomalis, Stephen, "Numerical Prefixes," 2007. http://phrontistery.info/numbers.html

Section 4.4

1. Exercise 42 from Problem 11 from May 2003 Course 1 Examination of the *Education and Examination Committee of the Society of Actuaries*. Reprinted by permission of the Society of Actuaries.

2. Exercise 43 from World Bank, World Development Indicators. http://www.google.com/publicdata?ds=wb-wdi&met=it_net_user&idim=country:USA&dl=en&hl=en&q=number+of+internet+users

3. Exercise 44 from http://esa.un.org/unpp/index.asp

4. Exercise 45 from U.S. Census Bureau, U.S. Interim Projections by Age, Sex, Race, and Hispanic Origin. http://www.census.gov/ipc/www/usinterimproj/.

5. Exercise 49 from Spratt, John et al., "Decelerating Growth and Human Breast Cancer," *Cancer*, Vol. 71, No. 6, 1993, pp. 2013–2019.

6. Exercise 50 from U.S. Vital Statistics, 1995.

7. Exercise 51 from Friedman, Simon H. and Jens O. Karlsson, "A Novel Paradigm,"

Nature, Vol. 385, No. 6616, Feb. 6, 1997, p. 480.

8. Exercise 52 from Prestrud, Pal and Kjell Nilssen, "Growth, Size, and Sexual Dimorphism in Arctic Foxes," *Journal of Mammalogy*, Vol. 76, No. 2, May 1995, pp. 522–530.

9. Exercise 53 from DeNise, R. and J. Brinks, "Genetic and Environmental Aspects of the Growth Curve Parameters in Beef Cows," *Journal of Animal Science*, Vol. 61, No. 6, 1985, pp. 1431–1440.

10. Exercise 54 from LaRosa, John C. et al., "The Cholesterol Facts: A Joint Statement by the American Heart Association and the National Heart, Lung, and Blood Institute," *Circulation*, Vol. 81, No. 5, May 1990, p. 1722.

11. Exercise 55 from Cisne, John L., "How Science Survived: Medieval Manuscripts' 'Demography' and Classic Texts' Extinction," *Science*, Vol. 307, Feb. 25, 2005, pp. 1305–1307.

12. Exercise 57 from Allen, I. E., and J. Seaman, "Staying the Course–Online Education in the United States, 2008," The Sloan Consortium, Nov. 2008.

13. Exercise 59 from Kevin Friedrich, Sharon, PA.

14. Exercise 60 from Schoen, Carl, "A New Empirical Model of the Temperature-Humidity Index," American Meteorological Society, Vol. 44, Sept. 2005, pp. 1413–1420.

15. Exercise 61 from Bennett, Jay, "Statistical Modeling in Track and Field," *Statistics in Sports*, Arnold, 1998, p. 179.

16. Exercise 63 from Osserman, Robert, "Mathematics of the Gateway Arch, *Notice of the AMS*, Vol. 57, No. 2, Feb. 2010, pp. 220–229.

Section 4.5

1. Page 240 from http://www.kbb.com/new-cars/Toyota/corolla/2010/resale-value.

2. Exercise 60 from Sharkey, I. et al., "Body Surface Area Estimation in Children Using Weight Alone: Application in Pediatric Oncology," *British Journal of Cancer*, Vol. 85, No. 1, 2001, pp. 23–28.

3. Exercise 61 from Zanoni, B., C. Garzaroli, S. Anselmi, and G. Rondinini, "Modeling the Growth of *Enterococcus faecium* in Bologna Sausage," *Applied and Environmental Microbiology*, Vol. 59, No. 10, Oct. 1993, pp. 3411–3417.

4. Exercise 62 from Miller, Michelle N. and John A. Byers, "Energetic Cost of Locomotor Play in Pronghorn Fawns," *Animal Behavior*, Vol. 41, 1991, pp. 1007–1013.

5. Exercise 63 from Pearl, R. and S. Parker, *Proc. Natl. Acad. Sci.*, Vol. 8, 1922, p. 212, quoted in *Elements of Mathematical Biology* by Alfred J. Lotka, Dover Publications, 1956, pp. 308–311.

6. Exercise 66 from "Health, United States, 2009," U.S. Census.

7. Exercise 67 from Bradley, Christopher, "Media Clips," *Mathematics Teacher*, Vol. 93, No. 4, April 2000, pp. 300–303.

8. Exercise 68 from Bender, Edward, *An Introduction to Mathematical Modeling,* New York: Wiley, 1978, p. 213.

Review Exercises

1. Exercise 63 from "Japanese University Entrance Examination Problems in Mathematics," edited by Ling-Erl Eileen T. Wu, published by the Mathematical Association of America, copyright 1993, pp. 18–19.
2. Exercise 64 from Maurer, Stephen B., "Hat Derivatives," *The College Mathematics Journal,* Vol. 33, No. 2, Jan. 2002, pp. 32–37.
3. Exercise 79 from U.S. Postal Service, Annual Report of the Postmaster General.
4. Exercise 80 from www.bls.gov/spotlight/2008/older_workers
5. Exercise 81 from stats.bls.gov/data/inflation_calculator.htm
6. Exercise 84 from Marshall, William H. and Tina Wylie Echeverria, "Characteristics of the Monkeyface Prickleback," *California Fish and Game,* Vol. 78, No. 2, Spring 1992.
7. Exercise 85 from Pestrud, Pal and Kjell Nilssen, "Growth, Size, and Sexual Dimorphism in Arctic Foxes," *Journal of Mammalogy,* Vol. 76, No. 2, May 1995, pp. 522–530.
8. Exercise 86 from U.S Census Bureau. 2004, *US Interim Projections by Age, Sex, Race and Hispanic Origin.*
9. Exercise 87 from http://www.wwindea.org
10. Exercise 90 from Lo Bello, Anthony and Maurice Weir, "Glottochronology: An Application of Calculus to Linguistics," *The UMAP Journal,* Vol. 3, No. 1, Spring 1982, pp. 85–99.
11. Exercise 91 from www-nrd.nhtsa.dot.gov/pdf/nrd-30/NCSA/RNotes/1998/AgeSex96.pdf.

Chapter 5

Section 5.1

1. Page 252 from Garbarino, S., L. Nobili, M. Beelke, F. Phy, and F. Ferrillo, "The Contribution Role of Sleepiness in Highway Vehicle Accidents," *Sleep,* Vol. 24, No. 2, 2001, pp. 203–206. © 2001 American Academy of Sleep Medicine. Reproduced with permission of the American Academy of Sleep Medicine via Copyright Clearance Center.
2. Exercise 42 from Berry, Andrew, "Root Average Equals Turning Point Average," *The Mathematics Teacher,* Vol. 99, No. 9, May 2006, p. 595.
3. Exercise 49 from 2009 OASDI Trustees Report, Social Security Administration, http://www. ssa.gov/OACT/TR/2009/LD_figIVB1.html.
4. Exercise 50 from Household Data Annual Averages, Table 1: Employment Status of the Civilian Noninstitutional Population, 1940 to Date, www.bls.gov/cps/cpsaat1.pdf.

5. Exercise 51 from National Arbor Day Foundation, 100 Arbor Ave., Nebraska City, NE 68410. Ad in *Chicago Tribune,* Feb. 4, 1996, Sec. 2, p. 11. Used with permission of the National Arbor Day Foundation.
6. Exercise 53 from Garriott, James C., ed., *Medicolegal Aspects of Alcohol Determination in Biological Specimens,* PSG Publishing Company, 1988, p. 57.
7. Exercise 56 from Stefanadis, C., J. Dernellis et al., "Assessment of Aortic Line of Elasticity Using Polynomial Regression Analysis," *Circulation,* Vol. 101, No. 15, April 18, 2000, pp. 1819–1825.
8. Exercise 57 from Reed, George and James Hill, "Measuring the Thermic Effect of Food," *American Journal of Clinical Nutrition,* Vol. 63, 1996, pp. 164–169.
9. Exercise 58 from Perotto, D., R. Cue, and A. Lee, "Comparison of Nonlinear Functions of Describing the Growth Curve of Three Genotypes of Dairy Cattle," *Canadian Journal of Animal Science,* Vol. 73, Dec. 1992, pp. 773–782.
10. Exercise 60 from Kristensen, Hans M., "Nuclear Weapons Status and Options Under a START Follow-On Agreement," Federation of American Scientists Presentation to Arm Control Association Briefing, April 27, 2009.
11. Exercise 61 from http://www.fle-online.com/Dyno.asp.

Section 5.2

1. Exercise 40 from Dubinsky, Ed, "Is Calculus Obsolete?" *Mathematics Teacher,* Vol. 88, No. 2, Feb. 1995, pp. 146–148.
2. Exercise 45 from currentenergy.lbl.gov/ny/index.php.
3. Exercise 50 from Household Data Annual Averages, Table 1: Employment Status of the Civilian Noninstitutional Population, 1940 to Date, www.bls.gov/cps/cpsaat1.pdf.
4. Exercise 52 from Mezzadra, C., R. Paciaroni, S. Vulich, E. Villarreal, and L. Melucci, "Estimation of Milk Consumption Curve Parameters for Different Genetic Groups of Bovine Calves," *Animal Production,* Vol. 49, 1989, pp. 83–87.
5. Exercise 53 from Schwartz, C. and Kris Hundertmark, "Reproductive Characteristics of Alaskan Moose," *Journal of Wildlife Management,* Vol. 57, No. 3, July 1993, pp. 454–468.
6. Exercise 54 from Reed, George and James Hill, "Measuring the Thermic Effects of Food," *American Journal of Clinical Nutrition,* Vol. 63, 1996, pp. 164–169.
7. Exercise 55 from Eagly, A. H. and K. Telaak, "Width of the Latitude of Acceptance as a Determinant of Attitude Change," *Journal of Personality and Social Psychology,* Vol. 23, 1972, pp. 388–397.

Section 5.3

1. Exercise 70 from http://www.tutor2u.net/business/marketing/products_lifecycle.asp.
2. Exercise 71 from 2009 OASDI Trustees Report, Social Security Administration, http://www.ssa.gov/OACT/TR/2009/LD_figIVB1.html.
3. Exercise 77 from Fudenberg, Drew and Jean Tirole, "Learning by Doing and Market Performance," *Bell Journal of Economics,* Vol. 14, 1983, pp. 522–530.
4. Exercise 80 from *The New York Times,* Aug. 29, 1993, p. E2.
5. Exercise 85 from Weymouth, F. W., H. C. McMillin, and Willis H. Rich, "Latitude and Relative Growth in the Razor Clam," *Journal of Experimental Biology,* Vol. 8, 1931, pp. 228–249.
6. Exercise 86 from Speer, John F. et al., "A Stochastic Numerical Model of Breast Cancer Growth That Simulates Clinical Data," *Cancer Research,* Vol. 44, Sept. 1984, pp. 4124–4130.
7. Exercise 87 from Song, A. and S. Eckhoff, "Optimum Popping Moisture Content for Popcorn Kernels of Different Sizes," *Cereal Chemistry,* Vol. 71, No. 5, 1994, pp. 458–460.
8. Exercise 88 from Kulesa, P. et al., "On a Model Mechanism for the Spatial Performing of Teeth Primordia in the Alligator," *Journal of Theoretical Biology,* Vol. 180, 1996, pp. 287–296.
9. Exercise 89 from *The New York Times,* Dec. 17, 1995, p. 49.
10. Exercise 92 was provided by Frederick Russell of College of Southern Maryland.
11. Exercise 95 was suggested by Larry Taylor of North Dakota State University.

Review Exercises

1. Exercise 64 from Monthly Energy Review April 2010, Table 9.4 Motor Gasoline Retail Prices, U.S. City Average, www.eia.doe.gov/mer/pdf/mer.pdf.
2. Exercise 65 from Meltzer, David E., "Age Dependence of Olympic Weightlifting," *Medicine and Science in Sports and Exercise,* Vol. 26, No. 8, Aug. 1994, p. 1053.
3. Exercise 66 from West, Geoffrey B., James H. Brown, and Brian J. Enquist, "A General Model for the Origin of Allometric Scaling Laws in Biology," *Science,* Vol. 276, April 4, 1997, pp. 122–126.
4. Exercise 67 from Davie, A. and D. Evans, "Blood Lactate Responses to Submaximal Field Exercise Tests in Thoroughbred Horses," *The Veterinary Journal,* Vol. 159, 2000, pp. 252–258.
5. Exercise 68 from Murray, J. D., *Mathematical Biology,* Springer–Verlag, 1989, p.163.
6. Exercise 69 from Pearl, R. and S. Parker, *Proc. Natl. Acad. Sci.,* Vol. 8, 1922, p. 212, quoted in Elements of Mathematical Biology by Alfred J. Lotka, Dover Publications, 1956, pp. 308–311.

7. Exercise 70 from Hurley, Peter J., "Red Cell Plasma Volumes in Normal Adults," *Journal of Nuclear Medicine*, Vol. 16, 1975, pp. 46–52, and Pearson, T. C. et al. "Interpretation of Measured Red Cell Mass and Plasma Volume in Adults," *British Journal of Haematology*, Vol. 89, 1995, pp. 748–756.

8. Exercise 71 from Courtis, S. A., "Maturation Units for the Measurement of Growth," *School and Society*, Vol. 30, 1929, pp. 683–690.

9. Exercise 72 from *Science*, Vol. 299, March 28, 2003, p. 1991.

10. Exercise 73 from Kristensen, Hans M., "Nuclear Weapons Status and Options Under a START Follow-on Agreement," Federation of American Scientists Presentation to Arms Control Association Briefing, April 27, 2009.

Chapter 6

Section 6.1

1. Page 308 from "Canada Spot Exchange Rate, Canadian $/US $," Federal Reserve. www.federalreserve.gov/releases/h10/hist/dat00_ca.txt.

2. Exercise 40 from Problem 19 from the May 2003 Course 1 Examination of the *Education and Examination Committee of the Society of Actuaries*. Reprinted by permission of the Society of Actuaries.

3. Exercises 41 and 42 from Bank Crime Statistics, Federal Bureau of Investigation. www.fbi.gov/publications.htm.

4. Exercise 53 from Harris, Edward F., Joseph D. Hicks, and Betsy D. Barcroft, "Tissue Contributions to Sex and Race: Differences in Tooth Crown Size of Deciduous Molars," *American Journal of Physical Anthropology*, Vol. 115, 2001, pp. 223–237.

5. Exercise 54 from Rowan, N., C. Johnstone, R. McLean, J. Anderson, and J. Clarke, "Prediction of Toxigenic Fungal Growth in Buildings by Using a Novel Modelling System," *Applied and Environmental Microbiology*, Vol. 65, No. 11, Nov. 1999, pp. 4814–4821.

Section 6.2

1. Page 313 from Biddle, Wayne, "Skeleton Alleged in the Stealth Bomber's Closet," *Science*, Vol. 244, May 12, 1989.

2. Page 317 from Cullen, Michael R., *Mathematics for the Biosciences*. Copyright © 1983 PWS Publishers. Reprinted by permission.

3. Exercise 38 from Ricker, W. E., "Stock and Recruitment," *Journal of the Fisheries Research Board of Canada*, Vol. 11, 1957, pp. 559–623.

4. Exercise 43 from Cook, R. M., A. Sinclair, and G. Stefánsson, "Potential Collapse of North Sea Cod Stocks," *Nature*, Vol. 385, Feb. 6, 1997, pp. 521–522.

Section 6.3

1. Page 327 from Gwartney, James D., Richard L. Stroup, Russell S. Sobel, *Economics: Private and Public Choice*, 9th ed., The Dryden Press, 2000, p. 510.

2. Example 3 from Wales, Terrence J., "Distilled Spirits and Interstate Consumption Efforts," *The American Economic Review*, Vol. 57, No. 4, 1968, pp. 853–863.

3. Example 4 from Hogarty, T. F. and K. G. Elzinga, "The Demand for Beer," *The Review of Economics and Statistics*, Vol. 54, No. 2, 1972, pp. 195–198.

4. Exercise 3 from the Uniform CPA Examination Questions and Unofficial Answers, Copyright © 1991 by the American Institute of Certified Public Accountants, Inc. Reprinted (or adapted) with permission.

5. Exercise 5 from http://www.investopedia.com/terms/g/giffen-good.asp.

6. Exercise 27 from Cooper, John, C., "Price Elasticity of Demand for Crude Oil: Estimates for 23 Countries," *OPEC Review*, March 2003, pp. 1–8.]

7. Exercise 28 from Kako, T., Gemma, M., and Ito, S., "Implications of the Minimum Access Rice Import on Supply and Demand Balance of Rice in Japan," *Agricultural Economics*, Vol. 16, 1997, pp. 193–204.

8. Exercise 29 from Atwood, Jeff, "Coding Horror," August 5, 2009. www.codinghorror.com.

9. Exercise 30 from http://support.sas.com/rnd/app/examples/ets/simpelast/.

10. Exercise 31 from Battersby, B. and E. Oczkowski, "An Econometric Analysis of the Demand for Domestic Air Travel in Australia," *International Journal of Transport Economics*, Vol. XXVIII, No. 2, June 2001, pp. 193–204.

11. Exercise 32 from http://my-forest.com/economics/index.html.

12. Exercise 33 from Gordon, Warren B., "The Calculus of Elasticity," *The AMATYC Review*, Vol. 26, No. 2, Spring 2006, pp. 53–55.

Section 6.4

1. Exercises 44 and 45 from Tanyeri-Aber, A. and Rosson, C., "Demand for Dairy Products in Mexico," Agricultural Economics, Vol. 16, 1997, pp. 6–76.

2. Exercise 46 from Gagliardi, L. and F. Rusconi, "Respiratory Rate and Body Mass in the First Three Years of Life," *Archives of Disease in Children*, Vol. 76, 1997, pp. 151–154.

5. Exercise 44 is based on an example in *A Concrete Approach to Mathematical Modelling* by Michael Mesterton-Gibbons, Wiley-Interscience, 1995, pp. 93–96.

6. Exercise 47 from http://www.ups.com/send/preparemailandpackages/measuringtips.htm.

Section 6.5

1. Exercise 17 from Wanderley, S., M. Costa-Neves, and R. Rega, "Relative Growth of the Brain in Human Fetuses: First Gestational Trimester," *Archives d'anatomie, d'histologie et d'embryologie*, Vol. 73, 1990, pp. 43–46.

2. Exercise 18 from Robbins, C., *Wildlife Feeding and Nutrition*, New York: Academic Press, 1983, p. 119.

3. Exercise 19 from Robbins, C., *Wildlife Feeding and Nutrition*, New York: Academic Press, 1983, p. 133.

4. Exercise 20 from Robbins, C., *Wildlife Feeding and Nutrition*, New York: Academic Press, 1983, p. 114.

Section 6.6

1. Exercise 23 from Garriott, James C. (ed.), *Medicolegal Aspects of Alcohol Determination in Biological Specimens*, PSG Publishing Company, 1988, p. 57.

2. Exercise 30 from Landon, D., C. Waite, R. Peterson, and L. Mech, "Evaluation of Age Determination Techniques for Gray Wolves," *Journal of Wildlife Management*, Vol. 62, No. 2, 1998, pp. 674–682.

3. Exercise 31 from Van Lunen, T. and D. Cole, "Growth and Body Composition of Highly Selected Boars and Gilts," *Animal Science*, Vol. 67, 1998, pp. 107–116.

Review Exercises

1. Exercise 53 from Tanyeri-Aber, A. and Rosson, C., "Demand for Dairy Products in Mexico," *Agricultural Economics*, Vol. 16, 1997, pp. 67–76.

2. Exercise 57 from Matsumoto, B., K. Nonaka, and M. Nakata, "A Genetic Study of Dentin Growth in the Mandibular Second and Third Molars of Male Mice," *Journal of Craniofacial Genetics and Developmental Biology*, Vol. 16, No. 3, July–Sept. 1996, pp. 137–147.

3. Exercise 58 from Voros, E., C. Robert, and A. Robert, "Age-Related Changes of the Human Skin Surface Microrelief," *Gerontology*, Vol. 36, 1990, pp. 276–285.

4. Exercise 65 from Redheffer, Ray, *Differential Equations: Theory and Applications*, Jones and Bartlett Publishers, 1991, pp. 107–108.

Extended Application

1. Based on "A Total Cost Model for a Training Program" by P. L. Goyal and S. K. Goyal, Faculty of Commerce and Administration, Concordia University.

3. Exercise 47 from Murray, J. D., *Mathematical Biology*, New York: Springer-Verlag, 1989, pp. 156–158.

4. Exercise 48 from Lotka, Alfred J., *Elements of Mathematical Biology*, Dover Publications, 1956, p. 313.

Chapter 7

Section 7.1

1. Exercise 57 from "US Wireless Quick Facts Year-End Figures," CTIA, International Association for the Wireless Telecommunications Industry. http://www.ctia.org/advocacy/research/index.cfm/AID/10323.
2. Exercise 61 from Dennis, Brian and Robert F. Costantino, "Analysis of Steady-State Populations with the Gamma Abundance Model: Application to *Tribolium,*" *Ecology,* Vol. 69, No. 4, Aug. 1988, pp. 1200–1213.
3. Exercise 65 from "Digest of Education Statistics," National Center for Education Statistics, Table 271. http://nces.gov/programs/digest;d08/tables/dt08_271.asp.
4. Exercise 66 from "Digest of Education Statistics," National Center for Education Statistics, Table 279. http://nces.gov/programs/digest;d08/tables/dt08_279.asp

Section 7.2

1. Page 374 from Quirin, Jim and Barry Cohen, *Chartmasters' Rock 100,* 5th ed. Copyright 1992 by Chartmasters. Reprinted by permission.
2. Exercise 43 from *National Transportation Statistics 2006*, Bureau of Transportation Statistics.
3. Exercise 44 from *Hospital Statistics,* American Hospital Association.

Section 7.3

1. Exercise 23 from MacLaskey, Michael, *All About Lawns,* ed. by Alice Mace, Ortho Information Services, © 1980, p.108.
2. Exercise 25 from U.S. Energy Information Administration, Monthly Energy Review June 2010, Table 10.1, "Renewable Energy Production and Consumption by Source." www.eia.doe.gov/mer/ppdf/mer.pdf.
3. Exercise 27 from Department of Environment, Food, and Rural Affairs, Foot and Mouth Disease, http://footandmouth.csl.gov.uk.
4. Exercise 28 from California Highway Patrol, Table 1A; "Fatal Collisions by Month 1999–2008." http://www.chp.ca.gov/switrs/#section1.
5. Exercises 29 and 30 from http://www.roadandtrack.com/tests/data-panel-archive.
6. Exercises 33 and 34 from Sustainable by Design, www.susdesign.com/windowheatgain.
7. Exercise 35 based on an example given by Stephen Monk of the University of Washington.
8. Exercise 38 from Wildbur, Peter, *Information Graphics,* Van Nostrand Reinhold, 1989, pp. 126–127.
9. Exercise 39 from *The New York Times,* Jan. 27, 2006, p.86.

Section 7.4

1. Exercise 62 from Gompertz, Benjamin, "On the Nature of the Function Expressive of the Law of Human Mortality," *Philosophical Transactions of the Royal Society of London,* 1825.
2. Exercise 66 from Jorgenson, Jon T., et al., "Effects of Population Density on Horn Development in Bighorn Rams," *Journal of Wildlife Management,* Vol. 62, No. 3, 1998, pp. 1011–1020.
3. Exercise 67 from Finke, M., "Energy Requirements of Adult Female Beagles," *Journal of Nutrition,* Vol. 124, 1994, pp. 2604s–2608s.
4. Exercise 68 from Nord, Gail and John Nord, "Sediment in Lake Coeur d'Alene, Idaho," *Mathematics Teacher,* Vol. 91, No. 4, April 1998, pp. 292–295.
5. Exercise 69 was originally contributed by Ralph DeMarr, University of New Mexico. It was updated with information from the U.S. Census Bureau. http://www.census.gov/popest/national/asrh/2008-nat-res.html.
6. Exercise 70 was originally contributed by Ralph DeMarr, University of New Mexico. It was updated with information from the U.S. Census Bureau, Current Population Reports, Table 680, P60-235, August 2008. http://www.census.gov/prod/2008pubs/p60-235.pdf.

Section 7.5

1. Exercise 38 from DeCicco, John and Marc Ross, "Improving Automotive Efficiency," *Scientific American,* Vol. 271, No. 6, Dec. 1994, p. 56. Copyright © 1994 by Scientific American, Inc. All rights reserved. Oil_and_Liquid_Fuels.
2. Exercise 38b from Table 4a, U.S. Crude Oil and Liquid Fuels Supply, Consumption, and Inventories, http://www.eia.doe.gov/steo.
3. Exercise 41 from Table H-4. "Gini Ratios for Households, by Race and Hispanic Origin of Householder: 1967 to 2008," http://www.census.gov/hhes/www/income/data/historical/inequality/index.html.

Section 7.6

1. Page 414 from Chodos, D. J. and A. R. DeSantos, *Basics of Bioavailability*, Upjohn Company, 1978.
2. Exercise 29 from Mezzadra, C., R. Paciaroni, S. Vulich, E. Villarreal, and L. Melucci, "Estimation of Milk Consumption Curve Parameters for Different Genetic Groups of Bovine Calves," *Animal Production,* Vol. 49, 1989, pp. 83–87.
3. Exercise 30 from Department of Environment, Food, and Rural Affairs, Foot and Mouth Disease, http://footandmouth.csl.gov.uk.
4. Exercise 34 from Mezzadra, C., R. Paciaroni, S. Vulich, E. Villarreal, and L. Melucci, "Estimation of Milk Consumption Curve Parameters for Different Genetic Groups of Bovine Calves, *Animal Production,* Vol. 49, 1989, pp. 83–87.

Review Exercises

1. Exercise 78 from Problem 7 from May 2003 Course 1 Examination of the *Education and Examination Committee of the Society of Actuaries.* Reprinted by permission of the Society of Actuaries.
2. Exercise 84 from Economic News Release. http://data.bls.gov/news.release/prod3.t02.htm.
3. Exercise 87 from "Annual U.S. Field Production of Crude Oil," U.S. Energy Information Administration http://www.eia.doe.gov/dnav/pet/hist/LeafHandler.ashx?n-petas-mcrfpus19f=a.
4. Exercise 88 from Problem 38 from May 2003 Course 1 Examination of the *Education and Examination Committee of the Society of Actuaries.* Reprinted by permission of the Society of Actuaries.
5. Exercise 91 from Hastings, Alan and Robert F. Costantino, "Oscillations in Population Numbers: Age-Dependent Cannibalism," *Journal of Animal Ecology,* Vol. 60, No. 2, June 1991, pp. 471–482.
6. Exercise 92 from Oliver, M. H. et al., "Material Undernutrition During the Periconceptual Period Increases Plasma Taurine Levels and Insulin Response to Glucose but not Arginine in the Late Gestation Fetal Sheep," *Endocrinology,* Vol 14, No. 10, 2001, pp. 4576–4579.
7. Exercise 93 from Freeze, Brian S. and Timothy J. Richards, "Lactation Curve Estimation for Use in Economic Optimization Models in the Dairy Industry," *Journal of Dairy Science,* Vol. 75, 1992, pp. 2984–2989.
8. Exercise 94 from Table IF "Property-Damage-Only Collisions by Month 1999–2008. The California Highway Patrol." http://www.chp.gov/switrs/#section1.

Extended Application

1. Page 422 from http://www.eia.gov/totalenergy/data/annual.
2. Page 423 from http://www.geotimes.org/nov02/feature_oil.html.

Chapter 8

Section 8.1

1. Exercise 38 from Sam Northshield, Plattsburgh State University.
2. Exercise 43 from Reed, George and James Hill, "Measuring the Thermic Effect of Food," *American Journal of Clinical Nutrition*, Vol. 63, 1996, pp. 164–169.
3. Exercise 44 from Hungate, R. E., "The Rumen Microbial Ecosystem," *Annual Review of Ecology and Systematics*, Vol. 6, 1975, pp. 39–66.

Section 8.2

1. Exercise 40 from http://www.nctm.org/wlme/wlme6/five.htm.
2. Exercise 43 from http://www.nctm.org/eresources/view_article.asp?article_id=6219&page=5.

Section 8.4

1. Exercise 49 from Murray, J. D., *Mathematical Biology*, Springer-Verlag, 1989, p. 642, 648.
2. Exercise 50 from Ludwig, Donald, "An Unusual Free Boundary Problem from the Theory of Optimal Harvesting," in *Lectures on Mathematics in the Life Sciences, Vol. 12: Some Mathematical Questions in Biology*, American Mathematical Society, 1979, pp. 173–209.

Extended Application

1. For a report on this case, see ASBCA No. 44791, 1994. For an introduction to learning curves see Heizer, Jay and Barry Render, *Operations Management*, Prentice-Hall, 2001, or Argote, Linda and Dennis Epple, "Learning Curves in Manufacturing," *Science*, Feb. 23, 1990.

Chapter 9

Section 9.1

1. Exercise 34 from Cobb, Charles W. and Paul H. Douglas, "A Theory of Production," *American Economic Review*, Vol. 18, No. 1, Supplement, March 1928, pp. 139–165.
2. Exercise 35 from Storesletten, Kjetil, "Sustaining Fiscal Policy Thorough Immigration," *Journal of Political Economy*, Vol. 108, No. 2, April 2000, pp. 300–323.
3. Exercise 38 from Harding, K. C. et al., "Mass-Dependent Energetics and Survival in Harbour Seal Pups," *Functional Ecology*, Vol. 19, No.1, Feb., 2005, pp. 129–135.
4. Exercise 39 from Haycock G.B., G.J. Schwartz, and D.H. Wisotsky, "Geometric Method for Measuring Body Surface Area: A Height Weight Formula Validated in Infants, Children and Adults," *The Journal of Pediatrics*, Vol. 93, No.1, 1978, pp. 62–66.
5. Exercise 40 from Alexander, R. McNeill, "How Dinosaurs Ran," *Scientific American*, Vol. 264, April 1991, p. 4.
6. Exercise 41 from Van Holt, T., D. Murphy, and L. Chapman, "Local and Landscape Predictors of Fish-assemblage Characteristics in the Great Swamp, New York," *Northeastern Naturalist*, Vol. 12, No. 3, 2006, pp. 353–374.
7. Exercise 42 from Chowell, F. and F. Sanchez, "Climate-based Descriptive Models of Dengue Fever: The 2002 Epidemic in Colima, Mexico," *Journal of Environmental Health*, Vol. 68, No. 10, June 2006, pp. 40–44.
8. Exercises 43 and 44 from Iverson, Aaron and Louis Iverson, "Spatial and Temporal Trends of Deer Harvest and Deer-Vehicle Accidents in Ohio," *Ohio Journal of Science*, 99, 1999, pp. 84–94.
9. Exercise 45 from Appleby, M., A. Lawrence, and A. Illius, "Influence of Neighbours on Stereotypic Behaviour of Tethered Sows," *Applied Animal Behaviour Science*, Vol. 24, 1989, pp. 137–146.

Section 9.2

1. Exercise 53 from Robbins, C., *Wildlife Feeding and Nutrition*, New York: Academic Press, 1983, p. 114.
2. Exercise 54 from Harding, K. C. et al., "Mass-Dependent Energetics and Survival in Harbour Seal Pups," *Functional Ecology*, Vol. 19, No. 1, Feb., 2005, pp. 129–135.
3. Exercise 55 from Haycock G.B., G.J. Schwartz, and D.H. Wisotsky, "Geometric Method for Measuring Body Surface Area: A Height Weight Formula Validated in Infants, Children and Adults," *The Journal of Pediatrics*, Vol. 93, No.1, 1978, pp. 62–66.
4. Exercise 56 from http://www.anaesthesiauk.com/article.aspx?articleid=251.
5. Exercise 58 from http://www.win.niddk.nih.gov/publications/tools.htm.
6. Exercise 60 from Westbrook, David, "The Mathematics of Scuba Diving," *The UMAP Journal*, Vol. 18, No. 2, 1997, pp. 2–19.
7. Exercise 61 from Bosch, William and L. Cobb, "Windchill," *The UMAP Journal*, Vol. 13, No. 3, 1990, pp. 481–489.
8. Exercise 62 from www.weather.com (May 9, 2000).
9. Exercise 66 from Fiore, Greg, "An Out-of-Math Experience: Einstein, Relativity, and the Developmental Mathematics Student," *Mathematics Teacher*, Vol. 93, No. 3, 2000, pp. 194–199.
10. Exercise 67 from Sanders, Mark and Ernest McCormick, *Human Factors in Engineering Design*, 7th ed., New York: McGraw-Hill, 1993, pp. 290–291.

Section 9.3

1. Exercise 32 from Problem 35 from May 2003 Course 1 Examination of the *Education and Examination Committee of the Society of Actuaries*. Reprinted by permission of the Society of Actuaries.
2. Exercise 40 from Grofman, Bernard, "A Preliminary Model of Jury Decision Making as a Function of Jury Size, Effective Jury Decision Rule, and Mean Juror Judgmental Competence," *Frontiers of Economics*, Vol. 3, 1980, pp. 98–110.
3. Exercise 41 from http://www.intel.com/technology/mooreslaw/.
4. Exercise 42 from Hindra, F. and Oon-Doo Baik, "Kinetics of Quality Changes During Food Frying," *Critical Reviews in Food Science and Nutrition*, Vol. 46, 2006, pp. 239–258.

Section 9.4

1. Exercise 41 from Duffin, R., E. Peterson, and C. Zener, *Geometric Programming: Theory and Application*, New York: Wiley, 1967. Copyright © 1967 John Wiley & Sons, Inc.
2. Exercise 42 from Owen, Guillermo et al., "Proving a Distribution-Free Generalization of the Condorcet Jury Theorem," *Mathematical Social Sciences*, Vol. 17, 1989, pp. 1–16.

Section 9.5

1. Page 502 from Colley, Susan Jane, "Calculus in the Brewery," *The College Mathematics Journal*, Vol. 25, No. 3, May 1994, p. 227.
2. Exercise 25 from Gotch, Frank, "Clinical Dialysis: Kinetic Modeling in Hemodialysis," *Clinical Dialysis*, 3rd ed., Norwalk: Appleton & Lange, 1995, pp. 156–186.
3. Exercise 26 from Fitzsimmons, N., S. Buskirk, and M. Smith, "Population History, Genetic Variability, and Horn Growth in Bighorn Sheep," *Conservation Biology*, Vol. 9, No. 2, April 1995, pp. 314–323.
4. Exercise 27 from Brown, J. P. and P. E. Sendak, "Association of Ring Shake in Eastern Hemlock with Tree Attributes," *Forest Products Journal*, Vol. 56, No. 10, October 2006, pp. 31–36.
5. Exercise 28 from Walker, Anita, "Mathematics Makes a Splash: Evaluating Hand Timing Systems," *The HiMAP Pull-Out Section*, Spring 1992, COMAP.

Review Exercises

1. Exercise 99 from Chumlea, W., S. Guo, C. Zellar et al., "Total Body Water Reference Values and Prediction Equations for Adults," *Kidney International*, Vol. 59, 2001, pp. 2250–2258.
2. Exercise 100 from Hayes, J., J. Stark, and K. Shearer, "Development and Test of a Whole-Lifetime Foraging and Bio-energetics Growth Model for Drift-Feeding Brown Trout," *Transactions of the American Fisheries Society*, Vol. 129, 2000, pp. 315–332.
3. Exercise 101 from *National Vital Statistics Reports*, Vol. 51, No. 3, December 19, 2002.

Extended Application

1. Page 523 from www.uspto.gov/patents. You can locate patents by number or carry out a text search of the full patent database.

Chapter 10

Section 10.1

1. Page 534 from Fisher, J. C. and R. H. Pry, "A Simple Substitution Model of Technological Change," *Technological Forecasting and Social Change*, Vol. 3, 1971–1972. Copyright © 1972 by Elsevier Science Publishing Co., Inc. Reprinted by permission of the publisher.

2. Page 534 from http://www.fabrics-manufacturers. com/consumption-statistics.html.
3. Exercise 40 from http://www.internetworldstats. com/emarketing.htm.
4. Exercise 41 from Problem 27 from May 2003 Course 1 Examination of the *Education and Examination Committee of the Society of Actuaries.* Reprinted by permission of the Society of Actuaries.
5. Exercise 43 from Specht, R. L., "Dark Island Heath (Ninety-Mile Plain, South Australia) V: The Water Relationships in Heath Vegetation and Pastures on the Makin Sand," *Australian Journal of Botany,* Vol. 5, No. 2, Sept. 1957, pp. 151–172.
6. Exercises 45 and 46 from Segal, Arthur C., "A Linear Diet Model," *The College Mathematics Journal,* Vol. 18, No. 1, Jan. 1987.
7. Exercise 47 from http://news.bbc.co.uk/2/hi/ uk_news/8083179.stm.
8. Exercise 48 from Population Division of the Dept. of Economic and Social Affairs of the UN Secretariat, *World Population Prospects: The 2008 Revision.*
9. Exercises 49 and 50 from http://www.census. gov/population/www/projections/usinterimproj/.
10. Exercise 52 from *The New York Times,* Nov. 17, 1996, p. 3.
11. Exercise 55 first appeared in the *American Mathematical Monthly,* Vol. 44, Dec. 1937.
12. Exercises 58 and 59 from Callas, Dennis and David J. Hildreth, "Snapshots of Applications in Mathematics," *The College Mathematics Journal,* Vol. 26, No. 2, March 1995.

Section 10.2

1. Example 4 from Andrews, Larry C., *Ordinary Differential Equations with Applications,* Scott, Foresman and Company, 1982, p. 79.
2. Exercise 23 from Hoppensteadt, F. C. and J. D. Murray, "Threshold Analysis of a Drug Use Epidemic Model," *Mathematical Biosciences,* Vol. 53, No. 1/2, Feb. 1981, pp. 79–87.
3. Exercise 24 from Anderson, Roy M., "The Persistence of Direct Life Cycle Infectious Diseases Within Populations of Hosts," in *Lectures on Mathematics in the Life Sciences, Vol. 12: Some Mathematical Questions in Biology,* American Mathematical Society, 1979, pp. 1–67.

Section 10.3

1. Exercise 33 from France, J., J. Kijkstra and M. S. Dhanoa, "Growth Functions and Their Application in Animal Science," *Annales de Zootechnie,* Vol. 45 (Supplement), 1996, pp. 165–174.
2. Exercise 34 from Thurstone, L. L., "The Learning Function," *The Journal of General Psychology,* Vol. 3, No. 4, Oct. 1930, pp. 469–493.

Section 10.4

1. Page 552 from Lotka, A. J., *Elements of Mathematical Biology,* Dover, 1956.

2. Page 556 from Andrews, Larry C., *Ordinary Differential Equations with Boundary Value Problems,* HarperCollins Publishers, Inc., 1991, pp. 85–86. Reprinted by permission of the author.
3. Exercise 13 from Bender, Edward A., *An Introduction to Mathematical Modeling.* Copyright © 1978 by John Wiley and Sons, Inc. Reprinted by permission.

Review Exercise

1. Exercises 64–66 from http://www.census. gov/prod/ www/abs/ma.html.
2. Exercise 67 from Southwick, Lawrence, Jr. and Stanley Zionts, "An Optimal-Control-Theory Approach to the Education-Investment Decision," *Operations Research,* Vol. 22, 1974, pp. 1156–1174.
3. Exercise 71 from *Harper's,* Oct. 1994, p. 13.

Extended Application

1. Page 564 from Bender, Edward A., *An Introduction to Mathematical Modeling.* Copyright © 1978 by John Wiley & Sons, Inc. Reprinted by permission from *An Introduction to Mathematical Modeling* by Edward A. Bender (Dover, 2000).

Chapter 11

Section 11.1

1. Example 4 from *National Vital Statistics Reports,* Vol. 54, No. 13, April 19, 2006, Table 6, p. 25.
2. Exercise 37 from Problem 34 from May 2003 Course 1 *Examination of the Education and Examination Committee of the Society of Actuaries.* Reprinted by permission of the Society of Actuaries.
3. Exercise 38 from Problem 40 from the 2005 Sample Exam P of the *Education and Examination Committee of the Society of Actuaries.* Reprinted by permission of the Society of Actuaries.
4. Exercise 41 from Dennis, Brian and Robert F. Costantino, "Analysis of Steady-State Populations with the Gamma Abundance Model: Application to *Tribolium,*" *Ecology,* Vol. 69, No. 4, Aug. 1988, pp. 1200–1213.
5. Exercise 42 from Karevia, Peter, "Experimental and Mathematical Analysis of Herbivore Movement: Quantifying the Influence of Plant Spacing and Quality on Foraging Discrimination," *Ecology Monographs,* Vol. 2, No. 3, Sept. 1982, pp. 261–282.
6. Exercise 43 from "Number of U.S. Facebook Users Over 35 Nearly Doubles in Last 60 Days," March 25, 2009, http://www. insidefacebook.com/2009/03/25/number-of-us-facebook-users-over-35-nearly-doubles-in-last-60-days/.
7. Exercises 46 and 47 from Wang, Jeen-Hwa and Chiao-Hui Kuo, "On the Frequency Distribution of Interoccurrence Times of

Earthquakes," *Journal of Seismology,* Vol. 2, 1998, pp. 351–358.
8. Exercise 48 from Traffic Safety Facts, 2008 Data, NHTA'S National Center for Statistics and Analysis, http://www-nrd.nhtsa.dot.gov/ pubs/811155.pdf.
9. Exercise 49 from http://www-nrd.nhtsa.dot. gov/pdf/nrd-30/NCSA/RNotes/1998/ AgeSex96. pdf.
10. Exercise 51 from Fatal Crashes by Time of Day and Day of Week, USA, Year: 2008. FARS Encyclopedia: Crashes–Time. http://www-fars.nhtsa.dot.gov/Crashes/ CrashesTime.aspx.

Section 11.2

1. Example 4 from National Center for Health Statistics, *Method for Constructing Complete Annual U.S. Life Tables,* Series 2, No. 129, Dec. 1999.
2. Exercise 23 from Problem 12 from May 2003 Course 1 Examination of the *Education and Examination Committee of the Society of Actuaries.* Reprinted by permission of the Society of Actuaries.
3. Exercise 27 from Problem 51 from the 2005 Sample Exam P of the *Education and Examination Committee of the Society of Actuaries.* Reprinted by permission of the Society of Actuaries.
4. Exercise 28 from Problem 53 from the 2005 Sample Exam P of the *Education and Examination Committee of the Society of Actuaries.* Reprinted by permission of the Society of Actuaries.
5. Exercise 29 from Problem 55 from the 2005 Sample Exam P of the *Education and Examination Committee of the Society of Actuaries.* Reprinted by permission of the Society of Actuaries.
6. Exercise 30 from Problem 68 from the 2005 Sample Exam P of the *Education and Examination Committee of the Society of Actuaries.* Reprinted by permission of the Society of Actuaries.
7. Exercise 34 from Kareiva, Peter, "Experimental and Mathematical Analyses of Herbivore Movement: Quantifying the Influence of Plant Spacing and Quality on Foraging Discrimination," *Ecology Monographs,* Vol. 2, No. 3, 1982, pp. 261–282.
8. Exercise 35 from Dennis, Brian and Robert F. Costantino, "Analysis of Steady-State Population with the Gamma Abundance Model: Application to *Tribolium,*" *Ecology,* Vol. 69, No. 4, Aug. 1988, pp. 1200–1213.

To view the complete source list, visit the Downloadable Student Resources site, www.pearsonhighered.com/mathstatsresources. The complete list is also available to qualified instructors within MyMathLab or through the Pearson Instructor Resource Center, www.pearsonhighered.com/irc.

KEY DEFINITIONS, THEOREMS, AND FORMULAS

3.1 Rules for Limits

Let a, A, and B be real numbers, and let f and g be functions such that

$$\lim_{x \to a} f(x) = A \qquad \text{and} \qquad \lim_{x \to a} g(x) = B.$$

1. If k is a constant, then $\lim_{x \to a} k = k$ and $\lim_{x \to a} [k \cdot f(x)] = k \cdot \lim_{x \to a} f(x) = k \cdot A$.

2. $\lim_{x \to a} [f(x) \pm g(x)] = \lim_{x \to a} f(x) \pm \lim_{x \to a} g(x) = A \pm B$

   (The limit of a sum or difference is the sum or difference of the limits.)

3. $\lim_{x \to a} [f(x) \cdot g(x)] = [\lim_{x \to a} f(x)] \cdot [\lim_{x \to a} g(x)] = A \cdot B$

   (The limit of a product is the product of the limits.)

4. $\lim_{x \to a} \dfrac{f(x)}{g(x)} = \dfrac{\lim_{x \to a} f(x)}{\lim_{x \to a} g(x)} = \dfrac{A}{B}$ if $B \neq 0$

   (The limit of a quotient is the quotient of the limits, provided the limit of the denominator is not zero.)

5. If $p(x)$ is a polynomial, then $\lim_{x \to a} p(x) = p(a)$.

6. For any real number k, $\lim_{x \to a} [f(x)]^k = [\lim_{x \to a} f(x)]^k = A^k$, provided this limit exists.

7. $\lim_{x \to a} f(x) = \lim_{x \to a} g(x)$ if $f(x) = g(x)$ for all $x \neq a$.

8. For any real number $b > 0$, $\lim_{x \to a} b^{f(x)} = b^{\lim_{x \to a} f(x)} = b^A$.

9. For any real number b such that $0 < b < 1$ or $1 < b$,
   $\lim_{x \to a} [\log_b f(x)] = \log_b [\lim_{x \to a} f(x)] = \log_b A$ if $A > 0$.

3.1 Limits at Infinity

For any positive real number n,

$$\lim_{x \to \infty} \frac{1}{x^n} = 0 \qquad \text{and} \qquad \lim_{x \to -\infty} \frac{1}{x^n} = 0.$$

3.3 Instantaneous Rate of Change

The instantaneous rate of change for a function f when $x = a$ is

$$\lim_{h \to 0} \frac{f(a + h) - f(a)}{h}, \qquad \text{provided this limit exists.}$$

3.4 Derivative

The derivative of the function f at x, written $f'(x)$, is defined as

$$f'(x) = \lim_{h \to 0} \frac{f(x + h) - f(x)}{h}, \qquad \text{provided this limit exists.}$$

Rules for Derivatives

The following rules for derivatives are valid when all the indicated derivatives exist.

4.1

Constant Rule If $f(x) = k$, where k is any real number, then $f'(x) = 0$.

4.1

Power Rule If $f(x) = x^n$ for any real number n, then $f'(x) = nx^{n-1}$.

4.1

Constant Times a Function Let k be a real number. Then the derivative of $f(x) = k \cdot g(x)$ is

$$f'(x) = k \cdot g'(x).$$

4.1

Sum or Difference Rule If $f(x) = u(x) \pm v(x)$, then

$$f'(x) = u'(x) \pm v'(x).$$

4.2 *Product Rule*   If $f(x) = u(x) \cdot v(x)$, then

$$f'(x) = u(x) \cdot v'(x) + v(x) \cdot u'(x).$$

4.2 *Quotient Rule*   If $f(x) = u(x)/v(x)$, and $v(x) \neq 0$, then

$$f'(x) = \frac{v(x) \cdot u'(x) - u(x) \cdot v'(x)}{[v(x)]^2}.$$

4.3 *Chain Rule*   If y is a function of u, say $y = f(u)$, and if u is a function of x, say $u = g(x)$, then $y = f(u) = f[g(x)]$, and

$$\frac{dy}{dx} = \frac{dy}{du} \cdot \frac{du}{dx}.$$

4.3 *Chain Rule (Alternate Form)*   If $y = f[g(x)]$, then $dy/dx = f'[g(x)] \cdot g'(x)$.

4.4 *Exponential Function*

$$\frac{d}{dx}[a^{g(x)}] = (\ln a)a^{g(x)}g'(x)$$

$$\frac{d}{dx}[e^{g(x)}] = e^{g(x)}g'(x)$$

4.5 *Logarithmic Function*

$$\frac{d}{dx}[\log_a|g(x)|] = \frac{1}{\ln a} \cdot \frac{g'(x)}{g(x)}$$

$$\frac{d}{dx}[\ln |g(x)|] = \frac{g'(x)}{g(x)}$$

5.2 First Derivative Test   Let c be a critical number for a function f. Suppose that f is continuous on (a, b) and differentiable on (a, b) except possibly at c, and that c is the only critical number for f in (a, b).

1. $f(c)$ is a relative maximum of f if the derivative $f'(x)$ is positive in the interval (a, c) and negative in the interval (c, b).

2. $f(c)$ is a relative minimum of f if the derivative $f'(x)$ is negative in the interval (a, c) and positive in the interval (c, b).

5.3 Second Derivative Test   Let f'' exist on some open interval containing c (except possibly at c itself), and let $f'(c) = 0$.

1. If $f''(c) > 0$, then $f(c)$ is a relative minimum.

2. If $f''(c) < 0$, then $f(c)$ is a relative maximum.

3. If $f''(c) = 0$ or $f''(c)$ does not exist, then the test gives no information about extrema, so use the first derivative test.

7.2 Substitution

Each of the following forms can be integrated using the substitution $u = f(x)$.

Form of the Integral   **Result**

1. $\displaystyle\int [f(x)]^n f'(x)\, dx,\quad n \neq -1$ $\qquad \displaystyle\int u^n\, du = \frac{u^{n+1}}{n+1} + C = \frac{[f(x)]^{n+1}}{n+1} + C$

2. $\displaystyle\int \frac{f'(x)}{f(x)}\, dx$ $\qquad \displaystyle\int \frac{1}{u}\, du = \ln |u| + C = \ln |f(x)| + C$

3. $\displaystyle\int e^{f(x)} f'(x)\, dx$ $\qquad \displaystyle\int e^u\, du = e^u + C = e^{f(x)} + C$

7.4 Fundamental Theorem of Calculus

Let f be continuous on the interval $[a, b]$, and let F be *any* antiderivative of f. Then

$$\int_a^b f(x)\, dx = F(b) - F(a) = F(x)\Big|_a^b.$$

7.6 Trapezoidal Rule

Let f be a continuous function on $[a, b]$ and let $[a, b]$ be divided into n equal subintervals by the points $a = x_0, x_1, x_2, \ldots, x_n = b$. Then, by the trapezoidal rule,

$$\int_a^b f(x)\, dx \approx \left(\frac{b-a}{n}\right)\left[\frac{1}{2}f(x_0) + f(x_1) + \cdots + f(x_{n-1}) + \frac{1}{2}f(x_n)\right].$$

7.6 Simpson's Rule

Let f be a continuous function on $[a, b]$ and let $[a, b]$ be divided into an even number n of equal subintervals by the points $a = x_0, x_1, x_2, \ldots, x_n = b$. Then, by Simpson's rule,

$$\int_a^b f(x)\, dx \approx \left(\frac{b-a}{3n}\right)[f(x_0) + 4f(x_1) + 2f(x_2) + 4f(x_3) + \cdots + 2f(x_{n-2}) + 4f(x_{n-1}) + f(x_n)].$$

8.1 Integration by Parts

If u and v are differentiable functions, then

$$\int u\, dv = uv - \int v\, du.$$

8.4 Improper Integrals

If f is continuous on the indicated interval and if the indicated limits exist, then

$$\int_a^\infty f(x)\, dx = \lim_{b \to \infty} \int_a^b f(x)\, dx,$$

$$\int_{-\infty}^b f(x)\, dx = \lim_{a \to -\infty} \int_a^b f(x)\, dx,$$

$$\int_{-\infty}^\infty f(x)\, dx = \int_{-\infty}^c f(x)\, dx + \int_c^\infty f(x)\, dx,$$

for real numbers a, b, and c, where c is arbitrarily chosen.

9.3 Test for Relative Extrema

For a function $z = f(x, y)$, let f_{xx}, f_{yy}, and f_{xy} all exist in a circular region contained in the xy-plane with center (a, b). Further, let

$$f_x(a, b) = 0 \qquad \text{and} \qquad f_y(a, b) = 0.$$

Define the number D by

$$D = f_{xx}(a, b) \cdot f_{yy}(a, b) - [f_{xy}(a, b)]^2.$$

Then

(a) $f(a, b)$ is a relative maximum if $D > 0$ and $f_{xx}(a, b) < 0$;

(b) $f(a, b)$ is a relative minimum if $D > 0$ and $f_{xx}(a, b) > 0$;